Discovering
Algebra
An Investigative Approach

Jerald Murdock

Ellen Kamischke

Eric Kamischke

DISCOVERING

MATHEMATICS™

Key Curriculum Press
Innovators in Mathematics Education

Project Editor
Joan Lewis

Project Administrator
Shannon Miller

Editors
Christian Aviles-Scott, Dan Bennett,
Curt Gebhard, Greer Lleuad, Crystal Mills,
Kent Turner, Pam Tyson

Editorial Assistants
James A. Browne, Heather Dever, Halo Golden,
Beth Masse, Susan Minarcin, Jason Taylor,
Lara Wysong

Mathematics Consultant and Writer
Larry Copes

Teacher Consultants
Kara Granger, Beth Schlesinger,
William Putnam

Accuracy Checker
Dudley Brooks

Editorial Production Manager
Deborah Cogan

Production Editor
Christine Osborne

Copy Editor
Erin Milnes

Production Director
Diana Jean Parks

Production Coordinator
Ann Rothenbuhler

Art Director and Cover Designer
Jill Kongabel

Text Designer
Jenny Somerville

Art Editor
Jason Luz

Art and Design Coordinator
Caroline Ayres

Technical Art
Precision Graphics

Photo Editor
Margee Robinson

Compositor and Prepress
TSI Graphics

Printing
Von Hoffmann Press

Executive Editor
Casey FitzSimons

Publisher
Steven Rasmussen

© 2002 by Key Curriculum Press. All rights reserved.

No part of this publication may be reproduced, stored in a retrieval system, or
transmitted, in any form or by any means, electronic, photocopying, recording,
or otherwise, without the prior written permission of the publisher.

®The Geometer's Sketchpad, Dynamic Geometry, and Key Curriculum Press
are registered trademarks of Key Curriculum Press. ™Sketchpad is a trademark of
Key Curriculum Press.

™Fathom Dynamic Statistics and the Fathom logo are trademarks of KCP
Technologies.

All other trademarks are held by their respective owners.

Key Curriculum Press
1150 65th Street
Emeryville, CA 94608
510-595-7000
editorial@keypress.com
http://www.keypress.com
Printed in the United States of America
10 9 8 7 6 5   06 05
ISBN 1-55953-472-9

Discovering Algebra Acknowledgments

Mathematical Content Reviewers

Bill Medigovich, San Francisco, California

Mary Jean Winter, Michigan State University, East Lansing, Michigan

Answer Checker

Jim Stetson

Editorial Consultants

John Allman, Steven Chanan, Cindy Clements, William Finzer, and Ladie Malek, Key Curriculum Press

Frank Aviles, San Francisco, California

Stacey Miceli, Chicago, Illinois

Leslie Nielsen, Issaquah Middle School, Issaquah, Washington

Equity Reviewers

Edward Castillo, Sonoma State University, Rohnert Park, California

Genevieve Lau, Skyline College, San Bruno, California

Charlene Morrow, Mount Holyoke College, South Hadley, Massachusetts

Arthur Powell, Rutgers University, Newark, New Jersey

William Yslas Velez, University of Arizona, Tucson, Arizona

Social Sciences and Humanities Reviewers

Ann Lawrence, Middletown, Connecticut

Karen Michalowicz, The Langley School, McLean, Virginia

Science Reviewers

Andrey Aristov, Loyola High School, Los Angeles, California

Matthew Weinstein, Macalester College, St. Paul, Minnesota

Laura Whitlock, Ph.D., Sonoma Center for Innovative Education in Science, Rohnert Park, California

Development Reviewers

Tina Barbieri, Palm Beach Day School, Palm Beach, Florida

Tom Beatini, Glen Rock Junior-Senior High School, Glen Rock, New Jersey

Marlys Brimmer, Ridgeview High School, Bakersfield, California

Martha Brown, Prince George's County Schools, Upper Marlboro, Maryland

Lois Burke, Charlottesville High School, Charlottesville, Virginia

Barbara Close, Palm Beach Day School, Palm Beach, Florida

Judy Cubillo, Northgate High School, Walnut Creek, California

Dave Damcke, Jefferson High School, Portland, Oregon

Pennie DeHoff, Tucson, Arizona

Larry Deis, El Molino High School, Forestville, California

Genie Dunn, Killian Senior High School, Miami, Florida

Patty Flowers, Humboldt High School, Humboldt, Tennessee

Claudia Gary, Amphitheater High School, Tucson, Arizona

Pamela Harris, Southwest Texas State University, San Marcos, Texas

Rachel Henning, Mims Elementary School, Mission, Texas

Murrel Hoover, South Charleston High School, South Charleston, West Virginia

Carolyn Jordan, Franklin High School, Stockton, California

Larry Lucas, Haskell Indian Nations University, Lawrence, Kansas

Bill Marthinsen, Piedmont High School, Piedmont, California

Jane Moore, Mariposa Middle School, Mariposa, California

John Oppedisano, Jefferson High School, Portland, Oregon

Luis Ortiz Franco, Chapman University, Orange, California

Marilyn Peak, Spencer County High School, Taylorsville, Kentucky

Debbie Preston, Keystone School, San Antonio, Texas

Darlene Pugh, Malta High School, Malta, Montana

Oran Pyle, Mariposa County High School, Mariposa, California

Beth Schlesinger, San Diego, California

Rick Shanley, Collegiate School, New York, New York

Phil Smith, Wayne High School, Fort Wayne, Indiana

Denise Tenanty, Masconomet High School, Topsfield, Massachusetts

Lisa Usher, San Pedro Math, Science, Technology Center, San Pedro, California

Dick Vinetz, Grant High School, Van Nuys, California

Field-test Participants

Jacqueline Abubakari, Arbor Middle School, Decatur, Georgia

Marjorie Ader, Platte Canyon High School, Bailey, Colorado

Doug Alford, South Orange Middle School, South Orange, New Jersey

Maelynn Anderson, Skyview High School, Vancouver, Washington

Jim Barys, Wachusett Regional High School, Holden, Massachusetts

Ellen Basile, Maplewood Middle School, Maplewood, New Jersey

Candice Beattys, South Orange and Maplewood Middle School, South Orange, New Jersey

Lanna Bell, The Lexington School, Lexington, Kentucky

Brian Boyd, Oakwood High School, Dayton, Ohio

Jeffrey Choppin, Benjamin Banneker Academic High School, Washington, D.C.

Jan Christianson, Hong Kong International School, Hong Kong

Mary Ann Clark, The Langley School, McLean, Virginia

Cel Cowan, Hamlin High School, Hamlin, Texas

Pat Cusick, Hong Kong International School, Hong Kong

Howard David, Amherst Regional High School, Amherst, Massachusetts

Carol DeCuzzi, Audubon Junior-Senior High School, Audubon, New Jersey

Cathy Doll, Henry County High School, New Castle, Kentucky

Bill Dolyniuk, Woodside Elementary School, Woodside, California

Marilyn Eglovitch, Maplewood Middle School, Maplewood, New Jersey

Janine Evans, Hong Kong International School, Hong Kong

Patty Flowers, Humboldt High School, Humboldt, Tennessee

Sy Friedman, Amherst Regional High School, Amherst, Massachusetts

Jennifer Gardner, Arrowsmith Academy, Berkeley, California

Janet Gibson, Denver City High School, Denver City, Texas

Steve Gile, Kingsburg High School, Kingsburg, California

Kara Granger, Northwestern High School, Hyattsville, Maryland

Kevin Harris, Pennridge Central Middle School, Perkasie, Pennsylvania

Carol Hattan, Skyview High School, Vancouver, Washington

Susan Heinrich, Hong Kong International School, Hong Kong

Edna Horton Flores, Kenwood Academy, Chicago, Illinois

Marilyn Howard, University School at Tulsa University, Tulsa, Oklahoma

Steven Isaak, Green Valley High School, Henderson, Nevada

Nancy Jameson, Rowland Hall-St. Mark's School, Salt Lake City, Utah

Mary Jensen, Friday Harbor High School, Friday Harbor, Washington

Patty Kincaid, Sheldon High School, Eugene, Oregon

Norman Krumpe, Oakwood High School, Dayton, Ohio

Bill Manchester, Amherst Regional High School, Amherst, Massachusetts

John Moran, Amherst Regional High School, Amherst, Massachusetts

Kristin Muldowney, Hong Kong International School, Hong Kong

David Mullins, Amherst Regional High School, Amherst, Massachusetts

Paul Myers, Woodward Academy, College Park, Georgia

Wayne Nirode, Troy High School, Troy, Ohio

Ken Nossardi, C. K. McClatchy High School, Sacramento, California

Paul Peelle, Amherst Regional High School, Amherst, Massachusetts

Doris Peim, South Orange Middle School, South Orange, New Jersey

Benita Pfeiffer, Hong Kong International School, Hong Kong

Ken Rohrs, Hong Kong International School, Hong Kong

Wakako Rohlich, Amherst Regional High School, Amherst, Massachusetts

Alison Ruebusch, Catlin-Gabel School, Portland, Oregon

Debi Rydeski, Columbia High School, Burbank, Washington

Rick Shanley, The Collegiate School, New York, New York

Rebecca Simpson, Green Mountain High School, Lakewood, Colorado

Janet Taylor, Hong Kong International School, Hong Kong

Denise Tenanty, Masconomet High School, Topsfield, Massachusetts

Tim Voegeli, Fairmont High School, Kettering, Ohio

Gladys Whitehead, Prince George's County Public Schools, Upper Marlboro, Maryland

Karla Wiggins, South Orange Middle School, South Orange, New Jersey

June Wilby, Amherst Regional High School, Amherst, Massachusetts

Cher Williams, Onalaska High School, Onalaska, Washington

Cheryl Wright, Amherst Regional High School, Amherst, Massachusetts

A Note from the Publisher

The mathematics we learn and teach in school has changed dramatically over the past few decades. The changes have been driven by new discoveries in mathematics and science, new research in education, changing societal needs, and the use of new technology in work and in education. The algebra you find in this book won't look exactly like the algebra you may have seen in older textbooks. It covers the same topics and includes lots of familiar equations, but the mathematics in *Discovering Algebra* also works with data from science, emphasizes techniques for data analysis, and uses technology tools such as graphing calculators to help visualize important concepts. We have included important new mathematics in this text to better prepare students for the educational and career opportunities they'll find in our fast-changing world.

For 30 years, Key Curriculum Press has developed mathematics materials for schools. By focusing on mathematics alone, our authors, our editors, and our consultants are able to keep pace with the changes in our field to produce curriculum that enables more students to succeed in school mathematics.

Over the years, in spite of the changes in mathematics and mathematics education, we have found one truth that has not changed: Students learn mathematics best when they understand the concepts behind it. With this in mind, Key Curriculum Press introduces our Discovering Mathematics series, beginning with this book, *Discovering Algebra: An Investigative Approach*. Through the investigations that are the heart of the series, students discover many important mathematical principles themselves. In the process, they come to value their own ability to succeed at mathematics; they realize that they can recreate their discoveries should they ever forget a fundamental concept; and they develop their abilities to continue discovering and learning about mathematics. And when they understand mathematics, they perform well on tests of any kind.

Many years of research, thoughtful work, and class testing have gone into the development of *Discovering Algebra*. The text has been written, piloted with students, rewritten, and piloted again. Over the course of five years, hundreds of teachers have used the trial editions of the text with many thousands of students. Thousands of pages of feedback from pilot teachers have been combined with professional, scientific, and mathematical reviews, along with results from standardized tests, to ensure that the book offers a sound conceptual framework and sufficient skill development for students.

If you are a student, as you work through this course we hope you gain knowledge for a lifetime. If you are a parent, we hope you enjoy watching your child develop mathematical power. If you are a teacher, we hope you find that *Discovering Algebra* makes a significant positive impact in your classroom. Regardless of who you are, the professional team at Key Curriculum Press wishes you success and urges you to continue your involvement and interest in mathematics and education.

Steven Rasmussen, President
Key Curriculum Press

Contents

CHAPTER	Fractions and Fractals	1

0

CHAPTER 5

Fitting a Line to Data   **250**

CHAPTER	Quadratic Models	531

10

Jerald Murdock

Ellen Kamischke

Eric Kamischke

Mathematics, like art, literature, philosophy, or music, is a journey toward finding beauty and meaning in our chaotic world through the discovery and celebration of patterns. Just as poetry is much more than the rhyming text on a greeting card, mathematics is more than exercises and symbols. The paths you and your students are embarking on are rich with mathematical investigations and applications that encourage exploration of ideas, data, patterns, and relationships.

Discovering Algebra helps you rethink algebra and redefine what a good mathematics teacher can do in today's classroom. This *Teacher's Edition* can help you guide and encourage better learning of important mathematics—mathematics that students can immediately make sense of and use. This curriculum is influenced by our experience with students, the National Council of Teachers of Mathematics (NCTM) *Principles and Standards for School Mathematics,* recent projects funded by the National Science Foundation (NSF), and insights of contemporary curriculum leaders.

The course is algebra, but students will frequently consider and represent patterns involving shape and size, or those found in data sets. Algebra blends with geometry, data analysis, discrete mathematics, and statistics.

The teaching-learning model actively engages students in both guided and open-ended mathematical explorations that help them make sense of their experiences. We believe this approach can help empower students mathematically so that they can participate in the world as creative, productive citizens.

As authors, we can trace the development of our innovative book to the mid–1980s. We were high school mathematics teachers working closely together at Interlochen Arts Academy in Interlochen, Michigan, where we wrote short units for computer graphing software that author Eric Kamischke had developed. Then graphing calculators became available for high school students.

During the spring of 1991, in a proposal to the NSF, we formalized our efforts: to develop instructional materials that would incorporate graphing calculators and provide materials and inservice help to other teachers. Interlochen's 1991–1995 NSF-supported (MDR9154410) Graphing Calculator Enhanced Algebra Project (GCEAP) proposed a major reevaluation of both algebra content and instructional strategies to reflect the way people do mathematics with technology.

Since 1993, we have had the good fortune of working with Key Curriculum Press, a leading-edge publisher that provides materials to meet needs not satisfied by traditional publishers. Key seeks authors who are also teachers to guarantee that materials will work when they reach the classroom. Key published our *Advanced Algebra Through Data Exploration* in 1998.

Discovering Algebra has been developed and shaped by what we learned as we wrote *Advanced Algebra* and as we worked with many teachers since the summer of 1992. We base this edition of *Discovering Algebra* on teacher suggestions from more than 60 teachers during the 1998–1999 nationwide field test, advice of educational consultants, and feedback on the Preliminary Edition.

Jerald Murdock
Ellen Kamischke
Eric Kamischke

You are about to embark on an exciting mathematical journey. The goal of your trip is to reach the point at which you have gathered the skills, tools, confidence, and mathematical power to participate fully as a productive citizen in a changing world. Your life will always be full of important decision-making situations, and your ability to use mathematics and algebra can help you make informed decisions. You need skills that can evolve and adapt to new situations. You need to be able to interpret and make decisions based on numerical information, and to find ways to solve problems that arise in real life, not just in textbooks. On this journey you will make connections between algebra and the world around you.

You're going to discover and learn much useful algebra along the way. Learning algebra is more than learning facts and theories and memorizing procedures. We hope you also discover the pleasure involved in mathematics and in learning "how to do mathematics." Success in algebra is a recognized gateway to many varied career opportunities.

With your teacher as a guide, you will learn algebra by doing mathematics. You will make sense of important algebraic concepts, learn essential algebraic skills, and discover how to use algebra. This requires a far bigger commitment than just "waiting for the teacher to show you" or studying "worked-out examples."

During this journey, successful learning will come from your personal involvement, which will often come about when you work with others in small groups. Talk about algebra, share ideas, and learn from and with your fellow group members. Work and communicate with others to strengthen your understanding of the mathematical concepts presented in this book. To improve your skills as a teammate, respect differences between group members, listen carefully when others are sharing, stay focused during the process, be responsible and respectful, and share your own ideas and suggestions.

Your graphing calculator is a tool that helps you explore new ideas and answer questions that come up along the way. With the calculator, you will be able to quickly manipulate large amounts of data so that you can see the overall picture. Throughout the text you will be referred to **Calculator Notes** that will provide useful information for using this tool. In your life and future career, technology is likely to play an important role. Learning to efficiently use your graphing calculator today, and being able to interpret its output, will prepare you to successfully use more advanced technology in your future.

The text itself will be a guidebook, leading you to explore questions and giving you the opportunity to ponder. Read the book carefully, with paper, pencil, and calculator close at hand. Work through the **Examples** and answer the questions that are asked along the way. Perform the **Investigations** as you travel through the course, being careful when making measurements and collecting data. Keep your data and calculations neat and accurate so that your work will be easier and the concepts clearer in the long run. Some **Exercises** require a great deal of thought. Don't give up. Make a solid attempt at each problem that is assigned. Sometimes you will need to fill in details later, after you discuss a problem in class or with your group.

Your notebook will serve as a record of your travels. In it you will record your notes, answers to questions in the text, and solutions for your homework problems. You may also want to keep a journal of your personal impressions along the way. You can place some of your especially notable accomplishments in a portfolio, which will serve as a "photo album" of the highlights of your trip. Collect pieces of work in your portfolio as you go, and refine the contents as you make progress on your journey. Each chapter ends with a feature called **Assessing What You've Learned.** This feature gives suggestions for organizing your notebook, writing in your journal, updating your portfolio, and other ways to reflect on what you have learned.

You should expect struggles, hard work, and occasional frustration. Yet, as your algebra skills grow, you'll overcome obstacles and be rewarded with a deeper understanding of mathematics, an increased confidence in your own problem-solving abilities, and the opportunity to be creative. Features called **Project, Improving Your ... Skills,** and **Take Another Look** will give you special opportunities to creatively apply and extend your learning. We hope that your journey through *Discovering Algebra* will be a meaningful and rewarding experience.

And now it is time to begin. You are about to discover some pretty fascinating things.

Features of the Student Edition

Lessons open with a smooth transition from prior learning and a motivational context that engages attention and keeps students moving forward.

Thought provoking **quotations** introduce some lessons.

Chapter openers have a refreshing fine arts or performing arts theme. Examples come from several cultures and all genres of art, historical and contemporary. A caption connects the opening image with the chapter content.

LESSON
8.3

Graphs of Real-World Situations

A picture is worth a thousand words.
NAPOLEON

Like pictures, graphs communicate a lot of information. So you need to be able to draw and make sense of graphs. In previous chapters you learned to interpret bar graphs, circle graphs, histograms, and box plots. Then you learned to graph data from recursive routines and equations. Some graphs were lines and others were curves.

In this lesson you will apply many of those ideas as you begin to explore the graphs of functions. You will learn how to draw and interpret graphs of some real-world situations.

Frida Kahlo (1907–1954), a Mexican artist, is well known for her fascinating self-portraits. After surviving a bus accident, she had 32 surgical operations, painting many works from her bed. The contrasts between *Self-Portrait with Changuito* (left) and *Self-Portrait with Dog Ixcmintli and Sun* (right) reflect differences in the output of her work at different times in her life.

AMPLE This graph shows the depth of the water in a leaky swimming pool. Tell what quantities are varying and how they are related. Give possible real-world events in your explanation.

olution The graph shows that the water level or depth changes over a 15-hour time period. At the beginning, when no time has passed, $t = 0$, and the water in the pool is 2 feet deep, so $d = 2$. During the first 6-hour interval ($0 \leq t \leq 6$), the water level drops. The leak seems to get worse as time passes. When $t = 6$ and $d = 1$, it seems that someone starts to refill the pool. The water level rises for the next 5 hours, during the interval $6 \leq t \leq 11$. At $t = 11$, the water reaches its highest level at just above 3 feet, so $d = 3$. At the

Functions

CHAPTER
6

Systems of Equations and Inequalities

Freshly painted umbrellas dry in the sun outside the Nagatsu factory in Kyushu, Japan. The sticks in their frames form intersecting lines like the graphs of linear equations.

OBJECTIVES

In this chapter you will
- learn to solve systems of linear equations
- solve systems using the substitution method
- solve systems using the elimination method
- solve systems using matrices
- graph inequalities in one and two variables
- solve systems of linear inequalities

The carefully written student text and the open page design with contemporary art will help students want to use their book as a guide to their learning. The reading level of the text is designed to facilitate comprehension for students below grade level in reading.

Objectives announce the aims of the chapter in language the students understand.

Investigations are a careful blend of guidance and discovery. Some are purely mathematical and others involve real-world situations to motivate the construction of new learning. Investigations allow students to work independently in groups, and they provide rich discussion opportunities for whole-class participation.

Students are told the **materials** they need for an investigation.

Examples are placed before or after investigations depending on where they will be most helpful. Some are presented in real-world contexts and others in purely mathematical form. All examples are fully worked out computationally and include reasons that justify the process. Notation conventions are clearly explained. Results are interpreted back into the real-world context.

LESSON

2.5

Circle Graphs and Relative Frequency Graphs

Circle graphs, like bar graphs, summarize data in categories. **Relative frequency graphs** also summarize data in categories, but instead of including the actual number for each category, they compare the number in that category to the total for all the categories. Relative frequency graphs show fractions or percents, not values.

Investigation
Circle Graphs and Bar Graphs

You will need
- graph paper
- a protractor
- a compass or circle template
- a ruler

The bar graph shows the approximate land area of the seven continents.

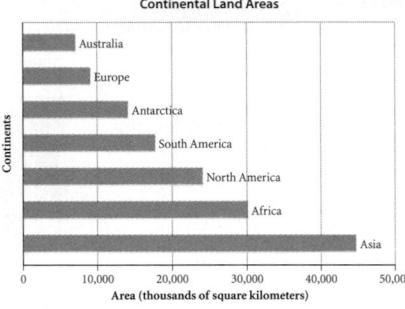

Continental Land Areas

Step 1   Determine from the bar graph the approximate area of each conti total land area.

Step 2   Convert the data in the bar graph to a circle graph. Use the fact th 360 degrees in a circle. Write proportions to find the number of d sector of the circle graph.

Step 3   Convert the data in the bar graph to a relative frequency circle gr showing the land area, the graph will show percents of total land a

Step 4   Convert the data in the bar graph to a relative frequency bar grap percents rather than land areas.

Step 5   Compare the graphs you made with the original graph. What adva there to each kind of graph?

Steps help guide students through an investigation. Horizontal bars indicate convenient places to pause and discuss progress.

The **visual side** of algebra comes through to students and appeals to visual learners as well as reinforcing symbolically expressed concepts for students of all learning styles.

EXAMPLE

Randy has been asked to create a graphical display showing the distribution of the library's collection in six categories. His boss has asked him to create two rough drafts. Together they will decide which one to finalize for the display.

Here is the data:

Library Collection

Category	Number of Items
Children's fiction	35,994
Children's nonfiction	28,106
Adult fiction	48,129
Adult nonfiction	69,834
Media	11,830
Other	5,766
Total	199,659

▶ **Solution**

Randy puts the number of items in each category into list L_1. He wants the calculator to determine in list L_2 the number of degrees needed for each sector. He writes a proportion to find the number of degrees in the sector for a particular category.

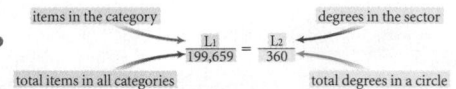

items in the category   degrees in the sector

$$\frac{L_1}{199,659} = \frac{L_2}{360}$$

total items in all categories   total degrees in a circle

By multiplying both sides of the proportion by 360, he finds the formula to enter into list L_2.

$$L_2 = L_1 \cdot \frac{360}{199,659}$$

His calculator quickly determines the number of degrees for each sector of the circle graph. Using a protractor to measure the angles, Randy creates the graph.

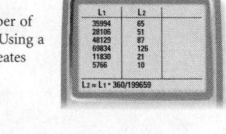

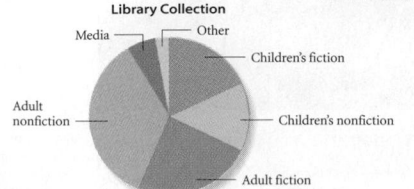

Library Collection

The exercise sets are developmental and begin with opportunities for self-testing of new material in **Practice Your Skills.** These first five or six exercises involve direct applications of what students learned through the investigation and examples.

Graphing calculator use is modeled in the examples. Students use calculators in nearly all investigations and in some of the exercises. Screen captures demonstrate the functionalities available to students, including tables, graphs, and operations on lists. By providing calculator notes with keystroke-level instructions in separate books, the page text remains uncluttered and the sense of the mathematics remains paramount over keystroke details. The calculator notes allow *Discovering Algebra* to be used with different calculators and with new calculators as they become available. Exercises requiring a graphing calculator are called out.

...rtical axis represents... ...put variable *x* represents altitude so the *x*-axis is la... ...e output variable *y* represents temperature so the *y*-axis is labeled temperature. What are the input and output variables in the investigation and in Example A?

EXERCISES

You will need your calculator for problems **2, 3, 5, 6, 9,** and **10.**

Practice Your Skills

1. Match the recursive routine in the first column with the equation in the second column.

a. 3 `ENTER`
Ans + 4 `ENTER` ; `ENTER` , . . .

i. $y = 4 - 3x$

b. 4 `ENTER`
Ans + 3 `ENTER` ; `ENTER` , . . .

ii. $y = 3 + 4x$

c. −3 `ENTER`
Ans − 4 `ENTER` ; `ENTER` , . . .

iii. $y = -3 - 4x$

d. 4 `ENTER`
Ans − 3 `ENTER` ; `ENTER` , . . .

iv. $y = 4 + 3x$

2. You can use the equation $d = 24 - 45t$ to model the distance from a destination for someone driving down the highway, where distance *d* is measured in miles and time *t* is measured in hours. Graph the equation and use the trace function to find the approximate time... each distance given in 3a and b.

Reason and Apply are more challenging problems that develop reasoning and transfer of knowledge. These problems often include several steps that build on each other. They may also require students to combine information from earlier chapters with what they just learned.

Reason and Apply

6. **APPLICATION** Louis is beginning a new exercise wor... shows him the calculator table with *x*-values showi... time in minutes. The Y₁-values are the total calorie... while running, and the Y₂-values are the number o... to burn.

a. Find how many calories Louis has burned befor... how many he burns per minute running, and th... wants to burn.

b. Write a recursive routine that will generate the v...

c. Use your recursive routine to write a linear equa... your equation generates the table values listed i...

d. Write a recursive routine that will generate the v...

e. Write an equation that generates the table value...

f. Graph the equations in Y₁ and Y₂ on your calcu... a time of up to 30 minutes. What is the real-wor... in Y₁?

g. Use the trace function to find the approximate coordinates of the point where the lines meet. What is the real-world meaning of this point?

7. Jo mows lawns after school. She finds that she can... the equation $P = -300 + 15N$ to calculate her pro...

a. Give some possible real-world meanings for the numbers −300 and 15 and the variable, *N*.

b. Invent two questions related to this situation an... then answer them.

c. Explain why the equation $P = 15N - 300$ provi... the same values as the equation $P = -300 + 15...$

d. Identify the variables.

Review

Review keeps previously learned skills from falling out of use, especially when they will be needed in upcoming lessons.

11. At a family picnic, your cousin tells you that he always has... how to compute percents. Write him a note explaining wha... these problems as examples of how to solve the different types of percent problems, with answers for each.

a. 8 is 15% of what number?

b. 15% of 18.95 is what number?

c. What percent of 64 is 326?

d. 10% of what number is 40?

12. **APPLICATION** Carl has been keeping a record of his gas purchases for his new car. Each time he buys gas, he fills the tank completely. Then he records the number of gallons he bought and the miles since the last fill-up. Here is his record:

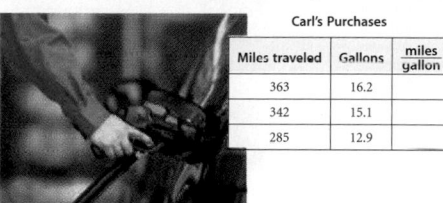

Carl's Purchases

Miles traveled	Gallons	miles/gallon
363	16.2	
342	15.1	
285	12.9	

Answers to most of the odd-numbered exercises can be found in the selected answers in the back of the student text.

a. Copy and complete the table by calculating the ratio of miles per gallon for each purchase.

b. What is the average rate of miles per gallon for Carl's new car so far?

c. The car's tank holds 17.1 gallons. To the nearest mile, how far should Carl be able to go without running out of gas?

d. Carl is planning a trip across the United States. He estimates that the trip will be 4230 miles. How many gallons of gas can Carl expect to buy?

13. Match each recursive routine to a graph. Explain how you made your decision and tell what assumptions you made.

a. 2.5 `ENTER`
Ans + 0.5 `ENTER` ; `ENTER` , . . .

b. 1.0 `ENTER`
Ans + 1.0 `ENTER` ; `ENTER` , . . .

c. 2.0 `ENTER`
Ans + 1.0 `ENTER` ; `ENTER` , . . .

d. 2.5 `ENTER`
Ans − 0.5 `ENTER` ; `ENTER` , . . .

As they read about and investigate patterns and relationships, students will learn to move easily between **multiple representations:** tables, graphs, recursive routines, equations, and words.

Graphs and other technical art are carefully designed for clarity and emphasis.

i.

ii.

iii.

iv.

(Graphs with axes labeled "Answer" (vertical) and "Stage" (horizontal), values 0 to 5)

Activity Day is a chance for students to solidify what they have learned by carrying out a fun investigation. No new material is introduced, so student groups have a chance to spend more time experimenting. And there is more time for groups to share their results and learn from what other students discovered.

The **Chapter Review** begins with a summary of the new mathematical ideas. The exercises are similar to those in the lesson and those that will appear on the chapter test. A more extensive **Mixed Review** is provided every four chapters.

Procedure notes help keep students working independently and simplify steps.

Small parts of pictures from the chapter remind students where an idea was first presented.

Boldface terms can be found in the glossary at the back of the student text.

LESSON
3.5

Activity Day

Variation with a Bicycle

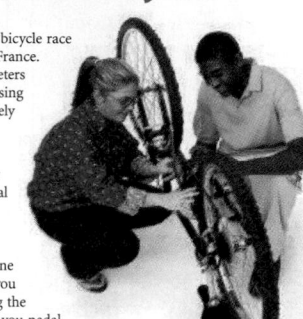

The Tour de France is a demanding bicycle race through Switzerland, Germany, and France. For 23 days, cyclists ride 3630 kilometers on steep mountain roads before crossing the finish line in Paris. The cyclists rely on their knowledge of gear shifting and bicycle speeds.

Many bicycles have several speeds or gears. In a low gear, it's easier to pedal uphill. In a high gear, it's harder to pedal, but you can go faster on flat surfaces and down hills. When you change gears, the chain shifts from one sprocket to another. In this activity you will discover the relationships among the bicycle's gears, the numbers of times you pedal, and the teeth on the sprockets.

Activity

The Wheels Go Round and Round

You will need
• a meterstick or metric tape measure
• bicycle

In Steps 1–5, you'll analyze the effect of the rear sprockets.

Procedure Note

Changing Gears

Each time you change gears on the bike, you may have to turn the bike right side up and rotate the pedal and crankshaft a few times until the gear change takes effect. Then you turn the bike upside down to start observing and recording data.

Rear sprocket assemblies

Tooth

Crankshaft

Front sprocket assemblies

Variation and Graphs

CHAPTER
5
REVIEW

In Chapter 4, you learned how to write equations in intercept form, $y = a + bx$. In this chapter, you learned how to calculate **slope** using the slope formula $b = \frac{y_2 - y_1}{x_2 - x_1}$. You also used the slope formula to derive another form for a linear equation—the **point-slope form**. The point-slope form, $y = y_1 + b(x - x_1)$, is the equation of a line through point (x_1, y_1) with slope b. You learned that this form is very useful in real-world situations when the starting value is not on the y-axis.

You continued to investigate equivalent forms of expressions and equations using tables and graphs. You used the distributive property of multiplication over addition and the **commutative** and **associative** properties of addition and multiplication to write point-slope equations in intercept form.

You discovered how to use the first and third quartiles from the five-number summaries of x- and y-values in a data set to write a linear model for data based on **Q-points**.

EXERCISES

You will need your calculator for problems **3**, **4**, and **9**.

1. The slope of the line between $(2, 10)$ and $(x_2, 4)$ is -3. Find the value of x_2. Show your work.

2. Give the slope and the y-intercept for each equation.
   a. $y = -4 - 3x$
   b. $2x + 7 = y$
   c. $38x - 10y = 24$

3. Line a and line b are shown on the graph at right. Name the slope and the y-intercept, and write the equation for each line. Check your equations by graphing.

4. Write each equation in the form requested. Check your answers by graphing.
   a. Write $y = 13.6(x - 1902) + 158.2$ in intercept form.
   b. Write $y = -5.2x + 15$ in point-slope form using $x = 10$ as the first coordinate of the point.

5. Consider the point-slope equation $y = -3.5 + 2(x + 4.5)$.
   a. Name the point used to write this equation.
   b. Write an equivalent equation in intercept form.
   c. Factor your answer to 5b and name the x-intercept.
   d. A point on the line has a y-coordinate of 16.5. Find the x-coordinate of this point and use this point to write an equivalent equation in point-slope form.
   e. Explain how you can verify that all four equations are equivalent.

CHAPTER 5 REVIEW **303**

Assessing What You've Learned helps
students capture their learning experiences
and identify gaps in understanding. It explains
the use of portfolios, journals, notebooks,
presentations, and performance assessment
as ways to assume responsibility for their own
learning. Students will find chapter-specific
suggestions for reinforcing their strengths
and addressing their weaknesses.

Connections highlight how the mathematics
applies to students' lives—in the workplace or in
the fields of science, history, technology, and the
arts. They enhance opportunities for visual learning.

History
CONNECTION

The Panama Canal allows
ships to cross the strip of land
between the Atlantic and
Pacific Oceans. Before the
canal was completed in 1913,
ships had to sail thousands of
miles around the dangerous
Cape Horn, even though only
50 miles separate the two
oceans.

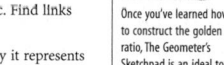

APTER 4 REVIEW • C

Assessing What You've Learned
GIVING A PRESENTATION

Making presentations is an important career skill. Most jobs require workers to
share information, to help orient new coworkers, or to represent the employer to
clients. Making a presentation to the class is a good way to develop your skill at
organizing and delivering your ideas clearly and in an interesting way. Most
teachers will tell you that they have learned more by trying to teach something than
they did simply by studying it in school.

Here are some suggestions to make your presentation go well:

▶ Work with a group. Acting as a panel member might make you less nervous than
giving a talk on your own. Be sure the role of each panel member is known so that
the work and the credit are equally shared.

▶ Choose the topic carefully. You can summarize the results of an i

Projects are open-ended
opportunities for students to
dig deeper and use their
strengths. Some specifically
recommend software options.

project

THE GOLDEN RATIO

In this project, you'll research the amazing number mathematicians call the
golden ratio. (There are plenty of books and web sites on the topic. Find links
from **www.keymath.com/DA** .) Your project should include

▶ Basic information on the golden ratio, such as its exact value, why it represents
a mathematically "ideal" ratio, and how to construct a golden rectangle. (Its
length-to-width ratio is the golden ratio.)

▶ Some history of the golden ratio, including its role in ancient Greek architecture.

▶ At least one other interesting mathematical fact about the golden ratio, such as
its relationship to the Fibonacci sequence or its own reciprocal.

▶ A report on where to find the
golden ratio in the environment,
architecture, or art. List items
and their measurements or
include prints from photographs,
art, or architecture on which you
have drawn the golden rectangle.

THE GEOMETER'S SKETCHPAD

Once you've learned how
to construct the golden
ratio, The Geometer's
Sketchpad is an ideal tool
for further exploration.
You can create a Custom
Tool for dividing segments
into the golden ratio, and
then use this tool to help
you construct the golden
rectangle or even the
golden spiral.

EW • CHAPTER 4 REVIEW • CHAPTER 4 REVIEW • CHAPTE

Take Another Look
gives students a
chance to approach
concepts from a
different perspective
by offereng a visual
slant or calling out
connections between
mathematical ideas.

TAKE ANOTHER LOOK

The picture at right is a **contour map** that reveals the
character of the terrain. All points on an **isometric line** are
the same height in feet above sea level. The graph below
shows how the hiker's walking speed changes as she covers
the terrain on the dotted-line trail shown on the map.

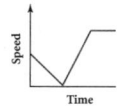

What quantities are changing in the graph
and in the map?

How does each display reveal rate of change?

How could you measure distance on each
display?

What would the graph sketch of this hike
look like if distance were plotted on the
vertical axis instead of speed?

What do these two displays tell you when you
study them together?

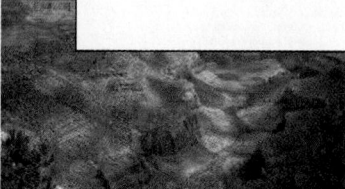
Sediment layers form contour lines in the Grand Canyon.

IMPROVING YOUR REASONING SKILLS

Did these plants grow at the same rate? If not, which plant was tallest on Day 4? Which
plant took the most time to reach 8 cm? Redraw the graphs so that you can compare
their growth rates more easily.

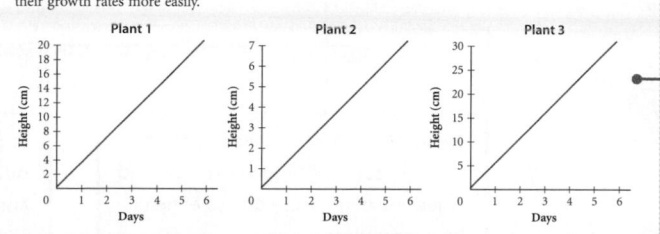

Improving Your Skills is a
puzzle-like feature that diverges
from the topic under study and
practices a particular skill.

Features of the Teacher's Edition

An **overview** of the chapter contents appears first in the interleaf pages that open each chapter of the Teacher's Edition.

Using This Chapter will tell you which lessons deal with topics that might not be required for your algebra curriculum.

Materials for the chapter are summarized in this list.

CHAPTER 3

Variation and Graphs

Overview

In Chapter 2, students used proportions to solve a variety of problems. In this chapter, they learn to use rates and variation to solve similar problems. The study of variation gives students experience with ratios and rates and lays the groundwork for graphing and solving linear equations.

Lesson 3.1 introduces rate as a special kind of ratio and shows how to use rates as a shortcut to solving some proportions. **Lesson 3.2** connects the concept of rate and unit conversion to direct variation. The investigation demonstrates how to use thinking about rates to shortcut setting up proportions. Students create a scatter plot of a real data set, draw a line through the points, and find an equation of the form $y = kx$. Learning to work with and graph simple equations is an important step in working with more generalized linear equations in later chapters.

Similar figures are presented in **Lesson 3.3**. Students use rates to make scale drawings and solve percent problems. **Lesson 3.4** concerns inverse relationships, in particular inverse variations. Students explore both direct and inverse variations through bicycle gear-shifting in the **Lesson 3.5** Activity Day.

The Mathematics

Direct Variation

In the proportion $\frac{x}{8} = \frac{3}{2}$, the 3 and the 2 represent different quantities, but you can think of their ratio as a constant number. If you know that number, you can solve the proportion simply by multiplying $\frac{3}{2}$ by 8.

In Chapter 2, the term *proportional* was used for the quantities x and y with the property that $\frac{y}{x} = k$, where k is a constant. In Chapter 3, students learn the term *directly proportional*. The unknown x can be determined by finding y.

The equation $\frac{y}{x} = k$ is equivalent to $y = kx$; this is called a *direct variation*. The constant k is called the

constant of variation. In general, the constant k direct variation can be positive or negative.

When k is positive, however, x and y increase decrease together. In that case, you can usually think of k as a rate. A *rate* is a ratio with 1 in the denominator. Some conversion factors from Chapter 2 are rates. Chapter 3 introduces many more. For example, at a constant speed (rate), distance traveled varies directly with time.

Dimensional analysis clarifies the role of rates. example, if x is time in hours and y is distance eled in kilometers, then the constant k is a rate kilometers per hour:

$$y \text{ kilometers} = \frac{k \text{ kilometers}}{\text{hour}} x \text{ hours}$$

Other examples of direct variation with a posi constant of variation are

• side length and perimeter of equilateral trian
• diameter and circumference of a circle
• height of an object and length of its shadow a given time of day
• length and weight of pipes of the same diame and composition
• sales and commission at a given rate
• speed and distance traveled over a given leng of time

Another example of directly proportional quan ties are lengths of corresponding sides of simil polygons. The ratio of proportionality is called *scale factor* or *scale*. Multiplying the length of c side by the scale factor gives the length of the c responding side.

Inverse Variation

When one quantity increases while another qua tity decreases, the quantities are *inversely relat* the special case in which the product of the two quantities is constant, they are *inversely propor tional*. They satisfy the *inverse variation* $y = \frac{k}{x}$. Again, k is called the *constant of variation*.

Examples of inverse variations are

• length and width of rectangles with a fixed area
• time and speed required to travel a given distance
• number of items and cost per item you can buy for a fixed amount
• volume and pressure of gas at a fixed temperature
• interest rate and amount of investment needed to maintain a fixed income
• number of workers and hours to do a fixed job
• radius and number of revolutions of a wheel to cover a particular distance
• measure of the central angle of a regular polygon and the number of sides
• speed of a gear and the number of its teeth

Graphs of Variations

Variations are equations, so they have graphs.

The graph of the direct variation $y = kx$ is a straight line through the origin, with slope k. For larger rates k, the value of y increases faster as values of x increase.

The graph of the inverse variation $y = \frac{k}{x}$ is the graph of $xy = k$: a hyperbola with the axes as asymptotes. Usually we have $x > 0$, and $y > 0$, so we consider just one branch of the hyperbola. Larger (positive) values of k result in a hyperbola that falls faster than one with a smaller value of k, indicating that y decreases faster as x increases.

Using This Chapter

If you skipped much of Chapter 2, you need to teach dimensional analysis as you present Lessons 3.1 and 3.2. Lesson 3.5 is optional. If you don't have time to do it, though, at least try to work with some inverse relationships, and show students the graph of an inverse variation as an example of a nonlinear graph. These ideas come up later. You might use examples such as the relationship between the force needed to open a door and the distance from the hinge where force is applied, or the relationship between weights balanced on a seesaw and their distances from the fulcrum.

Resources

Discovering Algebra Resources

Calculator Notes 1E, 1H, 1I, 3A, and 3B

Teaching and Worksheet Masters
 Lesson 3.2 Longest Ship Canals
 Lessons 3.2 and 3.4 Direct and Inverse Variation
 Lesson 3.3 Apartment Floor Plan

Assessment Resources A and B
 Quiz 1 (Lessons 3.1 and 3.2)
 Quiz 2 (Lessons 3.3 and 3.4)
 Chapter 3 Test
 Chapter 3 Constructive Assessment Options
 Chapters 1 to 3 Exam

More Practice Your Skills for Chapter 3

Condensed Lessons for Chapter 3

Other Resources

www.keymath.com/DA

Materials

• centimeter graph paper
• centimeter rulers
• 12-inch rulers
• nickels (9 per group)
• pencils
• tape
• meterstick or metric tape measure
• multiple-speed bicycles

Pacing Guide

	day 1	day 2	day 3	day 4	day 5	day 6	day 7	day 8	day 9	day 10
standard	3.1	3.2	quiz, 3.3	3.3	3.4	quiz, 3.5	3.5	review	assessment	
enriched	3.1	3.2	quiz, 3.3	3.3, project	3.4	quiz, 3.5	3.5	review, TAL	assessment	
block	3.1, 3.2	quiz, 3.3	3.4, project	3.5, review	assessment					

The Mathematics gives a survey of the mathematics of the chapter that places the mathematics in context and discusses mathematics topics in greater depth or at a higher level than is presented in the student text.

Pacing Guides let you see at a glance which lessons will require two days and when you might use a quiz or enrich your curriculum with a project.

To promote student research, this publisher **web site** has links to useful and interesting sources of data for chapter activities and information-packed sites about related topics.

The **Lesson Outline** helps you structure the class period and lets you know whether you should plan to spend two days on the lesson.

Planning helps you complete lesson plans, gather materials, and prepare worksheets, transparencies, or calculator notes.

Objectives for you summarize the objectives stated at the beginning of each chapter.

Teaching gives practical help in guiding the investigation, facilitating student sharing, evaluating student progress, and explaining new mathematical ideas.

One step is for classes experienced at investigating. It is an alternative to the more guided several-step investigation.

Help is keyed to the **steps** of the investigation.

The **National Council of Teachers of Mathematics Standards** addressed by this lesson are listed. The NCTM Principles as they apply to this book are discussed on page xvii.

The **Lesson Objectives** help you know what students should gain from the lesson.

Suggestions on how to use the fine-art image to introduce the chapter are given here as well as more information about the specific image or the area of the arts it represents. Questions for class discussion are also suggested.

Reproduced textbook page (LESSON 2.3)

LESSON

2.3

Proportions and Measurement Systems

Have you ever visited another country? If so, you needed to convert your money to theirs and perhaps some of your measurement units to theirs as well. Many countries use the units of the Système Internationale, or SI, known in the United States as the metric system.

So instead of selling gasoline by the gallon, they sell it by the liter. Distance signs are in kilometers rather than in miles, and vegetables are sold by the kilo (kilogram) rather than by the pound.

Vegetables are sold at a French market.

Investigation
Converting Centimeters to Inches

You will need
- a yardstick or tape measure
- a meterstick or metric tape measure

In this investigation you will find a ratio to help you convert inches to centimeters and centimeters to inches. Then you will use this ratio in a proportion to convert some measurements from the system standard in the United States to measurements in the metric system, and vice versa.

Step 1 Measure the length or width on each of six different-sized objects, such as a pencil, a book, your desk, or your calculator. For each object, record the inch measurement and the centimeter measurement in a table like this.

Inches to Centimeters

Object	Measurement in inches	Measurement in centimeters

Step 2 Enter the measurements in inches in your calculator's list L₁ and the measurements in centimeters in list L₂. In list L₃ enter the ratio of centimeters to inches, $\frac{L_2}{L_1}$, and let your calculator fill in the ratio values. [▶ See **Calculator Note 1I**. ◀]

Step 3 How do the ratios of centimeters to inches compare for the different measurements? If one of the ratios is much different from the others, recheck your measurements. Choose a single representative ratio of centimeters to inches.

Ratios should be about $\frac{2.54 \text{ cm}}{1 \text{ in.}}$.

NCTM STANDARDS

CONTENT	PROCESS
• Number	• Problem Solving
• Algebra	• Reasoning
Geometry	• Communication
• Measurement	• Connections
Data/Probability	• Representation

LESSON OBJECTIVES
- Review the English measurement system and the metric system
- Convert measurement units with conversion factors
- Convert measurement units with dimensional analysis

LESSON 2.3 · PLANNING

LESSON OUTLINE

One day:
- 25 min Investigation
- 10 min Sharing
- 5 min Example
- 5 min Closing
- 5 min Exercises

MATERIALS
- yardstick or tape measure (1 per group)
- meterstick or metric tape measure (1 per group)
- Calculator Note 1I

TEACHING

As they convert between different measurement units in this lesson, students use ratios called *conversion factors*. Products of several conversion factors can be set up to do more complicated conversions, in a process called *dimensional analysis*.

The name *Système Internationale* is French for the *International System* of measurement units. The SI is used in most science textbooks.

Guiding the Investigation

One step [Ask] "In as many ways as possible, decide how many centimeters are equivalent to an inch." As you circulate, encourage thinking beyond measuring the yardstick with the meter stick.

Step 1 Be sure that students understand how to read the inch markings. Often the fractional parts of an inch give students trouble.

LESSON 2.3 Proportions and Measurement Systems **105**

Reproduced textbook page (CHAPTER 2)

CHAPTER

2

CHAPTER 2 OBJECTIVES

- Learn to set up and solve proportions in which a variable represents an unknown number
- Use proportions to help make predictions from data gathered for the capture-recapture method
- Change measurement units through conversion factors and dimensional analysis
- Draw and interpret circle graphs and relative frequency graphs
- Determine observed and theoretical probabilities, and see how the former approximates the latter
- Use probabilities to describe patterns when outcomes are random

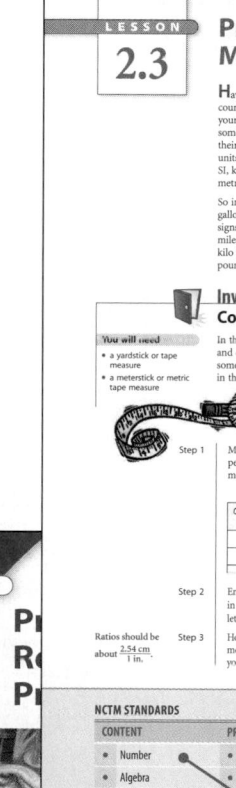

Murals are just one of the many art forms around us that come to life with the help of ratios and proportions. To plan a mural, the artist draws sketches on paper then uses ratio to enlarge the image to the size of the final work.

OBJECTIVES

In this chapter you will
- use proportional reasoning to understand problem situations
- create and interpret circle graphs
- explore probability

Most students have constructed pictures using enlargement grids; they capture the drawing fragment within the small square and transfer it to the large square. The enlarged drawing itself is called a *cartoon*. Ask, "If the artist first worked on paper that was about 18 inches by 24 inches, what did he or she have to multiply by to create a large drawing to transfer the image to the building?" [Reasonable answers are between 25 and 35.] Sketches for modern murals can be projected on a wall, which also involves an enlargement by a ratio.

You might also ask, "How are the real students like their images on the mural?" [The ratios of corresponding lengths are all the same. Or, the corresponding pairs of lengths are proportional. The images are about 3 times the height of a student. They look a lot bigger because they appear to have a volume of about 3³ times that of a student.] You might also ask, "What is the ratio of real students in the picture to images in the mural?" or "What is the ratio of children pictured on the mural to children?" [The ratio is 7 students to 4 images of children on the mural.]

Assessing Progress helps you evaluate student progress on skills from earlier lessons.

Vocabulary and other **language** points are explained.

Alerts warn of possible student errors or misunderstanding.

A key question will focus students' attention or check on their understanding.

SHARING IDEAS

If students use a variety of sizes of paper, they will come up with different systems of equations, but the solutions will be the same. Have any such variety represented in presentations of work.

Ask if there's another way to check the solutions. Students might suggest using a ruler to measure the paper clips and pennies.

Assessing Progress
Watch for the ability to follow directions, make careful measurements, and set up a **system of two linear equations.**

EXAMPLE B
This example shows how to eliminate a variable in a system of equations by multiplying the equations by different numbers.

[Language] UPC stands for *Universal Product Code.* It's the code that identifies each manufactured product and usually appears on the package as a bar code.

[Alert] Be sure students understand where the two equations come from. Using dimensional analysis might help.

[Ask] "How do you choose the numbers by which to multiply both sides of the equation?" [Students might use the coefficients themselves, opposites of coefficients, the products or least common multiples of coefficients.]

Part b of the solution illustrates how an arrow can mean *i*... Remind students that the ... arrow is more than a vagu... between steps in a process... required, again stress the ... logical correctness in check... a solution.

You might ask students to ... the same system by substit... and think about when one ... method is preferable to the ... A few more examples mig... them see that substitution ... preferable if the coefficie... variable in one equation is ...

326 CHAPTER 6 System...

The goal of the elimination method is to get one of the variables to have a coefficient of 0 when you add the two equations. Sometimes you must first multiply one or both of the equations by some convenient number before you combine them. If you start with additive inverses such as s and $-s$ in Example A, then you can simply add the equations.

EXAMPLE B The makers of FastBreak breakfast cereal offer a basketball in a promotional giveaway. There are two ways to get it. In Offer A you earn the basketball with 2 UPC symbols and $7. With Offer B you need 10 UPC symbols and $4. FastBreak has collected 1234 UPC symbols and $1,405. How many basketballs must it send out?

a. Write a system of equations that models this problem.

b. Use elimination to solve this system.

c. Check your solution.

▶ **Solution** It is helpful to organize the information in a table first. Let a represent the number of basketballs sent out in Offer A, and let b represent the number of basketballs sent out in Offer B.

Basketball Giveaway

	Offer A (per basketball)	Offer B (per basketball)	Total
UPC symbols	2	10	1234
Dollars	7	4	1405

a. These equations describe the number of UPC symbols and the amou... money collected:

UPC symbols $\qquad 2a + 10b = 1234$

dollars $\qquad 7a + 4b = 1405$

b. To eliminate a from the resulting equation you must make its coeffic... additive inverses, that is, numbers with opposite signs. If you multipl...

Eminently readable feature-by-feature lesson commentary appears in the wrap-around space at the point most relevant for helping you plan, teach, and guide students as they build understanding.

---

pennies in it, you might say there is "no mode" because no year occurs most often. How many modes does your data set have? What are they? Does your mode have to be a whole number? *Modes will vary but must be a whole number; the sample data has a mode of 1991.*

Step 5   Draw a square around the year corresponding to the mode(s) on the number line of your dot plot. Label each value "mode."

Step 6   *Mean of sample data is 1990.*

Step 6   Find the sum of the mint years of all your pennies and divide by the number of pennies. The result is called the mean. What is the mean of your data set?

Step 7   Show where the mean falls on your dot plot's number line. Draw an arrowhead under it and write the number you got in Step 6. Label it "mean."

Step 8   Now enter your data into a calculator list, and use your calculator to find the mean and the median. Are they the same as what you found using pencil and paper? [▶ See Calculator Notes 1A and 1B. Refer to Calculator Note 1J to check the settings on your calculator. ◀]

Save the dot plot you created. You will use it in Lesson 1.3.

Measures of Center

Mean
The mean is the sum of the data values divided by the number of data items. The result is often called the average.

Median
For an odd number of data items, the median is the middle value when the data values are listed in order. If there is an even number of data items, then the median is the average of the two middle values.

Mode
The mode is the data value that occurs most often. Data sets can have two modes (bimodal) or more. Some data sets have no mode.

Each measure of center has its advantages. The mean and the median may be quite different, and the mode, if it exists, may or may not be useful. You will have to decide which measure of center is most meaningful for each situation.

EXAMPLE This data set shows the number of people who attended a movie theater over a period of 16 days.

{14, 23, 10, 21, 7, 80, 32, 30, 92, 14, 26, 21, 38, 20, 35, 21}

a. Find the measures of center.

b. The theater's management wants to compare its attendance to that of other theaters in the area. Which measure of center best represents the data?

Step 4   "No mode" is relative to the size of the data set. If you have one thousand distinct data values, having five to ten modes may be worthwhile information. But if you have only 12 data values, having five modes is not very informative.

Step 6   Adding up these four-place numbers may challenge some students' patience. Encourage them to use only the last two places for the number of years since 1900 (or even 1970) rather than the full dates. Be sure they realize they can use the symbol -//- to indicate a "break" in the x-axis between extreme outliers and the rest of the pennies.

Step 7   [Link] You might mention that if the horizontal number line were a beam and the data points were weights, the balance point would be at the mean.

If you're saving Step 8 for a second day on this lesson, collect handwritten data for later entry into calculators.

Step 8   [Alert] Watch for confusion about the word *mode;* the mode on the graphing calculator (Calculator Note 1J) is not related to the statistical mode of a data set. [Language] The word *mode* has several meanings; *mean* also has several meanings.

SHARING IDEAS

As you ask groups to copy their dot plots onto transparencies, include some with an even number and others with an odd number of data points. If possible, include one whose data have outliers and one with bimodal data.

[ESL] Ask students to translate *mean, median,* and *mode* into their first language. In Spanish the words would be *promedia* (mean), *intermedio* (median), and *el más común* (mode).

Links are made to other parts of mathematics or to the real world.

Suggestions in **Sharing Ideas** help you guide students as they share and discuss investigation results.

You are prompted to help students who are new users of English.

NCTM STANDARDS

CONTENT	PROCESS
• Number	• Problem Solving
Algebra	• Reasoning
Geometry	Communication
Measurement	• Connections
• Data/Probability	• Representation

LESSON 1.2 Summarizing Data with Measures of Center   **45**

A problem from the chapter is suggested as a way to **review** all the new ideas.

In **Closing the Lesson** the new mathematical ideas are reviewed and related.

Building Understanding suggests how you might assign exercises and how to anticipate difficulties and offer help.

This chart indicates ways to use the exercises.

Helping with the Exercises indicates errors to watch for or ways to help students having difficulty with an **exercise.**

Annotations that appear in magenta on the student page or in wrap-around text give answers or suggestions for possible answers.

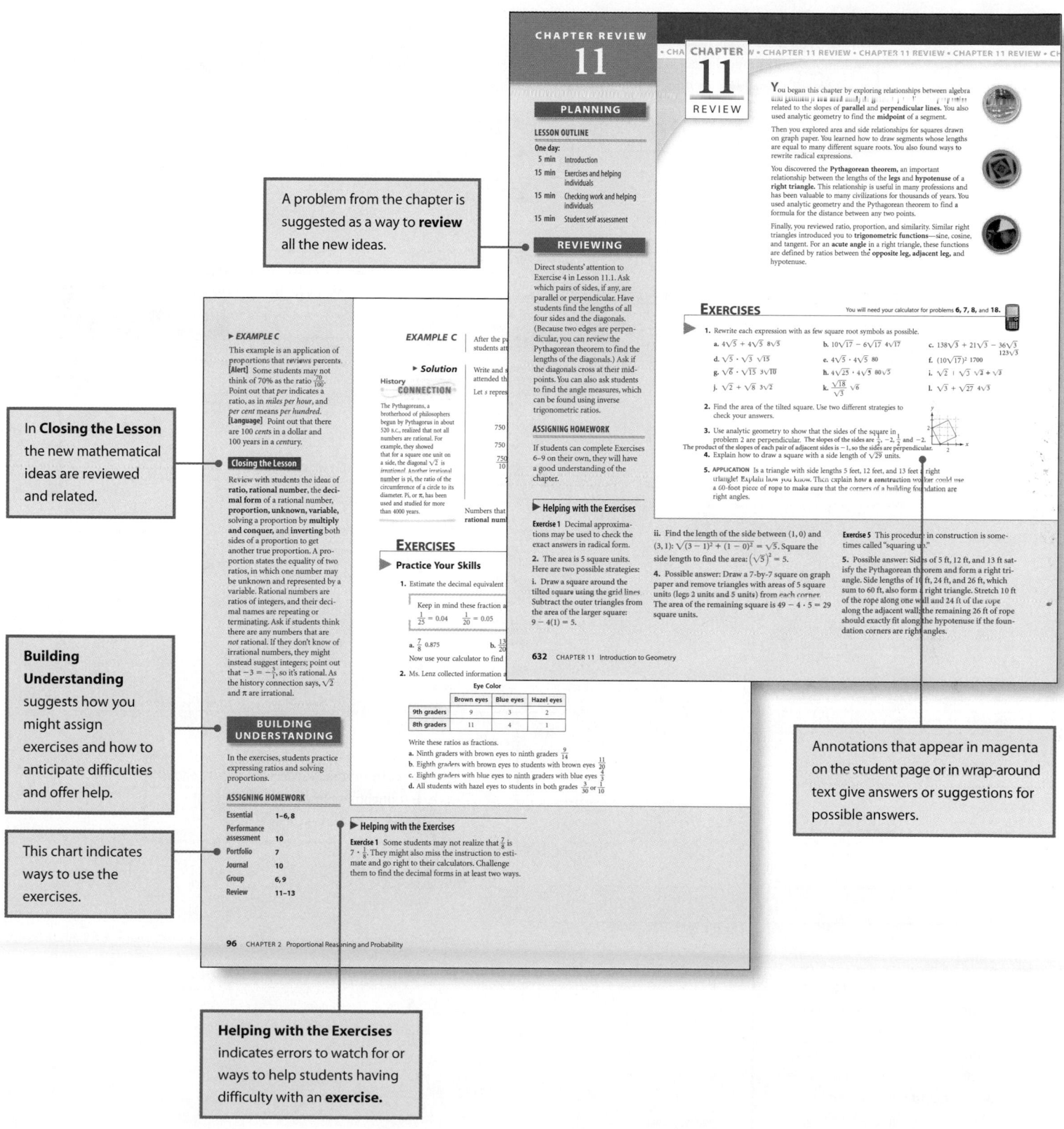

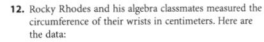

Additional information and answers are presented for Take Another Look, which offers a new perspective on one of the chapter concepts.

---

12. Rocky Rhodes and his algebra classmates measured the circumference of their wrists in centimeters. Here are the data:

15.2 14.7 13.8 17.3 18.2 17.6 14.6 13.5 16.5
15.8 17.3 16.8 15.7 16.2 16.4 18.4 14.2 16.4
15.8 16.2 17.3 15.7 14.9 15.5 17.1

a. Rocky made the stem plot shown here. Unfortunately, he was not paying attention when these plots were discussed in class. Write a note to Rocky telling him what he did incorrectly.

b. Make a correct stem plot of this data set.

c. What is the range of this data set?

13	8 5
14	7 6 2 9
15	2 8 7 8 7 5
16	5 8 2 4 4 2
17	3 6 3 3 1
18	2 4

Key
10 | 4   means 10.4 cm

project

BASEBALL RELATIONSHIPS

Baseball statistics, or "stats," often show relationships between variables. These relationships could be used to identify the best players on a particular Major League team, or in an entire league, and may influence the strategy of managers. These scatter plots show stats of 50 National League players from the 1996 season.

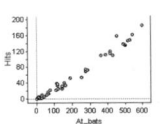

Find the stats of any one team for a particular season. Your data can come from library research, or you can use the links at **www.keymath.com/DA** . Briefly review all the stats and make at least two conjectures about relationships like "If a player appears in more games, he will have more times at bat." Use scatter plots to see if relationships truly exist between the variables. For each conjecture, your project should include

▸ A sentence stating the conjecture and a scatter plot to accompany it.

▸ An explanation of how the scatter plot either supports your conjecture or casts doubt on it. If a relationship does appear to exist, why do you think this is so?

Conclude your project by explaining how you think a baseball manager could use your graphs to strategize for a game.

Fathom
These scatter plots were made with Fathom. With Fathom, you can download baseball stats directly from the Internet. Scatter plots are easy to make with this software, which allows you to investigate relationships between many more pairs of variables.

▸ **Take Another Look**

The graph shows the change in the hiker's speed over time. The contour map shows the change in the hiker's elevation as she follows the dotted line. It also shows the horizontal distance she has traveled from her starting point.

The graph shows rate of change in speed as the steepness (slope) of a line. At first the speed is steadily decreasing, then it is increasing, then it remains constant. The contour map shows rate of change in elevation by the distance between the contour lines. When the lines are very close, the elevation is changing quickly.

If a scale were provided, you could infer distance from the graph by estimating the average speed up to a certain point and multiplying that by the time at that point. Students might refer to the formula distance equals rate (speed) times time ($d = rt$). You could measure distance on the contour map using the map scale.

Student sketches of the graph of (*distance, time*) should show three sections. In each the graph is increasing. In the first section, the graph is a curve that is concave down showing a steadily decreasing speed. The second section of the graph is concave up since the speed is increasing, and the third section is a steep straight line showing a constant, fast speed.

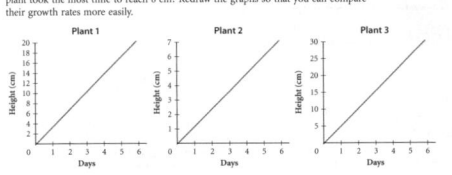

The graph and the contour map together show you why the speed of the hiker was decreasing at the beginning. She was climbing a hill. As she began to go down the hill, she kept moving faster and faster until she was running.

248   CHAPTER 4   Linear Equations

TAKE ANOTHER LOOK

The picture at right is a **contour map** that reveals the character of the terrain. All points on an **isometric line** are the same height in feet above sea level. The graph below shows how the hiker's walking speed changes as she covers the terrain on the dotted-line trail shown on the map.

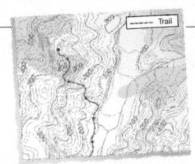

What quantities are changing in the graph and in the map?

How does each display reveal rate of change?

How could you measure distance on each display?

What would the graph sketch of this hike look like if distance were plotted on the vertical axis instead of speed?

What do these two displays tell you when you study them together?

Sediment layers form contour lines in the Grand Canyon.

IMPROVING YOUR REASONING SKILLS

Did these plants grow at the same rate? If not, which plant was tallest on Day 4? Which plant took the most time to reach 8 cm? Redraw the graphs so that you can compare their growth rates more easily.

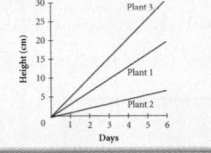

Plant 1   Plant 2   Plant 3

Both approaches and answers to each enrichment Improving Your Skills problem are given.

12a. Rocky made several takes. The data values nee organized in increasing o The key should show actu values from the data.

12b.
13	5 8
14	2 6 7 9
15	2 5 7 7 8
16	2 2 4 4 5
17	1 3 3 3 6
18	2 4

Key
13 | 5 means 13.5 cm

12c. 18.4 − 13.5 = 4.9 c

IMPROVING REASONING SKILLS

The differences in the vertical scale (height) indicate that the plants are growing at different rates. Plant 3 is the fastest growing; it is about 20 cm tall in 4 days. Plant 2 is slowest growing, it takes almost 6 days to reach 7 cm. This should suggest to students that the steepness of a line is relative to the scales on the axes. They have seen this many times on their calculators.

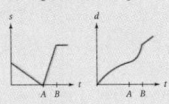

Supporting the project

MOTIVATION

Baseball fans often memorize the stats of their favorite players or teams. Managers analyze the stats and look for relationships among them.

OUTCOMES

▸ The conjecture refers to a relationship between two variables.
▸ The scatter plot includes data from the same two variables.
▸ The explanation uses good mathematical language.

▸ The explanation defends the claim about whether the scatter plot supports the conjecture or not.
• Several conjectures are explored.
• Writing and thinking are exceptionally good.

A short prompt is designed to get students interested in the project. Outcomes at two levels are listed. Outcomes you should expect of all projects are indicated with triangular bullets. Outcomes with round bullets will be found only in the best projects.

NCTM Principles and Standards 2000

As part of the Key Curriculum Press *Discovering Mathematics* series, *Discovering Algebra: An Investigative Approach* exemplifies the *Principles and Standards for School Mathematics* set forth by the National Council of Teachers of Mathematics (NCTM) in 2000.

The Equity Principle

Excellence in mathematics education requires equity—high expectations and strong support for all students. (*Principles and Standards for School Mathematics.* Reston, Virginia: National Council of Teachers of Mathematics, 2000. page 12.)

Discovering Algebra grew out of the belief that all students—not just a select few—are capable of learning mathematics. Research shows that most learning takes place while students are actively engaged in the learning process. That's why *Discovering Algebra* is structured around investigations and activities. Through these experiences, students who may have difficulty memorizing formulas or doing calculations come to understand algebra concepts, see relationships, and explain and apply them.

Different students, however, need a variety of experiences. Some need extra skills practice; *Discovering Algebra* provides that. Others learn from puzzles and different perspectives; *Discovering Algebra's* sections called Improving Your Reasoning Skills and Take Another Look provide such challenges. Some students face learning disabilities or language barriers; while others learn quickly and can explore further. This *Discovering Algebra Teacher's Edition* provides advice for accommodating such differences, partly through **[Alert]**, **[Language]**, and **[ESL]** prompts, and through **[Ask]** prompts that can help clarify or challenge.

The Curriculum Principle

A curriculum is more than a collection of activities: it must be coherent, focused on important mathematics, and well articulated across the grades. (NCTM *Principles and Standards,* page 14.)

The investigations and activities in *Discovering Algebra* are coherently organized and carefully crafted. They promote an intuitive understanding of algebra concepts and objects. Only after students come to understand a concept through experience are they introduced to the appropriate symbols and given opportunities to practice mechanics and problem solving.

Moreover, the investigations help students focus on a major mathematical idea in each lesson. The central idea is summarized in each lesson's closing in the *Discovering Algebra Teacher's Edition.*

This algebra course does not stand alone. *Discovering Algebra* is part of the Key Curriculum Press *Discovering Mathematics* series, which articulates a three-year high school curriculum that includes the most important ideas of algebra and geometry.

As part of this series, *Discovering Algebra* ties algebra to other mathematical topics. It introduces data analysis techniques; it enhances proportional reasoning in several contexts; it provides geometric investigations of topics from coordinate geometry; it revisits properties of arithmetic before generalizing them to algebraic principles; and it strengthens reasoning skills.

The *Principles* state (p. 15), " . . . teachers also need to be able to adjust [the curriculum] and take advantage of opportunities to move lessons in unanticipated directions." To help you maintain this flexibility, this *Discovering Algebra Teacher's Edition* includes ideas of the form "If students ask . . . then you might . . . ."

The Teaching Principle

Effective mathematics teaching requires understanding what students know and need to learn and then challenging and supporting them to learn it well. (NCTM *Principles and Standards,* page 16.)

The *Discovering Algebra Teacher's Edition* assumes that you are a creative professional, and don't just follow a script. It supports four primary aspects of professional teaching outlined in the *Principles:*

Your mathematical understanding is supported by the essay titled The Mathematics in each chapter interleaf of the *Discovering Algebra Teacher's Edition.* Each essay summarizes the chapter, places it in historical context, and relates new material to what your students already know.

The *Discovering Algebra Teacher's Edition* helps you **create an effective learning environment** by offering advice throughout. For example, later in this introduction you can find tested ideas about how to promote inquiry through cooperative learning.

While interacting with students, you may encounter confusion or questions that hint at a misunderstanding. The commentary in *Discovering Algebra Teacher's Edition* alerts you to common comprehension issues, suggests reasons why they arise, and proposes questions (with an **[Ask]** prompt) and alternative representations that you might use to help the students.

Occasional advice about **analysis and reflection** on your teaching helps save you time in engaging in these important activities.

The Learning Principle

Students must learn mathematics with understanding, actively building new knowledge from experience and prior knowledge. (NCTM Principles and Standards, page 20.)

To achieve deep understanding, students need to connect new ideas to old ones in personal ways. To make those connections, *Discovering Algebra* uses simple, hands-on investigations in age-appropriate contexts.

The investigations of *Discovering Algebra* require students to make sense of numbers, algebraic expressions, patterns, functions, solution procedures, and answers. They encourage the use of multiple representations—numeric, graphic, symbolic, and verbal. Thus, the investigations lead to deep understanding through active engagement. Working and communicating with their peers strengthens students' understanding of concepts and helps them build on their prior knowledge.

Moreover, *Discovering Algebra* takes students beyond learning algebra. First, it helps students learn to be resourceful, flexible problem solvers. Students develop ways of thinking about mathematics. They become open to new ideas, examining alternatives, dealing with real data and the uncertainties they introduce, thinking critically, and working independently.

Second, the text empowers students. The investigations ask students to develop their mathematical power by exploring, verbalizing and writing their reactions, using graphs to illustrate data, and drawing conclusions. Students are also empowered because the technology helps them hurdle some unproductive paper-and-pencil barriers. Many students feel more in control as they decide when and how to use appropriate technology to inform their mathematics conjectures by manipulating, investigating, and summarizing data. More students will thus enjoy a successful journey through one of the most important gateways to their future.

The Technology Principle

Technology is essential in teaching and learning mathematics; it influences the mathematics that is taught and enhances students' learning. (NCTM Principles and Standards, page 24.)

Technology is not a substitute for conceptual understanding, but it is one way to help deepen that understanding. Technological tools allow students to get beyond algorithmic barriers and focus on concepts. Graphing calculators and dynamic software are tools for exploring and investigating—tools that encourage students to ask "What if . . ." questions because it's easy to pursue them; tools that let students focus on decision making, reflection, reasoning, and problem solving; tools that erase artificial lines separating algebra, geometry, and precalculus.

Discovering Algebra was written to make powerful use of graphing calculators. For virtually every investigation students use calculators in addition to paper and pencil. Extensive calculator notes for various graphing calculators aid students as they learn to use the calculator. These notes also include help for students as they use motion sensors (Texas Instrument's CBL or CBR, or Casio's EA-100 probe-based data collector) to explore, graph, and make sense of time/distance/rate situations.

Suggestions for projects include some that students can do with The Geometer's Sketchpad or Fathom Dynamic Statistics software. To help students make connections, *Discovering Algebra* includes numerous exercises from situations all over the world. Links at www.keymath.com/DA lead students to Web pages related to those situations and help students collect and update data.

The Assessment Principle

Assessment should support the learning of important mathematics and furnish useful information to both teachers and students. (NCTM Principles and Standards, page 22.)

Often the focus of teaching is on final assessment: quizzes and tests with right or wrong answers. But also important is ongoing assessment of how students are thinking, so they can see how well they're learning and you can modify your teaching. *Discovering Algebra* supports both kinds of assessment.

During investigations you'll assess through observation as you move around the classroom and as students present ideas. In this *Discovering Algebra Teacher's Edition,* each lesson includes a section that lists what you can assess as you observe students while working or presenting ideas. The list includes concepts and skills from earlier lessons, as well as visualization, problem solving, and group work. On pages xxx–xxxi of this introduction you can find advice about assessing group work. This formative assessment leads to adjustments of your pedagogical plans and strategies.

Students can learn from self-assessment as well. Ideas for self-assessment appear at the end of each chapter of the student book. Ancillaries for the course include many resources—quizzes and tests (with answers) for skill assessment, and constructive assessments (with rubrics) to evaluate deeper understanding.

The Content and Process Standards

Each lesson addresses many of the NCTM content and process standards. At the beginning of each lesson, the *Discovering Algebra Teacher's Edition* lists the relevant standards.

Teaching with *Discovering Algebra*

In today's society, everyone, not just those who are good at mathematics, must understand algebraic concepts and be able to work with technology. Changes in society and changing expectations of employers have required changes in the algebra curriculum. *Discovering Algebra* answers that need for change. It teaches a range of skills more applicable to today; it works toward building technological understanding; it connects algebra to other areas of math; and it encourages an informal understanding of algebraic ideas.

Michael T. Battista of Kent State University writes that algebra teaching should

> focus on the basic skills of today, not those of 40 years ago. Problem solving, reasoning, justifying ideas, making sense of complex situations, and learning new ideas independently—not paper-and-pencil computation—are now critical skills for all Americans. In the Information Age and the Web era, obtaining the facts is not the problem; analyzing and making sense of them is. ("The Mathematical Miseducation of America's Youth," *The Phi Delta Kappan.* February, 1999).

Toward that end, *Discovering Algebra* can help you teach a practical blend of technology-related and paper-and-pencil problem-solving skills. With the assistance of graphing calculators, your students can study new and different algebraic topics and apply them in our technologically rich society.

Discovering Algebra uses technology, along with applications, to foster a deeper understanding of algebraic ideas. The explorations emphasize symbol sense, algebraic manipulations, and conceptual understandings. The investigative process encourages the use of multiple representations—numerical, graphic, symbolic, and verbal—to deepen understanding for all students and to serve a variety of learning styles. Explorations from multiple perspectives help students simplify and understand what were formerly difficult algebraic abstractions. Investigations actively engage students to make personal and meaningful connections to the mathematics they discover.

Discovering Algebra contains many connections to other areas of mathematics, especially geometry and statistics. Moreover, because lessons are based on cooperative learning, students have opportunities to share meaningful links to work they have completed in the past.

Traditional algebra teaches skills and ideas before examples and applications. The investigative approach works the other way. Interesting questions and simple hands-on investigations precede the introduction of formulas and symbolic representations. By providing meaningful contexts for students, the investigations motivate relevant algebraic concepts and processes. Moreover, these investigations are accessible. They use inexpensive and readily available materials, require little prerequisite technical knowledge, and follow simple procedures. Students can conduct them with a minimum of direction and intervention from you.

Teaching with *Discovering Algebra* decreases the time students spend on rote memorization, teacher exposition, and extended periods of paper-and-pencil drill. It changes the rules for what is expected of students and what they should expect of their teacher. Thus, teaching from *Discovering Algebra* requires untraditional thinking and behavior and an untraditional classroom. Success depends on your sensitivity, patience, enthusiasm, and determination.

Using Technology

Discovering Algebra makes extensive use of graphing calculators to capitalize on these strengths.

- Graphing calculators can generate many examples quickly, allowing students to focus on the meaning of and patterns in the results rather than on getting those results alone. Scatter plots and families of graphs make patterns easier to see, especially for students who tend to avoid more abstract mathematics.

- Students gain experience interpreting the output of technology, an important skill in today's world. *Discovering Algebra* helps students focus on the meaning of the numbers and expressions they enter into a calculator and of the results they generate. Finding meaning in numbers, variables, expressions, tables, and graphs is a year-long challenge for everyone. As the teacher, you will ask for the real-world meaning of such things as a term in a sequence, the slope of a line, the *y*-intercept, and possible outcomes.

- With a calculator, students can take on more realistic problems, which may require more difficult computations. Students can explore variables that actually vary and functions that describe real-world phenomena.

- The graphing calculator allows some students to hurdle error-plagued paper-and-pencil barriers. Even students who enter this course with poor computational and manipulation skills will find that, through their graphing calculator screens, they can visualize mathematical processes and concepts that previously eluded them.

- Many students will want to perform more than one calculation on their graphing calculators. They will ask, "But what if we changed just this one thing?" Routine exercises become open-ended ongoing mathematical explorations, initiated by students.

Cooperative Learning

Students with a wide range of backgrounds flourish in a course that catches their interest. *Discovering Algebra*'s emphasis on technology can help generate that interest. Having students work on investigations within small cooperative groups can engage them even more. The investigations in *Discovering Algebra* encourage students to plan together, brainstorm, determine and organize tasks, and communicate their individual and collective results.

If you haven't already experienced the power of teaching with cooperative groups, you might have several questions: Why use groups? How big should they be? How do I decide group membership? How often do they change? Do they require tables? How do I teach group cooperative skills? How should a class period be structured? How do I behave as a teacher?

Cooperative learning has many benefits:

- Students learn and practice the essential life skill of working with others. In a cooperative learning environment, the group cooperative skills are an integral part of the curriculum.

- Students are exposed to more ideas for solving problems. In solving challenging problems, "two heads are better than one."

- Students who are good in social situations can gain confidence in their mathematical abilities, even if they were once unsuccessful in mathematics.

- Students understand an idea more deeply if they have to articulate it for someone else.

- Students learn to solve more complex problems than they could if they didn't have a group to contribute different areas of expertise.

- Groups working in a supportive atmosphere can provide quicker feedback on ideas than a single teacher could offer.

- Some students will be more willing to contribute to a small group than to a full class, thus practicing oral communication skills.

Groups of two, three, or four are usually best. If you or your students are new to group work, you might start with pairs. Later you can move to groups of four, with groups of three as needed to make it come out evenly. Most investigations in *Discovering Algebra* are appropriate for groups of three or four.

How do you assign groups? At first, assign them randomly. As you get to know the students, think about who works well with whom. Mixing abilities usually works, but putting the strongest and weakest students together can frustrate both. Keep student groups together long enough for the members to get to know each other, but change group assignments for variety and for a healthy dynamic. You might reorganize the groups at the beginning of each chapter.

(continued)

(Cooperative Learning continued)

Appropriate furniture helps facilitate group behavior. Tables work well because they allow students plenty of workspace. But you can have an effective cooperative setting with desks also. Be sure that the desktops are brought together to form as close to a single surface as possible.

Discuss the reasons for working in groups. In the working world, people often work in groups, consult others for help, and rely on others to complete part of a large project. The ability to work with a group is a valuable skill. Try to counter some students' ideas that learning is competitive, that a course is about delivering information that only the teacher knows, or that stronger students have to slow down to teach weaker students.

Develop with your students some specific guidelines to follow for productive work. These might include the following: be considerate, listen without interrupting, ask questions when needed, help others in the group, and make sure everyone in the group understands the ideas well enough to present them to the class.

Appropriate assignments, such as the investigations in *Discovering Algebra,* are important for group work. Some assignments include problems that elicit many ideas, while others have tasks that group members can divide among themselves.

Hold each group fully accountable. End each group session with presentations to the class. You won't have time for all groups to present each time, but each student should be prepared.

Hold individuals responsible for group participation. Make part of the individual grade dependent on how that person contributes to the group. Several days a week, as you move around the room, carry a copy of the roll and record a plus or minus sign for participation. Make these marks part of each student's grade.

As the course progresses, you might extend your observation sheet to include such items as contributing ideas, asking questions, giving directions, actively listening, expressing support, encouraging members to participate, summarizing, talking through problems, and justifying viewpoints. Let your students know when you are looking for specific contribution skills, so that they can learn from the assessment.

Trust the group process. Give plenty of time for the group to correct mistakes. If a student asks you a question, turn it back to the group. If some students are causing behavior problems, try to facilitate a group solution: review the guidelines with the group or help students clarify their roles. In some cases you will need to call a student aside for a discussion of poor behavior. When you do so, also talk with the other group members individually about your expectations. Sometimes it helps to point out that the groups will change and that students will have a chance to work with a different set of classmates soon.

Promoting Inquiry

The investigations in *Discovering Algebra* encourage students to inquire about relevant ideas and issues beyond the bounds of the course. Students have legitimate opportunities to experiment, hypothesize, measure, analyze, test, talk, write, explain, and justify their ideas. In short, they engage in real mathematics.

Inquiry-based classes go beyond engaging students in activities. They place students in the role of researchers. The quality of students' investigations is linked to the quality of their own inquiries. The motivation for pursuing answers to their own questions is very strong. In fact, it conforms to the old educational maxim, "Don't answer questions that students haven't asked." Being flexible challenges you and the students in several ways. An inquiry-based classroom requires several things from both you and your students.

- First, it challenges everyone to develop skills of problem posing as well as problem solving. Encourage the posing of new problems during Sharing and Closing. You can foster problem posing by using question openers such as those listed in the box.

- Second, develop a list of inquiry questions that includes some specific to the lesson you're ready to do next. Use those suggested in this *Teacher's Edition* for that lesson and those based on inter-lesson connections summarized in The Mathematics section of the *Teacher's Edition* chapter introductions. What questions does each lesson answer? Where did these questions arise in previous lessons?

- Third, students must take responsibility for their own learning and see that learning is an active, not a passive, process. *Discovering Algebra's* cooperative investigations are chosen for their value as vehicles to promote activity and help students construct their own knowledge. The interaction, discussion, questions, suggestions, and ideas that students offer while working in groups can benefit all members. Some students want more from you than to help them help themselves. It is not the way they have played the learning game. But it is the way a work environment runs. Their grades in the course, like their evaluations in the workplace, must take into account how well they're learning to do research in a team, not just the results of that research.

- Fourth, you have to play the role of an experienced co-researcher rather than someone with all the answers. Don't give too many hints. Give encouragement on the basis of good thinking, not just for right answers. Treat right answers as discussion topics until the class, the research team, agrees on them. As soon as you acknowledge a right answer, you often shut off thinking about that problem, even if students don't understand the answer. You will find that if you provide answers and explanations too quickly, students may continue to expect and depend on your answers.

- Fifth, students must not conceive of mathematics as a collection of facts and procedures. Many mathematical investigations, such as those in *Discovering Algebra*, don't have just one answer, and rarely is there only one valid approach to a solution. Justifying ideas and problem solving become more important than the actual answers. The goal is that students experience mathematics as a process of finding and connecting ideas. Let students know that the thinking and problem-solving skills they develop can serve them in all aspects of their lives. They are learning more than algebra.

- Sixth, as you plan, be flexible in responding to students' ideas. Spend planning time thinking of how students might respond to the problem under investigation. This *Teacher's Edition* can help you anticipate places where students may need help and give you good questions that will keep students thinking.

Inquiry Question Openers

What happens if …?	What's the largest/smallest …?
What if not …?	What are the properties of …?
Why …?	What other …?
How many …?	How do you know …?
In general, …?	Is it always true that …?
What do we mean by …?	Is it possible …?
Is there a relationship …?	How can you …?
Under what conditions …?	Is there a similarity between …?

Teaching with Cooperative Groups

If you are new to the investigative approach, the change of role may be the most difficult part of learning to create an inquiry-based classroom. You may be comfortable with situations in which you are the center of attention, following a standard script. Now you need to be a leader at problem-solving. You will be making professional judgments constantly in response to student contributions.

Don't assume that an emphasis on process over product leads to "anything goes." In fact, sloppy thinking is no more acceptable now than before. Indeed, careful thinking is a major goal of the course. Trust that everything in mathematics can be justified logically and that almost all conflicts between intuition and logic can be resolved on the side of logic.

How can you make your classroom student-centered? You need to keep the focus of your classroom on the students while continuing to be a role model and facilitator. After you give groups a few minutes to settle down and get started (a good time to take attendance), move among groups, observing carefully, encouraging as necessary. As you observe the working groups, sit down if possible so that attention isn't drawn to you. Say little. Don't be too quick to jump in and correct errors. If the other students in the group don't catch a common error, have a student present that error later so that the entire class can learn from it. If students ask you a question, reflect it back to the group. During student presentations, sit down and watch. Keep in mind that if you take the stage or answer a question, many students will stop thinking about that question. In the material presented with each chapter, this *Teacher's Edition* gives suggestions about particular difficulties students might have and how to respond to them.

As you listen to functioning groups, plan how to make use of students' ideas during Sharing. Who should present what? What questions should be asked and what points can you make in response? Through watching the groups at work, you can decide which of their approaches should be shared with the whole class and in what order. For example, you might decide that a group that actually constructed the next stage of a fractal design and counted areas should show their picture first; a student who used the area formula for triangles should describe that formula next; and a pair that found a pattern in the numbers should present last. Look for individuals who seem to understand various key steps particularly well. Be sensitive to the fact that sometimes students with poor calculation skills have creative ideas in problem solving. Keep in mind the suggestions for Sharing from this *Discovering Algebra Teacher's Edition*.

To facilitate good group work, go to any groups that don't seem to be on task. If joining them doesn't refocus them, ask about their progress. If they think they've finished the task, look at their work, ask questions, make suggestions, and challenge them to extend it. If they say they're stuck, ask one of them to describe what they've done, and ask others for their ideas about it. If they're not cooperating, remind them of the group process guidelines. Praise good group work and good thinking, even if it's not yet "on the right track."

Structuring the Class Period

The order of events in a class period can vary with the lesson, with your class's growing experience in independent learning, and according to how students are responding. If they're lethargic, have them act out the investigation. If they're overly excited, channel the energy into their investigation.

This *Teacher's Edition* includes suggested times for these stages of each lesson:

Introduction (5–10 minutes): Set the context for the investigation. Pose the problem and be sure the terms are clear. Many lessons in *Discovering Algebra* include examples to help students learn techniques and see the steps involved in solving real-world problems.

Investigation (15–50 minutes): Students work in groups while you observe, encourage, and craft plans for the rest of the class period. For some of the longer investigations, the Pacing Guide recommends spending two days on the lesson. The section Using This Chapter in the teacher pages that begin each chapter can help you decide which lessons you can skip, depending on the requirements of your curriculum, to give more time for other lessons.

Sharing (10 minutes): Selected students report to the class. You lead the asking of questions, elaborate on students' ideas, and praise good work. Lead into further examples.

Closing (5 minutes): Remind students of the main mathematical ideas and where they arose in the lesson. If cooperative learning is new to your students, you might lead a quick discussion on how the groups are functioning, what they can improve, and what they do well. Or, ask students to write briefly about what they learned or what they're confused about.

Exercises (5–10 minutes): Assign and begin work on the exercises.

For each lesson, the *Discovering Algebra Teacher's Edition* includes a section marked **One step,** which offers an alternative to the several-step investigation and perhaps to some examples. You can use the one-step approach for classes that have already become experienced at investigations. Even if your class needs more guidance at the beginning of the course, read through these alternatives to keep in view your goal of helping your students become more independent investigators. This will help you become better at finding "teachable moments" in students' ideas and experiences.

Communicating

Over the last decade, teachers and researchers have become aware that the ability to read mathematics and other technical material and to effectively communicate ideas not only enhances students' understanding of the concepts, but improves their other language abilities as well.

The authors have taken care to make *Discovering Algebra* readable. Students will read the steps of an investigation, the real-world contexts of the exercises, and other parts of the text. Ideally you won't tell them what they can read in the textbook, and you'll point out that they can use the textbook as a resource.

As you embark on teaching the course, be sensitive to students' inexperience at reading mathematics and, perhaps, English. You can take steps to help. Ask a student to read a brief passage or instruction aloud. If one student in the class or a group reads an instruction or problem aloud, the others (especially auditory learners) might benefit from hearing it. Don't ask students to read aloud anything that they haven't first read silently. You might, for example, ask all students to read through the instructions to themselves and then ask a student in each group to read it aloud. You might suggest that a group rotate readers counterclockwise to be sure everyone gets a chance.

It is also helpful to ask students to paraphrase. After one student reads aloud, ask another student to restate the instruction or problem in other words. Once they read a passage, help struggling students by emphasizing main ideas and relating them to what students have learned.

(continued)

Assessing

Assessment is your way of getting feedback about how well your students are learning. Students want to know how well they are doing, too. The materials accompanying *Discovering Algebra* support a variety of methods for assessing both processes and results.

While you circulate among groups or watch students' presentations, engage in informal assessment of students' previous learning. Each lesson's Assessing Progress section in the *Discovering Algebra Teacher's Edition* provides ideas of what to look for. Also, during class, note how effectively the lesson is proceeding. This allows you to make adjustments as you go. Throughout the *Teacher's Edition* you can find ideas for what to look for. You can also assess group process skills as described in the Cooperative Learning section above. And because *Discovering Algebra* emphasizes mathematics as a journey, journals can be especially useful for assessment.

You will also use more intentionally focused activities to evaluate students' learning. *Discovering Algebra Assessment Resources A* and *B* together include two versions of quizzes, tests, and Constructive Assessment options for each chapter as well as unit tests and a final exam. If you feel that in-class examinations don't allow you to assess the depth of students' understanding, you can use the projects in *Discovering Algebra* for assessment. Or you might have students prepare portfolios of their best work. The *Teacher's Edition* lists exercises that you might assign for this purpose. At the end of each chapter of the student text, the section called Assessing What You've Learned contains suggestions for student self-assessment. These will give you and students additional feedback on their learning.

You can find more details about assessment in the front matter for *Assessment Resources A* and *Assessment Resources B*.

(Communicating continued)

The real-life contexts of the material will help motivate the reading and enhance critical thinking skills as students solve problems that relate to things they understand. If students are having trouble understanding an exercise they read, re-emphasize the mathematical concepts involved in solving the problems. It is crucial that students recognize the underlying mathematical concepts and how they apply to certain situations.

Students can often increase their comprehension of a chapter if they outline the chapter and list the objectives and main ideas in their notebooks.

Writing and speaking clearly are also important. The process of deciding what to say or write deepens one's understanding of the ideas. As the old saying goes, "Writing is nature's way of telling us what we don't understand." The same holds true for careful speaking.

Critique students' writing and speaking to deepen their understanding. Cultivate in your students the habit of reflecting on what they are saying and writing. Encourage them not to turn in their scratch paper for a homework assignment, but to write up a careful presentation of their answer and its justification. When you grade homework or critique a presentation, ask for clarification of vague expressions or murky logic.

Encourage students to use graphics in communicating, both orally and in writing. They can create overhead transparencies, use computer graphics programs, or use the linking software that accompanies graphing calculators to incorporate graphical illustrations into oral presentations and written reports.

Pacing Guide for an Enriched Class

Chapter 0

day 1	day 2	day 3	day 4	day 5	day 6	day 7	day 8	day 9	day 10
0.1	0.2	0.3	project	0.4	0.5	review, TAL	assessment		

Chapter 1

day 1	day 2	day 3	day 4	day 5	day 6	day 7	day 8	day 9	day 10
1.1	1.2	1.2	1.3	1.4	1.4, project	quiz, 1.5	1.5	1.6	1.6

day 11	day 12	day 13	day 14	day 15
1.7, project	1.8	1.8	review, TAL	assessment

Chapter 2

day 1	day 2	day 3	day 4	day 5	day 6	day 7	day 8	day 9	day 10
2.1	project	2.2	2.3	2.4	quiz, 2.5	2.6	2.7, project	review, TAL	assessment

Chapter 3

day 1	day 2	day 3	day 4	day 5	day 6	day 7	day 8	day 9	day 10
3.1	3.2	quiz, 3.3	3.3, project	3.4	quiz , 3.5	3.5	review, TAL	assessment	

Chapter 4

day 1	day 2	day 3	day 4	day 5	day 6	day 7	day 8	day 9	day 10
4.1	4.2	4.3	4.4, quiz	4.4	4.5	quiz	project	4.6	project

day 11	day 12	day 13	day 14	day 15
4.8	4.9	project	review, TAL	assessment

Chapter 5

day 1	day 2	day 3	day 4	day 5	day 6	day 7	day 8	day 9	day 10
5.1	project, 5.2	5.2	5.3	5.4	quiz, 5.4	5.5	quiz, 5.6	5.6, project	5.7

day 11	day 12	day 13	day 14
5.8	5.8	review, TAL	assessment

Chapter 6

day 1	day 2	day 3	day 4	day 5	day 6	day 7	day 8	day 9	day 10
6.1	6.2	6.2	6.3	6.3	6.4	6.4	quiz, 6.5	6.6, project	6.7

day 11	day 12
review, TAL	assessment

Pacing Guide for an Enriched Class (continued)

Chapter	day 1	day 2	day 3	day 4	day 5	day 6	day 7	day 8	day 9	day 10
7	7.1	7.2	7.2, project	quiz, 7.3	7.4	7.4	7.5	7.6	7.6	quiz, 7.7
	day 11	**day 12**	**day 13**	**day 14**						
	7.7, project	7.8	review, TAL	assessment						
8	**day 1**	**day 2**	**day 3**	**day 4**	**day 5**	**day 6**	**day 7**	**day 8**	**day 9**	**day 10**
	8.1	8.2, project	8.3	8.4	8.5	8.6	8.6	8.7	review, TAL	assessment
9	**day 1**	**day 2**	**day 3**	**day 4**	**day 5**	**day 6**	**day 7**	**day 8**	**day 9**	**day 10**
	9.1	9.1, project	9.2	quiz, 9.3	9.4	9.5	quiz, 9.6	9.7	project, quiz	review, TAL
	day 11									
	assessment									
10	**day 1**	**day 2**	**day 3**	**day 4**	**day 5**	**day 6**	**day 7**	**day 8**	**day 9**	**day 10**
	10.1	10.2	quiz, 10.3	10.3	10.4	10.5	quiz, project	10.6	10.7	10.8
	day 11	**day 12**								
	review, TAL	assessment								
11	**day 1**	**day 2**	**day 3**	**day 4**	**day 5**	**day 6**	**day 7**	**day 8**	**day 9**	**day 10**
	11.1	11.2	quiz, 11.3	11.4, project	11.5	11.5, project	quiz, 11.6	11.6	11.7	11.7
	day 11	**day 12**	**day 13**							
	11.8	review, TAL	assessment							

Materials List

Material	Quantity	Material	Quantity
beans, red	1/4 cup per group	paper cups	3 per group
beans, white	1 cup per group	patty paper	
blank transparencies	1 per group	pennies (not new)	100 per group
books (for analysis)	2 books per group on a variety of topics	plastic cups; 5-oz size	1 per group
books (for stacking)	4 per group	poster paper	1 per group
bucket or other object to pass	1 per class	protractors	1 per group
cables for linking calculators		rope—different thicknesses and lengths (around 1 meter each)	2 per group
centimeter graph paper		rulers	1 per student
centimeter rulers	2 per group	small boxes, *optional*	1 per group
colored pencils or pens		small round objects (coins, buttons, bottle caps)	8–10 per group
compasses or circle templates	1 per group	stopwatches (or clock with second hand)	1 per group
computer with CD-ROM to download data and programs, *optional*		straightedge	1 per person
dice	1 per student pair	string	several 24-cm lengths per group
electronic motion sensor, *optional*	1 per group	table and chair	1 each per group
empty coffee can	1 per group	tape	
graph paper		toothpicks	about 50 per group
large marbles	1 per group	toy figures	1 per group
metersticks	4 per group (or 4-meter marked rope, 1 per group)	transparency markers	1 per group
multiple-speed bicycle	1 per group or 1 for demonstration	uncooked spaghetti	1 box per group
nickels	9 per group	washers, *optional*	10 per group
packets of colored candies	1 per group	yardsticks	1 per group

0

Fractions and Fractals

Overview

In Chapter 0, students review fractions, integers, and exponents while investigating patterns in fractal designs. They are also getting used to the investigative approach and to working in groups.

In **Lesson 0.1**, fractions are reviewed in the context of comparing areas within fractal drawings. **Lessons 0.2 and 0.3** review exponents as a tool for counting features and segment lengths of a fractal at various stages in its generation. **Lesson 0.4** extends to algebraic expressions the notion of recursion, which will be used throughout the text. It also reviews operations with signed numbers. In **Lesson 0.5,** students practice making careful measurements.

The Mathematics

Fractions

To work successfully with variables and equations, students need confidence in working with fractions, both common and decimal, as well as integers. Students should understand equivalent fractions and their use in finding common denominators. They should realize that multiplying by a fraction less than 1 makes a quantity smaller, while dividing by that same fraction makes it larger. They need to know how fractions and integers relate to each other and to the number line.

The decimals displayed on a calculator usually conceal some of the patterns that common fractions reveal, so an ability to use and understand both forms of fractions is important.

Fractals and Recursion

A procedure is *iterative* if it does the same thing repeatedly in a systematic way. One example of iteration is squaring the whole numbers from 1 to 10.

A *recursive* procedure is a special kind of iterative procedure; to get the next result, it uses the result of the previous step. For example, a procedure that adds the whole numbers from 1 to 10 by first adding those from 1 to 9 and then adding 10 is recursive.

The fractals in this chapter are geometric shapes generated by recursive procedures. That is, they're formed by a procedure that uses each stage to produce the next one. For a true mathematical fractal, the recursive procedure is applied infinitely. Every fractal also has the property that a part of it is a shrunken version of the whole figure. That is, fractals are *self-similar*. Self-similarity is possible only because fractals are generated through infinitely many stages.

Many fractals were developed about 100 years ago, but not in an attempt to model reality. Rather, mathematicians were testing the implications of new mathematical definitions. Some mathematicians applied infinite processes to line segments to obtain what they called "curves" with unusual properties, such as passing through every point in a square. For example, the Koch curve of Lesson 0.3 was first proposed in 1906 by the Swedish mathematician Neils Fabian Helge von Koch as a curve on which every point is a "corner." (Koch is pronounced "KawCK," where *CK* is the sound made in Scottish or German by putting the tongue in the position of *k* while pronouncing *h*.)

Other mathematicians wanted nothing to do with these "monsters." As Charles Hermite wrote, "I turn away with fear and horror from this lamentable sore." This extreme case illustrates disagreement over mathematics.

Only about a quarter century ago, Benoit Mandelbrot of IBM was looking for a mathematical way to represent natural phenomena whose appearance is irregular due to fragmentation and turbulence. The squares and triangles of traditional mathematics just don't look like clouds and mountains, but these "monsters" do. He began calling them fractals because they were "fractured." He also found that in a very real sense, they had fractional dimensions. Indeed, fractal models have since been used in many areas, including anatomy (arteries and veins, lungs, the brain), topography (coastlines, watersheds), graphs of the stock market, and even linguistics (patterns of word use). Research is finding new areas where fractal models are useful.

Generating fractals is just one example of using a recursive procedure. Lesson 0.4 gives a glimpse of the many other types of recursion in mathematics. There students first encounter recursive sequences, which they'll learn more about as they study linear equations in Chapter 4 and exponential growth in Chapter 7.

Using This Chapter

You can use this chapter at the beginning of the school year even if your classes have not yet stabilized. Success with *Discovering Algebra* does not depend on detailed coverage of this chapter, so you may decide to use only the first three lessons or perhaps the last two.

Even if you cannot do every investigation, use as many as your time and situation permit. In this chapter, the investigations are especially designed to help students prepare for the rest of the course. Students less experienced in investigating and group work should engage in at least a part of each one.

Throughout this chapter only, you will notice balloons with hints for students on how to use the text successfully. Encourage students to read these.

Resources

Discovering Algebra Resources

Calculator Notes 0A, 0B, 0C, 0D, 0E, 0F, 0G

Teaching and Worksheet Masters
  Lesson 0.1 Connect the Dots
  Lesson 0.2 How Many?
  Lesson 0.3 How Long Is This Fractal?
  Lesson 0.4 Number Lines
  Lesson 0.5 A Chaotic Pattern?
  Lesson 0.5 Centimeter Rulers

Assessment Resources A and B
  Chapter 0 Test
  Chapter 0 Constructive Assessment Options

More Practice Your Skills for Chapter 0

Condensed Lessons for Chapter 0

Other Resources

Robert Devaney, Jonathan Choate, and Alice Foster. *Fractals: A Toolkit of Dynamics Activities.* Emeryville, California: Key Curriculum Press, 1998.

Robert Devaney, Jonathan Choate, and Alice Foster. *Iteration: A Toolkit of Dynamics Activities.* Emeryville, California: Key Curriculum Press, 1998.

Heinz-Otto Peitgen et al., *Fractals for the Classroom, Parts One and Two.* New York: Springer-Verlag, 1992.

Students may enjoy experimenting with fractal-drawing computer programs.

www.keymath.com/DA

Materials

- dice
- metric rulers
- blank transparencies
- transparency markers

Pacing Guide

	day 1	day 2	day 3	day 4	day 5	day 6	day 7	day 8	day 9	day 10
standard	0.1	0.1	0.2	0.3	0.3	0.4	0.5	review	assessment	
enriched	0.1	0.2	0.3	project	0.4	0.5	review, TAL	assessment		
block	0.1	0.2, 0.3	0.4	0.5, review	assessment					

0

Fractions and Fractals

You have probably seen designs like this—you may even have heard the word "fractal" used to describe them. Complex fractals are created by infinitely repeating simple processes; some are created with basic geometric shapes such as triangles or squares. With fractals, mathematicians and scientists can model the formation of clouds, the growth of trees, and human blood vessels.

OBJECTIVES

In this chapter you will
- review operations with fractions
- review operations with positive and negative numbers
- use exponents to represent repeated multiplication
- explore designs called fractals
- learn to use this book as a tool

- Review the conventional order of operations, midpoint of a line segment, the repeated multiplication model for exponents, and equivalent fractions

- Practice measuring skills and addition and multiplication of fractions, decimals, and signed numbers

- Learn to use the calculator with fractions, to evaluate expressions recursively, and to run programs

- Learn to correctly use the terms *congruent figures, exponent, line segment* and *curve, attractor, value of an expression* and *evaluating an expression*, and *vertex*

- Extend exponent concepts to include fractions as bases and use exponents to describe growth patterns

- Recognize a fractal design and describe the recursive procedure that created it

- Gain intuition about the concept of limit

- Become familiar with cooperative investigating

This fractal is generated by a much more complex procedure than those used for fractal designs in the chapter. You might ask students what patterns they see. They may see similar shapes, but the patterns aren't completely regular. Mention that fractals are generated by a regular process that produces apparent randomness, and that they'll be seeing such processes (called *chaotic*) in this chapter.

The fractal in this picture, called a *Julia set,* is generated by considering each point within a circle, moving it to another point by evaluating an equation at that point, then moving that to other point by taking the result and using it to again evaluate the equation, and so on, repeatedly. The starting point is colored according to how fast the successive points are moving away from it. (For this fractal, the operation on a point is given by considering the point as a complex number z and then moving it to the point $z^4 + z - 0.4 + 0.04i$.)

LESSON
0.1

The Same yet Smaller

A procedure that you do over and over, each time building on, is **recursive**. You'll see recursion used in many different ways throughout this book. In this lesson you'll draw a **fractal** design using a recursive procedure. After you draw the design, you'll work with fractions to examine its parts.

> Words in **bold** type are important mathematical terms. They may be new to you, so they will be explained in the text. You can also find a definition in the glossary.

> Investigations are a very important part of this course. Often you'll discover new concepts in an investigation, so be sure to take an active role.

Investigation
Connect the Dots

You will need

- a ruler
- the worksheet Connect the Dots

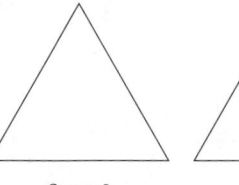 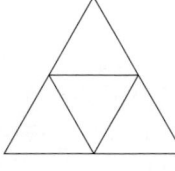 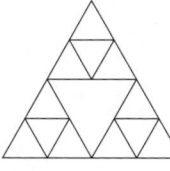 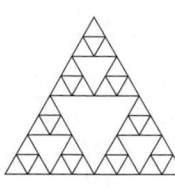

Stage 0   Stage 1   Stage 2   Stage 3

Step 1 Examine the figures above. The starting figure is the Stage 0 figure. To create the Stage 1 figure, you join the *midpoints* of the sides of the triangle. You can locate the midpoints by counting dots to find the middle of each side. The Stage 1 figure has three small upward-pointing triangles. See if you can find all three.

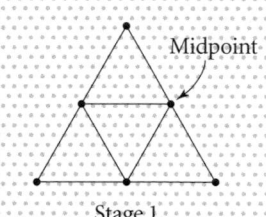

Midpoint

Stage 1

LESSON
0.1

PLANNING

LESSON OUTLINE

First day:

15 min	Getting started
30 min	Investigation group work
5 min	Closing

Second day:

15 min	Sharing by groups
10 min	Examples
10 min	Closing
15 min	Exercises

MATERIALS

- rulers (one per student)
- Connect the Dots or isometric dot paper (W, one per student)
- Calculator Note 0A

TEACHING

This lesson emphasizes equivalent fractions and arithmetic operations on fractions.

 Guiding the Investigation

After forming groups, talk a little bit about fractals. Then display the transparency Connect the Dots.

One step If your class is experienced at investigations, you might simply ask, "What's the sum of the areas of all upward-pointing triangles at the next stage of this figure?" and let them go to work. The students must figure out the generating rule, how to find the areas of the triangles, and how to calculate the desired sum. You can remind individuals or groups of the mathematical ideas being reviewed, handing out Calculator Note 0A and More Practice Your Skills sheets as needed.

If your class is less independent, here are some ideas for more guidance.

Step 1 [ESL] Be sure all students, especially ESL students, understand what a *midpoint* of a line segment is.

LESSON OBJECTIVES

- Review conventional order of operations
- Review midpoint of a line segment
- Review equivalent fractions
- Practice arithmetic operations on fractions
- Learn to use the calculator with fractions
- Recognize a pattern in a fractal design and describe the recursive procedure that created it
- Become familiar with congruent figures

Step 2 At Stage 2, line *segments* connect the midpoints of the sides of the three upward-pointing triangles that showed up at Stage 1. What do you notice when you compare Stage 1 and Stage 2?

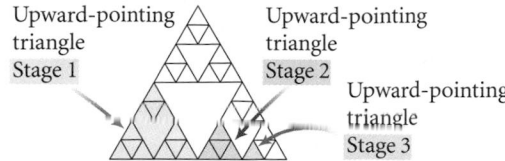

Upward-pointing triangle Stage 1

Upward-pointing triangle Stage 2

Upward-pointing triangle Stage 3

Step 3 How many new upward-pointing triangles are there in the Stage 3 figure? 27

Step 4 On your worksheet, create the Stage 4 figure. A blank triangle is provided. Connect the midpoints of the sides of the large triangle, and continue connecting the midpoints of the sides of each smaller upward-pointing triangle at every stage. How many small upward-pointing triangles are in the Stage 4 figure? 81

Stage 4

Step 5 What would happen if you continued to further stages? Describe any patterns you've noticed in drawing these figures.

You have been using a *recursive rule*. The rule is "Connect the midpoints of the sides of each upward-pointing triangle."

If you could continue this process forever, you would create a fractal called the *Sierpiński triangle*. At each stage the small upward-pointing triangles are *congruent*—the same shape and size.

> Words in *italic* are words you may have seen before or that you can probably figure out. If you need help with them, a definition is given in the glossary.

> This marker shows a convenient stopping place.

Step 6 If the Stage 0 figure has an area equal to 1, what is the area of one new upward-pointing triangle at Stage 1? $\frac{1}{4}$

Step 7 How many different ways are there to find the combined area of the smallest upward-pointing triangles at Stage 1? For example, you could write the *addition expression* $\frac{1}{4} + \frac{1}{4} + \frac{1}{4}$. Write at least two other expressions to find this area. Use as many different operations (like addition, subtraction, multiplication, or division) as you can. $3 \times \frac{1}{4}$ or $1 - \frac{1}{4}$ or $(1 \div 4) \times 3$

Step 8 What is the area of one of the smallest upward-pointing triangles at Stage 2? How do you know? $\frac{1}{4}$ of $\frac{1}{4}$ is $\frac{1}{16}$

Step 9 How many smallest upward-pointing triangles are there at Stage 2? What is the combined area of these triangles? 9; $\frac{9}{16}$

Step 10 Repeat Steps 8 and 9 for Stage 3. $\frac{1}{64}$ (or $\frac{1}{4}$ of $\frac{1}{16}$); 27; $\frac{27}{64}$

Step 11 If the Stage 0 figure has an area of 8, what is the combined area of

a. One smallest upward-pointing triangle at Stage 1, plus one smallest upward-pointing triangle at Stage 2? $2\frac{1}{2}$

NCTM STANDARDS

CONTENT		PROCESS	
●	Number	●	Problem Solving
	Algebra	●	Reasoning
●	Geometry		Communication
●	Measurement	●	Connections
	Data/Probability	●	Representation

Step 2 Possible answers: Each upward-pointing triangle in Stage 1 contains 3 upward-pointing triangles in Stage 2. The large downward-pointing triangle remains unchanged from Stage 1 to Stage 2. Stage 2 has 3 smaller versions of Stage 1.

Step 4 Be sure students understand how to count dots to find the midpoint of a side.

Some students may skip the actual construction of the Stage 4 figure. Encourage them, but be sure they don't dominate their groups and prevent the more visual and kinesthetic learners from creating the pattern for themselves.

Step 5 [Link] In geometry class, students will learn to prove that the new triangles formed at any stage are congruent.

Step 5 Possible answers: At each stage, 3 new upward-pointing triangles are formed in each upward-pointing triangle. Quickly the new triangles become very small and the number of new triangles becomes very large.

Step 7 Help students see a variety of ways to write the combined area. [Ask] "How much less than the whole are the three upward-pointing triangles at Stage 1?" Plan for several approaches to be presented later.

Step 8 [Alert] Watch for difficulties in calculating these areas. [Ask] "How does the smallest triangle at each stage relate to the smallest triangle at the previous stage?"

Step 11 [Alert] Watch for difficulties in determining the sizes of the triangles. Point out that the smallest triangle at Stage 1 is one-fourth the area of the Stage 0 triangle.

Step 12 Challenge students to be creative. For example, they might consider downward-pointing triangles. Many students dislike mathematics because they don't see its creative aspects.

SHARING IDEAS

Steps 5, 10, and 12 are good opportunities for students to present their work. Encourage all participation, but challenge students to justify and explain all answers.

[Language] If the opportunity arises, you might point out that when used with a fraction, the English word *of* often calls for *multiplying*.

You might also pose a "What if . . ." question. **[Ask]** "In this investigation the sides of the triangle were bisected. What if they were trisected instead?" In Example B, students will explore such a triangle.

Assessing Progress

By watching students contribute to groups or the whole class, you can assess their understanding of **equivalent fractions,** fraction **operations** by hand, and **midpoint** of a line segment.

▶ EXAMPLE A

This example allows more review of fraction addition and multiplication. The multiplication symbol for arithmetic, ×, is used here. Students will be introduced to the algebraic notation for multiplication, ( ) and ·, in the next lesson.

Here students are shown two ways to think about adding areas of the same size. Challenge students to use multiple approaches to solving a problem. Even if a question has only one right answer, there may be several legitimate ways to find that answer.

[Ask] "What if the area of the original is not 1, but 18?" Finding the area of the oregano plots in this case requires a further step of multiplying $\frac{2}{9}$ by 18.

b. Two smallest upward-pointing triangles at Stage 2, minus one smallest upward-pointing triangle at Stage 3? $\frac{7}{8}$

c. One smallest upward-pointing triangle at Stage 1, plus three smallest upward-pointing triangles at Stage 2, plus nine smallest upward-pointing triangles at Stage 3? $4\frac{5}{8}$

Step 12 Make up one problem like those in Step 11, and exchange it with a partner to solve.

> This marker means the investigation is done.

The Polish mathematician Waclaw Sierpiński created his triangle in 1916. But the word *fractal* wasn't used until nearly 60 years later, when Benoit Mandelbrot drew attention to recursion that occurs in nature. Trees, ferns, and even the coastlines of continents can be examined as real-life fractals.

EXAMPLE A

> Examples are important learning tools. Have your pencil in hand when you study the solution to an example. Try to do the problem before reading the solution. Work out any calculations in the solution so that you're sure you understand them.

Evan designed an herb garden. He divided each side of his garden into thirds and connected the points. He planted oregano in the labeled sections. If the whole garden has an area of 1, what is the area of one oregano section? What is the total area planted in oregano?

▶ Solution

> Often you will find questions in a solution. Try to answer these questions before you continue reading.

Because there are nine equal-size sections, each oregano section is one-ninth of the garden's area. To find the total area planted in oregano, you can either add $\frac{1}{9} + \frac{1}{9} = \frac{2}{9}$ or multiply $2 \times \frac{1}{9} = \frac{2}{9}$. So the oregano is planted in sections with a total area equal to $\frac{2}{9}$ of the garden. Can you explain how each expression represents the area?

Let's examine some features of Evan's garden in more detail. You can think of Evan's garden as a Stage 1 figure with six identical upward-pointing triangles that each have an area of $\frac{1}{9}$.

What is the area of one small upward-pointing triangle at Stage 2?

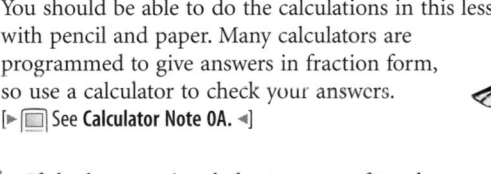

Stage 0   Stage 1   Stage 2

This feature will help you connect algebra to the people who continue to develop and use it.

History
CONNECTION

Benoit Mandelbrot (b 1924) first used the word *fractal* in 1975 to describe irregular patterns in nature. You can link to a biography of Mandelbrot at **www.keymath.com/DA** .

At Stage 2, nine smaller triangles are formed in each upward-pointing triangle from Stage 1. The shaded triangle in Stage 2 has an area that is $\frac{1}{9}$ of the Stage 1 shaded triangle. This equals $\frac{1}{9}$ of $\frac{1}{9}$, which you can write as $\frac{1}{9} \times \frac{1}{9}$ and is equal to $\frac{1}{81}$.

To find combined areas, you'll be adding, subtracting, and multiplying fractions. When there are more than two operations in an expression, it can be difficult to know where to start. To avoid confusion, all mathematicians have agreed to use the **order of operations.**

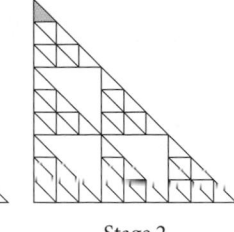

Order of Operations

1. Evaluate all expressions within parentheses.
2. Evaluate all powers.
3. Multiply and divide from left to right.
4. Add and subtract from left to right.

You should be able to do the calculations in this lesson with pencil and paper. Many calculators are programmed to give answers in fraction form, so use a calculator to check your answers.
[▶ 🖳 See **Calculator Note 0A.** ◀]

Go to the calculator notes whenever you see this icon. The calculator notes explain how to use your graphing calculator. You can get these notes from your teacher or log on to **www.keymath.com/DA** .

EXAMPLE B

If the largest triangle has an area of 1, what is the combined area of the shaded triangles?

▶ **Solution**

The area of the larger shaded triangle is $\frac{1}{9}$.

The area of each smaller triangle is $\frac{1}{9} \times \frac{1}{9}$ or $\frac{1}{81}$.

So the combined shaded area is
$\frac{1}{9} + \frac{1}{81} + \frac{1}{81} + \frac{1}{81}$.

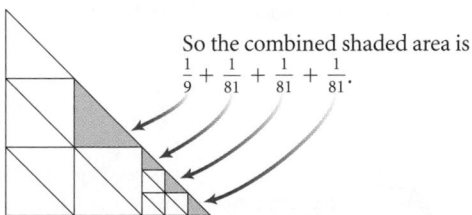

Notice that $\frac{1}{81} + \frac{1}{81} + \frac{1}{81} = \frac{3}{81}$, so the combined area is $\frac{1}{9} + \frac{3}{81}$.

ORDER OF OPERATIONS

Students need to understand order of operations whether they are doing arithmetic by hand or entering it into the calculator. The first lesson in Chapter 4 will cover this again.

The diver has not followed the correct order of operation in getting dressed. The trunks should have been put on before the wet suit.

▶ **EXAMPLE B**

Example B offers another opportunity to discuss adding and simplifying fractions. Before adding fractions, students must find a common denominator. **[Ask]** "Why is $\frac{12}{81}$ the same as $\frac{4}{27}$?" **[Language]** Make sure students can use the phrases *equivalent fractions* and *fraction reduced to lowest terms* in discussing this question.

Closing the Lesson

Mention the new mathematical ideas. Fractal designs are produced by **recursive** procedures; at each stage **congruent** figures are formed. The **calculator** can give answers as **fractions,** and patterns are often easier to see when the answers are fractions.

Invite students to look for patterns in clothing or architecture while at school and on their way home, and to try to see self-similarities as they look outdoors.

In these exercises, students are looking for patterns and practicing the skills of multiplying and dividing fractions, adding and subtracting fractions, using a common denominator, and reducing fractions.

ASSIGNING HOMEWORK

Essential	2–5, 7, 8, 11
Performance assessment	10
Portfolio	5
Journal	6, 8
Group	7, 9, 10
Review	12, 13

▶ Helping with the Exercises

Exercise 1 **[Alert]** Some students may need help finding the denominator for fractional parts of a figure, especially when the figure is not divided into parts of equal size.

Some students may still not understand equivalent fractions. Remind them that the value of a fraction does not change if the numerator and the denominator are multiplied or divided by the same number, as in $\frac{1}{4} = \frac{1 \times 4}{4 \times 4} = \frac{4}{16}$.

1c. $\frac{3}{5}$; $\frac{1}{25} + \frac{1}{25} + \frac{1}{25} + \frac{1}{25}$
$+ \frac{1}{25} + \frac{1}{25} + \frac{1}{25} + \frac{1}{25} + \frac{1}{25}$
$+ \frac{1}{25} + \frac{1}{25} + \frac{1}{25} + \frac{1}{25} + \frac{1}{25}$
$+ \frac{1}{25}$ or $15 \times \frac{1}{25}$ or $1 - \frac{10}{25} = \frac{3}{5}$

1d. $\frac{7}{625}$; $\frac{1}{625} + \frac{1}{625} + \frac{1}{625}$
$+ \frac{1}{625} + \frac{1}{625} + \frac{1}{625} + \frac{1}{625}$ or
$7 \times \frac{1}{625}$

To add fractions, you need *common denominators*. Since nine of the smallest triangles (with area of $\frac{1}{81}$) fit into a triangle with area of $\frac{1}{9}$, you can write $\frac{1}{9}$ as $\frac{9}{81}$. So you can rewrite the combined area as $\frac{9}{81} + \frac{3}{81}$, which equals $\frac{12}{81}$, or $\frac{4}{27}$ in *lowest terms*.

Think of another way to get the same answer. Check your method with a classmate to see if he or she agrees with you.

Notice that you're asked to rework a problem. Check your method and your answer by sharing them with a classmate. Working together is a powerful learning strategy.

Nature
● CONNECTION ●

The smallest leaves of a fern look very similar to the whole fern. This is an example of self-similarity in nature.

$\frac{1}{9} = \frac{9}{81}$

In the Sierpiński triangle, the design in any upward-pointing triangle looks just like any other upward-pointing triangle and just like the whole figure—they differ only in size. Objects like this are called **self-similar.** Self-similarity is an important feature of fractals, and you can find examples of self-similarity everywhere in nature.

EXERCISES

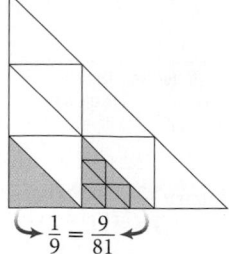

You will need your calculator for problems **1**, **2**, and **4.**

▶ Practice Your Skills

Do these calculations with paper and pencil. Check your work with a calculator.

In the Practice Your Skills problems, you will practice basic skills that you'll need to solve problems in the Reason and Apply section. Sometimes you'll review skills that you learned in earlier courses.

1. Find the total shaded area in each triangle. Write two expressions for each problem, one using addition and the other using multiplication. Assume that the area of each largest triangle is 1.

a.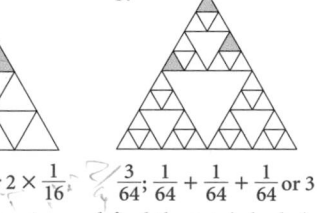
b.
c.
d.

$\frac{1}{8}$; $\frac{1}{16} + \frac{1}{16}$ or $2 \times \frac{1}{16}$; $\frac{3}{64}$; $\frac{1}{64} + \frac{1}{64} + \frac{1}{64}$ or $3 \times \frac{1}{64}$

2. Write an expression and find the total shaded area in each triangle. Assume that the area of each largest triangle is 1.

Sometimes it's easier and faster to do a calculation by hand than with a calculator.

a.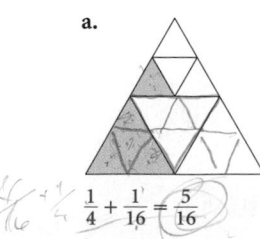
b.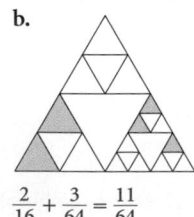
c.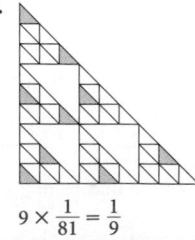
d.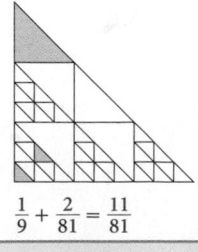

$\frac{1}{4} + \frac{1}{16} = \frac{5}{16}$

$\frac{2}{16} + \frac{3}{64} = \frac{11}{64}$

$9 \times \frac{1}{81} = \frac{1}{9}$

$\frac{1}{9} + \frac{2}{81} = \frac{11}{81}$

3. The first stages of a Sierpiński-like triangle are shown below.

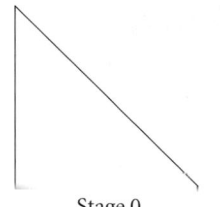 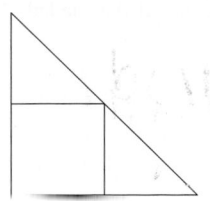 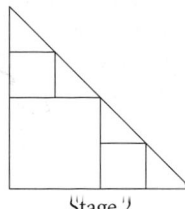 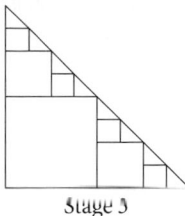

Stage 0   Stage 1   Stage 2   Stage 3

a. Draw Stage 4 of this pattern. You might find it easiest to start with a triangle that is about 8 cm or 4 in. along the bottom.

b. If the Stage 0 triangle has an area of 64, what is the area of the square at Stage 1? 32

c. At Stage 2, what is the area of the squares combined? 48

d. At Stage 3, what is the area of the squares combined? 56

4. Do each calculation, and check your results with a calculator. Set your calculator to give answers in fraction form.

a. $\frac{1}{3} + \frac{2}{9}$ $\frac{5}{9}$

b. $\frac{3}{4} + \frac{1}{2} + \frac{1}{3}$ $\frac{19}{12}$

c. $\frac{2}{5} \times \frac{3}{7}$ $\frac{6}{35}$

d. $2 - \frac{4}{9}$

                                                                   $\frac{14}{9}$ or $1\frac{5}{9}$

▶ Reason and Apply

5. Suppose the fractal design at right has an area equal to 1. Copy the figure four times and shade parts to show each area.

a. $\frac{1}{4}$ **b.** $\frac{3}{16}$ **c.** $\frac{5}{16}$ **d.** $1 - \frac{7}{16}$

6. You have been introduced to the Sierpiński triangle. What are some aspects of this triangle that make it a fractal?

7. Look at the Sierpiński-like pattern in the squares.

 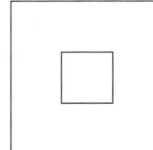 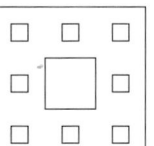

Stage 0   Stage 1   Stage 2

a. Describe in detail the recursive rule used to create this pattern.

b. Carefully draw the next stage of the pattern.

c. Suppose the Stage 0 figure represents a square carpet. The new squares drawn at each stage represent holes that have been cut out of the carpet. If the Stage 0 carpet has an area of 1, what is the total area of the holes at Stages 1 through 3?

d. What is the area of the remaining carpet at each stage?

> The exercises in this book may be different from what you're used to. There may be fewer problems, but you'll probably have to put more time into each one.

7a. Sample description: Divide each side of the square into thirds and connect those points with lines parallel to the sides. A square is formed in the middle. Erase everything except the center square. To get the next stage, do the same thing in all eight squares formed around the middle square.

7b.

Stage 3

7c. $\frac{1}{9}, \frac{17}{81}, \frac{217}{729}$ **7d.** $\frac{8}{9}, \frac{64}{81}, \frac{512}{729}$

Exercise 3 Students can use dot paper or start with edges of length 8 cm or 4 inches to make finding the midpoint easy.

3a.

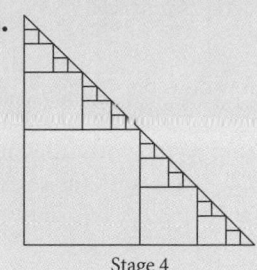

Stage 4

Exercise 4 [Ask] "For which parts of this problem, do you need to find a common denominator?"

5. Sample answers:

5a.

5b.

5c.

5d.

6. Answers should include creation by a repetitive process and smaller parts look like the whole.

Exercise 7 [Alert] Watch for inexact statements here. [ESL] Be sure students can use the term *parallel*.

[Ask] "If you want to subtract a fraction from 1, what might you change 1 into?" Any fraction whose numerator equals its denominator is equivalent to 1.

Exercise 8 Here the total is 8 units instead of 1. Answers will now be whole numbers. **[Alert]** In 8a, some students may still be thinking of fractions and write $\frac{1}{4}$ instead of 2.

[Ask] "How does dividing by 4 compare to multiplying by $\frac{1}{4}$?" Be prepared to convince students that they give the same answer; perhaps use a picture of a line segment.

11. Answers should be equivalent to

11a.

11b.

$\frac{3}{4} \times \frac{1}{4} \times \frac{1}{4} \times 32 = \frac{96}{64} = \frac{3}{2}$
$= 1\frac{1}{2}$

11c.

8. Suppose the area of the original large triangle at right is 8.

  a. Write a division expression to find the area of one of the shaded triangles. What is the area? $8 \div 4 = 2$

  b. What fraction of the total area is each shaded triangle? Use this fraction in a multiplication expression to find the area of one of the shaded triangles. $8 \times \frac{1}{4} = 2$

  c. What is the difference between dividing by 4 and multiplying by $\frac{1}{4}$?

  d. Write a multiplication expression using the fraction $\frac{3}{4}$ to find the combined shaded area. $8 \times \frac{3}{4} = 6$

Essentially, they are the same.

9. Suppose the original large triangle below has an area of 12.

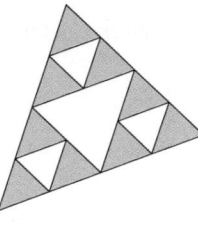

  a. What fraction of the area is shaded? $\frac{9}{16}$

  b. Find the combined area of the shaded triangles. Write two different expressions you could use to find this area.

9b. Sample answers:
$12 \times \left(\frac{9}{16}\right) = 6\frac{3}{4}$;
$(12 \div 16) \times 9 = 6.75$

10. Suppose the original large triangle at right has an area of 24.

  a. What fraction of the area is the shaded triangle at the top? $\frac{1}{9}$

  b. What fraction of the area is each smallest shaded triangle? $\frac{1}{81}$

  c. What is the total shaded area? Can you find two ways to calculate this area? $\left(\frac{4}{9} \times 24\right) + \left(\frac{3}{81} \times 24\right)$ or $24 \times \left(\frac{4}{9} + \frac{3}{81}\right) = 11\frac{5}{9}$

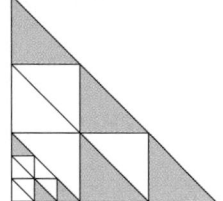

11. Rewrite each expression below using fractions. Then draw a Sierpiński triangle and shade the area described. In each case the Stage 0 triangle has an area of 32.

  a. One-fourth of one-fourth of 32 $\frac{1}{4} \times \frac{1}{4} \times 32 = \frac{32}{16} = 2$

  b. Three-fourths of one-fourth of one-fourth of 32

  c. One-half of one-half of one-fourth of 32
$\frac{1}{2} \times \frac{1}{2} \times \frac{1}{4} \times 32 = \frac{32}{16} = 2$

> You may need to refer back to examples or to work you did in an investigation as you work on a problem.

▶ Review

12. Assume the area of your desktop equals 1. Your math book rests on $\frac{1}{4}$ of your desktop, your calculator on $\frac{1}{16}$ of your desktop, and your scrap paper on $\frac{1}{32}$ of your desktop. What area is covered by these objects? Write an addition expression and then give your answer as a single fraction in lowest terms. $\frac{1}{4} + \frac{1}{16} + \frac{1}{32} = \frac{11}{32}$

13. Use the information from exercise 12 to find the area of your desk that is *not* covered by these materials. Write a subtraction expression and then give your answer as a single fraction in lowest terms. $1 - \frac{11}{32} = \frac{21}{32}$

LESSON
0.2

More and More

Did you notice that at each stage of a Sierpiński design, you have more to draw than in the previous stage? The new parts get smaller, but the number of them increases quickly. Let's examine these patterns more closely.

A strong positive mental attitude will create more miracles than any wonder drug.

PATRICIA NEAL

Investigation
How Many?

Explore how quickly the number of new triangles grows using multiplication repeatedly. Look for a pattern to help you *predict* the number of new triangles at each stage without counting them.

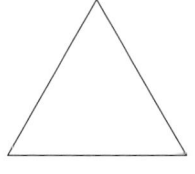 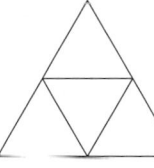 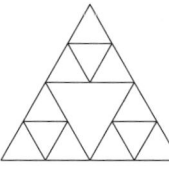 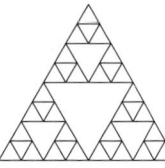 

| Stage 0 | Stage 1 | Stage 2 | Stage 3 | Stage 4 |

Step 1 Look at the fractal designs. Count the number of new upward-pointing triangles for Stages 0 to 4. Make a table like this to record your work. 1; 3; 9; 27; 81

Stage	Number of new upward-pointing triangles
0	1
1	

 Throughout this course you'll record results in a table. Tables provide a useful way to keep track of your work and see patterns develop.

Step 2 How does the number of new triangles compare to the number of new triangles at the previous stage? Each time the stage is increased by 1 the number of triangles is multiplied by 3.

Step 3 Using your answer to Step 2, find how many new upward-pointing triangles are at Stages 5, 6, and 7. 243; 729; 2187

Step 4 Explain how you could find the number of upward-pointing triangles at Stage 15 without counting. Use the answer for Stage 7, that is, 2187, and multiply it by 3 eight times.

At each stage, three new upward-pointing triangles are drawn in each of the upward-pointing triangles from the previous stage. How is this the same as repeatedly multiplying by 3? At each stage, there are 3 times the number of new upward-pointing triangles as the previous stage.

NCTM STANDARDS

CONTENT		PROCESS	
●	Number	●	Problem Solving
	Algebra	●	Reasoning
●	Geometry	●	Communication
	Measurement	●	Connections
	Data/Probability	●	Representation

LESSON OBJECTIVES

- Review the repeated multiplication model for exponents
- Learn a precise definition of exponent
- Generalize a growth pattern into a symbolic expression involving exponents

PLANNING

LESSON OUTLINE

One day:

10 min	Introduction
20 min	Investigation group work, sharing
5 min	Example
5 min	Closing
10 min	Exercises

MATERIALS

- How Many? (T)
- Calculator Note 0B

TEACHING

In this lesson students use exponents to model the rapidly changing numbers and sizes of parts between one stage of a fractal design and the next.

Guiding the Investigation

Once students are in their groups, put up the transparency How Many? Remind the class of the meanings of *recursive procedure* and *fractal* and elicit the idea that the number of parts in the Sierpiński triangle increases rapidly from one stage to the next.

One step [Ask] "How many upward-pointing triangles appear in Stages 0 to 4?" Write the numbers on the transparency. Ask, "How many upward-pointing triangles will the next stage have?" As you visit groups, remind students of exponential notation as needed and help them enter exponents on their calculators. If some students are proficient with the calculator, encourage peer teaching.

Step 1 Request that a few students show you how they found the numbers in their tables.

Step 2 Make sure students notice the word *new*.

Step 3 [Ask] "Why is the number being multiplied by 3 each time?" One answer is below Step 4 in the text.

SHARING IDEAS

While visiting groups, ask one student to copy onto a transparency the group's table from Step 1. Select other students to share their approaches to Steps 2 and 4. During reporting time, encourage all participation, but probe for explanations. Ask for a correct definition of *exponent*. Ask why exponents are useful to elicit the point that they save writing when a single number is being multiplied many times.

Assessing Progress

As you watch students contribute to groups or the whole class, look for understanding of **multiplication** and the **symbols** used to indicate multiplication as well as ideas that were new in the previous lesson: **recursive procedure, fractal, Sierpiński triangle, congruent figures,** and **order of operations.**

EXAMPLE

The example leads into a discussion of exponents. Call students' attention to the picture and question in the text.

[Language] The term *exponent* is defined in the student text, but some students may be more familiar with *raising a number to a power*.

Model good mathematical language to help students learn what terms mean. For example, use the word *factor* in context. Point out how the text answers your earlier

You can write the symbol for multiplication in different ways. For example, you can write 3×3 as $3 \cdot 3$ or $3(3)$. All three of these expressions have the same meaning. Each expression equals 9.

EXAMPLE | Describe how the number of new upward-pointing triangles is growing in this fractal.

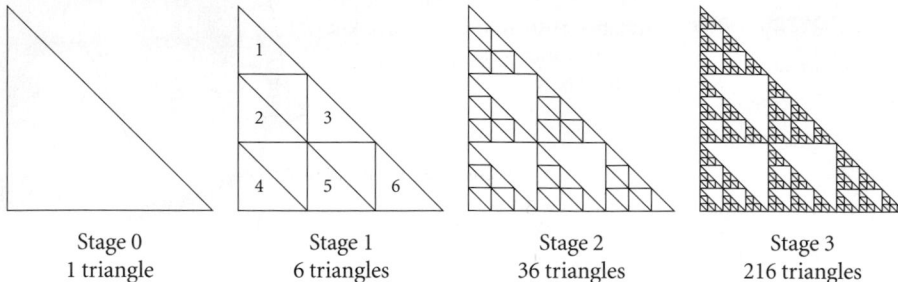

Stage 0	Stage 1	Stage 2	Stage 3
1 triangle	6 triangles	36 triangles	216 triangles

▶ **Solution** | At Stage 1, the six new upward-pointing triangles are numbered. At Stage 2, six new upward-pointing triangles are formed in each numbered Stage 1 triangle. At Stage 2, there are $6 \cdot 6$ or 36 new triangles. At Stage 3, six triangles are formed in each new upward-pointing Stage 2 triangle, so there are $36 \cdot 6$ or 216 new upward-pointing triangles.

Another way to look at the number of new upward-pointing triangles at each stage is shown in the table below.

	Number of new upward-pointing triangles		
Stage number	Total	Repeated multiplication	Exponent form
1	6	6	6^1
2	36	$6 \cdot 6$	6^2
3	216	$36 \cdot 6$ or $6 \cdot 6 \cdot 6$	6^3

The last number in each row of the table is a 6 followed by a small number. The small number, called an **exponent,** shows how many 6's are multiplied together. An exponent shows the number of **factors** of 6. What is the pattern between the stage number and the exponent?

Do you think the pattern applies to Stage 0? Put the number 6^0 into your calculator. [▶ 🖥 See **Calculator Note 0B** to learn how to enter exponents. ◀] Does the result fit the pattern?

How many upward-pointing triangles are there at Stage 4? According to the pattern, there should be 6^4. That's 1296 triangles! It is a lot easier to use the exponent pattern than to count all those triangles.

question about a correct definition of *exponent* by saying that the exponent tells "how many 6's are multiplied together." [Alert] The phrase "how many times you multiply a number by itself" is not strictly correct: 6^2 is not 6 multiplied by itself twice.

As part of the fractal design development, students are introduced to the concept of zero as an exponent. Detailed coverage of zero and negative exponents appears in Chapter 7.

Closing the Lesson

Review the careful definition of **exponent.** In 5^4, the exponent 4 indicates four factors of 5.

[Ask] "Why are exponents useful?" [They save a lot of multiplication and writing when the triangles become complicated.]

EXERCISES

You will need your calculator for problem **4.**

▶ Practice Your Skills

1. Write each multiplication expression in exponent form.

  a. $5 \times 5 \times 5 \times 5 \times 5$ 5^4 **b.** $7 \times 7 \times 7 \times 7 \times 7$ 7^5

  c. $3 \cdot 3 \cdot 3 \cdot 3 \cdot 3 \cdot 3 \cdot 3$ 3^7 **d.** $2(2)(2)$ 2^3

2. Rewrite each expression as a repeated multiplication three ways: using $\times$, $\cdot$, and parentheses.

  a. 3^4 **b.** 5^6 **c.** $\left(\dfrac{1}{2}\right)^3$

3. Write each number with an exponent other than 1. For example, $125 = 5^3$.

  a. 27 3^3 **b.** 32 2^5 **c.** 625 5^4 or 25^2 **d.** 343 7^3

4. Do the calculations. Check your results with a calculator.

  a. $\dfrac{2}{3} \cdot 12$ 8   **b.** $\dfrac{1}{3} + \dfrac{3}{5}$ $\dfrac{14}{15}$ **c.** $\dfrac{3}{4} - \dfrac{1}{8}$ $\dfrac{5}{8}$

  d. $5 - \dfrac{2}{7}$ $\dfrac{33}{7}$ or $4\dfrac{5}{7}$ **e.** $\dfrac{1}{4} \cdot \dfrac{1}{4} \cdot 8$ $\dfrac{1}{2}$ **f.** $\dfrac{3}{64} + \dfrac{3}{16} + \dfrac{3}{4}$ $\dfrac{63}{64}$

▶ Reason and Apply

5. Another type of fractal drawing is called a "tree." Study Stages 0 to 3 of this tree:

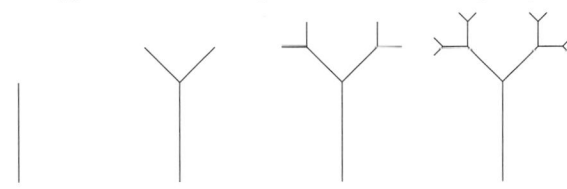

Stage 0   Stage 1   Stage 2   Stage 3

> Homework helps you reinforce what you've learned in the lesson and develops your understanding of new ideas.

  a. At Stage 1, two new branches are growing from the trunk. How many new branches are there at Stage 2? At Stage 3?   4 or 2^2; 8 or 2^3

  b. How many new branches are there at Stage 5? Write your answer in exponent form. 2^5

6. Another fractal tree pattern is shown below.

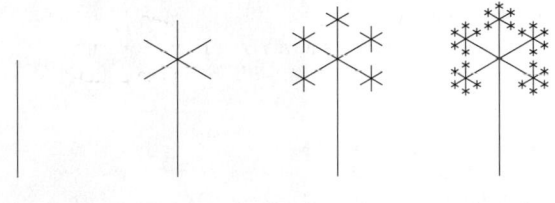

Stage 0   Stage 1   Stage 2   Stage 3

  a. At Stage 1, five new branches are growing. How many new branches are there at Stage 2?   25

2a. $3 \times 3 \times 3 \times 3$; $3 \cdot 3 \cdot 3 \cdot 3$; $3(3)(3)(3)$

2b. $5 \times 5 \times 5 \times 5 \times 5 \times 5$; $5 \cdot 5 \cdot 5 \cdot 5 \cdot 5 \cdot 5$; $5(5)(5)(5)(5)(5)$

2c. $\dfrac{1}{2} \times \dfrac{1}{2} \times \dfrac{1}{2}$; $\dfrac{1}{2} \cdot \dfrac{1}{2} \cdot \dfrac{1}{2}$; $\dfrac{1}{2}\left(\dfrac{1}{2}\right)\left(\dfrac{1}{2}\right)$

BUILDING UNDERSTANDING

In these exercises, students practice factoring and writing expressions with exponents. Then they use exponents to describe changes in fractal designs.

ASSIGNING HOMEWORK

Essential	1–3, 5, 8, 9
Performance assessment	9
Portfolio	6
Journal	7
Group	7, 9, 12
Review	10–12

▶ Helping with the Exercises

Exercise 1 [Language] You might use the term *expanded form* to describe these expressions.

Exercise 3 [Alert] Some students may need help finding the factors of a number. Suggest that they look for a factor—starting with 2, 3, 5, and 7—that will divide the number with no remainder. They should then keep dividing by that factor, counting the number of times they divide. When they can no longer divide by that factor, they should try another.

Exercise 3 Some students may suggest that 625 equals $(-5)^4$, 25^2, or $(-25)^2$. Encourage their creativity and praise their factual knowledge, but be sure they can find the expression with the smallest positive base.

Exercise 6 [Language] Some students might state that fractal designs become *larger* from stage to stage, for want of a better word. Begin to use the term *denser* to help students develop an intuition for limit. **[ESL]** Be sure all students understand the word *dense*. To model the word in another context, say that a city is more densely populated than the suburbs or the country.

To help make the idea of self-similarity more real to students, you might request that students bring in a self-similar object from nature or a written description of a self-similar object other than a fractal.

b. How many new branches are there at Stage 3? 125 or 5^3

c. How many new branches are there at Stage 5? Write your answer in exponent form. 5^5

7. At Stage 1 of this pattern, there is one square hole. At Stage 2, there are eight new square holes.

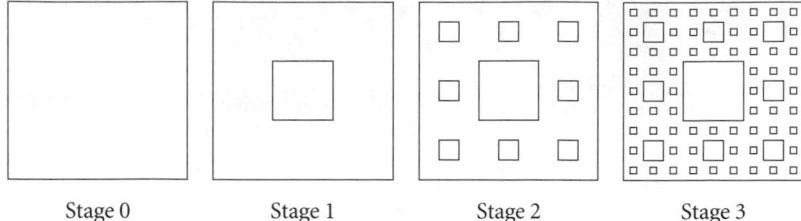

Stage 0   Stage 1   Stage 2   Stage 3

a. How many new square holes are there at Stage 3? 64

b. If you drew the Stage 4 figure, how many new square holes would you have to draw? 512

c. Write the answers to parts a and b in exponent form. 8^2; 8^3

d. How many new square holes would you have to draw in the Stage 7 figure? 262,144 or 8^6

e. Describe the pattern between the stage number and the exponent for these figures. The exponent is always or less than the stage numbe

f. Will the pattern you described in part e work for the Stage 1 figure? Why?
Yes, because $8^0 = 1$.

8. Study Stages 0 to 3 of this fractal "weed" pattern. At Stage 1, two new branches are created.

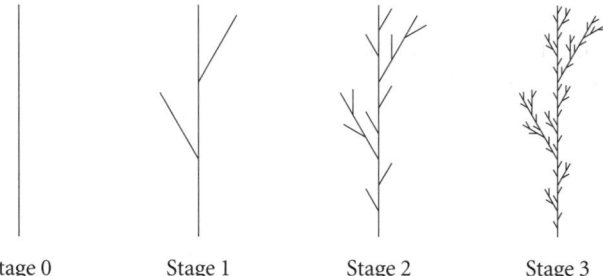

Stage 0   Stage 1   Stage 2   Stage 3

8c. Stage 3, $2 \cdot 5^2$; Stage 4, $2 \cdot 5^3$; Stage 5, $2 \cdot 5^4$

8d. The 2 is the number of new branches at Stage 1. The 5 is the number of smaller segments created in Stage 1. Then at each subsequent stage, new branches are added multiplying by 5 again.

a. How many new branches are created at Stage 2? 10

b. How many new branches are created at Stage 3? 50

c. You can write the expression $2 \cdot 5^1$ to represent the number of new branches in the Stage 2 figure. Write similar expressions to represent the number of new branches in Stages 3 to 5.

d. How do the 2 and the 5 in each expression relate to the figure?

Patterns like the "weed" in problem 8 can be used to create very realistic computer-generated plants. Graphic designers can use fractal routines to create realistic-looking trees and other natural features.

9. Look again at this familiar fractal design.

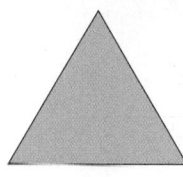 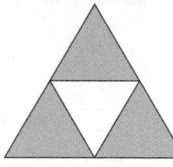 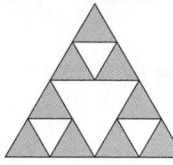 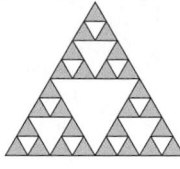

Stage 0   Stage 1   Stage 2   Stage 3

a. Make a table to calculate and record the area of one shaded triangle in each figure.

Stage number	Area of one shaded triangle	Total area of the shaded triangles
0	1	1
1		

b. Record the combined shaded area of each figure in your table.

c. Describe at least two patterns you discovered.

▶ Review

10. Ethan deposits $2 in a bank account on the first day, $4 on the second day, and $8 on the third day. He will continue to double the deposit each day. How much will he deposit on the eighth day? Write your answer as repeated multiplication separated by dots, in exponent form, and as a single number. $2 \cdot 2 \cdot 2 \cdot 2 \cdot 2 \cdot 2 \cdot 2 \cdot 2 = 2^8 = 256$ or $256

11. Write a word problem that illustrates $\frac{3}{4} \cdot \frac{1}{5}$, and find the answer.

12. The large triangles below each have an area of 1. Find the total shaded area in each.

a.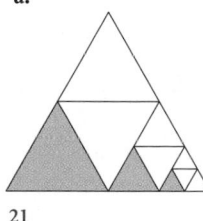

$\frac{21}{64}$

b.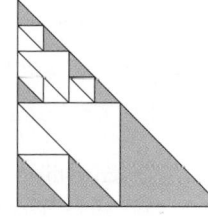

$\frac{29}{64}$

9b.

Stage	One triangle	Total area of the shaded triangles
1	$\frac{1}{4}$	$\frac{3}{4}$
2	$\frac{1}{16}$	$9 \cdot \frac{1}{16}$ or $\frac{3}{4} \cdot \frac{3}{4} = \frac{9}{16}$
3	$\frac{1}{64}$	$\frac{3}{4} \cdot \frac{9}{16} = \frac{27}{64}$

9c. The area of one shaded triangle is $\frac{1}{4}$ the area of one of the previous shaded triangles. The total area of the shaded triangles in each figure is $\frac{3}{4}$ the shaded area in the previous figure.

11. One possibility: The restaurant had $\frac{3}{4}$ of a pie left. Five people wanted pie. After cutting the pie into fifths, how much did each person get? $\frac{3}{20}$.

PLANNING

LESSON OUTLINE

One day:

25 min   Investigation

10 min   Sharing

5 min   Closing

10 min   Exercises

MATERIALS

- How Long Is This Fractal? (T), *optional*
- How Long Is This Fractal? (W, one per student), *optional*
- Calculator Note 0B

TEACHING

In this lesson the familiar idea of perimeter is extended to include the perimeter of a fractal.

Guiding the Investigation

Display the transparency How Long Is This Fractal?

Remind the class that fractals are useful in describing nature, especially irregular shapes such as coastlines. Point out three stages in the generation of the Koch curve, so-named for its inventor. (See page 1A for pronunciation of Koch.) Mathematicians of the early 1900s extended the notion of a curve to shapes like the Koch curve as they searched for one-dimensional figures that had unusual properties such as having a sharp turn at every point.

One step **[Ask]** "What's the total length of the Koch curve at Stage 4?" After students have produced satisfactory results, you might ask a similar question about Stage 17 of the fractal in the example.

The number of distinct scales of length of natural patterns is for all practical purposes infinite.

BENOIT MANDELBROT

Shorter yet Longer

In fractals like the Sierpiński triangle, new enclosed shapes are formed at each stage. Not all fractals are formed this way. One example is the *Koch curve*, which is not a smooth curve, but a set of connected line segments. It was introduced in 1906 by the Swedish mathematician Niels Fabian Helge von Koch. As you explore the Koch curve, you'll continue to work with exponents.

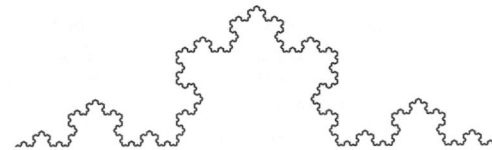

Investigation
How Long Is This Fractal?

Study how the Koch curve develops. One way to discover a fractal's recursive rule is to study what happens from Stage 0 to Stage 1. Once you know the rule, you can build, or generate, later stages of the figure.

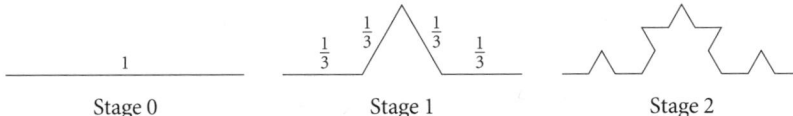

Stage 0   Stage 1   Stage 2

Step 1   Make and complete a table like this for Stages 0 to 2 of the Koch curve shown. How do the lengths change from stage to stage? If you don't see a pattern, try writing the total lengths in different forms.

Stage number	Number of segments	Length of each segment	Total length (Number of segments times length of segments)	
			Fraction form	Decimal form
0	1	1	1	1
1	4	$\frac{1}{3}$	$\frac{4}{3}$	1.33
2	16	$\frac{1}{9}$	$\frac{16}{9}$	1.78

Step 2 Remove the middle third of each segment, then create the other two sides of an equilateral triangle that would have that removed segment as a base.   **Step 2**

Step 3

Step 4 At Stage 3,   **Step 4** there are 64 segments. Each segment is $\frac{1}{27}$, so the length is $\frac{64}{27}$ or about 2.37.

Look at Stages 0 and 1. Describe the curve's recursive rule so that someone could recreate the curve from your description.

Predict the total length at Stage 3.
  Good estimate: "More than 2."
Find the length of each small segment at Stage 3 and the total length of the Stage 3 figure.

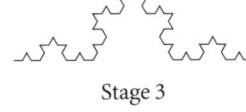

Stage 3

If the Koch curve were a piece of string, you could find its length by straightening the string and measuring it.

Step 1 Some students may change all improper fractions to mixed numbers. Suggest that the improper forms often make it easier to see patterns. If necessary, point out that though the individual pieces get shorter, the total length becomes longer.

LESSON OBJECTIVES

- Practice multiplication of fractions
- Extend exponent concepts to include fractions as bases
- Learn the concepts of line segment and curve
- Think about confined infinity to gain an informal intuition about the concept of limit

Step 5	Use exponents to rewrite your numbers in the column labeled *Total length in fraction form* for Stages 0 to 3. $1; \frac{4^1}{3^1}; \frac{4^2}{3^2}; \frac{4^3}{3^3}$
Step 6	Predict the Stage 4 lengths. Predictions will vary. Actual length is $\frac{4^4}{3^4}$ or $\frac{256}{81}$.
Step 7	Koch was attempting to create a "curve" that was nothing but corners. Do you think he succeeded? If the curve is formed recursively for many stages, what would happen to its length? Answers will vary. Students should notice that the length continues to increase.

Science
• CONNECTION •

Because a coastline, like a fractal curve, is winding and irregular, it is not possible to measure its length accurately. The Koch curve can be used as a model of a coastline. Understanding its structure helps geographers describe coastlines.

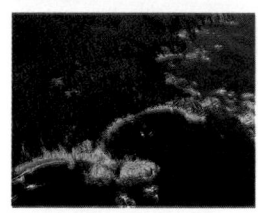

At later stages the Koch curve looks smoother and smoother. But, if you magnify a section at a later stage, it is just as jagged as at Stage 1. Mandelbrot named these figures *fractals* based on the Latin word *fractus,* meaning broken or irregular.

EXAMPLE

Look at the beginning stages of this fractal:

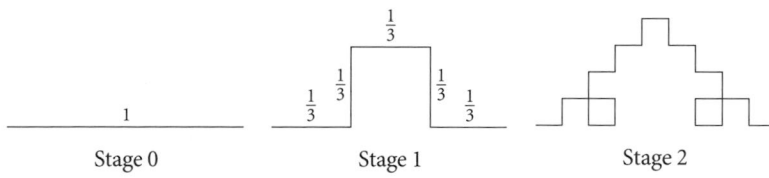

| Stage 0 | Stage 1 | Stage 2 |

a. Describe the fractal's recursive rule.

b. Find its length at Stage 2.

c. Write an expression for its length at Stage 17.

▶ **Solution**

a. You compare Stage 0 and Stage 1 to get the recursive rule. To get Stage 1, you divide the Stage 0 segment into thirds. Build a square on the middle third and remove the bottom. So the recursive rule is "To get to the next stage, divide each segment from the previous stage into thirds and build a bottomless square on the middle third."

NCTM STANDARDS

CONTENT	PROCESS
• Number	• Problem Solving
• Algebra	• Reasoning
• Geometry	Communication
• Measurement	• Connections
Data/Probability	• Representation

Step 6 For students who are finding this exploration easy, challenge them to write an expression for Stage 17 lengths. If all students can do that, you need not work through the example later.

SHARING IDEAS

A student with a good table in Step 1 might prepare a transparency for presentation, to be added to by a student with a good answer to Step 5. An imprecisely stated rule in Step 2 might help illustrate the amount of care needed in stating such rules. Someone could then show an answer to Step 6, and you could encourage students to develop their ideas about the questions in Step 7.

Ask students where else they've seen exponents used. They might mention units such as ft^2 on signs for office space, or scientific notation, which will come up in Chapter 7.

Encourage discussion about the paradoxical notion that the perimeter increases while the rectangular space containing the curve does not. The information on mathematicians who have disagreed about this possibility in the chapter interleaf, page 1A, might interest students who think there is no room for difference of opinion in the field.

Assessing Progress

While watching students contribute to groups or the whole class, note which students seem most and least comfortable with the principal ideas reviewed in this lesson: **fraction multiplication** and **exponents representing repeated multiplication.**

EXAMPLE

Refer students to the picture and questions in the text. **[Ask]** "What's another way to express the recursive rule given in the first part of the solution?" [The rule could be improved; it doesn't say whether the square goes out or in.]

You might use the term *for each* in the replacement: for each segment of the previous stage, there are five new segments, so you multiply the number of segments by 5. **[Link]** Students will use *for each* and multiplication in discrete mathematics when they learn the many uses of the Fundamental Counting Principle. (If for each of m ways of accomplishing a task there are n ways of accomplishing another task, then there are m times n ways of accomplishing the two tasks.)

Closing the Lesson

Say something briefly about each new mathematical idea. A **curve** can be **generated** from a **line segment** by a recursive rule. The changing lengths of curves can be represented by exponents on fractions.

BUILDING UNDERSTANDING

Calculators will be very important in determining some of the values, but remind students that they also need to be able to write the answers in fraction form. The patterns are not evident in the decimal form of the numbers.

ASSIGNING HOMEWORK

Essential	1–3, 5, 6
Performance assessment	7
Portfolio	6
Journal	8
Group	4, 8
Review	9–11

> Don't forget to think through the solution and answer any questions.

b. To find the length of the fractal at Stage 2, we'll start by looking at its length at Stage 1. The Stage 1 figure has 5 segments. Each segment is $\frac{1}{3}$ long. So the total length at Stage 1 is $5 \cdot \frac{1}{3}$. You can rewrite this as $\frac{5}{3}$.

At Stage 2, you replace each of the five Stage 1 segments with five new segments. So the Stage 2 figure has $5 \cdot 5$ or 5^2 segments.

Each Stage 1 segment is $\frac{1}{3}$ long, and each Stage 2 segment is $\frac{1}{3}$ of that. So each Stage 2 segment is $\frac{1}{3} \cdot \frac{1}{3}$ or $\left(\frac{1}{3}\right)^2$ long.

So, at Stage 2, there are 5^2 segments, each $\left(\frac{1}{3}\right)^2$ long. The total length at Stage 2 is $5^2 \cdot \left(\frac{1}{3}\right)^2$. You can rewrite this as $\left(\frac{5}{3}\right)^2$.

c. Do you see the connection between the stage number and the exponent? At each stage, you replace every segment from the previous stage with five new segments. The length of each new segment is $\frac{1}{3}$ the length of a segment at the previous stage. By Stage 17, you've done this 17 times. The Stage 17 figure is $\left(\frac{5}{3}\right)^{17}$ long.

Each segment is replaced with 5 segments. $\left(\dfrac{5}{3}\right)^{17}$ You've done this for 17 steps. Each new segment is $\frac{1}{3}$ the length of the old segment.

To compare total lengths, express each as a decimal rounded to the hundredth's place.

Stage number	Number of segments	Length of each segment	Total length (Number of segments times length of segments)	
			Fraction form	Decimal form
0	1	1	$1 \cdot 1$	1.00
1	$1 \cdot 5 = 5^1$	$1 \cdot \frac{1}{3} = \left(\frac{1}{3}\right)^1$	$5^1 \cdot \left(\frac{1}{3}\right)^1 = \left(\frac{5}{3}\right)^1$	1.67
2	$5 \cdot 5 = 5^2$	$\frac{1}{3} \cdot \frac{1}{3} = \left(\frac{1}{3}\right)^2$	$5^2 \cdot \left(\frac{1}{3}\right)^2 = \left(\frac{5}{3}\right)^2$	2.78
⋮	⋮	⋮	⋮	⋮
17	5^{17}	$\left(\frac{1}{3}\right)^{17}$	$5^{17} \cdot \left(\frac{1}{3}\right)^{17} = \left(\frac{5}{3}\right)^{17}$	5907.84

EXERCISES

You will need your calculator for problems **1, 3, 4, 6, 7,** and **8.**

▶ Practice Your Skills

1. Evaluate each expression. Write your answer as a fraction and as a decimal, rounded to the nearest hundredth. Remember, if the third digit to the right of the decimal is 5 or higher, round up.

   a. $\frac{5^3}{2^3}$ $\frac{125}{8}$; 15.63

   b. $\left(\frac{5}{3}\right)^2$ $\frac{25}{9}$; 2.78

   c. $\left(\frac{7}{3}\right)^4$ $\frac{2401}{81}$; 29.64

   d. $\left(\frac{9}{4}\right)^3$ $\frac{729}{64}$; 11.39

▶ Helping with the Exercises

Exercise 1 [Alert] Some students may need help rounding numbers. Be prepared to remind them of the rounding procedure.

2. The fractal from the example is shown below. How much longer is the figure at Stage 2 than at Stage 1? Use the table on the previous page and find your answer as a fraction and as a decimal rounded to the nearest hundredth. $\frac{5^2}{3^2} - \frac{5}{3} = \frac{10}{9}$ or $2.78 - 1.67 = 1.11$

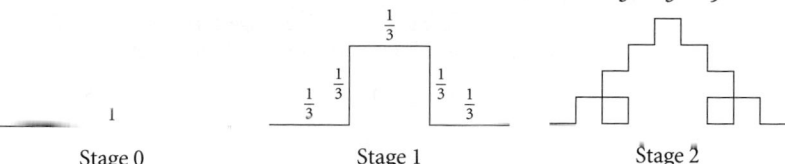

Stage 0   Stage 1   Stage 2

3. At what stage does the figure above first exceed a length of 10? $\left(\frac{5}{3}\right)^4 \approx 7.72$; $\left(\frac{5}{3}\right)^5 \approx 12.86$; Stage 5

4. Evaluate each expression and check your results with a calculator.

   a. $\frac{1}{5} + \frac{3}{4}$ $\frac{19}{20}$ **b.** $3^2 + 2^4$ 25   **c.** $\frac{2}{3} \cdot \left(\frac{6}{5}\right)^2$ $\frac{72}{75}$ or $\frac{24}{25}$ **d.** $4^3 - \frac{2}{5}$ $\frac{318}{5}$ or $63\frac{3}{5}$

▶ Reason and Apply

5. The Stage 0 figure below has a length of 1. At Stage 1, each segment has a length of $\frac{1}{4}$.

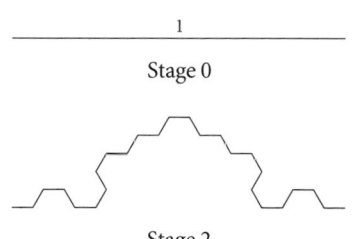

Stage 0

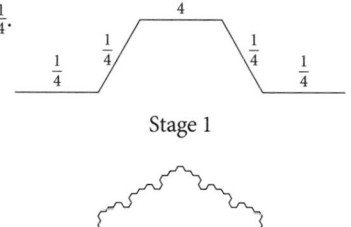

Stage 1

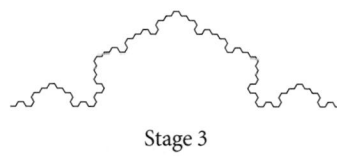

Stage 2   Stage 3

  a. Complete a table like the one below by calculating the lengths of the figure at Stages 2 and 3. Give each answer as a fraction in multiplication form, as a fraction in exponent form, and as a decimal rounded to the nearest hundredth. Try to figure out the total lengths at Stages 2 and 3 without counting.

  b. Which is the first stage to have a length greater than 3? A length greater than 10?

Stage number	Total length		
	Multiplication form	Exponent form	Decimal form
0	1	1^0	1
1	$5 \cdot \frac{1}{4} = \frac{5}{4}$	$\left(\frac{5}{4}\right)^1$	1.25
2	$5 \cdot 5 \cdot \frac{1}{4} \cdot \frac{1}{4} = \frac{25}{16}$	$5^2 \cdot \left(\frac{1}{4}\right)^2 = \left(\frac{5}{4}\right)^2$	1.56
3	$5 \cdot 5 \cdot 5 \cdot \frac{1}{4} \cdot \frac{1}{4} \cdot \frac{1}{4} = \frac{125}{64}$	$5^3 \cdot \left(\frac{1}{4}\right)^3 = \left(\frac{5}{4}\right)^3$	1.95

Exercise 2 The question "How much longer . . . ?" can be answered validly by either subtraction or division. As always, encourage different approaches, but point out both interpretations so that students can find the easier one to work with in Exercise 3.

Exercise 3 This problem is a reverse of problems asking for the length of a fractal design at a given stage. Reverse problems can help deepen understanding.

Calculator Note 0B shows how to raise numbers to powers. Though this lesson does not use the word *power* for *exponent,* the calculator note does. Students might ask how exponents are related to roots. Taking a root of a number is the opposite of raising the number to a power.

5b. Stage 5: $\left(\frac{5}{4}\right)^5 \approx 3.05$; Stage 11: $\left(\frac{5}{4}\right)^{11} \approx 11.64$

Exercise 6 [Alert] Watch for difficulties in handling exponents and fraction reduction at the same time. Suggest that it's easier to raise smaller numbers to powers than larger ones, so that it's usually better to do fraction reduction before exponentiation.

6. The Stage 0 figure below has a length of 1.

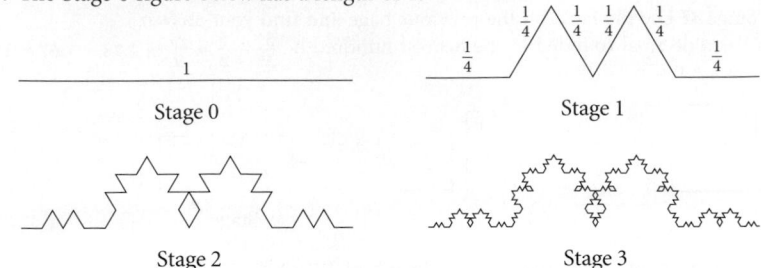

Stage 0   Stage 1

Stage 2   Stage 3

a. Complete a table like the one below by calculating the total length of the figure at each stage shown above. Give each answer as a fraction in expanded form, as a fraction in exponent form, and as a decimal rounded to the nearest hundredth.

Stage number	Total length		
	Expanded form	Exponent form	Decimal form
0	1	1^0	1
1	$6 \cdot \dfrac{1}{4} = \dfrac{6}{4} = \dfrac{3}{2}$	$6^1 \cdot \left(\dfrac{1}{4}\right)^1 = \left(\dfrac{6}{4}\right)^1 = \left(\dfrac{3}{2}\right)^1$	1.5
2	$6 \cdot 6 \cdot \dfrac{1}{4} \cdot \dfrac{1}{4} = \dfrac{36}{16} = \dfrac{9}{4}$	$6^2 \cdot \left(\dfrac{1}{4}\right)^2 = \left(\dfrac{6}{4}\right)^2 = \left(\dfrac{3}{2}\right)^2$	2.25
3	$6 \cdot 6 \cdot 6 \cdot \dfrac{1}{4} \cdot \dfrac{1}{4} \cdot \dfrac{1}{4} = \dfrac{216}{64} = \dfrac{27}{8}$	$6^3 \cdot \left(\dfrac{1}{4}\right)^3 = \left(\dfrac{6}{4}\right)^3 = \left(\dfrac{3}{2}\right)^3$	3.38

b. At what stage does the figure have a length of $\dfrac{243}{32}$? Stage 5

c. At what stage is the length closest to 100? Stage 11: $\left(\dfrac{6}{4}\right)^{11} \approx 86.50$

7. The Stage 0 figure below has a length of 1.

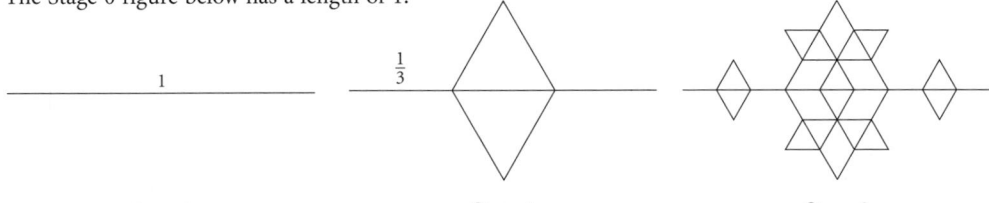

Stage 0   Stage 1   Stage 2

Stage 3   Stage 4

a. Complete a table like the one below by calculating the total length of each stage. Give each answer as a fraction in expanded form, as a fraction in exponent form, and as a decimal number rounded to the nearest hundredth. Figure out the lengths of Stages 3 and 4 without counting.

Stage number	Total length		
	Expanded form	Exponent form	Decimal form
0	1	1^0	1
1	$7 \cdot \frac{1}{3} = \frac{7}{3}$	$7^1 \cdot \left(\frac{1}{3}\right)^1 = \left(\frac{7}{3}\right)^1$	2.33
2	$7 \cdot 7 \cdot \frac{1}{3} \cdot \frac{1}{3} = \frac{49}{9}$	$7^2 \cdot \left(\frac{1}{3}\right)^2 = \left(\frac{7}{3}\right)^2$	5.44
3	$7 \cdot 7 \cdot 7 \cdot \frac{1}{3} \cdot \frac{1}{3} \cdot \frac{1}{3} = \frac{343}{27}$	$7^3 \cdot \left(\frac{1}{3}\right)^3 = \left(\frac{7}{3}\right)^3$	12.70
4	$7 \cdot 7 \cdot 7 \cdot 7 \cdot \frac{1}{3} \cdot \frac{1}{3} \cdot \frac{1}{3} \cdot \frac{1}{3} = \frac{2401}{81}$	$7^4 \cdot \left(\frac{1}{3}\right)^4 = \left(\frac{7}{3}\right)^4$	29.64

b. At what stage does the figure have a length of $\frac{16,807}{243}$? Stage 5

c. Will the figure ever have a length of 168? If so, at what stage? If not, why not?
No. Stage 6 has a length of slightly more than 161, and Stage 7 has a length of over 376.

IMPROVING YOUR REASONING SKILLS

As the Koch curve develops, segment length decreases as the number of segments increases.

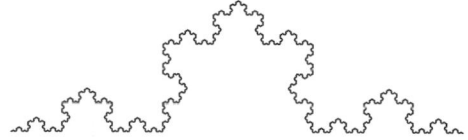

As you draw higher stages, the length of individual segments approaches, or gets closer and closer to, what number? What number does the number of line segments approach? Is it possible to draw the "finished" fractal? Why or why not? What would its total length be?

Now consider this pattern:

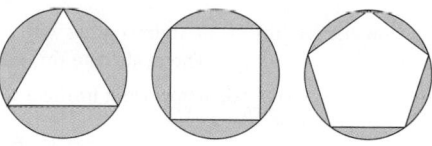

As you draw higher and higher stages, what does the length of each polygon side approach? What number does the number of sides approach? Is it possible to draw the "finished" polygon? What would it look like? What would the total perimeter of the polygon be? Is this pattern recursive? Is the result a fractal? Why or why not?

IMPROVING REASONING SKILLS

This activity builds on an analogy between forming fractals and approximating a circle with polygons. Like fractal designs, this figure lays some groundwork for the idea of limit. Students should realize that the polygon with "infinitely many sides" is the circle itself and that the area between the inscribed polygon and the circle approaches 0 as the number of sides increases.

Although the process shown for generating regular polygons isn't recursive (no stage builds on the previous one), it could be made so, for example, by repeatedly doubling the number of sides. Some students may believe that the resulting circle is a fractal because it's the result of an infinite sequence of recursive operations. Others may argue that it's not a fractal because it's not self-similar.

8. The figures below look a little more complicated because parts overlap. The Stage 0 figure has a length of 1.

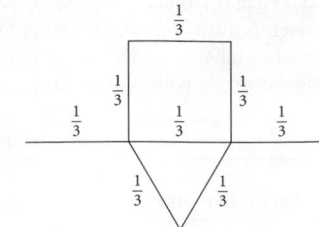

Stage 0

Stage 1

Stage 2

Stage 3

a. Complete a table like the one below by calculating the total length of the Stage 2 and Stage 3 figures shown above.

Stage number	Total length		
	Expanded form	Exponent form	Decimal form
0	1	1^0	1
1	$8 \cdot \frac{1}{3} = \frac{8}{3}$	$8^1 \cdot \left(\frac{1}{3}\right)^1 = \left(\frac{8}{3}\right)^1$	2.67
2	$8 \cdot 8 \cdot \frac{1}{3} \cdot \frac{1}{3} = \frac{64}{9}$	$8^2 \cdot \left(\frac{1}{3}\right)^2 = \left(\frac{8}{3}\right)^2$	7.11
3	$8 \cdot 8 \cdot 8 \cdot \frac{1}{3} \cdot \frac{1}{3} \cdot \frac{1}{3} = \frac{512}{27}$	$8^3 \cdot \left(\frac{1}{3}\right)^3 = \left(\frac{8}{3}\right)^3$	18.96

b. Look at how the lengths of the figures grow with each stage. Estimate how long the length will be in Stage 4. Then calculate this value. Estimates will vary; $\left(\frac{8}{3}\right)^4$ or about 50.

c. At what stage does your calculator begin to use a different notation for the length? After 23 stages, many calculators resort to scientific notation or get an overflow error.

▶ **Review**

Whenever possible it's a good idea to try to estimate your answer before calculating it. Estimating will help you determine whether or not your calculated answer is reasonable.

9. Write $\frac{14}{5}$ as a decimal. 2.8

10. What is $\frac{8}{3} - \frac{4}{9} \cdot \frac{3}{1}$? $1\frac{1}{3}$

11. Look at the fractal "cross" pattern below. At each stage, new line segments are drawn through the existing segments to create crosses.

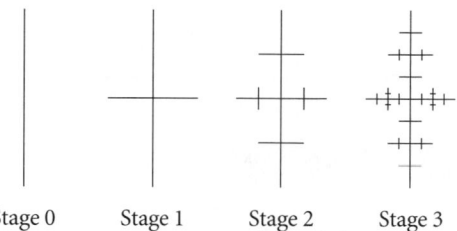

Stage 0   Stage 1   Stage 2   Stage 3

a. How many new segments are drawn in Stage 2? 4

b. How many new segments are drawn in Stage 3? 16

c. How many new segments would be drawn in Stage 4? 64

d. Use exponents to represent the number of new segments drawn in Stages 2 to 4. $4^1, 4^2, 4^3$

e. In general, how is the exponent related to the stage number for Stages 2 to 4? Does this rule apply to Stage 1? The exponent is one less than the stage number. For Stage 1, $4^0 = 1$

---

project

INVENT A FRACTAL!

Recursive procedures can produce surprising and even beautiful results. Consider these two fractals. (The top one was "invented" by student Andrew Riley!) Would you have expected that the Stage 1 figures would lead to the higher-stage figures?

Stage 0   Stage 1   Stage 4

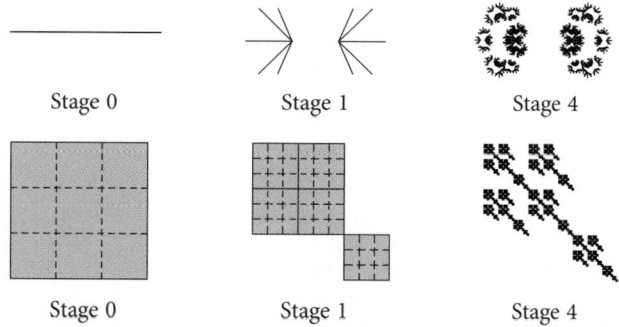

Stage 0   Stage 1   Stage 4

> **THE GEOMETER'S SKETCHPAD®**
>
> The Geometer's Sketchpad® was used to create these fractals. Sketchpad™ has several tools to help you create fractals. With Sketchpad, you can quickly and easily create Sierpiński's triangle, the Koch curve, and more. Learn how to use Sketchpad and create your own fractals!

Invent your own fractal. You can start with a line segment, as in the Koch curve or Andrew's fractal. Or try a two-dimensional shape like Stage 0 of Sierpiński's triangle or the square in the "kite" above. Your project should include

► A drawing of your fractal at Stages 0, 1, 2, and 3 (and possibly higher).

► A written description of the recursive rule that generates the fractal.

► A table that shows how one aspect of the figure changes. Consider area, length, or the number of holes or branches at each stage.

► A written explanation of how to continue your table for higher stages.

Invent a Fractal Project

This is the recursive process that created the first fractal pattern.

Divide a segment into thirds and remove the middle third. Call the four endpoints 1, 2, 3, and 4.

Rotate point 1 around point 2 by 45° and −45° and then rotate point 2 around point 1 by 45° and −45°. Do the analogous operations on the segment joining points 3 and 4.

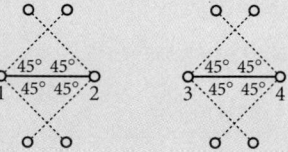

Now draw segments between point 2 and the four newly created points above and below it, and between point 3 and the four newly created points above and below it.

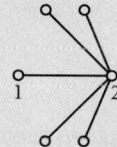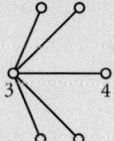

Then repeat the process on each of the ten segments shown above.

Supporting the project

MOTIVATION

To pique student interest, ask "What recursive rule did Andrew Riley use to create his fractal?" (See the side column for a complete description of Riley's process.)

OUTCOMES

► The written description is sufficient to produce the fractal illustrated.

► The table shows a clear pattern in the changing aspect.

► Correct mathematical language is used to explain how to continue the table.

• The fractal and table are extended to stages beyond Stage 3.

PLANNING

LESSON OUTLINE

One day:

5 min	Introduction
20 min	Investigation and sharing
10 min	Examples
5 min	Closing
10 min	Exercises

MATERIALS

- Number Lines (W, one per group)
- Number Lines (T)
- Calculator Notes 0C, 0D

TEACHING

In this lesson the notion of recursion is extended to algebraic expressions.

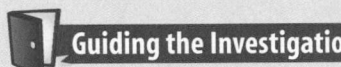 **Guiding the Investigation**

Give a worksheet to each group.

One step Assign each group of experienced investigators a different pair of expressions, one from Step 1 and one from Step 8 or the one from Example B. Summarize the instructions in the steps.

For new investigators, you might begin by working through the investigation steps for one expression as a whole class.

Steps 1 to 5 Draw a number line on the board and have a few volunteers mark their results. Because the results in this part all tend toward either infinity or negative infinity, it won't take long before they are headed well off the board. You can hand out Calculator Notes 0D and 0C as needed.

LESSON

0.4

Going Somewhere?

Leslie was playing miniature golf with her friends. First she hit the ball past the hole. Then she hit it back, but it went too far and missed again. She kept hitting the ball closer, but it still missed the hole. Finally she got so close that the ball fell in.

Some number processes also get closer and closer to a final target, until the result is so close that the number rounds off to the target value or answer. You'll explore processes like these while reviewing operations with positive and negative numbers.

 ## Investigation
A Strange Attraction

Step 1 Each member of your group takes one of these four expressions.

$$2 \cdot \square + 1 \qquad 3 \cdot \square - 4 \qquad -2 \cdot \square + 3 \qquad -3 \cdot \square - 1$$

Step 2 As a group, choose a starting number. Record your expression and starting number in a table like this.

Original Expression:	
Starting Number (at Stage 0):	
Stage number	Result
1	

Step 3 Put your starting number in the box, and do the computation. This process is called **evaluating the expression,** and the result is called the **value of the expression.** When doing the computation, be sure to follow the order of operations. Check your answer with a calculator, and record it in the table as your first result. [▷ 🖵 See **Calculator Note 0C** to learn about the difference between the negative key and the subtraction key. ◁]

Step 4 Take the result you got from Step 3, put it in the box in your expression, and evaluate your expression again. Place your new answer in the table as your second result.

Step 5 Continue this recursive process using your result from the previous stage. Evaluate your expression. Each time, record the new result in your table. Do this ten times.

[Alert] Watch for confusion: The first number to be entered in the calculator for recursion is not the first number in the expression.

[Language] Use the vocabulary *evaluating the expression* and *value of the expression* as you record results.
[Alert] Watch for use of the word *equation* instead of *expression*; every equation must have an equal sign.

If by chance a student suggests that you begin with a value that leads to convergence $\left(-1, 2, 1, \text{and } -\frac{1}{4}\right)$, follow through with the suggestion; then wonder aloud if you've found an expression for which all starting values lead to convergence, and ask for others to try.

Step 6 | Draw a number line and scale it so that you can show the first five results from your table. Plot the first result from your table, and draw an arrow to the next result to show how the value of the expression changes. For example,

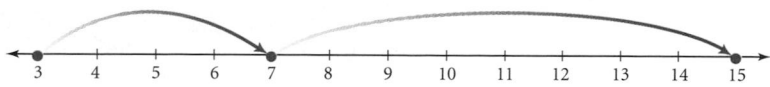

Step 7 | How do the results in your group compare?

Step 8 | Repeat Steps 1 to 6 with one of the expressions below.

$$0.5 \cdot \square - 3 \qquad 0.2 \cdot \square + 1 \qquad -0.5 \cdot \square + 3 \qquad -0.2 \cdot \square - 2$$

Step 9 | How do the results in your group compare? Do the results of these expressions differ from the results of your first expression? Students starting with the same expression will see that it tends toward the same specific value: $-6, 1.25, 2,$ or $-\frac{5}{3}$.

In this investigation, you explored what happens when you recursively evaluate an expression. First you selected a starting number to put into your expression, then you evaluated it. Then you put your result back into the same expression and evaluated it again. Calculators, like computers, are good tools for doing these repetitive operations.

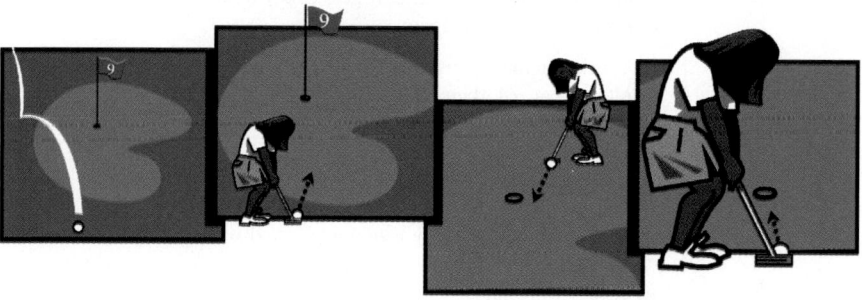

EXAMPLE A | What happens when you evaluate this expression recursively with different starting numbers?

$$0.5 \cdot \square - 2$$

▶ **Solution**

The starting number 1 gives

> Have your pencil and calculator in hand as you work through the solution to this example.

$$0.5 \cdot (1) - 2 = 0.5 - 2 = -1.5$$

$$0.5 \cdot (-1.5) - 2 = -0.75 - 2 = -2.75$$

$$0.5 \cdot (-2.75) - 2 = -1.375 - 2 = -3.375$$

$$0.5 \cdot (-3.375) - 2 = -1.6875 - 2 = -3.6875$$

$$0.5 \cdot (-3.6875) - 2 = -1.84375 - 2 = -3.84375$$

Steps 6 to 9 These steps give surprising results and are good as a small group activity.

You might demonstrate recursion on the calculator at this point. Sometimes recursion on a graphing calculator is called *homescreen iteration. Iteration* is a fancy word for repetition. Recursion is a special case of iteration.

Step 7 Students starting with the same expression will see the same patterns. The first two expressions will increase without bounds if students started with a number above -1 for the first or 2 for the second. Or those expressions will decrease toward negative infinity if they started with numbers below -1 or 2 respectively. The last two expressions will oscillate in ever-widening swings.

SHARING IDEAS

For each expression in Step 8, call on a different group to report its results. Make a table for the class. **[Ask]** "How might you predict from the expression the number that's being approached?" "Under what conditions will the process approach a number?"

Ask students to explain how the recursion is like the work they did with fractal drawing. Students may compare the attractor encountered in these later expressions with the confined infinity they saw before, thus deepening their intuitive feeling for mathematical limits.

Assessing Progress

As you listen to students contribute to groups or the whole class, check that they understand the concepts reviewed in this lesson: **operations on signed numbers, recursion,** and **order of operations.**

LESSON OBJECTIVES

- Practice arithmetic operations on signed numbers
- Work on recognizing patterns
- Extend the idea of recursion to expressions
- Learn to evaluate an expression recursively on the calculator
- Become familiar with the phrases *value of an expression* and *evaluating an expression*
- Learn the concept of an attractor

NCTM STANDARDS

CONTENT		PROCESS	
•	Number		Problem Solving
•	Algebra	•	Reasoning
	Geometry		Communication
	Measurement	•	Connections
	Data/Probability	•	Representation

► EXAMPLE A

This example is good to use if some students haven't grasped the details of a convergent case. Make sure students notice that the result of the previous iteration is inserted into the calculation for the next iteration. In this example any starting value will iterate toward a single value, -4.

► EXAMPLE B

This example shows an expression in which there are two attractors (-1 and 2) instead of just one. Some starting values lead to one of those attractors, and other starting values lead to the other.

Each result of the recursion seems to get closer to a certain number. If you continue the process a few more times, you'll get -3.9219, then -3.9609, then -3.9805. What do you think will happen after even more recursions?

Using 4 as a starting number in the same expression, you get

$$0.5 \cdot (4) - 2 = 2 - 2 = \boxed{0}$$

$$0.5 \cdot (0) - 2 = 0 - 2 = \boxed{-2}$$

$$0.5 \cdot (-2) - 2 = -1 - 2 = \boxed{-3}$$

$$0.5 \cdot (-3) - 2 = -1.5 - 2 = \boxed{-3.5}$$

$$0.5 \cdot (-3.5) - 2 = -1.75 - 2 = \boxed{-3.75}$$

Again the values seem to get closer to one number, perhaps -4. If any starting number that you try eventually gets closer and closer to -4, then -4 is called an **attractor** for this expression.

To check whether -4 is an attractor value, use it as the starting number. If your result stays at -4, then it is an attractor value.

Using -4 as the starting number gives

$$0.5 \cdot (-4) - 2 = -2 - 2 = -4$$

Since you get back exactly what you started with, -4 is an attractor value for the expression $0.5 \cdot \square - 2$.

Evaluating expressions recursively does not always lead to an attractor value. Some expressions continue to grow larger when evaluated recursively, while others are difficult, if not impossible, to recognize.

EXAMPLE B

What happens when you recursively evaluate this expression for different starting numbers?

$$\square^2 - 2$$

► Solution

Choosing 1 as a starting number:

$$(1)^2 - 2 = 1 - 2 = -1$$

$$(-1)^2 - 2 = 1 - 2 = -1$$

$$(-1)^2 - 2 = 1 - 2 = -1$$

The results are all -1's, so -1 is an attractor value for this expression. On a number line the results look like this:

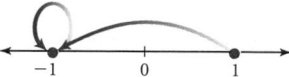

Choosing 3 as a starting number:

$$(3)^2 - 2 = 9 - 2 = 7$$
$$(7)^2 - 2 = 49 - 2 = 47$$
$$(47)^2 - 2 = 2209 - 2 = 2207$$

In this case the results get larger and larger.

Choosing -2 as a starting number:

$$(-2)^2 - 2 = 4 - 2 = 2$$
$$(2)^2 - 2 = 4 - 2 = 2$$

So 2 is another attractor value for this expression.

Choose any other starting number, and either you'll get a series of repeating -1's or 2's, or the values will get farther apart at each stage.

With enough practice you may be able to predict the attractor values for some simple expressions without actually doing any computations. But as you try this process with more complex expressions, the results are less predictable.

Remind students that as they have explored recursion on expressions, they have used **recursion on their calculator** and they **evaluated an expression** and found the **value of an expression.**

They have also learned a new mathematical term, **attractor.** Emphasize the closing paragraph in the student text.

Recursion on a calculator will be a useful skill for many of these exercises.

ASSIGNING HOMEWORK

Essential	1–5, 7, 8
Performance assessment	7
Portfolio	9, 10
Journal	4
Group	6, 8
Review	11, 12

▶ Helping with the Exercises

Exercise 4 Students may not know the term *absolute value* and may refer to "the number without its sign" in their answers.

4a. The sign of the answer is the sign of the number that is larger in absolute value. Subtract the number with the smaller absolute value from the number with the larger absolute value.

4b. Add the two numbers together. The result will be a negative number.

4c. Subtracting a negative number is the same as adding a positive number, so you just add the two numbers. The result will be a positive number.

4d. Subtracting a negative number is the same as adding a positive number, so the problem actually involves adding a negative number and a positive number.

4e. Multiply the two numbers. When you multiply two numbers with different signs, the result is a negative number.

4f. Multiply the two numbers. When you multiply two numbers with the same sign, the result is a positive number.

EXERCISES

You will need your calculator for problems **1, 2, 3, 5, 6, 7, 8,** and **9.**

▶ Practice Your Skills

1. Do each calculation and use a calculator to check your results. Then use a number line to illustrate your answer.

    a. $-4 + 7$ 3   **b.** $5 + (-8)$ -3

    c. $-2 - 5$ -7 **d.** $-6 - (-3)$ -3

2. Do each calculation and check your results on your calculator.

    a. $-2 \cdot 5$ -10 **b.** $6 \cdot (-4)$ -24 **c.** $-3 \cdot (-4)$ 12

    d. $-12 \div 3$ -4 **e.** $36 \div (-6)$ -6 **f.** $-50 \div (-5)$ 10

3. Do the following calculations. Remember, if there are no parentheses you must do multiplication or division before addition or subtraction. Check your results by entering the expression exactly as it is shown on your calculator.

    a. $5 \cdot -4 - 2 \cdot -6$ -8 **b.** $3 + -4 \cdot 7$ -25

    c. $-2 - 5 \cdot (6 + -3)$ -17 **d.** $(-3 - 5) \cdot -2 + 9 \cdot -3$ -11

4. Explain how to do each operation described below, and state whether the result is a positive or a negative number.

    a. adding a negative number and a positive number

    b. adding two negative numbers

    c. subtracting a negative number from a positive number

    d. subtracting a negative number from a negative number

    e. multiplying a negative number by a positive number

    f. multiplying two negative numbers

    g. dividing a positive number by a negative number

    h. dividing two negative numbers

▶ Reason and Apply

5. Pete Repeat was recursively evaluating this expression starting with 2.

$$-0.2 \cdot \square - 4$$

    a. Check his first two stages and explain what, if anything, he did wrong.

    $-0.2 \cdot 2 - 4 = 0.4 - 4 = -3.6$

    $-0.2 \cdot -3.6 - 4 = -0.72 - 4 = -4.72$

    b. Redo Pete's first two stages and do two more.

    c. Now do three recursions using Pete's expression and starting with -1. $-3.8, -3.24, -3.352$

    d. Do you think this expression has an attractor value? Explain.   Yes. The calculations seem to be approaching a value close to -3.3.

4g. Divide the two numbers. When you divide two numbers with different signs, the result is a negative number.

4h. Divide the two numbers. When you divide two numbers with the same sign, the result is a positive number.

Exercise 5 Here students check and then correct some common errors. Problems like this appear throughout the text and give students practice in looking for mistakes.

5a. In the first recursion, he should get $-0.2 \cdot 2 = -0.4$, not $+0.4$. His arithmetic when evaluating $0.4 - 4$ was correct. In the second recursion, he used the wrong value (-3.6 instead of -4.4) because of his previous error. His arithmetic was also incorrect because $-0.2 \cdot -3.6 = +0.72$, not -0.72. His arithmetic when evaluating $-0.72 - 4$ was correct.

5b. $-4.4, -3.12, -3.376, -3.3248$

6. To tell whether or not an expression has an attractor value, you often have to look at the results of several different starting values.

Starting value	2	−1	10
First recursion	−1.8	−2.1	−1
Second recursion	−2.18	−2.21	−2.1
Third recursion	−2.218	−2.221	−2.21

   a. Evaluate this expression for different starting values.

$$0.1 \cdot \square - 2$$

   Record the results for three recursions in a table like the one shown.

   b. Based on your table, do you think this expression reaches an attractor value in the long run? If so, what is it? If not, why not? Yes, about −2.222.

   c. If you found an attractor in 6b, use your calculator to see if substituting that value in the expression gives it back to you. Entering −2.222222222222 as a starting value returns the same value as an answer.

7. Investigate this expression.

$$-2 \cdot \square + 1$$

Starting value	2	−1	10
First recursion	−3	3	−19
Second recursion	7	−5	39
Third recursion	−13	11	−77

   a. Evaluate the expression for different starting values. Record your results in a table like the one shown.

   b. Based on your table, do you think this expression reaches an attractor in the long run? If so, what is it? If not, why not? No, the values get farther and farther apart.

8. Use a calculator to investigate the behavior of these expressions.
[▶ 🖳 See **Calculator Note 0D** for specific instructions. ◀]

   a. Use recursion to evaluate each expression many times, and record the attractor value you get after many recursions.

     i. $0.5 \cdot \square + 6$ 12   **ii.** $0.5 \cdot \square - 8$ −16   **iii.** $0.5 \cdot \square - 4$ −8

   b. Describe any connections you see between the numbers in the original expressions and their attractor values.

   c. Create an expression that has an attractor value of 6. One possibility is $0.5 \cdot \square + 3$.

9. Use a calculator to investigate the behavior of the expressions below.

     i. $0.2 \cdot \square + 6$ 7.5 or $\frac{15}{2}$ **ii.** $0.2 \cdot \square - 8$ −10   **iii.** $0.2 \cdot \square + 5$ 6.25

   a. Use recursion to evaluate each expression many times, and record its attractor value.

   b. Describe any connections you see between the numbers in the original expressions and the attractor values.

   c. Create an expression that has an attractor value of 2.25. One possibility is $0.2 \cdot \square + 1.8$.

10. How is the recursion process like drawing the Sierpiński triangle in Lesson 0.1? How is it like creating the Koch curve in Lesson 0.3? All involve repeating a process. Each time, the result becomes the starting value or figure for the next repetition.

▶ **Review**

11. What is $4 - 12 \div 4 \cdot \frac{1}{2} - 5^2$? −22.5

12. Find $(-3 \cdot -4) - (-4 \cdot 2)$. 20

Exercise 8 Students need Calculator Note 0D: Recursion

8b. When the coefficient of the box is 0.5, the attractor value is twice the constant. In general the attractor value is

$$\frac{constant\ term}{1 - coefficient\ of\ the\ box}$$

9b. When the coefficient is 0.2, the attractor value is 1.25 times the constant. In general the attractor value is

$$\frac{constant\ term}{1 - coefficient\ of\ the\ box}$$

LESSON

0.5

PLANNING

LESSON OUTLINE

One day:

35 min   Investigation with sharing

5 min   Closing

10 min   Exercises

MATERIALS

- dice (one per pair)
- centimeter rulers or rulers cut from the transparency Centimeter Rulers (one per pair)
- A Chaotic Pattern? (W, one per pair)
- transparencies and erasable transparency markers (one per pair)
- Calculator Notes 0E, 0F, 0G

TEACHING

In this lesson students discover that a fractal is an attractor of a random recursive process.

Guiding the Investigation

Every class will benefit from following the steps given in the text. Organize students into pairs. Give each pair of students the worksheet, a transparency, and a transparency marker.

Steps 1 to 4 You may want to demonstrate the first several steps in the Chaos game on the board or overhead projector, explaining the random element of rolling a fair die. Model careful measurement and care in remembering the location of the last dot marked. The first three or four dots might not fit the pattern, so after you have marked four points, erase the first three. Be sure students follow the directions to switch tasks after the first 20 points are marked.

Out of Chaos

If you looked at the results of 100 rolls of a die, would you expect to find a pattern in the numbers? You might expect each number to appear about one-sixth of the time. But you probably wouldn't expect to see a pattern in when, for example, a 5 appears. The 5 appears **randomly,** without order. You could not create a method to predict exactly when or how often a 5 appears. As you explore seemingly random patterns, you'll review some measurement and fraction ideas.

Nothing in nature is random.... A thing appears random only through the incompleteness of our knowledge.

BARUCH SPINOZA

Investigation
A Chaotic Pattern?

You will need

- a die
- a centimeter ruler
- a blank transparency and marker
- the worksheet A Chaotic Pattern?

What happens if you use a random process recursively to determine where you draw a point? Would you expect to see a pattern?

Work with a partner. One partner rolls the die. The other measures distance and marks points.

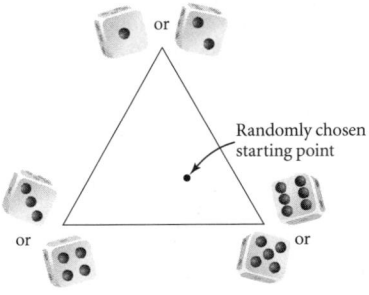

Randomly chosen starting point

For best results, measure as accurately as you can.

Step 1	Mark any point inside the triangle as your starting point.
Step 2	Roll the die.
Step 3	In centimeters, measure the distance from your starting point to the corner, or *vertex*, labeled with the number on the die. Take half of the distance, and place a small dot at this midpoint. This is your new starting point.
Step 4	Repeat Steps 2 and 3 until you've rolled the die 20 times. Then switch roles with your partner and repeat the process another 20 times.
Step 5	How is this process recursive? Answers should indicate that the result from one step becomes the input for the next step.
Step 6	Describe the arrangement of dots on your paper. Answers should mention that some areas show lots of dots while others remain empty. Some might mention Sierpiński's triangle.
Step 7	What would have happened if you had numbered the vertices of the triangle 1 and 3, 2 and 5, and 4 and 6? Answers should indicate that the result would be the same.

Step 2 [Language] *Die* is singular, *dice* is plural.

Step 3 [Language] Use the term *vertex* for a corner of the triangle and *vertices* for the plural.

These four steps take about 10 minutes to complete. Student results will look pretty chaotic at this point, although students may notice a lack of dots in the centers of their triangles. To keep students motivated, you might want to remind them that larger collections of data often show more patterns than

smaller sets. Hint that they will eventually see an unexpected pattern.

Step 5 The procedure is recursive: At each stage students are building on the point that results from the previous stage.

Step 8 | Place a transparency over your worksheet. Use a transparency marker and mark the vertices of the triangle. Carefully trace your dots onto the transparency.

Step 9 Since the **Step 9** | When you finish, place your transparency on an overhead projector. Align the position of a dot is used as | vertices of your triangle with the vertices of your classmates' triangles. This input for determining the next | allows you to see the results of many rolls of a die. Describe what happens when dot, the process is recursive. | you combine everyone's points. How is this like the result in other recursion processes? Is the result as random as you expected? Explain.

A *random* process can produce ordered-looking results while an orderly process can produce random-looking results. Mathematicians use the term *chaotic* to describe systematic, non-random processes that produce results that look random. Chaos helps scientists understand the turbulent flow of water, the mixing of chemicals, and the spread of an oil spill. They often use computers to do these calculations. Your calculator can repeat steps quickly, so you can use the calculator to plot thousands of points.

Step 10 | Enter the Chaos program into your calculator. [▶🖳 See **Calculator Note 0E** for the program. To learn how to link calculators, see **Calculator Note 0F.** To learn about how to enter a program, see **Calculator Note 0G.**◀]

The program randomly "chooses" one vertex of the triangle as a starting point. It "rolls" an imaginary die and plots a new point halfway to the vertex it chose. The program rolls the die 999 more times. It does this a lot faster than you can.

Step 11 | Run the program. Select an equilateral (equal-sided) triangle as your shape. When it "asks" for the fraction of the distance to move, enter $\frac{1}{2}$ or 0.5. It will take a while to plot all 1000 points, so be patient.

Step 12 | What do you see on your calculator screen, and how does it compare to your class's combined transparency image? Answers should mention Sierpiński's triangle and the similarity to the combined points for the class transparencies.

Most people are surprised that after plotting many points, a familiar figure appears. When an orderly result appears out of a random process like this one, the figure is a *strange attractor*. No matter where you start, the points "fall" toward this shape. Many fractal designs, like the Sierpiński triangles on your calculator screen, are also strange attractors. Accurate measurements are essential to seeing a strange attractor form. In the next example, practice your measurement skills with a centimeter ruler.

Science
CONNECTION

The growth and movement of an oil spill may appear random, but scientists can use chaos theory to predict its boundaries. This can aid restraint and cleanup. Learn more about the application of chaos theory with the Internet links at **www.keymath.com/DA** .

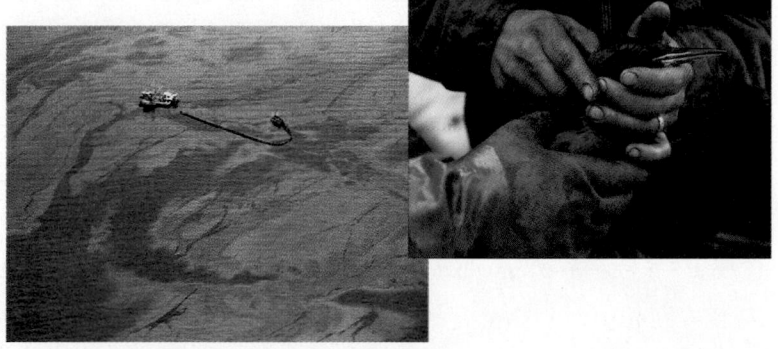

LESSON OBJECTIVES

- Practice multiplication with fractions and decimals
- Deepen understanding of recursion and attractors
- Practice careful measuring skills
- Review the meaning of a vertex of a triangle
- Learn to link calculators to transfer a program
- Learn to run a calculator program

NCTM STANDARDS

CONTENT		PROCESS	
	Number		Problem Solving
	Algebra	•	Reasoning
•	Geometry		Communication
•	Measurement	•	Connections
	Data/Probability	•	Representation

Step 9 As students add more layers of transparencies, they will begin to see a familiar pattern appear. It can be very exciting to see the Sierpiński triangle appearing out of the chaotic mass of dots. The Sierpiński triangle is an attractor for the Chaos game.

If a few dots don't fit the pattern, they were probably some of the first few dots a group marked.

Step 10 Students need Calculator Notes 0E, 0F, and 0G.

Step 11 You might challenge students to try various shapes and fractions and report their results.

SHARING IDEAS

Most reporting in this lesson takes place during the investigation. You need not push for deep understanding of the two terms defined here informally: *random* and *predict.* They will be used again later.

[Ask] "Did the game result in a random pattern?" [No, the process was random, but the overall result is predictable.]

You might mention that, technically, the Chaos game is misnamed; mathematical chaos is the production of seemingly random patterns by methods that are not random.

The random generation of the Sierpiński triangle gives experience with careful measurement and more practice at fraction multiplication and recursion.

Assessing Progress

To complete the investigation, students need to know what is meant by the **vertex** of a triangle and the **midpoint** of a segment. And they must be able to multiply fractions or decimals to find the midpoint of the segment between the vertex of the triangle and a point. They should understand that the Chaos game uses **recursion** and the resulting Sierpiński triangle is an **attractor.**

EXAMPLE

This example will be especially useful to students who are having difficulties with measurement and fractions. Using a marked number line might help some students.

Closing the Lesson

As needed, say something briefly about the importance of measuring carefully. Make sure there are no remaining questions about the new calculator skills of **transferring programs** between calculators and **running a calculator program.**

BUILDING UNDERSTANDING

Like the investigation, this set of exercises includes generating fractals by hand as well as with the calculator program.

ASSIGNING HOMEWORK

Essential	1–3, 6
Performance assessment	5
Portfolio	7
Journal	8
Group	4, 5
Review	9–11

▶ Helping with the Exercises

Exercise 1 [Alert] Watch for confusion about meaning of rounding to the nearest tenth of a centimeter.

Exercise 3 To students having difficulties with the level of abstraction, you might suggest assuming a distance such as 27 from *A* to *B* and then finding the distance of every point from *A*.

EXAMPLE

Find point *C* one-third of the way from *A* to *B*. Give the distance from *A* to *C* in centimeters.

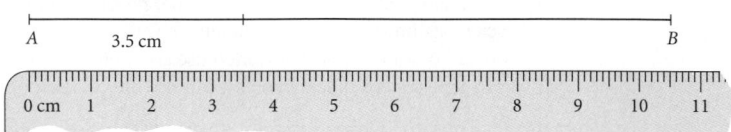

▶ **Solution**

Have your ruler handy so that you can check the measurements. Use your calculator to check the computations.

Measuring segment *AB* shows that it is about 10.5 cm long. (Check this.) Find one-third of this length.

$$\frac{1}{3} \cdot 10.5 = \frac{10.5}{3}$$ Multiply by $\frac{1}{3}$ or divide by 3.

$$= 3.5$$ Divide.

Place a point 3.5 cm from *A*.

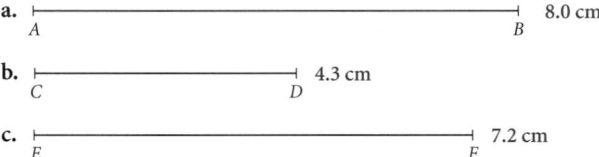

EXERCISES

You will need your calculator for problems **4, 5, 6, 7,** and **10.**

▶ Practice Your Skills

1. Estimate the length of each segment. Then measure and record the length to the nearest tenth of a centimeter.

   a. $\overline{A \quad\quad\quad\quad\quad B}$ 8.0 cm

   b. $\overline{C \quad\quad\quad D}$ 4.3 cm

   c. $\overline{E \quad\quad\quad\quad F}$ 7.2 cm

2. Draw a segment to fit each description.
   a. one-third of a segment 8.4 cm long 2.8-cm segment
   b. three-fourths of a segment 7.6 cm long 5.7-cm segment
   c. two-fifths of a segment 12.7 cm long about 5.1-cm segment

3. Mark two points on your paper. Label them *A* and *B*. Draw a segment between the two points.
   a. Mark a point two-thirds of the way from *A* to *B* and label it *C*.
   b. Mark a point two-thirds of the way from *C* to *B* and label it *D*.
   c. Mark a point two-thirds of the way from *D* to *A* and label it *E*.
   d. Which two points are closest together? Does it matter how long your original segment was? points *D* and *B*; no

4. Do these calculations. Check your results with a calculator.
   a. $-2 + 5 - (-7)$ 10 b. $(-3)^2 - (-2)^3$ 17 c. $\frac{3}{5} + \frac{-2}{3}$ $\frac{-1}{15}$ d. $-0.2 \cdot 20 + 15$ 11

3a–c. Answers should look proportional to this:

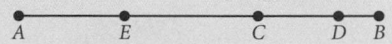

▶ Reason and Apply

[▶ 🖥] You'll need the program in **Calculator Note 0E** for these exercises.◀]

5. Draw a large right triangle on your paper. You can use the corner of a piece of paper or your book to help draw the right angle numbered 2. Number the vertices as shown at right.

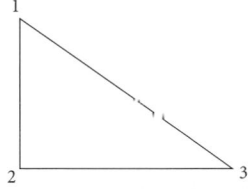

a. Choose a point anywhere inside your triangle. This is your starting point.

b. Write the numbers 1, 2, and 3 on three small pieces of paper. Put the pieces of paper in a cup or bowl and randomly select one to choose a vertex. Measure from the starting point to the chosen vertex. Mark a new point halfway from the starting point to the vertex.

c. Let the new point be your starting point, and repeat 5b at least 20 more times.

d. Describe any pattern you see forming. The resulting figure should resemble a right-angle Sierpiński triangle.

e. Run the Chaos program for problem 5 to see what happens when you plot 1000 points.

6. Draw a large square or rectangle on your paper. Number the vertices from 1 to 4 as shown at right.

a. Choose a point anywhere inside your figure. This is your starting point.

b. In order to choose a vertex to move toward, flip two different coins, such as a nickel and a penny. Use this scheme to determine the vertex:

Nickel	Penny	Vertex number
H	H	1
H	T	2
T	H	3
T	T	4

a–d. The resulting figure should resemble the Sierpiński carpet from Problem 7 in Exercises 0.1.

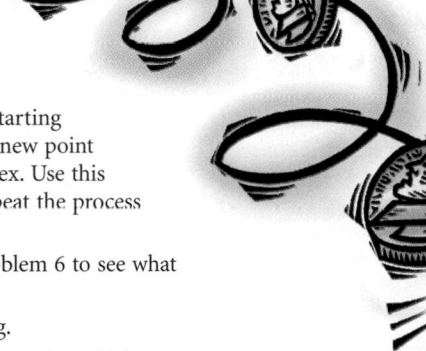

Measure the distance between your starting point and the chosen vertex. Mark a new point two-thirds of the distance to the vertex. Use this point as your new starting point. Repeat the process at least 20 more times.

c. Run the calculator simulation for problem 6 to see what happens when you plot 1000 points.

d. Describe any pattern you see forming.

e. How could you have used a die to determine which corner to move toward? What problems are there with using a die? Answers will vary. You could ignore any rolls of 5 or 6 and move toward the corners 1, 2, 3, or 4 on those rolls.

Exercise 5 Any triangle can be used to make a Sierpiński-like triangle in this way.

Exercise 7 The calculator allows entering either common fractions or decimal fractions at the "Enter a fraction" step.

7a. This game fills the entire square.

7b. This game creates a small Sierpiński triangle at each corner of the triangle.

7c. This game creates four small Sierpiński carpets, one at each corner of the square.

7d. This game creates a pattern like the Sierpiński triangle.

7. Experiment with the calculator program for each game description. For each, use the shape and fraction given. The program will start with a point inside the shape, randomly choose a vertex, and plot a point a fraction of the distance to the vertex. Describe your results and draw a sketch if possible.

   a. square, $\frac{1}{2}$ **b.** equilateral triangle, $\frac{2}{3}$

   c. square, $\frac{3}{4}$ **d.** right triangle, $\frac{2}{5}$

8. Suppose you are going to play a chaos game on a pentagon. Describe a process that would tell you which corner to move toward on each move. Possible answers: Use a die and ignore rolls of 6 or choose one of five playing cards to indicate the move. The answer should describe a process by which all corners are equally likely to be chosen.

▶ **Review**

9. Draw a segment that is 12 cm in length. Find and label a point that is two-thirds the distance from one of the endpoints. Point should divide segment into an 8-cm and a 4-cm segment.

10. Use a calculator to investigate the behavior of the expressions below.

   i. $-0.5 \cdot \boxed{} + 3$ 2   **ii.** $-0.5 \cdot \boxed{} + 6$ 4   **iii.** $-0.5 \cdot \boxed{} - 9$ −6

   a. Use recursion to evaluate each expression many times and record its attractor value.

   b. Describe any connections you see between the numbers in the original expressions and the attractor values. The attractor is two-thirds of the constant.

   c. Create an expression that has an attractor value of -10. One possibility is $-0.5 \cdot \boxed{} - 15$.

11. Look at the fractal below. The Stage 0 figure has a length of 1.

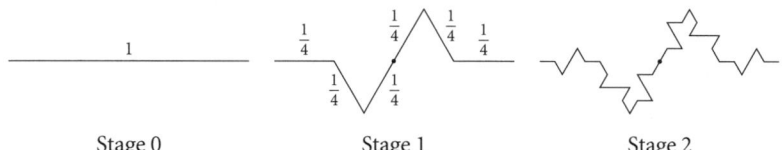

        Stage 0   Stage 1   Stage 2

Complete a table like the one below for Stages 0 to 2. Use any patterns you notice to extend the table for Stages 3 and 4.

Stage number	Total length		
	Multiplication form	Exponent form	Decimal form
0	1	1^0	1
1	$6 \cdot \frac{1}{4}$	$6^1 \cdot \left(\frac{1}{4}\right)^1$	1.5
2	$6 \cdot 6 \frac{1}{4} \cdot \frac{1}{4}$	$6^2 \cdot \left(\frac{1}{4}\right)^2 = \left(\frac{6}{4}\right)^2 = \frac{9}{4}$	2.25

11.

Stage number	Total Length		
	Multiplication form	Exponent form	Decimal form
3	$6 \cdot 6 \cdot 6 \cdot \left(\frac{1}{4}\right) \cdot \left(\frac{1}{4}\right) \cdot \left(\frac{1}{4}\right)$	$6^3 \cdot \left(\frac{1}{4}\right)^3 = \left(\frac{6}{4}\right)^3 = \frac{27}{8}$	3.38
4	$6 \cdot 6 \cdot 6 \cdot 6 \cdot \left(\frac{1}{4}\right) \cdot \left(\frac{1}{4}\right) \cdot \left(\frac{1}{4}\right) \cdot \left(\frac{1}{4}\right)$	$6^4 \cdot \left(\frac{1}{4}\right)^4 = \left(\frac{6}{4}\right)^4 = \frac{81}{16}$	5.06

CHAPTER 0 REVIEW

In this chapter you saw many instances of how you can start with a figure or a number, apply a mathematical rule, get a result, then apply the same rule to the result. This is called **recursion,** and it led you to find patterns in the results. When the recursive rule involved multiplication, you used an **exponent** as a shorthand way to show repeated multiplication.

Patterns in the results of recursion were often easier to see when you left them as common fractions. To add and subtract fractions, you needed a common denominator. You also needed to round decimals and measure lengths.

To **evaluate expressions** with any kind of numbers, you needed to know the **order of operations** that mathematicians use. The order is (1) evaluate what is in parentheses, (2) multiply out anything in exponent form, (3) multiply and divide as needed, and (4) add and subtract as needed. You used your knowledge of operations (add, subtract, multiply, divide, raise to a power) to write several expressions that gave the same number. Having an expression for the recursive rule helps you predict a value later in a sequence without figuring out all the values in between.

In this chapter you also got a peek at some mathematics that are new even to mathematicians, including **fractals** like the Sierpiński triangle, chaos, and strange attractors. You had to think about **random** processes and whether the long-term outcome of these processes was truly random.

EXERCISES

You will need your calculator for problems **1, 7,** and **8.**

1. Evaluate the expressions below.

a. 3^0 1

b. 3^1 3

c. 3^2 9

d. 3^3 27

e. 3^4 81

f. 3^8 6561

g. 3^{12} 531,441

h. $\left(\frac{1}{3}\right)^0$ 1

i. $\left(\frac{1}{3}\right)^1$ $\frac{1}{3}$

j. $\left(\frac{1}{3}\right)^2$ $\frac{1}{9}$

k. $\left(\frac{1}{3}\right)^3$ $\frac{1}{27}$

l. $\left(\frac{1}{3}\right)^4$ $\frac{1}{81}$

m. $\left(\frac{1}{3}\right)^8$ $\frac{1}{6561}$

n. $\left(\frac{1}{3}\right)^{12}$ $\frac{1}{531,441}$

2. Match equivalent expressions.

a. $\frac{1}{9} + \frac{1}{9} + \frac{1}{9}$ iii

b. $\frac{1}{9} + \frac{1}{9} + \frac{1}{3}$ v

c. $\frac{2}{9} + \frac{1}{9} + \frac{1}{27}$ ii

d. $\frac{4}{9} + \frac{2}{27} + \frac{3}{81}$ iv

e. $\frac{2}{81} + \frac{1}{3} + \frac{2}{27}$ i

i. $\frac{35}{81}$

ii. $\frac{10}{27}$

iii. $3 \times \frac{1}{9}$

iv. $\frac{12}{27} + \frac{2}{27} + \frac{1}{27}$

v. $2 \times \left(\frac{1}{9}\right) + \frac{1}{3}$

PLANNING

LESSON OUTLINE

One day:

5 min — Introduction

30 min — Helping individual students work on exercises

15 min — Student self assessment

REVIEWING

Direct students' attention to the table of lengths generated in the example on page 16. **[Ask]** "What expression could you evaluate recursively to generate the column of lengths?" If students try to use the last column, help them see that the patterns will be easier to find from the fourth column. You may need to remind students of how fractions and exponents are being used and that a recursive pattern needs a starting number (in this case, 1). The recursive rule is $\left(\frac{5}{3}\right)\square$. This might look simpler than the recursive expressions they worked with in Lesson 0.4, but it is different and therefore may be confusing. Be sure all students can use it to generate the sequence of lengths $1, \frac{5}{3}, \left(\frac{5}{3}\right)^2,$ $\left(\frac{5}{3}\right)^3,$ and so on.

As a challenge, ask them to consider the same fractal patterns but with the left segment $\left(\text{length } \frac{1}{3}\right)$ removed from Stage 0. Each length will be $\frac{1}{3}$ less than before: $\frac{2}{3}, \frac{5}{3} - \frac{1}{3}, \left(\frac{5}{3}\right)^2 - \frac{1}{3}, \left(\frac{5}{3}\right)^3 - \frac{1}{3}, \ldots$ The challenge is to get an expression that can be applied recursively so that each length is obtained from the previous length. As a hint, suggest that students first add $\frac{1}{3}$ to get the term back to that in the earlier sequence. Then they'll multiply that by $\frac{5}{3}$ and subtract $\frac{1}{3}$ from the total to get $\left(\square + \frac{1}{3}\right)\left(\frac{5}{3}\right) - \frac{1}{3}$. They can check

that this expression generates the sequence $\frac{2}{3}, \frac{5}{3} - \frac{1}{3}, \left(\frac{5}{3}\right)^2 - \frac{1}{3}, \left(\frac{5}{3}\right)^3 - \frac{1}{3}, \dots$ without simplifying to $\frac{2}{9} + \frac{5}{3}\,\square$, which requires students to understand that the distributive property can be used with expressions containing variables.

ASSIGNING HOMEWORK

If you plan to assess students individually, then they should work individually on these exercises. You can work more with students who have been having difficulties.

6a.

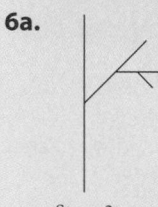

Stage 3

A branch is added at the midpoint of the newest segment equal to half its length at a 45° clockwise rotation.

6b.

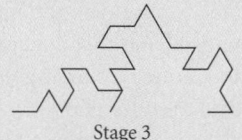

Stage 3

A "bottomless" equilateral triangle is built on the "right" half of segments.

6c.

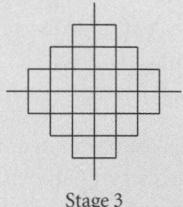

Stage 3

Each new segment is crossed at its midpoint by a centered perpendicular segment equal to it.

3. Evaluate these expressions.

a. $2 \times (24 + 12)$ 72 **b.** $2 + 24 \times 12$ 290 **c.** $2 - 24 + 12$ -10

d. $(2 + 24) \times 12$ 312 **e.** $(2 + 24) \div 12$ 2.1$\overline{6}$ **f.** $2 - (24 + 12)$ -34

4. Write a multiplication expression equivalent to each expression below in exponent form.

a. $\left(\frac{1}{3}\right)^3$ $\frac{1}{3} \times \frac{1}{3} \times \frac{1}{3}$ **b.** $\left(\frac{2}{3}\right)^4$ $\frac{2}{3} \times \frac{2}{3} \times \frac{2}{3} \times \frac{2}{3}$ **c.** $(1.2)^2$ 1.2×1.2 **d.** 16^5 $16 \times 16 \times 16 \times 16 \times 16$ **e.** 2^7 $2 \times 2 \times 2 \times 2 \times 2 \times 2 \times 2$

5. Write an addition expression that gives the combined total of shaded areas in each figure. Then evaluate the expression. The area of each figure is originally 1.

a. 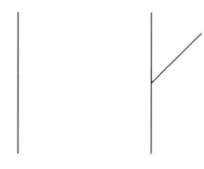 **b.** 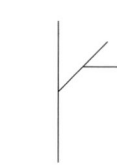 **c.**  **d.**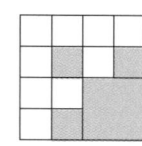

$\frac{1}{16} + \frac{1}{16} + \frac{1}{16} = \frac{3}{16}$ $\frac{1}{16} + \frac{1}{16} + \frac{1}{4} = \frac{3}{8}$ $\frac{1}{9} + \frac{1}{9} + \frac{1}{81} + \frac{1}{81} = \frac{20}{81}$ $\frac{1}{4} + \frac{1}{16} + \frac{1}{64} + \frac{1}{64} = \frac{11}{32}$

6. Draw the next stage of each fractal design below. Then describe the recursive rule for each pattern in words.

a.

Stage 0 Stage 1 Stage 2

b.

Stage 0 Stage 1 Stage 2

c.

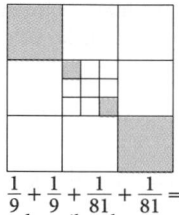

Stage 0 Stage 1 Stage 2

d.

Stage 0 Stage 1 Stage 2

7. Look at the figures below.

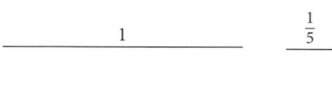  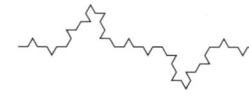

1 $\frac{1}{5}$

Stage 0 Stage 1 Stage 2

6d.

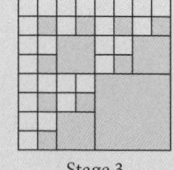

Stage 3

Each unshaded square is divided horizontally and vertically to create four congruent squares; the bottom-right square is shaded.

34 CHAPTER 0 Fractions and Fractals

a. Complete a table like the one below.

Stage number	Total length		
	Multiplication form	Exponent form	Decimal form
0	1	1	1
1	$7\left(\dfrac{1}{5}\right)$	$7^1\left(\dfrac{1}{5}\right)^1 = \left(\dfrac{7}{5}\right)^1 = \dfrac{7}{5}$	1.4
2	$7 \cdot 7 \cdot \left(\dfrac{1}{5}\right) \cdot \left(\dfrac{1}{5}\right)$	$7^2\left(\dfrac{1}{5}\right)^2 = \left(\dfrac{7}{5}\right)^2 = \dfrac{49}{25}$	1.96

b. If you were to draw Stage 20, what expression could you write with an exponent to represent the total length? Evaluate this expression using your calculator, and round the answer to the nearest hundredth. $\left(\dfrac{7}{5}\right)^{20} \approx 836.68$

8. Investigate the behavior of the expression below. Use recursion to evaluate the expression several times for different starting values. You may want to record your recursions in a table. Does the expression appear to have an attractor value?

$0.4 \cdot \boxed{} + 3$ The attractor is 5.

TAKE ANOTHER LOOK

- If a number gets larger when it is raised to a power, what kind of number is it?
- If a number gets smaller when it is raised to a power, what kind of number is it?
- What numbers stay the same when they are raised to a power?

To investigate these questions, choose positive and negative numbers, zero, and positive and negative fractions to put in the box and evaluate the expressions

$\boxed{}^3$ and $\boxed{}^4$

You may want to use a table like the one to the right to save your results.

You know that if the denominator of a fraction increases, the value of the fraction decreases. Why is that? Are there any exceptions?

Look again at your results for the expression $\boxed{}^3$.

What would the numerator of the fraction have to be so that

$\boxed{}$	$\boxed{}^3$
1	$1^3 = 1$
2	$2^3 = 8$

$\dfrac{\bigcirc}{\boxed{}^3}$ is smaller than $\boxed{}^3$? Greater than $\boxed{}^3$?

Now do the same thing with $\dfrac{\bigcirc}{\boxed{}^4}$.

Display your results in a chart.

Number in box	Exponent of 3 (or any odd integer greater than 1)	Exponent of 4 (or any even integer greater than zero)
Positive number greater than 1	bigger	bigger
Negative number less than -1	smaller (negative yet farther from zero)	bigger (becomes positive)
Zero or 1	stays the same	stays the same
-1	stays the same	bigger (becomes 1)
Positive fraction between 0 and 1	smaller (positive yet closer to zero)	smaller (positive yet closer to zero)
Negative fraction between 0 and -1	bigger (negative yet closer to zero)	bigger (becomes positive)

▶ **Take Another Look**

Some students may find the first three questions simple. They should work on them anyway because they'll find the rest of the activity easier if they're in the mode of recording results in a table.

The behavior of numbers raised to a power depends on the kinds of numbers involved, as described in the table at the bottom of this page.

The questions about increasing denominators give students an opportunity to think about numbers more broadly than they may be used to doing. The statement "If the denominator of a fraction increases, the value of the fraction decreases" is true for positive fractions written with both numerator and denominator positive. For negative fractions—written with only the numerator negative—increasing the denominator makes the magnitude of the fraction decrease, thereby moving it closer to zero. Because the fraction moves to the right along the number line, it becomes greater.

Students might find the questions about numerators most difficult because they must rely on generating data rather than using algebra. (Students will learn how to solve proportions in Chapter 2.)

For $\dfrac{\bigcirc}{\boxed{}^3}$ to be smaller than $\boxed{}^3$, the numerator would have to be less than $\boxed{}^6$. For $\dfrac{\bigcirc}{\boxed{}^3}$ to be greater than $\boxed{}^3$, the numerator would need to be greater than $\boxed{}^6$. Likewise, with $\dfrac{\bigcirc}{\boxed{}^4}$, the numerator would be compared to $\boxed{}^8$.

Students may find it surprising that the number in the box makes no difference.

From this chapter students have gained a familiarity with the investigative method, have improved facility with the graphing calculator, and have looked for and analyzed patterns. All these skills will be developed further throughout the course.

Students have reviewed fractions and exponents. For students who are having trouble adding or multiplying fractions, provide extra practice with the More Practice Your Skills worksheets.

The important new mathematical ideas are fractals and recursion. Because students will see recursion again, emphasize this idea as you review the chapter.

FACILITATING SELF ASSESSMENT

To help students complete the portfolio described in "Assessing What You've Learned," suggest that they consider for evaluation their work on Lesson 0.1, Exercise 5; Lesson 0.2, Exercise 6; Lesson 0.3, Exercise 6; Lesson 0.4, Exercise 9; and Lesson 0.5, Exercise 7.

TESTING

If many students missed much of this introductory chapter, you may choose not to give a test and simply regard the chapter as diagnostic.

You have several other options for assessing students' learning. You might use the chapter project, ask students to present the results of the computer program for different polygons, or evaluate their portfolios. (See "Assessing What You've Learned" at the very end of the chapter.)

Because investigative skills and group work are essential to *Discovering Algebra*, be sure to evaluate these skills as well as mathematical understanding.

Assessing What You've Learned
KEEPING A PORTFOLIO

If you look up "assess" in a dictionary, you'll find that it means to estimate or judge the value of something. The value you've gained by the end of a chapter is not what you studied, it's what you remember and what you've gained confidence in. There are ideas you may not remember, but you will be able to reconstruct them when you meet similar situations. That's mathematical confidence.

One way to hold on to the value you've gained is to start a portfolio. Like an artist's portfolio, a mathematics portfolio shows off what you can do. It also collects work that you found interesting or rewarding (even if it isn't a masterpiece!). It also reminds you of ideas worth pursuing. The fractal designs that you figured out or invented are worth collecting and showing. Your study of patterns in fractals is a rich example of investigative mathematics that is also a good reference for how to work with fractions and exponents.

Choose one or more pieces of your work for your portfolio. Your teacher may have specific suggestions. Document each piece with a paragraph that answers these questions:

► What is this piece an example of?
► Does it represent your best work? Why else did you choose it?
► What mathematics did you learn or gain confidence in with this work?
► How would you improve the piece if you wanted to redo it?

Portfolios are an ongoing and ever-changing display of your work and growth. As you finish other chapters, don't forget to update your portfolio with new work.

If you do want to give a test, you can use either Form A or Form B of the chapter test, or you can use Constructive Assessment items. Using the CD test generator, you can create an alternate version of the test or combine some items from Form A or Form B with Constructive Assessment items.

1

Data Exploration

Overview

In **Lesson 1.1,** students learn about bar graphs and dot plots. In **Lesson 1.2,** they see the measures of center—mean, median, and mode. **Lesson 1.3** introduces five-number summaries and box plots. Students will use these concepts again in this course. **Lesson 1.4** presents histograms and stem-and-leaf plots. On the activity day, **Lesson 1.5,** students state a conjecture on how two books differ and collect data to verify their conjecture. In **Lesson 1.6,** students graph scatter plots of two-variable data. The goals of **Lesson 1.7** are to practice estimation skills and understand the meaning of points on the coordinate plane in comparison with the graph of $y = x$. **Lesson 1.8** introduces matrices as an extension of data tables.

The Mathematics

A data set consists of specific information. If a data set is numerical, then several mathematical representations help summarize the data, display the data set, or make it easier to work with.

Statistics

Some representations of data are numbers, called *statistics.* Some statistics are measures of center; others are measures of spread.

The mean, median, and mode are measures of center. The *mean* (often called *average*) is the most commonly used, but *outliers* (extreme, atypical points) on one end of the data set can distort the mean's significance as a measure of center. The *median*, or middle value, is a better measure of center in that case. The *mode*, or modes, give the most commonly occurring value(s), the ones "most likely" to be picked at random.

Other useful statistics describe the spread of the data, or density of data points in various regions. The *range* is a single number giving the difference between the *maximum* value and the *minimum*. The first and third *quartiles* (boundaries of the quarters) are the medians of the two halves of the data set. Also useful in describing the spread is the *interquartile range*, or IQR, which is the difference between the third and first quartile. Perhaps the most useful way to measure spread is by a *five-number summary*, which lists the minimum, first quartile, median, third quartile, and maximum. It allows you to see the range and concentration of the data.

Graphs

Graphic displays make one- and two-variable data sets easier to understand.

For some one-variable data sets, it works best to group data into categories that aren't numerical. If a single picture or icon represents a specific number of data points in a category, the graph is called a *pictograph*. In contrast, a *bar graph* represents the data values in each category using a single bar, whose length is proportional to the number of points.

The best grouping in other one-variable data sets is by numbers, either single numbers or intervals of numbers. The analog of a bar graph is a *histogram*, in which bars over different intervals are adjacent. If the categories are single numbers, then a graph with rows or columns of dots, called a *dot plot*, might best represent the data. If the data set consists of numbers that are categorized by their first digit or several digits, a *stem-and-leaf plot* may be preferable, with the remaining digits of each number listed on the other side of a bar next to the first digits designating the category.

All of these graphs of one-variable data sets aid intuition about the measures of center and give some indication of spread. For a better indication of spread, a *box-and-whisker* plot parallels the five-number summary of the data.

Two-variable data sets consist of ordered pairs of data, which are described visually using coordinate geometry. A *scatter plot* is a graph of the points whose coordinates are given by the data set. For comparison (and later as a model), a scatter plot is often augmented by a line, such as $y = x$.

Matrices

Another way to display two-variable data is in a table. The mathematical abstraction of a table, a *matrix* (plural: *matrices*), is a useful tool for combining data sets. You can enter the data values from a table into a graphing calculator as a matrix, and the calculator can perform matrix operations such as multiplying a matrix by a constant and adding, subtracting, and multiplying matrices.

Using This Chapter

Exploring data gives many opportunities to use algebra and see its applications, but if data representations aren't part of your required curriculum, you could skip over most of these lessons now and return to them if you have time later. Include at least parts of Lessons 1.3 and 1.6, however, because students will need to understand scatter plots, five-number summaries, and graphing on the coordinate plane. Use Lesson 1.8 only if your curriculum requires that students study matrices. If you skip some lessons, you can present the associated calculator skills when they are required later.

Resources

Discovering Algebra Resources

Calculator Notes 1A, 1B, 1C, 1D, 1E, 1F, 1G, 1H, 1I, 1J, 1K, 1L, 1M, 1N, 1P, 2A (for TI-73 only 1Q, 1R)

Teaching and Worksheet Masters
  Lesson 1.1 Pulse Rate Sample Data
  Lesson 1.2 Pennies Sample Data
  Lesson 1.2 Measures of Center
  Lesson 1.3 Dot Plot for Pennies
  Lesson 1.3 Box Plot for Pennies
  Lesson 1.4 Hand-Span Sample Data
  Lesson 1.4 Communities Sample Data
  Lesson 1.6 Coordinate Plane
  Lesson 1.6 Distance from a Motion Sensor
  Lesson 1.7 Estimation Investigation
  Lesson 1.8 Pizza Prices

Assessment Resources A and B
  Quiz 1 (Lessons 1.1 and 1.2)
  Quiz 2 (Lessons 1.3 and 1.4)
  Quiz 3 (Lessons 1.6 to 1.8)
  Chapter 1 Test
  Chapter 1 Constructive Assessment Options

More Practice Your Skills for Chapter 1

Condensed Lessons for Chapter 1

Other Resources

Fathom™ Dynamic Statistics Software™
  (Key Curriculum Press)

www.keymath.com/DA

Materials

- watch or clock with second hand
- 250–300 pennies (not new)
- graph paper
- centimeter rulers
- a variety of books
- colored pencils or pens
- poster paper
- 80 small round objects (coins, buttons, bottle caps)
- tape
- metersticks or tape measures
- motion sensor (*optional*)
- computer and CD-ROM to download data
- protractors

Pacing Guide

	day 1	day 2	day 3	day 4	day 5	day 6	day 7	day 8	day 9	day 10
standard	1.1	1.2	1.2	1.3	1.4	1.4	quiz, 1.5	1.5	1.6	1.6
enriched	1.1	1.2	1.2	1.3	1.4	1.4, project	quiz, 1.5	1.5	1.6	1.6
block	1.1, 1.2	1.2, 1.3	1.4	quiz, 1.5	1.6	1.7, project	1.8, review	assessment		

	day 11	day 12	day 13	day 14	day 15	day 16	day 17	day 18	day 19	day 20
standard	1.7	1.8	1.8	review	assessment					
enriched	1.7, project	1.8	1.8	review, TAL	assessment					

1 Data Exploration

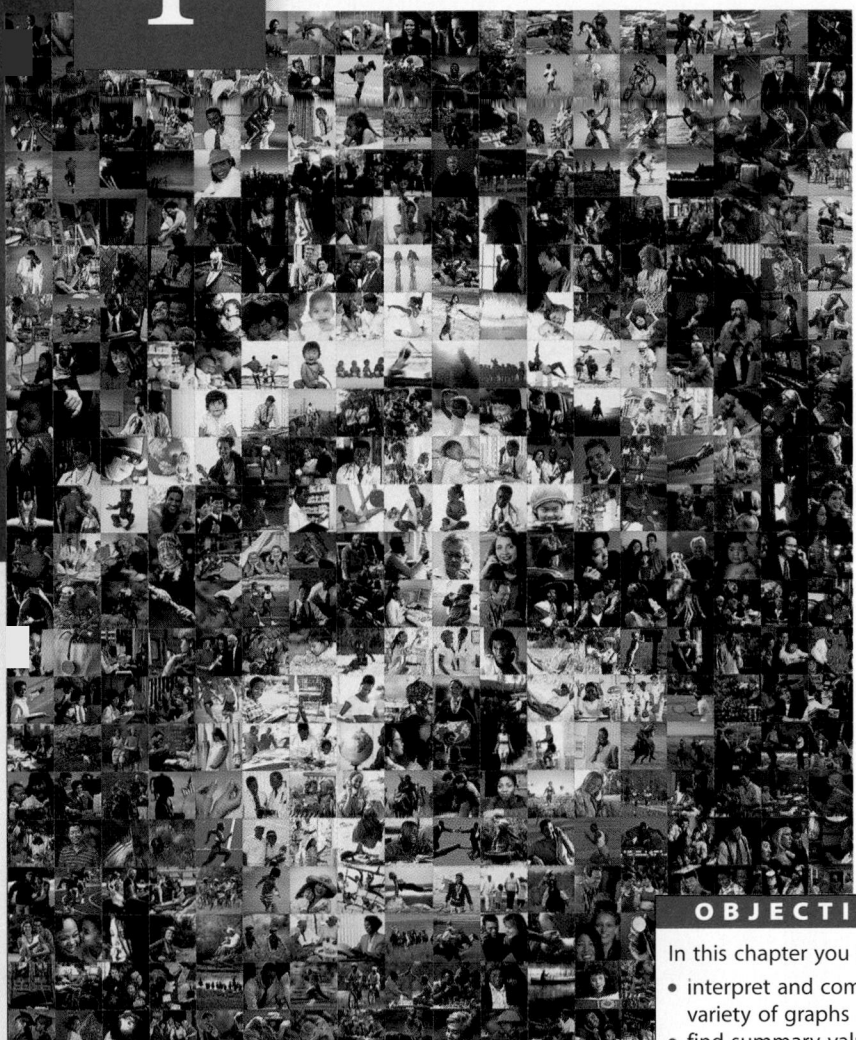

You are surrounded by information in many forms—in pictures, in graphs, in words, and in numbers. This information can influence what you eat, what you buy, and what you think of the world around you. This photo collage by Robert Silvers shows a lot of information.

OBJECTIVES

In this chapter you will
- interpret and compare a variety of graphs
- find summary values for a data set
- draw conclusions about a data set based on graphs and summary values
- review how to graph points on a plane
- organize and compute data with matrices

CHAPTER 1 OBJECTIVES

- Learn to calculate (on paper and by calculator), interpret, and compare statistics representing a one-variable data set: *minimum, maximum, range, mean, median, mode, quartiles, interquartile range, five-number summary, outliers*

- Learn to create (on paper and by calculator), interpret, and compare graphical displays of a one-variable data set: *pictograph, bar graph, dot plot, stem-and-leaf plot, histogram,* or *box plot*

- Learn to create (on paper and by calculator) and analyze scatter plot displays of a two-variable data set along with the line $y = x$

- Learn to represent two-variable data with a matrix; multiply the matrix by a number, add and subtract matrices, and multiply a matrix by a column matrix

- Gain experience making conjectures about a data set and testing them with appropriate statistical and graphical methods

- Use the terminology of statistics and graphs to present conclusions of a study of data

- Improve abilities to work with others on a mathematical problem

The picture illustrates how individual pieces of information can contribute to a larger meaning. You might ask, "What do you think this artist is trying to say about the population of the United States?" You could mention the census (taken every ten years) and ask students what attributes displayed in the photo might be counted in the census. You might even use the photo as a fictitious data set and have students summarize the data (such as by counting gender, ethnicity, or approximate ages) to characterize the artist's idea about the population.

LESSON
1.1

Bar Graphs and Dot Plots

PLANNING

LESSON OUTLINE

One day:

10 min	Getting settled, introduction
5 min	Example
15 min	Investigation
5 min	Sharing
5 min	Closing
10 min	Exercises

MATERIALS

- clock or watch with a second hand
- Pulse Rate Sample Data (W), *optional*
- Calculator Notes 1Q and 1R for TI-73 only

TEACHING

Mention the term *data* and how a set of data can be represented by a graph. **[Ask]** "What does the graph of CD data truly say about sales at any given store?" The fictional graph describing CDs depicts some data published by the Recording Industry Association of America. Students might visit the site (and others like it) to acquire realistic data for creating other graphs. They can find links to RIAA's original data at www.keymath.com/DA.

EXAMPLE

You may want to do this example with students lining up as a human pictograph, or as an activity in which students write their birthdays on sticky notes and build a pictograph of their birthday data.

Mention the term *pictograph*. Each picture or icon in a pictograph may represent several data points. **[ESL]** Make sure that students understand the word *category*.

I've always felt rock and roll was very, very wholesome music.

ARETHA FRANKLIN

This **pictograph** shows the number of CDs sold at Sheri's music store in one day. Can you tell just by looking which *type* sold the most? How many CDs of this type were sold? **Rock; 16**

This specific information, the kind Sheri may later use to make decisions, is sometimes called **data.** You use data every day when you answer questions like "Where is the cheapest place to buy a can of soda?" or "How long does it take to walk from class to the lunchroom?"

In this lesson you interpret and create graphs. Throughout the chapter, you'll learn more ways to organize and represent data.

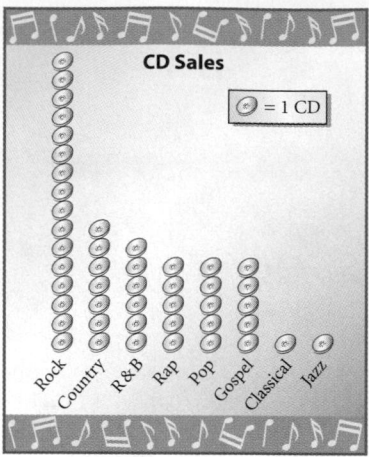

EXAMPLE

Joaquin's school posts a pictograph showing how many students celebrate their birthdays that month. Here is part of this pictograph. Create a table of data and a **bar graph** from the pictograph.

▶ **Solution**

This table lists the birthday data.

Number of Birthdays in Each Month

Jan	Feb	Mar	Apr
15	10	0	25

In the pictograph, there are three figures for January. Each figure represents five students. So 3 × 5 gives 15.

LESSON OBJECTIVES

- Create a pictograph, bar graph, and dot plot of a data set
- Calculate the minimum, maximum, and range of a data set
- Interpret pictographs, bar graphs, and dot plots
- Decide the appropriateness of bar graphs and dot plots for a given data set

NCTM STANDARDS

CONTENT		PROCESS	
●	Number	●	Problem Solving
	Algebra	●	Reasoning
	Geometry		Communication
	Measurement	●	Connections
●	Data/Probability	●	Representation

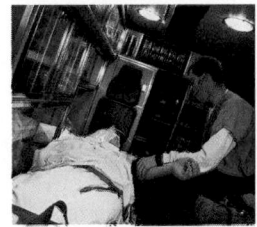

This bar graph shows the same data. The height of a bar shows the total in that **category,** in this case, a particular month. You use the *scale* on the *vertical axis* to measure the height of each bar. The vertical axis extends slightly past the greatest number of birthdays in any one month, so the data does not go beyond the scale.

Bar graphs gather data into categories and make it easy to present a lot of information in a compact form. In a bar graph you can quickly compare the quantities for each category.

In a **dot plot** each item of numerical data is shown above a number line or *horizontal axis*. Dot plots make it easy to see gaps and clusters in the data set, as well as how the data **spreads** along the axis.

In the investigation you'll gather and plot data about pulse rates. People's pulse rates vary, but a healthy person at rest usually has a pulse rate between certain values. A pulse that is too fast or too slow could tell a paramedic that a person needs immediate care.

Number of Birthdays in Each Month

Investigation
Picturing Pulse Rates

You will need

- a watch or clock with a second hand

Use the Procedure Note to learn how to take your pulse. Practice a few times to make sure you have an accurate reading.

Procedure Note

How to Take Your Pulse
1. Find the pulse in your neck.
2. Count the number of beats for 15 seconds.
3. Multiply the number of beats by 4 to get the number of beats per minute (bpm). This number is your pulse rate.

Step 1 | Start with pulse-rate data for 10 to 20 students.

Step 2 | Find the **minimum** (lowest) and **maximum** (highest) values in the pulse-rate data. The minimum and maximum describe the spread of the data. For example, you could say, "The pulse rates are between 56 and 96 bpm."

Based on your data, do you think a paramedic would consider a pulse rate of 80 bpm to be "normal"? What about a pulse rate of 36 bpm?
80 bpm is "normal," while 36 bpm is too low.

Step 3 | Construct a number line with the minimum value near the left end. Select a scale and label equal **intervals** until you reach the maximum.

Step 4 | Put a dot above the number line for each item of your data. When a data value occurs more than once, stack the dots.

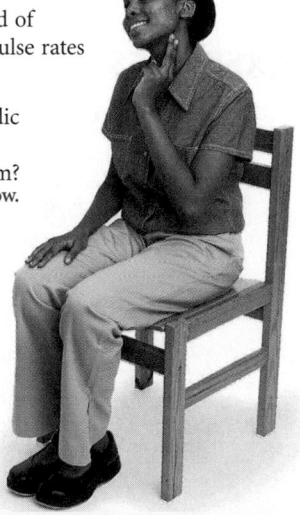

Step 3 [Alert] Some students may need to be reminded of how to draw a number line. Point out the one on the next page.

Step 4 You might remind students that they are creating a dot plot here. Dot plots are sometimes called "line plots" and can be made with X's instead of dots.

[Ask] "How does the number of people relate to the height of the bars?" or "Could the bars go sideways?"

You might start a list of new terms for this chapter and display them on a "wall of words."

Point out that the original data of names and birthdays is not recoverable from the pictograph.

If students are using TI-73 graphing calculators, they can use notes 1Q and 1R and draw calculator pictographs and bar graphs.

Guiding the Investigation

Mention that *beats per minute* is abbreviated as *bpm* in this lesson. Some students may know this abbreviation from music. You can also relate this to ratios and rates as $\frac{beats}{minute}$.

One step If your students are experienced investigators, use One step instead of Steps 1 to 6. Ask students to collect a data set of pulse rates and try to represent that set by a variety of graphs. Encourage a variety of representations. After a few students have shared their ideas, give a name to each standard representation. As students move into the main part of the investigation, mention that you'll be constructing a graph that's not a bar graph for data giving pulse rates. In general, try to avoid telling students things they can figure out.

Step 1 You can use the sample data, but most students will prefer actively collecting the data themselves. **[Alert]** Students who have trouble locating a pulse in the neck can try the wrist. You might also suggest that one person in a group serve as timer while the others count their pulses.

Coordinate the compiling of data into a class set. Without comment, skip students who show discomfort in volunteering; they may be generally self-conscious or have a medical condition. Record the data where everyone can see.

Step 5 The *Physicians Desk Reference* lists the normal resting pulse rate for adult men as 70 to 72 bpm and women as 78 to 82 bpm. **[Alert]** Some students may use the word *range* in the phrase, "the range is from 56 to 96 bpm." The maximum and minimum values help describe the spread of the data, but emphasize that in statistics the range is a single number—the maximum minus the minimum.

Step 6 **[ESL]** Many students have trouble pronouncing *statistics*. Don't insist on correct pronunciation.

SHARING IDEAS

Good steps to ask students to present: 4, 5, 6. Encourage presenters to use standard terminology, especially *dot plot*, *minimum*, *maximum*, and *range*.

Elicit the observation that all pulse rates are multiples of four. **[Ask]** "Why are the class data points multiples of 4?" [The one-minute data came from multiplying 15-second data by four.] This anticipates Exercise 5c.

Assessing Progress
Observe students' contributions to the lesson so you can assess their abilities to **order** and **multiply whole numbers** and their familiarity with the **number line.**

Closing the Lesson

This lesson includes quite a few mathematical ideas: data, pictograph, dot plot, bar graph, minimum, maximum, and range. Review these terms as you talk about the word *statistics*.

Here is an example for the data set {56, 60, 60, 68, 76, 76, 96}. Your line will probably have different minimum and maximum values.

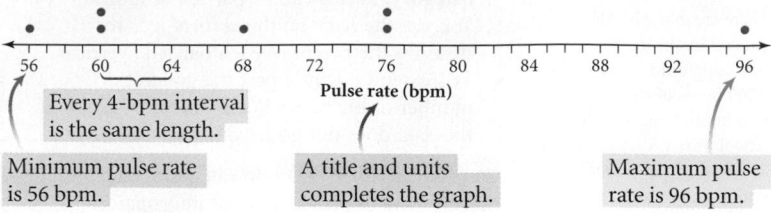

Every 4-bpm interval is the same length.

Pulse rate (bpm)

Minimum pulse rate is 56 bpm.

A title and units completes the graph.

Maximum pulse rate is 96 bpm.

The **range** of a data set is the difference between the maximum and minimum values. The data on the example graph have a range of 96–56 or 40 bpm.

Step 5 Most likely, the range will be greater than 12.

Step 5 What is the range of your data? Suppose a paramedic says normal pulse rates have a range of 12. Is this range more or less than your range? What information is the paramedic not telling you when she mentions the range of 12? The minimum and maximum.

Step 6 For your class data, are there data values between which a lot of points cluster? Are there any values that occur most often? What do you think these clusters would tell a paramedic? Do you think you could use your class data to say what is a "normal" pulse for all people? Why or why not?

Step 6 Clustering would help a paramedic identify a normal range. No, other factors, such as age, affect pulse. A single class's data is too small a sample.

History
CONNECTION

The word "statistics" was first used in the late eighteenth century to mean a branch of political science dealing with the collection of data about a state or country.

Statistics is a word used many ways. We sometimes refer to data we collect and the results we get as "statistics." For example, you could collect pulse-rate statistics from thousands of people and then determine a "normal" pulse-rate. The single value you calculate to be "normal" could also be called a statistic. You'll learn more about statistics and their usefulness in this chapter.

EXERCISES

▶ Practice Your Skills

1. Angelica has taken her pulse 11 times in the last six hours. Her results are 69, 92, 64, 78, 80, 82, 86, 93, 75, 80, and 80 beats per minute. Find the maximum, minimum, and range of the data. max: 93 bpm; min: 64 bpm; range: 93 bpm − 64 bpm = 29 bpm

2. The table shows the percentages of the most common elements found in the human body. Make a bar graph to display the data.

Elements in the Human Body

Oxygen	Carbon	Hydrogen	Nitrogen	Calcium	Phosphorus	Other
65%	18%	10%	3%	2%	1%	1%

MAKING THE CONNECTION

A data set can be represented either by statistics (minimum, maximum, range) or by graphs (pictographs, dot plots, bar graphs).

2.

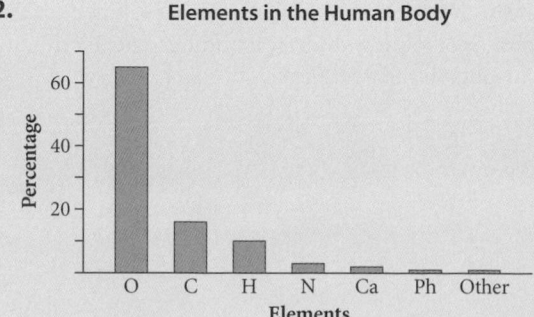

Elements in the Human Body

3. Use this bar graph to answer each question.

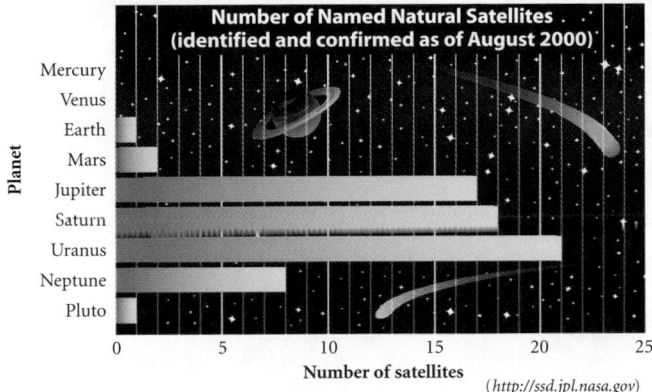

Number of Named Natural Satellites (identified and confirmed as of August 2000)

Planet (vertical axis): Mercury, Venus, Earth, Mars, Jupiter, Saturn, Uranus, Neptune, Pluto

Number of satellites (horizontal axis): 0, 5, 10, 15, 20, 25

(*http://ssd.jpl.nasa.gov*)

a. Which planet has the most satellites? Uranus (There are 21 named satellites.)

b. What does this graph tell you about Mercury and Venus? Mercury and Venus have no satellites.

c. How many more satellites does Jupiter have than Neptune? 9

d. How many times as many satellites does Saturn have than Mars? nine times

4. This table shows how long it takes the students in one of Mr. Matau's math classes to get to school.

a. Construct a dot plot to display the data. Your number line should show time in minutes.

b. How many students are in this class? 30

c. What is the combined time for Mr. Matau's students to travel to school? 241 minutes

d. What is the average travel time for Mr. Matau's students to get to school? $\frac{241}{30} \approx 8$ minutes

Travel Time to School

Time (min)	Number of students
1	2
3	2
5	6
6	1
8	6
10	7
12	3
14	2
15	1

▶ Reason and Apply

5. This graph is a dot plot of Angelica's pulse-rate data.

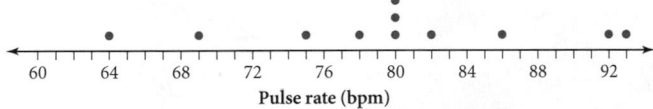

Pulse rate (bpm): 60, 64, 68, 72, 76, 80, 84, 88, 92

a. What pulse rate appeared most often? 80 bpm

b. What is the range of Angelica's data? 29 bpm

c. If your class followed the procedure directions for the investigation, your pulse rates should be multiples of four. Angelica's are not. How do you think she took her pulse? She counted her pulse rates for one full minute.

d. How would your data change if everyone in your class had taken his or her pulse for a full minute? How would the dot plot be different? Any whole number could occur, not just multiples of four.

e. Do you think medical professionals measure pulse rates for 1 minute or 15 seconds? Why? A full minute, sometimes longer, to ensure accuracy.

4a. **Travel Time to School**

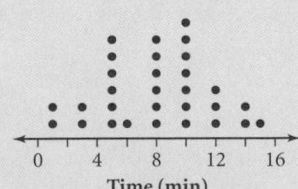

Time (min): 0, 4, 8, 12, 16

Exercise 5e Watch for different interpretations of the question "Why?": as asking either why "you think" something (one answer: "I've seen them") or why the professionals measure pulse rates the way they do (one answer: "to save time"). Encourage a range of thoughtful responses; the question is intended to keep students thinking.

BUILDING UNDERSTANDING

These exercises provide practice with finding the range of a data set and with reading and creating bar graphs and dot plots.

ASSIGNING HOMEWORK

Essential	1–6
Performance assessment	7
Portfolio	3
Journal	8
Group	6, 7
Review	9–11

▶ Helping with the Exercises

Exercise 3 **[Ask]** "Is there a reason for the order of categories on the horizontal axis?" [They are listed by increasing distance from sun.] **[ESL]** Some students might benefit from definitions of *horizontal* (like the horizon) and *vertical*.

[Ask] "How many more satellites does Neptune have than Mars?" [Six more, four times as many, or three times more are all correct answers.] Encourage a variety of legitimate responses. Show students that some mathematics questions have more than one answer and often answers can be arrived at in more than one way.

Exercise 4a Encourage students to place numbers on the vertical axis in a proportional way.

Exercise 4c Some students may simply add the times. Be sure they see why they must multiply times by numbers of students before adding.

Exercise 4d This question may be review. Those who don't remember what *average* means will be introduced to it as *mean* in Lesson 1.2.

Exercise 5b As needed, point out that the highest number marked is 92, not the maximum of 93.

Exercise 6 [Alert] Students may misinterpret the question 6a to mean "number of other kids (not people) in families in general," in which case the value of 0 might be legitimate. To distinguish between sets, suggest that students also consider the modes. A class of unusually tall students might legitimately switch 6b and 6c.

Exercise 7 [Alert] Students may have difficulties dealing with all the information at once. Be patient. One reason this is a good exercise for groups is that the groups are more likely to be self-correcting than an individual student, so that you don't have to intervene so much.

6. Each graph below displays information from a recent class survey. Determine which graph best represents each description:

    a. number of people living in students' homes iii
    b. students' heights in inches ii
    c. students' pulse rates in beats per minute iv
    d. number of working television sets in students' homes i

    i.
    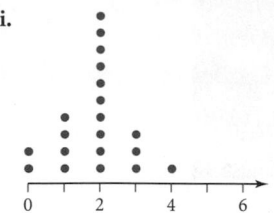

    ii.
    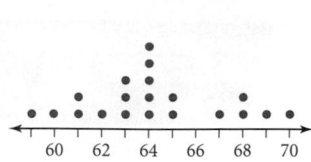

    iii.
    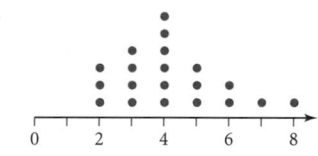

    iv.
    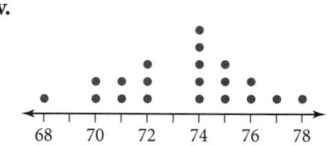

7. Reporter "Scoop" Presley of the school paper polled 20 students about their favorite type of music—classical, pop/rock, R&B, rap, or country. He delivered his story to his editor, Rose, just under the deadline. Rose discovered that Scoop, in his haste, had ripped the page with the bar graph showing his data. The vertical scale and one category were missing! Unfortunately, the only thing Scoop could remember was that three students had listed R&B as their favorite type. Reconstruct the graph so that it includes the vertical axis and the missing category with the correct count.

    3 in Classical

    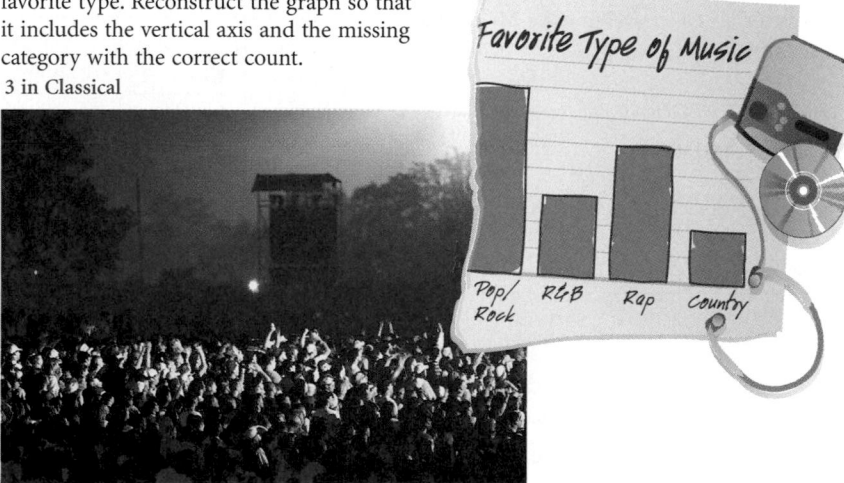

8. Suppose you collect information on how each person in your class gets to school. Would you use a bar graph or a dot plot to show the data? Explain why you think your choice would be the better graph for this information. A bar graph because the information falls into categories, is not numeric data, and cannot be scaled on a number line.

9. Rewrite each of these multiplication expressions using exponents.

   a. $10 \times 10 \times 10 \times 10$ 10^4

   b. $2 \cdot 2 \cdot 2 \cdot 5 \cdot 5 \cdot 5 \cdot 5 \cdot 5 \cdot 5$ $2^3 \cdot 5^6$

   c. $\dfrac{3^2(3^4)}{8(8)(8)}$ $\dfrac{3^6}{8^3}$

10. Use the order of operations to evaluate each expression.

   a. $7 + (3 \cdot 2) - 4$ 9   **b.** $8 + 2 - 4 \cdot 12 \div 16$ 7   **c.** $1 - 2 \cdot 3 + 4 \div 5$ -4.2

   d. $1 - (2 \cdot 3 + 4) \div 5$ -1 **e.** $1^2 \cdot 3 + (4 \div 5)$ 3.8

Exercise 10 Especially if you skipped Chapter 0, be ready to describe the conventional order of operations.

11. The early Egyptian *Ahmes Papyrus* (1650 B.C.) shows how to use a doubling method to divide 696 by 29. (George Joseph, *The Crest of the Peacock*, 2000, pp. 61–66)

Doubles of 29	58	116	**232**	**464**	928
Doubles of 1	2	4	**8**	**16**	32

Double the divisor (29) until you go past the dividend (696). Find doubles of 29 that sum to 696: $232 + 464 = 696$. Then sum the corresponding doubles of 1: $8 + 16 = 24$. So, 696 divided by 29 is 24.

   a. Divide 4050 by 225 with this method. 18;

Doubles of 225	**450**	900	1800	**3600**	7200
Doubles of 1	**2**	4	8	**16**	32

   b. Divide 57 by 6 with this method. (Hint: Use doubles and halves.) $9\frac{1}{2}$;

Doubles of 6	**6**	12	24	**48**	96		Half of 6	3
Doubles of 1	**1**	2	4	**8**	16		Half of 1	$\frac{1}{2}$

Exercise 11b Since doubles of 6 add to 54, a half of six must also be used.

IMPROVING YOUR **REASONING** SKILLS

Janet and JoAnn used the same data set of high and low temperatures for cities in a month in early spring. Which graph shows the low temperatures better? Is either graph better for showing the differences between high and low temperatures?

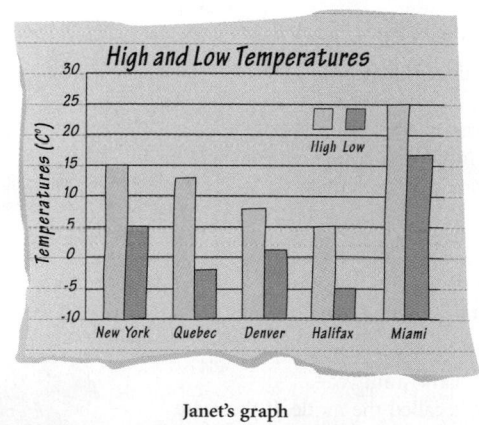

Janet's graph

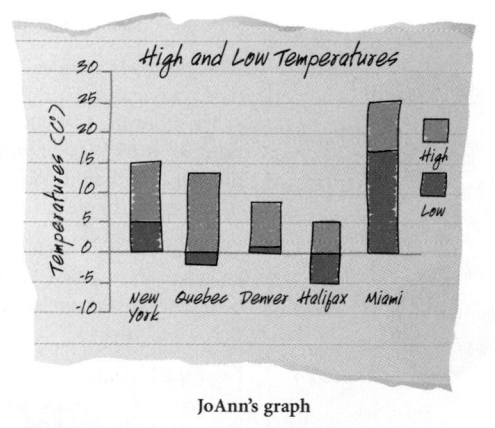

JoAnn's graph

IMPROVING **REASONING** SKILLS

Students may find that JoAnn's graph illustrates low temperatures better than Janet's, since negative values fall under the axis. JoAnn's choice of how to mark the graph makes it hard to compare the low temperatures. However, the side-by-side feature of Janet's graph allows easy comparison between high and low temperatures.

PLANNING

LESSON OUTLINE

One or two days:

20 min (or 40 min)	Investigation
15 min (or 25 min)	Sharing
5 min	Example
5 min (or 15 min)	Closing
5 min (or 15 min)	Exercises

MATERIALS

- 300 pennies (If you get pennies from a bank, ask for rolls that have been turned in, not new pennies.)
- Pennies Sample Data (W or T), *optional*
- Measures of Center (T), *optional*
- Calculator Notes 1A, 1B, 1J

TEACHING

In this investigation students learn the difference between the three measures of center.

Guiding the Investigation

As groups get started, give out pennies. Give an odd number to some groups and an even number to others. Then mention that some very useful statistics measure the center of a data set.

One step Ask students to measure the center of the set of dates of the pennies in as many ways as they can. As you circulate, encourage groups to think more broadly, and not only about the mean.

Students should read the introductory material in the text before beginning the investigation.

Step 1 [Alert] Make sure that students label their number line, title the plot, and show the units.

LESSON 1.2

Summarizing Data with Measures of Center

"Americans watch an average of four hours of television each day."

"Half of the participants polled had five or more people living in their home."

"When asked how many colleges they applied to, the most frequent answer given by graduating seniors was three."

Statements like these try to say what is typical. They summarize a lot of data with one number called a **measure of center.** The first statement uses the **mean,** or *average*. The second statement uses the **median,** or middle value. The third statement uses the **mode,** or most frequent value.

Investigation
Making "Cents" of the Center

You will need

- 20 to 30 pennies

In this investigation you learn to find the mean, median, and mode of a data set. You may have learned about these three measures of center in a past mathematics class.

Step 1 Sort your pennies by mint year. Make a dot plot of the years.

Step 2 Now put your pennies in a single line from oldest to newest. Find the median, or middle value. Does the median have to be a whole-number year? Why or why not? Would you get the same median if you arranged the pennies from newest to oldest?

Step 3 On your dot plot, circle the dot (or dots) that represent the median. Write the value you got in Step 2 beside the circled dot or dots, and label the value "median."

Step 4 Now stack pennies with the same mint year. The year of the tallest stack is called the mode. If there are two tall stacks, your data set is **bimodal.** If every stack had two

> **Procedure Note**
>
> **Finding the Median**
> If you have an odd number of pennies, the median is the year on the middle penny. If you have an even number of pennies, add the dates on the two pennies closest to the middle and divide by two.

Step 2 [Alert] Watch that students mark their number line with all years, not just the ones for which they have pennies.

Step 2 Medians will vary; the sample data (W) has a median of 1991. The median does not need to be a whole number. It may contain 0.5. Both orderings give the same median.

Step 3 Be sure that students don't simply call the median the average of the minimum and the maximum.

LESSON OBJECTIVES

- Calculate *mean, median,* and *mode* of a data set, noting *outliers*
- Enter data into calculator lists
- Given a list of data, use a calculator to find the mean and median
- Compare and interpret the three measures of center
- Explain the effects of outliers on the mean and median

pennies in it, you might say there is "no mode" because no year occurs most often. How many modes does your data set have? What are they? Does your mode have to be a whole number? **Modes will vary but must be a whole number; the sample data has a mode of 1991.**

Step 5 Draw a square around the year corresponding to the mode(s) on the number line of your dot plot. Label each value "mode."

Step 6 Mean of sample data is 1990.

Step 6 Find the sum of the mint years of all your pennies and divide by the number of pennies. The result is called the mean. What is the mean of your data set?

Step 7 Show where the mean falls on your dot plot's number line. Draw an arrowhead under it and write the number you got in Step 6. Label it "mean."

Step 8 Now enter your data into a calculator list, and use your calculator to find the mean and the median. Are they the same as what you found using pencil and paper? [▶ 🖥 See **Calculator Notes 1A and 1B.** Refer to **Calculator Note 1J** to check the settings on your calculator.◀]

Save the dot plot you created. You will use it in Lesson 1.3.

Measures of Center

Mean
The mean is the sum of the data values divided by the number of data items. The result is often called the average.

Median
For an odd number of data items, the median is the middle value when the data values are listed in order. If there is an even number of data items, then the median is the average of the two middle values.

Mode
The mode is the data value that occurs most often. Data sets can have two modes (bimodal) or more. Some data sets have no mode.

Each measure of center has its advantages. The mean and the median may be quite different, and the mode, if it exists, may or may not be useful. You will have to decide which measure is most meaningful for each situation.

EXAMPLE

This data set shows the number of people who attended a movie theater over a period of 16 days

{14, 23, 10, 21, 7, 80, 32, 30, 92, 14, 26, 21, 38, 20, 35, 21}

a. Find the measures of center.

b. The theater's management wants to compare its attendance to that of other theaters in the area. Which measure of center best represents the data?

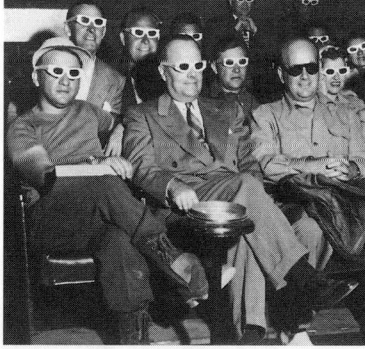

NCTM STANDARDS

CONTENT		PROCESS	
●	Number	●	Problem Solving
	Algebra	●	Reasoning
	Geometry		Communication
	Measurement	●	Connections
●	Data/Probability	●	Representation

Step 4 "No mode" is relative to the size of the data set. If you have one thousand distinct data values, having five to ten modes may be worthwhile information. But if you have only 12 data values, having five modes is not very informative.

Step 6 Adding up these four-place numbers may challenge some students' patience. Encourage them to use only the last two places for the number of years since 1900 (or even 1970) rather than the full dates. Be sure they realize they can use the symbol ─//─ to indicate a "break" in the *x*-axis between extreme outliers and the rest of the pennies.

Step 7 [Link] You might mention that if the horizontal number line were a beam and the data points were weights, the balance point would be at the mean.

If you're saving Step 8 for a second day on this lesson, collect handwritten data for later entry into calculators.

Step 8 [Alert] Watch for confusion about the word *mode;* the mode on the graphing calculator (Calculator Note 1J) is not related to the statistical mode of a data set. **[Language]** The word *mode* has several meanings; *mean* also has several meanings.

SHARING IDEAS

As you ask groups to copy their dot plots onto transparencies, include some with an even number and others with an odd number of data points. If possible, include one whose data have outliers and one with bimodal data.

[ESL] Ask students to translate *mean, median,* and *mode* into their first language. In Spanish the words would be *promedia* (mean), *intermedio* (median), and *el más común* (mode).

[Ask] "When are the mean, median, and mode close to being equal?" [With single-modal data without many outliers or with outliers symmetric about the mean.] Refer to the box "Measures of Center" in the text. Ask students to explain in their own words how to find each measure of center. You might point out that the word *mode* starts like *most;* the mode is the number that occurs most often.

If there are outliers in one or more data sets, mention the term as they show up. **[Ask]** "Is there ever a case in which outliers do *not* affect the mean?" [When they are equidistant on opposite sides of the mean.]

Assessing Progress

By observing students' contributions, you can assess their skills at **organizing data, ordering numbers, rounding, counting,** and **critical and creative thinking.**

MEASURES OF CENTER

[Ask] "Which is the best measure of center?" Encourage discussion that focuses on "For what purposes?" For example, the mean is good for summarizing heights or miles per gallon; the median is good for income or the cost of a home; the mode is good for shoe sizes if selling shoes.

[Ask] "What statistics besides measures of center might be used to describe the differences among these data sets?" You can review *maximum*, *minimum*, and *range*, and in a fast-moving class you might introduce the notion of quartiles to lead in to Lesson 1.3.

EXAMPLE

Talk through this example to help students understand how the mean can be affected by an outlier. You might comment on the mid-twentieth-century moviegoers who are wearing special glasses to view a three-dimensional movie.

▶ **Solution**

a. The mean is approximately 30 people.

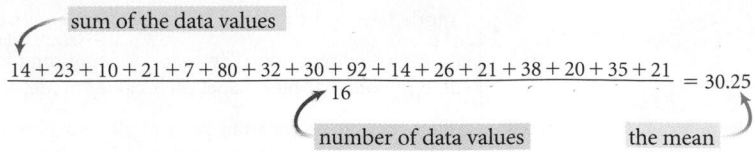

$$\frac{14 + 23 + 10 + 21 + 7 + 80 + 32 + 30 + 92 + 14 + 26 + 21 + 38 + 20 + 35 + 21}{16} = 30.25$$

sum of the data values — number of data values — the mean

The median is 22 people.

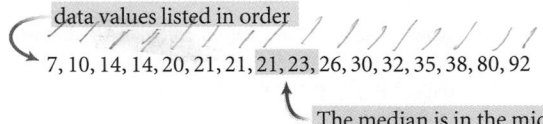

data values listed in order

7, 10, 14, 14, 20, 21, 21, 21, 23, 26, 30, 32, 35, 38, 80, 92

The median is in the middle, or halfway between 21 and 23.

The most frequent value, 21 people, is the only mode.

b. To determine which measure of center best summarizes the data, look for patterns in the data and look at the shape of the graph.

Attendance (number of people)

The dot plot clearly shows that, except for two items, the data are clustered between 7 and 38. The items with values 80 and 92 are far outside the range of most of the data and are called **outliers.**

Either the median, 22, or the mode, 21, could be used by the management to compare this theater's attendance to that of other theaters. The management could say, "Attendance was about 21 or 22 people per day over a 16-day period." The mean, 30, is too far to the right of most of the data to be the best measure of center. Yet the theater's management might prefer to use the mean of 30 in an advertisement. Why?

In the example, why is the mean so much larger than the median or the mode? This is because the mean is influenced by outliers in the data. To see how, recalculate the mean using 45 and 50, instead of 80 and 92, as the two greatest values. What would happen to the median and mode with this change? What happens if you remove these outliers and find the mean for the remaining 14 values? Using the mean to describe data that includes outliers can be misleading.

With outliers of 45 and 50, the mean is approximately 25; the median and mode are the same. With the outliers removed, the mean is approximately 20; the median is 21 and the mode stays the same.

Solution a The mean indicates that if the attendance were distributed evenly over the 16 days, approximately 30 people would have attended each day.

Solution b The median and mode are more appropriate for comparison purposes, but using the mean in an advertisement would make the theater seem more popular than it is.

If students use a calculator list for the theater data, it is easy for them to change values and recalculate the mean.

Closing the Lesson

The mathematical ideas in this lesson are the three primary measures of center (mean, median, mode). Mention that outliers can affect the mean but not the median.

Ask students to save their handwritten penny data and dot plots for use in Lesson 1.3.

EXERCISES

▶ Practice Your Skills

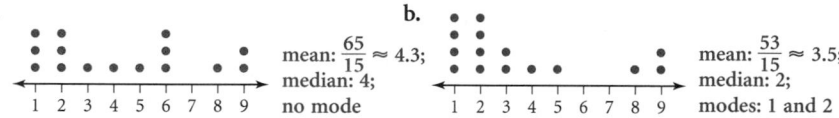

1. a. mean and median are 6; mode 5
   b. mean 5.1; median 5; modes 3, 8
   c. mean 10.25; median 9; no mode
   d. mean 17.5; median 20; mode 20

1. Find the mean, median, and mode for each data set.
   a. {1, 5, 7, 3, 5, 9, 6, 8, 10}   b. {6, 1, 3, 9, 2, 7, 3, 4, 8, 8}
   c. {12, 6, 11, 7, 18, 5, 2, 21}   d. {10, 10, 20, 20, 20, 25}

2. Find the mean, median, and mode for each dot plot.

   a.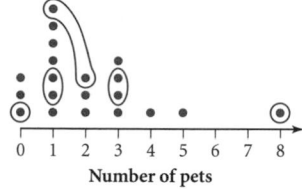
   mean: $\frac{65}{15} \approx 4.3$;
   median: 4;
   no mode

   b.
   mean: $\frac{53}{15} \approx 3.5$;
   median: 2;
   modes: 1 and 2

3. Students were asked how many pets they had. Their responses are shown in the dot plot below.

   Number of pets

   a. How many students were surveyed? 20
   b. What is the range of answers? 8
   c. What was the most common answer? 1

4. This graph gives the lift heights and vertical drops of the five tallest roller coasters at Cedar Point Amusement Park in Ohio.
   a. Find the mean, median, and mode for the lift heights. mean: 191.6 ft; median: 161 ft; no mode
   b. Find the mean, median, and mode for the vertical drops. mean: 181.2 ft; median: 155 ft; no mode

   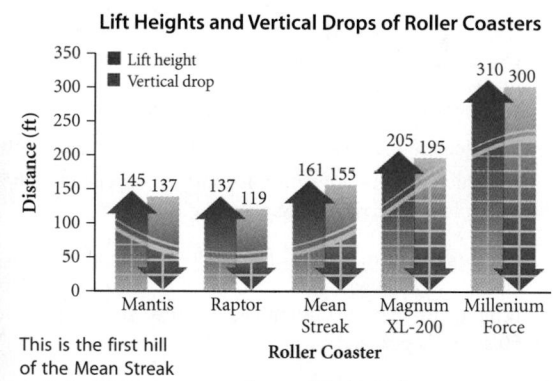

   Lift Heights and Vertical Drops of Roller Coasters

   ■ Lift height
   ■ Vertical drop

   Mantis 145 137; Raptor 137 119; Mean Streak 161 155; Magnum XL-200 205 195; Millenium Force 310 300

   Roller Coaster

   This is the first hill of the Mean Streak at Cedar Point.

   (www.cedarpoint.com)

5. If you purchase 16 grocery items at an average cost of $1.14, what is your grocery bill? Explain how you found the total bill. $18.24. If you multiply the mean by the number of items, you get the sum of the items.

BUILDING UNDERSTANDING

In these exercises students deepen their understanding of the three measures of center.

ASSIGNING HOMEWORK

Essential	1, 3, 5, 7, 9
Performance assessment	6
Portfolio	9, 10
Journal	8
Group	4, 10, 11
Review	12, 13

▶ Helping with the Exercises

Exercise 3 [Alert] Be sure that students realize that each dot represents one student, unlike in pictographs.

Exercise 4 [Alert] Graphs with double bars may be harder for some students to understand.

Exercise 5 This problem is the reverse of those that students have encountered before.
[Alert] Watch for difficulties in realizing that to go from the mean to the total requires multiplication.

6. Tsunamis are ocean waves caused by earthquakes. The heights of the 20 tallest tsunamis on record are listed in the table.

Tallest Tsunamis

Location of source	Year	Height (ft)	Location of source	Year	Height (ft)
Papua New Guinea	1998	45	Nankaido (Japan)	1946	20
Mindoro (Philippines)	1994	49	Kii (Japan)	1944	25
Sea of Japan	1983	49	Sanriku (Japan)	1933	93
Indonesia	1979	32	East Kamchatka (Russia)	1923	66
Celebes Sea	1976	98	South Kuril Island (Russia)	1918	39
Alaska	1964	105	Sanriku (Japan)	1896	98
Chile	1960	82	Sunda Strait (Indonesia)	1883	115
Aleutian Islands (Alaska)	1957	52	Chile	1877	75
Kamchatka (Russia)	1952	60	Chile	1868	69
Aleutian Islands (Alaska)	1946	105	Hawaii	1868	66

(*www.pmel.noaa.gov*)

(*The Hollow of the Deep-Sea Wave Off Kanagawa* by Katsushika Hokusai/Minneapolis Institute of Art Acc. No. 74.1.230)

a. Calculate the mean, median, and mode for the height data.

b. Which measure of center is most appropriate for the height data? Explain your reasoning.

7. The first three members of the stilt-walking relay team finished their laps of the race with a mean time of 53 seconds per lap. What mean time for the next two members will give an overall team mean of 50 seconds per lap?

8. Noah scored 88, 92, 85, 65, and 89 on five tests in his history class. Each test was worth 100 points. Noah's teacher usually uses the mean to calculate each student's overall score. Explain why the median is a better measure of center for Noah's work on the tests.

11a.

Highest-Paid Athletes

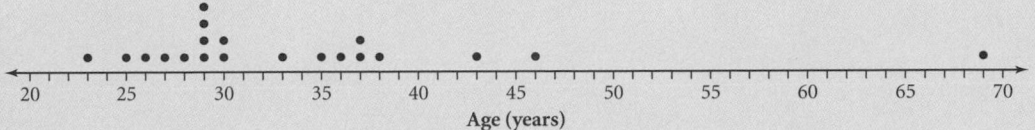

Age (years)

Exercise 6 In taking the median, some students may assume the heights are already in order because the years are in order.

6a. Mean: 67.15 feet; median: 66 feet; no mode.
6b. Either the mean or the median would be a good choice.

Exercise 7 This problem takes Exercise 5 another step. Students might solve it by multiplying and then subtracting, or by guess-and-check. Encourage them to try more than one approach.

7. The first three averaged 53 seconds each; together they took 159 seconds. The total time for the whole team must be 50(5) or 250 seconds. The two remaining members must have a total of 250 − 159 = 91 seconds. So the last two members must average 91 ÷ 2 = 45.5 seconds each.

8. The mean, 83.8, is lower than all but one of his scores. The median, 88, is more representative.

9. This table gives information about ten of the largest saltwater fish species in the world. The approximate mean weight of these fish is 1527.4 pounds.

a. Explain how to use the mean to find an approximate total weight for these ten fish. What is the total weight?

b. The median weight of these fish is about 1449 pounds. Assuming that no two weights are the same, what does the **median** tell you about the individual weights of the fish?

c. The range of weights is 1673 pounds, and the minimum weight is 991 pounds. What is the weight of the great white shark, the largest fish caught? 2664 pounds.

10. Create a set of data that fits each description.

a. The mean age of a family is 19 years, and the median age is 12 years. There are five people in the family. Sample answer: {8, 10, 12, 32, 33} years

b. Six students in the Mathematics Club compared their family sizes. The mode was five people, and the median was four people. Sample answer: {2, 3, 3, 5, 5, 5} people

c. The points scored by the varsity football team in the last seven games have a mean of 20, a median of 21, and a mode of 27 points. Sample answer: {7, 14, 20, 21, 24, 27, 27} points

11. This data set represents the ages of 20 of the highest-paid athletes in the United States as of December 31, 1998. (*www.infoplease.com*)

{35, 29, 29, 23, 46, 26, 25, 36, 69, 30, 28, 37, 37, 27, 38, 43, 29, 33, 30, 29}

a. Make a dot plot of the survey results.

b. Give the mean, median, and mode for the data. mean: 33.95; median: 30; mode: 29

c. Which measure of center best summarizes the data? Explain your reasoning.

Largest Saltwater Fish Species

Species	Location where caught
Swordfish	Chile
Bluefin tuna	Nova Scotia
Great white shark	South Australia
Atlantic blue marlin	Brazil
Greenland shark	Norway
Black marlin	Peru
Hammerhead shark	Florida
Tiger shark	California
Pacific blue marlin	Hawaii
Mako shark	Mauritius

(*The Top 10 of Everything 2001*, p. 41)

Review

12. Fifteen students gave their ages in months.

168   163   142   163   165   164   167   153   149   173   163   179   155   162   162

a. Would you use a bar graph, pictograph, or dot plot to display these data? Explain your reasoning.

b. Create the graph you chose in 12a.

13. Use this segment to measure or calculate in 13a–c.

a. What is the length in centimeters of the segment? 12 cm

b. Draw a segment that is $\frac{2}{3}$ as long as this segment. What is the length of your new segment? 8 cm

c. Draw a segment that is $\frac{1}{5}$ as long as the original segment. What is the length of this new segment? 2.4 cm

9a. Multiply the mean by 10; together they weigh approximately 15,274 pounds.

9b. Five of the fish caught weigh 1449 lbs or less, and five weigh 1449 lbs or more.

Exercise 10 This is a good problem for students to work on in a group that can stimulate ideas and be self-correcting. Some students may not know how to handle the freedom of creating their own set of numbers. Sample answers are shown. Encourage a variety of responses, including creative use of outliers, but watch for errors. If some students are completely baffled, suggest starting with a set of points all of which are the mean, then adjusting them to keep the mean the same but to change to the desired median, and finally adjusting them to achieve the desired mode without changing the other two measures of center.

11a. See page 48.

11c. Either the mode or the median is probably best. The mean is distorted by one extremely high value.

Exercise 13 Students may need help with this exercise if you skipped Chapter 0.

12a. A dot plot may be most appropriate for the numeric data. However, if each value was translated into years (divide by 12), you could make a bar graph or pictograph with ages as categories.

12b. Answers may vary.

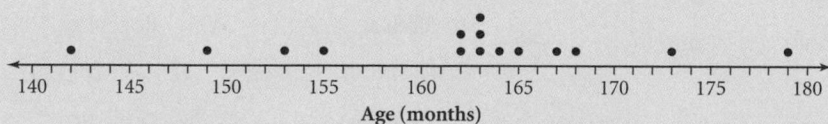

Student Ages

Age (months)

LESSON OUTLINE

One day:

10 min	Introduction and Example
20 min	Investigation
10 min	Sharing
5 min	Closing
5 min	Exercises

MATERIALS

- Dot Plot for Pennies (W or T), *optional*
- Box Plot for Pennies (T), *optional*
- Calculator Note 1C

TEACHING

Ask students to read the introduction. Point out that the five numbers include three medians: of the whole data set, of the upper half, and of the lower half.

EXAMPLE

The five-number summary represents the spread of a data set better than minimum, maximum, and range alone.

[Ask] "What are the measures of center of this data set?" [mean 675; median 416] "Do measures of center give you enough information to describe a data set?" Elicit the idea that measures of center don't describe the distribution, or spread, of data points.

Have students read the example to themselves. **[Ask]** "What do you observe in the data set?" Pool their observations. Add your own to the list as they seem to fit. Include in the list things listed at the bottom of page 50.

LESSON
1.3

To talk sense is to talk quantities. It is no use saying a nation is large—how large?
ALFRED NORTH WHITEHEAD

Five-Number Summaries and Box Plots

Michael Jordan of the Chicago Bulls scored the most points in the 1997–98 season in the National Basketball Association (NBA). In fact he scored almost three times as many points as the next highest scorer on his team. Does any measure of center give a complete description of how the Bulls scored as a team? A **five-number summary** could give a better picture. It uses five boundary points: the minimum and the maximum, the median (which divides the data set in half), the **first quartile** (the median of the first half), and the **third quartile** (the median of the second half).

Points Scored by Chicago Bulls Players Who Played Over 40 Games (1997–98 Season)

Chicago Bulls	Total points scored	Chicago Bulls	Total points scored
Michael Jordan	2357	Steve Kerr	376
Toni Kukoc	984	Dennis Rodman	375
Scottie Pippen	841	Randy Brown	288
Ron Harper	764	Jud Buechler	198
Luc Longley	663	Bill Wennington	167
Scott Burrell	416		

(*www.nba.com*)

Michael Jordan

EXAMPLE | Find the five-number summary for the number of points scored by the Chicago Bulls during the 1997–98 season (use the table above).

▶ Solution

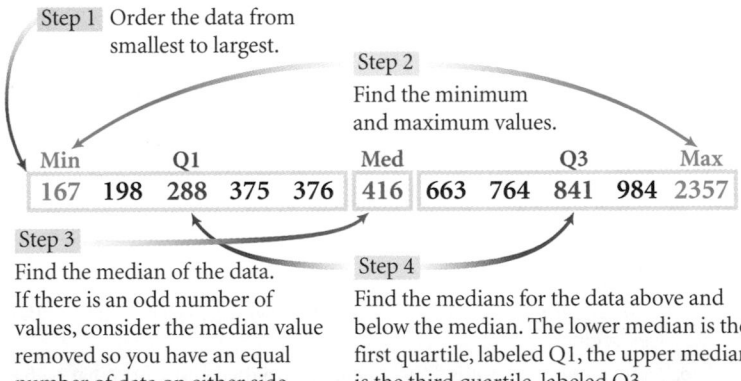

Step 1 Order the data from smallest to largest.

Step 2 Find the minimum and maximum values.

Min		Q1			Med			Q3		Max
167	198	288	375	376	416	663	764	841	984	2357

Step 3 Find the median of the data. If there is an odd number of values, consider the median value removed so you have an equal number of data on either side.

Step 4 Find the medians for the data above and below the median. The lower median is the first quartile, labeled Q1, the upper median is the third quartile, labeled Q3.

LESSON OBJECTIVES

- Calculate *quartiles, interquartile range,* and a *five-number summary* of a data set
- Create a *box plot* of a data set
- Given a list of data, use a calculator to graph a box plot
- Interpret box plots
- Decide the appropriateness of a box plot for a given data set
- Carefully define *outlier*

NCTM STANDARDS

CONTENT		PROCESS	
•	Number		Problem Solving
	Algebra	•	Reasoning
	Geometry	•	Communication
	Measurement	•	Connections
•	Data/Probability	•	Representation

The five-number summary is 167, 288, 416, 841, 2357. This is the minimum, first quartile, median, third quartile, and maximum in order of smallest to largest. The first quartile, median, and third quartile divide the data into four equal groups. Each of the four groups has the same number of values, in this case two.

A five-number summary helps you better understand the spread of the data along the number line. It also helps you compare different sets of data. A **box plot** is a visual way to show a five-number summary. This box plot shows the data for the Bulls.

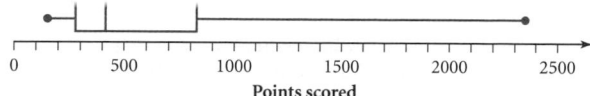

Notice how the box plot shows the spread of data and Michael Jordan as an extreme outlier. Can you see why this type of graph is sometimes called a **box-and-whisker plot**? Can you find the five-number summary values in this box plot? In the next investigation you'll see how to use the five-number summary to construct a box plot.

Investigation
Pennies in a Box

You will need

- your dot plot from Investigation: Making "Cents" of the Center

The illustrations are examples only. Your box plot should look different.

Step 1 Find the five-number summary values for your penny data.

Step 2 Place a clean sheet of paper over your dot plot and trace the number line. Using the same scale will help you compare your dot plot and box plot.

Step 3 Find the median value on your number line and draw a short vertical line segment just above it. Repeat this process for the first quartile and the third quartile.

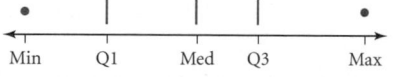

Step 4 Place dots above your number line to represent the minimum and maximum values from your dot plot.

This man is counting and bagging pennies at the U.S. Mint in Denver, Colorado.

- The mean is affected by an outlier (Jordan).
- The data are in order from largest to smallest.
- The median does not illustrate the wide spread of the data.
- There is no mode.
- The box plot shows the number-line position better than the table.

- You can't tell the number of data points from the box plot.
- Quartiles are not quarters, but the boundary values of the quarters.
- The five-number summary and box plot show the spread of data better than range does.

You might also show students a method for using their hand to represent a box plot:

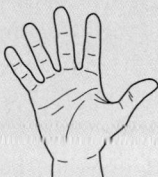

Open your hand. Then close your index, middle, and ring fingers to form the box.

Guiding the Investigation

After groups have gathered, be sure students have their dot plots from the pennies investigation in Lesson 1.2. You can provide a copy of the sample to any group that doesn't have its own data set.

One step If your class is experienced at investigations, ask them to figure out a way of graphing the five-number summary of their pennies data. During sharing time, introduce the idea of box plot and interquartile range and have them make box plots on calculators.

Steps 1 to 6 [ESL] Some students might benefit from making a flow chart of the process of making a box plot.

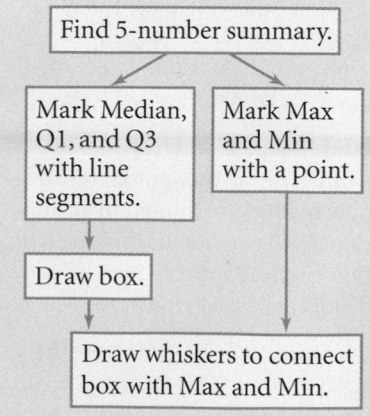

Step 6 [ESL] Be sure students understand the word *symmetric.* A box plot is symmetric when the first and last pairs of consecutive numbers in the five-number summary have the same difference, and the same for the second and third pairs. You might challenge students to come up with a data set that has a symmetric box plot.

Step 6 The five-number summary is clearly displayed by a box plot. The dot plot shows the actual data. Both show the spread.

Step 7 Calculator Note 1C gives instructions for making a box plot on a calculator. Don't clear out the penny list if it's still there from Lesson 1.2.

Step 8 Students should compare the relative size of the box and whiskers to determine whether the graphs are equivalent.

Step 10 Answers to 10a and the first part of 10b will depend on the data set. The fraction of pennies on each whisker will be exactly $\frac{1}{4}$ only if the number of data points is a multiple of 4. You might rephrase the question to ask about percents as well as fractions, as a review.

SHARING IDEAS

You can compare box plots if you ask all students to use the same length for a unit on their number line (such as 1 centimeter for 1 year). Ask several groups of students to put their box plots on transparencies, and then overlay them for comparison.

[Ask] "What if you had 10,000 pennies, and you wanted to show how many were minted in various years?" Discussion of this question can lead into the idea of histograms in Lesson 1.4.

Students might wonder how the IQR is used. **[Ask]** "Can we use the IQR to define outliers?" That question is answered in Exercise 10, so you need only encourage discussion here.

Step 5 Draw a rectangle with ends at the first and third quartiles. This is the "box." Finally, draw horizontal segments that extend from each end of the box to the minimum and maximum values. These are the "whiskers."

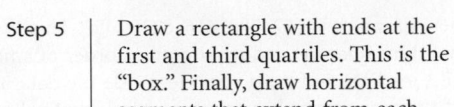

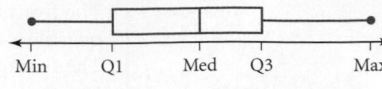

Step 6 Compare your dot plot and box plot. On which graph is it easier to locate the five-number summary? Which graph helps you to see the spread of data better?

Remember that the first quartile, median, and third quartile divide the data items into four equal groups. Although each section has the same number of data items, your boxes and whiskers may vary in length. Some box plots will be more *symmetric* than others. When would that happen?

Step 7 Clear any old data from your calculator and enter your penny mint years into list L1.

Step 8 Draw a calculator box plot. [▶ Follow the procedure outlined in **Calculator Note 1C.**◀] Does your calculator box plot look equivalent to the plot you drew by hand?

Step 9 Use the trace function on your calculator. What values are displayed as you trace the box plot? Are the five-number summary values the same as those you found before? **The five-number summary values shown should be the same as calculated before.**

The difference between the first quartile and third quartile is the **interquartile range,** or **IQR.** Like the range, the interquartile range helps describe the spread in the data.

Step 10 Complete this investigation by answering these questions:

a. What are the range and IQR of your data?

b. How many pennies fall between the first and third quartiles of the graph? What fraction of the total number of pennies is this number? Will this fraction always be the same? Explain. **about $\frac{1}{2}$**

c. Under what conditions will exactly $\frac{1}{4}$ of the pennies be in each whisker of the box plot? **The number of data points must be a multiple of 4.**

Box plots are a good way to compare two data sets. These box plots summarize the final test scores for two of Ms. Werner's algebra classes. Use what you have learned to compare these two graphs. Which class has the greater range of scores? Which has the greater IQR? In which class did the greatest fraction of students score above 80?

Ms. Werner's Algebra Test

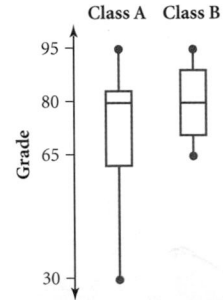

Notice that neither graph shows the number of students in the class nor the individual scores. If knowing each data value is important, then a box plot is not the best choice to display your data.

Assessing Progress

From your observations during the lesson, you can assess students' facility at **finding the median** and their understanding of **dot plots** and **spread.**

Closing the Lesson

Point out the two box plots showing test scores. Vertical plots, with low numbers on the bottom, foreshadow the coordinate system in Lesson 1.6. If you have time, ask students to write a comparison of the two box plots. Class A has the greatest range and the greatest IQR. In both classes, half the class scored above 80.

Mention the major ideas of this lesson: five-number summary, first quartile, third quartile, box (-and-whisker) plot, and interquartile range (IQR). All these ideas are designed to describe the spread of a data set—the density of its distribution along the number line.

EXERCISES

▶ **Practice Your Skills**

1. Find the five-number summary for each data set.

    a. {5, 5, 8, 10, 14, 16, 22, 23, 32, 32, 37, 37, 44, 45, 50} 5, 10, 23, 37, 50

    b. {10, 15, 20, 22, 25, 30, 30, 33, 34, 36, 37, 41, 47, 50} 10, 22, 31.5, 37, 50

    c. {44, 16, 42, 20, 25, 26, 14, 37, 26, 33, 40, 26, 47} 14, 22.5, 26, 41, 47

    d. {17, 43, 35, 34, 32, 21, 17, 16, 11, 9, 5, 5} 5, 10, 19, 34.5, 47

2. Sketch each graph below on your own paper.

    i.

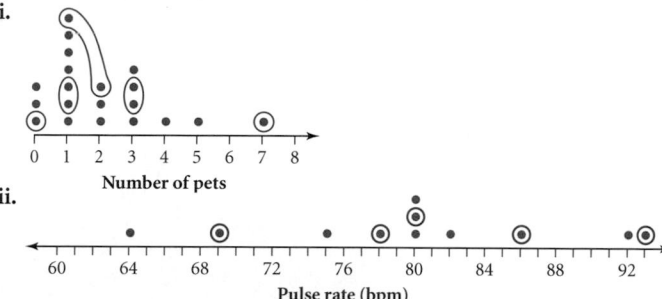

Number of pets

    ii.

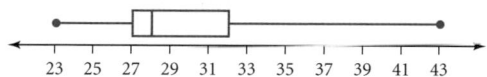

Pulse rate (bpm)

    a. Circle the points that represent the five-number summary values. If two data points are needed to calculate the median, first quartile, or third quartile, draw a circle around both points.

    b. List the five-number summary values for each data set. **i.** 0, 1, 1.5, 3, 7; **ii.** 64, 75, 80, 86, 93

3. Give the five-number summary and create a box plot for the listed values.

    {2, 6, 4, 9, 1, 6, 4, 7, 2, 8, 5, 6, 9, 3, 6, 7, 5, 4, 8}

4. Which data set matches this box plot? (More than one answer may be correct.) a and d

    23 25 27 29 31 33 35 37 39 41 43

    a. {23, 25, 26, 28, 28, 28, 28, 30, 31, 33, 41, 43}

    b. {23, 23, 24, 25, 26, 27, 29, 30, 31, 33, 41, 43}

    c. {23, 27, 28, 28, 33, 43}

    d. {23, 27, 28, 28, 29, 32, 43}

5. Check your vocabulary by answering these questions.

    a. How does the term *quartile* relate to how data values are grouped when using a five-number summary?

    b. What is the name for the difference between the minimum and maximum values in a five-number summary? The range.

    c. What is the name for the difference between the third quartile and first quartile in a five-number summary? The interquartile range, or IQR.

    d. How are outliers of a data set related to the whiskers of its box plot? Outliers are at or near the minimun and maximum values, which are the endpoints of the whiskers.

BUILDING UNDERSTANDING

In these exercises students practice moving between box plots and five-number summaries to deepen their understanding of both concepts.

ASSIGNING HOMEWORK

Essential	1–5, 8, 10
Performance assessment	9
Portfolio	8
Journal	5
Group	6, 7
Review	11, 12

▶ **Helping with the Exercises**

Exercise 3 Some students may have trouble deciding how to mark the axis. Point out that the maximum and minimum values determine the endpoints. It is not necessary to extend the number line beyond these points, nor is it wrong.

3. 1, 4, 6, 7, 9

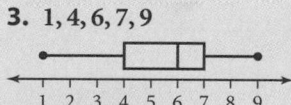

  1 2 3 4 5 6 7 8 9

5a. Quartiles are the boundaries dividing a data set into four groups, or quarters, with approximately the same number of values.

Exercise 6 Watch for misreading "mean score" as "median score." A good extension would be, "How much of this problem could he solve if he knew the median score instead of the mean?" [He could find the missing number only if it was between 23 and 29. It would be as far from the median as the median is from 27.]

7b. The mean for the Bulls is about 675 points, and the mean for the Raptors is about 715 points; the medians are 416 points and 644 points, respectively. The Chicago mean is much higher than its median because of Michael Jordan's total points. On the average, individual Toronto players scored more points than individual Chicago players.

7c. The median probably best represents the total-points-scored data for the Bulls. Students can justify choosing either the mean or the median for the Raptors. As a team owner, you might think the mean better reflects your team's talents.

7d. The lengths of the boxes are about the same, and the medians seem to divide the boxes into regions that look about the same length for the two teams.

▶ **Reason and Apply**

6. Stu had a mean score of 25.5 on four 30-point papers in English. He remembers three scores: 23, 29, and 27.

   a. Estimate the fourth score without actually calculating it. *The prediction should be less than the mean.*

   b. Check your estimate by calculating the fourth score. *23 points*

   c. What is the five-number summary for this situation? *23, 23, 25, 28, 29*

   d. Does it make sense to have a five-number summary for this data set? Explain why or why not. *Since there are only four values, a five-number summary may be inappropriate—the values themselves illustrate the spread of the data.*

7. **APPLICATION** The Toronto Raptors basketball team came in last in their division in the 1997–98 season.

Points Scored by Toronto Raptors Players Who Played Over 40 Games (1997–98 Season)

Toronto Raptors	Total points scored
Kevin Willis	1305
Doug Christie	1287
John Wallace	1147
Chauncey Billups	893
Charles Oakley	711
Dee Brown	658
Gary Trent	630
Reggie Slater	625
Tracy McGrady	451
Oliver Miller	401
Alvin Williams	324
John Thomas	151

(*www.nba.com*)

a. The five-number summary for the Chicago Bulls is 167, 288, 416, 841, 2357. (See the table and example at the beginning of this lesson.) Find the five-number summary for the Raptors. *151, 426, 644, 1020, 1305*

b. Find and compare the measures of center for the Bulls and the Raptors.

c. Decide which measure of center best describes each team. Explain your answer.

d. These box plots compare the points scored by the Chicago Bulls players to the points scored by the Toronto Raptors players. Compare the two teams' performance based on what you see in the graphs.

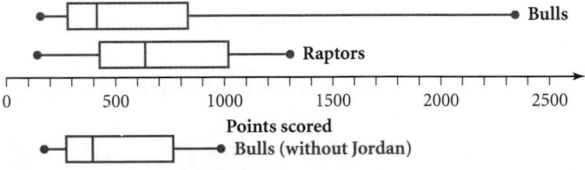

7e. Without Jordan, the range of the data is much smaller, and the length of the box is a little shorter. If Jordan's points scored are eliminated, the Raptors have the higher-scoring players. There is more variation in the number of points scored by individual Raptors than for individual Bulls.

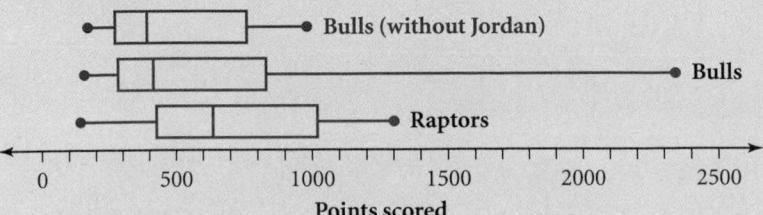

e. Remove Michael Jordan's points from the data table for the Chicago Bulls and make a new box plot. How does this new box plot compare to the original box plot for the Bulls? How does it compare to the box plot for the Raptors?

8. APPLICATION This table lists median weekly earnings of full-time workers by occupation and gender for 2000.

Median Weekly Earnings, 2000

Occupation	Men	Women
Managerial and professional specialty	999	697
Executive, administration, and managerial	995	684
Professional specialty	1001	708
Technical, sales, and admin. support	653	451
Technicians and related support	754	539
Sales occupations	683	379
Administrative support including clerical	552	455
Service occupations	405	313
Protective service	636	470
Precision production, craft, and repair	622	439
Mechanics and repairers	645	588
Operators, fabricators, and laborers	492	353
Machine operators, assemblers, and inspectors	498	353
Transportation and material moving	555	421
Handlers, equipment cleaners, helpers, and laborers	401	329
Farming, forestry, and fishing	342	288

(*http://stats.bls.gov*)

a. Make two box plots, one for men's salaries and one for women's salaries, above the same number line. Use them to compare the two data sets. Use the terms you have learned in this chapter.

b. What does the data tell you about women's and men's wages for the same type of work in 2000?

c. Do the box plots help you identify characteristics of the data better than the table does? Are there any aspects of the data that are better seen in the table?

d. How could you use the box plots to explain the slogan "Equal pay for equal work"?

During World War II many women took nontraditional jobs to support war industries. Some fought for and achieved equal pay for equal work.

Exercise 8 [Alert] Some students may miss the meaning of *median weekly earnings.* If needed, point out that the table shows weekly earnings. Students need to put the data in order before they can determine the summary values. **[ESL]** Be sure students understand that the slogan "Equal pay for equal work" states that men and women in the same job should get the same salary. Historically, women have earned less than men.

8a. For men, the mean salary is $639.56, and the five-number summary is 342, 495, 629, 718.5, 1001; for women, the mean salary is approximately $466.69, and the five-number summary is 288, 353, 445, 563.5, 708.

Median Weekly Earnings, 2000

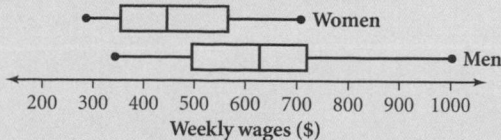

8b. Women received less pay than men for the same type of work.

8c. The box plots highlight the discrepancy in pay and spread of the data. However, dollar amount comparisons within a single profession are possible only in the table.

8d. Answers will vary. In 2000, gender equity did not exist for pay.

Exercise 9 [Language] The skin of a reticulated snake is covered with a pattern that resembles a net. *Reticulated* means netlike. **[Alert]** Some students may miss the fact that each data set has ten points. Even then, some may say that the length of the fifth-longest snake is at the median of the ten data points. Have them create a data set in which that's not the case. The box plots aren't comparable, because the markings on the *x*-axis refer to completely different quantities: lengths and speeds.

9d. The ten longest snakes vary in length from about 8 ft to 35 ft. About half of the snakes range in length from about 11 ft to 25 ft. Running speeds of the ten fastest mammals range from 42 mi/hr to 65 mi/hr. About half of the speeds are between 43 mi/hr and 50 mi/hr. The cheetah appears to run much faster than most other mammals.

9e. No, the units of these data sets are different.

Exercise 10 This essential problem defines *outlier*. Remind students that in Lesson 1.2 we saw that outliers are important in their effect on the mean. Sometimes statisticians disregard data points that meet this definition of *outlier*. If you used the term *limit* in Chapter 0, be sure students realize that its use in 10d is very different.

11. One possibility is a family with ages 4, 10, 14, 39, and 43. The 14 is fixed. The total of all ages must be 110 years.

---

9. These box plots display the recorded lengths of the ten longest snakes and the recorded running speeds for the ten fastest mammals.

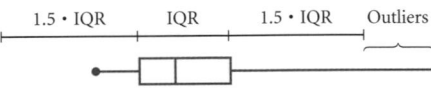

Longest Snakes   **Fastest Mammals**

Length (ft)   Speed (mph)

(*Factastic Book of 1001 Lists*, 1999, p. 41)   (*The Top Ten of Everything 2000*, p. 59)

a. The longest snake in the world is believed to be the reticulated python. What is the length of this snake? **35 feet**

b. The fifth-longest snake is the king cobra. Can you determine its length from the box plot? Explain. **More information is needed. The length is between 11 and 17.5 feet.**

c. The fastest mammal in the world is believed to be the cheetah. What is the fastest recorded speed for a cheetah? **65 mi/hr**

d. Explain what each box plot tells you about the spread of the data.

e. Could these two box plots be constructed above the same number line? Explain.

f. The fifth- and sixth-fastest mammals (Grant's gazelle and Thomson's gazelle) have been recorded at the same maximum speed. About how fast can they run? **about 47 mi/hr**

10. As a general rule, if the distance of a data point from the nearest end of the box is more than 1.5 times the length of the box (or IQR), then it qualifies as an outlier.

1.5 · IQR   IQR   1.5 · IQR   Outliers

a. The five-number summary for the number of points scored by the Chicago Bulls players is 167, 288, 416, 841, 2357. What is 1.5 times the interquartile range? **829.5**

b. What is the value of the first quartile less 1.5 times the interquartile range? **−541.5**

c. What is the value of the third quartile plus 1.5 times the interquartile range? **1670.5**

d. The values you found in 10b and 10c are the limits of outlier values. Identify any Chicago Bulls players who are outliers. **An outlier would have to score fewer than −541.5 points or more than 1670.5 points. Michael Jordan is an outlier.**

▶ **Review**

11. Create a data set for a family of five with a mean age of 22 years and a median age of 14.

12. The majority of pets in the United States are cats.

a. How many pet cats are there in the United States? Use the pictograph below. **66 million**

b. How many fewer dogs are there? **8 million**

c. If parakeets (11 million) were added to the pictograph below, how many pawprints would be drawn to represent parakeets? $5\frac{1}{2}$

Pets in the US

Cat 🐾🐾🐾🐾🐾 🐾🐾🐾🐾🐾 🐾🐾🐾🐾🐾 🐾🐾🐾🐾🐾 🐾🐾🐾🐾🐾 🐾🐾🐾🐾🐾 🐾🐾🐾
Dog 🐾🐾🐾🐾🐾 🐾🐾🐾🐾🐾 🐾🐾🐾🐾🐾 🐾🐾🐾🐾🐾 🐾🐾🐾🐾🐾 🐾🐾🐾🐾

🐾 = 2 million

(*The Top 10 of Everything 2001*, p. 45)

Histograms and Stem-and-Leaf Plots

This dot plot provides information about the amount of pocket money 16 students had with them on a given day. If you collect similar data for all the students in your school, you probably wouldn't want to make a dot plot because you would have too many dots. A box plot could be used to show the spread of the data set, but it wouldn't show whether you polled 16 or 600 students.

Pocket Money

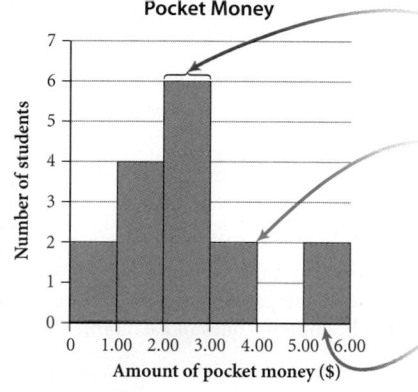

Amount of pocket money ($)

A **histogram** is related to a dot plot but more useful when you have a large data set. Histograms use columns to display how data are distributed and reveal clusters and gaps in the data. Unlike bar graphs, which use categories, the data for histograms must be numeric and ordered along the horizontal axis.

This histogram shows the same data as the dot plot.

Pocket Money

The width of each column represents an interval of $1.00. These intervals are also called **bins.**

The height of each bin shows the number of students whose money falls in that interval. This is the **frequency** of each bin.

Boundary values fall in the bin to the right. That is, this bin is $5.00 to $5.99.

All **bins** in a histogram have the same width, and the columns are drawn next to each other without any space between them. A gap between columns means that there is an interval with no data items that have those values. You can't name the individual values represented in a histogram, so a histogram summarizes data. You *can* determine the total number of data items. The sum of the bar heights in the histogram above tells you the total number of students.

NCTM STANDARDS

CONTENT	PROCESS
• Number	• Problem Solving
Algebra	• Reasoning
• Geometry	Communication
• Measurement	• Connections
• Data/Probability	• Representation

LESSON OBJECTIVES

- Create a histogram and a stem-and-leaf plot of a data set
- Given a list of data, use a calculator to graph a histogram
- Interpret histograms and stem-and-leaf plots
- Decide the appropriateness of a histogram and a stem-and-leaf plot for a given data set

PLANNING

LESSON OUTLINE

One day:

10 min	Introduction
20 min	Investigation
5 min	Sharing
5 min	Example
5 min	Closing
5 min	Exercises

MATERIALS

- graph paper
- centimeter rulers
- Hand-Span Sample Data (W, one copy per group), *optional*
- Communities Sample Data (W), *optional*
- Calculator Note 1D
- computer and CD-ROM to download data

TEACHING

Histograms are used for data with whole-number values, as well as for continuous, measured values.

INTRODUCTION

After forming groups, remind the class that in the previous lesson the box plot didn't show the actual numbers of pennies, as a dot plot would have. On the other hand, if they'd had lots of pennies, drawing a dot plot—with one dot per penny—would have been time-consuming. Ask for ideas of how to show the actual numbers in a large sample. Encourage a variety of suggestions.

Ask students to read through the introduction to themselves, and then have several students read it aloud. **[ESL]** Be sure students understand the word *bin*.

You might mention that the word *histogram* comes from the Greek words *histo,* which means "beam," and *gram,* which means "graph."

Point out that bins are named by the boundary values. The boundary item falls in the bin to the right. For discrete values, it is possible to list both ends of each bin, such as, 0–9, 10–19, 20–29. For continuous data, bins are designated by the smallest number and by the number all values in the bin are less than, such as 0–10, 10–20, 20–30. Though intervals are named differently depending on whether they represent discrete or continuous data, they are always labeled by boundary numbers.

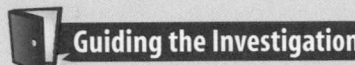

Guiding the Investigation

One step Ask each group to make a histogram of the hand spans of all the students in the class. Phrasing a problem so that it can be solved in several ways, or is subject to multiple interpretations, helps students understand more deeply the underlying ideas. As they work, remind them that bins need to be the same width, and let them experiment to find appropriate intervals. Pass out Calculator Note 1D to groups that finish their histograms quickly.

For classes following the steps of the investigation, you may want to take more time on the introduction. Point out that using such a long line for so few data items in this dot plot is not economical. Ask the expository questions. The histogram with $1 bins shows that most students have between $2 and $3. The histogram with 50¢ bins shows that most students have between $1 and $1.50, $2.00 and $2.50, or $2.50 and $3.00. The histogram with $2 bins shows that most students have between $2 and $4. Because the bin size groups the data, the tallest bars vary as the bin sizes change.

When you construct a histogram, you have to decide what bin width works best. These histograms show the same data set but use different bin widths. Use each of the three histograms presented to answer the question "Most students had pocket money between which values?" How do the bin widths of each histogram affect your answers?

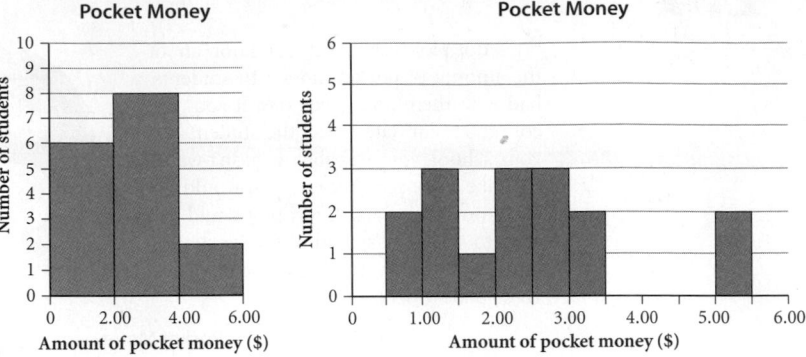

Too many bins may create an information overload. Too few bins may hide some features of the data set. As a general rule, try to have five to ten bins. Of course, there are exceptions.

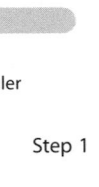

Investigation
Hand Spans

You will need
- graph paper
- a centimeter ruler

In this investigation you'll collect hand-span measurements and make a histogram. You'll organize the data using different bin widths and compare the results to a box plot.

Step 1	Measure your hand span in centimeters. Post your hand-span measurement in a classroom data table.
Step 2	Mark a zero point on your graph paper. Draw a horizontal axis to the right and a vertical axis up from this point.
Step 3	Scale the horizontal axis for the range of your data. Clearly divide this range into 5 to 10 equal bins. Label the boundary values of each bin.
Step 4	Count the data items that will fall into each bin. For example, in a bin from 20 cm to 22 cm, you would count all the items with values of 20.0, 20.5, 21.0, and 21.5. Items with a value of 22.0 are counted in the next bin.
Step 5	Scale the vertical axis for **frequency,** or count, of data items. Label it from zero to at least the largest bin count.
Step 6	Draw columns showing the correct frequency of the data items for each bin.

Step 1 Measurements should be made to the nearest half centimeter. Record all answers using one decimal place (for example, 20.0, 21.5). Have groups write their measurement data where everyone can see it.

Alternatively, give out the worksheet Hand-Span Sample Data.

Step 3 Bins are labeled by the smallest number that might be in the bin and the number that all values in the bin are less than, such as 20–22, 22–24. Because the hand-span measurements don't start at 0, students may want to use a broken horizontal axis.

Step 7 The calculator note instructs students to use a broken horizontal axis "Xmin equal to or slightly less than minimum of data." Suggest that they try using Xmin = 0 for comparison.

(Continued top of page 59)

Step 7 | Enter your hand-span measurements into list L₁ of your calculator. Create several versions of the histogram using different bin widths. [▶ 🖥 See **Calculator Note 1D** for instructions on creating a histogram.◀]

Step 8 | How did you select a bin width for your graph-paper histogram? Now that you have experimented with calculator bin widths, would you change the bin width of your paper graph? Write a paragraph explaining how to pick the "best" bin width.

Step 9 | Add a box plot of your hand-span data to both your graph-paper and calculator versions of the histogram. What information does the histogram provide that the box plot does not? Consider gaps in the data and the shape of the histogram.

A graph that is often more useful for small data sets is a **stem plot,** or **stem-and-leaf plot.** A stem plot, like a dot plot, displays each individual item. But, like a histogram, data values are grouped into intervals or bins. You need a **key** to interpret a stem plot.

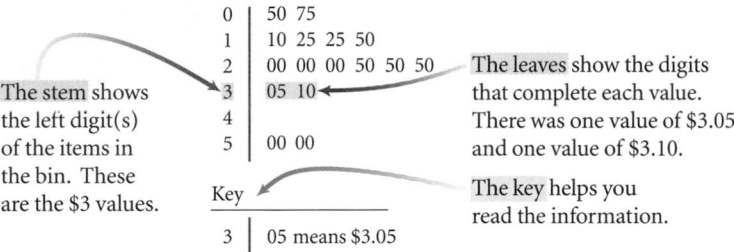

The stem shows the left digit(s) of the items in the bin. These are the $3 values.

The leaves show the digits that complete each value. There was one value of $3.05 and one value of $3.10.

The key helps you read the information.

Key

3 | 05 means $3.05

EXAMPLE | Use the stem-and-leaf plot to draw a histogram of the Canadian universities data.

▶ **Solution** | First, find the range of the data: 1995 − 1852 = 143 years. Then consider a "friendly" bin width. For a bin width of 10 years you'd need at least 15 bins, which is too many. For a bin width of 25 years you'd need 6 bins, which is more manageable.

You could start the first bin with the minimum value, 1852. However, it may be better to round down to 1850 so that each boundary will be a multiple of 25, and the centuries 1900 and 2000 will fall on boundaries.

Establishment Dates for Canadian Universities (1850–2000)

185	2 2 3 7
186	3 5
187	1 3 3 6 7 8
188	7 7
189	0 9
190	0 5 6 7 7 8
191	0 0 0 1 3 3 7 9
192	1 5
193	6
194	2 5 8
195	4 4 7 9
196	0 2 3 3 3 4 4 5 5 5 7 8 9 9
197	0 0 4 4 4 6 8 9
198	2
199	2 4 5

Key

197 | 3 means 1973

(www.aucc.ca)

Also suggest that when they have a histogram on the graphics screen, they use the calculator's trace function and note the endpoints of the intervals. This will demonstrate to them that boundary values are in the rightmost bin.

Step 8 For their paper histograms, students probably used the 5 to 10 bins, recommended in the instructions. For their calculator graphs, students should discover more subjective criteria, such as choosing "nice" boundary points and using a number of bins that accurately shows the distribution of the data. Paragraph write-ups will vary. Suggest that students exchange paragraphs and get peer feedback.

SHARING IDEAS

If students present to the full class as well as (or instead of) exchanging papers, they might show the results of Steps 6 and 9.

In discussing which histogram bin width is better, some students may say that the histogram with wider intervals contains more data. The larger area of the bars can mislead those who are having trouble with the abstraction of the graphic representation. The difficulty is so common, yet so hard for students to articulate, that you might mention it even if nobody else brings it up. **[Ask]** "Does this histogram represent more data than that one?"

As sharing takes place, you might want to compile a common list of students' observations of the important characteristics of histograms, such as

- Both axes of a histogram are marked with numbers, whereas one axis of a bar graph represents (usually) non-numeric categories.

- You can create a histogram on a calculator but not a bar graph.

- All intervals in a histogram should have the same width.

- The boundaries between intervals should be easy numbers to work with.

- 5 to 10 bins is usually about right.

- There should be enough intervals of the right size to show the distribution.

- Wider intervals obscure gaps and clusters in the data.

- Columns should be drawn next to each other with no space between them.

- A gap between columns indicates no data in that interval.

- A data value on the right boundary of an interval belongs in the next interval to the right.

EXAMPLE

For further experience with histograms, this example concerns turning a stem-and-leaf plot into a histogram. For a histogram with 16 bars, draw columns around the leaves.

Some students may need help to see that at least 15 bins are needed for widths of 10 years $\left(\frac{143}{10}\right.$ is more than $14\big)$ and 6 bins for a width of 25 years $\left(\frac{143}{25}\right.$ is more than $5\big)$.

As students explore different bin widths on their calculators, they should find that a bin width of 10 years looks erratic, with frequencies alternating between high and low. A bin width of 20, 25, or 30 years looks somewhat more uniform, with not too many bars. A bin width of 50 years shows an increasing trend but uses only 3 bins.

Assessing Progress

Your observations of students during this lesson can help you assess their skills at **counting, scaling,** and **organizing data,** and their understanding of **bar graphs.**

Closing the Lesson

Remind the class of the main ideas of this lesson: histogram and stem (-and-leaf) plot. Both graphical methods show measure of center and spread for data in numerical intervals, though neither shows spread as well as dot plots.

Count the frequency for each bin. You could use a table like this:

Bin	1850–1874	1875–1899	1900–1924	1925–1949	1950–1974	1975–1999
Frequency	9	7	15	5	23	7

Then create your histogram.

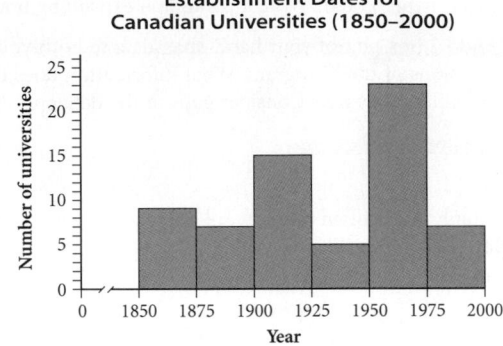

This histogram may or may not use the best bin width. Use your calculator to experiment with other bin widths for the data. Which bin width do you think highlights the spread of the data? Which do you feel highlights the clustering of data? Does one bin width show an increasing trend?

EXERCISES

You will need your calculator for problem **6.**

▶ Practice Your Skills

1. Marketing consultant Maive Wishnev surveyed people attending matinee and evening ballet performances. She made the two graphs below showing the ages of attendees whom she surveyed.

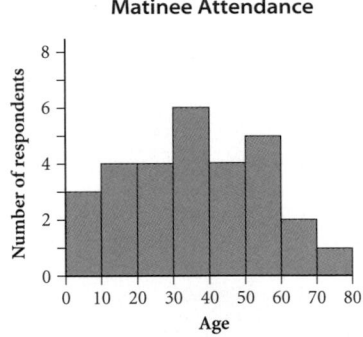
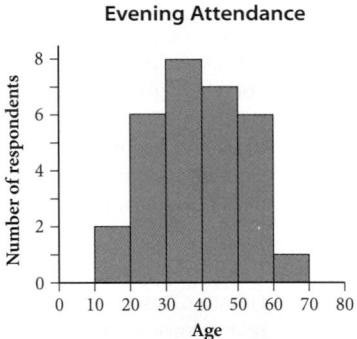

   a. How many people did she survey at each performance? matinee: 29; evening: 30
   b. At which performance did the ages of survey respondents vary more? matinee

BUILDING UNDERSTANDING

In these exercises students deepen their understanding of histograms and stem plots by connecting them with ideas presented earlier in the chapter.

ASSIGNING HOMEWORK

Essential	1–7, 10
Performance assessment	1, 7
Portfolio	6, 10
Journal	8, 9
Group	6, 8
Review	11, 12

c. How many children younger than 10 responded to the survey at the evening performance? None

d. What can you say about the number of 15-year-olds who were surveyed at the matinee? The number is less than or equal to 4.

2. Thirty students were asked at random to pick a number from 0 to 20. Here are the results:

{12, 7, 8, 3, 5, 7, 10, 13, 7, 10, 2, 1, 11, 12, 17, 4, 11, 7, 6, 18, 14, 17, 11, 9, 1, 12, 10, 12, 2, 15}

a. Construct two histograms for the data. Use different bin widths for each.

b. Do you notice any patterns in the data? What do the histograms tell you about the numbers that the students tend to choose? One observation is that students tended to choose numbers in the middle of the range.

c. Give the five-number summary for the data and construct a box plot.

d. Give the mode(s) for the data. 7 and 12

3. This box plot and histogram reflect the female life expectancy for several countries in 1999.

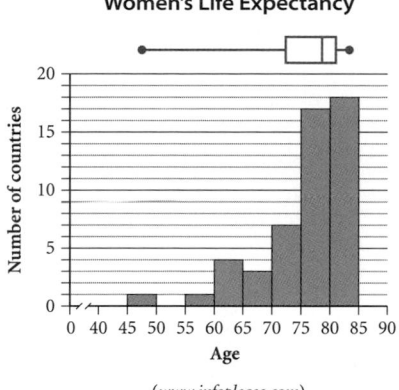

Women's Life Expectancy

(www.infoplease.com)

a. How many countries are represented? 51

b. The right whisker of the box plot is very short. What does this mean?

c. How many countries had female life expectancies of less than 60 years? 2

d. How can you tell that no country had a female life expectancy of greater than 85 years?

4. Redraw the histogram for Exercise 3, changing the bin width from 5 to 10.

5. Suppose some class members measure the lengths of their ring fingers. The measurements are 6.5, 6.5, 7.0, 6.0, 7.5, 7.0, 8.5, and 7.0 centimeters.

a. Identify the minimum, maximum, and range values of the data. minimum: 6.0 cm; maximum: 8.5 cm; range: 2.5 cm

b. Create a stem plot of these data values.

Exercise 5 On a stem-and-leaf plot, usually the first digits of numbers are on the left side of the vertical bar and later digits are on the right.

5b. Ring Finger Length

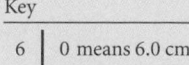

```
6 | 0  5  5
7 | 0  0  0  5
8 | 5
```

Key

6 | 0 means 6.0 cm

► **Helping with the Exercises**

Exercises 2c and 2d Some students may be surprised that they can't find the five-number summary from a histogram. Point out this disadvantage of histograms.

2a. Possible graphs:

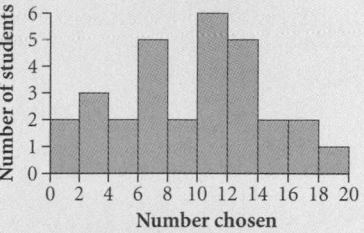

Random Numbers Selected Between 0 and 20

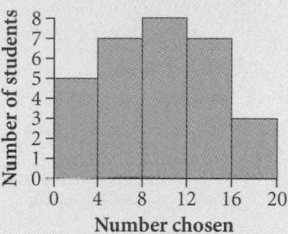

Random Numbers Selected Between 0 and 20

2c. 1, 6, 10, 12, 18

Random Numbers Selected Between 0 and 20

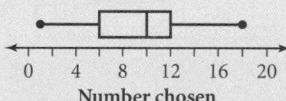

Exercise 3 The women pictured are from India, Mexico (Mayan), Kenya (Masai tribe), France, and Uighur (western China).

3b. Approximately $\frac{1}{4}$ of the countries had a female life expectancy between approximately 81 and 83 years.

3d. There are no bins to the right of 85 in the histogram. Also, the maximum point in the box plot is located at approximately 83 years.

4. Female Life Expectancy

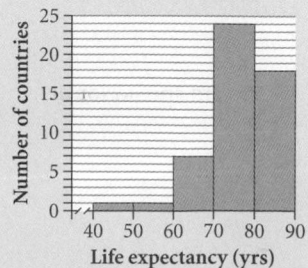

Exercise 6 If you have a computer, students might be able to save time by transferring the data from the CD to their calculators. Otherwise, have students work in pairs, one reading numbers and the other entering them, and then link their calculators to transfer the data.

Some students may be confused by having the lower left corner of the first bar away from the origin.

6b. Three models sold between 80,000 and 119,999 cars, inclusive.

6c. [80000, 440000, 40000, 0, 7, 1] or [0, 480000, 40000, 0, 7, 1]

6d.

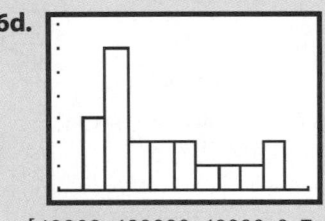

[40000, 480000, 40000, 0, 7, 1]

6e. An approximate five-number summary is 114000, 131000, 179000, 274000, 430000. The actual five-number summary is 113843, 131385.5, 179029.5, 274017.5, 429575.

Exercise 7 [ESL] Be sure students understand what a die is and that all students know that all outcomes of rolling a fair die are equally likely. Exercise 7d may challenge students who are weak at ratios. Encourage lots of guess-and-check as well as careful explanations.

▶ **Reason and Apply**

6. The histogram displays the passenger-car information listed in the table.

Top 20 Selling Passenger Cars in the United States in 1998

Car	Number sold	Car	Number sold
Toyota Camry	429,575	Chevrolet Lumina	177,631
Honda Accord	401,071	Ford Mustang	144,732
Ford Taurus	371,074	Nissan Altima	144,451
Honda Civic	334,562	Ford Contour	139,838
Ford Escort	291,936	Buick LeSabre	136,551
Chevrolet Cavalier	256,099	Buick Century	126,220
Toyota Corolla	250,501	Pontiac Grand Prix	122,915
Saturn	231,786	Dodge Neon	117,964
Chevrolet Malibu	223,703	Mercury Grand Marquis	114,162
Pontiac Grand Am	180,428	Nissan Maxima	113,843

(*2000 World Almanac and Book of Facts*, p. 716)

240,000 cars

a. What does 24 on the horizontal axis represent?

b. Explain the meaning of the first bin of this graph.

c. List calculator window values that would produce this graph. Give your answer in brackets like this: [Xmin, Xmax, Xscl, Ymin, Ymax, Yscl].

d. Graph this histogram on your calculator and on graph paper.

e. Use the table and your histogram to approximate the values of a five-number summary.

f. Graph a box plot on your calculator, and then sketch this plot above the histogram you drew in 6d.

7. Sketch what you think a histogram looks like for each situation below. Remember to label values and units on the axes. Answers will vary.

a. The outcomes when rolling a die 100 times.

b. The estimates of the height of the classroom ceiling made by 100 different students.

c. The ages of the next 100 people you meet in the school hallway.

d. The 100 data values used to make the box plot below. (Use a bin width of 1.)

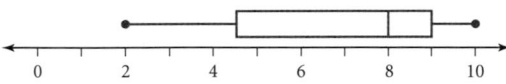

7a. The bar heights should be about the same, with about 16 or 17 in each of six bars.

7b. This graph will have a few short bins on the ends and several high ones around the actual height.

7c. This plot might be fairly flat except for several short bins on the high end.

7d. Fairly flat bins on the left getting taller as you go to the right, with the last two bins (8–9 and 9–10) both being 25 units tall.

8. Create a data set with eight data values for each graph.

a.

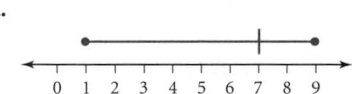

b.

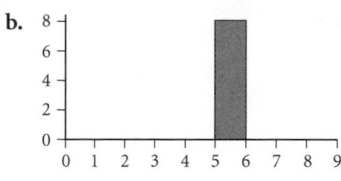

c.

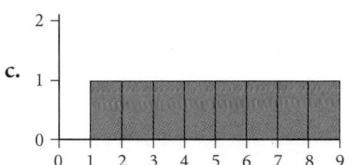

d.

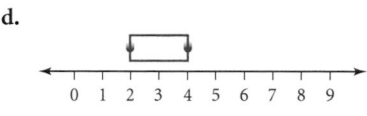

9. APPLICATION The histogram shows the results of an assessment on which 30 points were possible.

a. How would you assign the grades A–D?

b. Using your grading scheme, what grade would you assign the student represented by the bin farthest to the right?

c. Write a short description explaining why your grading scheme is a "fair" distribution of grades. Include comments about the measures of center, the variability, and the shape of the distribution.

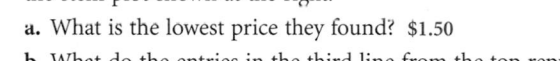

Assessment Results

Number of students vs *Number of points*

10. Chip and Dale, two algebra students, visited four different stores and recorded the prices on 1-pound bags of potato chips. They organized their data in the stem plot shown at the right.

a. What is the lowest price they found? $1.50

b. What do the entries in the third line from the top represent?

14 **c.** How many bags cost less than $2?

        Chips priced $1.75 and $1.79.

d. What is the most common price? $1.99

e. What is the range of prices for these chips? $1.09

Potato Chip Prices

15	0 0 9
16	9 9
17	5 9
18	5 9
19	9 9 9 9 9
20	9
21	5 9 9 9
22	5 9 9
23	9
24	
25	9

Key

25 | 9   means $2.59

Exercise 8a If students say "The box is missing," you might say that it has been "squeezed."

8a. {1, 7, 7, 7, 7, 7, 7, 9}. The Q1 value, the median, and the Q3 value are all 7.

Exercises 8b and 8c You might point out that data may include fractions.

8b. {5, 5, 5, 5, 5, 5, 5, 5}. Each value could be any number from 5 up to, but not including, 6.

8c. {1, 2, 3, 4, 5, 6, 7, 8}. These values could have any number of digits after the decimal point.

Exercise 8d Now the whiskers have been "squeezed."

8d. {2, 2, 2, 2, 2, 4, 4, 4} or {2, 2, 2, 4, 4, 4, 4, 4}. The minimum and Q1 are 2, Q3 and the maximum are 4, and the median is either 2 or 4; so it coincides with Q1 or Q3, respectively.

9a. Perhaps the top two get As, and the next seven (down to 15 points) get Bs. Those with 9 to 15 points would get Cs, and the bottom three would get Ds. Students who score in the center (mean, median, and mode) get a C grade.

9b. The outlier gets an A.

9c. Answers will vary.

▶ Review

11. Recreate on your paper this dot plot representing ages (in months) for students in an algebra class. Then add a box plot of this information to your graph.

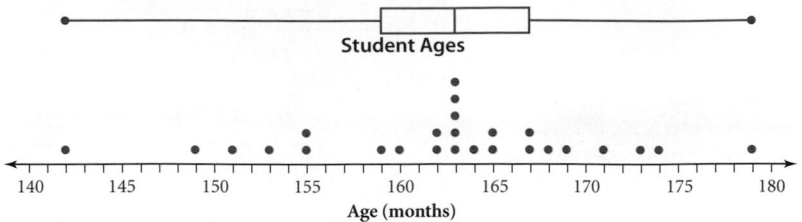

12. Create a data set of nine test scores with a five-number summary of 64, 72, 82, 82, 95 and a mean of 79. The underlined values are fixed. One possibility is {64, 70, 74, 80, 82, 82, 82, 82, 95}.

project

COMPARE COMMUNITIES

Compare Communities Project
Sample data sets are available on the worksheet Communities Sample Data or as part of the data supplied with Fathom™. If students have access to Fathom Version 2, they may be able to download entire data sets from the internet.

The U.S. Bureau of the Census collects data on people from all over the United States. Citizens and governments use these data to make informed decisions and to develop programs that best serve diverse communities. Here are two box plots that use census data to compare the ages of 50 people from two communities.

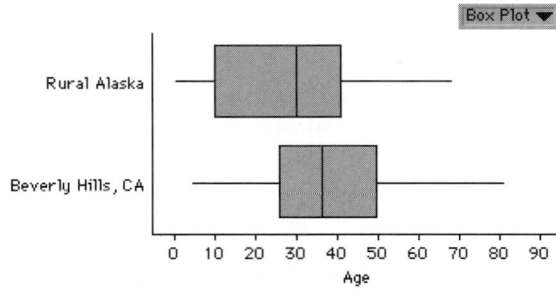

Fathom

These box plots were made with Fathom. You can use this software to work with many more people, attributes, or communities than you could by hand. Learn how to use Fathom to make a wide range of interesting graphs.

Use data provided with Fathom™ Dynamic Statistics™ software or get worksheets from your teacher listing official census data for many people in two communities. Compare the communities using three or more attributes such as gender, age, race, ancestry, marital status, education level, or income. Your project should include

▶ A graph of each attribute, for each community. You can use any type of graph from this chapter.

▶ A summary description of the attributes for each community. Compare and contrast the two communities using values like mean, median, or range.

▶ A written explanation of how you think each community could use your graphs to make decisions.

Supporting the project

MOTIVATION

"Can you identify the median age, range, and interquartile range in each?" "Based on the graphs, which community might benefit more from a preschool? From a senior center?" "What other information would help you make decisions like these?"

OUTCOMES

▶ Data are collected on individual people, not groups.
▶ The graphs are appropriate for the attributes chosen and are clearly drawn and labeled.
▶ The statistical terms are used accurately and the comparisons are valid.

▶ Good mathematical language is used in the explanation.
• The report included critical remarks on the census and decision-making procedures.

Activity Day

Exploring a Conjecture

data
measures of center
mean
median
mode
range
outlier

five-number summary
minimum
maximum
first quartile (Q1)
third quartile (Q3)
interquartile range (IQR)

pictograph
bar graph
dot plot
box-and-whisker plot
histogram
frequency
stem-and-leaf plot

Statistics is a branch of applied mathematics dedicated to collecting and analyzing numeric data. The **data analysis** that statisticians do is used in science, government, and social services like health care. In this chapter you have learned concepts fundamental to statistics: measures of center, summary values, and types of graphs to organize and display data. Terms you have learned are in the box to the left.

Each measure or graph tells part of the story. Yet having too much information for a data set might not be helpful. Statisticians, and other people who work with data, must choose which measures and graphs give the best picture for a particular situation. Carefully chosen statistics can be informative and persuasive. Poorly chosen statistics, ones that don't show important characteristics of the data set, can be accidentally or deliberately misleading.

Activity

The Conjecture

You will need

- two books
- graph paper
- colored pencils or pens
- poster paper

A **conjecture** is a statement that might be true but has not been proven. Your group's goal is to come up with a conjecture relating two things and to collect and analyze the numeric evidence to support your conjecture or cast doubt on it.

In this activity you'll review the measures and graphs you have learned. Along the way, you will be faced with questions that statisticians face every day.

Step 1 | Your group should select two books on different subjects or with different reading levels. Flip through the books, but do not examine them in depth. State a conjecture comparing these two books. Your conjecture should deal with a quantity that you can count or measure. For example, "The history book has more words per sentence than the math book."

NCTM STANDARDS

CONTENT	PROCESS
• Number	• Problem Solving
Algebra	• Reasoning
Geometry	• Communication
Measurement	Connections
• Data/Probability	• Representation

LESSON OBJECTIVES

- Review the statistics and graphs introduced so far
- Make conjectures about a data set and test them using appropriate statistical and graphical methods
- Use the terminology of statistics and graphs to present conclusions of a study of data

PLANNING

LESSON OUTLINE

One day:
25 min   Activity
20 min   Sharing
5 min   Closing

MATERIALS

- books on a variety of topics (academic and leisure) and of assorted page size, type size, number and type of illustrations, density of print, reading level, etc.
- graph paper
- materials for making posters
- Calculator Note 2A

TEACHING

The goal today is to deepen levels of understanding rather than to encounter new material.

G uiding the Activity

Step 1 Ask that students get your approval for their conjectures before going to Step 2. Some students may suggest conjectures that are not testable because they're not quantitative—they don't involve counting or measuring. Encourage all ideas but also a careful critique of the suggestions.

Examples of testable conjectures are "The first-grade book has words with fewer syllables than the ninth-grade book" and "The chemistry book and the biology book have the same number of words per page." If students are having trouble making any conjecture at all, you might suggest, "Describe this book. Does the same description fit the other book?"

Step 2 Elicit from students how they might get a "random sample" or a "typical sample" of data. One possibility is to create a random number routine to select pages from which to gather data. (Calculator Note 2A can help.)

Step 3 Point out any inconsistent data collection within a group. Group members should be measuring or counting in the same way with the same units. Every student should be gathering data.

Steps 4 to 7 Groups can divide the tasks for these steps. **[Alert]** Remind students to label their graphs. As needed, remind students that dot plots and stem plots show individual data items, that box plots help show the distribution but that they show no individual data items, and that histograms show how data sorts into bins.

Step 8 Provide students with clear guidelines for their reports or posters. They should analyze their displays to see what they can and can't learn. **[Alert]** Watch for weak support of conjectures. **[Ask]** "Could you have designed your data collection process differently?"

SHARING IDEAS

Ask at least one group to present its report orally, and ask the class to think about the extent to which the group justified its conclusions.

[Ask] "Could someone use the same data and reach a different conclusion?" Point out that selecting different measures or graphs might show different distributions of data. On the other hand, two truly random samples are likely to produce equivalent results.

[Ask] "How do your graphs demonstrate the truth of your conjecture?" Try to help students realize that there may not be a single best way to represent a data set.

Step 2   Decide how much data you'll need to convince yourself and your group that the conjecture is true or doubtful. Design a way to choose data to count or measure. For example, you might use your calculator to randomly select a page or a sentence. [▶ ▢ See **Calculator Note 2A** to generate random numbers.◀]

Step 3   Collect data from both books. Be consistent in your data collection, especially if more than one person is doing the collecting. Assign tasks to each member of your group.

Step 4   Find the measures of center, range, five-number summary, and IQR for each of the two data sets.

Step 5   Create a dot plot or stem-and-leaf plot for each set of data.

Step 6   Make box plots for both data sets above the same horizontal axis.

Step 7   Make a histogram for each data set.

Be sure that you have used descriptive units for all of your measures and clearly labeled your axes and plots before going on to the next step.

Step 8   Choose one or two of the measures and one pair of graphs that you feel give the best evidence for or against your conjecture. Prepare a brief report or a poster. Include

a. Your conjecture.

b. Tables showing all the data you collected.

c. The measures and graphs that seem to support or disprove your conjecture.

d. Your conclusion about your conjecture.

Could you have designed your data collection process differently? Would this change have provided more persuasive results?

Suppose another group used your data but chose a different selection of measures and graphs. Could they have reached a different conclusion about the truth or falsehood of your conjecture?

In the context of the report, mention the terms *sample* and *population*. Ask why they needed to use a sample to get this data set and why it matters how they got the sample.

Point out how statistics and graphs can be persuasive. Also point out any examples of how they can be misleading.

If students suspect that their conjecture is not adequately demonstrated, you might say that there are statistical tests, such as chi-square tests, that measure the significance of results by taking into account that different samples can yield different data.

Closing the Lesson

To conclude these lessons on one-variable data, you might discuss with students how statistics have become very important to society. For example, why might the first census have been taken? On a broader scale, what other fields of mathematics have been developed for governmental purposes?

Two-Variable Data

A **variable** is a trait or quantity whose value can change or vary. For example, birth dates will vary from person to person. In algebra, letters and symbols are used to represent variables. A person's birth date could be represented with the variable b, but the letter or symbol you choose doesn't matter. A data set that measures only one trait or quantity is called **one-variable data.** In Lessons 1.1 to 1.5, you learned to graph and summarize one-variable data.

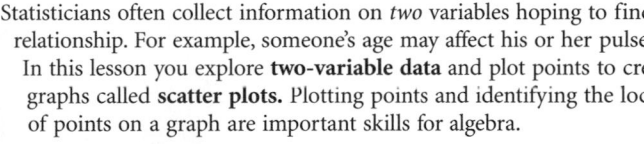

Statisticians often collect information on *two* variables hoping to find a relationship. For example, someone's age may affect his or her pulse rate. In this lesson you explore **two-variable data** and plot points to create graphs called **scatter plots.** Plotting points and identifying the location of points on a graph are important skills for algebra.

Two-variable graphs are constructed with two axes. Each axis shows possible values for one of the two variables. On the **coordinate plane,** the horizontal axis is used for values of the variable x, and the vertical axis is used for values of the variable y. The **origin** is the point where the x- and y-axes intersect. The axes divide the coordinate plane into four **quadrants.** Each point is identified with **coordinates** (x, y) that tell its horizontal distance x and vertical distance y from the origin. The horizontal distance of value x is always listed first, so (x, y) is called an **ordered pair.**

The coordinates tell you to move left 1 and up 2.

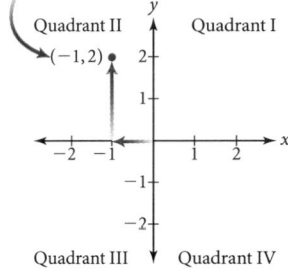

Investigation
Backed into a Corner

You will need

- 8 to 10 small round objects (coins, buttons, bottle caps)
- two centimeter rulers
- tape
- graph paper

In this investigation you'll collect two-variable data. A graph of these data points may help you to see a relationship between the two variables.

Step 1 | Tape one ruler to your desktop. Touch one vertex of the second ruler to the zero mark of the centimeter side of the first ruler and tape it in place to form an angle.

NCTM STANDARDS

CONTENT		PROCESS	
•	Number		Problem Solving
	Algebra	•	Reasoning
•	Geometry	•	Communication
•	Measurement	•	Connections
•	Data/Probability	•	Representation

LESSON OBJECTIVES

- On the coordinate plane, plot points with given coordinates, and determine the coordinates of plotted points
- Represent a two-variable data set with a scatter plot
- Given a pair of calculator lists, plot points whose coordinates are in those lists
- Interpret scatter plots

PLANNING

LESSON OUTLINE

First day:

25 min	Investigation with introduction
10 min	Sharing
15 min	Begin example

Second day:

25 min	Finish example
10 min	Closing
15 min	Exercises

MATERIALS

- 50 or more small, round, flat objects
- centimeter rulers (at least two per group)
- tape
- protractors (one per group)
- graph paper (one sheet per group)
- motion sensor, *optional*
- Calculator Notes 1E, 1F, 1G, 1K *(optional)*
- Coordinate Plane (T)
- Distance from a Motion Sensor (T), *optional*

TEACHING

The study of two-variable data begins with scatter plot representations and the use of letters for variables.

Guiding the Investigation

After students have gathered in their groups, talk through Steps 1 through 4.

Step 2
As needed, remind students that the diameter of a circle is the segment across the circle through the center.

Some students may have difficulty in measuring a diameter. Suggest that they move the object back and forth between two tick marks on the ruler to see the longest distance.

Step 3 [Link]
As a connection to a theorem of geometry, you might point out that the distance from the vertex is the same on both rulers.

Steps 5 to 7
Ask an experienced class to gather the data and represent it in a graph. Unsuccessful attempts at several kinds of representations will motivate learning about the coordinate plane and scatter plots—from you or each other.

Step 5
If your class needs more guidance, introduce points on the coordinate plane first, relying on the lesson introduction. You might project the coordinate plane transparency onto a hanging map or blueprint, or have several students stand on large coordinate axes on the floor. Point out the *x*- and *y*-axes. **[ESL]** Display the words *axis* and *axes,* and point out the singular and plural forms. Stress the importance of the order (first coordinate representing horizontal location, second coordinate vertical) when plotting points. **[Ask]** "What other words have the same root as *quadrant*?" [*Quad* means "four," as in *quadrangle, quadrilateral,* and *quadruplets.*]

Step 6 [ESL]
Explain that the word *scale* here refers to how numbers are marked on the axes. Some students may want to mark the axes with coordinates in a somewhat arbitrary way. Ask them to mark off a scale first. To do so, they can first mark the largest values of the two variables and then mark intermediate points proportionately. The desire to have 10–25 marks motivates the text's

---

Step 2 Measure the diameter of one of your round objects to the nearest tenth of a centimeter. Record this measurement in a table like this one.

Diameter (cm)	Distance from vertex (cm)

Step 3 Place the object inside your angle and push it toward the vertex until it fits snugly. Measure the point where the object touches the ruler that has zero at the vertex. This is the distance from the vertex. Enter this measurement, to the nearest tenth of a centimeter, in your table.

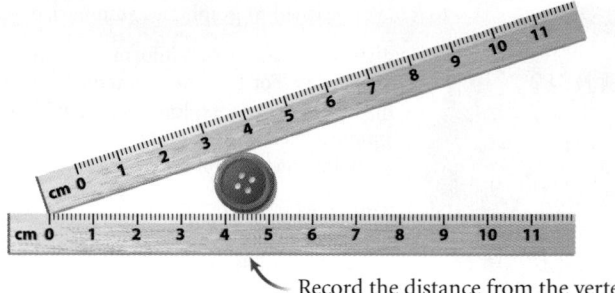

Record the distance from the vertex.

Step 4 Repeat Steps 2 and 3 for all of your round objects.

Step 5 Create a set of axes on your graph paper. Label the axes as shown.

Step 6 Scale the *x*-axis to fit all of your diameter values. For example, if the largest diameter was 5 cm, you might make each grid unit stand for 0.2 cm. Scale the *y*-axis to fit all of your distance-from-vertex values. For example, if the largest object touched the ruler at 30 cm, you might use 2 cm for each vertical grid unit.

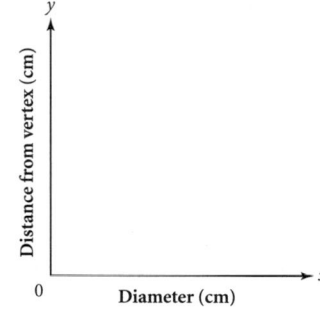

Step 7 Plot each piece of two-variable data from your table. Think of each row in your table as an ordered pair. Locate each point by first moving along the horizontal axis to the diameter measurement. Then move up vertically to the distance from vertex.

Step 8 Describe any patterns you see in the graph. Is there a relationship between your two variables? Graphs are linear. The larger the diameter the further the distance from the vertex of the angle. Smaller angles will result in a steeper slope.

Step 9 Enter the information from your table into two calculator lists. Make a scatter plot. The calculator display should look like the graph you drew by hand.
[▶ 🖳] See **Calculator Note 1E** to learn how to display this information on your calculator screen. ◀]

---

examples of 0.2 cm and 2 cm. Because the distance from the vertex will be much larger than the diameter, the scales on the *x*- and *y*-axes could then be different. This may be a new idea to some students.

Step 7 [Alert] Watch for confusion about which axis represents which variable. If asked, explain that any variable whose values you're trying to predict is (usually) plotted on the vertical axis. Don't belabor the point; this investigation is not unusual in that either variable might be considered the predictive variable.

Step 8 Students may say that the "points are straight." Point out that points have no length, and encourage use of the phrase "lie along a straight line" or "have a linear pattern." The slope of the line will vary according to the size of the angle. The larger the object, the farther it will be from the angle's vertex.

Step 9 One of the greatest difficulties students have in graphing on a calculator is setting the viewing window. If you skip the step of setting the window, the chances of seeing a graph are not very good. Help students work through Calculator Note 1E.

This officer is using a radar device that measures the speed of oncoming cars.

There are many ways to collect data. Have you ever seen a police officer measuring the speed of an approaching car? Have you wondered how technicians measure the speed of a baseball pitch? Did you know that satellites collect information about the earth from distances over 320 miles? Each of these measurements involves the use of remote sensors that collect data. In this course you may have the opportunity to work with portable sensor equipment.

EXAMPLE

This scatter plot shows how the distance from a motion sensor to a person varies over a period of 6 seconds. Describe where the person is in relation to the sensor at each second.

Distance From Motion Sensor

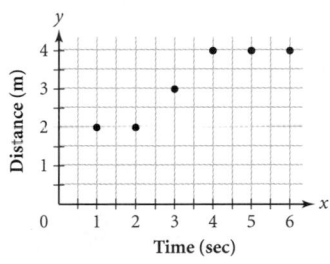

▶ **Solution**

The first point $(1, 2)$ shows that after 1 second the person was 2 meters away from the sensor.

The next point $(2, 2)$ indicates that after 2 seconds the person was still 2 meters away.

The point $(3, 3)$ means that after 3 seconds the person was 3 meters away.

The point $(4, 4)$ means that at 4 seconds the person was 4 meters away. He or she remained 4 meters away until 6 seconds had passed, as indicated by the points $(5, 4)$ and $(6, 4)$.

Negative values of time would indicate that events had occurred before the arbitrary starting point in time.

The graph in this example is a **first-quadrant graph** because all the values are positive. A lot of real-world data is described with only positive numbers, so first-quadrant graphs are very useful. However, you could graph the person's distance in front of the sensor as positive values and his or her distance *behind* the sensor as negative values. This would require more than one quadrant to show the data. If the graph showed negative values of time, how would you interpret this?

Time values might be negative if you were measuring time relative to a particular starting point—for example, time before noon and after noon.

Closing the Lesson

The mathematical ideas of this lesson are two-variable data, scatter plot, coordinate axes, and coordinates of a point. Two-variable data points are described by pairs of numbers. These can be represented visually by a scatter plot, in which each point has one of those pairs of data as its coordinates.

SHARING IDEAS

You might ask for several presentations of the results of Step 9, illustrating different slopes. **[Ask]** "Why do the graphs look different?" Encourage the class to see that smaller angles lead to greater distances from the vertex. As a consequence, the lines will be steeper.

Assessing Progress

By watching students' contributions, you can assess their skills with **careful measurement, organized data collection, entering data into lists, considering one variable at a time,** and using the **term** *straight line.*

EXAMPLE

If you have a second day for this lesson, the examples of motion provide opportunities for interpreting graphs in a context some students will see as relevant or fun.

If possible, use motion sensors attached to calculators. As you set up equipment, explain that remote sensors often use reflected pulses of high-pitched sound to measure distance. Try to arrange for each student to walk at least once and to operate the motion sensor. If motion sensors aren't available, use the transparency.

[Alert] Watch for the common misinterpretation of the graph as the *path* of the motion against time. This indicates that students are still having trouble assimilating the abstraction of the graph of distance. Repeatedly emphasize how the plot represents time and distance. It is not a map of north-south against east west. **[Ask]** "What other scenarios may give rise to the same distance values at seconds 4, 5, and 6?" [Both the sensor and the person may still be at a distance of 4 meters from each other, or the person may be walking in an arc around the sensor at a 4-meter radius, or both the person and the sensor may be moving in parallel 4 meters apart.] Encourage creativity.

In these exercises students practice both graphing from data and interpreting graphs.

ASSIGNING HOMEWORK

Essential	**1, 2, 4, 7, 10**
Performance assessment	**5, 7**
Portfolio	**8**
Journal	**6**
Group	**8, 10**
Review	**11, 12**

▶ Helping with the Exercises

Graph paper will be helpful for these exercises. If you're assigning Exercise 3 and students will not be working on it in the classroom, do the linking first.

Exercise 1 The viewing window being set up could be described as $-9 \leq x \leq 9$ and $-6 \leq y \leq 6$, but students may not be familiar with inequality symbols and you may not want to start that discussion now.

You may want to note that points are sometimes given a shorthand name with capital letters. Letters and symbols can represent constants as well as variables (π, $y = a + bx$).

Exercise 2 [Alert] Watch for confusion about quadrant names here; you may need to identify them again.

Exercise 3 Load this program into your calculator so that students can acquire it though linking.

Exercise 4 [Alert] Watch for confusion about this continuous graph. The curve contains infinitely many points and shows the walker's position at each moment.

EXERCISES

You will need your calculator for problems **3, 5, 6, 7,** and **9.**

▶ Practice Your Skills

1. Draw and label a coordinate plane so that the x-axis extends from −9 to 9 and the y-axis extends from −6 to 6. Represent each point below with a dot, and label the point using its letter name.

$A(-5, -3.5)$ $B(2.5, -5)$ $C(5, 0)$ $D(-1.5, 4)$ $E(0, 4.5)$

$F(2, -3)$ $G(-4, -1)$ $H(-5, 5)$ $I(4, 3)$ $J(0, 0)$

2. Sketch a coordinate plane. Label the axes and each of the four quadrants—I, II, III, and IV. Identify the axis or quadrant location of each point described.

   a. The first coordinate is positive, and the second coordinate is 0. **positive x-axis**

   b. The first coordinate is negative, and the second coordinate is positive. **quadrant II**

   c. Both coordinates are positive. **quadrant I**

   d. Both coordinates are negative. **quadrant III**

   e. The coordinates are (0, 0). **the origin**

   f. The first coordinate is 0, and the second coordinate is negative. **negative y-axis**

3. Use your calculator to practice identifying coordinates. The program POINTS will place a point randomly on the calculator screen. You identify the point by entering its coordinates to the nearest one-half. Run the program POINTS until you can easily name points in all four quadrants. [▶ 🖥 See **Calculator Note 1F.**◀]

▶ Reason and Apply

4. This graph pictures a walker's distance from a stationary motion sensor.

Motion Sensor Readings

These people are running and walking in a fund-raiser to support cancer research.

   a. How far away was the walker after 2 seconds? **about 2 m**

   b. At what time was the walker closest to the sensor? **about 5 sec**

   c. Approximately how far away was the walker after 10 seconds? **about 2.7 m**

   d. When, if ever, did the walker stop? **between 0 and 1 sec, and between 4.5 and 5.5 sec**

1.

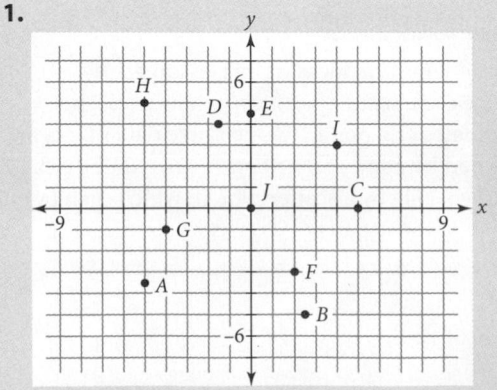

5. Look at this scatter plot.

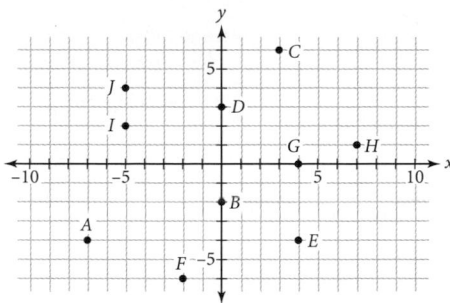

a. Name the (x, y) coordinates of each point pictured. $A(-7, -4), B(0, -2), C(3, 6), D(0, 3), E(4, -4),$ $F(-2, -6), G(4, 0), H(7, 1), I(-5, 2),$ and $J(-5, 4)$

b. Enter the x-coordinates of the points you named in 5a into list L_1 and the corresponding y-coordinates into list L_2. Set your graphing window to the values suggested in the pictured graph, and graph a scatter plot of the points. We call this the scatter plot of (L_1, L_2).

c. Which points are on an axis? Points *B*, *D*, and *G*

d. List the points in Quadrant I, Quadrant II, Quadrant III, and Quadrant IV.
I: *C* and *H* II: *I* and *J* III: *A* and *F* IV: *E*

6. Write a paragraph explaining how to make a calculator scatter plot, and identify point locations in the coordinate plane.

7. APPLICATION The graph below is created by connecting the points in a scatter plot as you move left to right. [▶🖳 See **Calculator Note 1G** to learn how to connect a scatter plot.◀]

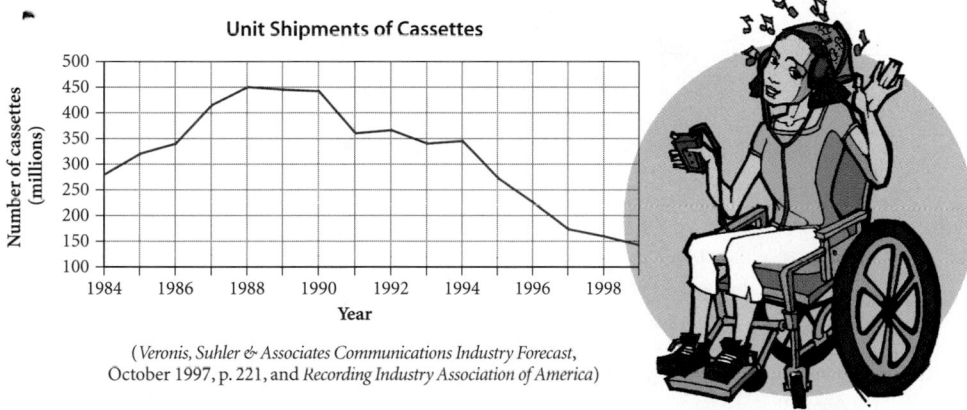

Unit Shipments of Cassettes

(*Veronis, Suhler & Associates Communications Industry Forecast*, October 1997, p. 221, and *Recording Industry Association of America*)

a. Approximate the 16 data points, for which the x-value is the year and the y-value is the number of cassettes, represented on this graph.

b. Name graphing window values and create this graph on your calculator screen. If necessary, adjust your coordinates from 7a so that your graph matches the graph shown above.

c. The graph shows a pattern or trend for shipments of music cassettes. Describe any patterns you see. What do you think happened in the 1980s that would cause the patterns in this graph?

Exercise 5 As always, watch for difficulties in specifying the viewing window. Try not to be too directive; students need to learn to do this for themselves. If necessary, refer students to Calculator Note 1E.

5b.

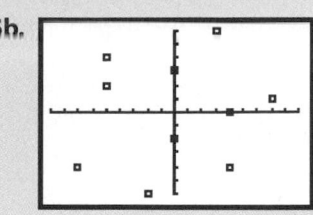

$[-9, 9, 1, -6, 6, 1]$

6. Location scenarios should include information such as *(positive, positive)* in Quadrant I, *(negative, positive)* in Quadrant II, *(negative, negative)* in Quadrant III, *(positive, negative)* in Quadrant IV, *(0, any number)* on y-axis, *(any number, 0)* on x-axis, and *(0, 0)* as the origin.

7a. Approximate answers (the second coordinates are in millions): (1984, 280), (1985, 320), (1986, 340), (1987, 415), (1988, 450), (1989, 445), (1990, 440), (1991, 360), (1992, 370), (1993, 340), (1994, 345), (1995, 273), (1996, 225), (1997, 173), (1998, 159), (1999, 142).

7b.

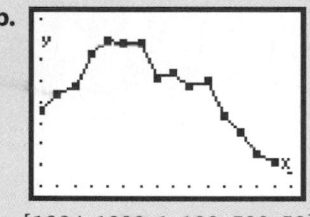

$[1984, 1999, 1, 100, 500, 50]$

7c. Answers will vary. Shipments increased until 1988 and then decreased. The introduction of compact discs may have influenced the decrease.

Exercise 8a In filling in the table, some students may jump to a pattern of 5, 10, 15, 20 and keep going, not noticing that the pattern breaks with 1996.

8b.

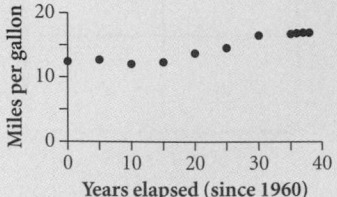

Average Miles per Gallon for All U.S. Automobiles (since 1960)

8c.

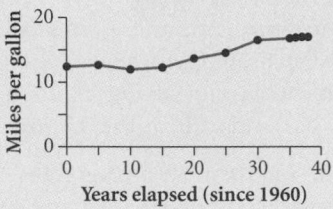

Average Miles per Gallon for All U.S. Automobiles (since 1960)

Exercise 8d [Alert] Watch for difficulties in understanding the meaning of a constant equation. The graph of equation $y = 16.7$ consists of exactly those points whose y-coordinate is 16.7. Thus it's a straight horizontal line.

8e.

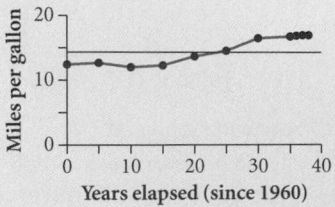

Average Miles per Gallon for All U.S. Automobiles (since 1960)

The horizontal line shows which data points are above the average of the averages and which are below it.

8f. Answers will vary.

---

8. The data in this table show the average miles per gallon (mpg) for all U.S. automobiles during the indicated years.

  a. Copy this table onto paper and calculate the years elapsed since 1960 to complete it.

  b. Graph a scatter plot on your paper of points whose x-value is years elapsed and whose y-value is miles per gallon. Carefully label and scale your axes. Give your graph a title.

  c. Connect each point in your scatter plot with a line segment from left to right.

  d. What is the mean mpg for these data? **14.6 mpg**

  e. Graph a horizontal line that starts on the y-axis at a height equal to your answer to 8d. What does this line tell you?

  f. Write a short descriptive statement about any pattern you see in these data and in your graph.

Average Miles per Gallon for All U.S. Automobiles

Year	Years elapsed	mpg
1960	0	12.4
1965	5	12.5
1970	10	12.0
1975	15	12.2
1980	20	13.3
1985	25	14.6
1990	30	16.4
1995	35	16.8
1996	36	16.9
1997	37	17.0
1998	38	17.0

(U.S. Department of Transportation)

9. The graph at right is a hexagon whose vertices are seven ordered pairs. Two of the points are (3, 0) and (1.5, 2.6). The hexagon is centered at the origin.

  a. What are the coordinates of the other points? $(-1.5, 2.6), (-3, 0), (-1.5, -2.6), (1.5, -2.6)$

  b. Create this connected graph on your calculator. Add a few more points and line segments to make a piece of calculator art. Identify the points you added. Answers will vary.

10. Xavier's dad braked suddenly to avoid hitting a squirrel as he drove Xavier to school. His speed during the trip to school is shown on the graph at right.

  a. At what time did Xavier's dad apply the brakes? **8:06**

  b. What was his fastest speed during the trip? **40 mi/hr**

  c. How long did it take Xavier to get to school? **12 min**

  d. Find one feature of this graph that you think is unrealistic. Answers will vary. That he never fully stopped during the trip to school is unusual.

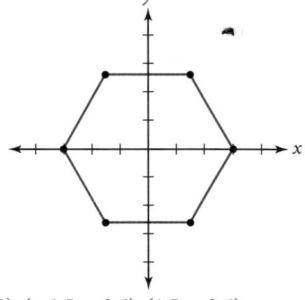

Xavier's Trip to School

Encourage students to be suspicious about conclusions drawn from taking a mean of means. For example, if the average test score in this class is 83 and in that class is 79, the average over the two classes would not be 81 unless the classes were the same size. In this exercise the mean of the means would give the average mileage of all cars over the years only if the number of gallons of gasoline used each year was the same.

Exercise 9 [Alert] Watch for difficulties in understanding that the points will be connected in the order in which they occur in the list. Encourage creativity in the art.

Exercise 10 Some students may find this exercise easy. Challenge them to find the distance traveled on the trip to school. To do so, they could take the average speed for each piece of the trip and multiply by the time taken for that piece. Or they could calculate the area under the graph.

Review

11. Create a data set with the specified number of items and the five-number summary values 5, 12, 15, 30, 47.

   a. 7   **b.** 10   **c.** 12

12. The table gives results for seventh-through ninth-grade students in the Third International Mathematics and Science Study.

   a. Find the five-number summary for this data set.

   b. Construct a box plot for these data.

   c. Between which five-number summary values is there the greatest spread of data? The least spread?

   d. What is the interquartile range?

   e. List the countries between the first quartile and the median. List those between the third quartile and the maximum, inclusive.

Results of the Third International Mathematics and Science Study (7th to 9th Grade)

Country	Mean score	Country	Mean score
Singapore	643	Sweden	519
Korea	607	New Zealand	508
Japan	605	England	506
Hong Kong	588	Norway	503
Belgium (Fl)	565	United States	500
Czech Republic	564	Latvia (LSS)	493
Slovak Republic	547	Iceland	487
Switzerland	545	Spain	487
France	538	Lithuania	477
Hungary	537	Cyprus	474
Russian Federation	535	Portugal	454
Canada	527	Iran, Islamic Republic	428
Ireland	527		

(*IEA Third International Mathematics and Science Study [TIMSS]*, 1994–95)

11a. The underlined values are fixed. One possibility is {5, 12, 14, 15, 20, 30, 47}.

11b. The underlined values are fixed. One possibility is {5, 10, 12, 13, 14, 16, 20, 30, 40, 47}.

11c. The underlined values are fixed. One possibility is {5, 10, 12, 12, 13, 14, 16, 20, 28, 32, 40, 47}.

Exercise 12 You or your students might want to research the data from 1999, to be released after this book goes to press. Check out the link from www.keymath.com/DA.

12a. 428, 490, 527, 555.5, 643

12b.

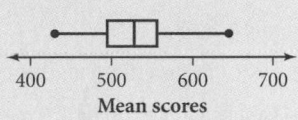
Result of TIMSS
400   500   600   700
Mean scores

12c. The greatest spread is between the third quartile and the maximum, or the right whisker. The least spread occurs between the median and the third quartile.

12d. 65.5 points

12e. Latvia, United States, Norway, England, New Zealand, Sweden; Czech Republic, Belgium, Hong Kong, Japan, Korea, Singapore

IMPROVING YOUR REASONING SKILLS

A **glyph** is a symbol that presents information nonverbally. These weather glyphs show data values for several variables in one symbol. How many variables can you identify? The diagram shows data for 12 hours starting at 12:00 noon. Which characteristics would you call categorical? Which are numerical? Is it possible to show all or part of the data in one or more of the graphs you learned to use in this chapter? How would you do that? Which graphs have the greatest advantages in this situation? Why?

You can learn more about weather symbols and research your local weather conditions by using the Internet links at **www.keymath.com/DA** .

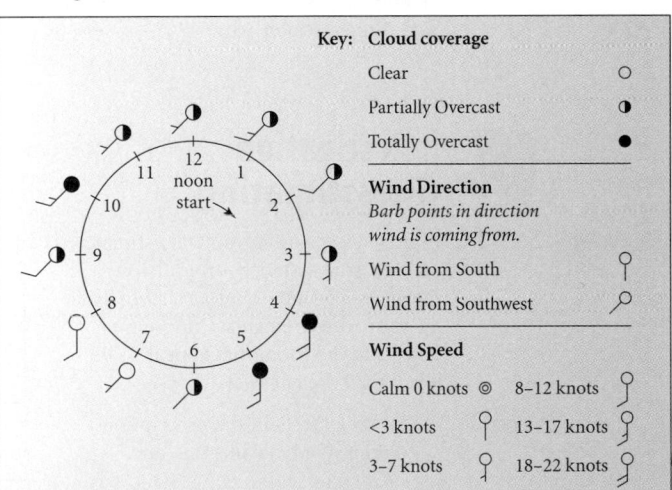

Key: **Cloud coverage**

Clear   ○
Partially Overcast   ◑
Totally Overcast   ●

Wind Direction
Barb points in direction wind is coming from.
Wind from South
Wind from Southwest

Wind Speed
Calm 0 knots ◎ 8–12 knots
<3 knots   13–17 knots
3–7 knots   18–22 knots

IMPROVING REASONING SKILLS

The weather glyphs show three variables—cloud coverage, wind direction, and wind speed. Time could be considered a fourth variable. Time is definitely a numerical variable, and wind direction is definitely categorical. Cloud coverage and wind speed, however, could be considered to be either numerical or categorical. Cloud coverage could be three categories, or it could be assigned a percent-age (0%, 50%, 100%) or a degree number (1, 2, 3). Wind speed is even more complex because, although a specific numeric wind speed would be measured (for example, 6 knots), each is then placed in a categorical bin (for example, 3–7 knots). Students could use an array of graphs to display the data but may be unable to find a single type of graph to display all the information.

Estimating

Statistics are no substitute for judgment.

HENRY CLAY

If you read Michael Crichton's book *Jurassic Park* or if you saw the movie, you may remember that the dinosaur population grows faster than expected. The characters estimate that there will be only 238 dinosaurs. But a computer inventory shows that there are actually 292 dinosaurs!

This skeleton of a Tyrannosaurus rex is at the Royal Tyrrell Museum in Alberta, Canada. Tyrannosaurs were among the largest carnivorous dinosaurs and measured up to 40 feet in length.

Dinosaurs in *Jurassic Park*

Species	Actual number	Estimated number	Species	Actual number	Estimated number
Tyrannosaurs	2	2	Hadrosaurs	11	11
Maiasaurs	22	21	Dilophosaurs	7	7
Stegosaurs	4	4	Pterosaurs	6	6
Triceratops	8	8	Hypsilophodontids	34	33
Procompsognathids	65	49	Euoplocephalids	16	16
Othnielia	23	16	Styracosaurs	18	18
Velociraptors	37	8	Microceratops	22	22
Apatosaurs	17	17	Total	292	238

(*Jurassic Park*, 1991, p. 164)

It is easy to see how the estimated numbers and the actual numbers compare by looking at this table. In this lesson you learn how to make efficient comparisons of data using a scatter plot.

Investigation
Guesstimating

You will need

- a meterstick, tape measure, or motion sensor

In this investigation, you will estimate and measure distances around your room. As a group, select a starting point for your measurements. Choose nine objects in the room that appear to be less than 5 meters away.

Step 1 | List the objects in the description column of a table like this one.

	x	y
Description	Actual distance (m)	Estimated distance (m)
(item 1)		
(item 2)		
(item 3)		
(item 4)		

LESSON OBJECTIVES

- Use a scatter plot to compare estimates with actual values
- Graph the equation $y = x$
- Judge the overall accuracy of a set of estimates from a scatter plot and the graph of $y = x$
- Develop the graphing practices of scaling and plotting carefully and labeling the graph and axes

NCTM STANDARDS

CONTENT	PROCESS
• Number	• Problem Solving
Algebra	• Reasoning
• Geometry	Communication
• Measurement	• Connections
• Data/Probability	• Representation

PLANNING

LESSON OUTLINE

One day:

35 min   Investigation

5 min   Closing

10 min   Exercises

MATERIALS

- tape measures, metersticks, or motion sensors
- graph paper
- Estimation Investigations (W), *optional*
- Calculator Notes 1H, 1I, 1K (*optional*)

TEACHING

USING THE QUOTE

Ask students how the quote opening the lesson relates to what they have learned in this chapter. Students should recognize that they have had to use judgment in selecting the best graphs and measures to use and that judgment is critical for analyzing the statistics they choose.

Guiding the Investigation

While students in groups are reading the opening example about dinosaurs, give out measuring devices and graph paper. Mention that the table is organized with the estimated column on the right to make a scatter plot of the data easier to understand.

To limit confusion, you can mark observation points on the floor, and ask groups to consider objects only in their region of the room.

One step Once a class experienced at investigations makes the table of data, ask them to compare a

(*Continued top of page 75*)

Step 2	Estimate the distances in meters or parts of a meter from your starting point to each object. If group members disagree, find the mean of your estimates. Record the estimates in your table.
Step 3	Measure the actual distances to each object and record them in the table.

Step 4	Draw coordinate axes and label actual distance on the x-axis and estimated distance on the y-axis. Use the same scale on both axes. Carefully plot your nine points. For each point, the actual distance is the x-value and the estimated distance is the y-value.
Step 5	Describe what this graph would look like if each of your estimates had been exactly the same as the actual measurement. How could you indicate this pattern on your graph? If all estimates were correct, they would lie on a diagonal line, specifically $y = x$.

Step 6	Make a calculator scatter plot of your data. Use your paper-and-pencil graph as a guide for setting a good graphing window.
Step 7	On your calculator, graph the line $y = x$. What does this line represent? Estimates that are correct [▶  See **Calculator Note 1H** to graph a scatter plot and an equation simultaneously.◀]
Step 8	What do you notice about the points for distances that were underestimated? What about points for distances that were overestimated? Underestimates fall under the line $y = x$; overestimates are above.
Step 9	How would you recognize the point for a distance that was estimated exactly the same as its actual measurement? Explain why this point would fall where it does.

Correct estimates will fall on the line $y = x$. In context the line $y = x$ represents *estimate = actual*.

Throughout this course you will create useful and informative graphs. Sometimes adding other elements to a graph as a basis for comparison can help you interpret your data. In the investigation, you added the line $y = x$ to your graph. How did this help you assess the accuracy of your estimates? Responses should include easily identifying overestimates and underestimates.

EXERCISES

You will need your calculator for problems **1, 4,** and **7.**

▶ Practice Your Skills

1. Enter the *Jurassic Park* data into your calculator. Put actual numbers into list L1 and estimated numbers into list L2 so that each (x, y) point has the form (*actual number, estimated number*).

   a. Identify the two variables in this situation.

   b. Graph a scatter plot of the data and record the window you used.

1a. Actual number of dinosaurs; estimated number of dinosaurs.

1b. The lower Ymin value allows trace numbers to appear on the calculator screen without interfering with plotted points.

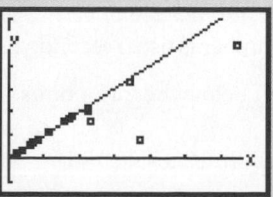

$[0, 70, 10, -10, 60, 10]$

scatter plot of the data with the line $y = x$ and to explain what it means for a point of the scatter plot to lie on the line.

Step 2 Show students a meterstick before they begin to estimate. Remind students to estimate and fill in the column *before* measuring. **[ESL]** Be sure students understand the word *estimate.* You can use terms like *near, close to,* and *more or less* to explain.

[Alert] Check that students don't put the estimate in the first rather than the second column.

Step 3 Be ready to help students who are having trouble measuring. Students using motion sensors will need Calculator Note 1K.

Step 5 If all estimates were the same as the actual measurements, the points plotted would have equal coordinates and thus lie on the line $y = x$.

Step 8 It is in these interpretations that the order of columns makes a difference. Overestimates will appear above the graph of $y = x$ and underestimates below.

SHARING IDEAS

Request at least two presentations of graphs. If a group did it backward from the instructions, include them.

[Ask] "Do you agree with the presenters' answers to the questions in Step 8?" At an appropriate time in the discussion, write the pair (*actual, estimated*) on the board. Students need to get used to seeing (x, y) pairs set out in this way. Point out that this defines x (and the x-axis) to stand for "actual," any y (and the y-axis) to stand for "estimate."

If a group used the pairs (*estimated, actual*), point out that information can still be gained from the graph. Now, however, the overestimates are below the line $y = x$ and the underestimates are above.

Another Investigation
As an extension, you can look up the ages of current political figures, artists, musicians, film personalities, writers, or perhaps even the principal of the school! Before showing these data to the students, have them guess the age of each person. Then form (*actual age, estimated age*) pairs that students can graph and analyze. You can find some examples on the Estimation Investigations worksheets. Cover the actual data written to the right of the problem before you copy the worksheet for students.

Assessing Progress
By observing students' contributions, you can assess their skills at **drawing axes** and **marking scales, plotting points,** and **making scatter plots,** on paper and on the calculator.

Closing the Lesson

If many students had difficulty using the line $y = x$, have the class read the last paragraph before the exercises. **[Ask]** "How does the reference line help you assess your estimates?"

Reinforce the new mathematical idea that the graph of the line $y = x$ consists of all points whose coordinates are equal.

BUILDING UNDERSTANDING

In these exercises, students compare estimated and actual values.

ASSIGNING HOMEWORK

Essential	1–3, 5, 7, 8
Performance assessment	3, 6
Portfolio	6
Journal	4, 9
Group	11
Review	11, 12

76 CHAPTER 1 Data Exploration

c. If x represents the actual number of dinosaurs and y the estimated number of dinosaurs, what equation represents the situation when the actual numbers equal the estimated numbers? Graph this equation on your calculator. $y = x$

d. Are any points of the scatter plot above the line you drew in 1c? What do these points represent? No. In no case is the estimated number of dinosaurs more than the actual number.

e. Are any points of the scatter plot below the line you drew in 1c? What do these points represent? Yes. The estimated count of five different species was less than the actual count.

2. Lucia and Malcolm each estimated the weights of five different items from a grocery store. Each of Lucia's estimates was too low. Each of Malcolm's was too high. The scatter plot at right shows the (*actual weight, estimated weight*) data collected. The line drawn shows when an *estimate* is the same as the *actual* measurement.

a. Which points represent Lucia's estimates?

b. Which points represent Malcolm's estimates?

Lucia and Malcolm's Estimates

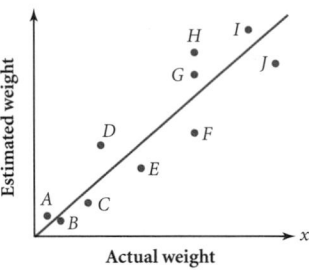

3. These points represent student estimates of temperatures in degrees Celsius for various samples of salt water. They are listed in the form (*actual temperature, estimated temperature*). Which points represent overestimates and which represent underestimates? Overestimates are B, C, D, E, and G. Underestimates are A, F, H, and I.

$A(27, 20)$	$B(-4, 2)$	$C(18, 22)$
$D(0, 3)$	$E(47, 60)$	$F(36, 28)$
$G(-2, 0)$	$H(33, 31)$	$I(-1, -2)$

4. This graph is a scatter plot of a person's distance from a motion sensor in a 5-second time period. The line is shown only as a guide.

a. Make a table of coordinates for the points pictured on this graph.

b. Describe how you would make a scatter plot of these data points on your calculator. Name the window values you would use.

c. What is the equation of the line pictured on the graph? $y = x$

d. Was the distance between the person and the sensor increasing or decreasing?

Distance from Motion Sensor

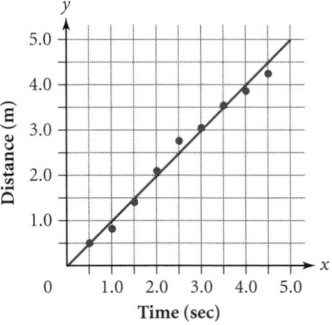

▶ **Helping with the Exercises**

Exercise 1 Be sure students understand that the ordered pair (*actual number, estimated number*) means that the estimated number is listed second.

2a. Lucia's estimates are all below the line, points B, C, E, F, and J.

2b. Malcolm's estimates are all above the line, points A, D, G, H, and I.

4a. Possible approximations are

Time	0.5	1.0	1.5	2.0	2.5	3.0	3.5	4.0	4.5
Distance	0.5	0.8	1.4	2.1	2.7	3.0	3.6	3.9	4.25

4b. Answers will vary. A minimum window is [0, 5, 1, 0, 5, 1].

4d. The distance is increasing at about 1 meter per second. The person may be walking away from the sensor.

▶ Reason and Apply

5. A group of students conducted an experiment by stretching a rubber band and letting it fly. They measured the amount of stretch (cm) in each trial and recorded it as x. The distance flown (cm) was measured and recorded as y. This scatter plot shows six trials.

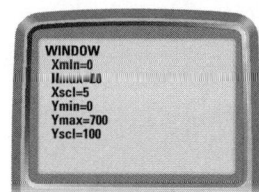 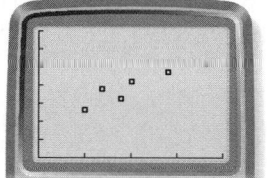

a. Describe any relationship you see. The more the rubber band is stretched, the farther it flies.

b. Based on the plot, how far might the rubber band fly if they stretch it 15 cm? Answers will vary between 400 and 600 cm.

c. How far should they stretch the rubber band if the target is at 400 cm? Answers will vary between 7 and 12 cm.

6. Copy this graph on your paper.

a. Plot a point that represents someone overestimating a $12 item by $4. Label it A. What are the coordinates of A? (12, 16)

b. Plot a point that represents someone underestimating an $18 item by $5. Label it B. What are the coordinates of B? (18, 13)

c. Plot and label the points $C(6, 8)$, $D(20, 25)$, and $E(26, 28)$. Describe each point as an overestimate or underestimate. How far off was each estimate?

d. Plot and label points F and G to represent two different estimates of an item priced at $16. Point F should be an underestimate of $3, and point G should be a perfect guess.

e. Where will all the points lie that represent an estimated price of $16? Describe your answer in words and show it on the graph.

f. Where will all the points lie that picture an actual price of $16? Describe your answer in words and show it on the graph.

g. If x represents the actual price and y represents the estimated price, where are all the points represented by the equation $y = x$? What do these points represent?

Estimated Prices vs. Actual Prices

7. Recall the *Jurassic Park* data on page 74. Enter the actual numbers and estimated numbers into two calculator lists. (You may still have the data in list L1 and list L2 from Exercise 1.) Use list L3 to calculate the estimated number minus the actual number for each dinosaur species. [▶ 🖵 See **Calculator Note 1I** for an explanation of how to use calculator lists in this way.◀]

a. What information does list L3 give you?

b. Use list L2 and list L3 to create a scatter plot of points in the form (*estimated number, estimated number − actual number*). Name the graphing window that provides a good display of your scatter plot.

c. How many points are below the x-axis? What do these points represent in general?

d. Name the coordinates of the point farthest from the x-axis. What do these coordinates tell you?

7a. It provides the differences between the estimated number of each species and the actual count. This helps identify over- and underestimates.

7b.

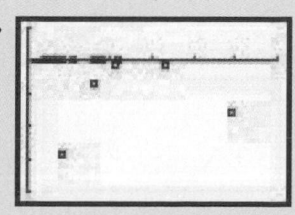

[0, 60, 10, −40, 10, 10]

7c. 5; underestimates

7d. (8, −29); the 8 is the estimated number of velociraptors; the number of velociraptors was underestimated by 29.

Exercise 5 [Alert] Watch for lack of understanding of how to determine the scale from the viewing window. Have them check Xmin, Xmax, Ymin, and Ymax for extreme points. Also, many students remain unaware that Xscl and Yscl determine where the hash marks are. Show students the effect each part of the window has on the graph.

Exercises 6f and 6g [Alert] For students having trouble concentrating on just one variable at a time, suggest that they make up and graph lots of points with the given attribute.

6c.

Estimated Prices vs. Actual Prices

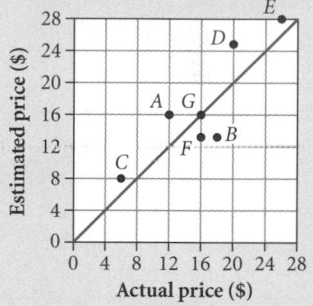

Point C is an overestimate of $2, point D an overestimate of $5, and Point E an overestimate of $2.

6d. See graph for 6c.

6e. On a horizontal line through the estimate $16 on the vertical axis.

6f. On a vertical line through the actual $16 on the horizontal axis.

6g. They are points on the line shown that bisects Quadrant I; these points indicate when *estimate = actual price*.

Exercise 7 Be ready with extra copies of Calculator Note 1I. The idea of performing operations on entire lists of numbers will be used repeatedly throughout the next few chapters.

If needed, suggest that students set their viewing windows by considering the maximum and minimum values in each list.

Exercise 8 If students are having difficulties with 8d, reiterate that the coordinates of points on the line $y = x$ are both the same.

Exercise 9 The apparent contradiction between 9a and 9b illustrates that an average of averages can be misleading. **[Ask]** "What would have to be true for a group of states to assure that the average of the state averages would be the same as the average score for all the students?" [The number of students taking the test in each state would need to be the same.]

9b. Those states in which students scored higher in mathematics had larger student populations taking the test. The national average is an average of all students, not an average of all state averages.

Exercise 10 **[Alert]** If students are confused about where slower rates are pictured, remind them that speed is distance over time— miles per hour or meters per second. You might point out that the rates are represented by the steepness of segments connecting the data points. This will come up in Chapters 4 and 5.

10a. About 1 m/sec, because the distances in meters are about equal to the times in seconds.

10b. Between 0.5 and 1.0, between 2.5 and 3.0, between 3.5 and 4.0, between 4.0 and 4.5.

10c. Between 1.0 and 1.5, between 1.5 and 2.0, between 2.0 and 2.5.

10d. Between 0 and 0.5, between 3.0 and 3.5.

Exercise 11 Watch for difficulties with 11a; the median might be on an edge of the box.

8. Draw the line for the equation $y = x$ on a coordinate grid with the x-axis labeled from -9 to 9 and the y-axis labeled from -6 to 6. Plot and label the points described.

   a. Point A with an x-coordinate of -4 and a y-coordinate 5 more than -4. $(-4, 1)$

   b. Point B with an x-coordinate of -2 and a y-coordinate 3 less than -2. $(-2, -5)$

   c. Point C with an x-coordinate of 1 and a y-coordinate 4 units above the line. $(1, 5)$

   d. Now plot several points with coordinates that are opposites (*inverses*) of each other, for example, $(-5, 5)$ or $(5, -5)$. Describe the pattern of these points. Write an equation to describe the pattern. Plotted points lie on the line that bisects Quadrant II and Quadrant IV. The equation $y = -x$ fits these points.

9. **APPLICATION** This graph shows the mean SAT verbal and mathematics scores for 50 states and the District of Columbia in 1999.

   a. Which statement is true: "More states had students with higher mathematics scores than verbal scores" or "More states had students with higher verbal scores than mathematics scores"?

   b. The national average mathematics score was slightly higher than the average verbal score. This may seem contradictory to 9a. Explain.

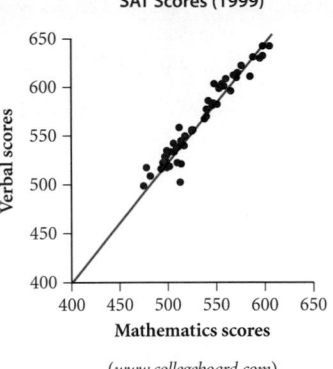

SAT Scores (1999)

(*www.collegeboard.com*)

10. Below is a scatter plot of points whose coordinates have the form (t, d). The variable t stands for time in seconds and d stands for distance in meters. The graph describes a person walking away from a motion sensor. The line $d = t$ is also graphed.

    a. Approximately how fast is the person moving? Explain how you know this.

    b. Name two different half-second intervals in which the person is moving more slowly than the rate you found in 10a.

    c. Name two different half-second intervals in which the person is moving faster than the rate you found in 10a.

    d. Name two different half-second intervals in which the person is moving at about the same rate you found in 10a.

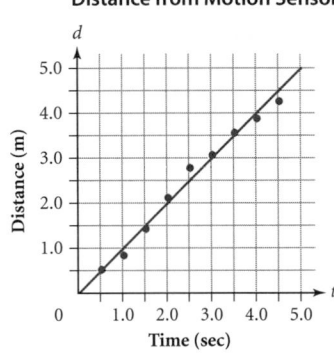

Distance from Motion Sensor

▶ Review

11. For each description, invent a seven-value data set so that all the values in the set are less than 10 and meet the conditions. Possible answers are shown.

    a. The box plot represents data with a median that is not inside the box. {1, 3, 3, 3, 4, 5, 6}

    b. The box plot represents data with an interquartile range of zero. {1, 4, 4, 4, 4, 4, 7}

    c. The box plot represents data with one outlier on the left. {1, 6, 6, 7, 7, 8, 9}

    d. The box plot is missing its right whisker. {1, 2, 3, 4, 5, 6, 6}

12. Rocky Rhodes and his algebra classmates measured the circumference of their wrists in centimeters. Here are the data:

15.2 14.7 13.8 17.3 18.2 17.6 14.6 13.5 16.5
15.8 17.3 16.8 15.7 16.2 16.4 18.4 14.2 16.4
15.8 16.2 17.3 15.7 14.9 15.5 17.1

```
13 | 8  5
14 | 7  6  2  9
15 | 2  8  7  8  7  5
16 | 5  8  2  4  4  2
17 | 3  6  3  3  1
18 | 2  4
```

Key

```
10 | 4   means 10.4 cm
```

a. Rocky made the stem plot shown here. Unfortunately, he was not paying attention when these plots were discussed in class. Write a note to Rocky telling him what he did incorrectly.

b. Make a correct stem plot of this data set.

c. What is the range of this data set?

12a. Rocky made several mistakes. The data values need to be organized in increasing order. The key should show actual values from the data.

12b.
```
13 | 5  8
14 | 2  6  7  9
15 | 2  5  7  7  8  8
16 | 2  2  4  4  5  8
17 | 1  3  3  3  6
18 | 2  4
```

Key

```
13 | 5   means 13.5 cm
```

12c. $18.4 - 13.5 = 4.9$ cm

project

BASEBALL RELATIONSHIPS

Baseball statistics, or "stats," often show relationships between variables. These relationships could be used to identify the best players on a particular Major League team, or in an entire league, and may influence the strategy of managers. These scatter plots show stats of 50 National League players from the 1996 season.

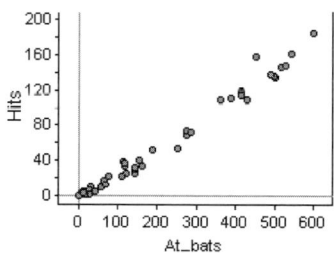

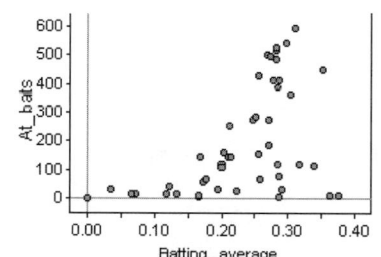

Fathom

These scatter plots were made with Fathom. With Fathom, you can download baseball stats directly from the Internet. Scatter plots are easy to make with this software, which allows you to investigate relationships between many more pairs of variables.

Find the stats of any one team for a particular season. Your data can come from library research, or you can use the links at **www.keymath.com/DA**. Briefly review all the stats and make at least two conjectures about relationships like "If a player appears in more games, he will have more times at bat." Use scatter plots to see if relationships truly exist between the variables. For each conjecture, your project should include

▶ A sentence stating the conjecture and a scatter plot to accompany it.

▶ An explanation of how the scatter plot either supports your conjecture or casts doubt on it. If a relationship does appear to exist, why do you think this is so?

Conclude your project by explaining how you think a baseball manager could use your graphs to strategize for a game.

Supporting the project

MOTIVATION

Baseball fans often memorize the stats of their favorite players or teams. Managers analyze the stats and look for relationships among them.

OUTCOMES

▶ The conjecture refers to a relationship between two variables.
▶ The scatter plot includes data from the same two variables.
▶ The explanation uses good mathematical language.

▶ The explanation defends the claim about whether the scatter plot supports the conjecture or not.
• Several conjectures are explored.
• Writing and thinking are exceptionally good.

LESSON OUTLINE

First day:

15 min	Example A
10 min	Example B
25 min	Exercises 1–10

Second day:

30 min	Investigation
10 min	Sharing
5 min	Closing
5 min	Exercises

MATERIALS

- Pizza Prices (T)
- Calculator Notes 1L, 1M, 1N, 1P

Tables are commonly used to represent two-variable data, and matrices are mathematical objects abstracting tables.

Organizing information is important in all areas of life as the opening quote indicates. Florynce Kennedy (1916–2000) was an outspoken civil rights and equal rights advocate.

There's no universal notation for giving the dimensions of a matrix; 6 by 2, 6 × 2, and (6, 2) are all common.

EXAMPLE A

[Alert] Watch for difficulties setting up the matrix. Some students may set up a 2 × 6 matrix, thinking that the first dimension is the "distance across" (the number of rows), as on a coordinate plane.

[Language] Use the term *entries* of a matrix.

Using Matrices to Organize and Combine Data

Don't agonize.

Organize.

FLORYNCE KENNEDY

Did you know that the average number of hours people work per week has been decreasing during the last 100 years? The table provides data for some countries. This table is a 6 × 2 (say "six by two"), because it has six rows of countries and two columns of years. Can you identify the entry in row 2, column 1? In which row and column is the entry 37.9? 51.7; row 5, column 2

Average Weekly Working Hours

	Country	1900	1998
row 1	Germany	51.6	29.0
row 2	France	51.7	31.7
row 3	Japan	51.7	38.3
row 4	Netherlands	52.0	30.8
row 5	United States	52.0	37.9
row 6	Britain	52.4	35.6

Rows are counted top to bottom.

This entry is in row 3, column 2.

column 1   column 2

Columns are counted left to right.

During the late 1800s and early 1900s, labor unions were formed to improve working conditions. These miners may have belonged to the American Federation of Labor, one of the first unions to include African American members.

A table is an easy way to organize data. A quicker way to organize data is to display it in a **matrix.** A matrix has rows and columns just like a table. To rewrite a table as a matrix, you simply use brackets to enclose the row and column entries. For the working-hour data, the **dimensions** of the matrix are the same as the table: 6 × 2.

In this lesson you represent different situations with matrices and explore some matrix calculations.

EXAMPLE A

a. Form a matrix from the table Average Weekly Working Hours.

b. Verify the row and column locations of the entries 31.7, 30.8, and 52.4 in matrix [A].

c. If the average person works 50 weeks per year, what were the average yearly working-hour totals for these countries in 1900 and 1998?

LESSON OBJECTIVES

- Use a matrix to represent a two-variable data set
- Set up a calculator matrix, enter data, and perform operations on matrices
- Multiply a matrix by a number
- Add and subtract matrices
- Multiply a matrix by an appropriate column matrix
- Understand the conditions under which two matrices can be added or multiplied

NCTM STANDARDS

CONTENT		PROCESS	
●	Number	●	Problem Solving
	Algebra	●	Reasoning
	Geometry		Communication
●	Measurement	●	Connections
●	Data/Probability	●	Representation

▶ Solution

a. Here is the 6 × 2 matrix. It is simply a table without labels.

$$[A] = \begin{bmatrix} 51.6 & 29.0 \\ 51.7 & 31.7 \\ 51.7 & 38.3 \\ 52.0 & 30.8 \\ 52.0 & 37.9 \\ 52.4 & 35.6 \end{bmatrix}$$

b. Enter the 6 × 2 matrix on your calculator. [▶ 🖥 See **Calculator Note 1L** to learn how to enter a matrix in your calculator.◀] The calculator display shows that 31.7 is located in row 2, column 2, of matrix A. By moving around in your editor, you see that the entry 30.8 is in row 4, column 2; 52.4 is in row 6, column 1.

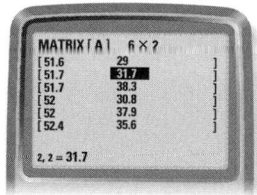

```
MATRIX [A]  6 X 2
[ 51.6   29    ]
[ 51.7   31.7  ]
[ 51.7   38.3  ]
[ 52     30.8  ]
[ 52     37.9  ]
[ 52.4   35.6  ]

2, 2 = 31.7
```

c. Use your calculator to multiply [A] by 50. [▶ 🖥 See **Calculator Note 1N** to learn how to multiply a matrix by a number.◀] This new matrix shows the result of multiplying each of the original entries by 50. The new entries are the average *yearly* working hours.

```
[A]
[[51.6   29   ]
 [51.7   31.7 ]
 [51.7   38.3 ]
 [52     30.8 ]
 [52     37.9 ]
 [52.4   35.6 ]]
```
× 50 =
```
50 * [A]
[[2580   1450 ]
 [2585   1585 ]
 [2585   1915 ]
 [2600   1540 ]
 [2600   1895 ]
 [2620   1780 ]]
```

As shown in the example, it is easy to operate on all the entries in a matrix with one calculation. In the next example you will learn how you can add or subtract two matrices to help you answer questions about a situation.

EXAMPLE B

Matrix [B] provides the costs of a medium pizza, medium salad, and medium drink at two different pizzerias: the Pizza Palace and Tony's Pizzeria. Matrix [C] provides the additional charge for large items at each pizzeria.

$$[B] = \begin{matrix} & \text{Pizza Palace} & \text{Tony's Pizzeria} \\ & \downarrow & \downarrow \\ \end{matrix} \begin{bmatrix} 8.90 & 9.10 \\ 2.35 & 2.65 \\ 1.50 & 1.60 \end{bmatrix} \begin{matrix} \leftarrow \text{pizza} \\ \leftarrow \text{salad} \\ \leftarrow \text{drink} \end{matrix}$$

$$[C] = \begin{bmatrix} 2.50 & 2.25 \\ 1.00 & 1.25 \\ 0.65 & 0.50 \end{bmatrix} \begin{matrix} \leftarrow \text{pizza} \\ \leftarrow \text{salad} \\ \leftarrow \text{drink} \end{matrix}$$

Write a matrix [D] displaying the costs at the Pizza Palace and at Tony's Pizzeria for a large pizza, large salad, and large drink.

EXAMPLE B

Have students use their calculators to try adding matrices of different dimensions. **[Ask]** "Why won't the calculator add the matrices?" [If two matrices don't have the same dimensions, not every value will have something to add to it.]

[ESL] Ask students if they understand what *corresponding entries* are. [They are elements in the same row and column in matrices with the same dimensions.]

You might point out that although any two matrices with the same dimensions can be added, context is also important. The matrix values being added must be logically summable.

If you wanted the price of a large pizza at the Pizza Palace, you would add the medium price and the additional charge.

$$8.90 + 2.50 = 11.40$$

The totals for matrix $[D]$ are found by adding all corresponding entries from matrix $[B]$ and matrix $[C]$. [► 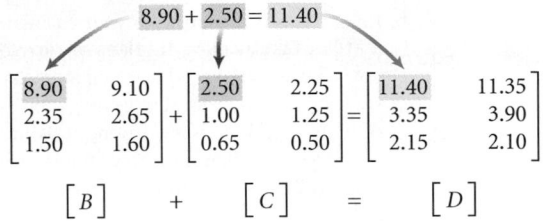 See **Calculator Note 1M** to learn how to use your calculator to add and subtract matrices.◄]

$$8.90 + 2.50 = 11.40$$

$$\begin{bmatrix} 8.90 & 9.10 \\ 2.35 & 2.65 \\ 1.50 & 1.60 \end{bmatrix} + \begin{bmatrix} 2.50 & 2.25 \\ 1.00 & 1.25 \\ 0.65 & 0.50 \end{bmatrix} = \begin{bmatrix} 11.40 & 11.35 \\ 3.35 & 3.90 \\ 2.15 & 2.10 \end{bmatrix}$$

$$[B] \qquad + \qquad [C] \qquad = \qquad [D]$$

Does the order in which you add these matrices make a difference? Take a moment to calculate $[B] + [C]$ and $[C] + [B]$ on your calculator.

If you try adding a 6×2 matrix and a 3×2 matrix with your calculator, you'll get an error message. This is because the matrices don't have the same dimensions. There is no way to match up corresponding entries to do the operations. In order for you to add or subtract matrices, each matrix must have the same number of rows or columns.

Can you multiply two matrices? In this investigation you will discover why matrix multiplication is more complicated.

Investigation
Row-by-Column Matrix Multiplication

Recall the table and matrix for large items at the Pizza Palace and Tony's Pizzeria from Example B.

Suppose you're in charge of ordering the food for a school club party.

$$[D] = \begin{bmatrix} 11.40 & 11.35 \\ 3.35 & 3.90 \\ 2.15 & 2.10 \end{bmatrix}$$

Large-Item Prices

	Pizza Palace	Tony's Pizzeria
Large pizza	$11.40	$11.35
Large salad	$3.35	$3.90
Large drink	$2.15	$2.10

Step 1 | What will be the total cost of 4 large pizzas, 5 large salads, and 10 large drinks at the Pizza Palace? $83.85

Step 2 | What will be the total cost for the same order at Tony's Pizzeria? $85.90

Step 3 | Describe how you calculated the total costs. Descriptions should include multiplying each item quantity by each cost and then adding all costs.

Guiding the Investigation

One step Give students a three-part question. First, give them the pizza data from the beginning of the investigation, on the investigation transparency Pizza Prices. Tell them that prices are all now 3% higher, and ask for a table of current prices. Then mention the word *matrix* and work through Example A. Ask them to use similar ideas to solve the pizza problem using their calculators. Then announce that you have orders for two club parties, and show a column matrix for each (also on the investigation transparency). Ask what the total order is, and then ask them to find the sum of the vectors on their calculators. Finally, ask how to find the cost of that total order from each store, and then let them find out how to make that calculation on their calculators.

As you see, calculating the food costs for your party requires multiplication and addition. Organizing your work like this is a first step to discovering how to multiply matrices.

$$\underset{\substack{\text{number}\quad\text{cost}}}{\overset{\text{pizza}}{\boxed{4}\cdot\boxed{11.40}}}+\underset{\substack{\text{number}\quad\text{cost}}}{\overset{\text{salad}}{\boxed{5}\cdot\boxed{3.35}}}+\underset{\substack{\text{number}\quad\text{cost}}}{\overset{\text{drink}}{\boxed{10}\cdot\boxed{2.15}}}=\overset{\text{total}}{\boxed{83.85}}$$

Step 4   Copy the calculation and replace each box with a number to find the total food cost for the same order at the Pizza Palace

Step 5   The row matrix [A] and column matrix [B] contain all the information you need to calculate the total food cost at the Pizza Palace.

$$[A] = [4 \quad 5 \quad 10] \quad [B] = \begin{bmatrix} 11.40 \\ 3.35 \\ 2.15 \end{bmatrix}$$

Explain what matrix [A] and matrix [B] represent.   [A] represents the food quantities; [B] represents the food costs.

Step 6   Enter [A] and [B] in your calculator and find their *product*, [A] · [B], or

$$[4 \quad 5 \quad 10] \cdot \begin{bmatrix} 11.40 \\ 3.35 \\ 2.15 \end{bmatrix}$$

[▶ 🖳 See **Calculator Note 1P** to learn how to multiply two matrices.◀] Explain in detail what you think the calculator does to find this answer.

Step 7   Repeat Step 4 to find the total food cost at Tony's Pizzeria.

Step 8   Write the product of a row matrix and a column matrix that calculates the total food cost at Tony's Pizzeria. Use your calculator to verify that your product matches your answer to Step 7.

Step 9   Explain why the number of columns in the first matrix must be the same as the number of rows in the second matrix in order to multiply them.

Step 10   Predict the answer to the matrix multiplication problem below. Use your calculator to verify your answer. What is its meaning in the real-world context?

$$[4 \quad 5 \quad 10] \cdot \begin{bmatrix} 11.40 & 11.35 \\ 3.35 & 3.90 \\ 2.15 & 2.10 \end{bmatrix} = ? \quad \begin{array}{l} [83.85 \quad 85.90] \text{ It represents the total cost at} \\ \text{the Pizza Palace and at Tony's Pizzeria.} \end{array}$$

Step 11   Explain how to calculate the matrix multiplication in Step 10 without using calculator matrices.   Explanations should include multiplying the row matrix by each column.

Step 12   Try this matrix multiplication:

$$\begin{bmatrix} 11.40 & 11.35 \\ 3.35 & 3.90 \\ 2.15 & 2.10 \end{bmatrix} \cdot [4 \quad 5 \quad 10] \quad \text{The dimensions do not allow multiplication.}$$

In general, do you think [A] · [B] = [B] · [A]? Explain.

Step 13   Write a short paragraph explaining how to multiply two matrices.

Step 5 You might mention that a matrix of just one row or column is often called a *vector*. Be sure students write their explanations of what the matrices represent.

Step 6 Explanations should include that each column entry in the first matrix is multiplied by the respective row entry in the second matrix, then everything is added.

Step 7 $\boxed{4}\cdot\boxed{11.35}+\boxed{5}\cdot\boxed{3.90}$ $+\boxed{10}\cdot\boxed{2.10}=\boxed{85.90}$

Step 8 $[4 \; 5 \; 10] \cdot \begin{bmatrix} 11.35 \\ 3.90 \\ 2.10 \end{bmatrix}$ $= [85.90]$

Step 9 Since each column entry is multiplied by a respective row entry, the number of columns and rows must be equal.

Step 10 Some students may write the answer vertically, but don't criticize. When they do it on the calculator, they will see it as a row.

Step 12 Even if both products [A] · [B] and [B] · [A] of two matrices are defined (that is, the matrices are square), the products are not usually equal. The products are the same only for a few special cases, such as when one of the matrices is the identity matrix $\begin{bmatrix} 1 & 0 & 0 \\ 0 & 1 & 0 \\ 0 & 0 & 1 \end{bmatrix}$.

Step 13 Students will need time to make the big jump from multiplying by a vector to multiplying by a larger matrix. Giving another example when you talk about Example C will help. **[Language]** Challenge students to use the language carefully: Each item in a row is multiplied by the *respective* item in the column, and the *products* are added.

Step 13 Answers should include the concept of multiplying rows by columns.

Good steps to ask students to present are 3, 10, and 12. **[Ask]** "When is matrix multiplication possible?" [When the number of columns of the first matrix equals the number of rows of the second matrix.]

Assessing Progress

By watching students contribute to groups or the whole class, you can assess their skills at **arithmetic with decimals** and keeping track of **two variables** at once.

EXAMPLE C

Talk through this third example if students are having difficulties in multiplying matrices.

Another Example

Example C shows a vector and a matrix, but two matrices can be multiplied as long as the inside dimensions are the same. If you plan to assign Review Exercise 10 in Lesson 2.4 or Exercise 15 in Lesson 6.2, show this example.

A 2 × 2 matrix can be multiplied by a 2 × 3. The inside dimensions are the same. The outside dimensions indicate that the answer will be 2 × 3.

$$\begin{bmatrix} 3 & -1 \\ 1 & 2 \end{bmatrix} \cdot \begin{bmatrix} 2 & 0 & 5 \\ 4 & 7 & -6 \end{bmatrix}$$

$$= \begin{bmatrix} 2 & -7 & 21 \\ 10 & 14 & -7 \end{bmatrix}$$

Closing the Lesson

Mention the important mathematical ideas new to this lesson: **matrix, dimension of matrix, addition of matrices,** and **multiplication of matrices.** Matrices are abstractions of tables representing data. Their dimensions are important in deciding whether they can be added or multiplied.

EXAMPLE C

Is it possible to multiply each pair of matrices? If so, what is the product? If not, why not?

a. $\begin{bmatrix} 2 & 3 & 4 \end{bmatrix} \cdot \begin{bmatrix} -1 \\ 1 \\ 5 \end{bmatrix}$

b. $\begin{bmatrix} 3 & -1 \end{bmatrix} \cdot \begin{bmatrix} 2 & 4 \\ 0 & 3 \\ 5 & -6 \end{bmatrix}$

c. $\begin{bmatrix} 3 & -1 \end{bmatrix} \cdot \begin{bmatrix} 2 & 0 & 5 \\ 4 & 7 & -6 \end{bmatrix}$

▶ **Solution**

a. Multiply the respective entries in the row matrix by those in the column matrix and then add the products.

$$\begin{bmatrix} 2 & 3 & 4 \end{bmatrix} \cdot \begin{bmatrix} -1 \\ 1 \\ 5 \end{bmatrix} = \begin{bmatrix} 21 \end{bmatrix}$$

$$(2 \cdot -1) + (3 \cdot 1) + (4 \cdot 5) = 21$$

b. This is not possible. There are not enough entries in the row matrix to match the three entries in each column.

c. Multiply the entries in the row matrix by the respective entries in each column. Each sum is a separate entry in the answer matrix.

Technology CONNECTION

The development of 3-D video games uses matrices to process the graphics.

You multiply a 1 × 2 matrix by a 2 × 3 matrix.

1 × 2, 2 × 3

The inside dimensions are the same so you can multiply. The 2 row entries match up with 2 column entries.

1 × 2, 2 × 3

The outside dimensions tell you the dimensions of your answer.

Multiply the row by each column.

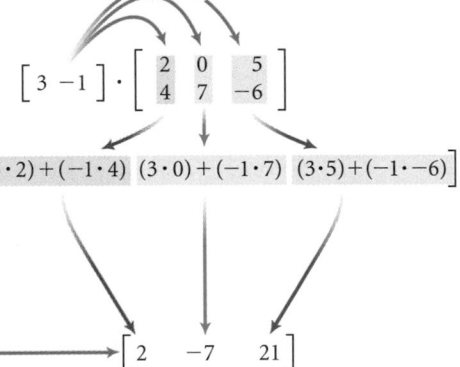

$$\begin{bmatrix} 3 & -1 \end{bmatrix} \cdot \begin{bmatrix} 2 & 0 & 5 \\ 4 & 7 & -6 \end{bmatrix}$$

$$\begin{bmatrix} (3 \cdot 2) + (-1 \cdot 4) & (3 \cdot 0) + (-1 \cdot 7) & (3 \cdot 5) + (-1 \cdot -6) \end{bmatrix}$$

$$\begin{bmatrix} 2 & -7 & 21 \end{bmatrix}$$

Matrix multiplication is probably too much work if all you want to do is plan a pizza party. But what if you had to manage the inventory (goods in stock) of a grocery store? Or a whole chain of grocery stores? If you needed to know the values of the inventory at each store, a matrix calculation carried out by a computer would be essential to your business.

EXERCISES

You will need your calculator for problems **4, 8,** and **12.**

Practice Your Skills

Use this information for problems 1–8. Troy Aikman, Randall Cunningham, and Steve Young have been top-performing quarterbacks in the National Football League throughout their careers. The rows in matrix [A] and matrix [B] show data for Aikman, Cunningham, and Young, in that order. The columns show the number of passing attempts, pass completions, touchdown passes, and interceptions, from left to right. Matrix [A] shows stats from 1992, and matrix [B] shows stats from 1998.

$$[A] = \begin{bmatrix} 473 & 302 & 23 & 14 \\ 384 & 233 & 19 & 11 \\ 402 & 268 & 25 & 7 \end{bmatrix} \quad [B] = \begin{bmatrix} 315 & 187 & 12 & 5 \\ 425 & 259 & 34 & 10 \\ 517 & 322 & 36 & 12 \end{bmatrix}$$

(*http://CNNSI.com*)

1. What does the entry in row 2, column 3, of matrix [A] tell you? **Randall Cunningham threw 19 touchdown passes in 1992.**

2. What does the entry in row 3, column 2, of matrix [B] tell you? **Steve Young made 322 pass completions in 1998.**

3. What are the dimensions of each matrix? **3 × 4**

4. Write a clear explanation of the procedure for entering [A] and [B] into your calculator. **Answers will vary.**

5. Find [A] + [B]. What is this matrix and what information does it provide?

6. Is [A] + [B] equal to [B] + [A]? Do you think this result is always true for matrix addition? Explain. **Yes; this result should always be true if the matrices have the same dimensions.**

7. Find [B] − [A]. What is this matrix and what information does it provide?

8. How can you use your calculator to find the average statistics for the three quarterbacks for these two seasons? Write a matrix expression that will give this average. $\left([A] + [B]\right) \cdot \left(\frac{1}{2}\right)$

Reason and Apply

9. Use matrices [A] and [B] to write the new matrices asked for in 9a–d.

$$[A] = \begin{bmatrix} 3 & -4 & 2.5 \\ -2 & 6 & 4 \end{bmatrix} \quad [B] = \begin{bmatrix} 5 & -1 & 2 \\ -4 & 3.5 & 1 \end{bmatrix}$$

a. [A] + [B]   **b.** −[A]   **c.** 3 · [B]   **d.** the average of [A] and [B]

10. Find the matrix [B] so that this equation is valid:

$$\begin{bmatrix} -2 & 0 \\ 6 & -11.6 \\ 4.25 & 7.5 \end{bmatrix} - [B] = \begin{bmatrix} 2.8 & 2.4 \\ 2.5 & -9.4 \\ 1 & 6 \end{bmatrix} \begin{bmatrix} -4.8 & -2.4 \\ 3.5 & -2.2 \\ 3.25 & 1.5 \end{bmatrix}$$

9a. $\begin{bmatrix} 8 & 5 & 4.5 \\ -6 & 9.5 & 5 \end{bmatrix}$

9b. $\begin{bmatrix} -3 & 4 & -2.5 \\ 2 & -6 & -4 \end{bmatrix}$

9c. $\begin{bmatrix} 15 & -3 & 6 \\ -12 & 10.5 & 3 \end{bmatrix}$

9d. $\begin{bmatrix} 4 & -2.5 & 2.25 \\ -3 & 4.75 & 2.5 \end{bmatrix}$

BUILDING UNDERSTANDING

In these exercises students manipulate matrices and use matrices to represent two-variable data.

ASSIGNING HOMEWORK

Essential	**1–3, 5, 6, 9, 12**
Performance assessment	**10, 12**
Portfolio	**12**
Journal	**4, 11**
Group	**12**
Review	**13, 14**

▶ Helping with the Exercises

Exercises 1 and 2 If needed, suggest that students label the rows and columns of each matrix.

Exercise 3 [Alert] Watch to be sure students don't switch the number of rows and columns.

5. $\begin{bmatrix} 788 & 489 & 35 & 19 \\ 809 & 492 & 53 & 21 \\ 919 & 590 & 61 & 19 \end{bmatrix}$

This matrix gives the totals from the two years.

Exercise 7 You might challenge some students to think about whether or not the result is the same as [A] − [B].

7. $\begin{bmatrix} -158 & -115 & -11 & -9 \\ 41 & 26 & 15 & -1 \\ 115 & 54 & 11 & 5 \end{bmatrix}$

This matrix gives the difference between the 1998 totals and the 1992 totals.

Exercise 8 [Language] Watch for confusion about the meaning of *average statistics*. Students will be calculating the mean of each pair of corresponding entries in the matrices.

Exercise 10 Most students will find this 3 × 2 matrix by thinking or guess-and-check.

10. $\begin{bmatrix} -4.8 & -2.4 \\ 3.5 & -2.2 \\ 3.25 & 1.5 \end{bmatrix}$

11. A possibility would be to find the labor costs of building an item using 16 hours billed at \$5.25 per hour and 30 hours billed at \$8.75 per hour. The product is

$$[5.25 \quad 8.75] \cdot \begin{bmatrix} 16 \\ 30 \end{bmatrix} = [346.50]$$

Exercise 12 Refer students to Example C on page 84. When two matrices are multiplied, the inside dimensions must be the same (3×3 times 3×1) and the outside dimensions tell the dimensions of the answer (3×1).

12a. Quantity $\begin{bmatrix} 74 & 25 & 37 \\ 32 & 38 & 16 \\ 120 & 52 & 34 \end{bmatrix}$

Profit $\begin{bmatrix} 0.90 \\ 1.25 \\ 2.15 \end{bmatrix}$

The number of columns in the quantity matrix must be the same as the number of rows in the profit matrix.

12b. Atlanta: 74(\$0.90) + 25(\$1.25) + 37(\$2.15) = \$177.40
Decatur: 32(\$0.90) + 38(\$1.25) + 16(\$2.15) = \$110.70
Athens: 120(\$0.90) + 52(\$1.25) + 34(\$2.15) = \$246.10

12c. $\begin{bmatrix} 177.40 \\ 110.70 \\ 246.10 \end{bmatrix}$; 3×1

12d. The answer matrix gives the profit at each location.

	Profit
Atlanta	\$177.40
Decatur	\$110.70
Athens	\$246.10

12e. The error message—ERR: DIM MISMATCH—means that the dimensions of the matrices in the product don't match.

11. Create a problem involving this matrix multiplication if the row matrix represents dollars and cents and the column matrix represents hours. Find the product without using your calculator's matrix menu.

$$[5.25 \quad 8.75] \cdot \begin{bmatrix} 16 \\ 30 \end{bmatrix}$$

12. **APPLICATION** Ms. Shurr owns three ice cream shops and wants to know how much she made on ice-cream cone sales for one day. The number of small, medium, and large cones sold at each location and the profit for each size are contained in the tables.

Number of Cones Sold

	S	M	L
Atlanta	74	25	37
Decatur	32	38	16
Athens	120	52	34

Cone Profit

	Profit
S	\$0.90
M	\$1.25
L	\$2.15

a. Write a quantity matrix that gives the number of each size sold at each location and a profit matrix that gives the profit for each size. What must be the same for each matrix?

b. Without using your calculator's matrix menu, find the profit from ice-cream cones for each location. Explain how you got your answer.

c. Check your answer to 12b by using your calculator to multiply the quantity matrix $[A]$, by the profit matrix $[B]$. What are the dimensions of the answer matrix?

d. What do the entries in the answer matrix tell you? Convert your matrix into a table with row and column headings so that Ms. Shurr can understand the information.

e. Try to calculate $[B] \cdot [A]$ on your calculator. What happens? What do you think the result means?

Review

13. Create a data set that fits the information.

a. Ten students were asked the number of times they had flown in an airplane. The range of data values was 7. The minimum was 0 and the mode was 2.

b. Eight students each measured the length of their right foot. The range of data values was 8.2 cm, and the maximum value was 30.4 cm. There was no mode.

14. Mr. Chin and Mrs. Shapiro had their classes collect data on the amount of change each student had in class on a particular day. The students graphed the data on the back-to-back stem plot at right.

Mrs. Shapiro's class		Mr. Chin's class
0 0 0 0	0	0 0 0 0 5 8
2 0	1	0 5 6
5 5 5	2	0 0 5 5 5 7
9 6 5 0 0	3	5 5
5	4	0 0 0 6
5 2 0 0	5	0 0 5 5 8
7 3 0 0	6	0 2 5 5
5 0	7	0 5 5 6
2 0 0	8	
4 1	9	
4 0 0	10	0 0 5
5 0	11	0
6 4 1	12	1 5 5
5 0	13	

Key

5 | 4 | 0 0 0 6 means 45¢ in Mrs. Shapiro's class; 40¢, 40¢, 40¢ and 46¢ in Mr. Chin's class

a. How many students are in each class? **Shapiro: 40; Chin: 41**

b. Find the range of the data in each class. **Shapiro: 1.35; Chin: 1.25**

c. How many students had more than \$1? **13**

d. What do the entries in the last row represent?

e. Without adding, make an educated guess which class has the most money altogether. Explain your thinking.

f. How much money does each class have? **Shapiro: \$24.96; Chin: \$20.34**

13a. The minimum in the data set must be 0, and the maximum must be 7. Also, the data value 2 must occur more frequently than any other. A sample solution is {0, 1, 2, 2, 2, 2, 3, 4, 6, 7}.

13b. The minimum must be 22.2, and the maximum must be 30.4. No values in the list should be repeated. A sample solution is {22.2, 24.5, 25.1, 26.2, 28.3, 28.7, 29.4, 30.4}.

14d. In Mrs. Shapiro's class, one student had \$1.35 and one student had \$1.30.

14e. Based on the stem plot, you might expect Mrs. Shapiro's class to have more money. There are more students with more than a dollar in her class and more with less than 50¢ in Mr. Chin's class.

CHAPTER 1 REVIEW

1

REVIEW

In this chapter you learned how statistical measures and graphs can help you organize and make sense of **data.** You explored several different kinds of graphs—**bar graphs, pictographs, dot plots, box plots, histograms,** and **stem plots**—that can be used to represent **one-variable** data.

You analyzed the strengths and weaknesses of each kind of graph to select the most appropriate one for a given situation. A bar graph displays data that can be grouped into **categories.** Numerical data can be individually shown with a dot plot. The **spread** of data is clearly displayed with a box plot built from the **five-number summary.** A histogram uses **bins** to show the **frequency** of data and is particularly useful for large sets of data. A stem plot also groups data into intervals but maintains the identity of each data value.

You can use the **measures of center** to describe a typical data value. In addition to the **mean, median,** and **mode,** statistical measures like **range, minimum, maximum, quartiles,** and **interquartile range** help you describe the spread of a data set and identify **outliers.**

You used the **coordinate plane** to compare estimates and actual values plotted on **two-variable** plots called **scatter plots.** Here, each variable is represented on a different axis, and an **ordered pair** shows the value of each variable for a single data item. You also analyzed scatter plots for situations involving the two variables *time* and *distance.* Scatter plots allowed you to find patterns in the data; sometimes these patterns could be written as an algebraic equation.

Lastly, you learned to use a **matrix** to organize data in rows and columns, very much like a table. You discovered ways to add, subtract, and multiply matrices and learned how the **dimensions** of matrices affect these computations. Computers and calculators can use matrices to calculate with large sets of data, making it easy to answer questions about the data.

EXERCISES

1. This data set gives the number of hours of use before each of 14 batteries required recharging: {40, 36, 27, 44, 40, 34, 42, 58, 36, 46, 52, 52, 38, 36}.
   a. Find the mean, median, and mode for the data set and explain how you found each measure.
   b. Find the five-number summary for the data set and make a box plot.

2. Seven students order onion rings. The mean number of onion rings they get is 16. The five-number summary is 9, 11, 16, 21, 22. How many onion rings might each student have been served? Possible answer: {9, 11, 14, 16, 19, 21, 22}.

1a. Mean: 41.5; divide the sum of the numbers by 14. Median: 40; list the numbers in ascending order and find the mean of the two middle numbers. Mode: 36; find the most frequently occurring number.

1b. 27, 36, 40, 46, 58

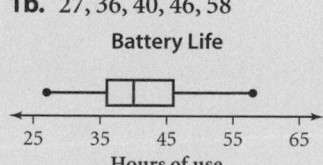

Battery Life

25 35 45 55 65
Hours of use

PLANNING

LESSON OUTLINE

5 min	Introduction
15 min	Exercises and helping individuals
15 min	Checking work and helping individuals
15 min	Student self assessment

REVIEWING

You might use the Jurassic Park data on page 74 to review. **[Ask]** "What statistics and graphs could be used to describe this data set?" Elicit these and other responses.

For the actual or for the estimated data as a one-variable data set,

- Find measures of center, measures of spread, and the five-number summary.

- Draw a box plot; identify any outliers, and describe how they affect the measures of center.

- Draw a dot plot or stem plot.

- Construct histograms with various bins.

- Divide the dinosaurs into groups by beginning letter of name, or use the length of the name, or put those whose names end in *saurus* in one group and the others in another group; then, construct a bar graph or a pictograph showing how many are in each group.

Comparing the estimated and actual data sets,

- Compare the box plots, dot plots, or histograms made for the actual or for the estimated.

Using the data set as two-variable data,

- Construct a scatter plot of the actual numbers versus the estimated numbers; decide from

the graph which species were overestimated.

· Divide the dinosaurs into groups and construct bar graphs or pictographs showing how many overestimates occurred in each category.

As students suggest graphs, talk about the strengths and weaknesses of the various graphs for representing the data and review how bar graphs and pictographs are used for nonnumerical categories, in contrast to dot plots, stem plots, and histograms.

ASSIGNING HOMEWORK

Students who complete Exercises 1 to 6 and 8 to 10 are reviewing all the graphical representations presented in the chapter. Assign Exercise 7 only if you covered matrices.

3a. See page 89.

3b. Greatest jump: from a master's degree to a doctorate; smallest difference: from not finishing high school to a high school diploma.

4a.

Top College Career Scorers

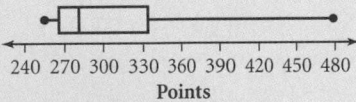

240 270 300 330 360 390 420 450 480
Points

4b. 479 points (Chamique Holdsclaw)

4c. Choices will vary. Mean: 311.8 points; median: 282.5 points; mode: 263 points.

5a. Mean: approximately 154; median: 121; there is no mode.

5b. Bin widths may vary.

Pages Read in Current Book

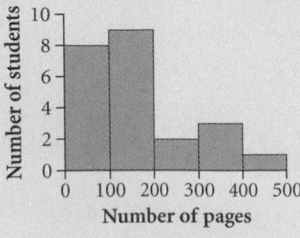

3. The table at right shows the mean annual wages earned by individuals with various levels of education in the United States in 1998.

a. Construct a bar graph for the data.

b. Between which two consecutive levels of education is there the greatest difference in mean annual wages? The smallest difference?

4. The table below shows the top ten scoring leaders in women's college basketball tournament history.

a. Construct a box plot for the data.

b. Are there are any outliers?

c. Which measure of center would you use to describe a typical value?

Mean Annual Wages, 1998

Level of education	Amount ($)
Did not finish high school	18,913
High school diploma only	25,257
Two-year degree (AA/AS)	33,765
Bachelor's degree (BA/BS)	45,390
Master's degree (MA/MS)	52,951
Doctorate degree	75,071

(U.S. Census 2000)

Leading Scorers in NCAA Tournament History (Women's Basketball, Through 1999–2000 Season)

Player, college	Total points
Chamique Holdsclaw, Tennessee	479
Bridgette Gordon, Tennessee	388
Cheryl Miller, USC	333
Janice Lawrence, Louisiana Tech	312
Penny Toler, Long Beach State	291
Dawn Staley, Virginia	274
Cindy Brown, Long Beach State	263
Venus Lacy, Louisiana Tech	263
Clarissa Davis, Texas	261
Janet Harris, Georgia	254

(www.infoplease.com)

5. Twenty-three students were asked how many pages they had read in a book currently assigned for class. Here are their responses: {24, 87, 158, 227, 437, 79, 93, 121, 111, 118, 12, 25, 284, 332, 181, 34, 54, 167, 300, 103, 128, 132, 345}.

a. Find the measures of center.

b. Construct histograms for two different bin widths.

c. Construct a box plot.

d. What do the histograms and the box plot tell you about this data set? Make one or two observations.

After an impressive college career, Chamique Holdsclaw went on to play professionally for the Washington Mystics.

Pages Read in Current Book

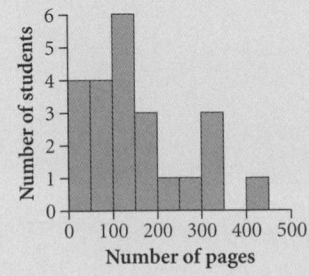

5c.

Pages Read in Current Book

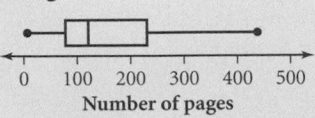

0 100 200 300 400 500
Number of pages

5d. Possible answer: Most of the students questioned had read fewer than 200 pages, with a fairly even distribution between 0 and 200.

6. Isabel made the estimates listed in the table at right for the year each item was invented.

  a. Create a scatter plot of data points with coordinates having the form (*actual year, estimated year*).

  b. Circle those points that picture an estimated year that is earlier than the actual year (underestimates).

  c. Define your variables and write the equation of a line that would represent all estimates being correct.

Invention Dates

Item	Actual year	Estimated year
Telephone	1876	1905
Color television	1928	1960
Video disk	1972	1980
Pacemaker	1952	1945
Motion picture	1893	1915
Ballpoint pen	1888	1935
Aspirin	1899	1917
Graphing calculator	1985	1980
Compact disc	1972	1990
Car radio	1929	1940

(*2000 World Almanac, pp. 609–610*)

7. APPLICATION The tables below show information for the Roxy Theater. The management is considering raising the admission prices.

Current Prices

	Matinee	Evening
Adult	$5.00	$8.00
Child	$3.50	$4.75
Senior	$3.50	$4.00

Price Increases

	Matinee	Evening
Adult	$0.50	$0.75
Child	$0.50	$0.25
Senior	$0.50	$0.25

Average Attendance

Adult	Child	Senior
43	81	37

  a. Convert each table to a matrix.

  b. Do a matrix calculation to find the new prices after the admission increase.

  c. Do a matrix calculation to find the total revenue of a matinee performance and an evening performance at the new prices.

8. The graph below shows Kayo's distance over time as she jogs straight down the street in front of her home. Point A is Kayo's starting point (her home).

  a. During which time period was Kayo jogging the fastest?

  b. Explain what the jogger might have been doing during the time interval between points B and C and between points D and E.

  c. Write a brief story for this graph using all five segments.

Jogger's Distance from Home

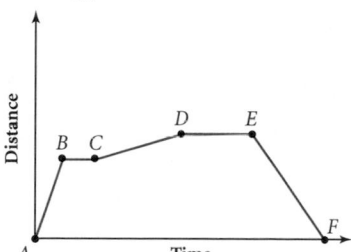

6a.

Invention Dates

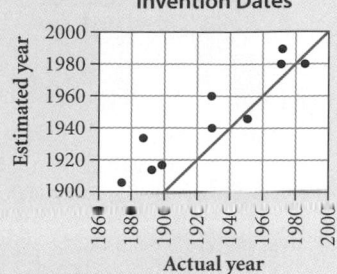

6b. (1952, 1945), (1985, 1980)

6c. $y = x$, where x represents actual year and y represents estimated year.

7a. $\begin{bmatrix} 5.00 & 8.00 \\ 3.50 & 4.75 \\ 3.50 & 4.00 \end{bmatrix} \begin{bmatrix} 0.50 & 0.75 \\ 0.50 & 0.25 \\ 0.50 & 0.25 \end{bmatrix}$

$[43 \ 81 \ 37]$

7b. $[A] + [B] = \begin{bmatrix} 5.50 & 8.75 \\ 4.00 & 5.00 \\ 4.00 & 4.25 \end{bmatrix}$

7c. $[C] \cdot ([A] + [B])$
$= [708.5 \ \ 938.5]$
matinee: $708.50
evening: $938.50

8a. between points A and B

8b. Kayo was not moving; perhaps she was resting.

8c. Possible answer: Kayo started out jogging fast but had to rest for a few minutes. Then she jogged much slower until she had to rest again. She finally got the energy to jog all the way home at a steady pace without stopping.

3a.

Mean Annual Wages, 1998

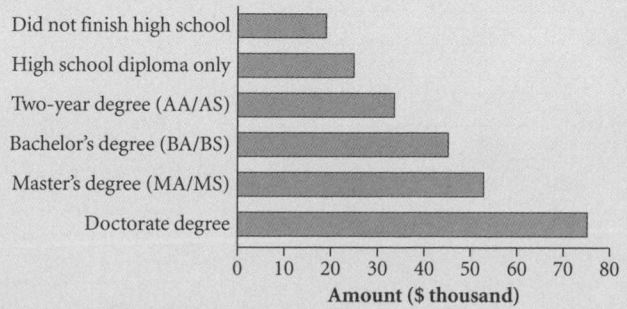

9d.

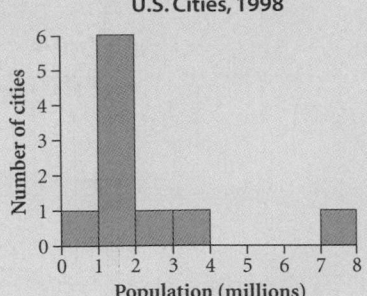

The Ten Most Populated U.S. Cities, 1998

9e. The bar graph helps show how each city compares with the others, since they remain identified by name. The stem plot shows distribution but also shows actual values. The histogram shows distribution and a definite clustering between 1 and 2 million, but it does not show individual cities or populations.

▶ **Take Another Look**

The picture graph uses both a broken vertical axis (starting at 15) and pictures with decreasing area to exaggerate the decline of green space. The normal bar graph, on the other hand, measures only acres and ignores the ratio to the increasing population, which the first bar takes into account. Therefore, both of these bar graphs are engineered to persuade. Answers will depend on whether students think acres per person or total acres is a better measure.

---

9. The table at right shows the approximate 1998 populations of the ten most populated cities in the United States.

   a. Write the approximate population of Chicago in 1998 as a decimal number. 2,820,000

   b. Create a bar graph of the data.

   c. Create a stem plot of the data.

   d. Create a histogram of the data.

   e. Each of the graphs you have created highlights different characteristics of the data. Briefly describe what features are unique to each graph.

10. This stem-and-leaf plot shows the number of minutes of sleep for eight students. In this plot the dots represent that the interval is divided in half at 50. Use the key to fully understand how to read this graph.

Minutes Sleeping

```
3 | 20
· | 60 90
4 | 00
· | 50 55 80 80
```

Key
```
3 | 20  means 320 minutes
· | 60  means 360 minutes
```

   a. What is the mean sleep amount?
   416.875 minutes
   b. What is the median sleep amount?
   425 minutes
   c. What is the mode sleep amount?
   480 minutes

The Ten Most Populated U.S. Cities, 1998

City	Population (millions)
Chicago	2.82
Dallas	1.08
Detroit	0.97
Houston	1.79
Los Angeles	3.60
New York	7.42
Philadelphia	1.44
Phoenix	1.20
San Antonio	1.11
San Diego	1.22

(www.census.gov)

TAKE ANOTHER LOOK

These two graphs display the same data set. Graph A is being presented by a citizen who argues that the city should use its budget surplus to buy land for a park. Graph B is being presented by another citizen who argues that a new park is not a high priority. Tell what position you would favor on this issue and what impact each graph has on your decision. Do you think either graph is deliberately misleading? What other information would you want to know?

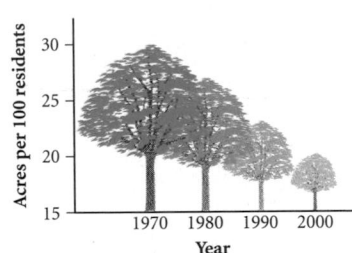

Greenspace Disappearing!

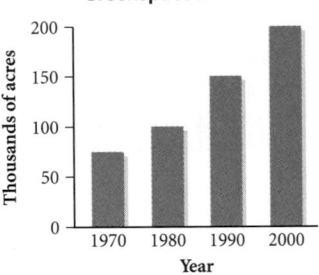

Greenspace in Our Town

Find another graphic display in a newspaper, magazine, or voter material that seems to be "engineered" to persuade the viewer to a particular point of view. Tell how the graph could be changed for a fairer presentation.

9b.

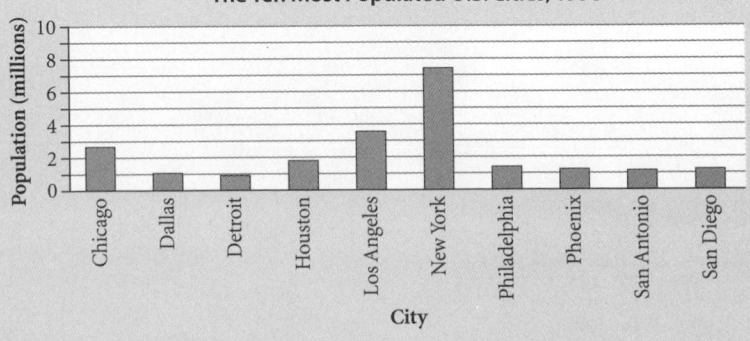

The Ten Most Populated U.S. Cities, 1998

9c. **The Ten Most Populated U.S. Cities, 1998**

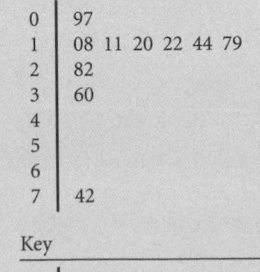

```
0 | 97
1 | 08 11 20 22 44 79
2 | 82
3 | 60
4 |
5 |
6 |
7 | 42
```

Key
```
2 | 82 means 2.82 million
```

Assessing What You've Learned

WRITE IN YOUR JOURNAL

Your course work in algebra will bring up many new ideas, and some are quite abstract. Sometimes you'll feel you have a good grip on these new ideas, other times less so.

Regular reflection on your confidence in mathematics generally, and your mastery of algebra skills in particular, will help you assess your strengths and weaknesses. Writing these reflections down will help you realize where you are having trouble, and perhaps you'll ask for help sooner. Likewise, realizing how much you know can boost your confidence. A good place to record these thoughts is in a journal—not a personal diary, but an informal collection of your feelings about what you're learning. Like a travel journal that others would find interesting to read and that will help recall the details of a trip, a mathematics journal is something your teacher will want to look at and something you'll look at again later.

Here are some questions to prompt your journal writing. Your teacher might give you other ideas, but you can write in your journal any time.

▶ How is what you are learning an extension of your previous mathematics courses? How is it completely new?
▶ What are your goals for your work in algebra? What steps can you take to achieve them?
▶ Can you see ways to apply what you are learning to your everyday life? To a future career?
▶ What ideas have you found hard to understand?

UPDATE YOUR PORTFOLIO At the end of Chapter 0, you may have started a portfolio. Now would be a good time to add one or more pieces of significant work from Chapter 1. You could choose an investigation, a homework problem, or your work on a Project or Take Another Look. It might be a good idea to include a sample of every type of graph you've learned to interpret or create.

Through this chapter, students have become more accustomed to working collaboratively and have improved their facility with graphing calculators. The principal new ideas discussed are representing data by statistics, by graphs, and by matrices.

FACILITATING SELF ASSESSMENT

Many of the exercises students have completed could be journal entries; they are marked Journal in Assigning Homework. Students also benefit from recording their thinking about their own ideas.

To help students complete the portfolio described in Assessing What You Have Learned, suggest that they consider for evaluation their work on Lesson 1.1, Exercise 3; Lesson 1.2, Exercise 9 and/or 10; Lesson 1.3, Exercise 8; Lesson 1.4, Exercise 6 and/or 10; Lesson 1.6, Exercise 8; Lesson 1.7, Exercise 6; and Lesson 1.8, Exercise 12.

TESTING

You can use either Form A or Form B of the chapter test, or you can use constructive assessment items from *Assessment Resources*. Using the CD test generator, you can create an alternate version of the test or combine some items from Form A or Form B with constructive assessment items.

2

Proportional Reasoning and Probability

Overview

In this chapter students review and extend previous work with proportions (**Lessons 2.1** and **2.2**) and percents (**Lesson 2.4**). In **Lesson 2.3,** students use conversion factors and dimensional analysis to convert units of measurement. **Lesson 2.5** uses proportions to create and interpret circle graphs and relative frequency bar graphs. In **Lessons 2.6** and **2.7,** probability concepts are explored using ratios.

Students who deeply understand proportions and their many applications will be mathematical thinkers for life. Tools like dimensional analysis will help them see and understand proportions in science and the world around them.

The Mathematics

Ratio

Traditionally, a *ratio* is a comparison of two quantities, such as "3 is to 2." The tradition dates back before Euclid, who defined a ratio as "a sort of relation in respect of size between two magnitudes of the same kind."

Euclid didn't think of these magnitudes as numbers; in fact, he treated the ratios of numbers (integers) separately from the ratios of magnitudes (lengths, areas, volumes). Today we think of a ratio as a fraction, a special kind of number.

A native speaker of a language recognizes nuances between words with the same dictionary definitions. Similarly, mathematicians use the word *ratio* slightly differently from *fraction*. That difference reflects the idea of a comparison. A fraction often measures a quantity, such as $\frac{2}{3}$ cup of sugar. Calling $\frac{2}{3}$ a ratio, though, emphasizes the comparison of 2 parts to 3 parts. More often, calling a fraction a ratio compares two quantities, such as 2 cups of sugar to 3 cups of flour. So in working with ratios, the units, also called *dimensions,* are important.

Because units are attached to the two numbers in a ratio, it's perhaps more natural to have a denominator of 1 in a ratio than in a pure fraction. Saying that 18 students is the same as $\frac{18}{1}$ students is true but contrived. If $\frac{18}{1}$ represents the ratio 18 students to 1 teacher, however, it makes more sense.

Since a ratio involves two units, it is sometimes very useful for converting between different units of measurement. Any conversion factor, such as $\frac{60 \text{ seconds}}{1 \text{ minute}}$, is a ratio. Using a product of several conversion factors, chosen so that their units cancel to effect a desired conversion, is called *dimensional analysis.* For example, to change x seconds to hours, multiply by 1 twice, using the ratios $\frac{1 \text{ minute}}{60 \text{ seconds}}$ and $\frac{1 \text{ hour}}{60 \text{ minutes}}$.

$$x \text{ \cancel{seconds}} \cdot \frac{1 \text{ \cancel{minute}}}{60 \text{ \cancel{seconds}}} \cdot \frac{1 \text{ hour}}{60 \text{ \cancel{minutes}}}$$

$$= x \cdot \frac{1}{3600} \text{ hours}$$

The probability of an outcome is also a ratio, whose numerator is the number of outcomes of the desired type and whose denominator is the total number of possible outcomes.

Proportion

A *proportion* is a statement that two ratios are equal. Four numbers are involved in a proportion. We say that corresponding pairs of these numbers are *proportional.*

In Chapter 1, we said, "Put numbers on the number line proportionally." That is

$$\frac{a - b}{a - c} = \frac{\textit{distance between points representing } a \textit{ and } b}{\textit{distance between points representing } a \textit{ and } c}$$

for any numbers a, b, and c, where $a \neq c$.

One application of proportions that many students find counterintuitive is the capture-recapture method. You can get a good estimate of the number of fish in a pond, for example, if you tag a sample of the fish, give the tagged fish time to redistribute themselves among the others, and then take a new sample.

Another use of proportions is in drawing relative frequency graphs, in which bars in a bar graph or sectors of a circle graph are labeled with fractions or percents rather than with the numbers giving the frequencies.

Representative Samples and Random Processes

A *representative sample* is one whose statistics are the same as those of the entire population. In the recapture phase of the capture-recapture investigation, the sample is representative if the portion of tagged fish in the sample is the same as that in the lake. One way to make a sample representative is to choose it in such a way that each member of the population has an equal probability of being chosen for the sample. Such a procedure is called a *random process*. Random processes are also used to determine experimental probabilities. Theoretical probabilities do not involve taking a sample. They are determined mathematically.

Using This Chapter

If your students understand ratios and proportions, you can skip some of the lessons. Be careful in making that decision, though; many students who can solve proportions are not skilled at setting them up and can't explain why their solution method works.

If students can use and understand proportions and percent and need little review of probability, you could do just one or two lessons, such as 2.2 and 2.3. If you think students have a pretty good grasp of proportions and percent and understand dimensional analysis, you might use probability to review proportions and do only Lessons 2.6 and 2.7. Or you might decide to extend Chapter 1 with a new kind of graph, the relative frequency graph of Lesson 2.5, as a review of proportions before students study rates in Chapter 3.

Resources

Discovering Algebra Resources

Calculator Notes 0A, 1I, 1J, 2A, 2B, 2C, 2D, (2E for TI-73 only)

Teaching and Worksheet Masters
    Lesson 2.2 Fish in the Lake Sample Data
    Lesson 2.5 Circle Graphs
    Lesson 2.5 Protractors

Assessment Resources A and B
    Quiz 1 (Lessons 2.1 to 2.3)
    Quiz 2 (Lessons 2.4 to 2.7)
    Chapter 2 Test
    Chapter 2 Constructive Assessment Options

More Practice Your Skills for Chapter 2

Condensed Lessons for Chapter 2

Other Resources

Mary Jean Winter and Ronald Carlson. *Probability Simulations.* Emeryville, California: Key Curriculum Press, 2000.

www.keymath.com/DA

Materials

- paper bags
- 3 to 4 pounds white beans
- $\frac{1}{2}$ pound red beans
- yardsticks and meter-sticks or tape measures
- graph paper
- rulers
- protractors
- compasses or circle templates
- packets of colored candies

Pacing Guide

	day 1	day 2	day 3	day 4	day 5	day 6	day 7	day 8	day 9	day 10
standard	2.1	2.2	2.3	2.4	quiz, 2.5	2.5	2.6	2.7	review	assessment
enriched	2.1	project	2.2	2.3	2.4	quiz, 2.5	2.6	2.7, project	review, TAL	assessment
block	2.1, 2.2	2.3, 2.4	quiz, 2.5	2.6, 2.7	review	assessment				

2 Proportional Reasoning and Probability

- Learn to set up and solve proportions in which a variable represents an unknown number

- Use proportions to help make predictions from data gathered for the capture-recapture method

- Change measurement units through conversion factors and dimensional analysis

- Draw and interpret circle graphs and relative frequency graphs

- Determine observed and theoretical probabilities, and see how the former approximates the latter

- Use probabilities to describe patterns when outcomes are random

Murals are just one of the many art forms around us that come to life with the help of ratios and proportions. To plan a mural, the artist draws sketches on paper then uses ratio to enlarge the image to the size of the final work.

OBJECTIVES

In this chapter you will

- use proportional reasoning to understand problem situations
- create and interpret circle graphs
- explore probability

Most students have constructed pictures using enlargement grids; they capture the drawing fragment within the small square and transfer it to the large square. The enlarged drawing itself is called a *cartoon*. Ask, "If the artist first worked on paper that was about 18 inches by 24 inches, what did he or she have to multiply by to create a large drawing to transfer the image to the building?" [Reasonable answers are between 25 and 35.] Sketches for modern murals can be projected on a wall, which also involves an enlargement by a ratio.

You might also ask, "How are the real students like their images on the mural?" [The ratios of corresponding lengths are all the same. Or, the corresponding pairs of lengths are proportional. The images are about 3 times the height of a student. They look a lot bigger because they appear to have a volume of about 3^3 times that of a student.] You might also ask, "What is the ratio of real students in the picture to images in the mural?" or "What is the ratio of children pictured on the mural to children?" [The ratio is 7 students to 4 images of children on the mural.]

Mathematics is not a way of hanging numbers on things so that quantitative answers to ordinary questions can be obtained. It is a language that allows one to think of extraordinary questions.

JAMES BULLOCK

Proportions

When you say, "I got 21 out of 24 questions correct on the last quiz," you are comparing two numbers. The **ratio** of your correct questions to the total number of questions is 21 to 24. You can write the ratio as 21 : 24 or as a fraction or decimal. The fraction bar means division, so these expressions are equivalent:

EXAMPLE A | Write the ratio 210 : 330 in several ways.

▶ **Solution** | $\dfrac{210}{330}$ or $\dfrac{7}{11}$ $210 \div 330$ or $7 \div 11$

To change a common fraction into a decimal fraction, divide the numerator by the denominator.

```
        .636363
330)210.000000
    198 0
     12 00
      9 90
      2 100
      1 980
        1200
         990
          . . .
```

When you divide 21 by 24, the decimal form of the quotient ends, or **terminates.** The ratio $\frac{21}{24}$ equals 0.875 exactly. But when you divide 210 by 330 or 7 by 11, you see a **repeating decimal** pattern 0.636363.... You can use a bar over the numerals that repeat to show a repeating decimal pattern $\frac{7}{11} = 0.\overline{63}$. [▶ 🖳 See **Calculator Note 0A** for more about converting fractions to decimals. ◀]

NCTM STANDARDS

CONTENT		PROCESS
• Number		Problem Solving
• Algebra		• Reasoning
Geometry		Communication
• Measurement		Connections
Data/Probability		• Representation

LESSON OBJECTIVES

- Rename fractions as decimal numbers
- Write ratios and proportions that express relationships in data
- Solve proportions by multiplying both sides of the equation by the same number
- Solve proportions by inverting both ratios
- Solve problems using proportions
- Review skills in working with percents

PLANNING

LESSON OUTLINE

One day:

10 min	Introduction, Example A
15 min	Investigation
5 min	Sharing
10 min	Examples B and C
5 min	Closing
5 min	Exercises

MATERIALS

- Calculator Note 0A

TEACHING

The method for solving proportions presented in this lesson is easy to explain and works with other kinds of equations. Students may have learned to solve proportions using cross-multiplication. Discourage them from using this method. **[Language]** Many students need to see and hear the vocabulary words *ratio*, *fraction*, *rational number*, *decimal*, *quotient*, and *divided by*. Create a display of the words and use them often as you talk with students about their work.

▶ EXAMPLE A

[ESL] Students who learned long division outside the United States may have different methods for doing it. Ask them to show the class how they do it.

Point out that it's not the rational number itself that repeats or terminates, but only its decimal name.

After Example C, numbers that can be written as the ratio of two integers are defined as rational numbers.

One step Pose the problem, "Write out descriptions of how to find the value of M as it replaces each whole number in the proportion $\frac{8}{19} = \frac{56}{133}$." As you circulate, bring out the ideas of a proportion as an equality of ratios, of rational numbers and their decimal equivalents, and of inverting the ratios of a proportion to solve it. Avoid answering questions yourself. Instead, call on students who might answer them or say, "Have you asked your group?"

Make sure students understand how a variable is used for an unknown number in a proportion.

Step 1 Some students may have trouble answering the question "Why can you do this?" They might not see that they're multiplying two equal numbers by the same thing. Point out that the left side is representing a number— the same number that the right side is representing.

Step 1 If both sides are equal, they remain equal when multiplied by the same number. $M = 8$.

Step 2 Some students might want to reduce the fraction before they do the multiplying. Doing so will make the arithmetic easier, so it could save them time if they are doing these without a calculator. After they multiply, they may want to divide out some factors; emphasize that this is "legal" because they're multiplying and dividing equal numbers by the same number. And when they "cancel," they are removing factors that are equivalent to 1.

Step 4 You could use this as a journal prompt.

Step 6 Explanations should include the phrase "first interchange the numerators and denominators of each fraction," and something about equivalent proportions. $k = 81$.

A **proportion** is an equation stating that two ratios are equal. For example, $\frac{2}{3} = \frac{8}{12}$ is a proportion. You can use the numbers 2, 3, 8, and 12 to write these true proportions:

$$\frac{2}{3} = \frac{8}{12} \qquad \frac{3}{2} = \frac{12}{8} \qquad \frac{3}{12} = \frac{2}{8} \qquad \frac{12}{3} = \frac{8}{2}$$

Do you agree that these are all true equations? One way to check that a proportion is true is by finding the decimal equivalent of each side. The statement $\frac{3}{8} = \frac{2}{12}$ is not true; 0.375 is not equal to $0.1\overline{6}$.

In algebra, a **variable** can stand for an unknown number, or for a set of numbers. In the proportion $\frac{2}{3} = \frac{M}{6}$, you can replace the letter M with any number, but only one number will make the proportion true. That number is unknown until the proportion is solved.

Investigation
Multiply and Conquer

You can easily guess the value of M in the proportion $\frac{2}{3} = \frac{M}{6}$. In this investigation, you examine ways to solve a proportion for an unknown number when guessing is not easy. It's hard to guess the value of M in the proportion $\frac{M}{19} = \frac{56}{133}$.

Step 1 Multiply both sides of the proportion $\frac{M}{19} = \frac{56}{133}$ by 19. Why can you do this? What does M equal?

Step 2 For each equation, choose a number to multiply both ratios by to solve the proportion for the unknown number. Then multiply and divide to find the missing value.

a. $\frac{21}{35} = \frac{Q}{20}$ $Q = 12$ **b.** $\frac{P}{12} = \frac{132}{176}$ $P = 9$

c. $\frac{L}{30} = \frac{30}{200}$ $L = 4.5$ **d.** $\frac{130}{78} = \frac{n}{15}$ $n = 25$

Step 3 Check that each proportion in Step 2 is true by replacing the variable with your answer.

Step 4 In each equation of Step 2, the variables are in the numerator. Write a brief explanation of one way to solve a proportion when one of the numerators is a variable. Answer should refer to multiplying both sides by the number under the variable.

Step 5 The proportions you solved in Step 2 have been changed by switching the numerators and denominators. That is, the ratio on each side has been *inverted*. Do the solutions from Step 2 also make these new proportions true? Yes.

a. $\frac{35}{21} = \frac{20}{Q}$ **b.** $\frac{12}{P} = \frac{176}{132}$ **c.** $\frac{30}{L} = \frac{200}{30}$ **d.** $\frac{78}{130} = \frac{15}{n}$

Step 6 How can you use what you just discovered to help you solve a proportion that has the variable in the denominator, such as $\frac{20}{135} = \frac{12}{k}$? Why does this work? Solve the equation.

Step 7 | There are many ways to solve proportions. Here are three student papers each answering the question "13 is 65% of what number?" What are the steps each student followed? What other methods can you use to solve proportions?

a.

$$\frac{65}{100} = \frac{13}{x}$$

$$\frac{100}{1} \cdot \frac{x}{1} \cdot \frac{65}{100} = \frac{13}{x} \cdot \frac{100}{1} \cdot \frac{x}{1}$$

$$\frac{65x}{65} = \frac{1300}{65}$$

$$x = 20$$

b.

$$\frac{65}{100} = \frac{13}{x}$$

$$\frac{\frac{13}{65}}{100} = \frac{13}{x}$$

$$20$$

$$20 = x$$

c.

$$\frac{65}{100} = \frac{13}{x}$$

$$\frac{100}{65} = \frac{x}{13}$$

$$\frac{13}{1} \cdot \frac{100}{65} = \frac{x}{13} \cdot \frac{13}{1}$$

$$20 = x$$

EXAMPLE B

Jennifer estimates that two out of every three students will attend the class party. She knows there are 750 students in her class. Set up and solve a proportion to help her estimate how many people will attend.

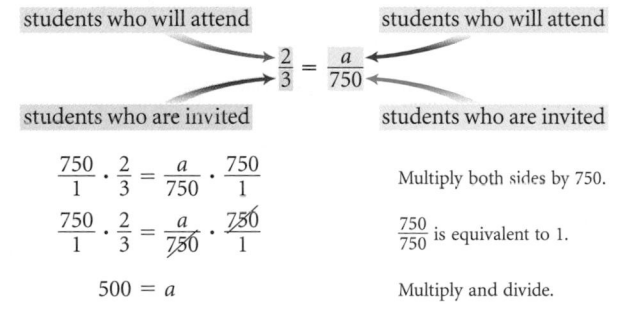

▶ Solution

To set up the proportion, be sure both ratios make the same comparison. Use a to represent the number of students who will attend.

students who will attend   students who will attend

$$\frac{2}{3} = \frac{a}{750}$$

students who are invited   students who are invited

$$\frac{750}{1} \cdot \frac{2}{3} = \frac{a}{750} \cdot \frac{750}{1}$$ Multiply both sides by 750.

$$\frac{750}{1} \cdot \frac{2}{3} = \frac{a}{750} \cdot \frac{750}{1}$$ $\frac{750}{750}$ is equivalent to 1.

$$500 = a$$ Multiply and divide.

Jennifer can estimate that about 500 students will attend the party.

Step 7 Consider assigning this for extra credit for especially capable or interested students rather than asking everyone to do it.

SHARING IDEAS

Have students share results of Step 2.

Using letters for unknowns may be new to students. **[Ask]** "What do the letters in the proportions represent?" Each represents a single unknown number. The letters are called *variables* even if they don't vary. In Chapter 1, each variable represented a variety of numbers.

Ask why the proportions in Step 5 have the same solutions. A mechanical answer is that you have inverted the fractions on both sides of the equation. Lead to deeper understanding by asking, "If $\frac{6}{B} = 2$, then why will $\frac{B}{6} = \frac{1}{2}$?" Encourage students to draw segments of these lengths— and to make up their own situations—to aid their understanding of proportions.

Assessing Progress

Assess students' abilities at multiplying, finding decimal forms of rational numbers, and inverting fractions.

▶ EXAMPLE B

This is an application of proportion. Some students may be able to solve the problem without proportions. Don't discourage them, but challenge them to set up and solve a proportion. **[Ask]** "Is 2 the part or the whole? 3? 750? The unknown?" Show how the proportion must be either

$$\frac{estimate\ part}{estimate\ whole} = \frac{students\ part}{students\ whole} \text{ or}$$

$$\frac{estimate\ part}{students\ part} = \frac{estimate\ whole}{students\ whole}.$$

This example is an application of proportions that reviews percents. **[Alert]** Some students may not think of 70% as the ratio $\frac{70}{100}$. Point out that *per* indicates a ratio, as in *miles per hour*, and *per cent* means *per hundred*. **[Language]** Point out that there are 100 *cent*s in a dollar and 100 years in a *cent*ury.

Closing the Lesson

Review with students the ideas of **ratio, rational number**, the **decimal form** of a rational number, **proportion, unknown, variable,** solving a proportion by **multiply and conquer,** and **inverting** both sides of a proportion to get another true proportion. A proportion states the equality of two ratios, in which one number may be unknown and represented by a variable. Rational numbers are ratios of integers, and their decimal names are repeating or terminating. Ask if students think there are any numbers that are *not* rational. If they don't know of irrational numbers, they might instead suggest integers; point out that $-3 = -\frac{3}{1}$, so it's rational. As the history connection says, $\sqrt{2}$ and π are irrational.

BUILDING UNDERSTANDING

In the exercises, students practice expressing ratios and solving proportions.

ASSIGNING HOMEWORK

Essential	1–6, 8
Performance assessment	10
Portfolio	7
Journal	10
Group	6, 9
Review	11–13

EXAMPLE C | After the party, Jennifer found out that 70% of the class attended. How many students attended?

► **Solution**

History

CONNECTION

The Pythagoreans, a brotherhood of philosophers begun by Pythagorus in about 520 B.C., realized that not all numbers are rational. For example, they showed that for a square one unit on a side, the diagonal $\sqrt{2}$ is *irrational*. Another irrational number is pi, the ratio of the circumference of a circle to its diameter. Pi, or π, has been used and studied for more than 4000 years.

Write and solve a proportion to answer the question "If 70 students out of 100 attended the party, how many students out of 750 attended?"

Let s represent the number of students who attended.

$\frac{70}{100} = \frac{s}{750}$	Write the proportion.
$750 \cdot \frac{70}{100} = \frac{s}{750} \cdot 750$	Multiply both sides by 750.
$750 \cdot \frac{70}{100} = s$	$\frac{750}{750}$ is equivalent to 1.
$\frac{750}{10} \cdot \frac{70}{10} = s$	100 is 10 · 10.
$75 \cdot 7 = s$	You use the fact that $\frac{10}{10} = 1$ twice.
$525 = s$	Multiply.

Numbers that can be written as the ratio of two integers are called **rational numbers.**

EXERCISES

You will need your calculator for problem **1.**

► Practice Your Skills

1. Estimate the decimal equivalent for each fraction.

> Keep in mind these fraction and decimal equivalents:
>
> $\frac{1}{25} = 0.04$ $\frac{1}{20} = 0.05$ $\frac{1}{8} = 0.125$ $\frac{1}{5} = 0.2$ $\frac{1}{2} = 0.5$ $\frac{2}{1} = 2.0$

a. $\frac{7}{8}$ 0.875   **b.** $\frac{13}{20}$ 0.65   **c.** $\frac{13}{5}$ 2.6   **d.** $\frac{52}{25}$ 2.08

Now use your calculator to find the decimal number.

2. Ms. Lenz collected information about the students in her class.

Eye Color

	Brown eyes	Blue eyes	Hazel eyes
9th graders	9	3	2
8th graders	11	4	1

Write these ratios as fractions.

a. Ninth graders with brown eyes to ninth graders $\frac{9}{14}$

b. Eighth graders with brown eyes to students with brown eyes $\frac{11}{20}$

c. Eighth graders with blue eyes to ninth graders with blue eyes $\frac{4}{3}$

d. All students with hazel eyes to students in both grades $\frac{3}{30}$ or $\frac{1}{10}$

► Helping with the Exercises

Exercise 1 Some students may not realize that $\frac{7}{8}$ is $7 \cdot \frac{1}{8}$. They might also miss the instruction to estimate and go right to their calculators. Challenge them to find the decimal forms in at least two ways.

3. Phrases such as miles per gallon, parts per million (ppm), and accidents per 1000 people indicate ratios. Write each ratio named below as a fraction. Use a number and a unit in both the numerator and the denominator.

   a. In 2000, the McLaren was the fastest car produced. Its top speed was recorded at 240 miles per hour.

   b. Pure capsaicin, a substance that makes hot peppers taste hot, is so strong that 10 ppm in water can make your tongue blister.

   c. In 2000, women owned approximately 350 of every thousand firms in the United States.

   d. The 2000 average income in Philadelphia, Pennsylvania, was approximately $35,500 per person. $\frac{35,500 \text{ dollars}}{1 \text{ person}}$

4. Write three other true proportions using the four integers in each proportion. **Sample answers:**

   a. $\frac{2}{5} = \frac{10}{25}$ $\frac{5}{2} = \frac{25}{10}$, $\frac{2}{10} = \frac{5}{25}$, $\frac{25}{5} = \frac{10}{2}$ b. $\frac{3}{7} = \frac{12}{28}$ $\frac{7}{3} = \frac{28}{12}$, $\frac{3}{12} = \frac{7}{28}$, $\frac{28}{7} = \frac{12}{3}$ c. $\frac{16}{20} = \frac{12}{15}$
   $\frac{20}{16} = \frac{15}{12}$, $\frac{15}{20} = \frac{12}{16}$, $\frac{16}{12} = \frac{20}{15}$

5. Find the value of the unknown number in each proportion.

   a. $\frac{24}{40} = \frac{T}{30}$ $T = 18$ b. $\frac{49}{56} = \frac{R}{32}$ $R = 28$ c. $\frac{52}{91} = \frac{42}{S}$ $S = 73.5$ d. $\frac{100}{30} = \frac{7}{x}$ $x = 2.1$

   e. $\frac{M}{16} = \frac{87}{232}$ $M = 6$ f. $\frac{6}{n} = \frac{62}{217}$ $n = 21$ g. $\frac{36}{15} = \frac{c}{13}$ $c = 31.2$ h. $\frac{220}{33} = \frac{60}{W}$ $W = 9$

▶ Reason and Apply

6. **APPLICATION** Write a proportion for each problem, and solve for the unknown number.

   a. Leaf-cutter ants that live in Central and South America weigh about 1.5 grams. One ant can carry a 4-gram piece of leaf that is about the size of a dime. If a person could carry proportionally as much as the leaf-cutter ant, how much could a 55-kilogram algebra student carry?

   b. The leaf-cutter ant is about 1.27 cm long and takes strides of 0.84 cm. If a person could take proportionally equivalent strides, what size strides would a 1.65-m–tall algebra student take?

   c. The 1.27-cm–long ants travel up to 0.4 km from home each day. If a person could travel a proportional distance, how far would a 1.65-m–tall person travel?

7. **APPLICATION** Jeremy has a job at the movie theater. His hourly wage is $7.38. Suppose 15% of his income is withheld for taxes and Social Security.

   a. What percent does Jeremy get to keep? 85%

   b. What is his hourly take-home wage? $\frac{t}{7.38} = \frac{85}{100}$, $t = \$6.27$

Exercise 3 When necessary, remind students that they can write any whole number as a ratio with denominator 1.

Exercise 6 The units can help students write the proportions. The top units can be the same on both sides, as well as the bottom units. But this is not the only way to write the proportion. In 6b, you can write the units of the proportion as $\frac{\text{gm creature}}{\text{gm load}} = \frac{\text{kg creature}}{\text{kg load}}$, comparing creature to load on both sides. Or you can write $\frac{\text{gm creature}}{\text{kg creature}} = \frac{\text{gm load}}{\text{kg load}}$, comparing gm with kg on each side.

This comparison can produce some interesting group discussion about the mighty little ant. The hypothetical algebra student in this problem weighs about 120 pounds and is 5 feet 5 inches tall. If he or she were proportionally as strong as an ant, the algebra student could carry 323 pounds, take strides of 3 feet 7 inches, and walk 32 miles a day.

[Extend] Suggest that students do Internet research on some small creature and predict the capabilities that people would have if proportionally able to jump as high, run as fast, lift as much, and so on.

Exercises 7 to 10 Don't discourage students who understand percents and fractions well enough to avoid proportions on these problems; rather, challenge them with the puzzle of representing the situations with proportions.

3a. $\frac{240 \text{ miles}}{1 \text{ hour}}$

3b. $\frac{10 \text{ parts capsaicin}}{1,000,000 \text{ parts water}}$ or $\frac{1 \text{ part capsaicin}}{100,000 \text{ parts water}}$

3c. $\frac{350 \text{ women-owned firms}}{1000 \text{ firms}}$ or $\frac{7 \text{ women-owned firms}}{20 \text{ firms}}$

6a. $\frac{4 \text{ g}}{1.5 \text{ g}} = \frac{x \text{ kg}}{55 \text{ kg}}$; $x = 146.\overline{6} \text{ kg}$

6b. $\frac{0.84 \text{ cm}}{1.27 \text{ cm}} = \frac{x \text{ m}}{1.65 \text{ m}}$; $x \approx 1.09 \text{ m}$

6c. $\frac{0.4 \text{ km}}{0.0127 \text{ m}} = \frac{x \text{ km}}{1.65 \text{ m}}$; $x \approx 52 \text{ km}$

9. $2\frac{2}{3}$ cups of water and $\frac{2}{3}$ cup of oatmeal; $6\frac{2}{3}$ cups water and $1\frac{2}{3}$ cups oatmeal

Exercise 10 [Link] Manipulating chemical formulas requires skill in algebra.

10b. You will need 3(470) or 1410 atoms of carbon and 6(470) or 2820 atoms of hydrogen.

10c. 500 molecules; use all the hydrogen atoms, 1500 atoms of carbon and 500 atoms of oxygen

8. In a resort area during the summer months, only one out of eight people is a year-round resident. The others are there on vacation. If the year-round population of the area is 3000, how many people are there in the summer? $\frac{1}{8} = \frac{3000}{P}$; $P = 24{,}000$

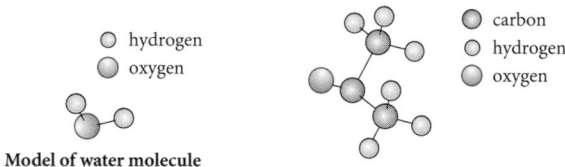

9. **APPLICATION** To make three servings of Irish porridge, you need 4 cups of water and 1 cup of oatmeal. How much of each ingredient will you need for two servings? for five servings?

10. **APPLICATION** When chemists write formulas for chemical compounds, they indicate how many atoms of each element combine to form a molecule of that compound. For instance, they write H_2O for water, which means there are two hydrogen atoms and one oxygen atom in each molecule of water. Acetone (or nail polish remover) has the formula C_3H_6O. The C stands for carbon.

○ hydrogen
○ oxygen

Model of water molecule

○ carbon
○ hydrogen
○ oxygen

Model of acetone molecule

a. How many of each atom are there in one molecule of acetone? 3 carbon, 6 hydrogen, 1 oxygen

b. How many atoms of carbon must combine with 470 atoms of oxygen to form acetone molecules? How many atoms of hydrogen are required?

c. How many acetone molecules can be formed from 3000 atoms of carbon, 3000 atoms of hydrogen, and 1000 atoms of oxygen?

▶ **Review**

11. In the dot plot below, circle the points that represent values for the five-number summary. If a value is actually the mean of two data points, draw a circle around the two points.

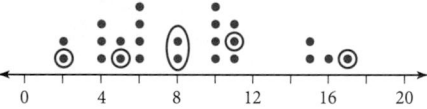

12. The *Forbes Celebrity 100* listed these ten people (and their incomes for 1999 in millions) among the celebrities that got the most media attention. Find the three measures of center for their incomes, and explain why they are so different.

George Lucas ($400)
Oprah Winfrey ($150)
Tom Clancy ($66)
Stephen King ($65)
Bruce Willis ($54)
Julia Roberts ($50)
Shania Twain ($48)
Tiger Woods ($47)
Mel Gibson ($45)
Jim Carrey ($45)

Tiger Woods

13. Use the order of operations to evaluate these expressions. Check your results on your calculator.
   a. $5 \cdot -4 + 8$ -12
   b. $-12 \div (7 - 4)$ -4
   c. $-3 - 6 \cdot 25 \div 30$ -8
   d. $18(-3) \div 81$ $-\frac{2}{3}$

THE GOLDEN RATIO

In this project, you'll research the amazing number mathematicians call the **golden ratio.** (There are plenty of books and web sites on the topic. Find links from **www.keymath.com/DA** .) Your project should include

▶ Basic information on the golden ratio, such as its exact value, why it represents a mathematically "ideal" ratio, and how to construct a golden rectangle. (Its length-to-width ratio is the golden ratio.)

▶ Some history of the golden ratio, including its role in ancient Greek architecture.

▶ At least one other interesting mathematical fact about the golden ratio, such as its relationship to the Fibonacci sequence or its own reciprocal.

▶ A report on where to find the golden ratio in the environment, architecture, or art. List items and their measurements or include prints from photographs, art, or architecture on which you have drawn the golden rectangle.

Once you've learned how to construct the golden ratio, The Geometer's Sketchpad is an ideal tool for further exploration. You can create a Custom Tool for dividing segments into the golden ratio, and then use this tool to help you construct the golden rectangle or even the golden spiral.

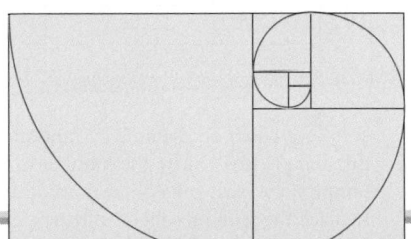

12. Mean: $97 million; median: $52 million; mode: $45 million. The very high incomes of George Lucas and Oprah Winfrey make the mean much higher than the median. The mode just happens to be the lowest income.

The Golden Ratio Project
The construction of the golden rectangle is shown. You can construct, with straightedge and compass, the longer edge of a golden rectangle by extending by 1 the hypotenuse of a right triangle with legs of length 1 and 2 and then bisecting the extension. Construct a perpendicular of length 1 at each end to get two more edges of the rectangle; join their other endpoints to complete the golden rectangle.

Examples of ratios of consecutive Fibonacci numbers (1, 1, 2, 3, 5, 8, 13, 21, 34, 55 . . .) include $\frac{55}{34} \approx 1.617647$; $\frac{233}{144} \approx 1.618056$.

Supporting the project

MOTIVATION

Find out about the ratio that was named the *golden ratio* during the golden age of Greece. The ratio is found in nature and in both ancient and modern architecture and art.

OUTCOMES

▶ The ratio, often symbolized by τ or ϕ, is $\frac{1 + \sqrt{5}}{2} \approx 1.618034$.

▶ It is exactly 1 more than its reciprocal $\left(\frac{1}{\tau} \approx 0.618034\right)$.

▶ A golden rectangle can be divided into a square and a golden rectangle.

▶ The ratio of consecutive numbers in the Fibonacci sequence gets closer and closer

to τ as you go further out in the sequence.

▶ The project should contain some history of τ along with examples.

• τ is a solution to the equation $x^2 - x - 1 = 0$.

• There are also golden triangles and a golden spiral.

LESSON

2.2

Capture-Recapture

The Universe is a grand book which cannot be read until one first learns to comprehend the language.... It is written in the language of mathematics.

GALILEO GALILEI

Wildlife biologists often need to know how many deer are in a national park or the size of the perch population in a large lake. It is impossible to count each deer or fish, so biologists use a method called "capture-recapture" that uses ratios to estimate the population.

To estimate a fish population, biologists first capture some of the fish and put tags on them. Then these tagged fish are returned to the lake to mingle with all of the untagged fish. The biologist allows time for the fish to mix thoroughly, captures another **sample,** and counts how many of the sample are tagged.

PLANNING

LESSON OUTLINE

One day:

20 min	Investigation
5 min	Sharing
10 min	Examples A, B, and C
5 min	Closing
10 min	Exercises

MATERIALS

- paper bag with about 1 cup of white beans (per group)
- less than $\frac{1}{4}$ cup red beans (per group)
- Fish in the Lake Sample Data (W or T), *optional*

TEACHING

Deepen understanding of ratio and proportion by using them in a new context: using the capture-recapture (sampling) method to estimate the size of population.

Guiding the Investigation

Instead of beans, you might use chocolate pretzels and regular pretzels, colored macaroni and uncolored macaroni, two different varieties of shaped snack crackers, and so on, with suitable containers. Objects used for tagged and untagged fish should be the same size and weight.

Steps 1 to 2 A good number to tag is 10%.

Step 2 If asked at this point, many students would say that they don't have enough information to predict the population of fish—from a lack of understanding of ratios rather than from concern about the quality of the sampling.

Investigation
Fish in the Lake

You will need

- a paper bag
- white beans
- red beans

In this investigation you'll **simulate** the capture-recapture method and examine how it works.

The bag represents a lake, the white beans are the untagged fish in the lake, and the red beans will replace white beans to represent tagged fish. Your objective is to estimate the total number of fish in the lake.

Step 1 Reach into the lake and remove a handful of fish to tag. Count and record the number of fish you removed. Replace these fish (white beans) with an equal number of tagged fish (red beans). Return the tagged fish to the lake. Set aside the extra beans.

Step 2 Allow the fish to mingle (seal the bag and shake it). Again remove a handful of fish, count them all, and count the number of tagged fish. In a table like this, record those counts and the ratio of tagged fish to total fish in the sample.

Tagging Simulation

Sample number	Number of tagged fish	Total number of fish	Ratio of tagged fish to total fish
1			
2			

You have taken one sample by randomly capturing some of the fish. You could use this sample to estimate the number of fish in the lake, but by taking several samples, you will get a better idea of the ratio of tagged fish to fish in the lake. Replace the fish, mix them, and repeat the sampling process four more times, filling in a row of your table each time.

LESSON OBJECTIVES

- Work with the idea of sample
- Become familiar with representative samples
- Understand the capture-recapture method

NCTM STANDARDS

CONTENT		PROCESS	
●	Number	●	Problem Solving
●	Algebra	●	Reasoning
	Geometry	●	Communication
	Measurement	●	Connections
●	Data/Probability	●	Representation

	Step 3	Choose one ratio to represent the five ratios. Explain how you decided this was a representative ratio.
	Step 4	If you mixed the fish well, should the fraction of tagged fish in a sample be nearly the same as the fraction of tagged fish in the whole lake? Why or why not?
	Step 5	Write and solve a proportion to find the number of fish in the lake. (About how many beans are in your bag?) Why is this method called capture-recapture? How accurate are predictions using this method? Why?

You can describe the results of capture-recapture situations using percents. Here are three kinds of percent problems—finding an unknown percent, finding an unknown total, and finding an unknown part. In each case, the percent equals the ratio of the part to the whole or total.

EXAMPLE A

Finding an Unknown Percent

In a capture-recapture process, 200 fish were tagged. From the recapture results, the game warden estimates that the lake contains 2500 fish. What percent of the fish were tagged?

largemouth bass

brook trout

yellow perch

▶ Solution

We know that the ratio of tagged fish (the part) to total fish in the lake (the whole) is 200 to 2500. This ratio is equivalent to the percent p of tagged fish in the sample.

$$\frac{p}{100} = \frac{200}{2500}$$ Write the proportion.

$$100 \cdot \frac{p}{100} = \frac{200}{2500} \cdot 100$$ Multiply both sides by 100.

$$p = 8$$ $\frac{100}{100} = 1$. Multiply and divide.

In the samples used for the estimate, 8% of the fish were tagged.

EXAMPLE B

Finding an Unknown Total

In a lake with 250 tagged fish, recapture results show that 11% of the fish are tagged. About how many fish are in the lake?

Step 3 Students may choose the mean or the median or simply one of their samples that is close to one of those values. Each choice is valid if the student can argue that it is representative.

Step 3 Students could choose mean, median, or mode. They could discard the largest and smallest ratio and take the mean of the other three.

Step 4 Yes, if the fish are evenly distributed in all samples and in the entire lake.

Step 5 The fact that the calculator gives an answer to six decimal places doesn't mean the answer is accurate to that many places. Frequent discussion of accuracy and of reporting an appropriate number of digits is important throughout the course. In general, the number of significant digits in the answer should be the same as the smallest number of significant digits that any number in the problem has. (Whole numbers from counting are considered to have infinitely many significant digits. Whole numbers arrived at through measurement have no significant digits to the right of the decimal point.)

Step 5 The proportion will have the ratio of red beans put into the bag to the total unknown number of beans on one side and the ratio determined by the experiment on the other. If several samples are made after thorough mixing, the estimates should be good.

SHARING IDEAS

Have one group present their results from Step 5. **[Ask]** "How representative are the samples?" and "What can we conclude if they are representative?" Elicit the idea that if a sample is representative, then the ratio of red to total beans in the sample should be the same as that ratio in the bag. Use, and ask students to use, the word *sample* repeatedly in discussing the capture-recapture method. In statistics, "population" is frequently used as the total regardless of whether the items are people, other living things, or inanimate objects.

Assessing Progress

Students should be using skills at solving proportions as well as multiplying and dividing. Assess the quality of group work as well.

▶ EXAMPLE A

This example is good for students who need a review of percents. Because the exact number of tagged fish is given, the answer is as good as the estimate of the number of fish in the lake—an estimate made from the experiment.

See page 715 for Step 5 sample data results.

Also a review of percent, this example is like the investigation. Unlike Example A, the answer is an estimate.

► **EXAMPLE C**

This example uses a capture-recapture experiment to review the third standard type of percent problem.

Closing the Lesson

The idea of a **sample,** the **representativeness** of a sample, and the **capture-recapture method** are the principal topics to mention in closing. The capture-recapture method uses proportions to make predictions about a whole population from a sample. The quality of a prediction is determined by how representative the sample is.

BUILDING UNDERSTANDING

Students work with ratios expressing percent and set up proportions to solve.

ASSIGNING HOMEWORK

Essential	1–5
Performance assessment	5
Portfolio	6
Journal	6, 11
Group	1, 2, 7
Review	8–11

► **Helping with the Exercises**

Exercise 1 As needed, keep referring students to the first sentence of the problem as a template.

► **Solution**

We can write 11% as the ratio $\frac{11}{100}$ or 11 parts to 100 (the whole). The variable will be the denominator because the unknown quantity is the whole—the total number of fish in the lake.

$$\frac{11}{100} = \frac{250}{f} \qquad \text{Write the proportion.}$$

$$\frac{100}{11} = \frac{f}{250} \qquad \text{Invert both ratios.}$$

$$250 \cdot \frac{100}{11} = \frac{f}{250} \cdot 250 \qquad \text{Multiply both sides by 250.}$$

$$2273 \approx f \qquad \text{Multiply and divide.}$$

There are approximately 2270 fish in the lake.

EXAMPLE C | **Finding an Unknown Part**

A lake is estimated to have 5000 fish after recapture experiments that showed 3% of the fish were tagged. How many fish were originally tagged?

► **Solution**

We can write 3% as the ratio $\frac{3}{100}$ or 3 parts to 100 (the whole). The variable is in the numerator because the unknown quantity is the number of tagged fish (the part).

$$\frac{t}{5000} = \frac{3}{100} \qquad \text{Write the proportion.}$$

$$5000 \cdot \frac{t}{5000} = \frac{3}{100} \cdot 5000 \qquad \text{Multiply both sides by 5000.}$$

$$t = 150 \qquad \text{Multiply and divide.}$$

About 150 fish were tagged.

In addition to making estimates of wildlife populations, you can use proportions to estimate quantities in many other everyday situations.

EXERCISES

► **Practice Your Skills**

1. The proportion $\frac{320}{235} = \frac{g}{100}$ represents the question "320 is what percent of 235?" Write each proportion as a percent question.

   a. $\frac{24}{w} = \frac{32}{100}$ b. $\frac{t}{450} = \frac{48}{100}$ c. $\frac{98}{117} = \frac{n}{100}$
   32% of what number is 24?   48% of 450 is what number?   What percent of 117 is 98?

2. You can write the question "What number is 15% of 120?" as the proportion   or 98 is what percent of 117? $\frac{x}{120} = \frac{15}{100}$. Write each question as a proportion.
   a. 125% of what number is 80?   a. $\frac{80}{d} = \frac{125}{100}$; b. $\frac{k}{46} = \frac{0.25}{100}$; c. $\frac{72}{470} = \frac{r}{100}$
   b. What number is 0.25% of 46?
   c. What percent of 470 is 72?

3. There are 1582 students attending the local high school. Seventeen percent of the students are twelfth graders. How many twelfth graders are there? 269

4. **APPLICATION** Write and solve a proportion for each situation.

   a. A biologist tagged 250 fish. Then she collected another sample of 75 fish, of which 5 were tagged. How many fish would she estimate are in the lake? $\frac{250}{f} = \frac{5}{75}; f = 3750$ fish

   b. A biologist estimated that there were 5500 fish in a lake in which 250 fish had been tagged. A ranger collected a sample in which there were 15 tagged fish. Approximately how many fish were in the sample the ranger collected? $\frac{250}{5500} = \frac{15}{s}; s = 330$ fish

▶ Reason and Apply

5. Marie and Richard played 47 games of backgammon last month. Marie's ratio of wins to losses was 28 to 19.

   a. Estimate the number of games you expect Marie to win if she and Richard play 12 more games. Explain your thinking. **Marie should win over half the games.**

   b. Write a proportion, and solve for Marie's expected number of wins if she and Richard play 12 more games.

   c. Write a proportion and solve it to determine how many games she and Richard will need to play before Richard can expect to win 30 games. $\frac{19}{47} = \frac{30}{G}; G \approx 74$ games

6. Jon opened a package of candies and counted them. He found 60 candies in the 1.69-ounce package.

   a. Estimate the number of candies in a 1-pound (16-ounce) bag. Explain your thinking for this estimate. **Slightly fewer than 600 pieces. Sixteen ounces is almost ten times 1.69 ounces.**

   b. Write the proportion and solve for the number of candies in a 1-pound package. Compare your approximation to your estimate in 6a.

   c. How much would 1 million candies weigh? Give your answer in pounds.

7. **APPLICATION** The year after a dictionary was published, two librarians notified the publisher of mistakes. One librarian found 43 errors, and the other found 62. The reprint editor noticed that 35 of the errors were mentioned by both librarians. He used the capture-recapture method to estimate the actual number of errors. How many errors are there likely to be? (Hint: The errors found by either librarian can be used as the number of tagged errors in the capture phase. The other librarian represents the recapture phase with the errors found by both being the tagged errors.) ≈ 76 errors

▶ Review

8. The ratio of ninth graders to eighth graders in the class is 5 to 3. Write these ratios as fractions.

   a. Ninth graders to eighth graders $\frac{5}{3}$

   b. Ninth graders to students in the class $\frac{5}{8}$

   c. Eighth graders to students in the class $\frac{3}{8}$

Exercises 5 and 6 Encourage students to estimate before they calculate and to make predictions based on the data provided.

5b. $\frac{28 \text{ games won by Marie}}{28 + 19 \text{ total games}} = \frac{M}{12}$;

$M = 7.15$ or 7 games

6b. $\frac{60 \text{ candy pieces}}{1.69 \text{ ounces}} = \frac{N \text{ pieces}}{16 \text{ ounces}}$

$16 \cdot \frac{60}{1.69} \approx 568$

About 568 candies in a 1-pound bag

6c. $\frac{1,000,000}{568} \approx 1761$ pounds

Exercise 7 [Alert] Many students may find this exercise difficult because they think of capturing as coming earlier in time than recapturing. Urge them to read and think about the hint. You might mention that similar methods are used by law enforcement agencies to estimate numbers of crimes that might go unreported. For example, if a defendant confesses to 50 petty thefts last year, only 40 of which are among the 900 reported, then the police might estimate that there were actually $\left(\frac{50}{40}\right)900 = 1125$ petty thefts last year. Encourage students to bring in other examples in which capture-recapture might help answer the question, "How do they know?"

Exercise 8 The assumption is made that everyone in the class is in either eighth grade or ninth grade.

9a. 1 sulfur atom, 2 hydrogen atoms, and 4 oxygen atoms

9b. 200 hydrogen atoms would combine with 100 sulfur atoms and 400 oxygen atoms.

9c. Use all 400 atoms of oxygen, 200 atoms of hydrogen, and 100 atoms of sulfur to make 100 molecules of sulfuric acid.

10c. Younger than 42, 44, 45, 66, 67, older than 69

10d.

Presidents' Ages at Inauguration

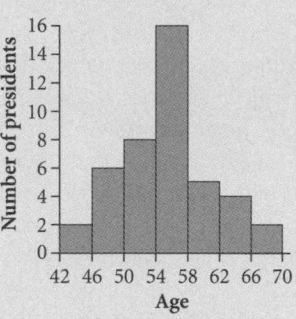

9. **APPLICATION** Sulfuric acid, a highly corrosive substance, is used in the manufacture of dyes, fertilizer, and medicine. Sulfuric acid is also used by artists for metal etching and in aqua tints. H_2SO_4 is the molecular formula for this substance. S stands for the sulfur atom. Use this information to answer each question.

   a. How many atoms of sulfur, hydrogen, and oxygen are in one sulfuric acid molecule?

   b. How many atoms of sulfur would it take to combine with 200 atoms of hydrogen? How many atoms of oxygen would it take to combine with 200 atoms of hydrogen?

   c. If 500 atoms of sulfur, 400 atoms of hydrogen, and 400 atoms of oxygen are combined, how many sulfuric acid molecules could be formed?

10. This histogram shows the ages of the first 43 presidents of the United States when they took office. **54 or 55**

   a. What is a good estimate of the median age?

   b. How many presidents were younger than 50 when they took office? **8**

   c. What ages do not represent the age of any president at inauguration?

   d. Redraw the histogram changing the interval width from 2 to 4.

11. On a group quiz, your group needs to calculate the answer to $12 - 2 \cdot 6 - 3$.
The three other group members came up with these answers:

   Marta   57

   Matt   −3

   Miguel   30

Who, if anyone, is correct? What would you say to the other group members to convince them? **Matt; by the order of operations, you multiply before you subtract.**

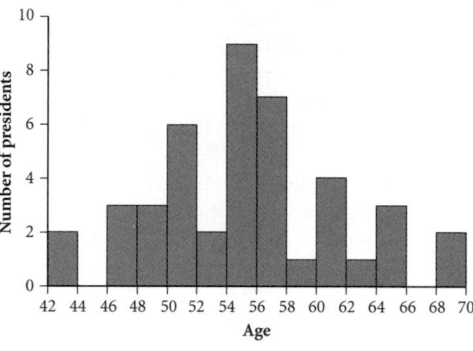

This untitled drypoint and aquatint is by the American artist Mary Cassatt (1844–1926).

Presidents' Ages at Inauguration

IMPROVING YOUR **REASONING** SKILLS

You have two containers of the same size; one contains juice and the other contains water. Remove one tablespoon of juice and put it into the water and stir. Then remove one tablespoon of the water and juice mixture and put it into the juice. Is there more water in the juice or more juice in the water?

IMPROVING **REASONING** SKILLS

If students are having difficulty figuring out that the percent of juice in the water is the same as the percent of the water in the juice, you might suggest that they think about particular amounts of liquid, such as 10 ounces of each with 1 ounce being transferred. Or produce some decks of playing cards. Give each pair of students a pile of ten red cards and a pile of ten black cards. Have one student pull any number of red cards out and mix them among the black cards. The other student pulls out the same number of cards from the other pile and puts them into the red pile. Keeping track of how many of each color are moving, and trying it with extreme cases, might deepen students' understanding of the ratios involved.

LESSON 2.3

Proportions and Measurement Systems

Have you ever visited another country? If so, you needed to convert your money to theirs and perhaps some of your measurement units to theirs as well. Many countries use the units of the Système Internationale, or SI, known in the United States as the metric system.

So instead of selling gasoline by the gallon, they sell it by the liter. Distance signs are in kilometers rather than in miles, and vegetables are sold by the kilo (kilogram) rather than by the pound.

Vegetables are sold at a French market.

 ## Investigation
Converting Centimeters to Inches

You will need
- a yardstick or tape measure
- a meterstick or metric tape measure

In this investigation you will find a ratio to help you convert inches to centimeters and centimeters to inches. Then you will use this ratio in a proportion to convert some measurements from the system standard in the United States to measurements in the metric system, and vice versa.

Step 1 Measure the length or width on each of six different-sized objects, such as a pencil, a book, your desk, or your calculator. For each object, record the inch measurement and the centimeter measurement in a table like this.

Inches to Centimeters

Object	Measurement in inches	Measurement in centimeters

Step 2 Enter the measurements in inches in your calculator's list L1 and the measurements in centimeters in list L2. In list L3 enter the ratio of centimeters to inches, $\frac{L2}{L1}$, and let your calculator fill in the ratio values. [▶ See **Calculator Note 1I.** ◀]

Ratios should be about $\frac{2.54 \text{ cm}}{1 \text{ in.}}$. **Step 3** How do the ratios of centimeters to inches compare for the different measurements? If one of the ratios is much different from the others, recheck your measurements. Choose a single representative ratio of centimeters to inches.

NCTM STANDARDS

CONTENT	PROCESS
• Number	• Problem Solving
• Algebra	• Reasoning
Geometry	• Communication
• Measurement	• Connections
Data/Probability	• Representation

LESSON OBJECTIVES

- Review the English measurement system and the metric system
- Convert measurement units with conversion factors
- Convert measurement units with dimensional analysis

PLANNING

LESSON OUTLINE

One day:
25 min   Investigation
10 min   Sharing
5 min   Example
5 min   Closing
5 min   Exercises

MATERIALS

- yardstick or tape measure (1 per group)
- meterstick or metric tape measure (1 per group)
- Calculator Note 1I

TEACHING

As they convert between different measurement units in this lesson, students use ratios called *conversion factors*. Products of several conversion factors can be set up to do more complicated conversions, in a process called *dimensional analysis*.

The name *Système Internationale* is French for the *International System* of measurement units. The SI is used in most science textbooks.

Guiding the Investigation

One step [Ask] "In as many ways as possible, decide how many centimeters are equivalent to an inch." As you circulate, encourage thinking beyond measuring the yardstick with the meter stick.

Step 1 Be sure that students understand how to read the inch markings. Often the fractional parts of an inch give students trouble.

Step 4 Some students will have trouble seeing a single number like 2.54 as a *ratio*. You'll need to remind them that the ratio is $\frac{2.54}{1}$. Because you're usually interested in knowing how many of one type of unit fit in 1 unit of a different type, most conversion factors have a 1 in the denominator.

Step 5 Be sure all students see that the more inches they have, the larger the number of centimeters, that the differential between the measurements is "steady" (that is, linear), and that it applies even to inch-centimeter comparisons that they did not examine.

Encourage students who can solve parts c and d without proportions, but challenge them to set up and solve proportions as a check.

SHARING IDEAS

Students might report results of Steps 3 and 5. As they do so, introduce the term *conversion factor*.

Point out that conversion factors can be used to convert within a measurement system as well as between systems. For example, to change yards to inches, you'd use the conversion factor 36 inches/yard. **[Ask]** "In which measurement system—SI or English—is it easier to convert between units?" [In the SI system, conversion factors within the system are all powers of 10.]

Assessing Progress
Assess students' skills at setting up and solving proportions, entering calculator data, measuring carefully, and factoring.

▶ **EXAMPLE**

This example develops the idea of dimensional analysis, which combines conversion factors into a product when no direct conversion factor is available.

Dimensional analysis requires deciding what conversion factors to multiply on the basis of how their units will "cancel" to leave the desired units. Some students

may need reminding of how to cancel factors when multiplying fractions. It may help to express 60 min as 60 · 1 min, for example, and show that $\frac{1\ min}{1\ min}$ "cancels" to a factor of 1. Suggest that students write the conversion factors in a way that makes clear which units will cancel, and check carefully for errors.

Some students will hesitate to insert units, either because they're not sure they have the freedom to do so or because they "don't want to make the problem harder." Show them how converting units in stages can help them keep track of the process.

Step 4 Choose a single representative ratio of centimeters to inches. Write a sentence that explains the meaning of this ratio. One inch is 2.54 cm.

Step 5 Using your ratio, set up a proportion and convert each length.

a. 215 centimeters = x inches 84.6 in.

b. 1 centimeter = x inches 0.4 in.

c. 1 inch = x centimeters 2.54 cm

d. How many centimeters high is a doorway that measures 80 inches? 203 cm

In the investigation you found a common ratio or **conversion factor** between inches and centimeters. Once you've determined the conversion factor, you can convert from one system to the other by solving a proportion. If your measurements in the investigation were very accurate, the mean and median of the ratios were very close to the conversion factor, 2.54 centimeters to 1 inch.

Some conversions require several steps. The example offers a strategy called **dimensional analysis** for doing more complicated conversions.

EXAMPLE A radio-controlled car traveled 30 feet across the classroom in 1.6 seconds. How fast was it traveling in miles per hour?

▶ **Solution**

Science
CONNECTION

Chemists often use dimensional analysis. For each chemical compound, 1 mole equals the compound's gram molecular weight. For water there are 18 grams per mole. If the density of water is 1 gram per milliliter, what is the volume of 1 mole of water in milliliters?

$$1\ mole \cdot \frac{18\ g}{1\ mole} \cdot \frac{1\ ml}{1\ g} = 18\ ml$$

Using the given information, you can write the speed as the ratio $\frac{30\ feet}{1.6\ seconds}$. Multiplying by 1 doesn't change the value of a number, so you can use conversion factors that you know (like $\frac{60\ minutes}{1\ hour}$) to create fractions with a value of 1. Then multiply your original ratio by those fractions to change the units.

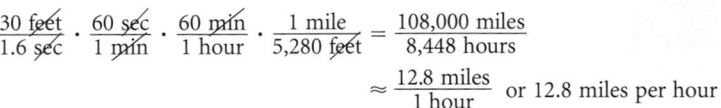

$$\frac{30\ \text{feet}}{1.6\ \text{sec}} \cdot \frac{60\ \text{sec}}{1\ \text{min}} \cdot \frac{60\ \text{min}}{1\ \text{hour}} \cdot \frac{1\ mile}{5{,}280\ \text{feet}} = \frac{108{,}000\ miles}{8{,}448\ hours}$$

$$\approx \frac{12.8\ miles}{1\ hour}\ \text{or 12.8 miles per hour}$$

Each of the fractions after the first one has a value of 1 because the numerator and denominator of each fraction are equivalent: 60 sec = 1 min, 60 min = 1 hr, and 1 mi = 5280 ft. The fractions equivalent to 1 are chosen so that when units cancel, the result is miles in the numerator and hours in the denominator.

Closing the Lesson

Conversion factors are ratios that allow the change from one unit to another. If a single conversion factor isn't available, convert by multiplying conversion factors in **dimensional analysis.**

Mention again that each conversion factor used in dimensional analysis really has a value of one. (For example, $\frac{60\ seconds}{minute}$ means $\frac{60\ seconds}{1\ minute}$ and both are equal to 1, so multiplying by the ratio doesn't change the value of an expression.)

EXERCISES

Practice Your Skills

1. Find the value of x in each proportion.

a. $\dfrac{1 \text{ meter}}{3.25 \text{ feet}} = \dfrac{15.2 \text{ meters}}{x \text{ feet}}$ $x = 49.4$

b. $\dfrac{1.6 \text{ kilometers}}{1 \text{ mile}} = \dfrac{x \text{ kilometers}}{25 \text{ miles}}$ $x = 40$

c. $\dfrac{0.926 \text{ meter}}{1 \text{ yard}} = \dfrac{200 \text{ meters}}{x \text{ yards}}$ $x \approx 216$

d. $\dfrac{1 \text{ kilometer}}{0.6 \text{ mile}} = \dfrac{x \text{ kilometers}}{350 \text{ miles}}$ $x = 583.\overline{3}$

2. Which ratio in problem 1a is used as the conversion factor? Use it to determine how many feet you would have to run in a 50-meter dash. $\dfrac{1 \text{ meter}}{3.25 \text{ feet}}$; 162.5 feet

3. Use dimensional analysis to change

a. 50 meters per second to kilometers per hour.

b. 0.025 day to seconds. $0.025 \text{ day} \cdot \dfrac{24 \text{ hr}}{1 \text{ day}} \cdot \dfrac{60 \text{ min}}{1 \text{ hr}} \cdot \dfrac{60 \text{ sec}}{1 \text{ min}} = 2160 \text{ sec}$

c. 1200 ounces to tons (16 oz = 1 lb; 2000 lb = 1 ton).

4. Write a proportion and answer each question using the conversion factor 1 ounce $\approx$ 28.4 grams.

a. How many grams does an 8-ounce portion of prime rib weigh? 227 g

b. If an ice-cream cone weighs 50 grams, how many ounces does it weigh? 1.76 oz

c. If a typical house cat weighs 160 ounces, how many grams does it weigh? 4544 g

d. How many ounces does a 100-gram package of cheese weigh? 3.52 oz

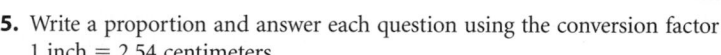

5. Write a proportion and answer each question using the conversion factor 1 inch = 2.54 centimeters.

a. A teacher is 62.5 inches tall. How many centimeters tall is she? 158.8 cm

b. A common ceiling height is 96 inches (8 feet). About how high is this in centimeters? 244 cm

c. The diameter of a CD is 12 centimeters. What is its diameter in inches? 4.72 in.

d. The radius of a typical soda can is 3.25 centimeters. What is its radius in inches? 1.28 in.

BUILDING UNDERSTANDING

Students practice applying conversion factors and dimensional analysis.

ASSIGNING HOMEWORK

Essential	1, 3–5, 7, 9
Performance assessment	6
Portfolio	8
Journal	11
Group	4
Review	11–13

▶ Helping with the Exercises

Exercise 4 Some students may be learning in their science classes that there is a fundamental difference between measures of weight (such as ounces, pounds, and tons) and measures of mass (such as grams). Weight depends on the pull of gravity whereas mass does not. The conversion factor 1 ounce $\approx$ 28.4 grams is true only where the pull of gravity is approximately equivalent to Earth's.

MAKING THE CONNECTION

In their science classes, students may be learning that distance, length, mass, time, and temperature are *fundamental quantities*, and that, for example, volume (length³) and speed (distance/time) are *derived* quantities. You might use that terminology also.

3a. $\dfrac{50 \text{ m}}{1 \text{ sec}} \cdot \dfrac{1 \text{ km}}{1000 \text{ m}} \cdot \dfrac{60 \text{ sec}}{1 \text{ min}} \cdot \dfrac{60 \text{ min}}{1 \text{ hr}} = 180 \text{ km/hr}$

3c. $1200 \text{ oz} \cdot \dfrac{1 \text{ lb}}{16 \text{ oz}} \cdot \dfrac{1 \text{ ton}}{2000 \text{ lb}} = 0.0375 \text{ ton}$

Reason and Apply

6. A group of students measured several objects around their school in both yards and meters. Use their data, shown in the table, to find a conversion factor between yards and meters.

Measurement in Yards and Meters

Yards	7	3.5	7.5	4.25	6.25	11
Meters	6.3	3.2	6.8	3.8	5.6	9.9

Use the conversion factor to answer these questions:

a. The length of a football field is 100 yards. How long is it in meters? 90 m

b. If it is 200 meters to the next freeway exit, how far away is it in yards? 222 yd

c. How many yards long is a 100-meter dash? 111 yd

d. How many meters of fabric should you buy if you need 15 yards? 13.5 m

7. a. Make a table like this showing the number of feet in lengths from 1 to 5 yards.

   b. For each additional yard in your table, how many more feet are there? **For each additional yard there are 3 more feet.**

   c. Write a proportion that you could use to convert the measurements between y yards and f feet. $\dfrac{f}{y} = \dfrac{3}{1}$

   d. Use the proportion you wrote to convert each measurement.

   450 feet **i.** 150 yards $= f$ feet   **ii.** 384 feet $= y$ yards 128 yards

Measurement in Yards and Feet

Yards	1	2	3	4	5
Feet	3	6	9	12	15

8. A rod is a unit of measure that was used many years ago.

   a. Using the table, find a common ratio that you can use to convert units between rods and feet. 1 rod = 16.5 feet

   b. Write a proportion using the ratio you found, and convert each measurement.

   57.75 feet **i.** 3.5 rods $= x$ feet   **ii.** 15 feet $= x$ rods 0.9 rod

Measurement in Rods and Feet

Rods	1.2	2	3	4
Feet	20	33	50	66

9. **APPLICATION** When mixed according to the directions, a 12-ounce can of lemonade concentrate becomes 64 ounces of lemonade.

   a. How many 12-ounce cans of concentrate are needed to make 120 servings if each serving is 8 ounces? **Fifteen 12-oz cans to make 960 oz.**

   b. How many ounces of concentrate are needed to make 1 ounce of lemonade? $\dfrac{12}{64}$ or 0.1875 oz

   c. Write a proportion that you can use to find the number of ounces of concentrate based on the number of ounces of lemonade wanted.

   d. Use the proportion you wrote to find the number of ounces of lemonade that can be made from a 16-ounce can of the same concentrate. $\dfrac{16}{L} = \dfrac{12}{64}; L \approx 85$ oz

10. Recipes in many international cookbooks use metric measurements. One cookie recipe calls for 120 milliliters of sugar. How much is this in our customary unit "cups"? (There are 1000 milliliters in a liter, 1.06 quarts in a liter, and 4 cups in a quart.)

9c. $\dfrac{\textit{number of ounces of concentrate}}{\textit{number of ounces of lemonade}} = \dfrac{12}{64}$

10. $120 \text{ ml} \cdot \dfrac{1 \text{ liter}}{1000 \text{ ml}} \cdot \dfrac{1.06 \text{ qt}}{1 \text{ liter}} \cdot \dfrac{4 \text{ cups}}{1 \text{ qt}} = 0.5088$ cup

or about $\frac{1}{2}$ cup

Exercises 7b and 9b These lay the foundation for the concept of linear growth and slope.

Exercise 8 Rods were used to measure land. This exercise can be used as a springboard to discuss other historical units of measure, such as *cubits* (the length of forearm from the elbow to the end of the middle finger, 17 to 21 inches) and *hands* (the width of a hand, about 4 inches).

► Review

11. The students in the mathematics and chess clubs worked together to raise funds for their respective groups. Together the clubs raised $480. There are 12 members in the Mathematics Club and only 8 in the Chess Club. How should the funds be divided between the two clubs? Explain your answer. **If the profits are divided in proportion to the number of students in the clubs, the Math Club would get $288, leaving $192 for the Chess Club.**

12. Draw a coordinate plane.

   a. Plot and label these points.
   $A(2, 1)$, $B(1, -4)$, $C(0, 4)$, $D(-1, 0)$, $E(-2, -1)$, $F(-3, 1)$, $G(4, 2)$

   b. Three of the points lie on a line. Name them and draw the line through them. **E A G**

   c. Name the point whose coordinates are the median of the x-values and the median of the y-values. Plot it and label it M. **$M(0, 1)$**

13. The box plot shows the length in centimeters of five members of the kingfisher family. Use this information to match each kingfisher to its length.

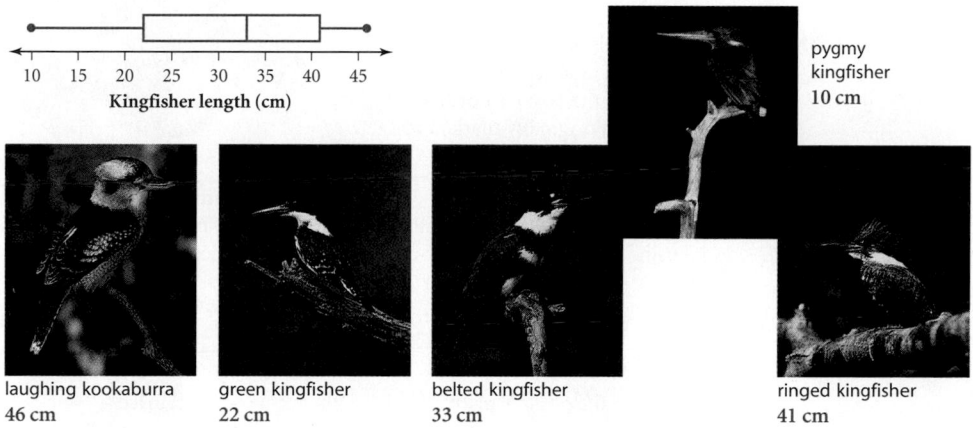

Kingfisher length (cm)

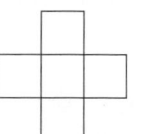
pygmy kingfisher
10 cm

laughing kookaburra
46 cm

green kingfisher
22 cm

belted kingfisher
33 cm

ringed kingfisher
41 cm

- These kingfishers range in size from the tiny pygmy kingfisher to the laughing kookaburra.
- The best known kingfisher, the belted kingfisher, breeds from Alaska to Florida. It is only 2.6 centimeters longer than the mean kingfisher length.
- The ringed kingfisher, a tropical bird, is much closer to the median length than the green kingfisher.

Exercise 11 This exercise has many possible answers. It can lead to a good group discussion.

12a.

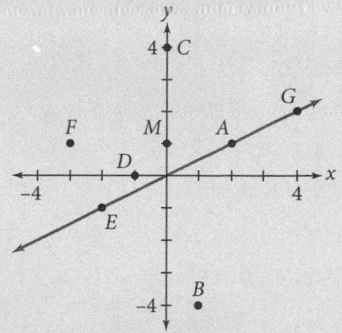

IMPROVING YOUR **VISUAL THINKING** SKILLS

A pentomino is made up of five squares joined along complete sides. The first pentomino can be folded into an open box. The second pentomino can't.

Draw all 12 unique pentominoes, and then identify those that can be folded into open boxes.

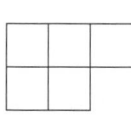

IMPROVING **VISUAL THINKING** SKILLS

Two pentominos are considered the same if either can be flipped over (reflected) or turned (rotated) to match the other. The 12 distinct, or unique, pentominos are often named after letters of the alphabet they resemble. Of these, the F, L, N, T, W, X, Y, and Z pentominos fold into an open box.

PLANNING

LESSON OUTLINE

One day:

5 min	Example A
5 min	Example B
20 min	Investigation
5 min	Sharing
5 min	Closing
10 min	Exercises

MATERIALS

- graph paper
- rulers
- Calculator Notes 1I and 1J

TEACHING

You can use proportions to calculate prices after discounts or taxes. You can use proportions with similar rectangles to find that the ratio of their areas is the square of the ratio of the lengths of their corresponding sides.

▶ **EXAMPLE A**

Previously students may have taken 35% of $34.99 (that is, 0.35 times 34.99) to find that the amount of discount is $12.25, and then subtracted this amount from the original price. By first finding the percent of the price retained (that is, 100% minus 35%), students can simplify the calculations. In fact, they might just multiply 0.65 by 34.99 without setting up a proportion.

LESSON

2.4

Be bold. If you're going to make a mistake, make a doozy, and don't be afraid to hit the ball.

BILLIE JEAN KING

Increasing and Decreasing

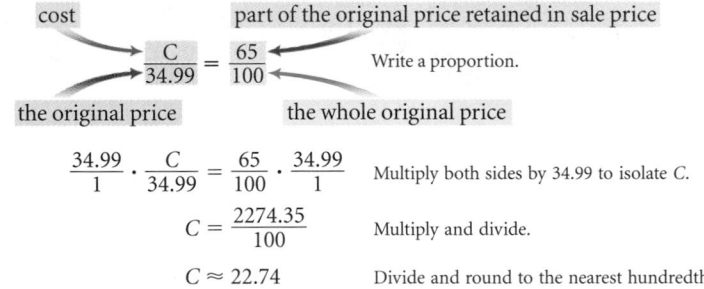

Emily sees a sign saying "Special! Every CD is marked down 25%." She wants to know the discounted price on a $17.99 CD. Nathan works as a bagger in a grocery store, where he makes $8.25 per hour. His boss announces that all employees are getting a 2.5% raise. Nathan wants to know his new hourly rate. $13.49; $8.46

These are different situations, but the pattern of reasoning to find solutions is the same for both. Example A uses a proportion to calculate a marked-down price.

EXAMPLE A | Birdbaths at the Feathered Friends store are marked down 35%. What is the cost of a birdbath that was originally marked $34.99?

▶ **Solution** | If an item is marked down 35%, then it must retain $100 - 35$ percent of its original price. That is, it will cost 65% of the original price. Set up a proportion using 65% and the ratio of cost, C, to original price.

cost   part of the original price retained in sale price

$$\frac{C}{34.99} = \frac{65}{100}$$ Write a proportion.

the original price   the whole original price

$$\frac{34.99}{1} \cdot \frac{C}{34.99} = \frac{65}{100} \cdot \frac{34.99}{1}$$ Multiply both sides by 34.99 to isolate C.

$$C = \frac{2274.35}{100}$$ Multiply and divide.

$$C \approx 22.74$$ Divide and round to the nearest hundredth.

So the cost after the 35% markdown is $22.74.

Percent-increase calculations use the same reasoning as percent decreases. If there is a 6% sales tax on an item, then the cost including tax will be 106% of the price. (Remember: When a quantity increases in size, whether it is a price or a population, the new number will be greater than 100% of the original number.)

EXAMPLE B | Suppose you work at the Feathered Friends store. It's your job to make a chart showing the cost of the eight varieties of birdseed including the 6% sales tax.

LESSON OBJECTIVES

- Calculate percent increase and decrease
- See that the ratio of areas of similar rectangles is the square of the ratio of their corresponding sides

NCTM STANDARDS

CONTENT		PROCESS	
•	Number	•	Problem Solving
•	Algebra	•	Reasoning
•	Geometry	•	Communication
•	Measurement	•	Connections
	Data/Probability	•	Representation

▶ **Solution**

All items are taxed at the same 6% rate, so the calculation will be the same for each birdseed variety. Whenever you have to repeat the same procedure over and over, it may be easier to let your calculator do the work for you in a list. You can enter the price of the feeds into list L1. In list L2 enter the correct formula for the cost including tax.

To find the formula to enter into list L2, write a proportion. Multiply both sides by L1.

$$\frac{L_2}{L_1} = \frac{106}{100}$$

$$L_2 = L_1 \cdot \frac{106}{100}$$

Price	Cost including 6% tax
8.25	
9.99	
10.75	
12.99	
15.99	
16.50	
17.95	
34.99	

The entries in list L2 will be the entries in list L1 times $\frac{106}{100}$, or 1.06. [▶ 🖳 See **Calculator Note 1I.** ◀] You can set the mode on the calculator to show only two decimal places for all numbers. [▶ 🖳 See **Calculator Note 1J.** ◀]

L1	L2
8.25	8.75
9.99	10.59
10.75	11.40
12.99	13.77
15.99	16.95
16.50	17.49
17.95	19.27
34.99	37.09

Investigation
Enlarging and Reducing

You will need
- graph paper
- a ruler

Graphic artists and designers frequently make reductions and enlargements of drawings. In this investigation you'll explore how the dimensions change when a picture is enlarged or reduced.

Step 1 On graph paper, carefully draw a rectangle and one of its diagonals. Record its length, width, diagonal length, perimeter, and area in a table like this one.

Rectangle Measurements

	Original dimensions	Enlarged dimensions	Reduced dimensions
Length			
Width			
Diagonal			
Perimeter			
Area			

Step 2 Suppose you are a designer and need to make a copy that increases the length and width by 20%. Draw this new rectangle with its diagonal. Record the new enlarged length, width, diagonal, perimeter, and area.

Step 3 Now suppose you need a copy that decreases the length and width of the original rectangle by 20%. Draw this new rectangle with its diagonal. Record its new length, width, diagonal, perimeter, and area.

▶ **EXAMPLE B**

Similarly, students can simplify calculations by finding a single multiplier for percent increase. A proportion can help find that multiplier. Then students can use calculator lists to calculate the after-tax cost of an entire list of items. Students who need practice multiplying might do some of these calculations by hand.

Guiding the Investigation

One step Give students the problem in Step 7. Welcome any disagreement in intuition; some controversy at the beginning can lead to deeper understanding. Urge students to consider a variety of cases, and require them to justify their thinking.

[ESL] Students may have trouble with the vocabulary. Display these words and talk about their meanings: *decrease, increase, dimensions, diagonal, perimeter,* and *area.*

Step 1 As needed, remind students that the area of a rectangle is the product of its length and width. Drawing a grid of boxes inside a rectangle might help them remember. **[Alert]** Many students believe that the perimeter and the area of a rectangle increase and decrease together— that you can't have two rectangles of equal perimeter with one larger in area. **[Ask]** "Can two rectangles with the same area have different perimeters?" [Yes, one example is a 3-by-4 rectangle with a perimeter of 14 and a 1-by-12 rectangle with a perimeter of 26.]

Step 4 Students may not notice that the ratio of areas is the square of the ratio of corresponding sides—one of the most important principles in mathematics.

Step 4 All the linear measurements increase by 1.2 or decrease by 0.8. The area increases by 1.44 or decreases by 0.64.

Step 5 Linear measurements increase by the same percent. Area increases by the percent squared.

Step 6 The area-ratio principle doesn't apply here, because the figures aren't similar. The percent increase is different in the two dimensions.

Step 6 Distorted. Everything is too long for the width.

Step 7 Encourage experimentation with tabletops of different dimensions to see that the area-ratio principle always holds.

Step 7 It will take 25 dominoes.
$$A_1 = lw;\ A_2 = \frac{1}{2}\,l \cdot \frac{1}{2}\,w = \frac{1}{4}\,lw$$

SHARING IDEAS

If no group has discerned the area-ratio principle as the square of side ratios, have several groups display their data and lead the class in finding the principle. Otherwise, have at least one presentation from a group that has found the principle. Ask for explanations.

Assessing Progress

Through your observations you can assess students' abilities to set up and solve **proportions,** to measure carefully, to organize data, and to convert between **percents** and decimals. You can also note their familiarity with **area** and **perimeter** of a rectangle.

Closing the Lesson

Summarize the ideas about calculating **percent increase** and **decrease** and the **area-ratio principle.**

Step 4 | Calculate the ratio of each new rectangle value to the corresponding original value. Describe how the ratios compare. Check your results with others in your group.

Step 5 | Describe how other aspects of the figure, such as the diagonal, perimeter, and area, are changed when you increase or decrease the length and width of a rectangle by the same percent.

Step 6 | Suppose the company that prints your school photos has a problem with its printing machines. It takes the original photo. But instead of increasing the length and width by 50% to make a large photo, it increases the length by 80% and the width by 30%. Describe what your large photo looks like.

Step 7 | A tabletop can be completely covered by 100 dominoes. If the length and width are reduced by half, will it take 50 dominoes to cover it? If not, will it take more or fewer dominoes? Explain your thinking.

EXERCISES

You will need your calculator for problems **3, 5,** and **8.**

▶ Practice Your Skills

1. If the length and width of a picture each have 150% of their value added to them, which of the statements are true and which are false?

F **a.** The picture is 50% wider and higher than it was before.

T **b.** The height and width of the picture are $2\frac{1}{2}$ times that of the original.

F **c.** The area of the picture is $2\frac{1}{2}$ times the area of the original.

T **d.** The area is more than 6 times the area of the original.

2. Write each percent change as a ratio comparing the result to the original quantity. For example, a 3% increase is $\frac{103}{100}$.

a. 8% increase $\frac{108}{100}$

b. 11% decrease $\frac{89}{100}$

c. 12.5% growth $\frac{1125}{1000}$ or $\frac{112.5}{100}$

d. $6\frac{1}{4}$% loss $\frac{9,375}{10,000}$ or $\frac{93.75}{100}$

e. x% increase $\frac{100 + x}{100}$

f. y% decrease $\frac{100 - y}{100}$

3. **APPLICATION** In 2000 the population of the United States was estimated to be 274,700,000. If the population grows by 1.1% per year, what would the population be in 2001? In 2002? In 2003? (Round to the nearest 1000 people.)

2001: 277,722,000
2002: 280,777,000
2003: 283,865,000

4. APPLICATION Justin wants to buy a computer. His mother will also use it, so she has agreed to pay 30% of the cost. The model Justin wants costs $2,649 including tax. How much will Justin need to pay? **$1,854.30**

5. This table shows the populations of the five most populous countries in the world in 2000.

Most Populous Countries in the World in 2000

Country	Population
China	1,261,832,000
India	1,014,004,000
United States	275,563,000
Indonesia	224,784,000
Brazil	172,860,000

a. The population of the world in 2000 was approximately 6,080,142,000. What percent of the world population was the total population of these five countries in 2000? **approximately 49%**

b. The population of India in 1990 was approximately 850,558,000. Approximately what percent increase does this indicate? **approximately 19.2%**

c. From 1990 to 2000, the population of China increased by roughly 10.8%. If the rate of population growth remains the same, what will the population be in 2010? **approximately 1,398,110,000**

▶ Reason and Apply

6. APPLICATION Raimy recently received a partial scholarship for college. Her first-year expenses, including tuition, textbooks, food, and housing, amount to $18,500. Her scholarship will cover 36% of her expenses. Raimy's parents say they can pay 35% of the balance. How much money will Raimy be responsible for? **$7696.**

7. APPLICATION Tamara works at a bookstore, where she earns $7.50 per hour.

a. Her employer is pleased with her work and gives her a 3.5% raise. What is her new hourly rate? **$7.76**

b. A few weeks later business drops off dramatically. The employer must reduce wages. He decreases Tamara's latest wage by 3.5%. What is her hourly rate now? **$7.49**

c. What is the final result of the two pay changes? Explain. **Her wage has dropped by $0.01 per hour because the increase was calculated as 3.5% of $7.50, but the decrease was based on $7.76.**

BUILDING UNDERSTANDING

These exercises give practice at calculating percent increase and decrease and using the area-ratio principle.

ASSIGNING HOMEWORK

Essential	1–6
Performance assessment	8
Portfolio	3
Journal	6
Group	5, 7
Review	9–11

▶ Helping with the Exercises

Exercise 3 This problem introduces the idea of recursively calculating the terms of a geometric sequence. This concept will be extended in Chapter 7.

Exercise 7 This problem shows that a 3.5% increase followed by a 3.5% decrease does not get you back to where you started. Ask students to keep all the decimal places of their calculation on calculators so they won't think the small difference is due to round-off error. This is not an easy concept for students to understand. It highlights that you must always know 3.5% of what, not just 3.5%. Get students having trouble to try it with 10% of 200 minus 10% of 220: $200 + 20 - 22$ is not back to 200. You might also draw a line segment, add 10% to it, and then point out that 10% of the longer one is more than the amount added.

8a. $3,115,200,000

8c. 20th Century Fox   31.38%
     MCA Universal   14.94%
     Sony   21.32%
     Time Warner   54.7%
     Viacom   1306.9%

8d. Revenue in 1993: $5,953.01 million, 2.5% increase

8. APPLICATION This table shows the revenues for six major film studios in 1992 and 1996.

  a. Write out $3,115.2 million as a whole number showing all the zeros.

  b. What was the increase in revenue for Walt Disney Company between 1992 and 1996? What percent is that of the 1992 profits? 7390.3 million; 237%

  c. Calculate the percent increase over the four-year period for the other studios.

  d. Time Warner reported a 50.9% increase in revenue from 1992 to 1993. What was the percent increase from 1993 to 1996?

  e. In 1994, Time Warner reported a 14.8% decrease in revenue from the previous year. What was its percent gain from 1994 to 1996? Revenue in 1994: $5,071.96 million, 20.3% increase

Film Studio Revenues

Film Studio	Revenue (in millions of dollars)	
	1992	1996
Walt Disney Company	3,115.2	10,505.0
20th Century Fox (News Corporation)	1,858.0	2,441.0
MCA Universal (Seagram)	4,709.4	5,413.0
Sony	2,475.3	3,003.0
Time Warner	3,945.0	6,103.0
Viacom	248.3	3,493.4

(*The Veronis, Shuler & Associates Communications Industry Report,* October 1997, p. 155)

Review

9. This histogram gives the height of the students in Mr. Moore's algebra class. Create a data set that corresponds to the graph.

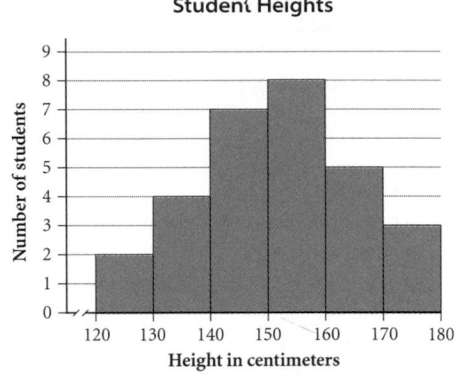

Student Heights

Only the first two digits of the values are known.
Possible answer: {124, 127, 131, 134, 134, 137, 140, 142, 144, 146, 147, 148, 148, 151, 152, 153, 155, 156, 158, 158, 158, 160, 163, 165, 166, 168, 174, 174, 177}

10. **APPLICATION** Ryan needs to be able to easily make larger or smaller portions of the pancake recipe used in the Pancake Company, where he is a chef. Here is a matrix he created of the ingredients:

$$\begin{bmatrix} 150 \\ 2 \\ 18 \\ 3.5 \\ 200 \\ 19 \\ 125 \end{bmatrix} \begin{matrix} \text{Cups flour} \\ \text{Cups salt} \\ \text{Cups sugar} \\ \text{Cups baking powder} \\ \text{Eggs} \\ \text{Cups melted butter} \\ \text{Cups milk} \end{matrix}$$

He wants to be able to make a half-size recipe, a recipe that is 50% larger than the original, and a double recipe.

a. What 1×3 matrix should Ryan multiply the ingredient matrix by to calculate how much of each ingredient to use in the different-size portions? [0.5 1.5 2]

b. Ryan can't remember if the order matters when he multiplies matrices. In what order should he multiply the original matrix and the 1×3 matrix to get a meaningful answer?

c. Calculate the resulting ingredient matrix, and label its rows and columns.

11. Draw a coordinate plane on graph paper. Label the axes and each of the four quadrants—I, II, III, and IV.

a. Mark points $W(5, 3)$ and $Q(-4, -2)$, and label them with their letter names. Sketch a rectangle that has diagonally opposite vertices W and Q. List the coordinates of the other two vertices, and identify the axis or quadrant location of each.

b. Sketch at least one other rectangle that has W and Q as vertices. List the coordinates of the other two vertices and identify their locations. **Possible answer: There are two squares and infinitely many rectangles that can be formed.**

IMPROVING YOUR REASONING SKILLS

Three children went camping with their parents, a dog, and a tin of cookies just for the children. They agreed to share the cookies equally.

The youngest child couldn't help thinking about the cookies, so alone she divided the cookies into three piles. There was one left over. She gave it to the dog, took her share, and left the rest of the cookies in the cookie tin.

A little later the middle child took the tin where he could be alone and divided the cookies into three piles. There was one left over. He gave it to the dog and took his share.

Not too long after that, the oldest child went alone to divide the cookies. When she made three piles, there was one left over, which the dog got. She took her share and put the rest back in the tin.

After dinner, the three children "officially" divided the contents of the cookie tin into three piles. There was one left over, which they gave to the dog.

What is the smallest number of cookies that the tin might have first contained?

Exercise 10 This review of Lesson 1.8 requires multiplication of matrices. As Example B in that lessons shows, multiplying a 7×1 matrix by a 1×3 matrix gives a 7×3 matrix.

10b. The ingredient matrix must be on the left, and the sizing matrix on the right. So you are multiplying a 7×1 matrix by a 1×3 matrix.

10c.

Half	50% larger	Double
75	225	300
1	3	4
9	27	36
1.75	5.25	7
100	300	400
9.5	28.5	38
62.5	187.5	250

11a.

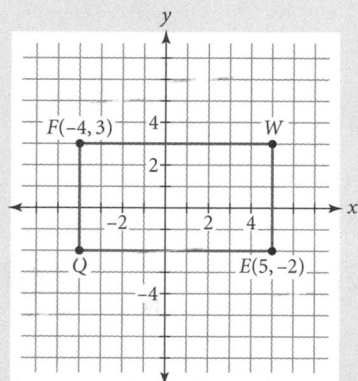

11b.

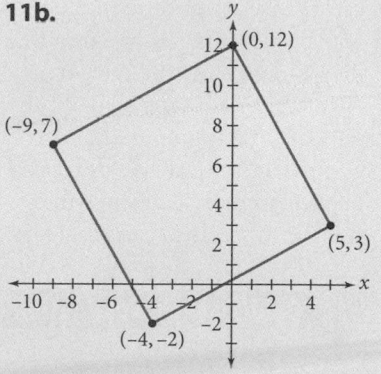

IMPROVING REASONING SKILLS

This problem can be solved in several ways. Working backward is probably the method most students will use. As they work backward, students will realize that some numbers won't work. For example, for each child and the dog to get 1 cookie at the end, there must have been 7 cookies when the last child gives 1 to the dog and takes one-third. But 7 is not two-thirds of any whole number, so each child must have gotten at least 2 in the end, and so on, up to 7, the smallest number that works.

Another approach is to create a function and put it into a graphing calculator. At the last split, the number of cookies each child will receive can be given by the formula $y = \frac{1}{3}\left(\frac{2}{3}\left(\frac{2}{3}\left(\frac{2}{3}(x - 1) - 1\right) - 1\right) - 1\right)$. Looking at the table for this equation, you see that 79 is the first value of x that gives an integral value for y (when $x = 79$, $y = 7$).

Circle Graphs and Relative Frequency Graphs

Circle graphs, like bar graphs, summarize data in categories. **Relative frequency graphs** also summarize data in categories, but instead of including the actual number for each category, they compare the number in that category to the total for all the categories. Relative frequency graphs show fractions or percents, not values.

PLANNING

LESSON OUTLINE

One day:

25 min   Investigation

5 min   Sharing

10 min   Example

5 min   Closing

5 min   Exercises

MATERIALS

- graph paper
- protractors
- compasses, circle templates, or Circle Graphs (W)
- rulers
- Calculator Note 2E for TI-73 only, *optional*
- Protractors (T), *optional*

TEACHING

Circle graphs, or pie charts, are good for showing the size of each category relative to the whole. Like histograms and bar graphs, they organize data by categories or intervals. With the help of proportions, circle graphs and bar graphs can be relabeled as *relative frequency graphs* that show fractions or percents in each category.

Guiding the Investigation

One step   Display a relative frequency circle graph and ask students to make one representing the data given by the bar graph in the investigation. As you circulate, encourage students to use calculator lists to convert frequencies to degrees all at once.

See page 715 for answers to Steps 2 and 4.

Investigation
Circle Graphs and Bar Graphs

You will need

- graph paper
- a protractor
- a compass or circle template
- a ruler

The bar graph shows the approximate land area of the seven continents.

Continental Land Areas

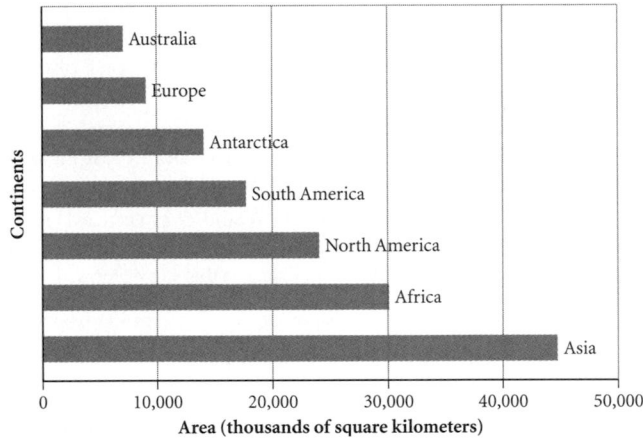

Step 1	Determine from the bar graph the approximate area of each continent and the total land area.
Step 2	Convert the data in the bar graph to a circle graph. Use the fact that there are 360 degrees in a circle. Write proportions to find the number of degrees in each sector of the circle graph.
Step 3	Convert the data in the bar graph to a relative frequency circle graph. Instead of showing the land area, the graph will show percents of total land area.
Step 4	Convert the data in the bar graph to a relative frequency bar graph that shows percents rather than land areas.
Step 5	Compare the graphs you made with the original graph. What advantages are there to each kind of graph?

Step 1 Approximate answers in millions of km²: Australia 7; Europe 9; Antarctica 14; South America 18; North America 24; Africa 30; Asia 45; Total 147.

Steps 3 and 4 As needed, point out that these steps require only changing the labels from numbers to percents.

Step 3   **Continental Land Areas**

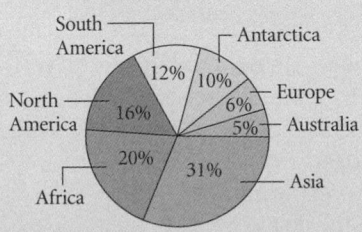

Step 5 Relative frequency graphs are good for showing ratios, but they don't report the amounts.

EXAMPLE

Randy has been asked to create a graphical display showing the distribution of the library's collection in six categories. His boss has asked him to create two rough drafts. Together they will decide which one to finalize for the display.

Here is the data:

Library Collection

Category	Number of Items
Children's fiction	35,994
Children's nonfiction	28,106
Adult fiction	48,129
Adult nonfiction	69,834
Media	11,830
Other	5,766
Total	199,659

▶ **Solution**

Randy puts the number of items in each category into list L1. He wants the calculator to determine in list L2 the number of degrees needed for each sector. He writes a proportion to find the number of degrees in the sector for a particular category.

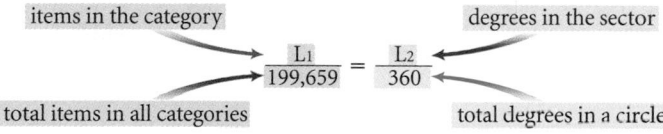

items in the category $\rightarrow$ degrees in the sector

$$\frac{L_1}{199,659} = \frac{L_2}{360}$$

total items in all categories — total degrees in a circle

By multiplying both sides of the proportion by 360, he finds the formula to enter into list L2.

$$L_2 = L_1 \cdot \frac{360}{199,659}$$

His calculator quickly determines the number of degrees for each sector of the circle graph. Using a protractor to measure the angles, Randy creates the graph.

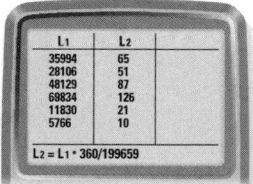

L1	L2
35994	65
28106	51
48129	87
69834	126
11830	21
5766	10

L2 = L1 * 360/199659

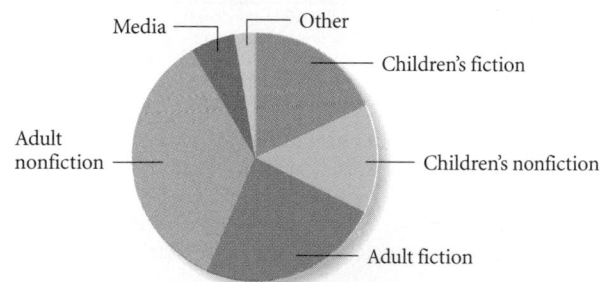

Library Collection

- Media
- Other
- Children's fiction
- Children's nonfiction
- Adult fiction
- Adult nonfiction

LESSON OBJECTIVES

- Learn about circle graphs
- Become familiar with relative frequency circle and bar graphs

NCTM STANDARDS

CONTENT	PROCESS
• Number	• Problem Solving
• Algebra	• Reasoning
• Geometry	Communication
• Measurement	• Connections
Data/Probability	• Representation

SHARING IDEAS

Select students to present the graphs of Steps 3 and 4. Lead a discussion of ideas about Step 5.

Some graphs display all data items, and others give only a visual summary.

[Ask] "Why do you think relative frequency graphs are called that?" [Language] Help students connect the phrase with the meanings of the words: *frequency* (how often) and *relative* (compared to the whole). Students will encounter the phrase *relative frequency* in the next lesson as well.

[Ask] "Which graphs in Chapter 1 were visual summaries and which displayed the actual data?" [Box plots are visual summaries, histograms summarize data in intervals (bins), and others display actual data.] [Ask] "Are circle graphs visual summaries, or do they display actual data?" [They display the data unless they are just relative frequency circle graphs.] [Ask] "Are relative frequency graphs visual summaries, or do they display actual data graphs." [Actual data values are not shown on relative frequency graphs.]

Assessing Progress

You can assess students' skills at solving proportions, working with ratios, careful measurement, drawing angles of given sizes, and seeing similarities and differences.

▶ **EXAMPLE**

This example shows how to use a calculator to convert data into degrees for a circle graph and into relative frequencies. Have students enter the lists in their own calculators. If there's time, have them apply the same techniques to the data of the investigation.

Review the important new ideas in this lesson: **circle graphs** and **relative frequency bar graphs.** If your class is using the TI-73, you might want to use Calculator Note 2E to show how to construct circle graphs on the calculator.

For a relative frequency graph, Randy finds the percent of the total represented by each category. He uses list L1 again, and the proportion

$$\frac{L1}{199{,}659} = \frac{L3}{100}$$

He solves for L3 and enters the formula that will give him the percent.

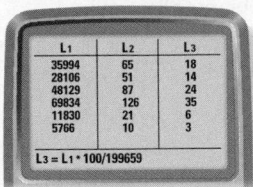

L₁	L₂	L₃
35994	65	18
28106	51	14
48129	87	24
69834	126	35
11830	21	6
5766	10	3

L3 = L1 * 100/199659

He makes a relative frequency circle graph by putting these percents in his circle graph. He uses the same percents to create a relative frequency bar graph.

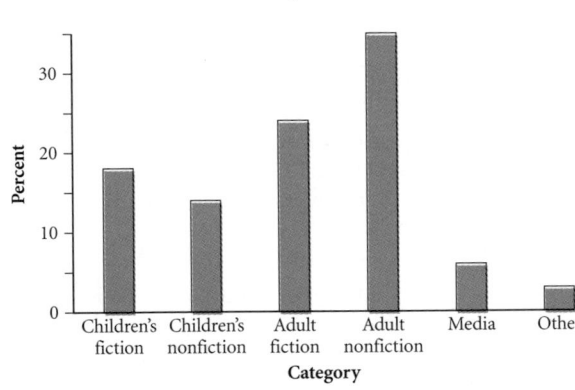

Randy and his boss decide to use the circle graph. They think it shows better how each category compares to the library's total collection.

In Chapter 1, you created box plots that gave a visual summary of the data. Like relative frequency graphs, they didn't contain the actual data values.

EXERCISES

Practice Your Skills

1. **APPLICATION** There are four basic blood types. The distribution of these types in the general population is shown in the relative frequency circle graph. In a city of 75,000 people, how many people with each blood type would you expect to find?

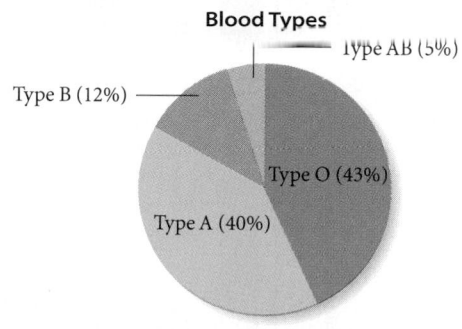

Blood Types

Type AB (5%)
Type B (12%)
Type O (43%)
Type A (40%)

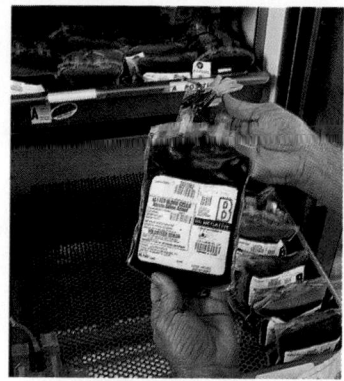

Type AB = 3750; Type B = 9000; Type A = 30,000; Type O = 32,250

2. Which data set matches the relative frequency circle graph at right? c

   a. {15, 18, 22, 25, 28} b. {20, 24, 30, 36, 45}
   c. {12, 18, 24, 30, 36} d. {9, 12, 18, 20, 24}

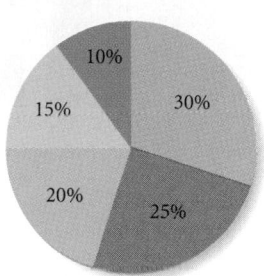

10%
30%
15%
20%
25%

3. In the bar graph of the library's collection created in the example, the bar for adult fiction represents 24%. Could there be a situation where all the bars represented 24%? Explain your thinking.
   No, the total height of all the bars must be 100%.

Reason and Apply

4. A manufacturer states that it produces colored candies according to the percents listed in this table. Create a circle graph to show this information.

Colored Candies Manufactured

Orange	Yellow	Blue	Red	Green	Brown
10%	20%	10%	20%	10%	30%

5. Chloe bought a small package of these candies and counted the number of each color. Her count is shown in the table at right.

Chloe's Colored Candies

Orange	Yellow	Blue	Red	Green	Brown
11	10	4	12	7	14

   a. Construct a relative frequency bar graph for Chloe's package of candies.

   b. Construct a relative frequency bar graph that shows on one graph both Chloe's small package of candies and the percents stated by the candy manufacturer. Use one color for the bar representing Chloe's candies and a different color for the bar representing the manufacturer's. Include a key showing what the two bar colors mean. What conclusions can you make?

BUILDING UNDERSTANDING

In these exercises, students practice constructing and interpreting circle graphs and relative frequency graphs.

ASSIGNING HOMEWORK

Essential	1, 2, 4, 6
Performance assessment	6
Portfolio	5
Journal	3
Group	5, 7, 10
Review	8–10

▶ Helping with the Exercises

Exercise 1 The circle graph shows visually the percents listed in the legend. The questions can be answered without reference to the graph itself.

4.

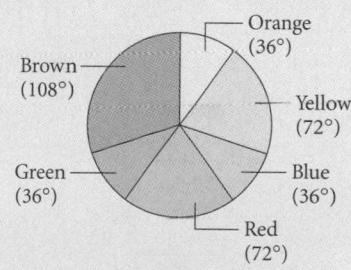

Orange (36°)
Brown (108°)
Yellow (72°)
Green (36°)
Blue (36°)
Red (72°)

Exercise 5 Help students understand that when two bar graphs are made on one set of axes, each bar graph has its own color bar.

5a.

Chloe's Candy Distribution

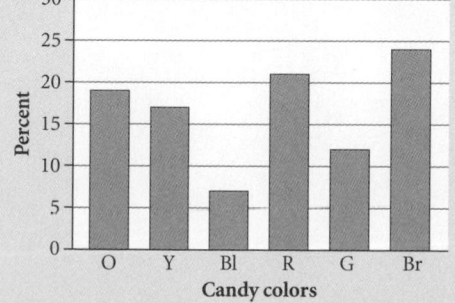

Percent / Candy colors
O Y Bl R G Br

5b. Comparing Chloe's Candy with the Manufacturer's

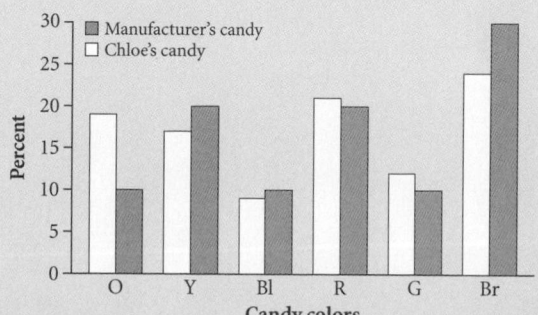

Manufacturer's candy
Chloe's candy

Percent / Candy colors
O Y Bl R G Br

Chloe's bag of candy had the same dominant color as the graph from the manufacturer, and her least color was one of the least manufactured. But the distributions are not very close.

Exercise 7b As needed, remind
students to use the techniques of
Lesson 2.4 for calculating percent
increase and decrease. Neither the
total nor the average of the percent changes of the classes will
give the percent change for the
school, because the classes have
different sizes. Remind students
that the average of averages of
parts doesn't usually give the
average of the whole. The average
of the changes (2%, −1.5%,
2.5%, −2%) is 0.5%, but since
there are different numbers of
students in each grade, the
change is 0.3%.

7c.

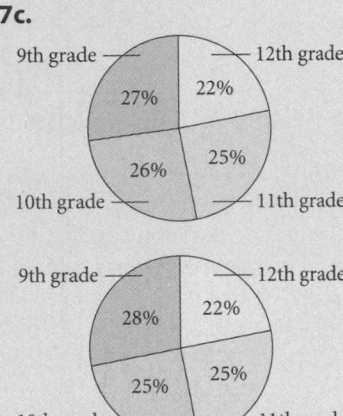

Proportionally, the ninth grade
increased 1% and the tenth
grade decreased 1%.

6. Match each bar graph with its corresponding relative frequency circle graph. Try to do this without calculating the actual percents.

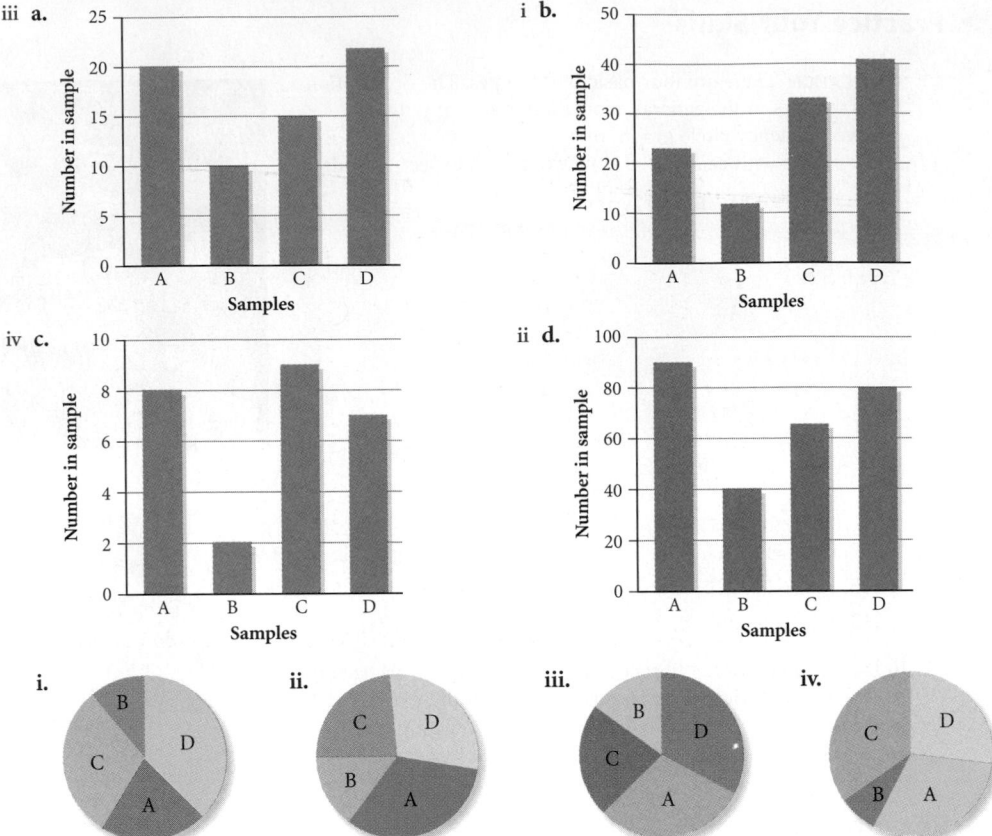

7. This table shows the number of students in each grade at a high school.

Class Size

Ninth grade	Tenth grade	Eleventh grade	Twelfth grade
185	175	166	150

a. What percent of the school is represented in each grade? **9th: 27%; 10th: 26%; 11th: 25%; 12th: 22%**

b. At semester break, the student population is counted again. The ninth grade has increased by 2%, the tenth grade has decreased by 1.5%, the eleventh grade has increased by 2.5%, and the twelfth grade has decreased by 2%. How many students are in each grade at the beginning of the second semester? By what percent has the total school population changed? What is the actual change in the number of students? **9th: 189; 10th: 172; 11th: 170; 12th: 147; 0.3%; 2 students**

c. Make a relative frequency circle graph for the situation at the beginning of the year and another circle graph for the situation at the beginning of the second semester. How has the distribution of students changed?

▶ Review

8. Evaluate these expressions. Check your results on your calculator.

a. $-3 \cdot -7 \cdot (8 + -5)$ 63
b. $(-5 - 7) \cdot -4 + 17$ 65
c. $6 + -2 \cdot (7 + 9)$ -26
d. $14 - (25 + 8 - 32)$ 13
e. $1 + 6 \cdot 5 - 11$ 20
f. $33 + 5 \cdot -8$ -7

9. Astrid works as an intern in a windmill park in
Holland. She has learned that the anemometer gives
off electrical pulses and that the pulses are counted
each second. The ratio of pulses per second to wind
speed in meters per second is always 4.5 to 1.

a. If the wind speed is 40 meters per second, how
many pulses per second should the anemometer
be giving off? **180 pulses per second**

b. If the anemometer is giving off 84 pulses per
second, what is the wind speed? **18.6 meters per second**

10. APPLICATION In 1999 there were 3142 counties in the
United States. Here is data on five counties:

Fastest-Growing Counties between 1990 and 1999

County	1990 population	1999 population	Change from 1990 to 1999	Percent growth
Douglas County, CO	60,391	156,860	96,469	159.7
Forsyth County, GA	44,083	96,686	52,603	119.3
Elbert County, CO	9,646	19,757	10,111	104.8
Park County, CO	7,174	14,218	7,044	98.2
Henry County, GA	58,741	113,443	54,702	93.1

(U.S. Bureau of the Census, 2000)

a. For each county, calculate the change in population from 1990 to 1999, and use
it to calculate the percent of growth. Which county had the largest percent of
growth in this time period? Douglas had the largest percent of growth.

b. Los Angeles County in California is
the largest county in the country. Its
population was 9,329,989 in 1999 and
8,863,164 in 1990. By what percent did
the population of Los Angeles County
grow during those nine years?

c. How does the growth of Los Angeles
County compare to the population
growth of the fastest-growing county?
Which do you think is a better
representation of the growth of a
county, the percent of change or the
actual number by which it grew?

San Fernando Valley, part of Los Angeles County

Exercise 9 An anemometer, like
the two pictured here, catches the
wind using cups or vanes and
records its speed.

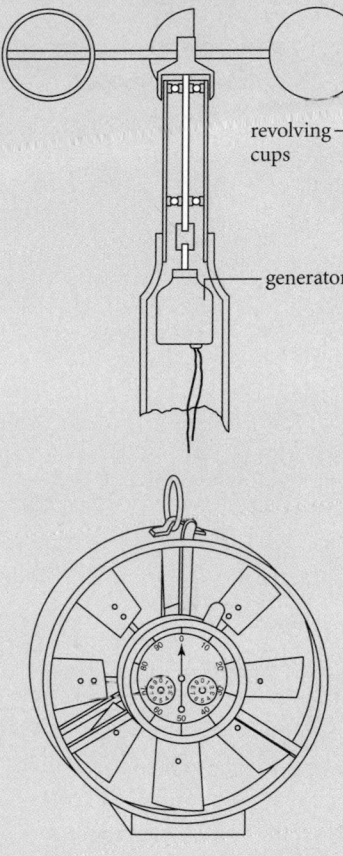

revolving cups

generator

10b. LA grew by 466,825, a
change of 5%.

10c. LA grew by over 4.8 times
more people than did Douglas,
but it increased by only 5.3%
rather than by 159.7%. The population growth of LA was more
manageable from the point of
view of the stress on the infra-
structure—roads, schools,
utilities.

PLANNING

LESSON OUTLINE

One day:

5 min	Example
25 min	Investigation
5 min	Sharing
5 min	Closing
10 min	Exercises

MATERIALS

- packets of colored candies (or small colored markers)
- paper bags
- protractors
- Calculator Notes 2A, 2B, and 2D

▶ **EXAMPLE**

Point out that to be useful for calculating probabilities about the squirrel population, the distribution must be accurate, or representative.

[Alert] Don't let students confuse the P(outcome) notation used for theoretical probability with multiplication.

LESSON

2.6

The theory of probability is at bottom only common sense reduced to calculation.

PIERRE SIMON DE LAPLACE

Probability Outcomes and Trials

Meteorologists know that out of ten days with a particular set of atmospheric conditions, it will probably rain on three of those days. When they see that set of conditions, they say, "The chance of rain is 30%."

The chance that something will happen is called its **probability.** The probability of an outcome is a ratio of the number of ways or times (in this case days) that outcome (rain) will occur, to the total number of ways or times (in this case days) under consideration. You can express that ratio as a percent.

The chance of rain can never be less than 0% or more than 100%, so a probability expressed as a decimal will always be between 0 and 1.

EXAMPLE

As part of Shandra's job with the forest service she tagged a total of 1470 squirrels last year. She tagged 820 black male squirrels, 100 black female squirrels, 380 gray male squirrels, and 170 gray female squirrels. If this distribution accurately reflects the squirrel population, what is the probability that the next squirrel she tags will be a gray male squirrel? A female squirrel? A red squirrel?

▶ **Solution**

The probability of the next squirrel tagged being a gray male squirrel can be expressed as the ratio

$$\frac{number\ of\ gray\ male\ squirrels\ tagged}{total\ number\ of\ squirrels\ tagged} = \frac{380}{1470} \approx 0.26$$

The probability of the next squirrel being female is

black female squirrels   gray female squirrels

$$\frac{number\ of\ female\ squirrels\ tagged}{total\ number\ of\ squirrels\ tagged} = \frac{100 + 170}{1470} \approx 0.18$$

During the last year Shandra hasn't tagged any red squirrels. Based on that information, the probability of a squirrel being red is 0.

$$\frac{number\ of\ red\ squirrels\ tagged}{total\ number\ of\ squirrels\ tagged} = \frac{0}{1470} = 0$$

LESSON OBJECTIVES

- Define the basic terms and concepts of probability
- Find observed probabilities through experimentation
- Calculate theoretical probabilities

NCTM STANDARDS

CONTENT		PROCESS	
●	Number		Problem Solving
●	Algebra	●	Reasoning
●	Geometry	●	Communication
●	Measurement	●	Connections
●	Data/Probability	●	Representation

In the example, each time a squirrel is tagged is a **trial.** Shandra conducted 1470 individual trials. Each possible result of a trial, is an **outcome.** Possible outcomes include a gray male squirrel, a black squirrel, or not a female squirrel.

An **observed probability** such as tagging a squirrel is based on experience or on collected data. Its usefulness is limited by the amount of data collected and the assumptions that conditions will remain the same.

The forest service can't know for sure the kinds and numbers of squirrels in the forest. If it did, it would use the known quantities to calculate the **theoretical probability** that the next squirrel to be tagged is gray. The notation P(gray squirrel) stands for the theoretical probability that the outcome of a trial will be "gray squirrel."

$$P(\text{gray squirrel}) = \frac{\text{number of gray squirrels in the forest}}{\text{total number of squirrels in the forest}}$$

Investigation
Candy Colors

You will need

- a packet of colored candies
- a paper bag

In the previous lesson you learned about *relative frequency*. This investigation will show you why the observed probability is sometimes also called the relative frequency. You will also work with known quantities to calculate theoretical probabilities.

Step 1 | Use a table like this one that lists the candy colors across the top to record the results of each trial.

	Experimental Outcomes					Total trials
	Red	Orange				
Tally						40
Experimental frequency						
Observed probability (relative frequency)						

Put the candies in the paper bag, then randomly select a candy by reaching into the bag without looking and removing a candy. Record the color as a tally mark, then replace the candy into the bag before the next person reaches in. Take turns removing, tallying the color, and replacing pieces of candy for a total of 40 trials. Your total for each color category is called its **experimental frequency.**

Record the experimental frequency for each outcome (color) in your table.

Step 2 **[ESL]** The word *draw* has several meanings in English. In this situation, to "draw from" the bag means "remove from" the bag.

Step 2 It is called *relative frequency* because the number of outcomes in each event is considered in relation to the total number of trials. To show the results as percents, divide the experimental frequency by 40 and write the answer as a percent. Total should be 100% or $\frac{40}{40}$ or 1.

Step 5 Sample answer: When there is a larger quantity of one color, the probability it will be drawn is higher. The total of all the probabilities should be 1.

SHARING IDEAS

Have students show (on transparencies) their tables from Steps 1 and 3. **[Ask]** "How would you describe the difference between observed and theoretical probabilities?" and "When are they the same?" [The observed probabilities get closer to the theoretical probabilities as the number of observations increases.] This idea provides more preparation for the concept of mathematical limit and leads into Lesson 2.7.

Assessing Progress

Look for skills at finding relative frequencies, organizing data collection, and contributing in a group.

Closing the Lesson

The principal ideas of this lesson are **trial, outcome, observed probability** (relative frequency), **theoretical probability,** and **equally likely outcomes.** Outcomes may be the result of trials and have observed probabilities, or they may be the result of theory and have theoretical probabilities, as symbolized by P(outcome). If there are *n* outcomes and they're all equally likely, the probability of each is $\frac{1}{n}$.

Step 2 From the experimental frequencies and the total number of trials (40), you can calculate the relative frequency or observed probability of each color. For instance, the observed probability of removing a red candy will be

$$\frac{\text{number of red candies drawn}}{\text{total number of trials}}$$

Record the observed probability in the bottom row of the table. Do you see why observed probability is also called relative frequency? How can you show these numbers as percents? What should the total be?

When all the candies are put into a bag, drawing one candy from the bag has several possible outcomes—the different colors listed on your table. Each individual candy is equally likely to be drawn, but some colors have a higher probability of being drawn than others.

Step 3 Make a second table.

	Outcomes					Total
	Red	Orange				
Number of candies counted						
Theoretical probability						

Dump out all the candies and count the number of candies of each color. Record this information in the top row.

Step 4 Use the known quantities in the first row to calculate the theoretical probability of drawing each candy color. For example,

$$P(R) = \frac{\text{number of red candies in the bag}}{\text{total number of candies in the bag}}$$

Record the results in the bottom row of the table.

Step 5 Is one color most likely to be drawn? Least likely? Explain the differences you found in the theoretical probabilities of drawing the different colors from the bag. What should their total be?

Step 6 Write a paragraph comparing your results for the theoretical probabilities you just calculated to the relative frequencies you calculated from your experiment.

In probability, an outcome is something that may or may not happen. When looking at the probability of a particular outcome, you must first ask yourself, "What outcomes are possible?" and "Are the outcomes equally likely?" If a packet of candy has exactly the same number of each color of candy, outcomes for each individual color, such as "green," are **equally likely.**

You will need your calculator for problem **10.**

Practice Your Skills

1. For each trial, list the possible outcomes.

 a. tossing a coin **heads or tails**

 b. rolling a die with faces numbered 1–6 **1, 2, 3, 4, 5, or 6**

 c. the sum when rolling 2 six-sided dice **2, 3, 4, 5, 6, 7, 8, 9, 10, 11, 12**

 d. spinning the pointer on a dial divided into sections A–E **A, B, C, D, or E**

2. The table below shows the distribution by fragrance of candles in a 20-candle assortment pack.

	Outcomes					
	Vanilla	Orange	Strawberry	Cinnamon	Winter	Total
Number	4	2	6	5	3	20
Theoretical probability	0.20	0.10	0.30	0.25	0.15	

 a. Copy the table and record in the bottom row the probability of selecting at random that type of candle.

 b. Suppose these 20 candles are put into a box. If you reach into the box without looking, what is the probability that you will pull out either a strawberry or a cinnamon candle? In other words, what is P(S or C)? **0.55**

 c. What is P(W or S or V)? **0.65**

 d. Suppose all 20-candle assortment packs made by this company have the same number of each type of candle listed above. If you empty ten assortment packs into a huge box, what is P(C) for the huge box? Explain why this is so.

3. One hundred tiny cubes were dropped onto a circle like the one at right, and all 100 cubes landed inside the circle. Twenty-seven cubes were completely or more than halfway inside the shaded region.

 a. Based on what happened, what is the observed probability of a cube landing in the shaded area?

 b. What is the theoretical probability in this situation? Explain your answer.

 90°

4. Igba-ita ("pitch and toss") is a favorite recreational game in Africa. In one version of Igba-ita, four cowrie shells are thrown in an effort to get a favorable outcome of all four up or all four down. Now coins are often used instead of cowrie shells and the name has changed to Igba-ego ("money toss"). Using four coins, what are the chances for an outcome in which all four land heads up or all four land tails up? $\frac{1}{8}$

 (Claudia Zaslovsky, *Africa Counts*, 1973, p. 113)

 You can learn about other cultural games with the links at **www.keymath.com/DA** .

BUILDING UNDERSTANDING

These exercises give students practice at calculating observed and theoretical probabilities.

ASSIGNING HOMEWORK

Essential	1–3, 5, 6
Performance assessment	7, 10
Portfolio	8
Journal	7
Group	4, 6, 9
Review	11–13

▶ Helping with the Exercises

Exercise 2b The notation P(S or C) is new. As needed, help students to see why the probabilities of the outcomes S and C must be added to find the probability of "either S or C."

2d. P(C) = 0.25. All probabilities would be the same as for the original pack because the ratios wouldn't change.

Exercise 3 Exercise 3 leads into Exercise 6, in which students calculate the area of a region by finding the portion of randomly distributed objects that fall inside the figure. Lesson 7 contains a similar exercise.

3a. $\frac{27}{100} = 0.27$

3b.
P(landing in shaded area) = 0.25 because the shaded area is $\frac{1}{4}$ of the circle.

Exercise 5 Encourage students to use a mixture of fraction, decimal, and percent labels for points.

Exercise 6 The calculator program GEOMPROB, explained in Calculator Note 2D, can be used by students who want to explore geometric probability further.

5. Draw and label a segment like the one below. Answers will vary. The probability of part 5f is 0 or very nearly 0; the probability of part 5g is 1.

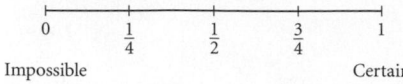

0        $\frac{1}{4}$ $\frac{1}{2}$ $\frac{3}{4}$ 1

Impossible   Certain

Plot and label points on your segment to represent the probability for each situation.

a. You will eat breakfast tomorrow morning.

b. It will rain or snow sometime during the next month in your hometown.

c. You will be absent from school fewer than five days this school year.

d. You will get an A on your next mathematics test.

e. The next person to walk in the door will be under 30 years old.

f. Next Monday every teacher at your school will give 100 free points to each student.

g. Earth will rotate once on its axis in the next 24 hours.

6. Suppose that 350 beans are randomly distributed on the rectangle shown at right and that 136 beans lie either totally inside the shaded region or more than halfway inside. Use this information to approximate the area of the shaded region. The program GEOMPROB in calculator note 2D can give more practice with geometric probability. Total area: $77 \cdot 63 = 4851; \frac{a}{4851} = \frac{136}{350}; a = 1885;$ the area is about 1900 square units.

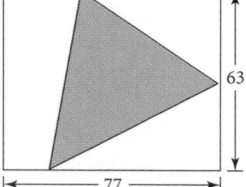

63

77

Reason and Apply

7a. Finding and counting a litter is a trial; an outcome may be having one cub (or two or three or four).

7b. No. If outcomes were equally likely, then the number of litters of each size would have been almost the same, with about nine litters of each size.

7. Dr. Lynn Rogers of the North American Bear Center does research on bear cub survival. He observed 35 litters in 1996. The distribution of cubs is shown in this table.

Bear Litter Study

Number of cubs	1	2	3	4
Number of litters	2	8	22	3

(*The North Bearing News,* July 1997)

a. Describe a trial for this situation. Name one outcome.

b. Is each outcome equally likely? Explain.

c. Based on the given information, what is the probability that a litter will be three cubs? $\frac{22}{35} \approx 0.63$

8. Twenty randomly chosen high school students were asked to estimate the percent of students in their school who are planning to attend college. Base your answers to the questions on their responses.

Student Responses

25	45	60	90
70	75	50	33
35	20	65	65
55	80	85	70
65	50	75	60

a. Draw a dot plot to organize the data.

b. What are the chances that the next student asked will give an estimate of at least 75%? $\frac{5}{20} = \frac{1}{4}$

c. If there are 4500 students in this high school, how many students do you think will give an estimate greater than or equal to 50%? $0.75 \cdot 4500 = 3375$

9. In the Wheel of Wealth game, contestants spin a large wheel like the one at right to see how much money each question is worth.

a. What is the probability that a contestant will have a question worth $500? $\frac{1}{6} \approx 17\%$

b. What is the probability that a contestant will have a question worth less than $500? $\frac{90 + 165}{360} = \frac{255}{360} \approx 71\%$

c. If one contestant spins the wheel and it lands in the $400 section, what is the probability that the next contestant will spin the wheel and have a question worth more than $400? $\frac{45 + 60}{360} = \frac{105}{360} \approx 29\%$

10. The Candy Coated Carob Company produces six different-colored candies with colors distributed as shown in the circle graph.

Carob Candy colors

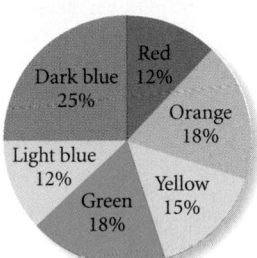

a. You could use the numbers between 1 and 100 to represent all the candies and choose numbers in this range to represent the percent of each color. For example, because P(red) = 12%, let the numbers from 1 to 12 represent a red candy. The next interval, which will represent orange, has to have 18 numbers because P(orange) = 18%. Therefore, let this interval be the numbers from 13 to 30. Identify intervals to represent the yellow, green, light-blue, and dark-blue candies. Make a table like this one and fill in the intervals.

	Outcomes						
	Red	Orange	Yellow	Green	Light blue	Dark blue	Total
Interval	1 to 12	13 to 30	31 to 45	46 to 63	64 to 75	76 to 100	
Number of candies							50
Probability							

Answers will vary. For some calculators a routine that works is randInt(1,100,50).

b. Enter a calculator routine that will generate a list of 50 random integers from 1 to 100. [▶ 🖥 See **Calculator Note 2A.** ◀]

c. Determine how many of each color you have in your collection. (You may want to sort, or order, your list first. [▶ 🖥 See **Calculator Note 2B.** ◀] Record the results in your table. Answers will vary. Sorting the list in ascending order is suggested.

d. Calculate the probability of selecting a particular color at random from your package. Answers will vary.

Exercise 9 To calculate the probabilities of the categories, students will need to find numerical data for each category by measuring the angles. Be prepared to lend protractors.

Exercise 10 To select categories with a calculator's random number generator, the categories must be labeled with intervals of numbers. Students can also use calculators to sort their lists.

8a.

```
                    •
          •   •     •   •
  •  •  ••  •  •  •  •  •  •  •
 ┗━┳━━┳━━┳━━┳━━┳━━┳━━┳━━┳━━
  10 20 30 40 50 60 70 80 90
        Student responses
```

Review

11. Give four pairs of coordinates that would create a shape like this when connected.

Answers could be any parallelogram with horizontal sides; one answer might be $(-4, 1), (-1, 3), (4, 3), (1, 1)$.

12a. $\dfrac{0.5 \text{ in.}}{1 \text{ mo}} \cdot \dfrac{12 \text{ mo}}{1 \text{ yr}} \cdot \dfrac{2.54 \text{ cm}}{1 \text{ in.}}$
$\cdot \dfrac{1 \text{ m}}{100 \text{ cm}} = 0.152 \dfrac{\text{m}}{\text{yr}}$

12. Michelle's hair grows $\frac{1}{2}$ inch each month.

  a. About how fast is that in meters per year? Use 2.54 cm = 1 in. and any other conversion factors you need.

  b. About how many years would it take for her hair to grow 1 meter? $\dfrac{1 \text{ yr}}{0.1523 \text{ m}} = \dfrac{x \text{ yrs}}{1 \text{ m}}$; about $6\frac{1}{2}$ yrs

Exercise 13 Solving this problem requires a good understanding of the game of baseball, which some students may not have.

13. APPLICATION The star hitter on the baseball team at City Community College was batting .375 before the start of a three-game series. During the three games, he came to the plate to bat eleven times. In these eleven plate appearances, he walked twice and had a sacrifice bunt. He either got a hit or made an out in his other plate appearances. If his batting average was the same at the end of the three-game series as at the beginning, how many hits did he get? (Note: Batting average is calculated by dividing hits by times at bat; sacrifice bunts and walks do not count as times at bat.) 3

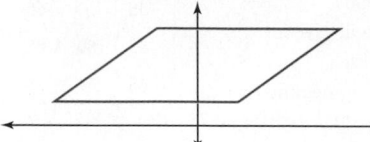

project

PROBABILITY, GENES, AND CHROMOSOMES

How are we or how are we not like our parents? In this project you will use probability ratios to describe how an individual's gender, eye color, color blindness, blood type, or other characteristic can be traced to his or her parents.

As you know, the number of girls and boys is not equal in every family. However, if a couple could have hundreds of children, over the long run, half would be girls and half boys. The same simple relationship does not hold for eye color. Parents who both have brown eyes may have a child with blue eyes. Or a father with blue eyes and a mother with brown eyes may never have a child with blue eyes.

Research the difference between the way gender is determined and the way eye color, or another human trait, is determined. Write a paper or develop a presentation that describes these probabilities.

Supporting the project

MOTIVATION

The first experiments to see how parents are like offspring were done on plants in the nineteenth century by Gregor Mendel before genes had been discovered. Now we are beginning to map the genetic codes of many forms of life, including humans. Students can focus on traits that they or their parents have. Students who are adopted or in foster care may prefer to study inheritance in plants.

OUTCOMES

▶ Clear explanations of how the gender of human children is determined
▶ Description of the heredity patterns of one or two other traits
• Additional information about the history and science of genetics
• Description of hereditary for more than three traits

Random Outcomes

*Prediction is difficult,
especially of the future.*

NIELS BOHR

Mario is an entomologist who
studies the behavior of bees. He
videotapes bees leaving a hive for
one day and counts 247 bees flying
east and 628 bees flying west. Can he
predict what the next bee will do?
Can he use the results of his study to
predict approximately how many of
the next 100 bees will fly east? Will
the videotape counts be the same if
he repeats the study a few days later? Is
there a pattern to the bees' flying
direction, or does it appear that the bees
fly randomly either east or west?

An outcome is **random** when you can't predict what will happen on the next trial. If
three bees always fly west after each eastbound bee, then this action is not random.
When a bee leaves a hive, he is following instinct and instructions from other bees.
To the bee, his actions are not random. But unless an observer can see a pattern
and predict the direction of the next bee, the pattern is random to the observer.

EXAMPLE

In the story, the entomologist cannot predict what the next bee will do. But he
can use the results of his study to predict approximately how many of the next
100 bees will fly east.

▶ Solution

The observed probability that a bee will fly west is

$$\frac{\textit{number of bees that flew west}}{\textit{all bees observed}} = \frac{628}{628 + 247} = \frac{628}{875}$$

The observed probability that a bee will fly east is

$$\frac{\textit{number of bees that flew east}}{\textit{all bees observed}} = \frac{247}{628 + 247} = \frac{247}{875}$$

Mario can calculate the *probability* of what the next bee will do, but he can't
predict its *actual direction*. From Mario's perspective the outcome is random.

The probability ratio $\frac{247}{875}$ is about 0.28, or 28%.

He can expect about 28 out of 100 bees to fly east. But this is a probability, not a
fact. He should not be surprised by 26 or 30 bees flying east. But if 50 or more
of the bees fly east, he might conclude that the conditions have changed and his
observations of yesterday no longer help him determine the probabilities for
today.

When you toss a coin, you cannot predict whether it will show heads or tails
because the outcome is random. You do know, however, that there are two equally
likely outcomes—heads or tails. Therefore, you know that the theoretical
probability of getting a head is $\frac{1}{2}$. What happens when you toss a coin many times?

PLANNING

LESSON OUTLINE

One day:

5 min	Example
25 min	Investigation
5 min	Sharing
5 min	Closing
10 min	Exercises

MATERIALS

• Calculator Notes 2A, 2B, 2C

TEACHING

[Language] *Random* is a relative
term. Traffic patterns are random
relative to an observer, even
though the drivers know where
they are going. Although the bees'
behavior is determined by
instinct and communication with
other bees, it is random relative to
a person looking at the beehive.

EXAMPLE

[Language] Ask students for
examples of predictions to help
explain the meaning of *predict*.

NCTM STANDARDS

CONTENT	PROCESS
• Number	Problem Solving
Algebra	• Reasoning
• Geometry	• Communication
• Measurement	• Connections
• Data/Probability	• Representation

LESSON OBJECTIVES

• Understand the difference between random and nonrandom
outcomes

• See how observed probabilities approach theoretical
probabilities as the number of trials increases

One step Challenge students to a similar coin-flipping experiment on a calculator as a way to show that the observed probabilities get closer to the theoretical probabilities as the number of trials increases.

Step 1 The TI-82 will allow only 99 random numbers to be stored in a list.

Step 2 [Alert] Although students can usually generate random integers one at a time on their calculators without much difficulty, some might be challenged by generating a whole list of them with one command. You may need to do lots of coaching.

Step 3 [Alert] Some students may be puzzled as to why the cumulative sum in list L3 gives the number of heads tossed so far. Stress that the word *total* in the title of the column means "total heads in this number of flips" rather than "total for the entire experiment."

Step 4 Similarly, the ratios being calculated concern only the number of flips given at this point in list L1.

Step 5 As needed, reassure students that the plotted data won't form a straight line. This may be the first *x-y* graph they've made other than that of $y = x$ in Chapter 1.

Step 6 The theoretical probability of 0.5 gives a horizontal straight line that the curve from Step 5 should approach.

Steps 7 and 8 By making this comparison as more and more trials are conducted, students should come to realize that observed probabilities approach theoretical probabilities as the number of trials increases. The graph provides a picture of a curve getting closer and closer to a line. Students again encounter the concept of limit.

Investigation
Calculator Coin Toss

In this investigation you will compare a theoretical probability with an observed probability from 100 trials. You will look at how the observed probability is related to the number of trials.

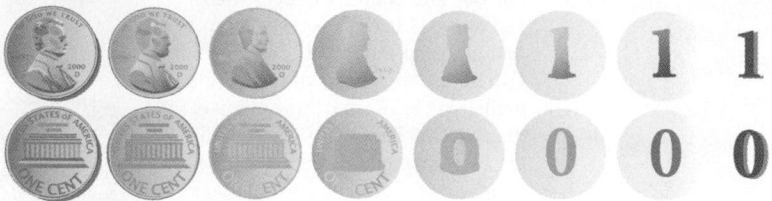

You could do this investigation by tossing coins 100 times, or you can use your calculator to simulate tossing many coins in a very short time.

Step 1 To number your tosses, enter the sequence of numbers from 1 to 100 into list L1 on your calculator. [▶️ See **Calculator Note 2B.** ◀]

Step 2 If a calculator randomly chooses 0 or 1, that's just like flipping a coin and getting tails or heads. Let 0 represent tails and 1 represent heads. Enter 100 randomly generated 0's and 1's into list L2. [▶️ See **Calculator Note 2A.** ◀]

Step 3 Display the cumulative sum of list L2 (number of heads) in list L3. [▶️ See **Calculator Note 2B.** ◀]

The table below shows an example in which the result of nine tosses was T, H, T, H, H, T, T, H, T. The numeral 1 in list L2 indicates heads. What does it mean if the eighth and ninth values in list L3 are both 4? **The ninth flip gave a 0 (tails) and added nothing to the cumulative sum of heads.**

Step 4 To calculate the ratio of heads to total number of tosses, enter $\frac{L3}{L1}$ in list L4. What does this ratio represent? **The observed probability of tossing a head**

Number of flips (L1)	Result of last flip (L2)	Total number of heads (L3)	Total heads Total tosses (L4)
1	0	0	0
2	1	1	0.50
3	0	1	0.33
4	1	2	0.50
5	1	3	0.60
6	0	3	0.50
7	0	3	0.43
8	1	4	0.50
9	0	4	0.44
⋮	⋮	⋮	⋮

SHARING IDEAS

Good results for groups to present are from Steps 4, 6, and 7. To help students appreciate the notion that more trials yield better results, ask them to imagine what the world would be like if that weren't the case. So much of what we learn is from estimating probabilities on the basis of trials!

Ask students to pretend they've flipped a fair coin seven times and gotten a tail each time. What's the probability that the eighth flip will also produce a tail? Some may say it's very small, because the probability of getting eight tails in a row is tiny. This reasoning is called the "Gambler's Fallacy." Others may say that the probability of getting a tail on the eighth flip is pretty high, because the coin is "running in tails." Actually, the coin has no memory of what has happened to it before; the eighth outcome is independent of all previous outcomes. So the probability of tails on the eighth flip is $\frac{1}{2}$.

Step 5	Create a scatter plot using list L1 as the x-values and list L4 as the y-values. Name an appropriate graphing window for this plot. **Possible answer:** [0, 100, 10, 0.2, 0.8, 0.1]
Step 6	Enter the theoretical probability of tossing a head in Y1 on the Y= screen. Graph your equation on the same screen as your scatter plot from Step 5.
Step 7	Compare your plot to that of other members of your group. Describe what appears to happen after 100 trials. What would you expect to see if you continued this experiment for 150 trials? Make a sketch of your predicted graph of 150 trials. Run the calculator simulation. [▶ ⬜ See **Calculator Note 2C.** ◀] Compare the results to your prediction.
Step 8	Explain what happens to the relationship between the theoretical probability and the observed probability as you do more and more trials. The larger the number of trials, the closer the observed probability will be to the theoretical probability.

If you tossed a coin many times, you would expect the ratio of the number of heads to the number of tosses to be close to $\frac{1}{2}$. The more times you toss the coin, the closer the ratio of heads to total tosses will be to $\frac{1}{2}$. With random events, patterns often emerge in the long run, but these patterns do not help predict a particular outcome.

When flipping a coin, you know what the theoretical probabilities are for heads and tails. However, in some situations you cannot calculate the theoretical probability of an outcome. After performing many trials, you can determine an observed probability based on your experimental results.

EXERCISES

You will need your calculator for problems **5** and **7**.

▶ Practice Your Skills

1. **APPLICATION** Suppose there are 180 twelfth graders in your school, and the school records show that 74 of them will be attending college outside their home state. You conduct a survey of 50 twelfth graders, and 15 tell you that they will be leaving the state to attend college. What is the theoretical probability that a random twelfth grader will be leaving the state to attend college? Based on your survey results, what is the observed probability? What could explain the difference?

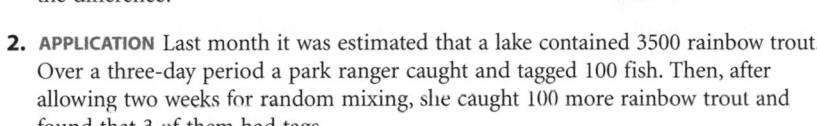

2. **APPLICATION** Last month it was estimated that a lake contained 3500 rainbow trout. Over a three-day period a park ranger caught and tagged 100 fish. Then, after allowing two weeks for random mixing, she caught 100 more rainbow trout and found that 3 of them had tags.

   a. What is the probability of catching a tagged trout? $\frac{100}{3500} \approx 0.0286$

   b. What assumptions must you make to answer 2a?

   c. Based on the number of tagged fish she caught two weeks later, what is the park ranger's observed probability? $\frac{3}{100} = 0.03$

---

1. Probability: $\frac{74}{180} \approx 0.411$; observed probability: $\frac{15}{50} = 0.30$. Possible answers: You can expect a wide variation in survey results. Perhaps your method of selecting students was not random. For example, your results could be biased because you talked only to students who were participating in after-school activities or only to students in a particular class. Perhaps the question was worded in such a way that students were biased in their response or reluctant to answer it honestly.

2b. You have to assume that the population is 3500, it remains stable (no fish die and no new fish hatch), and the fish are well mixed.

Assessing Progress

As students work through the investigation and present results, you can assess their abilities to enter and manipulate calculator lists, their understanding of **observed and theoretical probabilities,** and their intuition about limits.

Closing the Lesson

The primary idea of this lesson is that observed probabilities approach theoretical probabilities as the number of trials increases. Thus, probabilities can help predict the long-term results of many random outcomes.

BUILDING UNDERSTANDING

These exercises help strengthen the differences and relationships between observed and theoretical probabilities.

ASSIGNING HOMEWORK

Essential	1–4
Performance assessment	3
Portfolio	7
Journal	2, 4, 11
Group	5, 6
Review	8–11

Restaurant Workers

and full-time. Low stress work environment. Excellent benefits package. Call 555-7231, M-F, 9 to 5.

Many positions open at new restaurant chain in desirable downtown location, from entry level to management. Excellent benefits. Apply in person to Dave Lee, 2100 Buena Vista Avenue. No phone calls or emails please.

► **Helping with the Exercises**

Exercise 5 As needed, remind students how they took cumulative sums in the investigation. Although students' intuition may say correctly that the walk will probably stay close to the starting point, you might point out that the walk might land on any point if enough steps are taken.

3. Suppose 250 people have applied for 15 job openings at a chain of restaurants.

   a. What fraction of the applicants will get a job? $\frac{15}{250} = 0.06$

   $\frac{235}{250} = 0.94$ b. What fraction of the applicants will not get a job?

   c. Assuming all applicants are equally qualified and have the same chance of being hired, what is the probability that a randomly selected applicant will get a job? **0.06**

► **Reason and Apply**

4. If 25 randomly plotted points landed in the shaded region shown in the grid, about how many points do you estimate were plotted? $\frac{32}{126}x = 25; x = 98$

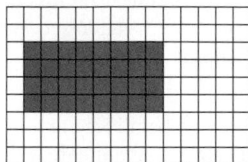

5. In a random walk, you move according to rules with each move being determined by a random process. The simplest type of random walk is a one-dimensional walk where each move is either one step forward or one step backward on a number line.

   a. Start at 0 on the number line and flip a coin to determine your move. Heads means you take one step forward to the next integer, and tails means you take one step backward to the previous integer. What sequence of six tosses will land you on the number-line locations $+1, +2, +1, +2, +3, +2$? **H, H, T, H, H, T**

   b. Explore a one-dimensional walk of 100 moves using a calculator routine that randomly generates $+1$ or -1.

   In list L1, generate random numbers with 1 representing a step forward and -1 representing a step backward. Describe what you need to do with list L1 to show your number-line location after every step. [► ▣ See **Calculator Notes 2A** and **2B.** ◄]

   c. Describe the results of your simulation. Is this what you expected?

   d. Would increasing the number of steps affect the results? Explain.
   With many trials you might be closer to 0.

5b. Find the cumulative sum of list L1.

```
2randInt(0,1,100
)-1→L1
{1 -1 1 1 -1 -1…
cumSum(L1)
{1 0 1 2 1 0 -1…
```

5c. After many steps, you may be close to 0.

6. A thumbtack can land "point up" or "point down."

    a. When you drop a thumbtack on a hard surface do you think the outcomes will be equally likely? If not, what would you predict for P(up)?

    b. Drop a thumbtack 100 times onto a hard surface, or drop 10 thumbtacks 10 times. Record the frequency of "point up" and "point down." What are your observed probabilities for the two responses?

    c. Make a prediction for the probabilities on a softer surface like a towel. Repeat the experiment over a towel. What are your observed probabilities? In one actual experiment, there was no change between a hard surface and a soft one.

7. APPLICATION A teacher would like to use her calculator to randomly assign her 24 students to 6 groups of 4 students each. Create a calculator routine to do this. One possible routine is randint(1,6). Assign students in order to groups 1 to 6, skip a number once that group is full.

▶ ## Review

8. The points listed here form a letter of the alphabet when they are connected in the proper order. What letter is it? **the letter P**

$(-1, 1)$	$(-1, -1)$	$(-1, -2)$	$(1, 1)$	$(2, 0)$	$(0, 1)$	$(-1, -4)$
$(0, -2)$	$(-1, 0)$	$(-1, -3)$	$(2, -1)$	$(1, -2)$	$(-1, -5)$	$(-1, -6)$

9. APPLICATION Zoe is an intern at Yellowstone National Park. One of her jobs is to estimate the chipmunk population in the campground areas. She starts by trapping 60 chipmunks, giving them a checkup, and banding their legs. A few weeks later, Zoe traps 84 chipmunks. Of these, 22 have bands on their legs. How many chipmunks should Zoe estimate are in the campgrounds? **229**

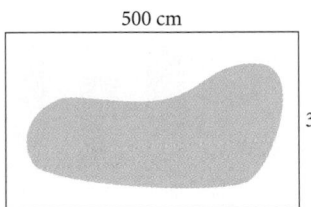

10. If a number has an exponent, write it in standard form. If a number is in standard form, write it with an exponent other than 1.

    a. 4^3 **64**   **b.** $\left(\dfrac{1}{6}\right)^2$ $\dfrac{1}{36}$ **c.** $\left(\dfrac{3}{4}\right)^2$ $\dfrac{9}{16}$

    d. 27 3^3 **e.** $\dfrac{1}{125}$ $\left(\dfrac{1}{5}\right)^3$ **f.** $\dfrac{4}{81}$ $\left(\dfrac{2}{9}\right)^2$ or $\dfrac{2^2}{3^4}$

11. Explain how to use probability to find the area of the irregular shape in the rectangle.

500 cm

300 cm

Answers will vary. You could cover the rectangle with a grid, count the squares in the shaded area, and compare that number to the total number of squares in the rectangle. Or you could cover the area with beans and compare the number of beans inside the shaded area to the total number of beans. Multiply the ratio $\dfrac{\text{number of beans in shaded area}}{\text{total number of beans}}$ times 150,000 cm^2.

Exercise 6 If you decide to do this problem in class, then use groups and consider appointing a safety monitor for each group.

6b. Answers will vary. Experience with a 9-mm-diameter head and a 6-mm point has indicated that the observed probability of "point up" is about 0.55.

Exercise 7 This exercise asks for a routine that has the teacher reject assignments to groups that are already full, rather than a routine that lists the assignment of each student.

CHAPTER

2

PLANNING

LESSON OUTLINE

5 min	Introduction
15 min	Exercises and helping individuals
15 min	Checking work and helping individuals
15 min	Student self assessment

REVIEWING

Display the proportion $\frac{x}{360} = \frac{9}{16}$. **[Ask]** "What problem about circle graphs could this proportion represent?" Encourage a variety of answers, submitting each for critique by the class. [One idea: the number of degrees in a sector representing $\frac{9}{16}$ of the total.] To review the multiply-and-conquer strategy, ask how to solve the problem. Then ask, "What problem about probability could this proportion represent?" [One possibility: x is the number of cartons of chocolate milk in a bin of 360 milk cartons if the probability of randomly choosing chocolate milk is $\frac{9}{16}$.] **[Ask]** "What problem about conversion of measurement units might this proportion represent?" Encourage imagination. [One idea: if you have a measurement unit that's $\frac{1}{16}$ of a circle, and an angle that's 9 of those units, x is the number of degrees in that angle.] Now invert the right side of the proportion to get $\frac{x}{360} = \frac{16}{9}$ and ask, "What's an increase/decrease problem that this proportion might represent?" [One idea: x is the amount charged by a store for an item for which it paid $360 and marked up by about 78%.]

Finally, invert both sides of the new proportion to get $\frac{360}{x} = \frac{9}{16}$. **[Ask]** "What's a capture-recapture situation that this proportion

In this chapter you explored and analyzed relationships among ratios, proportions, and percents. You used a **variable** to represent an unknown number, defined a proportion using the variable, and then solved the proportion to determine the value of the variable. You learned that a ratio of two integers is a **rational number** and that decimal representations of rational numbers either **terminate** or have a **repeating** pattern.

You can also use ratios as **conversion factors** to change from one unit of measure to another. You used **dimensional analysis** to convert units such as miles per hour to meters per second.

You constructed a relative frequency graph using your knowledge of ratios, proportions, and percents. Circle graphs and bar graphs can be **relative frequency graphs** that allow you to compare different categories in a data set proportionally.

Probability values for an **outcome** are ratios between 0 and 1 that compare the number of successful outcomes to the total number of trials. You learned that the more **samples** you select or **trials** you do in an experiment, the closer your **observed probability** will be to the **theoretical probability.** You investigated **random** outcomes. You saw that probability values can help you predict what will happen if you do many trials, but they will not help you predict the next outcome.

EXERCISES

1. Solve each proportion for the variable.

   a. $\frac{5}{12} = \frac{n}{21}$ $n = 8.75$ **b.** $\frac{15}{47} = \frac{27}{w}$ $w = 84.6$ **c.** $\frac{2.5}{3} = \frac{k}{6.2}$ $k = 5\frac{1}{6}$ or $5.1\overline{6}$

2. A group of 350 students were surveyed, and their eye colors are shown on the graph below right. Approximately how many students have each eye color?

3. Jeff can build 7 birdhouses in 5 hours. Write three different proportions that you could use to find out how long it would take him to build 30 birdhouses.

   Possible answers:

   $\frac{7 \text{ bhs}}{5 \text{ hrs}} = \frac{30 \text{ bhs}}{x \text{ hrs}}$; $\frac{7 \text{ bhs}}{30 \text{ bhs}} = \frac{5 \text{ hrs}}{x \text{ hrs}}$; $\frac{5 \text{ hrs}}{7 \text{ bhs}} = \frac{x \text{ hrs}}{30 \text{ bhs}}$; $\frac{30 \text{ bhs}}{7 \text{ bhs}} = \frac{x \text{ hrs}}{5 \text{ hrs}}$

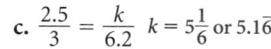
Survey of Eye Color

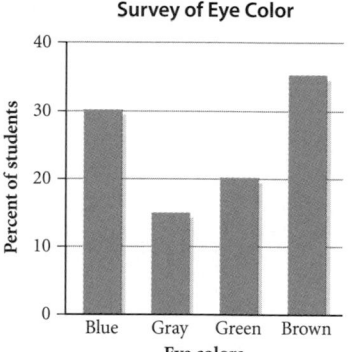

might represent?" [One possibility: x is the number of fish in a pond in which 360 were banded originally and in a recaptured sample of 16, 9 were banded.]

2. 105 students have blue eyes, approximately 52 or 53 have gray eyes, 70 have green eyes, and approximately 122 or 123 have brown eyes.

4. **APPLICATION** Tuition costs at several different colleges are listed in the table. Costs have been predicted to go up 3.7% for next year.

a. What will the costs be next year?

b. Explain how you can use the list feature of your calculator to quickly find the estimated cost next year for each school.

This year
$2,860
$3,580
$8,240
$9,460
$11,420
$22,500
$26,780

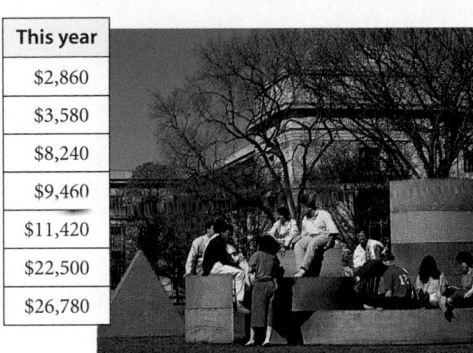

5. Plot the point (6, 3) on a graph.

a. List four other points where the *y*-coordinate is 50% of the *x*-coordinate. Plot them on the same graph.

b. Describe the pattern formed by the points.

6. In a fairy tale written by the Brothers Grimm, Rapunzel has hair that is about 20 ells in length (1 ell = 3.75 feet) by the time she is 12 years old. In the story, Rapunzel is held captive in a high tower with a locked door and only one window. From this window, she lets down her hair so that people can climb up.

a. Approximately how long was Rapunzel's hair in feet when she was 12 years old? 75 ft

b. If Rapunzel's hair grew at a constant rate from birth, approximately how many feet did her hair grow per month? 0.52 foot/month

7. Find the area of each shaded region. Then determine the probability of a random point landing in the shaded region of each figure.

a.
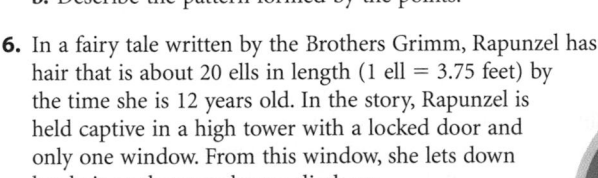

$12.5 \text{ cm}^2; \dfrac{12.5}{40} = 0.3125$

b.
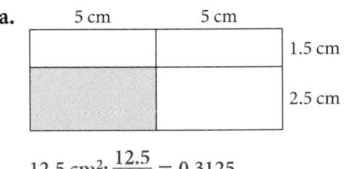

$32.5 \text{ cm}^2; \dfrac{32.5}{45} = 0.7\overline{2}$

8. **APPLICATION** This year the leftover candy at Sal's Candy Mart was marked down 15% after Halloween.

a. If bags of candy originally sold for $2.49, $1.89, and $3.29 before Halloween, what did they sell for after Halloween? $2.12; $1.61; $2.80

b. Write a proportion to calculate the sale price for any bag of leftover candy at Sal's when you know the original price. Use your proportion to verify your answers to 8a. Answers will vary; $\dfrac{85}{100} = \dfrac{\text{sale price}}{\text{original price}}$

c. Sal's original prices were twice the wholesale cost minus 1 cent. What were the wholesale prices? 1.25, 0.95, 1.65

d. What percent profit did he make on the marked-down candy? About 70%

4a. Answers are rounded to the nearest $10.

This year	Next year
$2,860	$2,970
$3,580	$3,710
$8,240	$8,540
$9,460	$9,810
$11,420	$11,840
$22,500	$23,330
$26,780	$27,770

4b. Enter this year's data into L_1. Define $L_2 = 1.037 \cdot L_1$.

5a. Possible points include (2, 1), (3, 1.5), (4, 2), (5, 2.5), (6, 3), (7, 3.5), (8, 4).

5b. All points appear to lie on a line.

▶ Take Another Look

The pattern is symmetric. The pictures for all fractions with denominators of 7 look the same, but they start on different digits. Students should observe that the digits of these repeating decimals are the same and repeat in the same order.

To draw other interesting patterns, choose another denominator that is a prime number.

If the repeating pattern is not too long, students can use their calculators to find the fraction corresponding to the pattern they draw. Here are examples of $\frac{1}{13} = 0.\overline{076923}$ and $\frac{1}{17} = 0.\overline{0588235294117647}$.

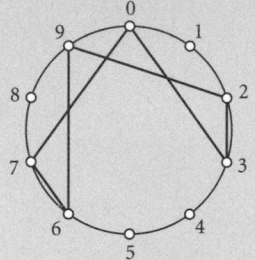

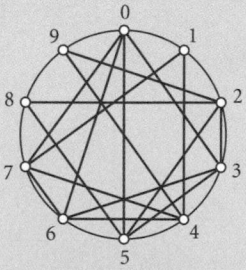

9. At a state political rally, someone said "We should raise test scores so that all students are above the state median." Analyze this statement. What is the probability of this happening? **Answers will vary. If the person is talking about the entire state, this cannot occur. All scores cannot be greater than the middle score. The probability is 0.**

10. Each year the Spanish club has a fund-raising raffle. First, second, and third prizes are $525, $125, and $25. The net gain is how much money you actually win if you deduct the cost of the ticket. Net gains and their respective probabilities are shown in the table.

Raffle Chances

Won	Net gain	Probability
$525	$500	1%
$125	$100	5%
$25	$0	10%
0	−$25	84%

a. What is the cost of one raffle ticket? **$25**

b. If you buy one ticket, what is the probability that you will win more than $50? **0.06, or 6%**

c. If only 100 tickets are sold, what would be the net winnings or losses for the group of 100 buyers? **One person is $500 ahead, 5 people are $100 ahead, 10 people will be even, and 84 people will be $25 behind. This is a net loss of $1100, or $11 per person.**

11. **APPLICATION** A thirteenth-century Chinese manuscript (*Shu-shu chiu-chang*) contains this problem: You are sold 1,534 shih of rice but find that millet is mixed with the rice. In a sample of 254 grains, you find 28 grains of millet. About how many shih of rice do you have? How much millet did you buy? (Ulrich Libbrecht, *Chinese Mathematics in the Thirteenth Century*, 1973, p. 79) **1365 shih rice; 169 shih millet**

TAKE ANOTHER LOOK

▶ You can make an interesting picture of some fractions using their decimal equivalents and a circle.

Calculate the decimal equivalent of $\frac{1}{7}$. On a circle like this one, create a pattern using the decimal equivalent of $\frac{1}{7}$. Start with your pencil at the point on the circle corresponding to the digit in the tenths place of the decimal equivalent. Use a ruler to draw a line segment from that point to the point corresponding to the digit in the hundredths place. Then draw a line from that point to the point corresponding to the digit in the thousandths place, then on to the ten-thousandths place, and so on. Continue until the pattern starts to repeat. Describe the pattern you have created.

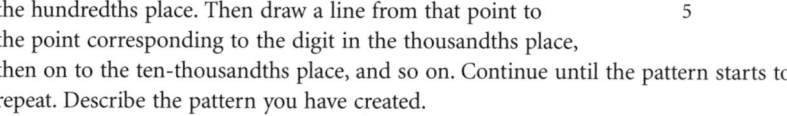

Repeat the process for the other fractions with a denominator of 7. What do you notice?

Explore and see if there are other fractions that create similar circle pictures. What do these fractions have in common?

Draw a symmetrical, repeating pattern. What fraction does it correspond to?

Assessing What You've Learned

ORGANIZE YOUR NOTEBOOK

You've been creating tables and answering questions as you do the investigations. You've been working the exercises and taking quizzes and tests. You've made notes on things that you want to remember from class discussions. Are those papers getting folded and stuffed into your book or mixed in with work from other classes? If so, it's not too late to get organized. Keeping a well-organized notebook is a habit that will improve your learning.

Your notebook should help you organize your work by lesson and chapter and give you room to summarize. Look through your work for a chapter and think about what you have learned. Write a short summary of the chapter. Include in the summary the new words you learned and things you learned about the graphing calculator. Write down questions you still have about the investigations, exercises, quizzes, or tests. Talk to classmates about your questions, or ask your teacher.

WRITE IN YOUR JOURNAL Add to your journal by expanding on a question from one of the investigations or exercises. Or use one of these prompts:

▶ It has been said that over half of the problems on the SAT Mathematics Test can be solved using proportional reasoning. Discuss why you think this might be true.

▶ Describe the progress you are making toward the goals you have set for yourself in this class. What things did you do and learn in this chapter that are helping you achieve those goals? What changes might you need to make to help keep you on track?

UPDATE YOUR PORTFOLIO Find the best work you have done in Chapter 2 to add to your portfolio. Choose at least one piece of work where you used proportions to solve a percent problem and at least one probability investigation or exercise. Choose one relative frequency graph you made. You might decide to put the graph with the graphs you selected for your Chapter 1 portfolio.

Use tests from Assessment Resources A and B to assess how well students have learned the thought processes related to ratios: setting up and solving proportions, the capture recapture method, converting units of measurement, percent increase and decrease, circle graphs and relative frequency graphs, and probability.

Ideally, students have also become more accustomed to working collaboratively and have improved their facility with graphing calculators.

FACILITATING SELF ASSESSMENT

This student self assessment focuses on organizing notes. Students need to see a reason for doing this, such as keeping track of questions or finding portfolio items easily. If you allow students to use their notebooks during tests, you'll be giving them a powerful reason for knowing where to find things.

Variation and Graphs

CHAPTER 3 OBJECTIVES

- Use dimensional analysis to choose appropriate rates that simplify solving proportions

- Use direct variation equations and their graphs to solve real-world problems

- Develop an intuitive understanding of slope and linear function

- Investigate the definition and properties of similar shapes, including those used in scale drawings

- Investigate inverse variation through real-world problems

- Learn the basic inverse variation equations

- Use inverse variation equations and their graphs to solve real-world problems

- Investigate the direct and inverse variations between gear selection and wheel speeds on a multispeed bicycle

Ratios and proportions describe many aspects of music composition and production. Most musicians today rely on electronic devices to generate, record, or perform music, and to amplify the sound of acoustic instruments. The individual levers and dials of a mixing board, for instance, create variations in the quality of sound.

OBJECTIVES

In this chapter you will
- learn what rates are and use them to make predictions
- study how quantities vary directly and inversely
- use equations and graphs to represent variation
- solve real-world problems using variation

The image pictures "turntabling," a modern music phenomenon. The disc jockey manipulates the record speed by hand, sometimes playing more than one recording at a time, sometimes at different speeds. The rate that a record spins is given in revolutions per minute (rpm). Long-playing records spin on their own at 33 rpm.

You might ask students where else rates appear in music. Beat is the number of strokes per measure of music. For example, there are three beats per measure in waltz time and four beats per measure in march time. Rhythm refers to the way those beats are typically divided into smaller time durations in the composition.

Rates are involved in mixing music tracks to determine the loudness (air-wave amplitude) and sound quality of each component (such as voice, instrument, or drums) for the proportional emphasis that the producer wants.

Sound intensity varies with the square of the distance from its source. Pitch is achieved by frequency of the air wave; higher overtones divide the period of the fundamental wave into shorter cycles.

LESSON
3.1

Using Rates

Measurement began our might.

WILLIAM BUTLER YEATS

Most people earn their income by working in a job. Yet the amount people are paid varies. Some workers have fixed yearly salaries while others are paid by the hour. How often a worker gets a paycheck varies too. One person may be paid weekly while another is paid monthly.

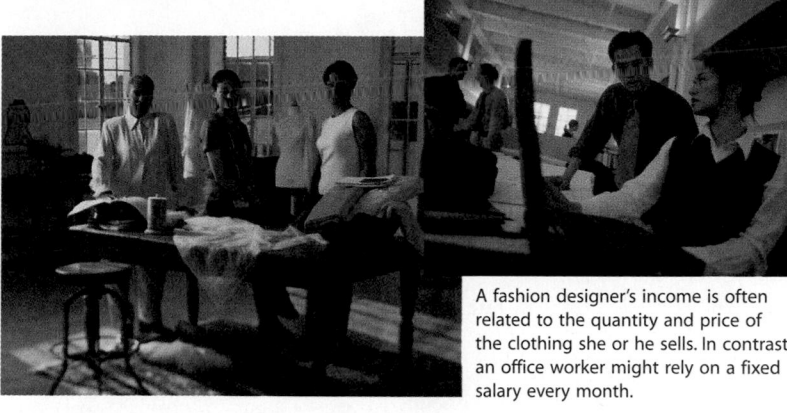

A fashion designer's income is often related to the quantity and price of the clothing she or he sells. In contrast, an office worker might rely on a fixed salary every month.

How could you compare the income of two workers based only on a paycheck? How could a worker find out how much she or he would earn for a particular time spent working? In this lesson you will learn about a type of ratio that is useful for answering questions like these.

Investigation
Off to Work We Go

You will need
- graph paper

Lacy works as a temporary employee. Last week, she earned $300 for 20 hours of work. In the first part of this investigation you will examine Lacy's pay and predict how much she will earn under different circumstances.

Step 1 $\frac{\$300}{20 \text{ hours}}$

Step 1 Write Lacy's pay as a ratio in the form
$$\frac{pay}{hours\ worked}$$

Step 2 $\frac{300}{20} = \frac{x}{40}$, $x = 600$. Since 40 is $2 \cdot 20$, use one step: $2 \cdot 300$. This works for any multiple of 20 hours—60, 80, 100, and so on.

Step 2 Using your ratio from Step 1, write and solve a proportion to find out what Lacy's pay would be for a week of full-time work (40 hours). Is there a shorter method to solve your proportion? For what other number of hours of work could you use it?

NCTM STANDARDS

CONTENT		PROCESS	
•	Number	•	Problem Solving
•	Algebra	•	Reasoning
	Geometry	•	Communication
•	Measurement	•	Connections
	Data/Probability	•	Representation

LESSON OBJECTIVES

- Learn what a rate is
- Get practice using rates to solve problems
- Gain experience in choosing appropriate rates for dimensional analysis

PLANNING

LESSON OUTLINE

One day:
5 min	Introduction
25 min	Investigation
5 min	Sharing
5 min	Example
5 min	Closing
5 min	Exercises

MATERIALS
- graph paper
- Calculator Notes 1A, 1L

TEACHING

This lesson introduces rates— ratios with a denominator of 1. They are often called *unit rates*. Using rates to make comparisons can be easier than setting up and solving proportions. You can represent rates graphically by the steepness of a line.

INTRODUCTION

Students may suggest looking at either yearly or hourly salary to compare salaries and predict how much will be earned over time.

Guiding the Investigation

[Language] Temp agencies provide short-term temporary employees to companies. Someone who works for a temp agency could be placed in a different position daily, weekly, or monthly. Temporary employees are paid by the temp agency, but their hourly salaries depend on the requirements of the job placement.

One step Give Lacy's and Joseph's pay information (from text before Step 7), and ask which of them makes more in a week. If any discussion arises of how many hours are in a workweek, emphasize the need to make assumptions in solving real-world problems. Keep coming back to the question of a rate—a ratio with 1 in the denominator. If students are doing well, ask how many hours Lacy must work to earn $1,000 for a trip.

Step 2 Remind students that writing the units first might help in setting up proportions.

Step 5 Some students may recognize that rather than multiplying each time you can find sequential values by adding $15 each time. The colon notation used for ratios, $15 : 1 hour, is rarely used for rates. Usually the rate magnitude is not written as part of the numerator; for instance, $15 \frac{\text{dollars}}{\text{hour}}$ is customary.

Step 8 Students might recognize that both Lacy's and Joseph's data go through the origin. Later, students will see that all graphs of direct variations pass through the origin.

SHARING IDEAS

Key steps to present are 6 and 9. Although it may be too early to use the word *slope*, elicit the idea that the steepness of the graph increases with an increase in pay rate.

[Ask] "What are some other rates?" [Some possibilities are miles per hour (mph), beats per minute (bpm), interest rates (sometimes worded as cents to the dollar), calling card or phone rates (cents per minute), and nutritional analyses (calories per serving). Some conversion factors can be thought of as rates (2.54 cm per inch), although they might also be reciprocals of rates (1 inch per 2.54 cm).] You could point out that some rates are constant (rate of pay will not

Step 3 $15   Step 3

Step 4 $\frac{15}{1}$; $\frac{15}{1} = \frac{x}{3}$, $x = 45$.   Step 4
Or multiply by 3.

                   Step 5
Step 5 $(1, 15), (2, 30), (3, 45),$ $(4, 60), (5, 75)$. The points lie on a line.

Step 3 Use your ratio from Step 1 to find out how much Lacy earns for 1 hour of work.

Step 4 Write Lacy's pay for 1 hour of work as a ratio. Then write and solve a proportion to find out how much Lacy would be paid for 3 hours of work. Find another way to calculate how much Lacy will be paid for 3 hours of work.

Step 5 Complete a table like this for 1 to 5 hours of work. Let x represent time worked in hours and y represent pay in dollars, and plot the points you get from the information in your table. Do you notice any patterns?

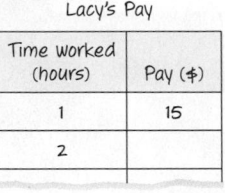

Lacy's Pay

Time worked (hours)	Pay ($)
1	15
2	

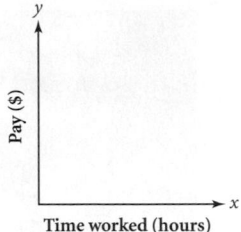

A **rate** is a ratio with 1 in the denominator. In Steps 3 and 4, you found Lacy's rate of pay. Here are three ways to write this rate.

$$\frac{\$15}{1 \text{ hour}} \qquad\qquad \$15 \text{ for each hour} \qquad\qquad \$15 \text{ per hour}$$

The word *per* means "for each." It means the number is a rate.

Step 6 Answers   Step 6 | How would you use the rate to find Lacy's pay for x hours?
should include that the denominator of 1 in the rate allows you to simply multiply by the number of hours worked.

Joseph is another temporary employee. He was paid $513 for 38 hours of work. In this part of the investigation you will compare Joseph's pay to Lacy's.

Step 7 $\frac{\$13.50}{1 \text{ hour}}$ Step 7

Step 8 $(1, 13.50),$ Step 8
$(2, 27), (3, 40.50), (4, 54),$ $(5, 67.50)$ These points also lie on a line. Since Lacy's rate of pay is higher, she will earn more than Joseph for any given number of hours worked.   Step 9

Step 9 Answers should include that rates have common denominators, so you only have to compare numerators to identify the greater or lesser rate.

Step 7 Find Joseph's rate of pay.

Step 8 Create a table for Joseph's pay for 1 to 5 hours. Plot the points on the same graph as Lacy's pay. How do Lacy's and Joseph's rates of pay affect what you see on the graph?

Step 9 Explain how a rate makes it convenient to compare two workers' pay.

fluctuate for each hour), whereas others are based on an average of values that do vary. For example, a player's points per game will vary for each game, but an average could be calculated and stated as a rate.

To emphasize that comparing rates is valid only if they have the same units, you may want to ask students to compare Lacy's and Joseph's pay if Joseph's were $\frac{\$13.50}{1 \text{ day}}$ and Lacy's were $\frac{2 \text{ British pounds}}{1 \text{ hour}}$.

You can encourage critical thinking by turning to the class for a critique of all answers, rather than confirming yourself that an answer is correct.

Assessing Progress
Through your observations, you can assess students' understanding of **ratios** and **proportions** and perhaps **conversion factors**.

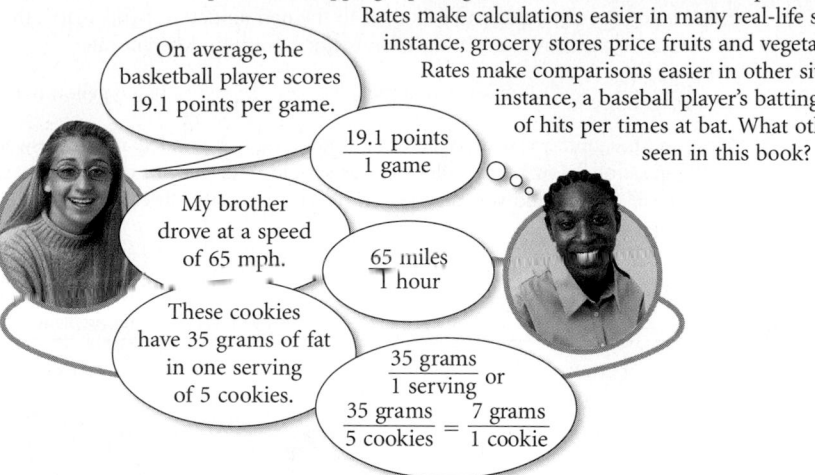

A rate of pay is just one example of a rate. Your weekly allowance, the cost per pound of shipping a package, and the number of cookies per box are all rates. Rates make calculations easier in many real-life situations. For instance, grocery stores price fruits and vegetables by the pound. Rates make comparisons easier in other situations. For instance, a baseball player's batting average is a rate of hits per times at bat. What other rates have you seen in this book?

On average, the basketball player scores 19.1 points per game.

$\frac{19.1 \text{ points}}{1 \text{ game}}$

My brother drove at a speed of 65 mph.

$\frac{65 \text{ miles}}{1 \text{ hour}}$

These cookies have 35 grams of fat in one serving of 5 cookies.

$\frac{35 \text{ grams}}{1 \text{ serving}}$ or $\frac{35 \text{ grams}}{5 \text{ cookies}} = \frac{7 \text{ grams}}{1 \text{ cookie}}$

EXAMPLE

Otto's car began the week with a full tank of gas. Otto drove to and from work, a total of 320 miles. He then filled his tank with 14.7 gallons of gas.

a. How many miles per gallon (mpg) does Otto's car get?

b. How far could Otto drive using only 5 gallons of gasoline?

c. When Otto's car recently broke down, he rented a newer model car. Otto drove the same distance, 320 miles, but it took only 11.5 gallons of gas to fill the tank. What was the rate of miles per gallon for the rental car? Which car can go more miles on 10 gallons of gas?

▶ Solution

Otto's car used 14.7 gallons to go 320 miles.

a. The ratio is $\frac{320 \text{ miles}}{14.7 \text{ gallon}}$. So you divide miles by gallons.

$$320 \div 14.7 \approx 21.8$$

Otto's car gets approximately 21.8 mpg or $\frac{21.8 \text{ miles}}{1 \text{ gallon}}$.

b. Multiply the rate by 5. You can use dimensional analysis to help verify the units in your answer.

$$\frac{21.8 \text{ miles}}{1 \text{ gallon}} \cdot \frac{5 \text{ gallons}}{1} = 109 \text{ miles}$$

c. $320 \div 11.5 \approx 27.8$

The rate of the rental car is approximately 27.8 mpg. This is greater than 21.8 mpg. The rental car can go farther than Otto's car on 1 gallon of gas, so it can also go farther than his car on 10 gallons of gas.

Consumer CONNECTION

Many factors influence the rate at which cars use gas, including size, age, and driving conditions. Advertisements for new cars often give the average mpg for city traffic (slow, congested) and highway traffic (fast, free flowing). These rates help consumers make an informed purchase.

Extension

Prepare students for the exercises. **[Ask]** "How many gallons of gas will your car use to go 200 miles?" [To solve this problem, students will probably find the number of gallons per mile and multiply it by 200.] Point out that students can use dimensional analysis to see which rate they need to use. Because they want to find a number of gallons, and they know a number of miles, they either need the $\frac{\text{gallons}}{\text{mile}}$ rate to multiply by a number of miles, or they need the $\frac{\text{miles}}{\text{gallon}}$ rate to divide into a number of miles. Students might need a review of division with fractions. **[Language]** Students may not know the term *reciprocal* for $\frac{\text{gallons}}{\text{mile}}$ and $\frac{\text{miles}}{\text{gallon}}$, but they may say that the ratios are inverted or flipped, or that the division has been reversed.

EXAMPLE

Some students may need you to elaborate on this situation. Since Otto began with a full tank, the missing 14.7 gallons (replaced when he fills up) was used to drive the 320 miles. **[ESL]** *Mileage* is not always the same as *miles*. It can mean "miles per gallon."

Some students might want to solve part c by calculating $\frac{\text{gallons}}{\text{mile}}$ and multiplying by 10. Encourage this approach (and skip the extension), but be sure the explanation in the student book comes up too.

You can help prepare students for later lessons by phrasing the rates in a variety of ways. "30 miles per gallon" is equivalent to "30 miles for each gallon" and "on each gallon you can drive 30 miles."

MAKING THE CONNECTION

Students could determine mileage information about the cars owned by people in their household.

As needed, reiterate that a **rate** is a ratio with 1 in the denominator. Also mention that **dimensional analysis** requires putting rates together so that the units "cancel," for instance, $\frac{1 \text{ mile}}{1 \text{ mile}} = 1$.

BUILDING UNDERSTANDING

Students work with rates, including reciprocal rates. Encourage them to write in the units and to use dimensional analysis.

ASSIGNING HOMEWORK

Essential	1–4, 7
Performance assessment	5
Portfolio	8
Journal	10
Group	6, 9
Review	11–14

▶ Helping with the Exercises

1a. $\frac{3 \text{ pounds}}{30 \text{ days}} = 0.1$ pound per day

1b. $\frac{5 \text{ pounds}}{45 \text{ days}} = 0.\overline{1}$ pound per day

1c. Crystal's cat

Exercise 4 Reciprocal rates are needed for the two parts.

Exercise 5 [Alert] Some students may be misled by the extra information about continents and countries. The rate needed for 5c is the reciprocal of the rate needed for 5a and 5b.

When you are finding a rate, consider which unit would be most helpful in the denominator based on the questions to be answered. In part a of the example, miles is in the numerator and gallons is in the denominator. If you switch the quantities in the numerator and denominator, you get a different rate.

$$\frac{320 \text{ miles}}{14.7 \text{ gallons}} \approx 21.8 \text{ miles per gallon} \qquad \frac{14.7 \text{ gallons}}{320 \text{ miles}} \approx 0.046 \text{ gallon per mile}$$

How are these rates related? Which rate would you use to calculate the answer to the question "How many gallons of gas should Otto put in his car to drive 200 miles?" Why would you choose that rate? Is it possible to calculate the answer using the other rate?

EXERCISES

You will need your calculator for problem **10.**

▶ Practice Your Skills

1. Tab and Crystal both own cats.

   a. Tab buys a 3-pound bag of cat food every 30 days. At what rate does his cat eat the food?

   b. Crystal buys a 5-pound bag of cat food every 45 days. At what rate does her cat eat the food?

   c. Whose cat, Tab's or Crystal's, eats more food per day?

2. If Ray drove 240 miles in 4 hours, what was his average rate of speed? **60 miles per hour**

3. Jocelyn's old car gets only 10 miles per gallon.

   a. What is the car's rate of gasoline use in gallons per mile? **0.1 gallon per mile**

   b. How many gallons of gas does the car need to go 220 miles? **22 gallons**

   c. How far could Jocelyn drive on 15 gallons of gas? **150 miles**

▶ Reason and Apply

4. Find a rate for each situation. Then use the rate to answer the question.

   a. Kerstin drove 350 miles last week and used 12.5 gallons of gas. How many gallons of gas will he use if he drives 520 miles this week? **18.72 gallons (0.036 gallon per mile)**

   b. Angelo drove 225 miles last week and used 10.7 gallons of gas. How far can he drive this week using 9 gallons of gas? **189 miles (21 miles per gallon)**

5. On his Man in Motion World Tour, Canadian Rick Hansen wheeled himself 24,901.55 miles to support spinal cord injury research, rehabilitation, and wheelchair sport. He covered 4 continents and 34 countries in two years, two months, and two days. Learn more about Rick's journey with the link at   .

China was one of the many countries through which Rick Hansen traveled during the Man in Motion World Tour.

Photo courtesy of The Rick Hansen Institute

a. Find Rick's average rate of travel in miles per day. (Assume there are 365 days in a year and 30.4 days in a month.) $\frac{24901.55 \text{ miles}}{(2 \cdot 365 + 2 \cdot 30.4 + 2) \text{ days}} \approx 31.4$ miles per day

b. How much farther would Rick have traveled if he had continued his journey for another $1\frac{1}{2}$ years? $\frac{31.4 \text{ miles}}{1 \text{ day}} \cdot \frac{(1.5 \cdot 365) \text{ days}}{1} \approx 17191.5$ miles

c. If Rick continued at this same rate, how many days would it take him to travel 60,000 miles? How many years is that? $\frac{31.4 \text{ miles}}{1 \text{ day}} = \frac{60,000 \text{ miles}}{t}$; $t \approx 1911$ days, or more than 5 years

6. APPLICATION Wynonna bought a prepaid calling card for $5. It allows her to make 80 minutes' worth of calls.

a. Estimate the prepaid calling card's cost per minute. Explain your thinking.

b. Calculate the actual cost per minute. 6.25¢

c. After purchasing her prepaid card, Wynonna made a 15-minute call to her friend Samson. How much is the card worth now? about $4.06

d. Wynonna's home telephone service provides a calling card that costs 8¢ per minute. Which calling card would you recommend she use? Why?

e. Wynonna has made plans for a 7-day vacation. She expects to make 30 minutes of calls each day. How many prepaid calling cards should she take along? 3

7. APPLICATION Marie and Tracy bought boxes of granola bars for their hiking trip. They noticed that the tags on the grocery-store shelf use rates.

CRUNCHY GRANOLA BARS
Unit Price
.42 per Bar $2.49
8776 657588 485857

CHEWY GRANOLA BARS
Unit Price
.25 per Ounce $2.99
95847 98564 162828

a. Each tag above uses two rates. Identify all four rates. $2.49 per box, 42¢ per bar, $2.99 per box, 25¢ per ounce

b. A box of Crunchy Granola Bars contains 6 bars. Is the price per bar correct? yes

c. A box of Chewy Granola Bars contains 8 bars. Use the information on the tag to find the number of ounces per bar. $\frac{1.495 \text{ ounces}}{1 \text{ bar}}$

d. A box of Crunchy Granola Bars weighs 10 ounces. What is the price per ounce? approximately 25¢

e. If Marie and Tracy like Crunchy Granola Bars as much as they like Chewy Granola Bars, which should they buy? Explain your answer.

8. APPLICATION Portia drove her new car 425 miles on 10.8 gallons of gasoline.

a. What is the car's rate of gasoline consumption in miles per gallon? approximately 39.4 mpg

b. If this is the typical mileage for Portia's car, how much gas will it take for a 750-mile vacation trip? 19 gallons

c. If gas costs $1.35 per gallon, how much will Portia spend on gas on her vacation? $25.65

d. The manufacturer advertised that the car would get 30 to 35 miles per gallon. How does Portia's mileage compare to the advertised estimates?

6a. More than 5¢ but less than 10¢; if the card were worth 50 minutes, $\frac{\$5.00}{50 \text{ minutes}} = \$0.10$ per minute; if the card were worth 100 minutes, $\frac{\$5.00}{100 \text{ minutes}} = \$0.05$ per minute.

6d. Possible answer: The prepaid calling card because it charges less per minute.

Exercise 7 [Alert] Some students may have difficulty with the dimensional analysis in 7c. Encourage them to try different units for ratios as they try to build ratios with units that will cancel, leaving *ounces* in the numerator and *bar* in the denominator.

7e. Possible answers: If comparing price per ounce, neither is a better value because both bars are 25¢ per ounce. If comparing price per bar, they should buy the Chewy Granola Bars because they are cheaper per bar. If Marie and Tracy prefer fewer, larger bars, they should buy Crunchy Granola Bars. If they prefer more, smaller bars, they should buy Chewy Granola Bars.

8d. Portia's gas mileage is more than 12% or 4.4 miles per gallon greater than the higher estimate. It is more than 31% greater than the lower estimate.

Exercise 9d Students must make the assumption that the food for the dogs costs the same amount per pound.

Exercise 10 Calculator Notes 1A and 1L are relevant here.

10a.

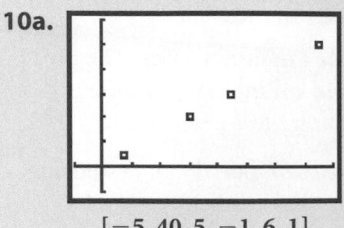

$[-5, 40, 5, -1, 6, 1]$

9. **APPLICATION** Chris' 85-pound black Labrador retriever, Tootsie, eats 40 pounds of dog food every 2 weeks. Each 40-pound bag costs $36.

   a. Write two ratios for this situation. Possible answers: $\dfrac{20 \text{ pounds of food}}{1 \text{ week}}$, $\dfrac{\$36}{1 \text{ bag of food}}$, $\dfrac{\$0.90}{1 \text{ pound of food}}$, $\dfrac{\$18}{1 \text{ week}}$.

   b. How much does it cost to feed Tootsie per year? **$936**

   c. If Cathy's 60-pound golden retriever, Clara, eats the same amount per pound of body weight as Tootsie, how much does Clara eat per week? $\dfrac{20}{85} = \dfrac{f}{60}$; **14 pounds of food per week**

   d. How much less does Cathy spend on dog food each year than Chris? **$280.80**

10. Emily bought a used scooter to use on short trips. She timed her travel on the scooter from home to her favorite places. Then she used a map to estimate the distances. She made this table:

Travel Time with Scooter

Location	Time (minutes)	Distance (miles)
Music store	15	2
Juice bar	3.75	0.5
Thrift store	22.5	3
Jeremy's house	37.5	5

   a. Enter the table values into your calculator in two lists. Use list L₁ for the times and list L₂ for the distances. Make a scatter plot of points with coordinates in the form (*time, distance*). What patterns do you notice? **Students should notice that the points are on a line.**

   b. In list L₃, calculate L₂ ÷ L₁. What pattern do you notice? Explain in words what the values in list L₃ represent. **The values in list L₃ should all be 0.13333, the rate of speed in miles per minute.**

   c. Find Emily's average speed on the scooter in miles per hour. Your findings in 10b should be helpful. $\dfrac{8 \text{ miles}}{1 \text{ hour}}$

▶ Review

11. Solve each proportion for x.

   a. $\dfrac{x}{3} = \dfrac{7}{5}$ $x = \dfrac{21}{5}$ or 4.2

   b. $\dfrac{2}{x} = \dfrac{9}{11}$ $x = \dfrac{22}{9}$ or $2.\overline{4}$

   c. $\dfrac{x}{c} = \dfrac{d}{e}$ $x = \dfrac{cd}{e}$

12. In 1997, there were an estimated 25,779,000 foreign-born people living in the United States. Of those people, 6,822,000 were born in Asian countries. (*www.census.gov*)

   a. What percent of foreign-born people in the United States were born in Asia? **approximately 26.5%**

   b. The Philippines was the birthplace of 16.6% of the Asian-born people living in the United States. How many people who lived in the United States in 1997 were born in the Philippines? **1,132,452 people**

13. Hannah bought a small bag of jelly beans to share with her friends, Asha, Annelise, Patrick, and Jamal. They counted the candies and got the results shown on the blue note paper.

   a. Draw a circle graph to show the proportion of the different colors of candies in the bag.

   b. Then the friends bought a large bag of jelly beans to share. They counted the number of each color in the large bag and got the results shown on the green note paper.
   Draw a relative frequency bar graph for the large bag of jelly beans.

   c. Compare the distribution of colors in the small bag to the distribution in the large bag.

   d. If the five friends share the jelly beans in the large bag, how many pieces should each one get? Will it be fair if each one gets all the jelly beans of one color?

Small Bag
12 yellow
13 purple
11 green
14 orange
10 red

Large Bag
78 yellow
96 purple
72 green
76 orange
104 red

85 pieces. No, because each color does not occur equally often.

14. This table shows information that Ms. Osborne collected about the students in her art class.

   a. Write the ratio of right-handed boys to right-handed girls. 10 : 9

   b. Write the ratio of left-handed boys to left-handed girls. 2 : 1

   c. Write the ratio of ambidextrous boys to boys in the class. 1 : 15

   d. Write the ratio of ambidextrous girls to girls in the class. 2 : 13

   e. Draw a relative frequency graph to represent this data set. Explain why you chose to draw either a circle graph or a bar graph.

Ms. Osborne's Class

	Right-handed	Left-handed	Ambidextrous
Boys	10	4	1
Girls	9	2	2

IMPROVING YOUR **REASONING** SKILLS

This problem is adapted from an ancient Chinese book, *The Nine Chapters of Mathematical Art*.

A city official was monitoring water use when he saw a woman washing dishes in the river. He asked, "Why are there so many dishes here?" She replied, "There was a dinner party in the house." His next question was "How many guests attended the party?" The woman did not know but replied, "Every two guests shared one dish for rice. Every three guests used one dish for broth. Every four guests used one dish for meat. And altogether there were sixty-five dishes used at the party." How many guests attended the party?

This is a detail from the 17th-century Chinese scroll painting *Landscapes of the Four Seasons* by Shen Shih-Ch'ing.

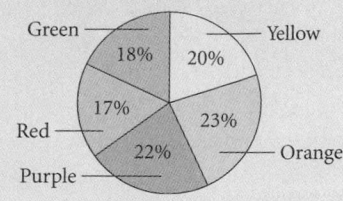

13a. Jelly Beans in Small Bag

Green — 18%
Yellow — 20%
Orange — 23%
Purple — 22%
Red — 17%

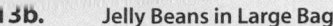

13b. Jelly Beans in Large Bag

(bar graph, Percent vs Color: Yellow, Orange, Purple, Red, Green)

13c. Small bag: orange occurs most frequently, red least frequently; large bag: red occurs most often, green least often

14e. Sample bar graph:

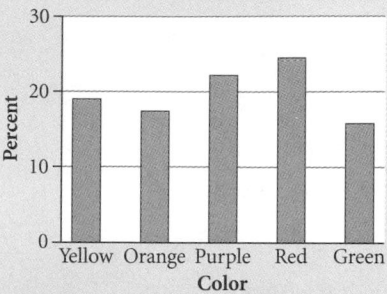

Hand Preference

(bar graph, Percent vs Right Handed, Left Handed, Ambidextrous; Boys and Girls)

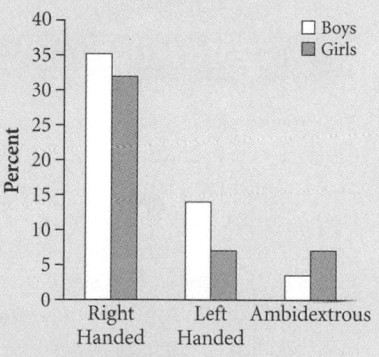

Reasoning Skills Answer

You could find the rate as about $1.083 \frac{\text{dishes}}{\text{guest}}$, which translates into $0.923 \frac{\text{guests}}{\text{dish}}$, and then multiply by 65 dishes. Or you can just use the ratio $\frac{13 \text{ dishes}}{12 \text{ guests}}$ and multiply the numerator and denominator by 5, to get $\frac{65 \text{ dishes}}{60 \text{ guests}}$. There were 60 guests.

IMPROVING **REASONING** SKILLS

Because you want the number of guests, and you know the number of dishes, students might think that you can find the rate $\frac{\text{guests}}{\text{dish}}$ and multiply by the number of dishes. Encourage them to try, but point out that the given rates of $2 \frac{\text{guests}}{\text{rice dish}}$, $3 \frac{\text{guests}}{\text{broth dish}}$, and $4 \frac{\text{guests}}{\text{meat dish}}$ have different units (kinds of dishes) in the denominators, so they can't be combined. The difference in units reflects the fact that the same guests will be sharing all three dishes. To get the same units in the denominator, use the reciprocals of the rates. The information translates into three ratios with the same denominators:

$$\frac{1 \text{ rice dish}}{2 \text{ guests}} \cdot \frac{1 \text{ broth dish}}{3 \text{ guests}} \cdot \frac{1 \text{ meat dish}}{4 \text{ guests}}$$

To find the total of dishes, get a common denominator of 12 guests and add:

$$\frac{6 + 4 + 3 \text{ dishes}}{12 \text{ guests}} = \frac{13 \text{ dishes}}{12 \text{ guests}}$$

Direct Variation

In Lesson 3.1, you worked with rates. Rates have a 1 in the denominator so they are convenient to calculate with. You also see patterns when you use a rate to make tables and graphs. In this investigation you'll use algebra to understand these patterns better.

PLANNING

LESSON OUTLINE

One day:

25 min	Investigation
10 min	Sharing
5 min	Example
5 min	Closing
5 min	Exercises

MATERIALS

- graph paper
- Calculator Notes 1E, 1H, 1I, 3A
- Longest Ship Canals (T), *optional*
- Direct and Inverse Variation (T), *optional*

TEACHING

The investigation in this lesson is the first step in finding a model to fit a scatter plot. Data points whose coordinates are directly proportional have an equation, a *direct variation*, of the form $y = kx$, whose graph is a straight line through the origin.

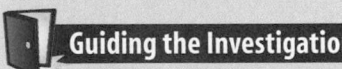

Guiding the Investigation

One step Prompt students to use a graph, a rate, an equation, and a calculator table to fill in the numbers missing from the table of lengths of ship canals.

Step 1 Two points have the same coordinates (80, 129) so students really plot only seven points. Again, remind students to mark off their axes in equal intervals.

See page 715 for graphs for Steps 1, 2, and 3.

Investigation
Ship Canals

You will need
- graph paper

In this investigation you will use data about canals to draw a graph and write an equation that states the relationship between miles and kilometers. You'll see several ways of finding the information that is missing from this table.

Longest Ship Canals

Canal	Length (miles)	Length (kilometers)
Albert (Belgium)	80	129
Alphonse XIII (Spain)	53	85
Houston (Texas)	50	81
Kiel (Germany)	62	99
Main-Danube (Germany)	106	171
Moscow-Volga (Russia)	80	129
Panama (Panama)	51	82
St. Lawrence Seaway (Canada/U.S.)	189	304
Suez (Egypt)	101	162
Trollhätte (Sweden)	54 or 55	87

(*The Top 10 of Everything 1998*, p. 57)

Step 1
Step 1 (189, 304) (106, 171) (80, 129) (80, 129) (62, 99) (53, 85) (51, 82) (50, 81)

Step 2
Step 2 A linear pattern or "rate." Find 101 miles on the *x*-axis and go up to the line; the *y*-value is the corresponding length in kilometers. Use a reverse process to approximate the length in miles of the Trollhätte Canal.

Step 1 Carefully draw and scale a pair of coordinate axes for the data in the table. Let *x* represent the length in miles and *y* represent the length in kilometers. Plot points for the first eight coordinate pairs.

Step 2 What pattern or shape do you see in your graph? Connect the points to illustrate this pattern. Explain how you could use your graph to approximate the length *in kilometers* of the Suez Canal and the length *in miles* of the Trollhätte Canal.

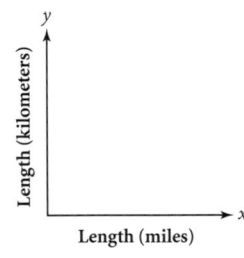

Step 3
Step 3 On your calculator, make a plot of the same points and compare it to your hand-drawn plot. Use list L₁ for lengths in miles and list L₂ for lengths in kilometers. [▶ See **Calculator Note 1E** to review this type of plot. ◀]

LESSON OBJECTIVES

- Learn the properties of a direct variation equation
- Graph a direct variation equation
- Read a direct variation graph to find missing values in the corresponding table
- Use a direct variation equation to extrapolate values from a given data set
- Develop intuitive understanding of the concepts of slope and linear equation

NCTM STANDARDS

CONTENT	PROCESS
• Number	• Problem Solving
• Algebra	• Reasoning
Geometry	• Communication
• Measurement	• Connections
• Data/Probability	• Representation

Step 4 The values from 1.5968 to 1.62 represent the ratio of km to mi.

Step 4 Use list L3 to calculate the ratio $\frac{L_2}{L_1}$. [▶ 🖳 See **Calculator Note 1I** to review using lists to calculate this way. ◀] Explain what the values in list L3 represent. If you round each value in list L3 to the nearest tenth, what do you get? 1.6

Step 5 161.6 kilometers; yes, use a proportion; divide 87 by the rate, or use a reciprocal rate; $t = 54.375$ miles.

Step 5 Use the rounded value you got in Step 4 to find the length in kilometers of the Suez Canal. Could you also use your result to find the length in miles of the Trollhätte Canal?

The number of kilometers is the same in every mile, so the value you found is called a **constant**.

Step 6 Multiply by the rate: $1.6 \cdot x$.

Step 6 How can you change x miles to y kilometers? Using variables, write an equation to show how miles and kilometers are related. $y = 1.6x$

Step 7 For the Suez Canal: $y = 1.6(101)$, $y = 161.6$. For the Trollhätte Canal: $87 = 1.6x$, $x = 54.375$.

Step 7 Use the equation you wrote in Step 6 to find the length in kilometers of the Suez Canal and the length in miles of the Trollhätte Canal. How is using this equation like using a rate?

Step 8 Graph your equation on your calculator. [▶ 🖳 See **Calculator Note 1H** to review graphing equations. ◀] Compare this graph to your hand-drawn graph. Why does the graph go through the origin?

Step 9 Trace the graph of your equation. [▶ 🖳 See **Calculator Note 1H** to review tracing equations. ◀] Approximate the length in kilometers of the Suez Canal by finding when x is approximately 101 miles. Trace the graph to approximate the length in miles of the Trollhätte Canal? How do these answers compare to the ones you got from your hand-drawn graph?

Step 10 Use the calculator's table function to find the missing lengths for the Suez Canal and the Trollhätte Canal. [▶ 🖳 See **Calculator Note 3A** to learn about the table function. ◀]

Step 11 In this investigation you used several ways to find missing values—approximating with a graph, calculating with a rate, solving an equation, and searching a table. Write several sentences explaining which of these methods you prefer and why.

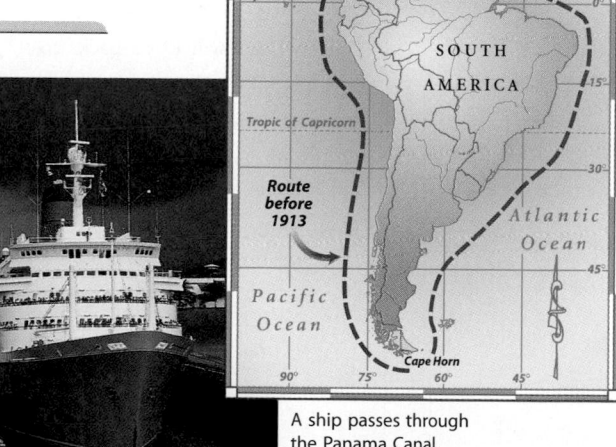

Panama Canal route

Equator

SOUTH AMERICA

Tropic of Capricorn

Route before 1913

Atlantic Ocean

Pacific Ocean

Cape Horn

A ship passes through the Panama Canal.

History CONNECTION

The Panama Canal allows ships to cross the strip of land between the Atlantic and Pacific Oceans. Before the canal was completed in 1913, ships had to sail thousands of miles around the dangerous Cape Horn, even though only 50 miles separate the two oceans.

Step 8 The calculator line should resemble the hand-drawn line. The line passes through the origin, 0 kilometers = 0 miles. The lines students drew may not pass through the origin; you could discuss factors that would cause the line to not fit perfectly (e.g., the rounded values in the table).

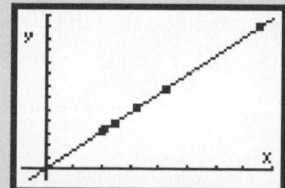

Step 9 For the Suez Canal, the y-value will be approximately 162 kilometers. For the Trollhätte Canal, using 87 kilometers for y results in an x-value of approximately 54 or 55 miles. These values should be close to those approximated from the hand-drawn graph.

Step 10 Using an x-value increase of 1, the table gives a y-value for the Suez Canal of exactly 161.6 kilometers. For the Trollhätte Canal, using an x-value increase of 0.01, the nearest y-value to 87 kilometers is 87.008, corresponding to an x-value of 54.38 miles.

Step 4 [Alert] If students accidentally enter kilometers in list L1 and miles in list L2, or if they calculate $\frac{L_1}{L_2}$, they will get a rate of approximately 0.62, representing the number of miles in a kilometer.

Students will see that all the values are approximately equal. [Ask] "Should they all be exactly equal?" "Why or why not?" [The ratios, representing the number of kilometers in one mile, should be equal. They may not be, because the numbers in the table have been rounded to whole numbers.] You may also want students to compare the ratio with the unit conversion factor of Lesson 2.3.

Step 5 To help students see the kilometer-to-mile conversion factor as a rate, explain it as "for every mile you drive, you pass over 1.6 kilometers." This is equivalent to saying, "There are 1.6 kilometers in every mile," as expressed in the language of Lesson 3.1.

Step 6 Other than proportions, this may be the first equation students have written themselves. [Alert] Many students have difficulties with the leap from the concrete discussion of real-world models to abstract variables like x and y. Use the rate in Step 4 and discuss what x and y represent.

Step 7 Solving the equation $87 = 1.6x$ is a new challenge. Encourage guess-and-check and working backward from 87 to x, as well as dividing both sides of the equation by 1.6.

Step 8 [Alert] Although students graphed the equation $y = x$ in Lesson 1.7, this more complicated equation may prove difficult for some. Be ready with calculator notes. Students should see that a line through the origin represents pairs of values that are the constant multiples of each other.

Have students show the four ways of filling in the numbers missing from the table: graph, rate, equation, and calculator table. Have them present ways they found of solving the equations in Step 7.

To encourage critical thinking, ask the class to critique each presentation thoroughly before you say anything about it. A critique includes praise as well as criticism. Listen carefully for how students' critiques reflect their own understanding of the ideas.

Introduce the term *direct variation* for the equation $y = kx$. Emphasize that the three letters have different meanings: x and y represent collections of numbers in the table, but k is a never-changing number called the *constant of variation*.

[Ask] "How does a direct variation relate to the idea of *proportional* from Chapter 2?" [The x- and y-values in a table are proportional when there's a constant by which you can multiply each x-value to get the corresponding y-value—you don't have to perform any other operation. That is, x and y are proportional when there's a direct variation relating them.] We also use the term *directly proportional* for this case. We'll look at inversely proportional quantities in Lesson 3.4.

Assessing Progress

Your observations can reveal students' ability to make a **scatter plot** and their understanding of **rates** and of graphs as collections of points.

▶ **EXAMPLE**

Use this example if students need more thinking about rates and dimensional analysis. **[Ask]** "How can you decide which unit should be in the numerator and which in the denominator of a rate?" [The unit in the denominator should be what you are changing or multiplying by].

Ratios, rates, and conversion factors are closely related. In this investigation you saw how to change the ratio $\frac{129 \text{ km}}{80 \text{ mi}}$ to a rate of approximately 1.6 kilometers per mile. You can also use that rate as a conversion factor between kilometers and miles. The numbers in the ratio vary, but the resulting rate remains the same, or constant. Kilometers and miles are **directly proportional**—there will always be the same number of kilometers in every mile. When two quantities vary in this way, they have a relationship called **direct variation**.

Direct Variation

An equation in the form $y = kx$ is a **direct variation.** The quantities represented by x and y are **directly proportional,** and k is the **constant of variation.**

You can represent any ratio, rate, or conversion factor with a direct variation. Using a direct variation equation or graph is an alternative to solving proportions. A direct variation equation can also help you organize calculations with rates.

EXAMPLE

A grocery store advertises a sale on soda.

a. Write a rate for the cost per six-pack.

b. Write an equation showing the relationship between the number of six-packs purchased and the cost.

c. How much will 15 six-packs cost?

d. Sol is stocking up for his restaurant. He bought $210 worth of soda. How many six-packs did he buy?

▶ **Solution**

a. The ratio given is $\frac{\$6.00}{4 \text{ six-packs}}$. This simplifies to a rate of $1.50 per six-pack.

b. Use x for the number of six-packs and y for the cost in dollars. Write a proportion.

$$\frac{y}{x} = \frac{1.50}{1}$$
y corresponds to 1.50 and x corresponds to 1.

$$x \cdot \frac{y}{x} = \frac{1.50}{1} \cdot x$$
To isolate y, multiply both sides by x.

$$y = 1.50x$$
Your result is a direct variation equation.

The solution to part b asks if every point on the graph of the equation makes sense in this situation. Assuming the six-packs are sold only as six-packs, the x-values are limited to positive integers—you couldn't buy 0.25 of a six-pack. In this situation, the graph is a continuous model of a discrete relationship.

The other proportion suggested in the solution also results in $y = 1.50x$. Your goal is to have students realize that the rate can be directly substituted for k

to arrive at an equation. Try to have students think about how the "per" (denominator) units of the rate will cancel with the units of x.

If students have graphed their equation and are trying to evaluate it at $x = 15$ by tracing, they might not find a point with x-coordinate exactly 15. They can get close by zooming in on the table.

The constant, k, is the rate $1.50 per six-pack. Does every point on the graph of this equation make sense in this situation?

Would you get the same equation if you started with the proportion $\frac{y}{x} = \frac{6.00}{4}$? Do you see another way to find the equation once you know the rate?

c. You can trace the graph to find the point where $x = 15$, or you can substitute 15 into the equation for x, the number of six-packs.

$$y = 1.50(15) = 22.50$$

Fifteen six-packs cost $22.50.

Cost of Soda

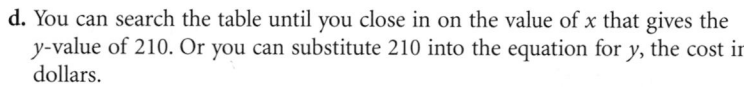

d. You can search the table until you close in on the value of x that gives the y-value of 210. Or you can substitute 210 into the equation for y, the cost in dollars.

$210 = 1.50x$	Substitute 210 for y into the original equation.
$\dfrac{210}{1.50} = \dfrac{1.50x}{1.50}$	To isolate x, divide both sides by 1.50.
$140 = x$	Simplify.

Sol purchased 140 six-packs for $210.

EXERCISES

You will need your calculator for problems **1, 2, 3, 6,** and **9.**

▶ Practice Your Skills

Let x represent distance in miles and y represent distance in kilometers. Enter the equation $y = 1.6x$ into your calculator. Use it for problems 1–3.

1. Trace the graph of $y = 1.6x$ to find each missing quantity. Adjust the window settings as you proceed.

  a. 25 miles ≈ $\boxed{40}$ kilometers

  b. 120 kilometers ≈ $\boxed{75}$ miles

2. Use the calculator table function to find the missing quantity.

  a. 55 miles ≈ $\boxed{88}$ kilometers

  b. 450 kilometers ≈ $\boxed{}$ miles
  281

3. Find the missing values in this table. Round each value to the nearest tenth.

Distance (miles)	Distance (kilometers)
2.8	4.5
7.8	12.5
650.0	1040.0
937.5	1500.0

Closing the Lesson

As needed, remind students that a **direct variation** is an equation, often written $y = kx$, that relates two **directly proportional** variables x and y using a **constant of variation** k.

BUILDING UNDERSTANDING

As students work with direct variations and their graphs, continue to stress the importance of dimensional analysis.

ASSIGNING HOMEWORK

Essential	1–4, 7
Performance assessment	10
Portfolio	6, 9
Journal	6d
Group	5, 8, 9
Review	11–13

▶ Helping with the Exercises

Exercise 1 to 2 Some students may prefer to use calculator tables. Others may prefer the graph.

Students can also use the first dynamic algebra exploration at www.keymath.com/DA to complete these exercises.

Exercise 3 Students might recognize that the data tables they are developing are input-output tables. The widely differing values for x should help them connect the input-output process with the infinite length of the line that represents the relationship.

Exercise 4 Remind students having difficulties with 4c and 4d that they can think of the right side as having a denominator of 1, so they might use their method of solving proportions.

Exercise 5 Some students might be confused by having two variables with names other than *x* and *y*. Remind them that variable names in Chapter 1 were not *x* and *y*.

6b.

[0, 80, 10, 0, 18, 1]

6c. A: $0.18; B: $0.21. These are the constants of variation in the equations, rounded to the nearest tenth.

Exercise 6d Encourage reference to the steepness of the lines. You might begin to use the word *slope*, but save its careful definition for Chapter 5.

6d. The steeper graph with the higher rate represents the more expensive market (Market B). The cheaper market is associated with the less steep graph.

---

4. Solve each equation for *x*.

   a. $14 = 3.5x$ 4
   b. $x = 45(0.62)$ 27.9
   c. $\frac{x}{7} = 0.375$ 2.625
   d. $\frac{12}{x} = 0.8$ 15

5. The equation $c = 1.25f$ shows the direct variation relationship between the length of fabric and its cost. The variable *f* represents the length of the fabric in yards, and *c* represents the cost in dollars. Use this equation to answer these questions.

   $3.13 a. How much does $2\frac{1}{2}$ yards of fabric cost?

   4 yards b. How much fabric can you buy for $5?

   c. What is the cost of each additional yard of fabric? $1.25

Christo (b 1935, Bulgaria) and Jeanne-Claude (b 1935, Morocco) are environmental sculptors who wrap large objects and buildings in fabric. This is the German Reichstag in 1995.

▶ **Reason and Apply**

6. **APPLICATION** Market A sells 7 ears of corn for $1.25. Market B sells a baker's dozen (13 ears) for $2.75.

   a. Copy and complete the tables below showing the cost of corn at each market.

   Market A

Ears	7	14	21	28	35	42
Cost	1.25	2.50	3.75	5.00	6.25	7.50

   Market B

Ears	13	26	39	52	65	78
Cost	2.75	5.50	8.25	11.00	13.75	16.50

   b. Let *x* represent the number of ears of corn and *y* represent cost. Find equations to describe the cost of corn at each market. Use your calculator to plot the information for each market on the same set of coordinate axes. Round the constants of variation to three decimal places. Market A: $y = 0.179x$; Market B: $y = 0.212x$

   c. If you wanted to buy only one ear of corn, how much would each market charge you? How do these prices relate to the equations you found in 6b?

   d. How can you tell from the graphs which market is the cheaper place to buy corn?

Why is 13 called a baker's dozen? In the 13th century, bakers began to form guilds to prevent dishonesty. To avoid the penalty for selling a loaf of bread that was too small, bakers began giving 13 whenever a customer asked for a dozen.

7. Bernard Lavery, a resident of the United Kingdom, has held several world records for growing giant vegetables. The graph shows the relationship between weight in kilograms and weight in pounds.

  a. Use the information in the graph to complete the table.

Bernard Lavery's Vegetables

Vegetable	Weight (kilograms)	Weight (pounds)
Cabbage	56	123
Summer squash	49	108
Zucchini	29	64
Kohlrabi	28	62
Celery	21	46
Radish	13	28
Cucumber	9	20
Brussels sprout	8	18
Carrot	5	11

(*The Top 10 of Everything 1998*, p. 98)

Relationship Between Kilograms and Pounds

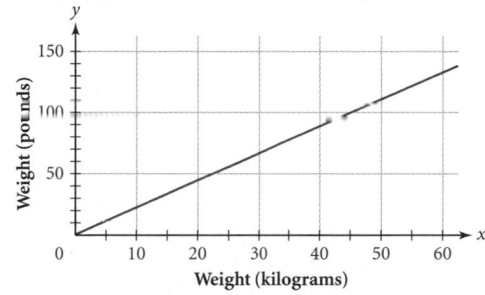

Weight (pounds) vs Weight (kilograms)

  b. Calculate the rate of pounds per kilogram for each vegetable entry from the table. Use the rate you think best represents the data as the constant of variation. Write an equation to represent the relationship between pounds and kilograms. $y = 2.2x$

  c. Use the equation you wrote in 7b to find the weight in kilograms for a pumpkin that weighs 6.5 pounds. 2.95 kilograms

  d. Use the equation you wrote in 7b to find the weight in pounds for an elephant that weighs 3600 kilograms. 7920 pounds

  e. How many kilograms are in 100 pounds? How many pounds are in 100 kilograms? 100 lb = 45.4$\overline{5}$ kg; 100 kg = 220 lb

8. As part of their homework assignment, Thu and Sabrina each found equations from a table of data relating miles and kilometers. One entry in the table paired 150 kilometers and 93 miles. From this pair of data values, Thu and Sabrina wrote different equations.

  a. Thu wrote the equation $y = 1.61x$. How did he get it? What does 1.61 represent? What do x and y represent?

  b. Sabrina wrote $y = 0.62x$ as her equation. How did she get it? What does 0.62 represent? What do x and y represent?

  c. Whose equation would you use to convert miles to kilometers?

  d. When would you use the other student's equation?
  Sabrina's equation may be more convenient when converting kilometers to miles.

Exercise 7a From the graph, students should be able to find the missing values to within 5 pounds or kilograms of the given answers.

Exercise 7b Students might check their conjectured equation by plotting it on a graph with the data points.

8a. Thu calculated the ratio $\frac{150}{93} = 1.61$. 1.61 is the rate of kilometers per mile. x represents miles and y represents kilometers.

8b. Sabrina calculated the ratio $\frac{93}{150} = 0.62$. Her constant is the rate of miles per kilometer. x represents kilometers and y represents miles.

8c. Thu's equation may be more convenient because you can directly multiply the number of miles by 1.61.

Exercise 9 The *exchange rate* between two monetary systems is the conversion factor.

You may want to take some class time for students to come up with a common list of ten items.

9a. A sample answer is to use U.S. coins and denominations in dollars. {100, 50, 20, 10, 5, 1, 0.50, 0.25, 0.10, 0.05, 0.01}

9b. Multiply the list by the exchange rate. For example, to convert to Japanese yen, multiply the list by 108.770. {10877, 5438.5, 2175.4, 1087.7, 543.85, 108.77, 54.385, 27.19, 10.88, 5.44, 1.09}

9c. Divide list L2 by the exchange rate to obtain the original values.

9d. Use dimensional analysis.

$$\frac{2119.150 \text{ liras}}{1 \text{ dollar}} \cdot \frac{1 \text{ dollar}}{2.140 \text{ marks}}$$

= 990.257 liras per mark.

10d.

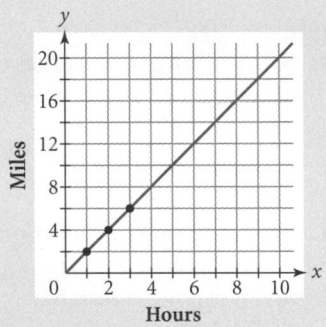

10e. 2 mi/hr. This represents the constant walking speed.

11c. 65 mph is 104 kilometers per hour. A speed limit sign might post 100 kilometers per hour.

---

9. APPLICATION If you're planning to travel to another country, you will need to learn about its monetary system. This table gives some exchange rates that tell how many of each monetary unit are equivalent to one U.S. dollar.

International Monetary Units

Country	Monetary unit	Exchange rates (units per American dollar)
Brazil	real	1.790
Germany	mark	2.140
Italy	lira	2119.150
Japan	yen	108.770
Mexico	peso	9.350
Pakistan	rupee	45.400
United Kingdom	pound	0.670

(Federal Reserve Bank of New York for August 2, 2000)

a. Make a list of ten items and the price of each item in U.S. dollars. Enter these prices into list L1 on your calculator.

b. Choose one of the countries in the table and convert the U.S. dollar amounts in your list to that country's monetary unit. Use list L2 to calculate these new values from list L1.

c. Using list L3, convert the values in list L2 back to the values in list L1.

d. Describe how you would convert marks to liras.

e. Check the business section of a newspaper for current exchange rates for five countries of your choice.

10. If you travel at a constant speed, the distance you travel is directly proportional to your travel time. Suppose you walk 3 miles in 1.5 hours.

a. How far would you walk in 1 hour? 2 miles

b. How far would you walk in 2 hours? 4 miles

c. How much time would it take you to walk 6 miles? 3 hours

d. Represent this situation with a graph.

e. What is the constant of variation in this situation, and what does it represent?

f. Define variables and write an equation that relates time to distance traveled.
$d = 2t$ where d is distance traveled in miles and t is travel time in hours.

▶ **Review**

11. U.S. speed limits are posted in miles per hour (mph). Germany's Autobahn has stretches where speed limits are posted at 130 kilometers per hour.

a. How many miles per hour is 130 kilometers per hour? 81.25 mph

b. How many kilometers per hour is 25 mph? 40 kilometers per hour

c. If the United States used the metric system, what speed limit do you think would be posted for 65 mph?

12. APPLICATION Cecile started a business entertaining at children's birthday parties. As part of the package, Cecile comes in costume and plays games with the children. She also makes balloon animals and paints each child's face. When she started the business, she charged $3.50 per child, but she is rethinking what her charges should be so that she will make a profit.

Exercise 12 [Language] Most of the rates in the problem contain the word *per*. But $12 an hour is also a rate—$12 per hour.

a. The average children's party takes about 3 hours. Cecile wants to make at least $12 an hour. What is the minimum number of children she should arrange to entertain at a party at her current rate? **11 children**

b. The balloons and face paint cost Cecile about 60¢ per child. What percent is that of the fee per child? **approximately 17%**

c. Cecile decided to raise her rates so that the cost of supplies for each child is only 10% of her fee. If the supplies for the party cost 60¢ per child, what should she charge per child? **$6 per child**

13. Ms. Zany sometimes plays the "homework game" with her class as a way to check that students have done their homework. The students sit at tables numbered 1 to 8. Ms. Zany spins a spinner numbered from 1 to 10. (Each number is equally likely to be spun.) If she spins the number of a table (1 to 8), she checks the homework of the students sitting at that table. If she spins a 9, she doesn't check anyone's homework. If she spins a 10, she checks everyone's homework.

a. What is the probability that your homework will get checked if you are sitting at table 5 in Ms. Zany's class? $\frac{2}{10} = \frac{1}{5}$

b. Does the probability of getting your homework checked depend on the number of your table? Why or why not? **No. All the tables are equally likely to have their table number spun; 9 or 10 being spun affects all tables equally.**

IMPROVING YOUR REASONING SKILLS

Can all unit conversion problems be modeled by direct variation? Consider this table of temperatures in degrees Fahrenheit and degrees Celsius.

Calculate the ratios of degrees Fahrenheit to degrees Celsius for each city. How do they compare to the kilometers-to-miles ratios you found in the investigation? Plot degrees Celsius on one axis and degrees Fahrenheit on the other axis. How is this graph similar to the other graphs in this lesson? How is it different? Water freezes at 0°C. Is zero also the temperature at which water freezes in degrees Fahrenheit? Is the relationship between degrees Fahrenheit and degrees Celsius a direct variation?

City	°C	°F
Athens	30	86
Barcelona	18	64
Buenos Aires	11	52
Cairo	35	95
Johannesburg	15	59
Phoenix	40	104
Rio de Janeiro	25	77

IMPROVING REASONING SKILLS

The quantities are not directly proportional because there's no constant multiplier for finding the temperature in one system from the corresponding temperature in the other system. Though the points on the graph lie in a straight line, the line does not go through the origin.

LESSON

3.3

Scale Drawings and Similar Figures

To design a building, an architect makes a scale drawing to show what the floor plan will look like. In the finished building the floor plan will look like the scale drawing, only larger. Maps of towns and cities are scale drawings that show how far and in which directions the roads go. Even school pictures come in large, medium, and small sizes. You can think of them as scale images of yourself. In all of these situations, a rate or **scale factor** relates the measurements of the drawing or image to the measurements of the real thing. When you create or interpret scale drawings, you use direct variation.

Objects in mirror are closer than they appear.

MANDATED BY FEDERAL LAW ON CONVEX MIRRORS ON AUTOMOBILES

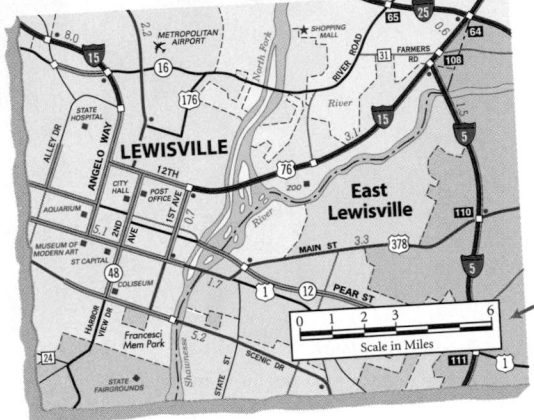

The scale of a map helps you determine actual measurements on the ground. Sometimes the scale is part of the map's legend.

Investigation
Floor Plans

You will need

• a centimeter ruler

This scale drawing shows the floor plan of an apartment. Use it to investigate how rate and direct variation apply to scale drawings.

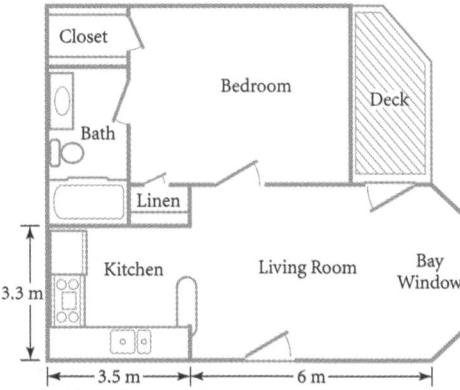

Step 5 $y = 1.5x$ where x is the drawing measurement in centimeters and y is the actual length in meters.

closet: 2 m × 1.5 m
bath: 2 m × 3.9 m
kitchen: 3.3 m × 3.5 m
linen: 1.5 m × 1 m
living room: (w/o bay window):
    6 m × 4.3 m

PLANNING

LESSON OUTLINE

One day:

5 min	Introduction
20 min	Investigation
5 min	Sharing
10 min	Examples
5 min	Closing
5 min	Exercises

MATERIALS

• Apartment Floor Plan (W), *optional*

• centimeter ruler

• white paper or centimeter graph paper

TEACHING

The study of similar figures in this lesson shows how rates and direct variations can provide a shortcut to setting up and solving proportions, especially with the help of dimensional analysis.

 Guiding the Investigation

One step Pose the problem, "Here is a floor plan of an apartment. The manager of the apartment complex wants to build a small model of this apartment to show apartment hunters. She wants a 6-meter wall of the apartment to be 10 centimeters in the model. How can she quickly determine from the floor plan how long to make each wall of the model?" As students work, encourage a variety of approaches. Ask to see ratios written with units. Ask to see an equation.

LESSON OBJECTIVES

• Create and interpret scale drawings

• Investigate the definition and properties of similar polygons

NCTM STANDARDS

CONTENT		PROCESS	
•	Number	•	Problem Solving
•	Algebra	•	Reasoning
•	Geometry		Communication
•	Measurement	•	Connections
	Data/Probability	•	Representation

Step 1

Actual (m)	Drawing (cm)
6.0	4.0
3.5	2.3
3.3	2.2

Step 1 Using a centimeter ruler, measure the three lengths given on the floor plan. Use a table to pair each actual length in meters with the scale drawing measurement in centimeters.

Step 2 Compare each pair of measurements in a ratio.

$$\frac{\textit{actual measurement in meters}}{\textit{scale drawing measurement in centimeters}}$$

Use your calculator to convert each ratio to a decimal. What units should you apply to each decimal number? How do the numbers compare? Explain your findings.

Step 2 Each ratio is approximately 1.5. The units are meters per centimeter. The ratios are a rate; if everything in the drawing is to scale, the measurements are proportional.

Step 3 State the scale for the drawing and explain how you got it. Then write this scale as a rate.

Step 3 Sample answer: 1 centimeter represents 1.5 meters, a rate of $\frac{1.5 \text{ m}}{1 \text{ cm}}$

Step 4 Measure the length and width of the bedroom in centimeters on the scale drawing. Use the scale you wrote in Step 3 to calculate the dimensions of the actual bedroom.

Step 4 approximately 3.7 cm by 2.9 cm; approximately 5.6 m by 4.4 m

Step 5 Write a direct variation equation to relate the scale drawing to the actual apartment. Use this equation to calculate the actual dimensions of the other rooms.

Step 6 The manager of the apartment complex wants a small model of this apartment to show apartment hunters. She wants the 6-meter wall of the apartment to be 10 centimeters in the model. Write an equation that she can use to convert from the scale drawing to the model.

Step 6 $y = 2.5x$, where x is the drawing's measurement and y is the model's measurement

Step 7 On a clean sheet of paper, draw an accurate floor plan for the model of the apartment. Think carefully about how to find the length of each wall and the angles where the walls meet. Compare your floor plan to the scale drawing in the book. Describe how they are alike and how they are different.

Step 7 Sample answer: The model has walls that are proportionally longer and the angles have stayed the same. It is an enlargement of the book's diagram.

Career
CONNECTION

There are several ways to do three-dimensional scale drawings. Some drawings show the object from the top, bottom, and each side. Others show the object with a perspective drawing. Architects use Computer Aided Design (CAD) software to help with these drawings.

Step 3 If students can correctly identify the rate of 1.5 m per (1) cm, but have trouble with the scale, they can simply reverse the numbers: 1 cm per 1.5 m. **[Alert]** Students should be clear that this scale is not a conversion factor between meters and centimeters. There are 1.5 m in the apartment for each cm in the scale drawing. If necessary, verbalize the rate as "1.5 m in the actual building to 1 cm in the drawing." If students write the scale as 1 cm = 1.5 m, be sure they understand that they are not writing an equation as they would in a conversion factor. **[Language]** You can introduce the term *legend*. A legend (on a map or scale drawing) lists and explains the symbols used. It usually includes a note about the scale and might also include symbols for kinds of landmarks or roads. The term *legend* is used in the project on page 162.

Step 5 This can be a great leap for students who are working with equations for the first time. It will help if they can put it into words. **[Ask]** "What do you do to get from the measurement on the drawing to the actual measurement? How might you represent the measurements on the drawing with a variable?"

Step 6 [Language] Here the term *model* means a three-dimensional representation, smaller than life-size.

Extension Question
[Ask] "The manager has bought a set of doll house furniture to help people visualize space in the model apartment. The back of the toy couch is 5 cm tall. The toy closely resembles a real couch with a back height of 1.3 m. Will the toys be the right size for the model apartment?" [Encourage variety in students' approaches. The model described in Step 6 has a scale of $\frac{6 \text{ m}}{10 \text{ cm}}$, which is 0.6 m per cm. The toy furniture has a scale of $\frac{1.3 \text{ m}}{5 \text{ cm}}$, or 0.26 m/cm. If students recall that the scale drawing in the book had a scale

of 1.5 m per cm, they may intu-itively reason that as the scaled object gets bigger, the scale factor gets smaller. Hence, the furniture will be too big for the model. Alternatively, students may use reciprocal rates $\left(\frac{cm}{m}\right)$ to see that the model uses 1.7 cm to show each meter, while the couch uses 3.8 cm for each meter. Hence the back of the toy couch would need to be only $1.7 \frac{cm}{m} \cdot 1.3$ m or 2.21 cm to fit the scale of the model—5 cm is too big.]

SHARING IDEAS

Students might present results of Steps 3, 5, and 6, and the exten-sion question. **[Link]** In your responses, introduce the geomet-ric term *similar figures* for those with directly proportional side lengths and congruent angles. **[Ask]** "How might it happen that the actual shapes of apartment rooms do not turn out to be simi-lar to the shapes in the drawing." [The wall thickness might be ignored in the drawing.]

[Ask] "How is a scale factor like a conversion factor and a rate?" [A scale factor is a kind of rate, as are conversion factors that have 1 in the denominator.] **[Ask]** "How do scale factors relate to directly proportional quantities and direct variation?" [The quantities in the diagram and their corresponding quantities in the room are directly proportional. The scale factor is the constant of variation.]

USING THE QUOTE

Refer to the opening quote, *Objects in mirror are closer than they appear*, which appears on all convex automobile mirrors. **[ESL]** Remind students that a convex mirror bulges outward, as opposed to a concave mirror, which caves inward. **[Ask]** "Is the figure in the mirror similar to the original figure?" [The image in the mirror is certainly a different size, but the shape is distorted.]

The floor plan you drew in the investigation should have the same *shape* as the scale drawing in the book—your floor plan is just bigger. It is the same shape because you made the angles the same. It is proportionally bigger because you used the same ratio to increase the length of every wall. Your drawing and the one in the book are **similar figures.** Similar polygons have sides that are proportional and angles that are *congruent*. Would the apartment's actual floor plan be similar to the scale drawing? What might happen during construction that would make the apartment *not* similar to the drawing?

EXAMPLE A | Here are two similar figures.

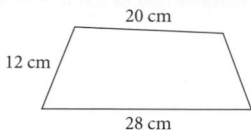

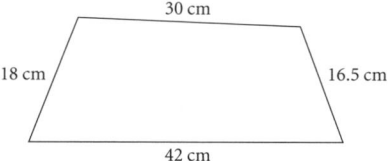

a. Write an equation to find the missing length.

b. Find the missing length on the smaller figure.

▶ **Solution** | **a.** Begin with a ratio using a pair of corresponding sides whose lengths are given, for example, 20 cm and 30 cm. Let x represent an unknown length from the smaller figure, and let y represent the corresponding unknown length from the larger figure.

$$\frac{y}{x} = \frac{30}{20}$$

Set up a proportion. y and 30 are both measurements from the large figure, and x and 20 are both measurements from the small figure.

$$x \cdot \frac{y}{x} = \frac{30}{20} \cdot x$$

To isolate y, multiply both sides by x.

$$y = 1.5x$$

Write as a direct variation.

So the equation $y = 1.5x$ can be used to find missing lengths. 1.5 is the scale factor.

b. The missing length on the smaller figure corresponds to 16.5 cm on the larger figure. Remember that in our equation, x represents a measurement on the smaller figure and y represents a measurement on the larger figure.

$$16.5 = 1.5x$$

Substitute 16.5 for y.

$$\frac{16.5}{1.5} = \frac{1.5x}{1.5}$$

To isolate x, divide both sides by 1.5.

$$x = 11$$

Reduce.

The missing length is 11 cm.

Assessing Progress

Through students' work and presentations, you can assess their skills at careful drawing and measuring, converting ratios to decimals, and deciding what fac-tors to multiply. Also look for the depth of understanding of **direct variations.**

▶ **EXAMPLE A**

This example will help students who need more practice with scale factors. **[ESL]** Be sure students understand that the word *congruent* means equal in measure. Stress that saying "Similar figures have the same shape" might be misinterpreted. Two figures could be thought of as having the same shape (rect-angular, for example) without being similar. Motivate the steps with a question. **[Ask]** "How will you find the length of the missing side? What information is useful? Can we use it to set up a proportion?"

EXAMPLE B

Sean and Jon are making a map of their school. After taking many measurements, they decided that 1 inch on their map should represent 20 feet on the actual school grounds. Write an equation to help them make all the conversions. Explain how a graph of this equation could help Sean and Jon make their map.

▶ **Solution**

In a direct variation, if you substitute a value for x, you only have to multiply to find y. Since their measurements are currently in feet, Sean and Jon should let x represent actual lengths in feet and y represent scale measurements in inches.

$$\frac{y}{x} = \frac{1}{20}$$ y corresponds to 1 inch and x to 20 feet.

$$x \cdot \frac{y}{x} = \frac{1}{20} \cdot x$$ To isolate y, multiply both sides by x.

$$y = \frac{1}{20}x$$ Write as a direct variation.

With a graph of this equation, Sean and Jon can trace along the line to read each measurement. For example, the point (50, 2.5) means that a distance of 50 feet on school grounds is 2.5 inches on the map. They can also use the table function to find map distances.

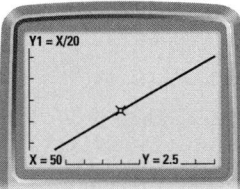

If you have to make several conversions, using equations, list calculations, graphs, and tables can save you time.

Career
CONNECTION

Many artists create very large works. Often the artist will begin with a smaller model or sketch and then apply properties of similarity to make the larger object. Maya Lin (shown below) used both sketches and a model to design the Vietnam Veterans Memorial in Washington, D.C.

▶ **EXAMPLE B**

This is a good example for students who haven't yet seen rates and direct variation as a shortcut to solving proportions. Encourage them to use dimensional analysis. Help them see that the variables x and y can represent whatever they choose, though usually y represents the amounts to be calculated from given values of x.

Closing the Lesson

As needed, remind students that a **scale factor** is a rate (with denominator 1) that describes how many times as large one object is as a **similar** object. To calculate a length on the first object, you use the scale factor as the constant in a direct variation.

BUILDING UNDERSTANDING

As students solve problems involving similar figures, encourage a variety of approaches, including proportions and rates. If you stress the use of dimensional analysis, students may come to see the equivalence of these approaches.

ASSIGNING HOMEWORK

Essential	1–4, 6, 7, 13, 14
Performance assessment	5, 8, 10
Portfolio	11
Journal	13
Group	3, 8, 9, 12
Review	15–18

▶ **Helping with the Exercises**

Exercise 3 [Alert] Students who have trouble with spatial orientation need help to see which sides correspond.

EXERCISES

You will need your calculator for problem **11.**

▶ **Practice Your Skills**

1. Use the equation $y = \frac{1}{20}x$ to find the unknown value.

   a. Find y if $x = 15$. $y = \frac{3}{4} = 0.75$

   b. Find y if $x = 40$. $y = 2$

   c. Find x if $y = 5$. $x = 100$

   d. Find x if $y = 5.4$. $x = 108$

2. Write two different proportions that you can use to find the length of the missing side in the similar triangles shown. Verify that both equations produce the same result. $\frac{x}{10} = \frac{15}{20}$ or $\frac{x}{15} = \frac{10}{20}$, $x = 7.5$ cm

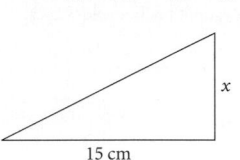
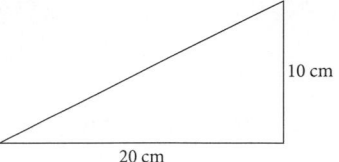

3. These polygons are similar:

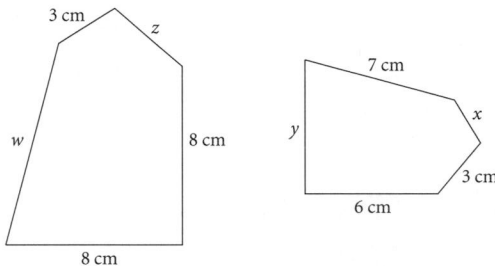

   a. Write an equation that relates lengths on the larger polygon to lengths on the smaller polygon. $l = \frac{8}{6}s$

   b. Use your equation to find the unknown side lengths. $w = 9.\overline{3}$ cm; $x = 2.25$ cm; $y = 6$ cm; $z = 4$ cm

4. **APPLICATION** Model train scales are given as $\frac{length\ of\ model\ in\ feet}{length\ of\ train\ in\ feet}$.

   An N-gauge model train has a scale of 1 : 160. An HO-gauge model train has a scale of 1 : 87.

   a. If an N-gauge caboose is 3 inches long, what is the length of the actual caboose? 40 ft

   b. What is the length of an HO scale model of the caboose in part a? 0.46 ft, or approximately $5\frac{1}{2}$ in.

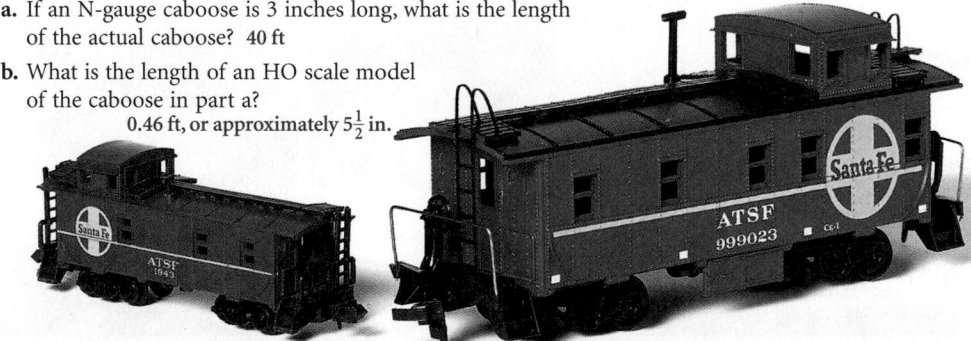

5. **APPLICATION** The scale on a map reads "1 inch = 15 miles."

    a. How far apart are two towns that are 2.8 inches apart on the map? 42 miles

    b. Suppose you know that Acme and Bates are 22 miles apart. How far apart should they be on the map? approximately 1.5 inches

    c. On the map, a distance on the highway is labeled as 47 miles. How long is the distance on the map? a little over 3 inches

    d. On the map, the distance across a large lake is 3.5 inches. How many miles across is the lake? 52.5 miles

▶ Reason and Apply

6. Find two pairs of rectangles that are similar. Explain how you know they are similar.

i.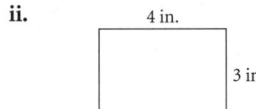

ii.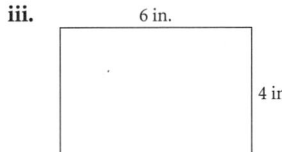

iii.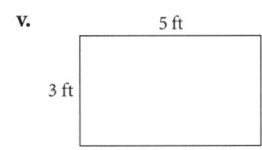

iv.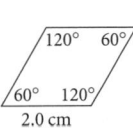

v.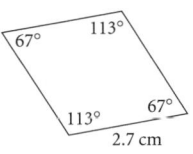

vi.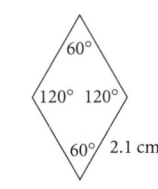

Rectangles i and iii are similar, and rectangles iv and vi are similar. In each pair, corresponding sides are proportional and corresponding angle measures are equivalent.

7. The polygons below are rhombuses (parallelograms with four equal sides). Find two pairs of rhombuses that are similar. Explain how you know they are similar.

i.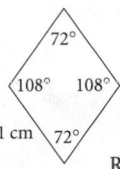

ii.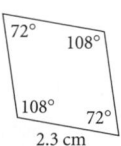

iii.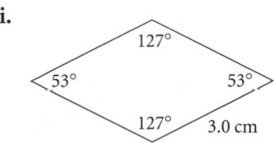

iv.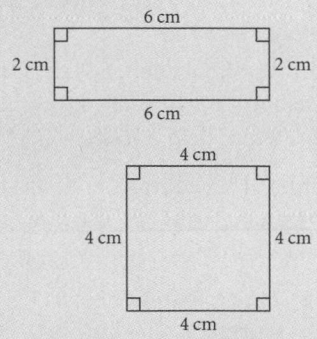

v.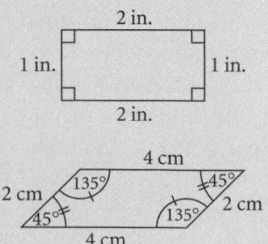

vi.

Rhombuses i and iii are similar, and rhombuses iv and v are similar. In each similar pair, corresponding sides are proportional and the angle measures are the same.

8. You know that similar figures must have proportional sides *and* congruent angles.

    a. Draw two polygons that have congruent angles but are *not* similar.

    b. Draw and label two polygons that have proportional sides but are *not* similar.

Exercise 5 Encourage students to convert the actual size to feet and compare it with a known size, such as the length of the classroom. **[Alert]** Be sure students realize that a scale written as "1 inch = 15 miles" is not an equation. The conversion "1 inch = 2.54 cm" is an equation.

Exercise 6 You may need to remind students that all angles of a rectangle are equal. **[Alert]** Comparing so many ratios might be difficult for some students. Suggest that they compare the first with all the others, then the second with those after it, the third with those after it, etc.

Exercise 7 Some students may have forgotten that all sides of a rhombus have the same length, so only the angles need to be compared to check for similarity.

Exercise 8 This exercise grows out of the two previous exercises.

8a. Answers will vary. One possibility is a nonsquare rectangle and a square.

8b. Answers will vary. One possibility is a rectangle and a parallelogram in which the ratio of the short side to the long side for each shape are the same.

Exercise 9 [Alert] Some students think of shadows as the three-dimensional space blocked from the sun. Stress that the "cast" shadow is a two-dimensional region on the ground. Even so, students may have trouble drawing the diagram they need, so you might want them to work in a group on this problem. If time and weather conditions permit, you might have students do this exercise outdoors.

Students can use the second dynamic algebra exploration at www.keymath.com/DA to extend this exercise.

9a.

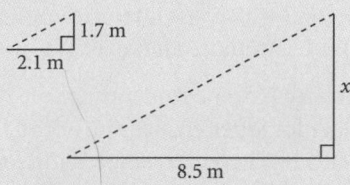

Exercise 10 Encourage students to fill in the units.

10a. x represents actual distance in miles and y represents map distance in inches.

10b. x represents map distance in inches and y represents actual distance in miles.

10c. Yes; it just depends which variable is substituted. 4.7 inches represents 58.75 miles.

Exercise 11 Encourage students to write in the units and use dimensional analysis.

11a. One possibility is 1 cm represents 250 km.

11b. Answers will vary. Using the scale of 1 cm = 250 km, an equation would be $y = \frac{1}{250}x$, where x is the actual river length in kilometers and y is the scale drawing length in centimeters. Approximations from a graph will vary based on window settings but should be around 26.7 cm, 25.8 cm, 25.2 cm, 23.9 cm, and 22.2 cm, respectively.

9. **APPLICATION** When objects block sunlight, they cast shadows, and similar triangles are formed. On a sunny day, Sunanda and Chloe measure the shadow of the school flagpole. It is 8.5 meters long. Chloe is 1.7 meters tall and her shadow is 2.1 meters long.

   a. Sketch and label your own diagram to represent this situation.

   b. Using the similar triangles in your diagram, write a proportion and find the height of the flagpole. $\frac{1.7}{2.1} = \frac{x}{8.5}$; $x \approx 6.88$ or 6.9 meters high

   c. Explain how you could use this method to find the height of a very tall tree. You could measure the length of the tree's shadow and write a proportion using a person's height and the length of his or her shadow.

10. **APPLICATION** Pablo and JoAnne are trying to find some actual distances from a map on which 1 inch represents 12.5 miles. Pablo gives the equation as

$$y = \frac{1}{12.5}x$$

and JoAnne writes the equation as

$$y = 12.5 \cdot x$$

   a. What does each variable represent in Pablo's equation?

   b. What does each variable represent in JoAnne's equation?

   c. Could you use either equation to convert a map distance of 4.7 inches to the actual distance in miles? How many miles does 4.7 inches represent?

11. **APPLICATION** For his social studies class, Tommy has to make scale drawings of the five longest rivers. Each drawing must fit on a piece of notebook paper, and all the drawings need the same scale so that the river lengths can be compared.

Five Longest Rivers

Name of river (location)	Length (km)
Nile (Tanzania/Uganda/Sudan/Egypt)	6670
Amazon (Peru/Brazil)	6448
Yangtze-Kiang (China)	6300
Mississippi-Missouri-Red Rock (United States)	5971
Yenisey-Angara-Selenga (Mongolia/Russia)	5540

(*The Top 10 of Everything 2001*, p. 20)

   a. Suggest a scale that Tommy could use. Be sure the longest river will fit on one piece of paper.

   b. Use your scale to write a direct variation equation. Graph the equation on your calculator. Use the graph to help Tommy convert each river length for his drawing.

Murchison Falls, Nile River

These two triangles are similar.

12. You know that similar figures have proportional sides. What about perimeter and area?

Side length	Perimeter	Area
6	24	36
4	16	16
3	12	9
2	8	4
8	32	64

   a. On graph paper draw five squares, each with the side length given in the table. Copy and complete the table.

   b. Choose any two squares from 12a. Compare the ratio of the side lengths to the ratio of the perimeters. Do the same comparison for a different pair of squares. Do you see a relationship? The ratio of the perimeters is the same as the ratio of the sides.

   c. Choose any two squares from 12a. Compare the ratio of the side lengths to the ratio of the areas. Do the same comparison for a different pair of squares. Do you see a relationship? The ratio of the areas is the square of the ratio of the sides.

   d. If you drew two squares such that the side length of one was five times the side length of the other, how would their perimeters compare? How would their areas compare? The perimeter would be 5 times as long. The area would be 25 times as great.

   e. Write at least two conjectures about the relationships among side lengths, perimeters, and areas of two different-size squares. Answers may include that similar figures have proportional sides and perimeter, yet area is a squared factor.

13. Percent problems are another type of direct variation.

   a. Thirty percent of the students in an algebra class have pets. If there are 24 students in the class, how many have pets? Set up a proportion for this problem. $\frac{30}{100} = \frac{x}{24}$; $x \approx 7$

   b. Rewrite the proportion from 13a using s for the number of students in the class and p for the number of students with pets. $\frac{30}{100} = \frac{p}{s}$

   c. Solve the proportion in 13b for the variable p. $p = \frac{30}{100}s$ or $p = 0.3s$

   d. Explain how to write a general direct variation equation for percent problems using these terms: total, part, percent. part = percent · total

14. If 2874 people in a town own bicycles and that number is 74% of the town's population, how many people are in the town? Use a direct variation equation to solve this problem.
$2874 = 0.74t$, $t = 3883.78$ or 3884 people in the town

Bicycles are the most popular form of transportation in Beijing, China.

Review

15. APPLICATION Two cats eat one 14-pound bag of cat food every six weeks. The bag costs $12.98.

   a. What is the cost of feeding both cats per day? $0.31 per day

   b. How many pounds of cat food will one cat eat per year? (Assume both cats eat the same amount.) $60\frac{2}{3}$ pounds of food per cat per year

   c. How much does it cost to feed both cats for one year, and how much will their owner spend on bags of cat food? It costs about $113, or 8.7 fourteen-pound bags. The owner will have to buy 9 bags and spend $116.82.

16. Find the five-number summary for this data set:

47	28	11	74	58	63	85
36	39	17	27	75	48	

11, 27.5, 47, 68.5, 85

Exercise 12 If students don't have graph paper, have them measure in centimeters so that figures will fit onto standard paper. As needed, remind students of the area-ratio principle found in the investigation of Lesson 2.4. Encourage students to draw diagrams for a visual understanding of this principle.

Exercise 13 Some students may see no need to set up a proportion because they can solve the problem in another way. Ask them to do so anyway to see how percent gives a rate (as indicated by the word *per*) and how a direct variation is a special case of a proportion.

Exercise 14 This exercise flows from Exercise 13.

Exercise 15 Students may think the two questions in 15c are the same. The first question asks for a product of the rate and the number of days in a year. The second asks students to realize that the owner must spend a little more to buy full bags of food.

This method has been known as "foil" (first, outside, inside, last). You might want to challenge students by having them try algebraic expressions like $(a + b)(a + b)$, where a is the tens digit of any number and b is the ones digit. The result here would be a formula for squaring any double-digit number. Students will see a geometric representation of this in Chapter 4. To help students develop critical thinking, **[Ask]** "Will it work for all pairs of numbers?" [For one thing, the formula needs to be modified for carrying if any product is larger than 10.]

17. A bag of fruit-flavored candies contains 8 banana, 8 orange, 6 lemon, 12 cherry, 10 grape, and 4 mango. If you select a candy at random, what is the probability that you will choose a mango-flavored candy from this package? P(mango flavored) $= \frac{4}{48}$ or $\frac{1}{12}$ or 0.083

18. An ancient Indian Sutra describes a formula for multiplication that can be summarized as "vertically and crosswise." (Śrī Bhārati Kṛṣṇa Tīrthajī, *Vedic Mathematics*, 1992, p. 39)

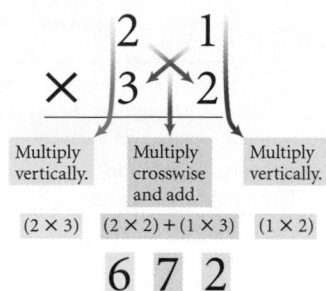

So, $21 \cdot 32 = 672$.

Use this method to multiply these numbers:

a. $12 \cdot 13$
$(1 \times 1), (1 \times 3) + (2 \times 1), (2 \times 3)$; 156

b. $21 \cdot 14$
$(2 \times 1), (2 \times 4) + (1 \times 1), (1 \times 4)$; 294

c. $42 \cdot 31$
$(4 \times 3), (4 \times 1) + (2 \times 3), (2 \times 1)$; 12, 10, 2; the 10 carries into the next place to give 1302.

MAKE YOUR OWN MAP

Maps are useful in planning trips, giving directions, and estimating distances. You can get maps at many stores, auto clubs, gas stations, and even on the Internet (see the resources at **www.keymath.com/DA**).

Apply the skills you've learned in this chapter to make a map of your trip to school. Whether you walk, bike, bus, or drive to school, your first step will be to measure the distance of your path to school, including how far you travel between changes in direction. Choose an appropriate scale based on the measurements you record.

Your hand-drawn map should include

▶ Your home and school.

▶ A legend to show the scale factor you used.

▶ The path you take to school including the actual measurements of each part of the route.

On your map, draw a straight line from your home to school. Use the legend to calculate this distance measured on the ground. Compare this to the distance of your current path. Can you follow this straight-line path? Is your current path to school the shortest?

Supporting the project

MOTIVATION

Making a map of the area between home and school might help you find a shorter or faster way to get to school each day.

OUTCOMES

▶ The map is a clearly labeled, accurate scale drawing with a reasonable scale.

▶ The map includes student's home and school and a legend showing at least the scale factor.

▶ A path from home to school is marked on the map, and each segment of the path is marked with its length.

• Ratio, proportion, and conversion factors were used.

• The legend contains more than a scale.

• The map has scale factors for both miles and kilometers.

• All buildings and streets are drawn to scale and conform to the definition of similarity.

Inverse Variation

In each relationship you have worked with in this chapter, if one quantity increased, so did the other. If one quantity decreased, so did the other. If the working hours increase, so does the pay. The shorter the trip, the less gas the car needs. These are direct relationships. Do all relationships between quantities work this way? Can two quantities be related so that increasing one causes the other to decrease?

*One person's constant is
another person's variable.*
SUSAN GERHART

Try opening your classroom door by pushing on it close to the hinge. Try it again farther from the hinge. Which way takes more force? As the distance from the hinge *increases*, the force needed to open the door *decreases*. This is an example of an *inverse* relationship.

Investigation
Seesaw Nickels

You will need

- a pencil
- a 12-inch ruler
- nine nickels
- tape

If a grown man and a small child sit on opposite ends of a seesaw, what happens? Would changing or moving the weight on one end of the seesaw affect the balance? You'll find out as you do the experiment in this investigation.

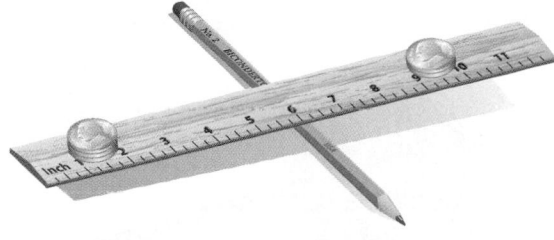

Step 1 | On a flat desk or table, try to balance the ruler across a pencil near the ruler's 6-inch mark.

Step 2 | Stack two nickels on the ruler so that they're centered 3 inches to the right of the pencil. You may need to tape them in place.

NCTM STANDARDS

CONTENT		PROCESS	
•	Number	•	Problem Solving
•	Algebra	•	Reasoning
	Geometry	•	Communication
•	Measurement		Connections
	Data/Probability	•	Representation

LESSON OBJECTIVES

- Investigate inverse variation through real-world problems
- Learn the basic inverse variation equations
- Graph inverse variation functions

PLANNING

LESSON OUTLINE

One day:

5 min   Introduction

20 min   Investigation

10 min   Sharing

5 min   Examples

5 min   Closing

5 min   Exercises

MATERIALS

- nickels (9 per group)
- 12-inch rulers
- tape
- pencils
- Direct and Inverse Variation (T), *optional*

TEACHING

Like direct variation, inverse variation (aided by dimensional analysis) can provide a shortcut to setting up and solving proportions. In direct-variation situations, the *ratios* of corresponding data values are constant. Each equals a constant *multiplied by* the other. This lesson presents inverse-variation situations, where the constants are the *products* of corresponding data values, in which each data value equals the constant *divided by* the other data value.

Guiding the Investigation

One step Pose the problem, "Simulate a seesaw with the ruler and pencil, and find patterns in how to make it balance with various numbers of nickels piled at various places on the two sides." Encourage students to change just one variable at a time and to express patterns as proportions.

[ESL] Be sure students know what a seesaw, sometimes called a *teeter-totter,* is.

You might demonstrate the procedure using a setup larger than a ruler and nickels, such as a length of 1-by-4 (inch) lumber and (full) cans of tuna. The cans don't slide off easily, and they stack well.

Step 2 As needed, remind students that the centers of the coins should be 3 inches from the center of the pencil. Any taping should be light so as not to affect the balance.

Step 3 As needed, remind students to measure to the center of the nickels.

Step 4 Be sure that each group uses a standard frame of reference to decide which nickels are "left" and "right."

Step 5 Students can easily get hung up on the logistics and miss the mathematics of the investigation. Some students may be satisfied to say something like, "The more nickels there are, the farther the nickel is from the center." Ask them to clarify which nickels they're talking about in each case. Encourage them to express the relationship more specifically in terms of variables.

Step 7 Some students may still have difficulties defining variables. Be sure they've written a sentence using the words, and then show them how to replace those words with letters and translate the sentence into an equation.

Step 7
(*left nickels*) · (*left distance*)
= (*right nickels*) · (*right distance*)

$$\frac{left\ nickels}{right\ distance} = \frac{right\ nickels}{left\ distance}$$

Step 3 Place one nickel on the left side of the ruler so that it balances the two right-side nickels. Be sure that the ruler stays centered over the pencil. How far from the pencil is this one nickel centered?

Step 3 The one nickel should balance at 6 inches from the pencil.

Step 4 Repeat Step 3 for two, three, four, and six nickels on the left side of the ruler. Measure to the nearest $\frac{1}{2}$ inch. Copy and complete this table.

Left side		Right side	
Number of nickels	Distance from pencil	Number of nickels	Distance from pencil
1	6	2	3
2	3	2	3
3	2	2	3
4	1.5	2	3
6	1	2	3

Step 5 As you increase the number of nickels on the left side, how does the distance from the balance point change? What relationships do you notice?

Step 5 Possible answer: As the number of nickels increases, the distance from the pencil decreases. The number multiplied by the distance equals 6.

Step 6 The same relationship holds—as the number of nickels increases, the distance from the pencil decreases—but this time the product is 9.

Step 6 Make a new table and repeat the investigation with three nickels stacked 3 inches to the right of center. Does the same relationship seem to hold true?

Step 7 Review the data in your tables. How does the number of nickels on the left and their distance from the pencil compare to the number of nickels on the right and their distance from the pencil? In each of your tables, do quantities remain constant? Write a sentence using the words *left nickels*, *right nickels*, *left distance*, and *right distance* to explain the relationship between the quantities in this investigation. Define variables and rewrite your sentence as an equation.

Step 8 Explain why you think this relationship between the number of nickels on the left side and the distance from the pencil is an *inverse* relationship.

Step 8 Possible answers: As one quantity increases, the other decreases. The "Right side" columns remained constant, and the product of the "Left side" values remained constant.

Step 6

Left side		Right side	
No.	Distance	No.	Distance
1	can't be done	3	3
2	4.5	3	3
3	3	3	3
4	2 or 2.5 (2.25)	3	3
6	1.5	3	3

SHARING IDEAS

During the group work, you might select groups to put their tables from Steps 4 and 6 onto transparencies for sharing. Have a student describe the numerical relationship from Step 5 and other students describe any other relationships they found. As needed, emphasize that the variables have an inverse relationship and that the number of nickels multiplied by the distance from the pencil remains constant. Be sure someone shows the results of Step 7, so you can write down the general equation of an inverse variation and point out the constant of variation.

Help students rephrase their ideas more clearly by putting yourself in the place of someone ignorant of what's going on and asking for clarification. You might also try asking "Do you mean . . ." and rephrasing what they're trying to say with clear terminology.

In the investigation you worked with a fundamental principal of seesaws, or levers. You probably discovered this equation:

$$(\textit{left nickels}) \cdot (\textit{left distance}) = (\textit{right nickels}) \cdot (\textit{right distance})$$

You could also write this relationship as a proportion.

$$\frac{\textit{left nickels}}{\textit{right nickels}} = \frac{\textit{right distance}}{\textit{left distance}} \quad \text{or} \quad \frac{\textit{left nickels}}{\textit{right distance}} = \frac{\textit{right nickels}}{\textit{left distance}}$$

Can you use what you know about solving proportion problems to show that all three of these equations are equivalent?

Look closely at the proportions above. How do they differ from the proportions you have written so far? When you wrote proportions for direct relationships, you had to make sure that the numerator and denominator of each ratio corresponded in the same way. In this inverse relationship, the proportion has ratios that correspond in the opposite (or inverse) way. The numerators and denominators seem to be flipped. These are called inverse proportions.

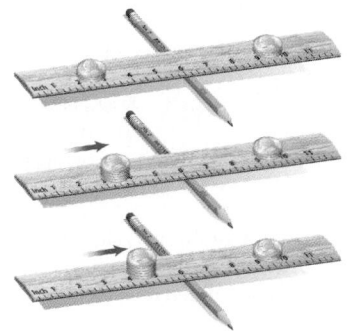

In the investigation, as the number of coins on the left side increased, the coins' distance from the pencil decreased. "Increasing" and "decreasing" show the inverse relationship.

EXAMPLE A

Tyline measured the force needed as she opened a door by pushing at various distances from the hinge. She collected the data shown in the table. Find an equation for this relationship. (A newton, abbreviated N, is the metric unit of force.)

Distance (cm)	Force (N)
40.0	20.9
45.0	18.0
50.0	16.1
55.0	14.8
60.0	13.3
65.0	12.3
70.0	11.6
75.0	10.7

▶ **Solution**

Enter the data into two calculator lists and graph points. The graph shows a curved pattern that is different from the graph of a direct relationship. If you study the values in the table, you can see that as distance increases, force decreases. The data pairs of this relationship might have a constant product like your data in the investigation.

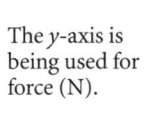
The y-axis is being used for force (N).

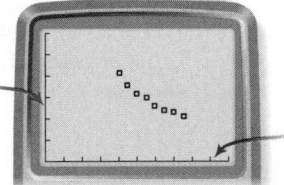

The x-axis is being used for distance from the hinge (cm).

Prepare students for the exercises with a question. [Ask] "How can you tell whether the variables are in direct or inverse variation or neither?" [If all the quotients of pairs are the same, there's a direct variation. If all the products are the same, there's an inverse variation.] Also discuss how to draw a quick diagram of the balancing situation, to prepare for Exercise 8.

You might encourage students to graph the pairs (*left distance*, *left nickels*) and try to describe the shape. This may be the first graph they've made that isn't a straight line. If this idea arises naturally here, you may be able to skip Example A.

Assessing Progress
Watch for skills at careful measurement, following directions, isolating variables in experimentation, and recording data.

Another Example
If you have two days for this lesson, you might dramatize another inverse relationship with the help of a 60-watt bulb. Show students the light and then ask them to close or cover their eyes as you move the lit bulb toward and away from them. [Ask] "What difference do you see in the light when it is close compared to what you see when it is far away?" [The distance away and the brightness are inversely related. Be careful not to say that they satisfy an inverse variation, though, since the brightness varies inversely as the *square* of the distance away.]

▶ **EXAMPLE A**

This example is good for students who have not yet seen that the product of variables in an inverse variation is constant. [Ask] "How would you size the viewing window to get this graph?" Reemphasize looking for maximum and minimum values of the two variables.

Calculate the products in another list. Their mean is approximately 810, so use that to represent the product. Let x represent distance and y represent force.

Distance (cm)	Force (N)	Force · Distance (N-cm)
40	20.9	836.0
45	18.0	810.0
50	16.1	805.0
55	14.8	814.0
60	13.3	798.0
65	12.3	799.5
70	11.6	812.0
75	10.7	802.5

Science

⦿ CONNECTION ⦿

Scientists use precise machines to measure the amount of force needed to pull or push. Manufacturers use these tools to test the strength of products like boxes. You can also measure force with simple tools like a spring scale. This box shows a certificate that gives the results from several force tests.

$$xy = 810 \qquad \text{The product of distance and force is 810.}$$

$$\frac{xy}{x} = \frac{810}{x} \qquad \text{Divide both sides by } x.$$

$$y = \frac{810}{x} \qquad \text{Now you have the } y= \text{ form so you can enter this equation in your calculator.}$$

Graph the equation. Does it go through all of the points? Why do you think the graph is not a perfect fit? It is a good practice to explore small changes to your equation's constant. A slightly different value might give an even better fit.

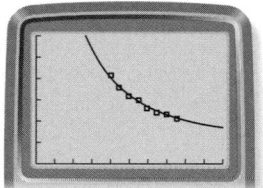

The two variables in the inverse relationships you have seen have a constant product, so you can write an **inverse variation** equation. You can represent the constant product with k just like you use k to represent the constant ratio of a direct variation. The graph of an inverse variation is always curved and will never cross the x- or y-axis. Why couldn't x or y be zero?

Inverse Variation

An equation in the form $y = \frac{k}{x}$ is an **inverse variation.** Quantities represented by x and y are **inversely proportional,** and k is the **constant of variation.**

EXAMPLE B

Ohm's law states that when the electromotive force E (in volts) is constant, the current i (in amperes) is inversely proportional to the resistance R (in ohms). A current of 24 amperes is flowing through a conductor whose resistance is 6 ohms.

Every knob or lever of this sound recording console regulates electric resistance in a current. The resistance varies directly with voltage and inversely with current.

a. What current flows in the system if the resistance increases to 8 ohms?

b. What is the resistance of the conductor if a current of 7.2 amperes is flowing?

▶ **Solution**

You can show the formula as

$$i = \frac{E}{R}$$

E is being used instead of k as the constant in the inverse variation equation.

$$i = \frac{144}{R}$$

The product of i and R is always the same, so use the product of 24 and 6 to get 144 for the value of E.

Use this equation to answer the questions.

a.
$$i = \frac{144}{8}$$

Substitute 8 for R.

$$i = 18$$

Divide.

The current flowing through a conductor with resistance 8 ohms is 18 amperes.

b.
$$7.2 = \frac{144}{R}$$

Substitute 7.2 for i.

$$\frac{1}{7.2} = \frac{R}{144}$$

To get R in the numerator, invert both ratios. The proportion is still true.

$$144 \cdot \frac{1}{7.2} = \frac{R}{144} \cdot 144$$

To isolate R, multiply both sides by 144.

$$20 = R$$

Multiply and divide.

The resistance of the conductor when a current of 7.2 amperes is flowing through it is 20 ohms.

▶ **EXAMPLE B**

Some students may have only a vague understanding of force, current, and resistance. Point out that, though the problem is about electricity, its flow is similar to water through a pipe. The more mineral deposit (resistance) there is in the pipe, the slower the flow (current), assuming constant water pressure (force). All electrical appliances, from lightbulbs to CD players, provide resistance to the flow of electricity. Some have dials, such as dimmers for a room light or volume controls on a radio, that increase or decrease resistance and thus change the flow.

If students are interested, you might mention that the resistance in a circuit is usually given by the formula $R = \frac{E}{i}$ where resistance R is measured in ohms, E is for electromotive force (measured in volts), and i is current in international amperes (amps).

In solving the proportion, students might be confused by the inversion if they don't realize that 7.2 in part b can be considered $\frac{7.2}{1}$.

Closing the Lesson

Remind the class that two positive data values have an **inverse relationship** when larger values of one correspond to smaller values of the other. In a special case, the variables satisfy an **inverse variation**, an equation of the form $y = \frac{k}{x}$, where k is the **constant of variation**. In this case, the product xy of the variables is constant.

Tell students that the next lesson will make use of multispeed bicycles. Ask for volunteers to bring in bicycles.

BUILDING UNDERSTANDING

Students practice working with inverse variations and their graphs. Continue to encourage the use of dimensional analysis. Allow students to use proportions, but encourage them to rewrite their equations as inverse variations.

ASSIGNING HOMEWORK

Essential	1–6, 8, 9
Performance assessment	7, 10
Portfolio	11, 12
Journal	5, 8
Group	11
Review	13, 14

► Helping with the Exercises

Exercise 2 Two quantities x and y are *inversely proportional* if they satisfy an inverse variation—that is, if their product is constant.

Exercise 5 Some students may be confused by the difference between an inverse relationship and inverse variation. This is an inverse relationship, but it's not an inverse variation because the products aren't all the same. In fact, the sums are the same, so the relationship is linear: $x + y = 3$.

5. This is not an inverse variation. The product of the quantities (*time spent watching TV, time spent doing homework*) is not a constant. Instead, the sum is a constant. This is a relationship of the form $x + y = k$, or $y = k - x$, not an inverse variation $xy = k$, or $y = \frac{k}{x}$.

EXERCISES

You will need your calculator for problems **3**, **10**, and **11**.

► Practice Your Skills

1. Rewrite each equation in $y=$ form.

   a. $xy = 15$ $y = \frac{15}{x}$ **b.** $xy = 35$ $y = \frac{35}{x}$ **c.** $xy = 3$ $y = \frac{3}{x}$

2. Two quantities, x and y, are inversely proportional. When $x = 3$, $y = 4$. Find the missing coordinates for the points below.

   a. $(4, y)$ $(4, 3)$ **b.** $(x, 2)$ $(6, 2)$ **c.** $(1, y)$ $(1, 12)$ **d.** $(x, 24)$ $(0.5, 24)$

3. Find five points that satisfy the inverse variation equation $y = \frac{20}{x}$. Graph the equation and the points to make sure the coordinates of your points are correct.
   Possible choices include $(4, 5), (2, 10), (5, 4), (10, 2),$ and $(2.5, 8)$.

4. **APPLICATION** The amount of time it takes to travel a given distance is inversely proportional to how fast you travel.

   a. How long would it take to travel 90 miles at 30 mph?   3 hr

   b. How long would it take to travel 90 miles at 45 mph?   2 hr

   c. How fast would you have to go to travel 90 miles in 1.5 hours?   60 mph

5. Henry noticed that the more television he watched, the less time he spent doing homework. One night he spent 1.5 hours watching TV and 1.5 hours doing homework. Another night he spent 2 hours watching TV and only 1 hour doing homework. To try to catch up, the next night he spent only a half hour watching TV and 2.5 hours doing homework. Is this an inverse variation? Explain why or why not.

► Reason and Apply

6. For each table of x- and y-values below, decide if the values show a direct variation, an inverse variation, or neither. Explain how you made your decision. If the values represent a direct or inverse variation, write an equation.

a.

x	y
2	12
8	3
4	6
3	8
6	4

b.

x	y
2	24
6	72
0	0
12	144
8	96

c.

x	y
4.5	2.0
0	9.0
3.0	3.0
9.0	0
6.0	1.5

d.

x	y
1.3	15.0
6.5	3.0
5.2	3.75
10.4	1.875
7.8	2.5

Exercise 6 This is the type of problem students might see on a standardized exam. As you may have mentioned during Sharing, they should check products and quotients, looking for constants. **[Alert]** In 6b students may be confused and think that since all other quotients are 12, $\frac{0}{0} = 12$. Similarly, in 6c the 0's might be confusing. All products are 9 except those involving 0.

6a. Inverse variation: $y = \frac{24}{x}$ or $xy = 24$

6b. Direct variation: $y = 12x$

6c. Neither; neither x nor y can be zero in an inverse variation; when $x = 0, y = 0$ in a direct variation.

6d. Inverse variation: $y = \frac{19.5}{x}$ or $xy = 19.5$

7. APPLICATION In Example A, you learned that the force in newtons needed to open a door is inversely proportional to the distance in centimeters from the hinge. For a heavy freezer door, the constant of variation is 935 N-cm.

  a. Find the force needed to open the door by pushing at points 15 cm, 10 cm, and 5 cm from the hinge. 62.3 N, 93.5 N, and 187 N.

  b. Describe what happens to the force needed to open the door as you push at points closer and closer to the hinge. How does the change in force needed compare as you go from 15 cm to 10 cm and from 10 cm to 5 cm?

  c. How is your answer to 7b shown on the graph of this equation? On the graph, the curve goes up very steeply near the y-axis.

8. Emily and her little brother Sid are playing on a seesaw. Sid weighs 65 pounds. The seesaw balances when Sid sits on the seat 4 feet from the center and Emily sits on the board $2\frac{1}{2}$ feet from the center.

  a. About how much does Emily weigh?

  b. Sid's friend Seogwan sits with Sid at the same end of the seesaw. They weigh about the same. Can Emily balance the seesaw with both Sid and Seogwan on it? If so, where should she sit? If not, explain why not.

9. To use a double-pan balance, you put the object to be weighed on one side and then put known weights on the other side until the pans balance.

  a. Explain why it is useful to have the balance point halfway between the two pans.

  b. Suppose the balance point is off-center, 15 cm from one pan and 20 cm from the other. There is an object in the pan closest to the center. The pans balance when 7 kg is placed in the other pan. What is the weight of the unknown object? $15 \cdot M = 20 \cdot 7; M \approx 9.3$ kg

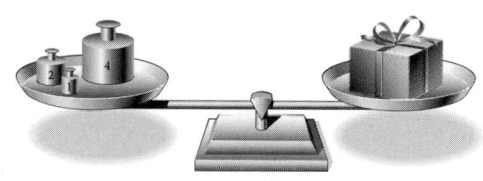

10. APPLICATION The student council wants to raise $10,000 to purchase computers. All students are encouraged to participate in a fund-raiser, but it is likely that some will not be able to.

  a. Pick at least four numbers to represent how many students might participate. Make a table showing how much each student will have to raise if each participant contributes the same amount. The table should include points like $(100, 100)$, $(200, 50)$, $(250, 40)$, $(400, 25)$.

  b. Plot the points represented by your table on a calculator graph. Find an equation to fit the points. $y = \dfrac{10,000}{x}$

  c. Suppose there are only 500 students in the school. How would this number of students affect your graph? Sketch a graph to show this limitation. The graph should stop at $x = 500$ because there are only that many students.

7b. As you move closer to the hinge, it takes more force to open the door. Moving from 15 cm to 10 cm needs an increase of about 31.2 N. Moving 5 cm closer requires an increase of 93.5 N. As you move closer, the force needed increases more rapidly. When you get very close to the hinge, the force needed becomes extremely large.

8a. 104 pounds

8b. $130 \cdot 4 = 104 \cdot D$; Emily needs to sit approximately 5 feet from the center. The seesaw is only 4 feet long from the center to the seat, so she can't balance the two boys as long as they stay on the seat. However, if the boys move and Emily sits on the seat, it can be done. $130 \cdot D = 104 \cdot 4$; $D = 3.2$ feet from the center.

Exercise 9 Some students might not have experience with a double-pan balance. You might borrow one from a science lab. Students will need to be familiar with pan balances for Chapter 4.

9a. If the balance point is at the center, then the weight of an unknown object will be exactly the same as the weight that balances it on the other side. If the balance point is off-center, you must know the two distances and do some calculation.

10b.

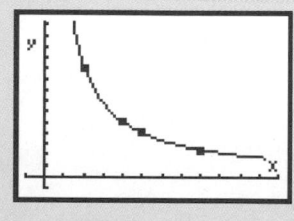

$[-50, 600, 50, -10, 140, 10]$

10c.

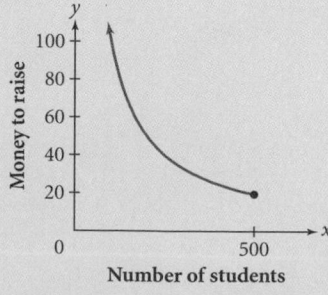

11a.

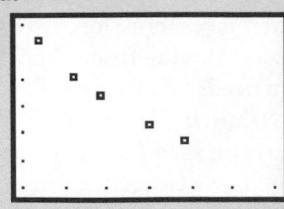

[400, 1000, 100, 30, 90, 10]

11b. Possible answer:

$$y = \frac{37{,}227.1}{x}$$

The constant 37,227.1 is the mean of the products of the frequencies and tube lengths.

Exercise 12 One atmosphere is the amount of air pressure at sea level, about 14.7 pounds per square inch.

12d. Possible answer: You would have to increase the volume of the container. If you kept the same volume, you would have to suck some of the air out of the container.

12e.

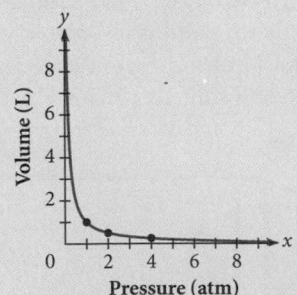

11. A tuning fork vibrates at a particular frequency to make the sound of a note in a musical scale. If you strike a tuning fork and place it over a hollow tube, the vibrating tuning fork will cause the air inside the tube to vibrate, and the sound will get louder. Skye found that if you put one end of the tube in water and raise and lower it, the loudness will vary. She used a set of tuning forks and, for each one, recorded the tube length that made the loudest sound.

Tuning Fork Experiment

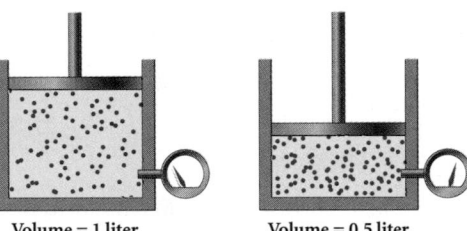

Note	Frequency (hertz)	Tube length (cm)
A_4	440.0	84.6
C_5	523.3	71.1
D_5	587.3	63.4
F_5	698.5	53.3
G_5	784.0	47.5

Answers will vary. On this graph, x represents frequency and y represents tube length.

a. Graph the data on your calculator and describe the relationship.

b. Find an equation to fit the data. Explain how you did this and what your variables and constants represent.

c. The last tuning fork in Skye's set is A_5 with a frequency of 880.0 hertz. What tube length should produce the loudest sound? $y = \frac{37{,}227.1}{x}; y \approx 42.3$ cm

12. APPLICATION To squeeze a given amount of air into a smaller and smaller volume, you have to apply more and more pressure. Boyle's law describes the inverse variation between the volume of a gas and the pressure exerted on it. Suppose you start with a 1-liter open container of air. If you put a plunger at the top of the container without applying any additional pressure, the pressure inside the container will be the same as the pressure outside the container, or 1 atmosphere (atm).

Volume = 1 liter   Volume = 0.5 liter

a. What will the pressure in atmospheres be if you push the plunger down until the volume of air is 0.5 liter? 2 atm

b. What will the pressure in atmospheres be if you push the plunger down until the volume of air is 0.25 liter? 4 atm

c. Suppose you exert enough pressure so that the pressure in the container is 10 atm. What will the volume of the air be? 0.1 liter

d. What would you have to do to make the pressure inside the container less than 1 atm?

e. Graph this relationship, with pressure (in atmospheres) on the horizontal axis and volume (in liters) on the vertical axis.

▶ Review

13. **APPLICATION** A CD is on sale for 15% off its normal price of $13.95. What is its sale price? Write a direct variation equation to solve this problem. $s = 0.85p$. **The sale price is $11.86.**

14. Calcium and phosphorus play important roles in building human bones. A healthy ratio of calcium to phosphorus is 5 to 3.

    a. If Mario's body contains 2.5 pounds of calcium, how much phosphorus should his body contain? **1.5 pounds**

    b. About 2% of an average woman's weight is calcium. Kyle weighs 130 pounds. How many pounds of calcium and phosphorus should her body contain?
    2.6 pounds of calcium and 1.56 pounds of phosphorus

project

FAMILIES OF RECTANGLES

On each set of axes below, two rectangles are drawn with a common vertex at the origin. On the left, the rectangles are similar. On the right, the rectangles have the same area. If you draw more rectangles following the same patterns and then connect their upper-right vertices, what kinds of curves will you get? Write an equation for each pattern.

> **THE GEOMETER'S SKETCHPAD**
>
> With The Geometer's Sketchpad, you can construct families of rectangles and other polygons. Commands like Trace and Locus can help you dynamically create curves.

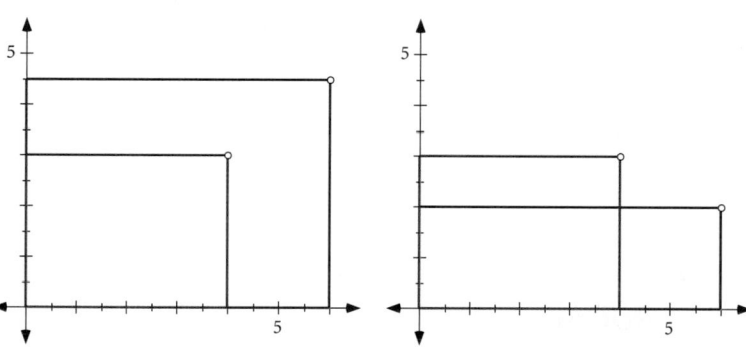

Explore at least four different families of rectangles or families of other shapes. Draw each family on its own set of axes and then describe the pattern in words. Connect corresponding vertices and see whether a curve is formed. (In mathematics, a straight line is actually considered a type of curve.) For each curve, write an equation or describe it in words. Summarize your findings in a paper or presentation.

Supporting the project

MOTIVATION

Are algebra and geometry unrelated fields of mathematics?

OUTCOMES

▶ Students can define simple families with patterns like "same area" or "same length."

▶ Students recognize the connected curves as direct variation and inverse variation.

▶ Students explore other families of rectangles or other polygons. These families may have other curves such as horizontal or vertical lines, or circles.

• Students write equations for the curves generated and recognize the connection to attributes of the polygons. (For example, when the area of rectangles is held constant, the equation is $xy = k$ or $lw = A$).

• Students show that geometric figures can be described with algebra and that algebra can derive from geometry.

PLANNING

LESSON OUTLINE

One day:

5 min	Introduction
30 min	Activity
10 min	Sharing
5 min	Closing

MATERIALS

- meterstick or metric tape measures
- multiple-speed bicycles with 12, 18, or 24 speeds, one per group. Each bike should have at least two sprockets in the front and several in the back.

TEACHING

Students who have ridden multi-speed bicycles probably know which gear combinations are easier than others, but few of them have thought about how direct and inverse variations can represent the relative efficiency of the gears. As an alternative, you might do the demonstration yourself with a single bicycle. If students bring in bicycles, warn administrators and security personnel.

G uiding the Activity

One step Pose the problem, "Experienced bicyclists often use terms such as *a 70-cm gear*, meaning a combination of sprockets for the chain that moves the bicycle 70 cm in one complete turn of the pedals. What centimeter gears are available on the bicycle you're working with today?" As you circulate, encourage systematic collection of data, careful measurement, and use of direct and inverse variation equations to express relationships among number of teeth and either number of revolutions of the rear wheel or distance traveled.

Activity Day

Variation with a Bicycle

The Tour de France is a demanding bicycle race through Switzerland, Germany, and France. For 23 days, cyclists ride 3630 kilometers on steep mountain roads before crossing the finish line in Paris. The cyclists rely on their knowledge of gear shifting and bicycle speeds.

Many bicycles have several speeds or gears. In a low gear, it's easier to pedal uphill. In a high gear, it's harder to pedal, but you can go faster on flat surfaces and down hills. When you change gears, the chain shifts from one sprocket to another. In this activity you will discover the relationships among the bicycle's gears, the numbers of times you pedal, and the teeth on the sprockets.

Activity
The Wheels Go Round and Round

You will need

- a meterstick or metric tape measure
- a multispeed bicycle

In Steps 1–5, you'll analyze the effect of the rear sprockets.

> ### Procedure Note
>
> **Changing Gears**
>
> Each time you change gears on the bike, you may have to turn the bike right side up and rotate the pedal and crankshaft a few times until the gear change takes effect. Then you turn the bike upside down to start observing and recording data.

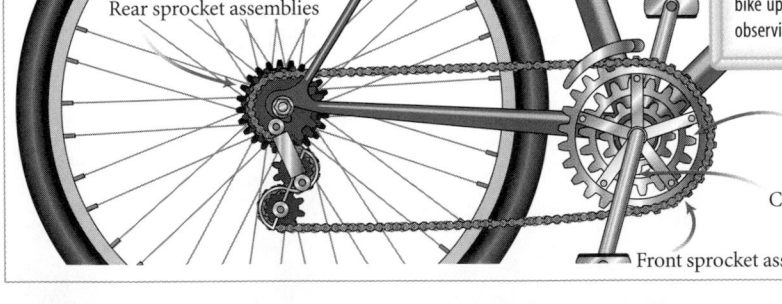

Rear sprocket assemblies

Tooth

Crankshaft

Front sprocket assemblies

LESSON OBJECTIVE

- Investigate the direct and inverse variations between gear selection and wheel speeds on a multispeed bicycle

NCTM STANDARDS

CONTENT		PROCESS	
•	Number		Problem Solving
•	Algebra	•	Reasoning
•	Geometry		Communication
•	Measurement	•	Connections
•	Data/Probability	•	Representation

| Step 1 | Shift the bicycle into its lower gear. |
| Step 2 | Count the number of teeth on the front and rear sprockets in use. Record your numbers in a table like this one: |

Number of teeth on front sprocket	Number of teeth on rear sprocket	Number of revolutions of rear wheel for one revolution of pedals

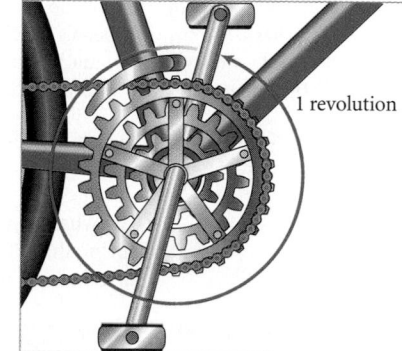

1 revolution

Step 3	Line up the air valve, or a chalk mark on the tire of the rear wheel, with part of the bicycle frame. This will be the "starting point." Rotate the pedal through one complete revolution and stop the wheel immediately. Estimate the number of wheel revolutions to the nearest tenth and enter it into the table.
Step 4	Shift gears so that the chain moves onto the next rear sprocket. Do not change the front sprocket. Repeat Steps 2 and 3. Record your data in a new row of your table. Repeat this process for each rear sprocket on your bicycle.
Step 5	Describe how the number of teeth on the rear sprocket affects how the wheel turns. What kind of variation is this? Plotting the data on your calculator may help you to see this relationship. Define variables and write an equation that relates the number of wheel revolutions to the number of teeth on the rear sprocket. Explain the meaning of the constant in this equation.

In Steps 6–10, you'll analyze the effect of the front sprockets.

Step 6	Shift the bicycle into its lowest gear again.
Step 7	Count and record the number of teeth on the sprockets in use in a second table.
Step 8	As you did in Step 3, record the number of wheel revolutions for one revolution of the pedal crankshaft.
Step 9	Keep the chain on the same rear sprocket and shift gears so that the chain is placed onto the next front sprocket. Repeat Steps 7 and 8. In the second table you should have one row of data for each front sprocket.
Step 10	Describe how the number of teeth on the front sprocket affects the turning of the wheel. What kind of variation models this relationship? Plot the data on your calculator to verify your answer. Define variables and write an equation that relates the number of teeth on the front sprocket to the number of wheel revolutions. What is the meaning of the constant in this equation?

Steps 2–4 Sample table for a bike with six rear sprockets:

Front	Rear	Rev.
30	28	1.1
30	24	1.3
30	21	1.4
30	18	1.7
30	16	1.9
30	14	2.1

Steps 7–9 Sample table for a bike with three front sprockets:

Front	Rear	Rev.
30	28	1.1
39	28	1.4
50	28	1.8

Step 1 Bicycles should be upside down. If needed, stabilize them with books or clothing. The lowest gear has the chain on the smallest front sprocket and the largest rear sprocket. You may need to point out the sprockets.

Step 2 Students may need to shift the chain off of these sprockets in order to see all the teeth.

Step 3 [Alert] Caution students to turn the pedal slowly in order to keep track of the revolutions. It will not be easy to count an exact number of revolutions of the rear wheel for one revolution of the pedals. Students will have to approximate this number as well as they can.

Step 5 The fewer the number of teeth on the rear sprocket, the more the wheel turns in one pedal rotation. As the size of the rear sprocket increases, the number of wheel revolutions decreases. Indirect variation models this behavior. If w is the number of wheel revolutions, r is the number of teeth on the rear sprocket, and k is the proportionality constant, then the equation is $w = \frac{k}{r}$. The constant k represents the number of teeth on the front sprocket.

Step 8 The pedal crankshaft is the metal bar connecting the pedals through the bicycle's frame. Tell students to rotate the pair of pedals once.

Step 10 Encourage students to check their conjectured equation by graphing it on the calculator and comparing it to the scatter plot.

Step 10 The higher the number of teeth on the front sprocket, the more turns the wheel makes. Direct variation models this relationship. If f is the number of teeth on the front sprocket, w is the number of wheel revolutions, and k is the proportionality constant, then the equation is $f = kw$. The constant k represents the number of teeth on the rear sprocket.

Step 11 When looking for the mathematical relationship, students should concentrate on gear combinations where the number of teeth on the rear sprocket is a multiple of the number of teeth on the front sprocket. In these cases, the rear wheel should make a whole number of rotations for one revolution of the pedals.

Step 11 The proportion is

$$\frac{number\ of\ front\ teeth}{number\ of\ rear\ teeth}$$

$$= \frac{number\ of\ wheel\ revolutions}{1\ crankshaft\ revolution}$$

or $\frac{f}{r} = \frac{w}{1}$. For a gear combination with 45 front sprocket teeth and 15 rear teeth, the equation should predict 3 wheel revolutions.

Step 12 [Alert] Some students may not realize that going faster means more revolutions of the rear wheel for one revolution of the pedals.

SHARING IDEAS

You might ask students to share their data and their results to Steps 5 and 10.

Keep asking them to use the terms *direct variation* and *inverse variation*.

If you have time, extend the problem with a question. **[Ask]** "How many different pairs of front and back sprockets could be considered?" This question can lead to a discussion of multiplication in a "for each" situation. One way to answer it is by multiplying the number of front sprockets by the number of back sprockets.

Assessing Progress
Watch for careful measurement and understanding of direct and inverse variation and rates.

Closing the Lesson

As needed, say that bicycle gears provide examples of both direct and inverse variations.

Now you'll see why gear shifting is such an important strategy in a bicycle race.

Step 11 Find a proportion relating the number of front teeth, rear teeth, wheel revolutions, and pedal revolutions. Use it to predict the number of wheel revolutions for a gear combination you have not tried yet. Test your prediction by doing the experiment on this gear combination.

Step 12 In a high gear, the rear wheel revolves more in one pedal rotation, making the bike travel farther and faster.

Step 13 A 70-cm diameter rim has a 220-cm circumference. This wheel makes 455 revolutions per kilometer.

Step 12 Explain why different gear ratios result in different numbers of rear wheel revolutions. Why is it possible to go faster in a high gear?

Step 13 Find the circumference of the rear wheel in centimeters. How far will the bicycle travel when the wheel makes one revolution? How many revolutions will it take to travel 1 kilometer without coasting?

Step 14 For the lowest and highest gear, how many times do you need to rotate the pedals for the bike to travel 1 kilometer? (Hint: Write a proportion or other equation involving the gear ratio and the number of revolutions of the pedals and the wheel.) **Step 14** A bicycle whose lowest front-to-back gear ratio is 30 : 28 has to be pedaled 425 times per kilometer. A high front-to-back gear ratio of 50 : 14 has to be pedaled 128 times per kilometer.

The wheels on Lance Armstrong's bicycle made roughly 1.6 million revolutions during the 2000 Tour de France. If he hadn't coasted or changed gears, he could have pedaled more than 1.5 million times.

Averaging 39 kilometers per hour, American Lance Armstrong became the 2000 Tour de France champion. He completed the race in 92 hours, 33 minutes, and 8 seconds in spite of his struggle with cancer.

CHAPTER 3 REVIEW

In this chapter you learned about variation. From your knowledge of ratios and proportions, you learned the concept of **rate.** You solved application problems with rates involving measurement units, pay, monetary conversions, and gasoline consumption.

You learned that quantities are **directly proportional** when an increase in one value leads to a proportional increase in another. These quantities form a **direct variation** when their *ratio* is constant. The constant ratio is called a **constant of variation.** When you graphed a direct relationship, you discovered a line that always passes through the origin.

For an **inverse variation,** the *product* of two quantities is constant. In this relationship, an increase in one variable causes a decrease in the other. The graph of the relationship is curved rather than straight, and the graph does not touch either axis.

You combined what you've learned about ratios, proportions, and direct variation to interpret scale drawings. You learned how to calculate the **scale factor** and how to use it in making your own drawings. You also investigated **similar figures** and learned how to find the length of missing sides using proportion and direct variation.

EXERCISES

You will need your calculator for problems **7** and **10.**

1. **APPLICATION** Nicholai's car burns 13.5 gallons of gasoline every 175 miles.
   a. What is the car's fuel consumption rate? approximately 13 miles per gallon or 0.077 gallon per mile
   b. At this rate, how far will the car go on 5 gallons of gas? 65 miles
   c. How many gallons does Nicholai's car need to go 100 miles? 7.7 gallons

2. **APPLICATION** Two dozen units in an apartment complex need to be painted. It takes 3 gallons of paint to cover each apartment.
   a. How many apartments can be painted with 36 gallons? 12 apartments
   b. How many gallons will it take to paint all 24 apartments? 72 gallons

3. On many packages the weight is given in both pounds and kilograms. The table shows the weights listed on a sample of items.

Kilograms	1.5	0.7	2.25	11.3	3.2	18.1	5.4
Pounds	3.3	1.5	5	25	7	40	12

   a. Use the information in the table to find an equation that relates weights in pounds and kilograms. Explain what the variables represent in your equation.
   b. Use your equation to calculate the number of kilograms in 30 pounds. approximately 13.6 kilograms
   c. Calculate the number of pounds in 25 kilograms. 55 pounds

3a. If *x* represents the weight in kilograms and *y* represents weight in pounds, one equation is $y = 2.2x$ where 2.2 is the data set's mean ratio of pounds to kilograms.

PLANNING

LESSON OUTLINE

One day:
10 min   Review
25 min   Exercises
15 min   Student self assessment

REVIEWING

As you talk about rates, always emphasize the units. **[Ask]** "What could be the units for each rate at the end of the first review paragraph?" [Examples are cm per inch, dollars per hour, dollars per franc, miles per gallon.] A constant of variation is a rate. It is the ratio of two directly proportional quantities or the product of two inversely proportional quantities. Dimensional analysis helps relate the units of the constant of variation and the units of the two quantities in a direct or indirect variation.

Use this problem to review and relate the ideas in the chapter.

Suppose that you're making a scale model with scale factor 3 m/cm. The actual figure and the model are **similar figures** because their corresponding sides are proportional. The longer the actual length is, the longer the corresponding length of your model will be: $a = 3m$, standing for *actual length = scale factor* times *model length*. Stress units: m in actual = *x* cm in model.

Mention that 3 m per cm is a **rate;** that *a* and *m* are directly proportional, which is equivalent to saying that *a* and *m* have a constant ratio equal to 3 and to saying that the equation $a = 3m$ is a **direct variation;** and that 3 is the **constant of variation.**

Now suppose you know that you're trying to model a length of 500 feet and you want to decide on a scale factor. The larger the scale factor, the shorter the length in the model. If f represents the scale factor in feet/inch and m the number of inches in the model, you now have $500 = fm$. Mention that the quantities f and m are **inversely proportional,** which is equivalent to saying that they have a **constant product** 500 and that the equation $fm = 500$ is an **inverse variation,** and that 500 is the **constant of variation.**

BUILDING UNDERSTANDING

As students work individually on these exercises, you can work more with individual students who have been having difficulties. The Mixed Review contains problems from Chapters 1 to 3.

Exercise 7 [Alert] Refer students to Example B of Lesson 3.4 for help in setting up the problem.

7b. One possibility: $y = \dfrac{45.5}{x}$ The constant 45.5 is the mean of the products.

Exercise 8 The rate in this exercise is the velocity.

8c. Inversely: $100 = vt$, or $t = \dfrac{100}{v}$

---

4. Consider this graph of a sunflower's height above ground.

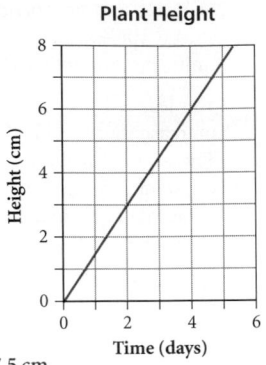

Plant Height

a. How tall was the sunflower after 5 days? about 7.5 cm

b. If the growth pattern continues, how many days will it take the plant to reach a height of 25 cm? approximately 17 days

c. Write an equation to represent the height of the plant after any number of days.
$H = 1.5 \cdot D$ where H represents height in centimeters and D represents time in days.

5. APPLICATION The scale on a map reads "1 inch = 21 miles."

a. How far apart on the map are two towns that are actually 47 miles apart? approximately 2.2 inches

b. A lake on the map is $\frac{3}{4}$-inch wide. How wide is the actual lake? 15.75 miles

6. Find the missing side lengths of the similar figures.

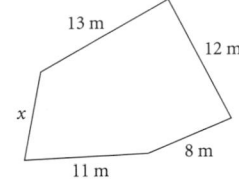

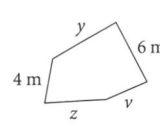

$x = 8$ m, $y = 6.5$ m, $z = 5.5$ m, and $v = 4$ m

7. Use the table of values to answer each question.

a. Is the data in the table related by a direct variation or an inverse variation? Explain. Because the product of the x- and y-values is approximately constant, it is an inverse relationship.

b. Find an equation to fit the data. You may use your calculator graph to see how well the equation fits the data.

c. Use your equation to predict the value of y when x is 32. $y = \dfrac{45.5}{32}, y \approx 1.4$

x	y
12	4
5	9
16	3
22	2
9	5
43	1

8. In the formula $d = vt$, d represents distance in miles, v represents rate in miles per hour (mph), and t represents time in hours. Use the word *directly* or *inversely* to complete each statement. Then write an equation for each.

a. If you travel at a constant rate of 50 mph, the distance you travel is ____ proportional to the time you travel. Directly: $d = 50t$

b. The distance you travel in exactly 1 hour is ____ proportional to your rate. Directly: $d = 1v$

c. The time it takes to travel 100 miles is ____ proportional to your rate.

9. Kris and Robbie are sitting on a 10-foot seesaw and have it perfectly balanced. At 100 pounds, Kris must sit 3.2 feet away from the center to balance Robbie, who weighs 80 pounds.

a. How far from the center is Robbie sitting? 4 feet

b. Robbie's 30-pound dog jumps into his lap. Can Robbie still balance the seesaw with Kris? How? Yes, Robbie can balance by sitting 2.9 feet from the center.

10. APPLICATION Boyle's law describes the inverse variation between the volume of a gas and the pressure exerted on it. In the experiment shown, a balloon with a volume of 1.75 liters is sealed in a bell jar with 1 atm of pressure. As air is pumped out of the jar, the pressure decreases, and the balloon expands to a larger volume.

Pressure = 1 atm

Pressure = 0.8 atm

Air pumped out

a. Find the volume under 0.8 atm of pressure. 2.1875 liters

b. Find the pressure when the balloon's volume is 0.75 liter. $2.\overline{3}$ atm

c. Write an equation that calculates the volume in liters from the pressure in atmospheres. $y = \dfrac{1.75}{x}$

d. Graph this relationship on your calculator, then sketch it on your own paper. Show on your graph the solutions to 10a and b.

MIXED REVIEW

11. APPLICATION Sonja bought a pair of 210-cm cross-country skis. Will they fit in her ski bag, which is $6\frac{1}{2}$ feet long? Why or why not? No, they won't fit. 210 centimeters is 6.89 feet.

12. Fifteen students counted the number of letters in their first and last names. Here is the data set.

6	15	8	12	8	17	9	7
13	15	14	9	16	15	10	

a. Make a histogram of the data with a bin width of 2.

b. If you selected one of the students at random, what is the probability that his or her name has 8 or 9 letters? $P(8 \text{ or } 9) = \dfrac{4}{15}$

13. Make a circle graph of these letter grades for an algebra class. Describe how you used the concept of direct variation in drawing the graph.

Grade	A	B	C	D	F
Number of students	8	10	12	6	4

The size and the angle of each piece of the circle graph vary directly with the percent of students who earned each grade.

14. Evaluate these expressions.

a. $-3 \cdot 8 - 5 \cdot 6$
-54

b. $(-2 - (-4)) \cdot 8 - 11$
5

c. $7 \cdot 8 + 4 \cdot (-12)$
8

d. $11 - 3 \cdot 9 - 2$
-18

10d.

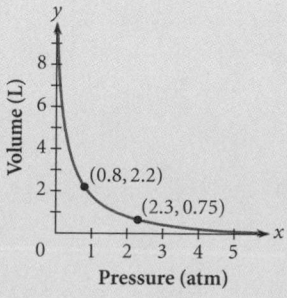

12a. Name Length

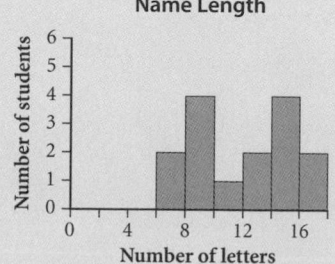

13. Algebra Grades

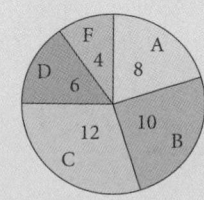

15c.

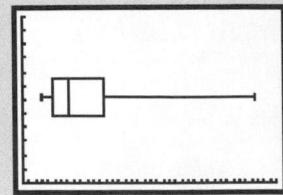

$[0, 4000, 100, 1, 12, 1]$

15. California has many popular national parks. This table shows the number of visitors in thousands to national parks in 1999.

a. Find the mean number of visitors. approximately 1100 thousand visitors

b. What is the five-number summary for the data? 354, 465, 740, 1272, 3494

c. Create a box plot for the data.

d. Identify any parks in California that are outliers in the numbers of visitors they had. Explain why they are outliers. Yosemite; the number of visitors exceeds 1272 by more than 1.5 (1272 − 465).

Park Attendance

National park	Visitors (thousands)
Channel Islands	607
Death Valley	1228
Joshua Tree	1316
Kings Canyon	560
Lassen Volcanic	354
Redwood	370
Sequoia	873
Yosemite	3494

(*U.S. National Park Service*)

Joshua Tree National Park, California

Lassen Volcanic National Park, California

16. A ball is randomly selected from a bin that contains balls numbered from 1 to 99.

a. What is the probability that the number is even? $\frac{49}{99}$

b. What is the probability that the number is divisible by three? $\frac{33}{99}$ or $\frac{1}{3}$

c. What is the probability that the number contains a 2? $\frac{19}{99}$

d. What is the probability that the number has only one digit? $\frac{9}{99}$ or $\frac{1}{11}$

17. Remember that Ohm's law states that electrical current is inversely proportional to the resistance. A current of 18 amperes is flowing through a conductor whose resistance is 4 ohms.

a. What is the current that flows through the system if the resistance is 8 ohms? 9 amperes

b. What is the resistance of the conductor if a current of 12 amperes is flowing? 6 ohms

18. APPLICATION Amber makes $6 an hour at a sandwich shop. She wants to know how many hours she needs to work to save $500 in her bank account. On her first paycheck, she notices that her net pay is about 75% of her gross pay.

a. How many hours must she work to earn $500 in gross pay? $\frac{500}{6} \approx 83.3$ hours

b. How many hours must she work to earn $500 in net pay? $\frac{500}{0.75 \cdot 6} \approx 111.1$ hours

TAKE ANOTHER LOOK

The equation $y = kx$ is a *general equation* because it stands for a whole family of equations such as $y = 2x$, $y = \frac{1}{4}x$, even $y = \pi x$. (Does $C = \pi d$ look more familiar?)

What might k be in each line of these graphs? (If you're stumped, choose a point on the line, then divide its y-coordinate by its x-coordinate. Remember, $k = \frac{y}{x}$ is equivalent to $y = kx$.)

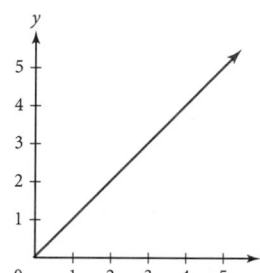 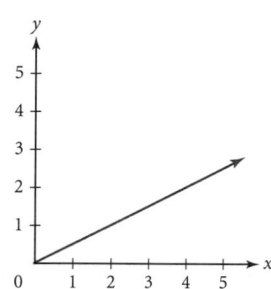 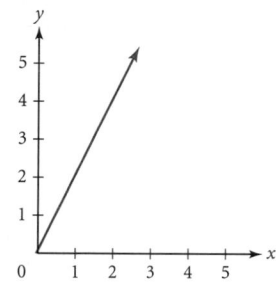

Most real-world quantities, like time and distance, are measured or counted in positive numbers. If two positive quantities vary directly, their graph $y = kx$ is in Quadrant I, where both x and y are positive. Because the quotient $\frac{y}{x}$ of two positive numbers x and y must be positive, the constant k is positive.

But the graph at right also shows a direct variation. What can you say about k for this graph?

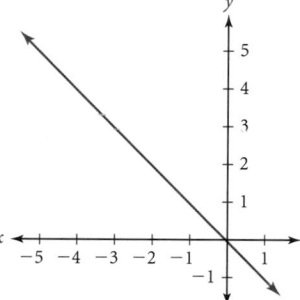

What relationship do you see between the lines in each of these situations shown below? Between the k-values?

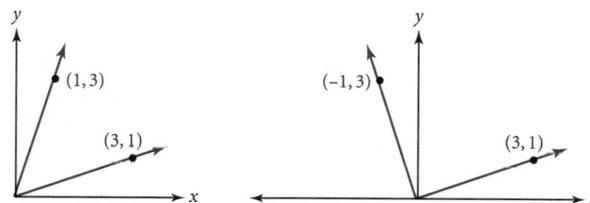

Finally, do you have a direct variation if $k = 0$? Why or why not?

► **Take Another Look**

Help students reflect on the idea that $C = \pi d$, $D = rt$, and $P = 4s$ are direct variations. Even $A = lw$ is a direct variation if l is held constant.

This discussion moves students closer to the concept of slope and forms a good habit of putting the change in y over the change in x. In the first three graphs:

$$k = 1, k < 1\left(= \frac{1}{2}\right), \text{and}$$
$$k > 1(= 2)$$

In the fourth graph, $k < 0(= -1)$ because the quotient $\frac{y}{x}$ is negative for every point on the graph.

In the last graph pair, the first set of lines is symmetric across $y = x$. Their k values $\left(3 \text{ and } \frac{1}{3}\right)$ are reciprocal. The second set of lines is perpendicular. Their k values $\left(-3 \text{ and } \frac{1}{3}\right)$ are negative reciprocals. The geometric relationships can be confirmed with similar triangles.

Students may say that $y = 0x$ is a direct variation because it fits the definition. Or they may say $y = 0$ is a constant function not a direct variation because there is no variation.

ASSESSING

FACILITATING
SELF ASSESSMENT

To help students complete the portfolio described in Assessing What You've Learned, suggest that they consider for evaluation their work on Lesson 3.1, Exercise 8; Lesson 3.2, Exercise 6 and/or 9; Lesson 3.3, Exercise 9 or 11; and Lesson 3.4, Exercise 11 or 12.

TESTING

You can use either Form A or Form B of the Chapter 3 test, or you can select four or five Constructive Assessment items for your chapter test. Using the CD test generator, you can create an alternate version of the test or combine some items from Form A or Form B with Constructive Assessment items.

You may also choose to make use of the Mixed Review as preparation for an exam that covers the material through Chapter 3. Two forms of the Chapter 1 to Chapter 3 Exam are available in Assessment Resources A and B.

Assessing What You've Learned
PERFORMANCE ASSESSMENT

This chapter has been about direct and inverse variation, so assessing what you've learned really means checking to see if you can tell how two quantities vary, what that variation looks like on a graph, and what the equation of the relationship is. If you have two direct variations using the same units, you should be able to compare them on the same graph and tell how their equations are different. Can you do one of the investigations in this chapter on your own? Can you get the equations for two variations from the word descriptions of the relationships? Showing that you can do tasks like these is sometimes called "performance assessment."

Review the Investigation "Off to Work We Go" in Lesson 3.1 about time and pay, and the Investigation "Floor Plans" in Lesson 3.3 about scale drawings. Reconstruct the graphs and drawings to see if these skills have become easier for you. Get help with any part of the process that you're not sure of. Review the Investigation "Seesaw Nickels" in Lesson 3.4 about the relationship between weight and distance from the balance point. Is it clear to you how inverse variation is different from direct variation? Graph at least one direct variation and one inverse variation for a classmate, a family member, or your teacher. Tell them how you can show the relationship in a table and in an equation as well as in a graph. Tell what the constant means in each equation.

ORGANIZE YOUR NOTEBOOK Make sure your notebook contains notes on each investigation from this chapter. Your notes should give the results of this investigation as well as some steps along the way. Be sure you have written down the definitions of new words such as *direct variation* and *inverse variation*. Are your notes complete enough that you could write a chapter summary from them?

WRITE IN YOUR JOURNAL Add to your journal using one of these prompts:
▶ Does graphing relationships help you understand how the quantities vary? Do you understand variation between quantities better when you look at a graph or when you read an equation?
▶ Tables of values, graphs, equations, and word descriptions are four ways to tell about a variation. What other mathematical ideas can you show in more than one way?

UPDATE YOUR PORTFOLIO Choose a variation that you studied in this chapter. Describe the relationship in words. Show it as a table of *x*- and *y*-values, as an equation, and as a graph. Be sure you have defined your variables and carefully labeled the graph. Add this work to your portfolio. Also, you may want to add your work from a project or the Activity Day.

4

Linear Equations

Overview

Chapter 4 begins to develop the concept of linear equations. In **Lesson 4.1**, students review the distributive property and the rules for order of operations. In **Lesson 4.2**, students work with "number tricks," series of operations that lead to a fixed number or back to the original number. They learn to solve equations by working backward to undo the operations of a number trick. In **Lessons 4.3** and **4.4**, students define recursive routines to answer questions and make predictions about linear relationships. In the Activity Day, **Lesson 4.5**, students deepen their understanding of starting value and rate of change by looking at and interpreting linear motion graphs. In **Lesson 4.6**, the intercept form of a line is formally introduced. In **Lesson 4.7**, students work with input-output tables, foreshadowing the study of functions. They use these tables to write equations in intercept form. **Lesson 4.8** gives students another way to solve equations— balancing. The final Activity Day, **Lesson 4.9**, gives students a chance to consolidate their understanding of modeling linear relationships with linear equations.

The Mathematics

Linearity

Having a **constant rate of change** is a primary characteristic of linearity. You start somewhere and advance by the same amount at each step. This kind of change is represented by a recursive sequence, easily generated on a calculator.

-15 (ENTER) Start with -15.
Ans $+10$ (ENTER) Add 10 to the answer.
(ENTER); (ENTER) Continue to add 10 to each answer.

A constant rate of change produces linear growth, though the values will be shrinking instead of growing if the rate of change is negative.

A second way of thinking about linearity is through equations that relate variables. Students begin to use, write, and make sense of $y = a + bx$, the intercept form of the equation of a line. Seeing and using multiple representations helps connect the recursive sequence start value with a and its

constant rate of change with b. The calculator steps for the recursive sequence above are equivalent to $y = -15 + 10x$, when the initial x-value is 0.

In Lesson 5.2, students will see the *point-slope form*, $y = y_1 + b(x - x_1)$. The traditional *slope-intercept form*, $y = mx + b$, is also mentioned here.

A third way of thinking about linearity is through **graphs.** Indeed, the term *linearity* comes from the fact that the associated graphs are (straight) lines. Students have seen linear graphs before—in data points, the graph of $y = x$ for comparing estimates with actual distances in Chapter 1, and direct variations $y = kx$ in Chapter 3.

The new forms of equations of a line indicate new ways of thinking of the graph. For example, the intercept form $y = a + bx$ allows students to graph by starting at point $(0, a)$ and moving vertically b units for each unit they move across from left to right. This process reflects the constant rate of change of linear growth and the recursive sequence. Later, in Chapter 9, students will discover the point-slope equation $y = y_1 + b(x - x_1)$ represents a vertical shift of y_1 and a horizontal shift of x_1.

Most data sets from real-world situations with a linear trend aren't exactly linear. Lesson 4.9 provides an activity for finding an equation that models a set of data points that lie close to but not on a straight line. Students will learn more about lines of fit in Chapter 5.

Working Backward

Many real-life situations call for predicting when linear growth will reach a certain value. Ways of making that prediction reflect the three ways of thinking about linearity—constant rate of change, equations that relate variables, and graphs.

From the **constant-rate-of-change** perspective, you can run the recursive routine until it reaches the desired output, counting input steps as you go. To mount a flagpole 75 feet up on a building, what floor would you go to if the building's basement floor is 15 feet below ground and its floors are 10 feet apart? Just run the sequence -15 (ENTER); Ans $+ 10$; (ENTER); (ENTER); ... until you get to 75 feet.

To undo those steps and get back to the number of floors, you can subtract -15 and divide by 10, the distance between floors.

This undoing process might be easier to see using an **equation** representation: If $10x - 15 = 75$, you can "get back to" x by adding 15 to 75 and then dividing by 10. The equation can also be solved using the metaphor of an equation as a pan balance. To keep it balanced, you do the same thing to both sides.

The third approach to linearity, **graphs,** also provides a means of working backward. You can graph the data points or the equation on a graphing calculator and then use the trace feature to approximate the input value for the desired output value. Using the calculator's table features is another way of approximating a solution by working backward. These two ways of working backward appear in Lesson 4.2.

Using This Chapter

Lesson 4.2 is important because solving equations by undoing is a powerful strategy, and it's easy to understand. Lesson 4.3 is essential since recursive sequences will be used throughout the book. If you must skip some lessons, you could skip the activity days, Lessons 4.5 and 4.9. No new material is presented, and these activities are both done later in the book. A walker investigation appears in Lesson 6.1 and rope tying is done again in Lesson 6.2.

Resources

Discovering Algebra Resources

Calculator Notes 0A, 0D, 1H, 3A, 4A, 4B, 4C, 4D, 4E

Teaching and Worksheet Masters
   Lesson 4.1 Cross-number Puzzle
   Lesson 4.1 Parentheses
   Lesson 4.1 Order of Operations
   Lesson 4.2 Undoing Operations
   Lesson 4.4 Elevator Table
   Lesson 4.4 On the Road Again Grid
   Lesson 4.4 On the Road Again Graph
   Lesson 4.7 Wind Chill
   Lesson 4.8 Pan Balance

Assessment Resources A and B
   Quiz 1 (Lessons 4.1 to 4.3)
   Quiz 2 (Lessons 4.4 and 4.5)
   Quiz 3 (Lessons 4.6 to 4.8)
   Chapter 4 Test
   Chapter 4 Constructive Assessment Options

More Practice Your Skills for Chapter 4

Condensed Lessons for Chapter 4

Other Resources

Rochelle Wilson Meyer and Walter Meyer. *Play It Again Sam: Recurrence Equations and Recursion in Mathematics and Computer Science.* Arlington, Massachusetts: Consortium for Mathematics and Its Applications (COMAP), 1990.

www.keymath.com/DA

Materials

- boxes of toothpicks
- graph paper
- colored pencils
- 4-meter measuring tapes, metersticks, or ropes
- motion sensors, *optional*
- stopwatches or watches with second hands
- 300 pennies or other markers
- paper cups (three per group)
- pieces of rope of different lengths (around 1 m) and thickness (two per group)

Pacing Guide

	day 1	day 2	day 3	day 4	day 5	day 6	day 7	day 8	day 9	day 10
standard	4.1	4.2	4.3	quiz	4.4	4.4	4.5	quiz	4.6	4.7
enriched	4.1	4.2	4.3	4.4, quiz	4.4	4.5	quiz	project	4.6	project
block	4.1, 4.2	4.3, quiz	4.4	4.5, 4.6	4.7, quiz	4.8, quiz	4.9, review	assessment		

	day 11	day 12	day 13	day 14	day 15	day 16	day 17	day 18	day 19	day 20
standard	4.8	quiz	4.9	review	assessment					
enriched	4.8	4.9	project	review, TAL	assessment					

Linear Equations

CHAPTER 4
OBJECTIVES

- Review or learn the rules governing order of operations and the distributive property

- Use calculator list operations to investigate the concepts of variables, terms, and expressions

- Simplify algebraic expressions that include multiplication and division, or addition and subtraction, by the same number

- Solve linear equations by undoing operations and by balancing

- Investigate recursive (arithmetic) sequences using the calculator

- Write a linear equation in intercept form given a recursion routine, a graph, or data

OBJECTIVES

In this chapter you will
- review the rules for order of operations
- describe number tricks using algebraic expressions
- write recursive routines
- study rate of change
- learn to write equations for lines
- use equations and tables to graph lines
- solve linear equations

Weavers repeat steps when they make baskets and mats, creating patterns of repeating shapes. This process is not unlike recursion. In the top photo, a mat weaver in Myanmar executes a traditional design with palm fronds. You see bowls crafted by Native American artisans in the bottom photo.

By keeping the amount of woven material constant, the weavers produce consistent patterns of lines and shapes. To make the mat in the chevron-like pattern, the weaver first lays the warp pieces lengthwise in a repeating pattern of two light and one dark. Then he weaves the woof (or weft) pieces horizontally with the same sequence of two light and one dark. Each woof piece alternately hops over three of the warp pieces and goes under three warp pieces. Students can

try this themselves with colored paper strips and write instructions for how to start the woof piece at the edge.

Coiled baskets begin with a spiral at the base, in contrast to baskets that begin with a wagon-wheel–like pattern at the base. As the basket diameter widens, the space widens between elements that cross the coil. Help students notice that some of the designs, when the baskets are viewed from top or

bottom, have reflective (mirror) symmetry, but others have radial symmetry.

Myanmar (Burma) is located between India and Thailand near the Bay of Bengal. The Native American baskets are a Pima coiled tray, an Apache coiled jar, two Yokuts coiled bowls, a Mono coiled bowl, and a Maidu coiled bowl.

Order of Operations and the Distributive Property

PLANNING

LESSON OUTLINE

One day:

5 min	Introduction
15 min	Example A, Investigation
10 min	Sharing, Distributive property
10 min	Examples B, C, D
5 min	Closing
5 min	Exercises

MATERIALS

- Cross-number Puzzle (W)
- Parentheses (W or T), *optional*
- Order of Operations (T), *optional*
- Calculator Notes 4A, 0A

TEACHING

This lesson reviews the conventional order of operations. Correctly using the order of operations is fundamental to evaluating algebraic expressions and solving equations. You can use the distributive property to make some mental arithmetic easier.

INTRODUCTION

If you skipped order of operations in Chapter 0, you'll need to spend more time on it here.

Ty's calculator is performing each operation as he enters the expression. Melinda's is following the order of operations. If possible, demonstrate with both kinds of calculators.

[Link] If students aren't seeing the need for the conventional order of operations, mention another common convention that helps avoid confusion: reading and writing from left to right.

Ty and Melinda both entered $4 + 6 \cdot 3$ into their calculators. Ty's simple, four-function calculator showed 30 when he pressed =. Melinda's graphing calculator showed 22 when she pressed ENTER. What did the two calculators do differently? Which calculator is correct? How do you know whether you should add first or multiply first? According to the order of operations, the correct answer is 22 because the rules tell you to multiply before adding.

Order of Operations

1. Evaluate expressions within parentheses or other grouping symbols.
2. Evaluate all powers.
3. Multiply and divide from left to right.
4. Add and subtract from left to right.

These rules ensure that everyone who correctly does a given calculation gets the same answer. A scientific or graphing calculator, which lets you finish entering an entire expression before it does the calculation, follows these rules. Understanding the rules will help you perform calculations correctly, with or without a calculator.

EXAMPLE A Evaluate the expression $\frac{18.7 + 11.3}{5}$ without a calculator. Then enter the expression into your calculator to see if you get the same answer.

▶ **Solution** The fraction bar is a grouping symbol, so the numbers in the numerator form a group.

$$\frac{18.7 + 11.3}{5} = \frac{30}{5} \qquad \text{Add the grouped numbers.}$$

$$= 6 \qquad \text{Now divide.}$$

In order for your calculator to recognize the grouping, you need to enter parentheses around the numerator: (18.7 + 11.3)/5 ENTER.

What's wrong with entering the expression as 18.7 + 11.3/5?

Does entering the keystrokes 18.7 + 11.3 ENTER /5 ENTER produce the correct result?

[Language] If you use the term *convention*, be sure students know that here it means an agreed-upon rule or "tradition," not a large meeting.

▶ **EXAMPLE A**

This example emphasizes that division comes before addition in the conventional order of operations. Encourage students to estimate the value of the fraction rather than calculate by hand. Remember to ask the two questions at the end of the example. Students should understand that, because calculators follow order of operation rules, 18.7 is added to the quotient $\frac{11.3}{5}$ unless the ENTER key or a set of parentheses forces the calculator to find the numerator first.

Investigation
Cross-number Puzzle

You will need

- the worksheet Cross-number Puzzle

Working the cross-number puzzle in this investigation will help you review and practice the rules for order of operations. Remember that you have to perform operations within the numerator or denominator of a fraction before you do the division indicated by the fraction bar. The square root symbol, $\sqrt{}$, can be used as a grouping symbol. Use parentheses where necessary. Part of your challenge is to figure out how to enter the entire expression into your calculator so that you get the correct answer without having to calculate part of the expression first.

[▶ ▭ See **Calculator Note 4A** to review the instant replay command. ◀]

Write your answer in fraction form or decimal form if the clue asks for it.

[▶ ▭ See **Calculator Note 0A** for help converting answers from decimal numbers to fractions and vice versa. ◀] Each negative sign, fraction bar, or decimal point occupies one square in the puzzle; commas are not part of the answer. For instance, an answer of 2508.5 needs six squares.

Be sure to work all the problems, even if an answer is entirely filled in by the time you get to its clue. That way you can check your work.

	¹1	0	6	2	²1	8	
	5					7	
	5				³3	6	
⁴5	5	7	⁵8			7	
6			⁶1	4	3	/	4	2
⁷7	9	⁸4	9			4	
	3			⁹1	0	
	3		¹⁰4			
¹¹1	1	4	8	0		
3	9		.			
1			5			
0			2			
¹²1	8	5	1	9	3	

Across

1. $\frac{2}{3}$ of 159,327

3. $\dfrac{-1 + 17^2}{4 + 2^2}$

4. $4835 - 541 + 1284$

6. $\dfrac{3 + 140}{3 \cdot 14}$ (fraction form)

7. $8075 - 3(42)$

9. $\sqrt{6^2 + 8^2}$

11. $\dfrac{740}{18.4 - 2.1 \cdot 9}$

12. 57^3

Down

1. $9(-7 + 180)$

2. $\left(\dfrac{9}{2}\right)\left(\dfrac{17}{5} + \dfrac{25}{4}\right)$ (fraction form)

4. $3 - 3(12 - 200)$

5. $9 \cdot 10^2 - 9^2$

8. $15 + 47(922)$

10. $25.9058 \cdot 20/4 - 89$ (decimal form)

11. $1284 - \dfrac{877}{0.2}$

In an expression like $3(4 + 2)$, you multiply 3 by the sum inside the parentheses. You can do this by finding the sum first:

$3(4 + 2) = 3(6)$ Add.

$\qquad\qquad = 18$ Multiply.

Or multiply 3 by each number in the sum and then add:

$3(4 + 2) = 3 \cdot 4 + 3 \cdot 2$ Multiply through the parentheses.

$\qquad\qquad = 12 + 6$ Evaluate.

$\qquad\qquad = 18$ Add.

LESSON OBJECTIVES

- Review or learn the rules governing order of operations
- Introduce the distributive property

NCTM STANDARDS

CONTENT	PROCESS
● Number	● Problem Solving
● Algebra	● Reasoning
Geometry	Communication
Measurement	● Connections
Data/Probability	● Representation

Guiding the Investigation

One step Give the puzzle on the Parentheses worksheet.

Insert as few parentheses as necessary into each expression so that, when that expression is entered into your calculator, it gives the same result as 5 * 13 − 5 * 4.

As you circulate, encourage students to write down patterns they see in how their calculators are performing. During Sharing, help them write rules describing the order of operations they've observed, and help them describe the distributive property of multiplication over addition and subtraction, being especially careful where subtraction is involved or where a minus sign, the factor −1, precedes the parentheses.

[ESL] Not all languages lend themselves to crossword puzzles. Be sure all students know how to use the Cross-number Puzzle clues. Numbered boxes start a number answer. When rows and columns intersect, the digit or symbol in that box works for the answers both across and down.

Encourage students to work in pairs. If students have weak arithmetic skills, you may need to break the puzzle into pieces and model how to solve it.

Another grouping symbol is a superscript exponent. So, by the order of operations, the expression 2^{3+5} means to calculate $3 + 5$, then raise 2 to the eighth power. On a calculator, parentheses must be used to show this grouping: $2\wedge(3 + 5)$.

[Alert] A potential source of confusion is -9^2 in 5 down. Does this mean $(-9)^2$, which is $+81$, or $-(9^2)$, which is -81? Again we have to rely on mathematical convention: Exponentiation has precedence over negation or subtraction, so the expression means $-(9^2)$. However, on some calculators, if the negative sign is used the parentheses are not needed.

Answers to Parentheses

a. $5 \cdot (13 - 4)$

b. $5 \cdot 3\wedge 2$

c. $5 \cdot 13 + 5 \cdot (-4)$

d. $(100 + 35) / (1 + 2)$

e. $(6 + 3) \cdot 5$

f. $5 + 5 \cdot 8$

g. $5 \cdot (1 + 8)$

h. $5 \cdot 3\wedge(1 + 1)$

i. $65 - 5 \cdot (3 + 1)$

j. $87 - 6 \cdot (10 - 3)$

k. $-3\wedge 2 + 54$

SHARING IDEAS

You might have 14 different students fill in spaces on a transparency of the puzzle. Point out the use of parentheses to denote multiplication of numbers. The multiplication symbol from arithmetic looks so much like the variable x that it can be confusing in algebra.

Try to arrange to have 1 down filled in last, and ask if the answer can be found in two ways. That discussion will lead into the distributive property.

Assessing Progress

Note arithmetic skills and careful organization of data, as well as comfort with order of operations.

The second method uses the **distributive property.** Think of "distributing" the number outside the parentheses to all the numbers inside. The figure at right shows a visual model of $3(4 + 2) = 18$.

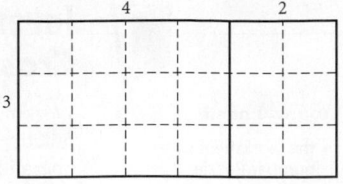

In an expression like $3(4 + 2)$, you multiply 3 by the sum inside the parentheses. You can do this two ways.

Method 1   Find the sum first:

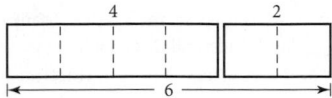

$4 + 2 = 6$

Then multiply 3 by the sum:

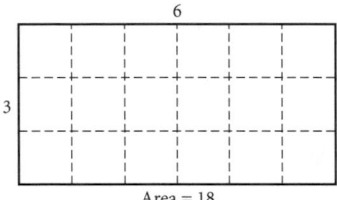

$3(6) = 18$

Area = 18

Method 2   Multiply 3 by each number in the sum:

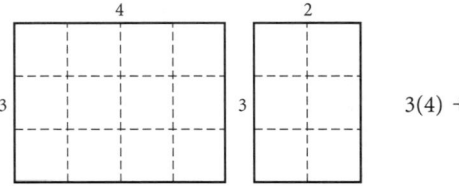

$3(4) + 3(2)$

Then add the products:

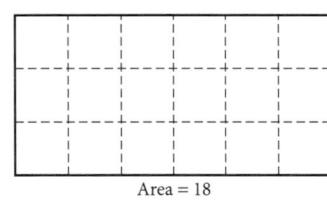

$12 + 6 = 18$

Area = 18

Distributive Property

For any values of a, b, and c, this equation is true:

$$a(b + c) = a(b) + a(c)$$

Distributive Property

Consider using algebra tiles to show several examples of the distributive property. Mention that a "distributed" sum or difference can be entered into a calculator without parentheses.

$4(2 + 3)$ $4 \cdot 2 + 4 \cdot 3$

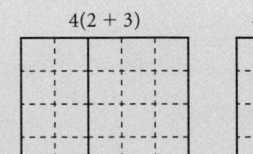 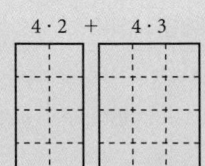

We say that a is "distributed through the parentheses." The values of a, b, and c can be 2, 5.6, $\frac{2}{3}$, -7, or 0, and the distributive property still holds true. Sometimes we call it "the distributive property of multiplication over addition" or "the distributive property over subtraction," depending on these values.

The distributive property is useful for doing mental math.

EXAMPLE B | Find these amounts without using your calculator.

a. four CDs at $14.95 each

b. a 15% tip on a $24 restaurant bill

▶ **Solution** | In each situation, use the distributive property.

a. Each CD is a nickel less than $15. So the cost is about $4 \cdot 15 = 60$. Subtract the four nickels (20 cents) to get $59.80.

Written out, the calculation looks like this:

$$4(15 - 0.05) = 4(15) - 4(0.05)$$ Distribute 4 over the subtraction $15 - 0.05$.

$$= 60 - 0.20$$ Multiply.

$$= 59.80$$ Subtract.

b. Think of 15% as 10% + 5%. Ten percent of $24 is $2.40. Five percent is half that, or $1.20. So the total tip is $3.60.

Written out, the calculation looks like this:

$$(0.10 + 0.05)24 = 0.10(24) + 0.05(24)$$ Distribute 24 over the addition $0.10 + 0.05$.

$$= 2.40 + 1.20$$ Multiply.

$$= 3.60$$ Add.

An **algebraic expression** is a symbolic representation of mathematical operations that can involve both numbers and variables. You can use the distributive property to rewrite some algebraic expressions without parentheses.

▶ **EXAMPLE B**

Even if students understand the distributive property, they may not have seen these applications or thought about how the short-cuts in mental arithmetic use the distributive property.

Reiterate that an algebraic expression is not an equation—that an equation must contain an equal sign.

This example is helpful for students who are having difficulties understanding the distributive property. It also gives practice in manipulating calculator lists. You might want to start with specific examples of ticket prices and then lead students to the abstract symbols.

Because the rectangular representation of distributivity is a little tricky when you have a variable, some students might be helped if you represent $4(T + 3)$ as four rows, each containing 1 circle (with a T in it) and 3 boxes (with a $+1$ in each). Counting down, then, they can see that the expression is equivalent to 4 circles and 12 boxes, representing $4T + 12$. The circle and box representation is used in Lesson 4.2.

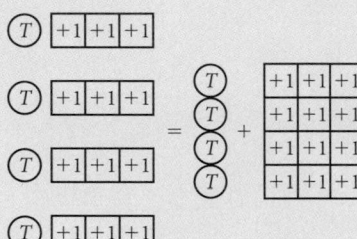

You might mention that mathematicians usually don't use symbols $\times$ or $\cdot$ or parentheses for multiplication by a quantity that's represented by a letter. **[Alert]** When students enter expressions into some calculators, they must press ⟨ × ⟩ wherever there is an implied multiplication.

Point out that in the solution there are four values in list L1, so substituting L1 for T really creates four expressions. When they are evaluated, there are four answers.

EXAMPLE C

The cost of tickets to a play varies depending on where the seats are located in the theater. The service charge for ordering tickets by phone is $3 per ticket, regardless of the ticket price.

This is the Balet Folklórico performing at the University of Guadalajara.

a. Let the variable T represent the cost of one ticket. Write two equivalent expressions for the cost of four tickets ordered by phone. One expression should use parentheses and the other should not.

b. Make up four ticket prices. Use your calculator to check that both of the expressions you wrote in part a give the same total prices for four tickets.

▶ **Solution**

a. One expression for the cost of four tickets is $4(T + 3)$. You can rewrite the expression without parentheses by distributing the 4 to get $4T + 12$.

b. Enter this list of ticket prices in list L1: $\{18, 30, 45, 65\}$. The calculator screen at right shows the list of prices stored in list L1; then it shows two expressions (using L1 in place of the variable T) evaluated. They give the same results. It costs $84 to order four $18 tickets by phone, $132 to order four $30 tickets, and so on.

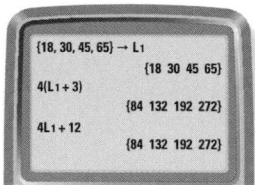

EXAMPLE D

Rewrite the expression $2 - 3(4 + L_1)$ in an equivalent form without parentheses. Enter and store a short list like $\{1, -3, -10, 5\}$ in list L1 to verify your result on your calculator.

▶ **Solution**

The expression $2 - 3(4 + L_1)$ is really a family of expressions because list L1 holds four values.

$$2 - 3(4 + L_1) = 2 + (-3)(4 + L_1)$$ Change from subtracting to adding a negative number.

$$= 2 + (-3)(4) + (-3)(L_1)$$ Distribute the -3 over the addition $4 + L_1$.

$$= 2 + (-12) + (-3L_1)$$ Multiply.

$$= -10 - 3L_1$$ Add.

Evaluating $-10 - 3L_1$ for each number stored in list L1 gives $-13, -1, 20,$ and -25.

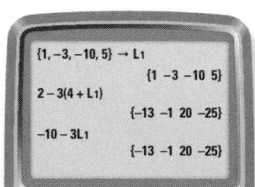

▶ **EXAMPLE D**

[Alert] When students see expressions such as $2 - 3(4 + x)$, they often forget to distribute the "$-$" in -3 to the second number in the parentheses and might write $-12 + 3x$ instead of $-12 - 3x$.

If this is your students' first experience with the distributive property, you may want to give more examples. If you haven't used the optional one-step puzzle, Parentheses, you might try it now.

Closing the Lesson

The **distributive property** allows shortcuts in doing mental arithmetic to solve real-world problems.

Practice Your Skills

1. Evaluate these expressions without a calculator. Use your calculator to check your answers.

   a. $\frac{-27 + 39}{4}$ 3
   b. $2 + (-3)(4)$ -10
   c. $\frac{3}{4} - \frac{1}{2}$ $\frac{1}{4}$
   d. $\frac{12}{4 - 6}$ -6

2. Check your understanding of the rules for order of operations

   a. Evaluate the expression $-4(13 - 2 \cdot 3^2)$ without using your calculator. 20

   b. Enter the expression into your calculator, and confirm both the correct order of operations you used and your computed answer.

3. Check your understanding of the distributive property.

   a. Write an expression equivalent to $2(L + W)$. $2L + 2W$

   b. Verify that your answer is equivalent to $2(L + W)$ by substituting 15 for L and 4 for W into both the original expression and your answer to 3a. $2(15 + 4) = 38; 2 \cdot 15 + 2 \cdot 4 = 38$

4. Describe two different procedures to find the value of the expression $8075 - 3(37 + 5)$. Use the distributive property to show that the two ways are equivalent.

5. **APPLICATION** Jasmine makes $7 per hour in her part-time job at the corner store. She worked 4 hours on Monday, 3 hours on Wednesday, and 5 hours on Saturday.

   a. How much did she earn for the week? $84

   b. Explain two ways you can answer the question in 5a. Why are the answers the same?

6. **APPLICATION** Use the distributive property to find the 15% tip amount for these food bills.

   a. $36
   b. $21 $3.15
   c. $48.50 $7.28
   $(10\% + 5\%) \cdot \$36 = \$3.60 + \$1.80 = \$5.40$

7. **APPLICATION** Find the 15% tip amount for these food bills by writing and solving proportions. For example, $\frac{15}{100} = \frac{x}{36}$.

   a. $36 $5.40
   b. $21 $3.15
   c. $48.50 $7.28

Reason and Apply

8. Peter and Seija evaluated the expression $37 + 8 \cdot \frac{6}{2}$. Peter said the answer was 135. Seija said it was 61. Who is correct? What error did the other person make?
   Seija. Peter incorrectly added before multiplying (45 times $\frac{6}{2}$ is 135).

9. In what order would you perform the operations to evaluate these expressions and get the correct answers?

   a. $9 + 16 \cdot 4.5 = 81$
   b. $18 \div 3 + 15 = 21$
   c. $3 - 4(-5 + 6^2) = -121$

BUILDING UNDERSTANDING

Students practice arithmetic operations and the distributive property.

ASSIGNING HOMEWORK

Essential	1–3, 5, 6, 8, 9, 12
Performance assessment	4, 13
Portfolio	14
Journal	8, 10, 11
Group	11, 14
Review	7, 15–17

▶ **Helping with the Exercises**

2b.
```
-4(13-2*3²)
                20
```

4. Procedure 1: Add 37 and 5 to get 42. Multiply 3 by 42 to get 126. Then subtract 126 from 8075 to get 7949. Procedure 2: The expression is equivalent to $8075 + (-3)(37 + 5)$. By the distributive property, this expression is equivalent to $8075 + (-3) \cdot 37 + (-3) \cdot 5 = 8075 - 111 - 15 = 7949$.

5b. You can multiply 7 by each day's hours and then add those products to find the week's total. Or you can find the total hours for the week first and multiply this result by $7 \cdot 4 + 7 \cdot 3 + 7 \cdot 5 = 7(4 + 3 + 5)$.

Exercises 6 and 7 [Ask] "Which method is easiest to do in your head?"

9a. First multiply 16 by 4.5, then add 9.

9b. First divide 18 by 3, then add 15.

9c. First square 6, then add -5, then multiply by 4 and subtract 124 from 3.

10. You can use your calculator's instant replay command when you need to evaluate an expression for several different sets of values. This command is helpful because it recalls previous entries one at a time and displays them on the screen.

[▶ 🖳 See **Calculator Note 4A** to review the instant replay command. ◀]

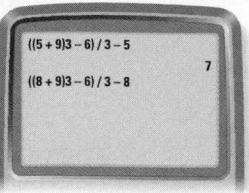

((5 + 9)3 − 6) / 3 − 5

((8 + 9)3 − 6) / 3 − 8
   7

a. Enter and evaluate the expression $((5 + 9)3 - 6)/3 - 5$. What is the answer? $((5 + 9)3 - 6)/3 - 5 = 7$

b. Use the instant replay function to get back to the original expression. Replace each 5 with 8 and reevaluate the expression. What's your answer this time? $((8 + 9)3 - 6)/3 - 8 = 7$

c. Go back to the last expression you entered. Use the insert key and replace each 8 with 25. Reevaluate the expression. Now what's your answer? $((25 + 9)3 - 6)/3 - 25 = 7$

d. Go back to the last expression you entered. Use the delete key and replace each 25 with 2. Reevaluate the expression. What answer did you get? $((2 + 9)3 - 6)/3 - 2 = 7$

e. Use the instant replay function repeatedly to return to the original expression in 10a. Replace each 5 with the same number of your choice. Reevaluate the expression once more. What's your answer? Expressions will vary; the answer is 7.

f. What did you notice about the answers to 10a–e? Why do you think this happened? Possible answer: The answer is always 7 because the original expression, $\frac{(5 + 9)3 - 6}{3} - 5$, can be reduced to $5 + 9 - 2 - 5$.

11a. Possible answers:
$(3 + 2)(5) - 7 = 18$. First add 3 and 2. Then multiply this sum by 5 and subtract 7. Or, $3(2) + 5 + 7 = 18$. Multiply 3 by 2 and add the product, 6, to 5 plus 7.

11b. $8 - 5(6 - 7) = 13$. First subtract 7 from 6. Then multiply this difference by 5. Then subtract this answer from 8.

Exercise 14 You could keep a running list of answers to this question. Students can return to it as they have extra time.

11. Insert operation signs, parentheses, or both into each string of numbers to create an expression equal to the answer given. Keep the numbers in the same order as they are written. Write an explanation of your answer, including information on the order in which you performed the operations.

a. 3 2 5 7 = 18   b. 8 5 6 7 = 13

12. Using a list in an expression is the same as writing as many expressions as there are numbers in the list. Use the distributive property to rewrite each family of expressions without parentheses. Then enter $\{-3, 4.5, 10\}$ into list L_1 to check answers.

a. $3(L_1 - 4.2)$ b. $14 + 3(L_1 - 4.2)$ c. $14 - 3(L_1 - 4.2)$
$3L_1 - 12.6; \{-21.6, 0.9, 17.4\}$ $3L_1 + 1.4; \{-7.6, 14.9, 31.4\}$ $-3L_1 + 26.6; \{35.6, 13.1, -3.4\}$

13. Write and then evaluate an expression that performs the following sequence of operations: Add 6 and 3, multiply by the square of 4, divide by the sum of 8 and 2, and then subtract 9. $\frac{(6 + 3) \cdot 4^2}{8 + 2} - 9 = 5.4$

14. This problem is sometimes called Einstein's problem: "Use the digits 1, 2, 3, 4, 5, 6, 7, 8, 9 and any combination of the operation signs $(+, -, \cdot, /)$ to write an expression that equals 100. Keep the numbers in consecutive order and do not use parentheses."

Here is one solution:

$123 - 4 - 5 - 6 - 7 + 8 - 9 = 100$

Your task is to find another one.

Other solutions: $1 + 2 + 3 + 4 + 5 + 6 + 7 + 8 \cdot 9 = 100$.
$12 + 34 + 5 \cdot 6 + 7 + 8 + 9 = 100$
$-1 + 2 + 3 + 4 \cdot 5 \cdot 6 - 7 - 8 - 9 = 100$

► Review

15. The table displays data for a bicyclist's distance from home during a four-hour bike ride.

 a. Make a scatter plot of the data.

15 mph **b.** Find the bicyclist's average speed.

$y = 15x$ **c.** Find an equation that models the data and graph it on the scatter plot.

 d. At what times might the bicyclist be riding downhill or pedaling uphill? Explain.

Downhill during the intervals of 1 to 2.25 hours away from home and 3.5 to 4 hours (the bike's velocity is above average) and uphill during the interval of 2.25 to 3.5 hours away from home (the bike's speed is slower than average).

Time (hr)	Distance (mi)
0	0
0.25	4
0.50	8
1.00	15
1.50	25
2.00	36
2.25	40
2.75	41
3.00	44
3.50	48
4.00	60

16. You are helping to design boxes for game balls that are 1 inch in diameter. Your supervisor wants the balls in one rectangular layer of rows and columns.

 a. Use a table to show all the ways to package 24 balls in rows and columns.

 b. Plot these points on a graph.

 c. Do the possible box dimensions represent direct or inverse variation? Explain how you know. Indirect variation, the product of the length and width equals a constant.

 d. Represent the situation with an equation. Does the equation have any limitations?

Learn about a career in packaging science with the links at **www.keymath.com/DA** .

17. APPLICATION A restaurant menu item is $12.95. You have a coupon for 20% off the price. Sales tax is 8% of the discounted amount. Find the amounts for the discount, tax, and a 15% tip that is based on the original amount. What is the total bill?
The 20% discount is $2.59. The 8% tax on 10.36 is $0.83. The tip amount on $12.95 is $1.94, so the total is $13.13.

Exercise 15 This exercise reviews Lessons 1.6 and 3.2.

15a.   **Distance Traveled**

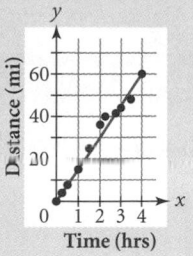

Exercise 16 This exercise reviews Lesson 3.4.

16a.

Length (in.)	Width (in.)
1	24
2	12
3	8
4	6
6	4
8	3
12	2
24	1

16b.   **Possible Boxes**

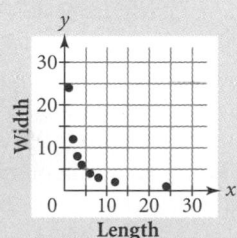

16d. $l \cdot w = 24$ or $w = \frac{24}{l}$. Yes, the situation requires the dimensions to be whole numbers of inches, so you can't have a box that is 16 inches by 1.5 inches even though it satisfies the equation.

Exercise 17 This exercise reviews Lesson 2.4.

LESSON

4.2

Writing Expressions and Undoing Operations

All change is not growth; all movement is not forward.

ELLEN GLASGOW

Try this trick: Think of a number from 1 to 25. Add 9 to it. Multiply the result by 3. Subtract 6 from the current answer. Divide this answer by 3. Now subtract your original number. No matter what number you chose, your final result should be 7! Surprised? Try the trick again, starting with a different number. Did you get 7 again? Did all your classmates also get 7? How does this number trick work? The illustration at right shows the sequence this number trick generates if your original number is 11. If you understand the rules for order of operations and a little algebra, you can determine why this trick works and design your own number tricks.

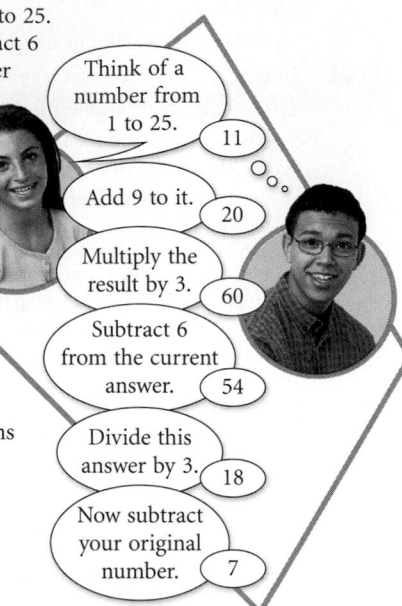

Think of a number from 1 to 25. — 11
Add 9 to it. — 20
Multiply the result by 3. — 60
Subtract 6 from the current answer. — 54
Divide this answer by 3. — 18
Now subtract your original number. — 7

Investigation
The Math Behind the Number Trick

Are you convinced that the number trick you just did works regardless of the number you start with? Would it still work if you chose a decimal number, a fraction, or a negative number? You can test the trick on several numbers at once by using a calculator list with several different starting numbers.

Step 1 Enter a list of at least four different numbers on the calculator home screen and store this list in list L1. In the example at right, the list is {20, 1.2, −5, 4}, but you should try different numbers. Perform the operations on your own starting numbers. The last operation is to subtract your original number.

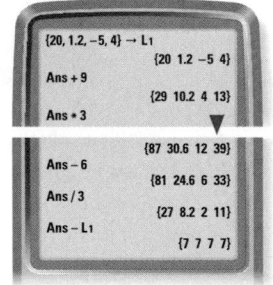

{20, 1.2, −5, 4} → L1
{20 1.2 −5 4}
Ans + 9
{29 10.2 4 13}
Ans * 3
{87 30.6 12 39}
Ans − 6
{81 24.6 6 33}
Ans / 3
{27 8.2 2 11}
Ans − L1
{7 7 7 7}

Step 2 Explain how the last operation is different from the others.

Step 3 Number tricks like this work because certain operations, such as multiplication and division, get "undone" in the course of the trick. Which step undoes Ans · 3?

Ans/3

PLANNING

LESSON OUTLINE

One day:

5 min	Introduction
20 min	Investigation
5 min	Sharing
10 min	Examples A, B, C
5 min	Closing
5 min	Exercises

MATERIALS

- Undoing Operations (W, T), *optional*

TEACHING

In this lesson, students continue to look for patterns in sequences and work with the order of operations. They also write expressions for number tricks and evaluate these expressions. The undoing and working backward strategy is introduced as a way to find the starting number in a number trick.

INTRODUCTION

Do the number trick with the whole class and ask several students for their answers.

Guiding the Investigation

One step Ask students to use an algebraic expression to express the number trick, to explain why it works, and then to make up their own trick. As you circulate, keep asking what cancels what and why, and suggest testing number tricks with calculator lists of several numbers at once.

Step 1 Encourage students to include decimals and negative numbers in their lists.

Step 2 You are subtracting your original number instead of a constant value.

LESSON OBJECTIVES

- Use calculator list operations to investigate the concepts of variables, terms, and expressions
- Rewrite algebraic expressions that include multiplication and division, or addition and subtraction, by the same number
- Build toward symbolic equation solving by working backward and undoing operations

NCTM STANDARDS

CONTENT	PROCESS
● Number	● Problem Solving
● Algebra	● Reasoning
Geometry	Communication
Measurement	● Connections
Data/Probability	● Representation

You can use the scheme below to help you figure out why any number trick works. The symbol $+1$ represents one positive unit. You can think of n as a variable or as a container for different unknown starting numbers.

Stage		Description
1	n	Pick a number.
2	n $+1$ $+1$ $+1$	Add 3.
3	n n $+1$ $+1$ $+1$ $+1$ $+1$ $+1$	Multiply by 2.
4	n n $+1$ $+1$	Subtract 4.
5	n $+1$	Divide by 2.
6	$+1$	Subtract the original number.
7	$+1$ $+1$ $+1$	Add 2.

Step 4 | Explain what happens at each stage. The descriptions are provided for Stages 1 and 6.

Step 5 | At which stage or stages will everyone's result be the same? Explain why this happens.

Step 5 At Stages 6 and 7; the original number has been subtracted.

Step 6 | Use your original list of starting numbers and record the results of this number trick at each stage.

Step 7 | Invent your own number trick that has at least five stages. Test it on your calculator with a list of at least four different numbers to make sure all the answers are the same. When you're convinced the number trick is working, try it on the other members of your group.

The math behind number tricks can help you understand the roles that variables and expressions play in algebra. A single expression can represent an entire number trick.

EXAMPLE A

Investigate this calculator routine to see if it is a number trick. Pick several numbers to store into list L₁ and use it as the variable x. For example, use $\{1, -2, 0, 0.3\}$ to apply the operations on four numbers at the same time.

a. Describe each stage of the routine in words.

b. List the sequence of numbers this routine generates if you let x equal -5.

x	ENTER
Ans $\cdot$ (-10)	ENTER
Ans $+$ 35	ENTER
Ans/(-10)	ENTER
Ans $-$ 14.5	ENTER
Ans $-$ x	ENTER

Step 3 Be sure students understand that the scheme isn't illustrating the number trick they've been working with.

Step 6 Sample sequences using $\{20, 1.2, -5, 4\}$ as starting numbers are

Stage	Starting Number			
1	20	1.2	-5	4
2	23	4.2	-2	7
3	46	8.4	-4	14
4	42	4.4	-8	10
5	21	2.2	-4	5
6	1	1	1	1
7	4	4	4	4

Step 7 Here's a sample number trick: Pick a number. Double it. Add 6. Divide by 2. Subtract 7. Subtract your starting number. Some students may have difficulty coming up with a trick that gives predictable results each time. Be sure they consult with their groups if they are having difficulty.

SHARING IDEAS

Have students or groups challenge each other with their number tricks. Introduce the idea of an equation and challenge students to describe their own number tricks with expressions. Ask if they can write an algebraic expression similar to those for number tricks, but one in which the answer depends on the starting number but is not the starting number. Then ask them to find a single number that solves the equation—a solution.

Assessing Progress

Look for familiarity with arithmetic operations and the distributive property, with handling calculator lists, and with algebraic expressions.

► EXAMPLE A

For students having difficulty representing number tricks with algebraic expressions, this example provides a transition through calculator lists.

[Alert] Students may see how dividing by −10 undoes the multiplication of x by −10 but not see that −10 is also being divided into 35 to get −3.5 (which, when added to −14.5, gives −18).

c. Write a sequence of expressions that shows each stage of this routine. Use x to represent your starting number.

d. Pick a new starting number and store it in x. Find the number this routine produces.

► Solution

a. The first column in the table describes the routine in words.

b. The second column shows the sequence of numbers generated by a starting number of −5.

c. The third column gives the sequence of expressions for the routine for a starting number x. Notice that the final expression contains all the information about the stages of the trick in a concise, symbolic form. You can also begin to see why the number trick works. The division by −10 "undoes" the multiplication by −10, and by subtracting x in the last stage you'll get the same result no matter what x is.

a. Description	b. Sequence	c. Expression
Pick the starting number.	−5	x
Multiply by −10.	50	$-10x$
Add 35.	85	$-10x + 35$
Divide by −10.	−8.5	$\dfrac{-10x + 35}{10}$
Subtract 14.5.	−23	$\dfrac{-10x + 35}{-10} - 14.5$
Subtract your original number.	−18	$\dfrac{-10x + 35}{-10} - 14.5 - x$

d. The routine always results in −18.

EXAMPLE B

Consider the expression

$$4\left(\frac{x + 7}{4} + 5\right) - x + 13$$

a. Write in words the number trick that the expression describes.

b. Test the number trick to be sure you get the same result no matter what number you choose.

c. Which operations that undo previous operations make this number trick work?

► Solution

a. Pick a number, x.
   Add 7.
   Divide the answer by 4.
   Add 5 to this result.
   Multiply the answer by 4.
   Subtract your original number.
   Then add 13.

b. One way to test the trick is to enter the expression into your calculator. Be sure to use parentheses for grouping symbols like the fraction bar:

$$4((x + 7)/4 + 5) - x + 13$$

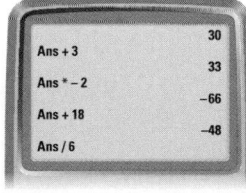

Press $\boxed{\text{ENTER}}$ to evaluate the expression for whatever value you now have stored in x.
Your answer should be 40. Store a different value for x, then use the instant replay function to recall the expression. Press $\boxed{\text{ENTER}}$ to evaluate the expression with your new x-value. You should get 40 again. Use a list to test the trick on several numbers at the same time.

c. The multiplication by 4 undoes the division by 4. You start with x and later you subtract x, so it doesn't matter what value you choose for x.

An **equation** is a statement that says the value of one number or algebraic expression is equal to the value of another number or algebraic expression. You can represent the number trick in Example B with the equation

$$4\left(\frac{x + 7}{4} + 5\right) - x + 13 = 40$$

The 40 on the right of the equal sign is the result of evaluating the expression on the left for some value of x. If x can be replaced by a number that makes the equation true, then that number is a **solution** to the equation. This is a very unusual equation, because it's true no matter what number x is. The equation has infinitely many solutions. That's what makes it a trick!

Of course, not every algebraic expression represents a number trick. In most algebraic expressions that you'll evaluate, your final result will depend on what your original number is.

EXAMPLE C

Consider the expression

$$\frac{18 - 2(x + 3)}{6}$$

a. Rewrite the expression by changing the subtraction to addition.

b. Find the value of the expression if x is 30. Write an equation that sets the expression equal to this result.

c. Solve the equation $\frac{18 + (-2)(x + 3)}{6} = 15$. That is, find the value of x that makes the equation true.

▶ Solution

a. Subtracting is the same as adding a negative quantity, so you can rewrite the expression as $\frac{18 + (-2)(x + 3)}{6}$.

b. One way to evaluate the expression is to perform its operations in order, as shown on the calculator screen.

The result of the final operation is -8.
The equation is

$$\frac{18 + (-2)(x + 3)}{6} = -8$$

▶ EXAMPLE B

This example gives a number trick with nested parentheses. Point out why the group within a group requires two sets of parentheses. You want the calculator, following the order of operations, to evaluate the innermost parentheses first.

[Alert] Some students may not understand that on a calculator a variable is the name of a storage location that holds a number. They might understand better if you ask them to type in x $\boxed{\text{ENTER}}$ before and after storing their own value in x.

Note that an equation is a statement that two expressions are equal. You might point out that an *expression* is like a noun phrase and an *equation* is like an entire sentence that says something about two noun phrases. If one of the expressions involves a variable, a *solution* to the equation is a number that replaces that variable to make a true statement.

▶ EXAMPLE C

This example presents an equation that has only one solution. Although the given solution uses the undoing approach, students could also find approximate solutions by graph tracing and by calculator tables.

You can simplify algebraic expressions if the same number is being both added and subtracted, or if it is being both multiplied and divided. An **equation** is a sentence claiming that two expressions are equal. A **solution** is a number that, when substituted for the variable, makes the equation a true statement.

c. You can solve the equation by working backward, **undoing** each operation you did in part b. Determine how to undo each operation and perform them on the number 15 in reverse order.

1. List the operations on x.

2. Start with the answer and work backward to undo each operation.

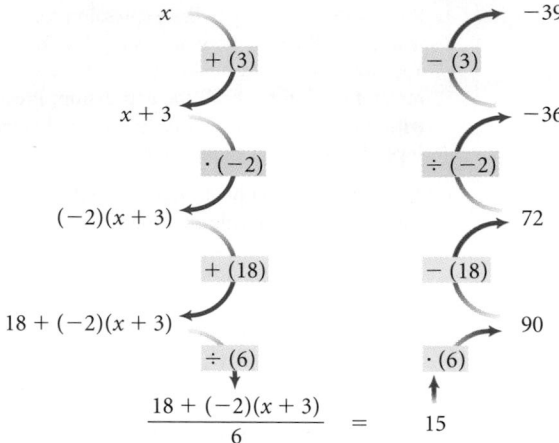

$$\frac{18 + (-2)(x + 3)}{6} = 15$$

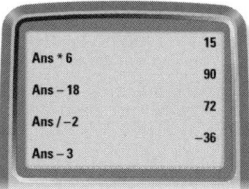

So $x = -39$. You may find it helpful to use this format when solving equations by undoing operations.

Equation: $\dfrac{18 + (-2)(x + 3)}{6} = 15$ Work backwards

Operations on x	Undo operations	
$+ (3)$	$- (3)$	$x = -39$
$\cdot (-2)$	$\div (-2)$	-36
$+ (18)$	$- (18)$	72
$\div (6)$	$\cdot (6)$	90
		15

EXERCISES

You will need your calculator for problems **1, 5,** and **8.**

▶ Practice Your Skills

1. Evaluate each expression without a calculator. Then check your result with your calculator.

   a. $-4 + (-8)$ -12

   b. $(-4)(-8)$ 32

   c. $-2(3 + 9)$ -24

   d. $5 + (-8)(-3)$ 33

   e. $(-3)(-5) + (-2)$ 13

   f. $\frac{-15}{3} + 8$ 3

   g. $\frac{23 - 3(4 - 9)}{-2}$ -19

   h. $\frac{-4(7 + (-8))}{8} - 6.5$ -6

   i. $\frac{6(2 \cdot 4 - 5) - 2}{-4}$ -4

2. Sondro entered $75 + 81 + 71 + 78 + 83/5$ to find his average score on five tests. He got 321.6. Is this his correct average? If not, what did he do wrong, and what is his correct average? No, this isn't correct. Sondro should have used parentheses:
$$(75 + 81 + 71 + 78 + 83)/5 = 77.6$$

3. Evaluate each expression if $x = 6$.

   a. $2x + 3$ 15

   b. $2(x + 3)$ 18

   c. $5x - 13$ 17

   d. $\frac{x + 9}{3}$ 5

4. For each equation identify the order of operations. Then work backward through the order of operations to find x.

   a. $\frac{x - 3}{2} = 6$ 15

   b. $3x + 7 = 22$ 5

   c. $\frac{x}{6} - 20 = -19$ 6

5. Marteen gave Juwan these instructions: Pick a number, add 3, multiply by 4, and subtract 5.

   a. Juwan began with 2 and got 15. He then entered the expression $2 + 3 \cdot 4 - 5$ into his calculator and got 9. What's wrong? Juwan forgot the parentheses.

   b. What expression should Juwan enter into his calculator so that he will get 15?
$$(2 + 3)4 - 5 \text{ or } 4(2 + 3) - 5.$$

6. Justine asked her group members to do this calculation: Pick a number, multiply by 5, and subtract 2. Quentin got 33 for an answer. Explain how Justine could determine what number Quentin picked. What number did Quentin pick? Work backward by adding 2 to 33 and then dividing by 5; 7

▶ Reason and Apply

7. Daxun, Lacy, Claudia, and Al are working on a number trick. Here are the number sequences their number trick generates:

Description	Daxun's sequence	Lacy's sequence	Claudia's sequence	Al's sequence
Pick the starting number.	14	-5	-8.6	x
Add 5.	19	0	-3.6	$x + 5$
Multiply by 4.	76	0	-14.4	$4(x + 5)$
Subtract 12.	64	-12	-26.4	$4(x + 5) - 12$
Divide by 4.	16	-3	-6.6	$\frac{4(x + 5) - 12}{4}$
Subtract the original number.	2	2	2	$\frac{4(x + 5) - 12}{4} - x$

BUILDING UNDERSTANDING

Students practice simplifying algebraic expressions and solving linear equations. They revisit both kinds of number tricks.

ASSIGNING HOMEWORK

Essential	2–4, 6–9, 11 or 13
Performance assessment	9, 14
Portfolio	11, 13
Journal	5, 10, 14
Group	7, 9, 11–16
Review	1, 17–19

▶ Helping with the Exercises

Exercises 7 and 9 Many students have a lot of fun with these essential exercises. Some, however, may find that keeping track of several columns is a challenge. Encourage them to work with other students.

a. Describe the stages of this number trick in the first column.

b. Complete Claudia's sequence.

c. Write a sequence of expressions for Al in the last column.

8. In the scheme below, the symbol +1 represents +1 and the symbol −1 represents −1. The symbol n represents the original number.

Stage		**Description**
1	n	Pick a number.
2	n −1 −1 −1	Subtract 3.
3	n n −1 −1 −1 −1 −1 −1	Multiply your result by 2.
4	n n −1 −1	Add 4.
5	n −1	Divide by 2.
6	−1	Subtract the original number.
7	+1 +1 +1	Add 4.

a. Explain what is happening as you move from one stage to the next. The explanation for Stage 6 is provided.

b. At which stage will everyone's result be the same? Explain. At Stages 6 and 7; the original number has been subtracted.

c. Verify that this trick works by using a calculator list and an answer routine.

d. Write an expression similar to the one shown in the solution to part c of Example A to represent this trick. $\dfrac{2(n-3)+4}{2} - n + 4$

9. Jo asked Jack and Nina to try two other number tricks that she had invented for homework. Their number sequences are shown in the tables. Use words to describe each stage of the number tricks.

a. Number Trick 1:

Description	Jack's sequence	Nina's sequence
Pick the starting number.	5	3
Multiply by 2.	10	6
Multiply by 3.	30	18
Add 6.	36	24
Divide by 3.	12	8
Subtract your original number.	7	5
Subtract your original number again.	2	2

b. Number Trick 2:

Description	Jack's sequence	Nina's sequence
Pick the starting number.	−10	10
Add 2.	−8	12
Multiply by 3.	−24	36
Add 9.	−15	45
Subtract 15.	−30	30
Multiply by 2.	−60	60
Divide by 6 (you should have your original number).	−10	10

10. Marcella wrote an expression for a number trick.

   a. Describe Marcella's number trick in words.

   b. Pick a number and use it to do the trick. What answer do you get?
Pick another number and do the trick again. What is the "trick"?
Solutions will vary. The trick always produces the original number.

Marcella's Trick
$\dfrac{4(x-5)+8}{2} - x + 6$

11. Write your own number trick with at least six stages.

   a. No matter what number you begin with, make the trick result in -4.

   b. Describe the process you used to create the trick. Sample answer: Start with -4 and make up operations, then work backward.

   c. Write an expression for your trick. $\dfrac{2(x-3)+10}{2} - x - 6$

12. The final answer to the number trick shown at right is 3. Starting with the final number, work backward, from the bottom to the top, undoing the operation at each step.

   a. What is the original number? 3

   b. How can you check that your answer to 12a is correct? Start with 3 and see if you get the answer 3.

   c. What is the original number if the final result is 15? 15

   d. What makes this sequence of operations a number trick?
The final result is always the original number no matter what number you choose.

x	_____
Ans $\cdot$ 8	_____
Ans $+$ 9	_____
Ans $/$ 4	_____
Ans $+$ 5.75	14
Ans $/$ 2	7
Ans $-$ 4	3

13. Write your own number trick so that, no matter what number you begin with, the result will always be that number.

14. The sequence of operations at right isn't a number trick. It will always give you a different final answer depending on the number you start with.

   a. What is the final value if you start with 18? 8.8

   b. What number did you start with if the final answer is 7.6? 15

   c. Describe how you got your answer to 14a. Undo the operations shown in reverse order.

   d. Let x represent the number you start with. Write an algebraic expression to represent this sequence of operations. $\dfrac{2(x+10)-12}{5}$

   e. Set the expression you got in 14d equal to zero. Then solve the equation for x. Check that your solution is correct by using the value you got for x as the starting number. Do you get zero again? $x = -4$; yes, sequence: $-4, 6, 12, 0, 0$

Ans $+$ 10	_____
Ans $\cdot$ 2	_____
Ans $-$ 12	_____
Ans $/$ 5	_____

15. Consider the expression

$$\frac{5(x+7)}{3}$$

   a. Find the value of the expression if $x = 8$. List the order you performed the operations. 25; add 7, multiply by 5, divide by 3.

   b. Solve the equation $\dfrac{5(x+7)}{3} = -18$ by undoing the sequence of operations in 15a. -17.8

16. Consider the expression

$$\frac{2.5(x-4.2)}{5} - 4.3$$

   a. Find the value of the expression if $x = 8$. Start with 8 and use the order of operations. -2.4

   b. Solve the equation $\dfrac{2.5(x-4.2)}{5} - 4.3 = 5.4$ by undoing the sequence of operations in 16a. $x = 23.6$

10a. Pick a number. Subtract 5. Multiply by 4. Add 8. Divide by 2. Subtract the original number. Add 6.

11a. Sample answer: Pick a number. Subtract 3. Multiply by 2. Add 10. Divide by 2. Subtract the original number. Subtract 6.

13. Sample answer: Pick a number. Add 5. Multiply by 4. Divide by 2. Subtract original number. Subtract 10.
$$\frac{4(x+5)}{2} - x - 10$$

Exercises 15 and 16 Suggest that students use the format of the Undoing Operations template.

15b.

Equation: $\dfrac{5(x+7)}{3} = -18$ Work backward

Operation on x	Undo operation	$x = -17.8$
$+ (7)$	$- (7)$	-10.8
$\cdot (5)$	$\div (5)$	-54
$\div (3)$	$\cdot (3)$	-18

16b.

Equation: $\dfrac{2.5x(x-4.2)}{5} - 4.3 = 5.4$ Work backward

Operation on x	Undo operation	$x = 23.6$
$- (4.2)$	$+ (4.2)$	19.4
$\cdot (2.5)$	$\div (2.5)$	48.5
$\div (5)$	$\cdot (5)$	9.7
$- (4.3)$	$+ (4.3)$	5.4

Exercise 17 Encourage students to write units and use dimensional analysis. This exercise reviews Lessons 2.3 and 3.1.

Exercise 18 This exercise reviews Lesson 4.1.

Exercise 19 This exercise reviews Lessons 0.1 and 3.2.

▶ **Review**

17. An HO-scale electric slot car set advertises that its cars travel in scale speeds of 200 miles per hour. If the HO scale factor is 1 : 87, find the car's actual speed in

   a. Feet per minute. 202.3 ft/minute

   b. Centimeters per second. 102.8 cm/sec

18. In the 15th century, Arab and Persian mathematicians popularized the use of an algebraic relations method of multiplication. These mathematicians multiplied two numbers, a and b, by using the relations method: $ab = 10b - (10-a)b$.

   Here's an example:

$$7 \times 8 = 10 \times 8 - (10 - 7)\times 8$$
$$= 10 \times 8 - 3 \times 8$$
$$= 80 - 24$$
$$= 56$$

   (Q. Mushtaq and A. L. Tan, *Mathematics: The Islamic Legacy,* 1990, p. 97)

   Does this method work with fractions? Decimals? Negative numbers? Pick two numbers and multiply them using this method. Discuss the advantages and disadvantages of using algebraic relations.

19. Natalie works in a shop that sells mixed nuts. Alice drops by and decides to buy a bag of mixed nuts with

$\frac{3}{4}$ cup of almonds

$\frac{2}{3}$ cup of cashews

$\frac{1}{2}$ cup of pecans

   a. How many cups of mixed nuts will there be in the bag? $1\frac{11}{12}$ cups

   b. Almonds cost $6.98 a cup, cashews cost $7.98 a cup, and pecans cost $4.98 a cup. What is the cost of Alice's bag of nuts? $13.05

18. Yes, this method works for fractions, decimals, and negative numbers. Sample student answer:

$$9 \times 6 = 10 \times 6 - (10 - 9) \times 6$$
$$= 10 \times 6 - 1 \times 6$$
$$= 60 - 6$$
$$= 54$$

This method works best when it converts multiplying large numbers to multiplying small numbers. It would not be useful for 3×4.

Recursive Sequences

A mathematician, like a painter or a poet, is a maker of patterns. If his patterns are more permanent than theirs, it is because they are made of ideas.

G. H. HARDY

The Empire State Building in New York City has 102 floors and is 1250 feet high. How high up are you when you reach the 80th floor? You can answer this question using a recursive sequence. In this lesson you will learn how to analyze geometric patterns, complete tables, and find missing values using numerical sequences.

A **recursive sequence** is an ordered list of numbers defined by a starting value and a rule. You generate the sequence by applying the rule to the starting value, then applying it to the resulting value, and repeating this process.

EXAMPLE A

The measurements in the table represent heights above and below ground at different floor levels in a 25-story building. Write a **recursive routine** that provides the sequence of heights −4, 9, 22, 35, . . . , 217, . . . that corresponds to the building floor numbers 0, 1, 2, ... Use this routine to find each missing value in the table.

Floor number	Basement (0)	1	2	3	4	. . .	10	. . .		. . .	25
Height (ft)	−4	9	22	35		. . .		. . .	217	. . .	

▶ **Solution**

The starting value is −4 because the basement is 4 feet below ground level. Each floor is 13 feet higher than the floor below it, so the rule for finding the next floor height is "add 13 to the current floor height."

The calculator screen shows how to enter this recursive routine on your calculator. Press −4 (ENTER) to start your number sequence. Press +13 (ENTER). The calculator automatically displays Ans +13 and computes the next value. Simply pressing (ENTER) again applies the rule for finding successive floor heights. [▶ ☐ See **Calculator Note 0D.** ◀]

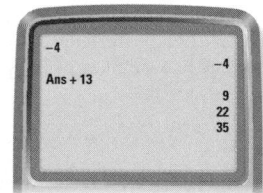

How high up is the 10th floor? Count the number of times you press (ENTER) until you reach 10. Which floor is at a height of 217 feet? Keep counting until you see that value on your calculator screen. What's the height of the 25th floor? Keep applying the rule by pressing (ENTER) and record the values in your table.

NCTM STANDARDS

CONTENT		PROCESS	
●	Number		Problem Solving
●	Algebra	●	Reasoning
●	Geometry		Communication
	Measurement		Connections
	Data/Probability	●	Representation

LESSON OBJECTIVES

• Review or become familiar with the concept of recursion

• Investigate recursive (arithmetic) sequences using the calculator

PLANNING

LESSON OUTLINE

One day:

5 min	Introduction, Example A
20 min	Investigation
10 min	Sharing, Example B
5 min	Closing
10 min	Exercises

MATERIALS

• Calculator Note 0D

• boxes of toothpicks (or have students draw line segments on paper)

TEACHING

In this lesson the idea of recursion, introduced in Chapter 0 with fractals and evaluation of expressions, is revisited with calculator home-screen iteration of recursive sequences.

INTRODUCTION

Students may say that they need only a proportion to determine the height of the 80th floor of the Empire State Building. However, they would be making some assumptions. Leave that question open until Exercise 6.

If you skipped Chapter 0, explain that while executing a *recursive procedure* you do the same thing repeatedly, at each step operating on the result of the previous step.

▶ **EXAMPLE A**

This example reminds students of how to use home-screen iteration. You may choose to do it as a class, especially if you skipped recursion in Chapter 0. **[ESL]** *Floor*, or *story*, means the level of a building usually beginning with 1. The *height of a floor* could mean the floor to ceiling distance of one

floor, but here it means the *height above ground* of that floor.

Some students might wonder if −4 is the first term of the sequence. Point out that sequences can have "zeroth" terms to make the counting easier.

The 10th floor is 126 feet high. 217 feet is the height of the 17th floor. The 25th floor is 321 feet high.

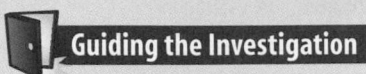

Guiding the Investigation

If you think your students will not handle toothpicks appropriately, you can have them draw line segments on paper.

One step Draw the three figures showing the growth of a triangle pattern or ask students to look at page 200. Point out how the perimeter is changing. Ask students to design their own toothpick pattern with a changing feature, like perimeter, and to write a recursive routine to make their calculator generate the sequence of numbers. As you interact, be sure students see the idea of a common difference between consecutive terms.

Step 1 If students do not use just 1 toothpick per side, their answers will be multiples of 3.

Step 2 Be sure that students extend their patterns of shapes horizontally. If they add shapes in several directions, their sequences will not be arithmetic.

The table is organized so that the calculated sequences are displayed vertically in columns. Calculators will also display the results of recursive routines vertically on the home screen.

Step 3 As needed, ask what the starting values for the number of toothpicks and for the perimeter are. Then ask about the rules for finding the next numbers.

Step 5 Be sure the pattern of squares remains a row.

Investigation
Recursive Toothpick Patterns

You will need
- a box of toothpicks

In this investigation you will learn to create and apply recursive sequences by modeling them with puzzle pieces made from toothpicks.

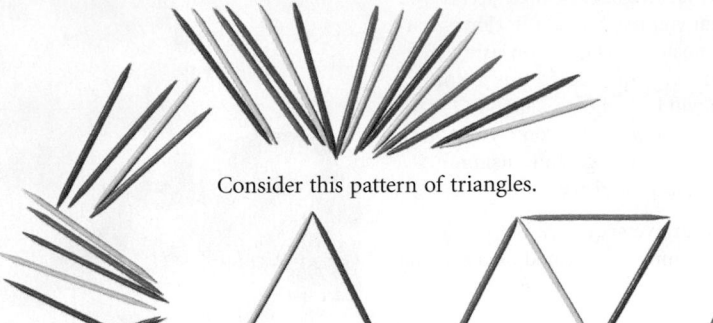

Consider this pattern of triangles.

Figure 1   Figure 2   Figure 3

See Step 2.   **Step 1**   Make Figures 1–3 of the pattern using as few toothpicks as possible. How many toothpicks does it take to reproduce each figure? How many toothpicks lie on the perimeter of each figure?

Step 2   Copy the table with enough rows for six figures of the pattern. Make Figures 4–6 from toothpicks by adding triangles in a row and complete the table.

Step 3 Toothpicks: add 2 to the previous number. Perimeter: add 1 to the previous perimeter. Press 3 ENTER and then Ans + 2 ENTER to find the successive numbers of toothpicks. For the perimeter, press 3 ENTER and then Ans + 1 ENTER.

	Number of toothpicks	Perimeter
Figure 1		
Figure 2		

Step 3   What is the rule for finding the number of toothpicks in each figure? What is the rule for finding the perimeter? Use your calculator to create recursive routines for these rules. Check that these routines generate the numbers in your table.

Step 4 Figure 10: 21   **Step 4**   Now make Figure 10 from toothpicks. Count the number of toothpicks and find the perimeter. Does your calculator routine give the same answers? Find the number of toothpicks and the perimeter for Figure 25.

toothpicks with a perimeter of 12. Figure 25: 51 toothpicks with a perimeter of 27.

Next, you'll see what sequences you can generate with a new pattern.

Step 5   Design a pattern using a row of squares, instead of triangles, with your toothpicks. Repeat Steps 1–4 and answer all the questions with the new design.

Step 6   Choose a unit of measurement and explain how to calculate the area of a square made from toothpicks. How does your choice of unit affect calculations for the areas of each figure?

Step 2

Toothpicks	Perimeter
3	3
5	4
7	5
9	6
11	7
13	8

Step 5

Toothpicks	Perimeter
4	4
7	6
10	8
13	10
16	12
19	14

Toothpicks: add 3 to the previous number. Perimeter: add 2 to the previous perimeter. Press 4 ENTER and then Ans + 3 ENTER to find the number of toothpicks. To find the perimeter, press 4 ENTER and then Ans + 2 ENTER. Figure 10: 31 toothpicks with 22 on the perimeter. Figure 25: 76 toothpicks with 52 on the perimeter.

Now you'll create your own puzzle piece from toothpicks. Add identical pieces in one direction to make the succeeding figures of your design.

Step 7 | Draw Figures 1–3 on your paper. Write recursive routines to generate number sequences for the number of toothpicks, perimeter, and area of each of six figures. Record these numbers in a table. Find the values for a figure made of ten puzzle pieces.

Step 8 | Write three questions about your pattern that require recursive sequences to answer. For example: What is the perimeter if the area is 20? Test your questions on your classmates.

You have been writing number sequences in table columns in this investigation. Remember that you can also display sequences as a list of numbers like this:

1, 3, 5, 7, . . .

Each number in the sequence is called a term. The three periods indicate that the numbers continue.

EXAMPLE B

Find the missing values in each sequence.

a. 7, 12, 17, ___ , 27, ___ , ___ , 42, ___ , 52

b. 5, 1, −3, ___ , −11, −15, ___ , ___ , −27, ___

c. −7, ___ , −29, ___ , −51, −62, ___ , −84, ___

d. 2, −4, 8, −16, 32, ___ , 128, −256, ___ , ___

How many hidden numbers can you find?

▶ **Solution**

a. The starting value is 7 and you add 5 each time to get the next number. The missing numbers are shown in red.

starting value

$+5$ $+5$ $+5$ $+5$ $+5$ $+5$ $+5$ $+5$ $+5$
7,   12,   17,   22,   27,   32,   37,   42,   47,   52

b. The starting value is 5 and you subtract 4 each time to get the next number. The missing numbers are shown in red.

starting value

-4 -4 -4 -4 -4 -4 -4 -4 -4
5,   1,   −3,   −7,   −11,   −15,   −19,   −23,   −27,   −31

c. The starting value is −7. The difference between the fifth and sixth terms shows you subtract 11 each time.

starting value

-11 -11 -11 -11 -11 -11 -11 -11
−7,   −18,   −29,   −40,   −51,   −62,   −73,   −84,   −95

Step 6 Students might measure a toothpick in centimeters, in inches, or as a unit 1 toothpick long. For a unit of 1 toothpick, the area is equal to the number of squares in the figure. Otherwise, the number of squares must be multiplied by the unit area of each square to calculate the area of the entire figure.

Step 7 Students may be reluctant to use figures whose areas are difficult to find in terms of the edges. Suggest that they can use the area of the first figure as 1 square unit even if their basic shape is not a square.

SHARING IDEAS

Ask several students to present their questions from Step 8. For one of them in which the perimeter increases by a different number from the number of toothpicks, ask why. [Usually one or more toothpicks in the perimeter at the previous step are no longer in the perimeter.]

You might begin to use the term *rate of change* to describe what's happening—for example, the amount being added to the perimeter is the rate of change of the perimeter, in toothpicks per stage.

If you ask some students to put up their sequences with terms missing and have the class guess the missing terms, you can anticipate Example B.

Assessing Progress

Watch for systematic data collection, familiarity with geometric shapes and terms such as *perimeter* and *area,* and ability to contribute to group work.

▶ **EXAMPLE B**

This example is good for students who may not have grasped the nature of a sequence with common differences of consecutive terms. Many students enjoy creating and solving problems like this. Encourage differing approaches to part c. You can find the common difference by looking down the list to the first consecutive pair, or by finding half the difference between the third term and the first term: $\frac{1}{2}(-29 - -7) = -11$. Numbers in the illustration include 0, 1, 2, 3, 4, 7, 8, and 11.

A dynamic algebra exploration at www.keymath.com/DA can be used to explore recursive sequences such as the ones in this example and in Example A and the investigation.

Closing the Lesson

You generate a **recursive sequence** on the calculator by entering a starting number and then designating how to operate on one number to get the next number in the sequence.

BUILDING UNDERSTANDING

Students practice writing recursive routines to generate sequences, most of which have a constant difference between consecutive terms.

ASSIGNING HOMEWORK

Essential	1–5, 7
Performance assessment	6, 8, 9
Portfolio	7
Journal	11, 12
Group	7, 10, 12, 13
Review	14–16

▶ Helping with the Exercises

3b. 5 ENTER Ans + 3 ENTER , ENTER , . . .

Exercise 5d For the first time, students see a recursive sequence defined by division.

5a. Start with 3, then apply the rule Ans + 6; 10th term = 57.

5b. Start with 1.7, then apply the rule Ans − 0.5; 10th term = −2.8.

5c. Start with −3, then apply the rule Ans · −2; 10th term = 1536.

5d. Start with 384, then apply the rule Ans/2 or Ans · 0.5; 10th term = 0.75.

d. Adding or subtracting numbers does not generate this sequence. Notice that the numbers double each time. Also, they switch between positive and negative signs. So the rule is to multiply by −2. Multiply 32 by −2 to get the first missing value of −64. The last missing values are 512 and −1024.

starting value

$$\downarrow \overset{\cdot(-2)}{\frown} \overset{\cdot(-2)}{\frown} \overset{\cdot(-2)}{\frown} \overset{\cdot(-2)}{\frown} \overset{\cdot(-2)}{\frown} \overset{\cdot(-2)}{\frown} \overset{\cdot(-2)}{\frown} \overset{\cdot(-2)}{\frown} \overset{\cdot(-2)}{\frown}$$

2, −4, 8, −16, 32, −64, 128, −256, 512, −1024

EXERCISES

You will need your calculator for problems **1–10, 12**, and **13**.

▶ Practice Your Skills

1. Evaluate each expression without using your calculator. Then check your result with your calculator.

  a. $-2(5 - 9) + 7$ 15

  b. $\dfrac{(-4)(-8)}{-2}$ −16

  c. $\dfrac{5 + (-6)(-5)}{-7}$ −5

2. Use the distributive property to rewrite each family of expressions without using parentheses. Enter {2, 5, 6} into list L₁ and {−3.6, −0.5, 12} into list L₂ on your calculator, and use them to check your answers numerically.

  a. $3(L_1 + 2)$
    $3L_1 + 6; \{12, 21, 24\}$

  b. $-4(6 - L_2)$
    $-24 + 4L_2; \{-38.4, -26, 24\}$

  c. $-7(L_1 - 3)$
    $-7L_1 + 21; \{7, -14, -21\}$

3. Consider the sequence of figures made from a row of pentagons.

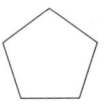 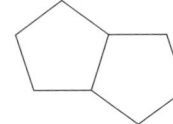 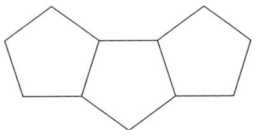

  Figure 1   Figure 2   Figure 3

  a. Copy and complete the table for five figures.

  b. Write a recursive routine to find the perimeter of each figure. Assume each side is 1 unit long.

  c. Find the perimeter of Figure 10. 32

  d. Which figure has a perimeter of 47? Figure 15

Figure #	Perimeter
1	5
2	8
3	11
4	14
5	17

4. Find the first six values generated by the recursive routine

  −14.2   ENTER   −14.2, −10.5, −6.8, −3.1, 0.6, 4.3

  Ans + 3.7   ENTER , ENTER , . . .

5. Write a recursive routine to generate each sequence. Then use your routine to find the 10th term of the sequence.

  a. 3, 9, 15, 21, . . .

  b. 1.7, 1.2, 0.7, 0.2, . . .

  c. −3, 6, −12, 24, . . .

  d. 384, 192, 96, 48, . . .

▶ Reason and Apply

6. **APPLICATION** In the Empire State Building the longest elevator shaft reaches the 86th floor, 1050 feet above ground level. Another elevator takes visitors to the observation area on the 102nd floor, 1224 feet above ground level.

   a. Write a recursive routine that gives the height above ground level for each of the first 86 floors. Tell what the starting value and the rule mean in terms of the building.

   b. Write a recursive routine that gives the heights of floors 86 through 102. Tell what the starting value and the rule mean in this routine.

   c. When you are 531 feet above ground level, what floor are you on?

   d. When you are on the 90th floor, how high up are you? When you are 1137 feet above ground level, what floor are you on? **1093.5 feet; 94th floor**

7. The diagram at right shows a sequence of gray and white squares each layered under the previous one.

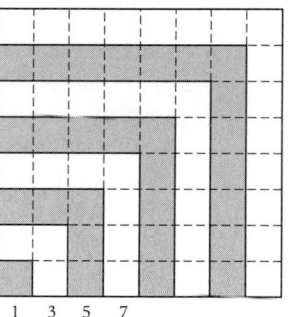
1 3 5 7

   a. Explain how the sequence 1, 3, 5, 7, . . . is related to the areas of these squares.

   b. Write a recursive routine that gives the sequence 1, 3, 5, 7, ...

   c. Use your routine to predict the number of additional unit squares you would need to enlarge this diagram by one additional row and column. Explain how you found your answers. **17, the value of the ninth term in the sequence**

   d. Use your routine to predict the 20th number in the sequence 1, 3, 5, 7, ... **39**

   e. The first term in the sequence is 1, and the second is 3. Which term is the number 95? Explain how you found your answer. **The 48th term is 95. Students might press (ENTER) 48 times or compute 2(48) − 1.**

8. Imagine a tilted L-shaped puzzle piece made from 8 toothpicks. Its area is 3 square units. Add puzzle pieces in the corner of each "L" to form successive figures of the design. In a second figure, the two pieces "share" two toothpicks so that there are 14 toothpicks instead of 16.

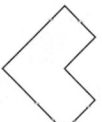   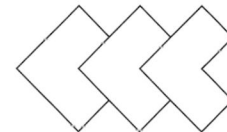

   Figure 1   Figure 2   Figure 3

   a. As you did in the investigation, make a table with enough columns and rows for the number of toothpicks, perimeter, and area of each of six figures.

   b. Write a recursive routine that will produce the number sequence in each column of the table.

   c. Find the number of toothpicks, perimeter, and area of Figure 10.

   d. Find the perimeter and area of the figure made from 152 toothpicks.

Exercise 6 Students may have difficulty calculating the heights of the floors. Encourage them to draw a picture to see that there are 16 floors, numbered 86 through 101, spanning a distance of 174 feet, and that the lower 85 floors cover a distance of 1050 feet. If students notice the height here of 1224 and the height in the introduction of 1250, say the 1224 is to the floor of the 102nd floor and 1250 includes the height of the antenna.

6a. The recursive routine is 0 (ENTER) and then Ans + 12.35 (ENTER). The starting value is 0, the height of ground level (the first floor). Add the average floor height for the next 85 floors: 12.35 feet.

6b. The recursive routine is 1050 (ENTER) and then Ans + 10.875 (ENTER). The starting value is the height of the 86th floor. Add 10.875, the average floor height of floors 86 through 101.

6c. When you are 531 feet high, you are 43 floors up and thus on the 44th floor.

7a. Possible explanation: The smallest square has an area of 1. The next larger white square has an area of 4, which is 3 more than the smallest square. The next larger gray square has an area of 9, which is 5 more than the 4-unit white square.

7b. The recursive routine is 1 (ENTER), Ans + 2 (ENTER), (ENTER), and so on.

8a. The table for six figures of the L-shaped puzzle pieces is

Figure	Toothpicks	Perimeter	Area
1	8	8	3
2	14	12	6
3	20	16	9
4	26	20	12
5	32	24	15
6	38	28	18

8b. To find the number of toothpicks, press 8 (ENTER) and then Ans + 6 (ENTER). To find the perimeter, press 8 (ENTER) and then Ans + 4 (ENTER). For the area, press 3 (ENTER) and then Ans + 3 (ENTER).

8c. Figure 10 has 62 toothpicks, a perimeter of 44, and an area of 30.

8d. Figure 25, made from 152 toothpicks, has a perimeter of 104 and an area of 75.

Exercise 9b In other words, let the
25th floor height be the first term,
the 24th floor height be the second term, and so on.

9b. Press 101 and then Ans − 4.
The 19th term represents the
height of the 7th floor. The
height is 29 meters.

10a. One routine is press −4
and then Ans + 12. Another is
press −4 and then Ans · −2.

10b. Two possible sequences are
{−16, −4, 8, 20, 32, 44, 56, ...},
and {2, −4, 8, −16, 32, −64,
128, ...}.

Exercise 11 If appropriate, ask
about multiples of 7 that aren't
positive. Mention non-negative
multiples (which include 0) as
well as negative multiples.

11c. Possible answer: There are
14 multiples between 100 and
200. There are also 14 multiples of
7 between 200 and 300. But there
are 15 between 300 and 400.

11d. Possible answer: The 4th
multiple of 7 is 4 · 7 = 28, the
5th multiple of 7 is 5 · 7 = 35,
and so on. Recursively, you start
with 7 and then continue
adding 7.

---

9. **APPLICATION** The table gives some floor heights in a building.

Floor	...	−1	0	1	2	...		...	25
Height (m)	...	−3	1	5	9	...	37	...	

a. How many meters are between the floors in this building? 4 meters

b. Write a recursive routine that will give the sequence of floor heights if you start
at the 25th floor and go to the basement (floor 0). Which term in your sequence
represents the height of the 7th floor? What is the height?

c. How many terms are in the sequence in 9b? 26 terms

d. Floor "−1" corresponds to the first level of the parking substructure under the
building. If there are five parking levels, how far underground is level 5? 19 meters

10. Consider the sequence __ , −4, 8, __ , 32, ...

a. Find two different recursive routines that could generate these numbers.

b. For each routine, what are the missing numbers? What are the next two numbers?

c. If you want to generate this number sequence with exactly one routine, what
more do you need? More numbers are needed to uniquely determine a recursive routine.

11. Positive multiples of 7 are generally listed as 7, 14, 21, 28, ...

a. If 7 is the 1st multiple of 7 and 14 is the 2nd multiple, then what is the 17th
multiple? 17 · 7 = 119

b. How many multiples of 7 are between 100 and 200? 14

c. Compare the number of multiples of 7 between 100 and 200 with the number
between 200 and 300. Does the answer make sense? Do all intervals of 100 have
this many multiples of 7? Explain.

d. Describe two different ways to generate a list containing multiples of 7.

12. Some babies gain an average of 1.5 pounds per month during the first 6 months
after birth.

a. Write a recursive routine that will generate a table of monthly weights for a baby
weighing 6.8 pounds at birth. The routine is press 6.8 (ENTER) and then Ans + 1.5 (ENTER); (ENTER) . . . .

b. Write a recursive routine that will generate a table of monthly weights for a baby
weighing 7.2 pounds at birth. The routine is press 7.2 (ENTER), and then Ans + 1.5 (ENTER); (ENTER) . . . .

c. How are the routines in 12a and 12b the same?
How are they different? The starting terms differ. The
rule itself is the same.

d. Copy and complete the table of data for this
situation.

Age (mo)	0	1	2	3	4	5	6
Weight of Baby A (lb)	6.8	8.3	9.8	11.3	12.8	14.3	15.8
Weight of Baby B (lb)	7.2	8.7	10.2	11.7	13.2	14.7	16.2

e. How are the table values for the two babies the
same? How do they differ?
Each baby always increases by 1.5 pounds. The starting
values are different. The difference between the babies'
weights is always 0.4 pound.

13. Write recursive routines to help you answer 13a–d.

   a. Find the 9th term of 1, 3, 9, 27, ... Press 1 (ENTER); Ans · 3 (ENTER); (ENTER) . . . . The 9th term is 6561.

   b. Find the 123rd term of 5, −5, 5, −5, ... Press 5 (ENTER); Ans · (−1) (ENTER); (ENTER) . . . . The 123rd term is 5.

   c. Find the term number of the first positive term of the sequence −16.2, −14.8, −13.4, −12, ... −16.2 (ENTER); Ans + 1.4 (ENTER); (ENTER) . . . . The 13th term is the first positive term.

   d. Which term is the first to be either greater than 100 or less than −100 in the sequence −1, 2, −4, 8, −16, . . . ? Press −1 (ENTER); Ans · (−2) (ENTER); (ENTER) . . . . The 8th term is the first to be greater than 100.

▶ Review

14. The table gives the normal monthly precipitation for three of the soggiest cities in the United States.

   a. Display the data in three box plots, one for each city, and use them to compare the precipitation for the three cities.

   b. What information do you lose by displaying the data in a box plot? What type of graph might be more helpful for displaying the data?

It's a rainy day in Portland, Oregon.

Precipitation for Three Cities

Month	Precipitation (in)		
	Portland	San Francisco	Seattle
January	5.4	4.1	5.4
February	3.9	3.0	4.0
March	3.6	3.1	3.8
April	2.4	1.3	2.5
May	2.1	0.3	1.8
June	1.5	0.2	1.6
July	0.7	0.0	0.9
August	1.1	0.1	1.2
September	1.8	0.3	1.9
October	2.7	1.3	3.3
November	5.3	3.2	5.7
December	6.1	3.1	6.0

(*2000 New York Times Almanac*, pp. 480–481)

15. **APPLICATION** Andy has a part-time job at the hardware store. He earns $7.25 an hour, but on Sundays and holidays he earns "time and a half." This means that for every hour he works on Sundays and holidays, he gets paid for 1.5 hours.

   a. Last week, Andy worked 8 hours a day Monday through Thursday, he had Friday off, and then he worked 6 hours on Sunday. What are his earnings for the week? $297.25

   b. Explain two ways that you can answer the question in 15a. Use the distributive property in one of your explanations. $4 \cdot 8(7.25) + 6 \cdot 1.5(7.25) = 232 + 65.25 = 297.25$
          or $7.25 (4 \cdot 8 + 6 \cdot 1.5) = 7.25(41) = 297.25$

16. Central High School is selling lottery tickets to raise money for a new sound system. There are 18 prizes for the lottery, and the students and teachers sold 2400 tickets.

   a. What fraction of the people who bought tickets will win a prize? $\frac{3}{400}$

   b. What fraction of the people who bought tickets will not win a prize? $\frac{397}{400}$

   c. What is the probability that someone who buys one ticket will win a prize? $\frac{3}{400}$

Exercise 14 [ESL] *Precipitation* is rain or snow. *Soggiest* means wettest. This exercise reviews Lessons 1.3 and 1.4.

Exercise 15 This exercise reviews Lesson 4.1.

Exercise 16 This exercise reviews Lesson 2.6.

14a. The top box plot is Portland, the middle is San Francisco, and the bottom is Seattle.

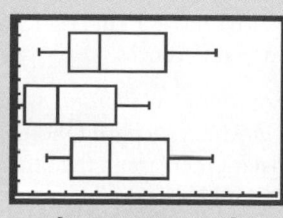

[0, 7, 0.5, 0, 12, 1]

San Francisco has the least precipitation and is the only city in which there is a month with no precipitation. One indicator that the weather is much better in San Francisco is that the month with no precipitation is not an outlier.

14b. You lose information about what time of year is soggiest. A bar graph or scatter plot would show trends over the months of the year more clearly.

PLANNING

LESSON OUTLINE

First day:

  5 min   Introduction and example

 45 min   Investigation

Second day:

15 min   Investigation

20 min   Sharing and closing

15 min   Exercises

MATERIALS

- Calculator Notes 0D, 4A, 4B
- Elevator Table (T), *optional*
- On the Road Again Grid (W)
- On the Road Again Graph (T)
- graph paper
- colored pencils

TEACHING

This lesson describes quantities generated by a recursive additive sequence (that is, with a constant rate of change) as having a *linear relationship* because their graph is a set of points that lie on a straight line. If the amount being added (the rate of change per step) is positive, the line is rising (increasing) from left to right, with larger rates of change giving steeper lines. If the rate of change is negative, the line is falling (decreasing) from left to right.

EXAMPLE

This example shows how you can enter a list of two starting values so that you can keep track of the term number in a sequence. The output is a list of paired values.

LESSON

4.4

Linear Plots

In this lesson you will learn that the starting value and the rule of a recursive sequence take on special meaning in certain real-world situations. When you add or subtract the same number each time in a recursive routine, consecutive terms in the sequence change by a constant amount. Using your calculator, you will see how these two important features of a sequence let you generate data for tables quickly. You will also plot points from these data sets and learn that the starting value and rule relate to specific characteristics of the graph.

In most sciences, one generation tears down what another has built, and what one has established, the next undoes. In mathematics alone, each generation builds a new story to the old structure.

HERMANN HANKEL

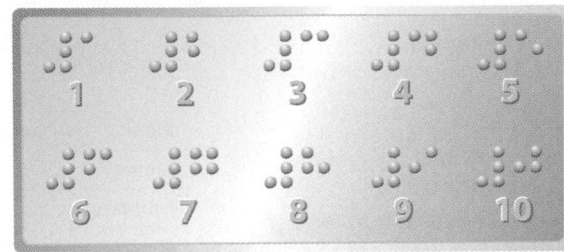

Many elevators use Braille symbols. This alphabet for the blind was developed by Louis Braille (1809–1852).

EXAMPLE

You walk into an elevator in the basement of a building. Its control panel displays "0" for the floor number. As you go up, the numbers increase one by one on the display, and the elevator rises 13 feet for each floor. The table shows the floor numbers and their heights above ground level.

Floor number	Height (ft)
0 (basement)	−4
1	9
2	22
3	35
4	48
. . .	. . .

a. Write recursive routines for the two number sequences in the table. Enter both routines into calculator lists.

b. Define variables and plot the data in the table for the first few floors of the building. Does it make sense to connect the points on the graph?

c. What is the highest floor with a height less than 200 feet? Is there a floor that is exactly 200 feet high?

▶ **Solution**

a. The starting value for the floor numbers is 0, and the rule is to add 1. The starting value for the height is −4, and the rule is to add 13. You can generate both number sequences on the calculator using lists.

Press {0, −4} and press **ENTER** to input both starting values at the same time. To use the rules to get the next term in the sequence, press {Ans(1) + 1, Ans(2) + 13} **ENTER**.
[▶ 🖥 See **Calculator Note 4B**. ◀]

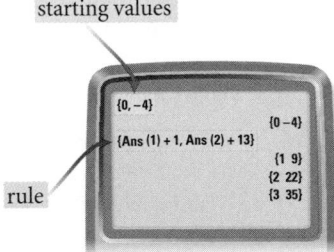

starting values

rule

```
{0, −4}
                    {0 −4}
{Ans (1) + 1, Ans (2) + 13}
                        {1  9}
                        {2  22}
                        {3  35}
```

In this routine, the calculator displays a new list of numbers horizontally every time you press **ENTER**. But the terms for the sequences of floor numbers and heights appear vertically aligned on the screen.

Be sure students understand the important difference between the use of braces and parentheses.

Emphasize the use of dimensions: $\frac{height}{floor}$. On the Elevator Table transparency, you might make marks between consecutive terms in the right column to show the common differences.

Ask why the term *linear relationship* was chosen for two quantities in which each unit increase of one results in a constant increase in the other. An answer to this question could wait until Sharing.

These commands tell the calculator to add 1 to the first term in the list and to add 13 to the second number. Press (ENTER) again to compute the next floor number and its corresponding height as the elevator rises.

b. Let x represent the floor number and y represent the floor's height in feet. Mark a scale from 0 to 5 on the x-axis and -10 to 80 on the y-axis. Plot the data from the table. The graph starts at $(0, \ 1)$ on the y-axis. The points appear to be in a line. It does not make sense to connect the points because it is not possible to have a decimal or fractional floor number.

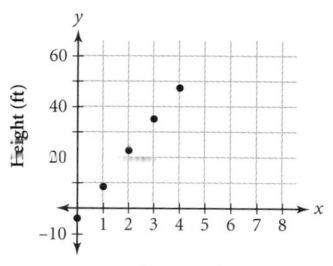

Floor number

c. The recursive routine generates the points $(0, -4)$, $(1, 9)$, $(2, 22)$, ..., $(15, 191)$, $(16, 204)$, ... The height of the 15th floor is 191 feet. The height of the 16th floor is 204 feet. So, the 15th floor is the highest floor with a height less than 200 feet. No floor is exactly 200 feet high.

Notice that to get to the next point on the graph from any given point, move right 1 unit on the x-axis and up 13 units on the y-axis. The points you plotted in the example showed a **linear relationship** between floor numbers and their heights. In what other graphs have you seen linear relationships?

Investigation
On the Road Again

You will need

- the worksheet On the Road Again Grid

A family wants to meet for a camping trip. Mom and Dad are at the campsite when they realize they forgot to pack the tent. Their son and daughter have just left their apartments and cannot be reached by phone. Will Mom and Dad get home before their son and daughter can call from the campsite? When and where will they pass each other on the highway? In this investigation you will learn how to use recursive sequences to answer questions like these.

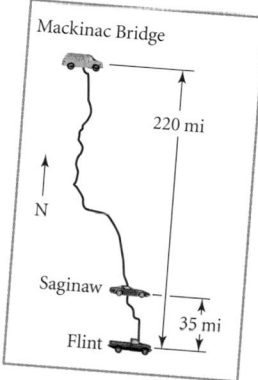

A green minivan starts at the Mackinac Bridge and heads south for Flint on Highway 75. At the same time, a red sports car leaves Saginaw and a blue pickup truck leaves Flint. The car and the pickup are heading for the bridge. The minivan travels 72 miles per hour. The pickup travels 66 mph. The sports car travels 48 mph.

LESSON OBJECTIVES

- Graph scatter plots of recursive sequences
- Continue to explore the connection between graphs and tables and how they can be used to solve problems
- Build toward an introduction of the intercept form of a line

NCTM STANDARDS

CONTENT	PROCESS
• Number	• Problem Solving
• Algebra	Reasoning
• Geometry	• Communication
• Measurement	• Connections
Data/Probability	• Representation

Guiding the Investigation

You can modify the context of the investigation to suit your geographical area by choosing cities and highways with approximately these distances. (If you're using a different geographical context, from that in the student text, make a transparency and worksheet like those provided.)

[ESL] *Heading for Flint* means traveling in the direction of Flint. *Heads south* means travels south.

One step Pose the investigation problem. Ask students to answer the questions in as many ways as possible. Encourage use of recursive calculator routines and graphs. Some students may say that "obviously" the minivan arrives before the pickup because it's going faster over the same distance. It may not be as obvious that the minivan will arrive before the sports car since the sports car has less distance to travel, but it's moving more slowly. Others may point out that the pickup will pass the sports car in less than 2 hours because the situation is equivalent to the sports car sitting still 35 miles up the road and the pickup traveling at 18 mph $(66 - 48)$. Encourage this kind of variety in approach.

Sometimes it helps to get six students to the front of the classroom with signs that say *green minivan, Mackinac Bridge,* etc., and have them "walk" through the problem before they begin working on it.

A dynamic algebra model at www.keymath.com/DA can help students see the graph change as the cars travel.

Step 1 Throughout the investigation, encourage the use of units and dimensional analysis.

If any students are concerned that the minivan is exceeding the legal speed limit by 2 mph, you might simply say that its speedometer is slightly inaccurate.

Step 2 As needed, discuss how the different directions the vehicles are traveling affect whether you add or subtract in the recursive routine. Also discuss the difference between miles apart on the highway and miles apart in a straight line. The distances are all miles apart on the highway. Be ready to remind students how to enter recursive data into calculator lists. (See Calculator Note 0D.)

Step 2 To input the starting values in a calculator list for the times of the minivan, pickup, and sports car, respectively, press {0, 220, 0, 35}. To apply the rule, press {Ans (1) + 1, Ans(2) − 1.2, Ans(3) + 1.1, Ans(4) + 0.8}. The starting values represent each vehicle's distance from Flint. The rule is to add or subtract the speed in miles per minute, depending on the vehicle's direction.

Step 3 Each student can choose a different car and generate the related sequence recursively. Remind students of the instant replay function [Calculator Note 4A]. To modify the recursive routines to use 10-minute intervals, you can recall the last entry and change it to read {Ans (1) + 10, Ans(2) − 12, Ans(3) + 11, Ans(4) + 8}.

Step 4 This is the most important phase of the investigation. Using different colors is important for differentiating the vehicles.

Step 5 The data will provide three linear patterns. Students should draw these three intersecting lines and justify connecting the points. This may be the first continuous graph students have constructed since Chapter 3. **[Alert]** Some students may interpret their coordinate graphs as the *paths* of the cars rather than indicating the distance from a fixed point over time.

See page 716 for answers to Step 3.

Step 1 minivan 1.2 mi/min; pickup 1.1 mi/min; sports car 0.8 mi/min

Step 1   Find each vehicle's average speed in miles per minute (mi/min).

Step 2   Write recursive routines to find each vehicle's distance from Flint at each minute. What are the real-world meanings of the starting value and the rule in each routine? Use calculator lists.

Step 3   Make a table to record the highway distance from Flint for each vehicle. After you complete the first few rows of data, change your recursive routines to use 10-minute intervals for up to 4 hours.

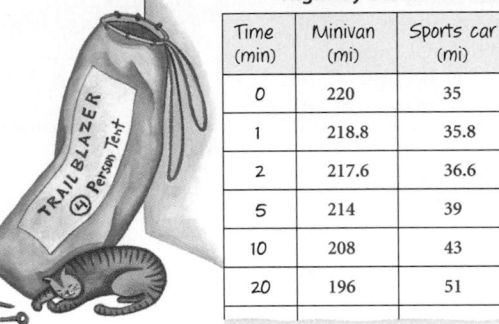

Highway Distance from Flint

Time (min)	Minivan (mi)	Sports car (mi)	Pickup (mi)
0	220	35	0
1	218.8	35.8	1.1
2	217.6	36.6	2.2
5	214	39	5.5
10	208	43	11
20	196	51	22

Procedure Note

After you enter the recursive routine in the calculator, press **ENTER** five or six times. Copy the data displayed on your calculator screen onto your table. Repeat this process.

Step 4 Let *x* represent the time in minutes since the vehicles started their trips. Let *y* represent the distance from Flint in highway miles.

Step 4   Define variables and plot the information from the table onto a graph. Mark and label each axis in 10-minute intervals, with time on the horizontal axis. Using a different color for each vehicle, plot its (*time, distance*) coordinates.

Step 5   On the graph, do the points for each vehicle seem to fall on a line? Does it make sense to connect each vehicle's points in a line? If so, draw the line. If not, explain why not. Yes. A line through the points for each vehicle represents every possible instant of time.

Step 6 On the *y*-axis representing each vehicle's initial distance from Flint. The rules affect the steepness and direction of each line.

Step 8 The minivan meets the sports car about 110 miles from Flint just after 90 minutes. The pickup is about 100 miles from Flint.

Use your graph and table to find the answers for Steps 6–10.

Step 6   Where does the starting value for each routine appear on the graph? How does the recursive rule for each routine affect the points plotted?

Step 7   Which line represents the minivan? How can you tell?

Step 8   Where are the vehicles when the minivan meets the first one headed north?

Step 9   How can you tell by looking at the graph whether the pickup or the sports car is traveling faster? When and where does the pickup pass the sports car?

Step 7 Be sure students go beyond an answer like "It's colored green." **[Ask]** "How did you know which line to color green?"

Step 7 The line going down from left to right. It starts 220 units above the origin on the *y*-axis and gets closer to the *x*-axis as time passes and the minivan gets closer to Flint.

Steps 8 and 9 [Alert] Students may have difficulty keeping in mind both variables being graphed. Some

students may say, or at least believe, something like "They're at the same place but not at the same time."

You can find very good approximations for the answers to the questions using either the table or the graph. Ask students to save their results if you are going to assign Exercise 11 in Lesson 6.2.

Step 9 The pickup is traveling faster because its line is steeper. It passes the sports car when their lines cross on the graph after roughly 115 minutes.

Step 10 The minivan arrives after approximately 185 minutes, the pickup 15 minutes later, and the sports car arrives 45 minutes later.

Step 11 The vehicles travel a constant speed, never speeding up, slowing down, or stopping.

Step 10 Which vehicle arrives at its destination first? How many minutes pass before the second and third vehicles arrive at their destinations? How can you tell by looking at the graph?

Step 11 What assumptions about the vehicles are you making when you answer the questions in the previous steps?

Step 12 Consider how to model this situation more realistically. What if the vehicles are traveling at different speeds? What if one vehicle runs out of gas or breaks down? What if one driver stops to get a bite to eat? What if the vehicles' speeds are not constant? Discuss how these questions affect the recursive routines, tables of data, and their graphs. **Step 12** If speeds are not constant, the points will not lie in a line and you would not be adding or subtracting the same number in the recursive routine. The lines would have horizontal pieces if drivers stop.

Assessing Progress
Your observations should help you assess students' skills at finding averages, writing **recursive routines,** plotting points, calculating **rates of change,** and interpreting **graphs.**

Closing the Lesson

Quantities generated by a **recursive additive sequence** (with a constant rate of change) have a **linear relationship** because their graph lies on a straight line. If the amount being added (the rate of change per step) is positive, the line is rising (increasing) from left to right, with larger rates of change giving steeper lines. If the rate of change is negative, the line is falling (decreasing) from left to right.

EXERCISES

You will need your calculator for problems **1, 4, 6, 7, 8,** and **10.**

▶ **Practice Your Skills**

1. Find out if each expression is positive or negative without using your calculator. Then check your answer with your calculator.

   a. $-35(44) + 23$ negative; -1517 **b.** $(-14)(-36) - 32$ positive; 472 **c.** $25 - \dfrac{152}{12}$ positive; $12.\overline{3}$

   d. $50 - 23(-12)$ positive; 326 **e.** $\dfrac{-12 - 38}{15}$ negative; $-3.\overline{3}$ **f.** $24(15 - 76)$ negative; -1464

2. List the terms of each number sequence of y-coordinates for the points shown on each graph. Then write a recursive routine to generate each sequence.

   a. 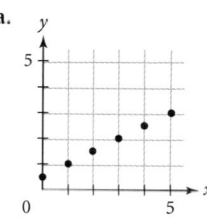 **b.** 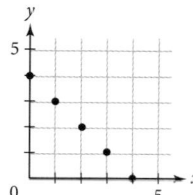 **c.** 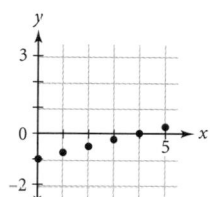 **d.**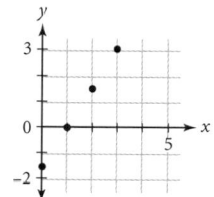

3. Make a table listing the coordinates of the points plotted in 2b and d.

4. Plot the first five points represented by each recursive routine in 4a and b on separate graphs.

   a. {0, 5} `ENTER`
      {Ans(1) + 1, Ans(2) + 7} `ENTER` ; `ENTER` , . . .
   b. {0, -3} `ENTER`
      {Ans(1) + 1, Ans(2) - 6} `ENTER` ; `ENTER` , . . .
   c. On which axis does each starting point lie? What is the x-coordinate of each starting point? the y-axis; 0
   d. As the x-value increases by 1, what happens to the y-coordinates of the points in each sequence in 4a and b? In 4a, the y-coordinates increase by 7. In 4b, the y-coordinates decrease by 6.

x	y
0	4
1	3
2	2
3	1
4	0

x	y
0	-1.5
1	0
2	1.5
3	3

BUILDING UNDERSTANDING

Students move among recursive sequences, linear relationships, and their graphs.

ASSIGNING HOMEWORK

Essential	2–4 , 7, 8
Performance assessment	9, 11
Portfolio	7
Journal	8, 9
Group	7, 10
Review	1, 5, 6, 12–15

▶ **Helping with the Exercises**

2a. {0.5, 1, 1.5, 2, 2.5, 3}; 0.5, Ans + 0.5

2b. {4, 3, 2, 1, 0}; 4, Ans − 1

2c. {−1, −0.75, −0.5, −0.25, 0, 0.25}; −1, Ans + 0.25

2d. {−1.5, 0, 1.5, 3}; −1.5, Ans + 1.5

Exercise 4 Be sure students note the instructions for 4a and 4b at the beginning of the exercise.

See page 716 for answers to Exercises 4a and 4b.

SHARING IDEAS

If some students had the idea that the graphs showed the paths of the vehicles, ask the class to critique that notion. This is a good place to ask questions beginning with "Are you saying . . . ?" in order to clarify ideas without having to tell much.

Have students present Steps 9 to 12. **[Ask]** "Are you saying that the relationship is linear because the rate of change is constant?" Elicit the idea that the amount being added at each step of the recursive routine is the rate of change.

If you have time, ask how linear relationships relate to directly proportional quantities and direct variation. [Directly proportional quantities have a linear relationship, but not necessarily vice versa. The graph of a direct variation is a straight line through the origin. The graph of a linear relationship between variables is also a straight line, but not necessarily through the origin.]

As a synonym for *steepness*, you might use the word *slope*, though its formal definition will not come until Chapter 5.

7a. $\{0, 272\}$ (ENTER);
$\{\text{Ans}(1) + 1, \text{Ans}(2) - 68\}$
(ENTER); (ENTER); (ENTER); (ENTER);
(ENTER) .

7b.

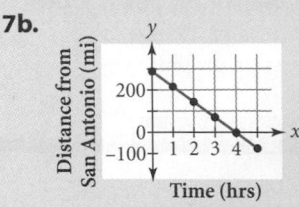

7c. The starting value is the point $(0, 272)$ on the graph.

7d. On the graph, you move right 1 unit and down 68 units to get from one point to the next. In the recursive routine, you add 1 to the first number and subtract 68 from the second number.

7e. This is a linear graph relating a distance to any time between 0 and 5 hours. The line represents the distances at all possible times; points only represent distances at certain times.

7f. The car is within 100 miles of San Antonio after 2.53 hours have elapsed. Explanations will vary. Graphically, it is the time after which the line crosses the horizontal line $y = 100$.

7g. The car takes 4 hours to reach San Antonio. Answers will vary. The answer is the fourth entry in the table. Graphically, it is where the line crosses the x-axis.

---

5. Consider the expression
$$\frac{4 - 5(x + 3)}{6}$$

  a. Use the order of operations to find the value of the expression if $x = 9$. $-9.\overline{3}$

  b. Set the expression equal to 12. Solve for x by undoing the sequence of operations.

6. The direct variation $y = 2.54x$ describes the relationship between two standard units of measurement where y represents centimeters and x represents inches.

  a. Write a recursive routine that would produce a table of values for any whole number of inches. Use a calculator list. $\{0, 0\}$ (ENTER) and $\{\text{Ans}(1) + 1, \text{Ans}(2) + 2.54\}$

  b. Use your routine to complete the missing values in this table.

Inches	Centimeters
0	0
1	2.54
2	5.08
14	35.56
17	43.18

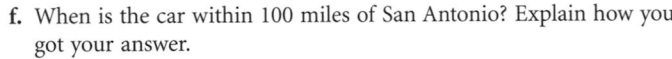

▶ Reason and Apply

7. APPLICATION A car is moving at a speed of 68 miles per hour from Dallas toward San Antonio. Dallas is 272 miles from San Antonio.

  a. Write a recursive routine to create a table of values relating time to distance from San Antonio for 0 to 5 hours in 1-hour intervals.

  b. Graph the information in your table.

  c. What is the connection between your plot and the starting value in your recursive routine?

  d. What is the connection between the coordinates of any two consecutive points in your plot and the rule of your recursive routine?

  e. Draw a line through the points of your plot. What is the real-world meaning of this line? What does the line represent that the points alone do not?

  f. When is the car within 100 miles of San Antonio? Explain how you got your answer.

  g. How long does it take the car to reach San Antonio? Explain how you got your answer.

8. APPLICATION A long-distance telephone carrier charges $1.38 for international calls of 1 minute or less and $0.36 for each additional minute.

  a. Write a recursive routine using calculator lists to find the cost of a 7-minute phone call.

  b. Without graphing the sequence, give a verbal description of the graph showing the costs for calls that last whole numbers of minutes. Include in your description all the important values you need in order to draw the graph.

5b. Equation:
$$\frac{4 - 5(x + 3)}{6} = 12$$

Operation on x	Undo operation	Work backward $x = -16.6$
$+ (3)$	$- (3)$	-13.6
$\cdot (-5)$	$\div (-5)$	68
$+ (4)$	$- (4)$	72
$\div (6)$	$\cdot (6)$	12

8a. Possible answer: $\{1, 1.38\}$ (ENTER); $\{\text{Ans}(1) + 1, \text{Ans}(2) + 0.36\}$ (ENTER); (ENTER); .... The recursive routine keeps track of time and cost for each minute. Apply the routine until you get $\{7, 3.54\}$. A 7-minute call costs $3.54.

8b. Possible answer: The graph should consist of points that lie on a line. It should include the point $(1, 1.38)$. Each subsequent point should be one unit to the right and $0.36 higher than the point before it.

9. These tables show the changing depths of two submarines as they come to the surface.

USS Alabama

Time (sec)	0	5	10	15	20	25	30
Depth (ft)	−38	−31	−24	−17	−10	−3	4

USS Dallas

Time (sec)	0	5	10	15	20	25	30
Depth (ft)	−48	−40	−32	−24	−16	−8	0

a. Graph the data from both tables on the same set of coordinate axes.

b. Describe what you found by graphing the data. How are the graphs the same? How are they different?

c. Does it make sense to draw a line through each set of points? Explain what these lines mean. Yes, each line means that any time in this range corresponds to depth below the surface.

d. What is the real-world meaning of the point (30, 4) for the USS *Alabama*?
The submarine's nose rises slightly above the water when surfacing.

10. Each geometric design is made from tiles arranged in a row.

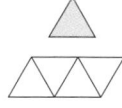 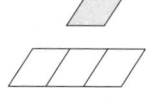 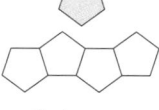 

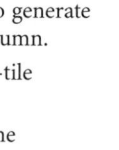

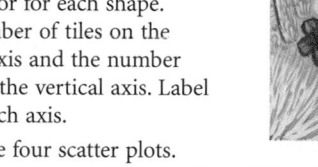

Triangle   Rhombus   Pentagon   Hexagon

a. Make a table like the one shown. Find the number of tile edges on the perimeter of each design, and fill in ten rows of the table. Look for patterns as you add more tiles.

Tile Edges on the Perimeter

Number of tiles	Triangle	Rhombus	Pentagon	Hexagon
1	3	4	5	6
2	4	6	8	10
3	5	8	11	14
4	6	10	14	18
...				
10	12	22	32	42

b. Write a recursive routine to generate the values in each table column.

c. Find the perimeter of a 50-tile design for each shape.

d. Draw four plots on the same coordinate axes using the information for designs of one to ten tiles of each shape. Use a different color for each shape. Put the number of tiles on the horizontal axis and the number of edges on the vertical axis. Label and scale each axis.

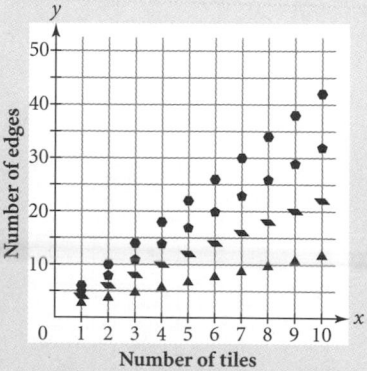

e. Compare the four scatter plots. How are they alike, and how are they different?

f. Would it make sense to draw a line through each set of points? Explain why or why not. No, because there must be a whole number of tiles and a whole number of edges.

10e. The points of each graph appear to lie on a line and each graph starts at 1. The graphs increase in steepness from the triangle tile to the hexagon tile.

Exercise 9 Discuss how measuring depth on the bow, bridge, or stern of a submarine as it is surfacing affects data.

9a.

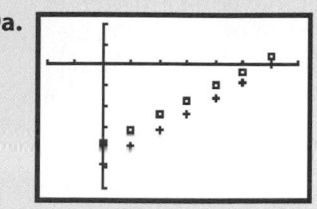

$[−10, 35, 5, −60, 20, 10]$

9b. The points for each submarine appear to lie on a line. The USS *Dallas* surfaces at a faster rate.

Exercise 10 Suggest that students calculate the perimeter by treating each edge as one unit long.

10b. Number of tiles: The starting value is 1; the rule is add 1.
Triangle: The starting value is 3; the rule is add 1.
Rhombus: The starting value is 4; the rule is add 2.
Pentagon: The starting value is 5; the rule is add 3.
Hexagon: The starting value is 6; the rule is add 4.
To generate the sequences for all tiles simultaneously, enter $\{1, 3, 4, 5, 6\}$ and $\{\text{Ans}(1) + 1, \text{Ans}(2) + 1, \text{Ans}(3) + 2, \text{Ans}(4) + 3, \text{Ans}(5) + 4\}$.

10c. triangle 52; rhombus 102; pentagon 152; hexagon 202

10d.

11a. Answers will vary. The graph starts at $(0, 5280)$. The points $(0, 5280)$, $(1, 4680)$, $(2, 4080)$, and $(3, 3480)$ will appear to lie on a line. From $(3, 3480)$ to $(8, -1520)$, the points will appear to lie on a steeper line. The bicyclist ends up 1520 feet past you.

11b.

Bicyclist

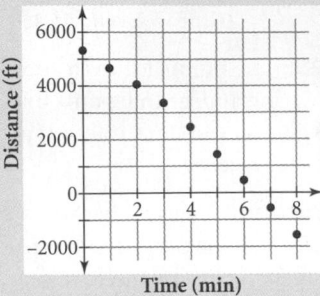

11c. Sample answer: Where on the graph does the bicyclist pass you? The answer is on the x-axis between 6 and 7 minutes.

13a. Subtract 32 from the temperature in Fahrenheit, multiply the difference by 5, and then divide by 9.

13b. $F = \dfrac{9C}{5} + 32$

$C = \dfrac{5(F - 32)}{9}$

Exercise 14 If students have difficulty converting from °F to °C, refer them back to Exercise 13.

14a. 0.118, about $\frac{1}{8}$ of a liter of water; 170.25, about 170 grams of flour

15a.

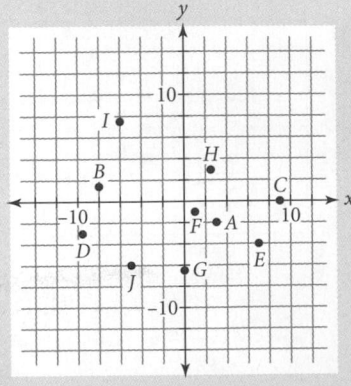

---

11. A bicyclist, 1 mile (5280 feet) away, pedals toward you at a rate of 600 feet per minute for 3 minutes. The bicyclist then pedals at the rate of 1000 feet per minute for the next 5 minutes.

  a. Describe what you think the plot of time against distance from you will look like.

  b. Graph the data using 1-minute intervals for your plot.

  c. Invent a question about the situation, and use your graph to answer the question.

▶ **Review**

12. Consider the expression

$$\frac{5.4 + 3.2(x - 2.8)}{1.2} - 2.3$$

12b. Order	Undo	Work backward $x = 3.4$
$- (2.8)$	$+ (2.8)$	0.6
$\cdot (3.2)$	$\div (3.2)$	1.92
$+ (5.4)$	$- (5.4)$	7.32
$\div (1.2)$	$\cdot (1.2)$	6.1
$(- 2.3)$	$+ (2.3)$	3.8

  a. Use the order of operations to find the value of the expression if $x = 7.2$. $13.9\overline{3}$

  b. Set the expression equal to 3.8. Solve for x by undoing the sequence of operations you listed in 12a. $x = 3.4$

13. Isaac has an easy time remembering how to convert from Celsius to Fahrenheit. He multiplies the temperature in Celsius by 9, divides the result by 5, and then adds 32.

  a. Write clear directions for converting Fahrenheit to Celsius.

  b. Write an expression for each conversion.

14. **APPLICATION** Karen is a U.S. exchange student to Austria. She wants to make her favorite pizza recipe for her host family, but she needs to convert the quantities to the metric system. Instead of using cups for flour and sugar, her host family measures dry ingredients in grams and liquid ingredients in liters. Karen has read that 4 cups of flour weigh 1 pound.

In her dictionary, Karen looks up conversion factors and finds that 1 ounce $\approx$ 28.4 grams, 1 pound $\approx$ 454 grams, and 1 cup $\approx$ 0.236 liter.

  a. Karen's recipe calls for $\frac{1}{2}$ cup water and $1\frac{1}{2}$ cups flour. Convert these quantities to metric units.

  b. Karen's recipe says to bake the pizza at 425°. Convert this temperature to degrees Celsius. 218.3, about 220° Celsius.

15. Draw and label a coordinate plane with each axis scaled from -10 to 10.

  a. Represent each point named with a dot, and label it using its letter name.

    $A(3, -2)$ $B(-8, 1.5)$ $C(9, 0)$ $D(-9.5, -3)$ $E(7, -4)$

    $F(1, -1)$ $G(0, -6.5)$ $H(2.5, 3)$ $I(-6, 7.5)$ $J(-5, -6)$

  b. List the points in Quadrant I, Quadrant II, Quadrant III, and Quadrant IV. Which points are on the x-axis? Which points are on the y-axis?

  c. Explain how to tell which quadrant a point will be in by looking at the coordinates. Explain how to tell if a point lies on one of the axes.

15b. Quad I H; Quad II B, I; Quad III D, J; Quad IV A, E, F; x-axis C; y-axis G.

15c. Sample answer: If the coordinates are both zero, then the point is on the origin. If the x-coordinate is zero, then the point is on the y-axis. If the y-coordinate is zero, then the point is on the x-axis.

If the first coordinate is positive, then the point will be in Quadrant I or IV. To tell which quadrant, look at the y-coordinate. If the y-coordinate is positive, the point is in Quadrant I. If the y-coordinate is negative, the point is in Quadrant IV.

If the first coordinate is negative, then the point will be in Quadrant II or III. To tell which quadrant, look at the y-coordinate again. If the y-coordinate is positive, the point is in Quadrant II. If the y-coordinate is negative, the point is in Quadrant III.

Time-Distance Relationships

Time-distance relationships are some of the most useful applications of algebra. You worked with this topic in the investigation and exercises of Lesson 4.4. Now you will explore this type of relationship in more depth by considering many walking scenarios. You'll learn the mathematics behind the starting position, speed, direction, and final position of a walker.

Activity
Walk the Line

You will need

- a 4-meter measuring tape or four metersticks per group
- a motion sensor
- a stopwatch or a watch that shows seconds

A **graph sketch** tells what quantities the axes represent but does not have specific number scales. Both of these graph sketches show that distance increases as time passes. Which graph sketch better represents a walk? Why?

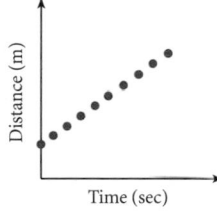

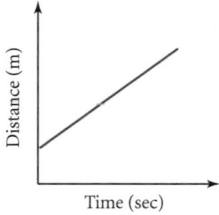

Imagine that you have a 4-meter measuring tape (or four metersticks) stretched out on the floor.

NCTM STANDARDS

CONTENT		PROCESS	
•	Number	•	Problem Solving
•	Algebra	•	Reasoning
	Geometry	•	Communication
•	Measurement	•	Connections
	Data/Probability	•	Representation

LESSON OBJECTIVES

- Explore time-distance relationships generally
- Write walking instructions or act out walks for a given graph
- Sketch graphs based on given walking instructions or table data
- Use an electronic data collection device, motion sensor, and graphing calculator to collect and graph data

PLANNING

LESSON OUTLINE

One day:
 5 min   Introduction
30 min   Activity
15 min   Sharing

MATERIALS

- Calculator Note 4C
- 4-meter measuring tapes, metersticks, or ropes
- motion sensors, *optional*
- stopwatches or watches with second hands

TEACHING

Becoming more familiar with distance-time graphs helps deepen students' understanding of graphs and rates of change.

INTRODUCTION

Post a plot with a large circle in the first quadrant. **[Ask]** "How would a walker walk to produce this graph?" After some students respond that the walker should simply walk in a circle, label the axes with time and distance. Help students see that a single walker can't produce such a graph without being at two places at one time. Ask for other examples of plots that are impossible to make with a walker and motion sensor.

The second graph sketch represents a walk better because walking is a continuous motion: At any given time there is a distance, not just at those times marked by dots in the first sketch. Thinking the *discrete* points represent footsteps, some students might choose the first sketch. Discuss the difference between discrete and continuous graphs.

(Elevator floors and polygonal designs were discrete, but highway distances were continuous.) **[Language]** The homonym for *discrete*, which means separate or individual, is *discreet*, which means delicately prudent or diplomatic.

Guiding the Activity

You might modify this activity to use finger walking and rulers. Use the same numbers but substitute inches for meters as the unit of measure.

Step 1 If some students need a hint, suggest that since the vertical axis measures distance, they can answer the question "How far does the walker start from the zero mark?" by considering the vertical axis. If students have trouble answering "How fast?" you can give a hint by pointing to graph a. **[Ask]** "How long did it take the walker to go 1 m from his or her starting point?" Some students may have trouble seeing that the walker ever walks toward the zero mark because the line proceeds away from the start. Ask them to look at graph b and read off "meters" at the start and finish. Graphs c and d can't be defined with just one rule. Graph e is an impossible walk because the walker can't be both 1 foot and 4 feet away at the same time.

Steps 7 and 8 Students will gain more if they are physically involved, even if minimally. If you have limited classroom space or equipment, you might have different groups demonstrate walks one at a time. You might select students to demonstrate two or three "walks" before the others start. However you do it, try to have each student take on each role—walker, director, recorder, and coach—for at least one walk. A group need not have a coach if there are too few students.

See page 716 for answers to Steps 1, 3, 4, and 5.

A motion sensor measures your distance from the tape's 0-mark as you walk, and it graphs the information. On the calculator graphs below, the horizontal axis shows time for 0–6 seconds and the vertical axis shows distance for 0–4 meters.

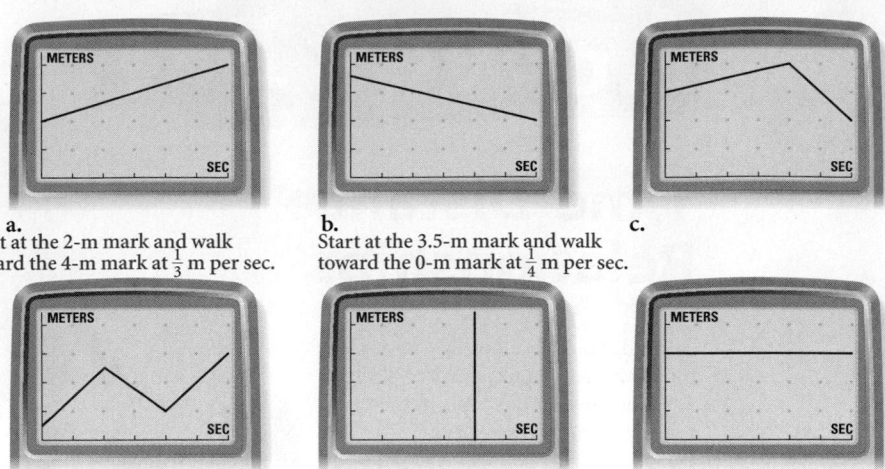

a. Start at the 2-m mark and walk toward the 4-m mark at $\frac{1}{3}$ m per sec.

b. Start at the 3.5-m mark and walk toward the 0-m mark at $\frac{1}{4}$ m per sec.

c.

d.

e. Not possible.

f. Start at the 3-m mark and stand still for 6 sec.

Step 1 Write a set of walking instructions, if possible, for each graph. Tell where the walk begins, how fast the person is walking, and whether the person is walking toward or away from the 0-mark.

Step 2 If it is not possible to write walking instructions for a graph or graphs, tell which one(s) and why it can't be done.

Step 2 Graph e because you cannot be at every possible meter mark at the same time.

Now you'll sketch graphs based on walking instructions or data.

Step 3 Graph a walk from each set of instructions.

a. Start at the 2.5-meter mark and stand still.

b. Start at the 3-meter mark and walk toward the 0-mark at a constant rate of 0.5 meters per second.

c. Start at the 0.5-meter mark and walk toward the 4-meter mark at 0.25 meters per second.

Table a

Time (sec)	Distance (m)
0	0.8
1	1.0
2	1.2
3	1.4
4	1.6
5	1.8
6	2.0

Table b

Time (sec)	Distance (m)
0	4.0
1	3.6
2	3.2
3	2.8
4	2.4
5	2.0
6	1.6

Step 4 Write a set of walking instructions based on the data in Tables a and b, and sketch graphs of the walks.

Step 5 Write recursive routines for both tables in Step 4. Explain how the starting values and rules show up in the walking instructions, graphs, and tables.

If you're not using a motion sensor, the recorder records the walker's distance from the 0-meter mark each second in a table. The timer begins timing when the director tells the walker to start and counts the seconds aloud.

If you are using a motion sensor, the recorder should hold the calculator and start collecting data on the director's command. Use the WALKER program and select Option 1: FREE FORM. (See Calculator Note 4C.) The timer should hold the motion sensor chest high, keeping it level and aimed directly at the walker. Timing begins when the director tells the walker to start. The coach makes helpful comments to guide the walker.

Now it's time for your group to act out each of the graphs and walking instructions. The object is not to match each graph exactly but to approximately match its shape. Your group will need a space about 4 meters long and 1.5 meters wide (13 feet by 5 feet). In this space, tape to the floor a 4-meter measuring tape or four meter sticks end-to-end.

Step 6   Now sketch a graph to represent these walking instructions. Start at the 0.5-meter mark and walk toward the 4-meter mark at 0.75 meters per second for 2 seconds, then walk toward the 0-meter mark at 0.5 meters per second for 2 seconds. Then walk to the 0.5-meter mark in the final 2 seconds.

Step 7   In your group, choose a walker, a director, a timer, a recorder, and a coach. Make a table to record distances from the 0-m mark for 0–6 seconds. [▶ 🖥 If your group is using a motion sensor see **Calculator Note 4C.** ◀]

Step 8 Students should discuss the difficulty of walking a constant speed, of recording the walker's exact position, and of changing speeds or direction at a specific instant in time. The director needs to give good directions.

Step 8   Choose three walks to act out: one from a graph, one from a table, and one from a set of instructions. After acting out each walk, discuss what you could have done to better model each situation. Then repeat the walk, rotating jobs.

Procedure Note

The walker follows the instructions the director gives. The timer begins timing when the director says to do so and counts the seconds out loud for the walker. The recorder records the walker's position for each second in the table. The coach guides the walker.

Step 6   The graph is

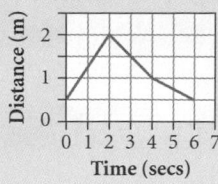

Walking Instructions

Step 8   The walker moves either backward or forward along the line, following the director's instructions. The director refers to the graph, table, or walking instructions from the group's chosen walks and gives directions to the walker based on this information.

Groups should discuss what they could have done to match more accurately the given graph, table, or instructions.

SHARING IDEAS

If groups have not been watching each other throughout the investigation, have a couple of them demonstrate the most interesting walks they did. Keep asking how to tell the speed from a graph, but don't force an answer yet.

Closing the Lesson

Distance-time graphs can be confusing because it's easy to think of them as descriptions of actual paths, as though they were overhead maps. They are not paths, but the changing distance from a fixed point over time. The starting point and the direction and steepness of the lines (indicating speed, or rate of change) are important in determining how the line models a walk.

project

PASCAL'S TRIANGLE

The first five rows of Pascal's triangle are shown.

```
              1
           1     1
        1     2     1
     1     3     3     1
  1     4     6     4     1
1     5    10    10     5     1
```

The triangle can be generated recursively. The sides of the triangle are 1's, and each number inside the triangle is the sum of the two diagonally above it.

Complete the next five rows of Pascal's triangle. Research its history and practical application. What is the connection between Sierpiński's triangle and Pascal's triangle? Can you find the sequence of triangular numbers in Pascal's triangle? What is its connection to the Fibonacci number sequence? Present your findings in a paper or a poster.

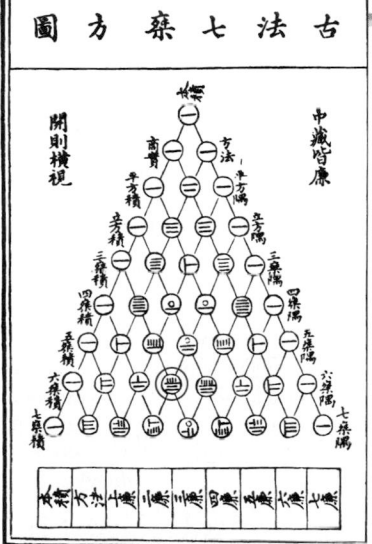

What became known as Pascal's triangle was first published in *Siyuan yujian xicao* by Zhu Shijie in 1303. This ancient version actually has one error. Can you find it?

The fourth number in row 8 should be 35.

Supporting the project

MOTIVATION

This pattern of numbers is named for a 17th-century mathematician. He used this pattern of numbers extensively in his study of probability. Was that the beginning of the history of this number pattern? (See page 716 for the next 4 rows.)

OUTCOMES

▶ The recursive rule includes a starting 1 and the rule: if a number is at the end of a row, it's 1, and if it's not, then it's the sum of the two numbers diagonally above it.

▶ Report includes history going back to the ancient Chinese civilization.

▶ The triangular numbers (1, 3, 6, 10, 15, ...) are in the third diagonal.

▶ The Fibonacci numbers (1, 1, 2, 3, 5, 8, ...) are sums of the numbers on diagonals described by "start with a 1 on the left, go over one-and-a-half numbers and up to the next row, follow that diagonal adding the numbers." (For example, $1 + 3 + 1 = 5$; $1 + 4 + 3 = 8$; $1 + 5 + 6 + 1 = 13$.)

▶ Sierpiński's triangle see page 716.

LESSON
4.6

Linear Equations and the Intercept Form

So far in this chapter you have used recursive routines, graphs, and tables to model linear relationships. In this lesson you will learn how to write linear equations from recursive routines. You'll begin to see some common characteristics of linear equations and their graphs, starting with the relationship between exercise and calorie consumption.

Different physical activities cause people to burn calories at different rates depending on many factors such as body type, height, age, and metabolism. Coaches and trainers consider these factors when suggesting workouts for their athletes.

PLANNING

LESSON OUTLINE

One day:

25 min	Investigation
5 min	Sharing
10 min	Examples
5 min	Closing
5 min	Exercises

MATERIALS

- Calculator Note 1H

TEACHING

In this lesson students make the transition from using recursive routines to writing linear equations in intercept form. The intercept form is like the well-known slope-intercept form except that the order is different: $y = a + bx$. This order emphasizes starting with the number a and adding the number b repeatedly (x times).

[Language] Define *metabolism* as the physical and chemical processes that maintain the body, for instance, turning food into energy.

 Guiding the Investigation

One step Pose the problem "Manisha starts her exercise routine by jogging to the gym. Her trainer says this activity burns 215 calories. At the gym she pedals a stationary bike, burning 3.8 calories per minute. How long will it take her to burn a total of 538 calories?" Encourage students to use recursive routines, scatter plots, linear equations, graphs, and calculator tables to solve the problem. Be sure sharing includes ideas about equations.

 ## Investigation
Working Out with Equations

Manisha starts her exercise routine by jogging to the gym. Her trainer says this activity burns 215 calories. Her workout at the gym is to pedal a stationary bike. This activity burns 3.8 calories per minute.

First you'll model this scenario with your calculator.

Step 1 {0, 215} (ENTER),
{Ans(1) + 1, Ans(2) + 3.8}
(ENTER); (ENTER)....

Step 1 Use calculator lists to write a recursive routine to find the total number of calories Manisha has burned after each minute she pedals the bike. Include the 215 calories she burned on her jog to the gym.

Step 2 Copy and complete the table using your recursive routine.

Step 3 291 calories; 60 minutes

Step 3 After 20 minutes of pedaling, how many calories has Manisha burned? How long did it take her to burn 443 total calories?

Manisha's Workout

Pedaling time (min)	Total calories burned
x	*y*
0	215
1	218.8
2	222.6
20	291
30	329
45	386
60	443

LESSON OBJECTIVES

- Write a linear equation in intercept form given a recursion routine, a graph, or data
- Learn the meaning of *y*-intercept for a linear equation in intercept form

NCTM STANDARDS

CONTENT		PROCESS	
•	Number	•	Problem Solving
•	Algebra	•	Reasoning
	Geometry	•	Communication
	Measurement	•	Connections
	Data/Probability	•	Representation

Next you'll learn to write an equation that gives the same values as the calculator routines.

Step 4 $215 + 3.8(20) = 291$

Step 4 Write an expression to find the total calories Manisha has burned after 20 minutes of pedaling. Check that your expression equals the value in the table.

Step 5 $215 + 3.8(38) = 359.4$ calories. You don't need to calculate the previous terms in the sequence or create a table to find the answer.

Step 5 Write and evaluate an expression to find the total calories Manisha has burned after pedaling 38 minutes. What are the advantages of this expression over a recursive routine?

Step 6 $y = 215 + 3.8x$

Step 6 Let x represent the pedaling time in minutes, and let y represent the total number of calories Manisha burns. Write an equation relating time to total calories burned.

Step 7 Sample checks: $215 + 3.8(1) = 218.8$; $215 + 3.8(60) = 443$

Step 7 Check that your equation produces the corresponding values in the table.

Now you'll explore the connections between the linear equation and its graph.

Step 8 Plot the points from your table on your calculator. Then enter your equation into the Y= menu. Graph your equation to check that it passes through the points. Give two reasons why drawing a line through the points realistically models this situation. [▶ 🖥 See **Calculator Note 1H** to review how to plot points and graph an equation. ◀]

Step 9 $538 = 215 + 3.8x$; $x = 85$. Check: $215 + 3.8(85) = 538$.

Step 9 Substitute 538 for y in your equation to find the elapsed time required for Manisha to burn a total of 538 calories. Explain your solution process. Check your result.

Step 10 How do the starting value and the rule of your recursive routine show up in your equation? How do the starting value and the rule of your recursive routine show up in your graph? When is the starting value of the recursive routine also the value where the graph crosses the y-axis?

The equation for Manisha's workout shows a linear relationship between the total calories burned and the number of minutes pedaling on the bike. You probably wrote this linear equation as

$$y = 215 + 3.8x \qquad \text{or} \qquad y = 3.8x + 215$$

The form $y = a + bx$ is the **intercept form.** The value of a is the **y-intercept,** which is the value of y when x is zero. The intercept gives the location where the graph crosses the y-axis. The number multiplied by x is b, which is called the **coefficient** of x.

Step 1 [**Language**] If students ask, tell them that a *calorie* is the amount of heat energy needed to warm 1 gram of water 1°C.

Step 4 As needed, suggest that students create an expression in words before they use symbols. Be sure they haven't forgotten to include the initial 215 calories.

Step 9 Allow any legitimate method for solving the equation, but insist on a good explanation. [**Alert**] Be especially wary of any student impulses to move numbers from one side of the equation to the other.

SHARING IDEAS

Ask several students to share their solution methods for Step 9.

At an appropriate time, introduce the term *intercept form* for the equation $y = a + bx$ and the related terms *y-intercept* and *coefficient of x*. [**Ask**] "Is the equation $y = bx + a$ equivalent to the intercept form?" [Yes] [**Ask**] "Are equations $y = ax + b$ and $y = mx + b$ (often called the *slope-intercept form*) also equivalent?" [Yes] Put in numbers for a, b, and m as needed. These questions may allow students to see that the letters a, b, and m represent constants in particular equations, whereas letters x and y represent variables. [**Ask**] "Is $y = a - bx$ equivalent to $y = bx - a$?" [No, $y = a - bx$ is equivalent to $y = -bx + a$. It is very important to keep the signs consistent.]

[**Ask**] "How are linear equations related to other equations you have seen in this course?" [The first, in Chapter 1, was $y = x$. It's a special case of a direct variation $y = kx$, with $k = 1$. And direct variations are special cases of the intercept form $a + bx$, with $a = 0$ and $b = k$.] Elicit the fact that the coefficient of x, no matter what it's called, gives the rate of change as well as the common difference between consecutive terms in the recursive sequence.

Step 8 The x-axis represents every instant of time, and the y-axis every fraction of calories burned. This graph models a continuous linear relationship.

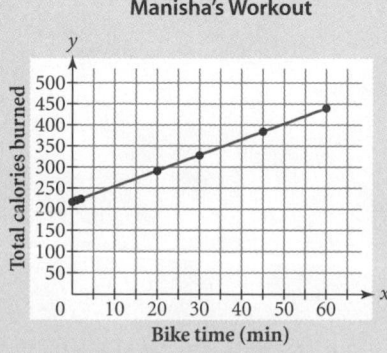

Manisha's Workout

Step 10 In the equation $y = 215 + 3.8x$, the starting value is 215 and the rule to add shows up as the coefficient of x. The starting value is the y-intercept. The rule add 3.8 shows the steepness of the line. The starting value of the recursive routine is the y-intercept only when the starting value of x is zero.

[Ask] "Which variables might be called input and output variables for the investigation?" [The input variable is the number of minutes spent exercising. The output variable is the number of calories burned.] See Lesson 8.3 for a discussion of independent and dependent variables.

Assessing Progress

Your observations should allow you to assess students' abilities with recursive routines, calculator lists, scatter plots, graphing equations, and undoing operations.

▶ *EXAMPLE A*

Besides revisiting the ideas of the investigation, this example shows how the graph of the linear equation $y = a + bx$ is a shift of the graph of $y = bx$. Translation will come up in Chapter 9.

Be sure students understand that the second column in the given table refers only to the calories burned while swimming, whereas the third column refers to the total number of calories burned, including those Sam burned before he started swimming.

Some students may say one graph is a vertical shift of the other. Other students, considering the graphs over quadrants other than the first, may see it as a horizontal shift. Encourage both viewpoints. Explore why they are equivalent and how the constants affect the amount of shift.

In the equation $y = 215 + 3.8x$, 215 is the value of a. It represents the 215 calories Manisha burned while jogging before her workout. The value of b is 3.8. It represents the rate her body burned calories while she was pedaling. What would happen if Manisha chose a different physical activity before pedaling on the stationary bike?

You can also think of direct variations in the form $y = kx$ as equations in intercept form. For instance, Sam's trainer tells him that swimming will burn 7.8 calories per minute. When the time spent swimming is 0, the number of calories burned is 0, so a is 0 and drops out of the equation. The number of calories burned is proportional to the time spent swimming, so you can write the equation $y = 7.8x$.

The constant of variation k is 7.8, the rate at which Sam's body burns calories while he is swimming. It plays the same role as b in $y = a + bx$.

EXAMPLE A

Suppose Sam has already burned 325 calories before he begins to swim for his workout. His swim will burn 7.8 calories per minute.

a. Create a table of values for the calories Sam will burn by swimming 60 minutes and the total calories he will burn after each minute of swimming.

b. Define variables and write an equation in intercept form to describe this relationship.

c. On the same set of axes, graph the equation for total calories burned and the direct variation equation for calories burned by swimming.

d. How are the graphs similar? How are they different?

▶ **Solution**

a. The total numbers of calories burned appear in the third column of the table. Each entry is 325 plus the corresponding entry in the middle column.

b. Let y represent the number of total calories burned, and let x represent the number of minutes Sam spends swimming.

$$y = 325 + 7.8x$$

Sam's Swim

Swimming time (min)	Calories burned by swimming	Total calories burned
0	0	325
1	7.8	332.8
2	15.6	340.6
20	156	481
30	234	559
45	351	676
60	468	793

c. The direct variation equation is $y = 7.8x$. Enter it into Y₁ on your calculator. Enter the equation $y = 325 + 7.8x$ into Y₂. Check to see that these equations give the same values as the table by looking at the calculator table.

d. The lower line shows the calories burned by swimming and is a direct variation. The upper line shows the total calories burned. It is 325 units above the first line because, at any particular time, Sam has burned 325 more calories. Both graphs have the same value of b, which is 7.8 calories per minute. The graphs are similar because both are lines with the same steepness. They are different because they have different y-intercepts.

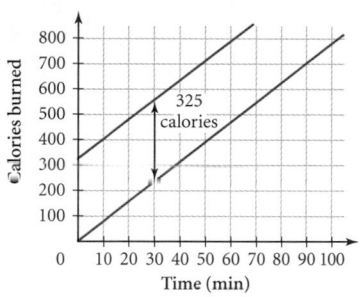

Can you tell what different values of a in the equation $y = a + bx$ will do to the graph?

EXAMPLE B

There is a linear relationship between the air temperature outside an airplane and the plane's altitude in meters. The temperature at sea level is 14.7°C. The temperature drops 7 degrees for every 1000-meter increase in altitude.

a. Define variables and write an equation in intercept form for this relationship.

b. Use your equation to evaluate expressions to complete a table of values for altitudes of 0, 1000, 2000, . . . , 6000 meters.

c. Explain the real-world meaning of the values of a and b in your equation.

d. Graph the relationship. Show how to find the temperature from the graph when the altitude is 4200 meters.

e. Use the graph to find a coordinate where the temperature is approximately 0°C. What is the real-world meaning of this coordinate?

f. Use the graph to find the altitude that gives a temperature of −10°C.

▶ **Solution**

a. If x represents altitude in meters and y represents the outside temperature in °C, the equation for the relationship is $y = 14.7 - 0.007x$.

b. Substitute the altitude values for x in the equation to get the y-values, which are the corresponding temperatures.

c. The real-world meaning of a is the temperature at altitude 0 meters, or sea level. The value of b, $\frac{-7}{1000}$ or −0.007, is the rate at which the temperature (in °C) changes for each meter increase in altitude.

Altitude (m) x	Temperature (°C) y
0	14.7
1000	7.7
2000	0.7
3000	−6.3
4000	−13.3
5000	−20.3
6000	−27.3

▶ **EXAMPLE B**

Use this example if your students are having difficulties connecting graphs to tables or relating the mathematical theory to the real world.

Ask how realistic the data are. Students may point out that the actual numbers depend on the time of year and the location. Part c also gives you the chance to connect the mathematics to the real-world context.

In part d, students may not be able to find an exact value for y corresponding to the x-value of 4200 by tracing unless they have a friendly window. (Friendly windows are discussed in detail in Chapter 8.) They can get such a window by using zoom decimal and then multiplying appropriately. The window $[0, 9400, 1000, -25, 15, 5]$ works well.

You can also find the value in part d by substituting 4200 for x in the equation $y = 14.7 - 0.007x$. This means the temperature is given by the expression $14.7 - 0.007 \cdot 4200$, which equals −14.7°C.

In part f, the two points on either side of $y = -10$ have x-values of about 3489 and 3574. Picking either of these values, or anything in between, is OK. To find the exact value, you can solve the equation $-10 = 14.7 - 0.007x$ by undoing the order of operations, giving approximately 3529.

Closing the Lesson

A linear equation in intercept form $y = a + bx$ reflects the recursive routine used to generate a sequence of data values with a constant rate of change. Such a routine begins with a and adds b repeatedly. (The value of either a or b may be negative.)

BUILDING UNDERSTANDING

Students practice writing, graphing, and exploring linear equations, primarily in intercept form.

ASSIGNING HOMEWORK

Essential	1, 2 or 3, 6–7, 10
Performance assessment	9, 10
Portfolio	6
Journal	7, 8
Group	7
Review	4, 5, 11–15

▶ Helping with the Exercises

Exercise 1 If students have difficulties relating recursive routines with the explicit equations, suggest that they make tables of data.

Exercises 2 and 3 These are the first equations in a while that don't use just x and y for variable names. Encourage students to write the dimensions of each number and variable, especially the rate.

2c. 24 represents the initial number of miles the driver is from his or her destination.

2d. 45 means the driver is driving at a speed of 45 miles per hour.

d. Trace on the graph to find the x-value 4200. The corresponding y-value is -14.7. At an altitude of 4200 meters the temperature is $-14.7°C$. The window shown is $[0, 7000, 1000, -25, 20, 5]$. [▶ See **Calculator Note 1H.** ◀]

e. Move the cursor until the y-value is 0. You will see that the x-value is 2100. (This point is where the graph crosses the x-axis.) It means that the plane's altitude is 2100 meters when the temperature is 0°C outside.

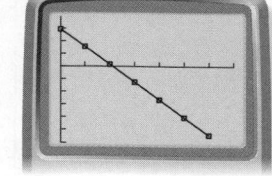

f. Move the cursor to a point on the graph where the y-value is approximately -10, and find that the corresponding x-value is approximately 3529. So the plane's altitude is about 3529 meters when the temperature is $-10°C$ outside.

In linear equations it is sometimes helpful to say which variable is the input variable and which is the output variable. The horizontal axis represents the input variable, and the vertical axis represents the output variable. In Example B, the input variable x represents altitude so the x-axis is labeled altitude, and the output variable y represents temperature so the y-axis is labeled temperature. What are the input and output variables in the investigation and in Example A?
The input variable is time, and the output variable is calories burned.

EXERCISES

You will need your calculator for problems **2, 3, 5, 6, 9,** and **10.**

▶ Practice Your Skills

1. Match the recursive routine in the first column with the equation in the second column.

ii **a.** 3 (ENTER)
   Ans + 4 (ENTER); (ENTER), . . .

iv **b.** 4 (ENTER)
   Ans + 3 (ENTER); (ENTER), . . .

iii **c.** −3 (ENTER)
   Ans − 4 (ENTER); (ENTER), . . .

i **d.** 4 (ENTER)
   Ans − 3 (ENTER); (ENTER), . . .

i. $y = 4 - 3x$

ii. $y = 3 + 4x$

iii. $y = -3 - 4x$

iv. $y = 4 + 3x$

2. You can use the equation $d = 24 - 45t$ to model the distance from a destination for someone driving down the highway, where distance d is measured in miles and time t is measured in hours. Graph the equation and use the trace function to find the approximate time for each distance given in 2a and b.

   a. $d = 16$ mi $t \approx 0.18$ hr   **b.** $d = 3$ mi $t \approx 0.47$ hr

   c. What is the real-world meaning of 24?

   d. What is the real-world meaning of 45?

Some rental cars have in-dash navigation systems.
© 2000 Hertz System, Inc. Hertz is a registered service mark and trademark of Hertz System, Inc.

3. You can use the equation $d = 4.7 + 2.8t$ to model a walk in which the distance from a motion sensor d is measured in feet and the time t is measured in seconds. Graph the equation and use the trace function to find the approximate distance from a motion sensor for each time value given in 3a and b.

   a. $t = 12$ sec $d \approx 38.3$ ft   **b.** $t = 7.4$ sec $d \approx 25.42$ ft

   c. What is the real-world meaning of 4.7?   **d.** What is the real-world meaning of 2.8?
The walker started 4.7 feet away from the motion sensor.   The walker was walking at a rate of 2.8 feet per second.

4. Undo the order of operations to find the x-value in each equation.

   a. $3(x - 5.2) + 7.8 = 14$ $x \approx 7.267$ **b.** $3.5\left(\dfrac{x - 8}{4}\right) = 2.8$ $x = 11.2$

5. Use the distributive property to rewrite each expression without parentheses. Use list $L_1 = \{-3.5, 2.5, 11\}$ and list $L_2 = \{-2.8, 4.2, 21\}$ to verify your answer.

   a. $-2(L_1 - 5)$ $-2L_1 + 10$ **b.** $-4(-L_1 + L_2)$ $4L_1 - 4L_2$

► Reason and Apply

6. APPLICATION Louis is beginning a new exercise workout. His trainer shows him the calculator table with x-values showing his workout time in minutes. The Y_1-values are the total calories Louis burned while running, and the Y_2-values are the number of calories he wants to burn.

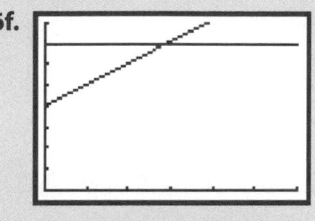

   a. Find how many calories Louis has burned before beginning to run, how many he burns per minute running, and the total calories he wants to burn.

   b. Write a recursive routine that will generate the values listed in Y_1. 400 (ENTER); Ans + 20.7 (ENTER)

   c. Use your recursive routine to write a linear equation in intercept form. Check that your equation generates the table values listed in Y_1. $Y_1 = 400 + 20.7x$

   d. Write a recursive routine that will generate the values listed in Y_2. 700 (ENTER); Ans + 0 (ENTER)

   e. Write an equation that generates the table values listed in Y_2. $Y_2 = 700 + 0x$ or $Y_2 = 700$

   f. Graph the equations in Y_1 and Y_2 on your calculator. Your window should show a time of up to 30 minutes. What is the real-world meaning of the y-intercept in Y_1?

   g. Use the trace function to find the approximate coordinates of the point where the lines meet. What is the real-world meaning of this point?

7. Jo mows lawns after school. She finds that she can use the equation $P = -300 + 15N$ to calculate her profit.

   a. Give some possible real-world meanings for the numbers -300 and 15 and the variable, N.

   b. Invent two questions related to this situation and then answer them.

   c. Explain why the equation $P = 15N - 300$ provides the same values as the equation $P = -300 + 15N$.

   d. Identify the variables.
The input variable is N for number of lawns, and the output variable is P for profit.

Exercise 6 In Example B, students moved the cursor to trace a graph and find the intersection of a line with the x-axis. In this exercise, they trace to find the intersection of a horizontal line $y = 700$ with $y = 400 + 20.7x$.

6a. Louis has burned 400 calories before beginning to run. His calorie-burning rate is 20.7 calories per minute, and he wants to burn 700 total calories.

6f.

[0, 30, 5, 0, 800, 100]

6g. The approximate coordinates of the point where the lines meet are (14.5, 700). This means that after 14.5 minutes of running, Louis will have burned off his desired total of 700 calories.

Exercise 7 Students may be confused with the variable names. Suggest that they write the appropriate dimensions.

7a. One possible scenario: Jo has an initial start-up cost of $300 for equipment and expenses. She makes $15 for every lawn she mows, N.

7b. Sample questions: How many lawns will Jo have to mow to break even? [Solve $-300 + 15N = 0$. Jo must mow 20 lawns.] How much profit will Jo earn if she mows 40 lawns? [Substitute 40 for N. $300.]

7c. Subtracting 300 from $15N$ is the same as adding $15N$ to -300.

Exercise 9 If students are having difficulties writing an equation, encourage them to generate the data with a recursive routine. As needed, point out that 12% is a rate.

9a. $y = 45 + 0.12x$, where x represents dollar amounts customers spend and y represents Manny's daily income in dollars.

9b.

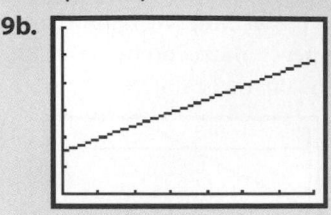

[0, 840, 120, 0, 180, 30]

Exercise 10d The shift of graphs is like that of Example A.

10a. $y = 3.8x$; 114 calories in Workout 1

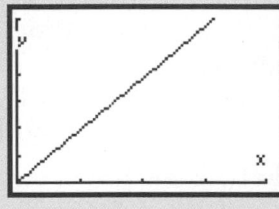

[0, 40, 10, 0, 120, 20]

10b. $y = 114 + 6.9x$; 103.5 calories in Workout 2; 217.5 total calories

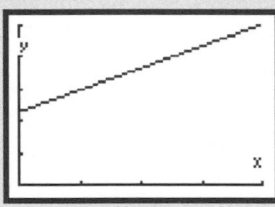

[0, 20, 5, 0, 250, 50]

10c. $y = 217.5 + 7.3x$; 146 calories in Workout 3; 363.5 total calories

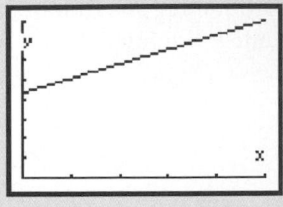

[0, 25, 5, 0, 400, 50]

10d. Since Paula burns 30(3.2) calories walking, 96 would be added to the value of a in each equation. In the graphs, each y-intercept (and line) would be shifted 96 units higher.

8. As part of a physics experiment, June threw an object off a cliff and measured how fast it was traveling downward. When the object left June's hand, it was traveling 5 meters per second, and it sped up as it fell. The table shows a partial list of the data she collected as the object fell.

Time (sec)	Speed (m/sec)
0	5.0
0.5	9.9
1.0	14.8
1.5	19.7

   $s = 5 + 9.8t$ or $s = 9.8t + 5$
   a. Write an equation to represent the speed of the object.
   b. What was the object's speed after 3 seconds? 34.4 m/sec
   c. If it were possible for the object to fall long enough, how many seconds would pass before it reached a speed of 83.4 meters per second? 8 sec
   d. What limitations do you think this equation has in modeling this situation? It doesn't account for air resistance and terminal speed.

9. **APPLICATION** Manny has a part-time job as a waiter. He makes $45 per day plus tips. He has calculated that his average tip is 12% of the total amount his customers spend on food and beverages.
   a. Define variables and write an equation in intercept form to represent Manny's daily income in terms of the amount his customers spend on food and beverages.
   b. Graph this relationship for food and beverage amounts between $0 and $900.
   c. Write and evaluate an expression to find how much Manny makes in one day if his customers spend $312 for food and beverages. $45 + 0.12 \cdot 312 = \$82.44$
   d. What amounts spent on food and beverages will give him a daily income between $105 and $120? between $500 and $625

10. **APPLICATION** Paula is cross-training for a triathlon in which she cycles, swims, and runs. Before designing an exercise program for Paula, her coach consults a table listing rates for calories burned in various activities.

Cross-training activity	Calories burned (per min)
Walking	3.2
Bicycling	3.8
Swimming	6.9
Jogging	7.3
Running	11.3

    a. Write and graph an equation in intercept form to find the number of calories burned for each minute of cycling. How many calories has Paula burned by completing Workout 1?
    b. After finishing Workout 1, Paula begins the swimming workout. Write and graph an equation to find the total number of calories she has burned so far during each minute of swimming. How many calories has she burned by completing Workout 2? How many calories has she burned by completing both workouts?
    c. Write and graph an equation to find the total number of calories Paula has burned during each minute of jogging. How many calories has she burned by completing Workout 3? How many calories has she burned after completing all three workouts?
    d. If Paula walks for half an hour before she begins her exercise program, how does that affect the equations and graphs in 10a–c?

    Paula's Exercise Program

    Workout 1   Bike for 30 minutes
    Workout 2   Swim for 15 minutes
    Workout 3   Jog for 20 minutes

▶ Review

11. At a family picnic, your cousin tells you that he always has a hard time remembering how to compute percents. Write him a note explaining what percent means. Use these problems as examples of how to solve the different types of percent problems, with answers for each.

a. 8 is 15% of what number? $\frac{8}{n} = \frac{15}{100}$, $n \approx 53.3$

b. 15% of 18.95 is what number? $\frac{15}{100} = \frac{n}{18.95}$, $n = 2.8$

c. What percent of 64 is 326? $\frac{P}{100} = \frac{326}{64}$, $p \approx 509.4$

d. 10% of what number is 40? $\frac{10}{100} = \frac{40}{n}$, $n = 400$

12. **APPLICATION** Carl has been keeping a record of his gas purchases for his new car. Each time he buys gas, he fills the tank completely. Then he records the number of gallons he bought and the miles since the last fill-up. Here is his record:

Carl's Purchases

Miles traveled	Gallons	$\frac{miles}{gallon}$
363	16.2	22.4
342	15.1	22.6
285	12.9	22.1

a. Copy and complete the table by calculating the ratio of miles per gallon for each purchase.

b. What is the average rate of miles per gallon for Carl's new car so far? 22.4 miles per gallon

c. The car's tank holds 17.1 gallons. To the nearest mile, how far should Carl be able to go without running out of gas? 383 miles

d. Carl is planning a trip across the United States. He estimates that the trip will be 4230 miles. How many gallons of gas can Carl expect to buy? approximately 189 gallons

13. Match each recursive routine to a graph. Explain how you made your decision and tell what assumptions you made.

ii **a.** 2.5 `ENTER`
Ans + 0.5 `ENTER` ; `ENTER` , . . .

iv **b.** 1.0 `ENTER`
Ans + 1.0 `ENTER` ; `ENTER` , . . .

iii **c.** 2.0 `ENTER`
Ans + 1.0 `ENTER` ; `ENTER` , . . .

i **d.** 2.5 `ENTER`
Ans − 0.5 `ENTER` ; `ENTER` , . . .

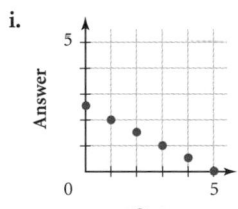

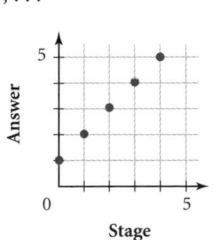

Exercise 11 Students can answer this question using proportions as they learned in Chapter 2, but allow other correct methods.

11. Partial answer: Write the percent as one ratio of a proportion. Put the part over the whole in the other ratio.

Exercise 12 This exercise reviews Lesson 3.1.

13. Sample explanation: I matched the rate of change to each graph. I assumed the starting value was the *y*-intercept.

14. Bjarne is training for a bicycle race by riding on stationary bicycles with time-distance readouts. He is riding at a constant speed. The graph shows his accumulated distance and time as he rides.

Distance Traveled

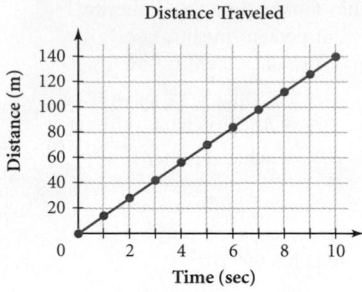

Time (sec)	Distance (m)
1	14
2	28
3	42
4	56
5	70
6	84
7	98
8	112
9	126
10	140

a. How fast is Bjarne bicycling?   14 meters per second

b. Copy the table and show the distance for each time.

c. Write a recursive routine for Bjarne's ride.   {0,0} [ENTER]   {Ans(1) + 1, Ans(2) + 14} [ENTER]

d. Looking at the graph, how do you know that Bjarne is neither slowing down nor speeding up during his ride?   The points lie on a line.

e. If Bjarne keeps up the same pace, how far will he ride in one hour?   50,400 meters, or 50.4 kilometers

Bicyclists race through San Luis Obispo, California.

15. Consider the expression $\frac{4(y - 8)}{3}$.

a. Find the value of the expression if $y = 5$. Make a table to show the order of operations.

b. Solve the equation $\frac{4(y - 8)}{3} = 8$, by undoing the sequence of operations. $y = 14$

15a. The expression equals -4.

Ans $-$ 8	-3
Ans $\cdot$ 4	-12
Ans/3	-4

Linear Equations and Rate of Change

How can it be that mathematics, being after all a product of human thought independent of experience, is so admirably adopted to the objects of reality?

ALBERT EINSTEIN

In this lesson you will continue to develop your skills with equations, graphs, and tables of data by exploring the role that the value of b plays in the equation

$$y = a + bx.$$

You have already studied the intercept form of a linear equation in several real-world situations. You have used them to relate calories to minutes spent exercising, floor heights to floor numbers, and distances to time. So, defining variables is an important part of writing equations. Depending on the context of an equation, its numbers take on different real-world meanings. Can you recall how these equations modeled each scenario?

Equation	Situation
$y = 215 + 3.8x$	calories burned in a workout
$y = 321 - 13x$	floor heights in a building
$y = -300 + 15x$	earnings from mowing lawns
$y = 45 + 0.12x$	income from restaurant tabs
$y = 220 - 1.2x$	distance a car is from Flint

Winds of 40 mph blow on North Michigan Ave. in 1955 Chicago.

In most linear equations, there are different output values for different input values. This happens when the coefficient of x is not zero. You'll explore how this coefficient relates input and output values in the examples and the investigation.

In addition to giving the actual temperature, weather reports often indicate the temperature you *feel* as a result of the wind chill factor. The wind makes it feel colder than it actually is. In the next example you will use recursive routines to answer some questions about wind chill.

PLANNING

LESSON OUTLINE

One day:

5 min	Introduction
5 min	Example A
20 min	Investigation
5 min	Sharing
5 min	Example B
5 min	Closing
5 min	Exercises

MATERIALS

- Calculator Note 4D
- Wind Chill (T), *optional*

TEACHING

Students get closer to the notion of slope by studying rates of change as they construct linear equations and their graphs from input-output tables. The concept of input and output variables is a precursor to the study of functions. (The input value is the *independent variable* and the output value is the *dependent variable*, though those terms aren't used until Lesson 8.3.)

[ESL] Input variables are those that are put in. Output variables are those that come out.

NCTM STANDARDS

CONTENT	PROCESS
• Number	Problem Solving
• Algebra	• Reasoning
• Geometry	• Communication
• Measurement	• Connections
Data/Probability	• Representation

LESSON OBJECTIVES

- Interpret equations in intercept form using input and output variables
- Explore the relationships among tables, scatter plots, recursive routines, and equations

This example introduces the notion of an input-output table and revisits the idea of rate of change. Students may be confused by the use of table entries as units rather than degrees. That is, the rate of change is 7° wind chill per table entry, or $\frac{7}{5}$° wind chill per degree temperature.

Wind chill for other wind speeds are given in the investigation (1.35 for 15 mph), Example B (1.55 for 35 mph) and Exercise 2 (1.6 for 40 mph). The faster the wind, the greater the wind chill factor.

 Guiding the Investigation

One step Show the Wind Chill transparency or refer to the table at the bottom of this page. **[Ask]** "What is the actual temperature if the weather report indicates a wind chill of 10.6 degrees at this wind speed?" As students work, encourage them to use recursive routines, calculate rates of change, and use equations and graphs.

Step 1 In real life, it's not always clear which variable denotes input and which denotes output. The convention is to put input values in the left column and output values on the right.

EXAMPLE A

The table relates the approximate wind chills for different actual temperatures when the wind speed is 20 miles per hour. Assume the wind chill is a linear relationship for temperatures between −5° and 35°.

Temperature (°F)	−5	0	5	10	15	20	25	30	35
Wind chill (°F)	−45	−38	−31			−10		4	11

a. What are the input and output variables?

b. Use calculator lists to write a recursive routine that generates the table values. What are the missing entries?

c. What is the change in temperature from one table entry to the next? What is the corresponding change in the wind chill?

► **Solution**

a. The input variable is the actual air temperature in °F. The output variable is the temperature you feel as a result of the wind chill factor.

b. The recursive routine to complete the missing table values is {−5, −45} ENTER and {Ans(1) + 5, Ans(2) + 7} ENTER. The calculator screen displays the missing entries.

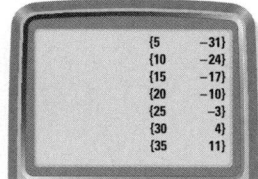

c. For every 5° increase in temperature, the wind chill increases 7°.

In Example A, the rate at which the wind chill drops can be calculated from the ratio $\frac{7}{5}$. In other words, it feels 1.4° colder for every 1° drop in air temperature. This number is the rate of change for a wind speed of 20 mph. The **rate of change** is equal to the ratio of the change in output values divided by the corresponding change in input values.

How does the rate of change differ with various wind speeds?

 Investigation
Wind Chill

In this investigation you'll use the relationship between temperature and wind chill to explore the concept of rate of change and its connections to tables, scatter plots, recursive routines, equations, and graphs.

The data in the table represent the approximate wind chill temperatures in degrees Fahrenheit for a wind speed of 15 miles per hour. Use this data set to complete each task.

Step 1 Let x be the input variable representing the temperature in °F, and let y be the output variable   **Step 1** representing wind chill in °F.

Define the input and output variables for this relationship.

Temperature (°F)	Wind chill (°F)
−5	−38
0	−31.25
1	−29.9
2	−28.55
5	−24.5
15	−11
35	16

| Step 2 | Plot the points and describe the viewing window you used. |

Step 3 Starting value: **Step 3**
{−5, −38} (ENTER)
Rule for one-degree **Step 4**
increment: {Ans(1) + 1,
Ans(2) + 1.35}

Write a recursive routine that gives the pairs of values listed in the table.

Copy the table. Complete the third and fourth columns of the table by recording the changes between consecutive input and output values. Then find the rate of change.

Input	Output	Change in input values	Change in output values	Rate of change	
−5	−38				
0	−31.25	5	6.75	$\frac{+6.75}{+5} =$	1.35
1	−29.9	1	1.35	$\frac{+1.35}{+1} = 1.35$	
2	−28.55	1	1.35	$\frac{+1.35}{+1} =$	1.35
5	−24.5	3	4.05	$\frac{+4.05}{+3} = 1.35$	
15	−11	10	13.5	$\frac{+13.5}{+10} =$	1.35
35	16	20	27	$\frac{+27}{+20} = 1.35$	

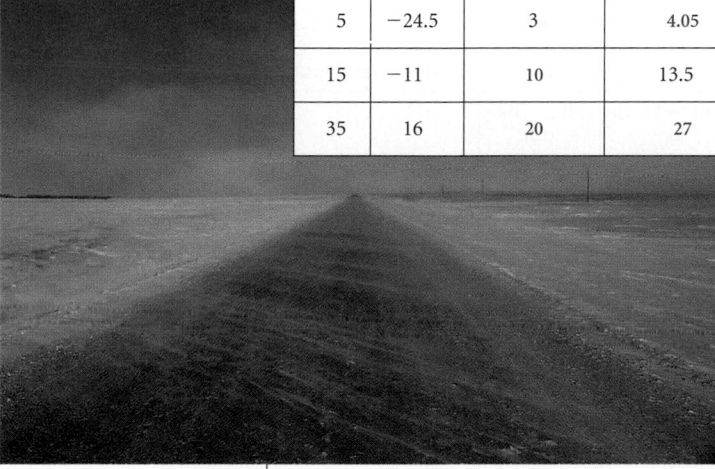

High wind speeds in Saskatchewan, Canada, drop temperatures below freezing.

Step 5
Step 5 $y = -31.25 + 1.35x.$
The rule appears as the coefficient of x.

Use your routine to write a linear equation in intercept form that relates wind chill to temperature. Note that the starting value, −38, is not the y-intercept. How does the rule of the routine appear in your equation?

Step 6
Graph the equation on the same set of axes as your scatter plot. Use the calculator table to check that your equation is correct. Does it make sense to draw a line through the points? Where does the y-intercept show up in your equation?

Step 7
What do you notice about the values for rate of change listed in your table? How does the rate of change show up in your equation? In your graph?

Step 8
Explain how to use the rate of change to find the actual temperature if the weather report indicates a wind chill of 10.6° with 15-mph winds.

The rate of change in the wind chill factor for 15-mph winds is 1.35. The rate for 20-mph winds is 1.4. What is the rate of change for a wind chill factor of 35 mph?

Step 2

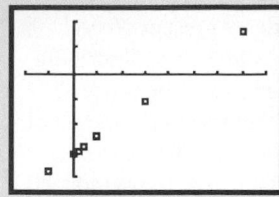

$[-10, 40, 5, -40, 20, 10]$

Step 5 If students are having difficulties, remind them that they can find the rate of change easily if they know output values for input values that are one unit apart. Also, elicit the idea that an input value of 0 gives the y-intercept as an output value.

Step 6 Yes, a line represents every possible temperature. The y-intercept shows up as the value of a in the equation. It is not the starting value of the routine.

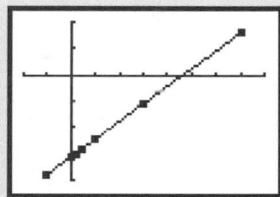

Step 7 The values for rate of change are all equivalent. The rate of change appears as b, the coefficient of x. In the graph, to go from one point to the next, you move right 1 unit and up 1.35 units, which is the rate of change.

Step 8 Encourage a variety of approaches: tracing, calculator tables, or working backward to solve the equation $10.6 = -31.25 + 1.35x.$

Step 8 Explanations will vary. Students can add 1.35° sixteen times to −11° in the table while adding 16° to 15°. They can also subtract 1.35° four times from 16° while subtracting 4° from 35°. The answer is 31°F.

Have students present several approaches to Step 8, describing their equations in the process. Mention that the constant rate of change is sometimes called the *wind chill factor* at this wind speed. Draw out the fact that a rate of change is a rate as defined in Chapter 3—that is, a ratio with denominator 1. The rate of change is the output change of each additional unit of input.

[Ask] "What do the equations have in common?" [Elicit the idea that the output variable is usually isolated on the left side, making it easy to enter functions in the calculator for graphing. The right side is like the recursive routine: There's a constant that corresponds to the starting value, and the rate of change is multiplied by the input variable.]

[Ask] "How is the rate of change represented on the graph?" As needed, draw a picture showing that, as the line moves one unit from left to right, it rises by the amount of the rate of change, so that the larger the rate of change, the steeper the line. You might use the word *slope* as a synonym for steepness of the line, representing the rate of change. The term *slope* will be defined in Lesson 5.1.

Assessing Progress

Observe students' skills at making a scatter plot, choosing a viewing window, writing a recursive routine, writing a linear equation from a recursive routine, and graphing a line. Also assess understanding of rate of change.

▶ *EXAMPLE B*

This example is a review of the investigation. Ask if it seems valid to look for a relationship between wind speed and wind chill factor. The values given in Example A, the investigation, and Example B appear to lie in straight lines. Actually, however, the wind chill factors level off at wind speeds greater than 45 mph.

EXAMPLE B The wind chill data for a wind speed of 35 miles per hour is shown in the table.

a. Add three columns to the table to record the change in input values, change in output values, and corresponding rate of change.

b. Use the table and a recursive routine to write a linear equation in intercept form.

c. What are the real-world meanings of the values for a and b in your equation?

Input	Output
Temperature (°F) (L_1)	Temperature (°F) (L_2)
−25	−89
−20	−81.25
10	−34.75
15	−27
25	−11.5
35	4

▶ **Solution**

a. You can create a table to find that the wind chill increases 7.75° for every 5° increase in temperature. The rate of change, or value of b, is 1.55.

Input	Output	Change in input values	Change in output values	Rate of change
Temperature (L_1)	Wind Chill (L_2)			
−25	−89			
−20	−81.25	5	7.75	$\frac{+7.75}{+5} = 1.55$
10	−34.75	30	46.5	$\frac{+46}{+30} = 1.55$
15	−27	5	7.75	$\frac{+7.75}{+5} = 1.55$
25	−11.5	10	15.5	$\frac{+15.5}{+10} = 1.55$
35	4	10	15.5	$\frac{+15.5}{+10} = 1.55$

The wind chill factor drops temperatures well below zero in Antarctica.

b. The recursive routine

$\{-25, -89\}$ `ENTER` ,

$\{Ans(1) + 5, Ans(2) + 7.75\}$ `ENTER` , `ENTER` , `ENTER` , `ENTER` , `ENTER`

gives the result $\{0, -50.25\}$. So the y-intercept, or value of a, is −50.25. Then the equation in intercept form is

$$y = -50.25 + 1.55x$$

where x represents the air temperature in °F, and y represents the wind chill temperature you feel.

Note that the starting value of the recursive routine is not the same as the value of the y-intercept in the equation.

c. The value of a is the wind chill you feel when the temperature is 0°F. The value of b states that it feels 1.55° colder for every 1° drop in air temperature.

Because the values for rate of change increase with wind speed, you feel colder when the wind blows faster. But actually, wind speeds greater than 45 mph result in very little additional cooling effect. Also, wind chill factors are usually reported only for temperatures less than 35°F. Think about what limitations a linear equation has when you are modeling these situations.

EXERCISES

You will need your calculator for problems **1, 4, 5, 10,** and **12.**

▶ Practice Your Skills

1. Copy and complete the table of output values for each equation.

a. $y = 50 + 2.5x$

Input x	Output y
20	100
−30	−25
16	90
15	87.5
−12.5	18.75

b. $L_2 = -5.2 - 10 \cdot L_1$

L_1 x	L_2 y
0	−5.2
−8	74.8
24	−245.2
−35	344.8
−5.2	46.8

2. Use the equation $w = -52 + 1.6t$ to approximate the wind chill temperatures for a wind speed of 40 miles per hour. Find the wind chill temperature w for each actual temperature t given in 2a and b.

a. $t = 32°$ $w = -0.8°$

b. $t = 12°$ $w = -32.8°$

c. What is the real-world meaning of 1.6?

d. What is the real-world meaning of −52?

3. Describe what the rate of change looks like in each graph.

a. the graph of a person walking at a steady rate toward a motion sensor

b. the graph of a person standing still The rate is not negative or positive. It is zero. The line is a horizontal line.

c. the graph of a person walking at a steady rate away from a motion sensor

d. the graph of one person walking at a steady rate faster than another person
The rate for the speedier walker will be greater than the rate for the person walking more slowly, so the graph for the speedier walker will be steeper than the graph for the slower walker.

Closing the Lesson

You can calculate the **rate of change** by a difference of output values divided by a difference of corresponding input values. The rate of change determines the **steepness** of the graph of the linear equation representing the data.

BUILDING UNDERSTANDING

Students work more with input and output variables, rates of change, and graphs of equations.

ASSIGNING HOMEWORK

Essential	1–3, 6–8
Performance assessment	5, 6
Portfolio	10
Journal	2, 8
Group	4, 5, 9
Review	11–14

▶ Helping with the Exercises

Exercise 2 If students are having difficulty, encourage them to measure or count how many units the line rises or falls as it moves one unit from left to right.

2c. The wind chill temperature changes by 1.6° for each 1° change in actual temperature.

2d. If the actual temperature is 0°, the wind chill temperature is −52°.

3a. The rate is negative, so the line goes from the upper left to the lower right.

3c. The rate is positive, so the line goes from the lower left to the upper right.

4. A sample:

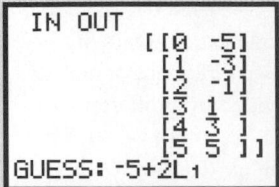

```
IN OUT
        [[0 -5]
         [1 -3]
         [2 -1]
         [3  1 ]
         [4  3 ]
         [5  5 ]]
GUESS: -5+2L₁
```

Exercise 6 The wind chill factor given here for 30 mph is not between those for 20 mph and 35 mph as might be expected. If students notice this, point out that the intercept is also not between the intercept for 20 mph and 35 mph. There are different formulas for wind chill.

[Language] Students may wonder how to answer the question in 6d about the difference in the graphs. Encourage them to use the words *discrete* and *continuous*.

6a. The input variable x is the temperature in °F, and the output variable y is the wind chill in °F.

6b. The rate of change is 1.6°. For every 5° increase in temperature, there is an 8° increase in wind chill.

6d. Both graphs show linear relationships with identical rates of change and y-intercepts. The graphs are different in that the points are discrete and the equation continuous.

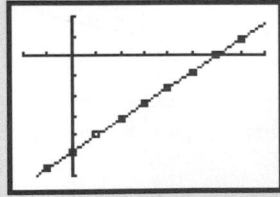

$[-10, 40, 5, -60, 20, 10]$

4. Use the "Easy" setting of the INOUT game on your calculator to produce four data tables. Copy each table and write the equation you used to match the data values in the table. [▶ 🖵 See **Calculator Note 4D** to learn how to run the program. ◀]

▶ Reason and Apply

5. Each table below shows a different input-output relationship.

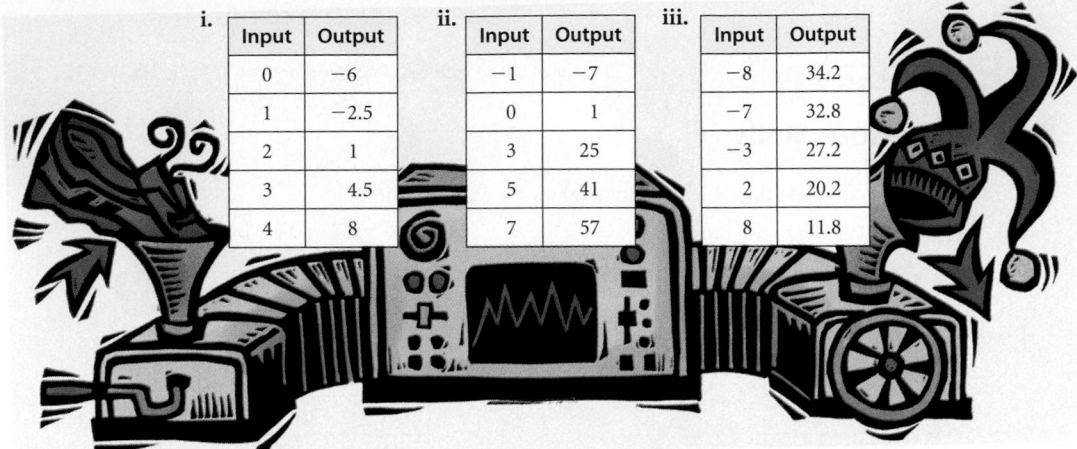

i.

Input	Output
0	−6
1	−2.5
2	1
3	4.5
4	8

ii.

Input	Output
−1	−7
0	1
3	25
5	41
7	57

iii.

Input	Output
−8	34.2
−7	32.8
−3	27.2
2	20.2
8	11.8

   a. Find the rate of change in each table. Explain how you found this value. **i.** 3.5; **ii.** 8; **iii.** −1.4

   b. For each table, find the output value that corresponds to an input value of 0. What is this value called? **i.** −6; **ii.** 1; **iii.** 23; the y-intercept

   c. Use your results from 5a and b to write an equation in intercept form for each table. **i.** $y = -6 + 3.5x$; **ii.** $y = 1 + 8x$; **iii.** $y = 23 - 1.4x$

   d. Use a calculator list of input values to check that each equation actually produces the output values shown in the table.

6. The wind chill temperatures for a wind speed of 30 mph are given in the table.

Temperature (°F)	−5	0	5	10	15	20	25	30	35
Wind chill (°F)	−56	−48	−40	−32	−24	−16	−8	0	8

   a. Define input and output variables and graph the data.

   b. Find the rate of change. Explain how you got your answer.

   c. Write an equation in intercept form. $y = -48 + 1.6x$

   d. Plot the points and graph the equation on the same set of axes. How are the graphs for the points and the equation similar? How are they different?

7. The equation $y = 35 + 0.8x$ gives the distance a sports car is from Flint after x minutes.

   a. How far is the sports car from Flint after 25 minutes? $35 + 0.8(25) = 55$ miles

   b. How long will it take until the sports car is 75 miles from Flint? Show how to find the solution using two different methods. 50 min. Students might use a graph or the undo method.

8. You can use the equation $7.3x = 200$ to describe a rectangle with an area of 200 square units like the one shown. What are the real-world meanings of the numbers and the variable in the equation? Solve the equation for x and explain the meaning of your solution.

200 square units	7.3 units

x units

9. The total area of the figure at right is 1584 square units. You can use the equation $1584 - 33x = 594$ to represent an area of 1584 square units less the area of $33x$ square units. The area remaining is 594 square units.

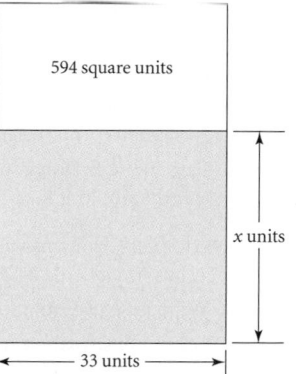

594 square units

x units

33 units

   a. What is the area of the shaded rectangle? 990 square units

   b. Write the equation you would use to find the height of the shaded rectangle. Possible answers: $33x = 990; x = \frac{990}{33}$

   c. Solve the equation you wrote in 9b to find the height of the shaded rectangle.   30 units

10. Use the "Medium" setting of the INOUT game on your calculator to produce four data tables. Copy each table and write the equation you used to match the data values in the table. [▶ 🖳 See **Calculator Note 4D.** ◀]

▶ Review

11. Show how you can solve these equations by using an undoing process. Check your results by substituting the solutions into the original equations.

   a. $-15 = -52 + 1.6x$ **b.** $7 - 3x = 52$

12. APPLICATION To plan a trip downtown, you compare the costs of three different parking lots. ABC Parking charges $5 for the first hour and $2 for each additional hour. Cozy Car charges $3 an hour, and The Corner Lot charges a $15 flat rate for a whole day.

   a. Make a table similar to the one shown. Write recursive routines to calculate the cost of parking up to 10 hours at each of the three lots.

Hours Parked	ABC Parking	Cozy Car	The Corner Lot
1	5	3	15
2	7	6	15
3	9	9	15

   b. Make three different scatter plots on the same pair of axes showing the parking rates at the three different lots. Use a different color for each parking lot. Put the hours on the horizontal axis and the cost on the vertical axis.

   c. Compare the three scatter plots. Under what conditions is each parking lot the best deal for your trip? Use the graph to explain.

   d. Would it make sense to draw a line through each set of points? Explain why or why not. No, since you have to pay for a whole hour for any fraction of the hour, the price of parking does not increase continuously.

11a.
-15	-15
Ans + 52	37
Ans/1.6	23.125
$52 + 1.6(23.125) = -15$	Check

11b.
52	52
Ans $-$ 7	45
Ans/-3	-15
$7 - 3(-15) = 52$	Check

12b.

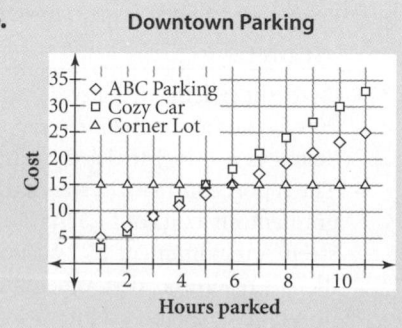

Downtown Parking

Hours parked

Exercise 8 This exercise gives an opportunity to emphasize that a direct variation is a special kind of linear equation.

8. Because height times width gives area, 7.3 and x represent the height and width, respectively. The number 200 represents the area of the rectangle in square units. The solution is about 27.4 units.

10. A sample:

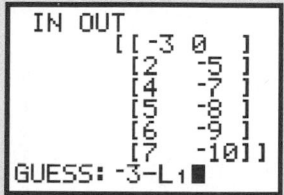

12a.

Hours	ABC	Cozy	Corner
4	11	12	15
5	13	15	15
6	15	18	15
7	17	21	15
8	19	24	15
9	21	27	15
10	23	30	15

ABC: {1,5},
{Ans(1) + 1, Ans(2) + 2}
Cozy: {1,3},
{Ans(1) + 1, Ans(2) + 3}
Corner: {1,15},
{Ans(1) + 1, Ans(2) + 0}

Exercise 12b On calculators without a color display, students can use different graph marks.

12c. For 3 hours or less, Cozy Car is the least expensive option because on the graph its line is lower than the lines of the others. For 3 to 5 hours, ABC Parking has the best price, since its graph has a lower cost in that time frame. For more than 6 hours, The Corner Lot is the least expensive option.

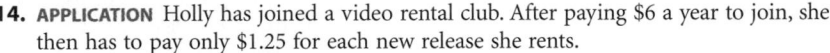

13. Today while Don was swimming, he started wondering how many lengths he would have to swim in order to swim different distances. At one end of the pool, he stopped, gasping for breath, and asked the lifeguard. She told him that 1 length of the pool is 25 yards and that 72 lengths is 1 mile. As he continued swimming, he wondered:

   a. Is 72 lengths really a mile? Exactly how many lengths would it take to swim a mile? **70.4 lengths**

   b. If it took him a total of 40 minutes to swim a mile, what was his average speed in feet per second? **2.2 feet per second**

   c. How many lengths would it take to swim a kilometer? **43.7 lengths**

   d. Last summer Don got to swim in a pool that was 25 meters long. How many lengths would it take to swim a kilometer there? How many for a mile? **40 lengths for a kilometer, 64 lengths for a mile**

14. APPLICATION Holly has joined a video rental club. After paying $6 a year to join, she then has to pay only $1.25 for each new release she rents.

   a. Write an equation in intercept form to represent Holly's cost for movie rentals. $y = 6 + 1.25x$

   b. Graph this situation for up to 60 movie rentals.

   c. Video Unlimited charges $60 for a year of unlimited movie rentals. How many movies would Holly have to rent for this to be a better deal? **44 movies**

14b.

[0, 65, 10, 0, 100, 10]

project

LEGAL LIMITS

To make a highway accessible to more vehicles, engineers reduce its steepness, also called its **gradient,** or simply grade. This highway was designed with switchbacks so the gradient would be small.

A gradient is the inclination of a roadway to the horizontal surface. Research the federal, state, and local standards for the allowable gradients of highways, streets, and railway routes.

Find out how gradients are expressed in engineering terms. Give the standards for roadway types designed for vehicles of various weights, speeds, and engine power in terms of rate of change. Describe the alternatives available to engineers to reduce the gradient of a route in hilly or mountainous terrain. What safety measures do they incorporate to minimize risk on steep grades? Bring pictures to illustrate a presentation about your research, showing how engineers have applied standards to roads and routes in your home area.

Supporting the project

MOTIVATION

Steep roads are harder to travel and harder to maintain. Engineers must consider steepness as they design roads. What guidelines and regulations do they follow?

OUTCOMES

▶ Gradient is defined. Stated as a percent, 20% means a rise of 20 ft every 100 ft. Stated as a ratio, 1 to 5 is the same as 20%.

▶ The report summarizes standards. A 20% gradient, an angle of 11.3°, is considered steep. A maximum of 15% sustained gradient is recommended. Where there is heavy snowfall, the maximum is 10%.

▶ The report includes techniques for reducing gradients and safety measures that can minimize risk on steep grades.

▶ The report includes relevant pictures of nearby roads.

• Gradients for roads are compared to gradients for railroads.

• The slope of a line is carefully defined.

Solving Equations Using the Balancing Method

Thinking in words slows you down and actually decreases comprehension in much the same way as walking a tightrope too slowly makes one lose one's balance.

LENORE FLEISCHER

In the previous two lessons, you learned about rate of change and the intercept form of a linear equation. In this lesson you'll learn symbolic methods to solve these equations. You've already seen the calculator methods of tracing on a graph and zooming in on a table. These methods usually give approximate solutions. Working backward to undo operations is a symbolic method that gives exact solutions. Another symbolic method that you can apply to solve all kinds of equations is the **balancing method.** In this lesson you'll investigate how to use the balancing method to solve linear equations. You'll discover that it's closely related to the method of working backward.

Investigation
Balancing Pennies

You will need
- pennies
- three paper cups

Here is a visual model of the equation $2x + 3 = 7$. A cup represents the variable x and pennies represent numbers. Assume that each cup has the same number of pennies in it and that the containers themselves are weightless.

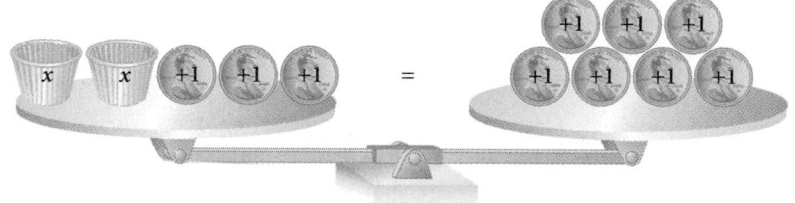

Step 1 2 Step 1
Sample explanation: Four pennies must be in the 2 cups, so there are 2 per cup.

How many pennies must be in each cup if the left side of the scale balances with the right side? Explain how you got your answer.

Your answer to Step 1 is the solution to the equation $2x + 3 = 7$. It's the number that can replace x to make the statement true. In Steps 2 and 3, you'll use pictures and equations to show stages that lead to the solution.

Step 2 $2x = 4$ Step 2

Redraw the picture above, but with three pennies removed from each side of the scale. Write the equation that your picture represents.

NCTM STANDARDS

CONTENT		PROCESS	
●	Number	●	Problem Solving
●	Algebra	●	Reasoning
	Geometry	●	Communication
●	Measurement	●	Connections
	Data/Probability	●	Representation

LESSON OBJECTIVES
- Learn the balancing method to solve equations by doing the same thing to both sides

PLANNING

LESSON OUTLINE

One day:
25 min	Investigation
5 min	Sharing
5 min	Example A
5 min	Example B
5 min	Closing
5 min	Exercises

MATERIALS
- Calculator Notes 3A, 4E
- Pan Balance (T), *optional*
- pennies (different amounts from 20 to 60 per group) or squares of construction paper or bingo chips or algebra tiles
- 10 washers, *optional*
- paper cups (three per group)

TEACHING

You can solve equations in which the unknown variable occurs only once by undoing operations. If the variable occurs more than once, however, other methods are needed. The balancing metaphor (doing the same thing to both sides) is useful for solving a variety of equations. In this lesson it's used for solving linear equations.

 Guiding the Investigation

One step Show the Pan Balance transparency. Pose the problem, "How could you use pennies and cups on a pan balance to represent and solve the equation $2x + 3 = 7$?" Be open to a variety of approaches. As needed, point out that an unbalanced scale doesn't tell as much as a balanced one. Remind students that to keep a balance they must do the same

thing to both sides. If there's time, have each group set up an equation for other groups to solve.

Step 1 If students ask, point out that the balance pictured is a pan balance, with all masses on each side concentrated on one point, as opposed to the beam balance of Chapter 3, in which the distribution of masses was important.

Steps 7 and 8 Some students may use undoing to solve the equation. Encourage them to see how working backward is used on one side of the balance to isolate the unknown.

SHARING IDEAS

At one station for groups to visit last during Steps 4 through 8, you might place 5 pennies under each of 2 cups, lay out 4 pennies and 7 washers on the same side, and put 10 pennies and 3 washers on the other side. Explain to visiting students that each washer represents -1. Then when the class is together, ask how they solved the equation $2x + 4 - 7 = 10 - 3$, or $2x - 3 = 7$. Elicit the idea that, just as they combined the 4 pennies with 4 washers and removed them, they could add 3 pennies to each side to "cancel out" the 3 washers remaining on the left. In other words, when you add a number to its opposite, you get 0, and you can remove 0's from anywhere on the balance without effect. (This wouldn't work on an actual pan balance, because the washers would have positive weights.)

If you don't have washers, ask during Sharing how you might model the equation $2x - 3 = 7$ on a pan balance. Some students may suggest that you could add 3 pennies to the 7 on the right. Point out that doing so models the equation $2x = 7 + 3$. **[Ask]** "Do the equations $2x - 3 = 7$ and $2x = 7 + 3$ have the same solution?"

Step 3 $x = 2$

Step 3 Redraw the picture, this one showing half of what was on each side of the scale in Step 2. There should be just one cup on the left side of the scale and the correct number of pennies on the right side needed to balance it. Write the equation that this picture represents. This is the solution to the original equation.

Now your group will create a pennies-and-cups equation for another group to solve.

Step 4 Divide the pennies into two equal piles. If you have one left over, put it aside. Draw a large equal sign (or form one with two pencils) and place the penny stacks on opposite sides of it.

Step 5 From the pile on one side of your equal sign, make three identical stacks, leaving at least a few pennies out of the stacks. Hide each stack under a paper cup. You should now have, on one side of your equal sign, 3 cups and some pennies.

Step 6 On the other side you should have a pile of pennies. On both sides of the equal sign you have the same number of pennies, but on one side some of the pennies are hidden under cups. You can think of the two sides of the equal sign as being the two sides of a balance scale. Write an equation for this setup, using x to represent the number of pennies hidden under one cup.

Step 7 Move to another group's setup. Look at their arrangement of pennies and cups, and write an equation for it. Solve the equation; that is, find how many pennies are under one cup without looking. When you're sure you know how many pennies are under each cup, you can look to check your answer.

Step 8 Write a brief description of how you solved the equation.

You could do problems like this on a balance scale as long as the weight of the cup is very small. But an actual balance scale can only model equations in which all the numbers involved are positive. Still, the idea of balancing equations can apply to equations involving negative numbers. Just remember, when you add any number to its opposite, you get 0. Think of negative and positive numbers as having opposite effects on a balance scale. You can remove 0 from either side of a balance-scale picture without affecting the balance. These three figures all represent 0:

$1 + (-1) = 0$

$-3 + 3 = 0$

$2x + (-2x) = 0$

EXAMPLE A | Draw a series of balance-scale pictures to solve the equation $6 = -2 + 4x$.

▶ **Solution** | The goal is to end up with a single x-cup on one side of the balance scale. One way to get rid of something on one side is to add its opposite to both sides.

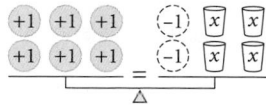

Here is the equation $6 = -2 + 4x$ solved by the balancing method:

Picture	Action taken	Equation
	Original equation.	$6 = -2 + 4x$
	Add 2 to both sides.	$6 + 2 = -2 + 2 + 4x$
	Remove the 0.	$8 = 4x$
	Divide both sides by 4.	$\dfrac{8}{4} = \dfrac{4x}{4}$
	Reduce.	$2 = x$ or $x = 2$

Balance-scale pictures can help you see what to do to solve an equation by the balancing method. But you won't need them once you get the idea of doing the same thing to both sides of an equation. And they're less useful if the numbers in the equation aren't "nice."

Assessing Progress

Observe how well students follow directions, work in groups, and understand the idea of balance.

▶ **EXAMPLE A**

This example continues the idea of working with negative numbers on a picture of a pan balance (not an actual balance).

► EXAMPLE B

This example illustrates how to solve a linear equation by four different methods. With an appropriate choice of window, the calculator methods can be made more exact but, in general, finding that window is more difficult than using the other methods.

EXAMPLE B Solve the equation $-31 = -50.25 + 1.55x$ using each method.

  a. the balancing method

  b. undoing operations

  c. tracing on a calculator graph

  d. zooming in on a calculator table

► **Solution** **a.** the balancing method

$$-31 = -50.25 + 1.55x \qquad \text{Original equation.}$$

$$-31 + 50.25 = -50.25 + 50.25 + 1.55x \qquad \text{Add 50.25 to both sides.}$$

$$19.25 = 1.55x \qquad \text{Evaluate and remove the 0.}$$

$$\frac{19.25}{1.55} = \frac{1.55x}{1.55} \qquad \text{Divide both sides by 1.55.}$$

$$12.42 \approx x, \text{ or } x \approx 12.42 \qquad \text{Reduce.}$$

b. undoing operations

Start with -31.

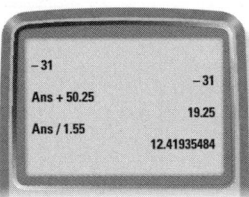

In parts a and b, if you convert the answer to a fraction, you get an exact solution of $\frac{385}{31}$.

c. tracing on a calculator graph

Enter the equation into Y1. Adjust your window settings and graph. Press TRACE and use the arrow keys to find the x-value for a y-value of -31. (See Example B in Lesson 4.6 to review this procedure.) The top screen shows that for a y-value of approximately -31.6 the x-value is 12.02.

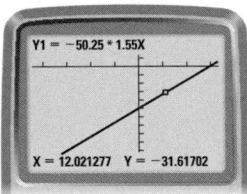 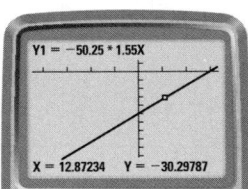

d. zooming in on a calculator table

To find a starting value for the table, use guess-and-check or a calculator graph to find an approximate answer. Then use the calculator table to find the answer to the desired accuracy. [► ▨ See **Calculator Note 3A** to review zooming in on a table. ◄]

Once you have determined a reasonable starting value, zoom in on a calculator table to find the answer using smaller and smaller values for the table increment.

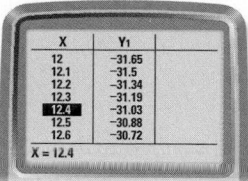 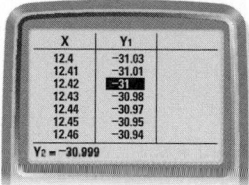

You can also check your answer by using substitution.

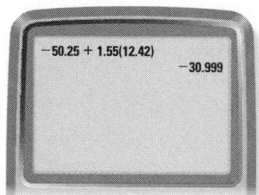

From Example B, you can see that each method has its advantages. The methods of balancing and undoing use the same process of working backward to get an exact solution. The two calculator methods are easy to use but usually give approximate solutions to the equation. You may prefer one method over others, depending on the equation you need to solve. If you are able to solve an equation using two or more different methods, you can check to see that each method gives the same result. With practice, you may develop symbolic solving methods of your own. Knowing a variety of methods, such as the balancing method and undoing, as well as the calculator methods, will improve your equation-solving skills, regardless of which method you prefer.

EXERCISES

You will need your calculator for problems **6, 8, 14,** and **15.**

▶ Practice Your Skills

1. Give the equation that each picture models.

a. $2x = 6$

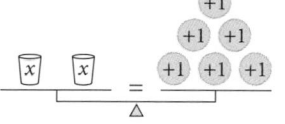

b. $x + 2 = 5$

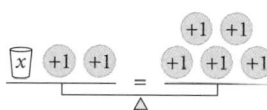

c. $2x - 1 = 3$

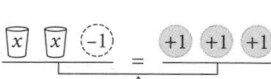

d. $2 = 2x - 3$

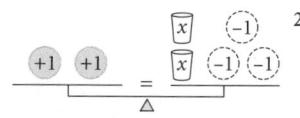

Closing the Lesson

Of the two exact methods of solving linear equations, the **balancing method**—doing the same thing to both sides of the equation—is more easily adapted to most other types of equations.

BUILDING UNDERSTANDING

Students practice solving equations by a variety of methods.

ASSIGNING HOMEWORK

Essential	**1–6, 9**
Performance assessment	**12**
Portfolio	**7, 12**
Journal	**6**
Group	**8, 10–11**
Review	**13–16**

▶ Helping with the Exercises

4. Possible answers:

4a. First multiply both sides by 10, then subtract 120 from both sides.

4b. After multiplying both sides by 3, multiply both sides by 100, then subtract 1200 from both sides, then divide both sides by 312.

Exercise 5 Most students will divide both sides of 5a by 12. For 5b, students may multiply both sides by 6, then subtract 12 from both sides, or they may subtract 2 from both sides, then multiply both sides by 6.

6b.

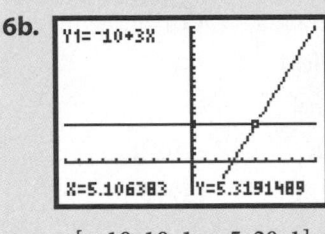

$[-10, 10, 1, -5, 20, 1]$

6c.

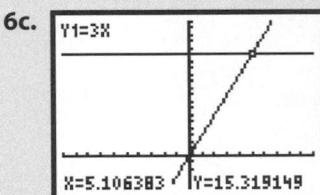

6d.

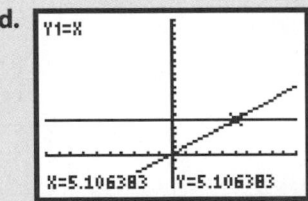

6e. Even though the lines are different in each graph, in all three graphs the x-coordinate of the intersection is the same: $x = 5$. This illustrates that transforming the equation by doing the same thing to both sides does not change the solution.

2. Copy and fill in the tables to solve the equations as in Example A.

Picture	Action taken	Equation
$\boxed{x}$ $\boxed{x}$ (−1) (−1) = +1 / +1 +1 +1	Original equation.	$2x - 2 = 4$
(+1) (+1) / $\boxed{x}$ $\boxed{x}$ (−1) (−1) = +1 +1 +1 / +1 +1 +1	Add 2 to both sides.	$2x - 2 + 2 = 4 + 2$
$\boxed{x}$ $\boxed{x}$ = +1 +1 +1 / +1 +1 +1	Remove 0 from left side.	$2x = 6$
$\boxed{x}$ $\boxed{x}$ = +1 +1 +1 / +1 +1 +1	Divide both sides by 2.	$\frac{2x}{2} = \frac{6}{2}$
$\boxed{x}$ = +1 +1 +1	Reduce.	$x = 3$

3. Give the next stage of the equation, matching the action taken, to reach the solution.

a. $0.1x + 12 = 2.2$ Original equation.

$0.1x + 12 - 12 = 2.2 - 12$ Subtract 12 from both sides.

$0.1x = -9.8$ Remove the 0 and subtract.

$x = -98$ Divide both sides by 0.1.

b. $\dfrac{12 + 3.12x}{3} = -100$ Original equation.

$12 + 3.12x = -300$ Multiply both sides by 3.

$-12 + 12 + 3.12x = -12 - 300$ Subtract 12 from both sides.

$3.12x = -312$ Remove the 0.

$x = -100$ Divide both sides by 3.12.

4. For each original equation in problem 3, tell how you could solve the equation using different steps or using the same steps in a different order.

5. Solve these equations. Tell what action you take at each stage.

    a. $144x = 12$ $x = \frac{1}{12}$ **b.** $\frac{1}{6}x + 2 = 8$ $x = 36$

Reason and Apply

6. In the solution to the equation $-10 + 3x = 17$ shown below, some of the steps are left out.

$$-10 + 3x = 5$$
$$3x = 15$$
$$x = 5$$

    a. Describe the steps that transform the original equation into the second equation and the second equation into the third (the solution). Add 10 to both sides, divide both sides by 3.

    b. Graph $Y_1 = -10 + 3x$ and $Y_2 = 5$, and trace to the lines' intersection. Write the coordinates of this point.

    c. Graph $Y_1 = 3x$ and $Y_2 = 15$, and trace to the lines' intersection. Write the coordinates of this point.

    d. Graph $Y_1 = x$ and $Y_2 = 5$, and trace to the lines' intersection. Write the coordinates of this point.

    e. What do you notice about your answers to 6b–d? Explain what this illustrates.

7. Solve the equation $4 - 1.2x = 12.4$ by using each method.

    a. balancing

    b. undoing

    c. tracing on a graph

    d. zooming in on a table

8. Write calculator routines that solve each equation by undoing. Give the solution to each one.

    a. $3 + 2x = 17$ **b.** $0.5x + 2.2 = 101.0$ **c.** $x + 307.2 = 2.1$

    d. $2(2x+2) = 7$ **e.** $\dfrac{4 + 0.01x}{6.2} - 6.2 = 0$

9. You can solve familiar formulas for a specific variable. For example, solving $A = lw$ for l you get

    $A = lw$ Original equation.

    $\dfrac{A}{w} = \dfrac{lw}{w}$ Divide both sides by w.

    $\dfrac{A}{w} = l$ Reduce.

You can also write $l = \frac{A}{w}$. Now try solving these formulas for the given variable.

    a. $C = 2\pi r$ for r $r = \dfrac{C}{2\pi}$ **b.** $A = \frac{1}{2}(hb)$ for h $h = \dfrac{2A}{b}$ **c.** $P = 2(l + w)$ for l $l = \dfrac{P}{2} - w$

    d. $P = 4s$ for s $s = \dfrac{P}{4}$ **e.** $d = rt$ for t $t = \dfrac{d}{r}$ **f.** $A = \frac{1}{2}h(a + b)$ for h

                                                                                     $h = \dfrac{2A}{a + b}$

7a. $4 - 1.2x = 12.4$ Original equation.

    $4 - 4 - 1.2x = 12.4 - 4$ Subtract 4 from both sides.

    $-1.2x = 8.4$ Subtract.

    $\dfrac{-1.2x}{-1.2} = \dfrac{8.4}{-1.2}$ Divide both sides by

                           -1.2.

    $x = -7$ Reduce.

7b. Start with 12.4.

                                    12.4

    Ans $- 4$ 8.4

    Ans$/-1.2$ -7

7c.

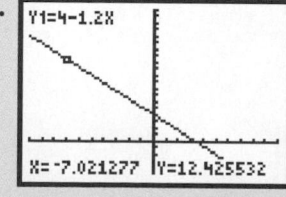

             $[-10, 10, 1, -5, 20, 1]$

Exercise 7 There are different ways to keep the equation balanced as it is solved. For example, the first step could be to multiply both sides by 10.

7d.

X	Y₁	
-3	7.6	
-4	8.8	
-5	10	
-6	11.2	
-7	12.4	
-8	13.6	
-9	14.8	

X = -7

8a. Start at 17.

  Ans $- 3$ 14

  Ans$/2$ 7

  $x = 7$

8b. Start at 101.0.

  Ans $- 2.2$ 98.8

  Ans$/0.5$ 197.6

  $x = 197.6$

8c. Start at 2.1.

  Ans $- 307.2$ -305.1

  $x = -305.1$

8d. Start at 7.

  Ans$/2$ 3.5

  Ans $- 2$ 1.5

  Ans$/2$ 0.75

  $x = 0.75$

8e. Start at 0.

  Ans $+ 6.2$ 6.2

  Ans $\cdot$ 6.2   38.44

  Ans $- 4$ 34.44

  Ans$/0.01$ 3444

  $x = 3444$

Exercise 9 As needed, remind students that lw means the product of l and w and that the answer gives the length of a rectangle in terms of its area and width. Although it's not necessary to be familiar with the other formulas in order to solve the equations, you might ask what the other formulas are used for. [a. circumference of a circle; b. area of a triangle; c. perimeter of a rectangle; d. perimeter of a square; e. distance traveled at a constant rate; f. area of a trapezoid] In part f, some students might distribute the h and then have trouble isolating h. As a hint, you can suggest treating $(a + b)$ as a single number.

12a. $y = 1 + \frac{1}{2}x$. The output value is half the input value plus 1.

x	y
0	1
1	1.5
2	2
3	2.5
4	3

12b. $y = -x$. The output value is the additive inverse (or negative) of the input value, or the sum of the input and the output value is 0.

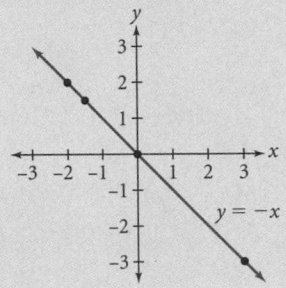

10. Tell what number, if multiplied by the given number, gives 1.

    a. 12 $\frac{1}{12}$

    b. $\frac{1}{6}$ 6

    c. 0.02   50

    d. $-\frac{1}{2}$ -2

11. Tell what number, if added to the given number, gives 0.

    a. $\frac{1}{5}$ $-\frac{1}{5}$

    b. 17 -17

    c. -2.3 2.3

    d. $-x$ x

12. You can represent linear relationships with a graph, a table of values, an equation, or a rule stated in words. Here are two linear relationships. Give all the other ways to show each relationship.

    a.

    b.

x	y
-2	2
-1.5	1.5
0	0
3	-3

▶ Review

13. **APPLICATION** Economy drapes for a certain size window cost \$90. They have shallow pleats, and the width of the fabric is $2\frac{1}{4}$ times the window width. Luxury drapes of the same fabric for the same size window have deeper pleats. The width of the fabric is 3 times the window width. What price should the store manager ask for the luxury drapes? $\frac{\$90}{2.25} = \frac{x}{3}, x = \$120$

14. Stella has decided to save to go on a trip at the end of her senior year of high school. She has decorated a glass jar and has put the \$350 that she has saved so far into the jar. The first day of every month, she is planning to put \$120 into the jar.

    a. Write a recursive routine that will generate a table showing the amount Stella will have in the jar each month for the next 12 months.   300   Ans + 120

    b. The trip Stella wants to go on will cost \$4,800. How many months will it take for her to save up enough to go on the trip?   38 months

    c. If Stella waited until September of her junior year to start saving, how much would she have to put in the jar every month in order to have saved enough by June of her senior year (assuming she still starts with \$350)?   \$205 per month for 22 months.

15. Run the LINES1 program on your calculator. [▶ ▣ See **Calculator Note 4E** to learn how to use the LINES program. ◀] Sketch a graph of the randomly generated line on your paper. Use the trace function to locate the y-intercept and to determine the rate of change. When the calculator says you have the correct equation, write it under the graph. Repeat this program until you get three correct equations in a row.

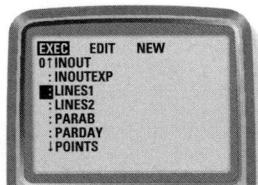

16. The local bagel store sells a baker's dozen of bagels for $6.49, while the grocery store down the street sells a bag of 6 bagels for $2.50.

a. Copy and complete the tables showing the cost of bagels at the two stores.

Bagel Store

Bagels	13	26	39	52	65	78
Cost	6.49	12.98	19.47	25.96	32.45	38.94

Grocery Store

Bagels	6	12	18	24	30	36	42	48	54	60
Cost	2.50	5.00	7.50	10.00	12.50	15.00	17.50	20.00	22.50	25.00

b. Graph the information for each market on the same coordinate axes. Put bagels on the horizontal axis and cost on the vertical axis.

c. Find equations to describe the cost of bagels at each store.

d. How much does one bagel cost at each store? How do these cost values relate to the equations you wrote in 15c?

e. Looking at the graphs, how can you tell which store is the cheaper place to buy bagels? The grocery store because its line is lower.

f. Bernie and Buffy decided to use a recursive routine to complete the tables. Bernie used this routine for the bagel store:

6.49 (ENTER)

Ans · 2 (ENTER)

Buffy says that this routine isn't correct, even though it gives the correct answer for 13 and 26 bagels. Explain to Bernie what is wrong with his recursive routine. What routine should he use?

Bernie's routine calculates each price by doubling the last. It works the first time, because if you buy twice as many bagels, you pay twice as much. But using Bernie's routine, if you buy 36 bagels at the bagel store, you pay $25.96 instead of $19.47, which amounts to paying four times as much as a single dozen instead of three times the price of a dozen. The routine should be: 6.49 (ENTER), Ans + 6.49, (ENTER), (ENTER), …

Exercise 16 As needed, remind students that a *baker's dozen* is 13.

16b.

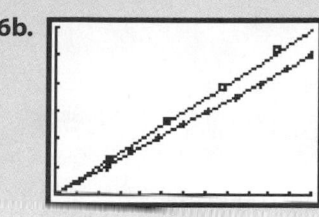

$[0, 72, 6, 0, 30, 5]$

The line with the square markers is the bagel store, and the line with the crosses is the grocery store.

16c. y represents cost; x represents number of bagels.

bagel store: $y = \dfrac{6.49}{13}x$
(or $y = 0.50x$)
grocery store: $y = \dfrac{2.50}{6}x$
(or $y = 0.42x$)

16d. Bagel store: about 50 cents per bagel; grocery store: about 42 cents per bagel. These are the coefficients of x or constants of variation in the equations.

PLANNING

LESSON OUTLINE

One day:

30 min   Activity

20 min   Improving Reasoning Skills

MATERIALS

- Pieces of rope of different lengths (around 1 m) and thickness (two per group)

- metersticks or measuring tapes (one per group)

TEACHING

Students experiment with slope and *y*-intercept of a line that fits real-world data.

Guiding the Activity

One step   Pose the problem, "Tie up to six knots in each rope, and predict the length of 10 m of rope after making 17 knots in each."

Step 1   [ESL]   The *length of the rope* after knots are tied means the distance from one end to the other when the rope is stretched tightly. If students want to use a knot other than an overhand knot, encourage the creativity, but ask them to make all the knots the same and to be sure there's enough rope to tie at least six of those knots.

Step 3   Be sure that students realize that the rate of change is negative.

Step 4   If students want to graph various lines to find the rate of change and *y*-intercept of one that seems to fit the data best, encourage them. This experimentation with slope and intercept can give them a good feel for both.

Activity Day

Modeling Data

Whenever measuring is involved in collecting data, you can expect some variation in the pattern of data points. Usually, you can't construct a mathematical model that fits the data exactly. But in general, the better a model fits, the more useful it is for making predictions or drawing conclusions from the data.

Activity
Tying Knots

You will need

- two pieces of rope of different lengths (around 1 m) and thickness

- a meterstick or a measuring tape

In this activity you'll explore the relationship between the number of knots in a rope and the length of the rope and write an equation to model the data.

Number of knots	Length of knotted rope (cm)
0	
1	
2	

Step 1   The length of rope should reduce the same amount for each knot tied.

Step 2   The data shows a linear relationship.

Choose one piece of rope and record its length in a table like the one shown. Tie 6 or 7 knots, remeasuring the rope after you tie each knot. As you measure, add data to complete a table like the one above.

Graph your data, plotting the number of knots on the *x*-axis and the length of the knotted rope on the *y*-axis. What pattern does the data seem to form?

What is the approximate rate of change for this data set? What is the real-world meaning of the rate of change? What factors have an effect on it?

Step 4   What is the *y*-intercept for the line that best models the data? What is its real-world meaning?   The *y*-intercept is the length of rope without any knots.

LESSON OBJECTIVES

- Find an equation that fits a set of real-world data

- Use a mathematical model to make predictions

NCTM STANDARDS

CONTENT		PROCESS	
•	Number	•	Problem Solving
•	Algebra	•	Reasoning
	Geometry		Communication
•	Measurement	•	Connections
	Data/Probability	•	Representation

Step 5 Graphs will vary. The line should go down from a positive y-intercept. It should pass through, or near, most of the points.

Step 5 Write an equation in intercept form for the line that you think best models the data. Graph your equation to check that it's a good fit.

Now you'll make predictions and draw some conclusions from your data using the line model as a summary of the data.

Step 6 Use your equation to predict the length of your rope with 7 knots. What is the difference between the actual measurement of your rope with 7 knots and the length you predicted using your equation?

Step 7 Use your equation to predict the length of a rope with 17 knots. Explain the problems you might have in making or believing your prediction.

Step 8 What is the maximum number of knots that you can tie with your piece of rope? Explain your answer.

Step 9 Does your graph cross the x-axis? Explain the real-world meaning, if any, of the x-value of the intersection point.

Step 10 Substitute a value for y into the equation. What question does the equation ask? What is the answer?

Step 10 The question it asks is: "How many knots are tied to produce a rope of length y?"

Step 11 Repeat Steps 1–5 using a different piece of rope. Graph the data on the same pair of axes.

Step 12 Different lengths of ropes will account for different y-intercepts. Different thicknesses will account for different rates of change. Both affect the value of the x-intercepts.

Step 12 Compare the graphs of the lines of fit for both ropes. Give reasons for the differences in their y-intercepts, in their x-intercepts, and in their rates of change.

IMPROVING YOUR **REASONING** SKILLS

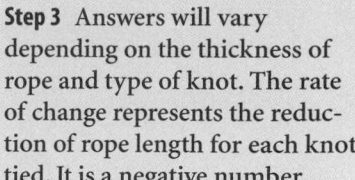

There are 100 students and 100 lockers in a school hallway. All of the lockers are closed. The first student walks down the hallway and opens every locker. A second student closes every even numbered locker. The third student goes to every third locker and opens it if it is closed, or closes it if it is open. This pattern repeats so that the nth student leaves every nth locker the opposite of how it was before. After all 100 students have opened or closed the lockers, how many lockers are left open?

Step 3 Answers will vary depending on the thickness of rope and type of knot. The rate of change represents the reduction of rope length for each knot tied. It is a negative number.

Step 5 As needed, explain that a line is a good fit to a data set if it goes pretty close to each data point and there are about as many points above the line as below the line. Students will see a more systematic way of finding lines of good fit in Lessons 5.6 and 5.7.

Step 7 The equation says that you will eventually have a rope length of zero if you tie enough knots.

Step 8 Answers will vary. Students will get to a point where they're tying knots on top of knots.

Step 9 The equation will cross the x-axis, but the points will not. It is not possible to have zero rope length no matter how many knots are tied. Even the knot itself accounts for some of the length.

Step 12 [Ask] "Does the thickness of the rope itself have any bearing on the results? Does the type of knot affect the results?"

SHARING IDEAS

Pick out any unusual approaches for sharing.

Assessing Progress

Observe students' understanding of how y-intercept and rate of change affect the graph of a line.

IMPROVING **REASONING** SKILLS

After accumulating data, many students will see the pattern that the lockers left open at the end correspond to perfect squares—1, 4, 9, 16, 25, 36, 49, 64, 81, and 100. If needed, explain that *left open* means open after all 100 students pass. Be sure they actually answer the question of how many lockers are left open (10). Ask students to explain.

[A locker changes once for each factor of its number. For example, locker number 24 is changed by students 1, 2, 3, 4, 6, 8, 12, and 24. So if a locker's number has an even number of factors, it is left closed. If a locker's number has an odd number of factors, it is left open. The numbers with an odd number of factors are the perfect squares.]

[Ask] "*Why* do perfect squares have an odd number of factors?" [Factors of numbers come in pairs: (1, 24), (2, 12), (3, 8), (4, 6). The square root of a perfect square is paired with itself—factors of 36: (1, 36), (2, 18), (3, 12), (4, 9), (6, 6). So the number of distinct factors is odd.]

CHAPTER REVIEW

4

CHAPTER

4

REVIEW

PLANNING

LESSON OUTLINE

One day:

5 min	Introduction
15 min	Exercises and helping individuals
15 min	Checking work and helping individuals
15 min	Student self assessment

REVIEWING

Direct students' attention to the table of floor heights in Example A of Lesson 4.3. **[Ask]** "What linear equation describes the floor heights?" [$y = -4 + 13x$; this is intercept form.] Then ask "What floor has height 282 feet?" To answer this, review how to set up the linear equation $-4 + 13x = 282$ and ask how to solve it. Bring out the ideas of generating the table recursively, undoing, and balancing. Also mention the distributive property and, if needed, review rules for order of operations.

ASSIGNING HOMEWORK

Assigning either the evens or the odds will give students a good review. They could work on the others in groups.

▶ Helping with the Exercises

Exercise 1 Reasons students give for each step will depend on their method of solving the equation.

You started this chapter by applying the rules for order of operations and learning how to enter numerical expressions into your calculator. You used the **distributive property** to write equivalent forms of **algebraic expressions.** You investigated **recursive sequences** by using their starting values and **rates of change** to write **recursive routines.** You saw how rates of change and starting values appear in plots.

In a walking investigation you observed, interpreted, and analyzed graphical representations of relationships between time and distance. What does the graph look like when you stand still? When you move away from or move toward the motion sensor? If you speed up or slow down? You identified real-world meanings of the **y-intercept** and the rate of change of a **linear relationship,** and used them to write the **equation** in the **intercept form** of $y = a + bx$. You learned the role of b, the coefficient of x. You explored relationships among graph sketches, tables, recursive rules, equations, and graphs.

Throughout the chapter you developed your equation-solving skills. You found **solutions** to equations by using an **undoing** process and by using a **balancing** process. You found approximate solutions by tracing calculator graphs and by zooming in on calculator tables. Finally, you learned how to model data that doesn't lie exactly on a line, and you used your model to predict inputs and outputs.

EXERCISES

You will need your calculator for problems **4, 6, 7,** and **8.**

1. Solve these equations. Give reasons for each step.

   a. $-x = 7$ $x = -7$

   b. $4.2 = -2x - 42.6$ $x = -23.4$

2. These tables represent linear relationships. For each relationship, give the rate of change, the recursive rule, the y-intercept, and the equation in intercept form.

   a.
x	y	
0	3	1
1	4	add 1
2	5	3

   $y = 3 + x$

   b.
x	y	
1	0.01	0.01
2	0.02	add 0.01
3	0.03	0

   $y = 0.01x$

   c.
x	y	
−2	1	2
0	5	add 2
3	11	5

   $y = 5 + 2x$

   d.
x	y	
−4	5	$-\frac{1}{2}$
12	−3	subtract $\frac{1}{2}$
2	2	3

   $y = 3 - \frac{1}{2}x$

3. Match these walking instructions with their graph sketches.

i. 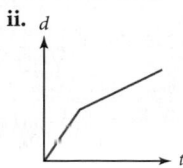 **ii.** 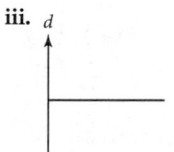 **iii.** *d*

a. The walker stands still. iii

b. The walker takes a few steps toward the zero mark, then walks away. i

c. The walker steps away from the zero mark, stops, then continues more slowly in the same direction. ii

4. Graph each equation on your calculator, and trace to find the approximate *y*-value for the given *x*-value.

a. $y = 1.21 - x$ when $x = 70.2$ $y = -68.99$

b. $y = 6.02 + 44.3x$ when $x = 96.7$ $y = 4289.83$

c. $y = -0.06 + 0.313x$ when $x = 0.64$ $y = 0.14032$

d. $y = 1183 - 2140x$ when $x = -111$ $y = 238{,}723$

5. Write the equations for linear relationships that have these characteristics.

a. The output value is equal to the input value. $y = x$

b. The output value is 3 less than the input value. $y = -3 + x$

c. The rate of change is 2.3 and the *y*-intercept is -4.3. $y = -4.3 + 2.3x$

d. The graph contains the points (1, 1), (2, 1), and (3, 1). $y = 1$

6. On a recent trip to Detroit, Tom started from home, which is 12 miles from Traverse City. After 4 hours he had traveled 220 miles.

a. Write a recursive routine to model Tom's distance from Traverse City during this trip. State at least two assumptions you're making.

b. Use your recursive routine to determine his distance from Traverse City for each hour during the first 5 hours of the trip.

c. What is the rate of change, and what does it mean in the context of this situation?

7. The profit for a small company depends on the number of bookcases it sells. One way to determine the profit is to use a recursive routine such as

{0, −850} ENTER
{Ans(1) + 1, Ans(2) + 70} ENTER ; ENTER ,...

a. Explain what the numbers and expressions 0, −850, Ans(1), Ans(1) + 1, Ans(2), and Ans(2) + 70 represent.

b. Make a plot of this situation.

c. When will the company begin to make a profit? Explain.

d. Explain the relationship between the values −850 and 70 and your graph.

e. Does it make sense to connect the points in the graph with a line? Explain why or why not. No, partial bookcases cannot be sold.

7b.

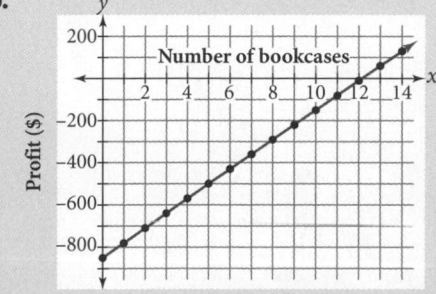

Exercise 6a If students don't know what to assume about how Tom's home is related to Traverse City and Detroit, tell them to assume something they can explain in 6b.

6a. The starting value is 12; Ans + 55.
Possible assumptions: Tom's home is 12 miles closer to Detroit than Traverse City. He travels at a constant speed. We are measuring highway distance.

6b.

Hours	0	1	2	3	4	5
Distance (mi)	12	67	122	177	232	287

6c. Tom traveled 55 miles each additional hour. The rate of change is 55 miles per hour.

7a. 0 represents no bookcases sold; −850 represents fixed overhead, such as startup costs; Ans(1) represents the previously calculated number of bookcases sold; Ans(1) + 1 represents the current number of bookcases sold, one more than the previous; Ans(2) represents the profit for the previous number of bookcases; Ans(2) + 70 represents the profit for the current number of bookcases—the company makes $70 more profit for each additional bookcase.

7c. Sample answer: The graph crosses the *x*-axis at approximately 12.1 and is positive after that. It shows you need to make at least 13 bookcases to make a profit.

7d. −850, the profit if the company makes 0 bookcases, is the *y*-intercept; 70, the amount of additional profit for each additional bookcase, is the rate of change; *y* goes up by $70 each time *x* goes up by 1 bookcase.

9e. $\dfrac{5.5 \text{ meters}}{0.6 \text{ meters/second}}$
$= 9.1\overline{6}$ seconds.
Approximately 9 seconds.

10a. Let v represent the value in dollars and y represent the number of years. $v = 5400 - 525y$

10b. The rate of change is -525. In each additional year, the value of the computer system decreases by $525.

10c. The y-intercept is 5400. The original value of the computer system is $5,400.

10d. The x-intercept is approximately 10.3. This means that the computer system no longer has value after approximately 10.3 years.

8. A single section and a double section of a log fence are shown.

a. How many additional logs are required each time the fence is increased by a single section? **3**

b. Copy and fill in the missing values in the table below.

Number of sections	1	2	3	4	...	30	...	50
Number of logs	4	7	10	13	...	91	...	151

c. Describe a recursive routine that relates the number of logs required to the number of sections. **4 (ENTER), Ans + 3 (ENTER), (ENTER), ...**

d. If each section is 3 meters long, what is the longest fence you can build with 217 logs? **216 meters**

9. The time-distance graph shows Carol walking at a steady rate. Her partner used a motion sensor to measure her distance from a given point.

a. According to the graph, how much time did Carol spend walking? **4 seconds**

b. Was Carol walking toward or away from the motion sensor? Explain your thinking. **Away; the distance is increasing.**

c. Approximately how far away from the motion sensor was she when she started walking? **approximately 0.5 meters**

d. If you know Carol is 2.9 meters away from the motion sensor after 4 seconds, how fast was she walking? $\dfrac{2.9 - 0.5}{4} = 0.6$ meter per second

e. If the equipment will measure distances only up to 6 meters, how many seconds of data can be collected if Carol continues walking at the same rate?

f. Looking only at the graph, how do you know that Carol was neither speeding up nor slowing down during her walk? **The graph is a straight line.**

10. Suppose a new small-business computer system costs $5,400. Every year its value drops by $525.

a. Define variables and write an equation modeling the value of the computer in any given year.

b. What is the rate of change, and what does it mean in the context of the problem?

c. What is the y-intercept, and what does it mean in the context of the problem?

d. What is the x-intercept, and what does it mean in the context of the problem?

11. For each table, write a formula for list L2 in terms of list L1.

a.

List 1	List 2
0	−5.7
1	−3.4
2	−1.1
3	1.2
4	3.5
5	5.8

$L_2 = -5.7 + 2.3 \cdot L_1$

b.

List 1	List 2
−3	19
−1	3
0	−5
2	−21
5	−45
6	−53

$L_2 = -5 - 8 \cdot L_1$

c.

List 1	List 2
3	13.5
−2	11
9	7.5
0	12
6	15
−5	9.5

$L_2 = 12 + 0.5 \cdot L_1$

12. Suppose Andrei and his younger brother are having a race. Because the younger brother can't run as fast, Andrei lets him start out 5 meters ahead. Andrei runs at a speed of 7.7 meters per second. His younger brother runs at 6.5 meters per second. The total length of the race is 50 meters.

   a. Write an equation to find how long it will take Andrei to finish the race. Solve the equation to find the time.

   b. Write an equation to find how long it will take Andrei's younger brother to finish the race. Solve the equation to find the time.

   c. Who wins the race? How far ahead was the winner at the time he crossed the finish line?

13. Consider the equation $2(x - 6) = -5$.

   a. Show two different ways you can solve the equation. Solution methods will vary; $x = 3.5$.

   b. Show how you can check your result by substituting it into the original equation.

$$2(3.5 - 6) = 2(-2.5) = -5$$

14. Solve each equation using the method of your choice. Then use a different method to verify your solution.

   a. $14x = 63$ $x = 4.5$

   b. $-4.5x = 18.6$ $x = -4.1\overline{3}$

   c. $8 = 6 + 3x$ $x = 0.\overline{6}$

   d. $5(x - 7) = 29$ $x = 12.8$

   e. $3(x - 5) + 8 = 12$ $x = 6.\overline{3}$

15. The equation $w = -38.3 + 1.4t$ approximates the wind chill temperatures in °F for a wind speed of 20 miles per hour. Find the actual temperature t for each given wind chill temperature w. Verify your results with another method.

   a. $w = -40°$ $t \approx -1.2°$

   b. $w = 15°$ $t \approx 38.1°$

   c. $w = -23°$ $t \approx 10.9°$

   d. $w = 0°$ $t \approx 27.4°$

12a. $50 = 7.7t$;

$t = \dfrac{50}{7.7} \approx 6.5$ seconds

12b. $50 = 5 + 6.5t$;

$t = \dfrac{50 - 5}{6.5} \approx 6.9$ seconds

12c. Andrei wins. When Andrei finishes, his younger brother is $50 - (5 + 6.5(6.5)) \approx 2.8$ meters from the finish line.

Exercise 14 Students can also check their results by substituting them into the equation.

▶ Take Another Look

The graph shows the change in the hiker's speed over time. The contour map shows the change in the hiker's elevation as she follows the dotted line. It also shows the horizontal distance she has traveled from her starting point.

The graph shows rate of change in speed as the steepness (slope) of a line. At first the speed is steadily decreasing, then it is increasing, then it remains constant. The contour map shows rate of change in elevation by the distance between the contour lines. When the lines are very close, the elevation is changing quickly.

If a scale were provided, you could infer distance from the graph by estimating the average speed up to a certain point and multiplying that by the time at that point. Students might refer to the formula distance equals rate (speed) times time ($d = rt$). You could measure distance on the contour map using the map scale.

Student sketches of the graph of (*distance, time*) should show three sections. In each the graph is increasing. In the first section, the graph is a curve that is concave down showing a steadily decreasing speed. The second section of the graph is concave up since the speed is increasing, and the third section is a steep straight line showing a constant, fast speed.

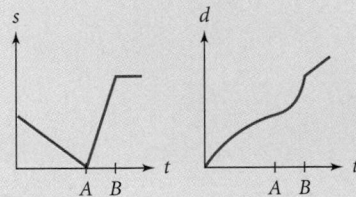

The graph and the contour map together show you why the speed of the hiker was decreasing at the beginning. She was climbing a hill. As she began to go down the hill, she kept moving faster and faster until she was running.

TAKE ANOTHER LOOK

▶ The picture at right is a **contour map** that reveals the character of the terrain. All points on an **isometric line** are the same height in feet above sea level. The graph below shows how the hiker's walking speed changes as she covers the terrain on the dotted-line trail shown on the map.

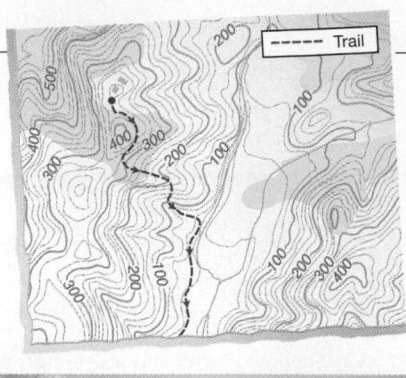

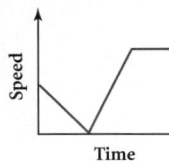

What quantities are changing in the graph and in the map?

How does each display reveal rate of change?

How could you measure distance on each display?

What would the graph sketch of this hike look like if distance were plotted on the vertical axis instead of speed?

What do these two displays tell you when you study them together?

Sediment layers form contour lines in the Grand Canyon.

IMPROVING YOUR REASONING SKILLS

Did these plants grow at the same rate? If not, which plant was tallest on Day 4? Which plant took the most time to reach 8 cm? Redraw the graphs so that you can compare their growth rates more easily.

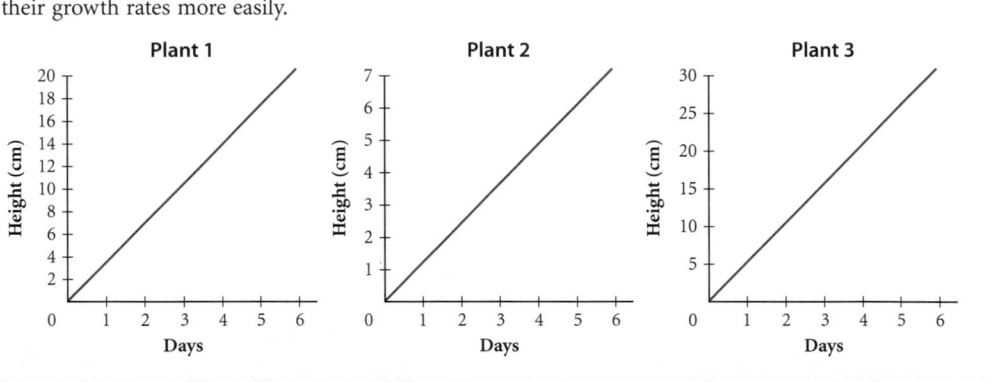

IMPROVING REASONING SKILLS

The differences in the vertical scale (height) indicate that the plants are growing at different rates. Plant 3 is the fastest growing; it is about 20 cm tall in 4 days. Plant 2 is slowest growing, it takes almost 6 days to reach 7 cm. This should suggest to students that the steepness of a line is relative to the scales on the axes. They have seen this many times on their calculators.

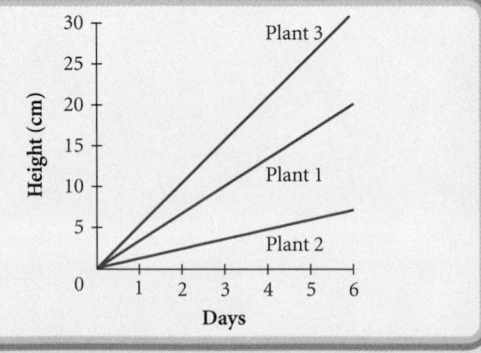

Assessing What You've Learned

GIVING A PRESENTATION

Making presentations is an important career skill. Most jobs require workers to share information, to help orient new coworkers, or to represent the employer to clients. Making a presentation to the class is a good way to develop your skill at organizing and delivering your ideas clearly and in an interesting way. Most teachers will tell you that they have learned more by trying to teach something than they did simply by studying it in school.

Here are some suggestions to make your presentation go well:

▶ Work with a group. Acting as a panel member might make you less nervous than giving a talk on your own. Be sure the role of each panel member is clear so that the work and the credit are equally shared.

▶ Choose the topic carefully. You can summarize the results of an investigation, do research for a project and present what you've learned and how it connects to the chapter, or give your own thinking on Take Another Look or Improving Your Reasoning Skills.

▶ Prepare thoroughly. Outline your presentation and think about what you have to say on each point. Decide how much detail to give, but don't try to memorize whole sentences. Illustrate your presentation with models, a poster, a handout, or overhead transparencies. Prepare these visual aids ahead of time and decide when to introduce them.

▶ Speak clearly. Practice talking loudly and clearly. Show your interest in the subject. Don't hide behind a poster or the projector. Look at the listeners when you talk.

Here are other ways to assess what you've learned:

ORGANIZE YOUR NOTEBOOK You might need to update it with examples of undoing and balancing to solve an equation, or with notes about how to trace a line or search a table to approximate the coordinates of the solution. Be sure you understand the meanings of important words like linear equation, rate of change, and intercept form.

WRITE IN YOUR JOURNAL What method for solving equations do you like better? Do you always remember to define variables before you graph or write an equation? How are you doing in algebra generally? What things don't you understand?

UPDATE YOUR PORTFOLIO Choose a piece of work you did in this chapter to add to your portfolio—your graph from the investigation On the Road Again (Lesson 4.4), your longest number trick with its undo operations, or your research on a project.

You can use either Form A or Form B of the chapter test, or you can use Constructive Assessment items. Using the CD test generator, you can create an alternate version of the test or combine some items from Form A or Form B with Constructive Assessment items.

FACILITATING SELF ASSESSMENT

During Sharing, students have had some opportunity to present ideas to the class. In a presentation, they will get practice at planning, preparing visual aids, and then presenting a topic of their choosing.

To help students complete the portfolio described in Assessing What You've Learned, suggest that they consider for evaluation their work on Lesson 4.1, Exercise 8; Lesson 4.2, Exercise 11 or 13; Lesson 4.3, Exercise 7 or 12; Lesson 4.4, Exercise 7; Lesson 4.6, Exercise 6; Lesson 4.7, Exercise 10; and Lesson 4.8, Exercise 7 or 12.

CHAPTER 5

Fitting a Line to Data

Overview

In the context of finding lines of fit, Chapter 5 emphasizes slope. **Lesson 5.1** presents a formula for determining slope using any two points on a line. Students learn about all four slope types: positive, negative, zero (horizontal), and undefined (vertical). In **Lesson 5.2,** students use their understanding of the intercept form from Chapter 4 to fit lines to data. **Lessons 5.3** and **5.4** introduce students to the point-slope form and how to apply it. In **Lesson 5.5,** students apply their understanding of the point-slope form to fit lines to data. **Lessons 5.6** and **5.7** establish and develop the method for determining lines of fit based on quartiles of the two data sets. This standardized procedure allows everyone to get the same equation to model a given set of data. In **Lesson 5.7,** students compare and evaluate methods of fitting lines to data. **Lesson 5.8** is an activity day for reviewing lines of fit.

The Mathematics

Slope

Previous chapters related the constant rate of change of a data set or equation to the steepness of a line graphing the data or equation. Now we see a definition of the slope of a line through two points, as the ratio of the difference in y-coordinates to the difference in corresponding x-coordinates, $\frac{change\ in\ y}{change\ in\ x}$ or $\frac{y_2 - y_1}{x_2 - x_1}$.

This formula gives a static way of thinking about slope. A dynamic concept can also be useful. Imagine that you're moving from left to right along the line. Page 255 describes this dynamic concept of slope and gives ideas on helping students understand dividing by zero.

Point-Slope Form

The point-slope form of an equation is often given as $\frac{y - y_1}{x - x_1} = m$ where m is the given slope and (x_1, y_1) is a point on the line. This form emphasizes that for "any point on the line" (x, y), the slope of the line segment between that point and the given point is the ratio $\frac{y - y_1}{x - x_1}$.

Using b instead of m to represent the slope, you can write the point-slope form as $y - y_1 = b(x - x_1)$, obtained from the previous form by multiplying both sides by the denominator. You can think of this equation as a translation of the direct variation $y = bx$ to the left by an amount x_1 and down by an amount y_1.

Neither of these forms lends itself to graphing on a graphing calculator, for which y must be expressed in terms of x. This difficulty is answered by the form the student text uses, $y = y_1 + b(x - x_1)$.

You can also see the equation $y = y_1 + b(x - x_1)$ as the result of translating the variation $y = bx$ to the right x_1 units and up y_1 units. When you slide up, you add to the y. But if you also slide the line to the right, the line now goes through the point (x_1, y_1) and the line will cross the y-axis below the y_1.

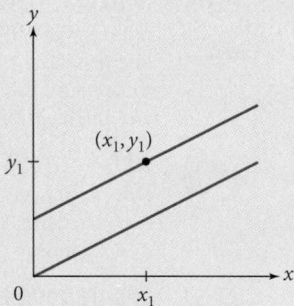

In practice, the point-slope form (in any version) is more useful than the intercept (or slope-intercept) form. Rarely do you know the y-intercept. If you need to find it, you can substitute 0 for x_1 in the point-slope form. So resist any impulses to change all linear equations into an intercept form.

Lines of Fit

A *line of fit,* or *regression line,* is a straight line that is intended to represent the points in a data set. You use it when you're trying to make a prediction based on a scatter plot that looks linear. You might use several methods of finding lines of fit. Very common is the *method of least squares,* in which the sum of squares of vertical distances between a line and the data points is minimized. This method is implemented on many graphing calculators. It will be studied in statistics courses and advanced algebra.

In *Discovering Algebra*, students find the slope b of the line between two representative points, graph the variation $y = bx$, and then translate that line to an appropriate y-intercept so that it appears to represent the data. This gives the intercept form, $y = a + bx$, of a line. Students also get the point-slope form by translating the line $y = bx$ to pass through one of the two points chosen to find the slope.

A third method presented in the student text, uses *Q-points*. For each of the four Q-points, the first coordinate is a first or third quartile of the x-coordinates of the data set, and the second coordinate is the first or third quartile of the set of y-coordinates: (x_{Q1}, y_{Q1}), (x_{Q3}, y_{Q1}), (x_{Q1}, y_{Q3}), and (x_{Q3}, y_{Q3}). These four points form a rectangle. The line of fit lies along one diagonal.

Another method of linear regression is implemented on calculators and might be an appropriate extension for students in this course. It's called the *median-median method*. In it, you divide each set of coordinates into thirds. You use the medians of the outer thirds to determine a line. Then you shift that line vertically $\frac{1}{3}$ of the way toward the point given by the medians of the middle thirds. (In practice, the middle thirds may have one point more or less than the outer thirds.)

Using This Chapter

The Activity Day is a very popular activity. Though no new material is presented, try to make time for it or combine it with the Chapter Review.

Resources

Discovering Algebra Resources

Calculator Notes 1C, 4E, 5A, 5B

Teaching and Worksheet Masters
  Lesson 5.1 The Four Slope Types
  Lesson 5.1 Coordinate Plane
  Lesson 5.4 Properties of Numbers
  Lesson 5.4 Properties of Equality

Assessment Resources *A* and *B*
  Quiz 1 (Lessons 5.1 to 5.3)
  Quiz 2 (Lessons 5.4 and 5.5)
  Quiz 3 (Lessons 5.6 and 5.7)
  Chapter 5 Test
  Chapter 5 Constructive Assessment Options

More Practice Your Skills for Chapter 5

Condensed Lessons for Chapter 5

Other Resources

Tim Erickson, *Data in Depth, Exploring Mathematics with Fathom!* Emeryville, California: Key Curriculum Press, 2000.

www.keymath.com/DA

Materials

- graph paper
- rulers
- uncooked spaghetti
- several books
- 5-oz plastic cups
- string
- pennies (100 per group)
- empty boxes (one per group), *optional*
- sharp scissors (or an awl)
- stopwatches
- bucket or other object to pass
- toy figures
- identical rubber bands (300)
- tape measures, metersticks, or yardsticks
- video camera, *optional*

Pacing Guide

	day 1	day 2	day 3	day 4	day 5	day 6	day 7	day 8	day 9	day 10
standard	5.1	5.2	5.2	5.3	5.4, quiz	5.4	5.5	quiz, 5.6	5.6	5.7
enriched	5.1	project, 5.2	5.2	5.3	5.4	quiz, 5.4	5.5	quiz, 5.6	5.6, project	5.7
block	5.1, 5.2	5.2, 5.3	5.4	5.5, quiz	5.6	5.7, project	5.8	review, assessment		

	day 11	day 12	day 13	day 14	day 15	day 16	day 17	day 18	day 19	day 20
standard	5.8	5.8	review	assessment						
enriched	5.8	5.8	review, TAL	assessment						

5 Fitting a Line to Data

CHAPTER 5 OBJECTIVES

- Learn how to calculate the slope of a line with slope triangles and the slope formula

- Learn about slopes of rising, falling, horizontal, and vertical lines

- Learn the point-slope form of an equation of a line

- Check the equivalence of linear equations by using algebraic properties

- Learn several approaches to finding a line that represents a set of real-world data points (eyeballing, representative points, Q-points)

- Evaluate the results of those approaches

Artists, like mathematicians, use lines to summarize their observations. An artist's data include contour, texture, color, shape, motion, and balance. The American artist Romaine Brooks (1874–1970) reduced her entire set of observations into the lines you see in this pencil sketch titled *Departure*.

OBJECTIVES

In this chapter you will

- define and calculate slope
- write an equation that fits a set of real-world data
- review the intercept form of a linear equation
- learn the point-slope form of a linear equation
- recognize equivalent equations written in different forms

You can ask the class to brainstorm other meanings of the word *line*. Some of these are boundary, course, limit, queue, series, and trend. Point out that, in colloquial use, a line is not always straight. In mathematics, *line* usually means a straight line or segment, whereas *curve* refers to a bending of a line.

A mathematical curve can be thought of geometrically, as a locus of points, or algebraically, as a set of points satisfying an equation. Curves that model numeric data are usually algebraic so that we can make predictions from the equation. Geometric curves can also be considered to be modeling data—data that are points rather than numbers.

In visual art, a line might be used in various ways. For example, an *outline* calls attention to the shape, as this artist has emphasized the shape the human figure assumes in this position. A *contour* gives the illusion of three-dimensionality; the continuation of the line of the lower leg into the interior of the foot shape gives the clue that the toe is farther from us than the heel. A *gesture* line is executed fast recording the path of the eye and depicting motion.

LESSON

5.1

A Formula for Slope

The nearest thing to nothing that anything can be and still be something is zero.

ANONYMOUS

You have seen that the steepness of a line can be a graphical representation of a real-world rate of change like a car's speed, the number of calories burned with exercise, or a constant relating two units of measure. Often you can estimate the rate of change of a linear relationship just by looking at a graph of the line. Can you tell which line in the graph matches which equation?

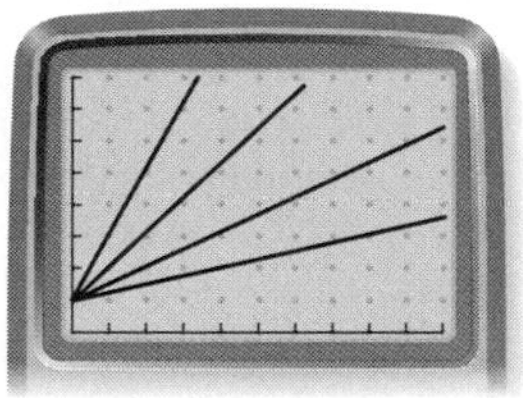

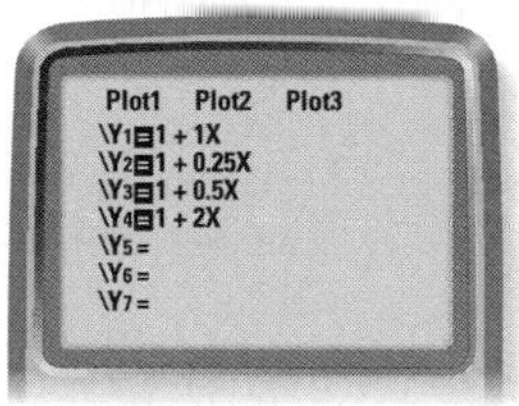

Slope is another word used to describe the steepness of a line or the rate of change of a linear relationship. In this investigation you will explore how to find the slope of a line from two points on the line.

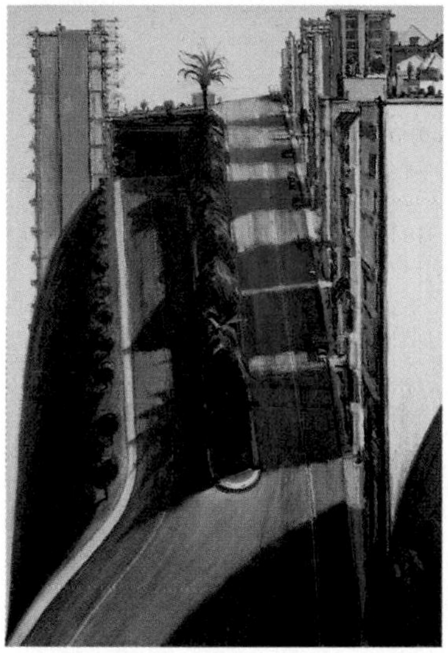

Wayne Thiebaud's oil painting *Urban Downgrade, 20th and Noe* (1981) is an artistic representation of the steepness, or slope, of a street in San Francisco, California. Thiebaud is an American artist born in 1920.

Investigation
Points and Slope

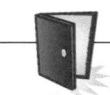

You will need
- graph paper

Hector recently signed up with a limited-usage Internet provider. He knows that there is a flat monthly charge and an hourly rate for the number of hours he is connected during the month. The table shows the amount of time he spent using the Internet for the first three months and the total fee he was charged.

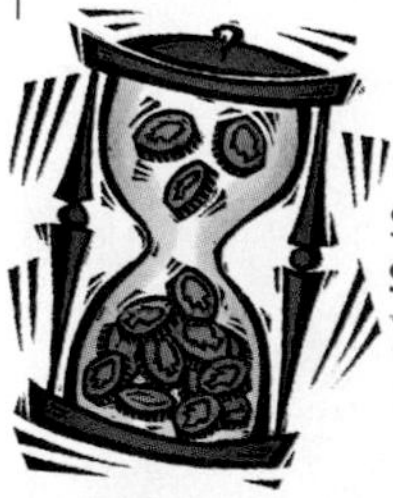

Internet Use

Month	Time (hr)	Total fee ($)
September	4	16.75
October	5	19.70
November	8	28.55

Step 1 Is there a linear relationship between the time in hours that Hector uses the Internet and his total fee in dollars? If so, why do you think such a relationship exists?

Step 1 Yes

Step 2 Use the numbers in the table to find the rate in dollars per hour. Explain in words how you calculated this rate.

Step 2 \$2.95 per hour

LESSON

5.1

PLANNING

LESSON OUTLINE

One day:

5 min	Introduction
20 min	Investigation
5 min	Sharing
10 min	Example and Visual Learning
5 min	Closing
5 min	Exercises

MATERIALS

- Calculator Note 4E
- graph paper
- rulers
- The Four Slope Types (T), *optional*
- Coordinate Plane (T from Chapter 1)

TEACHING

The steepness of the graph of line $y = a + bx$ is measured by the rate of change b, called the *slope* of the line.

INTRODUCTION

[Ask] "How are the lines on page 251 alike and how do they differ? How are these similarities and differences reflected in the equations?" [All the lines have a y-intercept of 1, but the slopes are different. The steepest line will match the equation with the largest rate of change. From bottom to top, the corresponding equations are Y2, Y3, Y1, and Y4.]

[ESL] A common idea of *slope* is a piece of ground that isn't horizontal, such as a ski slope. In mathematics, slope is a number measuring steepness.

NCTM STANDARDS

CONTENT		PROCESS	
•	Number	•	Problem Solving
•	Algebra	•	Reasoning
	Geometry	•	Communication
•	Measurement		Connections
	Data/Probability	•	Representation

LESSON OBJECTIVES

- Investigate and solve real-world problems that involve the slope of a line
- Learn how to calculate slopes with slope triangles and the slope formula
- Learn about slopes of rising, falling, horizontal, and vertical lines

One step Point out Hector's bill and ask students to write careful instructions for finding the slope of a line between any two given points. As students work, suggest that they use slope triangles.

Step 1 Students can decide the relationship is linear either by graphing or by finding that the rate of change (in $/hour) is constant.

Step 2 [Alert] Some students may not realize that there is a monthly charge. This relationship is not a direct variation and the intercept will not be zero.

Step 2 Explanations should include increase total fee and divide by the increase in time. If students select only the consecutive points $(4, 16, 75)$ and $(5, 19, 70)$, they may not be aware of the division by 1.

Step 3 Ask students to label their axes with their respective units of measure as well as to mark them numerically.

Step 3 The line should support the linear relationship students assumed in Step 1.

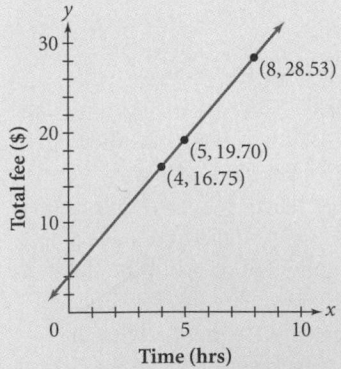

Step 4 [Alert] Many students may not understand how to find the lengths of the arrows. Suggest that they draw horizontal and vertical lines to mark equal lengths on the axes. Use the Coordinate Plane transparency to demonstrate this process.

Students may draw the horizontal arrow before the vertical arrow, and the slope triangle may lie below the line.

Step 3 Draw a pair of coordinate axes on graph paper. Use the x-axis for time in hours and the y-axis for total fee in dollars. Plot and label the three points the table of data represents. Draw a line through the three points. Does this line support your answer in Step 1?

Step 4 Choose two points on your graph. Use arrows to show how you could move from one point to the other using only one vertical move and one horizontal move. How long is each arrow? What is the real-world meaning of each length?

Step 5 The arrows show the change in total fee and time usage, which are divided to find the rate. The slope is $2.95/hour, as in Step 2.

Step 5 How do the arrow lengths relate to the hourly rate that you found in Step 2? Use the arrow lengths to find the hourly rate of change, or slope, for this situation. What units should you apply to the number?

In Step 4, you used arrows to show the vertical change and the horizontal change when you moved from one point to another. The right triangle you created is called a **slope triangle.**

Step 6 Any pair of points will give the same slope.

Step 6 Choose a different pair of points on your graph. Create a slope triangle between them and use it to find the slope of the line. How does this slope compare to your answers in Step 2 and Step 5?

Step 7 Subtracting corresponding coordinates gives the arrows' lengths. Using $(4, 16.75)$ and $(8, 28.55)$, a numerical expression would be $\frac{28.55 - 16.75}{8 - 4}$.

Step 7 Think about what you have done with your slope triangles. How could you use the coordinates of any two points to find the vertical change and the horizontal change of each arrow? Write a single numerical expression using the coordinates of two points to show how you can calculate slope.

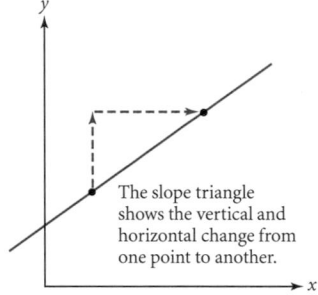

The slope triangle shows the vertical and horizontal change from one point to another.

Step 8 $\frac{y_2 - y_1}{x_2 - x_1}$ or $\frac{y_1 - y_2}{x_1 - x_2}$

Step 8 Write a symbolic algebraic rule for finding the slope between any two points (x_1, y_1) and (x_2, y_2). The subscripts mean that these are two distinct points of the form (x, y).

A slope triangle helps you visualize slope by showing you the vertical change and the horizontal change from one point to another. These changes are also called the "change in y" (vertical) and the "change in x" (horizontal). The example will help you see how to work with positive and negative numbers in slope calculations.

The steps going up this pyramid (located in the ruins of the ancient Mayan city of Chichen Itza) are like slope triangles that define the slope of the pyramid's face.

Step 4 Arrows should form a right triangle. With the points $(5, 19.70)$ and $(8, 28.55)$ as an example, the horizontal arrow should have a length of 3 hours. This length shows the real-world change in time usage between the 2 months.

Step 5 [Alert] Some students may divide the horizontal change by the vertical change. As a guide, have them think about the units of the rate: $ per hour, not hours per $.

Step 7 Students revisit what may have caused confusion in Step 4. When you think someone might not understand deeply enough, ask for explanations.

Step 8 This may be the first time that students have seen subscripted variables. Point out that they could have used (a, b) and (c, d), but (x_1, y_1) and (x_2, y_2) are in the form of coordinates.

EXAMPLE

Consider the line through the points (1, 7) and (6, 4).

a. Find the slope of the line.

b. Without graphing, verify that the point (4, 5.2) is also on that line.

c. Find the coordinates of another point on the same line.

▶ **Solution**

Plot the given points and draw the line between them.

a.

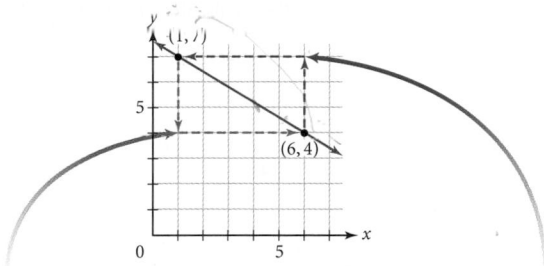

If you move from (1, 7) to (6, 4), the change in y is -3 (down 3) and the change in x is $+5$ (right 5). The slope is $\frac{-3}{+5}$.

If you move from (6, 4) to (1, 7), the change in y is $+3$ (up 3) and the change in x is -5 (left 5). The slope is $\frac{+3}{-5}$.

But $\frac{-3}{+5}$ is equivalent to $\frac{+3}{-5}$. You get the same slope, $-\frac{3}{5}$ or -0.6, no matter which point you start from. The slope triangles help you see this relationship more clearly.

Move to
(6, 4) from (1, 7).

$$\text{Slope} = \frac{4-7}{6-1} = \frac{-3}{5} = -\frac{3}{5} \quad \text{or}$$

Move to
(1, 7) from (6, 4).

$$\text{Slope} = \frac{7-4}{1-6} = \frac{3}{-5} = -\frac{3}{5}$$

b. The slope between any two points on the line will be the same. So, if the slope between the point (4, 5.2) and either of the original two points is -0.6, then the point is on the line. The slope between (4, 5.2) and (1, 7) is

$$\frac{7-5.2}{1-4} = \frac{1.8}{-3} = -\frac{1.8}{3} = -0.6$$

So the point (4, 5.2) is on the line.

SHARING IDEAS

Request that students present Steps 5 and 8. Ask why the ratio of differences gives the slope. Keep asking "Are you saying . . . ?" to clarify thinking and help all students understand these important but difficult ideas.

[Ask] "Is a linear equation a good model?" [The relationship applies only to whole numbers, so the points on the line between the whole numbers have no meaning.]

Mention various descriptions for slope. For example, "rise over run," "vertical change over horizontal change," or "the change in y over the change in x." Draw out the formula (as written at the bottom of page 254) and write the slope using deltas, $\frac{\Delta y}{\Delta x}$. Delta is the Greek letter "D," standing here for "difference." [Language] A *delta* also is defined as a triangular deposit built up at the mouth of a river.

If students worked on the project Legal Limits from Lesson 4.7, point out that slope is the same as gradient. Slope is usually given as a fraction and gradient as a percent. A 9% gradient is a slope of $\frac{9}{100}$.

[Ask] "If unlimited service is $21.95 per month, do you think Hector should change pricing plans or stay with limited usage?"

Assessing Progress

Note students' understanding of **linear relationships** and their abilities to find **rates** and **graph points** and to treat horizontal and vertical components separately.

EXAMPLE

This example provides a method for finding the slope of a line between two points. It also introduces the notion that a point C is on the same line as points A and B if and only if the slope of the segment AC is the same as the slope of the segment AB.

If students do not understand part a, **[Ask]** "Is there only one line through the points $(1, 7)$ and $(6, 4)$?" [Yes] **[Link]** In geometry, students will learn the postulate "Two points determine a line."

To find the slope when moving from $(1, 7)$ to $(6, 4)$, the coordinates of $(1, 7)$ will come last in the slope formula. Saying the move to $(6, 4)$ from $(1, 7)$ mentions the points in the order they appear in the slope formula.

Student confusion in part b may be due to a switch in the logic: If point A is on a line with slope b, and so is point B, then the slope of the segment AB is also b. Now students are being asked to use a partial converse: If point A is on a line with slope m and the slope of segment AB is also b, then point B is also on the line.

[Ask] "If you know the equation for the line (say $y = 7.6 - 0.6x$), how might you find another point on the line?" Encourage a variety of approaches: using the slope and $(0, 7.6)$, substituting a number for x and determining y, graphing, etc. Have students work on the last two questions for a few minutes to get $(1 - 5, 7 + 3)$ $= (-4, 10)$ and either $(1 + 1, 7 - 0.6) = (2, 6.4)$ or $(6 + 1, 4 - 0.6) = (7, 3.4)$.

VISUAL LEARNING

As needed, review division with signed numbers. Students may forget, for example, that the quotient of two negative numbers is a positive number.

Emphasize that either point can be called (x_1, y_1), as long as subtraction between corresponding coordinates takes place in the same order in the numerator and the denominator.

c. You can find the coordinates of another point by adding the change in x and the change in y from any slope triangle on the line to a known point.

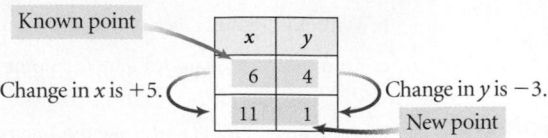

You could start with the point $(6, 4)$ and use

$$\frac{change\ in\ y}{change\ in\ x} = \frac{-3}{5}$$

This gives the new point $(6 + 5, 4 + (-3)) = (11, 1)$.

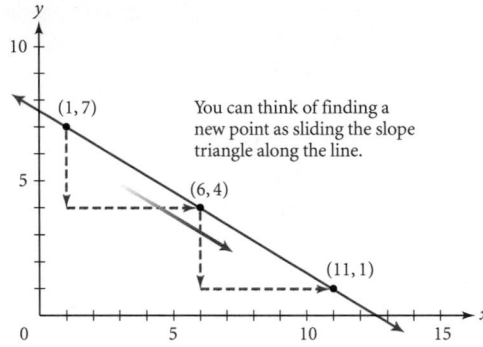

You can think of finding a new point as sliding the slope triangle along the line.

Try using point $(1, 7)$ and using

$$\frac{change\ in\ y}{change\ in\ x} = \frac{3}{-5} \quad (-4, 10)$$

to find another point. Try using either original point and using

$$\frac{change\ in\ y}{change\ in\ x} = \frac{-0.6}{1} \quad (7, 3.4) \text{ or } (2, 6.4)$$

to find another point.

Slope is an extremely important concept in mathematics and in applications like medicine and engineering that rely on mathematics. You may encounter different ways of describing slope—for example, rise over run or vertical change over horizontal change. But you can always calculate the slope using this formula.

History
• CONNECTION •

Slope is sometimes written $\frac{\Delta y}{\Delta x}$. The symbol Δ is the Greek capital letter delta. The use of Δ is linked to the history of calculus in the 18th century when it was used to mean "difference."

Slope Formula

The formula for the **slope** of the line passing through point 1 with coordinates (x_1, y_1) and point 2 with coordinates (x_2, y_2) is

$$slope = \frac{change\ in\ y}{change\ in\ x} = \frac{y_2 - y_1}{x_2 - x_1}$$

Visual learners will also be helped by a graphic display such as:

	x	y
	x_1	y_1
	x_2	y_2

$x_2 - x_1$ ⟨ ... ⟩ $y_2 - y_1$

Be sure all students now recognize that a is the intercept and b is the slope in $y = a + bx$.

A line that goes up from left to right has a positive slope. A line that goes down from left to right has a negative slope.

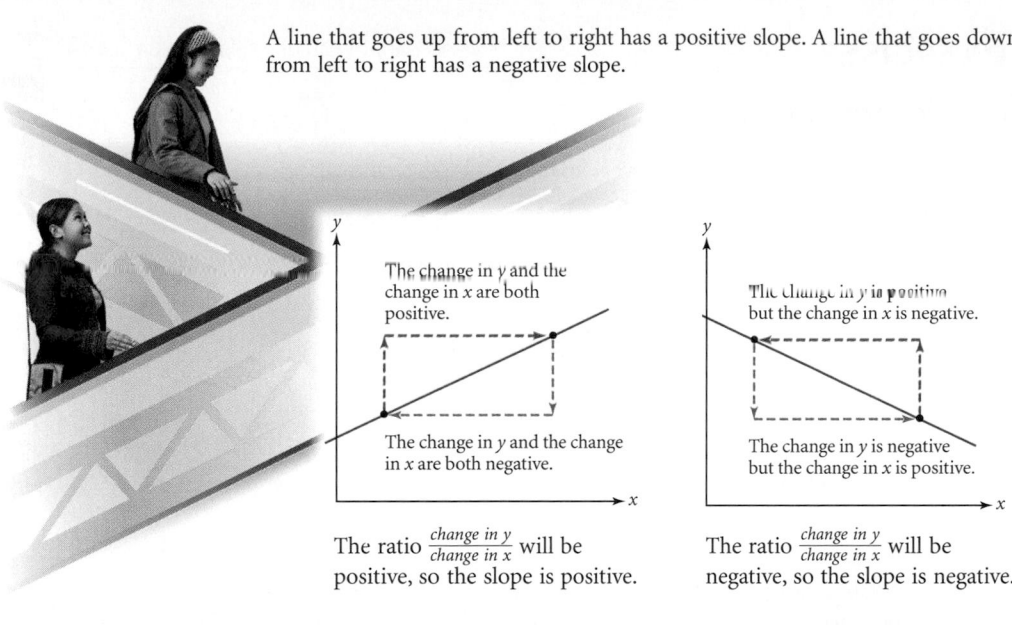

The change in y and the change in x are both positive.

The change in y and the change in x are both negative.

The ratio $\frac{change\ in\ y}{change\ in\ x}$ will be positive, so the slope is positive.

The change in y is positive but the change in x is negative.

The change in y is negative but the change in x is positive.

The ratio $\frac{change\ in\ y}{change\ in\ x}$ will be negative, so the slope is negative.

Horizontal lines have a slope of zero because they have no change in y. Vertical lines have no change in x. To calculate the slope of a vertical line, you would have to divide by zero, which is impossible—we say that the slope of a vertical line is undefined.

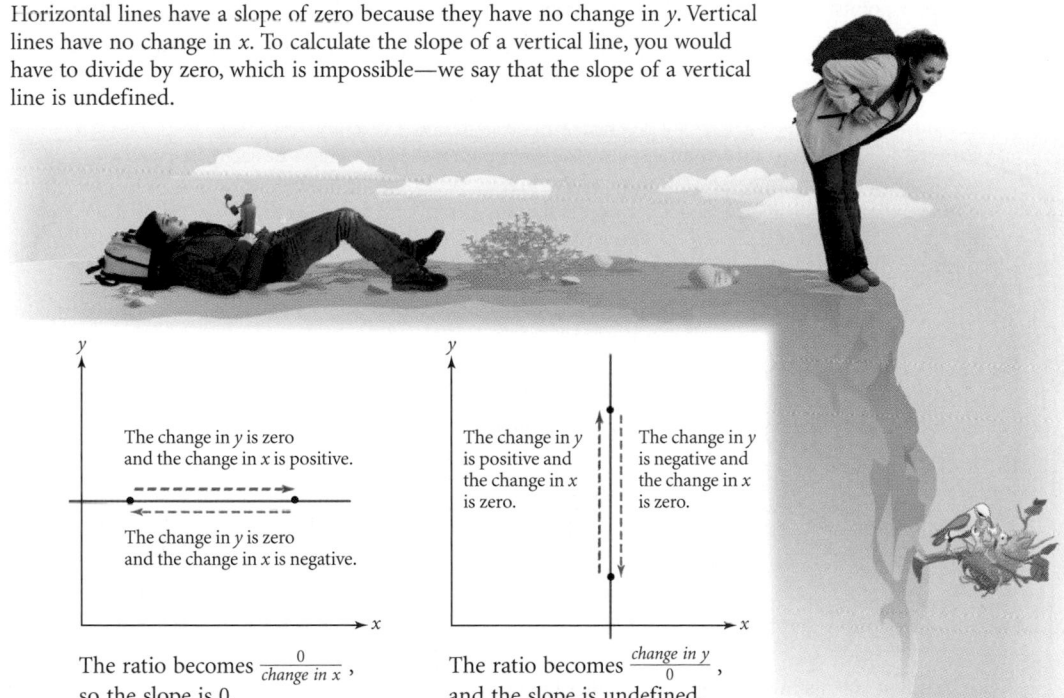

The change in y is zero and the change in x is positive.

The change in y is zero and the change in x is negative.

The ratio becomes $\frac{0}{change\ in\ x}$, so the slope is 0.

The change in y is positive and the change in x is zero.

The change in y is negative and the change in x is zero.

The ratio becomes $\frac{change\ in\ y}{0}$, and the slope is undefined.

As you work on the exercises, keep in mind that the slope of a line is the same as the rate of change of its equation. When a linear equation is written in intercept form, $y = a + bx$, which letter would represent the slope? *b*

Dynamic Concept of Slope

Imagine that you're moving from left to right along the line. What happens to your y-coordinate? If your y-coordinate is increasing, the slope is positive. You're climbing up the line. The larger the slope, the harder you have to work to make that climb. If your y-coordinate is decreasing, you're going downhill. The more negative the rate of change, the steeper the hill.

The dynamic approach can help students understand slopes of horizontal and vertical lines, which can be especially confusing. If your y-coordinate doesn't change, you're moving "on the level," so the line is horizontal. Its slope is 0. If, on the other hand, your motion can't be from left to right at all because the line is vertical, it's impossibly steep. The slope is undefined. If you calculate the slope of a vertical line by the static model, you're dividing by 0, the difference in x-coordinates of two points on the line.

A dynamic algebra exploration at www.keymath.com/DA can help students visualize what happens to the slope and the slope triangle as a line changes.

Dividing by Zero

If students ask why it is impossible to divide by 0, you might say, "If it were possible to divide 5 by 0, then the answer multiplied by 0 would equal 5." Or you might ask, "What number multiplied by 5 would equal 0?" If a student asks about the quotient $\frac{0}{0}$, say that $\frac{179}{541}$ (or any other number) multiplied by 0 equals 0. Any number could be the answer, so $\frac{0}{0}$ isn't defined. With any number that isn't 0, dividing by 0 isn't defined because no number can be the answer.

Closing the Lesson

As needed, say that the **slope of a line** can be calculated by finding two points on the line and dividing the vertical change between those points by the corresponding horizontal change. You can use the transparency the Four Slope Types.

BUILDING UNDERSTANDING

Students work with slopes to graph lines and to determine which points lie on given lines. They also relate slopes to rates of change in real-world problems.

ASSIGNING HOMEWORK

Essential	1–6, 9, 11
Performance assessment	8
Portfolio	7
Journal	10, 11c
Group	8
Review	12–15

▶ Helping with the Exercises

2a. 1.5; one possible point is (6, 10).

2b. −1.5; one possible point is (4, 2).

2c. 0; any point with a *y*-coordinate of 4 will be on the line.

2d. 1.875; one possible point is (17, 27).

Exercise 3 This exercise stresses that the slope of a line is constant and can be calculated from any two points. You might ask students to use slope triangles rather than memorize the formula.

Exercise 4 This program is a fun way for students to gain experience identifying the equations of lines.

Exercise 5 Be prepared to point out that the *x*-coordinates don't change on a vertical line.

5a. i. The *x*-values don't change, so the slope is undefined.

5a. ii. The *y*-values decrease as the *x*-values increase, so the slope is negative.

EXERCISES

You will need your calculator for problems **4** and **14.**

▶ Practice Your Skills

1. Find the slope of each line using a slope triangle or the slope formula.

a. 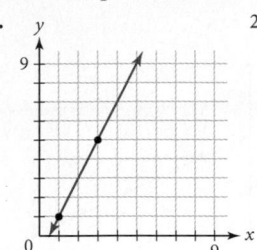   2   **b.** $\frac{2}{3}$ 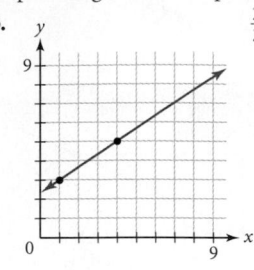   **c.** $-\frac{4}{3}$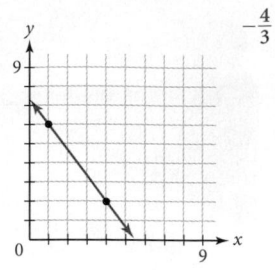

2. Find the slope of the line through each pair of points. Then name another point on the same line.

  a. (2, 4), (4, 7)   **b.** (6, −1), (2, 5)   **c.** (−2, 4), (8, 4)   **d.** (1, −3), (9, 12)

3. Given the slope of a line and one point on the line, name *two* other points on the same line. Then use the slope formula to check that the slope between each of the two new points and the given point is the same as the given slope. Possible answers:

  a. Slope $\frac{3}{1}$; point (0, 4)   (1, 7), (−1, 1)   **b.** Slope −5; point (2, 8)   (3, 3), (1, 13)

  c. Slope $-\frac{3}{4}$; point (8, 6)   (12, 3), (4, 9)   **d.** Slope 0.2; point (5, 7)   (6, 7.2), (4, 6.8)

4. Run the LINES1 program five times. On your own paper, sketch a graph of each randomly generated line. Find the slope of the line by counting the change in *y* and the change in *x* on the grid, or trace the line for two points to use in the slope formula. Then find the *y*-intercept and write the equation of the line in intercept form. [▶ 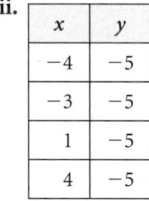 See **Calculator Note 4E** to learn how to use the LINES program. ◀] Answers will vary.

▶ Reason and Apply

5. Each table gives the coordinates of four points on a different line.

i.

x	y
4	−8
4	0
4	3
4	20

ii.

x	y
0	5
1	3
3	−1
4	−3

iii.

x	y
−4	−5
−3	−5
1	−5
4	−5

iv.

x	y
−4	−5
−2	−3.5
0	−2
4	1

  a. Without calculating, can you tell whether the slope of the line through each set of points is positive, negative, zero, or undefined? Explain how you can tell.

  b. Choose two points from each table and calculate the slope. Check that your answer is correct by calculating the slope with a different pair of points.

  c. Write an equation for each table of values.

5a. iii. The *y*-values don't change, so the slope is 0.

5a. iv. The *y*-values increase as the *x*-values increase, so the slope is positive.

5b. i. Using the points (4, 0) and (4, 3), the slope is $\frac{3-0}{4-4} = \frac{3}{0}$. Since you can't divide by 0, the slope is undefined.

5b. ii. Using the points (1, 3) and (4, −3), the slope is $\frac{-3-3}{4-1} = \frac{-6}{3} = -2$.

5b. iii. Using the points (−4, −5) and (−3, −5), the slope is $\frac{-5-(-5)}{-3-(-4)} = \frac{-5+5}{-3+4} = \frac{0}{1} = 0$.

5b. iv. Using the points (0, −2) and (4, 1), the slope is $\frac{1-(-2)}{4-0} = \frac{3}{4}$.

5c. i. $x = 4$; **ii.** $y = 5 - 2x$; **iii.** $y = -5$; **iv.** $y = -2 + \frac{3}{4}x$

6. Two lines have been graphed for you.

   a. How are the lines in the graph alike? How are they different?

   b. Which line matches the equation $y = -3 + \frac{2}{5}x$? line b

   c. What is the equation of the other line? $y = 1 + \frac{2}{5}x$

   d. How are the equations alike? How are they different?

7. APPLICATION Recall Hector's Internet use from the investigation. You probably found that his provider charges $2.95 per hour of use— that was the slope of the line you graphed.

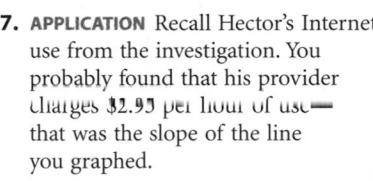

Internet Use

Month	Time (hr)	Total fee ($)
September	4	16.75
October	5	19.70
November	8	28.55

   a. Use the rate of change and the data in the table to find out how much the total fee is for 3 hours of use. How much is the total fee for 2 hours?

   b. Repeat the process in 7a to find out how much the total fee is for 0 hours of use. What is the real-world meaning of this number in this situation? (Look back at the investigation for help.) Continuing the process in 7a leads to (0, 4.95), or $4.95 for 0 hours. This is the flat monthly rate for Hector's Internet service.

   c. A mathematical model can be an equation, a graph, or a drawing that helps you better understand a real-world situation. Write a linear equation in intercept form that you can use to model this situation. $y = 4.95 + 2.95x$, where x is time in hours and y is total fee in dollars.

   d. Use your linear equation to find out how much the total fee is for 28 hours of use. Substitute 28 for x and solve for y: $y = 4.95 + 2.95(28) = 87.55$. $87.55 for 28 hours.

Many professions rely on models to understand real-world situations. For example, an architectural model like this one would be used to design a building. Mathematical models are another type of model that help people understand the real world.

Exercise 6 Students will often use the fact that lines are parallel when they have the same slope.

6a. The lines are parallel, so they have equal slope. The y-intercepts are different.

6d. The slope, $\frac{2}{5}$, is the same in each equation. The y-intercepts, -3 and 1, are different.

7a. Use the slope to move backward from (4, 16.75); $(4 - 1, 16.75 - 2.95) = (3, 13.80)$, or $13.80 for 3 hours; $(3 - 1, 13.80 - 2.95) = (2, 10.85)$, or $10.85 for 2 hours.

Exercise 7b This approach to finding the y-intercept by working backward is referred to in Lesson 7.3.

Exercise 7c Mathematical models have limitations. For example, the line contains points with negative x-values, but they don't make sense in the situations being modeled.

Exercise 8 Some students may be confused about how the height of the balloon can already be 14 m at time 0. If possible, ask the class to explain it. There can be several reasons. The timing need not start when the balloon starts rising. Or, the measurement might be to the top of the balloon, which might very well be 14 m high when the basket is still on the ground. In case the balloon has not started rising, the time-distance relationship would not be perfectly linear. To explain, you might note that when you start walking, you have some acceleration before you obtain any speed. Hence, the beginning of the graph would be slightly curved.

[Ask] "How can you tell whether the data are linear?" [Slope between pairs of points is constant.]

8b. m/min; the hot-air balloon rise at a rate of 30 m/min

Exercise 9 Some students may try measuring to find a numerical answer. Point out that when distances are represented by letters, students should express their answer in terms of those letters. [ESL] *In terms of* means *using*.

The slope can't be written using only those letters, however. A negation sign is needed because the slope is negative. Stress again that a negative slope means the line is falling *from left to right*. Any of the expressions

$y = e - \frac{a}{c}x, y = e + \frac{-a}{c}x,$ or $y = e + \frac{a}{-c}x$ are okay.

---

8. **APPLICATION** A hot-air balloonist gathered the data in this table.

Hot-Air Balloon Height

Time (min)	Height (m)
0	14
2.2	80
3.4	116
4	134
4.6	152

a. What is the slope of the line through these points? 30 m/min
b. What are the units of the slope? What is the real-world meaning of the slope?
c. Write a linear equation in intercept form to model this situation. $y = 14 + 30x$
d. What is the height of the balloon after 8 minutes? 254 meters
e. During what time interval is the height less than or equal to 500 meters? between 0 and 16.2 minutes

9. If a and c are the lengths of the vertical and horizontal segments and $(0, e)$ is the y-intercept, what is the equation of the line?

$y = e - \frac{a}{c}x$

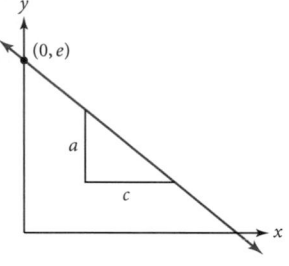

10. This line has a slope of 1. Graph it on your own paper.

a. Draw a slope triangle on your line. How do the change in y and the change in x compare?
b. Draw a line that is steeper than the given line. How do the change in y and the change in x compare? How does the numerical slope compare to that of the original line?
c. Draw a line that is less steep than the given line. How do the change in y and the change in x compare? How does the slope compare to the given line?
d. How would a line with a slope of -15 compare to your other lines? Explain your reasoning.

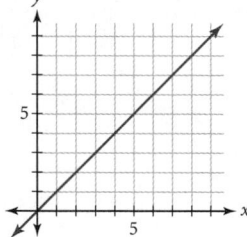

10a. The change in y and the change in x are the same for any slope triangle on the line.

10b. A steeper line would have a greater change in y than its change in x. Numerically the slope would be greater than 1.

10c. A less steep line would have a greater change in x than its change in y. Numerically the slope would be between 0 and 1.

10d. The line will decrease from left to right because the slope is negative; the line would be very steep because 15 is significantly greater than 1.

11. When you make a scatter plot of real-world data, you may see a linear pattern.

Exercise 11 This is a good exercise for leading up to Lesson 5.2.

a. Which line do you think "fits" each scatter plot? Think about slope and how the points are scattered. Explain how you chose your lines.

i. Line 2 is a better choice.

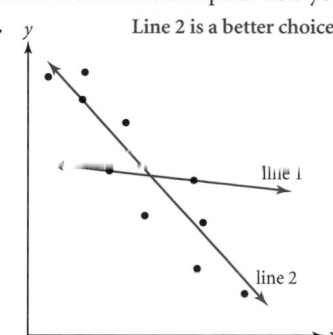

ii. Either line 3 or line 4 is a reasonable fit.

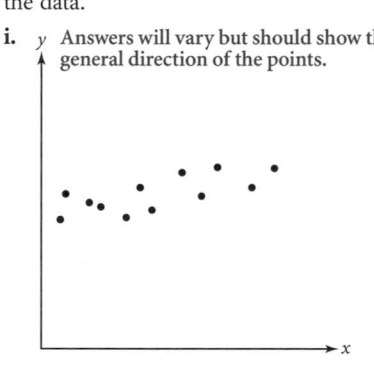

b. Sketch each scatter plot on your own paper. Then draw a line that you think fits the data.

i. *y* Answers will vary but should show the general direction of the points.

ii. *y* The actual points show no linear pattern, so a line is not an appropriate model.

c. List two features that you think are important for a line that fits data.
Answers will vary. The line should reflect the direction of the data, and about the same number of points should be above the line as below it.

▶ Review

12. Use the distributive property to rewrite each expression without using parentheses.

a. $3(x - 2)$ $3x - 6$

b. $-4(x - 5)$ $-4x + 20$

c. $-2(x + 8)$ $-2x - 16$

Exercise 12 This exercise reviews Lesson 4.1.

13. Calista has five brothers. The mean of her brothers' ages is 10 years, and the median is 6 years. Create a data set that could represent the brothers' ages. Is this the only possible answer? {3, 3, 6, 16, 22}; no

Exercise 13 This exercise reviews Lesson 1.2.

STEP RIGHT UP PROJECT

[ESL] The *tread* of a stair step is the horizontal part. The *riser* is the vertical piece connecting treads.

General information about treads and risers is available in home-improvement books.

---

14. Enter $\{-3, -1, 2, 8, 10\}$ into list L_1 on your calculator.

   a. Write a rule for list L_2 that adds 14 to each value in list L_1 and then multiplies the results by 2.5. What are the values in list L_2? $\ L_2 = 2.5(L_1 + 14);\ \{27.5, 32.5, 40, 55, 60\}$

   b. Write a different rule for list L_2 that produces the same results as the one you wrote in 14a but that doesn't use parentheses. Do you get the same results? $\ L_2 = 2.5L_1 + 35;$ yes

   c. Write a rule for list L_3 that works backward and undoes the operations in list L_2 to produce the values in list L_1. $\ L_3 = \frac{(L_2 - 35)}{2.5}$ or $L_3 = \frac{L_2}{2.5} - 14$

15. Convert each decimal number to a percent.

   a. 0.85 85%   **b.** 1.50 150%

   c. 0.065 6.5%   **d.** 1.07 107%

project

STEP RIGHT UP

How would it feel to climb a flight of stairs if every step was a little higher or lower than the previous one? The constant measure for treads and risers on most stairs keeps you from tripping. Have you noticed that the stairs outside some public buildings slow you down to a "ceremonial" pace? Or that little-used stairs to a cellar seem dangerously steep? Investigate the standards for stairs in various architectural settings and learn the reasons for their various slopes.

Your project should include

▶ Tread-and-riser data and slope calculations for several different stairways.

▶ The building codes or recommended standards in your area for home stairways. Is a range of slopes permitted? When are landings or railings required?

▶ Scale drawings for at least three different stairways.

After you've done your research, consider this question: Does a spiral staircase have a constant slope?

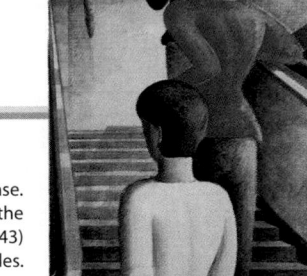

Slope triangles are like the steps of a staircase. This oil painting, *Bauhaus Stairway*, by the German artist Oskar Schlemmer (1888–1943) shows many slope triangles.

Bauhaus Stairway (1932) Oil on canvas, 63 7/8 × 45 in. The Museum of Modern Art, New York. Gift of Philip Johnson. Photograph © 2000 The Museum of Modern Art, New York

---

Supporting the project

MOTIVATION

If you were going to design a stairway for a child's play structure, would you use the same standards as for an adult house? If not, what would be the difference and why?

OUTCOMES

▶ The report is accurate on all required elements.

▶ The report on building codes cites sources.

▶ The report includes spiral staircases. To determine their slope, take measurements at the same distance from the center of the circle the staircase is based on.

▶ The scale for scale drawings of different stairways is marked.

• The report discusses safety.

• The paper lists reasonable factors for determining the slope within the allowable range, such as frequency of use or user age and agility.

Writing a Linear Equation to Fit Data

When you plot real-world data, you will often see a linear pattern. However, the points will rarely fall exactly on a line. How can you tell if a particular line is a good model for the data? One of the simplest ways is to ask yourself if the line shows the general direction of the data and if there are about the same number of points above the line as below the line. If so, then the line will appear to "fit" the data, and we call it a **line of fit.**

Sometimes one line will be a better model for your data than another. Each of these graphs shows a scatterplot of data points and possible lines of fit.

Although these birds are not in a line, can you visualize a line that shows the flocks' general direction?

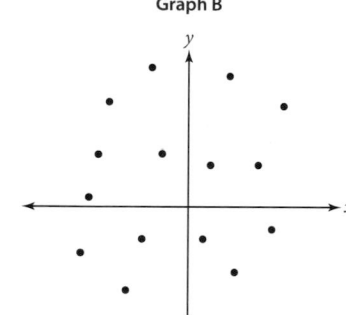

Graph A

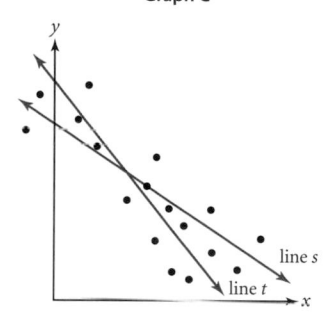

Graph B

Graph C

In Graph A, line *p* fits better because it shows the general direction of the data and there are the same number of points above the line as below the line. Although line *q* goes through several points, it does not show the direction of the data.

In Graph B, the data doesn't seem to have a pattern. No lines of fit are shown because you can't say that one line would fit the data better than another line would.

In Graph C, both lines show the general direction of the data and both lines have the same number of points above and below them. You could consider either line a line of fit. When making predictions, how would your calculations using the equation for line *s* differ from those using the equation for line *t*?

In the next investigation you will learn one method to find a possible line of fit.

PLANNING

LESSON OUTLINE

First day:

15 min   Introduction

35 min   Investigation (Steps 1 to 8)

Second day:

10 min   Investigation (Steps 9 to 12)

10 min   Sharing

20 min   Example

5 min   Closing

5 min   Exercises

MATERIALS

- graph paper (several sheets per group)
- a box of (uncooked) spaghetti, 10 inches long
- several books for each group
- 5-oz plastic cups (one per group)
- string
- pennies (100 per group)
- empty boxes (one per group), *optional*
- sharp scissors or awls (to punch cups for string)

TEACHING

A *line of fit* models a data set to allow predictions. In this lesson students work with slopes and intercepts to gain an intuitive feeling for finding lines of fit, otherwise known as *trend lines*.

NCTM STANDARDS

CONTENT	PROCESS
• Number	• Problem Solving
• Algebra	• Reasoning
Geometry	• Communication
Measurement	• Connections
• Data/Probability	• Representation

LESSON OBJECTIVES

- Draw a line that fits or models a set of points
- Write an intercept equation that fits a set of real-world data

Guiding the Investigation

Procedure note Using similar books allows students to make stacks of the same height. Facilitate cleanup by putting a box under each experiment to catch the broken spaghetti. Students should put a book on top of each end of the spaghetti.

One step Ask students to experiment with 1 to 6 strands of spaghetti and predict how many pennies 17 strands of spaghetti will support and predict how many spaghetti strands will be needed to support $5 worth of pennies. As you circulate, suggest that they lay out spaghetti strands on the graphs they create to help visualize lines of fit.

Step 1 The strings should be tied short to keep the cups off the base.

Step 2 As students devise their own tables for recording data, suggest that they think about the input and output variables in deciding how to title the columns.

[Language] The *maximum load* is the number of pennies in the cup *before* the one is added that breaks the beam.

Step 3 You may want to ask, "Would your results be the same if you moved the books closer or farther apart?" "Are your results with spread-out spaghetti strands different from those with bunched-up strands?"

Steps 4 to 7 These steps are to be worked on individually. Results will be shared with the group in Step 8.

Students might also make an accurate sketch on graph paper and then use a strand of spaghetti to visualize the line of fit. To get to an equation in intercept form, they might draw a slope triangle for the spaghetti and approximate where the spaghetti crosses the *y*-axis.

Investigation
Beam Strength

How strong do the beams in a ceiling have to be? How do bridge engineers select beams to support traffic? In this investigation you will collect data and find a linear model to determine the strength of various "beams" made of spaghetti.

You will need
- graph paper
- uncooked spaghetti
- several books
- a plastic cup
- string
- pennies

Steps 1 to 7 Answers will vary.

Step 1 Make two stacks of books of equal height. Punch holes on opposite sides of the cup and tie the string through the holes.

Step 2 Follow the procedure note for a beam made from one strand of spaghetti. Record the maximum load (the number of pennies) that this beam will support.

Step 3 Repeat Step 2 for beams made from two, three, four, five, and six strands of spaghetti.

Procedure Note

1. Hang your cup at the center of your spaghetti beam.
2. Support the beam between the stacks of books so that it overlaps each stack by about 1 inch. Put another book on each stack to hold the beam in place.
3. Put pennies in the cup, one at a time, until the beam breaks.

Step 4 Plot your data on your calculator. Let x represent the number of strands of spaghetti, and let y represent the maximum load. Sketch the plot on paper too.

Step 5 Use a strand of spaghetti to visualize a line that you think fits the data on your sketch. Choose two points on the line. Note the coordinates of these points. Calculate the slope of the line between the two points.

Step 6 Use the slope, b, that you found in Step 5 to graph the equation $y = bx$ on your calculator. Why is this line parallel to the direction the points indicate? Is the line too low or too high to fit the data?

Step 8 Students should see that everyone could come up with a different line of fit. Some lines may pass through data points, but it is not a requirement.

Step 7 Using the spaghetti strand on your sketch, estimate a good y-intercept, a, so that the equation $y = a + bx$ better fits your data. On your calculator, graph the equation $y = a + bx$ in place of $y = bx$. Adjust your estimate for a until you have a line of fit.

Step 8 In Step 5, everyone started with a visual model that went through two points. In your group, compare all final lines. Did everyone end up with the same line? Do you think a line of fit must go through at least two data points? Is any one line better than the others?

Step 4 When making a scatter plot on a calculator, a box or plus mark makes it easier to see a line of fit than does a dot mark. You might encourage students to use list L1 for input data and list L2 for output data whenever possible, to avoid having to specify lists each time they make a plot.

Step 7 You may want to write $y = bx$ as $y = 0 + bx$ to help students see the connection to $y = a + bx$ and the y-intercept of zero. As needed, remind them of the meaning of y-intercept.

Your line is a model for the relationship between the number of strands of spaghetti in the beam and the load in pennies that the beam can support.

Step 9 The additional number of pennies each additional strand of spaghetti can hold.

Step 9 Explain the real-world meaning of the slope of your line.

Step 10 Use your linear model to predict the number of spaghetti strands needed to support $5 worth of pennies.

Step 11 Use your model to predict the maximum loads for beams made of 10 and 17 strands of spaghetti.

Step 12 Errors in data collection or a poor line of fit.

Step 12 Some of your data points may be very close to your line, while others could be described as outliers. What could have caused these outliers?

Engineers conduct tests using procedures similar to the one you used in your investigation. The test results help them select the best materials and sizes for beams in buildings, bridges, and other forms of architecture.

Despite engineering tests, buildings can suffer damage during stress. This building in San Francisco, California, collapsed during an earthquake in October, 1989.

EXAMPLE

This table shows how many fat grams there are in some hamburgers sold by national chain restaurants.

This is not a real hamburger but a ceramic sculpture. Can you imagine how much total fat this burger would have if it were real?

Hamburger (1983) by David Gilhooly, Collection of Harry W. and Mary Margaret Anderson, Photo by M. Lee Fatheree

Nutrition Facts

Burger	Saturated fat (g)	Total fat (g)
Burger King "Big King"	18	43
Burger King "Double Cheeseburger with Bacon"	18	39
Burger King "Whopper Jr."	8	24
Burger King "Whopper with Cheese"	16	46
Hardee's "The Works"	12	30
Jack in the Box "Jumbo Jack with Cheese"	14	40
McDonald's "Arch Deluxe with Bacon"	12	34
McDonald's "Big Mac"	10	28
Wendy's "Big Bacon Classic"	12	30
Wendy's "Single with Everything"	7	20

(*Consumer Reports*, Dec. 1997, pp. 12–13)

a. Find a linear equation to model the data.

b. Tell the real-world meaning of the slope and intercept of your line.

c. Predict the total fat in a burger with 20 g of saturated fat.

d. Predict the saturated fat in a burger with 50 g of total fat.

SHARING IDEAS

Pick students who will present a variety of answers to Steps 10 and 11. (If students are putting their scatter plots on a calculator with an overhead projection panel, a spaghetti strand on the panel is good for showing a line of fit.) Point out that lines of fit that are slightly different on a graph can produce drastically different results when extended very far. **[Language]** You may want to introduce the term *extrapolate*, to estimate outside the observed range. Ask the class to decide which results they find most believable and why. Motivate the need for a nonsubjective, standard way to find a line of best fit, foreshadowing Lesson 5.6.

If the opportunity arises, you might ask how to get the equation by shifting the line through the origin not just up but also to the right, to go through one of the two points chosen. This could motivate Lesson 5.3.

You might say that a mathematical model or a line of fit is an idealization. Rarely will the data fit the model exactly. In fact, in the case of beams, the strength is proportional to the cross-sectional area (and thus to the square of the number of spaghetti strands) and inversely proportional to the amount of the beam between supports. So a linear model isn't a very good predictor beyond a few strands.

Assessing Progress

Watch for students' understanding of **input-output tables** and their skills at making scatter plots, collecting data systematically, working with a group, finding the **slope** and *y*-**intercept** of a line, writing the intercept form of the equation of a line, **evaluating an equation** at a point, and **solving a linear equation.**

EXAMPLE

This example is for students who had difficulties with the investigation.

The scatter plot shows 9 points for the 10 data points because one of the points, (12, 30), is a double point.

In the solution to part a, students may ask how to determine how far the line should be raised. Point out that they can use the graph to guess at the *y*-intercept. The line $y = 8.8 + 1.9x$ goes through the points used to find the slope, but that line has more data points above the line than below.

[Context] Saturated fat increases the amount of cholesterol in the blood, leading to increased risk of heart disease and hardening of the arteries. Unsaturated fat doesn't increase cholesterol. Students can find links to updated nutrition facts about fast food at www.keymath.com/DA.

Health
CONNECTION

Most foods contain three types of fats—saturated, polyunsaturated, and monounsaturated—in varying amounts. Saturated fat is a health concern because it raises blood cholesterol and increases the risk of heart disease.

▶ **Solution**

Draw a scatter plot of the data. Let *x* be the number of grams of saturated fat, and let *y* be the total number of grams of fat.

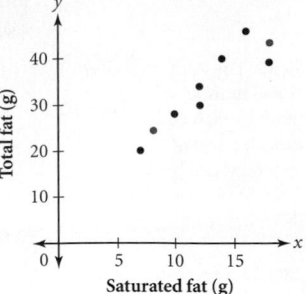

a. The scatter plot shows a linear pattern in the data. A line through the points (8, 24) and (18, 43) seems to show the direction of the data. Calculate the slope *b* of the line between these two points. Use (8, 24) as (x_1, y_1) and (18, 43) as (x_2, y_2).

$$b = \frac{y_2 - y_1}{x_2 - x_1} = \frac{43 - 24}{18 - 8} = \frac{19}{10} = 1.9$$

Substitute 1.9 for *b* in $y = bx$ to get

$$y = 1.9x$$

The equation $y = 1.9x$ shows the direction of the line, but it is too low.

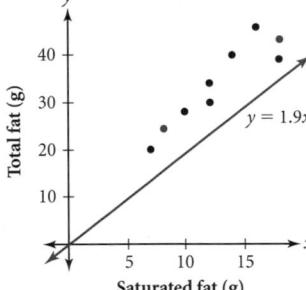

Estimate how far up to raise the line. The *y*-intercept is somewhere around 10. Adjust the intercept by tenths until you are satisfied. You may find that the equation

$$y = 9.4 + 1.9x$$

is a good model. Notice, however, that the line of fit doesn't have to go through any data points.

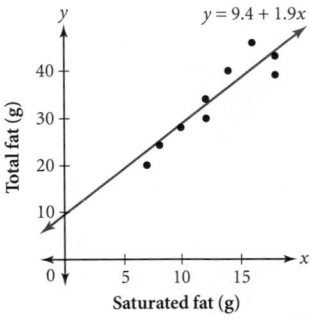

b. The *y*-intercept, 9.4, means that even without any saturated fat, a burger has about 9.4 grams of total fat. The slope, 1.9, means that for each additional gram of saturated fat there are an additional 1.9 grams of total fat.

c. Substitute 20 g of saturated fat for *x* in the equation.

$y = 9.4 + 1.9x$	Original equation.
$y = 9.4 + 1.9(20)$	Substitute 20 for *x*.
$y = 47.4$	Multiply and add.

The model predicts that there would be 47.4 g of total fat in a burger with 20 g of saturated fat.

d. Substitute 50 g of total fat for y in the equation.

$y = 9.4 + 1.9x$	Original equation.
$50 = 9.4 + 1.9x$	Substitute 50 for y.
$50 - 9.4 = 9.4 + 1.9x - 9.4$	Subtract 9.4 from both sides.
$40.6 = 1.9x$	Subtract.
$\dfrac{40.6}{1.9} = \dfrac{1.9x}{1.9}$	Divide both sides by 1.9.
$21.4 = x$	Reduce.

The model predicts that there would be about 21 g of saturated fat in a burger with 50 g of total fat.

Notice that you find the slope before the y-intercept when finding a line of fit. Because of the importance of slope, some mathematicians show it first. They use the **slope-intercept form** of a linear equation, often calling the slope m and the y-intercept b. This gives $y = mx + b$. Why is this equation equivalent to the intercept form that you have learned?

EXERCISES

You will need your calculator for problem **4.**

▶ Practice Your Skills

1. For each graph below, tell whether or not you think the line drawn is a good representation of the data. Explain your reasoning.

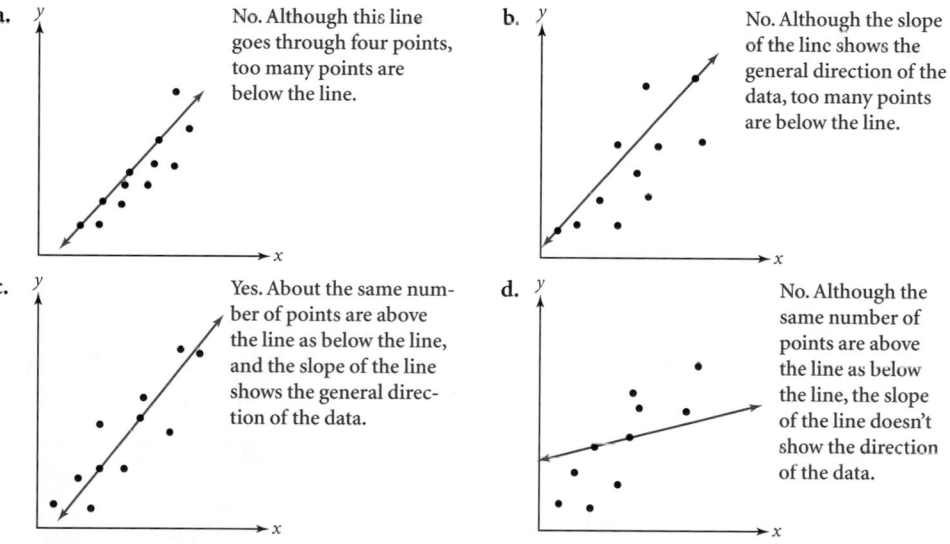

a. No. Although this line goes through four points, too many points are below the line.

b. No. Although the slope of the line shows the general direction of the data, too many points are below the line.

c. Yes. About the same number of points are above the line as below the line, and the slope of the line shows the general direction of the data.

d. No. Although the same number of points are above the line as below the line, the slope of the line doesn't show the direction of the data.

2. The line through the points (0, 5) and (4, 5) is horizontal. The equation of this line is $y = 5$ because the y-value of every point on it is 5. If a line goes through the points (2, −6) and (2, 8), what kind of line is it? What is its equation? vertical; $x = 2$

SLOPE-INTERCEPT FORM

[Ask] "Why can we now call the intercept b instead of a?" [The slope is whatever constant is multiplied by x, and the y-intercept is the constant being added to that product. You can represent those two constants with any pair of letters you choose, though some letters are more conventional to use than others.]

Closing the Lesson

You can use a line to **model** a set of data points for the sake of making predictions. The better the line fits the data, the better the predictions will be. If you have Fathom!™ software, you can use student data to create a scatter plot and then use Fathom's moveable line to picture possible **lines of fit.**

BUILDING UNDERSTANDING

Students acquire more practice with slopes and equations of lines, while gaining some experience with finding lines of fit.

ASSIGNING HOMEWORK

Essential	1–5, 10
Performance assessment	4
Portfolio	5, 7
Journal	1, 6, 8
Group	1, 5, 9
Review	11–13

▶ Helping with the Exercises

Exercise 2 The introduction of equations for vertical lines lays the groundwork for a later exercise in this lesson and for exercises through the rest of the student text.

Exercise 4 [Alert] Be sure students convert years to months.

4a. There is a linear pattern.

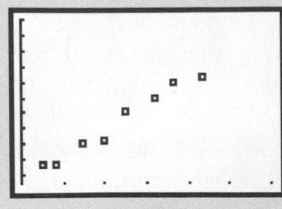

$[0, 36, 6, 200, 1200, 100]$

4b. Answers will vary. Using the points $(8, 376)$ and $(19, 684)$, the slope is 28.

4c. The slope represents the number of quarters Penny collects per month.

4d. $y = 28x$; the line needs to move up (the y-intercept needs to increase).

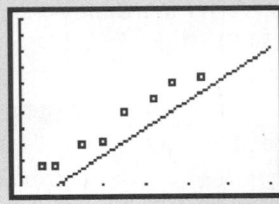

4e. A possible equation is $y = 152 + 28x$.

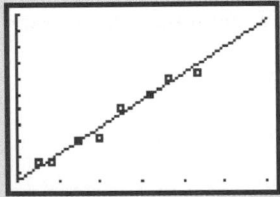

4f. The y-intercept represents the number of quarters Penny's grandmother gave her.

4g. Possible answer: 1160 quarters. The prediction may not be reliable because it extrapolates 10 months beyond the data.

---

3. Write the equation of the line in each graph.

a.

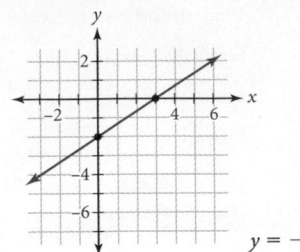

$y = -2 + \frac{2}{3}x$

b.

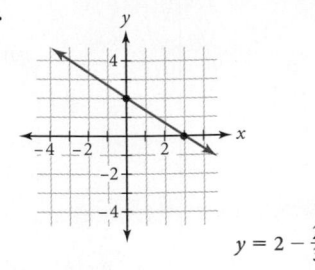

$y = 2 - \frac{2}{3}x$

c.

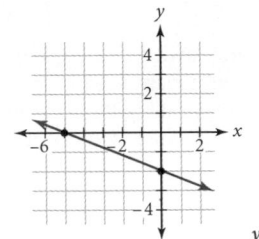

$y = -2 - \frac{2}{5}x$

d.

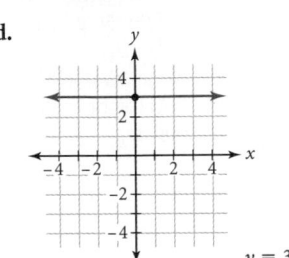

$y = 3$

4. On Penny's 15th birthday, her grandmother gave her a large jar of quarters. Penny decided to continue to save quarters in the jar. Every few months she counts her quarters and records the number in a table like this one. Predict how many quarters she'll have on her 18th birthday.

Penny's Savings

Number of months x	3	5	8	12	15	19	22	26
Number of quarters y	270	275	376	420	602	684	800	830

a. Make a scatter plot of the data on your calculator. Is there a pattern?

b. Select two points through which a line of fit would pass. Find the slope of the line between these points.

c. What is the real-world meaning of the slope?

d. Use the slope you found in 4b to write an equation of the form $y = bx$. Graph this line on the scatter plot. What do you need to do to this line to better fit the data?

e. Estimate the y-intercept and write an equation of the form $y = a + bx$. Graph this new line.

f. What is the real-world meaning of the y-intercept?

g. Use your equation to predict how many quarters Penny will have on her 18th birthday.

▶ Reason and Apply

5. Use the table to look for a relationship between a state's population and the number of members from that state in the House of Representatives.

Statistics for Some States

State	Estimated population (millions)	Number of members in House of Representatives	Number of members in Senate
Alabama	4.4	7	2
Indiana	5.9	10	2
Michigan	9.9	16	2
Mississippi	2.8	5	2
North Carolina	7.7	12	2
Oklahoma	3.4	6	2
Oregon	3.3	5	2
Tennessee	5.5	9	2
Utah	2.1	3	2
West Virginia	1.8	3	2

(*www.census.gov* and *www.house.gov*)

a. Which statement makes more sense: The population depends on the number of members in the House of Representatives, or the number of members in the House of Representatives depends on the population?

b. Based on your answer to 5a, define variables and make a scatter plot of the data.

c. Find the equation of a line of fit. What is the real-world meaning of the slope? What is the real-world meaning of the *y*-intercept?

d. California has an estimated population of 33 million. Use your equation to estimate the number of members California has in the House of Representatives.

e. Minnesota has eight members in the House of Representatives. Use your equation to estimate the population of Minnesota.

f. You might find that a direct variation equation of the form $y = bx$ fits your data. Is this a reasonable model for the data? Explain why or why not. The relationship should be a direct variation because it should go through the point $(0, 0)$. A state with no population would have no representatives.

6. Use the table in problem 5 to answer these questions.

a. Does the population of a state affect its number of members in the Senate? No. Each state has two senators regardless of its population.

b. Write an equation that models the number of senators from each state. Graph this equation on the same coordinate axes as 5c.

c. Describe the graph and explain why it looks this way.

The United States Constitution gives each state representation in the House of Representatives in proportion to its population. In the Senate, each state has equal representation regardless of size. This photo shows a joint session of both the House and Senate.

6b. $y = 2$, where *x* represents population in millions and *y* represents the number of senators.

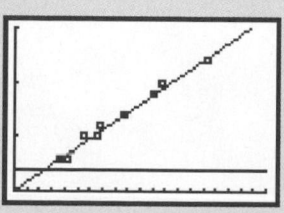

6c. The graph is a horizontal line because there's no change in *y*, the number of senators.

Exercise 5 You might encourage students to research the laws dictating the number of members in the House of Representatives. See www.keymath.com for links to sources. The first question gives an opportunity for more thinking about input and output variables.

5a. The number of representatives depends on the population.

5b. Let *x* represent population in millions, and let *y* represent the number of representatives.

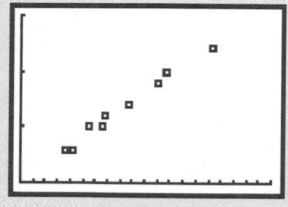

$[0, 10, 0.5, 0, 15, 5]$

5c. Answers will vary. Two possible points are $(3.3, 5)$ and $(7.7, 12)$. The slope between these two points is approximately 1.6. The equation $y = 1.6x$ appears to fit the data with a *y*-intercept of 0. The slope represents the number of representatives per 1 million people. The *y*-intercept means that a state with no population would have no representatives.

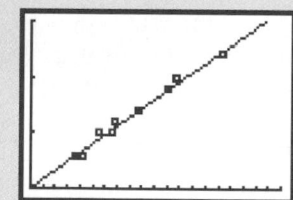

5d. The equation $y = 1.6x$ gives $y = 1.6(33) = 52.8$ or 53 representatives. On July 19, 2000, California had 52 representatives.

5e. The equation $y = 1.6x$ gives $8 = 1.6x$; $x = \dfrac{8}{1.6} = 5$; 5 million. The estimated population of Minnesota in July 1999 was 4.8 million.

Exercise 6a There is a constant relationship between the population and the number of senators.

Exercise 7 If the person walks at a truly constant rate, the data fit a line exactly. The slope and *y*-intercept of the line describe the data exactly.

8b. The *y*-intercept represents the start distance for the walk; the *x*-intercept represents the time elapsed when the walker reaches the detector.

8c. Answers will vary. Quadrant II could indicate walking before you started timing. Quadrant IV could indicate that the walker walks past you; the distances behind you are considered negative.

Exercise 9 Be sure students realize that no single line satisfies all four conditions. Students can slide a slope triangle to work backward and find the *y*-intercept. This technique foreshadows the beginning of Lesson 5.3.

Exercise 10 You may need to say that a "family of lines" is a collection of lines that share some property. If students haven't worked on Exercise 2, they may need help with the equation of a vertical line. In 10d, some students may have difficulty seeing that this equation describes a vertical line. In part, their difficulty may be in recognizing that *c* is a constant. Or they might not understand very deeply that the equation of a line gives a statement about both the *x*- and *y*-coordinates for points on that line. Pick a particular value of *c* and suggest that students name some points that satisfy that equation and draw the line connecting them. Then try it with another value or two.

7. Suppose your friend walks steadily away from you at a constant rate so that her distance at 2 seconds is 3.4 meters and her distance at 4.5 seconds is 4.4 meters. Let *x* represent time in seconds, and let *y* represent distance in meters.

$\dfrac{y_2 - y_1}{x_2 - x_1} = \dfrac{4.4 - 3.4}{4.5 - 2}$; the slope is 0.4 meter per second.

   a. What is the slope of the line that models this situation?

   The *y*-intercept is 2.6 meters; students can find this by working backward with the slope or by estimating from a graph.

   b. What is the *y*-intercept of this line? Explain how you found it.

   c. Write a linear equation in intercept form that models your friend's walk. $y = 2.6 + 0.4x$

8. Suppose this line represents a walking situation in which you're using a motion sensor to measure distance. The *x*-axis shows time and the *y*-axis shows distance from the sensor.

   a. Is the slope positive, negative, zero, or undefined? Explain your reasoning. The slope is negative because the distance decreases as the time increases.

   b. What is a real-world meaning for the *y*-intercept? For the *x*-intercept?

   c. If the line extended into Quadrant II, what could that mean? If the line extended into Quadrant IV?

9. Find the equation of a line that

   a. Has a positive slope and a negative *y*-intercept. Answers will vary. $y = -8 + 4x$ is one possibility.

   b. Has a negative slope and a *y*-intercept of zero. Answers will vary. $y = -2x$ is one possibility.

   c. Passes through the points $(1, 7)$ and $(4, 10)$. $y = 6 + x$

   d. Passes through the points $(-2, 10)$ and $(4, 10)$. $y = 10$

10. Each equation below represents a family of lines. Describe what the lines in each form have in common.

   a. $y = a + 3x$ All lines have a slope of 3; they are all parallel.

   b. $y = 5 + bx$ All lines cross the *y*-axis at 5; they radiate around the point $(0, 5)$.

   c. $y = a$ All lines are parallel to the *x*-axis, or horizontal.

   d. $x = c$ All lines are parallel to the *y*-axis, or vertical.

▶ Review

11. For each of these tables of *x*- and *y*-values, decide if the values indicate a direct variation, an inverse variation, or neither. Explain how you made your decision. If the values represent a direct or inverse variation, write an equation.

a.

x	y
−3	9
−1	1
−0.5	0.25
0.25	0.0625
7	49

neither

b.

x	y
−20	−5
−8	−12.5
2	50
10	10
25	4

inverse variation; $y = \dfrac{100}{x}$

c.

x	y
0	0
−6	15
8	−20
−12	30
4	−10

direct variation; $y = -2.5x$

d.

x	y
78	6
31.2	2.4
−145.6	−11.2
14.3	1.1
−44.2	−3.4

direct variation; $y = \dfrac{1}{13}x$

12. Show the steps to solve each equation. Then use your calculator to verify your solution.

a. $8 - 12m = 17$

$$8 - 12m = 17$$
$$-12m = 9$$
$$m = -0.75$$

b. $2r + 7 = -24$

$$2r + 7 = -24$$
$$2r = -31$$
$$r = -15.5$$

c. $-6 - 3w = 42$

$$-6 - 3w = 42$$
$$-3w = 48$$
$$w = -16$$

13. Give the mean and median for each data set.

a. {1, 2, 4, 7, 18, 20, 21, 21, 26, 31, 37, 45, 45, 47, 48} mean: 24.8$\overline{6}$; median: 21

b. {30, 32, 33, 35, 39, 41, 42, 47, 72, 74} mean: 44.5; median: 40

c. {107, 116, 120, 120, 138, 140, 145, 146, 147, 152, 155, 156, 179} mean: approximately 140.1; median: 145

d. {85, 91, 79, 86, 94, 90, 74, 87} mean: 85.75; median: 86.5

Exercise 11 This exercise reviews Chapter 3.

Exercise 12 Students might use either the undoing or the balancing method learned in Chapter 4.

Exercise 13 This exercise reviews Lesson 1.2.

IMPROVING YOUR **VISUAL THINKING** SKILLS

The traditional Japanese abacus, or "soroban," is still widely used today. Each column shows a different place value—1, 10, 100, 1000, and so on. The four lower beads are moved up to represent the digits from 1 to 4. The fifth bead is moved down to show the digit 5. The digits 6 to 9 are shown with a combination of lower and upper beads. The first abacus below shows the number 6053.

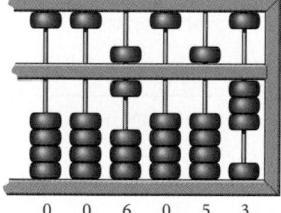 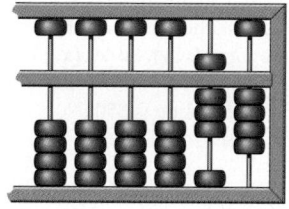 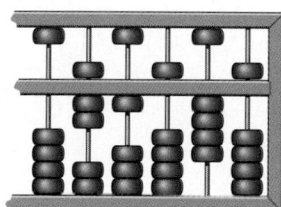

0 0 6 0 5 3

What numbers do the second and third abacuses show?

Sketch an abacus to show the number 27,059.

You can learn more about the abacus at **www.keymath.com/DA** .

IMPROVING **VISUAL THINKING** SKILLS

The second and third abacuses show 84 and 71,545 respectively. 27,059 would look like this:

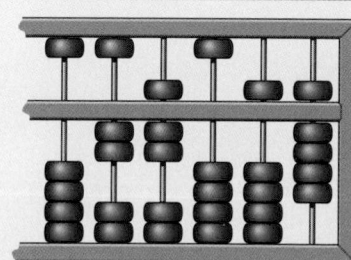

PLANNING

LESSON OUTLINE

One day:

10 min	Example
20 min	Investigation
5 min	Sharing
5 min	Closing
10 min	Exercises

MATERIALS

• Calculator Note 5A

TEACHING

This lesson shows that the slope-intercept form of the equation of a line can be found from two points without having to find the *y*-intercept.

INTRODUCTION

The reference to homework is to Lesson 5.1, Exercise 7b.

EXAMPLE

This example derives the point-slope form of the equation of a line. If the form happened to arise during Lesson 5.2, you may not need to spend much time on the example. Advise students who find it confusing that the investigation will make it clearer.

Encourage critical thinking by asking some questions. **[Ask]** "Do you think the situation is realistic? Do populations grow at a constant rate?" [Some do, but students may realize that population growth is often exponential up to a limit.]

Success breeds confidence.

BERYL MARKHAM

Point-Slope Form of a Linear Equation

So far you have worked with linear equations in intercept form, $y = a + bx$. When you know a line's slope and *y*-intercept, you can write its equation directly in intercept form. But what if you don't know the *y*-intercept? One method that you might remember from your homework is to work backward with the slope until you find the *y*-intercept. But you can also use the slope formula to find the equation of a line when you know the slope of the line and the coordinates of only one point on the line.

EXAMPLE Since the time Beth was born, the population of her town has increased at a rate of approximately 850 people per year. On Beth's 9th birthday the total population was nearly 307,650. If this rate of growth continues, what will be the population on Beth's 16th birthday?

▶ **Solution** Since the rate of change is approximately constant, a linear equation should model this population growth. Let *x* represent time in years since Beth's birth, and let *y* represent the population.

In the problem, you are given one point, (9, 307650). Any other point on the line will be in the form (x, y). So let (x, y) represent a second point on the line. You also know that the slope is 850. Now use the slope formula to find a linear equation.

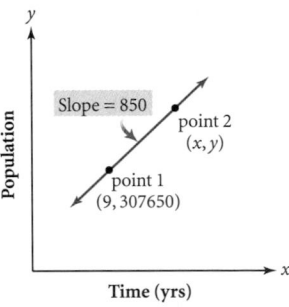

$$\frac{y_2 - y_1}{x_2 - x_1} = b$$ Slope formula.

$$\frac{y - 307{,}650}{x - 9} = 850$$ Substitute the coordinates of the point (9, 307650) for (x_1, y_1), and the slope 850 for *b*.

Since we only know one point, we use (x, y) to represent any point.

LESSON OBJECTIVES

• Learn the point-slope form of an equation of a line

• Write equations in point-slope form that model real-world data

NCTM STANDARDS

CONTENT		PROCESS	
•	Number		Problem Solving
•	Algebra	•	Reasoning
	Geometry		Communication
•	Measurement		Connections
	Data/Probability	•	Representation

$$(x - 9)\frac{y - 307{,}650}{(x - 9)} = 850(x - 9)$$ Multiply both sides by $(x - 9)$.

$$y - 307{,}650 = 850(x - 9)$$ Reduce the left side.

$$y - 307{,}650 + 307{,}650 = 307{,}650 + 850(x - 9)$$ Add 307,650 to both sides.

$$y = 307{,}650 + 850(x - 9)$$ Add.

The equation $y = 307{,}650 + 850(x - 9)$ is a linear equation that models the population growth. To find the population on Beth's 16th birthday, substitute 16 for x.

$$y = 307{,}650 + 850(x - 9)$$ Original equation.

$$y = 307{,}650 + 850(16 - 9)$$ Substitute 16 for x.

$$y = 313{,}600$$ Use order of operations.

The model equation predicts that the population on Beth's 16th birthday will be 313,600.

The equation $y = 307{,}650 + 850(x - 9)$ is a linear equation, but it is not in intercept form. This equation has its advantages too because you can clearly identify the slope and one point on the line. Do you see the slope of 850 and the point (9, 307650) within the equation? This form of a linear equation is appropriately called the **point-slope form.**

Point-Slope Form

If a line passes through the point (x_1, y_1) and has slope b, the **point-slope form** of the equation is

$$y = y_1 + b(x - x_1)$$

Investigation
The Point-Slope Form for Linear Equations

Silo and Jenny conducted an experiment in which Jenny walked at a constant rate. Unfortunately, Silo recorded only the data shown in this table.

Sorry Jenny,
had to motor, but this should
be enough info......
Peace, Silo

Elapsed time (sec) x	Distance to walker (m) y
3	4.6
6	2.8

Step 1 Find the slope of the line that represents this situation.

Step 2 Write a linear equation in point-slope form using the point (3, 4.6) and the slope you found in Step 1.

Step 3 Write another linear equation in point-slope form using the point (6, 2.8) and the slope you found in Step 1.

Step 1 −0.6 m/sec

Step 2 $y = 4.6 - 0.6(x - 3)$

Step 3 $y = 2.8 - 0.6(x - 6)$

[Alert] Some students may be confused about choosing (x, y) to represent any point on the line. They may not yet grasp the idea that the equation relates coordinates of exactly those points lying on the line. Help them keep in mind the goal of coming up with such an equation.

[Alert] A few students may be confused about multiplying both sides by $(x - 9)$. Remind them that they can consider $(x - 9)$ as a single number.

Students may want to simplify $y = 307{,}650 + 850(x - 9)$ to $y = 300{,}000 + 850x$. Although the equations are equivalent, the latter is not in point-slope form.

You might ask students to go through the derivation again, using x_1, y_1, and b instead of the numbers.

Emphasize that x_1, y_1, and b represent constants, whereas x and y represent variables. Also note that the coordinate x_1 is being subtracted from x.

 Guiding the Investigation

One step Direct students' attention to the Water Temperature table on page 272 and ask them to find a line of fit without finding the y-intercept.

Step 4 There appears to be only one line, which implies that the equations are equivalent.

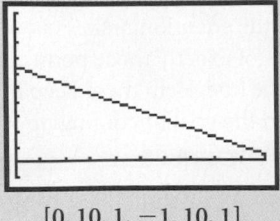

$$[0, 10, 1, -1, 10, 1]$$

Step 5 The Y_1- and Y_2-values are equivalent. Again, this implies that the two seemingly different equations are equivalent.

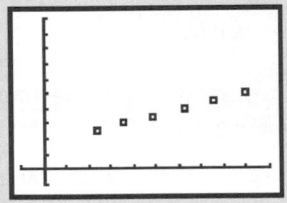

Step 6 Let x represent time in seconds, and let y represent temperature in degrees Celsius. The data set appears to have a linear pattern.

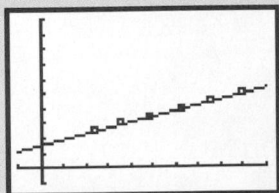

$$[-10, 100, 10, -10, 100, 10]$$

Steps 7 and 8 Each member of the group should be encouraged to select a different pair of points. Suggest that each group graph all of their lines on one calculator for easy comparison.

Step 8 Using the slope in Step 7, one possibility is $y = 35 + 0.38(x - 49)$.

| Step 4 | Enter the equation from Step 2 into Y_1 and the equation from Step 3 into Y_2 on your calculator, and graph both equations. What do you notice? |
| Step 5 | Look at a table of Y_1- and Y_2-values. What do you notice? What do you think the results mean? |

Now that you have some practice at writing point-slope equations, try using a point-slope equation to fit data.

The table shows how the temperature of a pot of water changed over time as it was heated.

Water Temperature

Time (sec) x	Temperature (°C) y
24	25
36	30
49	35
62	40
76	45
89	50

Step 6	Define variables and plot the data on your calculator. Describe any patterns you notice.
Step 7	Choose a pair of points from the data. Find the slope of the line between your two points.
Step 8	Write an equation in point-slope form for a line that passes through your two points. Graph the line. Does your equation fit the data?

Step 7 Answers will vary. Using $(49, 35)$ and $(62, 40)$, the slope is $\frac{5}{13}$.

| Step 9 | Compare your graph to those of other members of your group. Does one graph show a line that is a better fit than the others? Answers will vary. Since the data is in such a tight linear pattern, there may not appear to be one better line of fit—they will all be pretty good. |

If you look back at the investigation, you will notice that you found the point-slope form of a line even though you had only points to start with. This is possible because you can still use the point-slope form when you know two points on the line; there's just one additional step. What is it? You must calculate the slope using the two points.

EXERCISES

You will need your calculator for problems **3, 4, 5, 8, 9,** and **10.**

▶ **Practice Your Skills**

1. Name the slope and one point on the line that each point-slope equation represents.
   a. $y = 3 + 4(x - 5)$ $4; (5, 3)$
   b. $y = 1.9 + 2(x + 3.1)$ $2; (-3.1, 1.9)$
   c. $y = -3.47(x - 7) - 2$ $-3.47; (7, -2)$
   d. $y = 5 - 1.38(x - 2.5)$ $-1.38; (2.5, 5)$

2. Write an equation in point-slope form for a line, given its slope and one point that it passes through.
   a. Slope 3; point $(2, 5)$ $y = 5 + 3(x - 2)$
   b. Slope -5; point $(1, -4)$ $y = -4 - 5(x - 1)$

Step 9 [Ask] "How could you have wisely selected points in order to find the best fit line to begin with?" [Choose points neither close together nor too far apart that appear to lie on a line that passes near most of the data.]

SHARING IDEAS

Choose students to present several different equations that have the same graphs. **[Ask]** "Is there a way to tell that the equations are equivalent (have the same graphs) without actually graphing?" Encourage all ideas. You don't need to answer this question. Students will get more experience identifying equivalent equations in Lesson 5.4.

Assessing Progress

You can assess students' understanding of **input and output variables** and their abilities to find the **slope** of a line through two points and to **graph** data points and lines on a calculator.

3. A line passes through the points $(-2, -1)$ and $(5, 13)$.

   a. Find the slope of this line. 2

   b. Write an equation in point-slope form using the slope you found in 3a and the point $(-2, -1)$. $y = -1 + 2(x + 2)$

   c. Write an equation in point-slope form using the slope you found in 3a and the point $(5, 13)$. $y = 13 + 2(x - 5)$

   d. Verify that the equations in parts 3b and c are equivalent. Enter one equation into Y₁ and the other into Y₂ on your calculator, and compare their graphs and tables.
     The graphs coincide and the tables are identical.

4. APPLICATION This table shows a linear relationship between actual temperature and approximate wind chill temperature when the wind speed is 20 miles per hour.

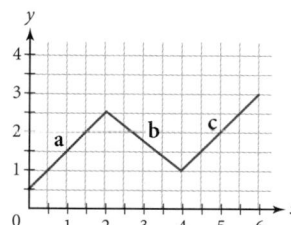

Wind Chill with Wind Speed of 20 mph

Temperature (°F) x	5	10	15	20	25
Wind chill (°F) y	−31	−24	−17	−10	−3

   a. Find the rate of change of the data (the slope of the line).

   b. Choose one point and write an equation in point-slope form to model the data.

   c. Choose another point and write another equation in point-slope form to model the data.

   d. Verify that the two equations in 4b and c are equivalent. Enter one equation into Y₁ and the other into Y₂ on your calculator, and compare their graphs and tables.

   e. What is the wind chill temperature when the actual temperature is 0°F? What does this represent in the graph? $-38°C$; this is the graph's *y*-intercept.

5. Play the BOWLING program at least four times. [▶☐ See **Calculator Note 5A** for instructions on how to play the game. ◀] Each time you play, write down any equations you try and how many points you score.

Reason and Apply

6. The graph at right is made up of linear segments **a**, **b**, and **c**. Write an equation in point-slope form for the line that contains each segment.

7. Look at quadrilateral *ABCD*.

   a. Write an equation in point-slope form for the line containing each segment in this quadrilateral. Check your equations by graphing them on your calculator.

   b. What is the same in the equations for the line through points *A* and *D* and the line through points *B* and *C*? What is different in these equations?

   c. What kind of figure does *ABCD* appear to be? Do the results from 7b have anything to do with this?
     ABCD appears to be a parallelogram because each pair of opposite sides is parallel. The equal slopes in 7b mean that the opposite sides are parallel.

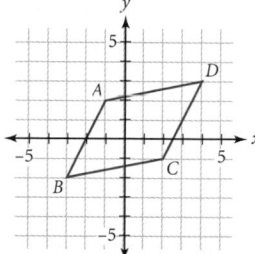

6a. $y = 0.5 + 1(x - 0)$ or $y = 2.5 + 1(x - 2)$
6b. $y = 2.5 - 0.75(x - 2)$ or $y = 1 - 0.75(x - 4)$
6c. $y = 1 + 1(x - 4)$ or $y = 3 + 1(x - 6)$

Exercise 7 [ESL] Students might not know the term *parallelogram*. Ask them to describe the figure if they cannot name it.

7a. $AD: y = 2 + 0.2(x + 1)$ or $y = 3 + 0.2(x - 4)$
$BC: y = -2 + 0.2(x + 3)$ or $y = -1 + 0.2(x - 2)$
$AB: y = 2 + 2(x + 1)$ or $y = -2 + 2(x + 3)$
$DC: y = 3 + 2(x - 4)$ or $y = -1 + 2(x - 2)$

7b. The slopes are the same; the coordinates of the points are different.

Closing the Lesson

As needed, say that if you know two points on a line, you can find an equation for a line without finding the *y*-intercept. The form is called the **point-slope form.**

DUILDING UNDERSTANDING

Students work with the point-slope form of linear equations to model real-world data.

ASSIGNING HOMEWORK

Essential	1–3, 8, 9
Performance assessment	4, 8
Portfolio	10
Journal	6, 7
Group	5, 8, 10
Review	11–13

▶ **Helping with the Exercises**

Exercise 1 [Alert] Some students may forget that the *x*-coordinate of the point is being subtracted from *x*. Therefore, in 1b, the *x*-coordinate of the point must be negative.

Exercises 3d and 4d Students can use distributivity as well as graphing or tables to check the equivalence of the lines. These exercises foreshadow Lesson 5.4.

4a. $\frac{7}{5} = 1.4$; the data are exactly linear, so any two points will give this slope.

4b. Answers will vary. Using the point $(5, -31)$, the equation is $y = -31 + 1.4(x - 5)$.

4c. Answers will vary. Using the point $(10, -24)$, the equation is $y = -24 + 1.4(x - 10)$.

4d. The two equations should give the same graphs and tables.

Exercise 8 Letters or packages weighing more than 13 oz are subject to a different rate schedule. Therefore, the possible x-values for these data are restricted to whole numbers from 1 to 13.

Research current postal rates through www.keymath.com/DA. The USPS web site states first-class rates in a quasi–slope-intercept form of "first ounce $0.34, each additional ounce $0.21."

Bring up the idea of step functions. **[Ask]** "How could you graph this relationship accurately?"

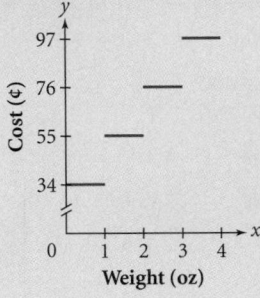

8a. The data appears linear.

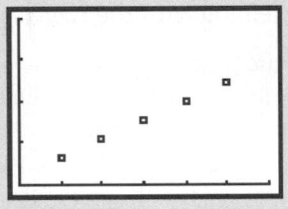

$$[0, 6, 1, 0, 2, 0.5]$$

8b. $0.21 per ounce; this is the cost per additional ounce after the first.

8c. Answers will vary. Using the point $(1, 0.34)$, the equation is $y = 0.34 + 0.21(x - 1)$.

8e. The rates are given for weights not exceeding the given weights, so a letter weighing 3.5 oz would cost the same as a 4-oz letter, or $0.97; a letter weighing 9.1 oz would cost the same as a 10-oz letter, or $2.23.

8f. Answers will vary. A continuous line includes points whose x-values are not whole numbers and whose y-values are not possible rates.

---

8. APPLICATION The table shows postal rates for first-class U.S. mail in the year 2001.

a. Make a scatter plot of the data. Describe any patterns you notice.

b. Find the slope of the line between any two points in the data. What is the real-world meaning of this slope?

c. Write a linear equation in point-slope form that models the data. Graph the equation to check that it fits your data points.

d. Use the equation you wrote in 8c to find the cost of mailing a 10-oz letter. $2.23

e. What would be the cost of mailing a 3.5-oz letter? A 9.1-oz letter? (Hint: Think about what the column header for the x-values means.)

f. The equation you found in 8c is useful for modeling this situation. Is the graph of this equation, a continuous line, a correct model for the situation? Explain why or why not.

9. APPLICATION The table below shows total fat grams and number of calories for some breakfast sandwiches sold by chain restaurants.

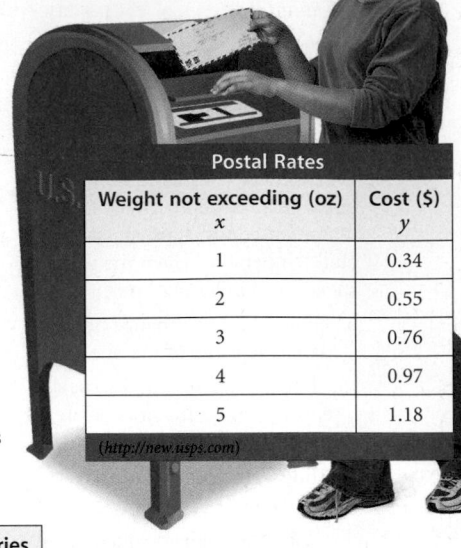

Postal Rates

Weight not exceeding (oz) x	Cost ($) y
1	0.34
2	0.55
3	0.76
4	0.97
5	1.18

(http://new.usps.com)

Nutrition Facts

Breakfast sandwich	Total fat (g) x	Calories y
Burger King Sausage, Egg, and Cheese Biscuit	43	620
Carl's Jr. Sunrise Sandwich	21	360
Hardee's Steak Biscuit	32	580
Jack in the Box Sourdough Breakfast Sandwich	24	450
McDonald's Sausage McMuffin with Egg	28	440
Subway Ham and Egg Breakfast Deli Sandwich	12	312
Taco Bell Country Breakfast Burrito	14	270
White Castle Sausage, Egg, and Cheese Breakfast Sandwich	25	340

(www.kenkuhl.com)

a. Make a scatter plot of the data. Describe any patterns you notice.

b. Select two points and find the equation of the line that passes through these two points in point-slope form. Graph the equation on the scatter plot.

c. According to your model, how many calories would you expect in a Hardee's Steak Biscuit with 32 grams of fat? In the graph of $y = 620 + 11.82 (x - 43)$ approximately 490 calories.

d. Does the actual data point representing the Hardee's Steak Biscuit lie above, on, or below the line you graphed in 9b? Explain what the point's location means.
Compared to the graph of $y = 620 + 11.82(x - 43)$, the point lies above the line. If a point lies above the line, then the sandwich has more calories than the model predicts.

Exercise 9g Some students may misinterpret this to mean that all fat-free foods have 112 calories. Warn them that many factors influence calories.

9a. The data appears linear.

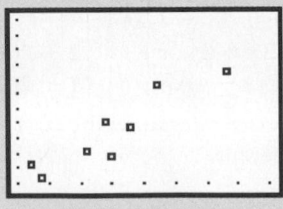

$$[10, 50, 5, 250, 800, 50]$$

9b. Answers will vary. Using the points $(21, 360)$ and $(43, 620)$, the equation is $y = 620 + 11.82(x - 43)$.

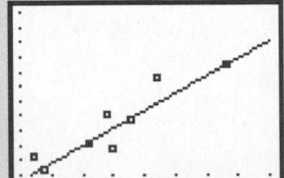

e. Check each breakfast sandwich to find if its data point falls above, on, or below your line. Answers will vary. Using $y = 620 + 11.82(x - 43)$ as a model, four points are above the line, two are on the line, and two are below the line.

f. Based on your results for 9d–e, how well does your line fit the data? Answers will vary.

g. If a sandwich has 0 grams of fat, how many calories does your equation predict? Does this answer make sense? Why or why not? Answers will vary. Using $y = 620 + 11.82(x - 43)$, approximately 112 calories; this makes sense because not all calories in food come from fat.

10. **APPLICATION** This table shows the amount of trash produced in the United States in 1980 and 1985.

a. Let x represent the year, and let y represent the amount of trash in millions of tons for that year. Write an equation in point-slope form for the line passing through these two points.

b. Plot the two data points and graph the equation you found in 10a.

c. In 1990, 196 million tons of trash were produced in the United States. Plot this data point on the same graph you made in 10b. Do you think the linear equation you found in 10a is a good model for these data? Explain why or why not.

U.S. Trash Production

Year	Amount of trash (million tons)
1980	152
1985	164

(*The Universal Almanac*, 1995)

This table shows more data about the amount of trash produced in the United States.

d. Add these data points to your graph. Adjust the window as necessary.

e. Do you think the linear equation found in 10a is a good model for this larger data set? Explain why or why not.

f. Find the equation of a better fitting line. You may find that you only need to adjust the slope value.

g. Use your new equation from 10f to predict the amount of trash produced in 2000. Answers will vary. Using $y = 152 + 3.4(x - 1980)$, 220 million tons.

U.S. Trash Production

Year	Amount of trash (million tons)
1960	88
1965	103
1970	122
1975	128

(*The Universal Almanac*, 1995)

Review

11. **APPLICATION** The volume of a gas is directly proportional to its temperature in kelvins (K). The volume of this gas is 3.50 liters at 280 K.

a. Find the volume of this gas when the temperature is 330 K. 4.125 liters

b. Find the temperature when the volume is 2.25 liters. 180 K

12. Insert operation signs, parentheses, or both into each string of numbers to create an expression equal to the answer given.

a. 1   2   3   4   5   = 1 $1(2 - 3)(4 - 5) = 1$ b. 1   2   3   4   5   = 3
c. 1   2   3   4   5   = 5 $(1 - 2)(3 - 4)5 = 5$ $1 - (2 - 3) - (4 - 5) = 3$

13. Find the slope of the line through the first two points given. Assume the third point is also on the line and find the missing coordinate.

a. $(-1, 5)$ and $(3, 1)$; $(5, \boxed{-1})$ -1 b. $(2, -5)$ and $(2, -2)$; $(\boxed{2}, 3)$ undefined
c. $(-10, 22)$ and $(-2, 2)$; $(\boxed{0}, -3)$ $-\frac{5}{2}$

Exercise 10 In 10e, a "good model" is one that allows accurate predictions. Part of the goal of 10g is to show that basing a model on a small set of data can lead to wild predictions. **[Ask]** "In what year does your equation predict that there were 0 million tons of trash?" [Using $y = 152 + 3.4(x - 1980)$, the year would be 1935.] "Is this possible?"

10a. $y = 152 + 2.4(x - 1980)$ or $y = 164 + 2.4(x - 1985)$

10b and 10c.

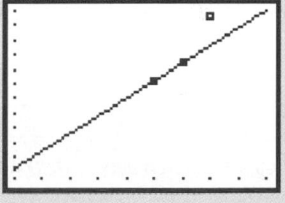

[1955, 2000, 5, 85, 200, 10]

The point (1990, 196) is not very close to the line, so the line isn't a good model for the data.

10d.

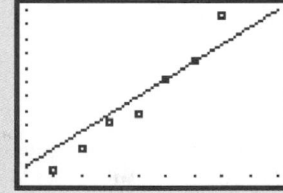

10e. The data is generally linear, but the line doesn't fit it very well. A line with a steeper slope would be a better fit.

10f. Answers will vary. $y = 152 + 3.4(x - 1980)$ gives a reasonable fit.

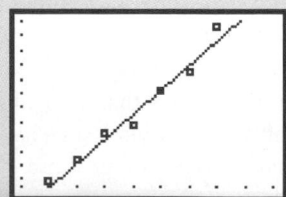

Exercise 12 This exercise reviews Lesson 4.1. As an extension, challenge students to find as many results as possible using the numbers 1, 2, 3, 4, and 5 or several ways to get the same result. Other answers:

a. $-1 - 2 + 3 - 4 + 5$

b. $1 \cdot (-2 \cdot 3) + 4 + 5$

c. $-1 \cdot 2 + (3 \cdot 4) - 5$

Exercise 13 This exercise reviews Lesson 5.1. Encourage a variety of approaches. Students might graph, step over one unit at a time, draw slope triangles, or make calculator tables.

Exercise 11 This exercise reviews Lesson 3.2. Remind students of direct variations $y = kx$. You might say that Kelvin units are the same size as Celsius degrees, but that 0 K is at about −273°C. It's called *absolute zero* because electrons at that temperature can't move. The word *degrees* is not used with the Kelvin scale.

LESSON

5.4

Equivalent Algebraic Equations

For the same line, there can be more than one equation in point-slope form. For example, you can write the equation for the line through the points $(1, 2)$ and $(2, 7)$ as $y = 2 + 5(x - 1)$ or $y = 7 + 5(x - 2)$ depending on which point you use. Actually, every line contains infinitely many points, so there are an infinite number of ways to write the equation in point-slope form. You can also write the equation of any line in intercept form, which looks different too. How can you tell when two equations actually represent the same line?

In this lesson you'll learn how to recognize equivalent equations by using mathematical properties and the rules for order of operations.

These self-portraits of the American pop artist Andy Warhol (1928–1987) are like equivalent equations. Each screen-printed image is the same as the next but Warhol's choice of colorization makes each look different.

Investigation
Equivalent Equations

Here are six different-looking equations in point-slope form.

- **a.** $y = 3 - 2(x - 1)$
- **b.** $y = -5 - 2(x - 5)$
- **c.** $y = 9 - 2(x + 2)$
- **d.** $y = 0 - 2(x - 2.5)$
- **e.** $y = 7 - 2(x + 1)$
- **f.** $y = -9 - 2(x - 7)$

Step 1 Answers will vary. Some students might say that they cannot tell whether the equations are the same or different. Others may try graphing them and see that they are equivalent. **Step 1**

Step 2 All equations become $y = 5 - 2x$. **Step 2**

Step 3 For all equations, the point-slope equation and the intercept equation should show identical **Step 3** graphs and tables, which means the same values satisfy both equations. **Step 4**

Do the six equations represent the same line or different lines? Explain your reasoning.

Divide these equations among the members of your group. Use the distributive property to rewrite the right side of each equation. You should get an equation in intercept form.

Enter your point-slope equation into Y_1, and enter your intercept equation into Y_2. Check that the two equations have the same calculator graph or table. How does this show that the equations are equivalent?

Now, as a group, compare your intercept equations. What do the results mean about the six equations?

PLANNING

LESSON OUTLINE

One day:

20 min	Investigation
5 min	Sharing
10 min	Examples
5 min	Closing
10 min	Exercises

MATERIALS

- Properties of Numbers (T), *optional*
- Properties of Equality (T), *optional*

TEACHING

This lesson shows that linear equations are equivalent if their graphs or tables are the same or if they can be symbolically manipulated into the same equation.

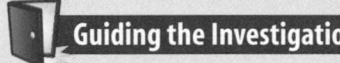

Guiding the Investigation

One step Point out the 15 equations listed on page 277 after Step 5 and ask students to categorize the lines by their equations. Encourage a variety of ways of checking for equivalence.

Step 1 [Alert] Watch for the claim that the lines are the same because their slopes are the same. Remind students that many different yet parallel lines can have the same slope. Students can compare the lines by graphing, looking at tables, or by distributing and simplifying.

Step 2 Depending on the size of the group, each student may need to work with more than one equation.

Step 4 All six point-slope equations transform into $y = 5 - 2x$.

LESSON OBJECTIVES

- Rewrite equations using the distributive property and grouping
- Rewrite equations to determine whether they are equivalent
- Write equivalent equations
- Review the distributive property and the commutative properties of addition and multiplication

NCTM STANDARDS

CONTENT		PROCESS	
•	Number		Problem Solving
•	Algebra	•	Reasoning
•	Geometry		Communication
	Measurement	•	Connections
	Data/Probability	•	Representation

Step 5 Answers will vary. You could graph the equations to see if they are the same line, or you could symbolically manipulate the point-slope form into intercept form.

Step 5 As a group, explain how you can tell that an equation in point-slope form is equivalent to one in intercept form. Think about how you can do this graphically and symbolically.

Here are fifteen equations. They represent only four different lines.

a. $y = 2(x - 2.5)$ **b.** $y = 18 + 2(x - 8)$

c. $y = 52 - 6(x + 8)$ **d.** $y = -6 + 2(x + 4)$

e. $y = 21 - 6(x + 4)$ **f.** $y = 14 - 6(x - 3)$

g. $y = -10 + 2(x + 6)$ **h.** $6x + y = 4$

i. $y = 11 + 2(x - 8)$ **j.** $12x + 2y = -6$

k. $y = 2(x - 4) + 10$ **l.** $y = 15 - 2(10 - x)$

m. $y = 7 + 2(x - 6)$ **n.** $y = -6(x + 0.5)$

o. $y = -6(x + 2) + 16$

Step 6 The intercept form for the equation of each line is given along with the letters of the equivalent equations.
$y = -5 + 2x$: a, m, i, l
$y = 2 + 2x$: b, d, g, k
$y = -3 - 6x$: e, j, n
$y = 4 - 6x$: c, f, h, o

Step 6 Test your answer to Step 5 by finding the intercept form of each equation and then grouping equivalent equations.

Step 7 As a group, explain how you can tell that two equations in point-slope form are equivalent. Students should recognize that intercept form is consistent and therefore transform each point-slope form to intercept form.

You have learned how to write linear equations in two different forms:

Intercept form $y = a + bx$

Point-slope form $y = y_1 + b(x - x_1)$

In the second part of the investigation, some of the equations had x and y on the same side, as in $12x + 2y = -6$. Equations like these are in **standard form.** Can you identify another equation in the investigation that is in standard form?

No matter what form you start with, you can always rewrite any linear equation in intercept form. Then it's easy to recognize equivalent equations. Let's review properties that help you change the form of an equation.

Step 6 Students might make four columns and place each equation in a column with equivalent equations. **[Alert]** Students may be confused by parts h and j. **[Ask]** "What's needed to get the equation in h (or j) into the same form as the others?" [For h, subtract 6x from both sides; for j, subtract 12x and divide by 2.] Students may forget to distribute the negative sign through part l and therefore may not be able to find an equivalent equation.

SHARING IDEAS

Ask for reports on Steps 5 and 6. If anyone disagrees about the report on Step 6, encourage discussion of the arguments rather than taking sides. If you announce a right answer, thinking about the problem might cease.

Introduce the term *standard form* to describe the equations in parts h and j. **[Ask]** "How would you generalize the standard form?" [The idea of generalizing may take some explaining. You might give the example of $y = a + bx$ as the generalized intercept form and elicit the idea that the standard form could be expressed as $ax + by = c$. Encourage representations that use other letters for the coefficients to communicate that the choice of letters is irrelevant.]

[Ask] "How many different equations in point-slope form might a line have?" [There are infinitely many because a line contains infinitely many points.]

Ask what steps students took in transforming equations. Solicit a generalization of each and write down the property names from page 278, or use the two transparencies to point out the properties.

In their work on the investigation and in their presentations, students will demonstrate their abilities to use the various properties of algebra, even if they don't yet know names for them.

EXAMPLE A

This example repeats the ideas of the investigation and cites the relevant properties.

[Alert] Some students may need to see a step between the last two steps of finding the intercept form

$$\frac{-2y}{-2} = \frac{2 - 6x}{-2}$$

For any values of a, b, and c, these properties are true:

Distributive Property

$a(b + c) = a(b) + a(c)$ Example: $6(-2 + 3) = 6(-2) + 6(3)$

Commutative Property of Addition

$a + b = b + a$ Example: $3 + 4 = 4 + 3$

Commutative Property of Multiplication

$ab = ba$ Example: $\frac{1}{2} \cdot \frac{3}{4} = \frac{3}{4} \cdot \frac{1}{2}$

Associative Property of Addition

$a + (b + c) = (a + b) + c$ Example: $2 + (1.5 + 3) = (2 + 1.5) + 3$

Associative Property of Multiplication

$a(bc) = (ab)c$ Example: $4\left(\frac{1}{3} \cdot 6.3\right) = \left(4 \cdot \frac{1}{3}\right)6.3$

There are also the properties that you have used to solve equations by balancing.

Properties of Equality

Given $a = b$, for any number c,

$a + c = b + c$ addition property of equality

$a - c = b - c$ subtraction property of equality

$ac = bc$ multiplication property of equality

$\dfrac{a}{c} = \dfrac{b}{c}\ (c \neq 0)$ division property of equality

EXAMPLE A | Is the equation $y = 2 + 3(x - 1)$ equivalent to $6x - 2y = 2$?

▶ **Solution** | Use the properties to rewrite each equation in intercept form.

$y = 2 + 3(x - 1)$ Original equation.

$y = 2 + 3x - 3$ Distributive property (distribute 3 over $x - 1$).

$y = 2 - 3 + 3x$ Commutative property (swap $3x$ and -3).

$y = -1 + 3x$ Subtract $2 - 3$.

So the intercept form of the first equation is $y = -1 + 3x$.

$6x - 2y = 2$ Original equation.

$-2y = 2 - 6x$ Subtraction property (subtract $6x$ from both sides).

$y = -1 + 3x$ Division property (divide both sides by -2).

The intercept form of the second equation is also $y = -1 + 3x$. So both are equivalent. You can also check that the intercept form and the point-slope form of the equation are equivalent by verifying that they produce the same line graph and have the same table of values. Unfortunately, you cannot enter the standard form into your calculator.

One of the authors, Jerald Murdock, works with two students.

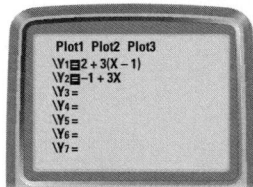
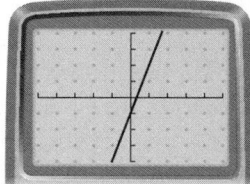
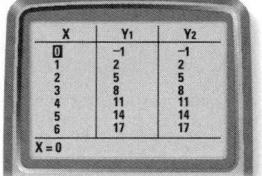

EXAMPLE B

Solve the equation $\frac{3x + 4}{6} - 5 = 7$. Identify the property of equality used in each step.

► **Solution**

$\frac{3x + 4}{6} - 5 = 7$	Original equation.
$\frac{3x + 4}{6} = 12$	Addition property (add 5 to both sides).
$3x + 4 = 72$	Multiplication property (multiply both sides by 6).
$3x = 68$	Subtraction property (subtract 4 from both sides).
$x = 22\frac{2}{3}$	Division property (divide both sides by 3).

EXERCISES

You will need your calculator for problems **1, 2, 9,** and **13.**

▶ **Practice Your Skills**

1. Is each pair of expressions equivalent? If they are not, change the second expression so that they are equivalent. Check your work on your calculator by comparing table values when you enter the equivalent expressions into Y_1 and Y_2.
   a. $3 - 3(x + 4)$ $3x - 9$ not equivalent; $-3x - 9$
   b. $5 + 2(x - 2)$ $2x + 1$ equivalent
   c. $5x - 3$ $2 + 5(x - 1)$ equivalent
   d. $-2x - 8$ $-2(x - 4)$ not equivalent; $-2(x + 4)$ or $2(-x - 4)$

EXAMPLE B

Depending on students' comfort level with solving equations, you may want to show more detail than in this solution. For example between the first two steps, show $\frac{3x + 4}{6} - 5 + 5 = 7 + 5$. **[Ask]** "Can the equation be solved a different way?" ["First multiply both sides by 6" is one possibility.]

Closing the Lesson

As needed, remind students that another form for a linear equation is the **standard form.** Linear equations are **equivalent** if their graphs are the same or if symbolic manipulation of one will give the other.

BUILDING UNDERSTANDING

Students practice using algebraic and equality properties to determine whether algebraic expressions are equivalent. They also encounter some simple factoring.

ASSIGNING HOMEWORK

Essential	**1, 2, 4, 6–9**
Performance assessment	**11, 12**
Portfolio	**11, 12**
Journal	**10**
Group	**3, 5**
Review	**3, 5, 11, 13–15**

▶ **Helping with the Exercises**

Exercise 1 This exercise gives yet another way to check equivalence of linear equations—besides graphing, algebraic manipulation, and calculator tables. Because the lists of several values agree does not prove that the equations are equivalent, but the chances are pretty good.

Exercise 2 Also encourage checks by substitution.

Exercise 3 Encourage variety. Students may solve 3a by dividing by 3 or multiplying by $\frac{1}{3}$. Similarly, 3b could involve division by -1 or multiplication by -1, and 3c could be solved by adding -15 or subtracting 15.

Exercise 4c Encourage different approaches. Students might solve by undoing. Or, if balancing, they might distribute first, add 12 first, or divide by 4 first.

If students check their work by starting with a claim of equality and working to a true statement, you might mention that the logic could be faulty, but don't emphasize it. Correct checking of solutions will come up in Chapter 6.

Exercise 6 Make sure you assign this problem. **[Alert]** If students don't remember how to find the GCF, remind them that a number m is a *factor* of a number n if n is the product of m and some number. For example, 3 is a factor of 15 because 15 is the product of 3 and 5. And 3 is a factor of $3x$ because $3x$ is the product of 3 and x. So 3 is a common factor of 15 and $3x$. Factoring is mentioned again in Exercise 5 of the Chapter Review.

Exercise 7c Some students may not recognize that $y = 5(2 + x)$ is the same as $y = 5(x + 2)$, which looks like the point-slope form except for the missing y-intercept.

2. Rewrite each equation in intercept form. Show your steps. Check your answer by using a calculator graph or table.

   a. $y = 14 + 3(x - 5)$ $y = -1 + 3x$

   b. $y = -5 - 2(x + 5)$ $y = -15 - 2x$

   c. $6x + 2y = 24$ $y = 12 - 3x$

3. Solve each equation by balancing and tell which property you used for each step.

   a. $3x = 12$ $x = 4$; division property

   b. $-x - 45 = 47$ $-x = 92$; addition property
   $x = -92$; multiplication property

   c. $x + 15 = 8$ $x = -7$; subtraction property

   d. $\frac{x}{4} = 28$
   $x = 112$; multiplication property

4. Solve each equation for x. Substitute your value into the original equation to check.

   a. $35 = 3(x + 8)$ $x = 3\frac{2}{3}$

   b. $\frac{15 - 3}{x - 4} = 10$ $x = 5.2$

   c. $4(2x - 5) - 12 = 16$ $x = 6$

5. An equation of a line is $y = 25 - 2(x + 5)$.

   a. Name the point used to write the point-slope equation. $(-5, 25)$

   b. Find x when y is 15. $x = 0$

▶ Reason and Apply

6. In the expression $3x + 15$, the *greatest common factor* (GCF) of both $3x$ and 15 is 3. You can write the expression $3x + 15$ as $3(x + 5)$. This process, called **factoring,** is the reverse of distributing. Rewrite each expression by factoring out the GCF that will leave 1 as the coefficient of x. Use the distributive property to check your work.

   a. $3x - 12$ $3(x - 4)$

   b. $-5x + 20$ $-5(x - 4)$

   c. $32 + 4x$ $4(8 + x)$

   d. $-7x - 28$ $-7(x + 4)$

7. Consider the equation $y = 10 + 5x$ in intercept form.

   a. Factor the right side of the equation. $y = 5(2 + x)$

   b. Use the commutative property of addition to swap the addends inside the parentheses. $y = 5(x + 2)$

   c. Your result should look similar to the point-slope form of the equation. What's missing? What is the value of this missing piece? The y_1 value is missing, which means it is zero; $y = 0 + 5(x + 2)$.

   d. What point could you use to write the point-slope equation in 7c? What is special about this point? $(-2, 0)$; this is the x-intercept.

8. In each set of three equations, two equations are equivalent. Find them and explain how you know they are equivalent.

   a. i. $y = 14 - 2(x - 5)$ Equations i and ii are equivalent.
       ii. $y = 30 - 2(x + 3)$
       iii. $y = -12 + 2(x - 5)$

   b. i. $y = -13 + 4(x + 2)$ Equations i and iii are equivalent.
       ii. $y = 10 + 3(x - 5)$
       iii. $y = -25 + 4(x + 5)$

   c. i. $y = 5 + 5(x - 8)$ Equations ii and iii are equivalent.
       ii. $y = 9 + 5(x + 8)$
       iii. $y = 94 + 5(x - 9)$

   d. i. $y = -16 + 6(x + 5)$ Equations i and iii are equivalent.
       ii. $y = 8 + 6(x - 5)$
       iii. $y = 44 + 6(x - 5)$

9. The equation $3x + 2y = 6$ is in standard form.

 a. Find x when y is zero. Write your answer in the form (x, y). What is the significance of this point? $x = 2$; the point $(2, 0)$ is the x-intercept.

 b. Find y when x is zero. Write your answer in the form (x, y). What is the significance of this point? $y = 3$; the point $(0, 3)$ is the y-intercept.

 c. On graph paper, plot the points you found in 9a and b and draw the line through these points.

 d. Find the slope of the line you drew in 9c and write a linear equation in intercept form. The slope is $-\frac{3}{2}$; $y = 3 - \frac{3}{2}x$.

 e. On your calculator, graph the equation you wrote in 9d. Compare this graph to the one you drew on paper. Is the intercept equation equivalent to the standard-form equation? Explain why or why not.

 f. Symbolically show that the equation $3x + 2y = 6$ is equivalent to your equation from 9d.

10. A line has the equation $y = 4 - 4.2x$.

 a. Find the y-coordinate of the point on this line whose x-coordinate is 2. $y = -4.4$

 b. Use the point you found in 10a to write an equation in point-slope form. $y = -4.4 - 4.2(x - 2)$

 c. Find the x-coordinate of the point whose y-coordinate is 6.1. $x = -0.5$

 d. Use the point you found in 10c to write a different point-slope equation. $y = 6.1 - 4.2(x + 0.5)$

 e. Show that the point-slope equations you wrote in 10b and d are equivalent to the original equation in intercept form. Explain your procedure. Answers will vary. You could rewrite each point-slope equation in slope-intercept form.

11. APPLICATION Dorine subscribes to an Internet service with a flat rate per month for up to 15 hours of use. For each hour over this limit, there is an additional per-hour fee. The table shows data about Dorine's first two bills.

Internet Use

Month	Logged on (hr)	Monthly fee ($)
January	20	15.20
February	23	17.75

 a. Define your variables and use the data in the table to write an equation in point-slope form that models Dorine's total fee. $y = 15.20 + 0.85(x - 20)$

 b. During March, Dorine was incorrectly charged $20 for being logged on for 25 hours. What should be her correct total fee? $19.45

 c. In April, Dorine was logged on for 14 hours. What was her total fee that month? Explain why you can't use your equation to answer this question. (Hint: Reread the problem carefully.)

 d. How many hours was Dorine logged on during a month when her fee was $23.70? 30 hours

9c.

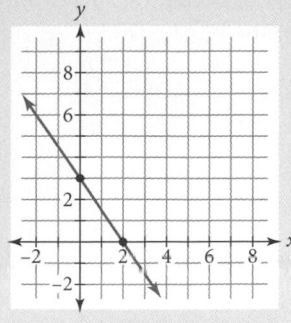

9e.

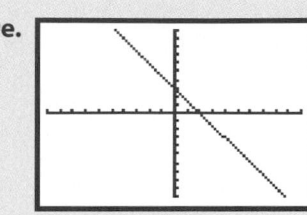

$[-10, 10, 1, -10, 10, 1]$

The two lines are the same; hence the equations are equivalent.

9f. See below.

11c. The equation is used to model the bill only when Dorine is logged on for more than 15 hours. Substituting 15 for x gives the flat rate of $10.95 for all amounts of time less than 15.

9f.

$3x + 2y = 6$	Original equation.
$2y = 6 - 3x$	Subtract $3x$ from both sides.
$y = 3 - \frac{3}{2}x$	Divide both sides by 2.

12a. The possible answers are
$y = 568 + 4.6(x - 5)$;
$y = 591 + 4.6(x - 10)$;
$y = 614 + 4.6(x - 15)$;
$y = 637 + 4.6(x - 20)$.

12c. The slope represents the number of calories burned per minute; the y-intercept represents the number of calories Avery burned from the time she went to sleep Friday night until she started hiking.

12d. Yes. It is equivalent to the slope-intercept equation $y = 545 + 4.6x$.

12e. The point $(60, 821)$ tells you that if Avery hikes for 60 minutes, she will have burned a cumulative total of 821 calories since she went to sleep Friday night.

Exercise 13 This exercise reviews Lesson 1.4.
[Language] *Compensation* includes wages and all other benefits.

13b. Germany is the country with the largest increase ($25.91), and Mexico is the country with the least increase ($0.04).

13c.

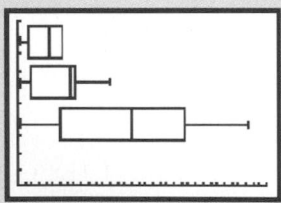

[0, 35, 1, 0, 10, 1]

Top box is 1975, middle 1985, lowest 1995.

Students should notice that there is a much larger range in data for 1995 than in 1975. The lowest compensation has not changed much, whereas the top end has moved considerably. The median has moved upward as well.

12. On Saturday morning, Avery took a hike in the hills near her house. The table shows the cumulative number of calories she burned from the time she went to sleep Friday night until she finished her hike.

a. Write a point-slope equation of a line that fits the data.

b. Rewrite your equation from 12a in intercept form.
$y = 545 + 4.6x$

c. What are the real-world meanings of the slope and the y-intercept in this situation?

d. Could you use the point-slope equation $y = 821 + 4.6(x - 60)$ to model this situation? Explain why or why not.

e. What is the real-world meaning of the point used to write the equation in 12d?

Avery's Hike

Time spent hiking (min)	Cumulative number of calories burned
5	568
10	591
15	614
20	637

▶ **Review**

13. The table shows hourly compensation costs in 15 countries for 1975, 1985, and 1995. Use the list commands on your calculator to do this statistical analysis.

a. Choose at least three countries and graph the hourly compensation costs for those countries over time. Write a paragraph describing the trends you notice and the conclusions you draw. Answers will vary.

b. Which of the 15 countries had the largest increase in compensation costs from 1975 to 1995? Which country had the least?

c. Create three box plots that compare the compensation costs for the three years. Write a brief paragraph analyzing your graph.

Hourly Compensation Costs (in U.S. dollars) for Production Workers

Country	1975	1985	1995
Australia	5.62	8.20	15.05
Canada	5.96	10.94	16.04
Denmark	6.28	8.13	24.07
France	4.52	7.52	20.01
Germany	6.31	9.53	32.22
Hong Kong	0.76	1.73	4.82
Israel	2.25	4.06	10.54
Italy	4.67	7.63	16.21
Japan	3.00	6.34	23.82
Luxembourg	6.50	7.81	23.35
Mexico	1.47	1.59	1.51
Spain	2.53	4.66	12.88
Sri Lanka	0.28	0.28	0.48
Taiwan	0.40	1.50	5.92
United States	6.36	13.01	17.19

(*2000 New York Times Almanac,* p. 515)

The production workers are inspecting automobile bodies at an American factory.

14. Plot the points $(4, 2)$, $(1, 3.5)$, and $(10, -1)$ on graph paper. These points are on the same line, or collinear, so you can draw a line through them.

  a. Draw a slope triangle between $(4, 2)$ and $(1, 3.5)$, and calculate the slope from the change in y and the change in x. $\frac{change\ in\ y}{change\ in\ x} = -\frac{1.5}{3} = -0.5$

  b. Draw another slope triangle between $(10, -1)$ and $(4, 2)$, and calculate the slope from the change in y and the change in x. $\frac{change\ in\ x}{change\ in\ y} = -\frac{3}{6} = -0.5$

  c. Compare the slope triangles and the slopes you calculated. What do you notice?

  d. What would happen if you made a slope triangle between $(10, -1)$ and $(1, 3.5)$?

15. Show how to solve the equation $3.8 - 0.2(z + 6.2) = 5.4$ by using an undoing process to write an expression for z. Check your answer by substituting it into the original equation. $z = \frac{3.8 + 5.4}{0.2} - 6.2; z = 39.8$

Exercise 14 Repeat that *collinear* means lying on the same line. This exercise reviews Lesson 5.1.

14c. Possible answers: The slope triangles are similar and the slopes are equal.

14d. Possible answers: You would get a larger yet similar triangle and a slope of -0.5.

Exercise 15 This exercise reviews Lesson 4.2.

IMPROVING YOUR **GEOMETRY** SKILLS

Think about triangles drawn on the coordinate plane.

Draw a triangle that satisfies each of these sets of conditions. If it's not possible, tell why not.

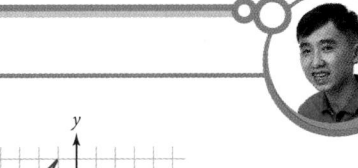

  1. a triangle with all three sides having positive slope

  2. a right triangle (one angle is 90°) with all three sides having negative slope

  3. an equilateral triangle (three equal sides) with one side having slope 0

  4. an obtuse triangle (one angle is greater than 90°) with one side having positive slope and the other two having negative slope

  5. an isosceles triangle (two equal sides) with all three sides having positive slope

  6. a right triangle with one side having undefined slope, one side having slope 0, and one side having slope 1

  7. a triangle with two sides having the same slope

IMPROVING GEOMETRY SKILLS

2. Imagine rotating a right angle about its vertex. Whenever one leg has a positive slope, the other has a negative slope, so this is not possible.

7. This is not possible because two lines with the same slope are parallel and never meet, but each pair of a triangle's edges must meet at a vertex.

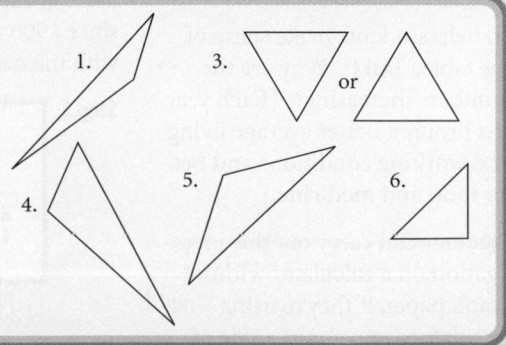

Say that you can use two points in a data set to find the **point-slope form of a linear equation** and then use it to make predictions.

BUILDING UNDERSTANDING

The exercises involve using the point-slope form of linear equations to fit data and make predictions.

ASSIGNING HOMEWORK

Essential	1–5, 8
Performance assessment	6
Portfolio	6
Journal	3, 5
Group	4–6
Review	7–10

▶ Helping with the Exercises

Exercise 3 Remind students that the *x-intercept* is the point where the line crosses the *x*-axis.

Exercise 4 Students might choose to use as input values the number of years since 1900 or 1950 or 1976.

4a. An equation could be $y = 341 + 1.5(x - 1982)$.

4b. All graphs should look approximately like this

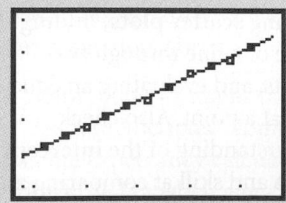

[1975, 2000, 5, 325, 375, 10]

4c. The equation $y = 341 + 1.5(x - 1982)$ gives 398 ppm.

4d. Using the equation in 4c, the *x*-intercept is about 1755. It represents the year when the concentration of CO_2 would

See page 287 for answers to Exercises 5d and 5e.

---

3. Graph each linear equation on your calculator and name the *x*-intercept. Make an observation about the *x*-intercept of any equation in the form $y = b(x - x_1)$.

   a. $y = 2(x - 3)$ 3   **b.** $y = \frac{1}{3}(x + 4)$ −4   **c.** $y = -1.5(x - 6)$ 6
   The *x*-intercept of $y = b(x - x_1)$ is at $x = x_1$.

4. **APPLICATION** Carbon dioxide is one of several greenhouse gases that is emitted into the atmosphere from a variety of sources, including automobiles. The table shows the concentration of carbon dioxide (CO_2) in the atmosphere measured from the top of Mauna Loa volcano in Hawaii each January. The concentration of CO_2 is measured in parts per million (ppm).

   a. Define variables and write an equation in point-slope form that models the data.

   b. Graph your equation to confirm that the line fits the data.

   c. Use your equation to predict what the concentration of CO_2 will be in 2020.

   d. What would be the *x*-intercept for your equation? Does its real-world meaning make sense? Explain why or why not.

CO₂ Concentration

Year	CO$_2$ (ppm)
1976	332
1978	335
1980	338
1982	341
1984	344
1986	346
1988	350
1990	354
1992	356
1994	358
1996	362

Mauna Loa is the largest and most active volcano on Earth. Research on Mauna Loa has revealed a great deal about global changes in the atmosphere.

(*Carbon Dioxide Information and Analysis Center*)

▶ Reason and Apply

5. **APPLICATION** Alex collected this table of data by using two thermometers simultaneously. Alex suspects that one or both of the thermometers are somewhat faulty.

   a. Graph the data.

   b. Write an equation in point-slope form that models Alex's data.

   c. Graph your equation to confirm that the line fits the data.

   d. The freezing point of water is 0°C, which is equivalent to 32°F. The boiling point of water is 100°C, which is equivalent to 212°F. Use this information to write another equation in point-slope form that models the true relationship between the Celsius and Fahrenheit temperature scales.

   e. Write the equations from 5b and d in intercept form. Are they equivalent?

   f. Do you think that Alex's thermometers are faulty? Explain why or why not. The difference could be a result of measurement error or faulty procedures.

Temperature Readings

Celsius (°C) x	Fahrenheit (°F) y
14.5	55.0
20.0	67.0
28.4	86.7
39.5	105.6
32.3	87.1
29.0	81.6
26.2	82.3
25.7	75.2
31.2	88.6

---

have been zero. This is not a reasonable conclusion because plants depend on CO_2 so there would have been some concentration of CO_2 for as long as there have been plants.

5a.

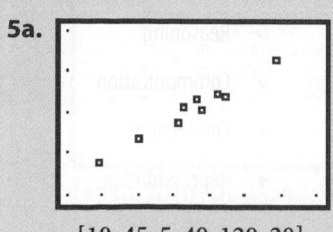

[10, 45, 5, 40, 120, 20]

5b. Using the points (20, 67) and (31.2, 88.6), the slope is approximately 1.9 and a possible equation is $y = 67 + 1.9(x - 20)$.

5c. One possible answer:

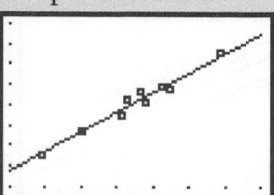

6. APPLICATION The table lists the concentration of dissolved oxygen (DO) in parts per million at various temperatures in degrees Celsius from a sample of lake water.

a. Graph the data.

b. Write an equation in point-slope form that models the data.

c. Graph your equation to confirm that the line fits the data.

d. Use your equation to predict the concentration of dissolved oxygen in parts per million when the water temperature is 2°C.

e. Use your equation to predict the water temperature in degrees Celsius when the concentration of dissolved oxygen is 11 ppm. Using $y = 11 - 0.6(x - 13)$, the temperature is 13°C.

Dissolved Oxygen

Temperature (°C) x	DO (ppm) y
17	8
15	9
13	11
16	10
11	14
13	11
10	14
8	14
6	16
7	13
8	14
4	17
5	15
9	13
6	16

► **Review**

7. Rewrite each equation in intercept form. Show the steps to make the conversion. Check your answer with a calculator table or graph.

a. $y = 3 + 2(x - 7)$ $y = -11 + 2x$ **b.** $y = -11 + 41x + 28$ $y = 17 + 41x$

c. $y = 5 - 6(x - 9)$ $y = 59 - 6x$ **d.** $y = 4(7 - x) - 19$ $y = 9 - 4x$

8. Give the five-number summary for each data set.

a. {1, 2, 4, 7, 18, 20, 21, 21, 26, 31, 37, 45, 45, 47, 48}   1, 7, 21, 45, 48

b. {30, 32, 33, 35, 39, 41, 42, 47, 72, 74}   30, 33, 40, 47, 74

c. {107, 116, 120, 120, 138, 140, 145, 146, 147, 152, 155, 156, 179}   107, 120, 145, 153.5, 179

d. {85, 91, 79, 86, 94, 90, 74, 87}   75, 82, 86.5, 90.5, 94

9. APPLICATION Bryan has bought a box of biscuits for his dog, Anchor. Anchor always gets three biscuits a day. At the start of the 10th day after opening the box, Bryan counts 106 biscuits left. Let x represent the number of days after opening the box, and let y represent the number of biscuits left.

a. In a graph of this situation, what is the slope? -3

b. Write a point-slope equation that models the situation. $y = 106 - 3(x - 10)$

c. When will the box be empty?

d. What is the real-world meaning of the y-intercept?

10. Consider this expression:

$$9\left(\frac{x - 11}{9} + 1\right) + 2$$

a. Write in words the number trick the expression describes.

b. Test the trick to be sure it works. Do you get the same result no matter what number you start with?   Check student work.

c. What operation(s) make this trick work, and how?

5d. $y = 32 + 1.8(x - 0)$ or $y = 212 + 1.8(x - 100)$

5e. The sample equation in 5b gives $y = 29 + 1.9x$; the equations in 5d both give $y = 32 + 1.8x$.

Exercise 6 [Ask] "Why does the graph of the data show only 12 points?" [Three are double points: $(6, 16)$, $(8, 14)$, $(13, 11)$.]

6a.

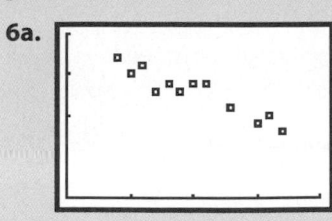

$[0, 20, 5, 0, 20, 5]$

6b. Using $(13, 11)$ and $(8, 14)$, the equation is $y = 11 - 0.6(x - 13)$ or $y = 14 - 0.6(x - 8)$.

6c. One possible answer:

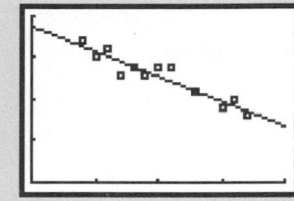

6d. Using $y = 11 - 0.6(x - 13)$, the concentration of dissolved oxygen is 17.6 ppm.

Exercise 7 The exercise reviews Chapter 4 and Lesson 5.4.

Exercise 8 This exercise presents an essential review of Lesson 1.3. In Lesson 5.6, students will need to be able to find five-number summaries.

Exercise 9 This exercise reviews Lesson 5.3.

9c. After 45 full days, there will be only one biscuit left, so the box will be empty in the middle of the 46th day.

9d. When the box was new, before Anchor had any biscuits, there were 136 biscuits.

Exercise 10 This exercise reviews Lesson 4.2.

10a. Start with a number. Subtract 11, divide by 9, then add 1. Multiply the result by 9 and then add 2. The result should be your original number.

10c. Dividing by 9 and multiplying by 9 undo each other. Adding 2 undoes the -2 that results inside the parentheses from $-11 + 9(1)$.

LESSON

5.6

More on Modeling

When you can measure what you are talking about and express it in numbers, you know something about it.

LORD KELVIN

Several times in this chapter you have found the equation of a representative line to fit data. Making, analyzing, and using predictions based on equation models is important in the real world. For this reason it is often helpful and even important that different people arrive at the same model for a given set of data. For this to happen, each person must get the same slope and *y*-intercept. To do that, they have to follow the same systematic method.

PLANNING

LESSON OUTLINE

First day:

50 min   Investigation (Steps 1 to 12)

Second day:

10 min   Investigation (Step 13)

10 min   Sharing

10 min   Example

 5 min   Closing

15 min   Exercises

MATERIALS

• Calculator Notes 1C, 5B

• Stopwatch or watch with second hand

• Bucket or other object to pass

• Graph paper

TEACHING

This lesson builds on the five-number summaries introduced in Lesson 1.3. One way to standardize the choice of two points through which a line of fit passes is to use the first and third quartiles of each data variable.

  Guiding the Investigation

Step 1 **[ESL]** A *bucket brigade* is a line of people passing a bucket, usually full of water to help put out a fire.

In place of a bucket, students can use any object they can pass. They might run each brigade several times and average the times. In this unusual situation, time is the output rather than the input variable. **[Ask]** "Which variable depends on which other variable?" [The time depends on how far the bucket needs to be passed.]

Investigation
Fire!!!!

You will need

• a stopwatch
• a bucket
• graph paper

In this investigation you will use a systematic method for finding a particular line of fit for data.

Procedure Note

Select a class member as timer. Everyone should line up single file. Your line might wrap around the room. Spread out so that there is an arm's length between two people.

Step 1 | Line up in a bucket brigade. Record the number of people in the line. (See the procedure note.) Starting at one end of the line, pass the bucket as quickly as you can to the other end. Record the total passing time from picking up the bucket to setting it down at the very end.

Step 2 | Now have one or two people sit down and close up the gaps in the line. Repeat the bucket passing. Record the new number of people and the new passing time.

Step 3 | Continue the bucket brigade until you have collected 10 data points in the form (*number of people, passing time in seconds*).

Step 2 Students who just sat down might do the timing, record the data, or begin to make scatter plots.

Step 4	Let x represent the number of people, and let y represent time in seconds. Plot your data on graph paper.
Step 5	List the five-number summary for the x-values and the five-number summary for the y-values.
Step 6	What are the first-quartile (Q1) and third-quartile (Q3) values for the x-values in your data set? What are the Q1- and Q3-values for the y-values in your data set?
Step 7	On your graph, draw a horizontal box plot just below the x-axis using the five-number summary for the x-values. Draw a vertical box plot next to the y-axis using the five-number summary for the y-values. A sample graph is shown. Your data and graph will look different based on the data that you collect.

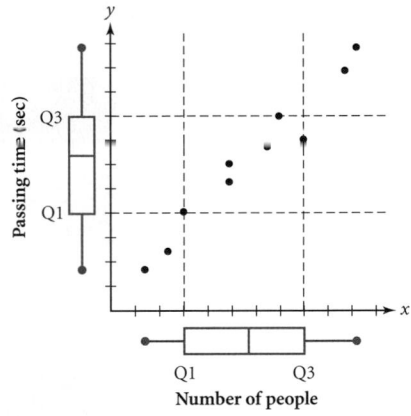

Step 8 Q-points may not be data points. Many factors influence this including the number of data points and the strength of the linear relationship (how close the points are to a line). However, everyone should get the same Q-points.

Step 8	Draw vertical lines from the Q1- and Q3-values on the x-axis box plot into the graph. Draw horizontal lines from the Q1- and Q3-values on the y-axis box plot into the graph. These lines should form a rectangle in the plot. Suppose we call the vertices of this rectangle **Q-points.** Do the Q-points have to be actual data points? Why or why not? Will everyone get the same Q-points?
Step 9	Draw the diagonal of this rectangle that shows the direction of the data. Extend this diagonal through the plot. Is the line a good fit for the data? Are any of the original data points on your line? If so, which ones?
Step 10	Find the coordinates of the two Q-points the line goes through and write a point-slope equation for the line.
Step 11	What are the real-world meanings of the slope and y-intercept of this model?
Step 12	What are the advantages and disadvantages of having a systematic procedure for finding a model for data?

| Step 13 | Use your calculator to plot the data points, draw the vertical and horizontal lines, and find a line of fit by this method. [▶ 🖳 See **Calculator Note 5B** for help on using the draw menu.◀] |

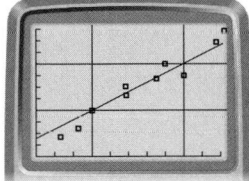

Step 5 Refer students to Lesson 1.3 or Calculator Note 1C to review five-number summaries.

Step 7 If students have difficulty making a vertical box plot, encourage them to make a horizontal one and then rotate it.

Step 11 The slope represents how long it takes one person to pass the bucket. The y-intercept is the time it takes for no people to pass the bucket (0), or it might represent the time to begin and end the experiment (lift the bucket and set it down at the end).

SHARING IDEAS

Ask what Q-points and equations different groups found. Be sure students see that this method always yields the same lines.

Have groups share their ideas about the questions in Steps 11 and 12. Step 12 foreshadows Lesson 5.7. **[Ask]** "Why are the Q-points better than extreme points?" [Points at the extremes of data tend to be less reliable because measurements and instruments are often least accurate there. Also, in the real world, some relationships have nonlinear end-behavior, such as a rubber band stretched to its breaking point.]

Ask when Q-points will be actual data points. See teacher comments in Chapter 1 (page 52) about when quartiles will be data points. Even if the quartiles are points in the one-variable data, the pairs of quartiles might not be points in the two-variable data.

LESSON OBJECTIVE

- Use quartiles to find an equation to fit a set of data

NCTM STANDARDS

CONTENT	PROCESS
● Number	● Problem Solving
● Algebra	● Reasoning
Geometry	● Communication
Measurement	● Connections
● Data/Probability	● Representation

From students' work on the investigation and their contributions to Sharing, you can assess their skills at recording data systematically, plotting points, finding **five-number summaries,** drawing **box plots,** and writing linear equations in **point-slope form.**

EXAMPLE

This example is good for students who didn't understand the investigation very well. **[Ask]** "Why is the slope negative?" [As a point moves along the line from left to right, its *y*-value decreases while its *x*-value increases. Hence either the numerator or denominator of the slope will be negative.]

This method of finding the line of fit based on Q-points is more direct than the methods you used in Lessons 5.2 and 5.5. It is more systematic, too, because everyone will get the same points and the points themselves relate to measures of center in the upper and lower halves of the data set. Points that have some distance between them, but are not at the extremes of the data, are probably more reliable for locating the line of fit.

These students are collecting water samples. Their samples can be analyzed for many things, including dissolved oxygen.

EXAMPLE

The table lists the concentration of dissolved oxygen (DO) in parts per million at various temperatures in degrees Celsius from a sample of lake water. Find a line of fit based on Q-points for the data and use it to predict the temperature for water with only 4 ppm dissolved oxygen.

Dissolved Oxygen

Temperature (°C) x	DO (ppm) y	Temperature (°C) x	DO (ppm) y
17	8	8	14
16	10	8	14
15	9	7	13
13	11	6	16
13	11	6	16
11	14	5	15
10	14	4	17
9	13		

▶ **Solution**

The five-number summaries are

For temperature (*x*-values): 4, 6, 9, 13, 17

For dissolved oxygen (*y*-values): 8, 11, 14, 15, 17

The first-quartile and third-quartile values are

For the x-values: Q1 = 6, Q3 = 13
For the y-values: Q1 = 11, Q3 = 15

A sketch of the scatter plot shows the appropriate Q-points are (6, 15) and (13, 11). Note that (6, 15) is not actually one of the data points but (13, 11) is.

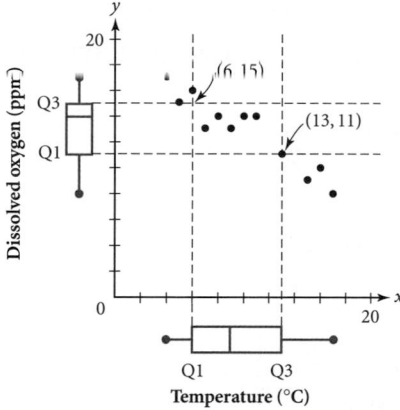

Calculating the slope between these two points you get

$$b = \frac{y_2 - y_1}{x_2 - x_1} = \frac{(11 - 15)}{(13 - 6)} = \frac{-4}{7} \approx -0.57$$

This means that if the temperature *rises* 1°C, the dissolved oxygen level *decreases* by 0.57 ppm. It also means that if the temperature *drops* 1°C, the dissolved oxygen level *increases* by 0.57 ppm.

Using the slope −0.57 and the coordinates of the point (6, 15) in the point-slope form gives

$$y = y_1 + (x - x_1)$$

$$y = 15 - 0.57(x \quad 6)$$

To find the temperature when the amount of dissolved oxygen is 4 ppm, substitute 4 for *y* in the equation and solve for *x*.

$y = 15 - 0.57(x - 6)$	Original equation.
$4 = 15 - 0.57(x - 6)$	Substitute 4 for *y*.
$-11 = -0.57(x - 6)$	Subtract 15 from both sides.
$19.3 \approx (x - 6)$	Divide both sides by −0.57.
$25.3 \approx x$	Add 6 to both sides.

At about 25°C, the water will have about 4 ppm dissolved oxygen.

If you go on to study statistics, you'll learn other systematic ways to find a line of fit for a data set, as well as how to find curves to model nonlinear patterns in data.

As needed, say that one way to standardize the choice of two points through which a line of fit passes is to use the first and third quartiles of each data variable.

ASSIGNING HOMEWORK

Essential	1–4, 7 or 8
Performance assessment	7, 8
Portfolio	4
Journal	3, 6, 12
Group	5, 9
Review	10–12

▶ **Helping with the Exercises**

Exercise 1 Either column of data
may be considered input or
output. Choosing the driving
distance as input would give
different equations. Different
procedures for rounding will
also lead to a variety of answers.

1c.

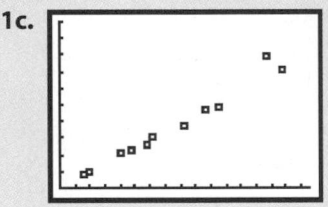

$[0, 1650, 100, 0, 2500, 250]$

1e.

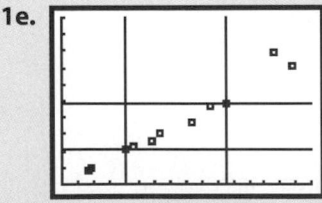

$(405, 514), (1052, 1194)$

1f.

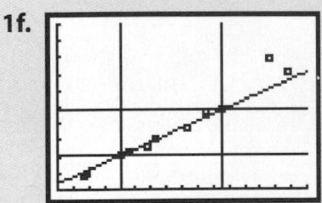

The slope is approximately 1.05;
$y = 1194 + 1.05(x - 1052)$ or
$y = 514 + 1.05(x - 405)$.

EXERCISES

You will need your calculator for problem **11.**

▶ **Practice Your Skills**

1. APPLICATION This table shows that the traveling distances between some cities
depend on how you travel.

Traveling Distances

From	To	Flying distance (mi)	Driving distance (mi)
Detroit, MI	Memphis, TN	623	756
St. Louis, MO	Minneapolis, MN	466	559
Dallas, TX	San Francisco, CA	1483	1765
Seattle, WA	Los Angeles, CA	959	1150
Washington, DC	Pittsburgh, PA	192	241
Philadelphia, PA	Indianapolis, IN	585	647
New Orleans, LA	Chicago, IL	833	947
Cleveland, OH	New York, NY	405	514
Birmingham, AL	Boston, MA	1052	1194
Denver, CO	Buffalo, NY	1370	1991
Kansas City, MO	Omaha, NE	166	204

a. What are the five-number summary values of the flying distances? 166, 405, 623, 1052, 1483

b. What are the five-number summary values of the driving distances? 204, 514, 756, 1194, 1991

c. Plot the data points. Let x represent flying distance in miles, and let y represent
driving distance in miles.

d. Will the slope of the line through these points be positive or negative? Explain
your reasoning. The slope will be positive because as the flying distance increases so does the driving distance.

e. Use the five-number summary values to draw a rectangle on the graph of the
data. Name the two Q-points you
should use for your line of fit.

f. Find the equation of the line and
graph the line with your data points.

g. The flying distance from Louisville,
Kentucky, to Miami, Florida, is
919 miles. Predict the driving distance
from Louisville to Miami. approximately
1054 miles

h. The driving distance from Phoenix,
Arizona, to Salt Lake City, Utah, is
651 miles. Predict the flying distance
from Phoenix to Salt Lake City.
approximately 535 or 536 miles

2. APPLICATION Let x represent total fat in grams, and let y represent saturated fat in grams. Use the model $y = 10 + 0.5(x - 28)$ to predict

  a. The number of saturated fat grams for a hamburger with a total of 32 grams of fat. **12 grams of saturated fat**

  b. The total number of fat grams for a hamburger with 15 grams of saturated fat.

                                          38 grams of total fat

3. In a few sentences, describe the differences in the procedures for finding the Q-points for these two data sets.

  a. 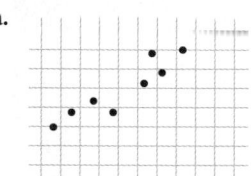   **b.**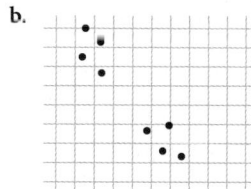

For the data set in 3a, the slope will be positive and you will choose the lower-left and upper-right corners of the rectangle for the Q-points. For the data set in 3b, the slope will be negative and you will choose the upper-left and lower-right corners of the rectangle.

▶ Reason and Apply

4. The table gives the winning times for the Olympic men's 400-meter dash.

  a. Define variables and find the line of fit based on Q-points for the data.

  b. Plot the data points and graph the equation of the model to verify that it is a good fit.

  c. What is the real-world meaning of the slope?

  d. Michael Johnson also won the 400-meter dash in the 2000 Olympic Games. Compare his actual winning time of 43.8 seconds with the winning time predicted by your model.

  e. Could you use this model to predict the winning time 100 years from now? Explain why or why not.

5. Create a data set that has Q-points at (4, 28) and (12, 47) so that only one of those two points is actually part of the data set.

6. Which linear equation below best fits the data at right? Explain your reasoning.

  i. $y = 1.3 + 0.18(x - 6)$

  ii. $y = 2.2 + 0.18(x - 6)$

  iii. $y = 1.3 - 0.18(x - 6)$

  iv. $y = 2.2 - 0.18(x - 6)$

Men's 400-meter Dash

Year	Champion	Country	Time (sec)
1952	George Rhoden	Jamaica	45.9
1956	Charles Jenkins	United States	46.7
1960	Otis Davis	United States	44.9
1964	Mike Larrabee	United States	45.1
1968	Lee Evans	United States	43.9
1972	Vincent Mathews	United States	44.7
1976	Alberto Juantorena	Cuba	44.3
1980	Viktor Markin	USSR	44.6
1984	Alonzo Babers	United States	44.3
1988	Steve Lewis	United States	43.9
1992	Quincy Watts	United States	43.5
1996	Michael Johnson	United States	43.5

(2000 World Almanac, p. 917)

Time (sec) x	Distance from motion sensor (m) y
2	2.8
6	2.2
8	1.7
9	1.5
11	1.3
14	0.9

Exercise 3 Students should realize that the procedures will be the same. The only difference is that they will choose a different set of points at diagonally opposite vertices of the rectangle as the Q-points needed to write the equation.

Exercise 4e This question shows the limitations of the model for making long-range predictions.

4a. The five-number summary for *years* is 1952, 1962, 1974, 1986, 1996. The five-number summary for *time* is 43.5, 43.9, 44.5, 45, 46.7. The Q-points are (1962, 45) and (1986, 43.9). The slope of the line between these two points is approximately -0.056. The possible equations are $y = 45 - 0.056(x - 1962)$ and $y = 43.9 - 0.056(x - 1986)$.

4b.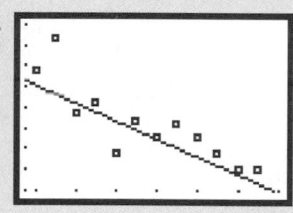

[1950, 2000, 8, 43, 47, 0.5]

4c. The slope means that each additional year the winning time decreases by 0.056 second.

4d. The prediction is 42.9 or 43.1 sec (depending on the equation used). This value is about 1 second less than his winning time.

4e. Answers will vary. However, there is a physical limit to how fast a runner can run. Eventually, the times will have to level off.

5. Answers will vary. One example is {(2, 22) (4, 30) (6, 28) (8, 35) (10, 42) (12, 47) (14, 53)}.

Exercise 6 Encourage lots of approaches here. Some students may find the Q-points and see which line goes through them. Others may count the number of data points on each side of a line or measure the vertical distance from data points to the line.

6. Reasons will vary. The data pattern has a negative slope, and the Q-points that lie on the line are (11, 1.3) and (6, 2.2). The point (6, 1.3) is not one of the Q-points used to draw the line, so equation iv is the correct equation.

Exercise 7 In this problem, as in the investigation, time is more reasonably an output than an input variable. **[Alert]** Some students may need help interpreting 2:00:45 as 45 seconds after 2:00.

7a. $y = 1.3 + 0.625(x - 4)$ or $y = 6.3 + 0.625(x - 12)$

7c. 36.3 seconds after 2:00, or approximately 2:00:36

8a. $y = 1.3 - 0.625(x - 92)$ or $y = 6.3 - 0.625(x - 84)$

8c. after 52.55 seconds, or at approximately 2:00:53

Exercise 10 This exercise reviews Lesson 3.2.

10. The size and cost are almost directly proportional; the 4-oz bottle costs $0.22 per oz, the 7.5-oz bottle costs $0.22 per oz, and the 18-oz bottle costs $0.2217 per oz. If you change the price of the 18-oz bottle to $3.96, then it also will cost exactly $0.22 per oz.

Exercise 11 This exercise reviews Lesson 4.4.

11a. Start with 370, then use the rule Ans − 54.

Time (hr)	Distance from Mt. Rushmore (mi)
0	370
1	316
2	262
3	208
4	154
5	100
6	46

11b.

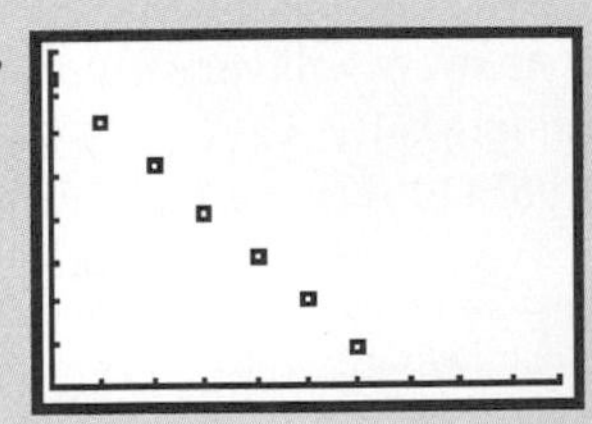

[0, 10, 1, 0, 400, 50]

11c.

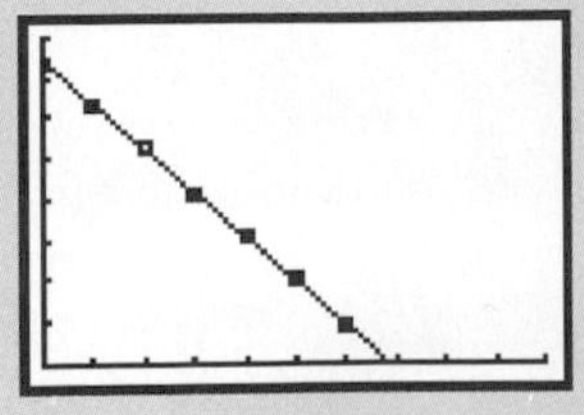

7. At 2:00 P.M., elevator A passes the second floor of the Empire State Building going up. The table shows the floors and the times in seconds after 2:00.

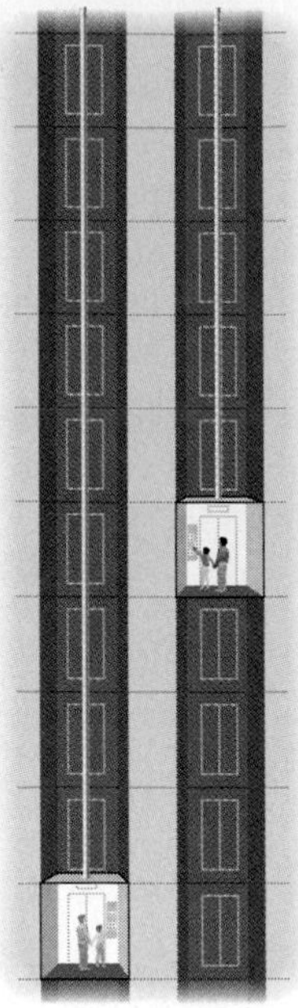

Floor x	2	4	6	8	10	12	14
Time after 2:00 (sec) y	0	1.3	2.5	3.8	5	6.3	7.5

a. What is the line of fit based on Q-points for the data?

b. Give a real-world meaning for the slope. The elevator is rising at a rate of 0.625 second per floor.

c. About what time will this elevator pass the 60th floor if it makes no stops?

d. Where will this elevator be at 2:00:45 if it makes no stops? almost at the 74th floor

8. At 2:00 P.M., elevator B passes the 94th floor of the same building going down. The table shows the floors and the times in seconds after 2:00.

Floor x	94	92	90	88	86	84	80
Time after 2:00 (sec) y	0	1.3	2.5	3.8	5	6.3	8.6

a. What is the line of fit based on Q-points for the data?

b. Give a real-world meaning for the slope. The elevator is moving down at 0.625 second per floor.

c. About what time will this elevator pass the 10th floor if it makes no stops?

d. Where will this elevator be at 2:00:34 if it makes no stops? between the 39th and 40th floors

9. Think about the elevators in problems 7 and 8.

a. Estimate when you expect that elevator A will pass elevator B if neither makes any stops. Estimates will vary.

b. Calculate the actual time. At 28.8 sec, or at about 2:00:29, the elevators will pass at the 48th floor.

Review

10. A 4-oz bottle of mustard costs $0.88, a 7.5-oz bottle costs $1.65, and an 18-oz bottle costs $3.99. Is the size of the mustard bottle directly proportional to the price? If so, show how you know. If not, suggest the change of one or two prices so that they will be directly proportional.

11. A car pulling a camper trailer is traveling from Sioux Falls, South Dakota, to Mt. Rushmore, which is near Rapid City, South Dakota. The car is traveling about 54 miles per hour, and it is about 370 miles from Sioux Falls to Mt. Rushmore.

a. Write a recursive routine to create a table of values in the form

(*time, distance from Mt. Rushmore*)

for the relationship from 0 to 6 hours.

b. Graph a scatter plot of your values using 1-hour time intervals.

c. Draw a line through the points of your scatter plot. What is the real-world meaning of this line? What does the line represent that the points alone do not?

d. What is the slope of the line? What is the real-world meaning of the slope?

The line represents the distance remaining at any time during the trip. With the line, you can see how far you are at any time, instead of just at the top of each hour.

11d. −54; the real-world meaning of the slope is that your distance from Mt. Rushmore decreases by 54 miles each hour.

e. When will the car be at the Wall Drug Store, which is 80 miles from Mt. Rushmore? Explain how you know.

f. When will the car arrive at Mt. Rushmore? Explain how you know.

12. Imagine that a classmate has been out of school for the past few days with the flu. Write him or her an e-mail describing how to convert an equation such as $y = 4 + 2(x - 3)$ from point-slope form to slope-intercept form. Be sure to include examples and explanations. End your note by telling your classmate how to find out if the two equations are equivalent. Answers will vary.

Wall Drug is a landmark in South Dakota. The store's fame began during the 1930s—the Great Depression—when it offered free ice water to travelers.

11e. Answers will vary. The car will reach the Wall Drug Store in the first half hour of the 5th hour of the trip. You can see this on the graph if you look at the line where it has a *y*-value of about 80.

11f. The car will reach Mt. Rushmore after almost 7 hours of travel. You can see this on the graph or in the table since after 7 hours, the car would be 8 miles too far if it kept going.

Exercise 12 Have students read and critique each other's e-mails. A good technique is to choose a few papers, delete the students' names, make overheads, and have each class critique the explanations from another class. This exercise reviews Lesson 5.4.

project

STATE OF THE STATES

Many characteristics of a state vary with the size of the state's population. Some of these relationships are linear. The more people who live in a state, the more houses, cars, schools, and prisoners there are. A lot of data about the states is available on the Internet. You can link to a very useful site through **www.keymath.com/DA** .

Here are two scatter plots that show a comparison of the population of a state to two different characteristics of the state—number of prisoners and median household income. Which scatter plot shows a linear pattern?

Fathom™

Fathom comes with many data sets that contain information about the states, and you can easily download more information from web sites.

Use Fathom's movable line to "eyeball fit" lines through points and read off the slope.

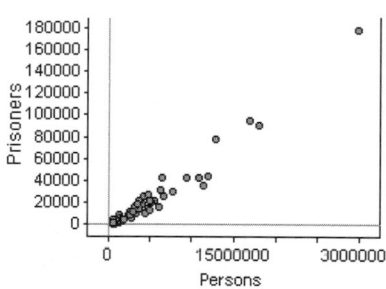

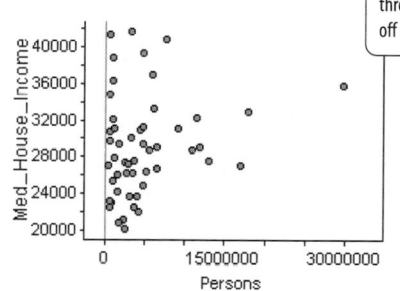

Your project should include

▶ Several scatter plots, investigating relationships between various pairs of characteristics for states.

▶ Lines of fit for your plotted data, their slopes and intercepts along with their real-world meanings (that is, if there appears to be a linear relationship).

▶ Explanations of why some relationships do not appear linear.

Supporting the project

MOTIVATION

The project gives students a chance to study data sets and decide how close to linear they are. **[Ask]** "What characteristics of a state are related to the population?" [One answer: The number of prisoners has a linear relationship to population.]

OUTCOMES

▶ The report includes several scatter plots of data.

▶ Lines of fit are graphed on the scatter plots.

▶ Slopes and intercepts of the lines of fit are included.

▶ There's a discussion of why the data are linear or nonlinear.

• The student explains how the lines of fit were determined.

• The student assesses the goodness of the fit.

• Predictions are made on the basis of the lines of fit.

LESSON OUTLINE

One day:

30 min	Investigation
5 min	Sharing
5 min	Closing
10 min	Exercises

MATERIALS

- graph paper
- uncooked spaghetti

TEACHING

In this lesson students practice their skills finding and evaluating lines of fit. They compare the results of the different methods when applied to the same data set.

Guiding the Investigation

One step Ask students to find lines of fit for the data in the tables in as many ways as possible and to decide which line fits best. Encourage a variety of approaches, at least reviewing all methods seen so far.

You might assign each group one of the two data tables, let them choose their own, or have each group work on both tables. Be aware that the points in Table 1 are exactly linear.

Step 1 It may not be clear to students how to assign input and output variables. Ask which quantity depends on which. **[Ask]** "Does age depend on the diameter of the pupil or vice versa?" [The diameter of the pupil depends on age.]

See page 717 for answers to Step 1.

Applications of Modeling

In Lesson 5.6, you learned a systematic method, using quartile values, to find a line of fit for data points that appear to have a linear pattern. In this lesson, you'll contrast that method with ways you've used before and evaluate your results.

Investigation
What's My Line?

You will need
- graph paper
- a strand of spaghetti

These tables show how the diameter of the pupil of your eye changes with age. Your objective is to arrive at three linear models—the first by "eyeballing" using a pencil or a strand of spaghetti, the second by choosing two data points that show the direction of the data, and the third by finding Q-points and drawing the line through them. Finally, you'll analyze your models and use them to make a prediction.

Procedure Note

Two people in your group should work with the data in Table 1, and two people should work with the data in Table 2. Work as a group to discuss the answers to the questions in Steps 8, 10, and 13.

Table 1

Age (yrs)	Diameter of pupil in daylight (mm)
20	4.7
30	4.3
40	3.9
50	3.5
60	3.1
70	2.7
80	2.3

Table 2

Age (yrs)	Diameter of pupil at night (mm)
20	8.0
30	7.0
40	6.0
50	5.0
60	4.1
70	3.2
80	2.5

(John Lord, *Sizes*, 1995, p. 120)

Step 2 Answers will vary. Students will need to use the y-intercept and the second point to find the slope—remind them that the x-coordinate for the y-intercept is zero.

First, you'll find a line of fit using an "eyeballing" method. Remember that the object of a linear model is to summarize or generalize the data.

Step 1 Plot the data on graph paper. Lay a piece of spaghetti on the plot so that it crosses the y-axis and follows the direction of the data. Try not to focus on the points themselves, but on the general direction of the "cloud" of points.

Step 2 Estimate the coordinate of the y-intercept. Locate a point with convenient coordinates along the strand. Use this information to write the equation of the line. Make a note of this equation.

Next, you'll find a line of fit by choosing "representative" data points.

Step 3 Make a scatter plot of the data on your calculator. Choose two data points that you think show the direction of the data.

LESSON OBJECTIVES

- Review various approaches to finding lines to fit sets of data
- Evaluate the results of those approaches

NCTM STANDARDS

CONTENT		PROCESS	
•	Number	•	Problem Solving
•	Algebra	•	Reasoning
	Geometry	•	Communication
	Measurement	•	Connections
•	Data/Probability	•	Representation

Step 4 | Use the two points to write a linear equation in point-slope form. Make a note of this equation.

Next, you'll find a line of fit using Q-points.

Step 5 | Use your calculator to get the five-number summaries of the x- and y-values. Draw a rectangle using the first- and third-quartile values for the x-values and the first- and third-quartile values for the y-values. Name the Q-points you should use for the data.

Step 6 Table 1:
$y = 4.3 - 0.04(x - 30)$ or
$y = 2.7 - 0.04(x - 70)$
Table 2:
$y = 7.0 - 0.095(x - 30)$ or
$y = 3.2 - 0.095(x - 70)$

Step 6 | Write the equation of the line of fit you can draw through your selected Q-points. Graph the equation to verify that it is the diagonal of the rectangle you drew on the plot. Make a note of this equation.

Step 8 The diameter of the pupil at night changes more drastically with age than does the diameter of the pupil in daylight.

Finally, you'll compare the lines and their characteristics and decide which method has given you the best-fitting line.

Step 7 | Compare the slopes of all three lines for each table. Do all these numbers have the same real-world meaning? What is it?

Step 9 The y-intercept on either graph represents the diameter of the pupil at birth.

Step 8 | What does the difference in slope values for Table 1 and Table 2 tell you about how your pupil diameter changes?

Step 10 Using the equations from Step 6, 25 years old for daylight; a little over 56 years old for night.

Step 9 | Compare the y-intercepts of all three lines for each table. Do they all have the same real-world meaning? If so, what is it?

Step 10 | At what ages would your models predict a pupil diameter of 4.5 mm? Show how to find this value symbolically.

Step 11 As you change the y-intercept, the y-coordinate of every point on the line increases by the same amount. For example, if the y-intercept changes from 4 to 6, the point (8, 9) moves to (8, 11).

Step 11 | What is the effect of a small change in the y-intercept when you use the model to predict a value in the middle of the data set?

Step 12 | What is the effect of a small change in the slope when you try to predict a y-value far from most of the given points?

Step 13 | Discuss the pros and cons of each procedure you used to find a line of fit. Which method do you like best and why?

> I like the "eyeballing" method because I can visualize the fit of the line before writing an equation.

> Using two points and the point-slope method works best for me. It's nice to have definite points.

> I prefer the certainty of using Q-points. I know my answer will be the same as anyone else's.

Step 12 An age "far from most of the given points" might be 120.

Step 12 A small change in the slope will have a magnifying change to points far out on the line. For example, if the equation $y = 1x$ is changed to $y = 5x$, the point (0.1, 0.1) only changes to (0.1, 0.5), but the point (100, 100) changes to (100, 500).

Step 2 Because students have found the y-intercept, they will probably want to use the intercept form for the equation. Try to be sure all students know how to do that.

Step 5 As needed, remind students to draw horizontal and vertical lines to help find the Q-points.

Step 5 Table 1:
x-values: 20, 30, 50, 70, 80
y-values: 2.3, 2.7, 3.5, 4.3, 4.7
Q-points: (30, 7.0) and (70, 2.7)

Table 2:
x-values: 20, 30, 50, 70, 80
y-values: 2.5, 3.2, 5.0, 7.0, 8.0
Q-points: (30, 7.0) and (70, 3.2)

Step 6 Although no particular form of the equation is requested, students will find it easiest to follow the precedent of Step 4 and find the point-slope form.

SHARING

After Step 6, have several groups graph their lines, using a single overhead transparency for each table of data and a different color for each group. You might lead a full-class discussion of Steps 7 through 13, or ask students to return to their groups. In a class discussion, be especially careful to treat all answers to questions equally, asking the class to critique them, whether or not you agree with them. (If some students seem uncomfortable participating in the class discussion, adjust your plans and return to groups.)

Ask if a line can be a reliable predictor if it's not the best fit. Elicit the idea that the quality of fit is determined by the accuracy of the prediction, so that it depends on the situation.

If your class is ready, you might introduce the median-median method for finding a line of fit. (See page 250B.) Students can check hand calculations with a calculator. Ask if this line is a better model of the data.

Assessing Progress

Watch for students' abilities to plot points, find slopes and y-intercepts, write equations of lines given y-intercept and a slope, write equations given points and slopes, find lines of fit using Q-points, figure out real-world meaning of a slope or y-intercept, and find the value of an expression at a point.

Closing the Lesson

A variety of approaches yield lines to model a set of data. The best of these will be a good predictor of data points not already found.

BUILDING UNDERSTANDING

Students practice making predictions from lines of fit found using Q-points.

ASSIGNING HOMEWORK

Essential	1, 4
Performance assessment	5
Portfolio	4
Journal	8
Group	2
Review	1–3, 6–8

Helping with the Exercises

4a. Let x represent years, and let y represent distance in meters. $y = 60.09 + 0.29(x - 1962)$ or $y = 68.16 + 0.29(x - 1990)$; the slope means the distance the discus is thrown increases by 0.29 m each year; the y-intercept is meaningless in this situation because it would indicate that a negative distance was the winning distance in year 0.

4b. 45.59 meters using $y = 60.09 + 0.29(x - 1962)$; the predicted distance is 0.33 m more than the actual distance.

As you study more about finding models for data, you will also learn more about methods you can use to tell how well a model fits data. In this course, the emphasis will be on finding a reasonable model—even though it may not be the best-fitting model—so that you can use it to make reliable predictions.

EXERCISES

You will need your calculator for problem **6**.

Practice Your Skills

1. The equation of a line in point-slope form is $y = 6 - 3(x - 6)$.
   a. Name the point on this line that was used to write the equation. $(6, 6)$
   b. Name the point on this line with an x-coordinate of 5. $(5, 9)$
   c. Using the point you named in 1b, write another equation of the line in point-slope form. $y = 9 - 3(x - 5)$
   d. Write the equation of the line in intercept form. $y = 24 - 3x$
   e. Find the coordinates of the x-intercept. $(8, 0)$

2. Solve each equation symbolically for x. Use another method to verify your solution.
   a. $3(x - 5) + 14 = 29$ $x = 10$
   b. $\dfrac{8 - 13}{x + 5} = 2$ $x = -7.5$
   c. $\dfrac{2(3 - x)}{4} - 8 = -7.75$ $x = 2.5$
   d. $11 + \dfrac{6(x + 5)}{9} = 42$ $x = 41.5$

3. Solve each equation for y.
   a. $2x + 5y = 18$
   $y = \dfrac{18 - 2x}{5}$ or $y = 3.6 - 0.4x$
   b. $5x - 2y = -12$
   $y = \dfrac{-12 - 5x}{-2}$ or $y = 6 + 2.5x$

Reason and Apply

4. **APPLICATION** This table shows winning distances for the Olympic men's discus throw.
   a. Define variables and find the line of fit based on Q-points for this data set. Give the real-world meanings of the slope and the y-intercept.
   b. In 1912, Armas Taipale of Finland threw the discus 45.21 m. What value does your model predict for that year? What is the difference in the two values?
   c. According to your model, what year might you expect the winning distance to pass 80 meters? Show how to find this value symbolically.

Men's Discus

Year	Champion	Country	Distance (m)
1952	Sim Iness	United States	55.03
1956	Al Oerter	United States	56.36
1960	Al Oerter	United States	59.18
1964	Al Oerter	United States	61.00
1968	Al Oerter	United States	64.78
1972	Ludvik Danek	Czechoslovakia	64.40
1976	Mac Wilkins	United States	67.50
1980	Viktor Rashchupkin	USSR	66.64
1984	Rolf Danneberg	West Germany	66.60
1988	Jürgen Schult	East Germany	68.82
1992	Romas Ubartas	Lithuania	65.12
1996	Lars Riedel	Germany	69.40
2000	Virgilijus Alelena	Lithuania	69.30

(*2000 World Almanac*, p. 919, and *www.olympics.com*)

4c. 2030 using $80 = 60.09 + 0.29(x - 1962)$; but the summer Olympics are held only every four years so either 2028 or not until 2032.

5. **APPLICATION** The table shows the timetable for the Coast Starlight train from Seattle to Los Angeles.

Coast Starlight

Location	Distance from Los Angeles (mi)	Arrival time	Elapsed time from Seattle (min)
Kelso, WA	1252	12:52	172
Vancouver, WA	1213	13:35	205
Salem, OR	1150	15:45	275
Eugene, OR	1079	17:07	357
Sacramento, CA	552	6:30	1160
Emeryville, CA	468	8:30	1280
Salinas, CA	355	12:01	1491
Santa Barbara, CA	103	18:17	1867

a. Define variables and give the line of fit based on Q-points for this data set. Give the real-world meaning of the slope.

b. While riding the train, you pass a sign that says you are 200 miles from Los Angeles. What length of time does your model predict you have traveled?

c. The train comes to a stop after 10 hours (600 minutes). According to your model, how far are you from Los Angeles? Show how to find this value symbolically.

Before 1971, when Amtrak created the Coast Starlight, passengers had to ride three different trains to go from Seattle to Los Angeles.

▶ Review

6. A music teacher is using a catalog to purchase new strings for instruments. The number of violin, viola, cello, and bass strings, the type of string, and the cost are listed in the matrices below.

$$[A] = \begin{array}{c} \\ \text{Violin} \\ \text{Viola} \\ \text{Cello} \\ \text{Bass} \end{array} \begin{array}{ccccc} E & A & D & G & C \\ \begin{bmatrix} 6 & 8 & 6 & 1 & 0 \\ 0 & 4 & 8 & 3 & 1 \\ 0 & 6 & 5 & 4 & 4 \\ 1 & 1 & 2 & 2 & 0 \end{bmatrix} \end{array} \qquad [B] = \begin{array}{c} E \\ A \\ D \\ G \\ C \end{array} \begin{array}{c} Cost \\ \begin{bmatrix} 8.50 \\ 12.25 \\ 16.50 \\ 18.75 \\ 22.25 \end{bmatrix} \end{array}$$

a. Without using your calculator's matrix menu, find the cost of strings for each section. Explain how you found your answer.

b. Verify that your answer to 6a is correct by using your calculator to multiply the quantity matrix by the cost matrix.

c. Label the rows and columns of your answer to 6a. Explain what the entries in the matrix tell you.

d. Do you get a meaningful answer if you multiply [B] · [A]? Why or why not?
   No, you can't multiply [B] · [A] because the inside dimensions don't match.

6a.
Violins' cost: 6($8.50) + 8($12.25) + 6($16.50) + 1($18.75) + 0($22.25) = $266.75

Violas' cost: 0($8.50) + 4($12.25) + 8($16.50) + 3($18.75) + 1($22.25) = $259.50

Cellos' cost: 0($8.50) + 6($12.25) + 5($16.50) + 4($18.75) + 4($22.25) = $320.00

Basses' cost: 1($8.50) + 1($12.25) + 2($16.50) + 2($18.75) + 0($22.25) = $91.25

6b.
$$\begin{bmatrix} 266.75 \\ 259.50 \\ 320.00 \\ 91.25 \end{bmatrix}$$

6c.
$$\begin{array}{c} \\ \text{Violin} \\ \text{Viola} \\ \text{Cello} \\ \text{Bass} \end{array} \begin{array}{c} Cost \\ \begin{bmatrix} 266.75 \\ 259.50 \\ 320.00 \\ 91.25 \end{bmatrix} \end{array}$$

Exercise 5 [Alert] Some students may not realize that columns 1 and 3 are extraneous. Either column 2 or 4 can be used for input data, and the other for output.

5. Let x represent distance from Los Angeles in miles, and let y represent elapsed time from Seattle in minutes.

5a. $y = 1385.5 - 1.49(x - 411.5)$ or $y = 240 - 1.49(x - 1181.5)$. The slope means the distance from Los Angeles decreases by 1 mile each 1.49 minutes.

5b. approximately 1701 min, or 28 hrs 21 min, by the first equation or approximately 1702 min, or 28 hrs 22 min, by the second equation

5c. approximately 939 miles by the first equation or 940 miles by the second equation

Exercise 6 [ESL] *Bass* is pronounced like *base*. Assign this exercise only if you have done Lesson 1.8.

Step 3 Students may want to proceed after one measurement. Ask them why they should take the mean of several measurements.

Step 8 Students can solve the equation by undoing or balancing. You'll want to guide them in selecting a location. Bleachers, a balcony, a stairwell, or a second-story window work well.

If possible, set up a video camera to film the final drops. Focus on the area near the floor or ground. Then view the videotape in slow motion, one frame at a time, to see just how close the object comes to the ground. You can actually measure a distance on the television screen and then compare the various distances to determine the best bungee jump.

SHARING IDEAS

You might ask the groups that made the best predictions to share with the class parts D to G of their report.

Assessing Progress

Assess students' abilities to work together, make careful measurements, gather data systematically, find the mean of a data set, make **scatter plots,** find equations of **lines of fit,** and **solve linear equations.**

Closing the Lesson

As needed, point out that after finding the best **line of fit,** students had to solve a **linear equation** to predict the number of rubber bands for a given distance.

Step 3 Repeat this jump several times and find a mean value for the distance. Record the number of rubber bands (2) and the mean distance the jumper falls in a table like this one.

Number of rubber bands	2	4	5	6
Distance fallen				

Step 4 Add one or two rubber bands to the bungee cord and repeat the experiment. Record this new information.

Step 5 Continue to make bungee cords of different lengths, and measure the distance your jumper falls until you have at least seven pairs of data. When using long cords, you may need to move to a higher place to measure the falls.

Step 6 Define variables and make a scatter plot of the information from your table.

Step 7 Find the equation of a line of fit for your data. You may use any procedure, but be able to justify why your equation is a reasonable fit.

Step 8 The test! Decide on a good location for all the groups to conduct final bungee jumps from a particular height. Use your equation to determine the number of rubber bands you need in the cord to give your jumper the greatest thrill—falling as close as possible to the ground without touching. When you have determined the number of rubber bands, make the bungee cord and wait your turn to test your prediction.

Step 9 Write a group report for this activity. Follow this outline to produce a neat, organized, thorough, and accurate report. Any reader of your report should not need to have watched the activity to know what is going on.

Report Outline

A. Overview — Tell what the investigation was about, its purpose or objective.

B. Data collection — Describe the data you collected and how you collected it.

C. Data table — Use labels and units.

D. Graph — Show all data points. Use labels and units. Show the line of fit.

E. Model — Define your variables and give the equation.

F. Procedure — Tell how you found this equation and why you used this method.

G. Calculations — Show how you decided how many rubber bands to use in the final jump.

H. Results — Describe what happened on the final jump.

I. Conclusion — What problems did you have? What worked really well? If you could repeat the whole experiment, what would you do to improve it?

CHAPTER 5 REVIEW

In Chapter 4, you learned how to write equations in intercept form, $y = a + bx$. In this chapter, you learned how to calculate **slope** using the slope formula $b = \frac{y_2 - y_1}{x_2 - x_1}$. You also used the slope formula to derive another form for a linear equation—the **point-slope form.** The point-slope form, $y = y_1 + b(x - x_1)$, is the equation of a line through point (x_1, y_1) with slope b. You learned that this form is very useful in real-world situations when the starting value is not on the y-axis.

You continued to investigate equivalent forms of expressions and equations using tables and graphs. You used the distributive property of multiplication over addition and the **commutative** and **associative** properties of addition and multiplication to write point-slope equations in intercept form.

You discovered how to use the first and third quartiles from the five-number summaries of x- and y-values in a data set to write a linear model for data based on **Q-points.**

EXERCISES

You will need your calculator for problems **3, 4,** and **9.**

1. The slope of the line between $(2, 10)$ and $(x_2, 4)$ is -3. Find the value of x_2. Show your work.

2. Give the slope and the y-intercept for each equation.

   a. $y = -4 - 3x$
   slope -3; y-intercept -4

   b. $2x + 7 = y$
   slope 2; y-intercept 7

   c. $38x - 10y = 24$
   slope 3.8; y-intercept -2.4

3. Line a and line b are shown on the graph at right. Name the slope and the y-intercept, and write the equation for each line. Check your equations by graphing. Line a has slope -1, y-intercept 1, and equation $y = 1 - x$. Line b has slope 2, y-intercept -2, and equation $y = -2 + 2x$.

4. Write each equation in the form requested. Check your answers by graphing.

   $$y = 13.6x - 25,709$$

   a. Write $y = 13.6(x - 1902) + 158.2$ in intercept form.

   b. Write $y = -5.2x + 15$ in point-slope form using $x = 10$ as the first coordinate of the point. $y = -37 - 5.2(x - 10)$

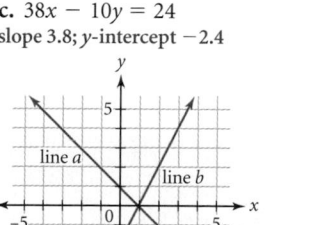

line a

line b

5. Consider the point-slope equation $y = -3.5 + 2(x + 4.5)$.

   a. Name the point used to write this equation. $(-4.5, -3.5)$

   b. Write an equivalent equation in intercept form. $y = 2x + 5.5$

   c. Factor your answer to 5b and name the x-intercept. $y = 2(x + 2.75)$; the x-intercept is -2.75.

   d. A point on the line has a y-coordinate of 16.5. Find the x-coordinate of this point and use this point to write an equivalent equation in point-slope form.

   e. Explain how you can verify that all four equations are equivalent.

PLANNING

LESSON OUTLINE

One day:

10 min Introduction

25 min Exercises

15 min Student self assessment

REVIEWING

Direct attention to the Nutrition Facts data from the example in Lesson 5.2. Ask the class how to predict the saturated fat in a burger with 50 g of total fat. Be sure students describe a variety of ways of finding a line of fit— eyeballing, finding a line through representative points, and using Q-points. Don't be satisfied with a list. Ask for demonstrations of techniques, emphasizing how to find the slope of a line with slope triangles and how to find the point-slope form of an equation. Show how different equations might arise from different choices of points, and review the algebraic properties used to check the equivalence of the equations and to solve equations.

ASSIGNING HOMEWORK

If students complete the odds on their own, they could do the evens in groups.

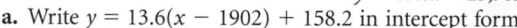

▶ **Helping with the Exercises**

1.
$$-3 = \frac{4 - 10}{x_2 - 2}$$
$$-3(x_2 - 2) = -6$$
$$x_2 - 2 = 2$$
$$x_2 = 4$$

Exercise 5 Some students may have missed the idea of factoring from exercise 6 in Lesson 5.4. Realizing that the x-intercept is the number subtracted from x might also be a challenge. **[Ask]** "What value of x makes y equal to 0?"

5d. The x-coordinate is 5.5; $y = 16.5 + 2(x - 5.5)$.

5e. Answers will vary. Possible methods are graphing, using a calculator table, or putting all equations in intercept form.

Exercise 6 You might ask students to give the reason for each step as a review of the algebraic properties.

7b. −1350; the car's value decreases by $1,350 each year.

7c. 12,600; Karl paid $12,600 for the car.

7d. $9\frac{1}{3}$; in $9\frac{1}{3}$ years the car will have no monetary value.

8b.

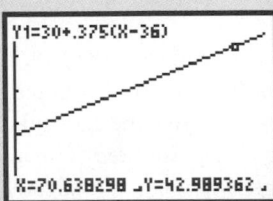

[0, 80, 10, 0, 50, 10]

9a. 1952, 1956, 1976, 1990, 2000; 1.67, 1.835, 1.93, 2.02, 2.05

Exercise 9 Rounding the numerical value of the slope will affect the answer to 9e.

---

6. Show all steps for a symbolic solution to each problem.

   a. $4 + 2.8x = 51$

   b. $38 - 0.35x = 27$

   c. $11 + 3(x - 8) = 41$

   d. $220 - 12.5(x - 6) = 470$

7. **APPLICATION** Suppose Karl bought a used car for $12,600. Each year its value is expected to decrease by $1,350.

   a. Write an equation modeling the value of the car over time. Let x represent the number of years Karl owns the car, and let y represent the value of the car in dollars. $y = 12600 - 1350x$

   b. What is the slope, and what does it mean in the context of the problem?

   c. What is the y-intercept, and what does it mean in the context of the problem?

   d. What is the x-intercept, and what does it mean in the context of the problem?

8. Recall the data about heating a pot of water from the investigation in Lesson 5.3. A possible linear model relating the time x to the temperature y is $y = 30 + 0.375(x - 36)$.

   a. What equation could you solve to find how long it would take before the pot of water reaches 43°C? $43 = 30 + 0.375(x - 36)$

   b. Find the approximate time indicated in 8a using a table or graph. $x \approx 71$ seconds

   c. Show a symbolic solution for your equation in 8a. $x = \frac{43 - 30}{0.375} + 36 = 70.\overline{6}$

9. **APPLICATION** The table gives the winning heights for the Olympic women's high jump.

   a. Find the five-number summaries for the year and height data.

Yelena Yelesina was the first Russian woman to win the Olympic high jump title.

Women's High Jump

Year	Champion	Country	Height (m)
1952	Esther Brand	South Africa	1.67
1956	Mildred McDaniel	United States	1.76
1960	Iolanda Balas	Romania	1.85
1964	Iolanda Balas	Romania	1.90
1968	Miloslava Rezkova	Czechoslovakia	1.82
1972	Ulrike Meyfarth	West Germany	1.92
1976	Rosemarie Ackerman	East Germany	1.93
1980	Sara Simeoni	Italy	1.97
1984	Ulrike Meyfarth	West Germany	2.02
1988	Louise Ritter	United States	2.03
1992	Heike Henkel	Germany	2.02
1996	Stefka Kostadinova	Bulgaria	2.05
2000	Yelena Yelesina	Russia	2.01

(*2000 World Almanac*, p. 920, and *www.olympics.com*)

6a. $4 + 2.8 = 51$

$2.8x = 51 - 4 = 47$

$x = \frac{47}{2.8} \approx 16.8$

6b. $38 - 0.35x = 27$

$-0.35x = 27 - 38 = -11$

$x = \frac{-11}{-0.35} \approx 31.4$

6c. $11 + 3(x - 8) = 41$

$3(x - 8) = 41 - 11 = 30$

$x - 8 = \frac{30}{3} = 10$

$x = 10 + 8 = 18$

6d. $220 - 12.5(x - 6) = 470$

$-12.5(x - 6) = 470 - 220 = 250$

$x - 6 = \frac{250}{-12.5} = -20$

$x = -20 + 6 = -14$

b. Name the Q-points for this data set. The Q-points are (1962, 1.835) and (1990, 2.020).

c. Write an equation for the line through the Q-points. $y = 1.835 + 0.007(x - 1962)$
or $y = 2.020 + 0.007(x - 1990)$

d. Graph the line and the data, and explain whether or not you think this line is a good model for the data pattern.

e. Predict the winning height for the year 2012. Using $y = 1.835 + 0.007(x - 1962)$, the prediction is 2.19 m.

10. APPLICATION This table shows the federal minimum hourly wage for 1974–1997.

a. Find the line of fit based on Q-points.

b. Give the real-world meaning of the slope.

c. Use your model to predict the minimum hourly wage for 2005.

d. Predict when the minimum hourly wage would have been $1.00.

11. Explain how to find the equation of a line when you know

a. The slope and the y-intercept.

b. Two points on that line. Answers will vary.

United States Minimum Wage

Year x	Hourly minimum y	Year x	Hourly minimum y
1974	$1.90	1980	$3.10
1975	$2.00	1981	$3.35
1976	$2.20	1990	$3.80
1977	$2.30	1991	$4.25
1978	$2.65	1996	$4.75
1979	$2.90	1997	$5.15

(*Bureau of Labor Statistics, U.S. Department of Labor*)

TAKE ANOTHER LOOK

Is rate of change the same as slope? For linear equations, you've seen that it is. But what about curves? You've studied inverse variations, whose equations have the form $y = \frac{k}{x}$. Let's look at the equation $y = \frac{12}{x}$ and its graph.

(3, 4) is a point on the curve. Let's choose another nearby point. Substituting 3.5 for x in the equation, you get $y \approx 3.4$. Using the points (3, 4) and (3.5, 3.4) in the formula for slope, you get

$$b = \frac{y_2 - y_1}{x_2 - x_1} = \frac{3.4 - 4}{3.5 - 3} = \frac{-0.6}{0.5} = -1.2$$

We can say that the *average* rate of change for $y = \frac{12}{x}$ on the interval $x = 3$ to $x = 3.5$ is -1.2. But -1.2 is not the "slope" of $y = \frac{12}{x}$. Instead, it is the slope of the *straight* line through the two points (3, 4) and (3.5, 3.4). Is the average rate of change on the x-interval from 3 to 3.25 the same as from 3.25 to 3.5?

Try points on the "wings" of the curve. For instance, (8, 1.5) is on the curve and so is (8.5, 1.4). Again, the y-coordinate is approximate. What is the average rate of change between these points? The x-interval is the same as for the points (3, 4) and (3.5, 3.4), but is the rate of change the same? What does this tell you? What *straight* line through (8, 1.5) has slope equal to the average rate of change on the interval $x = 8$ to $x = 8.5$?

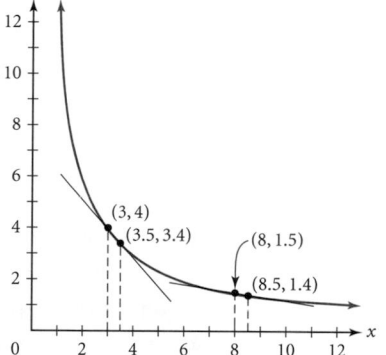

Exercise 10 [Link]
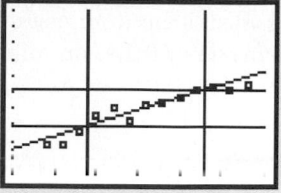

9d. Answers will vary. There are more points below the line than above the line.

[1940, 2000, 10, 1.5, 2.5, 0.1]

Exercise 10 [Link] You may want to have students do research on minimum wage. Some questions to research are, Why was a minimum-wage law enacted? Why doesn't the minimum wage change every year? Does the minimum-wage law apply to all professions?

10a. $y = 2.25 + 0.13(x - 1976.5)$ or $y = 4.025 + 0.13(x - 1990.5)$

10b. The slope means the minimum hourly wage increased approximately $0.13 per year.

10c. Using $y = 2.25 + 0.13(x - 1976.5)$, the prediction is $5.96.

10d. Using $y = 2.25 + 0.13(x - 1976.5)$, the prediction is 1967.

11. Possible answers:

11a. Written as $y = a + bx$, b is the slope and a is the y-intercept.

11b. If the points are (x_1, y_1) and (x_2, y_2), then the slope of the line is $\frac{y_2 - y_1}{x_2 - x_1} = b$. The equation is $y = y_1 + b(x - x_1)$.

▶ **Take Another Look**

The rate of change of a curve (other than a straight line) is not constant. In general, you can't find the slope of a curve at a point by finding the slope of a line between two points on the curve, no matter how close those points are together. The average rate of change over the x-interval from 3 to 3.25 will not be the same as from 3.25 to 3.5.

The rate of change between the points (3, 4) and (3.5, 3.4) is -1.2 and between (8, 1.5) and (8.5, 1.4) is -0.2. This tells us that the rate of change of the y-values is slower on the wings than at the portion of the graph nearest the origin. The line through (8, 1.5) and (8.5, 1.4) is $y = 1.5 - 0.2(x - 8)$.

ASSESSING

Choose three or four constructive assessment items from *Assessment Resources A or B*. Use one of the chapter tests, or create your own test.

FACILITATING SELF ASSESSMENT

To help students complete the portfolio described in Assessing What You've Learned, suggest that they consider for evaluation their work on Lesson 5.1, Exercise 7; Lesson 5.2, Exercises 5 and 7; Lesson 5.3, Exercise 10; Lesson 5.4, Exercises 11 and 12; Lesson 5.5, Exercise 6; Lesson 5.6, Exercise 4; and Lesson 5.7, Exercise 4.

Assessing What You've Learned

In each of the five chapters from Chapter 0 to Chapter 4, you were introduced to a different way to assess what you learned. Maybe you have tried all five ways—keeping a portfolio, writing in your journal, organizing your notebook, doing a performance assessment, and giving a presentation. Maybe you have tried just a couple of these methods. Probably, your teacher has adapted these ideas to suit the needs of your class.

By now, you should realize that assessment is not just giving and taking tests. In the working world, performance in some occupations can be measured in tests, but in all occupations, there is a need to communicate what you know to coworkers. In all jobs, workers demonstrate to their employer or to their clients, patients, or customers that they are skilled in their fields. They need to show they are creative and flexible enough to apply what they've learned in new situations. Assessing your own understanding and letting others assess what you know gives you practice in this important life skill. It also helps you develop good study habits, and that, in turn, will help you advance in school and give you the best possible opportunities in your work life. Keep that in mind as you try one or more of these suggestions.

 WRITE IN YOUR JOURNAL What have you enjoyed more in studying algebra—the numbers, symbols, graphs, and other abstract ways of describing relationships, or the concrete applications and examples that show how people use these ideas in the real world?

Do you find it interesting that a single linear relationship can be described in so many ways, or does that add confusion for you?

 ORGANIZE YOUR NOTEBOOK In your notebook, find examples of the different forms of a linear equation. Assign each form a color, and underline or highlight the equations of each type in the right color. Or create a table with the equation forms as the headings, and list several of each form in the columns.

 UPDATE YOUR PORTFOLIO Choose a piece of work from this chapter to add to your portfolio. Describe the work in a cover sheet, giving the objective, the result, and what you might have done differently.

 PERFORMANCE ASSESSMENT Show a classmate, a family member, or your teacher that you know how to solve an equation using the properties of equalities and other mathematical properties. Give your reasons for each step.

 GIVE A PRESENTATION Research a topic of interest to you that involves two kinds of numerical data. Present the data in a table, make a scatter plot, and describe the pattern of the points. If the data points show a linear pattern, tell how to find a line of fit for the data set and why that line is useful.

Systems of Equations and Inequalities

Overview

In this chapter students use systems of equations and inequalities to model and solve real-life problems. In **Lessons 6.1** through **6.4,** students learn five ways to solve a system of equations: tables, graphs, the substitution method, the elimination method, and row operations on matrices. Inequalities in one variable are introduced in **Lesson 6.5.** Students perform operations on inequalities and learn why multiplying an inequality by a negative number reverses the direction of the inequality. In **Lesson 6.6,** students learn how to graph inequalities in two variables and how to check whether given points are solutions. In **Lesson 6.7,** students graph and solve systems of inequalities.

The Mathematics

Systems of Linear Equations

A linear equation in two variables has infinitely many solutions. Each solution is an ordered pair of numbers. If you substitute those numbers for the two variables, the equation becomes a true statement. A *solution* to a system of two such equations is an ordered pair that is a solution to each equation.

A system may have zero solutions, one solution, or infinitely many solutions. In the first case, the equations are called *inconsistent*. In the last, they're called *redundant*.

A system of two linear equations in two variables can be solved using many methods.

Graphing. If you graph the line represented by each equation, the point of intersection will have coordinates that give a solution to (that is, *satisfy*) both equations. If the lines are parallel, there are no solutions. If the lines coincide, the system has infinitely many solutions.

Tables. To find a solution, if there is one, look for a pair of numbers that appears in the table of solution values for each equation.

Method of substitution. For an exact solution to a system, solve one equation for one of the variables in terms of the other variable, substitute the resulting expression for that variable in the other equation, solve that linear equation in the one remaining variable, substitute the solution into either original equation, and solve for the other variable.

Method of elimination. If the coefficients make it difficult to solve either equation for a variable, multiply one or both equations by a constant to get the opposite coefficients for the same variable in the two equations. Then add the equations, to eliminate that variable, leaving a single linear equation in the other variable. Substitute the solution to that equation into either original equation, which can then be solved for the other variable.

Row reduction of a matrix. Put the coefficients of the two equations into a matrix and then perform row operations that mimic the steps in the method of elimination. A procedure that systematically obtains a diagonal matrix from which the solution can be read easily is called *Gaussian elimination*. Calculators and computers solve systems by using variations on this method.

Linear Inequalities

An inequality is like an equation except that the equal sign is replaced by $<$ (less than), $>$ (greater than), $\leq$ (less than or equal to), or $\geq$ (greater than or equal to). Statements using $<$ or $>$ are called *strict inequalities*.

You can solve linear inequalities in one variable as you would solve linear equations by balancing, with the exception that if you multiply or divide by a negative number, you must switch the direction of the

inequality. The solution can be graphed as a ray on a number line. The endpoint of the ray is either an open circle (for strict inequalities) or a solid circle.

Solutions to an inequality in two variables can be graphed on a plane. First you make an equation by replacing the inequality symbol with an equal sign. Graph that equation. If the inequality is strict, draw a dashed line. Otherwise, use a solid line. The solutions to the inequality form a half-plane on one side of that line. Shade in the solution region.

The solutions to a system of inequalities in two variables are graphed as the intersection of the half-planes representing the solutions of the individual inequalities in the system. Such a graph provides a way of solving simple problems in the area of mathematics called *linear programming*. In that field, the inequalities are called *constraints*, and the solution set is called a *feasible region*. An example of a linear programming problem appears in Take Another Look on page 364.

Using This Chapter

If you prefer not to introduce all five methods for solving systems of equations, the most important methods are in the first two lessons. Lesson 6.4 depends on students having used matrices before.

Resources

Discovering Algebra Resources

Calculator Notes 5B, 6A, 6B, 6C, 6D, 8D

Teaching and Worksheet Masters
  Lesson 6.2 Rope Sample Data
  Lesson 6.4 Row Operations in a Matrix
  Lesson 6.5 Toe the Line
  Lesson 6.6 Graphing Inequalities Grids
  Lesson 6.6 Graphing Inequalities
  Lesson 6.7 Cereal Sales and Profit

Assessment Resources A and B
  Quiz 1 (Lessons 6.1 to 6.4)
  Quiz 2 (Lessons 6.5 to 6.7)
  Chapter 6 Test
  Chapter 6 Constructive Assessment Options

More Practice Your Skills for Chapter 6

Condensed Lessons for Chapter 6

Other Resources

Lǐ Yǎn and Dù Shírán. *Chinese Mathematics: A Concise History*. Oxford: Clarendon Press, 1987.

Waasen Institute. *Jinkōki*. Tokyo: Shyoscki Printing Co., 2000.

www.keymath.com/DA

Materials

- watches with second hands
- masking tape, 6-m ropes, or chalk
- motion sensors, *optional*
- 20 paper clips, all the same size
- 100 pennies
- 8.5-by-11 sheets of paper
- 2 ropes of different thickness, each 1 meter long (per group)
- meterstick or measuring tape
- 9-m-long thick rope, *optional*
- 10-m-long thin rope, *optional*
- 20–40-foot rope marked with whole numbers from −10 to 10 or paper or chalk number line with two markers, *optional*

Pacing Guide

	day 1	day 2	day 3	day 4	day 5	day 6	day 7	day 8	day 9	day 10
standard	6.1	6.2	6.2	6.3	6.3	6.4	6.4	quiz, 6.5	6.6	6.7
enriched	6.1	6.2	6.2	6.3	6.3	6.4	6.4	quiz, 6.5	6.6, project	6.7
block	6.1	6.2	6.3	6.4	6.5, 6.6	6.7, review	assessment			

	day 11	day 12	day 13	day 14	day 15	day 16	day 17	day 18	day 19	day 20
standard	quiz, review	assessment								
enriched	review, TAL	assessment								

6 Systems of Equations and Inequalities

Freshly painted umbrellas dry in the sun outside the Nagatsu factory in Kyushu, Japan. The sticks in their frames form intersecting lines like the graphs of linear equations.

CHAPTER 6 OBJECTIVES

- Model real-world situations with systems of two linear equations in two variables

- Approximate solutions to systems of two linear equations using tables and graphs, understanding how the relative position of the lines indicates the number of solutions to the system

- Solve systems of linear equations using substitution, elimination, and row operations on a matrix (Gaussian elimination)

- Model real-world situations with one-variable inequalities

- Solve one-variable inequalities, including use of the sign-change rule when multiplying or dividing both sides by a negative number

- Graph solutions to one-variable inequalities on a number line, showing whether or not they are strict inequalities

- Model real-world problems with two-variable inequalities and show their solutions as half-planes on the coordinate plane

- Model real-world problems with systems of two-variable inequalities and show the solutions as the intersection of two or more half-planes

OBJECTIVES

In this chapter you will
- learn to solve systems of linear equations
- solve systems using the substitution method
- solve systems using the elimination method
- solve systems using matrices
- graph inequalities in one and two variables
- solve systems of linear inequalities

This photo suggests many lines graphed on a plane, and the points of their intersection suggest solutions to systems of equations. As you look at this image with your students, you can look at it as a two-dimensional image or imagine the three-dimensional space in which the umbrellas exist. When looked at as a two-dimensional picture, the lines formed by the mechanism that opens and supports the umbrella cross the lines of the umbrella structure. What other lines intersect? What lines intersect in the three-dimensional space? At which points do several lines intersect?

Solving Systems of Equations

LESSON

6.1

PLANNING

LESSON OUTLINE

One day:

10 min	Example
20 min	Investigation
10 min	Sharing
5 min	Closing
5 min	Exercises

MATERIALS

- watch with second hand for each group
- 6-m or 6-yd path with 1-m or 1-yd marks on the floor or ground (masking tape or 6-m ropes with 1-m marks are good inside; use chalk or the yard markers on a football field outside)
- motion sensors, *optional* (1 or 2 per group)
- Calculator Notes 6A, 8D

TEACHING

A system of two linear equations can model some real-world problems. These systems can be solved using graphs or tables.

EXAMPLE

This example shows students how to solve a system of linear equations using a calculator graph or table. If you think your students will be confused by the graphs, consider doing the investigation first.

You might draw a diagram or ask two students to act out the situation to help students envision what's happening. **[Ask]** "Does Edna's graph slope upward because she's hiking uphill?" [No. Stress that the graph rises because her distance from the trailhead increases with time.]

In previous chapters you studied linear relationships in the contexts of elevators, wind chill, rope length, and walks. In this chapter you'll consider two or more linear equations together. A **system of equations** is a set of two or more equations that have variables in common. The common variables relate similar quantities. You can think of an equation as a condition imposed on one or more variables, and a system as several conditions imposed simultaneously.

When solving a system of equations, you look for a solution that makes each equation true. There are several strategies you can use. In this lesson you will solve systems using tables and graphs.

EXAMPLE

Edna leaves the trailhead at dawn to hike 12 miles toward the lake, where her friend Maria is camping. At the same time, Maria starts her hike toward the trailhead. Edna is walking uphill so she averages only 1.5 mi/hr, while Maria averages 2.5 mi/hr walking downhill. When and where will they meet?

a. Define variables for time and for distance from the trailhead.

b. Write a system of two equations to model this situation.

c. Solve this system by creating a table and finding the values for the variables that make both equations true. Then locate this solution on a graph.

d. Check your solution and explain its real-world meaning.

▶ **Solution**

a. Let x represent the time in hours. Both women hike the same amount of time. Let y represent the distance in miles from the trailhead. When Edna and Maria meet they will both be the same distance from the trailhead, although they will have hiked different distances.

LESSON OBJECTIVES

- Model real-world situations with systems of two equations
- Solve systems of two linear equations using tables and graphs
- Understand that the intersection of the two lines provides a solution to the system and thus to the real-world problem

NCTM STANDARDS

CONTENT		PROCESS	
	Number		Problem Solving
●	Algebra	●	Reasoning
	Geometry		Communication
●	Measurement	●	Connections
	Data/Probability	●	Representation

b. The system of equations that models this situation is grouped in a brace:

$$\begin{cases} y = 1.5x & \text{Edna's hike} \\ y = 12 - 2.5x & \text{Maria's hike} \end{cases}$$

Edna starts at the trailhead so she increases her distance from it as she hikes 1.5 mi/hr for x hours. Maria starts 12 miles from the trailhead and reduces her distance from it as she hikes 2.5 mi/hr for x hours.

c. Create a table from the equations. Fill in the times and calculate each distance. The table shows the x-value that gives equal y-values for both equations. When $x = 3$, both y-values are 4.5. So the solution is the ordered pair (3, 4.5). We say that these values "satisfy" both equations.

Hiking Times and Distances

x	$y = 1.5x$	$y = 12 - 2.5x$
0	0	12
1	1.5	9.5
2	3	7
3	4.5	4.5
4	6	2
5	7.5	-0.5

On the graph this solution is the point where the two lines intersect. You can use the trace function on your calculator to approximate the coordinates of the solution point, though sometimes you'll get an exact answer.

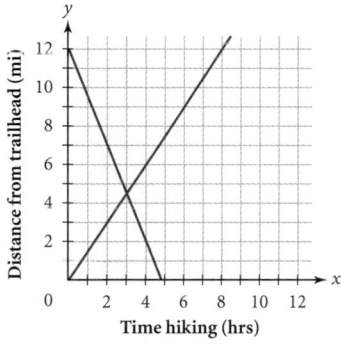

d. The coordinates (3, 4.5) must satisfy both equations.

Edna	Maria	
$y = 1.5x$	$y = 12 - 2.5x$	Original equations.
$4.5 \overset{?}{=} 1.5(3)$	$4.5 \overset{?}{=} 12 - 2.5(3)$	Substitute 3 for the time x and 4.5 for the distance y into both equations.
$4.5 = 4.5$	$4.5 = 4.5$	We get true statements, so (3, 4.5) is a solution for both equations.

So after hiking for 3 hours Edna and Maria meet on the trail 4.5 miles from the trailhead.

In part b of the solution, Maria's equation is written in intercept form. Edna's equation is a direct variation (also in intercept form with $y = 0$).

Encourage thinking about different ways to approach a problem. If a student asks, "Why not just divide the 12 miles by the speed of 4 mi/hr at which the two hikers are approaching each other?" point out that if the problem can be solved that way, the result is the same as if Maria didn't hike and Edna walked (or jogged) at 4 mi/hr. **[Ask]** "Are the situations the same?" [This would give the time they meet, but not the place.]

In part c of the solution, you may need to point out that the pair (x, y) is the solution. Also note the use of the word *satisfy*. It's used later, too.

Verifying Solutions
Each solution to a system should be checked in both of the original equations, and the check must be logically correct. A solution can be checked by evaluating both sides of the equation and showing they are equal.

The common faulty method, however, involves substituting in each side and then multiplying both sides of the equation by the same thing. This can yield inaccurate results unless you specify that you can't multiply the two sides by 0.

Illustrating Faulty Checking
You may illustrate why the faulty method is unreliable by checking whether $(1, 2)$ is a solution to the equation $\frac{2}{3}x + \frac{10}{3} = 2y$ and to the equation $3x = 4y$:

$$\frac{2}{3}(1) + \frac{10}{3} \overset{?}{=} 2(2) \qquad \text{Substitute.}$$

$$2 + 10 \overset{?}{=} 4(3) \qquad \text{Multiply by 3.}$$

$$12 = 12$$

This is a true statement, so $(1, 2)$ is a solution.

$$3(1) \overset{?}{=} 4(2) \qquad \text{Substitute.}$$

$$0(3) \overset{?}{=} 0(8) \qquad \text{Multiply by 0.}$$

$$0 = 0$$

This is a true statement, so $(1, 2)$ must be a solution.

It is true that $(1, 2)$ is a solution for the first equation because $\frac{2}{3}(1) + \frac{10}{3} = 4$ and $2(2) = 4$. However, $(1, 2)$ is not a solution to the second equation because $3(1) = 3$ and $4(2) = 8$, and 3 does not equal 8. Substitution shows $(1, 2)$ is not a solution.

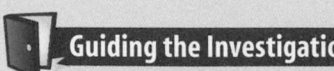
The example describes a situation in which two people walk in opposite directions, so the two slopes have opposite signs. In the step-by-step investigation, both walkers travel in the same direction, so their slopes will have the same sign.

One step Ask that one student in each group walk toward a point (the motion sensor, if you are using one) from 6 m away at a rate of 1 m/sec, while another student walks away from that point, beginning 2 m away, at 0.5 m/sec. The group is to represent the situation, and the meeting point, in as many ways as it can. Encourage students to use graphs and pairs of equations and to think about rates at which the two walkers are closing the distance between them. As groups accomplish this task, ask them to repeat the process, but this time with the first person beginning 0.5 m away from the given point and walking *away* from it at 1 m/sec.

Step 1 As needed, remind students that a graph sketch needs to have labels on the axes. Students may need clarification that the time is the number of seconds elapsed since the walkers started, and the distance is the number of meters the walkers are from one end of the path.

Step 2 In a group of three, one person will need to be both a walker and the timekeeper.

Step 3 Before students start recording data, they should practice the walks until they know how fast to move. Even with experience, they might want to do each walk three times and average the results.

See page 717 for answers to Steps 1, 4, and 6.

 ## Investigation
Where Will They Meet?

You will need
- two motion sensors
- measuring tape or chalk to make a 6-meter line segment

In this investigation you'll solve a system of simultaneous equations to find the time and distance at which two walkers meet.

Suppose that two people begin walking in the same direction at different average speeds. The faster walker starts behind the slower walker. When and where will the faster walker overtake the slower walker?

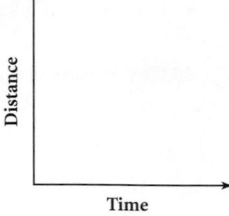

Step 1 | Draw a graph sketch showing both walks. Which line represents the faster walker?

Now act out the walk.

Step 2 | Mark a 6-meter segment at 1-meter intervals. In your group, designate walkers A and B, a timekeeper, and a recorder.

Step 3 | Practice these walks: Walker A starts at the 0.5-meter mark and walks toward the 6-meter mark at a speed of 1 m/sec. Walker B starts at the 2-meter mark and walks toward the 6-meter mark at 0.5 m/sec. When you can follow the walk instructions accurately, do the walk and record the solution.

> **Procedure Note**
> The timekeeper counts each second out loud. The recorder notes the time and position of the two walkers when they meet.

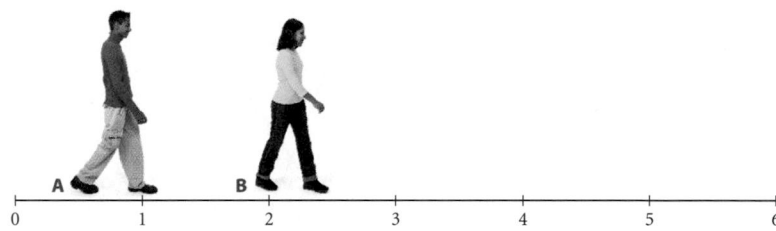

Step 4 | Graph points of the form (*time, distance*) for each walker. Then locate the point that you think represents the approximate time and distance from the zero mark when Walker A passes Walker B.

Next you'll model the walks with a system of equations.

Step 5 | Use the starting point and speed to write an equation for each walker.

Step 6 | Graph the two equations on the same set of axes as your data. Find the point of intersection of the two lines. Explain the real-world meaning of the coordinates of this point. Is the point you found by acting out the walks reasonably close to the point of intersection? (3, 3.5); after 3 seconds, both walkers are at the 3.5-meter mark.

If you are using one motion sensor, the timekeeper should stand at the 0-meter mark and hold the sensor to collect the data on each walker separately. When using two sensors, the recorder should hold the second sensor at the 6-meter mark and the two sensors should be started together. See Calculator Note 6A to learn how to merge the two data sets.

Step 4 Be sure students have the correct starting point for each graph: (0, 0.5) and (0, 2).

Step 5 Encourage students to use the terms *y-intercept* and *rate of change*.

Step 5 Let x represent time in seconds and y represent the meter mark number.

$y = 0.5 + x$ for Walker A
$y = 2 + 0.5x$ for Walker B

Step 7	Check that the coordinates of the point of intersection satisfy both of your equations.

Step 8 The line for Walker A will be steeper, and the solution point will move toward the y-axis.

Step 9 This graph shows two parallel lines. There is no solution point because the lines never meet.

Step 10 The two lines overlap into one on the graph. Every point on the line is a solution to the system of equations.

Next you'll consider what happens under different conditions.

Step 8 Suppose that Walker A walks even faster than 1 m/sec. How is the graph different? What happens to the point of intersection?

Step 9 Suppose that two people walk at the same speed and direction from different starting marks. What does this graph look like? What happens to the solution point?

Step 10 Suppose that two people walk at the same speed in the same direction from the same starting mark. What does this graph look like? How many points satisfy this system of equations?

Is it possible to draw two lines that intersect in two points? How many possible solutions do you think a linear system of two equations in two variables can have?

When you solve a system of two equations, you're finding a solution in the form (x, y) that makes both equations true. When you have a graph of two distinct linear equations the solution of the system is the point where the two lines intersect, if they cross at all. You can estimate these coordinates by tracing on the graph. To find the solution more precisely, zoom in on a table. In the next lesson you'll learn how to find the *exact* coordinates of the solution by working with the equations.

Dancers step between the parallel and intersecting sticks of a bamboo dance in Thailand.

EXERCISES

You will need your calculator for problems **3, 4, 6, 7, 10,** and **14.**

Practice Your Skills

1. Verify whether or not the given ordered pair is a solution to the system. If it is not a solution, explain why not.

 a. $(-15.6, 0.2)$

 $\begin{cases} y = 47 + 3x \\ y = 8 + 0.5x \end{cases}$

 b. $(-4, 23)$

 $\begin{cases} y = 15 - 2x \\ y = 12 + x \end{cases}$

 c. $(2, 12.3)$

 $\begin{cases} y = 4.5 + 5x \\ y = 2.3 + 5x \end{cases}$

1a. Yes, because $47 + 3(-15.6) = 0.2$ and $8 + 0.5(-15.6) = 0.2$.

1b. No, because $23 \neq 12 + (-4)$. The point satisfies only one of the equations.

1c. No, $12.3 \neq 4.5 + 5(2)$. Furthermore, the lines are parallel, so there is no solution to the system.

Step 7 Substitute $x = 3$ and $y = 3.5$ into both equations:

Walker A	Walker B
$3.5 = 0.5 + 3$	$3.5 = 2 + 0.5(3)$

Step 9 Encourage a variety of starting points.

Step 10 You might also ask students to consider a scenario in which two people walk in opposite directions, as in the example.

SHARING IDEAS

Have students present different responses to Steps 6, 8, 9, and 10. Ask why there is a variety. They might mention the need to accelerate from standing still and the difficulty of maintaining a constant speed.

Talk through the logic of verifying answers, as in Step 7. Be sure students do not get the notion that they can just substitute and then manipulate the equation to get one that is true. Instead, help students to see the validity of evaluating the expressions on each side of the equal sign and showing they are equal.

[Ask] "What are the disadvantages to the graphical method of solving a system of equations?" To motivate the rest of the chapter, elicit the ideas that the result might be only approximate and that the problem solver needs to construct graphs rather than just work with the equations.

Assessing Progress

You can assess understanding of time-distance relationships and graphs and how **slopes** of lines represent **rates of change.** You can also observe skills at plotting points, writing equations to represent motion with constant speed, and graphing **linear equations.**

Closing the Lesson

As needed, explain that some problem situations are modeled with a **system of two linear equations.** Solutions to these equations can be approximated by seeing where the equations' graphs intersect or by making tables.

BUILDING UNDERSTANDING

Students find and solve systems of equations modeling situations other than those involving time and distance.

ASSIGNING HOMEWORK

Essential	1–4, 6, 11
Performance assessment	8
Portfolio	8, 10
Journal	7, 11
Group	5, 6, 9
Review	12–16

▶ **Helping with the Exercises**

Exercise 1c The lines graphing the system are parallel, so the system has no solutions. The given point does lie on one of the lines, though, so accept an answer saying that the point isn't a solution because it doesn't lie on the other line.

Exercise 3 Be prepared to explain the list of numbers in the window settings. Visual learners might be helped with a display that shows [Xmin, Xmax, Xscl, Ymin, Ymax, Yscl]. These settings enable the calculator to provide exact solutions for these problems. Students will get approximate solutions if they graph in the standard viewing window (ZOOM 6: ZStandard). Friendly windows are introduced in Chapter 8. (See Calculator Note 8D.)

3a. $(8, 7)$

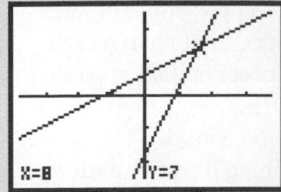

2. Match each graph of a system of equations with its corresponding table values. The tick marks on each graph are one unit apart.

	Graph of system	**Table values of system**

table iv **a.**
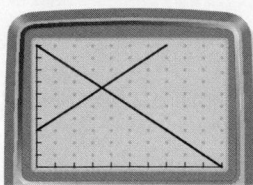

i.
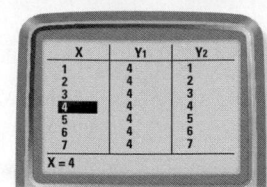

table iii **b.**
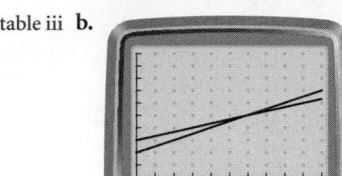

ii.
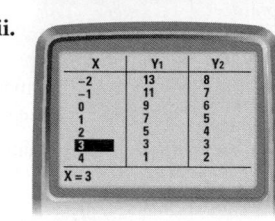

table i **c.**
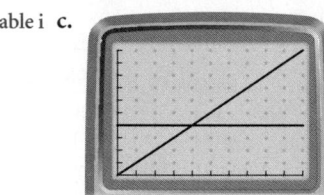

iii.

X	Y₁	Y₂
3	3.5	4
4	4	4.3333
5	4.5	4.6667
6	5	5
7	5.5	5.3333
8	6	5.6667
9	6.5	6
X = 6		

table ii **d.**
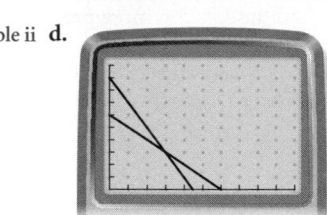

iv.

X	Y₁	Y₂
2	8	5
2.5	7.5	5.5
3	7	6
3.5	6.5	6.5
4	6	7
4.5	5.5	7.5
5	5	8
X = 3.5		

3. Graph each system on your calculator using the window given. Use the trace function to find the point of intersection. Do you think the calculator is giving you approximate or exact solutions?

a. $[-18.8, 18.8, 5, -12.4, 12.4, 5]$
$$\begin{cases} y = 3 + 0.5x \\ y = -9 + 2x \end{cases}$$

b. $[-4.7, 4.7, 1, -3.1, 3.1, 1]$
$$\begin{cases} y = 4x - 5.5 \\ y = -3x + 5 \end{cases}$$

4. Use the calculator table function to find the solution to each system of equations. (In 4b, you'll need to solve the equations for y first.)

a. $y = 7 + 2.5x$
    $y = 35.9 - 6x$

b. $2x + y = 9$
    $3x + y = 16.3$

5. Solve the equations for y. Evaluate each equation for $x = 1$. Substitute these values for x and y into their original equations. What does this tell you?

a. $4x + 2y = 6$

b. $2x - 5y = 20$

3b. $(1.5, 0.5)$

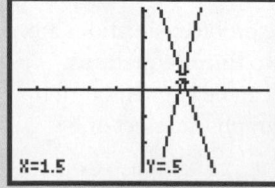

In this case, the calculator gives exact solutions that satisfy each system.

4a. $(3.4, 15.5)$

X	Y₁	Y₂
3	14.5	17.9
3.1	14.75	17.3
3.2	15	16.7
3.3	15.25	16.1
3.4	15.5	15.5
3.5	15.75	14.9
3.6	16	14.3
X = 3.4		

4b. $(7.3, -5.6)$

X	Y₁	Y₂
7	-5	-4.7
7.1	-5.2	-5
7.2	-5.4	-5.3
7.3	-5.6	-5.6
7.4	-5.8	-5.9
7.5	-6	-6.2
7.6	-6.2	-6.5
X = 7.3		

5a. $y = 3 - 2x$; $(1, 1)$: $4(1) + 2(1) = 6$

5b. $y = -4 + 0.4x$; $(1, -3.6)$: $2(1) - 5(-3.6) = 20$

A coordinate satisfies both forms of a linear equation.

Reason and Apply

6. **APPLICATION** Two friends start rival Internet companies in their homes. It costs Gizmo.kom $12,000 to set up the computers and buy the necessary office supplies. Advertisers pay Gizmo.kom $2.50 for each hit (each visit to the web site).

   a. Define variables and write an equation to describe the profits for Gizmo.kom.

   b. The profit equation for the rival company, Widget.kom, is $P = -5000 + 1.6N$. Explain possible real-world meanings of the numbers and variables in this equation, and tell why they're different from those in 6a.

   c. Use a calculator table to find the N-value that gives approximately equal P-values for both equations.

   d. Use your answer to 6c to select a viewing window, and graph both equations to display their intersection and all x- and y-intercepts.

   e. What are the coordinates of the intersection point of the two graphs? Explain how you found this point and how accurate you think it is.

   f. What is the real-world meaning of these coordinates?

7. **APPLICATION** After seeing her friends profit from their web sites in problem 6, Sally wants to start a third company, Gadget.kom, with the start-up costs of Widget.kom and the advertising rate of Gizmo.kom.

   a. What is Sally's profit equation? $P = -5000 + 2.5N$

   b. Graph the profit equations for Gadget.kom and Gizmo.kom.

   c. What does the graph tell you about Sally's profits?

   d. What can you learn about Sally's profits from the equations alone?

8. The total tuition for students at University College and State College consists of student fees plus costs per credit. Some classes have different credit values. The table shows the total tuition for programs with different numbers of credits at each college.

   a. Write a system of equations that represents the relationship between credit hours and total tuition for each college.

   b. Find the solution to this system of equations and check it. $(5, 175)$. Check: $175 = 25 + 30(5), 175 = 15 + 32(5)$.

   c. Which method did you use to solve this system? Why?

   d. What is the real-world meaning of the solution?

   e. When is it cheaper to attend University College? State College?

Total Tuition

Credits	University College ($)	State College ($)
1	55	47
3	115	111
6	205	207
9	295	303
10	325	335
12	385	399

7d. Sally pays less to start Gadget.kom, but she profits at the same rate as Gizmo.kom. Her profit will always be $7,000 more for the same N-values.

Exercise 8 Some students may find it easier to find the y-intercept if they insert a row for 0 credits into the table.

8a. $y = 25 + 30x$ where y is tuition for x credits at University College
$y = 15 + 32x$ where y is tuition for x credits at State College

8c. Answers will vary. The table is more accurate than tracing on the calculator graph.

8d. When a student takes 5 credit hours, the tuition at either college is $175.

8e. It is cheaper to attend University if taking more than 5 credits. Otherwise, it is cheaper to attend State.

Exercise 6 The k in .*kom* is a deliberate misspelling to avoid confusion with actual "dot com" companies. You may need to explain that profit is the difference between revenue ($2.50 multiplied by the number of hits) and cost (the start-up cost). **[Alert]** Some students may be uncomfortable with the use of variables other than x and y, especially for graphing. Help them convert to x and y. Some students also may need help in setting an appropriate viewing window for large numbers. **[Language]** The *start-up costs* for a business are the amount that must be spent to begin the business.

6a. Let P represent profit in dollars and N represent the number of hits.
$P = -12,000 + 2.5N$.

6b. P represents profit, N represents hits. Widget.kom's start-up cost is $5,000, and its advertisers pay $1.60 per hit. Because Widget.kom spent less in start-up costs, its web site might be less attractive to advertisers, hence, the lower rate.

6c. When $N \approx 7778, P \approx 7445$ in both equations.

6e. Use the table to find $(7778, 7445)$. Tracing on this graph is not precise.

6f. This intersection point indicates that for 7778 hits to their web sites, they make the same approximate profit of $7,445.

Exercise 7 [Alert] If students didn't hear the term *start-up costs* in connection with Exercise 6, they may not know what it means.

7c. Sally will always profit more than Gizmo.kom for the same number of web site hits. Because their lines never intersect, there is no solution to the system of equations, and their profits will never be equal.

See page 717 for answers to Exercises 6d and 7b.

9a. $d = 9 - t$ where d is drill team member's distance from end zone; $d = 3 + 0.5t$ where d is tuba player's distance from end zone

10a. The equations give winning times of 44.46 seconds and 44.456 seconds. The difference is 0.004.

10b. The equations give winning times of 42.948 seconds and 42.944 seconds. The difference is 0.004.

10c.

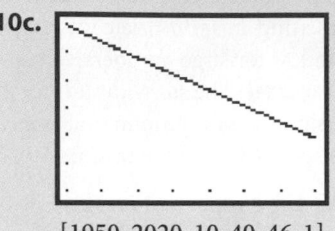

[1950, 2020, 10, 40, 46, 1]

10d. $\begin{cases} y = 150.948 - 0.054x \\ y = 150.944 - 0.054x \end{cases}$
The graph in 10c appears to show one line. However, the y-values are 0.004 apart. So the lines are parallel. The fact that the two equations that model the same data are slightly different is the result of the slope being rounded to -0.054. Exact value is $\frac{-13}{240}$.

11a. Because lines with different slopes always intersect, the y-intercept a can equal any number and b can be any number except -5.

11b. $a \neq 2$ and $b = -5$, same slope, different y-intercept, lines do not intersect.

11c. $a = 2$ and $b = -5$, same slope and y-intercept, lines overlap.

12a. *Miss B:* 5.769 minutes
Club C: 4.815 minutes

12b. *Miss B:* 24.81 gallons
Club C: 20.70 gallons

9. The high school band and drill team both practice on the football field. During one part of the routine, a drill team member walks from the 9-yard mark on the sideline at 1 yd/sec toward the 0-yard mark. At the same time, the tuba player walks from the 3-yard mark at 0.5 yd/sec in the opposite direction.

  a. Write a system of equations to describe this situation.

  b. Find the solution to this system and explain its meaning.
  (4, 5). After 4 seconds, the tuba player bumps into the drill team member at the 5-yard mark.

The marching band performs at half time during a football game at West Point.

10. In the equations $y = 45 - 0.054(x - 1962)$ and $y = 43.7 - 0.054(x - 1986)$, x represents the year and y represents the winning time in seconds. These are two ways to model the data for the men's 400-meter dash in the Olympics.

  a. Find the approximate winning time for the year 1972 given by each equation. What is the difference between the values?

  b. Find the approximate winning time for the year 2000 given by each equation. What is the difference between the values?

  c. Select an appropriate window and graph the two equations.

  d. Do you think these equations represent the same line? Explain your reasoning.

11. Consider the system of equations
$$\begin{cases} y = a + bx \\ y = 2 - 5x \end{cases}$$
Explain what values of a and b give this system

  a. exactly 1 solution   **b.** no solutions   **c.** infinitely many solutions

▶ **Review**

12. APPLICATION Hydroplanes are boats that move so fast that they skim the top of the water. The hydroplane, *Miss B*, qualified for the 2000 Columbia Cup race with a speed of 130.000 mi/hr. The hydroplane, *Club C*, qualified with a speed of 155.753 mi/hr. *(www.hydroracing.com, www.superior-racing.com, www.hydros.org)*

  a. How long will each hydroplane take to run a 5-lap race if one lap is 2.5 miles?

  b. Some boats limit the amount of fuel the motor burns to 4.3 gallons per minute. How much fuel will each boat use to run a 5-lap race?

Exercise 12 Assume that the boats maintain their qualifying speeds throughout the race. This exercise provides an excellent review of the concepts presented in Lesson 2.3 and shows the usefulness of dimensional analysis, but it is complex. If students have difficulty, encourage them to build the problem up slowly to the result in gallons.

$$\frac{\frac{4.3 \text{ gal}}{\text{min}} \cdot \frac{60 \text{ min}}{\text{hr}} \cdot \frac{2.5 \text{ mi}}{\text{lap}} \cdot 5 \text{ laps}}{\frac{155.753 \text{ mi}}{\text{hr}}}$$

$$\text{gal} = \frac{\text{gal}}{\text{min}}(\text{min})$$

$$= \frac{\text{gal}}{\text{min}} \cdot \frac{\text{min}}{\text{hr}}(\text{hr})$$

$$= \frac{\frac{\text{gal}}{\text{min}} \cdot \frac{\text{min}}{\text{hr}}(\text{mi})}{\frac{\text{mi}}{\text{hr}}}$$

$$= \frac{\frac{\text{gal}}{\text{min}} \cdot \frac{\text{min}}{\text{hr}} \cdot \frac{\text{mi}}{\text{lap}}(\text{laps})}{\frac{\text{mi}}{\text{hr}}}$$

c. Hydroplanes have a 50-gallon tank though generally only about 43 gallons are put in. The rest of the tank is filled with foam to prevent sloshing. How many miles can each hydroplane go on one 43-gallon tank of fuel?

d. Find each boat's fuel efficiency rate in miles per gallon.

13. Solve each equation using the method you like best. Then use a different method to check your solution.

a. $0.75x = 63.75$ 85

b. $18.86 = -2.3x$ -8.2

c. $6 = 12 - 2x$ 3

d. $9 = 6(x - 2)$ 3.5

e. $4(x + 5) - 8 = 18$ 1.5

14. APPLICATION At the civic center, Johanna volunteers to design a model of the town and railroad as it looked 150 years ago. The model needs to fit on a piece of plywood that is 4 feet wide. The region that Johanna needs to show is about 0.1 mile wide. The director suggested that she choose a model train size from the scales in the table.

a. Complete the table. It is helpful to use calculator lists for this process.

b. Which scale best fits Johanna's model?

c. Express the scale of her model in inches to feet?
1 inch : 11 ft is the exact ratio.

This hydroplane travels so fast that its image is blurred in the photo. Learn about hydroplane racing with a link at **www.keymath.com/DA** .

Train gauge (name of scale)	Scale (model:actual)	Represented width (ft)	Represented width (mi)
Z	1:220	880	0.167
N	1:160	640	0.12
HO	1:87	348	0.066
S	1:64	256	0.048
O	1:48	192	0.036
Maxi	1:32	128	0.024
G	1:24	96	0.018

15. Find each matrix sum and difference.

a. $\begin{bmatrix} 3 & -3 \\ -9 & 1 \end{bmatrix} + \begin{bmatrix} -2 & -8 \\ 3 & 7 \end{bmatrix}$ $\begin{bmatrix} 1 & -11 \\ -6 & 8 \end{bmatrix}$

b. $\begin{bmatrix} 5 & 0 \\ 2 & 7 \end{bmatrix} - \begin{bmatrix} -8 & 1 \\ -5 & -1 \end{bmatrix}$ $\begin{bmatrix} 13 & -1 \\ 7 & 8 \end{bmatrix}$

16. Solve each equation for y.

a. $y + 2 = 5x$ $y = 5x - 2$

b. $5y = 4 - 7x$ $y = 0.8 - 1.4x$

c. $2y - 6x = 3$ $y = 1.5 + 3x$

12c. *Miss B:* 21.67 miles
Club C: 25.96 miles
12d. *Miss B:* 0.504 mi/gal
Club C: 0.604 mi/gal

Exercise 13 This exercise reviews Lesson 4.8. Encourage students to check by substitution as well as by another method. Be sure the logic of the substitution is correct.

Exercise 14 This exercise reviews Lesson 3.3.

14a. To calculate all of these values easily, enter the denominators of the scales into list L1. At the top of list L2, enter $= L1 \cdot 4$. At the top of list L3, enter $= L2 / 5280$.

14b. The N scale best matches this model.

Exercise 15 This review of matrices foreshadows solving systems of equations with matrices in Lesson 6.4. Assign it only if you did Lesson 1.8 and presented the additional example of multiplying matrices.

Exercise 16 The phrase *combine like terms* is new. *Like terms* are terms with the same variable (if any). In the expression $4x + 5 - 3 + 2x$, $4x$ and $2x$ are like terms, as are 5 and -3. Pairs of like terms are combined by adding or subtracting the coefficients.

LESSON OUTLINE

First day:

15 min   Example A

20 min   Investigation

10 min   Sharing

5 min   Exercises

Second day:

15 min   Example B

5 min   Closing

30 min   Exercises

MATERIALS

- ropes of different thickness, each about 1 meter long, with ends sealed to prevent fraying (two per group)
- meterstick or measuring tape
- 9-m-long thick rope, *optional*
- 10-m-long thin rope, *optional*
- Rope Sample Data (W), *optional*

TEACHING

The substitution method can provide exact solutions to a system of linear equations. These solutions are not approximations, as is often true for solutions from tables or graphs.

One step Ask students to tie knots in the ropes and collect data as in Lesson 4.9. Then ask them to find the number of knots that make the thick and thin ropes the same length. They should not use graphing or tables. As you observe, encourage them to write equations next to each other to see how to substitute.

EXAMPLE A

This example introduces the substitution method for two equations in intercept form. You might have students make a diagram or act out the situation. Ask whether

Solving Systems of Equations Using Substitution

Graphing systems and comparing their table values are good ways to see solutions. However, it's not always easy to find a good graphing window or the right settings for a table. Also, the solutions you find are often only approximations. To find exact solutions, you'll need to work with the equations themselves. One way is called the **substitution method.**

EXAMPLE A

On a rural highway a police officer sees a motorist run a red light at 50 miles per hour, and begins pursuit. At the instant the police officer passes through the intersection at 60 mph, the motorist is 0.2 mile down the road. When and where will the officer catch the motorist?

a. Write a system of equations in two variables to model this situation.

b. Solve this system by the substitution method and check the solution.

c. Explain the real-world meaning of the solution.

▶ **Solution**

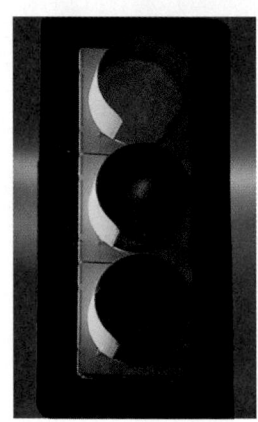

a. Let t represent the time in hours, with $t = 0$ being the instant the officer passes through the intersection. Let d represent the distance in miles from the intersection. Then the system of equations is

$$\begin{cases} d = 0.2 + 50t \\ d = 60t \end{cases}$$

The first equation represents the motorist, who is already 0.2 mile away when the timing begins. The second equation represents the officer.

b. When the officer catches the motorist, they will both be the same distance from the intersection. At this time, both equations will have the same d-value. So you can replace d in one equation with an equivalent expression for d that you find from the other equation. Substituting $60t$ for d into $d = 0.2 + 50t$ gives the new equation:

$$\begin{cases} d = 0.2 + 50t \\ d = 60t \end{cases} \longrightarrow 60t = 0.2 + 50t$$

There is now one equation to solve. Notice that the variable t occurs on both sides of the equal sign and d has been dropped out. Now you use the balancing method to find the solution.

$60t = 0.2 + 50t$	New equation.
$60t - 50t = 0.2 + 50t - 50t$	Subtract $50t$ from both sides of the equation.
$10t = 0.2$	Evaluate.
$\dfrac{10t}{10} = \dfrac{0.2}{10}$	Divide both sides of the equation by 10.
$t = 0.02$	Reduce and divide.

LESSON OBJECTIVES

- Understand the limitations of solving systems graphically
- Solve systems of linear equations using substitution

NCTM STANDARDS

CONTENT		PROCESS	
	Number	•	Problem Solving
•	Algebra	•	Reasoning
•	Geometry	•	Communication
•	Measurement	•	Connections
•	Data/Probability	•	Representation

To find the other half of the solution, substitute 0.02 for t into one of the original equations.

$$t = \boxed{(0.02)}$$

$$d = 0.2 + 50t \qquad d = 60t$$
$$d = 0.2 + 50(0.02) \quad \text{or} \quad d = 60(0.02)$$

If both equations give the same d-value, 1.2 in this case, then you have the correct solution.

c. The solution is the only ordered pair of values, (0.02, 1.2), that works in both equations. The police officer will catch the motorist 1.2 miles from the intersection in 0.02 hour, which is 1 minute 12 seconds after passing through the intersection.

The calculator screen shows the system of equations from the example in the window [0, 0.04, 0.01, −1, 3, 1]. It is difficult to guess at these window settings because the two lines have very similar slopes and close y-intercepts. But the substitution method helps you find the exact solution no matter how difficult it is to set windows or tables. Once you have the exact solution, it is much easier to find a good window to display it.

Investigation
All Tied Up

In this investigation you'll work with rope lengths and predict how many knots it would take in each rope to make a thicker rope the same length as a thinner one.

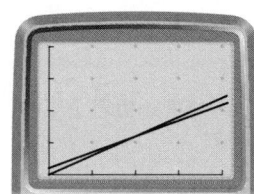

You will need

- two ropes of different thickness, both about 1 meter long
- a meterstick or measuring tape
- a 9-meter-long thin rope (optional)
- a 10-meter-long thick rope (optional)

First you'll collect data and write equations.

Step 1 Measure the length of the thinner rope without any knots. Then tie a knot and measure the length of the rope again. Continue tying knots until no more can be tied. Knots should be of the same kind, size, and tightness. Record the data for number of knots and length of rope in a table.

Step 2 Define variables and write an equation in intercept form to model the data you collected in Step 1. What are the slope and y-intercept, and how do they relate to the rope?

Step 3 Repeat Steps 1 and 2 for the thicker rope.

Step 4 Suppose you have a 9-meter-long thin rope and a 10-meter-long thick rope. Write a system of equations that gives the length of each rope depending on the number of knots tied.

Step 2 If x represents the number of knots and y represents the sample-rope length in cm, then the equation is $y = 100 - 6x$. The y-intercept, 100, is the length of rope in cm without knots, and the slope, −6, is the amount of thin rope in cm that each knot takes.

---

there is necessarily a solution to the problem. Because the patrol car is traveling faster than the motorist, a solution will exist. In fact, because the distance between them is 0.2 mi and closing at 10 mi/hr, some students might just point out that the time required will be $\frac{0.2}{10}$, or 0.02 hr.

[Alert] If students suggest substituting into the same equation, try it to show that all variables will drop out. They must use the other equation.

Some spatially challenged students may understand the substitution better if the two equations are written side-by-side rather than one under the other.

Guiding the Investigation

Steps 1 to 4 As an alternative to students' collecting data in Steps 1 and 3, you might pass out sample data. Assign each group the data from any one of these rope pairs: 1 and 4; 1 and 5; 2 and 4; 2 and 5; 3 and 4; 3 and 5. To move directly to Step 2, you can give them these equations, based on the sample data:

Type 1 Rope: $y = 89.9 - 4.2x$

Type 2 Rope: $y = 93.9 - 4.2x$

Type 3 Rope: $y = 100 - 6x$

Type 4 Rope: $y = 100 - 10.3x$

Type 5 Rope: $y = 97.8 - 13.6x$

The units are centimeters.

Step 4 Equations will vary because they use student-collected data.

Step 4 A sample system is
$$\begin{cases} y = 900 - 6x \\ y = 1000 - 10.3x \end{cases}$$

Step 1 Data will vary. A sample for the thin rope is

Number of knots	Length (cm)
0	100
1	94
2	88
3	81.3
4	75.7
5	69.9
6	63.5

Step 3 Data will vary. A sample for the thick rope is

Number of knots	Length (cm)
0	100
1	89.7
2	78.7
3	68.6
4	57.4
5	47.8
6	38.1

If x represents the number of knots and y represents the sample-rope length in cm, then the equation is $y = 100 - 10.3x$. The y-intercept, 100, is the length of rope without knots, and the slope, −10.3, is the amount of thick rope in cm that each knot takes.

Step 5 From sample:
$900 - 6x = 1000 - 10.3x$,
$4.3x = 100, x = 23\frac{11}{43}$. Round
to 23 knots.

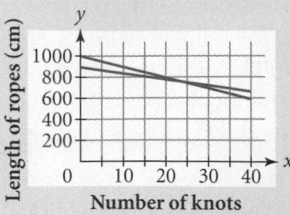

Step 7 The solution will be the number of knots needed in each rope for the two ropes to be the same length. Note that the number of knots must be an integer value. Students will need to round off their actual answers to get realistic values.

You may want to check the students' solutions as a class by having each half of the class tie knots in the optional longer ropes and then comparing the ropes.

SHARING IDEAS

Have students present several different systems for Steps 4 and 5. Ask students why the results differ. Then lead a discussion of ideas about Step 8. **[Ask]** "Solve the system $3x - 4y = 11$, $3x + 2y = -1$ in as many ways as you can." Not only does this get students thinking about solving equations in standard form, but it motivates solving by elimination in Lesson 6.3.

Assessing Progress

Watch for the ability to collect data systematically into tables, write linear equations in **intercept form,** find the **slope** and **y-intercept,** and decide the conditions under which a system has zero solutions, one solution, or infinitely many solutions.

EXAMPLE B

This example illustrates the substitution method for solving systems of equations in standard form. You may want to remind students what standard form means, in contrast to intercept form.

Step 6 Possible window is $[0, 40, 10, 0, 1000, 100]$. Estimated coordinates from sample: $(23, 760)$.

Step 7 At 23 knots, both ropes have nearly equal lengths of 760 cm.

Step 8 If two ropes have the same thickness, the slopes will be equal. If they have the same length, the y-intercepts are the same.

Next you'll analyze the system to find a meaningful solution.

Step 5 Solve this system of equations using the substitution method.

Step 6 Select an appropriate window setting and graph this system of equations. Estimate coordinates for the point of intersection to check your solution. Compare this solution with the one from Step 5.

Step 7 Explain the real-world meaning of the solution to the system of equations.

Step 8 What happens to the graph of the system if the two ropes have the same thickness? The same length?

So far in this chapter, you've seen equations only in intercept form. In other words, they are already solved for the output variable y. This form makes it easy to use the substitution method: You can simply set the two expressions in x equal to each other because they are both equal to y. Sometimes you have to put the equations into intercept form before substituting. In the next example, you'll have to change an equation in standard form to the intercept form.

EXAMPLE B   A molecule of hexane, C_6H_{14}, has six carbon atoms and fourteen hydrogen atoms. Its molecular weight in grams per mole, the sum of the atomic weights of carbon and hydrogen, is 86.178. The molecular weight of octane, C_8H_{18}, is 114.232 grams per mole. Octane has eight carbon atoms and eighteen hydrogen atoms per molecule. Find the atomic weights of carbon and hydrogen.

a. Define variables and write a system of linear equations in the standard form $ax + by = c$ for these molecular weights.

b. Put one equation in the intercept form $y = a + bx$ and substitute into the other equation to find a solution.

c. Check your solution in the original equations.

hexane

○ carbon
○ hydrogen

octane

► Solution   a. Let c represent the atomic weight of carbon in grams per mole. Let h represent the atomic weight of hydrogen in grams per mole. Since the molecular weight of the compounds is the sum of the atomic weights of carbon and hydrogen, we have the system

$$\begin{cases} 6c + 14h = 86.178 & \text{hexane's molecular weight} \\ 8c + 18h = 114.232 & \text{octane's molecular weight} \end{cases}$$

b. Use the balancing method to put one of the equations in intercept form. You can solve for either c or h.

Let's choose to solve the octane equation for c.

$8c + 18h = 114.232$	Original equation.
$8c + 18h - 18h = 114.232 - 18h$	Subtract $18h$ from both sides.
$8c = 114.232 - 18h$	Evaluate.
$\dfrac{8c}{8} = \dfrac{114.232}{8} - \dfrac{18h}{8}$	Divide every term on both sides by 8.
$c = 14.279 - 2.25h$	Reduce and divide.

Substitute this expression for c into the other equation, $6c + 14h = 86.178$, and solve for h:

$6c + 14h = 86.178$	Original equation.
$6(\mathbf{14.279 - 2.25h}) + 14h = 86.178$	Substitute $14.279 - 2.25h$ for c.
$85.674 - 13.5h + 14h = 86.178$	Distribute the 6 through the ( ).
$85.674 + 0.5h = 86.178$	Evaluate.
$85.674 - 85.674 + 0.5h = 86.178 - 85.674$	Subtract 85.674 from both sides.
$0.5h = 0.504$	Evaluate.
$\dfrac{0.5h}{0.5} = \dfrac{0.504}{0.5}$	Divide both sides by 0.5.
$h = 1.008$	Evaluate.

To find the corresponding c-value, substitute 1.008 for h in the equation.

$c = 14.279 - 2.25h$	The equation in intercept form.
$ = 14.279 - 2.25(1.008)$	Substitute 1.008 for h.
$ = 14.279 - 2.268$	Multiply.
$c = 12.011$	Subtract.

c. The solution to the system is (12.011, 1.008). So the atomic weight of carbon is 12.011 grams per mole and the atomic weight of hydrogen is 1.008 grams per mole. Check your answers by substituting them into the original equations.

$$8c + 18h = 114.232$$
$$8(12.011) + 18(1.008) \overset{?}{=} 114.232$$
$$96.088 + 18.144 \overset{?}{=} 114.232$$
$$114.232 = 114.232$$

$$6c + 14h = 86.178$$
$$6(12.011) + 14(1.008) \overset{?}{=} 86.178$$
$$72.066 + 14.112 \overset{?}{=} 86.178$$
$$86.178 = 86.178$$

Since we get true statements for both equations, the solution checks.

You also may want to review the chemical terms *mole* and *molecular weight*. One mole of a substance means 6.02×10^{23} molecules of that substance (just as one dozen eggs means twelve eggs). The molecular weight of a substance is the weight in grams of one mole of that substance.

Point out that the expression on the right side of the equation in part b can be written as $\frac{1}{8}(114.232 - 18h)$, and alert students to the common error of dividing the 8 into only one of the coefficients in the numerator. If students have worked with the exercises on factoring, they may be convinced by the alternative method of factoring the numerator as $8(14.279 - 2.25h)$ and then canceling the multiplication by 8 in the numerator with the division by 8 in the denominator.

You might have different groups solve the system in different ways: solving the same equation for h instead of c or solving the other equation for h or c.

Stress to students that *symbolic manipulation* means solving by hand manipulating the symbols to isolate the variables.

Challenge the class to solve these systems by substitution:

$$\begin{cases} -3x + y = 5 \\ -7x + y = 5 \end{cases} \qquad \begin{cases} x + 3y = 9 \\ y = 3x - 2 \end{cases}$$

The solutions are $(0, 5)$ and $(1.5, 2.5)$. Compare students' approaches.

The **substitution method** allows you to solve a system of linear equations exactly and without graphing or constructing a table. Following are the stages of this method:

1. Solve both equations for one variable and set the two equations equal. Or solve one equation for one variable and substitute that value in the other equation.

2. Solve the resulting equation for the other variable.

3. Substitute that solution for that variable in the first equation.

4. Solve for the second variable.

BUILDING UNDERSTANDING

Students practice solving systems of equations by substitution.

ASSIGNING HOMEWORK

Essential	1–4, 6, 7, 10
Performance assessment	9
Portfolio	12
Journal	8, 10
Group	5, 11
Review	13–15

▶ Helping with the Exercises

Exercise 2 Be sure students use good logic as they verify solutions, evaluating both sides and comparing them, and not multiplying both sides by the same number to get an equality. Even without substituting the value into either equation, students may recognize that both equations in 2c have the same slope, which means the lines are parallel and therefore cannot intersect.

There are many ways to solve systems using the substitution method. You can set expressions equal to one another, or solve for one of the variables and substitute the expression you get into the other equation. Both ways are examples of **symbolic manipulation,** which simply means that you are working with the properties you have used in the balancing method and "undoing" to keep sides of the equation equal. It does not matter which equation or variable you work with first, but you must always substitute the resulting expression into the *other* equation to find a solution. When you solve a system of equations using the substitution method, you can always find an exact solution, not just its approximate coordinates.

EXERCISES

You will need your calculator for problems **12** and **14**.

▶ Practice Your Skills

1. The system of equations

$$\begin{cases} d = 1.5t \\ d = 12 - 2.5t \end{cases}$$

describes the distance of two hikers, Edna and Maria, from the example in Lesson 6.1. By setting the expressions of the right sides of the equations equal to each other, you can find the time when Edna and Maria meet. Explain what happens in Stages 3 and 5 of the substitution process.

$d = 12 - 2.5t$	1. Original equation.
$1.5t = 12 - 2.5t$	2. Substitute $1.5t$ for d.
$1.5t + 2.5t = 12 - 2.5t + 2.5t$	3. Add $2.5t$ to both sides.
$4t = 12$	4. Evaluate.
$\dfrac{4t}{4} = \dfrac{12}{4}$	5. Divide both sides by 4.
$t = 3$	6. Reduce.

2. Check that each ordered pair is a solution to each system. If the pair is not a solution point, explain why not.

a. $(-2, 34)$
$$\begin{cases} y = 38 + 2x \\ y = -21 - 0.5x \end{cases}$$

b. $(4.25, 19.25)$
$$\begin{cases} y = 32 - 3x \\ y = 15 + x \end{cases}$$

c. $(2, 12.3)$
$$\begin{cases} y = 2.3 + 3.2x \\ y = 5.9 + 3.2x \end{cases}$$

3. Solve each equation by symbolic manipulation.

a. $14 + 2x = 4 - 3x$ **b.** $7 - 2y = -3 - y$

c. $5d = 9 + 2d$ **d.** $12 + t = 4t$

4. Solve the system of equations using the substitution method, and check your solution.

$$\begin{cases} y = 25 + 30x \\ y = 15 + 32x \end{cases}$$ $(5, 175)$. Check: $175 = 25 + 30(5)$ and $175 = 15 + 32(5)$.

2a. No, because the point satisfies only one equation.

2b. Yes, because $19.25 = 32 - 3(4.25)$ and $19.25 = 15 + 4.25$.

2c. No, because the point satisfies only one equation. Furthermore, the lines have the same slope, so they are parallel, and there is no solution.

3a. $2x + 3x = 4 - 14$
$\qquad 5x = -10$
$\qquad x = -2$

3b. $-2y + y = -3 - 7$
$\qquad -y = -10$
$\qquad y = 10$

3c. $5d - 2d = 9$
$\qquad 3d = 9$
$\qquad d = 3$

3d. $t - 4t = -12$
$\qquad -3t = -12$
$\qquad t = 4$

5. Substitute $4 - 3x$ for y. Then rewrite each expression in terms of one variable.

  a. $5x + 2y$ $5x + 2(4 - 3x) = 5x + 8 - 6x = -x + 8$ **b.** $7x - 2y$ $7x - 2(4 - 3x) = 7x - 8 + 6x = 13x - 8$

6. Solve each system of equations by substitution, and check your solution.

  a. $\begin{cases} y = 4 - 3x \\ y = 2x - 1 \end{cases}$ $(1, 1)$. Check: $1 = 4 - 3(1)$ and $1 = 2(1) - 1$.

  b. $\begin{cases} 2x - 2y = 4 \\ x + 3y = 1 \end{cases}$ $\left(\frac{7}{4}, -\frac{1}{4}\right)$ or $(1.75, -0.25)$.
  Check: $2(1.75) - 2(-0.25) = 4$ and $1.75 + 3(-0.25) = 1$.

▶ Reason and Apply

7. APPLICATION This system of equations models the profits of two home-based Internet companies.

$$\begin{cases} P = -12000 + 2.5N \\ P = -5000 + 1.6N \end{cases}$$

The variable P represents profit in dollars, and N represents hits to the company's web site.

  a. Use the substitution method to find an exact solution.

  b. Is an approximate or exact solution more meaningful in this model? The approximate solution, $N \approx 7778$ and $P \approx 7444$, is more meaningful because there cannot be a fractional number of web site hits.

8. Consider the system of equations

$$\begin{cases} P = -5000 + 2.5N \\ P = -12000 + 2.5N \end{cases}$$

from problem 6 in Lesson 6.1.

  a. What does the graph tell you about this system?

  b. How can you find the answer to 8a using the substitution method?

9. APPLICATION The manager of a movie theater wants to know the number of adults and children who go to the movies. The theater charges $8 for each adult ticket and $4 for each child ticket. At a showing where 200 tickets were sold, the theater collected $1304.

  a. Let the variable A represent the number of adult tickets and C represent the number of child tickets. Write an equation for the total number of tickets sold.

  b. Write an equation showing the total cost of the tickets.

  c. Use your equations from 9a and b to write a system whose solution represents the number of adult and child tickets sold. Solve this system by symbolic manipulation.

Exercise 8 As needed, point out that if students' calculations produce an equation of two non-equal constants, then the original system had no solution. If they get an equation like $0 = 0$, the two original equations were equivalent.

8a. There is no solution because the lines are parallel and slopes are equal.

8b. Answers will vary. One possible explanation: $-5,000 + 2.5N = -12,000 + 2.5N$. Subtracting $2.5N$ from both sides gives $-5,000 = -12,000$, which is never true. So this system has no solutions.

Exercise 9 This problem is a variation on an old puzzle, so a student might point out that it can be solved without use of a system of equations. Each of the 200 tickets brought in $4, making $800. The remainder of the ticket sales, $504, came from the adults, each of whom paid an additional $4. So there were $\frac{504}{4}$ adults, or 126. Encourage good thinking like this. Challenge these students to represent each step of their reasoning in a system of equations. Their method will motivate solving systems by elimination in Lesson 6.3.

9a. $A + C = 200$

9b. $8A + 4C = 1304$

9c. $A = 126$ and $C = 74$, so the theater sold 126 adult tickets and 74 child tickets.

7a. Answers will vary. A sample solution is

$-12,000 + 2.5N = -5,000 + 1.6N$ Set equations equal to each other.

$-12,000 + 0.9N = -5,000$ Subtract $1.6N$ from both sides.

      $0.9N = 7,000$ Add 12,000.

        $N = \dfrac{70,000}{9} = 7,777\frac{7}{9}$ Divide by 0.9.

        $P = -12,000 + 2.5\left(\dfrac{70,000}{9}\right) = 7,444\frac{4}{9}$

Exercise 11d Students may be confused by the phrase "twice as far." Ask them how far the sports car is from Flint at any time, and then ask what twice that distance would be.

11a. $\begin{cases} d = 35 + 0.8t \\ d = 1.1t \end{cases}$,

$1.1t = 35 + 0.8t$,

$\left(116\frac{2}{3}, 128\frac{1}{3}\right)$. The pickup passes the sports car roughly 128 miles from Flint after approximately 117 minutes.

11b. $\begin{cases} d = 220 - 1.2t \\ d = 1.1t \end{cases}$,

$220 - 1.2t = 1.1t$,

$\left(\frac{2200}{23}, \frac{2420}{23}\right) \approx (95.7, 105.2)$.

The minivan meets the pickup truck about 105 miles from Flint after approximately 96 minutes.

11c. $\begin{cases} d = 220 - 1.2t \\ d = 35 + 0.8t \end{cases}$,

$35 + 0.8t = 220 - 1.2t$,
$(92.5, 109)$. The minivan meets the sports car 109 miles from Flint after 92.5 minutes.

12a. Possible equations:
$y = 51.08 - 0.1215(x - 1972)$ or
$y = 48.83 - 0.1215(x - 1992)$.

---

10. Students in an algebra class did an experiment similar to the investigation Where Will They Meet? from Lesson 6.1. They had two walkers start at opposite ends of the marked length and walk toward each other. They wrote the system

$\begin{cases} d = 8.5 - 0.5t \\ d = 2.5 + 0.75t \end{cases}$

The first walker starts at the 8.5-m mark and walks toward the second walker at 0.5 m/sec. The second walker starts at the 2.5-m mark and walks toward the first walker at 0.75 m/sec.

a. What real-world information does the system tell you?

b. Use the substitution method to solve this system. $(4.8, 6.1)$

c. What is the real-world meaning of the solution you found in 10b? The two students passed each other at the 4.8-m mark after 6.1 sec.

11. The table at right gives the equations that model the three vehicles' distances in the investigation On the Road Again from Lesson 4.4.

The variable d represents the distance in miles from Flint and t represents time in minutes, with $t = 0$ being the instant all three vehicles start traveling.

For each event described in 11a–c, write a system of equations, solve using the substitution method, and explain the real-world meaning of your solution.

Distance from Flint

Equation	Vehicle
$d = 220 - 1.2t$	minivan
$d = 35 + 0.8t$	sports car
$d = 1.1t$	pickup truck

a. The pickup truck passes the sports car.

b. The minivan meets the pickup truck.

c. The minivan meets the sports car.

d. Write and solve an equation to find when the minivan is twice as far from Flint as the sports car. $220 - 1.2t = 2(35 + 0.8t)$; $t \approx 53.6$ min, minivan is about 156 mi, sports car is about 78 mi.

e. How do the solutions that you found using symbolic manipulation in 11a through d compare with those you found using recursive routines in the investigation in Lesson 4.4? The solutions found using substitution are exact (if not rounded off). Recursive routines sometimes give approximate answers because of their discreteness.

12. APPLICATION The table below shows the winning times for the women's 400-meter dash in the Olympics since 1964.

Women's 400-meter Dash

Year	Champion	Country	Time (sec)
1964	Betty Cuthbert	Australia	52.00
1968	Colette Besson	France	52.00
1972	Monika Zehrt	East Germany	51.08
1976	Irena-Szewinska Kirszenstein	Poland	49.29
1980	Marita Koch	East Germany	48.88
1984	Valerie Brisco-Hooks	United States	48.83
1988	Olga Bryzgina	U.S.S.R	48.65
1992	Marie-José Perec	France	48.83
1996	Marie-José Perec	France	48.25
2000	Cathy Freeman	Australia	49.11

(*2000 World Almanac*, p. 920, and *www.olympics.com*)

a. Find a line of fit based on Q-points for the data in the table.

Cathy Freeman won the gold medal in the 400-meter dash at the 2000 Olympics in Sydney, Australia.

b. A possible equation from the data for the men's 400-meter dash in Lesson 5.6 is $y = 45 - 0.054(x - 1962)$. Write and solve a system of equations whose solution tells you when the men and the women will have equal winning times for this Olympic event. *The approximate solution is (2070, 39).*

c. Select an appropriate window to graph this system and its solution.

d. Discuss the reasonableness of this model and the solution.

12c. Answers will vary. The window shown is [1960, 2100, 4, 30, 55, 5].

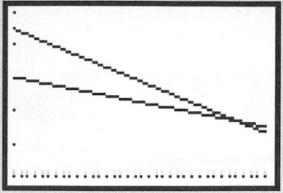

12d. Answers will vary. The exact solution for x is close to 2070, not an Olympic year. If the model holds true, the women's and men's times will be very close to 39 seconds in the 2068 or 2072 Olympics and the women's times will be faster than the men's times after that. It is probably not safe to extrapolate so far from known data. These equations imply that 0 seconds will eventually be a winning time.

▶ Review

13. You and your family are visiting Seattle and take the elevator to the observation deck of the Space Needle. The observation deck is 520 feet high while the needle itself is 605 feet high. The elevator travels at a constant speed, and it takes 43 seconds to travel from the base at 0 feet to the observation deck. **12.1 ft/sec**

a. What is the slope of the graph of this situation?

b. If the elevator could go all the way to the top, how long would it take to get there? **50 seconds**

c. If a rider got on the elevator at the restaurant at the 100-foot level, what equation models her ride to the observation deck?

14. Do each calculation by hand, and then check your results with a calculator. Express your answers as fractions.

a. $3 - \frac{5}{6}$ $2\frac{1}{6}$

b. $\frac{1}{4} + \frac{5}{12}$ $\frac{2}{3}$

c. $\frac{3}{4} \cdot \frac{2}{9}$ $\frac{1}{6}$

d. $\frac{1}{5} + \frac{2}{3} + \frac{3}{4}$ $\frac{97}{60}$ or $1\frac{37}{60}$

The Space Needle, shown here in the city skyline, was built for the 1962 Seattle World's Fair.

15. Match each matrix multiplication with its answer.

i **a.** $\begin{bmatrix} 8 & -2 \\ 1 & 9 \end{bmatrix} \times \begin{bmatrix} 3 & 8 \\ -1 & -4 \end{bmatrix}$

iii **b.** $\begin{bmatrix} 24 & -16 \\ -1 & -36 \end{bmatrix} \times \begin{bmatrix} 1 & 0 \\ 0 & 1 \end{bmatrix}$

ii **c.** $\begin{bmatrix} 6 & 8 \\ -7 & -1 \end{bmatrix} \times \begin{bmatrix} 2 \\ 3 \end{bmatrix}$

i. $\begin{bmatrix} 26 & 72 \\ -6 & -28 \end{bmatrix}$

ii. $\begin{bmatrix} 36 \\ -17 \end{bmatrix}$

iii. $\begin{bmatrix} 24 & -16 \\ -1 & -36 \end{bmatrix}$

Exercise 13 This exercise reviews Lesson 5.2.

13c. $y = 100 + 12.1x$, where x represents the time in seconds and y represents her height above ground level. To find out how long her ride to the observation deck is, solve the equation $520 = 100 + 12.1x$.

Exercise 14 This exercise reviews Lesson 0.1.

Exercise 15 The return to matrices helps prepare for Lesson 6.4. Assign this problem only if you did Lesson 1.8 and plan to do Lesson 6.4.

PLANNING

LESSON OUTLINE

First day:

10 min   Introduction and Example A

35 min   Investigation

Second day:

15 min   Investigation

10 min   Sharing

5 min   Example B

5 min   Closing

15 min   Exercises

MATERIALS

- Paper clips (4 per group)
- Pennies (15 for each group)
- 8.5-by-11 inch sheet of paper (1 for each group)
- Calculator Note 5B

TEACHING

The technique of elimination is useful for solving a system of equations in which no variable has the coefficient 1.

One step   Ask students to line up four paper clips along the long edge of a piece of paper and to fill in the rest of the length with pennies. Then do the same with two paper clips along the short edge, and write and solve a system of equations representing the situation. As you circulate, suggest that they double the second equation and combine it somehow with the first to eliminate a variable. Then have them use three instead of four paper clips along the long edge and try to solve the resulting equations by elimination.

LESSON OBJECTIVES

- Solve systems of linear equations in two variables by eliminating one variable
- See more examples in which two linear equations can model real-world situations

---

LESSON
6.3

Solving Systems of Equations Using Elimination

I happen to feel that the degree of a person's intelligence is directly reflected by the number of conflicting attitudes she can bring to bear on the same topic.

LISA ALTHER

When you add equal quantities to each side of an equation, the resulting equation has the same solution as the original.

$$
\begin{aligned}
y - 7 &= 12 \\
+\quad 7 &= 7 \\
\hline
y &= 19
\end{aligned}
$$

Original equations.

Add equal quantities to both sides.

The resulting equations are true and have the same solutions as the originals.

$$
\begin{aligned}
3x - 5y &= 9 \\
+\quad 5y &= 5y \\
\hline
3x &= 9 + 5y
\end{aligned}
$$

In the same way, when you add two quantities that are equal, c and d, to two other quantities that are equal, a and b, the resulting expressions are equal.

$$
\begin{aligned}
a &= b \\
+\ c &= d \\
\hline
a + c &= b + d
\end{aligned}
$$

Original equation.

Add equal quantities.

The resulting equation is true and has the same solutions as the originals.

The **elimination method** makes use of this fact to solve systems of linear equations.

EXAMPLE A   J. P. is thinking of two numbers, but he won't say what they are. He tells you that the sum of the two numbers is 163 and that their difference is 33. Find the two numbers.

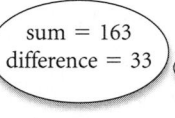
sum = 163
difference = 33

a. Write a system of equations for the sum and difference of these numbers.

b. Use the elimination method to solve this system.

▶ **Solution**   **a.** Let f and s represent the first and second numbers, respectively. Then the system is

$$
\begin{cases}
f + s = 163 \\
f - s = 33
\end{cases}
$$

The first equation describes the sum, and the second describes the difference.

b. Note that adding the equations eliminates the variable s. Then solve for f.

$$
\begin{aligned}
f + s &= 163 \\
f - s &= 33 \\
\hline
2f &= 196 \\
f &= 98
\end{aligned}
$$

Original equations.

Add.

Divide both sides by 2.

So the first number is 98. To find the second number, substitute 98 for f into one of the original equations:

$$98 + s = 163 \qquad \text{or} \qquad 98 - s = 33$$

NCTM STANDARDS

CONTENT		PROCESS	
	Number		Problem Solving
•	Algebra	•	Reasoning
	Geometry	•	Communication
•	Measurement		Connections
	Data/Probability	•	Representation

Either way, the second number is 65. Check that your solutions are correct.

$$f + s = 163 \qquad\qquad f - s = 33$$
$$98 + 65 \overset{?}{=} 163 \qquad\qquad 98 - 65 \overset{?}{=} 33$$
$$163 = 163 \qquad\qquad 33 = 33$$

Adding the two equations quickly leads to a solution because the resulting equation has only one variable. The other variable was eliminated! You won't always have coefficients that add to 0. You'll need another strategy for the elimination method to work.

Investigation
Paper Clips and Pennies

You will need

- three paper clips
- several pennies
- an 8.5-by-11-inch sheet of paper

In this investigation you'll create a system of equations by using paper clips and pennies as variables.

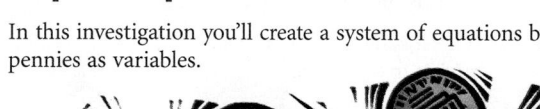

Steps 1 to 2 Sample answers are
For a regular paper clip:
$C + 13P = 11$
For a jumbo paper clip:
$C + 12P = 11$

Steps 3 to 4 Two samples are
For a regular paper clip:
$2C + 8P = 8.5$
For a jumbo paper clip: $2C + 6P = 8.5$

Step 6 The same equation results from either sample system: $-18P = -13.5$; $P = 0.75$. This makes sense because the penny's diameter is constant.

Step 9 Descriptions will vary. For example, you can eliminate P first and then substitute to find C. Or, instead of substituting to find the second value, you can start again and eliminate the other variable in the system.

Step 1 Lay one paper clip along the long side of the paper. Then add enough pennies to complete the 11-inch length.

Step 2 Use C for the length of one paper clip and P for the diameter of one penny. Write an equation in standard form showing your results.

Step 3 Now you'll write the other equation for the system. Lay two paper clips along the shorter edge of your paper, and then add pennies to complete the 8.5-inch length.

Step 4 Using the same variables as in Step 2, write an equation to record your results for the shorter side.

Step 5 In this system the equations from Steps 2 and 4 have different coefficients for each variable. What can you do to one equation so that the variable C is eliminated when you add both equations?

Step 6 Use your answer to Step 5 to set up the addition of two equations. Once you eliminate the variable C, use the balancing method to solve for P.

Step 7 Substitute the value for P into one of the original equations to find C.

Step 8 Check that your solution satisfies both equations.

Step 9 Describe at least one other way to solve this system by elimination.

Step 10 Explain the real-world meaning of the solution. Describe other experiments in measuring that you can solve using a system of equations.

Step 5 In both samples you need to multiply the top equation by (-2).

regular: $\begin{cases} C + 13P = 11 \\ 2C + 8P = 8.5 \end{cases} \rightarrow \begin{cases} -2C - 26P = -22 \\ 2C + 8P = 8.5 \end{cases}$

jumbo: $\begin{cases} C + 12P = 11 \\ 2C + 6P = 8.5 \end{cases} \rightarrow \begin{cases} -2C - 26P = -22 \\ 2C + 6P = 8.5 \end{cases}$

Step 7 regular: $C = 1.25$
jumbo: $C = 2$

Step 8 regular: $\begin{cases} 1.25 + 13(0.75) = 11 \\ 2(1.25) + 8(0.75) = 8.5 \end{cases}$

jumbo: $\begin{cases} 2 + 12(0.75) = 11 \\ 2(2) + 6(0.75) = 8.5 \end{cases}$

INTRODUCTION

You may want to demonstrate that the pairs of equations in the introduction have the same solutions: 19 for the two equations on the left, and points such as $(3, 0)$ and $\left(4, \frac{3}{5}\right)$ for the equations on the right.

EXAMPLE A

This example introduces the method of elimination. Be sure students understand the meanings of the variables and equations. You might place a plus sign between the two equations to emphasize that you're adding them.

Point out that, although substituting was used to solve for the second variable, the method of substitution isn't being used here, because different steps began the process of elimination. The solution to the system would be expressed algebraically as $(f, s) = (98, 65)$.

Be careful to write the check in a logically correct way.

Guiding the Investigation

Step 1 All groups should use the same size paper clips.

Step 3 Students may need to use clips and pennies that were used along the long edge.

Step 5 [Ask] "Why was C eliminated and not P?" [It's easier to eliminate C, but encourage a variety of approaches: multiplying one equation by -2 or dividing the other by -2, for example.]

Step 10 Answers will vary. Regular: $(0.75, 1.25)$. Jumbo: $(0.75, 2)$. The penny is 0.75 inch in diameter. The regular paper clip is 1.25 inches long, and the jumbo clip is 2 inches long. Linear systems in experiments like this one can measure walking and running strides, percent mixture in solutions, and city and highway fuel mileage.

If students use a variety of sizes of paper, they will come up with different systems of equations, but the solutions will be the same. Have any such variety represented in presentations of work.

Ask if there's another way to check the solutions. Students might suggest using a ruler to measure the paper clips and pennies.

Assessing Progress
Watch for the ability to follow directions, make careful measurements, and set up a **system of two linear equations.**

EXAMPLE B

This example shows how to eliminate a variable in a system of equations by multiplying the equations by different numbers.

[Language] UPC stands for *Universal Product Code.* It's the code that identifies each manufactured product and usually appears on the package as a bar code.

[Alert] Be sure students understand where the two equations come from. Using dimensional analysis might help.

[Ask] "How do you choose the numbers by which to multiply both sides of the equation?" [Students might use the coefficients themselves, opposites of coefficients, or the products or least common multiples of coefficients.]

Part b of the solution illustrates how an arrow can mean *implies.* Remind students that the implies arrow is more than a vague link between steps in a process. If required, again stress the need for logical correctness in checking a solution.

You might ask students to solve the same system by substitution and think about when one method is preferable to the other. A few more examples might help them see that substitution is preferable if the coefficient of one variable in one equation is 1.

The goal of the elimination method is to get one of the variables to have a coefficient of 0 when you add the two equations. Sometimes you must first multiply one or both of the equations by some convenient number before you combine them. If you start with additive inverses such as s and $-s$ in Example A, then you can simply add the equations.

EXAMPLE B

The makers of FastBreak breakfast cereal offer a basketball in a promotional giveaway. There are two ways to get it. In Offer A you earn the basketball with 2 UPC symbols and $7. With Offer B you need 10 UPC symbols and $4. FastBreak has collected 1234 UPC symbols and $1,405. How many basketballs must it send out?

a. Write a system of equations that models this problem.

b. Use elimination to solve this system.

c. Check your solution.

▶ **Solution**

It is helpful to organize the information in a table first. Let a represent the number of basketballs sent out in Offer A, and let b represent the number of basketballs sent out in Offer B.

Basketball Giveaway

	Offer A (per basketball)	Offer B (per basketball)	Total
UPC symbols	2	10	1234
Dollars	7	4	1405

a. These equations describe the number of UPC symbols and the amount of money collected:

UPC symbols $\quad 2a + 10b = 1234$

dollars $\qquad\quad 7a + 4b = 1405$

b. To eliminate a from the resulting equation you must make its coefficients additive inverses, that is, numbers with opposite signs. If you multiply the UPC equation by 7 and the dollars equation by -2, then you get two new equations set up for elimination.

$$7(2a + 10b) = 7(1234) \rightarrow \quad 14a + 70b = \ 8638 \quad \text{Multiply both sides by 7.}$$
$$-2(7a + 4b) = -2(1405) \rightarrow \underline{-14a - 8b = -2810} \quad \text{Multiply both sides by } -2.$$
$$62b = \ 5828 \quad \text{Add the equations.}$$
$$\frac{62b}{62} = \frac{5828}{62} \quad \text{Divide both sides by 62.}$$
$$b = \quad 94 \quad \text{Reduce.}$$

To find the value of a, you could substitute 94 for b in one of your original equations and solve for a, as you did in the previous lesson. Or you could go back to the original equations and use elimination on b. If you multiply the UPC equation by -2 and the dollars equation by 5, then you get two equations set up to eliminate b.

$$-2(2a + 10b) = -2(1234) \rightarrow -4a - 20b = -2468 \qquad \text{Multiply both sides by } -2.$$
$$5(7a + 4b) = 5(1405) \quad \rightarrow \quad \underline{35a + 20b = \quad 7025} \qquad \text{Multiply both sides by } 5.$$
$$31a = \quad 4557 \qquad \text{Add the equations.}$$
$$\frac{31a}{31} = \frac{4557}{31} \qquad \text{Divide both sides by } 31.$$
$$a = \quad 147 \qquad \text{Reduce.}$$

c. Substitute 147 for a and 94 for b in the original equations.

$$2a + 10b \overset{?}{=} 1234$$
$$2(147) + 10(94) \overset{?}{=} 1234$$
$$294 + 940 \overset{?}{=} 1234$$
$$1234 = 1234$$

$$7a + 4b = 1405$$
$$7(147) + 4(94) \overset{?}{=} 1405$$
$$1029 + 376 \overset{?}{=} 1405$$
$$1405 = 1405$$

So there are 147 basketballs from Offer A and 94 from Offer B. FastBreak must send out 241 basketballs.

Since there is no single right order to the steps in solving a system of equations, you can start by choosing a variable that's easy to eliminate. You can use both elimination and substitution if that's easiest. Always check your solution in the original system.

EXERCISES

You will need your calculator for problem **9.**

▶ **Practice Your Skills**

1. Consider the equation $5x + 2y = 10$.
  a. Solve the equation for y and sketch the graph.
  b. Multiply the equation $5x + 2y = 10$ by 3, and then solve for y. How does the graph of this equation compare with the graph of the original equation? Explain your answer.

2. Use the equation $5x - 2y = 10$ to find the missing coordinates of each point.
  a. $(6, a)$ $(6, 10)$
  b. $(-4, b)$ $(-4, -15)$
  c. $(c, 25)$ $(12, 25)$
  d. $(d, -5)$ $(0, -5)$

3. Solve each system of equations by elimination. Show your work.
  a. $\begin{cases} 6x + 5y = -20 \\ -6x - 10y = 25 \end{cases}$
  b. $\begin{cases} 5x - 4y = 23 \\ 7x + 8y = 5 \end{cases}$

1b. $y = \dfrac{30 - 15x}{6}$ or $y = \dfrac{30}{6} - \dfrac{15x}{6}$

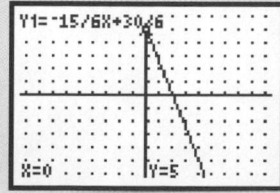

The graph is the same as the graph for 1a. Both equations are equivalent to
$$y = 5 - \frac{5}{2}x.$$

3a. You can simply add the equations as they are to eliminate the x-terms: $-5y = 5, y = -1$; $6x = -15, x = -2.5$. The solution is $(-2.5, -1)$.

3b. You can multiply the first equation by 2 to eliminate the y-terms: $17x = 51, x = 3$; $8y = -16, y = -2$. The solution is $(3, -2)$.

Closing the Lesson

As needed, point out that the **method of elimination** is a fourth method of solving a system of equations, along with **substitution, graphing,** and **tables.** The last two often give only approximations, but elimination and substitution give exact solutions.

BUILDING UNDERSTANDING

Students practice the elimination method for solving systems of equations and deepen their understanding of the graphical meaning of a solution to a system.

ASSIGNING HOMEWORK

Essential	1–4, 7, 9
Performance assessment	6, 10, 12
Portfolio	9, 13
Journal	8, 11
Group	5, 9
Review	14–17

▶ **Helping with the Exercises**

Exercise 1 Students may be surprised to see that when they multiply both sides of an equation by the same number the graph stays the same.

1a. $y = \dfrac{10 - 5x}{2}$ or $y = \dfrac{10}{2} - \dfrac{5x}{2}$

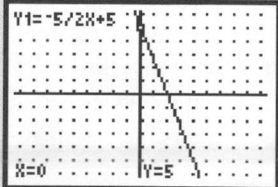

Exercise 4 Allow for a variety of approaches to solving this system.

4c. They are the same line, $y = -2.4 + 0.4x$.

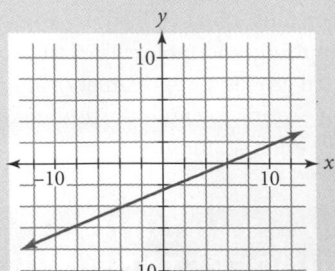

4d. It does not change the graph.

5a. Multiply the first equation by -5 and the second equation by 3, or multiply the first equation by 5 and the second equation by -3.

5b. Multiply the first equation by -8 and the second equation by 7, or multiply the first equation by 8 and the second equation by -7.

8.

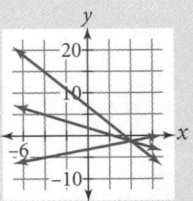

Exercise 9 Students continue to explore the connections between the elimination method and the graphs representing the equations, and they discover that all the equations intersect in a single point.

In 9c, students are asked to draw a vertical line on their calculators. If they don't remember how, refer them to Calculator Note 5B.

---

4. Consider this system of equations:
$$\begin{cases} 2x - 5y = 12 \\ 6x - 15y = 36 \end{cases}$$

   a. By what number can you multiply which equation to eliminate the x-term when you combine the equations by addition? Do this multiplication. Multiply the first equation by -3.

   b. What is the sum of these equations? $0 = 0$

   c. Solve each equation for y and graph the system.

   d. How does multiplying both sides of an equation affect the graph?

5. Consider this system of equations:
$$\begin{cases} 3x + 7y = -8 \\ 5x + 8y = -6 \end{cases}$$

   In 5a and b, tell how you can eliminate each variable when you combine the equations by addition.

   a. the x-term   **b.** the y-term

Reason and Apply

6. List the different ways you have learned to solve the system. Then choose one method and find the solution.
$$\begin{cases} 3x + 7y = -8 \\ 5x + 8y = -6 \end{cases}$$ (1) You can solve for y and graph, then look for the point where the lines cross; (2) create tables and zoom in to where the y-values are equal; (3) solve one equation for y and substitute into the other; or (4) multiply the equations and add them to eliminate x or y. The solution is $(2, -2)$.

7. Solve each system using the elimination method.

   a. $\begin{cases} 2x + y = 10 \\ 5x - y = 18 \end{cases}$ $(4, 2)$ **b.** $\begin{cases} 3x + 5y = 4 \\ 3x + 7y = 2 \end{cases}$ $(3, -1)$ **c.** $\begin{cases} 2x + 9y = -15 \\ 5x + 9y = -24 \end{cases}$ $(-3, -1)$

8. In 8a–c, solve each equation for y and sketch a graph of the result on the same set of axes.

   a. $x - 2y = 6$ $y = -3 + 0.5x$

   b. $3x + 4y = 8$ $y = 2 - 0.75x$

   c. Graph the equation you get from adding the original two equations in 8a and b. $y = 7 - 2x$

   d. What does the graph tell you? Adding equations does not change the solution point.

9. Refer to this system from Example A to answer each question.
$$\begin{cases} x + y = 163 \\ x - y = 33 \end{cases}$$

   a. Solve each equation for y and enter each equation into your calculator. Use the window [0, 150, 10, 0, 150, 10] to graph this system. $y = 163 - x$ and $y = -33 + x$

   b. Use the elimination method to find the y-value of the solution. Enter the resulting equation into Y$_3$ and add it to your graph from 9a.

   c. Use elimination to find the x-value of the solution. Draw a vertical line on the graph to represent the equation you found in 9b.

   d. Describe what you notice about the four lines on your screen and explain why this happens. The four lines intersect at the same point, $(98, 65)$. The solution to the system must satisfy all the equations—the original equations in the system and any new equations created by combining pairs of equations.

9b. $2y = 130; y = 65$

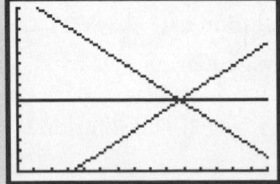

9c. $2x = 196; x = 98$

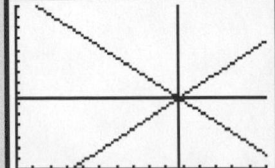

10. Part of Adam's homework paper is missing. If (5, 2) is the only solution to the system shown, write a possible equation that completes the system.

$2x + y = 12$
$4x$

Answers vary. Substitute (5, 2) for x and y in $4x + ay = b$ to get $20 + 2a = b$. One choice is $4x - 3y = 14$.

11. Anisha turned in this quiz in her algebra class.

Anisha _____ Score _____

Solve this system:
$y = x - 5$
$3y + 2x = 5$

Solution:
$3(x - 6) + 2x = 0$
$3x - 15 + 2x = 5$
$5x = 20$
$x = 4$

a. What method did she use? substitution

b. Is her solution correct? Her solution is half right. Anisha didn't find the value $y = -1$ when $x = 4$.

12. The school's photographer took pictures of couples at this year's prom. She charged $3.25 for wallet-size pictures and $10.50 for portrait-size pictures.

a. Write a system of equations representing the fact that Crystal and Dan bought a total of 10 pictures for $61.50.

b. Solve this system and explain what your answer means. They bought 6 wallet-size pictures and 4 portrait-size.

13. **APPLICATION** Automobile companies advertise two rates for fuel mileage. City mileage is the rate of fuel consumption for driving in stop-and-go traffic. Highway mileage is the rate for driving at higher speeds for long periods of time.

Cynthia's new car gets 17 mi/gal in the city and 25 mi/gal on the highway. She drove 220 miles on 11 gallons of gas.

a. Define variables and write a system of equations for the gallons burned at each mileage rate.

b. Solve this system and explain the meaning of the solution. (6.875, 4.125); 6.875 gallons in the city, 4.125 gallons on the highway

c. Find the number of city miles and highway miles Cynthia drove.

d. Check your answers.

▶ **Review**

14. For each pair of fractions, name a fraction that lies between them. Answers will vary. Samples are

a. $\frac{1}{2}$ and $\frac{3}{4}$ $\frac{5}{8}$

b. $\frac{2}{3}$ and $\frac{7}{8}$ $\frac{3}{4}$

c. $-\frac{1}{4}$ and $-\frac{1}{5}$ $-\frac{9}{40}$

d. $\frac{7}{11}$ and $\frac{5}{6}$ $\frac{2}{3}$

e. Describe a strategy for naming a fraction between any two fractions.

Exercise 10 If students are stuck, you might want to suggest that they write the second equation in the form $4x + ay = b$. For any value of a, there's a value of b that gives an equation satisfied by (5, 2).

Ask students how to tell graphically that there are infinitely many possibilities for the second equation. (Infinitely many lines pass through the given solution point.) Try to get them talking and writing about their ideas. Ask if only the line with slope 4 is a possibility. If all the students think so, suggest that, given that the coefficient of y can be any number, there's not enough information to determine the slope. Welcome challenges to both ideas, because the discussion can bring out misconceptions and help deepen understanding of how slope is calculated.

Exercise 14 This exercise reviews Lesson 0.1.

12a. $\begin{cases} w + p = 10 \\ 3.25w + 10.50p = 61.50 \end{cases}$

13a. Let c represent gallons burned in the city and h represent gallons burned on the highway.
$\begin{cases} c + h = 11 \\ 17c + 25h = 220 \end{cases}$

13c. $\frac{17 \text{ mi}}{\text{gal}} \cdot 6.875 \text{ gal} \approx 117 \text{ city mi}$,

$\frac{25 \text{ mi}}{\text{gal}} \cdot 4.125 \text{ gal} \approx 103 \text{ hwy mi}$

13d. Check: $\begin{cases} 6.875 + 4.125 = 11 \\ 17(6.875) + 25(4.125) = 220 \end{cases}$
and $117 + 103 = 220$.

14e. Sample answer: Find a common denominator, select a new numerator between the other two, and reduce.

Exercise 15 This exercise reviews Lesson 4.7.

15b. $T = 95.2 - 0.004E$. The slope is the rate of change in temperature with elevation. The y-intercept (in this case, T-intercept) is the temperature that day at sea level in the same area.

Exercise 16 This exercise reviews Lesson 5.3.

Exercise 17 This exercise reviews Lesson 2.6.

17a. $P\,(\text{top}) = \dfrac{400}{769}$ or 0.52 roughly.

$P\,(\text{bottom}) = \dfrac{369}{769}$ or 0.48 roughly.

---

15. **APPLICATION** When you go up a mountain, the temperature drops about 4 degrees Fahrenheit for every 1000 feet you ascend.

a. While climbing a trail on Mt. McKinley in Alaska, Marsha intended to record the elevation and temperature at three locations. Complete the table for her.

Marsha's Climb

	Elevation (ft)	Temperature (°F)
Start	4,300	78
Rest station	7,800	64
Highest point	11,900	47.6

This mountain climber is ascending Mt. McKinley in Denali National Park, Alaska.

b. Write an equation to model the relationship between elevation and temperature. Explain the meaning of the slope and y-intercept.

c. Mt. McKinley is 20,320 feet high. On the day Marsha was climbing, how cold was it at the summit? At the summit the temperature was 13.9 degrees Fahrenheit.

16. Write an equation in point-slope form using the given information.
   a. A line whose slope is -2 that passes through the point $(5, -3)$. $y = -3 - 2(x - 5)$
   b. A line whose slope is 2.5 that passes through the point $(-3, 7)$. $y = 7 + 2.5(x + 3)$

17. Students are randomly assigned to a locker in a school where even numbered lockers are on the bottom and odd numbered lockers are on top. There is a total of 800 lockers. However, due to construction mess, the bottom row of lockers is blocked off under lockers numbered 201 to 261. Those lockers have been removed from the system. The odd numbered lockers are still available.

   a. What is the probability of being assigned a locker on the top row? What is the probability of being assigned a locker on the bottom row?

   b. In the situation described, what is a trial? What is an outcome?
   A trial is assigning a locker. An outcome is whether a locker is in the top row or the bottom row.

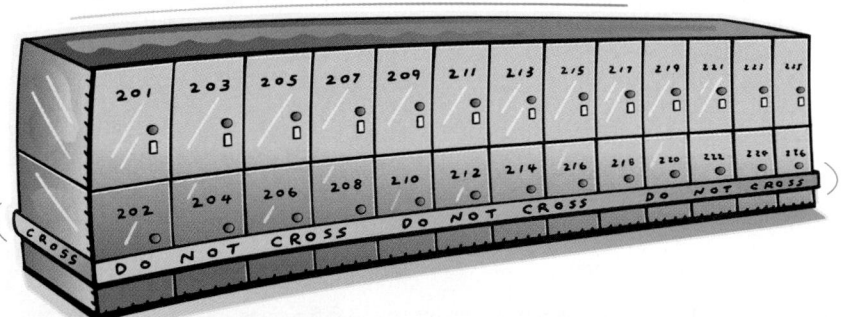

Solving Systems of Equations Using Matrices

The essence of mathematics is not to make simple things complicated but to make complicated things simple.

STANLEY GUDDER

In Lesson 1.8, you learned how to enter, display, and use matrices to organize and analyze data. In this lesson you will use matrices to solve systems of equations.

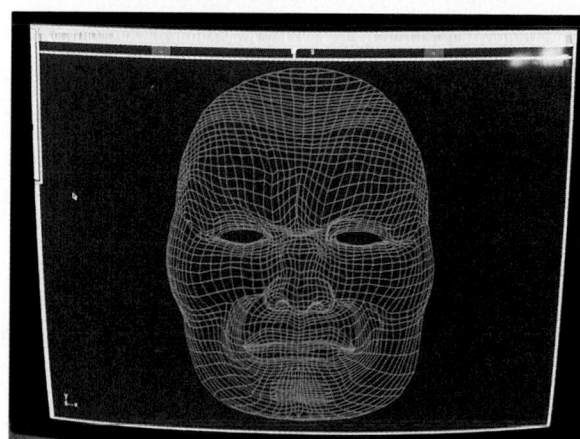

Software that renders 3-D computer-generated images uses matrices to organize data. This program graphs thousands of points and lines to draw the contours of a person's face.

If you look only at the numerals in a system of equations in standard form $ax + by = c$—that is, the coefficients of both variables and the constant terms—you have a matrix with two rows and three columns. If you have a system with both equations in standard form $ax + by = c$, you can write a matrix for the system:

$$\begin{cases} 5x + 3y = -1 \\ 2x - 6y = 50 \end{cases} \qquad \begin{bmatrix} 5 & 3 & -1 \\ 2 & -6 & 50 \end{bmatrix}$$

The numerals in the first equation match the numerals in the first row, and the numerals in the second equation match the numerals in the second row. But what does the solution look like in a matrix? The solution to the system above is $(4, -7)$, or $x = 4$ and $y = -7$. You want the rows of the solution matrix to represent the equations. So you can rewrite each equation to get the numerals for each row of the solution matrix:

$$\begin{array}{ccccc} x = 4 & & x + 0y = 4 & & \begin{bmatrix} 1 & 0 & 4 \\ 0 & 1 & -7 \end{bmatrix} \\ y = -7 & \rightarrow & 0x + y = -7 & \rightarrow & \end{array}$$

PLANNING

LESSON OUTLINE

First day:

| 20 min | Introduction and Example A |
| 30 min | Investigation |

Second day:

10 min	Investigation
5 min	Sharing
10 min	Example B
5 min	Closing
20 min	Exercises

MATERIALS

- Row Operations in a Matrix (T), *optional*
- Calculator Note 6B

TEACHING

Other methods for solving systems of equations employ matrices. Variations on the **Gaussian elimination method** are used by computers handling many linear equations in many variables.

This lesson is optional. Omit it unless you taught Lesson 1.8 on matrices.

Note that the text uses *numerals* instead of *numbers* to characterize coefficients and constants in the original equations that are transferred to the matrix. Students need to realize that x and y are numbers or sets of numbers.

NCTM STANDARDS

CONTENT	PROCESS
Number	• Problem Solving
• Algebra	• Reasoning
Geometry	Communication
Measurement	Connections
Data/Probability	• Representation

LESSON OBJECTIVES

- Use the method of Gaussian elimination for solving systems of linear equations
- Use the calculator to solve systems of linear equations

One step Show students how to represent a system like $2x + y = 11, 6x - 5y = 9$ as a matrix of coefficients. Then challenge them to change the numbers in the matrix to indicate a solution to the system derived by elimination. As groups finish, suggest that they try the same approach on a system that has infinitely many or no solutions such as $y = 3x + 4$ and $2y - 6x = 8$ or $y = 2x + 1$ and $y - 2x = 3$. During Sharing, lead students to formalize the row operations and the diagonalization procedure.

INTRODUCTION

As needed, point out how the row operations mimic the operations on equations used in the previous lesson.

Implies Arrows

One bad habit some algebra students can begin to develop is writing a long string of equalities, such as

$$3x + 5 = 8 = 3x = 3 = x = 1$$

Among other inaccuracies, this statement claims that $8 = 1$. The equal signs have two different meanings. Some show equations, but the ones between equations mean *implies*. If indeed the solution is to be written on one line, have students write the word *implies* or use its abbreviation, an arrow:

$$3x + 5 = 8 \Rightarrow 3x = 3 \Rightarrow x = 1$$

But stress that implies arrows must be used carefully in mathematical statements. When the student text uses a regular arrow, it often means "next you do this." The first of the arrows used near the bottom of page 331 could be written as an implies arrow, but the second one should not be.

EXAMPLE A

This example illustrates how a system can be solved using row operations on a matrix.

In the elimination method, you combined equations and multiplied them by numbers. In much the same way, you can modify the rows of a matrix by performing **row operations** on each number in those rows.

Row Operations in a Matrix

► Multiply (or divide) all numbers in a row by a nonzero number.
► Add all numbers in a row to corresponding numbers in another row.
► Add a multiple of the numbers in one row to the corresponding numbers in another row.
► Exchange two rows.

You can do these operations on the rows of a matrix to change the starting matrix into the solution matrix. The general strategy is to get a diagonal of 1's in the matrix with 0's above and below, like this:

$$\begin{bmatrix} 1 & 0 & a \\ 0 & 1 & b \end{bmatrix}$$

The ordered pair (a, b) is the solution, if one exists, to the system.

EXAMPLE A Solve this system of equations using matrices:
$$\begin{cases} x - 2y = 3 \\ 3x + y = 23 \end{cases}$$

► Solution Copy the numerals from each equation into each row of the matrix. Then use row operations to transform it into the solution matrix.

$$\begin{cases} x - 2y = 3 \\ 3x + y = 23 \end{cases} \longrightarrow \begin{bmatrix} 1 & -2 & 3 \\ 3 & 1 & 23 \end{bmatrix}$$

Add -3 times row 1 to row 2.

$$\begin{array}{llrrr} -3 \text{ times row 1} & \rightarrow & -3 & 6 & -9 \\ + \text{ row 2} & \rightarrow & 3 & 1 & 23 \\ \hline \text{New row 2} & \rightarrow & 0 & 7 & 14 \end{array} \quad \begin{bmatrix} 1 & -2 & 3 \\ 0 & 7 & 14 \end{bmatrix}$$

Divide row 2 by 7.

$$\begin{bmatrix} 1 & -2 & 3 \\ 0 & 1 & 2 \end{bmatrix}$$

Add 2 times row 2 to row 1.

$$\begin{array}{llrrr} 2 \text{ times row 2} & \rightarrow & 0 & 2 & 4 \\ + \text{ row 1} & \rightarrow & +1 & -2 & 3 \\ \hline \text{New row 1} & \rightarrow & 1 & 0 & 7 \end{array} \quad \begin{bmatrix} 1 & 0 & 7 \\ 0 & 1 & 2 \end{bmatrix}$$

Use the solution matrix to write the equations:

$$\begin{bmatrix} 1 & 0 & 7 \\ 0 & 1 & 2 \end{bmatrix} \rightarrow \begin{matrix} 1x + 0y = 7 \\ 0x + 1y = 2 \end{matrix} \quad \begin{matrix} \text{or} \\ \text{or} \end{matrix} \quad \begin{matrix} x = 7 \\ y = 2 \end{matrix}$$

The solution to the system is (7, 2).

Investigation
Diagonalization

In this investigation you will see how to combine row operations in your solution process.

Consider the system of equations

$$\begin{cases} 2x + y = 11 \\ 6x - 5y = 9 \end{cases}$$

Step 1 $\begin{bmatrix} 2 & 1 & 11 \\ 6 & -5 & 9 \end{bmatrix}$ **Step 1** | Write the matrix for this system. What does the first row contain? The second row?

Step 2 Add −3 times row 1 to row 2 and record **Step 2** the sum in row 2 to get $\begin{bmatrix} 2 & 1 & 11 \\ 0 & -8 & -24 \end{bmatrix}$ **Step 3** | Describe how to use row operations to get a 0 as the first entry in the second row. Write this matrix.

Step 3 Divide row 2 by −8 to get $\begin{bmatrix} 2 & 1 & 11 \\ 0 & 1 & 3 \end{bmatrix}$ **Step 4** | Next, get a 1 as the second number in the second row of your matrix from Step 2.

Step 5 | Use row operations on the matrix from Step 3 to get a 0 as the second number in row 1.

Step 5 | Next, get a 1 as the first number of row 1 of your matrix from Step 4. Tell what this matrix means, and give the solution to the system.

Step 6 Students can **Step 6** | Check your solution by solving the system with another method.
solve with a graph, a table, substitution, or elimination.

Look at the first three rules for Row Operations in a Matrix. How do they correspond to steps in the elimination process?

History
CONNECTION

German mathematician Carl Friedrich Gauss (1777–1855) made many contributions to mathematics. He developed the elementary row operations on matrices. In his honor the process of solving systems with matrices is sometimes called "Gaussian elimination."

[Language] A diagonalized matrix has the form

$$\begin{bmatrix} 1 & 0 & a \\ 0 & 1 & b \end{bmatrix}$$

Step 2 [Alert] Students often want to subtract 6 from all entries in row 2. Ask them to write out the corresponding equations to see that they've represented subtracting $6x$ and $6y$ from one side but 6 from the other side.

Step 3 [Alert] Some students may try to add 9 to all entries in row 2.

Step 4 Add −1 times row 2 to row 1 and write the answer in row 1:

$$\begin{bmatrix} 2 & 1 & 11 \\ 0 & 1 & 3 \end{bmatrix} \rightarrow \begin{bmatrix} 2 & 0 & 8 \\ 0 & 1 & 3 \end{bmatrix}$$

Step 5 Divide row 1 by 2 to get

$$\begin{bmatrix} 1 & 0 & 4 \\ 0 & 1 & 3 \end{bmatrix}$$

It means $x = 4$ and $y = 3$, so the solution is (4, 3).

SHARING IDEAS

Point out the quotation opening the lesson. Ask in what sense using matrices is simplifying.

Have a group that used elimination in Step 6 present their check. Ask the class to compare it to the matrix method.

Assessing Progress

You can assess students' familiarity with matrix-related terms such as **row, column,** and **entry** and with their understanding of the **elimination method** and how to check a solution to a system of equations. You also will be able to tell how deeply they understand which systems have no solutions and which have infinitely many solutions.

EXAMPLE B

This example provides another illustration of solving a system of equations using row operations. It resembles an old puzzle, so some students might suggest a non-algebraic solution. (Each of the 3247 people paid $3, bringing in $9741. The rest of the revenue, $4532, came from the adults, each of whom paid an additional $2. So there must have been 2266 adults.) Praise good thinking like this. Challenge students to represent that reasoning by row operations on a matrix. They'll get a sequence like that in the example, except that they'll eliminate the 3 in the second row.

Notice that in each solution, A and C are verified by showing that both sides of the equation equal the same thing. At no time are both sides multiplied or divided by the same thing.

Closing the Lesson

We now have seen five ways of solving systems of equations. Two of them—graphing and tables—often give only approximations. For exact answers, use substitution or elimination. Using **matrices** will produce exact solutions if no rounding is required.

Matrices are useful for solving systems involving large numbers. Here is another example.

EXAMPLE B   On Friday, 3247 people attended the county fair. The entrance fee for an adult was $5, and for a child 12 or under the fee was $3. The fair collected a total of $14,273. How many of the total attendees were adults and how many were children?

▶ **Solution**   Use A for the number of adults attending the fair and C for the number of children attending. Use these variables to write a system of equations and solve it using matrices. The attendance is the number of adults and children at the fair. So the first equation is $A + C = 3247$. The fair collected $5A$ dollars for A adults and $3C$ dollars for C children in attendance. The total collected is $5A + 3C$, so the second equation is $5A + 3C = 14273$.

With one equation describing attendance at the fair, and another describing ticket money collected, the system is

$$\begin{cases} A + C = 3247 \\ 5A + 3C = 14273 \end{cases} \longrightarrow \begin{bmatrix} 1 & 1 & 3247 \\ 5 & 3 & 14273 \end{bmatrix}$$

Use row operations to find the solution.

Add -5 times row 1 to row 2 to get new row 2. $\begin{bmatrix} 1 & 1 & 3247 \\ 0 & -2 & -1962 \end{bmatrix}$

Divide row 2 by -2. $\begin{bmatrix} 1 & 1 & 3247 \\ 0 & 1 & 981 \end{bmatrix}$

Add -1 times row 2 to row 1 to get new row 1. $\begin{bmatrix} 1 & 0 & 2266 \\ 0 & 1 & 981 \end{bmatrix}$

The final matrix shows that $A = 2266$ and $C = 981$. So there were 2266 adults and 981 children at the fair on Friday.

To check this solution, substitute 2266 for A and 981 for C into the original equations.

$$A + C = 3247 \qquad\qquad 5A + 3C = 14273$$
$$2266 + 981 \overset{?}{=} 3247 \qquad 5(2266) + 3(981) \overset{?}{=} 14273$$
$$3247 = 3247 \qquad\qquad 11330 + 2943 \overset{?}{=} 14273$$
$$14273 = 14273$$

These are true statements, so the solution checks. With row operations on matrices, you now have five methods to solve systems of linear equations. Like elimination and substitution, row operations on matrices give exact solutions. With practice, you will develop a sense of when it is easiest to use each solution method. The form of the equation often makes some methods easier to use than others. If an equation is solved for y, then it is easiest to use the substitution method. If the equations are in standard form, then it is easiest to solve by elimination or by using matrices.

EXERCISES

You will need your calculator for problem **8.**

Practice Your Skills

1. Write a system of equations whose matrix is

a. $\begin{bmatrix} 2 & 1.5 & 12.75 \\ -3 & 4 & 9 \end{bmatrix}$
b. $\begin{bmatrix} \frac{1}{2} & 0 & \frac{1}{2} \\ -1 & 2 & 0 \end{bmatrix} \begin{cases} \frac{1}{2}x = \frac{1}{2} \\ -x + 2y = 0 \end{cases}$
c. $\begin{bmatrix} 2 & 3 & 1 \\ 0 & 2 & 0 \end{bmatrix} \begin{cases} 2x + 3y = 1 \\ 2y = 0 \end{cases}$

2. Write the matrix for each system.

a. $\begin{cases} x + 4y = 3 \\ -x + 2y = 9 \end{cases} \begin{bmatrix} 1 & 4 & 3 \\ -1 & 2 & 9 \end{bmatrix}$
b. $\begin{cases} 7x - y = 3 \\ 0.1x - 2.1y = 3 \end{cases} \begin{bmatrix} 7 & -1 & 3 \\ 0.1 & -2.1 & 3 \end{bmatrix}$
c. $\begin{cases} x + y = 3 \\ x + y = 6 \end{cases} \begin{bmatrix} 1 & 1 & 3 \\ 1 & 1 & 6 \end{bmatrix}$

3. Write each solution matrix as an ordered pair.

a. $\begin{bmatrix} 1 & 0 & 8.5 \\ 0 & 1 & 2.8 \end{bmatrix}$ $(8.5, 2.8)$
b. $\begin{bmatrix} 1 & 0 & \frac{1}{2} \\ 0 & 1 & \frac{13}{16} \end{bmatrix}$ $\left(\frac{1}{2}, \frac{13}{16}\right)$
c. $\begin{bmatrix} 1 & 0 & 0 \\ 0 & 1 & 0 \end{bmatrix}$ $(0, 0)$

4. Use row operations to transform the matrix $\begin{bmatrix} 4.2 & 0 & 12.6 \\ 0 & -1 & 5.25 \end{bmatrix}$ into the form
$\begin{bmatrix} 1 & 0 & a \\ 0 & 1 & b \end{bmatrix}$

Write the solution as an ordered pair.

5. Consider the system
$\begin{cases} y = 7 - 3x \\ y = 11 - 2(x - 5) \end{cases}$

a. Convert each equation to the standard form $ax + by = c$. $\begin{cases} 3x + y = 7 \\ 2x + y = 21 \end{cases}$

b. Write a matrix for the system. $\begin{bmatrix} 3 & 1 & 7 \\ 2 & 1 & 21 \end{bmatrix}$

Reason and Apply

6. Give the missing description or matrix for each step of the process below. Give the solution as an ordered pair.

Description	Matrix
The matrix for $\begin{cases} 3x + 2y = 28.9 \\ 8x + 5y = 74.6 \end{cases}$	$\begin{bmatrix} 3 & 2 & 28.9 \\ 8 & 5 & 74.6 \end{bmatrix}$
Add 8 times row 1 to -3 times row 2 and put the result in row 2.	$\begin{bmatrix} 3 & 2 & 28.9 \\ 0 & 1 & 7.4 \end{bmatrix}$
Add -2 times row 2 to row 1 and put the result in row 1.	$\begin{bmatrix} 3 & 0 & 14.1 \\ 0 & 1 & 7.4 \end{bmatrix}$
Divide row 1 by 3. The solution is $(4.7, 7.4)$.	$\begin{bmatrix} 1 & 0 & 4.7 \\ 0 & 1 & 7.4 \end{bmatrix}$

BUILDING UNDERSTANDING

Students work with matrix representations of systems of equations and use row operations to solve those systems.

ASSIGNING HOMEWORK

Essential	1–6, 9, 10
Performance assessment	9, 10
Portfolio	9, 10, 11
Journal	7
Group	8, 11, 15
Review	12–14

▶ Helping with the Exercises

1a. $\begin{cases} 2x + 1.5y = 12.75 \\ -3x + 4y = 9 \end{cases}$

4. Divide row 1 by 4.2:
$\begin{bmatrix} 1 & 0 & 3 \\ 0 & -1 & 5.25 \end{bmatrix}$

Multiply row 2 by -1:
$\begin{bmatrix} 1 & 0 & 3 \\ 0 & 1 & -5.25 \end{bmatrix}$

Solution: $(3, -5.25)$

Exercise 6 If students have difficulty keeping track of this calculation, encourage them to write out intermediate steps.

8 times row 1	24	16	231.2
-3 times row 2	-24	-15	-223.8
Sum (new row 2)	0	1	7.4

Exercise 7 Systems of equations show the power of algebra, which was developed to make arithmetical reasoning easier. Algebra is not needed for this problem:

Each day Sal prepares a large basket of self-serve tortilla chips in his restaurant. One day, his 45 patrons ate 10.8 kg of chips. Sal knows from experience that each child eats 0.2 kg of chips and each adult eats 0.3 kg. How many adults and how many children patronized his restaurant that day?

In this case, you can think of dividing the 10.8 kg into portions of 0.2 kg each. There will be 54 of these portions. There were only 45 patrons, though, so 1.8 kg went to adults. Each adult eats 0.1 kg more than a child, so the 1.8 kg were distributed among 18 adults. The rest of the 45 patrons (27) must have been children.

However, for Exercise 7, Sal knows that on Monday, 40 adults and 15 children ate a total of 10.8 kg of chips, and on Tuesday, 35 adults and 422 children ate 12.3 kg of chips. Without using algebra, the thinking needed to solve this problem would be quite difficult.

Exercise 8 You may want to skip the calculator method if your students are using TI-73 or TI-82 calculators.

8.
```
[A]
    [[1 0 -1]
     [0 1  1 ]]
```

Exercise 9 If students have difficulty deciding what the equations should look like, ask what they want to find (to help them determine the variables) and what they know (to set up the equations).

9a. See page 337.

9c. See page 337.

---

7. APPLICATION Each day Sal prepares a large basket of self-serve tortilla chips in his restaurant. On Monday, 40 adult patrons and 15 child patrons ate 10.8 kg of chips. On Tuesday, 35 adult patrons and 22 child patrons ate 12.29 kg of chips. Sal wants to know whether adults or children eat more chips on average.

  a. Organize the information into a table.

  b. Define variables and write the system of equations showing the average amount of chips the adults and the children eat each day.

  c. Write a matrix for the system. $\begin{bmatrix} 40 & 15 & 10.8 \\ 35 & 22 & 12.29 \end{bmatrix}$

  d. Solve the system by transforming the matrix into the solution matrix $\begin{bmatrix} 1 & 0 & a \\ 0 & 1 & b \end{bmatrix}$.

  e. Write a sentence that describes the real-world meaning of the solution to the system. **Each adult ate an average of about 0.15 kg (150 g) of chips, and each child ate an average of 0.32 kg (320 g) of chips.**

8. Your calculator probably has built-in row operations to transform a matrix into its solution form. Transform this matrix using row operations on your calculator.
[▶ ▢ See **Calculator Note 6B.** ◀]

$$\begin{bmatrix} 8 & 7 & -1 \\ 3 & -1 & -4 \end{bmatrix}$$

9. APPLICATION Zoe must ship 532 tubas and 284 kettledrums from her warehouse to a store across the country. A truck rental company offers two sizes of trucks. A small truck will hold 5 tubas and 7 kettledrums. A large truck will hold 12 tubas and 4 kettledrums. If she wants to fill each truck so that the cargo won't shift, how many small and large trucks should she rent?

  a. Define variables and write a system of equations to find the number of small trucks and the number of large trucks Zoe needs to ship the instruments. Write one equation for each instrument.

$\begin{bmatrix} 5 & 12 & 532 \\ 7 & 4 & 284 \end{bmatrix}$ **b.** Write a matrix that represents the system.

  c. Perform row operations to transform the matrix into a solution matrix.

  d. Write a sentence describing the real-world meaning of the solution. **Zoe should order 20 small trucks and 36 large trucks.**

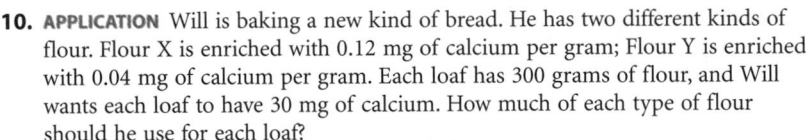

10. APPLICATION Will is baking a new kind of bread. He has two different kinds of flour. Flour X is enriched with 0.12 mg of calcium per gram; Flour Y is enriched with 0.04 mg of calcium per gram. Each loaf has 300 grams of flour, and Will wants each loaf to have 30 mg of calcium. How much of each type of flour should he use for each loaf?

  a. Write a system of equations that relates the number of grams of Flour X and the number of grams of Flour Y. (Hint: Write one equation for the total grams of flour and one for the total grams of calcium.) $\begin{cases} X + Y = 300 \\ 0.12X + 0.04Y = 30 \end{cases}$

  b. Write a matrix for the system. **10b.** $\begin{bmatrix} 1 & 1 & 300 \\ 0.12 & 0.04 & 30 \end{bmatrix}$ **10c.** $\begin{bmatrix} 1 & 0 & 225 \\ 0 & 1 & 75 \end{bmatrix}$

  c. Find the solution matrix.

  d. Explain the real-world meaning of the solution. **Will should mix 225 grams of Flour X with 75 grams of Flour Y.**

7a.

	Adults	Children	Total (kg)
Monday	40	15	10.8
Tuesday	35	22	12.29

7b. Let x represent the average weight of chips an adult eats and y represent the average weight of chips a child eats. The system is
$$\begin{cases} 40x + 15y = 10.8 \\ 35x + 22y = 12.29 \end{cases}$$

7d. Add -35 times row 1 to 40 times row 2:
$$\begin{bmatrix} 40 & 15 & 10.8 \\ 0 & 355 & 113.6 \end{bmatrix}$$

Divide row 2 by 355: $\begin{bmatrix} 40 & 15 & 10.8 \\ 0 & 1 & 0.32 \end{bmatrix}$

Add -15 times row 2 to row 1:
$$\begin{bmatrix} 40 & 0 & 6 \\ 0 & 1 & 0.32 \end{bmatrix}$$

Divide row 1 by 40: $\begin{bmatrix} 1 & 0 & 0.15 \\ 0 & 1 & 0.32 \end{bmatrix}$

11. APPLICATION On Monday a group of students started on a three-day bicycle tour covering a total of 286 km. On Tuesday they cycled 7 km less than on Monday. On Wednesday they traveled 24 km less than on Tuesday.

   a. Write a system of three linear equations representing this trip. Use m, t, and w to represent the distances in kilometers they cycled on Monday, Tuesday, and Wednesday, respectively. Write each equation in the form $am + bt + cw = d$.

   b. Write a 3×4 matrix to model this system of equations. Describe what the rows and columns of your matrix represent.

   c. List and describe a sequence of matrix row operations that will produce a matrix of the form   Sequence of row operations will vary. Solution matrix is

$$\begin{bmatrix} 1 & 0 & 0 & ? \\ 0 & 1 & 0 & ? \\ 0 & 0 & 1 & ? \end{bmatrix} \quad \begin{bmatrix} 1 & 0 & 0 & 108 \\ 0 & 1 & 0 & 101 \\ 0 & 0 & 1 & 77 \end{bmatrix}$$

   d. What is the solution to this problem?
   They cycled 108 km on Monday, 101 km on Tuesday, and 77 km on Wednesday.

▶ Review

12. These matrices show the cost of a 1-day ticket and a 3-day ticket for an adult, a teen, and a child at two amusement parks, Tivoli and Hill.

1-day ticket

	Tivoli	Hill
Adult	31	28
Teen	26	24
Child	21	16

3-day ticket

	Tivoli	Hill
Adult	72	65
Teen	55	55
Child	45	35

   a. Write a matrix equation displaying the difference in cost between a 3-day ticket and a 1-day ticket.

   b. Which type of ticket is the better deal and why?

   c. Which type of ticket should you buy if you are in the park for only 2 days?

13. Write a recursive sequence for the y-coordinates of the points shown on each graph. On each graph one tick mark represents 1 unit.

   a. $4, \text{Ans} - 0.5$ **b.** $-3, \text{Ans} + 2$

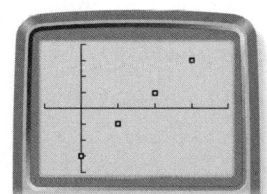

   c. $1/2, \text{Ans} - 1$ **d.** $0, \text{Ans} + 1$

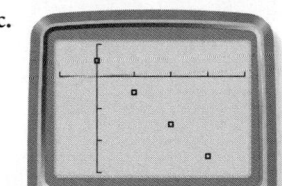 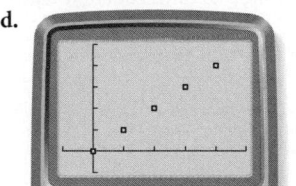

9a. Let x represent the number of small trucks and y represent the number of large trucks. The system is $\begin{cases} 5x + 12y = 532 \\ 7x + 4y = 284 \end{cases}$

9c. Solution steps will vary. $\begin{bmatrix} 5 & 12 & 532 \\ 7 & 4 & 284 \end{bmatrix} \rightarrow$

$$\begin{bmatrix} -16 & 0 & -320 \\ 0 & -64 & -2304 \end{bmatrix} \rightarrow \begin{bmatrix} 1 & 0 & 20 \\ 0 & 1 & 36 \end{bmatrix}$$

Exercise 11 This problem asks students to write a system of equations using three variables. Be sure students realize they need to use zero as place holders for missing variables when they write a system in matrix form.

11a. $\begin{cases} m + t + w = 286 \\ m - t = 7 \\ t - w = 24 \end{cases}$

11b. $\begin{bmatrix} 1 & 1 & 1 & 286 \\ 1 & -1 & 0 & 7 \\ 0 & 1 & -1 & 24 \end{bmatrix}$

The rows represent each equation. The columns represent the coefficients of each variable and the constants.

Exercise 12 This exercise reviews Lesson 1.8.

12a. $\begin{bmatrix} 72 & 65 \\ 55 & 55 \\ 45 & 35 \end{bmatrix} - \begin{bmatrix} 31 & 28 \\ 26 & 24 \\ 21 & 16 \end{bmatrix}$

$= \begin{bmatrix} 41 & 37 \\ 29 & 31 \\ 24 & 19 \end{bmatrix}$

12b. If you are planning to be in the park for 3 days, then the 3-day ticket is a much better deal. If you bought three 1-day tickets, the cost would be

$\begin{bmatrix} 93 & 84 \\ 78 & 72 \\ 21 & 48 \end{bmatrix}$

12c. If you are going to be in the park for 2 days, the cost would be $\begin{bmatrix} 62 & 56 \\ 52 & 48 \\ 42 & 32 \end{bmatrix}$

This is less than the cost of the 3-day ticket, so if you are going for only 2 days, you should buy two 1-day tickets.

Exercise 13 This exercise reviews Lesson 4.4.

Exercise 14 This exercise reviews
Lesson 5.4. Assign it as enrichment for students who enjoy a
challenge.

Exercise 15 This historical problem shows another way to solve a
system of equations. It may confuse students who are just learning the regular row operations.

14. At the Coffee Stop you can buy a mug for $25 and then pay only $0.75 per
hot drink.

   a. What is the slope of the equation that models the total cost of refills? What is the
   real-world meaning of the slope? **Slope 0.75. The slope is the cost per drink once you've bought the mug.**

   b. Use the point (33, 49.75) to write an equation in point-slope form that models
   this situation. $y = 49.75 + 0.75(x - 33)$

   c. Rewrite your equation in intercept form. What is the real-world meaning of the
   y-intercept? $y = 25 + 0.75x$. **The y-intercept is the cost of buying the mug.**

15. Over 2000 years ago the Chinese developed column equation matrices as a method
to solve linear equations. The numerals of each equation are arranged in columns
instead of rows. Use the biancheng (translated as "multiply throughout") and zhichu
(translated as "direct reduction") rules of operation to solve the system.

Rules of Operation

1. Biancheng: Multiply the numerals of the left column by the numeral at the
top of the right column.

2. Zhichu: Subtract the right column from the resulting left column
repeatedly until you get a 0 at the top.

For example, represent this system as a column equation matrix.

$$\begin{cases} 2x + y = 11 \\ 6x - 5y = 9 \end{cases} \rightarrow \begin{bmatrix} 2 & 6 \\ 1 & -5 \\ 11 & 9 \end{bmatrix}$$

Biancheng: Multiply the first column by 6 (highest top row numeral).

$$2(6) \rightarrow 12$$
$$1(6) \rightarrow 6$$
$$11(6) \rightarrow 66$$

Zhichu: Subtract the right column from the left column twice.

$$12 - 6 - 6 \rightarrow 0$$
$$6 - -5 - -5 \rightarrow 16$$
$$66 - 9 - 9 \rightarrow 48$$

Write a new equation and solve for y.

$$16y = 48 \qquad \text{or} \qquad y = 3$$

Substitute and solve for x.

$$2x + 3 = 11 \qquad \text{or} \qquad x = 4$$

Use a Chinese column equation matrix to solve the system

$$\begin{cases} x - 2y = 3 \\ 3x + y = 23 \end{cases}$$

$$\begin{bmatrix} 1 & 3 \\ -2 & 1 \\ 3 & 23 \end{bmatrix} \rightarrow \begin{matrix} 3 & -3 \\ -6 & -1 \\ 9 & -23 \end{matrix} \rightarrow \begin{matrix} 0 \\ -7 \\ -14 \end{matrix}$$

$$-7y = -14, y = 2 \text{ and } x = 7$$

(Jean-Claude Martzloff, *A History of Chinese Mathematics*, 1997, pp. 252–254; Lǐ Yǎn and Dù Shírán, *Chinese Mathematics, a Concise History*, 1987, pp. 46–48)

Inequalities in One Variable

Some material may be inappropriate for children under 13.

DESCRIPTION OF PG-13 RATING, MOTION PICTURE ASSOCIATION OF AMERICA

Drink at least six glasses of water a day. Store milk at temperatures below 40°F. Eat snacks with fewer than 20 calories. Spend at most $10 for a gift. These are a few examples of inequalities in everyday life. In this lesson you will analyze situations involving inequalities in one variable and learn how to find and graph their solutions.

An **inequality** is a statement that one quantity is less than or greater than another. You write inequalities using these symbols:

| less than | < | less than or equal to | ≤ |
| greater than | > | greater than or equal to | ≥ |

Sometimes you need to translate everyday language into the phrases you see in the table above. Here are some examples.

History CONNECTION

Thomas Harriot (1560–1621) introduced the symbols of inequality < and >. Pierre Bouguer first used the symbols ≤ and ≥ about a century later.
(Florian Cajori, *A History of Mathematics*, 1985)

Everyday phrase	Translation	Inequality
at least six glasses	The number of glasses is greater than or equal to 6.	$g \geq 6$
below 40°	The temperature is less than 40°.	$t < 40$
fewer than 20 calories	The number of calories is less than 20.	$c < 20$
at most $10	The price of the gift is less than or equal to $10.	$p \leq 10$
between 35°and 120°	35° is less than the temperature and the temperature is less than 120°.	$35 < t < 120$

You solve inequalities very much like you solve equations. You use the same strategies—adding or subtracting the same quantity to both sides, multiplying both sides by the same number or expression, and so on. However, there is one exception you need to remember when solving inequalities. You will explore this exception in the investigation.

PLANNING

LESSON OUTLINE

One day:

5 min	Introduction
20 min	Investigation
5 min	Sharing
10 min	Examples
10 min	Exercises

MATERIALS

- Toe the Line (W, T), *optional*
- rope marked with whole numbers from −10 to 10 about 1 or 2 feet apart or paper or chalk number line with two markers, *optional* (one per group)
- Calculator Note 6D

TEACHING

Inequalities help model problems for situations described by words like *greater than, less than, no more than, at least,* and so on. Students gain understanding kinesthetically of how multiplying or dividing by a negative number reverses the direction of the inequality. Otherwise, solving inequalities is much like solving equations by balancing.

INTRODUCTION

You might ask the class for other phrases modeled by inequalities, such as *no more than, above, over,* and *more than.* Point out that the larger value is at the larger side of the inequality symbol.

NCTM STANDARDS

CONTENT	PROCESS
• Number	Problem Solving
• Algebra	• Reasoning
• Geometry	• Communication
Measurement	• Connections
Data/Probability	• Representation

LESSON OBJECTIVES

- Write and solve one-variable inequalities and interpret the results based on real-world situations
- Graph solutions to one-variable inequalities on a number line, showing whether or not they are strict inequalities
- Learn the sign-change rule for multiplying or dividing both sides of a one-variable inequality by a negative number

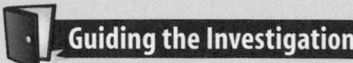
One step Ask walkers to follow the direction in Step 1. Each group should write inequalities to describe the result of each step and then look for patterns. Encourage them to note when the direction of inequality changes.

Step 1 In a group consisting of only three members, one can serve as both recorder and announcer. If time is a problem, use ropes prepared ahead of time. If space is also a problem, use a paper number line and move two markers instead of two people.

Step 2 You may want to discuss the term *relative position*. It means the position of the walkers in relation to each other. The one on the right is greater. Relative position does not indicate how far apart they are.

Step 4 Correct students who use the term *equation* to refer to an inequality.

Steps 5–9 Students can use the dynamic sketch on inequalities at www.keymath.com/DA as they analyze how addition, subtraction, multiplication, and division affect an inequality.

Investigation
Toe the Line

Procedure Note

The announcer calls out operations for Walkers A and B. The walkers perform operations on their numbers by walking to the resulting values on the number line. The recorder logs the position of each walker after each operation.

In this investigation you will analyze properties of inequalities and discover some interesting results.

First you'll act out operations on a number line.

You will need
- chalk or measuring tape to mark a segment

Step 1 In your group choose an announcer, a recorder, and two walkers. The two walkers make a number line on the ground with marks from −10 to 10. The announcer and recorder make a table with these column headings and twelve rows. The operations to use as row headings are Add 2. Subtract 3. Add −2. Subtract −4. Multiply by 2. Subtract 7. Multiply by −3. Add 5. Divide by −4. Subtract 2. Multiply by −1.

Operation	Walker A's position	Inequality symbol	Walker B's position
Starting number	2		4

Step 2 As a trial, act out the first operation in the table: Walker A simply stands at 2 on the number line, and Walker B stands at 4.

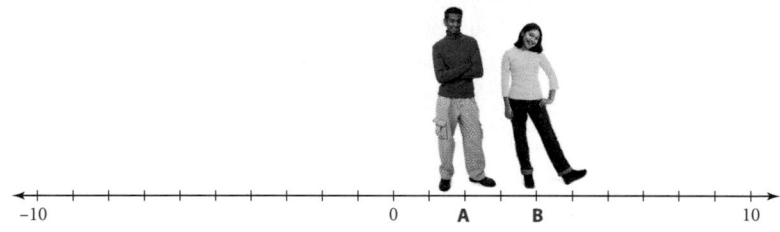

Enter the inequality symbol in the table that describes the relative position of Walkers A and B on the number line. Be sure you have written a true inequality.

Step 3 Call out the operations. After the walkers calculate their new numbers, record the operation and walkers' positions in the next row.

Step 4 As a group, discuss which inequality symbol to enter into each cell of the third column.

Step 5 The walkers' positions shift right and left but maintain the same distance apart. The inequality symbol doesn't change.

Step 6 The walkers' positions stretch from side to side, but the walkers do not switch relative positions. The inequality doesn't change.

Next you'll analyze what each operation does to the inequality.

Step 5 What happens to the walkers' relative positions on the number line when the operation adds or subtracts a positive number? A negative number? Does anything happen to the direction of the inequality symbol?

Step 6 What happens to the walkers' relative positions on the number line when the operation multiplies or divides by a positive number? Does anything happen to the inequality symbol?

Steps 1 to 4

Operation	A's position	Inequality symbol	B's position
Start	2	<	4
Add 2	4	<	6
Subtract 3	1	<	3
Add −2	−1	<	1
Subtract −4	3	<	5
Multiply by 2	6	<	10

Operation	A's position	Inequality symbol	B's position
Subtract 7	−1	<	3
Multiply by −3	3	>	−9
Add 5	8	>	−4
Divide by −4	−2	<	1
Subtract 2	−4	<	−1
Multiply by −1	4	>	1

Step 7 The walkers switch relative positions. The inequality symbol reverses directions.

Step 8 Multiplying or dividing by a negative number reverses the direction of the inequality. All elementary operations—add, subtract, multiply, and divide—with positive numbers maintain the direction of the inequality. Adding and subtracting negative numbers also preserve the inequality. Multiplying by a number greater than 1 or dividing by a fraction between −1 and 1 increases the distance between walkers.

Step 7 What happens to the walkers' relative positions on the number line when the operation multiplies or divides by a negative number? Does the inequality symbol change directions?

Step 8 Which operations on an inequality reverse the inequality symbol? Does it make any difference which numbers you use? Consider fractions and decimals as well as integers.

Step 9 Check your findings about the effects of adding, subtracting, multiplying, and dividing by the same number on both sides of an inequality by creating your own table of operations and walkers' positions.

In square dancing a caller tells the dancers which steps to take. Their maneuvers depend on their relative positions.

This example will show you how to graph solutions to inequalities.

EXAMPLE A

Graph each inequality on a number line.

a. $t > 5$

b. $x \leq -1$

c. $-2 \leq x < 4$

▶ Solution

a. Any number greater than 5 satisfies this inequality. So $5.0001 > 5$, $7\frac{1}{2} > 5$, and $1,000,000 > 5$ are all true statements. You show this by drawing an arrow through the values that are greater than 5.

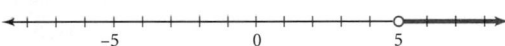

The open circle at 5 excludes 5 from the solutions because $5 > 5$ is not a true statement.

b. The inequality $x \leq -1$ reads, "x is less than or equal to -1." The solid circle at -1 includes the value -1 in the solutions because $-1 \leq -1$ is a true statement.

EXAMPLE A

In this example, students see how to graph the solution of an inequality on a number line. Remind students that an open circle is used to represent the endpoint of a strict inequality and that a filled circle is used to represent the endpoint of an inequality that includes equality. Be sure students understand that the ray indicates an unbounded set of numbers.

Step 7 Students may recall that the change in direction of the inequality occurred when the two walkers passed each other.

Step 8 As needed, note that multiplying by -1 reflects a point about zero. You may want students to circle rows in the table where the inequality switches directions.

Step 9 Encourage students to continue using decimals and other fractions. If time permits, explore the effects of operations such as square roots and exponents.

SHARING IDEAS

Note the opening quotation, and ask how it relates to the lesson. To bring out the strictness of the inequality, ask if 13-year-old kids are included in the advice.

Ask students what patterns they saw in Step 9. Watch for confusion about the relative position of two negative numbers—that is, the one closer to 0 is larger though it has the smaller magnitude.

Inequalities with $<$ or $>$ are *strict* inequalities. Entertain the common but infrequently articulated question of why you'd ever need any other kind of inequality. After all, if two numbers are equal, you can use an equal sign. Students may get ideas about this question from the table in the introduction.

[Ask] "How might inequalities be graphed?" Encourage all participation, not just "right answers." Through repeatedly asking what makes one graphing method better than another, you may be able to get enough ideas from students to make Example A unnecessary.

Assessing Progress
Watch for comfort with **negative integers** and operations on them, especially subtraction of negatives. Assess familiarity with the **number line,** the ability to follow systematic instructions, and the willingness to look for **patterns in data.**

EXAMPLE B

This example goes through the complete modeling process: representing a real-life problem with an inequality, solving the inequality, and interpreting the solution in the original context.

Writing the inequality may be the hardest step. Be sure students see the expression $8 - 0.25x$ as the amount of sleep she will get. **[Ask]** "Why does the inequality include $0.25x$ instead of $15x$?" [The units of time need to be the same for every number in the expression.] As students read through the solution, be sure they notice that the direction of the inequality sign reverses.

Once students have seen the solution to the inequality, complete the modeling process by asking what solution it gives to the original problem. Although the solution to the inequality includes all numbers greater than 12, only whole numbers greater than 12 make sense as solutions to the original problem.

Closing the Lesson

Inequalities help model problems for situations described by phrases such as *greater than, less than, no more than,* and *at least.* Solving inequalities is much like solving equations, except that multiplying or dividing by a negative number reverses the direction of the inequality. The **solutions** to an inequality can be represented by a **ray,** illustrating an unbounded set of numbers in the solution. If an open circle marks the ray's endpoint, the endpoint is not included in the solution. A solid circle indicates inclusion of the endpoint.

---

c. This statement is a **compound inequality.** It says that -2 is less than or equal to x and that x is less than 4. So the graph includes all values that are greater than or equal to -2 and less than 4. The solid circle at -2 includes -2 in the solutions because $-2 \leq -2$ is true. The open circle at 4 excludes 4 from the solutions because $4 < 4$ is not true.

When you graph inequalities, always label 0 on the number line as a point of reference.

EXAMPLE B

Erin says, "I lose 15 minutes of sleep every time the dog barks. Last night I got less than 5 hours of sleep. I usually sleep 8 hours." Find the number of times Erin woke up.

To solve the problem, let x represent the number of times Erin woke up, and write an inequality.

Graph your solutions.

Solve the inequality.

▶ **Solution**

The number of hours Erin slept is 8 hours, minus $\frac{1}{4}$ hour times x, the number of times she woke up. The total is less than 5 hours. So the inequality is $8 - 0.25x < 5$.

Solve the inequality for x. Remember to reverse the inequality symbol if you multiply or divide by a negative number.

$8 - 0.25x < 5$	Original inequality.
$8 - 0.25x - 8 < 5 - 8$	Subtract 8 from both sides of the inequality.
$-0.25x < -3$	Evaluate.
$\dfrac{-0.25x}{-0.25} > \dfrac{-3}{-0.25}$	Divide both sides by -0.25, and reverse the inequality symbol.
$x > 12$	Divide.

The dog woke her up more than 12 times. The graph of the solutions is

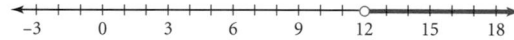

Any number to the right of 12 on the number line satisfies the inequality. The open circle at 12 shows that 12 itself is not a solution.

Working with inequalities is very much like working with equations. An equation shows a balance between two quantities, but an inequality shows an imbalance. The important thing to remember is that multiplying and dividing both sides of an equation by a negative number tips the scales in the opposite direction.

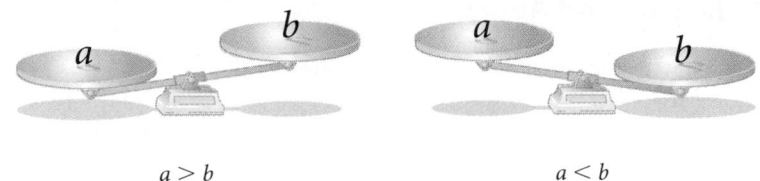

$a > b$ $a < b$

EXERCISES

You will need your calculator for problem **14.**

▶ **Practice Your Skills**

1. Tell what operation on the first inequality gives the second one, and give the answer using the correct inequality symbol.

   a. $3 < 7$ Multiply by 4; $12 < 28$
   $4 \cdot 3 \;\square\; 7 \cdot 4$

   b. $5 \le 12$ Multiply by -3; $-15 \ge -36$
   $-3 \cdot 5 \;\square\; 12 \cdot -3$

   c. $-4 \ge x$ Add -10; $-14 \ge x - 10$
   $-4 + -10 \;\square\; x + -10$

   d. $b + 3 > 15$ Subtract 8; $b - 5 > 7$
   $b + 3 - 8 \;\square\; 15 - 8$

   e. $24d < 32$ Divide by 3; $8d < 10\frac{2}{3}$
   $\frac{24d}{3} \;\square\; \frac{32}{3}$

   f. $24x \le 32$ Divide by -3; $-8x \ge -10\frac{2}{3}$
   $\frac{24x}{-3} \;\square\; \frac{32}{-3}$

2. Find three values of the variable that satisfy each inequality.

   a. $5 + 2a > 21$
   b. $7 - 3b < 28$ Values must be > -7.
   c. $-11.6 + 2.5c < 8.2$ Values must be < 7.92.
   d. $4.7 - 3.25d > -25.3$

3. Give the inequality graphed on each number line.

   a.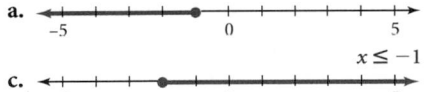
   $x \le -1$

   b.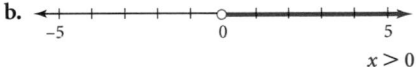
   $x > 0$

   c.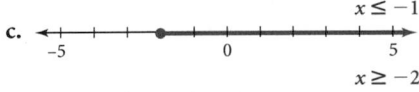
   $x \ge -2$

   d.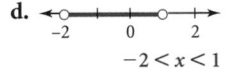
   $-2 < x < 1$

   e.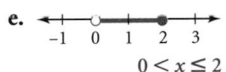
   $0 < x \le 2$

4. Translate each phrase into symbols.

   a. 3 is more than x $3 > x$
   b. y is at least -2 $y \ge -2$
   c. z is no more than 12 $z \le 12$
   d. n is not greater than 7 $n \le 7$

BUILDING UNDERSTANDING

Students practice setting up and solving inequalities and graphing their solutions.

ASSIGNING HOMEWORK

Essential	1–4, 6, 7, 11, 12
Performance assessment	8, 9
Portfolio	8
Journal	12
Group	10, 12
Review	5, 13–15

▶ **Helping with the Exercises**

Exercise 2 You might ask students to make up real-world problems that are modeled by the given inequalities.

2a. Answers will vary, but the values must be > 8.

2d. Values must be $< \frac{120}{13}$ or $9\frac{3}{13} \approx 9.2308$

Exercise 3 Exercises 3d and 3e introduce students to compound inequalities. As needed, explain to students that, because the x is written between two inequality symbols, x is greater than some number *and* x is less than some other number.

Exercise 4 [ESL] Some students may not know how the phrases *more than, no more than,* and *not greater than* correspond to phrases *greater than* and *less than or equal to* used earlier in the lesson.

5. Solve each equation for y.

   a. $3x + 4y = 5.2$ $y = \frac{5.2 - 3x}{4} = 1.3 - 0.75x$ **b.** $3(y - 5) = 2x$ $y = \frac{2x}{3} + 5$

▶ Reason and Apply

6. Solve each inequality and show your work.

   a. $4.1 + 3.2x > 18$ $x > 4.34375 = \frac{139}{32}$ **b.** $7.2 - 2.1b < 4.4$ $b > 1.\overline{3}$

   c. $7 - 2(x - 3) \geq 25$ $x \leq -6$ **d.** $11.5 + 4.5(x + 1.8) \leq x$ $x \leq -5.6$

7. Solve each inequality and graph the solutions on a number line.

   a. $3x - 2 \leq 7$

   b. $4 - x > 6$

   c. $3 + 2x \geq -3$

   d. $10 \leq 2(5 - 3x)$

8. Ezra received \$50 from his grandparents for his birthday. He makes \$7.50 each week for odd jobs he does around the neighborhood. Since his birthday, he has saved more than enough to buy the \$120 gift he wants to buy for his parents' 20th wedding anniversary. How many weeks ago was his birthday?

9. For each graph, tell what operation moves the two points in the inequality to their new positions. Write the new inequality, stating the position of the red dot first.

   a. $1 < 2$ Add 3 to both sides; $4 < 5$.

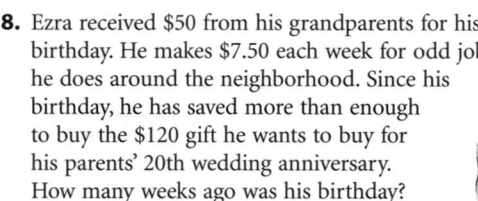

   b. $6 > 2$ Divide both sides by 2 (or multiply by 0.5); $3 > 1$.

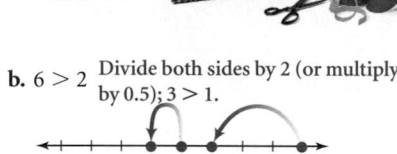

   c. $-1 < 1$ Multiply both sides by -3; $3 > -3$.

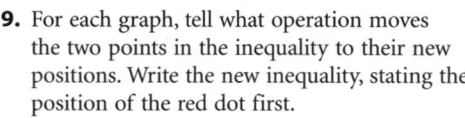
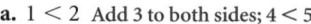
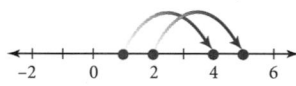

   d. $0 < 3$ Multiply both sides by 2; $0 < 6$.

10. Tell whether each inequality is true or false for the given value.

   a. $x - 14 < 9$, $x = 5$ $-9 < 9$ is true.

   b. $3x \geq 51$, $x = 7$ $21 \geq 51$ is false.

   c. $2x - 3 < 7$, $x = 5$ $7 < 7$ is false.

   d. $4(x - 6) \geq 18$, $x = 12$ $24 \geq 18$ is true.

11. Solve each inequality. Explain the meaning of the result. On a number line, graph the values of x that make the original inequality true.

   a. $2x - 3 > 5x - 3x + 3$

   b. $-2.2(5x + 3) \geq -11x - 15$

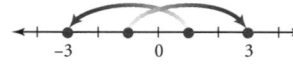

7a. $x \leq 3$

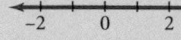

7b. $x < -2$

7c. $x \geq -3$

7d. $0 \geq x$ or $x \leq 0$

Exercise 8 A student who's very comfortable with dimensional analysis might go directly to the inequality $7.5t > 70$. Be encouraging.

8. $50 + 7.5w > 120$; $w > 9.\overline{3}$; Ezra has been saving for at least 10 weeks.

Exercise 9 Here's another way to visualize operations on inequalities. You may want to model a problem like this one to be sure students understand that the directions of the arrows indicate movement of the two numbers relative to the number line.

11a. The variable x drops out of the inequality, leaving $-3 > 3$, which is never true. So the original inequality is not true for any number x. You can't draw a graph to represent this situation on a number line.

11b. The variable x drops out of the inequality, leaving $-6.6 \geq -15$, which is always true. So the original inequality is true for any number x. The graph would be a line with arrows on both ends.

12. You read the inequality symbols $<$, $\leq$, $>$, and $\geq$ as "is less than," "is less than or equal to," "is greater than," and "is greater than or equal to," respectively. But you describe everyday situations with different expressions. Identify the variable in each statement and give the inequality to describe each situation.

a. I'll spend no more than $30 on CDs this month.

b. You must be at least 48 inches tall to go on this ride.

c. Three or more people make a carpool. $p \geq 3$ (p for people in carpool)

d. No one under age 17 will be admitted without a parent or guardian. $a \geq 17$ (a for age of person admitted)

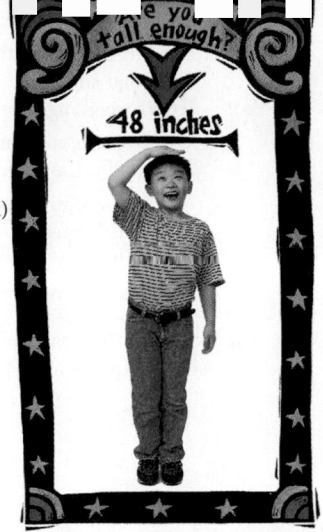

▶ **Review**

13. List the order in which you would perform these operations to get the correct answer.

a. $72 - 12 \cdot 3.2 = 33.6$ b. $2 + 1.5(3 - 5^2) = -31$

c. $21 \div 7 - 6 \div 2 = 0$

Divide 21 by 7 to get 3 and divide 6 by 2 to get 3. Subtract 3 from 3 to get 0.

Exercise 12 **[ESL]** Again, be sure all students know the meaning of the English phrases.

12a. $d \leq 30$ (d for dollars spent on CDs)

12b. $h \geq 48$ (h for height of riders)

Exercise 13 This exercise reviews Lesson 4.1.

13a. Multiply 12 by 3.2 to get 38.4. Subtract 38.4 from 72 to get 33.6.

13b. Square 5 to get 25. Subtract 25 from 3 to get -22. Multiply -22 by 1.5 to get -33. Add -33 to 2 to get -31.

IMPROVING YOUR **VISUAL THINKING** SKILLS

In this chapter you have seen three possible outcomes for a system of two equations in two variables. If one solution exists, it is the point of intersection. If no solution exists, the lines are parallel and there is no point of intersection. If infinitely many solutions exist, the two lines overlap.

But what do the solutions look like in a system of three linear equations in three unknowns? An equation like $3x + 2y = 12$ is a line, but an equation in three variables is a plane. Consider the graph of $3x + 2y + 6z = 12$. Imagine the x-axis coming out of the page. The shaded triangle indicates the part of the solution plane whose coordinates are all positive. The complete plane is infinite.

If you have two more planar equations, you have a system of three equations in three variables. There will be three planes on the graph. So the solutions to this system are where the planes intersect, if they do at all. Visualize how three planes could intersect to answer these questions.

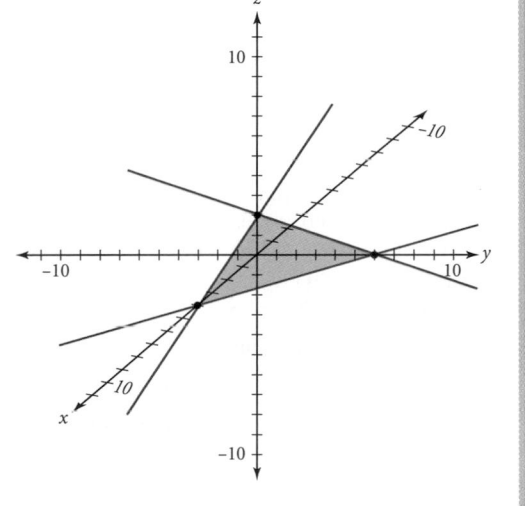

▶ Can three planes intersect in one point? How many solutions will this system have?

▶ If a system has infinitely many solutions, are all three equations the same plane?

▶ If the system has no solutions, are the planes parallel?

IMPROVING **VISUAL THINKING** SKILLS

If two planes aren't parallel, they intersect in a line. If three planes all intersect, the intersection of each pair is a line. Those lines might intersect in a point, which means the system will have one solution. Or the three lines might be the same line, in which case the system has infinitely many solutions even though the planes aren't the same. Or the lines of intersection of pairs of planes

might all be parallel, extending the edges of a triangular prism. Then the system will have no solution, but the planes themselves won't be parallel. If two of the planes are parallel, there will also be no solution.

Exercise 14 This exercise reviews
Lesson 4.4.

14a. 0.34 ENTER
Ans + 0.21 ENTER, ENTER, . . .

Weight (oz)	Rate ($)
1	0.34
2	0.55
3	0.76
4	0.97
5	1.18
6	1.39
7	1.60
8	1.81
9	2.02
10	2.23
11	2.44

14c. A line would mean that the cost would pass through each amount between the different increments. For example, if a package weighed 0.5 ounce, you would pay $0.17. However, the cost increases discretely. Instead, draw segments for each integral ounce. Note the open and closed circles.

Postage Costs

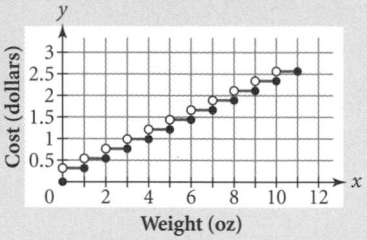

Exercise 15 This exercise reviews
Lesson 4.1.

See page 718 for answer to
Exercise 14b.

14. The table shows the 2001 U.S. postal rates for letters, flats, and parcels.

a. Use a recursive routine to create a table that shows the cost of sending letters weighing from 0 to 11 ounces.

b. Use 1-ounce units on the horizontal axis to plot the postal costs.

c. Kasey has drawn a line through the points on her graph. What real-world meaning does this line have? Is a line useful in this situation? Why or why not?

d. What is the cost of sending a 10.5-ounce parcel? $2.44

U.S. Postal Rates

Weight	Rate
First ounce or fraction of an ounce	$0.34
Each additional ounce or fraction	$0.21

(*http://new.usps.com*)

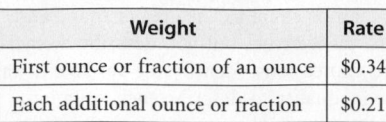

15. Use the distributive property to rewrite each expression without using parentheses.

a. $-2(x + 8)$ $-2x - 16$ **b.** $4(0.75 - y)$ $3 - 4y$

c. $-(z - 5)$ $-z + 5$

project

TEMPERATURES

Temperatures for your city vary depending on the time of day, season, and its location. Weather reports give the daily high and low temperatures and often compare them with the record temperatures in the past 100 years.

Research the range of temperatures for your geographic area. What are the record highs and lows? What are the record temperatures for a specific day, say, your birthday? How do the altitude and location of your area affect these temperatures?

Compare your results to temperatures on the moon. Research the temperatures of other planets such as Venus, Mars, and Pluto. What factors affect these data sets? Are the temperatures given in degrees Fahrenheit or degrees Celsius? Be sure to convert all data to the same units before comparing. Describe your findings with inequalities and graphs in a paper or give a presentation.

Your project should include

► Your hometown high and low temperatures.

► Algebraic expressions with compound inequalities.

► Clearly labeled graphs.

Some people think it may be possible to live on another planet or moon someday. Based on your findings, what do you think?

This view from the Apollo II spacecraft shows Earth above the Lunar terrain.

Supporting the project

MOTIVATION

How do record temperatures in your geographic area compare with those on other planets or on the moon?

OUTCOMES

► Record highs and lows for the area overall as well as temperatures for a specific day are all given in the same units.

► The report comments on how the altitude and geographic location affect the temperatures.

► Temperatures for the moon or for other planets are given and compared.

► Factors influencing extraterrestrial temperatures are given.

► The paper or presentation includes clearly labeled graphs as well as inequalities, including compound inequalities.

► Claims about the possibilities of living on another planet or on the moon are consistent with the data given.

LESSON
6.6

Graphing Inequalities in Two Variables

In Lesson 6.5, you learned to graph inequalities in one variable on a number line. However, some situations, such as the number of points a football team scores by touchdowns and field goals, require more than one variable. In this lesson you will learn to graph inequalities in two variables on the coordinate plane.

You have graphed many equations like $y = 1 + 0.5x$. In the following investigation you will learn how to graph inequalities such as $y < 1 + 0.5x$ and $y > 1 + 0.5x$.

 ## Investigation
Graphing Inequalities

You will need

• the worksheet Graphing Inequalities Grids

First you'll make a graph from one of four statements.

i. $y \,\square\, 1 + 0.5x$

ii. $y \,\square\, -1 - 2x$

iii. $y \,\square\, 1 - 0.5x$

iv. $y \,\square\, 1 - 2x$

Step 2 For $(-3, 3)$,
i. $3 > 1 + 0.5(-3)$
ii. $3 < -1 - 2(-3)$
iii. $3 > 1 - 0.5(-3)$
iv. $3 < 1 - 2(-3)$

Step 1 | Each member of the group should choose a different statement from above.

Step 2 | Use the coordinates of each point shown with a circle to test the statement you selected. Fill in each circle with the relational symbol, $<$, $>$, or $=$, that makes the statement true. For example, to test the upper left point in statement i, substitute $(-3, 3)$ for (x, y) as follows:

NCTM STANDARDS

CONTENT		PROCESS
	Number	• Problem Solving
•	Algebra	• Reasoning
	Geometry	• Communication
	Measurement	• Connections
•	Data/Probability	• Representation

LESSON OBJECTIVES

• Solve two-variable inequalities for y
• Graph inequalities on the coordinate plane and show the solutions as the intersection of two half-planes

PLANNING

LESSON OUTLINE

One day:
20 min   Investigation
 5 min   Sharing
 5 min   Example
 5 min   Closing
15 min   Exercises

MATERIALS

• Calculator Notes 6C, 6D
• Graphing Inequalities Grids (W)
• Graphing Inequalities (T), *optional*

TEACHING

Solutions to inequalities involving two variables can be visualized as points in half-planes.

Guiding the Investigation

One step Ask students to divide the statements i, ii, iii, and iv among group members. For each one, fill in each circle of the grid worksheet with the symbol $<$, $>$, or $=$ that should go into the box to make a true statement when that point's coordinates are substituted for x and y. Ask students to look for patterns. As you circulate, encourage students to check points with fractional coordinates, to redraw the graphs without the circles, and to use dotted lines as boundaries of strict inequalities.

Step 1 If a group has only three members, have them reserve statement iv until they've finished the others. In a group with two members, each could take two statements, such as i and iii for one and ii and iv for the other.

Step 3 Graphs vary depending on statement chosen.

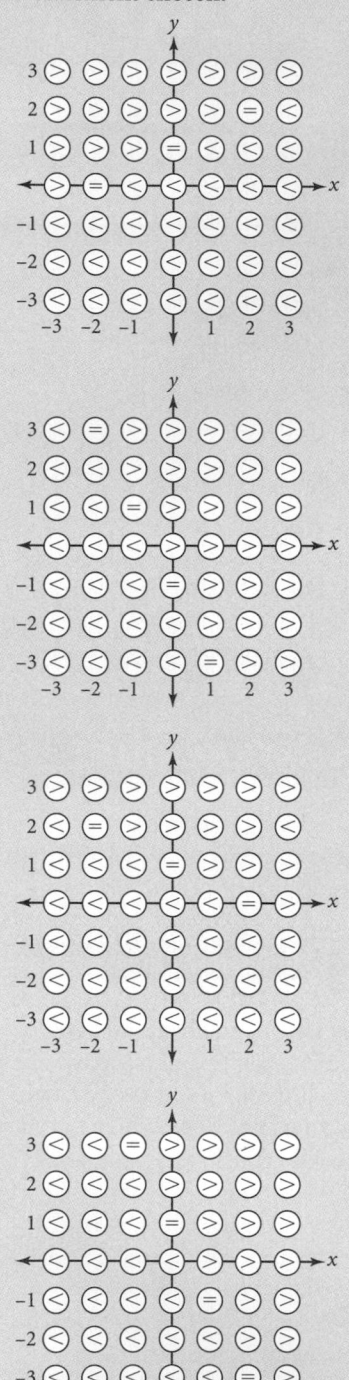

$$y \,\square\, 1 + 0.5x \qquad \text{Original statement.}$$
$$3 \,\square\, 1 + 0.5(-3) \qquad \text{Substitute 3 for } y \text{ and } -3 \text{ for } x.$$
$$3 > -0.5 \qquad \text{Evaluate and choose the symbol.}$$

Place a ">" in the upper left circle because this symbol makes statement i true.

Step 3 Repeat Step 2 for your statement. Work down each column to fill all 49 circles with one of the three symbols.

Next you'll analyze the results of your graph.

Step 4 The circles filled with equal signs form a line. The "greater than" symbol, >, is shaded above the line, and the "less than" symbol, <, is shaded below the line.

Step 4 What do you notice about the circles filled with the equal sign? Tell any other patterns you see.

Step 5 Test a point with fractional or decimal coordinates that is not represented by a circle on the grid. Compare your result with the symbols on the same side of the line of equal signs as your point.

Step 5 Coordinates vary. The symbol will be the same as the symbols on the same side of the line of equal signs.

Step 6 Draw a set of *xy*-axes, with scales from −3 to 3 on each axis. Under the graph write your statement with the less than symbol <. Shade the region of points that makes your statement true. If the points on the line make an inequality true, draw a solid line through them. If not, draw a dashed line. Repeat this step for each of the remaining symbols (>, ≤, ≥, =).

Step 7 The graphs for the symbols =, ≤, and ≥ require a solid line because points on the line satisfy the relationship. The strict inequalities, < and >, require a dashed line.

Finally, you'll draw general conclusions by comparing graphs in your group.

Step 7 Compare your graphs with those of others in your group. What graphs require a solid line? A dashed line?

Step 8 What graphs require shading? Shading above the line? Below the line?

Step 9 Discuss how to check the graph of an inequality with one point.

The graph of the solutions to a single inequality is called a **half-plane** because it includes all the points in the coordinate plane that fall on one side of the boundary line.

EXAMPLE Graph the inequality $2x - 3y > 3$, and check to see whether each point is part of the solution.

i. $(3, -2)$ **ii.** $(3, 1)$ **iii.** $(-1, 2)$ **iv.** $(-2, -3)$

► Solution To graph the inequality, first solve it for *y*:

$$2x - 3y > 3 \qquad \text{Original inequality.}$$
$$-3y > 3 - 2x \qquad \text{Subtract } 2x \text{ from both sides.}$$
$$y < -1 + \frac{2}{3}x \qquad \text{Divide both sides by } -3 \text{ and reverse the inequality symbol.}$$

Step 7 [Ask] "How are broken lines and solid lines related to open circles and filled circles?" [Broken lines, like open circles, represent strict inequalities. Solid lines, like filled circles, include equality.]

Step 8 The graphs for the symbols <, >, ≤, and ≥ require shading. For > and ≥, shade above the line. Shade below the line for < and ≤.

Step 9 Answers will vary. Plug into the inequality the coordinates from one point on one side of the line, say (0, 0). If the point satisfies the inequality, shade that side of the line. If not, shade the other side.

SHARING IDEAS

Ask students to describe the rules they derived for deciding which half-plane to shade, and prompt the class to critique them. Elicit the idea that the point (0, 0) is the easiest point to check and that doing so

will give the desired information unless the line representing the equation passes through the origin.

Assessing Progress

From your observations, you can assess students' understanding of **inequality symbols** and the number line, especially the fact that larger negative numbers are closer to 0. Also look for the ability to collect data systematically, to find patterns, and to work with a group.

Graph the line $y = -1 + \frac{2}{3}x$ with a dashed line to indicate that points on the line are not part of the solution to the inequality. Because the inequality in y is less than the expression in x on the right side, shade the region *below* the line. Points in this region will have y-values that are less than the expression in x.

If you plot the given points, you'll see that the points that satisfy the inequality lie in the shaded part of the plane.

To check numerically whether the given points satisfy the inequality, substitute the x- and y-values from each given coordinate pair for x and y in the inequality $2x - 3y > 3$, and enter the inequality into your calculator. When you press $\boxed{\text{ENTER}}$ you'll see 1 if the inequality is true or 0 if the inequality is false, as shown on the calculator screen.

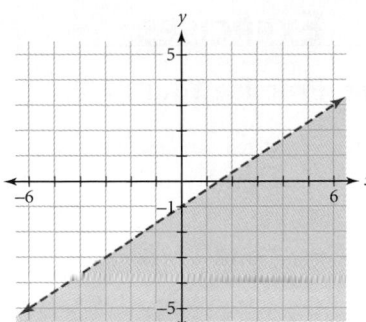

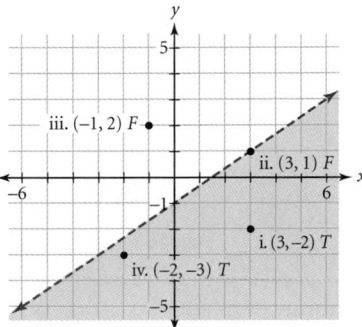

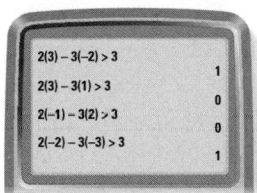

i. $2(3) - 3(-2) > 3$	$\longrightarrow$	$12 > 3$	$\longrightarrow$	True
ii. $2(3) - 3(1) > 3$	$\longrightarrow$	$3 > 3$	$\longrightarrow$	False
iii. $2(-1) - 3(2) > 3$	$\longrightarrow$	$-8 > 3$	$\longrightarrow$	False
iv. $2(-2) - 3(-3) > 3$	$\longrightarrow$	$5 > 3$	$\longrightarrow$	True

[▶ 🖳 See **Calculator Note 6C** to test inequalities. ◀]

Graphing Inequalities

▶ Draw a broken or dashed line on the boundary for inequalities with $>$ or $<$.
▶ Draw a solid line on the boundary for inequalities with $\geq$ or $\leq$.
▶ To graph inequalities of the form $y <$ or $y \leq$ shade below the boundary line.
▶ To graph inequalities of the form $y >$ or $y \geq$ shade above the boundary line.

[▶ 🖳 See **Calculator Note 6D** to graph inequalities in two variables on your calculator. ◀]

EXAMPLE

This example is good for students who had difficulties understanding the investigation. It also shows how calculators can evaluate the truth or falsity of statements.

You may need to remind students that points *satisfy* the inequality if they make the inequality true. They may check the potential solutions by hand as well as with a calculator.

Closing the Lesson

To visualize solutions to **inequalities involving two variables,** you can graph the line that is represented by the equation and then shade in the side of that line (**half-plane**) on which the solutions to the inequality lie.

BUILDING UNDERSTANDING

Students practice working with inequalities in two variables.

ASSIGNING HOMEWORK

Essential	1–3, 4 or 5, 6–9
Performance assessment	6, 7
Portfolio	8, 10
Journal	13
Group	10, 11
Review	12–14

▶ Helping with the Exercises

Exercise 1 If students are having difficulties, suggest that they turn the inequality into an equation and graph the line that is the boundary for the solutions to the inequality. Some may want to solve for y to have the equation in intercept form.

Exercise 3 This exercise reviews graphing inequalities on a number line. You may want to ask students to discuss the difference between inequalities graphed on a number line and those graphed on a coordinate plane.
[Ask] "How would you graph an inequality like $x < 5$ on coordinate axes?" [Shade the half-plane to the left of the vertical line $x = n$ if $x < n$ or the right of the line if $x > n$.]

3a.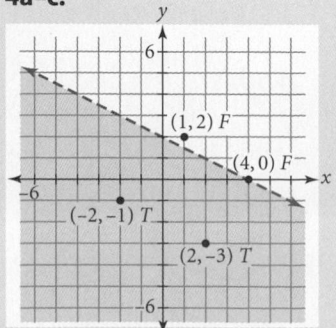

3b.

3c.

3d.

Exercises 4 to 6 [Alert] Check that students are using a broken line for strict inequalities and a solid line for others.

▶ Practice Your Skills

1. Match each graph with an inequality.

  a. $y \le 3 + 2x$ iii   **b.** $y \le 2 + 3x$ ii   **c.** $2x + 3y \le 6$ i   **d.** $2x + 3y \ge 6$ iv

i. 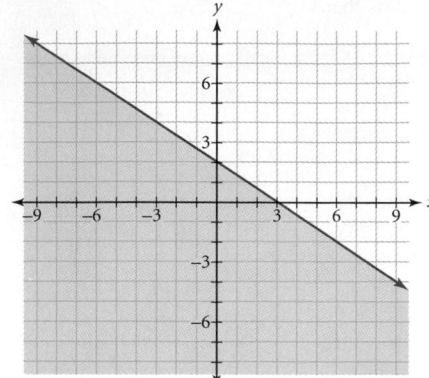   **ii.**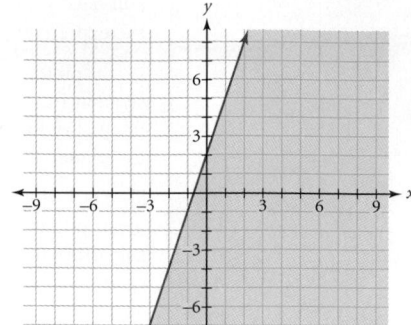

iii. 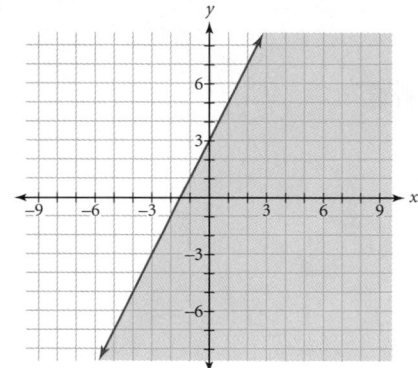   **iv.**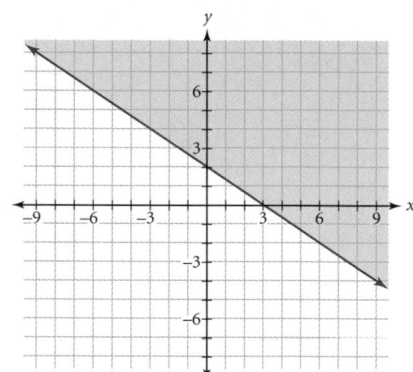

2. Solve each inequality for y.

  a. $84x + 7y \ge 70$ $y \ge -12x + 10$ **b.** $4.8x - 0.12y < 7.2$ $y > 40x - 60$

3. Sketch each inequality on a number line.

  a. $x \le -5$ **b.** $x > 2.5$ **c.** $-3 \le x \le 3$ **d.** $-1 \le x < 2$

4. Consider the inequality $y < 2 - 0.5x$.

  a. Graph the boundary line for the inequality on axes with scales from -6 to 6 on each axis.

  b. Determine whether each given point satisfies $y < 2 - 0.5x$. Plot the point on the graph you drew in 4a. Label the point T (true) if it is part of the solution or F (false) if it is not part of the solution region.

    i. $(1, 2)$ **ii.** $(4, 0)$ **iii.** $(2, -3)$ **iv.** $(-2, -1)$

  c. Use your results from 4b to shade the half-plane that represents the inequality.

4a–c.

5. Consider the inequality $y \geq 1 + 2x$.

    a. Graph the boundary line for the inequality on axes with scales from -6 to 6 on each axis.

    b. Determine whether each given point satisfies $y \geq 1 + 2x$. Plot the point on the graph you drew in 5a, and label the point T (true) if it is part of the solution or F (false) if it is not part of the solution.

      i. $(-2, 2)$ **ii.** $(3, 2)$ **iii.** $(-1, -1)$ **iv.** $(-4, -3)$

    c. Use your results from 5b to shade the half-plane that represents the inequality.

Reason and Apply

6. Sketch each inequality.

    a. $y \leq -3 + x$

    b. $y > -2 - 1.5x$

    c. $2x - y \geq 4$

7. Write the inequality for each graph.

    a.

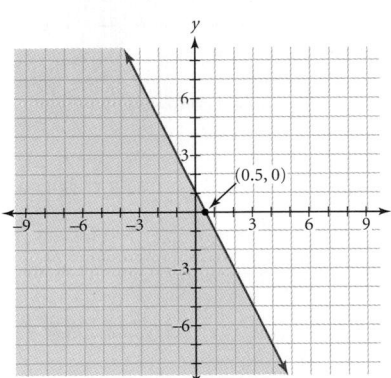

$y \leq 1 - 2x$

    b.

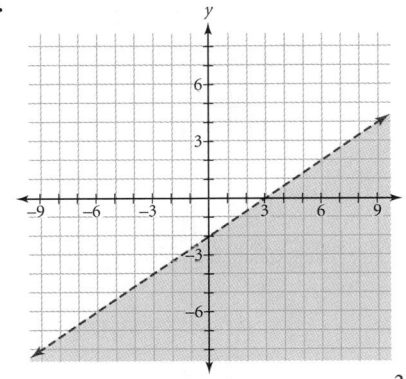

$y < -2 + \frac{2}{3}x$

    c.

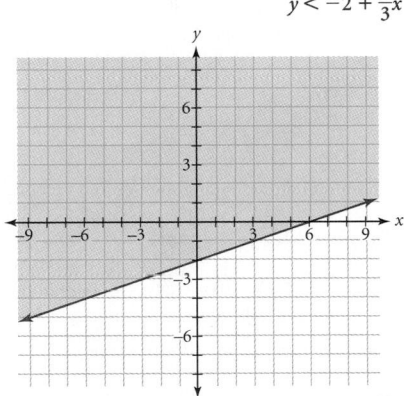

$y > 1 - 0.5x$

    d.

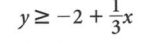

$y \geq -2 + \frac{1}{3}x$

5a–c.

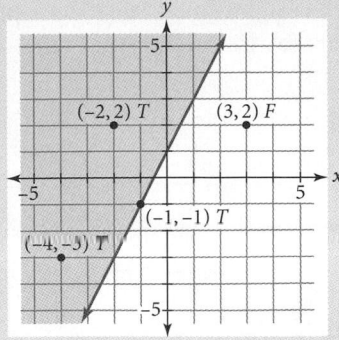

6a.

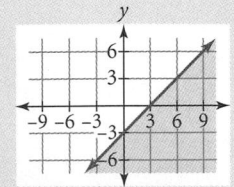

6b.

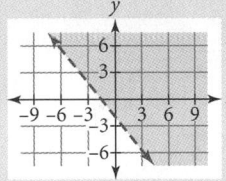

6c.

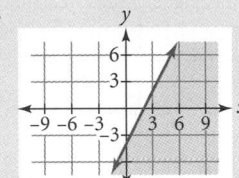

Exercise 8 Note that this full exercise requires 80 checks. You might want to assign only some of the parts.

9a.

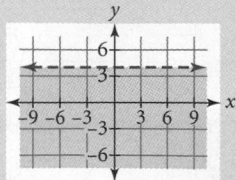

9b.

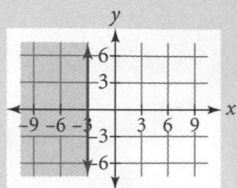

9c.

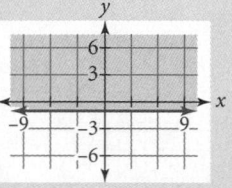

9d.

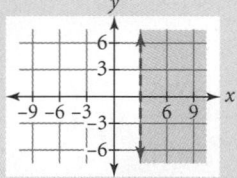

8. Which of these points lie in the shaded region of each graph?

$A(4, 3)$	$B(1, -2)$	$C(5, -4)$	$D(2, 0)$	$E(0, 5)$
$F(4, -7)$	$G(-2, 3)$	$H(1, 8)$	$I(-1, -4)$	$J(3, 11)$

i.   *A, C, D, F, J*

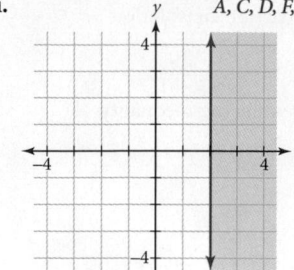

ii.   *B, C, D, F, I*

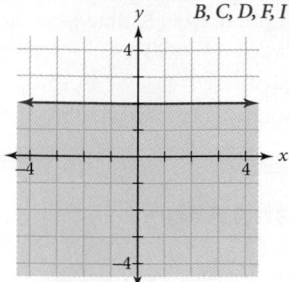

iii.   *B, C, D, F, G, I*

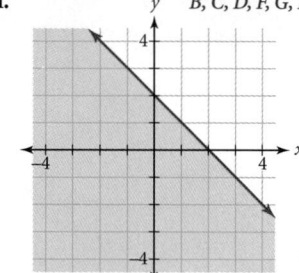

iv.   *A, E, H, J*

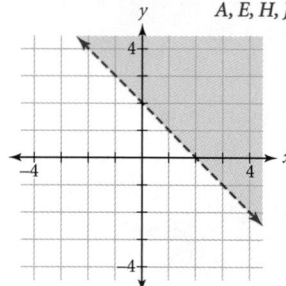

v.   *A, E, G, H, J*

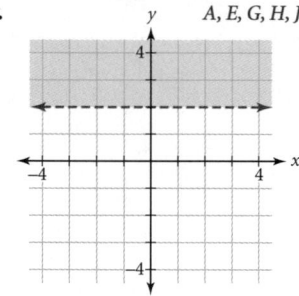

vi.   *B, E, G, H, I*

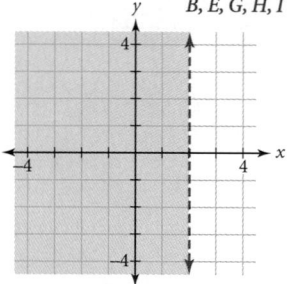

vii.   *A, D, E, G, H, J*

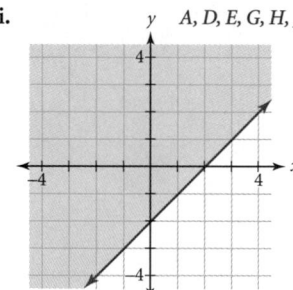

viii.   *B, C, F, I*

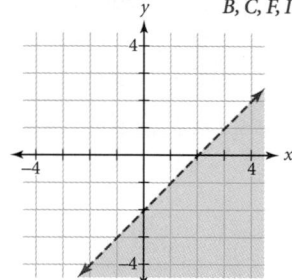

9. Sketch each inequality on coordinate axes.

   a. $y < 4$ **b.** $x \leq -3$ **c.** $y \geq -1$ **d.** $x > 3$

10. APPLICATION The total number of points from a combination of one-point free throws, *F*, and two-point shots, *S*, is less than 84 points.

    a. Write an inequality to represent this situation. $F + 2S < 84$

10b.
$F + 2S = 84$

    b. Write the equation for the boundary line of this situation.

    c. Graph this inequality. Label the horizontal axis two-point shots and the vertical axis free throws. Show the scale you used on the axes.

10d.
Possible
answer:
(0.50),
(10, 30),
(25, 0)

    d. On your graph, indicate three possible combinations of free throws and two-point shots that give a point total of 50. Name the coordinates of these points.

11. Graph the inequalities in problems 4 and 5 on your calculator. [▶🖳 See **Calculator Note 6D** to review graphing inequalities on your calculator. ◀]

Raul Acosta plays wheelchair basketball for the Eastern Paralyzed Veterans Association in New Jersey. The sport started in 1946 and is governed worldwide by the International Wheelchair Basketball Federation.

▶ **Review**

12. In social studies, Zach studies minimum wages of the past 60 years. He finds this set of data on the Internet.

    a. Graph the data from the table on the same set of axes. Use one color for minimum wage and another for 1998 dollars.

    b. Which is better represented by a line, the hourly minimum wage or the dollar value? minimum wage

12c.
Q points:
(1956, 1.00)
(1981, 3.35)

    c. Find the line of fit based on Q-points for the data points of the form (*year*, *1998 dollars*) that are best modeled by a line. $y = -182.864 + 0.094x$

    d. Graph the equation in 12c to verify that it is a good fit.

    e. What is the real-world meaning of the slope? How does it compare with the 1998 dollars graph?

13. Ellie was talking with her grandmother about a trip she took this summer. Ellie made the trip in 2.5 hours traveling at 65 mph. Ellie's grandmother remembers that she made the same trip in about 6 hours when she was Ellie's age.

    a. What speed was Ellie's grandmother traveling when she made the trip? about 27 mph

    b. Explain how this is an application of inverse variation.

14. Solve each equation for *y*.

    a. $7x - 3y = 22$ $y = \frac{7}{3}x - \frac{22}{3}$

    b. $5x + 4y = -12$ $y = -\frac{5}{4}x - 3$

Year	Minimum wage	1998 Equivalent dollars
1938	0.25	2.89
1939	0.30	3.52
1945	0.40	3.62
1950	0.75	5.07
1956	1.00	5.99
1961	1.25	6.27
1967	1.40	6.83
1968	1.60	7.49
1974	2.00	6.61
1975	2.10	6.36
1976	2.30	6.59
1978	2.65	6.63
1979	2.90	6.51
1980	3.10	6.13
1981	3.35	6.01
1990	3.50	4.74
1991	4.25	5.09
1996	4.75	4.93
1997	5.15	5.23

(www.dol.gov)

12a, d.

Minimum Wage

Exercise 10 Be sure students go through the entire modeling cycle for this real-world problem. In particular, they should consider how their solution to the inequality applies to the original situation. Though all points below the line are shaded to represent the solution set of the inequality, the only meaningful solutions to the problem are points representing non-negative integer values for free throws and two-point shots.

10c.

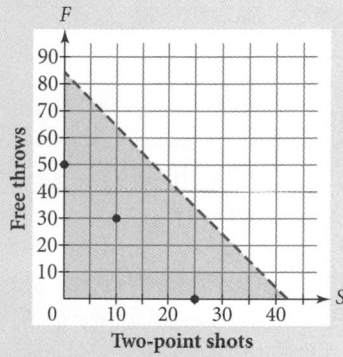

11.

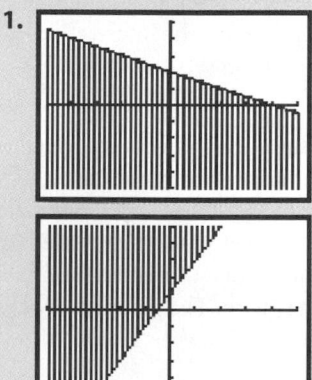

Exercise 12 This exercise is a follow-up to problem 10 in the Chapter 5 Chapter Review.

12e. The minimum wage increases 9¢ every year on average. But actual dollar value has decreased since 1978 and was highest in 1968.

Exercise 13 This exercise reviews Lesson 3.6.

13b. Since $d = r \cdot t$, and the distance was the same for both Ellie and her grandmother, you can set these products equal to each other. If you let $r =$ Ellie's grandmother's speed, then $2.5(65) = 6r$.

Exercise 14 This exercise reviews Lesson 4.6.

PLANNING

LESSON OUTLINE

One day:

15 min	Examples
20 min	Investigation
5 min	Sharing
5 min	Closing
5 min	Exercises

MATERIALS

- Cereal Sales and Profit (T)
- Calculator Note 6D

TEACHING

The solution set for a system of inequalities is the intersection of the solution sets for the individual inequalities.

One step Tell students that a cereal company is letting the buyer of each box of cereal enter a drawing for a $1,000 scholarship. One scholarship will be given away each month. The company makes a profit of between $0.47 and $1.10 on each box of cereal, depending on how the cereal is priced at different locations. If the company sells 2000 boxes in a month, will it make enough to cover the $1,000 scholarship? Ask students to answer the question in as many ways as they can. As you circulate, be sure that one way they answer is through a system of inequalities.

EXAMPLE A

This example shows how to solve an abstract system of inequalities.

[Ask] "In how many points can two half-planes intersect?" [Infinitely many if they intersect and zero if they don't.]

LESSON

6.7

All mathematical truths are relative, conditional.

CHARLES PROTEUS
STEINMETZ

Systems of Inequalities

You learned that the solution to a system of two linear equations, if there is exactly one solution, it is the coordinates of the point where the two lines intersect. In this lesson you'll learn about **systems of inequalities** and their solutions. Unlike the graphs of linear equations, the graphs of linear inequalities don't intersect in a single point, as you'll see in the examples and investigation in this lesson.

Translucent sheets of blue, red, and yellow intersect to form overlapping regions of new colors—orange, green, and purple.

EXAMPLE A

Graph the system of inequalities

$$\begin{cases} y \le -2 + \frac{3}{2}x \\ y > 1 - x \end{cases}$$

Graph the boundary lines and shade the half-planes. Indicate the solution area as the darkest region.

▶ **Solution**

First, determine if the boundary lines are solid or dashed. Graph $y = -2 + \frac{3}{2}x$ with a solid line because points on the line satisfy the inequality. Graph $y = 1 - x$ with a dotted line because its points do not satisfy the inequality.

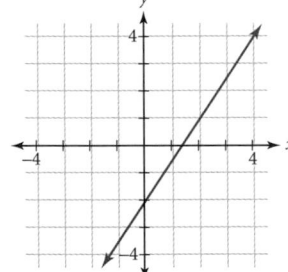 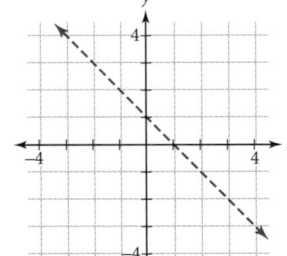 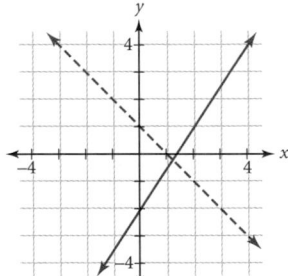

Shade the half-plane below the solid line $y = -2 + \frac{3}{2}x$ because its inequality has the "less than or equal to" symbol, $\le$. Shade above the dotted line $y = 1 - x$ because its inequality has the "greater than" symbol, $>$. Use different colors or patterns to distinguish each area shaded.

LESSON OBJECTIVES

- Solve systems of inequalities by graphing
- Interpret the mathematical solutions in terms of the problem context

NCTM STANDARDS

CONTENT		PROCESS	
	Number	•	Problem Solving
•	Algebra	•	Reasoning
•	Geometry	•	Communication
	Measurement	•	Connections
•	Data/Probability	•	Representation

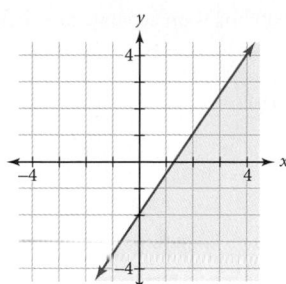 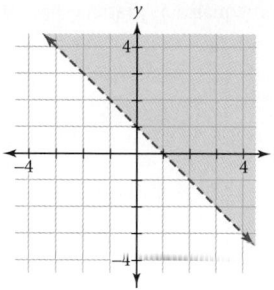 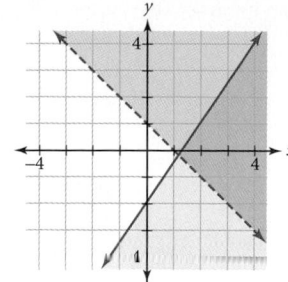

Each shaded area indicates the region of points that satisfy each inequality. The overlapping area bounded by $y \leq -2 + \frac{3}{2}x$ and $y > 1 - x$ satisfies both. The points that lie in both half-planes are the solutions to the system of inequalities.

EXAMPLE B

A cereal company is including a chance to win a $1,000 scholarship in each box of cereal. In this promotional campaign, it will give away one scholarship each month, regardless of the number of boxes sold. Because the cereal is priced differently at various locations, the profit from a single box is between $0.47 and $1.10. Graph the expected profit, given the initial cost of the scholarship, for up to 5000 boxes sold in a month. Show the solution region on a graph. Is it possible to sell 3000 boxes and make a profit of $1,000?

▶ **Solution**

Write a system of inequalities to model this situation. The lowest profit per box is $0.47. So $0.47x$ is the profit when x boxes are sold. Subtract $1,000 for the scholarship given each month. So the profit y is at least $0.47x - 1000$ dollars for x boxes sold. This is given by the inequality

$$y \geq -1000 + 0.47x$$

Likewise, if the maximum profit is $1.10 per box, then the profit is at most $1.1x - 1000$ dollars. So the second inequality is

$$y \leq -1000 + 1.1x$$

The profit during each month is given by the system

$$\begin{cases} y \geq -1000 + 0.47x \\ y \leq -1000 + 1.1x \end{cases}$$

Analogously, a system of linear *equations* has zero solutions, one solution, or infinitely many solutions because two lines intersect in zero points, one point, or infinitely many points.

Students enjoy watching their calculators graph a system of inequalities. If you have a projection panel, you might demonstrate.

Students may wonder if the intersection of two half-planes is a "quarter-plane." Since the area of the intersection is infinite, it isn't really smaller than a half-plane in any measurable sense.

EXAMPLE B

This example provides a real-world application of a system of inequalities. If students don't understand how to set up the inequalities, ask what they want to find (to help them determine the variables) and what they know (to write the inequalities).

You might ask why anyone would want to know if it's possible to sell 3000 boxes and make $1,000. For example, the marketing staff might want to know if the $1,000 scholarship will be paid for by selling a projected number of boxes.

Pick various points in the solution region. Ask why they are solutions to each inequality and then what real-world meaning they have. You might generate some good discussion if you pick a point with fractional coordinates. Students will need to realize that x is a number of boxes sold, so it must be an integer. The variable y is in dollars, so it can be fractional to an extent.

The transparency master shows the two graphs separately. If you cut the transparency apart, you can lay one graph on top of the other and get a result that looks like the third graph.

Each inequality is graphed for up to 5000 boxes on separate axes below.

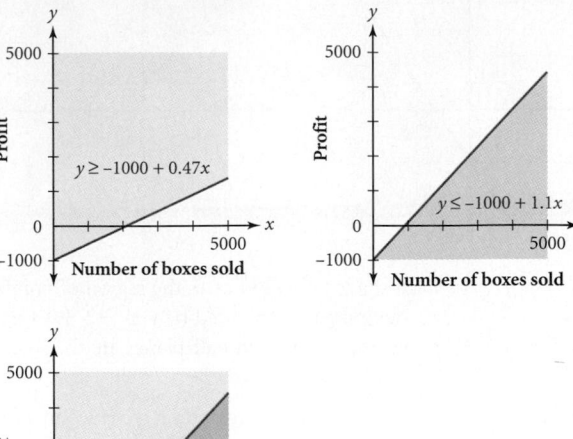

$y \geq -1000 + 0.47x$

Profit

Number of boxes sold

$y \leq -1000 + 1.1x$

Profit

Number of boxes sold

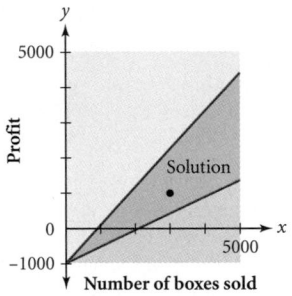

Profit

Solution

Number of boxes sold

The possible profits are in the region where the two half-planes overlap.

[▶ 🖳 See **Calculator Note 6D** to graph systems of inequalities on your calculator. ◀]

Profit

Solution

Number of boxes sold

To see if it is possible to make $1000 when 3000 boxes are sold, plot the point (3000, 1000) on the graph. Since the point is in the solution region, the coordinates satisfy both inequalities.

$$y \geq -1000 + 0.47x \qquad\qquad y \leq -1000 + 1.1x$$
$$1000 \;\square\; -1000 + 0.47(3000) \quad \text{and} \quad 1000 \;\square\; -1000 + 1.1(3000)$$
$$1000 \geq 410 \qquad\qquad\qquad 1000 \leq 2300$$

Both inequalities are true, so it is possible to sell 3000 boxes and make $1000.

The inequalities in a system are often called **constraints.** In Example B, the inequalities model constraints on the possible profits in the situation. In the following investigation you'll learn about another application that has constraints.

Step 6 Answers will vary. Yes, (0, 0) satisfies the system. This means the envelope has no length or width. Minimum and maximum lengths and widths could be added as constraints. For example, the Postal Service lists $11\frac{1}{2}$ inches and $6\frac{1}{8}$ inches as the maximum length and width for an envelope with a 34¢ stamp. So a sample system is

$$\begin{cases} l \leq 2.5w \\ l \geq 1.3w \\ w \leq 6.125 \\ l \leq 11.5 \end{cases}$$

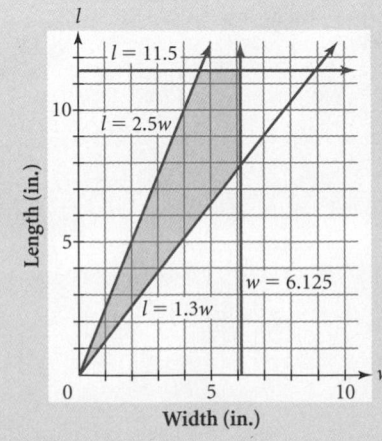

$l = 11.5$

$l = 2.5w$

Length (in.)

$w = 6.125$

$l = 1.3w$

Width (in.)

Investigation
A "Typical" Envelope

The U.S. Postal Service imposes several constraints on the acceptable sizes for an envelope. One constraint is that the ratio of length to width must be less than or equal to 2.5, and another is that this ratio must be greater than or equal to 1.3.

Step 1 Let l represent the length in inches and w represent the width in inches. $\frac{l}{w} \leq 2.5, \frac{l}{w} \geq 1.3$

Step 2 $\begin{cases} l \leq 2.5w \\ l \geq 1.3w \end{cases}$ Inequality is not reversed because w takes on only positive values.

Step 4 Answers will vary. Check by substituting coordinates from the overlapping regions to see if they satisfy both inequalities.

Step 1 Define variables and write an inequality for each constraint.

Step 2 Solve each inequality for the variable representing length. Decide whether or not you have to reverse directions on the inequality symbols. Then write a system of inequalities to describe the Postal Service's constraints on envelope sizes.

Step 3 Decide on appropriate scales for each axis and label a set of axes. Decide if you should draw the boundaries of the system with solid or dashed lines. Graph each inequality on the same set of axes. Shade each half-plane with a different color or pattern.

Step 4 Where on the graph are the solutions to the system of inequalities? Discuss how to check that your answer is correct.

Step 5 Decide if each envelope satisfies the constraints by locating the corresponding point on your graph.

a. 5 in. by 8 in. yes

b. 3 in. by 3 in. no

c. 2.5 in. by 7.5 in. no

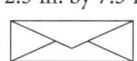

d. 5.5 in. by 7.5 in. yes

Step 6 Do the coordinates of the origin satisfy this system of inequalities? Explain the real-world meaning of this point. What constraints can you add to more realistically model the Postal Service's acceptable envelope sizes? How do these additions affect the graph?

With enough constraints the solution to a system of inequalities might resemble a geometric shape or polygon. No matter how small the region, there are infinitely many points that satisfy the system. Even a line may contain solutions to a system of inequalities. Of course, if no solutions exist, there is no solution region at all.

Step 3

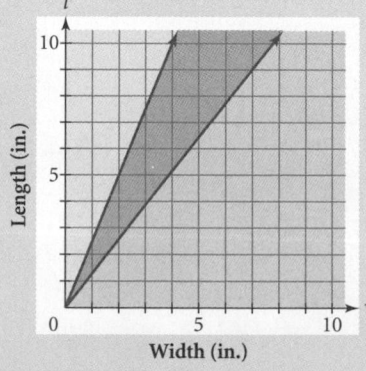

Step 5

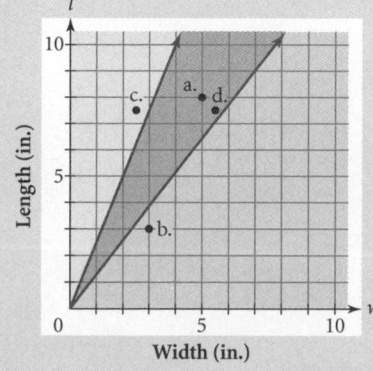

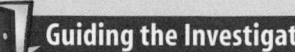

Guiding the Investigation

If you have a variety of envelopes that meet postal regulations, you might display them or even have students measure them, report their ratios, and try to guess the legal constraints.

Step 2 You may need to remind students to multiply each side of the inequality by w. Work on the habit of thinking about the sign of the multiplier of an inequality. Here, if nothing were known about w, you'd have to consider the case of negative w as well as positive w. Because w represents the width of an envelope, though, w is positive, so the direction of the inequality isn't changed.

Step 3 It is often good for students to graph on a larger region than is needed. If they graph on a calculator, though, they may not be able to shade differently.

Step 6 For information about other constraints, visit www.keymath.com/DA and see the link to the U.S. Postal Service. Tell students that, though the units in this ratio cancel, the Postal Service uses "inches" as its unit of measurement.

SHARING IDEAS

Have students briefly share answers to Step 5 and then discuss how well the solution to the system applies to the real-world problem. If you or students have looked up more constraints, bring those out.

Assessing Progress

Students will show their abilities at graphing the solution set for an **inequality in two variables** and their understanding of the need to reverse the direction of an inequality when multiplying by a negative number.

See page 356 for answers to Step 6.

Closing the Lesson

You can find the solutions for a **system of inequalities** by finding the half-planes giving the solutions for the individual inequalities and then looking at their intersection.

BUILDING UNDERSTANDING

Students work with systems of inequalities.

ASSIGNING HOMEWORK

Essential	1–5, 8, 10
Performance assessment	8, 12
Portfolio	8, 11
Journal	7
Group	7, 11, 12
Review	13–16

▶ Helping with the Exercises

Exercise 2 Students might check whether or not pairs give solutions by substituting in the inequalities rather than graphing.

2a. Yes, $(1, 2)$ satisfies both inequalities.

2b. No, only one inequality is satisfied.

3b.

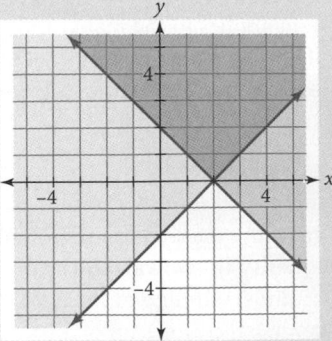

▶ Practice Your Skills

1. Match each system of inequalities with its graph.

a. $\begin{cases} y < 3 \\ x \geq 2 \end{cases}$ iii

b. $\begin{cases} y > 2 + x \\ y > 1 - x \end{cases}$ i

c. $\begin{cases} 2x - y \leq 6 \\ 3x + 2y \geq 12 \end{cases}$ ii

i.

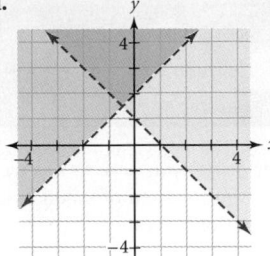

ii.

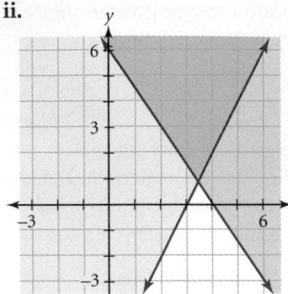

iii.

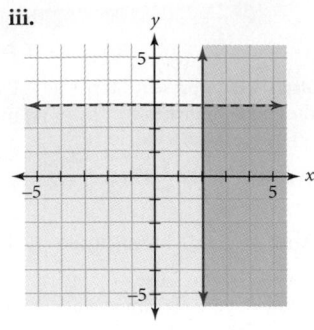

2. Here is the graph of this system of inequalities:

$$\begin{cases} y > x \\ y > 2 - \dfrac{1}{2}x \end{cases}$$

Is each point listed a solution to the system? Explain why or why not.

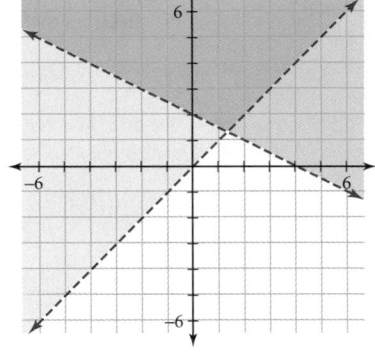

a. $(1, 2)$

b. $(3, 2)$

c. $\left(\dfrac{4}{3}, \dfrac{4}{3}\right)$

No, for both inequalities $\dfrac{4}{3} > \dfrac{4}{3}$ is not true.

d. $(5, -3)$

No, neither inequality is satisfied.

3. Consider these two inequalities together as a system.

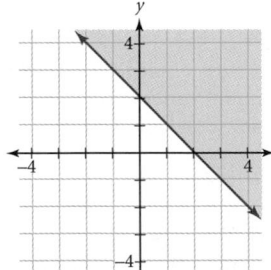

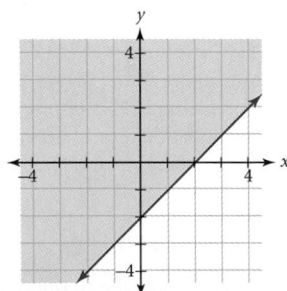

a. Name the inequality pictured in each graph. $y \geq -x + 2; y \geq x - 2$

b. Sketch a graph showing the solution to this system.

4. Sketch a graph showing the solution to each system.

a. $y \leq 2$
  $x < 2$

b. $x + y \leq 4$
  $x - y \leq 4$

5. Write a system of inequalities for the solution shown on the graph.

$$\begin{cases} y > 2 - x \\ y < 2 \\ x < 3 \end{cases}$$

▶ Reason and Apply

6. APPLICATION The cereal company from Example B decides to raise the scholarship amount to $1,250. It also lowers the cereal's price so that the expected profit from a single box is between $0.40 and $1.00.

a. Write the inequalities to represent this new situation. $y \geq -1250 + 0.40x;\ y \leq -1250 + 1.00x;\ x \geq 0$

b. Graph the expected revenue for up to 5000 boxes sold in a month.

7. On Kids' Night, every adult admitted into a restaurant must be escorted by at least one child. The restaurant has a maximum seating capacity of 75 people.

a. Write a system of inequalities to represent the constraints in this situation.

b. Graph the solution. Is it possible for 50 children to escort 10 adults to the restaurant?

c. Why would the restaurant reconsider the rules for Kids' Night? Add a new constraint to address these concerns. Draw a graph of the new solution.

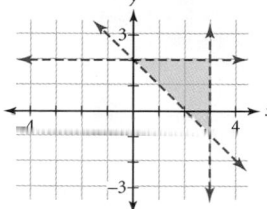

8. APPLICATION The American College of Sports Medicine considers age as one factor when it recommends low and high heart rates during workout sessions. For safe and efficient training, your heart rate should be between 55% and 90% of the maximum heart rate level. The maximum heart rate is calculated by subtracting a person's age from 220 beats per minute.

a. Define variables and write an equation relating age and maximum heart rate during workouts.

b. Write a system of inequalities to represent the recommended high and low heart rates during a workout.

c. Graph the solution to show a region of safe and efficient heart rates for people of any age.

d. What constraints should you add to limit your region to show the safe and efficient heart rates for people between the ages of 14 and 40? $a \geq 14$ and $a \leq 40$

e. Graph the new solution for 8d.

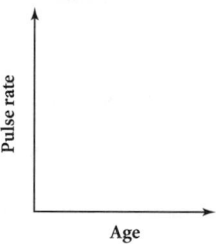

Pulse rate

Age

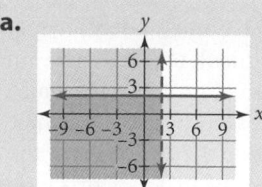

4a.

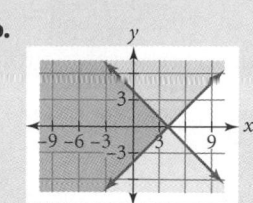

4b.

Exercise 6 Don't mark students wrong if they don't include the inequality $x \geq 0$. It is assumed but not stated in the example.

6b.

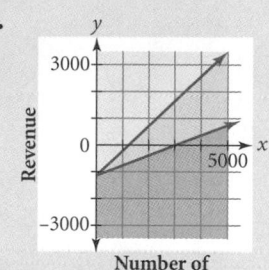

Revenue

Number of

Exercise 7 A very common error is to translate "every adult must be escorted by at least one child" as "$A \geq 1C$" or "$A \geq C$." Rather than just announcing that the inequality is backward, try to engage students in a conversation about how A and C represent numbers of adults and children, not just the words *adult* and *child*. Students may not list the inequalities $A \geq 0$ and $C \geq 0$.

7a. $\begin{cases} A \leq C \\ A + C \leq 75 \\ A \geq 0 \\ C \geq 0 \end{cases}$

7b.

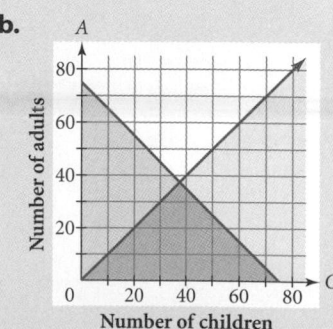

Number of adults

Number of children

All the points in the dark-shaded triangular region satisfy the two inequalities. The point (50, 10) represents the situation in which 50 children escort 10 adults.

7c. Answers will vary. It is possible to have all children and no adults at the restaurant. One possible additional constraint is that there must be at least one adult per five children, or $A \geq \frac{1}{5}C$. The solution for this set of constraints is the triangular region bounded by $A \leq C$, $A + C \leq 75$, and $A \geq 0.2C$.

8a. $r = 220 - a$, where a represents age in years and r represents the heart rate in beats per minute.

8b. $\begin{cases} r \leq 0.90(220 - a) \\ r \geq 0.55(220 - a) \end{cases}$ or $\begin{cases} r \leq 198 - 0.90a \\ r \geq 121 - 0.55a \end{cases}$

8c.

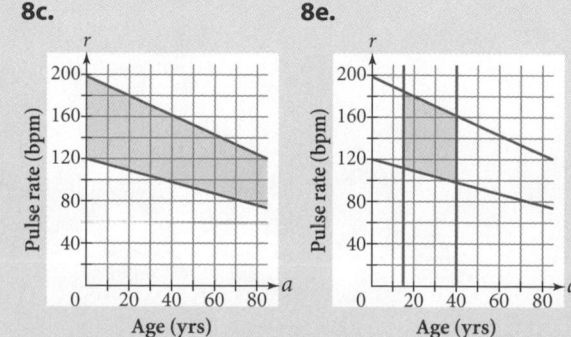

Pulse rate (bpm)

Age (yrs)

8e.

Pulse rate (bpm)

Age (yrs)

Exercise 10 Some students may find it difficult to work backward from the region to the inequalities, especially because they first have to find the equations of the lines. Suggest that they first write down a list of the steps and then follow their plan.

10. $AB: y \leq \frac{2}{3}x + \frac{5}{3}$;

$BC: y \leq -\frac{3}{5}x + \frac{59}{5}$;

$AC: y \geq \frac{1}{11}x + \frac{31}{11}$

11. The region is a pentagon.

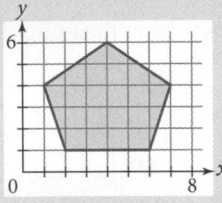

Exercise 13 This exercise reviews Lesson 2.3.

9. Write two inequalities that describe the shaded area below. Assume that the boundaries are solid lines and that each grid mark represents 1 unit.

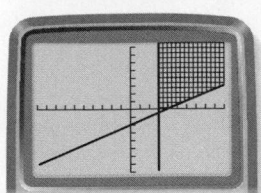

$x \geq 3$ and $y \geq -2 + \frac{1}{2}x$

10. Write a system of inequalities to describe the shaded area on the graph at right. Write each slope as a fraction.

11. Graph this system of inequalities on the same set of axes. Describe the shape of the region.

$$\begin{cases} y \leq 4 + \frac{2}{3}(x-1) \\ y \leq 6 - \frac{2}{3}(x-4) \\ y \geq -17 + 3x \\ y \geq 1 \\ y \geq 7 - 3x \end{cases}$$

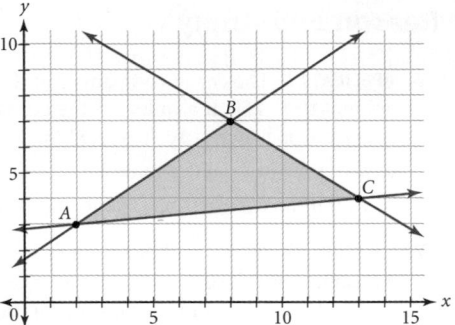

12. Write a system of inequalities that defines each shaded region in the parallelogram on the graph at right.

Region 1: $\begin{cases} y \geq 3 \\ y \geq x - 2 \\ y \leq \frac{1}{3}x + \frac{8}{3} \end{cases}$ Region 2: $\begin{cases} y \leq 3 \\ y \leq x - 2 \\ y \geq \frac{1}{3}x \end{cases}$

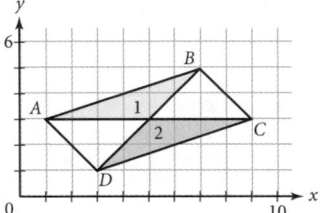

▶ **Review**

13. **APPLICATION** Manuel has a sales job at a local furniture store. Once a year, on Employees' Day, every item in the store is 15% off regular price. In addition, salespeople get to take home 25% commission on the items they sell as a bonus.

   a. A loft bed with a built-in desk and closet usually costs $839. What will it cost on Employee's Day? $713.15

   b. At the end of the day, Manuel's bonus is $239.45. How many dollars worth of merchandise did he sell? $957.80

14. Hugh went for a hike in Colorado that went by windmills that generate electricity. He decided to figure out how tall they are by using proportions. Hugh knows that he is about 6 feet tall, and he measured his shadow as being 4 feet long. The shadow cast by the windmill was about 175 feet long. How tall is the windmill? about 263 feet tall

Exercise 14 This exercise reviews Lesson 3.3.

Exercise 15 This exercise reviews Lesson 4.2.

15. Think about this number trick.

  a. Layla got a final number of 4. What was her original number? 4

  b. Robert got a final answer of 10. What was his original number? 10

  c. Write an algebraic expression to represent this sequence of operations. Let x represent the number you start with.

x	___
Ans · 3	___
Ans + 12	___
Ans ÷ 5	___
Ans − 1.4	___
Ans · 10	___
Ans − 10	___
Ans ÷ 6	___

15c. $\dfrac{10\left(\dfrac{3x + 12}{5} - 1.4\right) - 10}{6}$, which simplifies to x

16. Solve each system of equations by using a symbolic method. Check that your solutions are correct.

  a. $\begin{cases} y = 4x - 3 \\ y = 2x + 9 \end{cases}$
  $x = 6, y = 21$

  b. $\begin{cases} 3x - 4y = -2 \\ -2x + 3y = 1 \end{cases}$
  $x = -2, y = -1$

IMPROVING YOUR REASONING SKILLS

Suppose 9 crows each make 9 caws 9 times throughout the day. How many total caws are there?

Suppose 99 crows make 99 caws 99 separate times in one day. Now how many caws are there?

Answer the question again for 999 crows making 999 caws 999 times. If you continue this pattern of problems, at what number does your calculator round the answer? What is the exact number of caws in this case?

Write the answers to the first three questions and look for a pattern. Use it to find how many caws there are when the number is 99,999. With 86,400 seconds in a day, this means that each crow makes more than one caw per second every hour!

IMPROVING REASONING SKILLS

The calculators referred to in the calculator notes will use scientific notation for numerals with more than ten digits. Some more powerful calculators, such as the TI-89, will hold many more digits. At 9,999 crows, the TI-83 begins rounding.

Crows	Caws
9	729
99	970,299
999	997,002,999
9,999	999,700,029,999
99,999	999,970,000,299,999

CHAPTER REVIEW

6

•CHA **CHAPTER** • CHAPTER 6 REVIEW • CHAPTER 6 REVIEW • CHAPTER 6 REVIEW • CH.

6

REVIEW

PLANNING

LESSON OUTLINE

5 min	Introduction
15 min	Exercises and helping individuals
15 min	Checking work and helping individuals
15 min	Student self assessment

REVIEWING

Have students reexamine Example B of Lesson 6.3. Ask them to solve the system of equations by eliminating b instead of a and then again by substitution and by matrix row operations. Then change the problem so that the cereal makers haven't yet started the promotion but, based on experience, predict that they will receive fewer than 1600 UPC symbols and less than \$1,500. How many basketballs should they be prepared to send out? Students should graph the two inequalities and find the solution set in the first quadrant. As students examine the corners of the solution set, they see how many basketballs must be sent out for each of these four extreme cases and that the largest number they'll need is 271.

ASSIGNING HOMEWORK

Exercises 1 through 5 review the mechanics of solving systems of equations. Exercises 6 through 8 review inequalities and systems of inequalities. Exercise 9 is an application that you might want students to complete in groups. Assign Exercise 10 only if you are covering matrices.

1. See page 363.

3. See page 363.

In this chapter you learned to model many situations with a **system of equations** in two variables. You learned that systems of linear equations can have 0, 1, or infinitely many solutions. You used tables, used graphs, and solved symbolically to find the solutions of systems. You discovered that the methods of **elimination, substitution,** and **row operations** on a matrix allow you to find exact solutions to problems, not just the approximations of graphs and tables.

Then you analyzed situations involving **inequalities** and discovered how to find their solutions using graphs, tables, and **symbolic manipulation.** The graph of an inequality in one variable is a part of a number line, and the graph of a linear inequality in two variables is a shaded **half-plane** that contains points whose coordinates make the inequality true. You learned that a **compound inequality** is the combination of two inequalities.

You discovered how to use inequalities to define **constraints** that limit the solution possibilities in real-world applications. You learned how to graph a **system of inequalities.**

EXERCISES

1. Lines a and b at right form a system of equations. Write the equations of the lines and find the point of intersection.

2. Find the point where the graphs of the equations intersect. Check your answer.

$$\begin{cases} 3x - 2y = 10 \\ x + 2y = 6 \end{cases}$$ The lines meet at the point $(4, 1)$; the equations $3(4) - 2(1) = 10$ and $(4) + 2(1) = 6$ are both true.

3. Graph this system of equations, and find the solution point.

$$\begin{cases} y = 5 - 0.5(x - 3) \\ y = -4 + 1.5(x + 2) \end{cases}$$

4. Show the steps involved in solving this system symbolically by the substitution method. Explain each step.

$$\begin{cases} y = 16 + 4.3(x - 5) \\ y = -7 + 4.2x \end{cases}$$

5. Complete each sentence.

a. A system of two linear equations has no solution if . . .

b. A system of two linear equations has infinitely many solutions if . . .

c. A system of two linear equations has exactly one solution if . . .

6. Name the inequality that each graph represents.

a.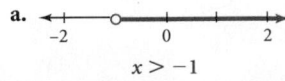
$x > -1$

b.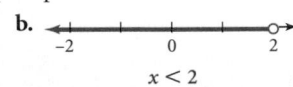
$x < 2$

c.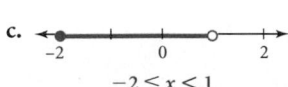
$-2 \leq x < 1$

4.
$16 + 4.3(x - 5) = -7 + 4.2x$	Set the right sides of the two equations equal to each other.
$16 + 4.3x - 21.5 = -7 + 4.2x$	Apply the distributive property.
$-5.5 + 4.3x = -7 + 4.2x$	Subtract.
$0.1x = -1.5$	Add $-4.2x$ and 5.5 to both sides.
$x = -15$	Divide both sides by 0.1.
$y = -7 + 4.2(-15)$	Substitute -15 for x and find y.
$y = -70$	Multiply and add.

The solution is $x = -15$ and $y = -70$.

5a–c. See page 363.

7. Solve the inequality $5 \le 2 - 3x$ for x and graph the solution on a number line.

8. Write a system of inequalities to describe this shaded area.

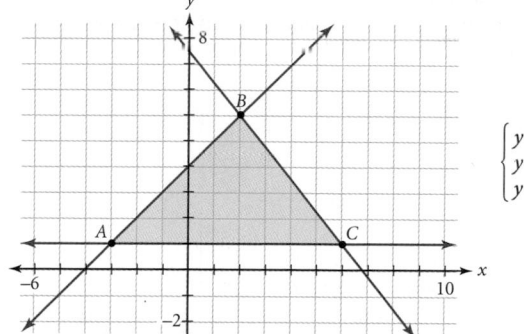

$$\begin{cases} y \le x + 4 \\ y \le -1.25x + 8.5 \\ y \ge 1 \end{cases}$$

9. **APPLICATION** Harold cuts lawns after school. He has a problem on Wednesdays when he cuts Mr. Fleming's lawn. His lawn mower has two speeds—at the higher speed he can get the job done quickly, but he always runs out of gas; at the lower speed he has plenty of gas, but it seems to take forever to get the job done. So he has collected this information.

- On Monday he cut a 15-meter-by-12-meter lawn at the higher speed in 18 minutes. He used a half tank of gas, or 0.6 liter.
- On Tuesday he cut a 20-meter-by-14-meter lawn at the lower speed in 40 minutes. He used a half tank of gas.
- Mr. Fleming's lawn measures 22 m by 18 m.

a. At the higher speed, how many square meters of lawn can Harold cut per minute? 10 sq. m/min

b. At the lower speed, how many square meters of lawn can Harold cut per minute? 7 sq. m/min

c. If Harold decides to cut Mr. Fleming's lawn using the higher speed for 10 minutes and the lower speed for 8 minutes, will he finish the job? No. He will cut 156 sq. m, and the lawn measures 396 sq. m.

d. Let h represent the number of minutes cutting at higher speed, and let l represent the number of minutes cutting at lower speed. Write an equation that models completion of Mr. Fleming's lawn. $10h + 7l = 396$

e. At the higher speed, how much gas does the lawn mower use in liters per minute? $\frac{1}{30}$ liters per min

f. At the lower speed, how much gas does the lawn mower use in liters per minute? $\frac{3}{200}$ liters per min

g. If Harold starts with a full tank, will he have enough gas to use the higher speed for 10 minutes and the lower speed for 8 minutes? Yes. He will use $\frac{34}{75}$ liter, and the tank holds 1.2 liters.

h. Write an equation in terms of h and l that has Harold use all of his gas. $\frac{h}{30} + \frac{3l}{200} = 1.2$

i. Using the equations from 9d and h, solve the system. $l = 14.4$ min, $h = 29.52$ min

j. What is the real-world meaning of the solution? If Harold cuts for 29.52 minutes at the higher speed and 14.4 minutes at the lower speed, he will finish Mr. Fleming's lawn and use one full tank of gas.

10. Use row operations to find the solution matrix for this system.

$$\begin{cases} 7x + 3y = -45 \\ x + 6y = -51 \end{cases} \quad \begin{bmatrix} 1 & 0 & -3 \\ 0 & 1 & -8 \end{bmatrix}$$

---

1. line a: $y = 1 - x$;

line b: $y = 3 + \frac{5}{2}x$;

intersection: $\left(-\frac{4}{7}, \frac{11}{7}\right)$

3. The point of intersection is $(3.75, 4.625)$.

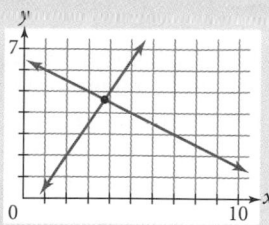

5a. ...the slopes are the same, but the intercepts are different (the lines are parallel).

5b. ...the slopes are the same and the intercepts are the same (the lines overlap).

5c. ...the slopes are different (the lines intersect in a single point).

7. $x \le -1$

▶ Take Another Look

Linear programming problems like this one deal with many more variables and inequalities. An application might require that you maximize or minimize a linear function of the two variables, using only points in this region. The largest and smallest values of the function occur along the boundary of the region, usually at a corner. These problems are a major application of mathematics today.

$$\text{System:} \begin{cases} x \le 6000 \\ y \le 8000 \\ x + y \le 10{,}000 \\ x \ge 0 \\ y \ge 0 \end{cases}$$

The profit is given by $15x + 10y$. At (6000, 4000), the profit is the maximum, $130,000.

TAKE ANOTHER LOOK

▶ Businesses use systems of equations and inequalities to determine how to maximize profits. A process called **linear programming** applies the concepts of constraints, points of intersection, and algebraic expressions to solve this very real application problem. Here is one example.

A company manufactures scooters and skateboards. The factory has the capacity to make at most 6000 scooters in one day, and the factory can make at most 8000 skateboards in one day. However, the factory can produce a combination of no more than 10,000 scooters and skateboards together. Define variables and write a system of three inequalities to describe these constraints. Label a set of axes and graph the solution. This is called a **feasible region.** What do the points in this shaded region represent? Find the points of intersection at the corners of this region.

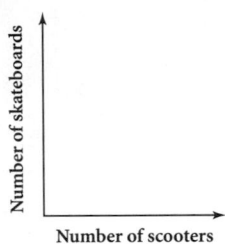

The company makes a profit of $15 per scooter and $10 per skateboard. How many of each should the company make to maximize its profits? To answer this question, use the variables defined earlier to write an expression to find the total profit the company makes from scooters and skateboards. Then substitute the coordinates of several points from the feasible region including the points of intersection. For example, if the company makes 5000 scooters and 5000 skateboards, substitute 5000 for x and 5000 for y into your expression to find the profit. Which point gave you the greatest profit? Be sure to try the points of intersection.

Professional skateboarder Tony Hawk performs a trick at the X Games.

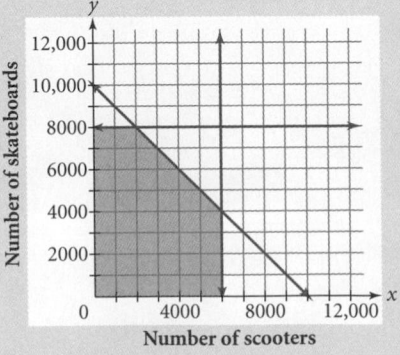

Assessing What You've Learned

 ORGANIZE YOUR NOTEBOOK Update your notebook with an example, investigation, or exercise that demonstrates each solution method for a system of equations. Add one problem that demonstrates each concept—inequalities in one variable, inequalities in two variables, and systems of inequalities.

 WRITE IN YOUR JOURNAL Add to your journal by answering one of these prompts:

▶ You have learned five methods to find a solution for a system of equations. Which method do you like best? Which one is the most challenging to you? What are the advantages and disadvantages of each method?

▶ You have now studied more than half of this book. What mathematical skills in the previous chapters were most crucial to your success in this chapter? Which concepts are your strengths and weaknesses?

▶ Describe in writing the difference between an inequality in one variable and an inequality in two variables. How do the graphs of the solutions differ? Compare these to the graph of a system of inequalities.

 UPDATE YOUR PORTFOLIO Choose your best graph of a system of inequalities from this chapter to add to your portfolio. Redraw the graph with a clearly labeled set of axes. Use color to highlight each inequality and its half-plane. Indicate the solution region with a visually pleasing design or pattern.

 GIVE A PRESENTATION Write your own word problem for a system of equations or inequalities. Choose a setting that is meaningful to you or that you wish to know more about, and write a problem to model the situation. It can be about winning times for Olympic events, the point where two objects meet while traveling, percent mixture problems, or something new you created. Solve the problem using one of the methods you learned in this chapter. Make a poster of the problem and its solution, and present it to the class. Work with a partner or in a group.

ASSESSING

For written assessment, use items from Constructive Assessment for this chapter or one of the chapter tests from Assessment Resources. Or create your own test using the CD-ROM test generator, omitting matrices if you are not covering them.

FACILITATING SELF ASSESSMENT

To help students complete the portfolio described in Assessing What You've Learned, suggest that they consider for evaluation their work in Lesson 6.1, Exercises 8 and 10; Lesson 6.2, Exercise 12; Lesson 6.3, Exercises 9 and 13; Lesson 6.4, Exercises 9 to 11; Lesson 6.5, Exercise 8; Lesson 6.6, Exercises 8 and 10; and Lesson 6.7, Exercises 8 and 11.

Exponents and Exponential Models

Overview

In **Lesson 7.1,** students model exponential growth through recursion with constant multipliers. In **Lesson 7.2,** students discover how to find exponential equations to model the real-life growth being simulated by the recursive routines. Students review the multiplication property of exponents in **Lesson 7.3,** where they are also introduced to the power properties. In **Lesson 7.4,** the multiplication property of exponents is used to introduce and develop scientific notation for large numbers. In **Lesson 7.5,** students review the division property of exponents. Then they combine exponential expressions involving both multiplication and division. **Lesson 7.6** introduces students to nonpositive exponents as they learn how to use scientific notation for very small numbers. In **Lesson 7.7,** students fit an exponential model to a set of data. **Lesson 7.8** is an activity day in which students do one of two experiments related to exponential decay.

The Mathematics

Exponential change

Up to this point in *Discovering Algebra,* the phenomena studied have exhibited a constant rate of change. That is, for every unit of change in *x,* the change in *y* has been constant. Values can be generated on a calculator by a recursive procedure involving *addition* of a constant at each step.

To model many real-life phenomena, however, values are often best generated on a calculator by recursively *multiplying* by a constant rather than by adding. Instead of consecutive terms having a constant *difference,* they have a constant *ratio.* (A series of terms generated by multiplying by a constant is known as a *geometric sequence,* though this terminology isn't used.) The rate of change is not constant, but rather is changing in a regular way. The most common examples of these phenomena are financial investments, populations, heating, cooling, and radioactive decay.

A constant rate of change can be represented by an equation of the form $y = a + bx$. The constant b is added to a repeatedly, x times. In contrast, with the multiplicative model, the equation is $y = ab^x$. The constant b is multiplying a repeatedly, x times. In this case, the rate of change is not constant, and b isn't the rate of change. In the exponential model, b represents the *constant multiplier* or *growth factor* (even when b is less than 1, in which case the values of y are shrinking rather than growing).

The graph of $y = a + bx$ is a straight line, so the equation represents what is called *linear change.* Because the variable x is an exponent in $y = ab^x$, this kind of change is called *exponential change.* The slope of the graph of $y = ab^x$ is always changing.

For most phenomena exhibiting exponential change, we don't know the constant multiplier directly, but rather we know what portion of y is being added on or removed for each unit change in x. For example, a financial investment earns interest at an annual rate of 5%. Or carbon-14 atoms decay at a rate of 35% every 1000 years. In these cases, the constant multiplier b is best written in terms of its relationship to 1. In the case of growth, we write $y = a(1 + r)^x$ and call r the *growth rate.* In the case of decay, we write $y = a(1 - r)^x$ and call r the *decay rate.* In actuality, r is not a rate because it has no units. The quantities $(1 + r)^x$ and $(1 - r)^x$ are dimensionless numbers being multiplied by the initial value a, which has the dimensions.

As a quick way to summarize exponential decay, especially of radioactive substances, scientists often refer to the *half-life* of a particle. The half-life is the amount of change in x that is required to halve the value of y. The fact that the half-life is independent of the initial value of y (it takes just as long to reduce y from 1000 to 500 as from 100 to 50) is an important feature of exponential change.

Going backward

Just as we often want to solve for x the linear equation $y = a + bx$ in order to predict when growth

will reach a certain point, we frequently want to solve for x the exponential equation $y = ab^x$. That is, we want to find the value of the exponent. There are several ways to do this. To get an approximate value of x, calculator graphs and tables are fine. So is the guess-and-check method—trying various values of x and gradually homing in on one that gives the desired value for y. For a more exact solution, logarithms can be used.

Scientific notation and significant digits

The note on significant digits on page 389 will help you guide students in determining the digits to display in scientific notation or after a decimal point.

Using This Chapter

If your students are already proficient at using the rules of exponents and scientific notation, you can skim or skip Lessons 7.3 through 7.5. If your class is familiar with nonpositive exponents, you can skip Lesson 7.6. On the other hand, if they're very weak at exponents and you skipped Chapter 0, you might want to do Lessons 0.2 and 0.3 before embarking on this chapter.

Resources

Discovering Algebra Resources

Calculator Notes 2A, 4B, 7A, 7B, 7C, 7D

Teaching and Worksheet Masters
 Lesson 7.2 Growth of the Koch Curve
 Lessons 7.3, 7.5, and 7.6 Properties of Exponents
 Lesson 7.7 100 Grid
 Lesson 7.7 Radioactive Decay Sample Data
 Lesson 7.7 Moore's Law Sample Data
 Lesson 7.8 Bounce Sample Data
 Lesson 7.8 Pendulum Sample Data

Assessment Resources A and B
 Quiz 1 (Lessons 7.1 and 7.2)
 Quiz 2 (Lessons 7.3 to 7.6)
 Quiz 3 (Lesson 7.7)
 Chapter 7 Test
 Chapter 7 Constructive Assessment Options
 Chapters 4 to 7 Exam

More Practice Your Skills for Chapter 7

Condensed Lessons for Chapter 7

Other Resources

Powers of Ten. Santa Monica, California: Pyramid Film and Video, 1984.

Philip and Phylis Morrison. *Powers of Ten.* New York: Scientific American Books, 1982.

Scott Beall. *Functional Melodies.* Emeryville, California: Key Curriculum Press, 2000.

Gary Asp et al. *Graphic Algebra.* Emeryville, California: Key Curriculum Press, 2000.

www.keymath.com/DA

Materials

- graph paper
- paper plates
- protractors
- small counters
- balls
- metersticks
- empty soda cans
- string
- motion sensors (CBL, CBR, or EA-100), *optional*

Pacing Guide

	day 1	day 2	day 3	day 4	day 5	day 6	day 7	day 8	day 9	day 10
standard	7.1	7.2	7.2	quiz, 7.3	7.4	7.4	7.5	7.6	7.6	quiz, 7.7
enriched	7.1	7.2	7.2, project	quiz, 7.3	7.4	7.4	7.5	7.6	7.6	quiz, 7.7
block	7.1, 7.2	7.2, 7.3	7.4	quiz, 7.5	7.6	7.7, quiz	7.8, review	assessment		

	day 11	day 12	day 13	day 14	day 15	day 16	day 17	day 18	day 19	day 20
standard	7.7	7.8	quiz, review	assessment						
enriched	7.7, project	7.8	review, TAL	assessment						

This "Chinese Horse" is part of a prehistoric cave painting in Lascaux, France. Scientific methods that use equations with exponents have determined that parts of the Lascaux cave paintings are more than 15,000 years old. For archaeologists, dating ancient artifacts helps them understand how civilizations evolved. Drawings and pieces of art help them understand what existed at that time and what was important to the civilization. You will see that exponents are useful in many other real-world settings too.

CHAPTER 7 OBJECTIVES

- Write the exponential form of a sequence generated recursively by a constant multiplier

- Review or learn the multiplication, division, and power properties of exponents

- Rewrite an expression with exponents as an expression with the opposite of those exponents

- Move between scientific notation (by hand and on calculators) and standard notation for numbers

- Write exponential equations that model real-world growth and decay data

OBJECTIVES

In this chapter you will

- write recursive routines for nonlinear sequences
- learn an equation for exponential growth or decrease
- use properties of exponents to rewrite expressions
- write numbers in scientific notation
- model real-world data with exponential equations

A standard procedure for determining age is by carbon dating (as discussed in the chapter), but many of the paints used in Lascaux are metal-based and don't contain carbon. The claim that the paintings are 15,000 years old are based on carbon dating of paintings done in charcoal and on other artifacts at the site. Non-carbon-based methods indicate that the caves are about 17,000 years old. Students can find more details about these dating methods and take a virtual tour of the site at http://www.culture.fr/culture/arcnat/lascaux/en/datation.htm. Students might also want to investigate the deterioration of the cultural treasures elsewhere, such as Italy or Egypt.

Knowledge of ancient art has made many modern artists concerned about the durability of the materials they use. Deterioration due to exposure and the growth of organisms in organic fibers and pigments is usually exponential. Students may be interested in the archival methods and materials for preserving family photographs or books. Will the latest film you saw be around in 15,000 years?

Recursive Routines

Slow buds the pink dawn
*   like a rose*
From out night's gray and
*   cloudy sheath*
Softly and still it grows
*   and grows*
Petal by petal, leaf by leaf
SUSAN COOLIDGE

Have you ever noticed that it doesn't take very long for a cup of steaming hot chocolate to cool to sipping temperature? If so, then you've also noticed that it stays about the same temperature for a long time. Have you ever left food in your locker? It might look fine for several days, then suddenly some mold appears and a few days later it's covered with mold. The same mathematical principle describes both of these situations. Yet these patterns are different from the linear patterns you saw in rising elevators and shortening ropes—you modeled those situations with repeated addition or subtraction. Now you'll investigate a different type of pattern, a pattern seen in a population that increases very rapidly.

Investigation
Bugs, Bugs, Everywhere Bugs

You will need
• graph paper

Imagine that a bug population has invaded your classroom. One day you notice 16 bugs. Every day new bugs hatch, increasing the population by 50% each week. So, in the first week the population increases by 8 bugs.

Step 1 In a table like this one, record the total number of bugs at the end of each week for 4 weeks.

Bug Invasion

Weeks elapsed	Total number of bugs	Increase in number of bugs (rate of change per week)	Ratio of this week's total to last week's total
Start (0)	16	/////	/////
1	24	8	$\frac{24}{16} = \frac{3}{2} = 1.5$
2	36	12	$\frac{36}{24} = \frac{3}{2} = 1.5$
3	54	18	$\frac{54}{36} = \frac{3}{2} = 1.5$
4	81	27	$\frac{81}{54} = \frac{3}{2} = 1.5$

Step 2 The increase in the number of bugs each week is the population's rate of change per week. Calculate each rate of change and record it in your table. Does the rate of change show a linear pattern? Why or why not?

Step 3 Let x represent the number of weeks elapsed, and let y represent the total number of bugs. Graph the data using (0, 16) for the first point. Connect the points with line segments and describe how the slope changes from point to point.

NCTM STANDARDS

CONTENT	PROCESS
• Number	Problem Solving
• Algebra	• Reasoning
Geometry	Communication
Measurement	• Connections
• Data/Probability	• Representation

LESSON OBJECTIVES

• Begin to investigate geometric sequences using recursive routines

• See examples of growth and decay that can be modeled recursively

PLANNING

LESSON OUTLINE

One day:
30 min Investigation
 5 min Sharing
 5 min Example
 5 min Closing
 5 min Exercises

MATERIALS
• graph paper
• counters, *optional*
• Calculator Note 4B

TEACHING

In some situations values change through multiplying by a constant rather than through adding a constant.

Guiding the Investigation

One step "You have a colony of 16 bugs, and the number is increasing by 50% each week. Use your calculators to generate recursively the colony's population for four weeks and to graph the data points. Then decide a good recursive routine to fit the data." Be sure at least one group thinks about multiplication and not just addition. While groups share a variety of ideas, elicit the notion of the constant multiplier.

Groups might model the growth using counters. Or you may work out the first stage or two with the class.

See page 718 for answers to Steps 2 and 3.

Step 2 Remind students that a rate of change has dimensions. **[Ask]** "What are the units for the rate of change?" [bugs per week].

[Alert] If students think the pattern is linear because the rate of change shows a regular pattern, ask what the constant difference is.

Step 3 Students can also be challenged to graph rate of change (y) versus weeks (x), or to find the rate of change of the rate of change. The rate of change of exponential growth is also exponential.

Step 6 {0, 16} sets the starting value (when 0 weeks have elapsed) at 16. {Ans (1) + 1, Ans (2) * 1.5} increases the number of weeks elapsed by 1 and multiplies the population value by 1.5.

Step 7 Students will get decimal values for weeks 5 to 8. **[Ask]** "What does a decimal mean when you are talking about population?" [A fraction of a bug cannot exist.] **[Alert]** If students have trouble entering this recursive routine, they may be entering parentheses for braces or vice versa. Remind them to use the 2nd key to make braces around the list {0, 16} and around the entire recursive expression.

Step 7

```
        {1 24}
        {2 36}
        {3 54}
        {4 81}
      {5 121.5}
    {6 182.25}
   {7 273.375}
  {8 410.0625}
```

SHARING IDEAS

Point out the quotation that opens the lesson. Susan Coolidge (1835–1905), also known as Sarah Chauncey Woolsey, was an author of children's literature. This stanza creates an image of growth, a central topic of this chapter.

Ask a student or group to present the table.

Step 4 The constant ratio of $\frac{3}{2}$ indicates that the bug population multiplies by 1.5 each week. Recursive routines with repeated addition create linear patterns. But here the rate of change between two successive bug populations changes every week. The new values result from multiplication by a constant amount.

Step 8 After 20 weeks: 53,204 bugs. After 30 weeks: 3,068,017 bugs. Possible answer: Natural factors—decreasing resources, death, migration—would cause the population to level off.

Step 4 Calculate the ratio of the number of bugs each week to the number of bugs the previous week and record it in the table. For example, divide the population after 1 week has elapsed by the population when 0 weeks have elapsed. Repeat this process to complete your table. How do these ratios compare? Explain what the ratios tell you about the bug population growth.

Step 5 What is the *constant multiplier* for the bug population? $\frac{3}{2}$ or 1.5

Step 6 Model the population growth by writing a recursive routine that shows the growing number of bugs. [▶ 🖳 See **Calculator Note 4B** to review recursive routines. ◀] Describe what each part of this calculator command does.

Step 7 By pressing (ENTER) a few times, check that your recursive routine gives the sequence of values in your table (in the column "Total number of bugs"). Use the routine to find the bug population at the end of weeks 5 to 8.

Step 8 What is the bug population after 20 weeks have elapsed? After 30 weeks have elapsed? What happens in the long run?

In the investigation, you found that repeated multiplication is the key to growth of the bug population. Populations of people, animals, and even bacteria show similar growth patterns. Many decreasing patterns, like cooling liquids and decay of substances, can also be described with repeated multiplication.

EXAMPLE

Maria has saved $10,000 from her part-time job and wants to invest it for college. She is considering two options. Plan A guarantees a payment, or return, of $550 each year. Plan B grows by 5% each year. With each plan, what would Maria's new balance be after 5 years? After 10 years?

▶ **Solution**

With plan A, Maria's investment would grow by $550 each year.

Year	Current balance	+	Return	=	New balance
1	10,000	+	550	=	10,550
2	10,550	+	550	=	11,100
3	11,100	+	550	=	11,650

A recursive routine to do this on your calculator is

{0, 10000} (ENTER)

{Ans(1)+1, Ans(2)+550} (ENTER)

(ENTER) , (ENTER) , . . .

After 5 years the new balance is $12,750. After 10 years it is $15,500.

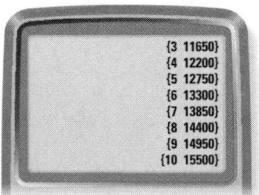

```
{3 11650}
{4 12200}
{5 12750}
{6 13300}
{7 13850}
{8 14400}
{9 14950}
{10 15500}
```

[Ask] "How does the population growth of the bugs differ from linear growth?" "Does it make sense that the more bugs there are, the more will be added each week?" [Yes, there are more bugs reproducing.]

See whether the class can think of other situations in which growth takes place in a "the more there are, the more you get" way. This is a good opportunity to practice asking students to explain their reasoning without acknowledging the correctness of their answers right away. If someone mentions financial investments, you can go right into the example.

Assessing Progress
The investigation allows you to assess students' understanding of 50% as $\frac{1}{2}$ and their ability to work with **decimals,** follow directions systematically, calculate **ratios,** and enter a **recursive sequence.**

With plan B, money earns *interest* each year. The amount of interest is 5% of the current balance. To find the new balance at the end of the first year, add the interest to the current balance. Notice that there is a factor of 10,000 in both the current balance and the interest. You can apply the distributive property to write the expression for the new balance in **factored form.**

Year	Current balance	+	Interest (balance × interest rate)	=	New balance (factored form)
1	10,000	+	10,000 × 0.05	=	10,000(1 + 0.05), or 10,500
2	10,500	+	10,500 × 0.05	=	10,500(1 + 0.05), or 11,025
3	11,025	+	11,025 × 0.05	=	11,025(1 + 0.05), or about 11,576

In the first year the balance grows by $500 to $10,500. To find the new balance for the next year, you need to add 5% of $10,500 to the current $10,500 balance in the account.

Each year the balance grows by 5%. To find each new balance, you use the constant multiplier 1 + 0.05, or 1.05.

You can generate the sequence of balances from year to year on your calculator using this recursive routine:

{0, 10000} (ENTER)

{Ans(1) + 1, Ans(2) · (1 + 0.05)} (ENTER)

(ENTER), (ENTER), . . .

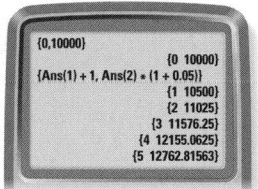

The calculator screen shows the sequence of new balances after the first 5 years. Notice that the balance grows by a larger amount each year. That's because each year you're finding a percent of a larger current balance than in the previous year. After 5 years the new balance is about $12,763. After 10 years it is about $16,289.

A graph illustrates how the investment plans compare. Given enough time, the balance from plan B, which is growing by a constant percent, will always outgrow the balance from plan A, which has only a constant amount added to it. After 20 years you see an even more significant difference: $26,533 compared to $21,000.

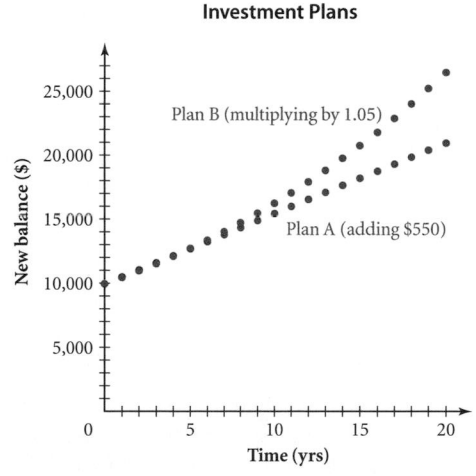

Investment Plans

Plan B (multiplying by 1.05)

Plan A (adding $550)

► **EXAMPLE**

This example reinforces the concepts introduced in the investigation and shows how to write exponential growth in factored form. It also compares the results over time for linear and exponential growth. Be sure students understand that an exponential model represents much faster growth than a linear model.

The situation described is not very realistic, because interest is almost always compounded more frequently than once per year. Let students know that the student text is simplifying matters to help understanding.

The starting balance is often called the *principal* and represented as *P.* Students can think of 1.05 as 105%, or 100% + 5% (principal + interest).

[Alert] Watch for difficulties with the factoring. Merely having students check it by distributing doesn't answer the question, "How could I see how to do that?" Point out the occurrence of the same number in each term of the sum and ask what would result if you divided each term by that number.

[Ask] How do banks actually round dollar amounts for balances? How does this compare to what your calculator does? [A calculator will round up if the decimal represents more than half a cent, but many banks ignore fractions of a cent.]

You might point out that plan A is actually more profitable in the first two years. **[Ask]** "Which plan should Maria choose?" Encourage answers that are based on different assumptions.

Students can use the dynamic algebra exploration on recursive routines with repeated multiplication at www.keymath.com/DA to explore this and similar examples.

Closing the Lesson

When values change through multiplying by a constant rather than through adding a constant, the constant is called a **constant multiplier.** The rate of change is not constant but increases as the amount increases. Examples include growth of populations and the growth of an interest-earning financial investment.

BUILDING UNDERSTANDING

Students practice recursive routines with constant multipliers.

ASSIGNING HOMEWORK

Essential	**1–4, 6 or 7, 8, 10**
Performance assessment	**5, 9**
Portfolio	**10**
Journal	**11**
Group	**6, 10**
Review	**12–14**

▶ Helping with the Exercises

Exercise 1 The decay in 1b is represented by having a constant multiplier less than 1.

Exercise 2 In this sequence resulting from a negative multiplier, the terms are neither decreasing nor increasing—they go back and forth between negative and positive.

It is helpful to think of a constant multiplier, like 1.05 in the example, as a sum. The plus sign in $1 + 0.05$ shows that the pattern increases and 0.05 is the percent growth per year, written as a decimal. When a balance or population decreases, say, by 15% during a given time period, you write the constant multiplier as a difference, for example, $1 - 0.15$. The subtraction sign shows that the pattern decreases and 0.15 is the percent decrease per time period, written as a decimal.

Constant multipliers can be positive or negative. These two sequences have the same starting value, but one has a multiplier of 2 and the other has a multiplier of -2.

$$3, 6, 12, 24, 48, \ldots$$
$$3, -6, 12, -24, 48, \ldots$$

How does the negative multiplier affect the sequence? The negative multiplier changes the sign on every other term. Every other term is the result of multiplying two negatives and the other terms result from a negative and a positive.

EXERCISES

You will need your calculator for problems **2, 3, 5, 6, 7, 9,** and **10.**

▶ Practice Your Skills

1. Give the starting value and constant multiplier for each sequence. Then find the 7th term of the sequence.

   a. 16, 20, 25, 31.25, . . .
   starting value: 16; multiplier: 1.25; 7th term: 61.035

   b. 27, 18, 12, 8, . . .
   starting value: 27; multiplier: $\frac{2}{3}$ or $0.\overline{6}$; 7th term: $2.\overline{370}$ or $\frac{64}{27}$

2. Use a recursive routine to find the first six terms of a sequence that starts with 100 and has a constant multiplier of -1.6. Start with 100, then apply the rule Ans · -1.6; first six terms are 100, $-160, 256, -409.6, 655.36, -1048.576.$

3. Use a recursive routine to find the first five terms of a sequence that starts with 72 and increases by 40% with each term. (Hint: Identify the constant multiplier first.)
   Start with 72, then apply the rule Ans · $(1 + 0.40)$; first five terms are 72, 100.8, 141.12, 197.568, and 276.5952.

4. Use the distributive property to rewrite each expression in an equivalent form. For example, you can write $500(1 + 0.05)$ as $500 + 500(0.05)$.

   a. $75 + 75(0.02)$ $75(1 + 0.02)$ or $75(1.02)$

   b. $1000 - 1000(0.18)$ $1000(1 - 0.18)$ or $1000(0.82)$

   c. $P + Pr$ $P(1 + r)$

   d. $75(1 - 0.02)$ $75 - 75 \cdot 0.02$

   e. $80(1 - 0.24)$ $80 - 80 \cdot 0.24$

   f. $A(1 - r)$ $A - A \cdot r$

5. You may remember from Chapter 0 that the geometric pattern below is the beginning of a fractal called the *Sierpiński triangle.* In this example, the beginning triangle, Stage 0, has a total shaded area of 32 square units. Write a recursive routine that generates the sequence of shaded areas in the pattern. Then use your routine to find the shaded area in Stage 2.

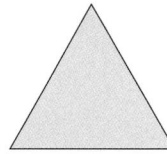
Stage 0
Area = 32 square units

Stage 1
Area = 24 square units

Stage 2

Start with 32, then apply the rule Ans · 0.75; Stage 2 has a shaded area of 18 square units.

Exercise 4 *P* is commonly used in applications to mean *principal.* In 4f, *A* could mean *amount.* In both 4c and 4f, *r* could mean *rate.* If students try to write answers as single numbers, point out that they haven't factored or distributed.

Exercise 5 Students who have done Chapter 0 can draw Stage 3 (or higher) and calculate the shaded area of this figure. **[Ask]** "In Chapter 0, what mathematical operation did you use to express areas and lengths of fractals?"

Reason and Apply

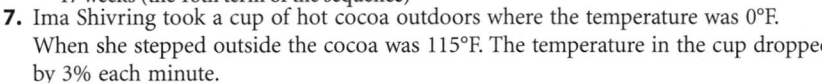

6. APPLICATION Toward the end of the year, to make room for next year's models, a car dealer may decide to drop prices on this year's models. Imagine a car that has a sticker price of $20,000. The dealer lowers the price by 4% each week until the car sells.

   a. Write a recursive routine to generate the sequence of decreasing prices.

   b. Find the 5th term and explain what your answer means in this situation.

   c. If the cost of the car to the dealer was $10,000, how many weeks will pass before its sale price will produce no profit for the dealer?
   17 weeks (the 18th term of the sequence)

7. Ima Shivring took a cup of hot cocoa outdoors where the temperature was 0°F. When she stepped outside the cocoa was 115°F. The temperature in the cup dropped by 3% each minute.

   a. Write a recursive routine to generate the sequence representing the temperature of the cocoa each minute. Start with 115, then apply the rule Ans · (1 − 0.03).

   b. How many minutes does it take for the cocoa to cool to less than 80°F? **12 minutes**

8. Recall the six expressions in problem 4. Imagine that each expression represents the value of an antique increasing or decreasing in value per year. For each expression, identify whether it represents an increasing or decreasing situation, give the starting value, and give the percent of increase or decrease per year.

9. APPLICATION Health care expenditures in the United States exceeded $1 trillion in the mid-1990s and are expected to exceed $2 trillion before 2010. Many elderly and disabled persons rely on Medicare benefits to help cover health care costs. According to the *1998 Wall Street Journal Almanac*, Medicare expenditures were only $6.9 billion in 1970.

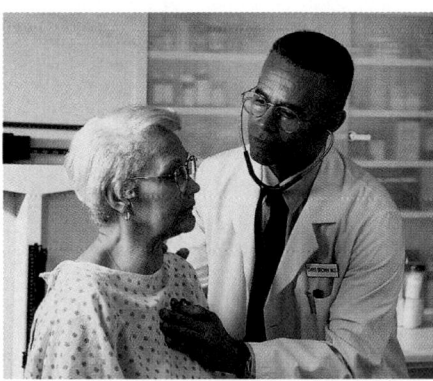

   a. Assume Medicare spending has increased by 14.2% per year since 1970. Write a recursive routine to generate the sequence of increasing Medicare spending.

   b. Use your recursive routine to find the missing table values. Round to the nearest $0.1 billion.

Medicare Spending

Year	1970	1975	1980	1985	1990	1995	2000	2005
Elapsed time (yrs) x	0	5	10	15	20	25	30	35
Spending ($ billion) y	6.9	13.4	26.0	50.6	98.2	190.8	370.5	719.7

6a. Start with 20,000, then apply the rule Ans · (1 − 0.04).

6b. 5th term: 16,986.93; $16,982.93 is the selling price of the car after four price reductions.

Exercise 6c Students can solve problems like this by using graphs, calculator tables, or guess-and-check.

8a. increasing; 75; 2%

8b. decreasing; 1000; 18%

8c. increasing; P; $(100 \cdot r)$%

8d. decreasing; 75; 2%

8e. decreasing; 80; 24%

8f. decreasing; A; $(100 \cdot r)$%

Exercise 9 [ESL] In the United States, a *billion* is a thousand millions. In other English-speaking countries, a billion is a million millions. "Round to the nearest 0.1 billion" means, in this text, round to the nearest hundred million.

9a. Start with 6.9, then apply the rule Ans · (1 + 0.142).

9c.

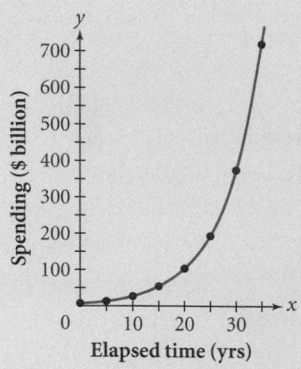

c. Plot the data points from your table and draw a smooth curve through them.

d. What does the shape of the curve suggest about Medicare spending? Do you think this is a realistic model? Answers will vary. The graph implies a smooth, ever-increasing amount of Medicare spending, which is probably not realistic.

10. APPLICATION The advertisement for a super-duper bouncing ball says it rebounds to 85% of the height from which it is dropped.

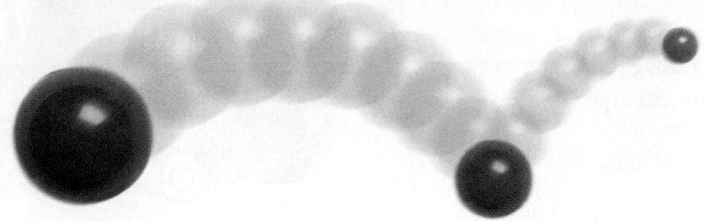

a. If the ball is dropped from a starting height of 2 m, how high should it rebound on the first bounce? 1.7 m

b. Write a recursive routine to generate the sequence of heights for the ball when it is dropped from a height of 2 m. Start with 2, then apply the rule Ans · 0.85.

c. How high should the ball rebound on the sixth bounce? approximately 0.75 m

d. If the ball is dropped from a height of 10 ft, how high should it rebound on the tenth bounce? (Hint: Modify your recursive routine in 10b.) approximately 1.97 feet

e. When the ball is dropped from a height of 10 ft, how many times will it bounce before the rebound height is less than 0.5 ft? 19 bounces

f. A collection of super-duper bouncing balls was tested. Each ball was dropped from a height of 2 m. The table shows the height of the first rebound for eight different balls. Do you think the advertisement's claim that the ball rebounds to 85% of the original height is fair? Explain your thinking.

Balls Dropped from 2 m

Ball number	1	2	3	4	5	6	7	8
Height of rebound (m)	1.68	1.67	1.69	1.78	1.64	1.68	1.66	1.8

Answers will vary. The average is 1.7 m, or 85% of 2 meters, and the median and mode are both 1.68 m, or 84% of 2 m. However, only two of the balls tested met or exceeded this height.

11. Grace manages a local charity. A wealthy benefactor has offered two options for making a donation over the next year. One option is to give $50 now and $25 each month after that. The second option is to give $1 now and twice that amount next month; each month afterward the benefactor would give twice the amount given the month before.

a. Determine how much Grace's charity would receive each month under each option. Use a table to show the values over the course of one year.

b. Use another table to record the total amount Grace's charity will have received after each month.

c. Let x represent the number of the month (1 to 12), and let y represent the total amount Grace received after each month. On the same coordinate axes, graph the data for both options. How do the graphs compare?

d. Which option should Grace choose? Why?

11c. The graph of the first plan is linear. The graph of the second is not; its slope increases between consecutive points.

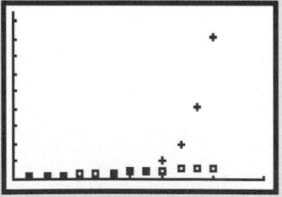

[0, 15, 3, 0, 4750, 500]

11d. Possible answer: Grace's charity will receive more total money with option 2. It will also give the charity more income later in the year when the budget may be tighter.

11a.

	Jan	Feb	Mar	Apr	May	June	July	Aug	Sep	Oct	Nov	Dec
Option 1	$50	$25	$25	$25	$25	$25	$25	$25	$25	$25	$25	$25
Option 2	$1	$2	$4	$8	$16	$32	$64	$128	$256	$512	$1,024	$2,048

11b.

	Jan	Feb	Mar	Apr	May	June	July	Aug	Sep	Oct	Nov	Dec
Option 1	$50	$75	$100	$125	$150	$175	$200	$225	$250	$275	$300	$325
Option 2	$1	$3	$7	$15	$31	$63	$127	$255	$511	$1,023	$2,047	$4,095

▶ Review

12. Write an equation in point-slope form for a line with slope -1.2 that goes through the point $(600, 0)$. Find the y-intercept. $y = -1.2(x - 600)$; y-intercept: $(0, 720)$

13. Match the recursive routine to the equation.
    a. $y = 3x + 7$ i
    b. $y = -3x + 7$ iii
    c. $y = 7x + 3$ ii
    d. $y = -7x + 3$ iv

    i. Start with 7, then apply the rule Ans + 3.
    ii. Start with 3, then apply the rule Ans + 7.
    iii. Start with 7, then apply the rule Ans − 3.
    iv. Start with 3, then apply the rule Ans − 7.

14. **APPLICATION** A long-distance phone company offers two calling plans. The first plan costs $12 per month and offers 60 minutes free per month; additional minutes cost 5¢ per minute. The second plan costs only $10 a month and offers 50 minutes free per month; additional minutes cost 9¢ per minute.

    a. Define variables and write an equation for the first plan if you use it for 60 minutes or less. Let x represent minutes of use and y represent cost; $y = 12$.

    b. Write an equation for the first plan if you use it for more than 60 minutes. $y = 12 + 0.05(x - 60)$

    c. Write two equations for the second plan similar to those you wrote in 14a and b. Explain what each equation represents.

    d. Sydney frequently calls her cousin in Australia from her hometown in Kentucky and talks for a long time, generally about 150 minutes per month. How much would each plan cost for her? Which plan should she choose?

    e. Louis makes only a few long-distance calls since most of his friends and family members live nearby. How much would each plan cost for Louis if he averages 55 minutes of long-distance calls per month? Which plan should he choose?

    f. For how many minutes of use will the cost of the plans be the same? How can you decide which of these two long-distance plans is better for a new subscriber? The plans cost the same for 87.5 minutes of use. Assuming the phone company charges only for whole minutes, a person who uses more than 87 minutes should choose the first plan. If she uses 87 minutes or less, then the second plan is better.

Exercise 12 This exercise reviews Lesson 5.3.

Exercise 13 This exercise reviews Lesson 4.6.

Exercise 14 This exercise reviews Lesson 6.2.

14c. $y = 10$ for 50 minutes or less of use; $y = 10 + 0.09(x - 50)$ for more than 50 minutes of use

14d. First plan: $16.50; second plan: $19.00. She should sign up for the first plan.

14e. First plan: $12.00 (he pays only the flat rate of $12.00); second plan: $10.45. He should sign up for the second plan.

IMPROVING YOUR **GEOMETRY** SKILLS

Use 16 toothpicks to make this pattern. Then remove 4 toothpicks so that you have 4 congruent triangles.

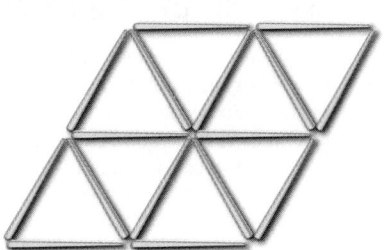

IMPROVING **GEOMETRY** SKILLS

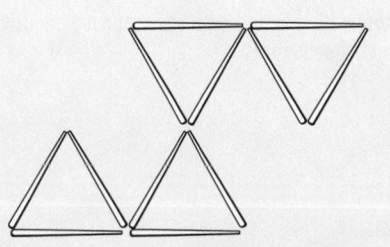

PLANNING

LESSON OUTLINE

First day:

40 min   Investigation

10 min   Sharing

Second day:

30 min   Examples

 5 min   Closing

15 min   Exercises

MATERIALS

- Growth of the Koch Curve (W)
- Calculator Note 7A

TEACHING

Values changing through multiplication by a constant can be represented using powers of the constant multiplier.

One step Remind students of the Koch curve from Chapter 0, or introduce it. Ask them to find the constant multiplier for the length when moving from one stage of the curve to the next. Then have them use that multiplier to write expressions for the length of the curve at each of several stages. As you circulate, remind students to use exponents. During Sharing, ask for other examples of exponential growth and have the class make up a problem about a savings account.

LESSON

7.2

Exponential Equations

Recursive routines are useful for seeing how a sequence develops and generating the first few terms. But, as you learned in Chapter 4, if you're looking for the 50th term, you'll have to do many calculations to find your answer. For most of the sequences in Chapter 4, you found that the graphs of the points formed a linear pattern, so you learned how to write the equation of a line.

Recursive routines with a constant multiplier create a different type of increasing or decreasing pattern. In this lesson you'll discover the connection between these recursive routines and exponents. Then, with a new type of equation, you'll be able to find any term in a sequence based on a constant multiplier without having to find all the terms before it.

This sculpture, *Door to Door* (1995), was created by Filipino artist José Tence Ruiz (b 1956) from wood, cardboard, and other materials. The decreasing size of the boxes suggests an exponential pattern.

Investigation
Growth of the Koch Curve

You will need

- the worksheet Growth of the Koch Curve

In this investigation you will look for patterns in the growth of a fractal. You may remember the *Koch curve* from Chapter 0. Here you will think about the relationship between the length of the Koch curve and the repeated multiplication you studied in Lesson 7.1. Stage 0 of the Koch curve is already drawn on the worksheet. It is a segment 27 units long.

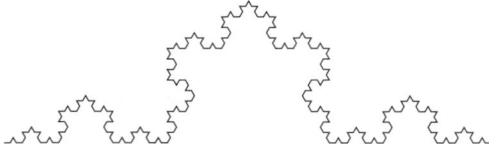

LESSON OBJECTIVES

- Explore exponential growth and decay patterns
- Discover the connection between recursive and exponential forms of geometric sequences

NCTM STANDARDS

CONTENT		PROCESS	
●	Number		Problem Solving
●	Algebra	●	Reasoning
●	Geometry	●	Communication
	Measurement	●	Connections
	Data/Probability	●	Representation

Step 1 The figure should look like the one shown in the Student Edition, where each segment is 9 units long.

Step 1 Draw the Stage 1 figure below the Stage 0 figure. The first segment is drawn for you on the worksheet. As shown here, the Stage 1 figure has four segments, each $\frac{1}{3}$ the length of the Stage 0 segment.

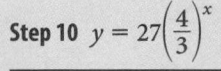
Stage 1

Step 2 Determine the total length at Stage 1 and record it in a table like this.

Stage number	Total length (units)	Ratio of this stage's length to previous stage's length
0	27	
1	36	$\frac{4}{3} = 1.\overline{3}$
2	48	$\frac{4}{3} = 1.\overline{3}$
3	64	$\frac{4}{3} = 1.\overline{3}$

Step 3 Draw the Stage 2 and Stage 3 figures of the fractal. Again, the first segment for each stage is drawn for you. Record the total length at each stage.

Step 4 See table for ratios; $\frac{4}{3}$, or $1.\overline{3}$.

Step 4 Find the ratio of the total length at any stage to the total length at the previous stage. What is the constant multiplier?

Step 5 Use your constant multiplier from Step 4 to predict the total lengths of this fractal at Stages 4 and 5.

Step 6 twice; $27\left(\frac{4}{3}\right)\left(\frac{4}{3}\right) = 27\left(\frac{4}{3}\right)^2$

Step 6 How many times do you multiply the original length at Stage 0 by the constant multiplier to get the length at Stage 2? Write an expression that calculates the length at Stage 2.

Step 7 three times; $27\left(\frac{4}{3}\right)\left(\frac{4}{3}\right)\left(\frac{4}{3}\right) = 27\left(\frac{4}{3}\right)^3$

Step 7 How many times do you multiply the length at Stage 0 by the constant multiplier to get the length at Stage 3? Write an expression that calculates the length at Stage 3.

Step 8 If your expressions in Steps 6 and 7 do not use exponents, rewrite them so that they do.

Step 9 Stage 5: $27\left(\frac{4}{3}\right)^5 = 113.\overline{7}$

Step 9 Use an exponent to write an expression that predicts the total length of the Stage 5 figure. Evaluate this expression using your calculator. Is the result the same as you predicted in Step 5?

Step 10 Let x represent the stage number, and let y represent the total length. Write an equation to model the total length of this fractal at any stage. Graph your equation and check that the calculator table contains the same values as your table.

Step 11 Possible answer: The length of the Koch curve rapidly increases.

Step 11 What does the graph tell you about the growth of the Koch curve?

Step 3

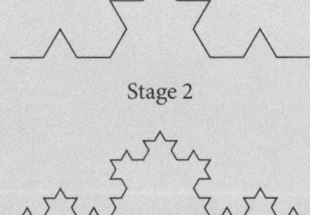

Stage 2

Stage 3

Step 10 $y = 27\left(\frac{4}{3}\right)^x$

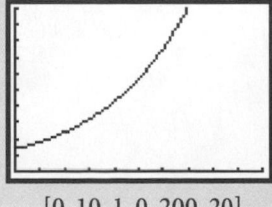

[0, 10, 1, 0, 200, 20]

Guiding the Investigation

Step 1 Students who did not do Chapter 0 may need help in seeing how to generate one stage from another. Have them write a rule. Students may describe something like, "Remove the middle third and put a bottomless triangle on it." Alternatively, students can look back at page 14 in Lesson 0.3 to see further stages, or use the worksheet "How Long Is This Fractal?" from Lesson 0.3. Note that the Stage 0 Koch curve in Chapter 0 does not have length 27.

Step 5 Predictions will vary. Exact answers are Stage 4:
$27\left(\frac{4}{3}\right)\left(\frac{4}{3}\right)\left(\frac{4}{3}\right)\left(\frac{4}{3}\right) = 85.\overline{3}$

Stage 5:
$27\left(\frac{4}{3}\right)\left(\frac{4}{3}\right)\left(\frac{4}{3}\right)\left(\frac{4}{3}\right)\left(\frac{4}{3}\right) = 113.\overline{7}$

Step 8 Some students may need to be reminded what exponents are, especially if you skipped Chapter 0. Remember not to say that the exponent gives "the number of times the base is multiplied by itself," because if you multiply the number $\frac{4}{3}$ by itself two times, you'll get $\frac{4}{3} \times \frac{4}{3} \times \frac{4}{3}$, which is $\left(\frac{4}{3}\right)^3$. Rather, say that the exponent gives the number of times the base is a factor in a product.

Step 10 [Ask] "Are all values of x and y meaningful in this equation?" [Only whole numbers can represent stage numbers and lengths of the first stage of these curves.]

[Alert] Be sure students understand the difference between the fractal itself and the graph that represents its growth.

Step 11 Have students consider the slope of the line between consecutive points. They should find that the rate of change is always increasing.

SHARING IDEAS

Select students to share their
ideas about Steps 10 and 11.

Assessing Progress
Watch for students' familiarity
with **fractions** and **exponents**.

▶ **EXAMPLE A**

This example reviews the meaning of exponents as repeated multiplication.

When the exponent must be a
whole number, the variable n is
sometimes used instead of x.
Students saw this in Chapter 0.

A recursive routine that uses a constant multiplier represents a pattern that
increases or decreases by a constant ratio or a constant percent. Because exponents
are another way of writing repeated multiplication, you can use exponents to model
these patterns. In the investigation you discovered how to calculate the length of
the Koch curve at any stage by using this equation:

$$y = 27\left(\frac{4}{3}\right)^x$$

Stage number

Total length   Starting length   Constant multiplier

Equations like this are called **exponential equations** because a variable, in this case
x, appears in the exponent.

When you write out a repeated multiplication expression to show each factor, it is
written in **expanded form.** When you show a repeated multiplication expression
with an exponent, it is in **exponential form** and the factor being multiplied is
called the **base.**

Expanded form   **Exponential form**

$$27\left(\frac{4}{3}\right)\left(\frac{4}{3}\right)\left(\frac{4}{3}\right) \qquad = \qquad 27\left(\frac{4}{3}\right)^3$$

The exponent
means there
are three such
factors of $\frac{4}{3}$.

There are three
factors of $\frac{4}{3}$.

The base means you are
multiplying factors of $\frac{4}{3}$.

EXAMPLE A | Write each expression in exponential form.

a. $(5)(5)(5)(5)(5)(5)$

b. $3(3)(2)(2)(2)(2)(2)(2)(2)(2)(2)$

c. The balance of a savings account that was opened 7 years ago with $200
earning 2.5% interest per year.

▶ **Solution** | **a.** 5^6

b. There are two factors of 3 and nine factors of 2, so you write

$$3^2 \cdot 2^9$$

You can't combine 3^2 and 2^9 any further because they have different bases.

c. There will be 7 factors of $(1 + 0.025)$ multiplied by the starting value of $200,
so you write

$$200(1 + 0.025)^7$$

EXAMPLE B

Seth deposits $200 in a savings account. The account pays 5% annual interest. Assuming that he makes no more deposits and no withdrawals, calculate his new balance after 10 years.

▶ **Solution**

The interest represents a 5% rate of growth per year, so the constant multiplier is $(1 + 0.05)$. Now find an equation that you can use to find the new balance after any number of years by considering these yearly calculations and results.

	Expanded form	Exponential form	New balance
Starting balance:	$200		= $200.00
After 1 year:	$200(1 + 0.05)	= $200(1 + 0.05)^1$	= $210.00
After 2 years:	$200(1 + 0.05)(1 + 0.05)	= $200(1 + 0.05)^2$	= $220.50
After 3 years:	$200(1 + 0.05)(1 + 0.05)(1 + 0.05)	= $200(1 + 0.05)^3$	= $231.53
After x years:	$200(1 + 0.05)(1 + 0.05) \ldots (1 + 0.05)	= $200(1 + 0.05)^x$	

You can now use the equation $y = 200(1 + 0.05)^x$, where x represents time in years and y represents the new balance in dollars, to find the new balance after 10 years.

$y = 200(1 + 0.5)^x$	Original equation.
$y = 200(1 + 0.05)^{10}$	Substitute 10 for x.
$y \approx 325.78$	Use your calculator to evaluate the exponential expression.

The new balance after 10 years would be $325.78.

Amounts that increase by a constant percent, like the savings account in the example, have **exponential growth.**

Exponential Growth

Any constant percent growth can be modeled by the exponential equation

$$y = A(1 + r)^x$$

A is the starting value, r is the rate of growth written as a positive decimal or fraction, x is the number of time periods elapsed, and y is the final value.

You can model amounts that decrease by a constant percent with a similar equation. What would need to change in the exponential equation to show a constant percent decrease?

▶ **EXAMPLE B**

The example is useful for students who didn't come up with an equation to graph during the investigation. It is also good for refining students' skills at "chunking," seeing a collection of symbols as a single entity. Some students have difficulty seeing the $(1 + 0.05)$ as a single quantity that's being multiplied by itself repeatedly. Temporarily rewriting the number as 1.05 might help, but the form $(1 + 0.05)$ emphasizes the fact that the multiplier is more than 1 and the account's value is increasing by 5%.

Students may ask why the constant multiplier is $1 + 0.05$ rather than just 0.05. Suggest that they look at $200(0.05)^x$ for several values of x. Then ask them to explain why the 1 is part of the expression.

All dollar values are rounded to the nearest cent. Students' calculators may show different numbers of decimal places, depending on how their mode is set.

Help students distinguish between the growth rate r and the rate of change. The number r is constant and has no units. The rate of change is the amount of increase in y for each unit of increase in x, and it has units.

Closing the Lesson

When values are increasing through multiplication by a constant that's more than 1, we say that we have **exponential growth** and represent the values by the equation $y = A(1 + r)^x$. You might want to make a bulletin board display of the exponential growth equation.

BUILDING
UNDERSTANDING

Students make connections
between recursive routines,
explicit definitions, calculator
lists, and tables of values.

ASSIGNING HOMEWORK

Essential	2–5, 9, 12
Performance assessment	8, 10
Portfolio	9, 12
Journal	6, 7, 13, 14
Group	12
Review	1, 11, 16

EXERCISES

You will need your calculator for problems **2, 5, 7, 8, 9, 11, 12,** and **13.**

▶ **Practice Your Skills**

1. Rewrite each expression with exponents.
   a. $(7)(7)(7)(7)(7)(7)(7)(7)$ 7^8
   b. $(3)(3)(3)(3)(5)(5)(5)(5)(5)$ $3^4 \cdot 5^5$
   c. $(1 + 0.12)(1 + 0.12)(1 + 0.12)(1 + 0.12)$ $(1 + 0.12)^4$

2. A bacteria culture grows at a rate of 20% each day. There are 450 bacteria today. How many will there be
   a. Tomorrow? $450(1 + 0.2) = 540$ bacteria
   b. One week from now?
   $450(1 + 0.2)^7 \approx 1612$ bacteria

A technician puts bacteria in several petri dishes of agar. Agar is a gelatin-like substance made from algae. The agar holds the bacteria in place on the petri dish and provides nutrients for growth of the bacteria.

3. Match each equation with a table of values.
   a. $y = 4(2)^x$ ii
   b. $y = 4(0.5)^x$ iii
   c. $y = 2(4)^x$ iv
   d. $y = 2(0.25)^x$ i

   i.

x	y
0	2
1	0.5
2	0.12
3	0.03

   ii.

x	y
0	4
1	8
2	16
3	32

   iii.

x	y
0	4
1	2
2	1
3	0.5

   iv.

x	y
0	2
1	8
2	32
3	128

4. Match each recursive routine with the equation that gives the same values.
   iv **a.** 1.05 (ENTER)
       Ans · (0.95) (ENTER)

       i. $y = 0.95(1.05)^x$

   ii **b.** 1.05 (ENTER)
       Ans + Ans · 0.05 (ENTER)

       ii. $y = 1.05(1 + 0.05)^x$

   i **c.** 0.95 (ENTER)
       Ans · (1 + 0.05) (ENTER)

       iii. $y = 0.95(0.95)^x$

   iii **d.** 0.95 (ENTER)
       Ans · (1 − 0.05) (ENTER)

       iv. $y = 1.05(1 − 0.05)^x$

5. For each table, find the value of the constants A and r such that $y = A \cdot r^x$.
(Hint: Enter your equation into Y_1 on your calculator. Then see if a table of values matches the table in the book.)

a.

x	y
0	1.2
1	2.4
2	4.8
3	9.6
4	19.2

$y = 1.2 \cdot 2^x$

b.

x	y
0	500
2	20
3	4
5	0.16
7	0.0064

$y = 500 \cdot 0.2^x$

c.

x	y
3	8
1	50
5	1.28
2	20
7	0.2048

$y = 125 \cdot 0.4^x$

6. The equation $y = 500(1 + 0.04)^x$ models the amount of money in a savings account that earns annual interest. Explain what each number and variable in this expression means.

7. Run the calculator program INOUTEXP and play the easy-level game five times. Each time you play, write down the input and output values you were given and the exponential equation that models those values. [▶ 🖳 See **Calculator Note 7A** for instructions on running the program INOUTEXP. ◀] Students will self-check their work as they run the program.

▶ Reason and Apply

8. **APPLICATION** A credit card account is essentially a loan. A constant percent interest is added to the balance. Stanley buys $100 worth of groceries with his credit card. The balance then grows by 4.5% interest each month. How much will he owe if he makes no payments in 4 months? Write the expression you used to do this calculation in expanded form and also in exponential form.
$100(1 + 0.045)(1 + 0.045)(1 + 0.045)(1 + 0.045) = 100(1 + 0.045)^4$; approximately $119.25

9. **APPLICATION** Phil purchases a used truck for $11,500. The value of the truck is expected to decrease by 20% each year. (A decrease in monetary value over time is sometimes called *depreciation*.)

a. Find the truck's value after 1 year.

b. Write a recursive routine that generates the value of the truck after each year.

c. Create a table showing the value of the truck when Phil purchases it and after each of the next 4 years.

d. Write an equation in the form $y = A(1 - r)^x$ to calculate the value, y, of the truck after x years.

e. Graph the equation from 9d, showing the value of the truck up to an age of 10 years.

Many people, like these ranch hands in Montana, rely on a truck for work and leisure.

9c.

Time elapsed (yrs)	0	1	2	3	4
Value ($)	11,500	9,200	7,360	5,888	4,710.40

▶ **Helping with the Exercises**

6. The initial deposit is $500. The account earns 4% interest per year; thus the constant multiplier is $(1 + 0.04)$. The variable x represents the number of years since the initial deposit. The variable y represents the balance after x years.

Exercise 7 In this game, students are deriving only the right-hand expression $A(1 + r)^x$. Yet on paper, they should include the $y =$.

Exercise 9 Make sure students are aware that in actuality depreciation is not always a constant rate—a car may depreciate 20% one year and 30% another. Students doing Exercise 13 will see that some goods, such as artwork and antiques, gain in value, or **appreciate**.

9a. $9,200

9b. Start with 11,500, then apply the rule Ans $\cdot (1 - 0.2)$.

9d. $y = 11,500(1 - 0.2)^x$

9e.

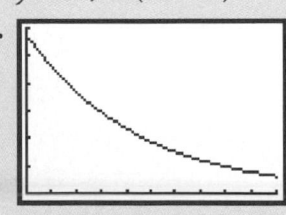

[0, 10, 1, 0, 12000, 2000]

12a. The number of layers doubles with each fold.

12b. Estimates will vary.

12c. Methods will vary. Eight folds gives 256 layers (512 pages) and nine folds gives 512 layers (1024 pages).

Exercise 13 This is the first time that students may notice that in $y = A(1 + r)^x$ the value of x can be fractional. The graph of the equation is continuous, not discrete.

Students may not know how to convert a decimal number of years to years and months. They can multiply 3.08 years by 12 months per year to get approximately 37 months, or 3 years and 1 month.

If students use trace to find for the intersection of the graphs to answer 13b, the width of the pixels on the screen may prevent them from finding the exact position.

13b.

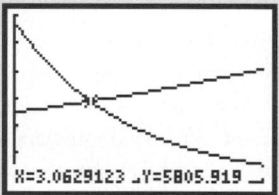

[0, 10, 1, 0, 12000, 2000]

The intersection point represents the time and the value of both cars when their value will be the same. By tracing the graph shown, students should see that both cars will be worth approximately $5,800 after a little less than 3 years 1 month.

10. Draw a "starting" line segment 2 cm long on a sheet of paper.

    a. Draw a segment 3 times as long as the starting segment. How long is this segment? **6 cm**

    b. Draw a segment 3 times as long as the segment in 10a. How long is this segment? **18 cm**

    c. Use the starting length and an exponent to write an expression that gives the length in centimeters of the next segment you would draw. **$2(3)^3$**

    d. Use the starting length and an exponent to write an expression that gives the length in centimeters of the longest segment you could draw on a 100-meter soccer field. **$2(3)^7$**

11. Run the calculator program INOUTEXP and play the medium- or difficult-level game five times. Each time you play, write down the input and output values you were given and the exponential equation that models those values. [▶ 🖳 See **Calculator Note 7A** for instructions on running the program INOUTEXP. ◀] Students will self-check their work as they run the program.

12. Fold a sheet of paper in half. You should have two layers. Fold it in half again so that there are four layers. Do this as many times as you can. Make a table and record the number of folds and number of layers.

    a. As you fold the paper in half each time, what happens to the number of layers?

    b. Estimate the number of folds you would have to make before you have about the same number of layers as the number of sheets in this textbook.

    c. Calculate the answer for 12b. You may use a recursive routine, the graph or table of an equation, or a trial-and-error method.

Origami is the Japanese art of paper folding.

13. **APPLICATION** Phil's friend Shawna buys an antique car for $5,000. She estimates that it will increase in value (*appreciate*) by 5% a year.

    a. Write an equation to calculate the value, y, of Shawna's car after x years. **$y = 5000(1 + 0.05)^x$**

    b. Simultaneously graph the equation in 13a and the equation you found in 9d. Where do the two graphs intersect? What is the meaning of this point of intersection?

14. Invent a situation that could be modeled by each equation below. Sketch a graph of each equation, and describe similarities and differences between the two models.

$$y = 400 + 20x$$
$$y = 400(1 + 0.05)^x$$

15. Consider the recursive routine

    {0, 100} **ENTER**

    {Ans(1)+1, Ans(2) · (1−0.035)} **ENTER**

    a. Invent a situation that this routine could model.

    b. Create a problem related to your situation. Carefully describe the meaning of the numbers in your problem.

    c. Use an exponential equation to solve your problem.

Answers will vary. Eventually 15c should use $y = 100(1 − 0.035)^x$.

14. Possible answer: The first equation could model a principal of $400 to which $20 is added each time period; the second equation could model a starting bank balance of $400, with 5% interest added to the balance each time period. Both models have the same starting value, 400. In both models, $y = 420$ when $x = 1$. For x greater than 1, y increases much more quickly in the second model.

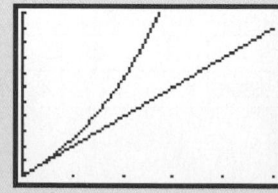

[0, 50, 10, 400, 1500, 100]

► Review

16. Look at this "step" pattern. In the first figure, which has one step, each side of the block is 1 cm long.

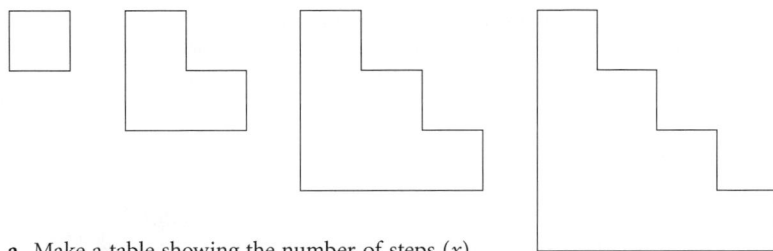

a. Make a table showing the number of steps (x) and the perimeter (y) of each figure.

b. On a graph, plot the coordinates your table represents.

c. Write an equation that relates the perimeter of these figures to the number of steps. $y = 4x$

d. Use your equation to predict the perimeter of a figure with 47 steps. **188 cm**

e. Is there a figure with a perimeter of 74 cm? If so, how many steps does it have? If not, why not? **A perimeter of 74 cm is not possible because it would require 18.5 steps.**

project

AUTOMOBILE DEPRECIATION

Cars usually lose value as they get older. Dealers and buyers may rely on books or Internet resources to help them find out how much a used car is worth. But many people don't understand what type of math is used to make these judgments.

Choose a model of automobile that has been manufactured for several years. Research the new-car value now. Then research how much the same model would be worth now if it were manufactured last year, the year before that, and so on. Does your data show a pattern? Can you write an equation that models your data?

Your project should include

▸ Data for at least 10 consecutive years.

▸ A scatter plot comparing age and value.

▸ The rate of change (if your data appears linear) or the rate of depreciation as a percent (if your data looks exponential).

▸ An equation that fits your data.

▸ A summary of your procedures and findings; include how you collected your data and how well your equation fits the data.

You might want to ask a local auto dealership how it determines a car's value. Does it use the same rate of depreciation for all cars? And how do special features, like a custom stereo, affect the value?

Exercise 16 This exercise reviews Lesson 4.3.

16a.

Number of steps x	1	2	3	4
Perimeter (cm) y	4	8	12	16

16b.

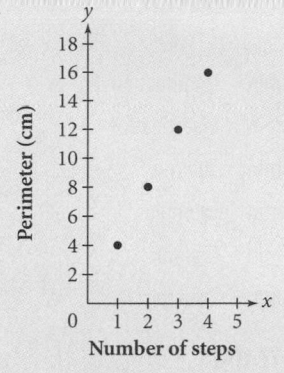

Student Data for Project
Data should be collected from *Kelly Blue Book* or reliable on-line resources with consideration to features of the car (should be similar for all years) and mileage (should increase as age increases). Students may tend to notice a linear pattern, depending on the actual rate of depreciation and how far back the data goes. If factors such as mileage and features are not closely considered, the data may appear to have no pattern.

Supporting the project

MOTIVATION

The cost of driving a car is more than gas and repairs. It also includes the depreciation, the difference between what was paid for the car and what it can be sold for.

OUTCOMES

▸ The data and scatter plot correctly illustrate the findings.

▸ The report includes a summary of procedures and findings, including information about how the data were found.

▸ An equation in the form $y = a + bx$ or $y = A(1 - r)^x$ should be stated and shown to be a good model of the data.

▸ The report includes the rate of depreciation.

• Appropriate adjustments are made to the equations to achieve a good fit.

• The report includes extended research on the equations actually employed in the used-car industry and how they are applied.

LESSON

7.3

Multiplication and Exponents

Growth for the sake of growth is the ideology of the cancer cell.

EDWARD ABBEY

Social Science CONNECTION

The U.S. Bureau of the Census only collects population information every 10 years. It uses mathematical models, like exponential equations, to make population predictions between census years.

In Lesson 7.2, you learned that the exponential expression $200(1 + 0.05)^3$ can model a situation with a starting value of 200 and a rate of growth of 5% over three time periods. How would you change the expression to model five time periods, seven time periods, or more? In this lesson you will explore that question and discover how the answer is related to a rule for showing multiplication with exponents.

Every year, the population of the United States increases. This photo shows Grand Central Station in New York City, which is the most populated U.S. city.

Suppose the population of a town is 12,800 and the town's population grows at a rate of 2.5% each year.

An expression for the population 3 years from now is $12{,}800(1 + 0.025)^3$. To represent one more year, you can write the expression $12{,}800(1 + 0.025)^4$. You can also think about the growth from 3 years to 4 years recursively. Because the rate of growth is constant, multiply the expression for 3 years by one more constant multiplier to get $12{,}800(1 + 0.025)^3 \cdot (1 + 0.025)^1$.

This means that

$$12{,}800(1 + 0.025)^3 \cdot (1 + 0.025)^1 = 12{,}800(1 + 0.025)^4$$

Both methods make sense and both evaluate to the same result.

So you can advance exponential growth one time period either by multiplying the previous amount by the base (the constant multiplier) or by increasing the exponent by one. Every time you increase by one the number of times the base is used as a factor, the exponent increases by one. But what happens when you want to advance the growth by more than one time period? In the next investigation, you will discover a shortcut for multiplying exponential expressions.

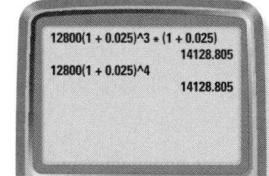

```
12800(1 + 0.025)^3 * (1 + 0.025)
                           14128.805
12800(1 + 0.025)^4
                           14128.805
```

PLANNING

LESSON OUTLINE

One day:

5 min	Introduction
15 min	Investigation
5 min	Sharing
15 min	Examples
5 min	Closing
5 min	Exercises

MATERIALS

- Properties of Exponents (T), *optional*
- Calculator Note 7B

TEACHING

Exponential growth models motivate rules for exponents when multiplying quantities with the same bases and when raising a power to a power.

INTRODUCTION

To estimate a population between actual counts, the growth rate is assumed to be the same between the counts. Theoretically, the graph of a population is a step function, because the population is counted by whole numbers. Because the steps would be so small in the graph, the best model assumes that the population takes on non-integer values and grows continuously. In estimating the population, students should round to the nearest integer.

[Alert] Some students may have forgotten that $(1 + 0.025)^1$ equals $1 + 0.025$. It's sometimes useful to write a number as itself to the first power when trying to see or demonstrate a pattern.

LESSON OBJECTIVES

- Review or learn the multiplication property of exponents
- Review or learn the power properties of exponents

NCTM STANDARDS

CONTENT		PROCESS
●	Number	Problem Solving
●	Algebra	● Reasoning
	Geometry	Communication
	Measurement	● Connections
	Data/Probability	● Representation

Investigation
Moving Ahead

Step 1

Rewrite each product below in expanded form and then rewrite it in exponential form with a single base.

a. $3^4 \cdot 3^2$ $(3 \cdot 3 \cdot 3 \cdot 3) \cdot (3 \cdot 3) = 3^6$

b. $x^3 \cdot x^5$ $(x \cdot x \cdot x) \cdot (x \cdot x \cdot x \cdot x \cdot x) = x^8$

c. $(1 + 0.05)^2 \cdot (1 + 0.05)^4$

d. $10^3 \cdot 10^6$ $(10 \cdot 10 \cdot 10) \cdot (10 \cdot 10 \cdot 10 \cdot 10 \cdot 10 \cdot 10) = 10^9$

Step 1c $(1 + 0.05) \cdot$
$(1 \cdot 0.05) \cdot (1 + 0.05) \cdot$
$(1 + 0.05) \cdot (1 + 0.05) \cdot$
$(1 + 0.05) = (1 + 0.05)^6$

Step 2 Sample answer:
You add the original
exponents to get the exponent on the final expression.

Step 2

Compare the exponents in each final expression you got in Step 1 to the exponents in the original product. Describe a way to find the exponents in the final expression without using expanded form.

Step 3 $m + n$

Step 3

Generalize your observations in Step 2 by filling in the blank.

$$b^m \cdot b^n = b^{\square}$$

Step 4

Step 4 a. the population after
3 more weeks; $16(1 + 0.5)^8$

Apply what you have discovered about multiplying expressions with exponents.

a. The number of bugs in a colony after 5 weeks is $16(1 + 0.5)^5$. What does the expression $16(1 + 0.5)^5 \cdot (1 + 0.5)^3$ mean in this situation? Rewrite the expression with a single exponent.

All ants live in colonies.

b. the value of the truck after
2 more years; $11,500(1 - 0.2)^9$

b. The depreciating value of a truck after 7 years is $11,500(1 - 0.2)^7$. What does the expression $11,500(1 - 0.2)^7 \cdot (1 - 0.2)^2$ mean in this situation? Rewrite the expression with a single exponent.

c. the growth after m more
time periods

c. The expression $A(1 + r)^n$ can model n time periods of exponential growth. What does the expression $A(1 + r)^{n+m}$ model?

Step 5 Sample
answer: You add to the exponent to indicate multiplying
by more time.

Step 5

How does looking ahead in time with an exponential model relate to multiplying expressions with exponents?

SHARING IDEAS

Have students share and critique answers to Step 4c.

Draw students' attention to the quotation that opens the lesson. Growth and reproduction of cells is one of many situations that can be modeled by exponential equations.

Ask how good they think predictions will be if made from exponential models of a population. What factors are not taken into account with an exponential

model of the town's growth or the growing cancer cells mentioned in the introduction? Generally, exponential models neglect to consider limits on resources (such as the availability of food, jobs, or land) and external interventions (such as immigration or treatments that kill cancer cells).

Assessing Progress

Look for understanding of **constant multipliers** and of **exponents** as showing repeated multiplication.

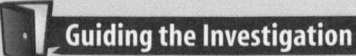

Guiding the Investigation

One step Say that the number of bugs in a colony has now been growing exponentially for 5 weeks and has reached a value of $16(1 + 0.5)^5$. Assuming that the growth will continue at the same rate, what will the population be 3 weeks from now? As you circulate, ask students to write an expression for that population in at least two ways—$16(1 + 0.5)^5(1 + 0.5)^3$ and $16(1 + 0.5)^8$. **[Ask]** "Can you always add exponents when multiplying?" [Bring out the necessity of having the same base.] Ask students to make up and solve some similar problems with which to challenge each other.

Step 1 Having students go through the "fully expanded form" will help them make sense of and remember the multiplication property of exponents. Students may have difficulty seeing $1 + 0.05$, the base, as a "chunk."

Step 2 You may want to extend this instruction to include "Write an explanation of this pattern for a friend who is absent today."

Step 3 [Ask] "What would happen in something like a^3b^2?" [Exponents can't be added if the bases differ. If the bases are the same, we say they are *like bases.*]

Step 4 The real-world contexts help students develop understanding of the multiplication rule for exponents. The contexts should also help students make the connection with the exponential growth and decay models.

▶ EXAMPLE A

This examples stresses that exponents are added only if multiplying powers of the same base. If two factors are raised to the same power, they can be multiplied and the product raised to that same power.

▶ EXAMPLE B

This example derives the power properties of exponents. *Power* is another word for *exponent*, but the terms are used slightly differently. In the expression x^2, the number 2 is the *exponent on x* or the *power on x*. We say that the entire expression is a *power of x*, and that the number 27 is a *power of 3*.

If students get confused when doing a problem, suggest rewriting the expression in expanded form and then combining like factors with an exponent.

Another Example

Write $3^4 \cdot 2^2$ with a single exponent. It is not 6^6. But both exponents are even numbers, so you can regroup $3 \cdot 3 \cdot 3 \cdot 3 \cdot 2 \cdot 2$ as $(3 \cdot 3 \cdot 2)(3 \cdot 3 \cdot 2)$, which is 18^2.

This investigation has helped you to discover the **multiplication property of exponents.**

> ### Multiplication Property of Exponents
>
> For any nonzero value of b and any integer values of m and n,
>
> $$b^m \cdot b^n = b^{m+n}$$

This property is very handy in rewriting exponential expressions. However, you can add exponents to multiply numbers only when the bases are the same.

EXAMPLE A Cal and Al got different answers when asked to write $3^4 \cdot 2^2$ in another exponential form. Who was right and why?

Cal

$3^4 \cdot 2^2$ is in simplest exponential form because the numbers have different bases.

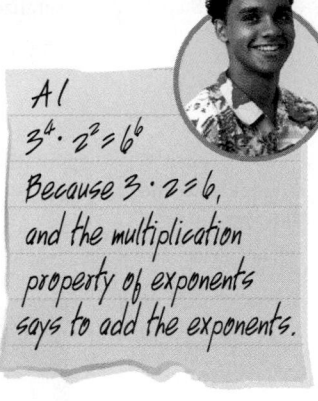

Al

$3^4 \cdot 2^2 = 6^6$

Because $3 \cdot 2 = 6$, and the multiplication property of exponents says to add the exponents.

▶ **Solution** Rewrite the original expression in expanded form

$$3 \cdot 3 \cdot 3 \cdot 3 \cdot 2 \cdot 2 = 3^4 \cdot 2^2$$

4 factors of 3   2 factors of 2

The factors are not all the same, so the multiplication property of exponents does not allow you to write this expression with a single exponent. Cal was right. Use your calculator to check that $3^4 \cdot 2^2$ and 6^6 are not equivalent.

EXAMPLE B Rewrite each expression with a single exponent.

a. $(4^5)^2$

b. $(x^3)^4$

c. $(5^m)^n$

d. $(xy)^3$

► Solution

a. Here, a number with an exponent has another exponent. You can say that 4^5 is **raised to the power** of 2. Begin by writing $(4^5)^2$ as two factors of 4^5.

$$(4^5)^2 = 4^5 \cdot 4^5 = 4^{5+5} = 4^{10}$$

There is a total of $5 \cdot 2$ or 10 factors of 4.

b. $(x^3)^4 = x^3 \cdot x^3 \cdot x^3 \cdot x^3 = x^{3+3+3+3} = x^{12}$

There is a total of $3 \cdot 4$ or 12 factors of x.

c. Based on parts a and b, when you raise an exponential expression to a power, you multiply the exponents.

$$(5^m)^n = 5^{mn}$$

d. Here, a product is raised to a power. Begin by writing $(xy)^3$ as 3 factors of xy.

$$(xy)^3 = xy \cdot xy \cdot xy = x \cdot x \cdot x \cdot y \cdot y \cdot y = x^3 y^3$$

Do you remember which property allows you to write $xy \cdot xy \cdot xy$ as $x \cdot x \cdot x \cdot y \cdot y \cdot y$?

This example has illustrated two more properties of exponents.

Power Properties of Exponents

For any nonzero values of a and b and any integer values of m and n,

$$(b^m)^n = b^{mn}$$
$$(ab)^n = a^n b^n$$

EXERCISES

You will need your calculator for problems **1, 8, 9, 12,** and **13.**

► Practice Your Skills

1. Use the properties of exponents to rewrite each expression. Use your calculator to check that your expression is equivalent to the original expression. [► 🖵 See **Calculator Note 7B** to learn how to check equivalent expressions. ◄]

   a. $(5)(x)(x)(x)(x)$ **b.** $3x^4 \cdot 5x^6$ **c.** $4x^7 \cdot 2x^3$ **d.** $(-2x^2)(x^2 + x^4)$
        $5x^4$ $15x^{10}$ $8x^{10}$ $-2x^4 - 2x^6$

2. Write each expression in expanded form. Then rewrite the product in exponential form.

   a. $3^5 \cdot 3^8$ **b.** $7^3 \cdot 7^4$ **c.** $x^6 \cdot x^2$ **d.** $y^8 y^5$ **e.** $x^2 y^4 \cdot xy^3$

3. Rewrite each expression with a single exponent.

   a. $(3^5)^8$ 3^{40} **b.** $(7^3)^4$ 7^{12} **c.** $(x^6)^2$ x^{12} **d.** $(y^8)^5$ y^{40}

4. Use the properties of exponents to rewrite each expression.

   a. $(rt)^2$ $r^2 t^2$ **b.** $(x^2 y)^3$ $x^6 y^3$ **c.** $(4x)^5$ $1024x^5$ **d.** $(2x^4 y^2 z^5)^3$ $8x^{12} y^6 z^{15}$

Exercises 1 and 8 Students can begin to check their work by substituting numbers for x and y and evaluating.

Exercise 2 Especially if this is the first time students have been asked to multiply exponential expressions with more than one variable, be sure they're not adding exponents for the factors with unlike bases.

2a. $(3 \cdot 3 \cdot 3 \cdot 3 \cdot 3)(3 \cdot 3 \cdot 3 \cdot 3 \cdot 3 \cdot 3 \cdot 3 \cdot 3) = 3^{13}$

2b. $(7 \cdot 7 \cdot 7)(7 \cdot 7 \cdot 7 \cdot 7) = 7^7$

2c. $(x \cdot x \cdot x \cdot x \cdot x \cdot x)(x \cdot x) = x^8$

2d. $(y \cdot y \cdot y \cdot y \cdot y \cdot y \cdot y \cdot y)(y \cdot y \cdot y \cdot y \cdot y) = y^{13}$

2e. $(x \cdot x \cdot y \cdot y \cdot y \cdot y)(x \cdot y \cdot y \cdot y) = (x \cdot x \cdot x)$ $(y \cdot y \cdot y \cdot y \cdot y \cdot y \cdot y) = x^3 y^7$

Exercises 2, 4, and 6 If students are not sure how to handle variables without exponents, suggest that they rewrite such variables with the exponent 1.

Closing the Lesson

When **multiplying powers** with like bases, the exponents are added. When **raising a power to a power,** the powers are multiplied. Some students may find it helpful to remember the properties of exponents as "one operation simpler." That is, multiplication means they need to add exponents, while powers means they multiply exponents. The second power property, $(ab)^n = a^n b^n$, is sometimes called *distributivity of exponentiation over multiplication.*

Exponentiation and multiplication are analogous in several ways. Exponentiation is repeated multiplication, and multiplication is repeated addition. Exponentiation takes precedence over multiplication in the order of operations, as multiplication takes precedence over addition. And, just as multiplication distributes over addition, the second power property of exponents says that exponentiation distributes over multiplication. You might show the analogies through examples.

BUILDING UNDERSTANDING

Students practice using the properties of exponents.

ASSIGNING HOMEWORK

Essential	1–7, 13
Performance assessment	6, 8, 12
Portfolio	11, 13
Journal	5, 9, 11
Group	10, 12, 14
Review	15–17

► Helping with the Exercises

Exercise 1 [Alert] Some students may propose that $x^2 + x^4 = x^6$. Suggest that they write out the expression in expanded form.

Exercises 5 and 8 Students may not recall that the conventional order of operations has exponentiation before multiplication. That is, ab^c means $a(b^c)$ rather than $(ab)^c$. ("Powers have more power.")

7. a, d, and g: 27,521.40084
b and f: 11,711.2344
c and h: 1060.32
e: 129,350.5839

9. Enclose the −5 in parentheses.

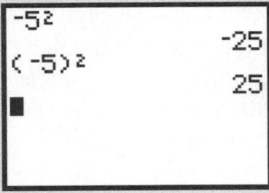

11d. Exponents are added when you multiply two exponential expressions with the same base. Exponents are multiplied when an exponential expression is raised to a power. An exponent is distributed when a product is raised to a power.

12a. $500(1 + 0.015)^6$; $546.72

▶ Reason and Apply

5. An algebra class had this problem on a quiz: "Find the value of $2x^2$ when $x = 3$."
Two students reasoned differently.

Student 1 Two times three is six. Six squared is thirty-six.
Student 2 Three squared is nine. Two times nine is eighteen.

Who was correct? Explain why.
Student 2 was correct. $2x^2 = 2 \cdot x \cdot x$; substitute $x = 3$: $2 \cdot 3 \cdot 3 = 18$

6. Match expressions from this list that are equivalent but written in different exponential forms. There can be multiple matches.

a. $(4x^4)(3x)$ **b.** $(8x^2)(3x^2)$ **c.** $(12x)(4x)$ **d.** $(6x^3)(2x^2)$

e. $12x^6$ **f.** $24x^4$ **g.** $12x^5$ **h.** $48x^2$

a, d, and g; b and f; c and h; e has no match.

7. Evaluate each expression in problem 6 using an x-value of 4.7.

8. Use the properties of exponents to rewrite each expression. Use your calculator to check that your expression is equivalent to the original expression. [▶ ⬚ See **Calculator Note 7B** to learn how to check equivalent expressions. ◀]

a. $3x^2 \cdot 2x^4$ $6x^6$ **b.** $5x^2y^3 \cdot 4x^4y^5$ $20x^6y^8$ **c.** $2x^2 \cdot 3x^3y^4$ $6x^5y^4$ **d.** $x^3 \cdot 4x^4$ $4x^7$

9. Cal and Al's teacher asked them, "What do you get when you square negative 5?" Al said, "Negative five times negative five is positive twenty-five." Cal replied, "My calculator says negative twenty-five. Doesn't my calculator know how to do exponents?" Experiment with your calculator and see if you can find a way for Cal to get the correct answer.

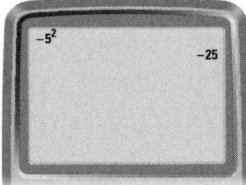

10. Evaluate $2x^2 + 3x + 1$ for each x-value.

a. $x = 3$ 28   **b.** $x = 5$ 66   **c.** $x = -2$ 3   **d.** $x = 0$ 1

11. The properties you learned in this section involve adding and multiplying exponents, and applying an exponent to more than one factor. Possible answers:

a. Write and solve a problem that requires adding exponents. $x^3 \cdot x^5 = x^8$

b. Write and solve a problem that requires multiplying exponents. $(x^3)^5 = x^{15}$

c. Write and solve a problem that requires applying an exponent to two factors. $(3x)^5 = 3^5x^5 = 243x^5$

d. Write a few sentences describing when to add exponents, when to multiply exponents, and when to apply an exponent to more than one factor.

12. **APPLICATION** Lara buys a $500 sofa at a furniture store. She buys the sofa with a new credit card that charges 1.5% interest per month, with an offer for "no payments for a year."

a. What balance will Lara's credit card bill show after 6 months? Write an exponential expression and evaluate it.

b. How much much total interest will be added after 6 months? **$46.72**

c. What balance will Lara's credit card bill show after 12 months? Write an exponential expression and evaluate it. **$500(1 + 0.015)^{12}$; $597.81**

d. How much more interest will be added between 6 and 12 months? **$51.09**

e. Explain why more interest builds up between 6 and 12 months than between 0 and 6 months. **Answers will vary. The increase is greater between 6 and 12 months because the interest each month is a percent of a greater current balance.**

13. Use the distributive property and the properties of exponents to write an equivalent expression without parentheses. Use your calculator to check your answers, as you did in problem 1.

a. $x(x^3 + x^4)$ $x^4 + x^5$

b. $(-2x^2)(x^2 + x^4)$ $-2x^4 - 2x^6$

c. $2.5x^{4.2}(6.8x^{3.3} + 3.4x^{4.2})$ $17x^{7.5} + 8.5x^{8.4}$

14. Write an equivalent expression in the form $a \cdot b^n$.

a. $3x \cdot 5x^3$ $15x^4$

b. $x \cdot x^5$ x^6

c. $2x^3 \cdot 2x^3$ $4x^6$

d. $3.5(x + 0.15)^4 \cdot (x + 0.15)^2$ $3.5(x + 0.15)^6$

e. $(2x^3)^3$ $8x^9$

f. $[3(x + 0.05)^3]^2$ $9(x + 0.05)^6$

15. Jack Frost started a snow-shoveling business. He spent $47 on a new shovel and gloves. Jack plans to charge $4.50 for every sidewalk he shovels.

a. Write an expression for Jack's profit from shoveling x sidewalks. (Hint: Don't forget his expenses.)

b. Write and solve an inequality to find how many sidewalks Jack must shovel before he makes enough money to pay for his equipment.

c. How many sidewalks must Jack shovel before he makes enough money to buy a $100 used lawn mower for his summer business? Write and solve an inequality to find out.
$4.5x - 47 > 100$; $x > 32.\overline{6}$; he must shovel 33 sidewalks to pay for his expenses and buy a lawn mower.

16. Find the equation of the line that passes through (2.2, 4.7) and (6.8, −3.9).
$y \approx 8.81 - 1.87x$

17. Solve each system.

a. $\begin{cases} y = 7.3 + 2.5(x - 8) \\ y = 4.4 - 1.5(x - 2.9) \end{cases}$
(5.3625, 0.70625)

b. $\begin{cases} 2x + 5y = 10 \\ 3x - 3y = 7 \end{cases}$
approximately (3.095, 0.762)

Exercise 13 [Alert] Some students may be confused by decimal exponents. Assure them that the multiplication property of exponents still applies.

Exercise 15 As an alternative solution for 15b, students can graph the equation $y = 4.5x - 47$, where y represents profit. The x-intercept is Jack's "break even" point of approximately 11 sidewalks. The point (33, 101.5) represents 33 sidewalks needed to make more than $100 profit. This exercise reviews Lesson 6.5.

15a. $4.5x - 47$

15b. $4.5x - 47 > 0$; $x > 10.\overline{4}$; he must shovel 11 sidewalks to pay for his equipment.

Exercise 16 This exercise reviews Lesson 5.3, but it asks for a line through points with non-integer coordinates.

Exercise 17 This exercise reviews Lesson 6.3, but each solution is a point with non-integer coordinates.

LESSON 7.3 Multiplication and Exponents **387**

PLANNING

LESSON OUTLINE

One day:

5 min	Introduction
20 min	Investigation
5 min	Sharing
5 min	Example
5 min	Closing
10 min	Exercises

MATERIALS

• Calculator Note 7C

TEACHING

This lesson introduces students to scientific notation, both as it's written by hand and as it's represented on a calculator.

One step Give the problem from the example and let students struggle for a while writing all of the 0's. Then call them together and ask for ideas about simpler notation. If none of them suggests scientific notation, introduce it, have them solve the problem, and let them experiment with scientific notation mode on their calculators to see how to work with it there.

Step 1 b. no; should be 3.2×10^6
c. no; only exponent should be on 10; should be 1.6×10^7

INTRODUCTION

The 75,000 genes in all cells are alike, so not all of the 3.75×10^{18} genes in the body are different. The scientific notation assumes that the number of genes given has three significant digits. See page 390 of this teacher's edition for a discussion of significant digits. Page 390 of the student book has a picture of a slide rule.

Scientific Notation for Large Numbers

In fact, everything that can be known has number, for it is not possible to conceive of or to know anything that has not.

PHILOLAUS

Did you know that there are approximately 75,000 genes in each human cell and more than 50 trillion cells in the human body? This means that $75,000 \cdot 50,000,000,000,000$ is a low estimate of the number of genes in your body!

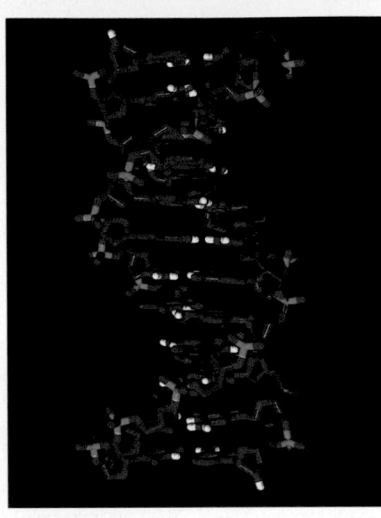

This is a computer model of a DNA strand. Many strands of DNA combine to form the genetic information in each cell.

Whether you use paper and pencil, an old-fashioned slide rule, or your calculator, exponents are useful when you work with very large numbers. For example, instead of writing 3,750,000,000,000,000,000 genes, scientists write this number more compactly as 3.75×10^{18}. This compact method of writing numbers is called **scientific notation.** You will learn how to use this notation for large numbers—numbers far from 0 on a number line.

Investigation
A Scientific Quandary

Consider these two lists of numbers.

In scientific notation	Not in scientific notation
3.4×10^5	27×10^4
7.04×10^3	$120,000,000$
6.023×10^{17}	42.682×10^{29}
8×10^1	4.2×12^6
1.6×10^2	$4^2 \times 10^2$

Step 1 Classify each of these numbers as in scientific notation or not. If a number is not in scientific notation, tell why not.

Step 2 Possible answer: The number is factored so that the first factor is a number between 1 and 10 and the second factor is a power of 10.

a. 4.7×10^3 yes
b. 32×10^5
c. $2^4 \times 10^6$
d. 1.107×10^{13} yes
e. 0.28×10^{11} no; should be 2.8×10^{10}

Step 2 Define what it means for a number to be in scientific notation.

LESSON OBJECTIVES

• Write in scientific notation numbers far from 0

• Rewrite in standard notation numbers that are in scientific notation

• Learn how calculators represent scientific notation

NCTM STANDARDS

CONTENT		PROCESS	
•	Number		Problem Solving
•	Algebra		Reasoning
	Geometry	•	Communication
•	Measurement	•	Connections
	Data/Probability	•	Representation

Step 6 Answers will vary. The exponent on 10 is the number of digits following the first digit in the original number. The digits before the 10 show the significant digits in the original number.

Step 5 If the number is negative in standard notation, the decimal number will likewise be negative in scientific notation and the minus sign will appear before the digits factor.

Step 7 Possible answer: Write the digits 415 with one digit before the decimal point: 4.15. Determine how many places the decimal point needs to move to have 4.15 become 415,000,000 and make this the exponent on 10. The scientific notation is 4.15×10^8.

Step 8 Possible answer: Move the decimal point in 6.4 five places to the right as represented in 10^5. The standard notation is 640,000.

Step 3 · **Step 4** · **Step 5** · **Step 6** · **Step 7** · **Step 8**

Use your calculator's scientific notation mode to help you figure out how to convert standard notation to scientific notation and vice versa.

Set your calculator to scientific notation mode. [▶ 🖥 See **Calculator Note 7C**. ◀]

Enter the number 5000 and press ⎡ENTER⎤. Your calculator will display its version of 5×10^3. Use a table to record the standard notation for this number, 5000, and the equivalent scientific notation.

Repeat Step 4 for these numbers:

a. 250   **b.** 5,530
c. 14,000   **d.** 7,000,000
e. 18   **f.** −470,000

In scientific notation, how is the exponent on the 10 related to the number in standard notation? How are the digits before the 10 related to the number in standard notation? If the number in standard notation is negative, how does that show up in scientific notation?

Write a set of instructions for converting 415,000,000 from standard notation to scientific notation.

Write a set of instructions for converting 6.4×10^5 from scientific notation to standard notation.

Physicist Suzanne Willis repairs a particle detector at Fermi National Accelerator Lab in Batavia, Illinois. When working with the physics of atomic particles, physicists need scientific notation to write quantities such as 2 trillion electron volts.

A number in scientific notation has the form $a \times 10^n$ where $1 \le a < 10$ or $-10 < a \le -1$ and n is an integer. In other words, the number is written as a number with one nonzero digit to the left of the decimal point multiplied by a power of 10. The number of digits to the right of the decimal point in a depends on the degree of accuracy you want to show.

EXAMPLE

Meredith is doing a report on stars and wants an estimate for the total number of stars in the universe. She reads that astronomers estimate there are at least 125 billion galaxies in the universe. An encyclopedia says that the Milky Way, Earth's galaxy, is estimated to contain more than 100 billion stars. Estimate the total number of stars in the universe. Give your answer in scientific notation.

Maria Mitchell (1818–1889) was the first professional woman astronomer in the United States.

[Ask] "What real-world quantities could be negative?" [Negative quantities might represent temperatures, locations or velocity relative to a fixed point, time before a given time, electrical charges, or acceleration (deceleration).]

Point out the quote introducing the lesson. Philolaus was a philosopher from 475 B.C. who lived in what is now southern Italy. Ask students if they agree that everything that can be known for sure can be counted or measured. Scientific notation provides convenient names for numbers far from 0. To motivate Lesson 7.6, you might challenge students to think about how to use scientific notation to describe numbers close to 0.

Assessing Progress
As you observe, you can assess students' skills at seeing patterns, following instructions, applying careful thinking, and working with groups.

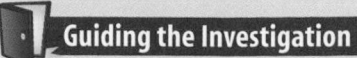

Guiding the Investigation

Step 1 Scientific notation commonly uses × instead of · to indicate multiplication by a power of ten, because typically numbers rather than letters are being multiplied.

If students aren't familiar with the fact that multiplying by 10 appends zeros or moves the decimal point, you may want to show them a list like this:

$4 \times 10 = 4 \times 10^1 = 40$
$4 \times 10 \times 10 = 4 \times 10^2 = 400$
$4 \times 10 \times 10 \times 10 = 4 \times 10^3 = 4000$

Step 3 *Standard notation* means using numerals and decimal points, with no exponents. You may want to have students also set the number of decimal places from floating to fixed 3. Then 4.7×10^8 will be 4.700E8.

Step 4 Many students should do this step and the next by hand before pulling out their calculators. Calculator displays will vary. Some use an E (in place of 10) followed by the exponent. Help students translate their display to written scientific notation.

Step 6 [Alert] Some students may need more experience with the number of digits in various powers of 10.

SHARING IDEAS

Ask students to share their instructions for Steps 7 and 8. Writing precise instructions is very difficult for many students, so be sympathetic. On the other hand, be sure that weaknesses in the instructions are noted. Work toward the definition that follows Step 8. If appropriate, you might elicit the fact that the condition on a can be written as $1 \le |a| < 10$.

See page 718 for answers to Steps 4 and 5.

► EXAMPLE

Have students check the solution with calculators. The Improving Your Reasoning Skills in Lesson 7.6 will introduce engineering notation, an alternative to scientific notation.

Significant Digits

Significant digits allow you to communicate the degree of rounding in a measurement. For example, if your measuring device showed you only that the length of an object was between 3 and 4 cm, and closer to 3 cm, then you'd report the length as 3 cm. If your device showed you that the length was between 3.0 and 3.1 cm, though, and closer to 3.0, you'd report the length as 3.0 cm.

But using significant digits to report the accuracy of a measurement can be tricky for whole numbers ending in 0. For example, what if your measurement gave you a value between 1200 and 1210, but closer to 1200? You would write 1200. But isn't that also what you'd write if your measurement were between 1200 and 1201, but closer to 1200? Or if your measurement were between 1200 and 1300, but closer to 1200? Just writing 1200 doesn't communicate the number of significant digits.

Here's where scientific notation comes to the rescue. To communicate that 1200 has only two significant digits, you'd write 1.2×10^3. If the 1200 has three significant digits, you'd write 1.20×10^3. And to say that it has four significant digits, you'd write 1.200×10^3.
[ESL] If you haven't said so before, remind students that in the United States the term *billion* means a thousand millions.

► Solution

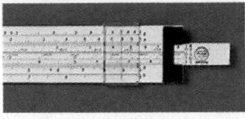

History
CONNECTION

A slide rule is a mechanical device that uses a scale related to exponential notation. Slide rules were widely used for calculating with large numbers until electronic calculators became readily available in the 1970s.

One billion is 1,000,000,000, or 10^9. Write the numbers in the example using powers of 10 and multiply them.

$(125 \times 10^9)(100 \times 10^9)$	125 billion (galaxies) times 100 billion (stars per galaxy).
$125 \times 100 \times 10^9 \times 10^9$	Regroup using the associative and commutative properties of multiplication.
$125 \times 10^2 \times 10^9 \times 10^9$	Express 100 as 10^2.
125×10^{20}	Use the multiplication property of exponents.

Since 125 is greater than 10, the answer is not yet in scientific notation.

$1.25 \times 10^2 \times 10^{20}$	Convert 125 to scientific notation.
1.25×10^{22}	Use the multiplication property of exponents.

So the universe contains more than 1.25×10^{22} stars.

Notice in this example that you used exponential expressions that were not in scientific notation. Numbers like 125 billion, 100×10^{18}, or 0.03×10^{12} can come up in calculations and sometimes these numbers make comparisons easier. Scientific notation is one of several ways to write large numbers consistently.

EXERCISES

You will need your calculator for problems **5, 8, 10,** and **14.**

► Practice Your Skills

1. Write each number in scientific notation.
   a. 34,000,000,000 3.4×10^{10} b. −2,100,000 -2.1×10^6 c. 10,060 1.006×10^4

2. Write each number in standard notation.
   a. 7.4×10^4 74,000   b. -2.134×10^6 −2,134,000   c. 4.01×10^3 4010

3. Use the properties of exponents to rewrite each expression.
   a. $3x^5(4x)$ $12x^6$ b. $y^8(7y^8)$ $7y^{16}$
   c. $b^4(2b^2 + b)$ $2b^6 + b^5$ d. $2x(5x^3 - 3x)$ $10x^4 - 6x^2$

4. Use the properties of exponents to rewrite each expression.
   a. $3x^2 \cdot 4x^3$ $12x^5$ b. $(3y^3)^4$ $81y^{12}$
   c. $2x^3(5x^4)^2$ $50x^{11}$ d. $(3m^2n^3)^3$ $27m^6n^9$

5. Owen insists on reading his calculator's display as "three point five to the seventh." Bethany tells him that he should read it as "three point five times ten to the seventh." He says, "They are the same thing. Why say all those extra words?" Write Owen and Bethany's expressions in expanded form and evaluate each to show Owen why they are not the same thing.
$3.5 \times 10^7 = 3.5 \cdot 10 \cdot 10 \cdot 10 \cdot 10 \cdot 10 \cdot 10 \cdot 10 = 35,000,000$;
$3.5^7 = 3.5 \cdot 3.5 \cdot 3.5 \cdot 3.5 \cdot 3.5 \cdot 3.5 \cdot 3.5 = 6433.9296875$

Closing the Lesson

Scientific notation provides a way of naming numbers in a consistent way and is especially useful for numbers that are very close to or far from 0. When written, the notation consists of a number between 1 and 10 (possibly 1 but not 10), or between −1 and −10, multiplied by a power of 10. Calculators use other representations.

▶ Reason and Apply

6. There are approximately 5.58×10^{21} atoms in a gram of silver. How many atoms are there in 3 kilograms of silver? Express your answer in scientific notation. 1.674×10^{25} atoms

7. Because the number of molecules in a given amount of a compound is usually a very large number, scientists often work with a quantity called a mole. One mole is about 6.02×10^{23} molecules.

    a. A liter of water has about 55.5 moles of H_2O. How many molecules is this? Write your answer in scientific notation.

    b. How many molecules are in 6.02×10^{23} moles of a compound? Write your answer in scientific notation. 3.3411×10^{25}

approximately 3.6×10^{47}

8. Write each number in scientific notation. How does your calculator show each answer? Answers will vary based on model of calculator used.

    a. 250 2.5×10^2; 2.5E2

    b. 7,420,000,000,000 7.42×10^{12}; 7.42E12

    c. −18 -1.8×10^1; −1.8E1

The number of molecules in one mole is called Avogadro's number. The number is named after the Italian chemist and physicist Amadeo Avogadro (1776–1856).

9. Cal and Al were assigned this multiplication problem for homework:

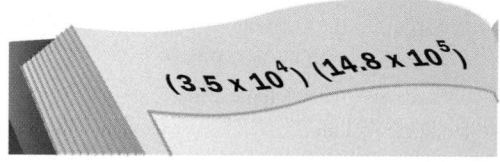

$(3.5 \times 10^4)(14.8 \times 10^5)$

Cal got an answer of 51.8×10^9, and Al got 5.18×10^{10}.

    a. Are Cal's and Al's answers equivalent? Explain why or why not.

    b. Whose answer is in scientific notation? Al's answer

    c. Find another exponential expression equivalent to Cal's and Al's answers. Possible answer: 518×10^8

    d. Explain how you can rewrite a number, such as 432.5×10^3, in scientific notation.

10. Consider these multiplication expressions.

    i. $(2 \times 10^5)(3 \times 10^8)$ 6×10^{13} **ii.** $(4.1 \times 10^3)(2 \times 10^5)$ 8.2×10^8

    a. Set your calculator in scientific notation mode and multiply each expression.

    b. Explain how you could do the multiplication in 10a without using a calculator.

    c. Find the product $(4 \times 10^5)(6 \times 10^7)$ and write it in scientific notation without using your calculator. $(4 \times 10^5)(6 \times 10^7) = 4 \times 6 \times 10^5 \times 10^7 = 24 \times 10^{12} = 2.4 \times 10^1 \times 10^{12} = 2.4 \times 10^{13}$

11. Americans make almost 2 billion telephone calls each day. *(www.britannica.com)*

    a. Write this number in standard notation and in scientific notation. 2,000,000,000; 2×10^9

    b. How many phone calls do Americans make in one year? (Assume that there are 365 days in a year.) Write your answer in scientific notation. 7.3×10^{11} calls per year

BUILDING UNDERSTANDING

The exercises provide practice with exponents and scientific notation.

ASSIGNING HOMEWORK

Essential	1, 2, 5, 8–10
Performance assessment	6, 11–13
Portfolio	7, 14
Journal	5, 9, 10
Group	14
Review	3, 4, 14, 15

▶ Helping with the Exercises

Exercise 6 Some students may want to use their calculators here. They don't need to use scientific notation mode. They can enter 5.58*10^23, for example, or on some calculators they can enter 5.58 EE23.

Exercise 9 Neither answer is better because the problem didn't ask for a result in scientific notation.

9a. Yes, because they are both equal to 51,800,000,000.

9d. Rewrite the digits before the 10 in scientific notation, then use the multiplication property of exponents to add the exponents on the 10's. In this case, $4.325 \times 10^2 \times 10^3 = 4.325 \times 10^5$.

10b. Regroup, multiply the numbers, and multiply the powers of 10 by adding the exponents.

Exercises 12 and 13 Encourage the use of dimensional analysis.

12a. 1.5×10^6 cells per hour

Exercise 13 This exercise, which reviews Lesson 7.3, can be used to foreshadow division with exponents in Lesson 7.5. That is, $10^{17} \div 10^4 = 10^{13}$.

The exercise is from *Graphic Algebra* (Key Curriculum Press), a good resource for other problems.

Exercise 14 Because the equation was a fit of data, it might not provide the exact population at any point. In particular, it is a little higher than the actual population now.

14a. 3.8 is the population (in millions) in 1900; 0.017 is the annual growth rate; t is the elapsed time in years since 1900; P is the population (in millions) t years after 1900.

14b. Answers will vary depending on the current year; $0 \le t \le$ (*current year* − 1900)

14c.

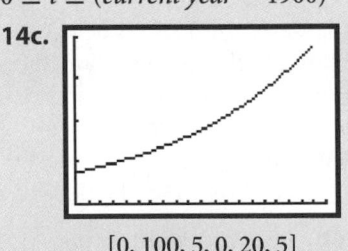

[0, 100, 5, 0, 20, 5]

Exercise 15 This exercise reviews Lesson 6.6.

15.

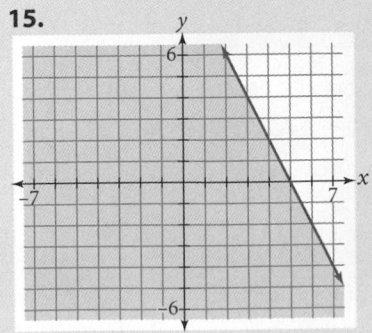

---

12. On average a person sheds 1 million dead skin cells every 40 minutes. (*The World in One Day*, 1997, p. 16)

   a. How many dead skin cells does a person shed in an hour? Write your answer in scientific notation.

   b. How many dead skin cells does a person shed in a year? (Assume that there are 365 days in a year.) Write your answer in scientific notation. 1.314×10^{10} cells per year

13. A *light-year* is the distance light can travel in one year. This distance is approximately 9,460 billion kilometers. The Milky Way galaxy is estimated to be about 100,000 light-years in diameter.

   a. Write both distances in scientific notation. 9.46×10^{12} km; 1.0×10^5 light-years

   b. Find the diameter of the Milky Way in kilometers. Use scientific notation. 9.46×10^{17} km

   c. Scientists estimate the diameter of the earth is greater than 1.27×10^4 km. How many times larger is the diameter of the Milky Way? $\dfrac{(9.46 \times 10^{17})}{(1.27 \times 10^4)} = 7.45 \times 10^{13}$

Dead skin cells are one of the components of dust.

▶ Review

14. **APPLICATION** The exponential equation $P = 3.8(1 + 0.017)^t$ approximates Australia's annual population (in millions) since 1900.

   a. Explain the real-world meaning of each number and variable in the equation.

   b. What interval of t-values will give information up to the current year?

   c. Graph $P = 3.8(1 + 0.017)^t$ over the time interval you named in 14b.

   d. What population does the model predict for the year 1950? approximately 8.8 million

   e. Use your answer from 14d to predict today's population.
   Answers will vary depending on the current year; $P = 3.8(1 + 0.017)^{50} \cdot (1 + 0.017)^{(current\ year - 1950)}$

15. Graph $y \le -2(x - 5)$.

IMPROVING YOUR **REASONING** SKILLS

The *Jinkōki* (2000, Wasan Institute, p. 146) tells this ancient Japanese problem:

> A breeding pair of rats produced 12 baby rats (6 female and 6 male) in January. There were 14 rats at that time. In February, each female-male pair of rats again bred 12 baby rats. The total number of rats was then 98. In this way, each month, the parents, their children, their grandchildren, and so forth, breed 12 baby rats each. How many rats would there be at the end of one year?

Solve this problem using an exponential model. If you use your calculator, you will get an answer in scientific notation that doesn't show all the digits of the answer. Can you devise a way to find the "missing" digits?

IMPROVING **REASONING** SKILLS

One way to solve this problem is to concentrate on pairs of rats rather than individual rats. Every month, each pair produces 6 more pairs, making a total of 7 pairs at the end of the month for each pair at the beginning of the month. Therefore, there's a constant multiplier of 7, and an exponential model for the number of pairs is 7^x. Doubling 7^{12}, to find number of individual rats after 12 months, gives you 27,682,574,402. On the TI-83 Plus, the number is shown in scientific notation as 2.76825744E10—the last two digits are not displayed. If students subtract 27000000000, the result is 682574402 and the final two digits are revealed. (Instead of typing in 27000000000 to subtract, they can enter 2.7EE10.)

Looking Back with Exponents

The eye that directs a needle in the delicate meshes of embroidery, will equally well bisect a star with the spider web of the micrometer.

MARIA MITCHELL

You've learned that looking ahead in time to predict future growth with an exponential model is related to the multiplication property of exponents. In this lesson you'll discover a rule for dividing expressions with exponents. Then you'll see how dividing expressions with exponents is like looking *back* in time.

PLANNING

LESSON OUTLINE

One day:

20 min	Investigation
5 min	Sharing
10 min	Examples
5 min	Closing
10 min	Exercises

MATERIALS

- Properties of Exponents (T), *optional*

Investigation
The Division Property of Exponents

Step 1

Write the numerator and the denominator of each quotient in expanded form. Then reduce to eliminate common factors. Rewrite the factors that remain with exponents.

a. $\dfrac{5^9}{5^6}$ **b.** $\dfrac{3^3 \cdot 5^3}{3 \cdot 5^2}$ **c.** $\dfrac{4^4 x^6}{4^2 x^3}$

Step 2 Descriptions Step 2 should include subtracting the exponent in the denominator from the exponents in the numerator.

Step 3
$$5^{(15-11)}\left(1 + \frac{0.08}{12}\right)^{(24-18)}$$
$$= 5^4\left(1 + \frac{0.08}{12}\right)^6$$

Step 2

Compare the exponents in each final expression you got in Step 1 to the exponents in the original quotient. Describe a way to find the exponents in the final expression without using expanded form.

Step 3

Use your method from Step 2 to rewrite this expression so that it is not a fraction. You can leave $\frac{0.08}{12}$ as a fraction.

$$\frac{5^{15}\left(1 + \dfrac{0.08}{12}\right)^{24}}{5^{11}\left(1 + \dfrac{0.08}{12}\right)^{18}}$$

Recall that exponential growth is related to repeated multiplication. When you look ahead in time you multiply by more constant multipliers, or increase the exponent. To look back in time you will need to undo some of the constant multipliers, or divide.

TEACHING

Real-world situations give insight into why exponents are subtracted when powers of like bases are divided.

 Guiding the Investigation

One step Pose the problem, "Six years ago Anne paid $18,500 for a van for her flower delivery service. Its value has been depreciating at a rate of 9% per year, so the van is currently worth $18,500(1 − 0.09)^6$ dollars. How much was it worth two years ago?" As you circulate, encourage students to write the value in at least two different ways: $\dfrac{18,500(1 - 0.09)^6}{(1 - 0.09)^2}$ and $18,500(1 - 0.09)^4$. Ask students if they think they can always subtract exponents when dividing, bring out the idea of having like bases, and ask them to make up and solve some related problems with which to challenge each other.

Step 1 Encourage students to write in exponents of 1. Doing so will aid in devising a rule in Step 2.

See page 718 for answers to Step 1.

NCTM STANDARDS

CONTENT	PROCESS
● Number	● Problem Solving
● Algebra	● Reasoning
Geometry	● Communication
Measurement	● Connections
Data/Probability	● Representation

LESSON OBJECTIVE

- Review or learn the division property of exponents

Step 3 Division by a power of 5 as well as the other power might confuse some students. Remind them that division is the opposite of multiplication, so that they can think of dividing 5^{15} by 5^{11}.

Step 4 If students are having difficulties expressing their ideas in parts a and b, suggest that they look at the wording in parts c and d.

SHARING IDEAS

Ask students to share their ideas about Step 2. Try to get them to formulate the Division Property of Exponents. Students can look on page 395, or you might want to display the division property shown on the Properties of Exponents transparency.
[Ask] "Why must b not equal 0?"
[Division by zero is undefined.]
[Ask] "Why doesn't the statement exclude m or n from being zero?" Getting students to conjecture about the value of b^0 motivates Lesson 7.6.

Assessing Progress

Look for understanding that **exponents** represent repeated multiplication, understanding that the fraction bar represents **division,** and ability to see "chunks."

▶ EXAMPLE A

This example is good for extending the division property beyond powers on a single base. If students are having trouble, keep stressing that they can write out the terms in expanded form, such as

$$\frac{6x^9}{5x^4} = \frac{6 \cdot x \cdot x \cdot x \cdot x \cdot x \cdot x \cdot x \cdot x \cdot x}{5 \cdot x \cdot x \cdot x \cdot x},$$

and then eliminate factors equivalent to 1, in this case getting $\frac{6x^5}{5}$.

Step 4 a. the balance
3 years prior; $500(1 + 0.04)^4$

b. the balance 5 years prior;
$21,300(1 - 0.12)^4$

c. $\dfrac{32(1 + 0.50)^5}{(1 + 0.50)^2}$

$= 32(1 + 0.50)^3$

d. $A(1 + r)^{(n-m)}$

Step 5 Dividing by $(1 + r)^n$ represents looking back n time periods.

Apply what you have discovered about dividing expressions with exponents.

a. After 7 years the balance in a savings account is $500(1 + 0.04)^7$. What does the expression $\frac{500(1 + 0.04)^7}{(1 + 0.04)^3}$ mean in this situation? Rewrite this expression with a single exponent.

b. After 9 years of depreciation, the value of a car is $21,300(1 - 0.12)^9$. What does the expression $\frac{21,300(1 - 0.12)^9}{(1 - 0.12)^5}$ mean in this situation? Rewrite this expression with a single exponent.

c. After 5 weeks the population of a bug colony is $32(1 + 0.50)^5$. Write a division expression to show the population 2 weeks earlier. Rewrite your expression with a single exponent.

d. The expression $A(1 + r)^n$ can model n time periods of exponential growth. What expression models the growth m time periods earlier?

How does looking back in time with an exponential model relate to dividing expressions with exponents?

Expanded form helps you understand many properties of exponents. It also helps you understand how the properties work together.

EXAMPLE A

Use the properties of exponents to rewrite each expression.

a. $\dfrac{6x^9}{5x^4}$ **b.** $\dfrac{(3x^2)(8x^4)}{-4x^3}$ **c.** $\dfrac{7.5 \times 10^8}{1.5 \times 10^3}$

▶ Solution

a. Use expanded form and reduce.

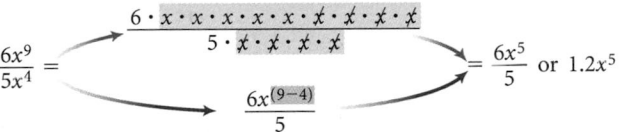

$$\frac{6x^9}{5x^4} = \frac{6 \cdot x \cdot x \cdot x \cdot x \cdot x \cdot \cancel{x} \cdot \cancel{x} \cdot \cancel{x} \cdot \cancel{x}}{5 \cdot \cancel{x} \cdot \cancel{x} \cdot \cancel{x} \cdot \cancel{x}} = \frac{6x^5}{5} \text{ or } 1.2x^5$$

$$\frac{6x^{(9-4)}}{5}$$

In expanded form, 4 factors of x are removed in the numerator and denominator. That leaves $9 - 4$, or 5, factors of x in the numerator.

b. Use expanded form and reduce.

$$\frac{(3x^2)(8x^4)}{-4x^3} = \frac{3 \cdot 8 \cdot x^2 \cdot x^4}{-4 \cdot x^3} = \frac{3 \cdot 8}{-4} \cdot \frac{x \cdot x \cdot x \cdot \cancel{x} \cdot \cancel{x} \cdot \cancel{x}}{\cancel{x} \cdot \cancel{x} \cdot \cancel{x}} = -6x^3$$

$$\frac{3 \cdot 8}{-4} \cdot x^{(2+4)-3}$$

In expanded form, 2 factors of x are combined with 4 factors of x in the numerator. Then 3 factors of x are removed in the numerator and denominator. That leaves $(2 + 4) - 3$, or 3, factors of x in the numerator.

c.

$$\frac{7.5 \times 10^8}{1.5 \times 10^3} = \frac{7.5}{1.5} \times \frac{10 \cdot 10 \cdot 10 \cdot 10 \cdot 10 \cdot \cancel{10} \cdot \cancel{10} \cdot \cancel{10}}{\cancel{10} \cdot \cancel{10} \cdot \cancel{10}} = 5.0 \times 10^5$$

$$\frac{7.5}{1.5} \times 10^{(8-3)}$$

So, division involving scientific notation can be done just like any other expression with exponents.

The investigation and example have introduced the **division property of exponents.**

Division Property of Exponents

For any nonzero value of b and any integer values of m and n,

$$\frac{b^n}{b^m} = b^{n-m}$$

The division property of exponents lets you divide expressions with exponents simply by subtracting the exponents.

EXAMPLE B

Six years ago, Anne bought a van for $18,500 for her flower delivery service. Based on the prices of similar used vans, she estimates a rate of depreciation of 9% per year.

a. How much is the van worth now?

b. How much was it worth last year?

c. How much was it worth 2 years ago?

▶ Solution

a. Right now the value of the van has been decreasing for 6 years. The original price was $18,500, and the rate of depreciation as a decimal is 0.09.

$$A(1 - r)^x = 18{,}500(1 - 0.09)^6 \approx 10{,}505.58$$

The van is currently worth $10,505.58.

b. A year ago the van was 5 years old. One approach is to use 5 as the exponent.

$$18{,}500(1 - 0.09)^5 \approx 11{,}544.59$$

Another approach is to undo the multiplication in part a by using division.

$$\frac{18{,}500(1 - 0.12)^6}{(1 - 0.12)} = 18{,}500(1 - 0.12)^5$$

The numerator on the left side of this equation represents the starting value multiplied by 6 factors of the constant multiplier $(1 - 0.09)$. Dividing by the constant multiplier once leaves you with an expression representing 5 years of exponential depreciation. Either way, the exponent is decreased by 1. The van was worth $11,544.59 last year.

▶ EXAMPLE B

This example provides another case of a decreasing exponential situation, especially useful for students who had difficulties with Step 4b of the investigation.

Closing the Lesson

Considering the value of an exponentially growing quantity at one time and at an earlier time can show that, when **powers** of like bases **are divided**, the **exponents are subtracted.**

BUILDING UNDERSTANDING

Students practice subtracting exponents while dividing powers of like bases.

ASSIGNING HOMEWORK

Essential	1–8
Performance assessment	4, 9, 11–13
Portfolio	8, 10
Journal	3
Group	10
Review	14, 15

▶ Helping with the Exercises

Exercise 2a If students write this as 1^8, urge them to write out the 7's and then remove values of 1. There will be eight 7's left.

Exercise 3 This exercise gives students a chance to confront a common error. **[Ask]** "Why can't you divide the common bases?" Encourage students to write the numerator and denominator in expanded form and then remove values equivalent to 1. To challenge students who understand this very well, **[Ask]** "Are there any conditions under which you could divide the common bases?" [If the same bases have the same power, then the ratio is 1. Dividing the bases to get a base of 1 will yield that ratio.]

3. Possible answer: $\frac{3^6}{3^2}$ means here are 6 factors of 3 in the numerator and 2 factors of 3 in the denominator. So there are 2 factors of 1, or $\frac{3}{3}$, in the entire expression, leaving 4 factors of 3 in the numerator, or 3^4.

Exercise 4 At this point students can solve the equation by using graphing or tables. Students will see in Lesson 7.6 that the equation can also be solved using negative exponents: $A = 10,000(1 + 0.1)^{-20}$.

---

c. To find the value 2 years ago, decrease the exponent in part a by 2.

$$18,500(1 - 0.09)^{(6-2)} = 18,500(1 - 0.09)^4 \approx 12,686.37$$

Subtracting 2 from the exponent gives the same result as undoing two multiplications. The van was worth \$12,686.37 two years ago.

EXERCISES

You will need your calculator for problems **4, 6, 8, 9, 10, 11, 12, 13, 14,** and **15.**

▶ Practice Your Skills

1. Eliminate factors equivalent to 1 and rewrite the right side of this equation.

$$\frac{x^5 y^4}{x^2 y^3} = \frac{x \cdot x \cdot x \cdot x \cdot x \cdot y \cdot y \cdot y \cdot y}{x \cdot x \cdot y \cdot y \cdot y} \quad x^3 y$$

2. Use the properties of exponents to rewrite each expression.

a. $\frac{7^{12}}{7^4}$ 7^8 **b.** $\frac{x^{11}}{x^5}$ x^6 **c.** $\frac{12x^5}{3x^2}$ $4x^3$ **d.** $\frac{7x^6 y^3}{14x^3 y}$ $0.5x^3 y^2$

3. Cal says that $\frac{3^6}{3^2}$ equals 1^4 because you divide the 3's and subtract the exponents. Al knows Cal is incorrect, but he doesn't know how to explain it. Write an explanation so that Cal will understand why he is wrong and how to get the correct answer.

4. APPLICATION Webster owns a set of antique dining-room furniture that has been in his family for many years. The historical society tells him that furniture similar to his has been appreciating in value at 10% per year for the last 20 years and that his furniture could be worth \$10,000 now.

a. Which letter in the equation $y = A(1 + r)^x$ could represent the value of the furniture 20 years ago when it started appreciating?

b. Substitute the other given information into the equation $y = A(1 + r)^x$.

c. Solve your equation in 4b to find out how much Webster's furniture was worth 20 years ago. Show your work.

5. Use the properties of exponents to rewrite each expression.

a. $(2x)^3 \cdot (3x^2)^4$ $648x^{11}$ **b.** $\frac{(5x)^7}{(5x)^5}$ $25x^2$ **c.** $\frac{(2x)^5}{-8x^3}$ $-4x^2$ **d.** $(4x^2 y^5) \cdot (-3xy^3)^3$ $-108x^5 y^{14}$

▶ Reason and Apply

6. The earth is 1.5×10^{11} meters from the sun. Light travels at a speed of 3×10^8 meters per second. How long does it take light to travel from the sun to the earth? Answer to the nearest minute. It takes 500 seconds, or approximately 8 minutes.

4a. *A* represents the starting value.

4b. $10,000 = A(1 + 0.1)^{20}$

4c. $10,000 = A(1 + 0.1)^{20}$

$\dfrac{10,000}{(1 + 0.1)^{20}} = A$

$1486.43 \approx A$

The furniture was worth about \$1,486 twenty years ago.

Exercises 6 and 7 Encourage the use of dimensional analysis in setting up the quotient. If students use a calculator to do the calculation, point out that the calculation can also be done with the division property of exponents.

7. **APPLICATION** **Population density** is the number of people per square mile. That is, if the population of a country were spread out evenly across an entire nation, the population density would be the number of people in each square mile.

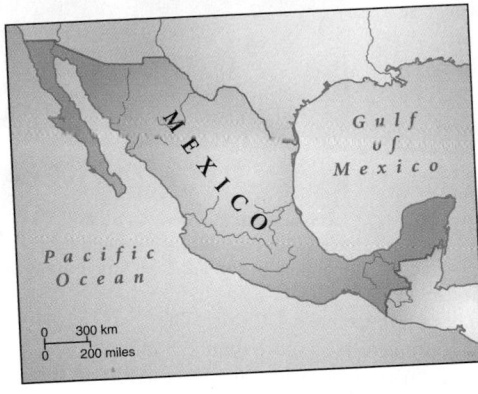

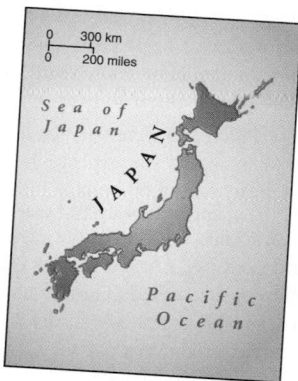

a. In 2000, the population of Mexico was about 1.0×10^8. Mexico has a land area of about 7.6×10^5 square miles. What was the population density of Mexico in 2000? *(2001 World Almanac, p. 822)* **about 132 people per square mile**

b. In 2000, the population of Japan was about 1.3×10^8. Japan has a land area of about 1.5×10^5 square miles. What was the population density of Japan in 2000? *(2001 World Almanac, p. 803)* **about 867 people per square mile**

c. How did the population densities of Mexico and Japan compare in 2000?
The population of Japan was about 6.6 times denser than that of Mexico.

8. **APPLICATION** Eight months ago, Tori's parents put $5,000 into a savings account that earns 0.25% interest per month. Now, her dentist has suggested that she get braces.

a. If Tori's parents use the money in their savings account, how much do they have? $5,100.88

b. If Tori's dentist had suggested braces 3 months ago, how much money would have been in her parents' savings account? $5,062.81

c. Tori's dentist says she can probably wait up to 2 months before having the braces fitted. How much will be in her parents' savings account if she waits? $5,126.42

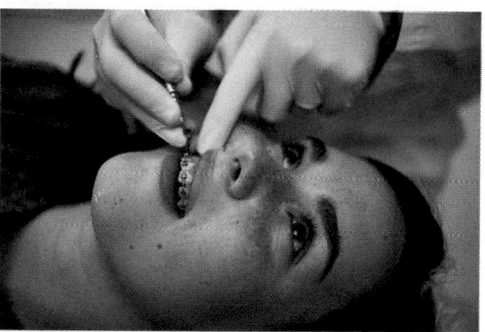

Orthodontic treatment can cost between $4,000 and $6,000 depending on the extent of the procedure. It is estimated that 5 million people were treated by orthodontists in the U.S. in 2000.

9. **APPLICATION** During its early stages, a disease can spread exponentially as those already infected come in contact with others. Assume that the number of people infected by a disease approximately triples every day. At one point in time there are 864 people who are infected. How many days earlier had fewer than 20 people been infected? Show at least two different methods for solving this problem.

10. The population of a city has been growing at a rate of 2% for the last 5 years. The population is now 120,000. Find the population 5 years ago.

Exercise 9 Students could also use the equation $20 = 864(3)^x$, thus previewing negative exponents from Lesson 7.6.

9. four days earlier

Method 1: Use a recursive routine 864 (ENTER), Ans/3 (ENTER), (ENTER), (ENTER), (ENTER)

Method 2: Use an equation: $y = 864\left(\frac{1}{3}\right)^x$; look at the table to find x when y is less than 20.

10. approximately 108,688

Exercise 11 Encourage students to use dimensional analysis. As needed, remind them that bpm means *beats per minute*.

Exercises 12 and 13 Urge students to use dimensional analysis carefully. Students may not know that 1 ton = 2000 pounds and 1 pound = 16 ounces.

Exercise 14 This problem reviews Lesson 2.3.

Exercise 15 This problem reviews Lesson 3.3 and requires division of exponential expressions.

15a. (Answers recorded to tenths.) Mercury: 4.1 cm; Venus: 10 cm; Earth: 10.5 cm; Mars: 5.6 cm; Jupiter: 117.3 cm; Saturn: 94.7 cm; Uranus: 69.3 cm; Neptune: 41.3 cm; Pluto: 2 cm; Sun: 1152 cm

11. **APPLICATION** In the course of a mammal's lifetime its heart beats about 800 million times, regardless of the mammal's size or weight. (This excludes humans.)

    a. An elephant's heart beats approximately 25 times a minute. How many years would you expect an elephant to live? Use scientific notation to calculate your answer. approximately 61 years

    b. A pygmy shrew's heart beats approximately 1150 times a minute. How many years would you expect a pygmy shrew to live? approximately 1.3 years

    c. If this relationship were true for humans, how many years would you expect a human being with a heart rate of 60 bpm to live?
    approximately 25.4 years

Pygmy shrews may be the world's smallest mammal, as small as 5 cm from nose to tail.

12. More than 57,000 tons of cotton are produced in the world each day. It takes about 8 ounces of cotton to make a T-shirt. The population of the United States in 2000 was estimated to be more than 275 million. If all the available cotton were used to make T-shirts, how many T-shirts could have been manufactured every day for each person in the United States in 2000? Write your answer in scientific notation. (*http://cotton.net*) approximately 8.3×10^{-1} T-shirt per person

13. Each day bees sip the nectar from approximately 3 trillion flowers to make 3300 tons of honey. How many flowers does it take to make 8 ounces of honey? Write your answer in scientific notation. (*The World in One Day*, 1997, p. 21) approximately 2.272×10^5 flowers

▶ Review

14. On his birthday Jon figured out that he was 441,504,000 seconds old. Find Jon's age in years. (Assume that there are 365 days per year.) 14 years

15. Halley is doing a report on the solar system and wants to make models of the sun and the planets showing relative size. She decides that Pluto, the smallest planet, should have a model diameter of 2 cm.

    a. Using the table, find the diameters of the other models she would have to make.

    b. What advice would you give Halley on her project?

    Possible answer: Halley should make her models much smaller because Jupiter is 1 m in diameter and the sun is greater than 11 m in diameter; it would be better to leave the sun out of her models altogether; if she makes Pluto with a diameter of 0.2 cm, Jupiter will be only about 12 cm in diameter.

Size of Planets and Sun

Planet	Diameter (mi)
Mercury	3.1×10^3
Venus	7.5×10^3
Earth	7.9×10^3
Mars	4.2×10^3
Jupiter	8.8×10^4
Saturn	7.1×10^4
Uranus	5.2×10^4
Neptune	3.1×10^4
Pluto	1.5×10^3
Sun	8.64×10^5

Zero and Negative Exponents

It is not knowledge which is dangerous, but the poor use of it.

HROTSWITHA

$\mathbf{H}$ave you noticed that so far in this chapter the exponents have been positive integers? In this lesson you will learn what a zero or a negative integer means as an exponent.

Investigation
More Exponents

Step 1 Use the division property of exponents to rewrite each of these expressions with a single exponent.

a. $\dfrac{y^7}{y^2}$ y^5 b. $\dfrac{3^2}{3^4}$ 3^{-2} c. $\dfrac{7^4}{7^4}$ 7^0 d. $\dfrac{2}{2^5}$ 2^{-4} e. $\dfrac{x^3}{x^6}$ x^{-3}

f. $\dfrac{z^8}{z}$ z^7 g. $\dfrac{2^3}{2^3}$ 2^0 h. $\dfrac{x^5}{x^5}$ x^0 i. $\dfrac{m^6}{m^3}$ m^3 j. $\dfrac{5^3}{5^5}$ 5^{-2}

Some of your answers in Step 1 should have positive exponents, some should have negative exponents, and some should have an exponent of zero.

Step 2 How can you tell what type of exponent will result simply by looking at the original expression?

Step 3 Go back to the expressions in Step 1 that resulted in a negative exponent. Write each in expanded form. Then reduce them. b. $\dfrac{1}{9}$ d. $\dfrac{1}{16}$ e. $\dfrac{1}{x^3}$ j. $\dfrac{1}{25}$

Step 4 Compare your answers from Step 3 and Step 1. Tell what a base raised to a negative exponent means.

Step 5 Go back to the expressions in Step 1 that resulted in an exponent of zero. Write each in expanded form. Then reduce them. c, g, and h.; 1

Step 6 Compare your answers from Step 5 and Step 1. Tell what a base raised to an exponent of zero means. Possible answer: A base raised to an exponent of zero is always equal to 1.

Step 2
Possible answer: Expressions resulting in positive exponents have a larger exponent in the numerator; expressions resulting in negative exponents have larger exponents in the denominator; and expressions resulting in exponents of zero have equal exponents in the numerator and denominator.

NCTM STANDARDS

CONTENT	PROCESS
• Number	
	• Problem Solving
• Algebra	• Reasoning
Geometry	• Communication
• Measurement	• Connections
Data/Probability	• Representation

LESSON OBJECTIVES

• Investigate the meaning of nonpositive exponents
• Write a number with a negative exponent in a form that has a positive exponent and write a number with a positive exponent in a form that has a negative exponent
• Write in scientific notation numbers close to 0

PLANNING

LESSON OUTLINE

First day:
40 min   Investigation
10 min   Sharing
Second day:
25 min   Examples
5 min   Closing
15 min   Exercises

MATERIALS

• Properties of Exponents (T), *optional*

TEACHING

Negative exponents are used in scientific notation for numbers very close to 0. They provide an alternative way to solve problems involving division of powers.

  Guiding the Investigation

One step Pose this problem, "How many atoms are on the point of a pin? Assume that the diameter of a pin point is 0.0010 cm, and that each atom has a diameter of 0.00000001 cm. How would you represent these numbers with scientific notation, and how many atoms are on the point of a pin?" As students work, remind them of how the positive exponent in scientific notation is one fewer than the number of digits to the left of the decimal point. After they get the quotient $\dfrac{1.0 \times 10^{-3}}{10^{-8}}$, you may need to remind them how to subtract a negative number from another negative.

Step 4 Possible answer: A base raised to a negative exponent means the same thing as one over the same base raised to the same exponent with the sign changed to positive.

Step 7 Some students may see that a negative exponent on a term in the numerator moves that term to the denominator with a positive exponent but not see that the process works in reverse. Suggest that they write the denominator in part b as a fraction and then multiply the numerator and denominator of the complex fraction by 3^8. The same idea applies to y^{-5} in part c.

Step 8 [Ask] "Do your findings extend to rewriting an expression with positive exponents as a fraction with negative exponents?"

SHARING IDEAS

Ask for ideas from Step 8. Work with the class to come up with a precise method. **[Ask]** "Why does the method work?" [One justification might be through patterns, as found in the table in the student text.]

Have the class critique the boxed definition in the student text. Ask why b must be "nonzero." Help students see that if b were 0, then you'd have $0 = \frac{1}{0^n}$, but division by 0 is undefined. When students realize the importance of the phrase "for any value of n," they may then raise questions about the case of $n = 0$. In that case, both sides of the equation are 1.

Elicit the question of the value of 0^0. Arguments for two different values can be made: zero to any other power is zero, but any other number to the zero power is one. Mathematicians say 0^0 is *indeterminate*, implying that it's undefined. If students are interested, they might graph $y = x^x$ on their calculators and see what value it gets close to as x approaches 0.

Step 7

Step 8 Possible answer: An expression with an exponent can be moved between the numerator and denominator of a fraction as long as the sign of the exponent is changed with each move.

Step 7 Use what you have learned about negative exponents to rewrite each of these expressions with positive exponents and only one fraction bar.

a. $\dfrac{5^{-2}}{1}$ $\dfrac{1}{5^2}$

b. $\dfrac{1}{3^{-8}}$ $\dfrac{1}{\frac{1}{3^8}} = \dfrac{3^8}{1}$

c. $\dfrac{4x^{-2}}{z^2 y^{-5}}$ $\dfrac{4y^5}{z^2 x^2}$

Step 8 In one or two sentences, explain how to rewrite a fraction with a negative exponent in the numerator or denominator as a fraction with positive exponents.

This table supports what you have learned about negative exponents and exponents of zero. To go down either column of the table, you divide by 3. Notice that each time you divide, the exponent decreases by 1. (Likewise, to go up either column of the table, you multiply by 3 and the exponent increases by 1.) In order to continue the pattern, 3^0 must have the value 1. As the exponents become negative, the base 3 appears in the denominator with a positive exponent.

$3^1 \div 3 = \dfrac{3^1}{3^1} = 3^{(1-1)} = 3^0$

$3^{-1} \div 3 = \dfrac{3^{-1}}{3^1} = 3^{(-1-1)} = 3^{-2}$

Exponential form	Fraction form
3^3	27
3^2	9
3^1	3
3^0	1
3^{-1}	$\dfrac{1}{3}$
3^{-2}	$\dfrac{1}{9}$
3^{-3}	$\dfrac{1}{27}$

$3 \div 3 = \dfrac{3}{3} = 1$

$\dfrac{1}{3} \div 3 = \dfrac{1}{3} \cdot \dfrac{1}{3} = \dfrac{1}{3^2} = \dfrac{1}{9}$

Negative Exponents and Exponents of Zero

For any nonzero value of b and for any integer value of n,

$$b^{-n} = \frac{1}{b^n} \quad \text{and} \quad \frac{1}{b^{-n}} = b^n$$

$$b^0 = 1$$

EXAMPLE A Use the properties of exponents to rewrite each expression withou fraction bar.

a. $\dfrac{3^5}{4^7}$

b. $\dfrac{25}{x^8}$

c. $\dfrac{5^{-3}}{2^{-8}}$

d. $\dfrac{3(17)^8}{17^8}$

Note the quotation introducing the lesson. Hrotswitha of Gansersheim (A.D. 935–1000) is the first recorded female German writer. She included some mathematics in her plays. She was also a Benedictine nun. You could use this quote to warn against the "dangers" of misapplying the various properties of exponents, though she undoubtedly had in mind a deeper social meaning.

Assessing Progress
You can assess understanding of **exponents** as indicating repeated multiplication, understanding of **subtraction of exponents** when dividing like bases, and abilities to subtract a larger integer from a smaller one.

▶ **Solution**

a. $\dfrac{3^5}{4^7} = 3^5 \cdot \dfrac{1}{4^7}$ Think of the original expression as having two separate factors.

      $= 3^5 \cdot 4^{-7}$ Use the definition of negative exponents.

b. $\dfrac{25}{x^8} = 25 \cdot \dfrac{1}{x^8} = 25 \cdot x^{-8} = 25x^{-8}$

c. $\dfrac{5^{-3}}{2^{-8}} = 5^{-3} \cdot \dfrac{1}{2^{-8}} = 5^{-3} \cdot 2^8$

d. $\dfrac{3(17)^8}{17^8} = 3 \cdot 17^0$ Use the division property of exponents.

      $= 3 \cdot 1$ Use the definition of an exponent of zero.

      $= 3$ Multiply.

You can also use negative exponents to look back in time with increasing or decreasing exponential situations.

EXAMPLE B

Solomon bought a used car for $5,600. He estimates that it has been decreasing in value by 15% each year.

a. If his estimate of the rate of depreciation is correct, how much was the car worth 3 years ago?

b. If the car is 7 years old, what was the original price of the car?

▶ **Solution**

a. You can solve this problem by considering $5,600 to be the starting value and then looking back 3 years.

      $y = A(1 - r)^x$ The general form of the equation.

      $y = 5{,}600(1 - 0.15)^{-3}$ Substitute the given information in the equation. -3 means you look back 3 years.

      $y \approx 9{,}118.66$

The value of the car 3 years ago was approximately $9,118.66.

▶ **EXAMPLE A**

This example is useful for students who didn't understand the rule derived in the investigation, especially the case of a negative exponent in the denominator.

▶ **EXAMPLE B**

This example shows a new way to solve a type of problem from the last lesson, that is, using negative exponents.

You may want to show that the solution to the equation $5600 = A(1 - 0.15)^3$ is equivalent to $A = \dfrac{5600}{(1 - 0.15)^3} = 5600(1 - 0.15)^{-3}$.

When time (in the exponent) is negative, the reference is to the time before some fixed point in time, usually the present. You might extend the example by asking students to use negative exponents to rework some problems from Lesson 7.5.

From this point on in the chapter, the term *small number* refers to numbers close to 0, primarily small positive numbers.

[Language] The term *pi meson* is pronounced pie′ may′ zon.

The gravitational constant is popularly known as *G*. Newton's law of universal gravitation states that every particle in the universe attracts every other particle with a force directly proportional to the product of the masses of the two particles and inversely proportional to the square of the distance between their centers of mass. In symbols, $F = \dfrac{Gm_1m_2}{s^2}$, where *F* is the force of attraction in newtons, m_1 and m_2 are the masses in kg, and *s* is the distance in meters (m). For units to cancel in dimensional analysis, *G* is in $\dfrac{\text{newton-m}^2}{\text{kg}^2}$.

Because the student text refers to moving decimal places, if you haven't done so already you may want to review the effects on the decimal point of multiplying and dividing by 10. Multiplying by 10 moves the decimal to the right one place, and dividing by 10 moves the decimal to the left one place. Some students remember this with a number line—negatives to the left, positives to the right.

Exponents can be decimals or other fractions as well as integers. **[Ask]** "What have you seen already in this chapter that indicates the existence of fractional and decimal exponents?" [The graphs of equations of the form $y = A(1 + r)^x$ are smooth curves, indicating that *x* can have noninteger values.]

Closing the Lesson

As needed, say that negative exponents are used in scientific notation for numbers very close to 0 and provide an alternative way of solving problems involving division of powers.

---

b. The original price is the value of the car 7 years ago.

$$y = 5{,}600(1 - 0.15)^{-7}$$

$$y \approx 17{,}468.50$$

The original price was approximately $17,468.50.

You can also use negative exponents to write numbers close to 0 in scientific notation. Just as positive powers of 10 help you rewrite numbers with lots of zeros, negative powers of 10 help you rewrite numbers with lots of zeros between the decimal point and a nonzero digit.

EXAMPLE C

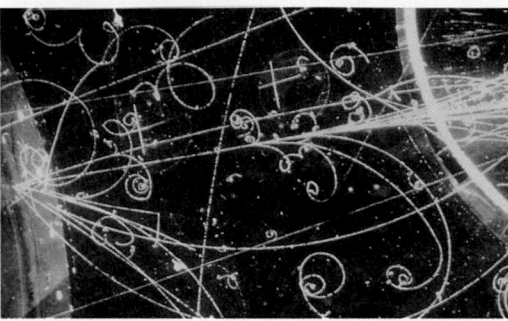

These particle tracks show the paths of particles like protons, electrons, and mesons during a nuclear reaction.

Convert each number to standard notation from scientific notation, or vice versa.

a. A pi meson, an unstable particle released in a nuclear reaction, "lives" only 0.000000026 second.

b. The number 6.67×10^{-11} is the gravitational constant in the metric system used to calculate the gravitational attraction between two objects that have given masses and are a given distance apart.

c. The mass of an electron is 9.1×10^{-31} kilogram.

► **Solution**

a. $0.000000026 = \dfrac{2.6}{100{,}000{,}000} = \dfrac{2.6}{10^8} = 2.6 \times 10^{-8}$

Notice that the decimal point in the original number was moved to the right eight places to get a number between 1 and 10, in this case, 2.6. To undo that, you must multiply 2.6 by 10^{-8}.

b. $6.67 \times 10^{-11} = \dfrac{6.67}{10^{11}} = \dfrac{6.67}{100{,}000{,}000{,}000} = 0.0000000000667$

Multiplying 6.67 by 10^{-11} moves the decimal point 11 places to the left, requiring 10 zeros after the decimal point—the first move of the decimal point changes 6.67 to 0.667.

c. Generalize the method in part b. To write 9.1×10^{-31} in standard notation you move the decimal point 31 places to the left, requiring 30 zeros after the decimal point.

$$9.1 \times 10^{-31} = 0.00000000000000000000000000000091$$

EXERCISES

You will need your calculator for problems **4, 7, 8, 12,** and **15.**

▶ Practice Your Skills

1. Rewrite each expression using only positive exponents.
   a. 2^{-3} $\frac{1}{2^3}$
   b. 5^{-2} $\frac{1}{5^2}$
   c. 1.35×10^{-4} $\frac{1.35}{10^4}$

2. Insert the appropriate symbol ($<$, $=$, or $>$) between each pair of numbers.
   a. 6.35×10^5 $\boxed{=}$ 63.5×10^4
   b. -5.24×10^{-7} $\boxed{<}$ -5.2×10^{-7}
   c. 2.674×10^{-5} $\boxed{>}$ 2.674×10^{-6}
   d. -2.7×10^{-4} $\boxed{>}$ -2.8×10^{-3}

3. Find the exponent of 10 that you need for scientific notation.
   a. $0.0000412 = 4.12 \times 10^{\square}$ -5 **b.** $46 \times 10^{-5} = 4.6 \times 10^{\square}$ -4 **c.** $0.00046 = 4.6 \times 10^{\square}$ -4

4. The population of a town is currently 45,647. It has been growing at a rate of about 2.8% per year.
   a. Write an expression of the form $45{,}647(1 + 0.028)^x$ for the current population. $45{,}647(1 + 0.028)^0$
   b. What does the expression $45{,}647(1 + 0.028)^{-12}$ represent in this situation? the population 12 years ago
   c. Write and evaluate an expression for the population 8 years ago. $45{,}647(1 + 0.028)^{-8} \approx 36{,}599$
   d. Write expressions without negative exponents that are equivalent to the exponential expressions from 4b and c. $\frac{45{,}647}{(1 + 0.028)^{12}}; \frac{45{,}647}{(1 + 0.028)^{8}}$

5. Juan says that 6^{-3} is the same as -6^3. Write an explanation of how Juan should interpret 6^{-3}, then show him how each expression results in a different value.
   Possible answer: Negative exponents mean to use a reciprocal with the exponent positive. $6^{-3} = \frac{1}{6^3} = \frac{1}{216}$, $-6^3 = -216$

▶ Reason and Apply

6. Use the properties of exponents to rewrite each expression without negative exponents.
   a. $(2x^3)^2(3x^4)$ $12x^{10}$
   b. $(5x^4)^2(2x^2)$ $50x^{10}$
   c. $3(2x)^3(3x)^{-2}$ $\frac{8x}{3}$
   d. $\left(\frac{2x^4}{3x}\right)^3$ $\frac{8x^9}{27}$

7. **APPLICATION** Suppose the annual rate of inflation is about 4%. This means that the cost of an item increases by about 4% each year. Write and evaluate an exponential expression to find the answers to these questions.
   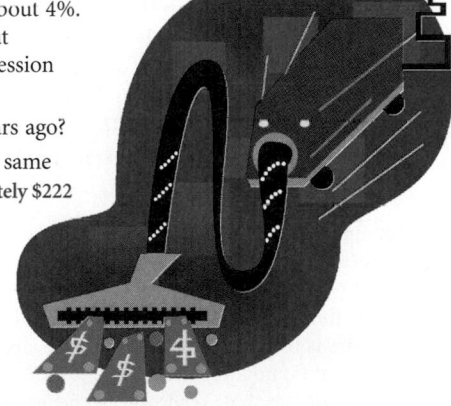
   a. If a piano costs $3,500 today, what did it cost 4 years ago?
   b. If a vacuum cleaner costs $250 today, what did the same model cost 3 years ago? $250(1 + 0.04)^{-3}$; approximately $222
   c. If tickets to a college basketball game cost $25 today, what did they cost 5 years ago?
   d. The median price of a house in the Midwest United States in June 2000 was $126,800. What was the median price 30 years ago?
   (*National Association of Realtors*)
   $126{,}800(1 + 0.04)^{-30}$; approximately $39,095

BUILDING UNDERSTANDING

Students practice working with nonpositive exponents.

ASSIGNING HOMEWORK

Essential	1, 4, 6, 7, 9
Performance assessment	8, 13
Portfolio	12
Journal	5
Group	9, 10–12
Review	14, 15

▶ Helping with the Exercises

Exercise 2 [Alert] This exercise tests students' number sense because they have to identify the order relationship between pairs of numbers. If students have difficulties with parts b and d, suggest that they treat the numbers as positive first and then multiply through by -1, changing the direction of the inequality.

Exercise 3 [Alert] Students need to be very careful: As they decrease 46 to 4.6, they add 1 to the negative exponent.

Exercise 4 Even students who would immediately recognize that $(1 + 0.028)^0 = 1$ might be confused by needing to go backward.

Exercise 7 This exercise is about inflation, so the constant multiplier is more than 1, but the situation calls for thinking back in time, so the exponent is negative. **[Alert]** Students might write a quotient instead of using a negative exponent.

7a. $3500(1 + 0.04)^{-4}$; approximately $2,992

7c. $25(1 + 0.04)^{-5}$; approximately $21

8a. approximately 1.2×10^{-3} square mile per person

8. **APPLICATION** The population of Japan in 2000 was about 1.3×10^8. Japan has a land area of about 1.5×10^5 square miles. (*2001 World Almanac*, p. 803)

    a. On average, how much land in square miles is there per person? (Note: This is a different problem from the one you may have solved in Lesson 7.5.)

    b. Convert your answer from 8a to square feet per person. approximately 3.35×10^4 square feet per person (or 3.22×10^4 square feet per person if the answer from 8a is used without rounding)

9. Decide whether each statement is true or false. Use expanded form to show either that the statement is true or what the correct statement should be.

    a. $(2^3)^2 = 2^6$

    b. $(3^3)^4 = 3^7$

    c. $(10^{-2})^4 = -10^8$

    d. $(5^{-3})^{-4} = 5^{12}$

10. A large ball of string originally held 1 mile of string. Abigail cut off a piece of string one-tenth of that length. Barbara then cut a piece of string that was one-tenth as long as the piece Abigail had cut. Cruz came along and cut a piece that was one-tenth the length of what Barbara had cut.

    a. Write each length of string in miles in scientific notation.

    b. If the process continues, how long a piece will the next person, Damien, cut off? 1×10^{-4} mile

    c. Do any of the people have a piece of string too short to use as a shoelace? (Hint: Would you use inches, feet, or miles to measure a shoelace?)

10a. Original: 1×10^0; Abigail: 1×10^{-1}; Barbara: 1×10^{-2}; Cruz: 1×10^{-3}

10c. Convert to inches; Damien's string is too short (6.3 in.).

11. Suppose $36(1 + 0.5)^4$ represents the number of bacteria cells in a sample after 4 hours of growth at a rate of 50% per hour. Write an exponential expression for the number of cells 6 hours earlier.
$$36(1 + 0.5)^{(4-6)} = 36(1 + 0.5)^{-2}$$

12c. Possible answer: In the savings account, interest is added at 6 months, so the interest earns interest. The one-year interest is $(1 + 0.025)^2 = 1.050625$; that is more than 5%.

12. **APPLICATION** Camila received a $1,200 prize for one of her essays. She decides to invest $1,000 of it for college. Her bank offers two options. The first is a regular savings account that pays 2.5% interest every 6 months. The second is a certificate of deposit that pays 5% interest each year.

    a. With the savings account, how much would Camila have after 1 year? After 2 years? $1,050.63 and $1,103.81

    b. With the certificate of deposit, how much would Camila have after 1 year? After 2 years? $1,050 and $1,102.50

    c. Explain why you get different results for 12a and b.

13. Enough chocolate is produced in the world each day to make 600,000,000 chocolate bars. If you use an estimate of 6,000,000,000 people as the world population, how many chocolate bars is this per person? Write your answer in scientific notation. (*The World in One Day*, 1997, p. 20)
1×10^{-1} or $\frac{1}{10}$ of a chocolate bar per person

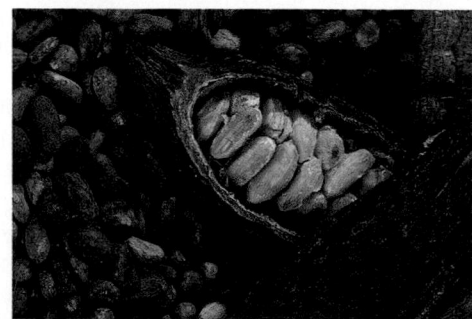
A major ingredient of chocolate is cocoa. Cocoa beans grow in large pods.

9a. true; $(2^3)^2 = 2^3 \cdot 2^3 = 2 \cdot 2 \cdot 2 \cdot 2 \cdot 2 \cdot 2 = 2^6$

9b. false; $(3^3)^4 = 3^3 \cdot 3^3 \cdot 3^3 \cdot 3^3 = 3 \cdot 3 \cdot 3 \cdot 3 \cdot 3 \cdot 3 \cdot 3 \cdot 3 \cdot 3 \cdot 3 \cdot 3 \cdot 3 = 3^{12}$

9c. false; $(10^{-2})^4 = \left(\frac{1}{10^2}\right)^4 = \left(\frac{1}{10 \cdot 10}\right)\left(\frac{1}{10 \cdot 10}\right)\left(\frac{1}{10 \cdot 10}\right)\left(\frac{1}{10 \cdot 10}\right) = \frac{1}{10^8} = 10^{-8}$

9d. true; $(5^{-3})^{-4} = \left(\frac{1}{5^3}\right)^{-4} = \frac{1}{\left(\frac{1}{5^3}\right)^4} = \frac{1}{\left(\frac{1}{5 \cdot 5 \cdot 5}\right)\left(\frac{1}{5 \cdot 5 \cdot 5}\right)\left(\frac{1}{5 \cdot 5 \cdot 5}\right)\left(\frac{1}{5 \cdot 5 \cdot 5}\right)} = \frac{1}{\frac{1}{5^{12}}} = \frac{1}{5^{-12}} = 5^{12}$

Review

14. Find the solution for each system.

    a. $\begin{cases} y = 3x - 5 \\ y = -2.5x + 9 \end{cases}$ Approximate coordinates are $(2.55, 2.64)$.

    b. $\begin{cases} y = 2(x - 4) + 15 \\ y = 15(x + 5) - 12 \end{cases}$ Approximate coordinates are $(-4.31, -1.62)$.

15. Set your calculator in scientific notation mode for this problem.

    a. Use your calculator to do each division.

       i. $\dfrac{8 \times 10^8}{2 \times 10^3}$ 4×10^5 ii. $\dfrac{9.3 \times 10^{13}}{3 \times 10^3}$ 3.1×10^{10} iii. $\dfrac{4.84 \times 10^9}{4 \times 10^4}$ 1.21×10^5 iv. $\dfrac{6.2 \times 10^4}{3.1 \times 10^8}$ 2×10^{-4}

    b. Describe how you could do the calculations in 15a without using a calculator.

    c. Find the answer to the quotient $\dfrac{4.8 \times 10^7}{8 \times 10^2}$ without using your calculator.

       0.6×10^5 or 6×10^4 in scientific notation

IMPROVING YOUR REASONING SKILLS

You have learned about scientific notation in this chapter. There is another convention for writing numbers called **engineering notation.**

Engineering notation	Not in engineering notation
2.5×10^9	2500×10^3
630×10^{-3}	630×10^{-2}
12×10^0	1.5×10^5
400×10^3	0.4×10^6
10.8×10^6	1.08×10^7

1. Write a definition for engineering notation based on the numbers in the lists. If your calculator has an engineering notation mode, you can enter more numbers to help support your definition.

2. Convert these numbers to engineering notation.
   a. 78,000,000
   b. 9,450
   c. 130,000,000,000
   d. 0.0034
   e. 0.31
   f. 1.4×10^8

3. You may have seen these symbols used as shorthand for numbers:

   n ("nano," or times $\frac{1}{1,000,000,000}$),
   μ ("micro," or times $\frac{1}{1,000,000}$),
   k ("kilo," or times 1,000),
   M ("mega," or times 1,000,000),
   G ("giga," or times 1,000,000,000).

   Explain how engineering notation is related to these symbols.

This tool, a micrometer, is used to accurately measure very small distances. Measurements made with it may be recorded in engineering notation.

Exercise 14 This exercise reviews Lesson 6.3.

Exercise 15 This exercise reviews Lesson 7.4. **[Alert]** Students may forget to use parentheses when entering the denominators into their calculators. For example, in 15a, they may enter 8 * 10^8/2 * 10^3 instead of 8*10^8/(2*10^3). Remind them that the numerator is divided by each factor in the denominator. In fact, they can avoid parentheses by entering 8*10^8/2/10^3. Be sure they explain their reasoning in 15c.

15b. Possible answer: Divide the coefficients of the powers of 10, and then divide the powers of 10 (subtract the exponents).

IMPROVING REASONING SKILLS

1. If students are having difficulties, suggest they get a hint from part 3. A possible definition of engineering notation that models the definition for scientific notation: $a \times 10^n$, where $1 \le a < 1000$ or $-1000 < a \le -1$ and n is a multiple of 3. As needed, point out that 0 is a multiple of 3, being 3×0.

2. a. 78×10^6 d. 3.4×10^{-3}
   b. 9.45×10^3 e. 310×10^{-3}
   c. 130×10^9 f. 140×10^6

   Answers that append 0's to the first parts are acceptable. (For example, the first may be 78.0×10^6 or 78.00×10^6.)

3. Engineering notation is related to the groups of three digits separated by commas in decimal numbers. That is, $n = 10^{-9}$, $\mu = 10^{-6}$, $k = 10^3$, $M = 10^6$, $G = 10^9$.

   Students may be interested in expanding this to a project by researching careers and applications that use engineering notation.

LESSON

7.7

Fitting Exponential Models to Data

In broken mathematics
We estimate our prize
Vast—in its fading ratio
To our penurious eyes!

EMILY DICKINSON

Victoria Julian has been collecting data on changes in median house prices in her area over the past 10 years. She plans to buy a house 5 years from now and wants to know how much money she needs to save each month toward the down payment. How can she make an intelligent prediction of what a house might cost in the future? What assumptions will she have to make?

Charming Mock Tudor on 1/2 acre

6 bedrms, 3 baths, stone fireplace, marquetry floors throughout, 2-car garage, huge landscaped lot. Best offer.

Sale pending

Lovely Westside Bungalow
Newly remodeled, 2 bedroom, 1.5 bath, off-street parking, laundry, porch, hdwd floors and fireplace, charming back patio

Affordable Brownstone
• 3 floors
• 3 bed/2 bath
• spacious
• charming details
• street parking
• needs only minor repairs
• walk to shops and park
• accepting offers starting Wed

In the real world, situations like population growth, price inflation, and the decay of substances often tend to approximate an exponential pattern over time. With an appropriate exponential model, you can sometimes predict what might happen in the future.

In Chapter 5, you learned about fitting linear models to data. In this lesson you'll learn how to find an exponential model to fit data.

Investigation
Radioactive Decay

You will need

- a paper plate
- a protractor
- a supply of small counters

The particles that make up an atom of some elements, like uranium, are unstable. Over a period of time specific to the element, the particles will change so that the atom eventually will become a different element. This process is called **radioactive decay.**

In this investigation, your counters represent atoms of a radioactive substance. Draw an angle from the center of your plate, as illustrated. Counters that fall inside the angle represent atoms that have decayed.

PLANNING

LESSON OUTLINE

First day:

| 5 min | Introduction |
| 45 min | Investigation |

Second day:

15 min	Sharing
10 min	Example
5 min	Closing
15 min	Exercises

MATERIALS

- paper plates
- protractors
- small flat counters that don't roll, such as lentils, split peas, popcorn kernels, or flat candies (100 per group)
- Calculator Note 2A
- 100 Grid (W or T), *optional*
- Radioactive Decay Sample Data (W), *optional*
- Moore's Law Sample Data (W), *optional*

TEACHING

Exponential models can help make predictions in many real-world situations.

INTRODUCTION

To make an intelligent prediction, Victoria must assume that prices will continue to increase at the same rate.

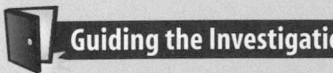

Guiding the Investigation

One step Have students generate the data as in the investigation. Then ask them to find an equation that models the data. As you

(Continued top of page 407)

LESSON OBJECTIVE

- Write exponential equations that model real-world growth and decay data

NCTM STANDARDS

CONTENT		PROCESS	
•	Number	•	Problem Solving
•	Algebra	•	Reasoning
•	Geometry		Communication
•	Measurement	•	Connections
•	Data/Probability	•	Representation

Step 1	Count the number of counters. Record this in a table as the number of "atoms" after 0 years of decay. Pick up all of the counters.
Step 2	Drop the counters on the plate. Count and remove the counters that fall inside the angle—these atoms have decayed. Subtract from the previous value and record the number remaining after 1 year of decay. Pick up the remaining counters.
Step 3	Repeat Step 2 until you have fewer than ten atoms that have not decayed. Each drop will represent another year of decay. Record the number of atoms remaining each time.

Procedure Note

Create a procedure for dropping counters randomly and evenly on the plate. Practice your method until you think you have a good technique. Make a plan for handling counters that fall on the lines of your angle and those that miss the plate—they need to be accounted for too.

circulate, encourage groups to find ratios and come up with an exponential equation. Also ask how their angles relate to those ratios.

To avoid using materials in Steps 1 through 3, have each group start with a 100 grid as their starting collection of atoms. Then they use a calculator to generate 15 random numbers to represent decayed atoms (see Calculator Note 2A). Duplicates may occur. They cross those numbers off the grid and record how many numbers are not crossed off. They generate another 15 random numbers, cross off new ones, and again count the remaining numbers. They repeat until fewer than ten numbers remain. If you use this method, skip Step 11.

With the plates, students should use protractors to make an angle that's less than 90° but not too small. They can approximate the center by using a ruler to find the midpoints of several diameters.

Step 2 To have the counters fall randomly, students should avoid aiming.

An acceptable plan would be to count counters on the line as being within the angle, and count those that fall outside the plate proportionately for inside or outside the angle.

Students may get a kinesthetic feeling for decay by eating candy counters.

Step 3 Rather than counting the remaining counters each time, students may find it easier to count the number decayed and subtract from the previous amount to find the number remaining.

Step 5 Suggest that students add a column of data for recording the ratios.

Step 5 See table of Step 1 for sample data. The ratios should be approximately the same.

Step 6 Answers will vary. Students could give reasons for selecting the mean, median, or another value. In this sample data, the mean is 0.802.

Step 7 For this sample data, using 0.802, the rate of decay is 19.8% per year.

Step 8 For this sample, $y = 201(1 - 0.198)^x$.

Step 4	Let x represent elapsed time in years, and let y represent the number of atoms remaining. Make a scatter plot of the data. What do you notice about the graph?
Step 5	Calculate the ratios of atoms remaining between successive years. That is, divide the number of atoms after 1 year by the number of atoms after 0 years; then divide the number of atoms after 2 years by the number of atoms after 1 year; and so on. How do the ratios compare?
Step 6	Choose one representative ratio. Explain how and why you made your choice.
Step 7	At what rate did your atoms decay?
Step 8	Write an exponential equation that models the relationship between time elapsed and the number of atoms remaining.
Step 9	Graph the equation with the scatter plot. How well does it fit the data?
Step 10	If the equation does not fit well, which values could you try to adjust to give a better fit? Record your final equation when you are satisfied.

Step 11 The ratio of the angle measure to 360° should be approximately the same as r. In the sample data, $\frac{68}{360} \approx 0.19$, which is close to the r-value used in Step 8.

Step 11	Measure the angle on your plate. Describe a connection between your angle and the numbers in your equation.
Step 12	Create a table of x- and y-values on your calculator. Is the starting value at year 0 the same as the data you collected? Are any of the subsequent values the same as your data? Explain why you could find differences.

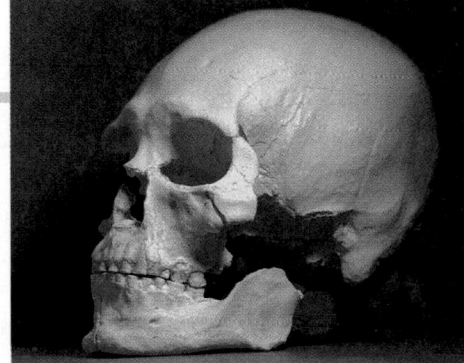

Archaeologists can approximate the age of artifacts with *carbon dating*. This process uses the rate of radioactive decay of carbon-14. Carbon is found in all living things, so the amount left in a bone, for example, is an indicator of the bone's age. This is a plastic casting of a skull found in 1997 in Richland, Washington. Carbon dating has dated the skull as 9200 years old.

Step 7 You may need to encourage students to look at their constant multiplier in the form of $(1 - r)$. That is, a ratio of $\frac{3}{4}$ would be $(1 - 0.25)$. The decay rate r, like a growth rate, is technically not a rate because it has no units.

Step 11 [Ask] "What is the ratio of your angle to the whole plate?"

Step 12 You may want to discuss theoretical value versus observed data. The starting value of the best fit might not actually be the starting value of the data.

See pages 718–719 for answers to Steps 1, 2, 3, 4, 9, 10, and 12.

Have several groups present their data, scatter plots, equations, and graphs for class critique and suggestions. In the equation $y = A(1 - r)^x$, the number A is the y-intercept of the graph, so students might adjust A to shift the graph vertically. For graphs of exponential equations, vertical shifts can also be thought of as horizontal shifts. To stretch the graph, students will tinker with the decay rate r.

Point out the opening quotation. It is the second stanza of an untitled poem from about 1859. The first stanza is

As by the dead we love to sit
Become so wondrous dear
As for the lost we grapple
Tho' all the rest are here.

[Language] *Penurious* means extremely stingy or poor. One interpretation is that the relationship between what we want and what we see that we currently have is an inverse relationship—the more we want, the less we see value in what we have. Another interpretation is that what we want decays exponentially with respect to how well we see what we have.

Ask if anyone can explain radioactive decay and its uses. Generally, a neutron of an atom is stable only when paired with a proton. Elements with larger nuclei (larger than that of bismuth) have too many neutrons and not enough protons for such pairing, so the neutrons gradually turn into protons and electrons. This transformation is the basis of radioactive decay.

Because carbon-14 atoms occur in all living things, knowing about their decay is extremely useful in determining the age of very old remains of life. It takes about 5730 years for half of their excess neutrons to decay. Therefore, if one bone has half the amount of carbon-14 as a similar bone, the first bone is 5730 years older.

The steps of finding an equation in this investigation provide a good method for finding an exponential equation that models data that displays an exponential pattern, either increasing or decreasing. These situations are often generated recursively by multiplying by a constant ratio. Thinking of the constant multiplier in the form $1 + r$ or $1 - r$ leads to these familiar equations:

$$y = A(1 + r)^x$$
$$y = A(1 - r)^x$$

You can then fine-tune the fit of your model by slightly adjusting the values of A and r.

EXAMPLE

Every musical note has an associated frequency measured in hertz (Hz), or vibrations per second. The table shows the approximate frequencies of the notes in the octave from middle C up to the next C on a piano. (In this scale, E# is the same as F and B# is the same as C.)

Piano Notes

Note name	Note above middle C	Frequency (Hz)
Middle C	0	262
C#	1	277
D	2	294
D#	3	311
E	4	330
F	5	349
F#	6	370
G	7	392
G#	8	415
A	9	440
A#	10	466
B	11	494
C above middle C	12	523

The arrangement of strings in a piano shows an exponential-like curve.

a. Find a model that fits the data.

b. Use the model to find the frequency of the note two octaves above middle C (note 24).

c. Find the note with a frequency of 600 Hz.

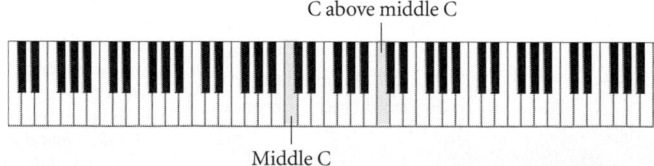
C above middle C

Middle C

[Ask] "Would there ever be 0 atoms remaining?" [In the exponential equation, the value of y will never decrease all the way to 0. The counters and even protons are discrete units, however, so they will all decay eventually.]

You may want to have the students generalize the steps they used in getting an equation:

1. Find the ratios between successive y-values.

2. Select a representative ratio, possibly a mean or median.

3. Think of ratio in the form of $(1 - r)$ if the values are decreasing and $(1 + r)$ if they're increasing.

4. Use the starting value to then write an equation in the form $y = A(1 - r)^x$ or $y = A(1 + r)^x$.

5. Adjust the values of A and r to make a better fit.

▶ **Solution**

a. Let x represent the note number above middle C, and let y represent the frequency. A scatter plot shows the exponential-like pattern. To find the exponential model, first calculate the ratios between successive data points. The mean of the ratios is 1.0593. So the frequency of the notes increases by about 5.93% each time you move up one note on the keyboard. The starting frequency is 262 Hz. So an equation is

$$y = 262(1 + 0.0593)^x$$

The graph shows a very good fit.

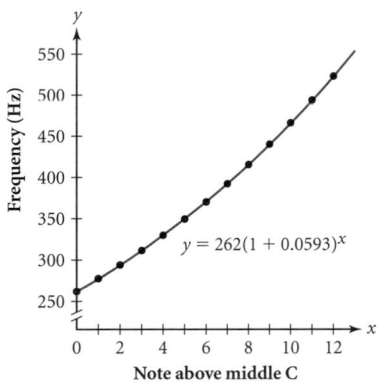

b. To find the frequency of the C two octaves above middle C (note 24), substitute 24 for x in the model.

$$y = 262(1 + 0.0593)^{24} \approx 1044$$

By this model, the frequency of note 24 is 1044 Hz.

c. To find the note with a frequency of 600 Hz, substitute 600 for y in the model.

$$600 = 262(1 + 0.0593)^x$$

Enter $262(1 + 0.0593)^x$ into Y₁ and 600 into Y₂ on your calculator. Graph both equations and trace to approximate the intersection point. Or you could look at a table to see where Y₁ = Y₂.

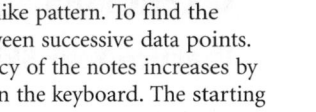

Both the graph and the table show an x-value between 14 and 15. The 14th note above middle C is a D and the 15th note is a D#. Since the piano notes correspond only to whole numbers, you cannot make a note with a frequency of 600 Hz on this piano.

If you wanted to find the frequency of notes below middle C, you would need to use negative values for x. The frequencies found using this equation will be fairly accurate because the data fits the equation so well. If the piano were very out of tune, the equation probably would not fit so nicely, and the model might be less valuable for predicting.

Music

CONNECTION

Before the 17th century, there were many ways to tune an instrument. The most popular, developed by the ancient Greek philosopher Pythagoras, used different tuning ratios between each pair of adjacent notes. This made some scales, like the scale of C, sound good but others, like the scale of A-flat, sound bad. Modern Western tuning now uses *even temperament*, based on an equal tuning ratio between adjacent notes, which leads to an exponential model.

Assessing Progress

Watch for skills at handling geometric tools (ruler, protractor), collecting data, making a **scatter plot,** and **graphing exponential equations.**

▶ **EXAMPLE**

This example shows exponential growth in a different context. You might ask students with a musical background to elaborate on the notation. The symbol # in the table means *sharp* or $\frac{1}{2}$ step above the note; C# is $\frac{1}{2}$ note above C. *Functional Melodies* (Key Curriculum Press) has additional activities involving the Pythagorean and even-tempered scales.

The solution in part c brings in a new way of solving an equation: graphing the two sides of the equation and finding the intersection point of the graphs.

Closing the Lesson

At the heart of finding the equation modeling exponential growth or decay is finding a ratio to use as a constant multiplier.

ASSIGNING HOMEWORK

Essential	1–3, 5, 6, 10
Performance assessment	9, 11
Portfolio	5–8
Journal	12
Group	5, 6
Review	4, 13

▶ **Helping with the Exercises**

Exercise 1e Some students may not realize that a number that's more than 1 is more than 100%.

Exercise 2d This question brings in the idea of limit, as seen in Chapter 0. **[Ask]** "Why do powers of positive numbers less than 1 get smaller and smaller?" [Multiplying a positive number by a positive number less than 1 yields a smaller positive number.]

5a. The ratios are 0.957, 0.956, 0.965, 0.964, 0.963, 0.961, 0.959, 0.958, 0.971, and 0.955.

5e.
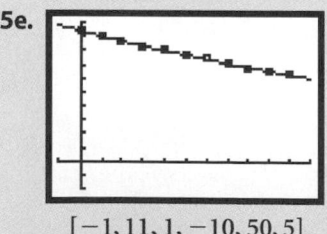
$[-1, 11, 1, -10, 50, 5]$

Adjustments to A or r are not necessary—the fit is good as is.

Exercise 5f Ask if the temperature will ever drop to freezing (0°C). Theoretically, an exponential expression will never actually get to 0, though in real life water freezes.

5f. 55 minutes
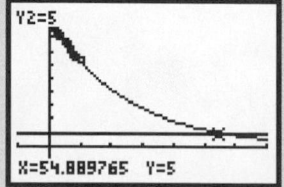
$[-10, 70, 10, -10, 50, 5]$

EXERCISES

You will need your calculator for problems **2, 5, 6, 7, 8, 10, 11,** and **12.**

▶ **Practice Your Skills**

1. Rewrite each value as either $1 + r$ or $1 - r$. Then state the rate of increase or decrease as a percent.

   a. 1.15 $1 + 0.15$; rate of increase: 15%   **b.** 1.08 $1 + 0.08$; rate of increase: 8%   **c.** 0.76
   $1 - 0.24$; rate of decrease: 24%
   d. 0.998   **e.** 2.5
   $1 - 0.002$; rate of decrease: 0.2% $1 + 1.5$; rate of increase: 150%

2. Use the equation $y = 47(1 - 0.12)^x$ to answer each question.

   a. Does this equation model an increasing or decreasing pattern? Decreasing

   b. What is the rate of increase or decrease? 12%

   c. What is the y-value when x is 13? $y \approx 8.92$

   d. What happens to the y-values as the x-values get very large? The y-values approach 0.

3. Write an equation to model the growth of an initial deposit of \$250 in a savings account that pays 4.25% annual interest. Let B represent the balance in the account, and let t represent the number of years the money has been in the account. $B = 250(1 + 0.0425)^t$

4. Use the properties of exponents to rewrite each expression with only positive exponents.

   a. $4x^3 \cdot (3x^5)^3$ $108x^{18}$ **b.** $\dfrac{60x^8y^4}{15x^3y}$ $4x^5y^3$ **c.** $3^2 \cdot 2^3$ 72   **d.** $\dfrac{(8x^3)^2}{(4x^2)^3}$ 1

   e. $x^{-3}y^4$ $\dfrac{y^4}{x^3}$ **f.** $(2x)^{-3}$ $\dfrac{1}{8x^3}$ **g.** $2x^{-3}$ $\dfrac{2}{x^3}$ **h.** $\dfrac{2x^{-4}}{(3y^2)^{-3}}$ $\dfrac{54y^6}{x^4}$

▶ **Reason and Apply**

5. Mya placed a cup of hot water in a freezer. Then she recorded the temperature of the water each minute.

   Water Temperature

Time (min) x	0	1	2	3	4	5	6	7	8	9	10
Temperature (°C) y	47	45	43	41.5	40	38.5	37	35.5	34	33	31.5

   a. Find the ratios between successive temperatures.

   b. Find the mean of the ratios in 5a. $0.9609 \approx 0.96$

   c. Write the ratio from 5b in the form $1 - r$. $1 - 0.04$

   d. Use your answer from 5c and the starting temperature to write an equation in the form $y = A(1 - r)^x$. $y = 47(1 - 0.04)^x$

   e. Graph your equation with a scatter plot of the data. Adjust the values of A or r until you get a satisfactory fit.

   f. Use your equation to predict how long it will take for the water temperature to drop below 5°C.

6. In science class Phylis used a light sensor to measure the intensity of light (in lumens per square meter, or lux) that passes through layers of colored plastic. The table below shows her readings.

Light Experiment

Number of layers	0	1	2	3	4	5	6
Intensity of light (lux)	431	316	233	174	128	98	73

a. Write an exponential equation to model Phylis's data. Let x represent the number of layers, and let y represent the intensity of light in lux.

b. What does your r-value represent?

c. If Phylis's sensor cannot register readings below 30 lux, how many layers can she add before the sensor stops registering?

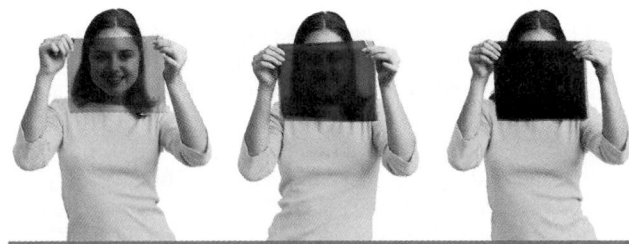

7. **APPLICATION** Recall Victoria from the opening of this lesson. She has collected this table of data on median house prices for her area.

a. Define variables and find an exponential equation to model Victoria's data.

b. Victoria plans to buy a house five years from now. What median price should she expect then?

c. Victoria plans to make a down payment of 10% of the purchase price. Based on your answer to 7b, how much money will she need for her down payment?

d. If Victoria saves the same amount each year for the next 5 years (without interest), how much will she need to save each month for her down payment?

Median House Prices

Year	Years since 1990	Median price ($)
1990	0	85,000
1991	1	95,000
1992	2	95,400
1993	3	101,250
1994	4	107,000
1995	5	114,000
1996	6	120,250
1997	7	127,580
1998	8	135,500
1999	9	144,000

8. The equation $y = 262(1 + 0.0593)^x$ models the frequency in hertz of various notes on the piano, with middle C considered as note 0. The average human ear can detect frequencies between 20 and 20,000 hertz. If a piano keyboard were extended, the highest and lowest notes audible to the average human ear would be how far above and below middle C?

9. Very small amounts of time much less than a second have special names. Some of these names may be familiar to you, such as a millisecond, or 0.001 second. Have you heard of a nanosecond or a microsecond? A nanosecond is 1×10^{-9} second, and a microsecond is 1×10^{-6} second. How many nanoseconds are in a microsecond?
1000 nanoseconds per microsecond

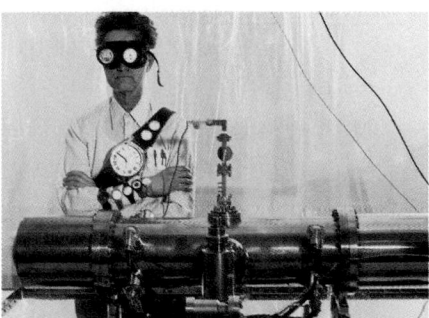

This is Jim Gray, keeper of the NBS-4 atomic clock. Much more accurate than mechanical clocks or watches, some atomic clocks may gain or lose less than a microsecond each year.

6a. Possible answer: $y = 431(1 - 0.26)^x$, where 0.26 is derived from the mean ratio of about 0.74.

6b. With each layer of plastic, the amount of light is reduced 26%.

6c. With 9 layers, the reading would be below 30.

Exercise 7 This data set is artificial. You might encourage students to research similar data for their own town or make their own projections.

7a. Possible answer: Let x represent years since 1990, and let y represent median price in dollars. An equation is $y = 85,000(1 + 0.06)^x$, where 0.06 is derived from the mean ratio of about 1.06.

7b. Answers will vary depending on the current year; let $x = (current\ year + 5 - 1990)$.

7c. Answers will vary depending on 7b; the answer should be the result of 7b multiplied by 0.10.

7d. Answers will vary depending on 7c; the answer should be the result of 7c divided by 60.

Exercise 8 A standard keyboard has 88 keys. **[Language]** *Audible* means capable of being heard.

8. Note 75 above middle C would be the highest audible note; note -44 (44 notes below middle C) would be the lowest audible note.

Exercise 9 Students may recall *nano* and *micro* from the Lesson 7.6 Improving Your Reasoning Skills problem about engineering notation. As needed, help with subtracting a negative number.

10. Suppose that on Sunday you see 32 mosquitoes in your room. On Monday you count 48 mosquitoes. On Tuesday there are 72 mosquitoes. Assume that the population will continue to grow exponentially.

    a. What is the percent rate of growth? 50%

    b. Write an equation that models the number of mosquitoes, y, after x days. $y = 32(1 + 0.5)^x$

    c. Graph your equation and use it to find the number of mosquitoes after 5 days. After 2 weeks. After 4 weeks. 243 mosquitoes; 9,342 mosquitoes; 2,727,126 mosquitoes

    d. Name at least one real-life factor that would cause the population of mosquitoes not to grow exponentially. Answers will vary. Possibilities include lack of resources and overcrowding.

Exercise 11 "Magic pot" stories have developed in the folklore of several cultures, including India, Italy, and China. Titles include *The Magic Porridge Pot, The Magic Pasta Pot,* and simply *The Magic Pot. The Sorcerer's Apprentice* has a similar theme.

11b. $y = 2(1 + 0.5)^{0.5} \approx 2.45$; approximately 2.45 liters

11c. approximately 115 liters

11. Many stories in children's literature involve magic pots. An Italian variation goes something like this: A woman puts a pot of water on the stove to boil. She says some special words, and the pot is filled with pasta. Then she says another set of special words, and the pot stops filling up.

    Suppose someone overhears the first words, takes the pot, and starts it in its pasta-creating mode. Two liters of pasta are created. Then the pot continues to create more pasta because the impostor doesn't know the second set of words. The volume continues to increase 50% per minute.

    a. Write an equation that models the amount of pasta in liters, y, after x minutes. $y = 2(1 + 0.5)^x$

    b. How much pasta will there be after 30 seconds?

    c. How much pasta will there be after 10 minutes?

    d. How long, to the nearest second, will it be until the entire house, which can hold 450,000 liters, is full of pasta? After about 30.4, or 30 minutes 24 seconds

12. In this problem you will explore the equation $y = 10(1 - 0.25)^x$.

    a. Find y for some large positive values of x, such as 100, 500, and 1000. What happens to y as x gets larger and larger? y gets closer and closer to 0.

    b. The calculator will say y is 0 when x equals 10,000. Is this correct? Explain why or why not. No, because y can never equal 0. The number is just smaller than the calculator is able to represent.

    c. Find y for some large negative values of x, such as -100, -500, and -1000. What happens to y as x moves farther and farther from 0 in the negative direction? y approaches infinity.

▶ Review

13. These polygons are similar. Find the lengths of the three unknown sides. $x = 3$ cm; $y = 7.2$ cm; $z = 9$ cm

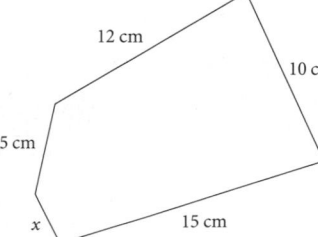
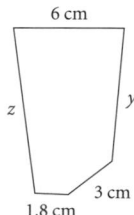

14. One of the most famous formulas in science is

$$E = mc^2$$

The formula describes the relationship between mass (*m*, measured in kilograms) and energy (*E*, measured in joules) and shows how they can be converted from one to the other. The variable *c* is the speed of light, 3×10^8 meters per second. How much energy could be created from a 5-kilogram bowling ball? Express your answer in scientific notation.

$5(3 \times 10^8)^2 = 5(9 \times 10^{16}) = 45 \times 10^{16}$ or 4.5×10^{17} joules

James Joule (1818–1889) was one of the first scientists to study how energy was related to heat. At the time of his experiments, many scientists thought heat was a gas that seeped in and out of objects. The SI (metric) unit of energy was named in his honor.

Exercise 14 A *joule* is the SI unit for the amount of work done by a force of one newton acting over a distance of one meter. It is equivalent to 1 watt second, so a 100-watt bulb burning for an hour produces 360,000 joules of energy.

project

MOORE'S LAW

In 1965 Gordon Moore, the co-founder of Intel Corporation, observed that the number of transistors on a computer chip doubled approximately every 2 years. Since a computer processor's speed and power are proportional to the number of transistors on it, computers should get twice as powerful every 2 years.

Has "Moore's Law" come true since 1965? Research technical specifications for various computer processors and find an exponential model that relates time and number of transistors. You can research data in magazines or at **www.keymath.com/DA**. How many years or months has it taken for computers to double in power? At what rate has the power of computers increased each year?

Your project should include

▶ A scatter plot of your data.

▶ An exponential equation that models the data and an explanation of each number and variable in your equation.

▶ A report summarizing your findings.

You may want to research news items that give recent projections and see if computer chip manufacturers are continuing to meet or exceed Moore's Law. You may also want to research other theories on computer production that examine variables such as purchase price or equipment required for production.

Fathom

With Fathom you can easily graph an exponential equation through data points. You can use a slider to make small adjustments in your equation until it fits. You can graph multiple models each with its own slider to compare different exponential equations.

Other Laws
Machrone's Law states that the machine you want always costs $5,000 (for example, the purchase price of transistors is cut in half every 2 years, so as the number of transistors doubles, the price stays the same). Rock's Law states that the cost of equipment to build semiconductors doubles every 4 years.

Supporting the project

MOTIVATION

This project challenges by requiring evaluation of data whose *x*-values are not likely to be sequential. (Students may use Moore's Law Sample Data.)

OUTCOMES

▶ The report gives evidence that the relationship between time and number of transistors is roughly exponential as Moore's Law predicts.

▶ Analysis (for the sample data) shows that a good model is $y = 2300(1 + 0.38)^x$, where *x* is time in years since 1971 and *y* is number of transistors.

▶ The report shows that, by the same model, the number of transistors doubles in a little over 2 years.

• The presentation may show an alternative model: $y = 2300(2)^{\frac{x}{2.15}}$. Here, the base of 2 represents doubling, and dividing *x* by 2.15 means that the number of transistors will double in 2.15 years.

PLANNING

LESSON OUTLINE

One day:

5 min	Introduction
35 min	Activity
10 min	Sharing

MATERIALS

- balls
- metersticks
- motion sensor, *optional*
- soda cans half-filled with water
- string
- Bounce Sample Data (W), *optional*
- Pendulum Sample Data (W), *optional*
- Calculator Note 7D

TEACHING

Exponential decay equations can model processes that slow down.

Guiding the Activity

It is not mandatory to have all of these materials. The motion sensor can be supplanted with careful low-tech data collection. The bouncing ball is needed only for Experiment 1, and the string and can are needed only for Experiment 2. In place of tying a string around the pull tab of a soda can, you might tie the string around the neck of a water bottle.

If materials in general are problematic, try the 100-grid alternative explained in Lesson 7.7.

LESSON
7.8

Activity Day

Decreasing Exponential Models and Half-Life

In Lesson 7.7, you learned that data can sometimes be modeled using the exponential equation $y = A(1 - r)^x$. In this lesson you will do an experiment, write an equation that models the decreasing exponential pattern, and find the **half-life**—the amount of time needed for a substance or activity to decrease to one-half its starting value. To find the half-life, approximate the value of x that makes y equal to $\frac{1}{2} \cdot A$.

In the previous investigation, if your plate was marked with a 72° angle and you started with 200 "atoms," a model for the data could be $y = 200(1 - 0.20)^x$. This is because the ratio of the angle to the whole plate is $\frac{72}{360}$, or 0.20. To determine the half-life of your atoms, you would need to find out how many drops you would expect to do before you had 100 atoms remaining. Hence, you could solve the equation $100 = 200(1 - 0.20)^x$ for x using a graph or a calculator table. The x-value in this situation is approximately 3, which means your atoms have a half-life of about 3 years.

Technology
CONNECTION

You can see simulations of atomic half-life with a link at **www.keymath.com/DA** .

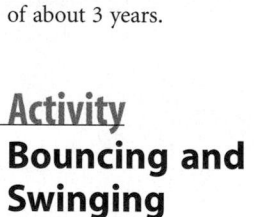

Activity
Bouncing and Swinging

You will need

- a motion sensor
- a meterstick
- a ball
- string
- a soda can half-filled with water

There are two experiments described in this activity. Each group should choose at least one, collect and analyze data, and prepare a presentation of results.

LESSON OBJECTIVE

- Write exponential equations that model real-world decay data

NCTM STANDARDS

CONTENT		PROCESS	
●	Number		Problem Solving
●	Algebra	●	Reasoning
	Geometry		Communication
●	Measurement	●	Connections
●	Data/Probability	●	Representation

Step 1 | Select one of these two experiments.

Experiment 1: Ball Bounce

You will drop a ball from a height of about 1 meter and measure its rebound height for at least 6 bounces. You can collect data "by eye" using a meterstick, or you can use a motion sensor. [▶ 🖳 See **Calculator Note 7D** for a program to use with your motion sensor. ◀] If you use a motion sensor, hold it $\frac{1}{2}$ meter above the ball and collect data for about 8 seconds; trace the resulting scatter plot of data points to find the maximum rebound heights.

Experiment 2: Pendulum Swing

Make a pendulum with a soda can half-filled with water tied to at least one meter of string—use the pull tab on the can to connect it to the string. Pull the can back about $\frac{1}{2}$ meter from its resting position and then release it. Measure how far the can swings from the resting position for several swings. You can collect data "by eye" using a meterstick (you may have to collect data for every fifth swing in this case), or you can use a motion sensor. [▶ 🖳 See **Calculator Note 7D** for a program to use with your motion sensor. ◀] If you use a motion sensor, position it 1 meter from the can along the path of the swing; the program will collect the maximum distance from the resting position for 30 swings.

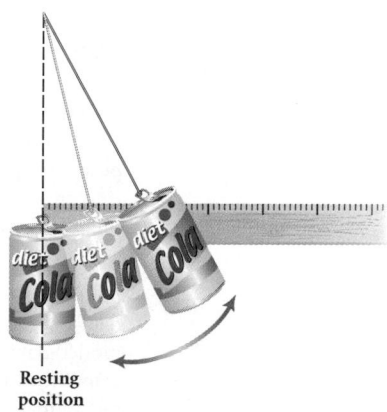

Resting position

Step 2 | Set up your experiment and collect data. Based on your results, you might want to modify your setup and repeat your data collection.

Step 3 The scatter plot should show an exponential pattern.

Step 3 | Define variables and make a scatter plot of your data on your calculator. (If you used a motion sensor, you should have this already.) Sketch the plot on paper. Does the graph show an exponential pattern?

Step 4 | Find an equation of the form $y = A(1 - r)^x$ that models your data. Graph this equation with your scatter plot and adjust the values if a better fit is needed.

Step 5 One method is to graph $Y_2 = \frac{1}{2} \cdot A$ and find the intersection with $Y_1 = A(1 - r)^x$. Using sample data half-lives are:
Exp. 1: about 2 bounces
Exp. 2: about 17 swings

Step 5 | Find the half-life of your data. Explain what the half-life means for the situation in your experiment. (Read p. 414 to review the calculation of half-life.)

Step 6 | Find the y-value after 1 half-life, 2 half-lives, and 3 half-lives. How do these values compare? With each consecutive half-life, the value of y will be $\frac{1}{2}$ the previous value of y.

Step 7 | Write a summary of your results. Include descriptions of how you found your exponential model, what the rate r means in your equation, and how you found the half-life. You might want to include ways you could improve your setup and data collection.

In the real world, eventually your ball will stop bouncing or your pendulum will stop swinging. Your exponential model, however, will never reach a y-value of zero. Remember that any mathematical model is, at best, an approximation and will therefore have limitations.

Step 3 The location of the pendulum bob is harmonic, but its maximum distance from center is roughly exponential.

Step 4 Students who collect pendulum data by eye will need to account for collecting data every fifth swing. One option is $y = A(1 + r)^{\frac{x}{5}}$, where x is the number of swings.

Step 4 For sample data:
Exp. 1: $y = 1(1 - 0.33)^x$

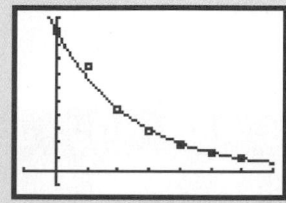

$[-1, 7, 1, -0.1, 1.1, 0.1]$

Exp. 2: $y = 0.50(1 - 0.04)^x$

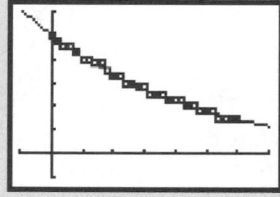

$[-5, 35, 5, -0.1, 0.6, 0.1]$

Step 6 You might want to introduce the equation $y = A\left(\frac{1}{2}\right)^{\frac{x}{t}}$ where t is the half-life. Students can see that the graph of this equation is similar to that of their equation in the form $y = A(1 - r)^x$. Ask them to think about why the graphs are the same. [Looking at special cases, when $x = t$, the quantity A is multiplied by $\frac{1}{2}$. When $x = 2t$, $\left(\frac{1}{2}\right)A$ is multiplied by $\frac{1}{2}$, to make $\frac{1}{4}A$. And so on. Symbolically, by definition of half-life, $b^t = \frac{1}{2}$, so raising both sides to the $\frac{x}{t}$ power gives $b^x = \left(\frac{1}{2}\right)^{\frac{x}{t}}$.]

ASSESSING PROGRESS

Watch for ability to collect data systematically, to define variables, to make a scatter plot, to find common ratios and write an appropriate exponential decay equation, and to find the half-life.

SHARING IDEAS

Groups can share the summaries from Step 7. The class can discuss the reasons for the differences in data and equations that model the data.

Closing the Lesson

As needed, point out that the term *exponential decay* refers to slowing processes other than radioactive decay.

PLANNING

LESSON OUTLINE

One day:

10 min Introduction

15 min Exercises

10 min Checking work

15 min Student self assessment

REVIEWING

Refer students to Exercise 13 of Lesson 7.2. **[Ask]** "What is the equation asked for in 13a?" [$y = 5000(1 + 0.05)^x$] "By this model, how much will the car be worth in 5 years?" [$6,381] "How much will it be worth 3 years after that?" [$7,387] "What about 2 years before that?" [After 6 years from the time he purchased the car, it will be worth $6,700.] Review the addition of exponents when multiplying powers of the same base and the subtraction of exponents when dividing powers of the same base. Then ask how much the car was worth 5 years ago. Model the problem using negative exponents, and discuss the fact that a model is not always accurate. **[Ask]** "What would the car be worth 5 years from now if, contrary to Sean's hopes, it depreciates at 7% per year instead of appreciating?" [$3,478]

ASSIGNING HOMEWORK

You might assign the even problems (2 to 10) for homework and allow students to work on the odd problems in class while you take time to work individually with students who have questions. Assign the mixed review before giving a test that covers Chapters 4 to 7.

7 REVIEW

You started this chapter by creating sequences that increase or decrease when you multiply each term by a constant factor. Repeated multiplication causes the rate of change between successive terms to increase or decrease. So the graphs of these sequences curve, getting steeper and steeper or less and less steep. You then discovered that **exponential equations** model these sequences, in which the constant multiplier is the **base** and the number of the term in the sequence is the exponent.

By writing exponential expressions in both **expanded form** and **exponential form**, you learned the **multiplication, division,** and **power properties of exponents,** and you explored the meanings of zero and negative exponents. You applied these properties to **scientific notation,** a way to express numbers with powers of 10.

When modeling data, you can often use an equation to make predictions. You now have two kinds of models for real-world data—a linear equation and an exponential equation. You can model many real-world quantities that increase as **exponential growth** with an equation in the form $y = A(1 + r)^x$. You can model many quantities that decrease, like **radioactive decay,** with an equation in the form $y = A(1 - r)^x$.

EXERCISES

You will need your calculator for problems **2, 3, 5, 7, 8, 10, 13, 19,** and **20.**

1. Write each number in exponential form with a base of 3.

   a. 81 3^4

   b. 27 3^3

   c. 9 3^2

   d. $\frac{1}{3}$ 3^{-1}

   e. $\frac{1}{9}$ 3^{-2}

   f. 1 3^0

2. Use the properties of exponents to rewrite each expression. Your final answer should have only positive exponents. Use calculator tables to check that your expression is equivalent to the original equation.

   a. $\frac{x \cdot x \cdot x}{x}$ x^2

   b. $2x^{-1}$ $\frac{2}{x}$

   c. $\frac{6.273x^8}{5.1x^3}$ $1.23x^5$

   d. 3^{-x} $\frac{1}{3^x}$

   e. $3x^0$ 3

   f. $x^{2.1} \cdot x^{5.6}$ $x^{7.7}$

   g. $(3^4)^x$ 3^{4x}

   h. $\frac{1}{x^{-2}}$ x^2

3. Consider this exponential equation:

   $y = 300(1 - 0.15)^x$

   a. Invent a real-world situation that you can model with this equation. Give the meaning of 300 and of 0.15 in your situation.

   b. What would the inequality $75 \leq 300(1 - 0.15)^x$ mean for your situation in 3a?

   c. Find all integer values of x such that $75 \leq 300(1 - 0.15)^x$.

▶ **Helping with the Exercises**

Exercise 2 In 2f students manipulate non-integer exponents for the first time.

Exercise 3 This exercise reviews inequality along with exponential equations.

3a. Possible answer: A $300 microwave depreciates at a rate of 15% per year.

3b. The years (x) for which the depreciating value of the microwave is at least $75

3c. Answers will vary given the context of 3a. $x \leq 8$ or $0 \leq x \leq 8$ (Some integers may be excluded by the real-life situation.)

4. Proaga says, "Three to the power of zero must be zero. An exponent tells you how many times to multiply the base, and if you multiply zero times you would have nothing!" Give her a convincing argument as to why 3^0 equals 1.

4. Answers will vary. Possible answer: $\frac{3^x}{3^x} = 3^{(x-x)} = 3^0$. Any number divided by itself is 1.

5. For each table, find the value of the constants A and r such that $y = A(1 + r)^x$ or $y = A(1 - r)^x$. Then use your equations to find the missing values.

a. $y = 200(1 + 0.4)^x$

x	y
0	200
1	280
2	392
3	548.8
4	768.32
5	1075.648
6	1505.9072

b. $y = 850(1 - 0.15)^x$

x	y
−2	1176.4706
−1	1000.0000
0	850
1	722.5
2	614.125
3	522.00625
4	443.7053

6. Convert each number from scientific notation to standard notation, or vice versa.

a. -2.4×10^6 $-2,400,000$

b. 3.25×10^{-4} 0.000325

c. $37,140,000,000$ 3.714×10^{10}

d. 0.00000008011 8.011×10^{-8}

7. A person blinks about 9365 times a day. Each blink lasts about 0.15 second. If one person lives 72 years, how many years will be spent with his or her eyes closed while blinking? Write your answer in scientific notation. $\approx 1.17 \times 10^0$ years

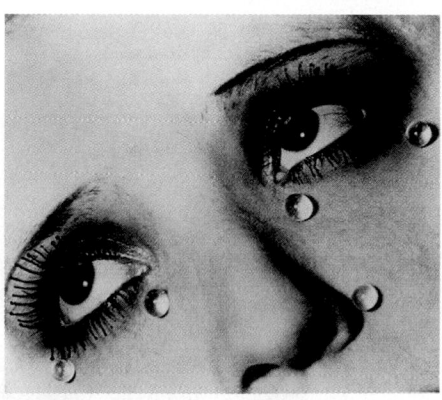

One of the purposes of blinking is to spread tears over the eye. The American photographer Man Ray (1890–1976) is well-known for this photo titled *Glass Tear*.

Exercise 7 Encourage students to use dimensional analysis.

8. APPLICATION In 1995, a can of soda cost 75¢ in a vending machine. If prices increase about 3% per year, in what year will the cost first exceed $2? after 34 years, or in the year 2029

Exercise 8 Encourage a variety of approaches, including recursion. This exercise provides a good chance to ask students about reasonable values of y. The equation $y = 0.75(1 + 0.03)^x$ can yield any positive real value of y. But if, for example, the smallest denomination of coin accepted by the machine is a nickel, then only multiples of 0.05 fit the real-world situation.

9a. False; 3 to the power of 3 is not 9; $27x^6$

9b. False; you can't use the multiplication property of exponents if the bases are different; $3^2 \cdot 2^3$ or 72

9c. False; the exponent -2 applies only to the x; $\dfrac{2}{x^2}$

9d. False; the power property of exponents says to multiply exponents; $\dfrac{x^6}{y^9}$

Exercise 10 As mentioned in Lesson 7.8, pendulum motion is fundamentally harmonic. But the aspect that this problem simulates can be roughly modeled with an exponential equation. However, that doesn't mean it is truly exponential.

10a. Possible answer: $y = 80(1 - 0.17)^x$, where x is the time elapsed in minutes and y is the maximum distance in centimeters; $(1 - 0.17)$ is derived from the mean ratio of approximately 0.83.

Exercises 11 to 19 Can be used to review for a final exam.

Exercise 11 This exercise reviews Chapter 6.

11a. Let t be the number of T-shirts, and let s be the number of sweatshirts.
$t + s = 12$
$6t + 10s = 88$

Exercise 12 This exercise reviews Lesson 1.4. **[Language]** An *entomologist* is one who studies or collects insects.

12a. Praying Mantis Length

```
 1 | 7
 2 | 1  2  6  6
 3 | 4
 4 | 8
 5 | 3  3  3  4  6  6
 6 | 2
 7 |
 8 | 2
 9 | 4  8
10 |
11 |
12 | 1
```

Key
1 | 7 means 1.7 cm

9. Classify each equation as true or false. If false, explain why and change the right side of the equation to make it true.

   a. $(3x^2)^3 = 9x^6$ **b.** $3^2 \cdot 2^3 = 6^5$ **c.** $2x^{-2} = \dfrac{1}{2x^2}$ **d.** $\left(\dfrac{x^2}{y^3}\right)^3 = \dfrac{x^5}{y^6}$

10. **APPLICATION** A pendulum is pulled back 80 centimeters horizontally from its resting position and then released. The maximum distance of the swings from the resting position is recorded for 5 minutes.

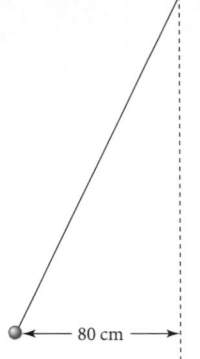

Pendulum Swings

Time elapsed (min)	0	1	2	3	4	5
Maximum distance from the resting position (cm)	80	66	55	46	38	32

80 cm

   a. Define variables and write an equation that models the maximum distance of the swing after each minute.

   b. What is the maximum distance from the resting position after 9 minutes? approximately 15.0 cm

   c. After how many minutes will the maximum distance from the resting position be less than 5 centimeters? 15 min

MIXED REVIEW

11. Three sisters went shopping for T-shirts and sweatshirts at an outlet store. They paid $6 for each T-shirt and $10 for each sweatshirt. They bought 12 shirts in all and the total cost was $88.

   a. Define variables and write a system of equations to represent this situation.

   b. Solve the system symbolically. How many shirts of each kind did the sisters buy? 8 T-shirts and 4 sweatshirts

12. Mr. Lee's science class received a collection of praying mantises from a local entomologist. The students have been measuring the mantises' lengths in centimeters.

Praying Mantis Length (cm)

5.6, 9.4, 1.7, 3.4, 5.3, 6.2, 8.2, 2.1, 5.3, 2.6, 5.6, 2.6, 5.4, 12.1, 5.3, 2.2, 4.8, 9.8

   a. Organize the data in a stem plot.

   b. What is the range of the data? 10.4 cm

   c. What are the measures of center? Which do you think best represents the data? Explain your thinking.

Praying mantises use their front legs to capture other insects.

12c. mean: approximately 5.4 cm; median: 5.3 cm; mode: 5.3 cm. Choice and explanations will vary.

13. Write a recursive routine to generate each sequence. Then use your routine to find the 10th term of the sequence.

   a. 21, 17, 13, 9, . . .   **b.** −5, 15, −45, 135, . . .   **c.** 2, 9, 16, 23, . . .

14. APPLICATION Chad polled the 9th graders and 12th graders at his school to ask about their plans for post–high school education. He got the following results.

Post–High School Plans

Plans	Number of 9th graders	Number of 12th graders
College or university	126	212
Junior college	88	122
Technical school	64	98
Travel	92	142
Work full time	132	78
Undecided	260	46
Total	762	698

   a. Make two circle graphs to represent the data that Chad collected.

   b. What is the percent of increase or decrease in students planning to attend a college or university from 9th to 12th grade? **approximately 68% increase**

   c. What is the percent of increase or decrease in students planning to work full time from 9th to 12th grade? **approximately 41% decrease**

   d. Marta attends another high school in Chad's town. She wants to use Chad's information to predict how many 12th graders from a class of 520 will travel after graduation. Explain how she can do this.

15. Angie has some guests visiting from Italy, and they are planning to drive from her house to Washington, D.C. Angie wants her friends to understand how far they will need to drive.

   a. Write a direct variation equation to convert miles to kilometers.

   b. Angie's house is about 250 miles from Washington, D.C. How many kilometers will her friends have to drive? **400 km**

   c. The hotel her friends are staying at says it is 2 miles from the Washington Monument. How far is that in kilometers? **3.2 km**

   d. The Washington Monument is taller than 555 feet. How tall is this in meters? **approximately 168 m**

The Washington Monument was built between 1848 and 1884. It was the tallest structure until the Eiffel Tower was built in 1889.

Exercise 13 This exercise reviews Lesson 4.3.

13a. Start with 21, then apply the rule Ans − 4; 10th term = −15.

13b. Start with −5, then apply the rule Ans · (−3); 10th term = 98,415.

13c. Start with 2, then apply the rule Ans + 7; 10th term = 65.

Exercise 14 This exercise reviews Lesson 2.6. Students may want to discuss whether it is appropriate to use the data from one school to make predictions about another. What assumptions are being made?

14a.

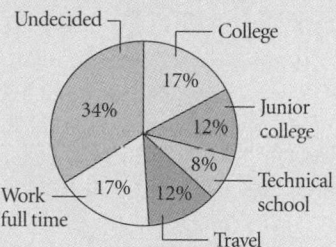

Plans for Ninth Graders

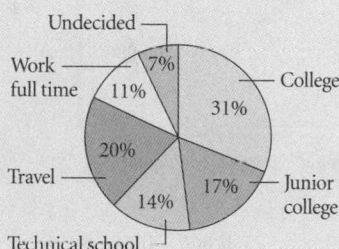

Plans for Twelfth Graders

14d. Answers will vary. Possible answer: In Chad's school, 142 out of 698 12th graders want to travel, or approximately 20%; for Marta, 0.20 · 520 gives 104 12th graders or $\frac{x}{520} = \frac{142}{698}$, so $x \approx 106$ 12th graders.

Exercise 15 This exercise reviews Lesson 3.2.

15a. $y = 1.6x$, where x is a measurement in miles and y is a measurement in kilometers

Exercise 16 This exercise reviews Lesson 6.5.

16. right triangle

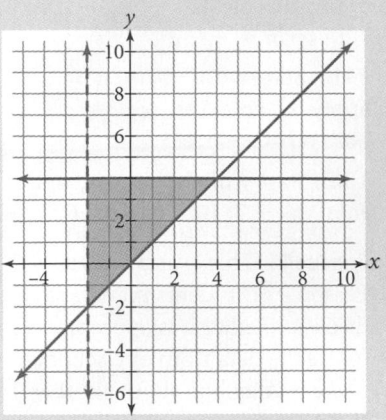

Exercise 18 This exercise reviews Lesson 5.1.

Exercise 19 This exercise reviews Lesson 5.7.

19. Answers will vary depending on the method used. The following possible answers used the Q-point method and a decimal approximation of the slope.

19a. $y = 96 - 5.7(x - 5)$

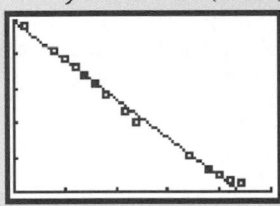

$[0, 25, 5, 0, 125, 25]$

19d. The y-intercept would become 200; $y = 200 - 5.7x$.

16. Sketch a graph showing these three inequalities on the same coordinate axes. What shape do you get?
$$\begin{cases} x > -2 \\ y \ge x \\ y \le 4 \end{cases}$$

17. Solve each equation using the method of your choice. Then use a different method to verify your answer.
   a. $-2(4-d) + 3 = -13$ -4
   b. $42 - 7(d - 8) = 7$ 13
   c. $0.5(d - 2) - 3 = -10$ -12

18. Find the slope and y-intercept of the line through each pair of points.
   a. $(3, 2), (1, -3)$ slope: $\frac{5}{2}$; y-intercept: $-\frac{11}{2}$
   b. $(-1, 4), (-1, 8)$ slope: undefined; y-intercept: none
   c. $(-11, 3), (-9, 2)$ slope: $-\frac{1}{2}$; y-intercept: $-\frac{5}{2}$

19. Does someone use the same amount of soap each day when he or she showers? Rex Boggs of Glenmore State High School in Queensland, Australia, decided to find out. He collected data for three weeks.

   a. Make a scatter plot and find a linear equation that fits the data. Use any method that you prefer.
   b. Write your equation in intercept form. $y = 124.5 - 5.7x$
   c. Using your equation, after how many days would Rex's soap weigh 34 grams? approximately 16 days
   d. How would your equation in 19b change if Rex starts with a family size soap bar that weighs 200 grams? Write an equation for a family size soap bar. (Assume the soap is used at the same rate.)
   e. If Rex starts with a family size soap bar, how much will it weigh after 20 days have elapsed? 86 grams

Soap Usage

Number of days elapsed	Weight of bar of soap (grams)
0	124
1	121
4	103
5	96
6	90
7	84
8	78
9	71
11	58
12	50
17	27
19	16
20	12
21	8
22	6

(*www.maths.uq.oz.au*)

Exercise 20 This exercise reviews Lesson 7.6.

20. APPLICATION When Anton started his career as an assistant manager of a tutoring center, his salary was $18,500 per year. He was told that he would get a 2.25% raise every year. Anton has now been with the company for 3 years.

   a. Find Anton's current salary in whole dollars. $19,777

   b. What will Anton's salary be in another 5 years? $22,104

   c. Anton's coworker Kobra has been with the company for 10 years. She can't remember her starting salary but knows that she got the same 2.25% raise for each of her first 8 years and then got a 3.5% raise during each of the last 2 years. She now makes $23,039. Write and solve an equation to find Kobra's starting salary. $y = 23{,}039(1 + 0.0225)^{-8}(1 + 0.035)^{-2}$; approximately $18,000

TAKE ANOTHER LOOK

▶

Scientific notation gives scientists and mathematicians one way to express extremely large and extremely small numbers. Sometimes scientists focus on only the power of 10 to describe size or quantity, calling this the **order of magnitude.**

Consider that the average distance from the earth to the sun is 9.29×10^7 miles. Unless a scientist is going to calculate with this figure, she may simply say the distance in miles from the earth to the sun is *on the order of 10^7*. By stating only the power of 10, what range of values is the scientist including?

Order of magnitude is also used to compare numbers. Suppose a sample of bacteria grows from several hundred to several thousand cells overnight. How many times larger is the sample now? A scientist may say the number of cells in the sample *increased by one order of magnitude,* because $\frac{10^3}{10^2}$ equals 10^1. What would the scientist say when the sample grows from several hundred cells to several hundred thousand cells? What fraction of cells would remain if the sample *decreased* by two orders of magnitude? (Note: The units must be equal to compare orders of magnitude.)

Think about the relative size of our universe as you answer these questions:

   1. Explain what it means for the typical size of a cell in meters to be on the order of 10^{-6}.

   2. Explain what it means for the length of a cow in meters to be on the order of 10^0.

   3. The distance in meters from the earth to the nearest star (other than the sun) is on the order of 10^{17}. Is it correct to compare the distance from the earth to the sun and the distance from the earth to the nearest star as an increase by 10 orders of magnitude, because $\frac{10^{17}}{10^7}$ equals 10^{10}?

   4. The diameter in meters of the Milky Way galaxy is 10^{20}. Describe the increase in order of magnitude between the size of a cell and the size of the galaxy.

When something increases 100%, should it be described as an increase in order of magnitude? Give an example to support your conclusion.

▶ **Take Another Look**

Students may better understand order of magnitude if they relate it to maximum place value and can describe orders of magnitude with words. (For example, 10^7 means the maximum place value is the ten millions.)

By stating the order of 10^7, the scientist is including a range of values at least 10,000,000 and less than 100,000,000.

If a sample grows from several hundred cells to several thousand cells, it increases roughly 10 times.

If the sample grows from several thousand cells to several hundred thousand cells, it has increased by two orders of magnitude. If the cells decrease by two orders of magnitude, about $\frac{1}{100}$ remain.

1. The size of a cell includes a range of values at least 0.000001 and less than 0.00001.

2. The length of a cow is at least 1 and less than 10 meters.

3. This is incorrect because the units are not equivalent (meters versus miles).

4. This is an increase by 26 orders of magnitude.

An increase by 100% does not represent an increase in order of magnitude. An increase of 100% means the quantity doubles, whereas an increase in an order of magnitude means that the quantity is multiplied by about 10.

Use assessment of students' group investigation skills, projects, and self assessment along with a written test to measure students' progress. For written assessment, use two or three constructive assessment items, use Form A or B of the chapter test, or use the test generator to construct a written test based on the lessons your class completed.

FACILITATING SELF ASSESSMENT

To help students complete the portfolio described in Assessing What You've Learned, suggest that they consider for evaluation their work on Lesson 7.1, Exercise 10; Lesson 7.2, Exercises 9, 12; Lesson 7.3, Exercises 11, 13; Lesson 7.4, Exercises 7, 14; Lesson 7.5, Exercises 8, 10; Lesson 7.6, Exercise 12; and Lesson 7.7, Exercises 5 to 7.

Assessing What You've Learned

WRITE IN YOUR JOURNAL Add to your journal by considering one of these prompts.

▶ Why is scientific notation convenient for writing extremely large or extremely small numbers? Are there numbers that you find to be less convenient to write in scientific notation? Does scientific notation help you to understand why our standard number system is called a "base 10" system?

▶ Compare and contrast linear and exponential data. How do the graphs differ? If you weren't specifically told to find either a linear or exponential equation to fit a graph of data, how would you decide which to try? How do the methods of fitting linear and exponential models compare?

PERFORMANCE ASSESSMENT Show a classmate, a family member, or your teacher that you know how to find an exponential model in the form $y = A(1 + r)^x$. You may want to go back and use the data sets from Lesson 7.7 or Lesson 7.8, or use data that you have collected from a project. Explain why you think the data is exponential, and when and why you would want to adjust the value of A or r.

GIVE A PRESENTATION Review the properties of exponents that you learned in this chapter. Think about the techniques you have used to remember these properties, or ask your peers, teachers, or family members how they remember these properties. Prepare a presentation for your class and demonstrate the memory methods you have learned. Your presentation will help your classmates remember the properties of exponents too!

Functions

Overview

Secret codes highlight **Lesson 8.1.** Students learn that unambiguous encoding requires a code that's a function. Functions are defined in **Lesson 8.2** as students learn about properties and geometric representations of functions. In **Lesson 8.3,** to learn the difference between independent and dependent variables, students construct and interpret simple graphs that describe real-world situations. **Lesson 8.4** focuses on function notation and its relationship to input and output variables. In **Lesson 8.5,** students learn to make graphs of real-world situations, and they also describe, read, and interpret these graphs and functions using the terms *linear, nonlinear, increasing, decreasing, discrete,* and *continuous.* In **Lesson 8.6,** students learn several ways to define absolute value and then construct and interpret graphs of absolute value functions. **Lesson 8.7** features the squaring and square root functions; students graph parabolas and relate the squaring function to finding the area of a square.

The Mathematics

Functions

Originally, calculus was developed to study geometric figures as described by equations. The notion of a function evolved to put calculus on a firm logical footing. Leonhard Euler, who introduced the modern function notation in the middle of the 1700s, thought of functions as algebraic expressions corresponding to curves.

As work progressed, the geometry of calculus gradually became less important, and the notion of function became more abstract. Although several different approaches are still used today, the most common one is that a function is a set of ordered pairs of objects (such as letters, numbers, or figures), in which no first element of an ordered pair can correspond to more than one second element. When you think of functions of numbers, you can envision a two-column table, perhaps infinite, with no number occurring twice in the first column. If you think of the ordered pairs of numbers as representing points on the plane, no vertical line will pass through the graph more than once.

Although graphing functions helps us understand them, there are many graphs of geometric figures (for example, circles) that don't pass the vertical line test. The importance of the function concept is evident in graphing calculators, which most easily graph functions rather than these other figures.

Squares and Square Roots

The *square* of a number is the product of that number multiplied by itself. The term comes from the geometric square, whose area is the product of the length of one edge multiplied by itself. The notation for the square of number x is x^2. Most calculators have a key to find the square of a number. You can also type in x^2. A number x and its opposite $(-x)$ have the same square.

Each number has only one square, so $y = x^2$ is a function. Its graph is a curve known as a *parabola.* The parabola is one of the *conic sections,* so-named because they are formed by the intersection of an infinite cone with a plane.

To go from the area of a square to the length of one edge, you use the *square root function,* symbolized by $\sqrt{\ }$. This function gives the *non-negative* number whose square is the given number. For example, $\sqrt{4} = 2$.

Both 2 and -2 have a square of 4, and they are both considered *square roots* of 4. The non-negative value is called the *principal square root,* and it is the value returned by the square root function.

Absolute Value

Absolute values arise from a need to talk about distances, which are never negative. If we want the *difference* between numbers a and b, we just subtract: $a - b$. If we want the *distance* between them, we don't want a negative number. If $a - b$ is negative, we want $b - a$ instead, because it's positive. We refer to the distance as the *absolute value* of $a - b$, represented $|a - b|$. It's the same as $|b - a|$.

The absolute value of a single number x, then, is its distance from 0. If $x < 0$, we say that $|x|$ is $-x$, which will be a positive number. If $x \geq 0$, then $|x|$ is simply x itself.

Because the square root function returns the non-negative square root, we can say that $\sqrt{x^2} = |x|$, whether x is positive or negative.

Using This Chapter

Because students need to understand the definition of function, to be familiar with the absolute value and squaring functions, and to have experience relating situations to graphs, you will want to do all the lessons in this chapter. You might want to use several of the one-step investigations to give students more opportunity to explore.

Resources

Discovering Algebra Resources

Calculator Notes 1E, 1H, 2B, 8A, 8B, 8C, 8D

Teaching and Worksheet Masters
  Lesson 8.1 Coding Grid
  Lesson 8.1 Letter-Number Chart
  Lesson 8.1 TFDSFU DPEFT

Lesson 8.2 Function or Not?
Lesson 8.3 Water Depth
Lesson 8.4 A Graphic Message
Lesson 8.5 Real-World Situations
Lesson 8.6 Pulse Rate Sample Data 2

Assessment Resources A and B
  Quiz 1 (Lessons 8.1 and 8.2)
  Quiz 2 (Lessons 8.3 to 8.5)
  Quiz 3 (Lessons 8.6 and 8.7)
  Chapter 8 Test
  Chapter 8 Constructive Assessment Options

More Practice Your Skills for Chapter 8

Condensed Lessons for Chapter 8

Other Resources

Green Globs and Graphing Equations. New York: Sunburst Communications, 1996. (Both Macintosh and Windows versions are available.)

www.keymath.com/DA

Materials

• large sheets of paper
• markers

Pacing Guide

	day 1	day 2	day 3	day 4	day 5	day 6	day 7	day 8	day 9	day 10
standard	8.1	8.2	quiz, 8.3	8.4	8.5	8.6	8.6	8.7	review	assessment
enriched	8.1	8.2, project	8.3	8.4	8.5	8.6	8.6	8.7	review, TAL	assessment
block	8.1, 8.2	8.3, 8.4	quiz, 8.5	8.6	8.7, quiz, review	assessment				

Functions

The musician in *The Lute Player* by an unknown artist called the Master of the Half Figures plays her lute while reading sheet music. When music is composed or transcribed it is written on a staff as notes in standard notation or as numbers in tablature. Playing music from notation and writing notation from music are very much like the relationships between input and output in mathematical functions.

OBJECTIVES

In this chapter you will
- learn the mathematics of code breaking
- use the vertical line test
- graph functions of real-world situations
- learn about function notation and vocabulary
- learn the absolute value and squaring functions

CHAPTER 8 OBJECTIVES

- Investigate the concept, definition, notation, properties, and graphs of functions

- Learn the terminology of independent and dependent variables

- Construct and interpret graphs and functions that describe real-world situations

- Evaluate functions by substitution, by using the graphing calculator, and by using graphs

- Describe, read, and interpret graphs of real-world situations using the terms *linear, nonlinear, increasing, decreasing, rate of change, continuous,* and *discrete*

- Learn to work with absolute values and with the absolute value function and its graph

- Learn to work with the squaring and square root functions and the parabolic graph of the squaring function.

Relationships between musical notation and musical pitch can be modeled mathematically. A piano player reading standard musical notation (input), plays the key that corresponds to each written note (output). Discuss with students if playing the piano from sheet music is an example of a relationship in which each input has only one output. [It is.] Ask if translating piano playing to standard musical notation is that same kind of relationship. [Yes, there is exactly one musical notation for each piano key.]

Other stringed instruments differ from the piano in this respect. The same pitch may be played in different ways. For example, on a guitar a high F can be played on the first string at the first fret, on the second string at the sixth fret, or on the third string at the ninth fret. Ask if guitar playing from music written in standard notation is a relationship with a single output for each input. [No, this is not a *function,* a word students will learn the meaning of in this chapter.]

In tablature, however, each method of producing a musical pitch has its own musical notation. Lute tablature defines a function rule that tells which notes to play on which strings. A player reading lute tablature (input) has only one choice of how to play the note (output). Discuss the relationship of writing tablature (output) from a lutist's playing (input). [This is also a function.]

LESSON

8.1

PLANNING

LESSON OUTLINE

First day:

50 min   Investigation

Second day:

5 min   Investigation

20 min   Sharing

15 min   Example

5 min   Closing

5 min   Exercises

MATERIALS

- Coding Grid (W)
- TFDSFU DPEFT (T), *optional*
- Letter-Number Chart (W or T)

TEACHING

Schemes for writing secret codes provide an introduction to mathematical functions. Take advantage of the fact that many students are intrigued by secret codes.

 Guiding the Investigation

One step Use the TFDSFU DPEFT transparency to show how to encode a word, and then ask each pair to create its own code with a regular pattern on the Coding Grid worksheets and to encode a message for other pairs to decode. As you circulate, bring out the idea that for coding the pattern should have only one output for each input, and for decoding the reverse should be the case. The thinking processes of developing codes and trying to decode are more important than any answers the students get. During Sharing, introduce the term *function*.

Secret Codes

The study of secret codes is called **cryptography.** Early examples of codes go back 4000 years to Egypt. Writing messages in code plays an important role in history and in technology. Today you can find applications of codes at ATMs, in communications, and on the Internet.

Cryptography is an intellectual battle between the code-maker and the code-breaker.
SIMON SINGH

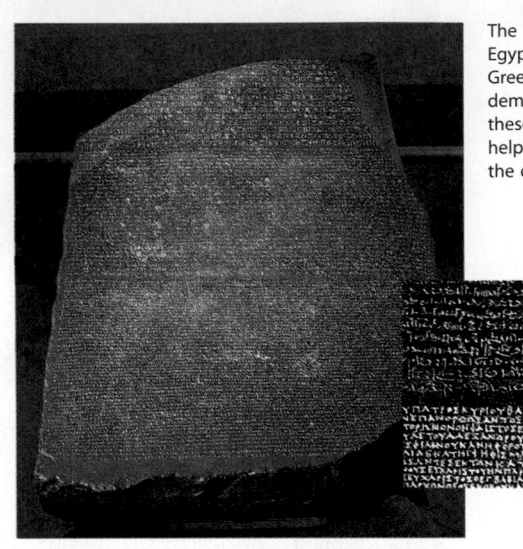

The Rosetta Stone, found near Rashid, Egypt, in 1799, bears inscriptions in Greek, Egyptian hieroglyphics, and demotic (everyday) Egyptian. Having these three versions of the same text helped language researchers "break the code" of hieroglyphics.

In this investigation you will learn some of the mathematics behind secret codes.

 ## Investigation
TFDSFU DPEFT

You will need

- the worksheet Coding Grid

The table below shows that the letter A is coded into the letter Q, and the letter B is coded into the letter R, and so on. It also shows that the letter U is coded into the letter K. This code is an example of a letter-shift code. Can you see why? How would you use the code to write a message?

Original input	A	B	C	D	E	F	G	H	I	J	K	L	M	N	O	P	Q	R	S	T	U	V	W	X	Y	Z
Coded output	Q	R	S	T	U	V	W	X	Y	Z	A	B	C	D	E	F	G	H	I	J	K	L	M	N	O	P

LESSON OBJECTIVE

- Investigate the concept of function through secret codes

NCTM STANDARDS

CONTENT		PROCESS	
	Number	●	Problem Solving
●	Algebra	●	Reasoning
	Geometry	●	Communication
	Measurement	●	Connections
	Data/Probability	●	Representation

You can also represent the code with a grid. Note that the input letters run across (horizontally). To code a letter, look for the shaded square directly above it. Then find the coded output by looking across to the letters that run up (vertically).

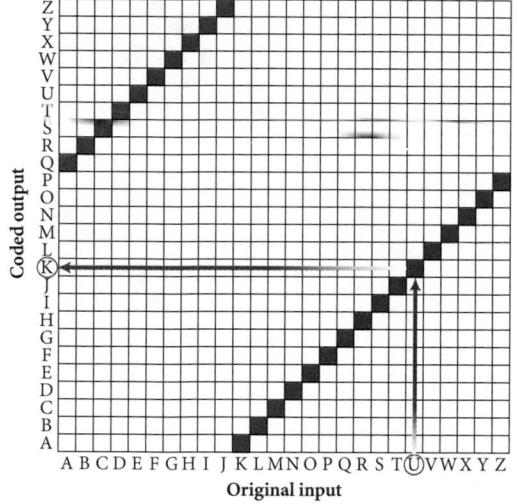

Coded output (vertical axis)
Original input (horizontal axis)

Step 1 Sample answers:

input: MATH;
output: CQJX

input: NUMBER;
output: DKCRUH

input: LETTER;
output: BUJJUH

Step 2 One rule:
Add 16 to the letter's
position in the alphabet. If the new position is greater than 26, subtract 26 so that K becomes A. Other rule: Subtract 10 from the letter's position in the alphabet. If the new position is less than 1, add 26 so that J becomes Z.

Step 1 Use the coding grid to write a two-word or three-word message.

Step 2 Think of the letters A through Z as the numbers 1 through 26. Write two different rules to describe this code. Write one rule that adds a constant number to a letter's position in the alphabet. Write a second rule that subtracts a constant number from the letter's position. Explain how to code letters that are shifted to a position which is less than 1 or greater than 26.

Step 3 Exchange your coded message with a partner. Use this grid or one of your rules to decode each other's messages.

Next you'll invent your own letter-shift code.

Steps 4–7 All possible letter-shift codes should appear on the grid as two parallel diagonal lines sloping upward. Possible codes might have different intercepts. Neither line segment will be directly above the other or across from the other at any point.

Step 4 Create a new code by writing a rule that shifts letters a certain specified number of places. Put the code on a grid like the one shown above. Do not allow your partner to see this grid.

Step 5 Use your new grid to code the same message you wrote in Step 1.

Step 6 Exchange your newly coded message with your partner. Use it along with the message in the first code to try to figure out each other's new codes. Write a rule or create a coding grid to represent your partner's new code.

Step 7 Compare your grid to your classmates' new grids. In what ways are the grids the same? How are they different? For any one grid, how many coded outputs are possible for one input letter? How many ways are there to decode any one letter in a coded message?

For this investigation, students are in pairs.

[ESL] Explain *code* and *decode* in terms of doing and undoing.

[Language] Other words for coding and decoding are *encryption* and *decryption*.

Step 1 If necessary, use the transparency to demonstrate how to encode a word. In making sense of graphs of equations, students have probably thought in terms of moving from the origin to the right (or left) and then up (or down). When considering the graph of a function, it's good to think of beginning with input on the *x*-axis, moving up (or down) to the graphed line or curve, and then moving horizontally to the *y*-axis for the corresponding output. Coding with a grid helps establish this way of thinking.

Watch for inappropriate words.

Step 2 Pass out or display the Letter-Number Chart. These will be useful again in Lesson 8.4 and the chapter review. All coding rules for this shift require adding or subtracting 26 if the shift produces a number outside the range of 1 to 26. This is sometimes called *wrapping*.

Step 4 Give students the worksheet to make grids for their codes.

Step 7 The grid scheme helps students see how various codes are similar and different. Provided that students have followed directions and written the alphabet in order, all code graphs should consist of two linear segments with positive slopes. The graph should have no more than one point (one shaded square) in any one vertical or horizontal line.

Step 8 Students may be confused because there's more than one way to encode most messages. Ask them what characteristic of the graph makes it confusing, to motivate Step 10.

Step 12 Codes created here may look somewhat random or have some other pattern. However, they should all have only one coded letter value for each original letter.

SHARING IDEAS

Have selected students present their code grids and answers to the questions in Step 12. This would be a good chance for students to shine who are more comfortable with language than mathematics. As the idea arises of one output for each input, introduce the terms *function, domain,* and *range*.

Begin writing out complete sentences on the board and insist that students do also. For example, they should write "If the input is 1, then the output is 4," rather than some shorthand. The inconvenience of doing so will motivate function notation in Lesson 8.4.

Draw attention to the quotation introducing the lesson, and ask what the difference is between breaking a code and decoding. Ask how difficult it would be to break each of the codes they created in the investigation—that is, to figure out the corresponding grids. The advantage to less organized grids is their security against code breakers. The disadvantage is that more information has to be transmitted for decoders to work and that this information may be intercepted. Modern mathematics has led to methods in which no information must be transmitted for decoding but whose complexity makes cracking virtually impossible. The quote is from a journalist who specializes in mathematics and science. He has written books on codes and one called *Fermat's Last Theorem*.

Here is another new code.

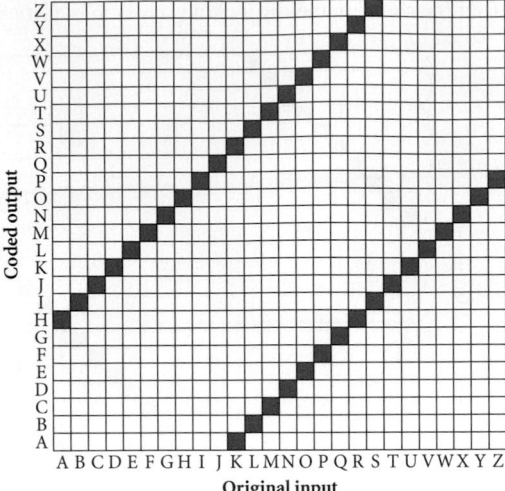

Step 8 Answers will vary. Each letter between K and S has two possible codes. So MOP can be coded eight different ways—CEF, CEW, CVF, CVW, TEF, TEW, TVF, TVW.

Step 9 There may be several ways to decode one message. HI can be decoded into AB, RS, AS, or RB.

Step 10 This coding grid's overlapping lines allow eight ways to code FUNCTION. The N and O (letters between K and S), each can be coded as two different letters.

Step 11 The grid in Step 1 is easier for decoding. Each coded letter between H and P corresponds to two input letters.

Step 12 Coding grids will be similar to the grid in Step 1 in that no two shaded squares will be in the same column or row.

Step 8

Step 9

Step 10

Step 11

Step 12

Use the grid above to send a new two- or three-word message to your partner. Exchange and decode each other's messages.

Did your partner successfully decode your message? Why or why not?

How is the new grid in Step 8 different from the grid in Step 1? Code the word FUNCTION to help you answer this question.

Which grid makes it easier to decode messages? Which coded output letters are difficult to decode into their original input letters?

Create a new coding scheme by shading squares that don't touch each other on the grid. Make the grid so that there is exactly one output for each input. How is it similar to the grid in Step 1? How is it different?

Letter-shift codes are relationships. Codes that have exactly one output letter for every input letter are examples of **functions.** The set of values that are inputs for a function is called its **domain.** In the investigation, the domain is the set of all letters of the alphabet. The **range** of a function is the set of all its possible output values for these codes. The range happens to be all the letters of the alphabet as well. But often the domain contains many values different from those in the range. Here is an example.

Domain	A	B	C	D	E	F	G	H	I	J	K	L	M
Range	65	66	67	68	69	70	71	72	73	74	75	76	77

Domain	N	O	P	Q	R	S	T	U	V	W	X	Y	Z
Range	78	79	80	81	82	83	84	85	86	87	88	89	90

Computers store letters as numbers. In the preceding example, the letter A is coded as the number 65, B as 66, and so on. In this case, the domain is the letters of the alphabet, but the range is the set of whole numbers from 65 through 90. The table on the opposite page shows how each letter is represented in this code.

Notice that each letter in the domain matches no more than one number in the range. This is what makes the code a function.

EXAMPLE

Tell whether or not each table of values represents a function. Give the domain and range of each relationship.

Table A

Input	Output
1	2
2	4
3	6

Table B

Input	1	0	1
Output	1	2	5

Table C

Input	1	2	3	4	5	6
Output	0	0	0	0	0	0

▶ **Solution**

To be a function, each input must have exactly one output. It is helpful to use arrows to show which input value matches which output value.

Table A

Each input value matches one output value. So this relationship is a function. The domain is {1, 2, 3}, and the range is {2, 4, 6}.

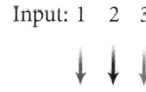

Table B

The input value 1 has two outputs, 1 and 5. This relationship is not a function because there is an input value with more than one output value. The domain is {0, 1}, and the range is {1, 2, 5}.

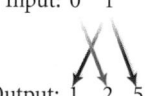

Table C

Each input value matches exactly one output value, 0. So this relationship is a function, even though all the inputs have the same output. The domain is {1, 2, 3, 4, 5, 6}, and the range is {0}.

You can represent a function with a table, a graph, an equation, symbols, a diagram, or even a written rule or description. Many of the relationships you have studied in this book are functions. You will revisit some of them as you learn more about functions in this chapter.

History CONNECTION

Alan Turing (1912–1954) was an English mathematician and pioneer in computing theory. During World War II, he led the team that cracked the German codes of the notorious Enigma machine. Learn more about Turing with the links at **www.keymath.com/DA** .

Closing the Lesson

A **function** is a relationship between two sets, called the **domain** and **range** of the function, such that every element of the domain is associated with one but only one element of the range. The domain can be thought of as the set of input values and the range as the set of output values.

A function might be defined through a table, arrow diagram, formula, equation, written description, or graph. Many functions are best defined through a table rather than with a formula or equation.

Have students look at the table of computer codes following the investigation. Ask for a description of the set of input values, and use the term *domain*. Do the same thing for the output values and *range*. This code is called ASCII (pronounced ASK-ee), which stands for American Standard Code for Information Interchange. Special characters, symbols, and numbers are represented by numbers 0 through 64 in ASCII, so the alphabet begins with 65.

Making the Connection

The History Connection refers to the Enigma machine. This machine used rotors to perform a different shift for each letter of the code, according to a prearranged keyword. For example, if the keyword was ENCODE, the first letter of the message would shift by 5 (because E is the fifth letter in the alphabet), the second letter would shift by 14 (because N is the fourteenth letter), and so on. The Enigma machine could handle very long "keywords" that were not really words at all. Because the Allies were very careful in how they responded to information gained by decoding German messages, the Germans were not aware that their code had been cracked and didn't make their methods more secure.

Assessing Progress

Through your observations, you can assess students' abilities to describe a mathematical procedure in English, to follow directions, and to work with a partner.

▶ **EXAMPLE**

This example isolates the ideas of function, domain, and range. Assume that the entire domain and entire range are listed in the table.

BUILDING UNDERSTANDING

Students work with coding and functions. Ask that they write out their homework in complete sentences, to begin to motivate function notation in Lesson 8.4. Writing this way is a good tool for acquiring a deeper understanding of ideas.

ASSIGNING HOMEWORK

Essential	1–7
Performance assessment	6, 10, 11
Portfolio	7, 8
Journal	10, 12
Group	7, 9, 13, 14
Review	15, 16

▶ Helping with the Exercises

Exercise 3 For the first time, students are asked in this exercise to crack a code. If they're having difficulties, say that code crackers often guess important words that might be in the message.

Exercise 4 Some students might say that the range and domain consist of all letters because they're all listed along the grid's edges. Ask which letters can actually be input values and which can be output values. You might ask about vertical lines through the graph, foreshadowing Lesson 8.2.

4a. letters with shaded squares—A, B, C, E, G, H, I, K, L, M, N, O, Q, U, V, Y, Z

4b. all 26 letters of the alphabet

EXERCISES

You will need your calculator for problems **7** and **8**.

▶ Practice Your Skills

1. Use this table to code each word.

Input	A	B	C	D	E	F	G	H	I	J	K	L	M	N	O	P	Q	R	S	T	U	V	W	X	Y	Z
Coded output	B	C	D	E	F	G	H	I	J	K	L	M	N	O	P	Q	R	S	T	U	V	W	X	Y	Z	A

a. RANGE SBOHF b. DOMAIN EPNBJO c. TABLE UBCMF d. GRAPH HSBQI

2. Use the grid at right to decode each word.
a. SXZED INPUT
b. YEDZED OUTPUT
c. BOVKDSYXCRSZ RELATIONSHIP
d. BEVO RULE

3. The title of the investigation, TFDSFU DPEFT, is the output of a one–letter-shift code.
a. Decode TFDSFU DPEFT. SECRET CODES
b. Write the rule or create the coding grid for the code. The coding scheme is a letter-shift of +1.

4. Use the coding grid below to answer 4a–c.
a. What are the possible input values?
b. What are the possible output values?
c. Is this code a function? Explain why or why not. No, the letters B, E, G, I, K, M, and Q each have more than one output.

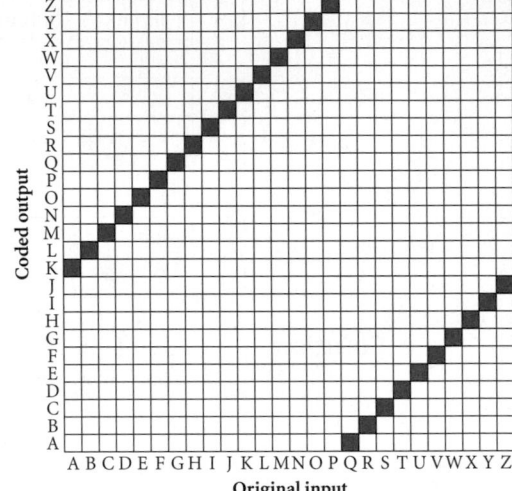

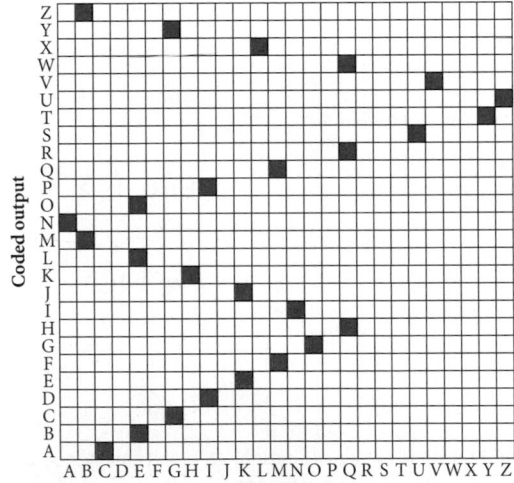

---

5. The table converts standard time to military time.

Standard time (A.M.)	1:00	2:00	3:00	4:00	5:00	6:00	7:00	8:00	9:00	10:00	11:00	12:00
Military time	0100	0200	0300	0400	0500	0600	0700	0800	0900	1000	1100	1200

Standard time (P.M.)	1:00	2:00	3:00	4:00	5:00	6:00	7:00	8:00	9:00	10:00	11:00	12:00
Military time	1300	1400	1500	1600	1700	1800	1900	2000	2100	2200	2300	2400

a. Describe the input. {1:00, 2:00, 3:00, 4:00, 5:00, 6:00, 7:00, 8:00, 9:00, 10:00, 11:00, 12:00}

b. Describe the output.

c. Does the table represent a function? Explain why or why not. It is not a function because each standard time designation has two military time designations. If students distinguish A.M. from P.M. times, then it is a function.

▶ Reason and Apply

6. Use the letter-shift grid at right to

  a. Find the output when the input is W. G

  b. Find the input when the output is W. M

  c. Code a Q. A

  d. Decode a K. A

7. APPLICATION Think of the letters A through Z as the numbers 1 through 26.

  a. Enter the position for each letter in the word FUNCTIONS into list L_1. Use your calculator to shift the letters 9 places to the right in the alphabet. Store the results in list L_2.

  b. What must you do to some of these numbers before coding them all into letters? What are the numbers in list L_2 after you do this?

  c. Use the results from 7b to code the word FUNCTIONS. ODWLCRXWB

  d. Plot pairs of the form (*input position, output position*) for this code.

  e. If you design a different letter-shift code, what letter-shift values should you avoid so that FUNCTIONS is not coded as the same word? Avoid any multiple of 26, such as 0, ±26, ±52, and so on.

8. Sylvana creates a code that doubles the position number of each letter in the alphabet. Then she subtracts 26 from the new positions that do not correspond to a letter in the alphabet. She stores the input values in list L_1 and the output values in list L_2.

  a. What numbers are in list L_1? $L_1 = \{1, 2, \ldots, 26\}$

  b. What numbers are in list L_2? $L_2 = \{2, 4, \ldots, 26, 2, 4, 6, 8, \ldots, 26\}$

  c. Plot Sylvana's code.

  d. Will she have difficulty coding or decoding messages?

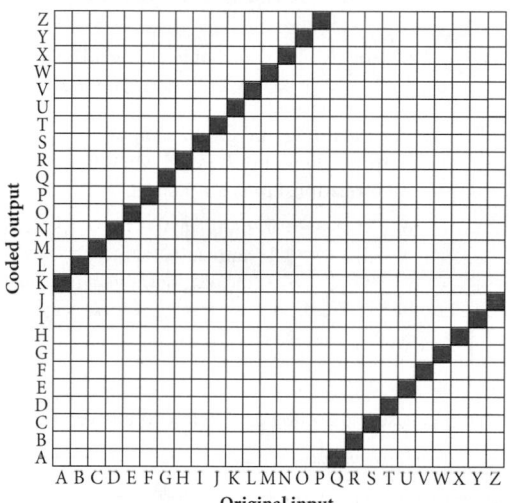

Coded output / Original input grid (A B C D E F G H I J K L M N O P Q R S T U V W X Y Z)

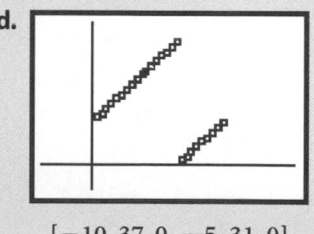

8c.

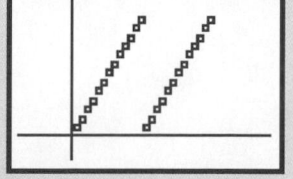

8d. Answers will vary. The graphs show that the 26 letters of the alphabet are coded into 13 letters. Sylvana will have difficulty decoding because each coded letter has two possible inputs.

Exercise 5 Students who see ambiguity in this exercise should use a condition to answer. If 1 A.M. is considered different from 1 P.M., then the table represents a function. If not, then an input of 1 has two outputs, 1 and 13, and so it is not a function.

5b. output: {0100, 0200, 0300, 0400, 0500, 0600, 0700, 0800, 0900, 1000, 1100, 1200, 1300, 1400, 1500, 1600, 1700, 1800, 1900, 2000, 2100, 2200, 2300, 2400}

Exercise 7b Students may be able to use their calculators to do any necessary wrapping, using the instruction $(L_1 + 9 > 26) * (L_1 + 9 - 26) + (L_1 + 9 \le 26) * (L_1 + 9) \to L_2$. Challenge students to understand the syntax of the statement: if $L_1 + 9 > 26$, that factor will be 1 for true giving $L_1 + 9 - 26 + 0 \to L_2$; if $L_1 + 9 \le 26$, the first term will be 0 for false and the second term will have a factor of 1 giving $0 + L_1 + 9 \to L_2$.

7a. $L_1 = \{6, 21, 14, 3, 20, 9, 15, 14, 19\}$
$L_2 = \{15, 30, 23, 12, 29, 18, 24, 23, 28\}$

7b. You have to subtract 26 from the numbers greater than 26. The new list L_2 is {15, 4, 23, 12, 3, 18, 24, 23, 2}.

7d.

[−10, 37, 0, −5, 31, 0]

Exercise 9 There are 54 possibilities, but only one is a word. Students might write all possible decoded letters next to the coded one to aid in their solution.

10a. No, there are two coding choices for C and two decoding choices for B.

10b. If every input letter matches only one output letter, that makes coding easier.

10c. Every output letter on the vertical axis should match only one input letter on the horizontal axis.

12a. Double the position of the letter and add 1. If the result is greater than 26, subtract 26. This number is the position of the coded letter.

---

9. Use the coding grid at right to decode CEOKEQC into a word. **ALGEBRA**

10. Here is a corner of a coding grid.

a. Does each input letter code to a single output? Does each output letter decode to a single input?

b. If this code were a function, which would be made easier, coding or decoding?

c. How would you change this grid to make the other part of coding in 10b easier?

11. For each diagram, give the input values and output values and then tell whether or not each relationship is a function.

a. 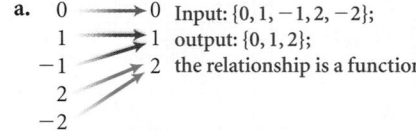 Input: {0, 1, −1, 2, −2}; output: {0, 1, 2}; the relationship is a function.

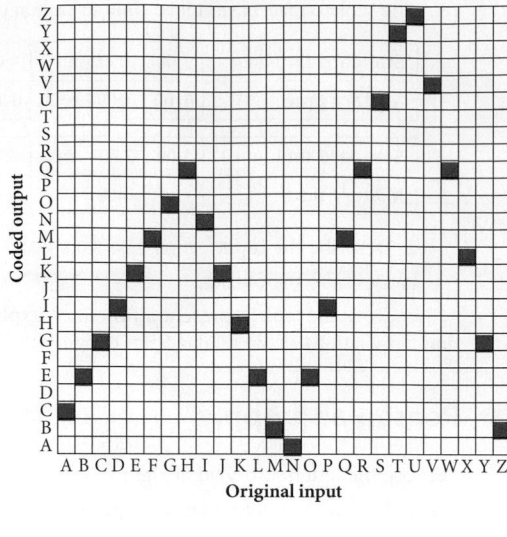

Coded output

Original input

b. 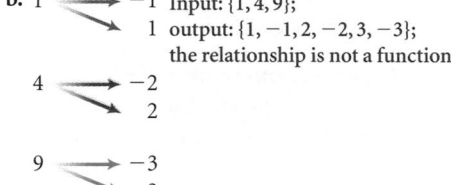 Input: {1, 4, 9}; output: {1, −1, 2, −2, 3, −3}; the relationship is not a function.

12. Use the coding grid below to answer 12a–c.

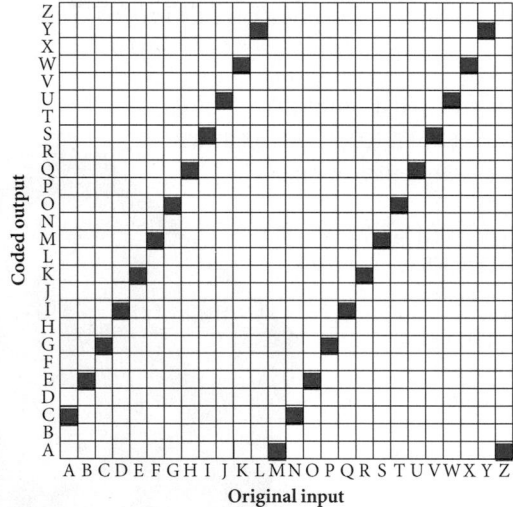

Coded output

Original input

a. Write a rule for this coding grid.

b. Code the word CODE. **GEIK**

c. Can you decode the word SPY? Explain why or why not. You cannot decode SPY because the letter P is not in the range.

13. The grid at right shows an ancient Hebrew code called "atbash."

(*http://all.net/books/ip/Chap2-1.html*)

a. Create a rule for the atbash code.

b. Is this code a function? Explain why or why not.

c. Use the atbash code to code your name.

14. If you know that SHOFJEWHQFXO is the study of coding and decoding, what is the rule for breaking this code? What is the original message? Subtract 10 from the letter's position, or add 16, so that your result is between 1 and 26, inclusive. The original message is CRYPTOGRAPHY.

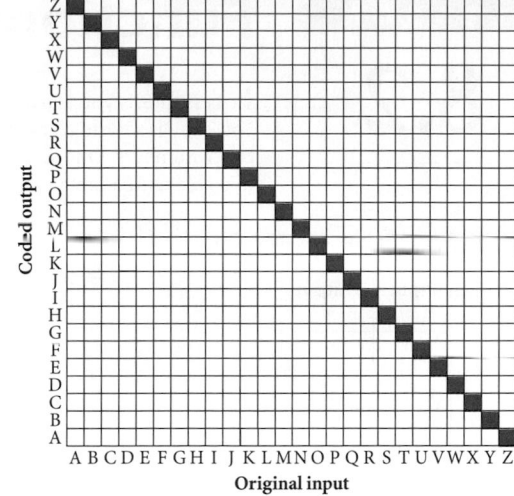

Coded output (vertical axis): Z Y X W V U T S R Q P O N M L K J I H G F E D C B A

Original input (horizontal axis): A B C D E F G H I J K L M N O P Q R S T U V W X Y Z

13a. Subtract the input letter's position from 27 to get the output letter's position.

13b. Yes, each input matches no more than one output.

13c. Sample: LISA codes into ORHZ.

▶ **Review**

15. If possible, perform the indicated operation.

a. $(4b^3)(7b^3)$ $28b^6$

b. $4a^3 + 7a^3$ $11a^3$

c. $\frac{7c^3}{4c^3}$ 1.75

d. $4(7d^3)^3$ $1372d^3$

e. $2a^3 + 3b^2$ Not possible

f. $(2x^3)(2x^2)$ $4x^5$

16. If 1 calorie is 4.1868 joules, then how many calories is 470 joules?

$\frac{1\text{ cal}}{4.1868\text{ J}} = \frac{x\text{ cal}}{470\text{ J}}$, so $x = \frac{470}{4.1868} \approx 112$ calories

Exercise 15 This exercise reviews Chapter 7.

Exercise 16 This exercise reviews Lesson 2.3.

COMPUTER NUMBER SYSTEMS

A computer stores alphanumeric symbols—letters, digits, and special characters—as a sequence of 1's and 0's in its memory. These numbers are called **binary numbers,** or base 2 numbers, because they contain only two digits—1 and 0. The number system that people use is a base 10 decimal system because it contains the ten digits from 0 through 9. How does a computer store 10 numerical digits, 26 letters, and several other characters using only two digits?

Research the binary number system and its use in computer memory. Are there other number systems that computers also use? How do computers convert letters into numbers? Is there a standard code that most systems follow?

Your project should include

▶ Sample conversions of base 10 numbers to binary numbers, and vice versa.

▶ A table that shows how to code letters and special characters.

Supporting the project

MOTIVATION

Dramatic growth in the size of computer memory enables the technology behind all digital media, such as the Internet, CDs and MP3s, DVDs, video games, and cell phones.

OUTCOMES

▶ A sample showing conversions between bases 2 and 10 is clearly explained.

▶ The report includes an ASCII table.

• The hexadecimal (base 16) system is mentioned and its relation to base 2 is explained.

• The ASCII table includes decimal, binary, and hexadecimal representations for each letter or symbol.

• Use of the binary system in computers is put in the context of the history of computers and programming.

Functions and Graphs

In Lesson 8.1, you learned that you can write rules for some of the coding grids. You can also write rules, often in the form of equations, to transform numbers into other numbers. One simple example is "Add one to each number." You can represent this rule with a table, an equation, a graph, or even a diagram.

PLANNING

LESSON OUTLINE

One day:

20 min	Investigation
5 min	Sharing
15 min	Example
5 min	Closing
5 min	Exercises

MATERIALS

- Calculator Note 1H
- Function or Not? (T), *optional*

TEACHING

The vertical line test is one method to determine whether or not a graph represents a function. Continue to use, and insist that students use, complete sentences when writing up the mathematics.

One step Show the Function or Not? transparency. Ask students which graphs represent functions, reviewing the fact that a function has only one output value for every input value. Then ask what the graphs of functions have in common that distinguish them from graphs of non-functions. As students think together, encourage them as necessary to use straightedges such as rulers to help explain their ideas to each other.

Table

Input x	Output y
7	8
−47	−46
10.28	11.28
x	x + 1

Equation

$y = x + 1$

Graph

Diagram

Domain		Range
−47	⟶	−46
7	⟶	8
10.28	⟶	11.28

This rule turns 7 into 8, −47 into −46, 10.28 into 11.28, and x into x + 1.

When you explored relationships in previous chapters, you used recursive routines, graphs, and equations to relate input and output data. To tell whether a relationship between input and output data is also a function, there is a test that you can apply to the relationship's graph on the xy-plane.

The Spanish painter Pablo Picasso (1881–1973) was one of the originators of the art movement Cubism. Cubists were interested in creating a new visual language, translating realism into a different way of seeing.

This painting is titled *Guitare et Journal*.

LESSON OBJECTIVES

- Learn a definition of function
- Learn about properties and geometric representations of functions

NCTM STANDARDS

CONTENT		PROCESS	
	Number		Problem Solving
•	Algebra	•	Reasoning
	Geometry		Communication
	Measurement		Connections
	Data/Probability	•	Representation

Investigation
Testing for Functions

Here are three relationships, each in different form.

Relationship A

Input x	Output y
3	7
31	63
4.7	10.4
0	1
−11	−21
51	103

$y = 2x + 1$

Relationship B

$y = 47(1.10)^x$

x	y
−10	18.1
−8	21.9
−6	26.5
−4	32.1
−2	38.8
0	47.0
2	56.9
4	68.8
6	83.3

Relationship C

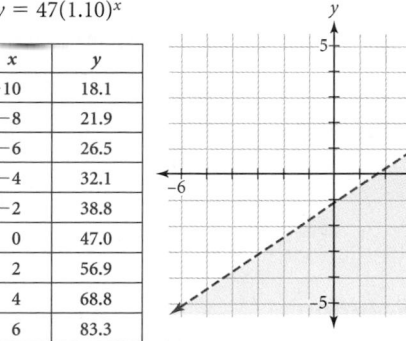

Step 1 | The table in Relationship A gives a few input and output values for a function. Make a graph of the complete function and write an equation. Use these three different expressions of the same relationship to answer Steps 2–5.

Step 2 | Enter possible input values into list L₁ and the corresponding output values into list L₂. Select an appropriate window and plot the points on your calculator. What is the complete domain and range of the graph?

Step 3 | Next, enter an equation for the relationship into Y₁. Graph the equation onto your point plot. What set of values can be entered into x? What is the resulting set of values for y?

Step 4 | Examine your table and graph. Does each input value have exactly one output value?

Step 5 | Move a vertical line, such as the edge of a ruler, from side to side on your graph. Count the number of intersections of your vertical line at each x-value to see if the graph represents a function. How does this result help you tell whether the relationship is a function?

Step 6 | Consider the equation in Relationship B. Express it as a graph and as a table. Use these three different expressions of Relationship B to again answer Steps 2–5.

Step 7 | Consider the graph in Relationship C. Express this relationship one other way and repeat Steps 2–5.

A function is a relationship between input and output values. Each input has exactly one output. The **vertical line test** helps you determine if a relationship is a function. If all possible vertical lines cross the graph once or not at all, then the graph represents a function. The graph does not represent a function if you can draw even one vertical line that crosses the graph two or more times.

Step 2 Relationship A
Input: all numbers
Output: all numbers

Step 3 For Relationship C students can enter the equation for the boundary line and then shade the inequality.

Step 4 Relationship A, yes; Relationship B, yes; Relationship C, no

Step 5 If all positions of the vertical line cause it to cross the graph in at most one point, then it is a function.

[Ask] "Is the inequality x < 5 a function?" [Some students may think of the graph on a number line, which is a horizontal line and passes the vertical line test. Remind students that we're dealing with two variables, and that the two-variable equivalent of x < 5 is 0y + x < 5. The graph of this inequality is a half-plane, which does not pass the vertical line test.]

If there's time, look back at the table for Relationship A and ask whether it represents a linear function.

Assessing Progress

You can assess students' abilities to create a table, graph, or equation representation of a function from one of the other forms and to plot points and graph functions on a calculator.

 Guiding the Investigation

Step 2 In Relationships A and C, students may limit their list of possible input values to those in the table or graph. That's acceptable. On the other hand, all real numbers are possible input values into the equation describing Relationship B. For any equation or inequality, a complete table is impossible.

Step 3 Relationship C is not an equation, but an inequality. If students enter an equation, it will probably be for the boundary line of the region representing the inequality.

Step 5 If you or a student is demonstrating this on a calculator projection panel, a piece of spaghetti makes a good straightedge.

Step 7 Several values of the output variable y are associated with any one value of the input variable x, so inequalities do not represent functions. Complete table is not possible.

Inequality:

$$y < -1 + \frac{2}{3}x$$

or

$$-2x + 3y < -3$$

Input: all numbers
Output: all numbers

SHARING IDEAS

Ask students to share their ideas on how to tell whether or not a graph represents a function. Connect the vertical line test to the definition of function: for each input value, there's only one output value.

[Ask] "What does function y = x + 1 do to numbers like 7?" [adds 1] "What does it do to t?" [t + 1] "to a + 5?" [a + 6]

See page 719 for answers to Steps 1 and 6.

Part b Remind students that the slope-intercept form of a line's equation is similar to the intercept form, but the constant term is last.

Writing the equation $x = 6$ as $0y + x = 6$ emphasizes that no matter what value is substituted for y the first term will be 0. **[Ask]** "Is this true for $y = 0$?" [Some students may hesitate, not being sure about multiplication by 0, before they say, "Yes."] Set an example for raising thoughtful questions. Another good question here is "What about horizontal lines? Are they graphs of functions?"

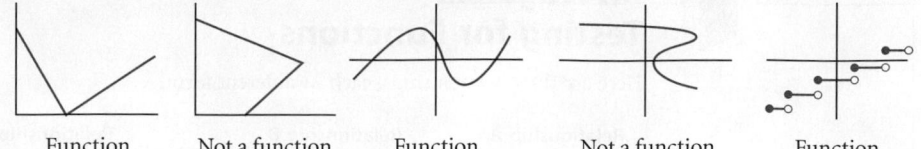

Function   Not a function   Function   Not a function   Function

You have learned many forms of linear equations. In the example, you will see whether or not all lines represent functions.

EXAMPLE

Name the form of each linear equation and use a graph to explain why it is or is not a function.

a. $y = 1 - 3x$ **b.** $y = 0.5x + 2$ **c.** $y = \frac{3}{4}x$ **d.** $2x + 3y = 6$

e. $y = 5 + 2(x - 8)$ **f.** $y = 7$ **g.** $x = 9$

► Solution

Each equation is written in one of the forms you have learned in this course. If you graph the equations, you can see that all of them except the graph for part g pass the vertical line test. So all the equations represent functions except for the one in part g.

a. This equation is in the intercept form $y = a + bx$.

b. This equation is in the slope-intercept form $y = mx + b$.

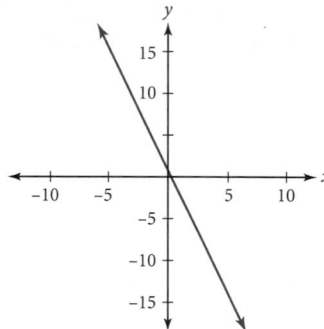
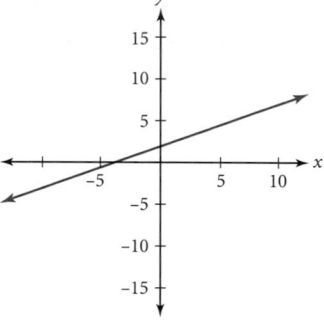

c. This equation is a direct variation in the form $y = kx$.

d. This equation is in the standard form $ax + by = c$.

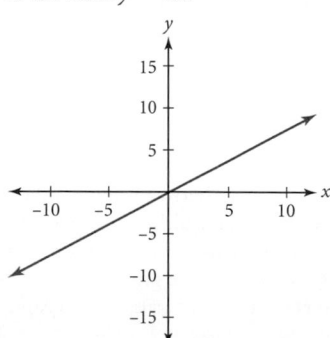
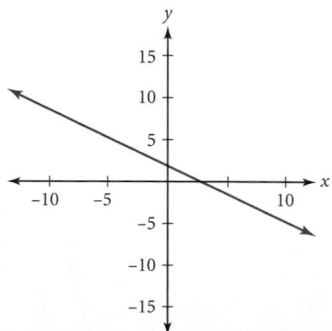

e. This equation is in the point-slope form $y = y_1 + b(x - x_1)$.

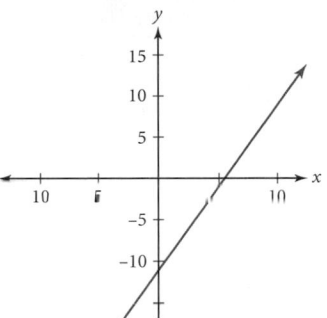

f. This equation is a horizontal line in the form $y = k$.

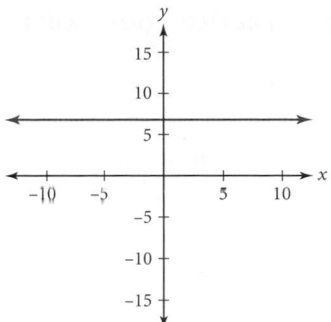

g. This equation is a vertical line in the form $x = k$.

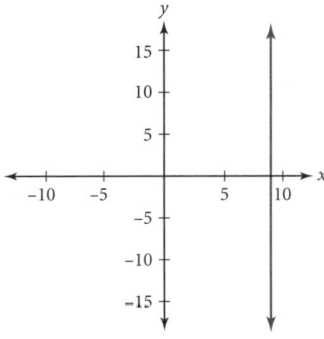

The graph of $x = 9$ fails the vertical line test. In fact, all vertical lines fail the vertical line test. If you rewrite the equation as $0y + x = 9$, you'll see that you can match infinitely many output values of y to one input value of x. (No matter what value you substitute for y, the factor 0 will drop the y-term out of the equation.) So this equation does not represent a function.

Carpenters use a tool called a level to determine if support beams are truly vertical.

As you work more with functions, you will be able to tell if a relationship is a function without having to consider its graph on the xy-plane. If the graph is shown, use the vertical line test. Otherwise, see if there is more than one output value for any single input value.

Closing the Lesson

Using rectangular coordinates a graph represents a function if and only if every vertical line through the graph intersects it at no more than one point.

[Link] In later courses students will encounter graphs using polar coordinates where functional relationships do not pass the vertical line test. Continue to emphasize that a function is a relationship such that every input produces one output.

Students practice determining whether graphs represent functions. Remind students to use complete sentences in writing out their assignment.

ASSIGNING HOMEWORK

Essential	1–4, 6, 11
Performance assessment	9, 10, 12
Portfolio	10
Journal	6, 8, 14
Group	5–8, 13
Review	15, 16

▶ Helping with the Exercises

2.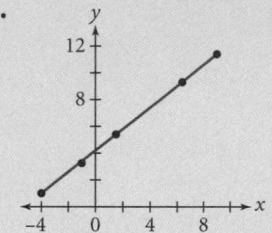

3.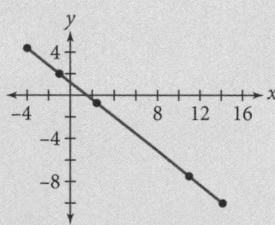

Exercise 4 This exercise depends on the previous three exercises. Be sure students justify their answers.

Exercise 5 [Ask] "Will every set of walking instructions represent a function?"

EXERCISES

You will need your calculator for problem **15.**

▶ Practice Your Skills

1. Use the equations to find the missing entries in each table.

a. $y = 4.2 + 0.8x$

Input x	Output y
−4	1
−1	3.4
1.5	5.4
6.4	9.32
9	11.4

b. $y = 1.2 - 0.8x$

Domain x	Range y
−4	4.4
−1	2
2.4	−0.72
11	−7.6
14	−10

2. On the same set of axes, plot the points in the table and graph the equation in problem 1a.

3. On the same set of axes, plot the points in the table and graph the equation in problem 1b.

4. Use the tables and graphs in problems 1–3 to tell whether or not the relationships in problem 1 are functions. Answers will vary. In the table, every input value produces exactly one output value. Both graphs in problems 2 and 3 pass the vertical line test. Both rules are functions.

▶ Reason and Apply

5. The graph at right describes another student's distance from you. What are the walking instructions for the graph? Does it represent a function?

6. Find whether or not each graph below represents a function. Does it pass the vertical line test?

a.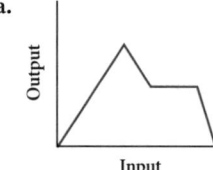

b.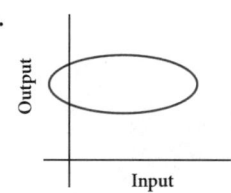

c.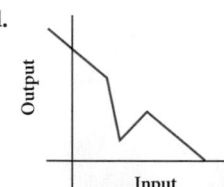

d.

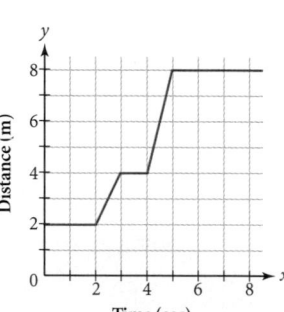

5. Sample answer: Start at the 2-m mark and stand still for 2 sec. Walk toward the 4-m mark at 2 m/sec for 1 sec. Stand still for another second. Walk toward the 8-m mark at 4 m/sec for 1 sec. Then stand still for 3 sec. Yes, the graph represents a function.

6a. yes

6b. No; many input values have two different output values.

6c. No; there is a vertical segment. All the points on the vertical segment have the same input value but different output values.

6d. yes

7. Does each relationship of the form (*input, output*) represent a function? If the relationship does not represent a function, find an example of one input that has two or more outputs.

  a. (*city, area code*)
  b. (*person, birth date*)
  c. (*last name, first name*)
  d. (*state, capital*)

8. The graphs of seven walks showing distance from a motion sensor are shown.

  i. 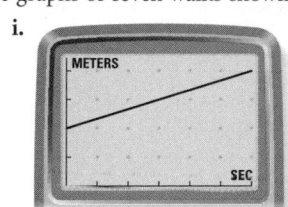   **ii.** 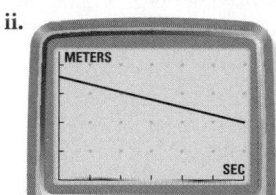   **iii.**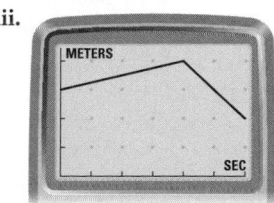

  iv. 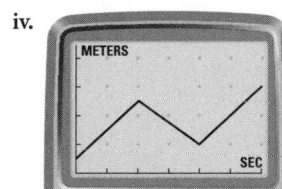   **v.** 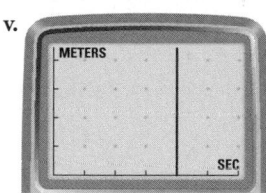   **vi.**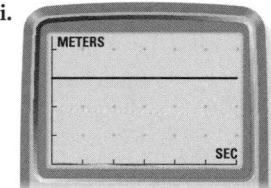

  vii.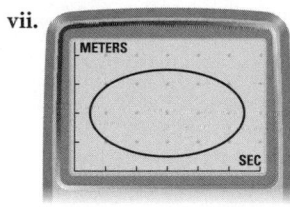

  a. Which graphs represent functions? graphs i, ii, iii, iv, and vi
  b. For which graphs is it not possible to write walking instructions? graphs v and vii
  c. What conclusion can you make?

Exercise 7 You might give another example. **[Ask]** "Is the relation (*musical note name, pitch*) a function?" [It depends. If several different notes are all called C, then the note name will have many different pitches, so it won't be a function. If the different C's have different names, like C′ and C″, then it will be a function.] You can also extend this exercise and give a preview of the Take Another Look at the end of the chapter by looking at inverse functions. **[Ask]** "If we interchange the input and output values, do we still have a function?" [(*area code, city*) no; (*birth date, person*) no; (*first name, last name*) no; (*capital, state*) yes.]

7a. No; Los Angeles, for example, has more than one area code (213, 310, ...).

7b. Yes; each person has only one birth date.

7c. No; the same last name will correspond to many different first names.

7d. Yes; each state has only one capital.

8c. Sample conclusion: It is not possible to walk a graph that does not represent a function.

<table>
</table>

Exercise 9 [Ask] "How can the table in part a be changed to make it a function?"

9a. Not a function; the input 3 has two different output values, 10 and 8.

9b. A function; each *x*-value corresponds to only one *y*-value.

9c. A function; each *x*-value corresponds to only one *y*-value.

Exercises 10 and 11 Both noncontinuous and continuous functions are acceptable. You might have students show each other a variety of answers.

Exercise 12 If students have difficulties filling in the tables, especially in 12c, encourage them to solve the equations for *y* in terms of *x*. In part c, *x* is a function of *y*, though *y* is not a function of *x*.

12a.

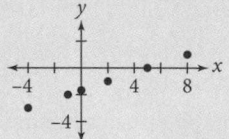

12b.

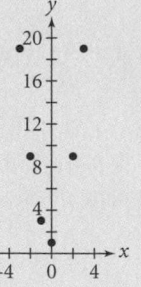

12c.

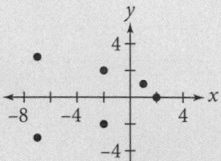

12d.

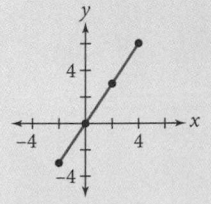

438 CHAPTER 8 Functions

9. Find whether or not each table of *x*- and *y*-values represents a function. Explain your reasoning.

a.
Input x	Output y
0	5
1	7
3	10
7	9
5	7
4	5
3	8

b.
Input x	Output y
3	7
4	9
8	4
5	5
9	3
11	9
7	6

c.
Input x	Output y
2	8
3	11
5	12
7	3
9	5
8	7
4	11

10. On graph paper, draw a graph that is a function and has these three properties:
   ▶ Domain of *x*-values satisfying $-3 \le x \le 5$
   ▶ Range of *y*-values satisfying $-4 \le y \le 4$
   ▶ Includes the points $(-2, 3)$ and $(3, -2)$

   Graphs must pass the vertical line test and pass through the points $(-2, 3)$ and $(3, -2)$.

11. On graph paper, draw a graph that is *not* a function and has these three properties:
   ▶ Domain of *x*-values satisfying $-3 \le x \le 5$
   ▶ Range of *y*-values satisfying $-4 \le y \le 4$
   ▶ Includes the points $(-2, 3)$ and $(3, -2)$

   Graphs will not pass a vertical line test, but they should include points $(-2, 3)$ and $(3, -2)$ and have the correct domain and range. Graphs of inequalities are possible.

12. Complete the table of values for each equation. Let *x* represent the input values, and let *y* represent the output values. Graph the points and find whether or not the equation describes a function. Explain your reasoning.

   a. $x - 3y = 5$ The graph is a line. This is a function; each *x*-value is paired with only one *y*-value.

x	2	8	−4	−1	0	5
y	−1	1	−3	−2	$-\frac{5}{3}$	0

   b. $y = 2x^2 + 1$ This is a function; each *x*-value is paired with only one *y*-value.

x	−2	3	0	−3	−1	±2
y	9	19	1	19	3	9

   c. $x + y^2 = 2$ The equation does not represent a function because there are two different *y*-values for many *x*-values.

x	−7	1	−2	−7	−2	2
y	±3	1	−2	−3	±2	0

   d. $x + 2y = 4x$ Tables will vary. Sample: $(0, 0), (2, 3), (-2, -3), (4, 6)$. The graph is a line and represents a function since each *x*-value is paired with only one *y*-value.

x					
y					

13. Identify all numbers in the domain and range of each graph.

a.

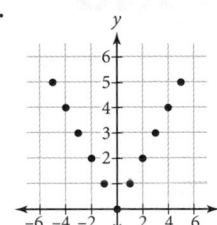

b.

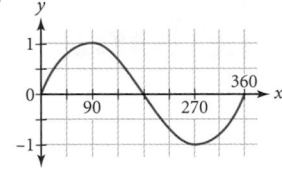

c.

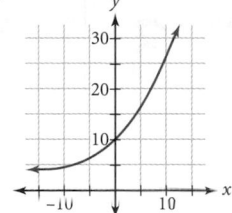

Exercise 13 The arrows on graph c indicate that the input values extend indefinitely. The output may or may not become 0 or negative.

13a. domain: $\{-5, -4, -3, -2, -1, 0, 1, 2, 3, 4, 5\}$;
range: $\{0, 1, 2, 3, 4, 5\}$

13b. domain: $0 \leq x \leq 360$;
range: $-1 \leq y \leq 1$

13c. domain: all numbers x;
range: all numbers y

14. Which letters of the alphabet pass the vertical line test?

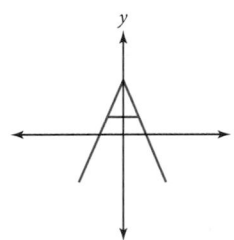

Some letters can represent functions, depending on the typeface. In some typefaces, graphs of the letters V and W are functions because they pass the vertical line test. Most letters are not functions because as graphs they fail the vertical line test.

▶ Review

15. If x represents actual temperature and y represents wind chill temperature, the equation

$$y = -52 + 1.6x$$

approximates the wind chill temperatures for a wind speed of 40 miles per hour. Enter this equation into Y₁ on your calculator and find the requested x- and y-values.

a. What x-value gives a y-value of $-15°$? Explain how you use the calculator table function to find this answer.

b. Enter

$$y = -15$$

into Y₂ on your calculator. Graph both equations. Explain how to use the graph to answer 15a.

16. Show how you can use an undoing process to solve these equations. (Hint for 16b: First, invert both fractions.)

a. $\dfrac{4(x - 7) - 8}{3} = 20$

b. $\dfrac{4.5}{x - 3} = \dfrac{2}{3}$

Exercise 15 This exercise reviews Lesson 6.1. Have copies of Calculator Note 1H available for students who may need to review graphing two equations.

15a. When $x = 23.125, y = -15$. Answers will vary. Zoom in on the table by changing the start values and the table increments (ΔTbl).

15b. Answers will vary. The lines intersect at the solution point.

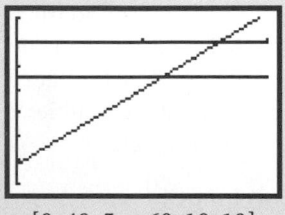

$[0, 40, 5, -60, 10, 10]$

Exercise 16 This exercise reviews Lesson 4.2.

16a. Equation:
$$\dfrac{4(x - 7) - 8}{3} = 20$$

Operation on x	Undo operation	Work backward
		$x = 24$
$- (7)$	$+ (7)$	17
$\cdot (4)$	$\div (4)$	68
$- (8)$	$+ (8)$	60
$\div (3)$	$\cdot (3)$	20

16b. Equation: $\dfrac{x - 3}{4.5} = \dfrac{3}{2}$

Operation on x	Undo operation	Work backward
		$x = 9.75$
$- (3)$	$+ (3)$	6.75
$\div (4.5)$	$\cdot (4.5)$	$\dfrac{3}{2}$

One day:

5 min	Example
20 min	Investigation
10 min	Sharing
5 min	Closing
10 min	Exercises

MATERIALS

• Water Depth (T), *optional*

TEACHING

A graph can be thought of as telling a story about how variables change values. Continue to model, and have students model, the use of complete sentences when writing about the mathematics.

Some students will be more comfortable with the terms *x-axis* and *y-axis* than with *horizontal axis* and *vertical axis*.

One step Ask students for examples of graphs that tell stories. They might mention graphs giving walking directions. Have each group choose a scenario from Step 6 of the investigation and draw a graph telling that story. If students run into difficulties, refer them to the example.

▶ EXAMPLE

This example illustrates how a graph can tell a story. Use the Water Depth transparency as you ask students about the various time periods. Be sure they understand that the different rates are depicted as slopes. Identify the variables to emphasize further the concepts of independent (input) and dependent (output) variables and domain and range.

LESSON

8.3

Graphs of Real-World Situations

A picture is worth a thousand words.

NAPOLEON

Like pictures, graphs communicate a lot of information. So you need to be able to draw and make sense of graphs. In previous chapters you learned to interpret bar graphs, circle graphs, histograms, and box plots. Then you learned to graph data from recursive routines and equations. Some graphs were lines and others were curves.

In this lesson you will apply many of those ideas as you begin to explore the graphs of functions. You will learn how to draw and interpret graphs of some real-world situations.

Frida Kahlo (1907–1954), a Mexican artist, is well known for her fascinating self-portraits. After surviving a bus accident, she had 32 surgical operations, painting many works from her bed. The contrasts between *Self-Portrait with Changuito* (left) and *Self-Portrait with Dog Ixcmintli and Sun* (right) reflect differences in the output of her work at different times in her life.

EXAMPLE This graph shows the depth of the water in a leaky swimming pool. Tell what quantities are varying and how they are related. Give possible real-world events in your explanation.

▶ **Solution** The graph shows that the water level or depth changes over a 15-hour time period. At the beginning, when no time has passed, $t = 0$, and the water in the pool is 2 feet deep, so $d = 2$. During the first 6-hour interval ($0 \leq t \leq 6$), the water level drops. The leak seems to get worse as time passes. When $t = 6$ and $d = 1$, it seems that someone starts to refill the pool. The water level rises for the next 5 hours, during the interval $6 \leq t \leq 11$. At $t = 11$, the water reaches its highest level at just above 3 feet, so $d = 3$. At the

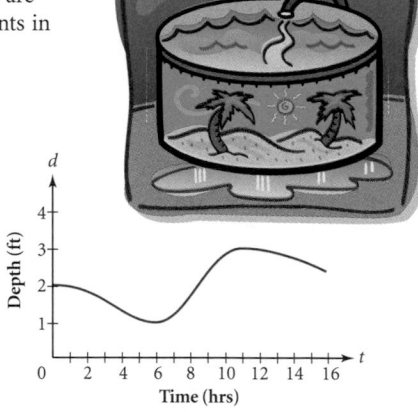

LESSON OBJECTIVES

• Construct and interpret graphs that describe real-world situations

• Learn the terminology of independent and dependent variables

NCTM STANDARDS

CONTENT		PROCESS
	Number	• Problem Solving
•	Algebra	• Reasoning
	Geometry	• Communication
	Measurement	• Connections
	Data/Probability	• Representation

11-hour mark, the in-flowing water must have been turned off. The pool still has a leak, so the water level starts to drop again.

In the example the depth of the water is a function of time. That is, the depth depends on how much time has passed. So, in this case, depth is called the **dependent variable.** Time is the **independent variable.** When you draw a graph, put the independent variable on the horizontal axis and put the dependent variable on the vertical axis.

On the graph of this function, you can see domain values that are possible for the independent variable. In the example the domain is the set of all instants of time from 0 through 15 hours. You express this interval as $0 \leq x \leq 15$, where x is the independent variable representing time.

You can also see the values that are possible for the dependent variable. In the example the range is the set of all numbers from 1 through about 3.3. You express this as $1 \leq y \leq 3.3$, where y is the dependent variable representing the depth of the water. Notice that the lowest value for the range does not have to be the starting value when x is zero.

Investigation
Identify the Variables

The people in the school gymnasium before a volleyball game consist of the players, coaches, and the people working the event. Slowly the fans arrive for the match. Just before the first game, the people are coming in as fast as the tickets can be sold. After the match is over, most of the parents and fans leave. Then more students arrive for the after-game dance. Most of the students leave after an hour. The people that remain are the ones who have been working at the gym all night long.

Step 1 independent variable x: time in hours

Step 2 Sample answer: From 0 through 6 including fractions

Step 3 dependent variable y: number of people

Step 4 Sample answer: From less than 10 to several hundred: positive whole numbers only

Step 1 Define an independent variable for this situation. Decide what units of measurement you will use for this variable.

Step 2 What are reasonable values for the domain? Are they positive or negative numbers? Whole numbers or decimals?

Step 3 Define the dependent variable for this situation. What are the units of measurement?

Step 4 What are reasonable values for the range? Are they positive or negative numbers? Whole numbers or decimals?

Step 5 Draw and label a graph for this situation.

You might want to revisit this example after students do the investigation. You need not insist on distinctions between strict and nonstrict inequalities. Help students to find the correct direction of the inequality sign.

For review, you might ask whether every line is the graph of some function. Be patient. Wait at least 40 seconds before answering the question yourself. Accept all answers and have the class critique them. Vertical lines are not graphs of functions, but all other lines are. Some students may think that horizontal lines are not. Constant functions, such as $f(x) = -3$, which give the same output value for every input value, may not be very interesting, but they're functions.

 Guiding the Investigation

Step 2 Times in seconds might be negative if, for example, time 0 is taken to be the beginning of the first game.

Step 5 Sample graph:

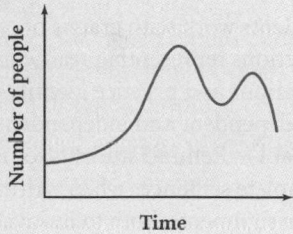

LESSON OUTLINE

One day:

25 min	Investigation
10 min	Sharing
10 min	Example
5 min	Closing
5 min	Exercises

MATERIALS

- Calculator Notes 8A, 8B
- A Graphic Message (W)

A commonly used notation simplifies the writing of accurate statements about the behavior of functions. Inject as much energy as you can into the class session. At this point in the course, students may be sagging.

One step Ask that students write out complete sentences describing what the function given by $y = 2x + 4$ does to numbers -5, -1, $\frac{-1}{2}$, 0, $\frac{2}{3}$, 3, and π. They might write "The function given by $y = 2x + 4$ has value 8 when x is 3" or "If $x = 3$, then $y = 8$." Then introduce the shorthand $f(x) = 2x + 4$ and $f(3) = 8$. Point out that the parentheses don't mean multiplication. **[Ask]** "How do you find $f(3)$ from the graph of $f(x)$?" [Go up from 3 on the x-axis and over on the y-axis to 8.] Finally, ask them to evaluate and decode the expressions in Step 2 and 3 of the investigation.

INTRODUCTION

Remind students of the inconvenience of writing out accurate statements about the behavior of functions, and show them the standard function notation. As you look at the graph with students, be sure they see $f(6) = 2$. When $f(x) = 1$, x is 4 or about 9.3.

Function Notation

Every function defines a relationship between an input (independent) variable and an output (dependent) variable. **Function notation** uses parentheses to name the input or independent variable for the function. For instance, $y = f(x)$, which you read as "y equals f of x," says "y is a function of x" or "y depends on x." (In function notation, the parentheses do *not* mean multiplication.)

The theory that has had the greatest development in recent times is without any doubt the theory of functions.

VITO VOLTERRA

Albert Einstein writes mathematical notation in a lecture to scientists in 1931.

You can show some functions with an equation. For example, the equation $y = 2x + 4$ represents a function, so you can write it as $f(x) = 2x + 4$. The notation $f(3)$ tells you to substitute 3 for x in the equation $y = 2x + 4$. So $f(3) = 2(3) + 4$. The value of $f(x)$ when $x = 3$ is 8. By itself, f is the name of the function. In this case, its rule is $2x + 4$.

Not all functions are expressed as equations. The graph shows a new function, $f(x)$. No rule or equation is given, but you can still use function notation to find output values. For example, on the graph below, the point at $x = 4$ has the coordinates $(4, f(4))$ or $(4, 1)$. The value of y when x is 4 is $f(4)$. So $f(4) = 1$. Check that $f(2)$ is 4. What is the value of $f(6)$? Can you find two x-values for which $f(x) = 1$?

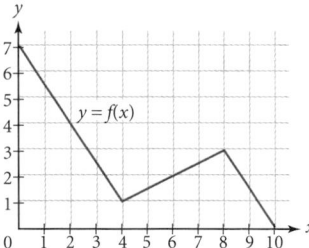

In the next investigation you will learn more about using function notation with graphs.

Investigation
A Graphic Message

You will need

• the worksheet
  A Graphic Message

Step 1
domain: $0 \leq x \leq 26$;
range: $0 \leq y \leq 20$

Step 2

Step 3

Step 4
Step 4 Step 2: EULER;
Step 3: LEONHARD

In this investigation you will apply function notation to learn the identity of the mathematician who introduced functions.

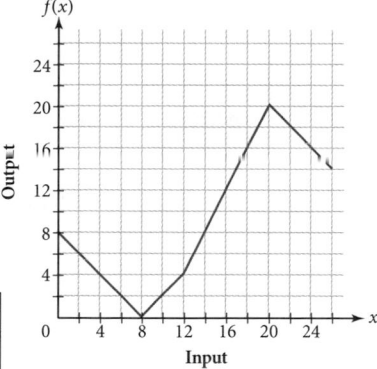

Step 1 Describe the domain and range of the function f in the graph.

Step 2 Use the graph to find each function value in the table. Then do the indicated operations.

Notation	Value
$f(3)$	5
$f(18) + f(3)$	21
$f(5) \cdot f(4)$	12
$f(15) \div f(6)$	5
$f(20) - f(10)$	18

Step 3 Use the rules for the order of operations to evaluate the expressions that involve function values. Do the operations inside parentheses first. Then find the function values before doing the remaining operations. Write your answers in a table.

Notation	Value
$f(0) + f(1) - 3$	12
$5 \cdot f(9)$	5
x when $f(x) = 10$	15
$f(9 + 8)$	14
$\dfrac{f(17) + f(10)}{2}$	8
$f(8 \cdot 3) - 5 \cdot f(11)$	1
$f(4 \cdot 5 - 1)$	18
$f(12)$	4

Step 4 Think of the numbers 1 through 26 as the letters A through Z. Find the letters that match your answers to Step 2 to learn the mathematician's last name. Find the letters that match your answers to Step 3 to discover the first name.

The mathematician whose name you decoded invented much of the mathematical notation in use today. In the example that follows, you will practice function notation with an equation.

Guiding the Investigation

The purpose of this investigation is to help students become more adept with function notation. Reinforce the ideas that input variables are called *independent variables* and output variables are known as *dependent variables*. Also emphasize that in the form $f(x) = y$ or $y = f(x)$, the variable x is independent and y is dependent.

Steps 2 and 3 With all of the arithmetic here, students may need reminding that the parentheses in function notation do not indicate multiplication.

Step 4 Have students consult the Letter-Number Chart you gave them in Lesson 8.1 or copies posted around the room.

SHARING IDEAS

Leonhard Euler's last name is pronounced "OIL-er." From Switzerland, he was the greatest eighteenth-century mathematician. Euler's works are still being compiled and are expected to fill 81 volumes. He completed much of his mathematical work after he became totally blind.

Point out the quotation opening the lesson. The power of functions has come from their notation. Functions make calculus concepts such as continuity and limits easier to define and talk about. By focusing on graphs that pass the vertical line test (and thus ignoring many other common figures, such as circles), mathematicians have been able to use the function notation to develop the concept of function into a very sophisticated one.

Assessing Progress
Through your observation of group work, you can assess students' understanding of **domain** and **range** and their abilities to associate numbers with letters for decoding.

LESSON OBJECTIVES

• Learn function notation

• Evaluate functions by substitution, by using the graphs made by hand, and on the graphing calculator

• Write and interpret functions that describe real-world data

NCTM STANDARDS

CONTENT		PROCESS	
	Number		Problem Solving
•	Algebra	•	Reasoning
	Geometry		Communication
	Measurement		Connections
	Data/Probability	•	Representation

This example gives students more practice with function notation. Using the calculator to check answers will reinforce the idea that the number inside the parentheses denotes the *x*-value (input) that is substituted into the function expression to get the *y*-value (output). Have Calculator Note 8A available for students. You might extend the example to ask about $f(-40)$. A temperature of -40 degrees C is the same as -40 degrees F.

Most TI models allow you to evaluate functions by using the notation $Y_1(x)$. However, Casio calculators interpret the parentheses as multiplication between the value in Y_1 and the value currently stored in *x*. Consult the calculator manual to see how to enter and evaluate functions on your calculator.

Closing the Lesson

Using notation such as $f(x) = 5x - 4$ to describe a function allows saying "$f(3) = 5(3) - 4 = 11$" instead of "the value of $y = 5x - 4$ is 11 when $x = 3$."

BUILDING UNDERSTANDING

Students practice using function notation and evaluating functions.

ASSIGNING HOMEWORK

Essential	1–5, 9, 12, 13
Performance assessment	6, 14
Portfolio	7
Journal	9, 11
Group	9, 10
Review	8, 11, 12, 15–17

▶ **Helping with the Exercises**

Exercise 1 [Alert] The name of one of these functions is *g* rather than *f*.

EXAMPLE You can use the function $f(x) = \frac{9}{5}x + 32$ to find the temperature $f(x)$ in degrees Fahrenheit for any given temperature *x* in degrees Celsius. Find the specified value.

a. $f(15)$ **b.** $f(-10)$

c. $f(5)$ **d.** *x* when $f(x) = -4$

▶ **Solution** In each case, substitute the value in parentheses for *x* in the function.

a. $f(15) = \frac{9}{5}(15) + 32$
$f(15) = 27 + 32$
$f(15) = 59$

b. $f(-10) = \frac{9}{5}(-10) + 32$
$f(-10) = -18 + 32$
$f(-10) = 14$

c. $f(5) = \frac{9}{5}(5) + 32$
$f(5) = 9 + 32$
$f(5) = 41$

d. $-4 = \frac{9}{5}x + 32$
$-36 = \frac{9}{5}x$
$-20 = x$

[▶ 🖳 See **Calculator Note 8A** to learn how to evaluate functions on your calculator. ◀]

Some calculators use the notation $Y_1(x)$ instead of $f(x)$. The function depends on the equation you have entered into Y_1. Other calculators allow you to directly define the function *f* as an expression.

EXERCISES

You will need your calculator for problems **1, 2,** and **9.**

▶ **Practice Your Skills**

1. Find each function value for $f(x) = 3x + 2$ and $g(x) = x^2 - 1$ without using your calculator. Then enter the equation for $f(x)$ into Y_1 and the equation for $g(x)$ into Y_2. Use function notation on your calculator to check your answers. [▶ 🖳 See **Calculator Note 8A** to review function notation on your calculator. ◀]

   a. $f(3)$ **b.** $f(-4)$ **c.** $g(5)$ **d.** $g(-3)$

2. Find the *y*-coordinate corresponding to each *x*-coordinate if the functions are $f(x) = -2x - 5$ and $g(x) = 3.75(2.5)^x$. Check your answers with your calculator.

   a. $f(6)$ $-2(6) - 5 = -17$ **b.** $f(0)$ $-2(0) - 5 = -5$ **c.** $g(2)$ $3.75(2.5)^2 = 23.4375$ **d.** $g(-2)$ $3.75(2.5)^{-2} = 0.6$

3. Use the graph of $y = f(x)$ at right to answer each question.

   a. What is the value of $f(4)$? $f(4) = 0$
   b. What is the value of $f(6)$? $f(6) = 4$
   c. For what value or values does $f(x) = 2$? $f(2) = 2, f(5) = 2$
   d. For what value or values does $f(x) = 1$?
   e. How many *x*-values make the statement $f(x) = 0.5$ true? Three
   f. For what *x*-values is $f(x)$ greater than 2? $x > 5$
   g. What are the domain and range shown on the graph?

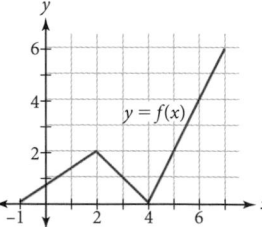

1a. $3(3) + 2 = 11, Y_2(3) = 11$

1b. $3(-4) + 2 = -10, Y_2(3) = -10$

1c. $(5)^2 - 1 = 24, Y_2(5) = 24$

1d. $(-3)^2 - 1 = 8, Y_2(-3) = 8$

Exercise 3f Students may find this confusing. Remind them that the $f(x)$ values are vertical heights in the graph.

3d. $f(0.5) = 1, f(3) = 1, f(4.5) = 1$

3g. domain: $-1 \leq x \leq 7$; range: $0 \leq y \leq 6$

4. APPLICATION The graph of the function $y = f(x)$ below shows the temperature y outside at different times x over a 24-hour period.

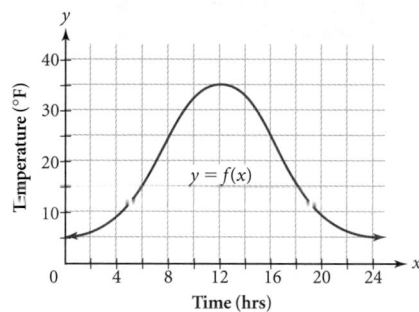

a. What are the dependent and independent variables?

b. What are the domain and range shown on the graph? domain: $0 \le x \le 24$; range: $5 \le y \le 35$

c. Use function notation to represent the temperature at 10 hours. $f(10)$

d. Use function notation to represent the time at which the temperature is 10°F. $f(x) = 10$

5. Use function notation to write the equation for a line through each pair of points.

a. $(0, 5)$ and $(1, 12)$ $f(x) = 7x + 5$

b. $(1, 5)$ and $(2, 12)$ $f(x) = 5 + 7(x - 1)$

 Reason and Apply

6. The function $f(x)$ gives the lake level over the past year, with x measured in days and y, that is, the $f(x)$ values, measured in inches above last year's mean height.

a. What is the real-world meaning of $f(60)$?

b. What is the real-world meaning of $f(x) = -3$?

c. What is an interpretation of $f(x) = f(150)$?

7. The graph shows part of the function $f(x) = 500(0.80)^x$.

a. What is the dependent variable and what are its units?

b. What is the independent variable and what are its units?

c. What part of the domain is pictured? What is the domain of the function? $0 \le x \le 10$; all real numbers x

d. What part of the range is pictured? What is the range of the function? $53 < y \le 500, y > 0$

e. What is $f(0)$ for this graph? 500

f. Find the value of x when $f(x) = 200$. about 4 hours

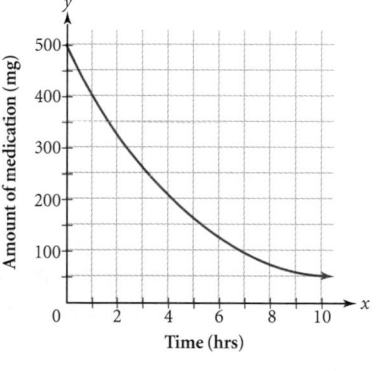

8. APPLICATION A bacteria population decreases at the rate of 8.5% an hour. There are 650 bacteria present at the start. The time it takes for the population to decrease to half its original size is called its half-life.

a. Write an equation and graph the function that describes this population decay.

b. Graph the line that represents half the starting amount of bacteria. Find the point of intersection of this line with the population decay function.

c. What is the real-world meaning of your answer in 8b? After 7.803 hours, there are 325 bacteria present.

4a. The dependent variable, y, is temperature in degrees Fahrenheit; the independent variable, x, is the time in hours.

6a. the level of the lake on the 60th day of the year

6b. At a certain time, the level was 3 inches below last year's mean.

Exercise 6c Students may find this difficult. Encourage them to think about it, taking it apart. **[Ask]** "What's $f(150)$? What if you rewrite the equation with that number in place of $f(150)$?" [$f(150)$ is the height of the lake on day 150.] **[Ask]** "Does function f have two output values for some input value, making it not a function after all?" [No, it has two input values that give the same output value, but there is only one output value for an input value.]

6c. On certain days the level was the same as it was on day 150.

Exercise 7d At $x = 10, y > 53$, but looking at the graph students might say the range pictured is between 50 and 500. Students may need to think to realize that the range includes all positive numbers. Remind them that x can be negative.

7a. amount of medication in milligrams

7b. time in hours

Exercise 8 In 8b students are asked to graph the horizontal line $y = 325$. This exercise reviews Lesson 7.7.

8a. $f(x) = 650(1 - 0.085)^x$

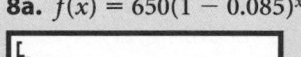

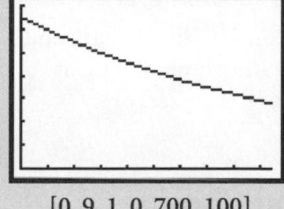

$[0, 9, 1, 0, 700, 100]$

8b. The point of intersection is $(7.8, 325)$. So the population is 325 after 7.8 hours.

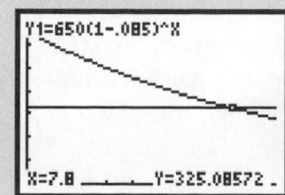

Exercise 9 Students need Calculator Note 8B for this exercise. Part b foreshadows Lesson 8.7. Some students may claim that $\sqrt{}$ is not a function, because positive numbers have more than one square root. For example, if the input value is 4, is the output value 2 or -2? Praise such thinking. Suggest they use their calculators to see that the symbol $\sqrt{}$ means the positive square root, so it designates a function.

9. Answers will vary.

9a. Domain: all real numbers, range: non-negative numbers, same input always gives same output—function.

9b. Domain: all non-negative numbers or lists containing non-negative numbers. Range: non-negative numbers or lists of non-negative numbers. Same input always gives same output—function.

9c. Domain: lists of any size of real numbers. Range: all real numbers. Same input always gives same output—function.

9d. Two answers are possible. (1) This command does not take an input, but it gives one output, such as .471359732. It will give a different output each time. It is not a function. (2) This command takes one input, the seed. It gives one output that depends on the seed. It will always give the same output for the same seed. It is a function.

11a. $f(x)$: independent variable x is time in seconds; dependent variable y is height in meters. $g(x)$: independent variable x is time in seconds; dependent variable y is velocity in meters per second.

11b. For $f(x)$: domain $0 \le x < 3.2$; range $0 \le y \le 50$. For $g(x)$: domain $0 \le x < 3.2$; range $-31 \le y \le 0$.

9. Many of the commands in your calculator are programmed as functions. Try each command several times with a variety of inputs. Describe the allowable input and corresponding output of each command. If you think the command is a function, describe its domain and range. [▶ 🖳 See **Calculator Note 8B** to access and use these commands. ◀]

   a. the square (x^2) command
   b. the square root $\left(\sqrt{}\right)$ command
   c. the sum of a list command
   d. the random command

10. Use the function $f(x) = \frac{5}{9}(x - 32)$ to convert temperatures in degrees Fahrenheit (x-values) to temperatures in degrees Celsius ($f(x)$-values) and vice versa.

   a. 72°F $f(72) \approx 22.2°C$
   b. $-10°F$ $f(-10) \approx -23.3°C$
   c. 20°C $f(x) = 20; x = 68°F$
   d. $-5°C$ $f(x) = -5; x = 23°F$

11. The graphs of $f(x)$ and $g(x)$ below show two different aspects of an object dropped straight down from the Tower of Pisa.

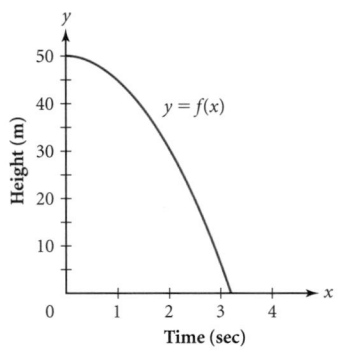 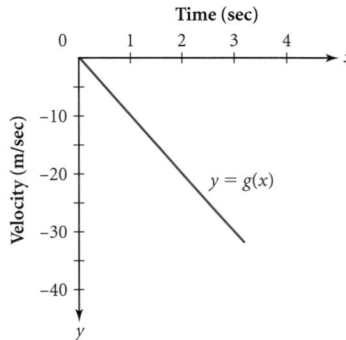

Answer these questions for the graph of each function.
   a. What are the dependent and independent variables?
   b. What are the domain and range?
   c. Describe a real-world sequence of events for each graph.
   d. About how far does the ball drop in the 1st second (from $x = 0$ to $x = 1$)?
   e. About how far does the ball drop in the 2nd second (from $x = 1$ to $x = 2$)?
   f. At what speed does the object hit the ground?

12. Could this set of ordered pairs represent a function? If so, what are its domain and range values?

   $(-2, 3), (3, -2), (1, 3), (0, -2)$

Yes, it could represent a function even though different inputs have the same output. Domain: $\{-2, 0, 1, 3\}$; range: $\{-2, 3\}$.

11c. Answers will vary. For the graph of $f(x)$, the ball is dropped from an initial height of 50 meters. It hits the ground after about 3.2 seconds. At the moment the ball is dropped, its velocity is 0 m/sec. For the graph of $g(x)$, the velocity starts at 0 m/sec and changes at a constant rate, becoming more and more negative.

11d. In the 1st second, the ball falls about 5 m, from 50 m at $x = 0$ to about 45 m at $x = 1$.

11e. In the 2nd second, the ball falls about 15 m, from about 45 m at $x = 1$ to about 30 m at $x = 2$.

11f. For the graph of $f(x)$, the ball hits the ground after about 3.2 seconds. For the graph of $g(x)$, at $x \approx 3.2$ the velocity is about -31 m/sec, or $g(x) \approx -31$.

Exercises 12 and 13 These exercises review Lesson 8.2.

13. Could this set of ordered pairs represent a function? Explain your reasoning.

(3, −2), (−2, 3), (3, 1), (−2, 0)

14. Use the graph of $f(x)$ at right to evaluate each expression. Write your answers as a number sequence. Then think of the numbers 1 through 26 as the letters A through Z to decode a message.

a. $f(8) + 6$ 6

b. $f(20) + 1$ 21

c. the sum of two x-values that give $f(x) = 8$ 14

d. $f(0) − 4$ 4

e. $f(7)$ 1

f. x when x is an integer and $f(x) = 15$ 25

g. $f(18) + f(5)$ 19

h. (the sum of two x-values that give $f(x) = 16$) ÷ 42 1

i. (x when $f(x) = 12$) − 8 8

j. $\dfrac{f(25)}{3}$ 5

k. $f(7) + f(8)$ 1

l. the largest domain value − the largest range value − 2 4

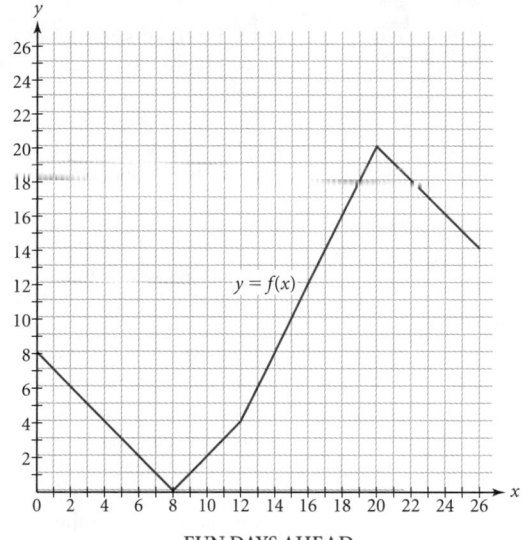

$y = f(x)$

FUN DAYS AHEAD

Review

15. Write each expression in exponential form without using negative exponents.

a. $(2^3)^{-3}$ $\dfrac{1}{2^9}$

b. $(5^2)^5$ 5^{10}

c. $(2^4 3^2)^3$ $2^{12} 3^6$

d. $(3^2 5^3)^{-4}$ $\dfrac{1}{3^8 5^{12}}$

16. Find the slope for the line through each pair of points.

a. (1, 3) and (−2, 6) −1

b. (−4, −5) and (7, 0) $\dfrac{5}{11}$

c. (−3, 6) and (9, 6) 0

17. Solve each equation.

a. $2x − 5 = 7x + 15$ $x = −4$

b. $3(x + 6) = 12 − 5x$ $x = −0.75$

c. $\dfrac{7(8 − x)}{4} = x + 3$ $x = 4$

13. No, the input −2 has two different output values, 3 and 0, and the input 3 has two different outputs, 1 and −2.

Exercise 15 This exercise reviews Chapter 7.

Exercise 16 This exercise reviews Lesson 5.1.

Exercise 17 This exercise reviews Chapter 4.

PLANNING

LESSON OUTLINE

One day:

5 min	Introduction
30 min	Investigation
5 min	Sharing
5 min	Example and Closing
5 min	Exercises

MATERIALS

- Large sheets of paper and markers, one set per group
- Real-World Situations (W), *optional*

TEACHING

The ideas of increasing and decreasing functions and rates of change are useful in representing real-world situations with functions.

One step Assign each group one of the situations on the Real-World Situations worksheet (also on page 454 of the student text) and ask them to represent the situation with a graph, drawn on a large sheet of paper. As you circulate, plant the idea (if necessary) in one or two groups that their graphs may need to be discrete. Post the sheets of paper, and ask the groups to come up with words describing the graphs so that they can be compared and contrasted. During the discussion, focus on the terms *linear, nonlinear, increasing, decreasing, rate of change, continuous,* and *discrete.*

INTRODUCTION

Be careful that your language doesn't equate "the graph" with "the function." The graph is a picture, but the function is a relationship between sets.

Of all the passions, the passion of learning is that which contributes the most to our happiness. It is, of all, that which will make us the least dependent on others.

EMILIE DU CHATELET

Interpreting Graphs

News reports and advertisements use graphs. So do science articles and political debates. To "read" a graph, you have to understand how the quantities in the graph relate to each other, how they make the graph go up or down or level off.

The function values in a graph can change at a constant rate or at a varying rate as the *x*-values of a function increase steadily. A function is **linear** if, as *x* increases at a constant rate, the function values change at a constant rate. Graphs A and D on the opposite page show linear functions. Here is another linear function. What is the constant rate of change for the function values in the graph below?

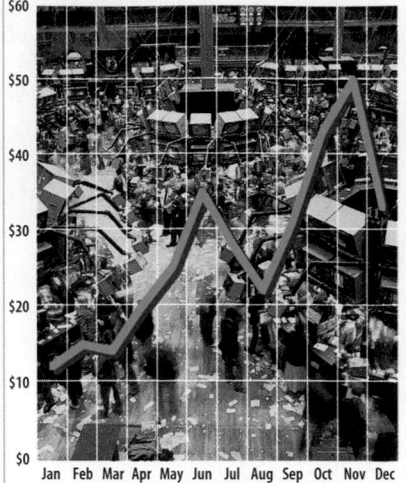

Traders on the floor of the New York Stock Exchange use graphs to show stock prices.

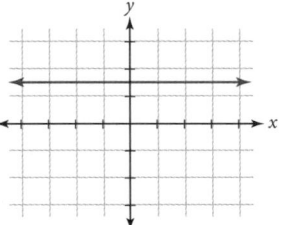

A function is **nonlinear** if, as *x* changes at a constant rate, the function values change at a varying rate. Graphs B, C, E, and F on the opposite page show nonlinear functions. Here is another nonlinear function.

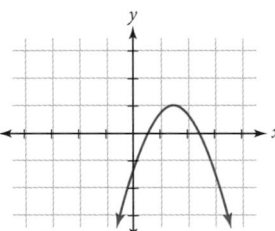

The function shown in the graph above rises, peaks, and then falls.

Functions with graphs like those in C and F of investigation Steps 1 and 2 are sometimes called *piece-wise linear,* because their graphs consist of line segments or rays. The first graph on page 452 of the student text is a straight line, so it represents a linear function. Its rate of change is the slope of the graph, 0.

Many graphs do not show the whole domain and the whole range of the function. Often, the function whose graph you're looking at has limitless domain and range. The graph below cannot show the whole domain of the function, but it *can* show the whole range.

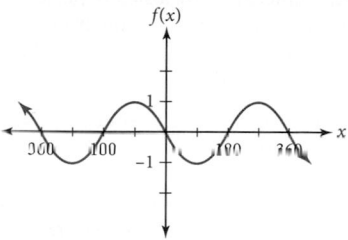

In the investigation that follows, you'll discover another aspect of functions and use their graphs to describe real-world situations.

Investigation
Matching Up

First, you'll consider the concepts of increasing and decreasing functions.

Step 1 | These are graphs of *increasing functions*. What do the three graphs have in common? How would you describe the rate of change in each?

Graph A

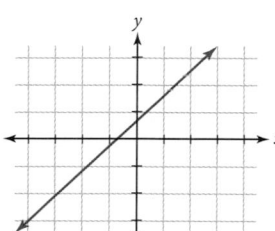

Graph B

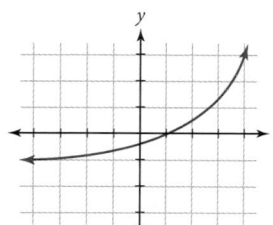

Graph C

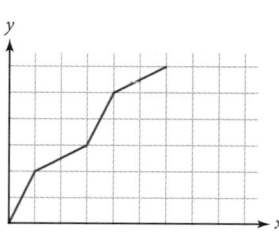

Step 2 | These are graphs of *decreasing functions*. What do these three graphs have in common? How are they different from those in Step 1? How would you describe the rate of change in these graphs?

Graph D

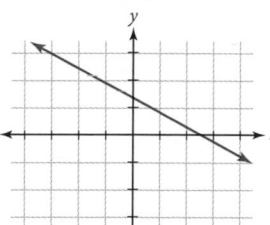

Graph E

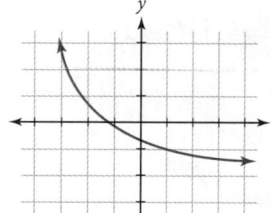

Graph F

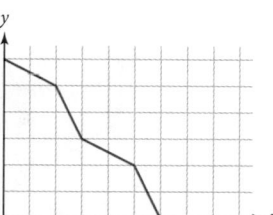

You can review the meaning of *range*. **[Ask]** "Assuming each grid mark represents 1 unit, what is the range for each function graphed?" [Possible answers: The first graph on page 452: $y = \frac{3}{2}$; the second graph: $y \leq 1$. Graphs A and D on page 453: all numbers; Graph B: $y > -1$; Graphs C and F: $0 \leq y \leq 6$; Graph E $y > -\frac{3}{2}$.]

Guiding the Investigation

Step 1 Sample answer: All three graphs rise from left to right. The function values increase as the x-values increase. The rate of change can be constant or vary, but is positive.

Step 2 Sample answer: All three graphs fall from left to right. Unlike in Step 1, the function values decrease as the x-values increase. The rate of change can be constant or vary, but is negative.

LESSON OBJECTIVES

- Describe, read, and interpret graphs of real-world situations using the terms *linear, nonlinear, increasing, decreasing, rate of change, continuous,* and *discrete*
- Make appropriate graphs of real-world situations

NCTM STANDARDS

CONTENT		PROCESS	
	Number		Problem Solving
●	Algebra	●	Reasoning
	Geometry	●	Communication
●	Measurement	●	Connections
	Data/Probability	●	Representation

Step 3 Situation A: time is independent, number of deer is dependent. Situation B: time is independent, hours of daylight is dependent. Situation C: width is independent, area is dependent. Situation D: time is independent, temperature difference is dependent.

Step 4 Encourage students to create graph descriptions that are more than a repetition of the situation in the description in the student text.

Steps 4–5 Sample description: Situation A matches Graph 5, which is increasing with a faster and faster rate of change, then slows to become linear with a small constant rate of change. Situation B matches Graph 3, which starts with a slow rate of change that speeds up, levels off and peaks, then decreases quickly at first then more slowly and levels off again. Situation C matches Graph 1, which increases with a steep rate of change that slows, levels off and peaks, then decreases, at first with a slow rate of change, and then faster and faster. Situation D matches Graph 4, which decreases, first at a fast rate of change, and then more slowly.

SHARING IDEAS

As students share their descriptions, clarify that "the function increases" means that the *function values* increase. Because we always read a graph from left to right, "as *x* increases" is often an assumed condition when using the terms *increasing function* and *decreasing function*.

[Language] Model correct use of both the adjectives *increasing* and *decreasing* as well as the verbs *increase* and *decrease*. It might be helpful to show students that they already know "*x* increases" as you move right on the *x*-axis and that "*y* increases" as you go up on the *y*-axis.

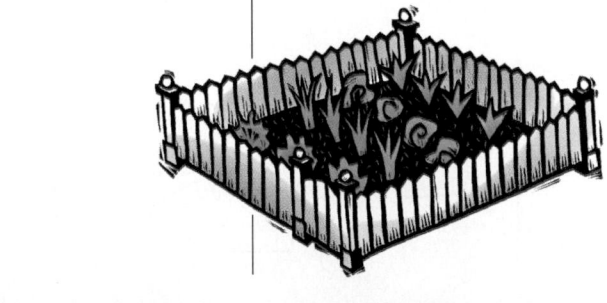

In Steps 3–5 you'll use this vocabulary to find and describe four of the graphs that match each of these real-world situations.

Situation A During the first few years, the number of deer on the island increased by a steady percentage. As food became less plentiful, the growth rate started slowing down. Now, the number of births and deaths is about the same.

Situation B In the Northern Hemisphere the amount of daylight increases slowly from January through February, faster until mid-May, and then slowly until the maximum in June. Then it decreases slowly though July, faster from August until mid-November, and then slowly until the year's end.

Situation C If you have a fixed amount of fencing, the width of your rectangular garden determines its area. If the width is very short, the garden won't have much area. As the width increases, the area also increases. The area increases more slowly until it reaches a maximum. As the width continues to increase, the area becomes smaller more quickly until it is zero.

Situation D Your cup of tea is very hot. The difference between the tea temperature and the room temperature decreases quickly at first as the tea starts to cool to room temperature. But when the two temperatures are close together, the cooling rate slows down. It actually takes a long time for the tea to finally reach room temperature.

Step 3 In Situation A decide which quantities are varying. Also decide which variable is independent and which is dependent.

Step 4 Match and describe the graph that best fits the situation. Write a description of the function and its graph using the words *linear, nonlinear, increasing, decreasing, rate of change, maximum* or *greatest value,* and *minimum* or *least value.* Tell why you think the graph and your description match the situation.

Step 5 Repeat Steps 3–4 for the other three situations.

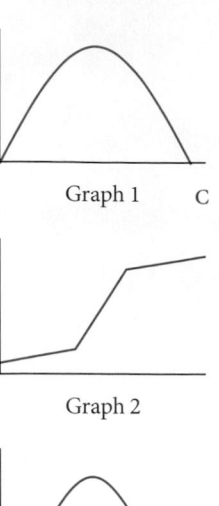

Graph 1   C

Graph 2

Graph 3   B

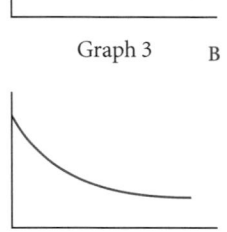

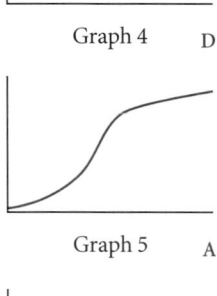

Graph 4   D

Graph 5   A

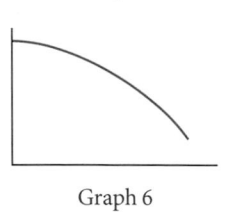

Graph 6

In the investigation you learned how to describe real-world situations with graphs and some function vocabulary. A function is **increasing** when the variables change in the same way—that is, the *y*-values *grow* when reading the graph from left to right. A function is **decreasing** when the variables change in different directions—that is, the *y*-values *drop* when reading the graph from left to right.

Situations C and D in the investigation represent **continuous** functions because there are no breaks in the domain or range. Many functions that are not continuous involve quantities that are counted or measured in whole numbers—for instance, people, cars, or stories of a building. In the investigation you have already seen two functions like this—the number of deer in Situation A and the number of days in Situation B. These are called **discrete** functions. When graphing the amount of daylight for every day of the year, the graph should really be a set of 365 points, as in the graph below. There is no value for day 47.35. Likewise, there may not be a day with exactly 11 hours 1 minute of daylight. But it's easier to draw this relationship as a smooth curve than to plot 365 points.

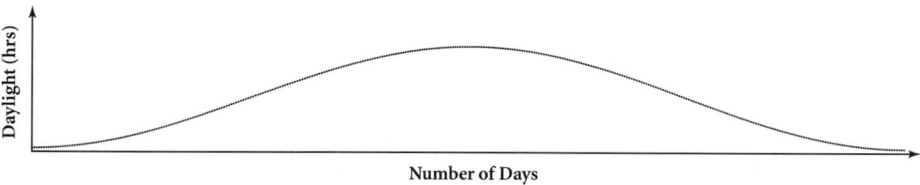

Sometimes it is useful to name a part of the domain for which a function has a certain characteristic.

EXAMPLE

Describe this graph, telling how the quantities in the graph relate to each other.

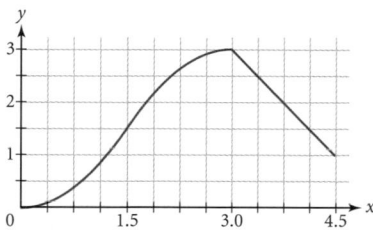

▶ Solution

Use the intervals marked on the *x*-axis to help you discuss where the function is increasing or decreasing and where it is linear or nonlinear.

On the interval $0 \leq x < 3.0$, the function is nonlinear and increasing. As *x* increases steadily, *y* changes at a varying rate, so the graph is nonlinear. When read from left to right, the graph rises. So the *y*-values grow and the function is increasing.

On the interval $3.0 \leq x < 4.5$, the function appears linear and is decreasing. As both *x* and *y* appear to change at a constant rate on the graph, the function is linear. When read from left to right, the graph falls. So the *y*-values drop and the function is decreasing.

[Ask] "How do the first two situations differ from the last two?" Be open to lots of ideas. As needed, bring out the idea that a *continuous function* has no breaks in its domain or range, as in Situations C and D.

Some students may argue that Situation B is best represented by a continuous rather than a discrete function. Point out that it is always important to interpret the mathematical model in terms of the original situation. Although Situations A and B might be considered discrete, they may well be modeled with continuous functions.

[Ask] "What kind of growth is represented in Situations A and D?" [Situation A begins as exponential growth, and Situation D shows exponential decay.]

Situation B is trigonometric (a sinusoidal cure). Be sure students understand what the graph looks like, because it's a model for Exercise 10.

Assessing Progress
Watch for students' ability to see **patterns,** interpret graphs, and identify **dependent** and **independent** variables.

▶ EXAMPLE

This example is good for students who were having difficulties describing graphs in the investigation. For the interval $3.0 \leq x < 4.5$, the solution excludes both endpoints. The function is said to be *stationary* at 3.0.

Closing the Lesson

A function increases if its *y*-values get larger as its *x*-values increase. Graphically, its graph rises from left to right. A function decreases if its *y*-values get smaller as its *x*-values increase. Its graph falls from left to right. A function's rate of change increases if the slope of the graph becomes larger, and decreases if the slope becomes smaller.

BUILDING UNDERSTANDING

Students practice describing graphs and making graphs to represent real-world situations.

ASSIGNING HOMEWORK

Essential	1, 3, 6, 8, 9, 11
Performance assessment	4, 5
Portfolio	6, 8
Journal	3, 7, 10
Group	5, 9
Review	2, 11–15

▶ **Helping with the Exercises**

Exercises 1 and 4 Here, "slower and slower" means slopes becoming closer to 0, with a graph becoming flatter. "Faster and faster" means slopes farther from 0, with a graph becoming steeper.

1a.
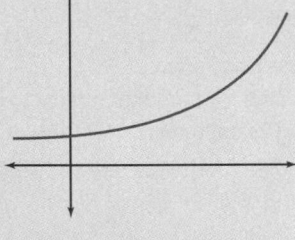

1b.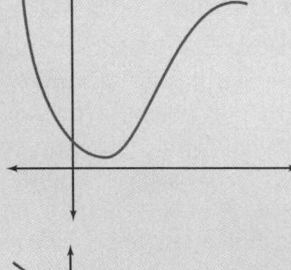

1c.

EXERCISES

▶ **Practice Your Skills**

1. Sketch a graph of a continuous function to fit each description.
   a. always increasing with a faster and faster rate of change
   b. decreasing with a slower and slower rate of change, then increasing with a slower and slower rate of change
   c. linear and decreasing
   d. decreasing with a faster and faster rate of change

2. Write an inequality for each interval in 2a–e. Include the least point in each interval and exclude the greatest point in each interval.

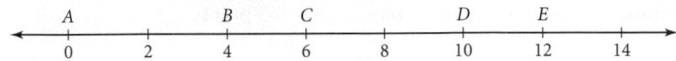

   a. A to B $0 \leq x < 4$ b. B to C $4 \leq x < 6$ c. B to D $4 \leq x < 10$ d. C to E $6 \leq x < 12$ e. A to E $0 \leq x < 12$

3. Describe each of these discrete function graphs using the words *increasing, decreasing, linear, nonlinear,* and *rate of change.*

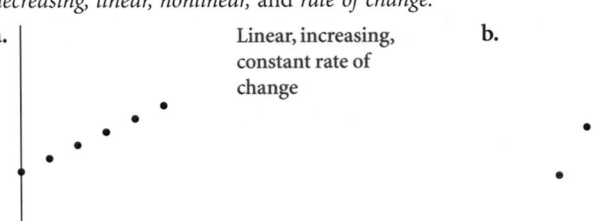

   a. Linear, increasing, constant rate of change
   b. Nonlinear, increasing, fast rate of change, then slowing down

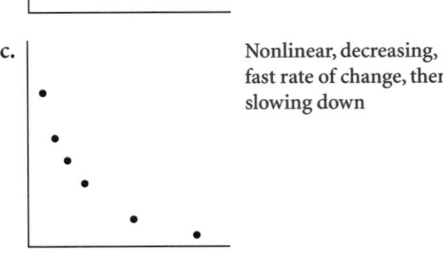
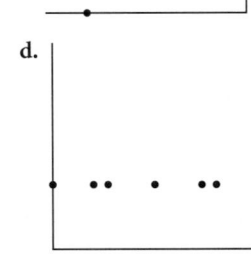

   c. Nonlinear, decreasing, fast rate of change, then slowing down
   d. Linear, neither increasing nor decreasing, zero rate of change

4. Sketch a discrete function graph to fit each description.
   a. always increasing with a slower and slower rate of change
   b. linear with a constant rate of change equal to zero
   c. linear and decreasing
   d. decreasing with a faster and faster rate of change

1d.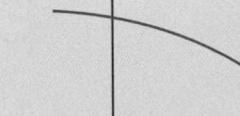

4a.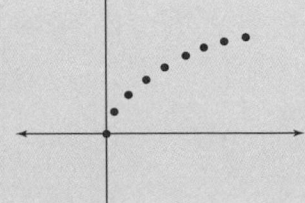

4b.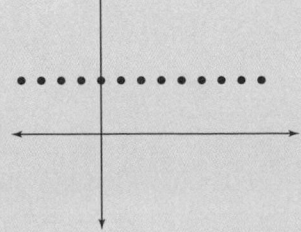

▶ Reason and Apply

5. A turtle crawls steadily from its pond across the lawn. Then a small dog picks up the turtle and runs with it across the lawn. The dog slows down and finally drops the turtle. The **turtle rests for a few** minutes after this excitement. Then a young boy comes along, picks up the turtle, and slowly carries it back to the pond. Which of the graphs describes the turtle's distance from the pond? Graph B

Graph A

Distance
Time

Graph B

Distance
Time

Graph C

Distance
Time

Graph D

Distance
Time

6. The graphs show the distance of a person from a fixed point for a 4-second interval. Answer both questions for each graph.

   i. Is the person moving toward or away from the point?
   ii. Is the person speeding up, slowing down, or moving at a constant speed?

a.
Distance (m)
Time (sec)
moving away, speeding up

b.
Distance (m)
Time (sec)
moving toward, speeding up

c.
Distance (m)
Time (sec)
moving away, constant speed

d.
Distance (m)
Time (sec)
moving toward, slowing down

e.
Distance (m)
Time (sec)
moving toward, constant speed

f.
Distance (m)
Time (sec)
moving away, slowing down

Exercise 5 [Alert] Some students may still be having difficulty representing motion on a graph. Be patient. Even if a student asks a question you just answered publicly, be grateful that the student is trying to make sense of the idea. Often people don't hear answers to questions they haven't asked.

Exercise 6 [Alert] Some students may think that the point is at the origin or y-intercept. Remind them that the graph is not the path of the moving person, and that the height of the graph represents the distance of that person from the fixed point.

4c.

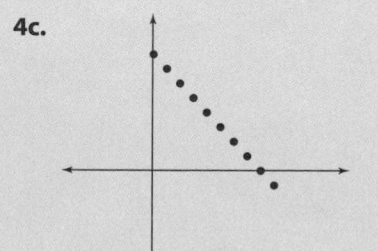

4d.

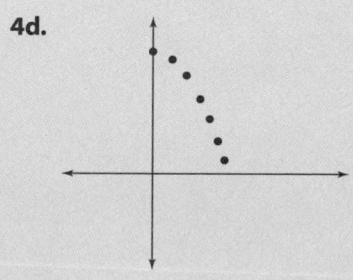

7. Answers will vary. Decreasing graphs like b, d, and e deserve the most attention. Probably d is the most realistic choice. As too many students get involved, the amount of time won't change by much. In fact, some students may sketch a graph that begins to increase after too many students get involved.

Exercise 8 As before, students can describe them as linear or nonlinear, as increasing or decreasing, and as having an increasing or decreasing rate of change.

8a. The graph is nonlinear, increasing with a faster and faster rate of change.

8b. The graph is nonlinear, decreasing with a faster and faster rate of change.

8c. The graph is linear and increasing with a constant rate of change.

8d. The graph is nonlinear, decreasing with a slower and slower rate of change.

8e. The graph is linear and decreasing with a constant rate of change.

8f. The graph is nonlinear, increasing with a slower and slower rate of change.

7. Which of the graphs most realistically shows the relationship between the number of volunteers and the amount of time required to clean up the trash on the school's campus? Explain your choice. If none of the choices seems correct, sketch your own graph to answer the question.

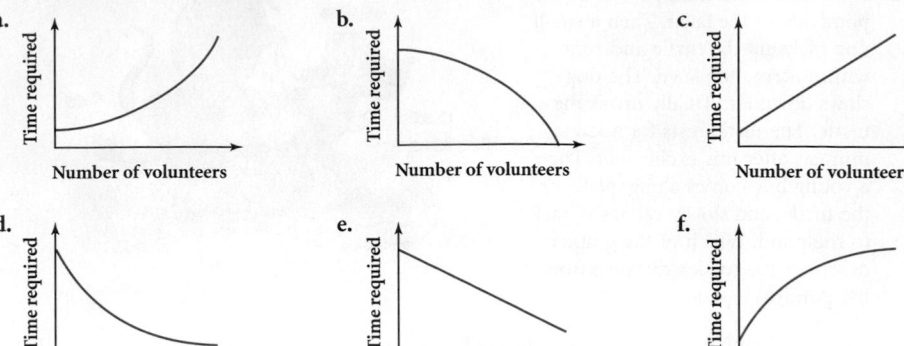

8. Describe the six graphs in problem 7.

9. APPLICATION This graph shows Anne's blood pressure level during a morning at school. Give the points or intervals when her blood pressure

a. Reached its highest level. about 11:00 A.M.

b. Was rising the fastest. between 10:10 and 10:40 A.M.

c. Was decreasing. between 9:00 and 9:45 A.M. and then again after 11:00 A.M.

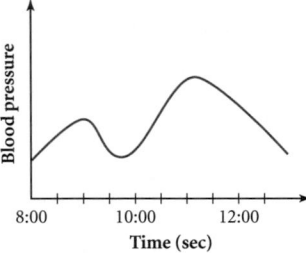

10. This graph shows the air temperature in a 24-hour period from midnight to midnight. Write a description of this graph, giving the intervals as the temperature changed.

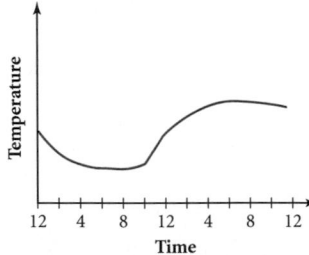

The temperature dropped slowly from midnight until 6 A.M. and was fairly constant through the morning. At about 10 A.M., the temperature began a dramatic rise for 2 hours. Then the temperature continued to rise more slowly for another 7 hours until 7 P.M. The temperature dropped only slightly from 7 P.M. until midnight.

Review

11. For each relationship, identify the independent and dependent variables.

    a. the mass of a spherical lollipop and the number of times it has been licked

    b. the number of scoops in an ice cream cone and the cost of the cone scoops, independent; cost, dependent

    c. the distance a rubber band will fly and the amount you stretch it before you release it

    d. the number of coins you flip and the number of heads

12. Use these equations for 12a–c.

    i. $4x + 2y = 16$ ii. $4x - 2y = 16$

    a. Solve these equations for y. (Show your work.)

    b. Rewrite the equations using $f(x)$ notation.

    c. Find the value of $f(-1)$ in both equations.

13. Consider the equation $y = -12.4 - 2.5(x + 5.4)$.

    a. Write it in intercept form. $y = -25.9 - 2.5x$

    b. Name the slope and y-intercept of the equation in 13a. The slope is -2.5 and the y-intercept is -25.9.

14. Solve the equation $740 = 16.8x + 405$ for x. When finished, check that you are correct. Approximately 19.94. The exact answer, $\frac{1675}{84}$, makes the equation true.

15. The top 10 grossing movies at the end of the 20th century are shown in the table. Find the five-number summary. The five-number summary is $293, 307, 343.5, 431, 601$.

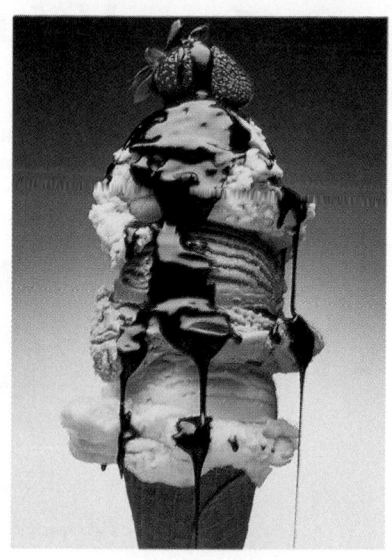

Rank	Total gross (millions of dollars)	Movie
1	601	Titanic
2	461	Star Wars
3	431	The Phantom Menace
4	400	E.T.
5	357	Jurassic Park
6	330	Forrest Gump
7	313	The Lion King
8	307	Return of the Jedi
9	306	Independence Day
10	293	The Sixth Sense

(*http://movieweb.com*)

Exercise 11 This exercise reviews Lesson 8.4.

11a. licks, independent; mass, dependent

11c. amount of stretch, independent; flight distance, dependent

11d. number of coins, independent; number of heads, dependent

Exercise 12 This exercise reviews Chapter 4 and Lesson 8.4.

12a. i. $y = \dfrac{16 - 4x}{2} = 8 - 2x$

12a. ii. $y = \dfrac{16 - 4x}{-2} = -8 + 2x$

12b. i. $f(x) = 8 - 2x$
     ii. $f(x) = -8 + 2x$

12c. i. $f(-1) = 10$
     ii. $f(-1) = -10$

Exercise 13 This exercise reviews Lesson 5.4.

Exercise 14 This exercise reviews Chapter 4. A proof would consist of a check using careful logic.

PLANNING

LESSON OUTLINE

First day:

15 min   Example A

25 min   Investigation

10 min   Sharing

Second day:

30 min   Example B

 5 min   Closing

15 min   Exercises

MATERIALS

- Calculator Notes 1E, 2B, 8C, 8D
- Pulse Rate Sample Data 2 (T), *optional*

TEACHING

The absolute value function is very useful for converting differences between values into distances, which can't be negative.

One step Remind students of how the spread of a data set was measured in Chapter 1: the range, the interquartile range, and the five-number summary. Ask them to generate and plot a set of data (as in Step 1 of the investigation) and to use differences from the mean to devise a method for describing the spread. Some groups will find the mean of differences, but others will see that differences cancel each other and may instead use distances, which are not negative. You might plant ideas in some groups to insure variety. During Sharing have several measures presented and critiqued (perhaps differences first), and introduce the idea of absolute value.

▶ **EXAMPLE A**

This example gives students practice at finding the absolute values of a variety of integers.

LESSON 8.6 Defining the Absolute Value Function

Cal and Al both live 3.2 miles from school, but in opposite directions. If you assign the number 0 to the school, you can show that Cal and Al live in opposite directions from it by assigning $+3.2$ to Al's house and -3.2 to Cal's apartment. For both Cal and Al, the distance from school is 3.2 miles.

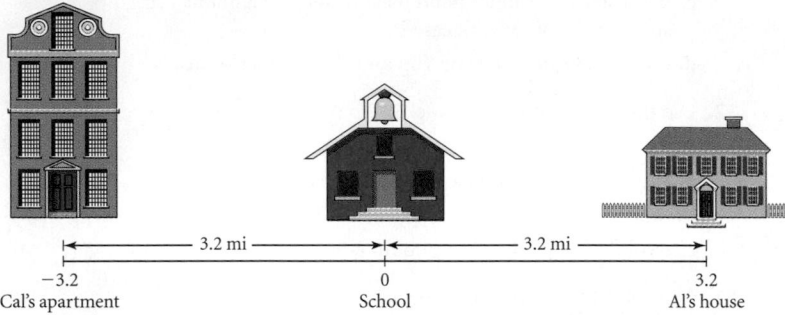

|← 3.2 mi →|← 3.2 mi →|
-3.2 0   3.2
Cal's apartment   School   Al's house

The **absolute value** of a number is its size, or magnitude, regardless of whether the number is positive or negative. In this lesson you'll develop a mathematical understanding of the absolute value function and its graph.

A way to visualize the absolute value of a number is to picture its distance from zero on a number line. A signed number and its opposite both have the same magnitude, or absolute value, because they're both the same distance from zero. For example, 3.2 and -3.2 are both 3.2 units from zero on a number line, so they both have an absolute value of 3.2.

Use the notation $|x|$ when you write absolute value expressions. So you would write $|3.2| = 3.2$ and $|-3.2| = 3.2$.

EXAMPLE A

Evaluate each expression.

a. $|5| + |-5|$ **b.** $2|-17| + 3$ **c.** $\dfrac{|-4|}{|4|}$

d. $|-6| - |-6|$ **e.** $-|8|$ **f.** $|0|$

▶ **Solution**

Substitute the magnitude of the number for each absolute value.

a. $|5| + |-5| = 5 + 5 = 10$ **b.** $2|-17| + 3 = 2(17) + 3 = 37$

c. $\dfrac{|-4|}{|4|} = \dfrac{4}{4} = 1$ **d.** $|-6| - |-6| = 6 - 6 = 0$

e. $-|8| = -8$ **f.** $|0| = 0$

Absolute value is useful for answering questions about distance, pulse rates, test scores, and other data values that lie on opposite sides of a central point such as a mean. The distance or difference of a data point from the mean of its data set is called its *deviation from the mean*.

LESSON OBJECTIVES

- Investigate the concept of absolute value
- Construct and interpret graphs of absolute value functions
- Learn the piecewise definition of the absolute value function
- Evaluate expressions containing absolute values

NCTM STANDARDS

CONTENT		PROCESS	
	Number		Problem Solving
●	Algebra	●	Reasoning
	Geometry	●	Communication
●	Measurement		Connections
●	Data/Probability	●	Representation

Investigation
Deviations from the Mean

In the investigation you will learn how the absolute value function tells how much an item of data or a whole set of data deviates from the mean.

Step 1 | Collect at least 10 pulse rates from your class. Record the data in a table and enter the numbers into list L₁ on your calculator.

Step 2 Deviations, or positive and negative differences, from the mean.

Step 2 Find the difference between each data point and the mean of the data in list L₁. [▶ ▣ See **Calculator Note 2B** to review finding the mean of a list. ◀] Record these numbers into a second column of your table and enter them into list L₂. What do these numbers represent?

Step 3 These numbers have the same magnitudes, but all are positive.

Step 3 Make a dot plot of the list L₁ data and note the distance from each data point to the mean. Record your results in a third column and enter them into list L₃. How are these entries different from those in list L₂? How are they alike?

Step 4 Answers will vary depending on the data. Any number can be in the domain. Only positive numbers and zero (non-negative numbers) will be in the range.

Step 4 Next, plot points of the form (L₂, L₃). [▶ ▣ See **Calculator Note 1E** to review scatter plots. ◀] What numbers are in the domain and range of the graph?

Step 5

Step 5 Use the trace function on your calculator and use the arrow keys to scan through each data point. Which input numbers are unchanged as output numbers?

Step 5 Positive domain values are unchanged.

Step 6

Step 6 Which input numbers are changed and how?

Step 6 Negative domain values are changed to their positive opposites.

Step 7

Step 7 Does it make sense to connect these points with a continuous graph? Why or why not?

Step 7 Yes, because a data point can be any decimal distance from the mean.

Step 8

Step 8 How does this graph compare to the graph of Y₁ = abs(x) on your calculator? [▶ ▣ See **Calculator Note 8C** to access the abs command. ◀]

Step 8 The graph of Y₁ = abs(x) passes through all the points on the graph.

Step 9 Find the mean of the deviations stored in list L₂. Compare it to the mean of the distances stored in list L₃. Which do you think is a better measure of the spread of the data?

Step 9 The mean in list L₃, the absolute value of the deviations from the mean, is the better measure. Otherwise, negative numbers in the deviation reduce the value of the mean of the deviations to zero. By definition of *mean* the mean difference will be zero.

Step 10 In your own words, write the rule for the function you graphed in Step 8. What number is output as y when the input x is positive or equal to zero? What number is output when x is negative? How can you use operations to change these numbers?

Step 10 Sample explanation: If x is positive or zero, output the same number as y. If x is negative, output the positive number by multiplying x by -1.

Despite deviations from each other in appearance, each impersonator clearly portrays Elvis Presley. This photo was taken at Graceland, the late singer's home in Memphis, Tennessee.

Guiding the Investigation

[Language] Remind students that a difference is the result of subtraction and may be negative, whereas a distance is never negative. A deviation is a difference, and might be negative. Model correct usage of both the noun (*deviation*) and the verb.

Step 1 To get enough data, you may choose to have the class do this investigation together. If students work in groups, they must collect data from at least 10 people. Or you can assign the data collection as part of the previous night's homework. To save time, you may display the transparency Pulse Rate Sample Data 2 or pass out copies of the data as a worksheet.

Step 2 Some students may need help breaking this down into the two parts: finding the mean, and then entering the expression L₁ − (*mean*) into list L₂. You might want to demonstrate part of this process for a smaller data set. You also might want students to do this and the next two steps by hand before working with the graphing calculator.

Step 3 Students may need to be reminded of what a dot plot is. See Lesson 1.1. Be sure that the distances recorded are never negative.

Step 10 Students may have difficulty jumping from "just make it positive" or "take away the negation sign" to "output the opposite if the input is negative."

Students might think that $-x$ is always a negative number. Remind them that $-x$ is "the opposite of x." Use that term rather than "the negative of x" or "negative x." You might say that the graph of $y = |x|$ is called *piecewise linear*.

SHARING IDEAS

After presentation of results, you might mention that a more standard measure of spread is the *standard deviation*. To calculate the standard deviation, squares of the differences are taken (instead of the absolute value) to make positive quantities, the squares are added, and then the square root of the sum is divided by the number of data points.

Remind students that the square root of a sum is not the sum of square roots. **[Ask]** "Is $\sqrt{x^2}$ the same as $|x|$?" Don't answer the question; it previews Lesson 8.7.

Assessing Progress

Assess students' skills at gathering **data** systematically, entering **calculator lists,** and **plotting points.** Also see how well they remember **dot plots.**

▶ EXAMPLE B

Method 3 Ask students to find examples of absolute-value equations with just one solution or no solutions. The equation $|x| + 5 = 5$ has only the solution $x = 0$. The equation $|x| + 5 = 4.9$ has no solutions, because a solution would imply that $|x|$ would be negative.

▶ ANOTHER EXAMPLE

You might want to ask students to generate the graph of the (*distance from the east goal line, yard line on the football field*). **[Ask]** "What are the domain and range of this function?" [Domain: $0 \leq$ *distance from the east goal line* ≤ 100, range: $0 \leq$ *yard line on the football field* ≤ 50.] **[Ask]** "How does the graph of this function compare to that of the absolute-value function?"

The **absolute value function** is defined by two rules. The first rule says to output the same number when the input value is positive or zero. The second rule says to output the opposite number when the input is negative. You express these rules like this:

$$|x| = \begin{cases} x & \text{if } x \geq 0 \\ -x & \text{if } x < 0 \end{cases}$$

For instance, if x is 3, then $|x|$ is also 3. On the other hand, if x is -3, then multiply by -1 to get 3 again. So there are two solutions to the equation $|x| = 3$.

EXAMPLE B | Solve the equation $12 = |x| + 7$.

▶ **Solution**
There are three ways to solve this equation. The first two are calculator methods that often give only approximate solutions.

Method 1: Looking at a Graph

Set y equal to both sides of the equation. You know from your work solving systems that you can enter $Y_1 = 12$ and $Y_2 = \text{abs}(x) + 7$ on your calculator. The x-coordinates of the two points of intersection, which represent the solutions, are $x = 5$ and $x = -5$. Note that the viewing window must be large enough for you to see both solutions.

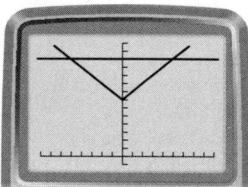

$[-8, 8, 1, -1, 14, 1]$

Method 2: Looking at a Table

A table shows these same solutions. You will have to scroll through the table to find both solutions.

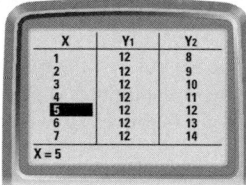 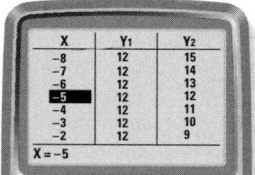

To be guaranteed of finding an exact solution, choose the next method.

Method 3: Solving Symbolically

The process for solving this equation symbolically is a bit different from that for solving most other equations because there is no function for "undoing" the absolute value—and you have to remember that there are two solutions.

$12 = |x| + 7$	The original equation.
$12 - 7 = |x| + 7 - 7$	Subtract 7 from each side of the equation.
$5 = |x|$	Find two numbers whose absolute value is 5.
$x = 5 \text{ or } x = -5$	The two solutions.

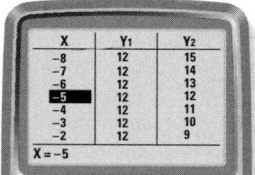

50

Yard line

0   50   100 x

Distance from east goal line (yds)

Whatever method you use to solve an absolute value equation, you always have to be sure that you are finding all possible solutions. In general, an absolute value equation has two solutions, one solution, or no solution.

If you're not sure how many solutions an equation should have, look at the graph of the situation first and then decide which method you want to use to solve the equation.

EXERCISES

You will need your calculator for problems **1, 3, 5,** and **12.**

▶ Practice Your Skills

1. Find the value of each expression without using a calculator. Check your results with your calculator. [▶ 🖥 See **Calculator Note 8C.** ◀]

   a. $|-7|$ 7   b. $|0.5|$ 0.5   c. $|-7 + 2|$ 5   d. $|-7| + |2|$ 9

   e. $-|5|$ −5   f. $-|-5|$ −5   g. $|-4| \cdot |3|$ 12   h. $\dfrac{|-6|}{|2|}$ 3

2. What x-values satisfy the equation $|x| = 10$? 10 and −10

3. Evaluate both sides of each statement to determine whether to replace the box with =, <, or >. Use your calculator to check your answers.

   a. $|5| + |7|$ ☐ $|5 + 7|$ 12 = 12   b. $|-5| \cdot |8|$ ☐ $|-40|$ 40 = 40

   c. $|-12 - 3|$ ☐ $|-12| - |3|$ 15 > 9   d. $|-2 + 11|$ ☐ $|-2| + |11|$ 9 < 13

   e. $\dfrac{|36|}{|-9|}$ ☐ $\left|\dfrac{36}{-9}\right|$ 4 = 4   f. $|4|^{|-2|}$ ☐ $|4^{-2}|$ 16 > $\dfrac{1}{16}$

4. Consider the functions $f(x) = 3x - 5$ and $g(x) = |x - 3|$. Find each value.

   a. $f(5)$ 10   b. $f(-2.5)$ −12.5
   c. $g(-5)$ 8   d. $g(1)$ 2

5. Plot the points determined by the function $y = |x|$ (or Y1 = abs(x)) using the domain {−4, −1.5, 0, 1.2, 3, 4.75}. Use a friendly window. [▶ 🖥 See **Calculator Note 8D** to learn about friendly windows. ◀] Use your graph, table, or equation to evaluate $|4.75|$ and $|-1.5|$.

▶ Reason and Apply

6. Create this graph on graph paper: When $x \geq 0$, graph the line $y = x$. When $x < 0$, graph the line $y = -x$. What single function has this same graph? $y = |x|$, the absolute value function

7. Solve this system of equations:
   $$\begin{cases} y = |x| \\ y = 2.85 \end{cases}$$

5.

x	−4	−1.5	0	1.2	3	4.75
y	4	1.5	0	1.2	3	4.75

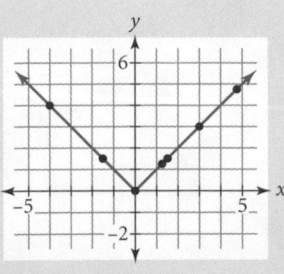

7. The solutions are (2.85, 2.85) and (−2.85, 2.85).

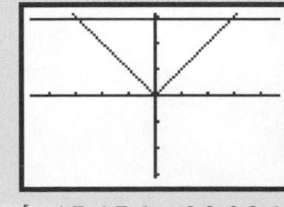

$[-4.7, 4.7, 1, -3.2, 3.2, 1]$

Closing the Lesson

The **absolute value function** converts differences between values into distances, which can't be negative, by outputting the opposite of negative numbers (and the number itself if it's not negative). It can be used to describe the **spread** of a data set by converting **deviations,** or differences, from the mean into distances and then finding the mean of the distances.

BUILDING UNDERSTANDING

Students find absolute values and work with the absolute-value function and its graph.

ASSIGNING HOMEWORK

Essential	1–8, 11, 13
Performance assessment	4, 11, 12
Portfolio	13
Journal	9, 13
Group	10, 11, 13
Review	14, 15

▶ Helping with the Exercises

Exercise 2 Be sure students see both values. They might think of x as positive in the expression $|x|$.

Exercise 3 Some students may need to write down the values they get for both sides of the statement. A good form for writing up part c, for example, is

$|-12 - 3| = |-15| = 15$
$|-12| - |3| = 12 - 3 = 9$
$15 > 9$

Exercise 6 This is another example of a piecewise linear graph.

6.

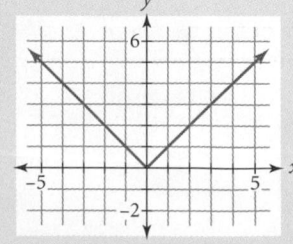

Exercise 12 Encourage students to consider two cases: when x is negative and when it's not.

12a. Note that when $x < 0$, the graph lies on the x-axis.

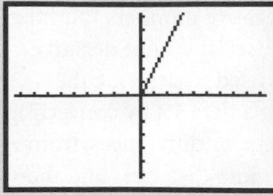

$[-9.4, 9.4, 1, -6.2, 6.2, 1]$

12b. Note that when $x = 0$, the function is undefined.

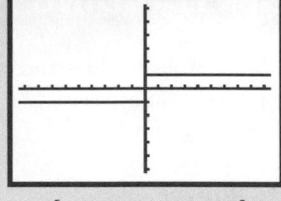

$[-9, 9, 1, -6, 6, 1]$

Exercise 13 Some students might reason that the range is greater in the Spider Lake data, so there is more variation from the mean. This approach works for this data set, but challenge students to find a data set where it doesn't work. **[Ask]** "Find a pair of data sets of equal size in which one has a larger range and the other has more deviation from the mean." [An example is $\{1, 5, 5, 5, 5, 5, 9\}$ and $\{2, 2, 2, 5, 8, 8, 8\}$.]

8. If possible, solve each equation for x.
   a. $|x| = 12$ $x = -12$ or $x = 12$
   b. $10 = |x| + 4$ $x = -6$ or $x = 6$
   c. $10 = 2|x| + 6$ $x = 2$ or $x = -2$
   d. $4 = 2(|x| + 2)$ $x = 0$

9. Describe the walk represented by this calculator graph. Each mark on the x-axis represents 1 second and each mark on the y-axis represents 1 meter from a motion sensor.

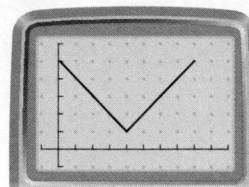

The walker starts 5 meters away from the motion sensor and walks toward the motion sensor at a rate of 1 meter per second for 4 seconds and then walks away from the motion sensor at the same rate.

10. The graph in Example B shows two solutions for x.
   a. Replace $Y_1 = 12$ with a horizontal line that gives exactly one solution for x. $Y_1 = 7$
   b. Replace $Y_1 = 12$ with a horizontal line that gives no solution for x. $Y_1 =$ any number less than 7

11. In 11a–d, identify which function, $f(x)$, $g(x)$, or $h(x)$, is used in each (*input, output*) pair.

$$f(x) = 7 + 4x \qquad g(x) = |x| + 6 \qquad h(x) = 18(1 + 0.5)^x$$

   a. (5, 11) $g(5) = |5| + 6 = 11$
   b. (1, 27) $h(1) = 18(1.5) = 27$
   c. (−2, 8)
       $g(-2) = |-2| + 6 = 8; h(-2) = 18(1.5)^{-2} = 8$
   d. (3, 19) $f(3) = 7 + 4 \cdot 3 = 19$

12. Predict the graph of each equation. Sketch your prediction and then check your answer with your calculator.
   a. $y = x + |x|$
   b. $y = \dfrac{x}{|x|}$

13. **APPLICATION** The table shows the weights of fish caught by wildlife biologists in Spider Lake and Doll Lake. In which lake did the fish weights vary more from the mean? Explain how you arrived at your answer.

Weights of Fish

Spider Lake (lb)	Doll Lake (lb)
1.2	0.9
2.1	1.1
0.8	1.6
1.4	1.9
2.7	2.1
1.0	1.4
0.4	1.4
2.4	2.2

Find the mean of the absolute values of the deviations for each lake. Spider Lake is 0.675 lbs, and Doll Lake is 0.375 lbs. The fish weights varied more in Spider Lake.

Exercise 14 This problem reviews Lessons 6.3 and 6.2.

Exercise 15 This problem reviews Lesson 6.5.

▶ Review

14. Solve each system of equations using the method of your choice. For each, tell which method you chose and why.

    a. $\begin{cases} -2x + 3y = 12 \\ 4x - 3y = -21 \end{cases}$ $(-4.5, 1)$

    b. $\begin{cases} 5x + y = 12 \\ 2x - 3y = 15 \end{cases}$ $(3, -3)$

15. Solve each inequality and graph the solution on a number line.

    a. $-2 < 6x + 8$ $-1\frac{2}{3} < x \text{ or } x > -1\frac{2}{3}$
    b. $3(2 - x) + 4 \geq 13$ $x \leq -1$
    c. $-0.5 \geq -1.5x + 2(x - 4)$ $15 \geq x \text{ or } x \leq 15$

15a. (number line: open circle at −3, marks at −3, 0, 5)

15b. (number line: closed circle at −2, marks at −2, 0, 2)

15c. (number line: closed circle at 15, marks at −15, 0, 15)

IMPROVING YOUR REASONING SKILLS

Consider the table of the squares of numbers between 0 and 50 that end in 5.

Number	5	15	25	35	45
Square	25	225	625	1225	2025

Do you notice a pattern that helps you mentally calculate these kinds of square numbers quickly? Can you square 65 in your head? When you think you have discovered the pattern, check your results with a calculator. Then try reversing the process to find the square root of 7225.

Practice this pattern and then race someone using a calculator to see who is quicker at computing a square number ending in 5. Will this pattern work for all numbers ending in 5? Why or why not? Are there numbers that make this pattern too difficult to use?

IMPROVING REASONING SKILLS

To square a positive integer that ends in 5, take the number before the 5 and multiply it by one more than itself. Put 25 on the end of that product. For example, to square 115, find that $(11)(12) = 132$ and append 25 to get 13225.

To find the square root of 7225, find consecutive positive integers (8 and 9) whose product is 72. Put a 5 after the smaller of these integers, to get $\sqrt{7225} = 85$. This procedure will be successful whenever the number preceding the 25 is the product of two consecutive integers.

LESSON
8.7

PLANNING

LESSON OUTLINE

One day:

5 min	Introduction
20 min	Investigation
5 min	Sharing
10 min	Example
5 min	Closing
5 min	Exercises

MATERIALS

- Calculator Notes 1E, 8B

TEACHING

The relationship between numbers and their squares (or lengths of geometric squares and their areas) is a function whose graph is a parabola. There's also a function that relates the areas of squares to the lengths of their sides.

One step Ask whether there's a function that relates the length of the side of a square to its area. After consideration of congruent squares, students will probably decide that the squaring function qualifies. Ask them to graph the function and describe its shape, using terms from Lesson 8.5. Encourage them to look at its mirror symmetry and to note the point that's the square of only one number. Then ask whether there's a function to go from the square's area to the length of an edge. The idea of the square root function ($\sqrt{\ }$) should come out. Help students describe its graph.

INTRODUCTION

As needed, remind students that the opposite of a number x, written $-x$, is the result of multiplying x by -1.

LESSON
8.7

Squares, Squaring, and Parabolas

Think of a number between 1 and 10. Multiply it by itself. What number did you get? Try it again with the opposite of your number. Did you get the same result? This number is called the **square** of a number. The process of multiplying a number by itself is called **squaring.** The square of a number x is "x squared," and you write it as x to the power of 2, or x^2. When squaring numbers on your calculator, remember the order of operations. Try entering -3^2 and $(-3)^2$ on your calculator. Which result is the square of -3?

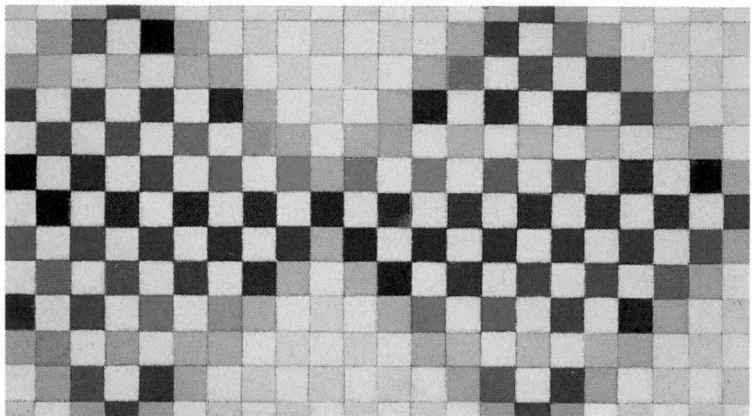

Do you think that the rule for squaring is a function? In order to answer this question, you will graph the relationship between numbers and their squares.

The mathematical process for the squaring takes its name from the application of finding a square's area. From the Latin *quadrare*, which means to make square, we also have the word "quadratic" to describe x^2.

Charmion Von Wiegand, *Advancing Magic Squares*, ca. 1958.
The National Museum of Women in the Arts, Washington, D.C.

Investigation
Graphing a Parabola

In this investigation you will explore connections between any number x and its square by graphing the coordinate pairs (x, x^2).

Step 1 | Make a table with column headings like the one shown. Put the numbers -10 through 10 in the first column, and then enter these numbers into list L1 on your calculator.

Number (x)	Square (x²)

The parabolic shape of the SETI (Search for Extraterrestrial Intelligence) radio telescope at Harvard University in Massachusetts collects radio signals from space.

Refer students to the list of the conventional order of operations in Lesson 0.1. Expressions in parentheses are evaluated first, so $(-3)^2 = 9$. But powers are evaluated before multiplication and therefore before taking opposites. So $-3^2 = -9$.

Step 1 [Ask] "What might the square of a number represent?" At least the idea of the area of a square should come up.

Steps 1–2 number column: $\{-10, -9, -8, \ldots, -1, 0, 1, 2, \ldots, 8, 9, 10\}$;
square column: $\{100, 81, 64, \ldots, 1, 0, 1, 4, \ldots, 64, 81, 100\}$

Step 3 The output values are non-negative integers. Both the positive and negative numbers match a positive square number.

Step 2

Without a calculator, find the square of each number and place it in the second column. Check your results by squaring list L₁ with the x^2 key. [▶ ☐ See **Calculator Note 8B.** ◀] Store these numbers in list L₂.

Step 3

How do the squares of numbers and their opposites compare? What is the relationship between the positive numbers and their squares? Between the negative numbers and their squares?

Step 4

Choose an appropriate window and plot points of the form (L₁, L₂). [▶ ☐ See **Calculator Note 1E** to review scatter plots. ◀] Graph Y₁ = x^2 on the same set of axes. What relationship does this graph show?

Step 5 yes; domain: all numbers; range: $y \geq 0$

Step 5

Is the graph of $y = x^2$ the graph of a function? If so, describe the domain and range. If not, explain why not.

The graph of $y = x^2$ is called a **parabola.** In later chapters you will learn how to create other parabolas based on variations of this basic equation.

Step 6 Quadrants I and II and the origin $(0, 0)$

Step 6

The points of the parabola for $y = x^2$ are in what quadrants?

Step 7 The output value 0 has only one input value, which is also 0. The point $(0, 0)$ is at the bottom of the parabola.

Step 7

What makes the point $(0, 0)$ on your curve unique? Where is this point on the parabola?

Step 8

Draw a vertical line through the point you found in Step 7. How is this line like a mirror?

Step 9

Compare your parabola with the graph of the absolute value function, $y = |x|$. How are they alike and how are they different?

Step 10

Which x- and y-values in your parabola could represent side lengths and areas of squares?

In the investigation you learned that on the graph of $y = x^2$ two different input values can have the same output. For instance, the square of -3 and the square of 3 are both equal to 9. What happens when you try to "undo" the squaring? If you want to find a number whose square is 9, is 3 or -3 the answer? Or are both the answer? You will learn about a function that undoes squaring in the example.

EXAMPLE

Find the side of the square whose area is 6.25 square centimeters.

▶ **Solution**

Let x represent the side of the square in centimeters. To find it, solve the equation $x^2 = 6.25$. There are three ways to do this.

6.25 cm² | x

x

Method 1: Graph

Graph the line $y = 6.25$ and the parabola $y = x^2$. The graphs intersect at two points. Use the trace key on your calculator to find the x-values of the intersections.

Step 2 As needed, remind students that a negative times a negative is positive.

Step 3 Students may say that the square of every number equals the square of its opposite or the square of its absolute value.

Step 5 After students determine that the graph is that of a function (by the vertical line test), ask "Does it make sense that the relationship between the length of an edge of a square and the area of that square is a function?" That is, do all congruent squares have the same area?

Step 7 You might introduce the term *vertex* to describe the turning point.

Step 8 Mirror symmetry, or reflective symmetry, is often called simply *symmetry*.

SHARING IDEAS

Arrange for sharing any unusual ideas. Answers to Steps 9 and 10 might be especially interesting.

As needed, mention the terms *parabola* and its *vertex*. As a follow-up to Step 10, you might ask how to find the length of a square's edge, given its area. If only one idea arises, ask for more. You can thus get into the ideas of the example.

Assessing Progress

Check for students' ability to enter and operate on calculator lists, to make scatter plots, and to apply the vertical line test.

▶ **EXAMPLE**

The ordered pair (x, x^2) here represents (*side of a square, square's area*). Because of the context of the problem, only positive values of x are reasonable. Both the domain and the range of the function are non-negative numbers.

See pages 719–720 for answers to Steps 4, 8, 9, and 10.

LESSON OBJECTIVES

- Learn about the squaring and square root functions
- Graph parabolas
- Compare the squaring function to other functions
- Relate the squaring function to finding the area of a square

NCTM STANDARDS

CONTENT		PROCESS	
●	Number	●	Problem Solving
●	Algebra	●	Reasoning
●	Geometry	●	Communication
	Measurement	●	Connections
	Data/Probability	●	Representation

Method 3 [Alert] This skips the step after $\sqrt{x^2} = \sqrt{6.25}$, that is, $|x| = 2.5$. It will help many students if you insert this step and the reason for it: "The absolute value of x is the positive square root of x^2, and 2.5 is the positive square root of 6.25."

Students frequently are confused between the square roots of a number and the square root function (whose output is only the non-negative square root, often called the *principal square root*). The square root symbol refers to the output of the square root function. So it's incorrect, for example, to say $\sqrt{4} = -2$. If you want to refer to both square roots, you use the notation $\pm\sqrt{}$ precisely because $\sqrt{}$ alone means only the positive root.

Closing the Lesson

The **squaring function** has a **parabolic graph.** Although every positive number has two square roots, the **square root function** ($\sqrt{}$) gives only the positive square root.

BUILDING UNDERSTANDING

Students practice using squares and square roots.

ASSIGNING HOMEWORK

Essential	1–6, 8, 10
Performance assessment	10, 11
Portfolio	9
Journal	2, 12
Group	9–11
Review	7, 13, 14

▶ Helping with the Exercises

Exercise 1 If students are having difficulties, suggest that they draw pictures.

Method 2: Calculator Table

You can also see these x-values in a calculator table. Adjust the table settings and zoom in on the table until you find the row in which both Y_1 and Y_2 equal 6.25. Remember to look for two solutions, one positive and one negative.

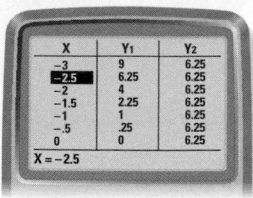 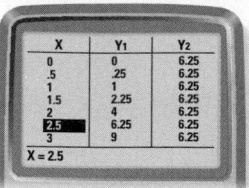

Method 3: Solve Symbolically

$$x^2 = 6.25$$
$$\sqrt{x^2} = \sqrt{6.25}$$
$$x = 2.5 \text{ or } x = -2.5$$

The original equation.

To solve for x, take the **square root** of both sides.

There are two solutions.

The equation has two solutions, but because the side of a square must be positive, the only realistic solution is 2.5 cm.

Your calculator will not give the negative solution when you press the square root key. The **square root function**, ($f(x) = \sqrt{x}$), undoes squaring, giving only the positive solution.

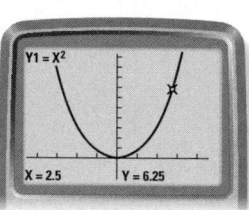

You can learn about the use of parabolas in the real world with the Internet links at **www.keymath.com/DA** .

EXERCISES

You will need your calculator for problem **5.**

▶ Practice Your Skills

1. Complete the table by filling in the missing values for the side, perimeter, and area of each square.

2. Look at the table of squares in the investigation Graphing a Parabola. Use values from this table to explain why $y = x^2$ is nonlinear.

3. Solve each equation for x.
   a. $|x| = 6$ $x = \pm 6$
   b. $x^2 = 36$ $x = \pm 6$
   c. $|x| = 3.8$ $x = \pm 3.8$
   d. $x^2 = 14.44$ $x = \pm 3.8$

Side (cm)	Perimeter (cm)	Area (sq cm)
1	4	1
2	8	4
3	12	9
4	16	16
14	56	196
15	60	225
21	84	441
25.2	100.8	635.04
47	188	2209

2. Answers will vary. One explanation is that the slope between $x = 0$ and $x = 1$ is 1, but the slope from $x = 1$ to $x = 2$ is 3. The rate of change varies.

Exercise 3 You may want to introduce the notation $\pm\sqrt{}$ here.

4. Solve each equation, if possible.

    a. $4.7 = |x| - 2.8$
        $x = 7.5$ or -7.5

    b. $-41 = x^2 - 28$
        No real solution

    c. $11 = x^2 - 14$
        $x = 5$ or -5

▶ Reason and Apply

5. For what values of x is $|x| \geq x^2$? To check your answer, graph $Y_1 = |x|$ and $Y_2 = x^2$ on the same set of axes.

6. For what values of y does the equation $y = x^2$ have

    a. No real solutions? $y < 0$
    b. Only one solution? $y = 0$
    c. Two solutions? $y > 0$

7. Graph the functions $f(x) = 3x - 5$ and $g(x) = |x - 3|$. What do the two graphs tell you about the equation $3x - 5 = |x - 3|$?

8. Solve each equation symbolically.

    a. $5 = |x| - 3$ $x = 8$ or -8
    b. $-4 = x^2 - 8$ $x = 2$ or -2
    c. $4 = 2|x| + 6$

9. Find the sum of each set of numbers in 9a–c.

    a. the first five odd positive integers $1 + 3 + 5 + 7 + 9 = 25$, or 5^2.

    b. the first 15 odd positive integers $1 + 3 + 5 + 7 + 9 + \ldots + 29 = 225$, or 15^2.

    c. the first n odd positive integers   The sum of the first n positive odd integers is n^2.

    d. Use the diagram to explain the connection among the sum of odd integers, square numbers, and these square figures.

  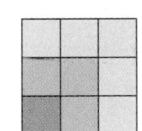 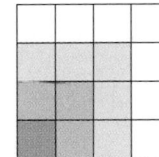

10. Write an equation for the function represented in each table. Check your answers.

    a.

x	-3	-1	0	1	4	6
y	14	10	8	6	0	-4

    b.

x	-3	-1	0	1	4	6
y	9	1	0	1	16	36

    c.

x	-3	-1	0	1	4	6
y	3	1	0	1	4	6

10a. $y = 8 - 2x$

10b. $y = x^2$

10c. $y = |x|$

11. This 4-by-4 grid contains squares of different sizes.

    a. How many of each size square are there? Include overlapping squares.

    b. How many total squares would a 3-by-3 grid contain? A 2-by-2 grid? A 1-by-1 grid?

    c. Find a pattern to determine how many squares an n-by-n grid contains. Use your pattern to predict the number of squares in a 5-by-5 grid.

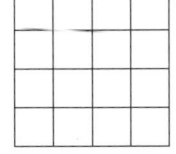

12. Explain why the equation $x^2 = -4$ has no solutions. (Hint: Why is it impossible for the product of a number multiplied by itself to be negative?)

Exercise 10 Students might find it helpful to graph the data and to think of input and output values. Part c is a good example of a case that might prove difficult because of its simplicity.

Exercise 11 Don't expect students to find a closed-form expression for the sum of squares of positive integers up to n. That formula is $\dfrac{n(n+1)(2n+1)}{6}$.

11a. 16: 1-by-1, 9: 2-by-2, 4: 3-by-3, and 1: 4-by-4

11b. A 3-by-3 grid has 14 squares: 9: 1-by-1, 4: 2-by-2, and 1: 3-by-3. A 2-by-2 grid has 5 squares: 4: 1-by-1 and 1: 2-by-2. A 1-by-1 grid has 1 square: 1-by-1.

11c. Answers will vary. One possible response: There are n^2 1-by-1 squares, $(n-1)^2$ 2-by-2 squares, $(n-2)^2$ 3-by-3 squares, and so on. There would be 55 squares in a 5-by-5 grid.

Exercise 4 Be sure students explain their answers, especially why there's no solution to the equation in 4b.

Exercise 5 Some students may need help in interpreting inequality on the graph.

5. $-1 \leq x \leq 1$. Good graphing window answers will vary. One possibility is a TI-83 friendly window like $[-2.35, 2.35, 1, 0, 3.1, 1]$.

Exercise 6 You may want to discuss the difference between real and imaginary numbers. Imaginary numbers are square roots of negative numbers. They were originally investigated because they were useful in finding solutions to cubic equations (equations that involve x^3).

7.

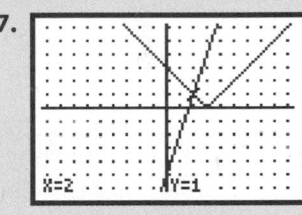

$[-9, 9, 1, -6, 6, 1]$

Answers will vary. They tell us that there is exactly one solution and that the x-value that makes the equation true is $x = 2$.

Exercise 8 Again, be sure students include good explanations of why the equation in 8c has no solution.

8c. No solution. You get $-1 = |x|$, but no number can have a negative absolute value. (The graphs of $y = 4$ and $y = 2|x| + 6$ do not intersect.)

9d. Each larger square is created by adding a border of small squares on two sides. The number of little squares added on each time is the next odd number. The resulting figure is the next square number.

See page 720 for answers to Exercise 12.

Exercise 13 This exercise reviews Lesson 7.2.

Exercise 14 Watch for difficulties with 14g, in which two variables are involved and there's a negative exponent in the denominator. Be sure students understand that they are free to use several steps in rewriting it. This exercise reviews Chapter 7.

▶ **Review**

13. The table shows exponential data.

    a. What equation in the form $y = ab^x$ can you use to model the data in the table? $y = 400(0.75)^x$

    b. Use your equation to find the missing values.

x	y
0	400
4	126.5625
3	168.75
1	300
≈ −3.19	1000

14. Use properties of exponents to find an equivalent expression if possible of the form ax^n. Use positive exponents.

   a. $24x^6 \cdot 2x^3$ $48x^9$

   b. $(-15x^4)(-2x^4)$ $30x^8$

   c. $\dfrac{72x^{11}}{3x^2}$ $24x^9$

   d. $4x^2(2.5x^4)^3$ $62.5x^{14}$

   e. $\dfrac{15x^5}{-6x^2}$ $-2.5x^3$

   f. $(-3x^3)(4x^4)^2$ $-48x^{11}$

   g. $\dfrac{42x^{-6}y^2}{7y^{-4}}$ $\dfrac{6y^6}{x^6}$

   h. $3(5xy^2)^3$ $375x^3y^6$

IMPROVING YOUR **VISUAL THINKING** SKILLS

Square numbers are so named because they result from the geometric application of finding the area of a square. A square of side length 3 has an area equal to 3^2, or 9. You can represent 9, and other perfect square numbers, with diagrams like this:

 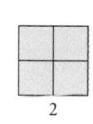 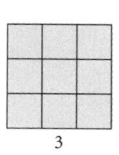 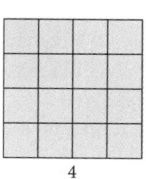 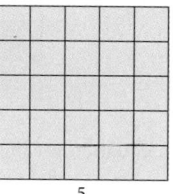

1   2   3   4   5

What numbers result when you represent them with cubes instead of squares? Use sugar cubes to make these shapes:

 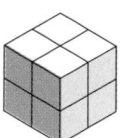 

How many sugar cubes does it take to make each figure? What is the relationship between the side length (measured in sugar cubes) and the total number of cubes needed for each figure? If you double the side length of a cubic figure, how many times larger is its resulting volume? If you triple the side length?

IMPROVING **VISUAL THINKING** SKILLS

The cubes consist of 1, 8, and 27 sugar cubes. These are their volumes, measured in cubic units. The volume of a cube of length n is $n \cdot n \cdot n$, which is n^3 or n *cubed*.

The last two questions lead to a profound result, so let students play with them. If the length of one edge is doubled, the volume is multiplied by 8, which is 2^3. If the length of an edge is tripled, the volume is multiplied by 27, which is 3^3. In general, if the length is multiplied by k, the volume is multiplied by k^3. In even more generality, if all lengths in an n-dimensional figure are multiplied by k, the "size" (either length, area, volume, or the equivalent for n dimensions) is multiplied by k^n.

CHAPTER 8 REVIEW

In this chapter you used functions to describe real-world relationships. You began by designing and decoding secret messages. You discovered that the easiest way to code is to use a **function**—it codes each input into a single output.

You investigated functions represented by rules, equations, tables, and graphs. You learned to tell whether a relationship is a function by applying the **vertical line test.** On a graph, this means that no vertical line can intersect the graph of a function at more than one point.

You learned how to use function notation $f(x)$ and some new vocabulary—**independent variable, dependent variable, domain,** and **range.** You learned when a function is **increasing** or **decreasing, linear** or **nonlinear,** and the difference between a **discrete** and **continuous** graph. You explored the **absolute value function,** $f(x) = |x|$, and the **squaring function,** $f(x) = x^2$, and their graphs. You learned that these two functions can have zero, one, or two solutions. You learned how to graph a **parabola.** You also used the **square root function** to undo the squaring function and get only the positive square root.

EXERCISES

You will need your calculator for problem **10.**

1. Answer each question for the graph of $f(x)$.
   a. What is the domain of the function?
   b. What is the range of the function?
   c. What is $f(3)$? 1
   d. For what values of x does $f(x) = 1$?
      at $x = -1$ or 3

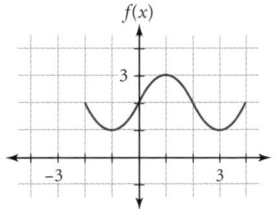

2. Which of these tables of x- and y-values represent functions? Explain your answers.

a.

x	y
0	5
1	7
3	10
7	9
5	7
4	5
2	8

b.

x	y
3	7
4	9
8	4
3	5
9	3
11	9
7	6

c.

x	y
2	8
3	11
5	12
7	3
9	5
8	7
4	11

PLANNING

LESSON OUTLINE

One day:

10 min	Introduction
15 min	Exercises and helping individuals
10 min	Checking work and helping individuals
15 min	Student self assessment

REVIEWING

Ask students to critique this coding scheme: translate each letter to an integer 1 through 26, square it, subtract 26 repeatedly until you get a number between 1 and 26, and then translate back to a letter. (If there's time, students might make a coding grid.) Because every input value has a single output value, the method involves a function. The method would lead to difficulties in decoding, however, because more than one input value is taken to the same output value. In fact, the integers n and $26 - n$ will be paired with the same output, just as integers and their opposites have the same squares. **[Ask]** "If n and $26 - n$ are opposites, does the absolute value of a number equal the positive square root of the number squared, as it does without wrapping?" [It does if the absolute value and "positive" are defined to be an integer from 1 to 13 and 14 to 26 are considered their opposites.]

ASSIGNING HOMEWORK

You might let the students work on the even-numbered exercises in groups in class and do the other exercises individually as homework. Groups could share and discuss their answers to exercise 10.

1a. $-2 \le x \le 4$

1b. $1 \le f(x) \le 3$

2a. A function; each x-value corresponds to only one y-value.

2b. Not a function; the input $x = 3$ has two different output values, 5 and 7.

2c. A function; each x-value corresponds to only one y-value.

▶ Helping with the Exercises

3. The graph is a horizontal line segment at 0.5 meters per second.

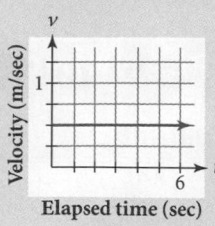

Elapsed time (sec)

Exercise 4 Archimedes (ca. 287–212 B.C.), a Greek philosopher and inventor, wrote about volume, pi, and spirals.

Descartes, René (1596–1650), a French mathematician and philosopher, invented the coordinate system.

Hypatia (ca. A.D. 370–415), was a woman Greek philosopher and mathematician.

Euclid (ca. 300 B.C.), was a Greek geometer whose *Elements* was the chief source of geometric reasoning until the 19th century.

4d. This code shifts 20 spaces forward, or 6 spaces back, in the alphabet.

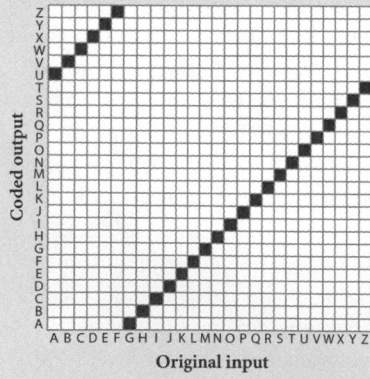

Original input

See page 720 for answers to Exercises 6b, 7a, and 7b.

472 CHAPTER 8 Functions

3. The graph at right shows the relationship between an object's distance from a motion sensor in meters and time in seconds. Sketch a graph to represent the velocity of this object dependent on time *t*.

Elapsed time (sec)

4. In a letter-shift code, ARCHIMEDES codes into ULWBCGYXYM. Use this information to determine the names of famous mathematicians in 4a–c.

a. XYMWULNYM DESCARTES

b. BSJUNCU HYPATIA

c. YOWFCX EUCLID

d. Create a grid and state a rule for this code.

5. The graph below shows the velocities of three girls inline skating over a given time interval. Assume that they start at the same place at the same time.

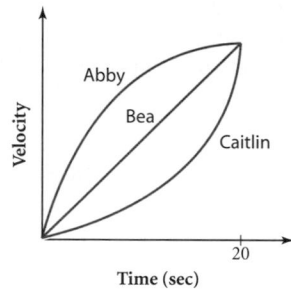

a. Create a story about these three girls that explains the graph.

b. Are Caitlin and Bea ever in front of Abby? Explain.

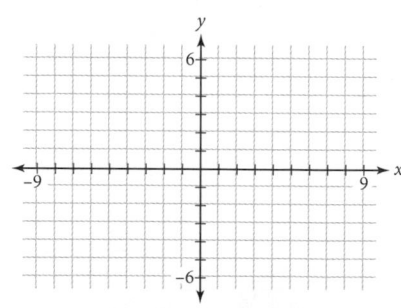

6. APPLICATION A recent catalog price for tennis balls was $4.25 for a can with three balls. The shipping charge per order was $1.00.

a. Write an equation that you can use to project the costs for ordering different numbers of cans. $y = 4.25x + 1.00$

b. Draw a graph showing this relationship.

c. How does raising the shipping charge by 50¢ affect the graph? It shifts the graph up 0.50 units on the y-axis.

d. What equation models the cost situation in 6c?
$$y = 4.25x + 1.50$$

7. Draw graphs that fit these descriptions:

a. a function that has a domain of $-5 \le x \le 1$, a range of $-4 \le y \le 4$, and $f(-2) = 1$

b. a relationship that is not a function and that has inputs on the interval $-6 \le x \le 4$ and outputs on the interval $0 \le y \le 5$

Exercise 5 Students may say that the three girls reach the same point after 20 seconds, but the graph shows only that they reach the same velocity. A good question is whether or not they do in fact arrive at the same point. In fact, Caitlin's average velocity is less than Bea's, which is less than Abby's, so they won't be at the same point. Actually, the distances traveled are given by the areas between the curves and the horizontal axis (in velocity · time units, which are distance units), so they differ considerably.

5a. Stories will vary. At the 20-second mark, each girl is moving at the same velocity. Bea's velocity increased steadily in a linear fashion. Caitlin's velocity increased very slowly at first and then became faster and faster. Abby's velocity increased very quickly at first and then increased at a slower rate.

5b. No. Because Abby starts out moving faster than both Bea and Caitlin, even when she slows down to their speed, she stays ahead.

8. Cody's code multiplies each letter's position by 2. Complete a table like the one shown. If a number is greater than 26, subtract 26 from it so that it represents a letter of the alphabet. Is the code a function? Is the rule for decoding a function?

A	B	C	D	E
1	2	3	4	5
2	4	6		

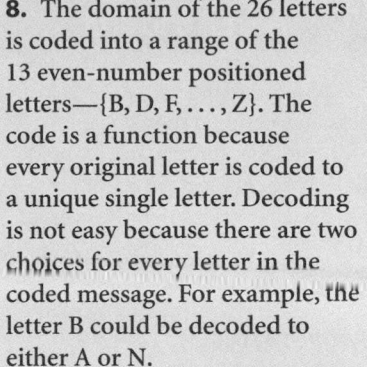

9. Consider the function $f(x) = |x|$.
   a. What is $f(-3)$? $f(-3) = |-3| = 3$
   b. What is $f(2)$? $f(2) = |2| = 2$
   c. For what x-value(s) is $f(x) = 10$? $x = 10$ or -10

10. Use your calculator for 10a–c.
    a. Graph the functions $y = \sqrt{x}$ and $y = x^2$ in a friendly window.
    b. Compare the graphs. How are they similar? How are they different?
    c. Explain why the graph of $y = \sqrt{x}$ has only one "branch."
    d. Sketch the graph of $y^2 = x$. Is this the graph of a function? Explain why or why not.

TAKE ANOTHER LOOK

You learned to solve linear equations by "undoing" the order of operations in them. You learned to code and decode secret messages. Both are examples of reversing the order of a process, or finding an **inverse.**

How do you find the inverse of a function? The equation $y = \frac{9}{5}x + 32$ converts temperatures from x in degrees Celsius to y in degrees Fahrenheit.

If you want to make a graph showing how to convert temperature from degrees Fahrenheit to degrees Celsius, you can swap the two variables in the equation and solve for y.

$$y = \frac{9}{5}x + 32 \qquad \text{The original equation.}$$
$$x = \frac{9}{5}y + 32 \qquad \text{Interchange } x \text{ and } y.$$
$$x - 32 = \frac{9}{5}y \qquad \text{Subtract 32 from both sides.}$$
$$5(x - 32) = 9y \qquad \text{Multiply both sides by 5.}$$

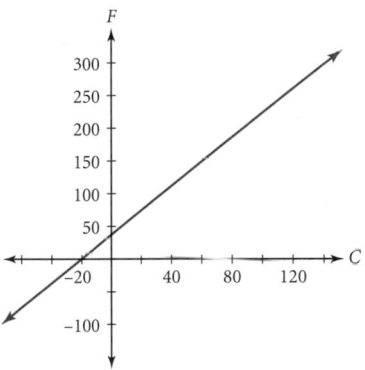

Take Another Look

The inverse of the equation that converts temperatures from degrees Celsius to degrees Fahrenheit is a function. Encourage exploration of several functions, both algebraically and with graphs. A function has an inverse that's a function if its graph passes the horizontal line test: no horizontal line crosses the graph at more than one point. Equivalently, some students may say that the graph "doesn't turn

around." That is, it's either always increasing or always decreasing. Students should also realize that the inverse of a constant function or a step function will not be a function.

Often a function is restricted to a domain over which it's just increasing or decreasing in order to have an inverse function. For example, when $y = x^2$ is restricted to the domain of positive values for x, it has the inverse function $y = \sqrt{x}$

8. The domain of the 26 letters is coded into a range of the 13 even-number positioned letters—{B, D, F, . . . , Z}. The code is a function because every original letter is coded to a unique single letter. Decoding is not easy because there are two choices for every letter in the coded message. For example, the letter B could be decoded to either A or N.

Exercise 10 Students may think that the graph in 10d is like that of 10c. Point out that 10d includes negative values of y. In this case, x is a function of y, but y is not a function of x. To do 10d on a calculator, enter $\sqrt{x}$ into list L1 and $-\sqrt{x}$ into list L2.

10a.

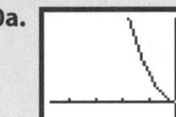
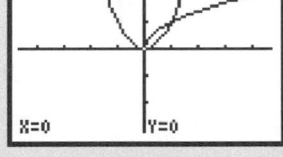

$$[-4.7, 4.7, 1, -3.1, 3.1, 1]$$

10b. The graph of $y = \sqrt{x}$ looks like half of the graph of $y = x^2$ lying on its side.

10c. The graph of $y = \sqrt{x}$ has only one branch because it gives only positive solutions.

10d. This equation does not represent a function, because a given input can have two different outputs. For example, if $x = 4, y = 2$ or $y = -2$.

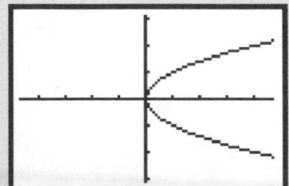

If you have given the quizzes for this chapter, you may decide to skip the chapter test and instead have students turn in a portfolio. Or use the Assessment Resources or the test generator to create a written assessment.

FACILITATING SELF ASSESSMENT

To help students complete the portfolio described in Assessing What You've Learned, suggest that they consider for evaluation their work on Lesson 8.1, Exercises 7 and 8; Lesson 8.2, Exercise 10; Lesson 8.3, Exercise 13; Lesson 8.4, Exercise 7; Lesson 8.5, Exercises 6, 8; Lesson 8.6, Exercise 13; and Lesson 8.7, Exercise 9.

$$\frac{5}{9}(x - 32) = y \qquad \text{Divide both sides by 9.}$$

$$y = \frac{5}{9}(x - 32) \qquad \text{Isolate } y \text{ on the left side.}$$

Note that after the switch x represents degrees in Fahrenheit and y represents degrees in Celsius. The domain of the inverse is the range of the original function and vice-versa. Is this a function? Does each input in °F give exactly one output in °C?

How can you tell from the graph of a function whether or not it has an inverse that's a function? Does every linear function in the form $y = a + bx$ have an inverse function? Does the squaring function or the absolute value function have an inverse function? Look for patterns in the graphs of these functions and others. Can you restrict the values on the domain of a function so that the inverse is a function?

Learn more about inverse functions with the Internet links at www.keymath.com/DA .

Assessing What You've Learned

 ORGANIZE YOUR NOTEBOOK Update your notebook with an example, investigation, or exercise that best demonstrates the concept of a function. Also, add one problem that illustrates the absolute value function and one that shows the squaring function.

 WRITE IN YOUR JOURNAL Add to your journal by answering one of these prompts.
  ► You have seen many forms of equations—direct and indirect variation, linear relationships, and exponential modeling. Do you think all equations represent functions? Can you represent all functions as equations?
  ► Is $y = f(x)$ function notation helpful to you or do you find it challenging? When do parentheses ( ) mean multiplication and when do they show the independent variable in function notation?

 UPDATE YOUR PORTFOLIO Choose your best explanation of a graph of a real-world situation from this chapter to add to your portfolio. Identify the independent and dependent variables of the situation. Describe all possible domain and range values as shown in the graph. Discuss whether the graph should be continuous or discrete.

 GIVE A PRESENTATION Create your own code for making secret messages. Explain the rule for your code with a grid or an equation or both. Is your code a function? Is it simple to code? Is it hard to decode? How does the concept of functions apply to code making and code breaking?

Transformations

Overview

This chapter introduces students to four basic transformations. Students examine vertical and horizontal translations in **Lessons 9.1** and **9.2.** They study these transformations by changing coordinates of polygons in the plane and then by modifying equations. They learn about reflections in **Lesson 9.3** and dilations in **Lesson 9.4.** In **Lesson 9.5,** students perform experiments to generate data that can be modeled by (transformed) linear and exponential functions. In **Lesson 9.6,** students study rational functions by transforming the basic function $y = \frac{1}{x}$. **Lesson 9.7,** concludes the chapter with the study of how matrices can be used to represent transformations, including some rotations, of geometric figures.

The Mathematics

Transformations

This chapter includes the study of four ways of transforming geometric figures and equations: translating (shifting), reflecting (across a line), dilating (stretching and shrinking), and, briefly, rotating around the origin. Many students are intrigued by the application of transformations to computer graphics, so you might tie each lesson to that topic.

The chapter considers transformations from two points of view. Students first consider moving polygons by working with the coordinates of their vertices, perhaps in calculator lists or, later, in a matrix. But nonpolygonal figures must be treated differently. Students study those that can be represented by equations to see the effects of various kinds of transformations on those equations.

Students often don't understand the relationship between the two points of view and resort to memorization. For example, they may not understand why you accomplish the same horizontal translation by *adding* to the x-coordinates of a polygon's vertices but *subtracting* from the variable x in an equation. Similarly, the operations involved in vertical translations and in dilations seem backward.

For example, if you're trying to dilate (stretch or shrink) a polygon vertically by a factor of 3, you multiply the y-coordinates of its vertices by 3. But if you're trying to dilate vertically the graph of $y = x^2$, you replace y with $\frac{y}{3}$ (and then solve for y as needed to get $y = 3x^2$.) Why?

To find the equation of the transformed graph, the strategy is to use the equation of the original graph. The x and y in the transformed equation will be coordinates of points lying on the transformed graph. So, if the transformation stretches everything vertically by 3, then points with coordinates $\left(x, \frac{y}{3}\right)$ lie on the original graph and therefore satisfy the original equation, meaning that $\frac{y}{3} = x^2$.

Similarly, if $y = |x|$ is translated 2 units to the right, to find the equation relating the coordinates (x, y) of the transformed graph, you use the fact that point $(x - 2, y)$ lies on the original graph, so that $y = |x - 2|$.

Double duty

You might encourage students who enjoy looking for relationships to explore how different transformations of some functions yield the same result.

The parent linear function $y = x$ has the interesting property that horizontal transformations are also vertical transformations. The graph of $y = x + a$ can be thought of as a shift of $y = x$ either a units to the left or a units upward.

Horizontal translations of the parent exponential functions are vertical stretches. For example, $y = 3^{x+2}$ is the same as $y = 3^2 3^x = (9)3^x$.

Vertical stretches of $y = x^2$ are horizontal shrinks. For example, $y = 16x^2$ can also be thought of as $y = (4x)^2$. And vertical stretches of $y = \frac{1}{x}$ are also horizontal stretches. For example, $y = \frac{4}{x}$ is the same as $y = \frac{1}{\frac{1}{4}x}$.

Matrices of transformations

In real-life computer graphics, points are moved by multiplying by matrices. For example, if a point is represented by $\begin{bmatrix} x \\ y \end{bmatrix}$, then multiplying it on the left

by the matrix $\begin{bmatrix} a & b \\ c & d \end{bmatrix}$ produces point $\begin{bmatrix} ax + by \\ cx + dy \end{bmatrix}$.

If b and c are 0 and d is 1, this transformation is a horizontal dilation or reflection.

Multiplying by any 2×2 matrix always takes the origin to itself, so such a matrix can't represent a translation. Computer graphics programs instead use 3 coordinates for points in the plane and 3×3 matrices for transformations. They represent each

point as $\begin{bmatrix} x \\ y \\ 1 \end{bmatrix}$. Multiplying on the left by matrix

$\begin{bmatrix} 2 & 0 & 0 \\ 0 & -1 & 0 \\ 0 & 0 & 1 \end{bmatrix}$ yields $\begin{bmatrix} 2x \\ -y \\ 1 \end{bmatrix}$, a horizontal dilation and

vertical reflection. Multiplying by $\begin{bmatrix} 1 & 0 & 3 \\ 0 & 1 & 2 \\ 0 & 0 & 1 \end{bmatrix}$ gives

$\begin{bmatrix} x + 3 \\ y - 2 \\ 1 \end{bmatrix}$, representing a translation.

Using This Chapter

The activities of Lesson 9.5 apply the ideas of previous lessons rather than introducing new material. Lesson 9.6 may be skipped if rational functions are not required core material for your school district. And use Lesson 9.7 only if your class is studying matrices.

Resources

Discovering Algebra Resources

Calculator Notes 1G, 1P, 4C, 8D, 9A, 9B, 9C, 9D

Teaching and Worksheet Masters
   Lesson 9.1 Moving a Quadrilateral
   Lesson 9.2 Moving Absolute Values
   Lesson 9.3 Flipping a Letter
   Lesson 9.4 Stretching a Polygon
   Lesson 9.5 The Rolling Marble

Assessment Resources A and B
   Quiz 1 (Lessons 9.1 and 9.2)
   Quiz 2 (Lessons 9.3 to 9.5)
   Quiz 3 (Lessons 9.6 and 9.7)
   Chapter 9 Test
   Chapter 9 Constructive Assessment Options

More Practice Your Skills for Chapter 9

Condensed Lessons for Chapter 9

Other Resources

Ronald J. Carlson and Mary Jean Winter. *Transforming Functions to Fit Data*. Emeryville, CA: Key Curriculum Press, 1998.

www.keymath.com/DA

Materials

- graph paper
- patty paper, *optional*
- large marbles
- poster board
- metersticks or yardsticks
- tape
- paper cups
- sheets of paper
- books
- tables and chairs
- motion sensor
- cables for linking calculators
- stopwatch or watch with second hand

Pacing Guide

	day 1	day 2	day 3	day 4	day 5	day 6	day 7	day 8	day 9	day 10
standard	9.1	9.1	9.2	quiz, 9.3	9.4	9.5	9.5, quiz	9.6	9.7, quiz	review
enriched	9.1	9.1, project	9.2	quiz, 9.3	9.4	9.5	quiz, 9.6	9.7	project, quiz	review, TAL
block	9.1	9.2, 9.3	quiz, 9.4	9.5, 9.6	quiz, 9.7	review, assessment				

	day 11	day 12	day 13	day 14	day 15	day 16	day 17	day 18	day 19	day 20
standard	assessment									
enriched	assessment									

Transformations

The Dome of the Rock is a famous site in Jerusalem. Built in the 7th century, it is well known for its beautiful tile work. Moving a small design left, right, up, or down could create some of the large patterns you see. Flipping or turning a design could create yet other patterns. Moving, flipping, or turning a design is important in creating many art forms. As you will see, changes like these are equally important in mathematics.

CHAPTER 9 OBJECTIVES

- Learn how to graph polygons on the graphing calculator

- Write list definitions to describe transformations (combinations of translations, reflections, dilations) of points

- Write equations of transformed function graphs

- Explore the concept of a parent function and its family

- Model real-world data with transformed equations

- Investigate rational functions of the form $f(x) = \frac{a}{x}$ and explore basic transformations of this function

- Model real-world data using rational functions

- Use matrices to represent transformations of polygons

OBJECTIVES

In this chapter you will
- move a polygon by changing its vertices' coordinates
- learn to change, or transform, graphs by moving, flipping, shrinking, or stretching
- write a new equation to describe the changed, or transformed, graph
- model real-world data with equations of transformations
- use matrices to transform the vertices of a polygon

Challenge students to find the smallest unit that can be repeated in the row of tilted squares at the top. That unit is analogous to a parent, and every other incidence of it can be thought of as a child of the parent. There is more than one way to visualize the parent unit. For instance, students can trap a light-colored diamond in a square and call that the parent, or they can visualize a square that captures a bow-tie shaped figure.

A design that repeats to fill the plane is called a *tessellation*. The patterns pictured here are based on squares, but tessellations can be derived from any plane-filling figure. The parent unit, called a *tessera*, is translated, reflected, or rotated in several directions.

You might ask students to think of other art forms that use transformations. Examples include fabric designs and some forms of cubism and pop art.

You may want to make a distinction between symmetry and transformation. Complete sections of the Dome of the Rock facade could be described as having reflective symmetry (one half of the section looks like the other half). The designer of the facade may have used a reflection transformation to create the second half.

LESSON OUTLINE

First day:

5 min	Introduction
30 min	Investigation
15 min	Example

Second day:

20 min	Investigation
10 min	Sharing
5 min	Closing
15 min	Exercises

MATERIALS

* graph paper, *optional*
* Calculator Notes 1G, 8D, 9A
* Moving a Quadrilateral (W), *optional*

TEACHING

Our study of geometric transformations begins with translations.

INTRODUCTION

In computer animation each individual point is sometimes called an *articulated variable* or "avar," a combination of real-life motion (articulated) and mathematics (variable).

 Guiding the Investigation

One step Give each group a copy of the Moving a Quadrilateral worksheet. Ask them to graph the darker quadrilateral on their calculators (referring them to Calculator Note 1G as needed) and then to express vertex coordinates for the lighter quadrilateral in terms of the vertex coordinates for the darker figure. If groups have difficulties, suggest that they try a horizontal or vertical shift.

Ask one member of each group to do the investigation on graph paper.

LESSON
9.1

If one is lucky, a solitary fantasy can totally transform one million realities.

MAYA ANGELOU

Translating Points

In computer animation, many individual points define each figure. You animate a figure by moving these points around the screen, little by little, through a series of frames. When you see the frames one after the other, the entire figure appears to move.

In mathematics, changing or moving a figure is called a **transformation.** So, every frame of an animation is a transformation of the one before it.

This computer-animated motorcycle was created with software called Maya. The "skeleton" of the motorcycle would look like the face in the photo on page 331. The software allows an animator to move the motorcycle by transforming the points of the underlying skeleton.

 ## Investigation
Figures in Motion

In this investigation you will learn how to move a polygon around the coordinate plane. You will first see what happens when you change the *y*-coordinates of the vertices.

Step 1 (2, 1), (2, 4), (6, 1)

Step 1 Name the coordinates of the vertices of this triangle.

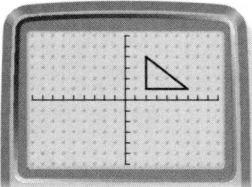

Step 2 Enter the *x*-coordinates of the vertices into list L₁ and the corresponding *y*-coordinates of the vertices into list L₂. Enter the first coordinate pair again at the end of each list. Graph the triangle by connecting the vertices.
[▶🖥] See **Calculator Note 1G** to review connected graphs. ◀

Step 2 The connected graphs shown in this chapter use the small dot mark instead of the larger box.

Steps 2 and 3

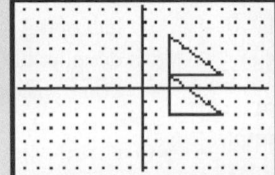

> **Procedure Note**
>
> For this investigation, use a friendly window with a factor of 2.
> [▶🖥] See **Calculator Note 8D** to review friendly windows. ◀

LESSON OBJECTIVES

* Learn how to graph polygons on the graphing calculator
* Write list definitions to describe translations of points

Step 3

Define list L3 and list L4 as follows

$$L_3 = L_1$$
$$L_4 = L_2 - 3$$

Graph a second triangle using list L3 for the *x*-coordinates of the vertices and list L4 for the *y*-coordinates of the vertices.

Step 4 $(2, -2), (2, 1),$
and $(6, -2)$; the triangle moves down 3 units; the *y*-coordinates are reduced by 3.

Step 4

Name the coordinates of the vertices of the new triangle. Tell how the original triangle has moved. How did the coordinates of the vertices change?

Step 5

Repeat Steps 3 and 4 with these definitions.

a. $L_3 = L_1$
$L_4 = L_2 + 2$
$(2, 3), (2, 6),$ and $(6, 3)$; the triangle moves up 2 units; the *y*-coordinates are increased by 2.

b. $L_3 = L_1$
$L_4 = L_2 - 1$
$(2, 0), (2, 3),$ and $(6, 0)$; the triangle moves down 1 unit; the *y*-coordinates are reduced by 1.

Step 6

Write definitions for list L3 and list L4 in terms of list L1 and list L2 to create each graph below. Check your definitions by graphing on your calculator.

a.

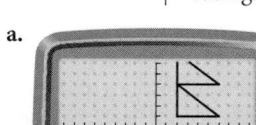

$L_3 = L_1;\ L_4 = L_2 + 3$

b.

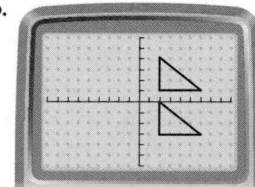

$L_3 = L_1;\ L_4 = L_2 - 4$

c.

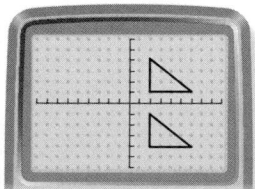

$L_3 = L_1;\ L_4 = L_2 - 5$

Next, you will include changes to the *x*-coordinates too.

Step 7 $(1, 2), (2, -2),$
$(-3, -1), (-2, 1)$

Step 7

Name the coordinates of the vertices of this quadrilateral.

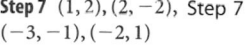

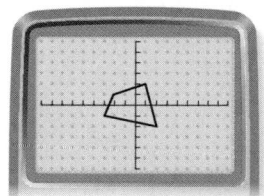

Step 8

Graph the quadrilateral using list L1 for the *x*-coordinates of the vertices and list L2 for the *y*-coordinates of the vertices.

Step 10 $(-2, 2), (-1, -2),$
$(-6, -1),$ and $(-5, 1)$; the quadrilateral moves left 3 units; the *x*-coordinates are reduced by 3.

Step 9

Define list L3 and list L4 as follows

$$L_3 = L_1 - 3$$
$$L_4 = L_2$$

Graph a second quadrilateral using list L3 for the *x*-coordinates of the vertices and list L4 for the *y*-coordinates of the vertices.

Step 11a $(3, 2), (4, -2)$
$(-1, -1),$ and $(0, 1)$; the quadrilateral moves right 2 units; the *x*-coordinates are increased by 2.

Step 10

Name the coordinates of the vertices of this new quadrilateral. Describe how the original quadrilateral moved. How did the coordinates of the vertices change?

Step 11b $(0, 5), (1, 1),$
$(-4, 2),$ and $(-3, 4)$; the quadrilateral moves left 1 unit and up 3 units; the *x*-coordinates are reduced by 1 and the *y*-coordinates are increased by 3.

Step 11

Repeat Steps 9 and 10 with these definitions.

a. $L_3 = L_1 + 2$
$L_4 = L_2$

b. $L_3 = L_1 - 1$
$L_4 = L_2 + 3$

NCTM STANDARDS

CONTENT	PROCESS
● Number	Problem Solving
● Algebra	● Reasoning
● Geometry	● Communication
Measurement	Connections
Data/Probability	● Representation

Step 11a

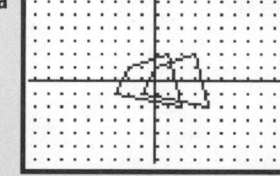

Step 11b

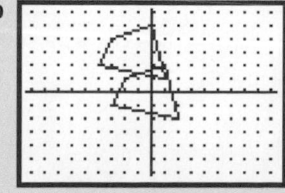

Step 3 This definition can take place in the STAT EDIT menu, or using the STO key: $L_1 \rightarrow L_3$ (ENTER) and $L_2 - 3 \rightarrow L_4$ (ENTER).

Step 5 Students will not need to keep entering $L_3 = L_1$, but they should be reminded that the *x*-coordinates are not changing.

Step 5a

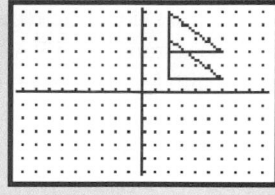

Step 5b

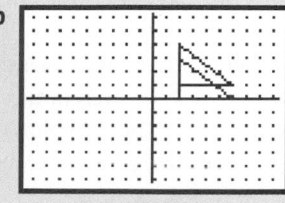

Step 6 [ESL] Some students may not realize that the phrase "in terms of" means "using."

Step 7 As needed, remind students that a quadrilateral is a four-sided figure. The calculator will produce a strange figure if students enter the coordinates in an order other than clockwise or counterclockwise around the figure. Let them learn from their error rather than trying to prevent it. Also, they must enter the first coordinate again at the end to close the figure.

Steps 8 and 9

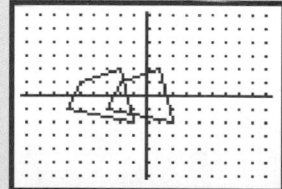

Steps 9 and 11 Ask any group you may be observing when they begin this step if they can predict the result before going to the calculator. Students need not enter $L_4 = L_2$ more than once, but they should be aware that the *y*-coordinates are not changing.

Step 13 Sample answer: Adding
to or subtracting from the
x-coordinate moves the figure
right or left, respectively. Adding
to or subtracting from the
y-coordinate moves the figure
up or down, respectively.

SHARING IDEAS

Ask several students to report
their ideas from Step 13. As
always, aim for variety to enrich
the discussion. Practice watching
students' faces and body language
and responding appropriately.
Students may get the idea that you
read minds! Or, if you're smooth
enough, you can have the satisfac-
tion of their not even noticing that
you asked for their opinion right
when they had an idea.

Remind the class that they have
been looking at transformations of
the geometric figures. Draw their
attention to the quotation at the
beginning of the lesson and ask
how that use of the word *transform*
is similar to and different from the
mathematical meaning. Introduce
the term *translation* for the kind
of transformation they've been
observing, and mention that the
resulting figure is called the *image*
of the original under the transformation. Ask how a geometric
translation like this is similar to
and different from a translation
between languages. **[Language]**
Transformation and *translation*
are defined here as nouns, but
they have verb forms also:
"Translate the point."

Assessing Progress

Through observing groups and
presentations, you can assess
students' ability to name **coordinates** of points, enter data into
calculator lists, and operate on
calculator lists.

▶ EXAMPLE

This example shows how the variables *x* and *y* are used to define
the translation. It is good for
students who haven't yet grasped
how to translate a point in two
directions, as in Step 11b.

Step 12 | Write definitions for list L₃ and list L₄ in terms of list L₁ and list L₂ to create each graph below. Check your definitions by graphing on your calculator.

a.

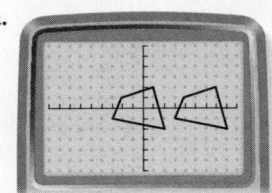

$L_3 = L_1 + 6; \; L_4 = L_2$

b.

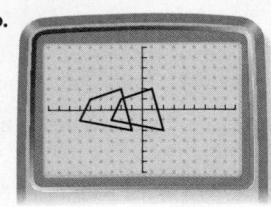

$L_3 = L_1 - 3; \; L_4 = L_2$

c.

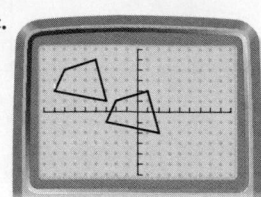

$L_3 = L_1 - 5; \; L_4 = L_2 + 3$

Step 13 | Summarize what you have learned about moving a figure around the coordinate plane.

In the investigation, each new polygon is the result of transforming the original polygon by moving it left, right, up, or down, or combinations of these movements. The figure that results from a transformation is called the **image** of the original figure. Transformations that move a figure horizontally, vertically, or both are called **translations.** You can define the translation of a point simply by adding to or subtracting from its coordinates.

EXAMPLE   Sketch the image of this figure after a translation right 4 units and down 3 units. Define the coordinates of any point in the image using (*x*, *y*) as the coordinates of any point in the original figure.

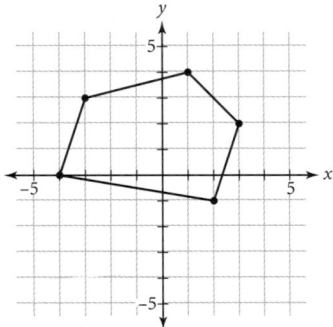

▶ Solution   Translate every point right 4 units and down 3 units. For example, move the vertex at (1, 4) to (5, 1). This is the same as adding 4 to the *x*-coordinate and subtracting 3 from the *y*-coordinate. That is, (1 + 4, 4 − 3) gives (5, 1).

A definition for any point in the image is

$$(x + 4, y - 3)$$

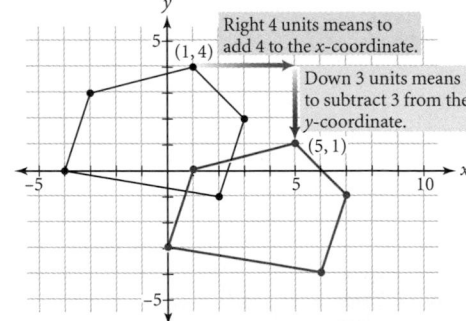

Right 4 units means to add 4 to the *x*-coordinate.

Down 3 units means to subtract 3 from the *y*-coordinate.

On your calculator, you can put the coordinates of the vertices of the original pentagon into list L1 and list L2. Then define list L3 as L3 = L1 + 4 and list L4 as L4 = L2 − 3.

Graphing confirms that your definition works.

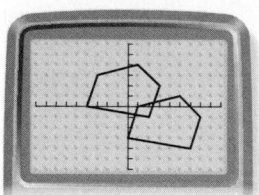

Science
●—CONNECTION—●

Many scientists support the theory of plate tectonics. According to this theory, the continents of the world were, at one time, together as a single continent. The German geophysicist and meteorologist Alfred Wegener (1880–1930) called this mass of land Pangaea. Over thousands of years, the individual continents drifted (or translated) to their current locations.

EXERCISES

You will need your calculator for problem **7.**

▶ Practice Your Skills

1. Name the coordinates of the vertices of each figure.

a.

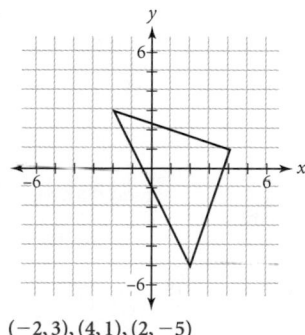

$(-2, 3), (4, 1), (2, -5)$

b.

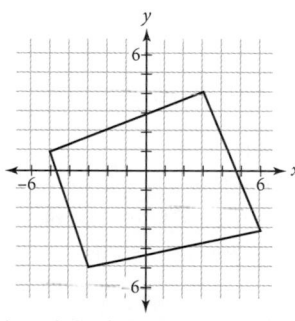

$(-5, 1), (3, 4), (6, -3), (-3, -5)$

2. The x-coordinates of the vertices of a triangle are entered into list L1. The y-coordinates are entered into list L2. Describe the transformation for each definition.

a. $L_3 = L_1 - 5$
$L_4 = L_2$
a translation left 5 units

b. $L_3 = L_1 + 1$
$L_4 = L_2 + 2$
a translation right 1 unit and up 2 units

MAKING THE CONNECTION

[Ask] "How is translation on a sphere, representing the earth, like and different from translations on a plane?" [For the image to be congruent to the original (as is the case for continents), different parts of the figure must actually move different distances.]

Closing the Lesson

One kind of **transformation** of a geometric figure is a **translation,** in which the figure is moved in the plane without changing size, shape, or orientation. To translate to the right horizontally, you add a positive number to every point's x-coordinate. To translate upward, you add a positive number to every point's y-coordinate. Adding a negative number (that is, subtracting) translates in the opposite direction.

BUILDING UNDERSTANDING

Students work with translations of points and figures.

ASSIGNING HOMEWORK

Essential	1, 2, 4, 5, 7
Performance assessment	3, 8
Portfolio	10
Journal	9
Group	6
Review	11–13

4a.

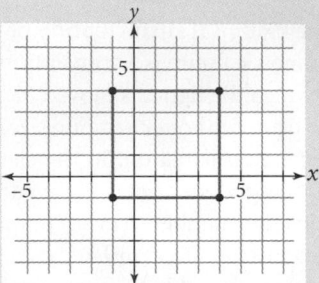

6a. Possible answer: Enter the x-coordinates into list L_1 and the y-coordinates into list L_2. $L_1 = \{2, 5, 1, 2\}$ and $L_2 = \{-1, 0, 2, -1\}$. Then make a connected graph.

3. The red triangle at right is the image of the black triangle after a transformation.

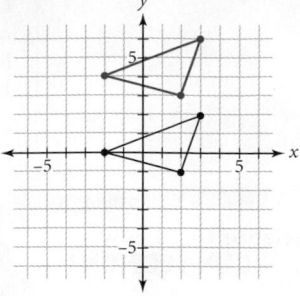

  a. Describe the transformation. **a translation up 4 units**

  b. Tell how the x-coordinates of the vertices change between the original figure and the image. **The x-coordinates are unchanged.**

  c. Tell how the y-coordinates of the vertices change.
  The y-coordinates are increased by 4.

4. Consider the square at right.

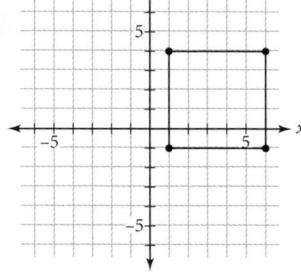

  a. Sketch the image of the figure after a translation left 2 units.

  b. Define the coordinates of any point in the image using (x, y) as the coordinates of any point in the original figure. $(x - 2, y)$

5. The "spider" in the upper left has its x-coordinates in list L_1 and its y-coordinates in list L_2.

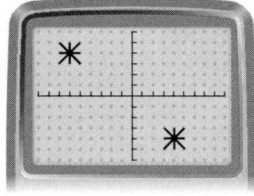

$[-9.4, 9.4, 1, -6.2, 6.2, 1]$

  a. Describe the transformation to its image in the lower right. **a translation right 10 units and down 8 units**

  b. Write definitions for list L_3 and list L_4 in terms of list L_1 and list L_2. $L_3 = L_1 + 10, L_4 = L_2 - 8$

  c. How would your answer to 5b change if the "spider" in the lower right were the original figure and the figure in the upper left were the image?
  The signs would change: $L_3 = L_1 - 10, L_4 = L_2 + 8.$

► **Reason and Apply**

6. Consider the triangle on the calculator screen at right.

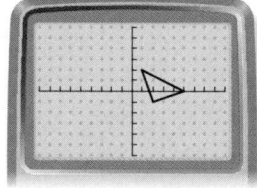

$[-9.4, 9.4, 1, -6.2, 6.2, 1]$

  a. Describe how to graph this triangle on your calculator.

  b. For each graph below, describe the transformation made to the original triangle.

  i.

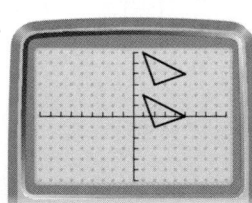

  a translation up 4 units

  ii.

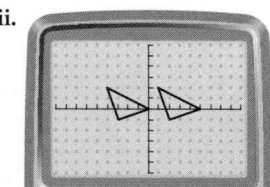

  a translation left 5 units

  iii.

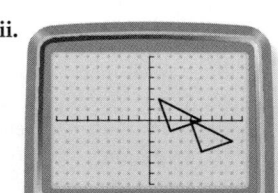

  a translation right 3 units and down 2 units

7. The coordinates of the vertices of a triangle are (2, 1), (4, 3), and (3, 0). Sketch the image that results from each definition. Use calculator lists to check your work.

a. $(x, y + 3)$ **b.** $(x - 2, y)$ **c.** $(x + 3, y - 1)$

8. If the triangle at right is the original figure, name the coordinates of the vertices of the image after

a. A translation up 4 units. (1, 5), (5, 5), (1, 10)

b. A translation left 7 units. $(-6, 1), (-2, 1), (-6, 6)$

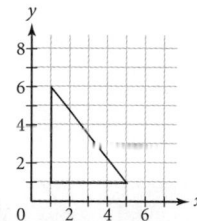

9. Lisa is designing a computer animation program. She has a set of coordinates for the arrow in the lower left. Now she wants the arrow to move to the upper right position.

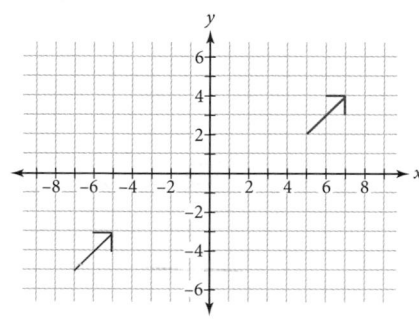

a. Describe the transformation that moves the arrow to the upper right.

b. Define the coordinates of any point in the image using (x, y) as the coordinates of any point in the original figure. $(x + 12, y + 7)$

c. Lisa decides that a single move is too sudden. She thinks that moving the arrow little by little, in 20 frames, would look better. How should she define the coordinates of any point in each new image using (x, y) as the coordinates of any point in the figure in the previous frame? $\left(x + \frac{12}{20}, y + \frac{7}{20}\right) = (x + 0.6, y + 0.35)$

10. Nick is also designing a computer animation program. His program first draws an N by connecting the points (7, 1), (7, 2), (8, 1), and (8, 2). Then, in each subsequent frame, the previous N is erased and an image is drawn whose coordinates are defined by $(x - 0.25, y + 0.05)$. The program uses recursion to do this over and over again.

What will be the coordinates of the N in the

a. 10th new frame?
(4.5, 1.5), (4.5, 2.5),
(5.5, 1.5), (5.5, 2.5)

b. 25th new frame?
(0.75, 2.25), (0.75, 3.25),
(1.75, 2.25), (1.75, 3.25)

c. 40th new frame?
$(-3, 3), (-3, 4),$
$(-2, 3), (-2, 4)$

▶ **Review**

11. Use $f(x) = 2 + 3x$ to find

a. $f(5)$ 17   **b.** $f(-4)$ -10 **c.** $f(x + 2)$ $8 + 3x$ **d.** $f(2x - 1)$ $-1 + 6x$

12. Solve each equation.

a. $5 = -3 + 2x$ $x = 4$ **b.** $-4 = -8 + 3(x - 2)$ $x = 3.\overline{3}$ **c.** $7 + 2x = 3 + x$ $x = -4$

7a.

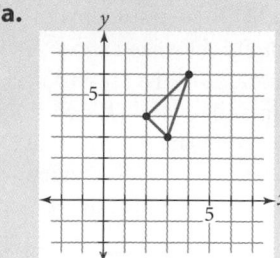

7b.

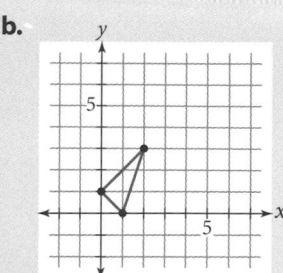

7c.

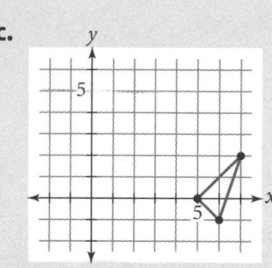

9a. a translation right 12 units and up 7 units

Exercise 10 The original figure is frame 0. Hence, by the 10th frame, the definition for the translation $(x - 0.25, y + 0.05)$ has been used 10 times.

Exercise 11 Parts c and d may be the first time students evaluate a function for a variable expression. You might want to describe it as "substituting $x + 2$ for x." They will do this again in Exercise 1 of Lesson 9.2. This exercise reviews Lesson 8.4.

Exercise 12 This exercise reviews Lesson 4.8.

Exercise 13 To help students in 13d, you might write an exponential equation from a graph. Give the points $(0, 1)$ and $(1, 1.5)$ so they can calculate the constant multiplier. This exercise reviews Lessons 4.6, 5.5, 7.2, and 8.6.

13. Find an equation for each graph.

a.

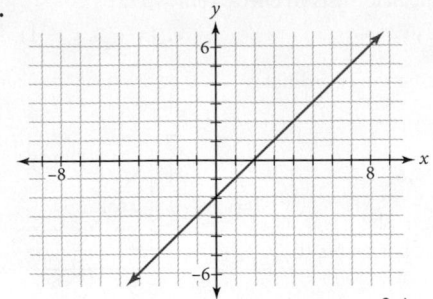

$$y = -2 + x$$

b.

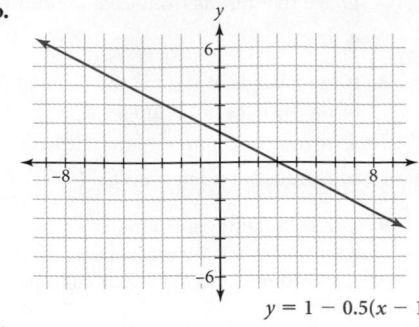

$$y = 1 - 0.5(x - 1)$$

c.

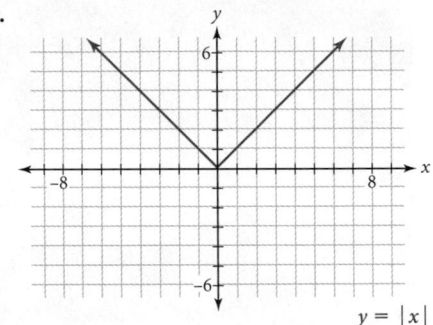

$$y = |x|$$

d.

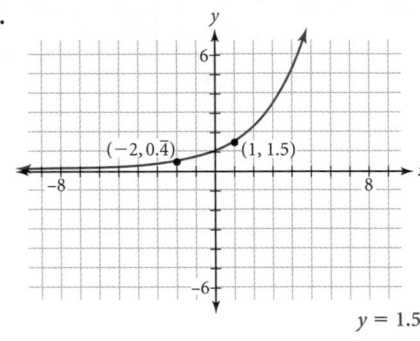

$$y = 1.5^x$$

Alternative Project
For students for whom this project is inaccessible (they don't have access to a computer or calculator, or they find the logic of programming languages confusing), you might want to suggest a research paper on professional animation software.

project

ANIMATING WITH TRANSFORMATIONS

As you've learned in this lesson, you can use transformations to create computer animation. Programs use mathematics to transform the points of a figure little by little. For example, Lisa's arrow in problem 9 appears to move because it makes 20 very small translations.

Now it's your turn to be the computer animator. Use a computer programming language to create an animation of any figure you choose. You can even use your calculator. [▶ 🖳 See **Calculator Note 9A** for a calculator program that moves Lisa's arrow. ◀]

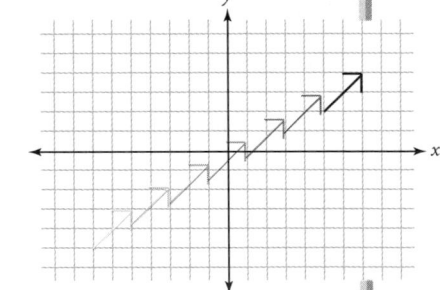

Your project should include

▶ The steps of your program.

▶ An explanation of what each step of the program does.

▶ A description of the transformations used.

▶ A sketch of your original figure and the final image.

As you learn about other transformations in this chapter, you can try including them in your project too. You might also want to research the programming languages and software that professional animators use.

Supporting the project

MOTIVATION

Many students who think they can't write computer programs become very excited when they discover how to move a figure across the screen of a graphing calculator, so don't be too quick to give in to protests.

OUTCOMES

▶ The program does not have any obvious errors (such as unclosed loops) that would keep it from running.

▶ Each step is described in easy-to-understand terms using proper terminology. ("This step stores the coordinates of the vertices." "This step produces a translation to the right 0.5 unit and up 0.4 unit.")

▶ The sketch shows accurately the original and final figure on a coordinate grid.

• Student uses elaborate figures or many different transformations.

• Student has researched and learned a new computer programming language.

• Student has used internet animation languages like Java, Shockwave, or Flash.

LESSON 9.2

LESSON
9.2

Translating Graphs

There are infinitely many linear and exponential functions. In the previous chapters, you wrote many of them "from scratch" using points, the y-intercept, the slope, the starting value, or the constant multiplier.

Poetry is what gets lost in translation.
ROBERT FROST

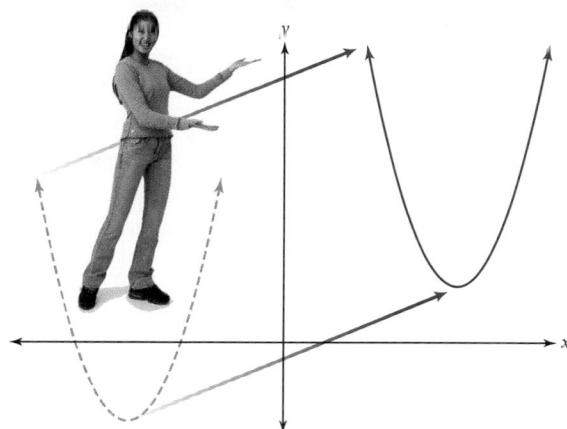

There are also infinitely many absolute value and squaring functions. But rather than starting from scratch, you can transform $y = |x|$ and $y = x^2$ to create many different equations. In this investigation you will use what you know about translating points to translate functions. If you discover any unexpected transformations along the way, make a note so that you can use them later in the chapter.

Investigation
Translations of Functions

> **Procedure Note**
> For this investigation, use a friendly window with a factor of 2.

First you'll transform the absolute value function by making changes to x.

Step 1 | Enter $y = |x|$ into Y1 and graph it on your calculator.

Step 2 | If you replace x with $x - 3$ in the function $y = |x|$, you get $y = |x - 3|$. Enter $y = |x - 3|$ into Y2 and graph it.

Step 3 | Think of the graph of $y = |x|$ as the original figure and the graph of $y = |x - 3|$ as its image. How have you transformed the graph of $y = |x|$?

The **vertex** of an absolute value graph is the point where the function changes from decreasing to increasing or from increasing to decreasing.

Step 4 | Name the coordinates of the vertex of the graph of $y = |x|$. Name the coordinates of the vertex of the graph of $y = |x - 3|$. Do these two points help verify the transformation you found in Step 3?

Step 3 a translation right 3 units

Step 4 $(0,0)$; $(3,0)$; yes, the point is translated right 3 units.

NCTM STANDARDS

CONTENT	PROCESS
Number	Problem Solving
• Algebra	• Reasoning
• Geometry	• Communication
Measurement	• Connections
Data/Probability	• Representation

LESSON OBJECTIVES

- Write equations to describe translations of the absolute value and squaring functions
- Explore the concept of a family of functions

LESSON OUTLINE

One day:

25 min	Investigation
5 min	Sharing
10 min	Examples
5 min	Closing
5 min	Exercises

MATERIALS

- Calculator Note 9B
- Moving Absolute Values (W), *optional*

TEACHING

To transform graphs that aren't polygons, you can't just connect images of vertices. Rather, we need to see how the equations of these graphs are affected by transformations.

Students have learned to write linear equations in intercept form with the constant term first: $y = a + bx$. Nonlinear equations will be written with the constant term last, as in $y = (x - 4)^2 + 2$, $y = |x - 7| + 2.5$, and $y = 3^{x+1} - 2$.

Guiding the Investigation

One step Give each group the Moving Absolute Values worksheet and ask them to alter the calculator function $Y1 = abs(x)$ to create the lighter graph. Let them experiment with adding and subtracting in various ways.

Step 1 You may need to remind students of the meaning of absolute value.

See page 720 for answers to Steps 1 and 2.

Step 4 Students may be confused by the fact that, to translate a point to the right, you add to the x-coordinate, but in an equation, you ultimately subtract from x. Challenge them to think about reasons to discuss during Sharing.

Step 6 Suggest to students having trouble that they look at individual points.

Steps 7 and 8

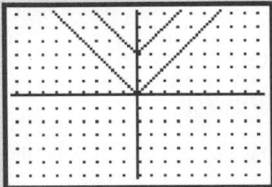

Step 13 Sample answer: To translate $h(h > 0)$ units to the right, replace x with $x - h$. To translate h units to the left replace x with $x + h$. When y is replaced by $y + h$, the graph is translated up h units. When y is replaced with $y - h$, the graph is translated down h units. Another sample answer: To translate the graph horizontally, subtract from x in the function (subtracting a positive number translates right; subtracting a negative number translates left). To translate the graph vertically, add to the entire function (adding a positive number translates up; adding a negative number translates down).

SHARING IDEAS

The quotation introducing the lesson, as well as the Science Connection, can be used to discuss the multiple meanings of *translation*. Robert Frost (1874–1963) was an American poet.

Have three or more groups report on the parts of Step 12.

Step 5
$y = |x - (-4)|$ or
$y = |x + 4|$; replace x with $x + 4$.

Step 5 Find a function for Y2 that will translate the graph of $y = |x|$ left 4 units. What is the function? In the equation $y = |x|$, what did you replace x with to get your new function?

Step 6 Write a function for Y2 to create each graph below. Check your work by graphing both Y1 and Y2.

a.
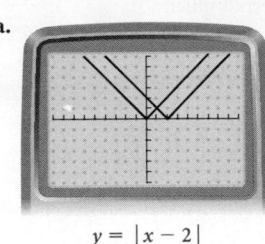
$y = |x - 2|$

b.
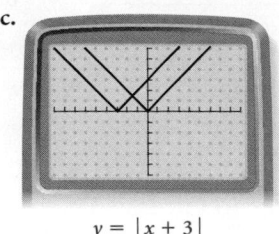
$y = |x - 5|$

c.
$y = |x + 3|$

Next, you'll transform the absolute value function by making changes to y.

Step 7 Clear all of the functions in your Y= menu. Enter $y = |x|$ into Y1 and graph it.

Step 8 If you replace y with $y - 3$ in the function $y = |x|$, you get $y - 3 = |x|$. Solve for y and you get $y = |x| + 3$. Enter $y = |x| + 3$ into Y2 and graph it.

Step 9 a translation up 3 units

Step 9 Think of the graph of $y = |x|$ as the original figure and the graph of $y = |x| + 3$ as its image. How have you transformed the graph of $y = |x|$?

Step 10 $(0, 0)$; $(0, 3)$; yes, the point is translated up 3 units.

Step 10 Name the coordinates of the vertex of the graph of $y = |x|$. Name the coordinates of the vertex of the graph of $y = |x| + 3$. Do these two points help verify the transformation you found in Step 9?

Step 11
$y = |x| - 3$; replace y with $y - (-3)$ or $y + 3$.

Step 11 Find a function for Y2 that will translate the graph of $y = |x|$ down 3 units. What is the function? In the function of $y = |x|$, what did you replace y with to get your new function?

Step 12 Write a function for Y2 to create each graph below. Check your work by graphing both Y1 and Y2.

a.
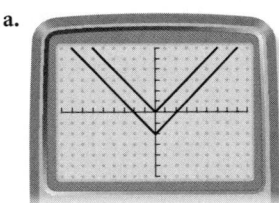
$y = |x| - 2$

b.
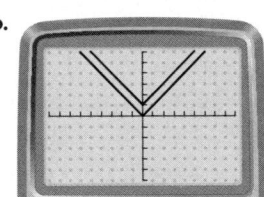
$y = |x| + 1$

c.
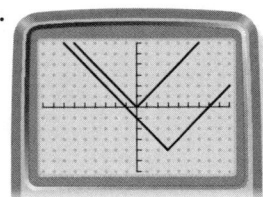
$y = |x - 3| - 4$

Step 13 Summarize what you have learned about translating the absolute value graph vertically and horizontally.

Remind students of Step 8: $y = |x| + 3$ is equivalent to $y - 3 = |x|$; either form is a translation 3 units up from $y = |x|$. Ask if they can explain why subtracting a positive number from the y-coordinate of a point translates that point *downward*, but subtracting a positive number from the variable y in $y = |x|$ translates the graph *upward*. Through discussion, bring out the idea that point (x, y) satisfies the equation of the shift if $(x, y - 3)$ lies on the original graph and thus satisfies $y - 3 = |x|$. As

usual, the longer you can wait before explaining, the more students will think about the ideas.

Ask how the vertex of an absolute value function is like and unlike the vertex of a polygon. Answers are unimportant. Your goal is to help students become more familiar with the term *vertex*. **[Language]** Point out that the plural of *vertex* is *vertices* and that the singular of *vertices* is not *verticie*.

Anni Albers (1899–1994), a German-American artist, used many transformations of a single triangle to create this serigraph. Can you find some translations?

Anni Albers, *Untitled*, ca. 1969. The National Museum of Women in the Arts, Washington, D.C.

Parent Functions
Introduce the terms *parent function* and *family of functions* and ask what equation would represent a translation of the parent function $y = x^2$ two units to the right and four units down. This leads into Example A.

You may want to use and display more formal notation for translations, such as $y = f(x - h) + k$, where h represents a horizontal translation and k represents a vertical translation. The constants h and k will be used in Chapter 10.

Assessing Progress
Watch for students' understanding of **translation** and their abilities to graph functions.

The most basic form of a function is often called a **parent function.** By transforming the graph of a parent function, you can create infinitely many new functions, or a **family of functions.** Functions like $y = |x - 3|$ and $y = |x| + 3$ are members of the absolute value family of functions with $y = |x|$ as the parent. Other families of functions include the linear family with $y = x$ as the parent, the squaring family with $y = x^2$ as the parent, and the base-3 exponential family with $y = 3^x$ as the parent.

Learning how to create a family of functions will help you to see relationships between equations and graphs. The translations you learned in the investigation apply to any function.

Science
CONNECTION

Earthquakes often translate the earth's crust along a *fault.* You can see faults most easily when buildings and other structures are translated too. These cable car tracks were bent by a fault during the great 1906 earthquake in San Francisco, California. Learn more about earthquakes and faults with the links at www.keymath.com/DA .

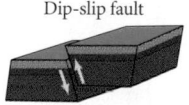
Dip-slip fault

Strike-slip fault

► **EXAMPLE A**

This example shows how to translate the squaring parent function in two directions. As needed, remind students of the meaning of squaring.

EXAMPLE A

The graph of the parent function $y = x^2$ is shown in black. Its image after a transformation is shown in red. Describe the transformation. Then write an equation for the image.

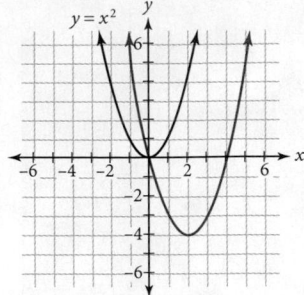

► **Solution**

The vertex of a parabola is the point where the squaring function changes from decreasing to increasing or increasing to decreasing. The vertex of the graph of $y = x^2$ is $(0, 0)$. The vertex of the image is $(2, -4)$. So the graph of $y = x^2$ is translated right 2 units and down 4 units to create the red image. You can check this with any other point. For example, the image of the point $(2, 4)$ is $(4, 0)$, which is also a translation right 2 units and down 4 units.

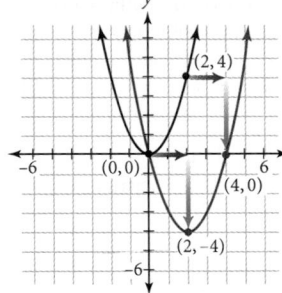

Every point on $y = x^2$ is translated right 2 units and down 4 units.

The equation of the image is
$$y - (-4) = (x - 2)^2$$
or
$$y = (x - 2)^2 - 4$$

Use the translation to write an equation for the red image.

$y = x^2$ Equation of the original parabola.

$y = (x - 2)^2$ Replace x with $x - 2$ to translate the graph right 2 units.

$y - (-4) = (x - 2)^2$ Replace y with $y - (-4)$, or $y + 4$, to translate the graph down 4 units.

$y = (x - 2)^2 - 4$ Solve for y.

The equation of the image is $y = (x - 2)^2 - 4$. You can graph this on your calculator to check your work.

In the next example you'll see how to translate an exponential function. Later you will use these skills to fit a function to a set of data.

EXAMPLE B

The starting number of bacteria in a culture dish is unknown, but the number grows by approximately 30% each hour. After 4 hours there are 94 bacteria present. Write an equation to model this situation. Then find the starting number of bacteria.

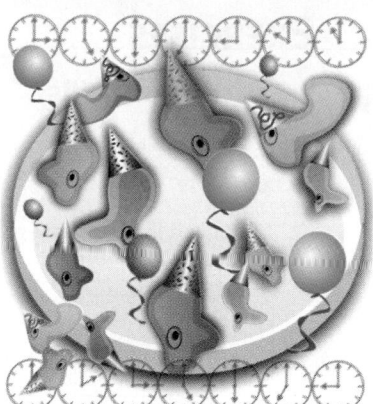

▶ **Solution**

Since the starting number is not known, suppose it was 94 bacteria. Then the function $y = 94(1 + 0.30)^x$ would be a correct model, in which x represents time elapsed in hours and y represents the number of bacteria.

However, there were 94 bacteria after 4 hours, not at 0 hours. So translate the point $(0, 94)$ right 4 units to $(4, 94)$. To translate the whole graph right 4 units, replace x with $x - 4$ in the function. You get

$$y = 94(1 + 0.30)^{(x-4)}$$

The graph shows how the new function translates every point in the graph right 4 units.

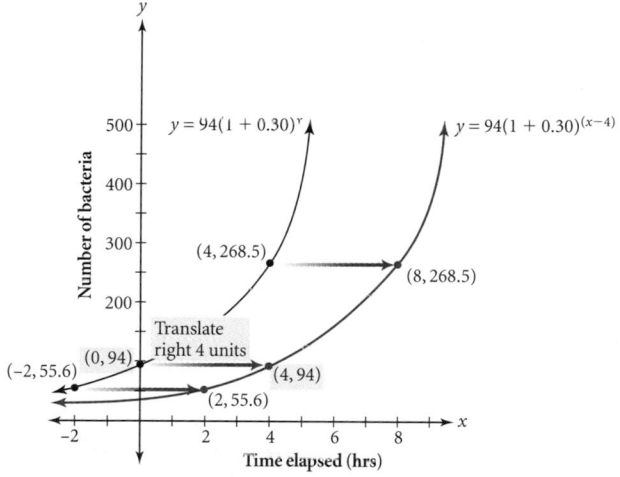

To find the starting value, substitute 0 for x in the new function.

$$94(1 + 0.30)^{(0-4)} = 94(1 + 0.30)^{-4} \approx 33$$

The starting number of bacteria was approximately 33 bacteria.

Using the starting value you found in the example, you could now write the function $y = 33(1 + 0.30)^x$. Can you use properties of exponents to show that $y = 94(1 + 0.30)^{(x-4)}$ is approximately equivalent to $y = 33(1 + 0.30)^x$? Do you think these functions would be considered members of the same family of functions?

EXAMPLE B

Here an exponential function is translated horizontally to model a real-world situation. Some students may think the graph is a translation down rather than to the right. If they try vertical lines between points, they will see there is not a uniform translation vertically.

If students are asking why subtraction in the equation is the same as adding to the coordinate, point out that any point (x, y) on the graph of $y = 94(1 + 0.30)^x$ has coordinates $(x, 94(1 + 0.30)^x)$. Then direct attention to the equation $y = 94(1 + 0.30)^{(x-4)}$ of the translated function. **[Ask]** "In terms of x, what are the coordinates of the point $(x + 4, y)$—that is, of a point 4 units to the right of the original graph?" "What does $x - 4$ in the equation do?"

Ask whether students think the function $y = 33(1 + 0.30)^x$ is of the same family as the original function. Have students develop the algebraic derivation requested in the student text:

$$y = 94(1 + 0.30)^{(x-4)}$$
$$= 94(1 + 0.30)^{-4}(1 + 0.30)^x$$
$$\approx 33(1 + 0.30)^x$$

From one point of view, the expression for the transformed graph is an equivalent expression for the shift of the first function, so the functions would appear to be part of the same family. On the other hand, students may argue that the functions are not of the same family because they have different starting values. Don't answer the question yet. It motivates Lesson 9.4, in which students will see that both functions are dilations of the parent $y = (1 + 0.30)^x$.

Students can use dynamic graphs at www.keymath.com/DA to explore translations of functions. The dynamic exploration closely follows the investigation and this example.

Closing the Lesson

Numbers are added to coordinates of points to indicate translations up or to the right (adding negative numbers for translations in the opposite directions), but those same numbers are subtracted from the variables to transform an equation of the graph similarly. The strategy is to build on the equation of the original function to create the equation of the transformed function.

Basic functions such as $y = x^2$, $y = |x|$, $y = x$, and $y = 3^x$ are called **parent functions,** and all transformations of these functions make up **families of functions.** The origin is a **vertex** of $y = x^2$ and also of $y = |x|$. A family member that is the image of the parent function under a transformation has as its vertex the image of the origin under the same transformation.

BUILDING UNDERSTANDING

Students practice working with functions of and real-world applications of translated graphs.

ASSIGNING HOMEWORK

Essential	1–4, 6, 9, 12, 13
Performance assessment	5, 8
Portfolio	10, 11
Journal	6, 7
Group	2, 10–13
Review	14, 15

▶ Helping with the Exercises

Exercise 1 You might ask what the parent function is and how it's been translated. [Parent function $y = |x|$ has been translated 4 units to the left and 1 unit upward.]

4a. a translation of $y = |x|$ right 1.5 units and down 2.5 units

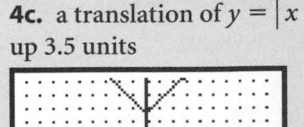

4b. a translation of $y = x^2$ left 3 units

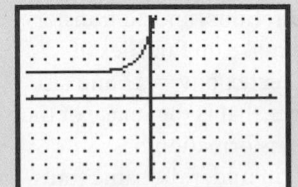

4c. a translation of $y = |x|$ up 3.5 units

EXERCISES

You will need your calculator for problems **4, 8, 10,** and **12.**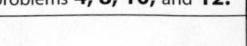

▶ Practice Your Skills

1. Use $f(x) = 2|x + 4| + 1$ to find
    a. $f(5)$ 19
    b. $f(-6)$ 5
    c. $f(-2) + 3$ 8
    d. $f(x + 2)$ $2|x + 6| + 1$

2. List L1 and list L2 contain coordinates for three points on the graph of $f(x)$. List L3 and list L4 contain coordinates for the three points after a transformation of f.

L1 x	L2 y		L3 x	L4 y
−1	3		7	−1
3	5		11	1
2	4		10	0

    a. Write definitions for list L3 and list L4 in terms of list L1 and list L2. L3 = L1 + 8; L4 = L2 − 4
    b. Describe the transformation. a translation right 8 units and down 4 units

3. Give the coordinates of the vertex for each graph.

  a.

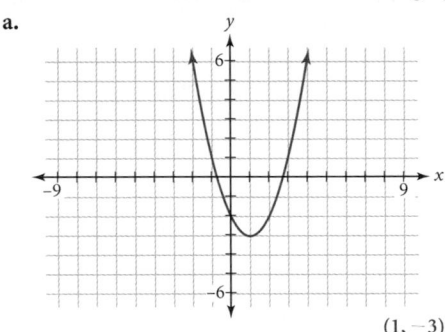

$(1, -3)$

  b.

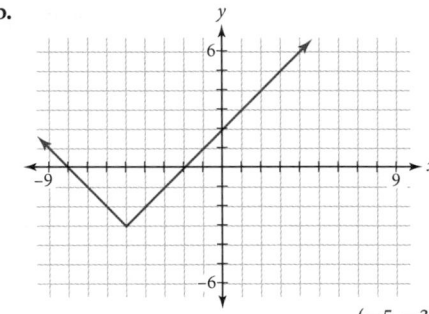

$(-5, -3)$

  c.

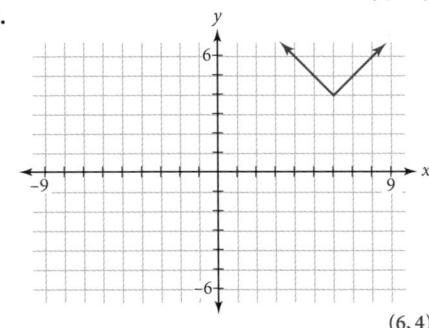

$(6, 4)$

  d.

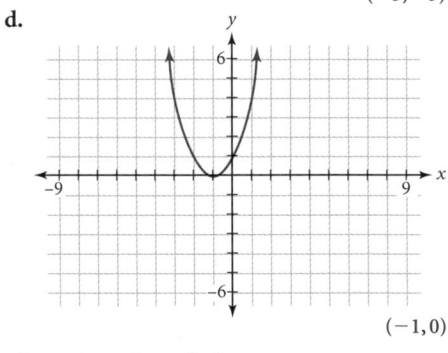

$(-1, 0)$

4. Graph each equation and describe the graph as a transformation of $y = |x|$, $y = x^2$, or $y = 3^x$.
    a. $y = |x - 1.5| - 2.5$
    b. $y = (x + 3)^2$
    c. $y = |x| + 3.5$
    d. $y = 3^{(x+1)} + 2$

4d. a translation of $y = 3^x$ left 1 unit and up 2 units

6a. a translation of $y = x^2$ right 1 unit and down 3 units; $y = (x - 1)^2 - 3$

6b. a translation of $y = |x|$ left 5 units and down 3 units; $y = |x + 5| - 3$

6c. a translation of $y = |x|$ right 6 units and up 4 units; $y = |x - 6| + 4$

6d. a translation of $y = x^2$ left 1 unit; $y = (x + 1)^2$

5. Write an equation for each of these transformations.

   a. Translate the graph of $y = x^2$ down 2 units. $y = x^2 - 2$

   b. Translate the graph of $y = 4^x$ right 5 units. $y = 4^{(x-5)}$

   c. Translate the graph of $y = |x|$ left 4 units and up 1 unit. $y = |x + 4| + 1$

▶ Reason and Apply

6. Describe each graph in problem 3 as a transformation of $y = |x|$ or $y = x^2$. Then write its equation.

7. This graph shows Beth's distance from her teacher as she turns in her test.

   a. What are the input and output variables?

   b. What are the units of the variables?

   c. What are the domain and range shown in the graph?

   d. Describe the situation.

   e. Write a function that models this situation.
$$d = |t - 2| + 1$$

Beth's Walk

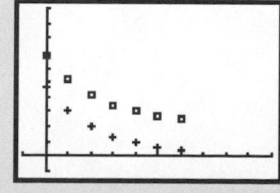

8. Graph $Y_1(x) = abs(x)$ on your calculator. Predict what each graph will look like. Check by comparing the graphs on your calculator.

   [▶ 🖳 See **Calculator Note 9B** for specific instructions for your calculator. ◀]

   a. $Y_2(x) = Y_1(x) - 4$ **b.** $Y_2(x) = Y_1(x - 4)$

9. Describe how the graph of $y = x^2$ will be transformed if you replace

   a. x with $(x - 3)$ a translation right 3 units   **b.** x with $(x + 2)$ a translation left 2 units

   c. y with $(y + 2)$ a translation down 2 units   **d.** y with $(y - 3)$ a translation up 3 units

10. Recall that an exponential equation in the form $y = A(1 - r)^x$ models some decreasing patterns. As you increase the value of x, the **long-run value** of y gets closer and closer to zero. Some situations, however, do not decrease all the way to zero. For example, as a cup of hot chocolate cools, the coolest it can get is room temperature. The long-run value will not be 0°C. Consider this table of data.

Time (min)	0	1	2	3	4	5	6
Temperature (°C)	68	52	41	34	30	27	25

   a. Define variables and make a scatter plot of the data. What type of function would fit the data?

   b. Find the ratio of each temperature to the previous temperature. Do these ratios support your answer to 10a?

   Assume the temperature of the room in this situation is 21°C. That means the long-run value of this data will also be 21°C.

   c. Make a new table by subtracting 21 from each temperature. Then make a scatter plot of the changed data. How have the points been transformed? What will be the long-run value?

   d. For your data in 10c, find the ratios of temperatures between successive readings. How do the ratios compare? What is the mean of these ratios?

10b. Ratios to the nearest thousandth: 0.765, 0.788, 0.829, 0.882, 0.9, 0.926; the ratios are not approximately constant and do not support an exponential function.

10c.

Time (min)	0	1	2	3	4	5	6
Temperature (°C)	47	31	20	13	9	6	4

A translation down 21 units; the long-run value will now be 0°C.

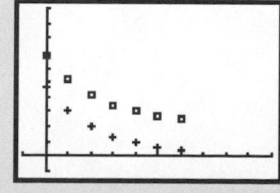

$[-1, 10, 1, -10, 100, 10]$

10d. Ratios to the nearest thousandth: 0.660, 0.645, 0.65, 0.692, 0.667, 0.667; the ratios are approximately constant; the mean is approximately 0.66.

7a. The input variable is t, time, the output variable is d, distance.

7b. Time is in seconds and distance is in meters.

7c. domain: $0 \le t \le 5$; range: $1 \le d \le 4$

7d. Possible answer: Beth starts 3 m from her teacher and walks toward the teacher at 1 m/sec for 2 sec. When she turns in the test, she is 1 m from the teacher. Beth then turns and walks away from the teacher at 1 m/sec for 3 sec.

Exercise 8 Calculator syntax varies. For example, some calculators interpret $Y_1(x - 4)$ as composition of functions, while others interpret it as multiplication. Calculator Note 9B is therefore mandatory for this exercise. You might want to challenge students to determine whether their calculators do composition or multiplication. They might try simple functions. For example, if $Y_1 = x$ and $Y_2 = Y_1(x)$, identical graphs would indicate composition.

8a. a translation down 4 units

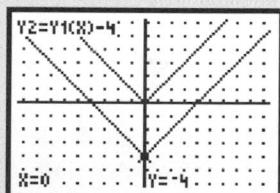

8b. a translation right 4 units

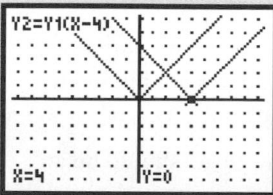

10a. Let x represent time in minutes, and let y represent temperature in degrees Celsius. The scatter plot suggests an exponential function.

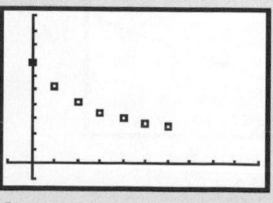

$[-1, 10, 1, -10, 100, 10]$

10g. $y = 47(1 - 0.34)^x + 21$

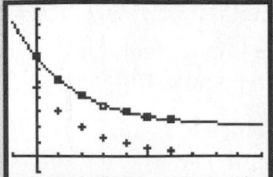

Exercise 11 Students will probably have more success with this exercise if they've worked on Exercise 10 first. In 11h, the equation can be solved using graphs, guess-and-check, or calculator tables. Logarithms aren't necessary.

11d. Let x represent time in hours, and let y represent temperature in degrees Celsius; $y = 221(1 - 0.15)^x$.

11e. $y = 221(1 - 0.15)^x + 20$; the vertical translation is needed to make the long-run value 20°C.

11f. $y = 221(1 - 0.15)^{(x-5)} + 20$; the horizontal translation is needed because the first temperature was measured 5 hours after the bowl was removed from the kiln.

11g. The temperature of the bowl immediately after it was removed from the kiln was approximately 518°C.

Exercise 12 Students might define x to be the number of years since 1999 [and get $y = 6.0(1 + 0.013)^x$] or as the number of years since 1990 [for $y = 6.0(1 + 0.013)^{x-9}$].

12a. The year is the input variable, x, and population in billions is the output variable, y.

12b. The graph should be an increasing exponential function passing through the point (1999, 6.0).

12d. $y = 6.0(1 + 0.013)^{(x-1999)}$

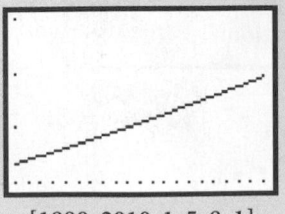

[1990, 2010, 1, 5, 8, 1]

---

e. Write an exponential equation in the form $y = A(1 - r)^x$ that models the data in 10c. $y = 47(1 - 0.34)^x$

f. In 10c you subtracted 21 from each temperature. What transformation takes the original data back to the original data? **a translation up 21 units**

g. Your equation in 10e models translated data. Change that equation so that it models the original data. Check the fit by graphing on your calculator.

11. APPLICATION Clay works for a company that manufactures pottery. He has designed a new bowl and needs to know how long it takes to cool after being removed from the kiln. He removes a sample bowl from the kiln. After 5 hours of cooling, the temperature of the bowl is 241°C. Clay calls this time 0. After 1 more hour, the temperature of the bowl is 208°C. Room temperature is 20°C.

a. What will be the long-run value for the temperature of the bowl? **20°C**

b. Transform the temperatures so that the long-run value will be 0°C. **221°C; 188°C**

c. Find the rate of cooling per hour using the temperatures in 11b. **approximately 15% per hour**

d. Define variables and write an exponential function that models the temperatures in 11b.

e. Change your function in 11d to show a translation up 20 units. Why do you need this translation?

f. Change your function in 11e to show a translation right 5 units. Why do you need this translation?

g. Use your function in 11f to find $f(0)$. What is the real-world meaning of this value?

h. The company's safety regulations say that a piece of pottery cannot be handled until it is at most 25°C. How long after the bowl is removed from the kiln can it be handled? **To make sure the temperature is 25°C or less, round up to 29 hours.**

12. APPLICATION In 1999, the world population was estimated to be 6.0 billion, with an annual growth rate of 1.3%. (*2000 World Almanac*, p. 878)

a. Define input and output variables for this situation.

b. Without finding an equation, sketch a graph of this situation for 1990 to 2010.

c. What one point on the graph do you know for sure? **(1999, 6.0)**

d. Write a function that models this situation. Graph your function on your calculator and name an appropriate window.

e. Use your graph to estimate the population to the nearest tenth of a billion in 1990 and 2010. (Assume a constant growth rate during this period.) **1990: 5.3 billion; 2010: 6.9 billion**

13. The graph of a linear equation of the form $y = bx$ passes through $(0, 0)$.

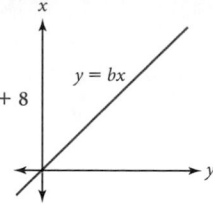

a. Suppose the graph of $y = bx$ is translated right 4 units and up 8 units. Name a point on the new graph. $(4, 8)$

b. Write an equation for the line in 12a after the transformation. $y = b(x - 4) + 8$

c. Suppose the graph of $y = bx$ is translated horizontally H units and vertically V units. Name a point on the new graph. (H, V)

d. Write an equation for the line in 12c after the transformation.
$$y = b(x - H) + V$$

▶ **Review**

14. Drew's teacher gives skill-building quizzes at the start of each class.

a. On Monday, Drew got 77 problems correct out of 85. What is her percent correct? 90.6%

b. On Tuesday, Drew got 100% on a quiz that had only 10 problems. Estimate her percent correct for the two-day total. Answers will vary.

c. Calculate her percent correct for the two-day total. 91.6%

15. Solve each system of equations.

a. $\begin{cases} y = 5 + 2x \\ y = 8 - 2x \end{cases}$
$(0.75, 6.5)$

b. $\begin{cases} y = -2 + 3(x - 4) \\ y = 3 + 5(x - 2) \end{cases}$
$(-3.5, -24.5)$

c. $\begin{cases} 2x + 7y = 13 \\ 5x - 14y = 1 \end{cases}$
$(3, 1)$

IMPROVING YOUR **VISUAL THINKING** SKILLS

Tammy and Jose are working on problem 13a on this page. They each decide to graph a linear equation of the form $y = bx$ to help visualize the question. They translate their graphs right 4 units and up 8 units. Their results are surprisingly different.

Jose's Graph

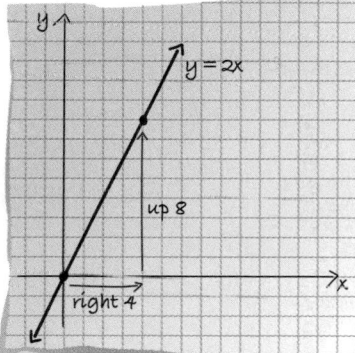

Tammy's Graph

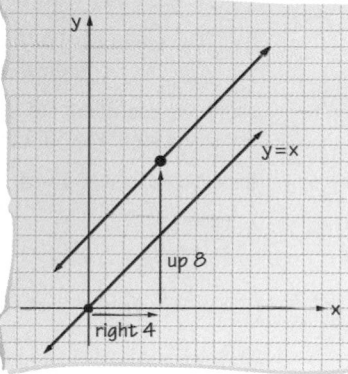

Why did Jose get the same graph after the translation?

If the graph of an equation of the form $y = bx$ is translated horizontally H units and vertically V units, when would you get the same graph after the translation?

IMPROVING **VISUAL THINKING** SKILLS

The vertical shift divided by the horizontal shift is the same as the slope of Jose's line, so the line was in effect shifted along itself to lie on top of itself. You might ask what shift would have shifted Tammy's line onto itself. Students may be surprised to discover that any translation by the same amount horizontally and vertically will leave Tammy's line unchanged. A translation horizontally H units and vertically V units will leave the line with slope $\frac{V}{H}$ unchanged. Ask if this could happen to figures other than lines, to help students appreciate the importance of lines' constant slope.

Exercise 13 It could surprise students that vertical translations of $y = bx$ are the same as horizontal translations. For example, the vertical upward shift given by $y - 3 = 2x$ is the same as $y = 2(x + 1.5)$, which can be seen as a translation to the left. This property is a characteristic of straight lines.

Exercise 14 This exercise reviews Lesson 2.2.

Exercise 15 This exercise reviews Lesson 6.3.

LESSON OUTLINE

One day:

20 min	Investigation
5 min	Sharing
10 min	Example
5 min	Closing
10 min	Exercises

MATERIALS

- graph paper
- patty paper, *optional*
- Calculator Note 9B
- Flipping a Letter (T), *optional*

TEACHING

Another kind of transformation is the reflection. Here we look at reflections across the axes.

 Guiding the Investigation

One step Show the Flipping a Letter transparency. Ask students what operations they should apply to the coordinates of the corners of the darker letter to achieve the lighter letter. Students might use their calculators to test their theories. As groups finish, have them experiment to reflect the graph of a function of their choice across the *x*-axis and the *y*-axis.

Steps 2 and 3

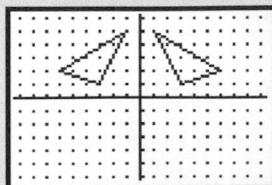

$[-9.4, 9.4, 1, -6.2, 6.2, 1]$

See page 720 for answers to Steps 6, 7, 8, and 10.

Reflecting Points and Graphs

The art of a people is a true mirror to their minds.

JAWAHARLAL NEHRU

Translations move points and graphs around the coordinate plane. Have you noticed that the image of the translation always looks like the original figure? Although the image of a translation moves, it doesn't flip, turn, or change size. To get these changes, you need other types of transformations.

Investigation
Flipping Graphs

In this investigation you will explore the relationships between the graph of an equation and its image when you flip it two different ways.

Step 1 $(1, 5), (3, 1),$ and $(6, 2)$

Step 4 $(-1, 5), (-3, 1),$ and $(-6, 2)$; the triangle is flipped over the *y*-axis; the *x*-coordinates of the new triangle become negative.

Step 5a $(1, -5),$ $(3, -1),$ and $(6, -2)$; the triangle is flipped over the *x*-axis; the *y*-coordinates of the new triangle become negative.

Step 1   Name the coordinates of the vertices of this triangle.

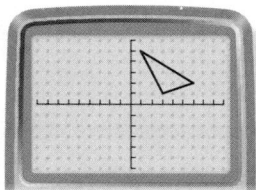

Procedure Note

For this investigation, use a friendly window with a factor of 2.

Step 2   Graph the triangle on your calculator. Use list L1 for the *x*-coordinates of the vertices and list L2 for the *y*-coordinates of the vertices.

Step 3   Define list L3 and list L4 as follows

$$L_3 = -L_1$$
$$L_4 = L_2$$

Graph a second triangle using list L3 for the *x*-coordinates of the vertices and list L4 for the *y*-coordinates of the vertices.

Step 4   Name the coordinates of the vertices of the new triangle. Describe the transformation. How did the coordinates of the vertices change?

Step 5   Repeat Steps 3 and 4 with these definitions.

a. $L_3 = L_1$
$L_4 = -L_2$

b. $L_3 = -L_1$
$L_4 = -L_2$

$(-1, -5), (-3, -1),$ and $(-6, -2)$; the triangle flips across the *x*- and *y*-axes; the signs change for both the *x*- and *y*-coordinates.

Next, you'll see if what you have learned about flipping points is true for the graphs of functions.

Step 6   Graph $y = 2^x$ on your calculator.

Step 7   Replace *x* with $-x$ in the function. Graph this second function. Describe how the second graph is related to the graph of $y = 2^x$. $y = 2^{-x}$; a flip across the *y*-axis.

Steps 3 to 5 Each time students go back to Step 3, they are to work with the original picture. They might use graph paper, which can be folded and held up to the light to see both the line of reflection and the congruence between the reflected image and the original. The translucence of patty paper allows one to see when polygons are on top of each other.

Step 5 Some students may recognize that in part b a reflection across the *x*-axis and then across the *y*-axis is equivalent to a 180-degree rotation about the origin.

Step 5a

Step 5b

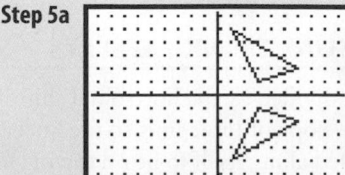

Step 8 $y = -2^x$; a flip across the x-axis.

Step 9

Step 9a $y = (-x - 1)^2$ flips across the y-axis; $y = -(x - 1)^2$ flips across the x-axis.

Step 10

Step 8 Now replace y with $-y$ in the function $y = 2^x$ and solve for y. Graph this third function. Describe how its graph is related to the graph of $y = 2^x$.

Step 9 Repeat Steps 6–8 using these functions. Make a note of anything unusual that you find.

Step 9b $y = -|x|$ flips across the y-axis although it appears unchanged due to the vertical symmetry of the graph;

a. $y = (x - 1)^2$

b. $y = |x|$

c. $y = x$

$y = -|x|$ flips across the x-axis.

Step 9c $y = -x$ can be considered a flip either across the x-axis or across the y-axis due to diagonal symmetry of the graph.

Step 10 Summarize what you have learned about flipping graphs.

A transformation that flips a figure to create a mirror image is called a **reflection**. A point is **reflected across the x-axis** when you change the sign of its y-coordinate. A point is **reflected across the y-axis** when you change the sign of its x-coordinate. You saw both types of reflections in the investigation. Similar reflections result when you change the sign of x or y in a function.

You can combine reflections with other transformations. Sometimes, different combinations will give the same result.

EXAMPLE A

The graph of a parent function is shown in black. Its image after a transformation is shown in red. Describe the transformation and then write a function for the image.

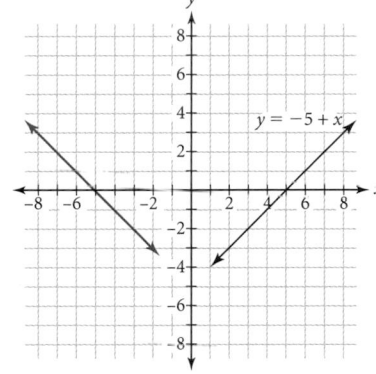

$y = -5 + x$

▶ Solution

This is a reflection across the y-axis. The image is produced by replacing each x-value in the original function with $-x$.

$$y = -5 + (-x)$$
or
$$y = -5 - x$$

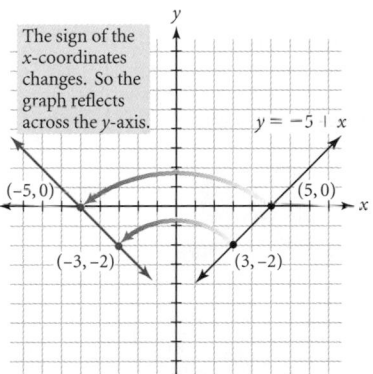

The sign of the x-coordinates changes. So the graph reflects across the y-axis.

$y = -5 + x$

$(-5, 0)$ $(5, 0)$

$(-3, -2)$ $(3, -2)$

LESSON OBJECTIVES

- Explore reflections and combinations of reflections and translations
- Write equations and list definitions that describe these transformations

NCTM STANDARDS

CONTENT	PROCESS
Number	Problem Solving
• Algebra	• Reasoning
• Geometry	Communication
Measurement	• Connections
Data/Probability	• Representation

Step 8 Encourage students to write out all the details.

SHARING IDEAS

Groups could report their ideas from Steps 9 and 10. Students may be surprised that the reflections of $y = x$ across the two axes are the same. Ask whether that's the case for other graphs. [It's true only for figures that are rotated onto themselves by a 180-degree rotation around the origin.] Be sure the term *reflection* is used.

A mathematical *reflection* across a line moves a figure to where it would *appear* to be if the line were a mirror. The term *flipping* for a reflection indicates that you can think of the figure as leaving the plane and being flipped over the line of reflection.

Ask whether the opening quotation uses the idea of a mirror in the same way as a mathematical reflection. Nehru (1889–1964), a leader in India's liberation from British rule, went on to become India's first prime minister.

You may want to say that a reflection across the y-axis is a *horizontal reflection* because the image is flipped left to right (or right to left). Similarly, a reflection across the x-axis is a *vertical reflection* because the image is flipped top to bottom (or bottom to top). A reflection can take place across any line. In Take Another Look, students will see reflections across the line $y = x$.

To write the equation of a translation, students went backward from what they did to coordinates. Some may notice that with reflections, they seem to be doing the same thing to both the variable and the coordinate: taking the opposite. **[Ask]** "Why does translation seem different from reflection in this respect?" [Elicit the idea that going backward from taking the opposite is taking the opposite.]

Watch for abilities to read coordinates of points, enter coordinates into lists, graph polygons on a calculator, graph equations on a calculator, and recognize translations.

▶ *EXAMPLE A*

This example illustrates going from the graph to the transformation.

▶ *EXAMPLE B*

Some students may prefer to see $x + 4$ rewritten as $x - (-4)$.

Extension Questions

If you have time, ask extension questions. **[Ask]** "What effect does a horizontal or vertical reflection have on the slope of a straight line? [It negates the slope.] "What one point does not change due to the reflection?" [The intercept with the line of reflection.] "What order of transformations, applied to the parent absolute value function, produces $y = -\left(\left|-x - 4\right| + 1\right)$?" [One answer: a reflection across the x-axis, followed by a translation right 4 units and up 1 unit, followed by a reflection across the y-axis.]

The two halves of the graphs in Example C, $y = f(x)$ and $y = f(x - 8) - 6$, are mirror images. The drawing is *symmetric*. **[Ask]** "Are there any other graphs you have learned about that are mirror images of themselves?" [Parabolas, absolute value functions, horizontal and vertical lines] "Is it possible for a *function's* graph to show no apparent change after a reflection across the x-axis?" [For this to be the case, the top half of the graph would have to be a mirror image of the bottom half. Unless the function were $f(x) = 0$, the graph would then fail the vertical line test.] You might also ask, "When

can translations be replaced by reflections?" [Every translation can be accomplished through two reflections, though not necessarily across axes. If the graph is appropriately symmetric, the translation can be accomplished with just one reflection.] These ideas will preview Exercise 11 and Take Another Look.

EXAMPLE B

The graph of a parent function is shown in black. Its image after a transformation is shown in red. Describe the transformation and then write a function for the image.

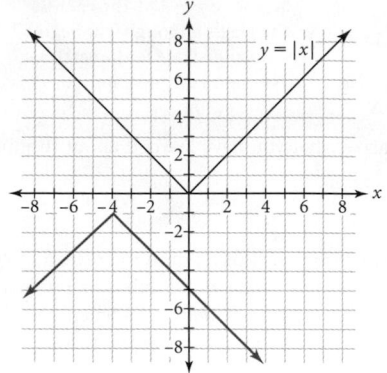

▶ **Solution**

There are several ways to think about this transformation. The order of the transformations will create different, yet equivalent, equations.

Here is one possible solution. Reflect the graph of the function across the x-axis, then translate it left 4 units and down 1 unit. To write the equation of the image, change the original function in the same order.

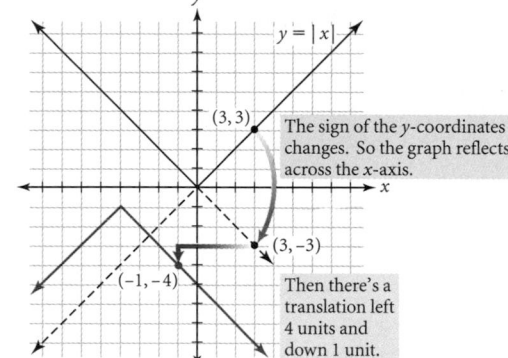

The sign of the y-coordinates changes. So the graph reflects across the x-axis.

Then there's a translation left 4 units and down 1 unit.

Technology
● **CONNECTION** ●

Many computer applications allow you to import and transform clip art. Most have commands like "reflect vertically" or "reflect horizontally." Since clip art doesn't normally have an x- or y-axis, these commands reflect the picture by flipping it top to bottom or left to right.

$$y = |x| \qquad \text{Original equation.}$$
$$-y = |x| \qquad \text{Replace } y \text{ with } -y \text{ to reflect across the } x\text{-axis.}$$
$$y = -|x| \qquad \text{Solve for } y.$$
$$y = -|x + 4| \qquad \text{Translate left 4 units.}$$
$$y = -|x + 4| - 1 \qquad \text{Translate down 1 unit.}$$

A function for the image is $y = -|x + 4| - 1$.

EXAMPLE C

The graph of a parent function is shown in black. Its image after a transformation is shown in red. Describe the transformation and then write a function for the image.

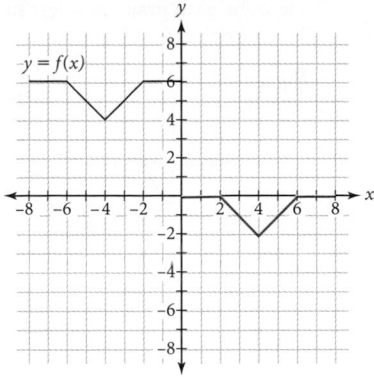

▶ **Solution**

As in Example B, you can think of this transformation in several ways. One solution is to translate right 8 units and down 6 units, as shown in the graph on the left below. That gives the function $y = f(x - 8) - 6$.

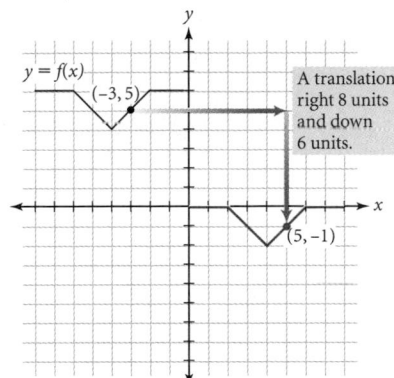

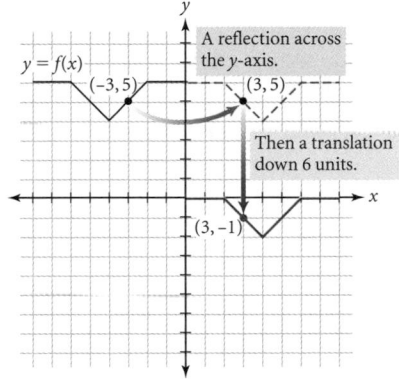

Another solution is to reflect the graph across the y-axis and then translate down 6 units, as shown in the graph on the right above. That gives the function $y = f(-x) - 6$.

In the investigation you probably saw no change when you reflected the graph of $y = |x|$ across the y-axis. In Example C, a reflection across the y-axis has the same result as a horizontal translation. Do you notice anything special about these graphs that could explain these strange results?

EXERCISES

You will need your calculator for problems **4, 5,** and **12.**

▶ **Practice Your Skills**

1. Use $f(x) = 0.5(x - 3)^2 - 3$ to find

   a. $f(5)$ -1

   b. $f(-6)$ 37.5

   c. $4 \cdot f(2)$ -10

   d. $f(-x)$ $0.5(-x - 3)^2 - 3$

   e. $-f(x)$ $-0.5(x - 3)^2 + 3$

Closing the Lesson

Taking the opposite of a y-coordinate or the variable y results in a **reflection** across the x-axis. Similarly, taking the opposite of an x-coordinate or variable results in a reflection across the y-axis.

BUILDING UNDERSTANDING

Students work with combinations of reflections and translations.

ASSIGNING HOMEWORK

Essential	1, 4–6, 12
Performance assessment	2, 7
Portfolio	8, 9
Journal	10
Group	3, 11
Review	12–14

Exercise 2 Now that students know about reflections, they might insert reflections to give a variety of transformations. They can check their work on a graphing calculator.

Exercise 3 Some students may be confused by being asked to identify the transformation of a function that is already a transformation of a parent function.

Exercise 4 **[Alert]** As in Exercise 8 of Lesson 9.2, make sure students know how their calculators behave. On some calculators, $Y_1(-x)$ will multiply rather than give a composition of functions.

4a. a reflection across the y-axis

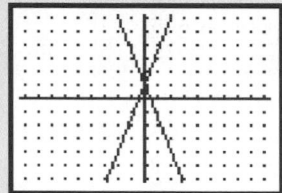

4b. a reflection across the x-axis

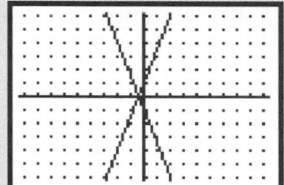

2. Describe each graph as a transformation of $y = |x|$ or $y = x^2$. Then write its equation.

a.
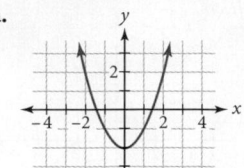
a translation of the graph of $y = x^2$ down 2 units; $y = x^2 - 2$

b.
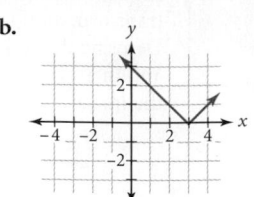
a translation of the graph of $y = |x|$ right 3 units; $y = |x - 3|$

c.
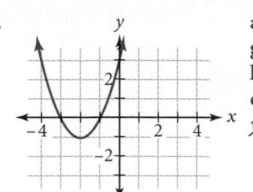
a translation of the graph of $y = x^2$ left 2 units and down 1 unit; $y = (x + 2)^2 - 1$

d.
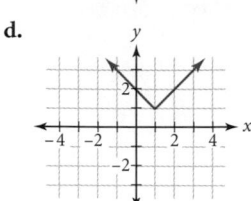
a translation of the graph of $y = |x|$ right 1 unit and up 1 unit; $y = |x - 1| + 1$

3. Describe each graph below as a transformation of $y = |x + 3|$, shown at right.

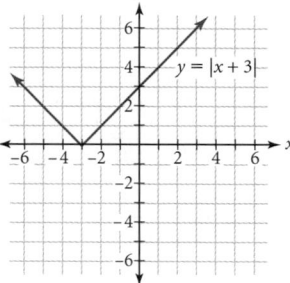

$y = |x + 3|$

a.
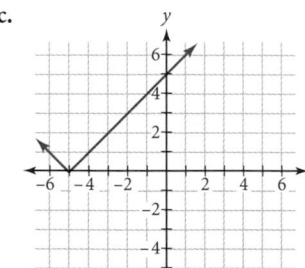
a reflection across the x-axis

b.
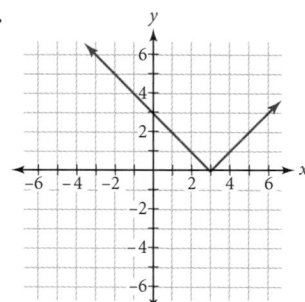
a translation right 6 units or a reflection across the y-axis

c.
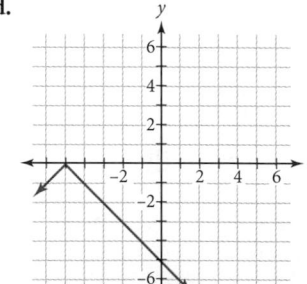
a translation left 2 units

d.
a translation left 2 units and a reflection across the x-axis

4. Graph $Y_1(x) = 1 + 2.5x$ on your calculator. Predict what each graph will look like. Check by comparing graphs on your calculator. [▶ ☐ See **Calculator Note 9B** for specific instructions for your calculator. ◀]

    a. $Y_2(x) = Y_1(-x)$

    b. $Y_2(x) = -Y_1(x)$

5. Describe the graph of each function below as a transformation of the graph of the parent function $y = x^2$. Check your answers by graphing on your calculator.

    a. $y = x^2$

    b. $y = -(x + 3)^2$

    c. $y = -x^2 + 3$

    d. $y = (-x)^2 + 3$

▶ **Reason and Apply**

6. Consider the triangle at right.

    a. Describe how you can graph this triangle on your calculator.

    b. How could you make these graphs?

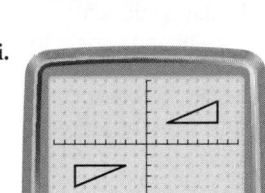

$[-9.4, 9.4, 1, -6.2, 6.2, 1]$

    i. 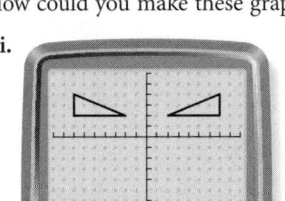   Define L₃ = −L₁ and L₄ = L₂.

    ii. Define L₃ = −L₁ and L₄ = −L₂.

    iii. 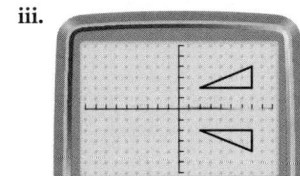   Define L₃ = L₁ and L₄ = −L₂.

    iv. 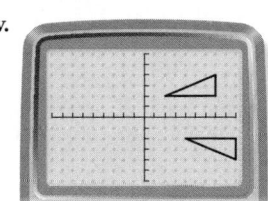   Define L₃ = L₁ + 2 and L₄ = −L₂.

7. The points in this table form a star when you connect them in order. Describe the transformation that results when you redefine the points as

    a. $(-x, y)$ a reflection across the y-axis

    b. $(x - 8, -y)$

    c. $(x + 2, y - 4)$ a translation right 2 units and down 4 units

    d. (y, x) (Hint: Try graphing this.) a reflection across the line $y = x$

x	y
6.0	2.0
2.4	3.2
4.6	0.1
4.6	3.9
2.4	0.8
6.0	2.0

Exercise 5 Some students may need to be reminded that the order of operations calls for $-x^2$ to mean $-(x^2)$ rather than $(-x)^2$.

Order is crucial in 5c. A translation followed by a reflection would be represented by the equation $y = -(x^2 + 3)$.

In 5d, because $y = x^2$ has vertical symmetry, the resulting graph will appear only to have been translated. You may want to ask students why $(-x)^2 + 3$ would be equivalent to $x^2 + 3$, to review $(-x)^2 = x^2$.

5a. a reflection across the x-axis

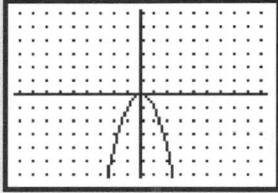

$[-9.4, 9.4, 1, -6.2, 6.2, 1]$

5b. a translation left 3 units and a reflection across the x-axis

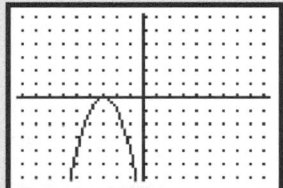

5c. a reflection across the x-axis followed by a translation up 3 units

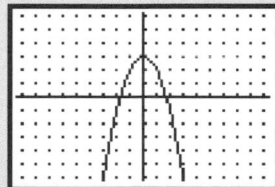

5d. a reflection across the y-axis and a translation up 3 units

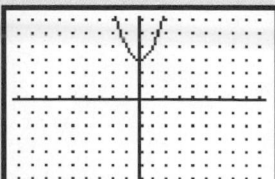

6a. Possible answer: Enter the x-coordinates into list L₁ and the y-coordinates into list L₂. L₁ = {2, 7, 7, 2} and L₂ = {2, 2, 4, 2}. Then make a connected graph.

Exercise 7 This exercise might be better done on paper than on a calculator. If students do use a calculator, they must reenter the first point as the last point to complete the star. If students have difficulties describing the transformations, you might suggest that they add columns (for $-x$ or $x - 8$, for example) to the table.

7b. a translation left 8 units and a reflection across the x-axis

Exercise 8 You might want to ask students to check their work by graphing. In 8b, the answer of $y = -|x - 3| + 3.5$ assumes a translation right by 3, followed by a reflection across the *y*-axis and then a translation up 3.5. An alternative is a translation right 3 and down 3.5, followed by a reflection across the *x*-axis. This gives $y = -\left(|x - 3| - 3.5\right)$.

Exercise 9 This exercise lays the groundwork for the problem referred to in Reviewing at the end of the chapter.

8. Anthony and Cheryl are using a motion sensor for a "walker" investigation.

a. This graph shows data that Cheryl collected when Anthony walked. Write an equation that models his walk. $y = |x - 4| + 1$

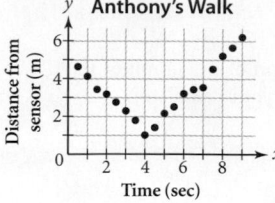

Anthony's Walk

b. Here is a description of Cheryl's walk.

Begin at a distance of 0.5 meter from the sensor. Walk away from the sensor at 1 meter per second for 3 seconds. Then walk toward the sensor at the same rate for 3 seconds.

Write an equation to model her walk. $y = -|x - 3| + 3.5$

c. Give the domain and range for the function that models Cheryl's walk.

domain: $0 \leq x \leq 6$; range: $0.5 \leq y \leq 3.5$

9. Bo is designing a computer animation program. She wants the star on the left to move to the position of the star on the right using 11 frames. She also wants the star to flip top to bottom in each frame. Define the coordinates of each image based on the coordinates of the previous figure.

$(x + 1, -y)$

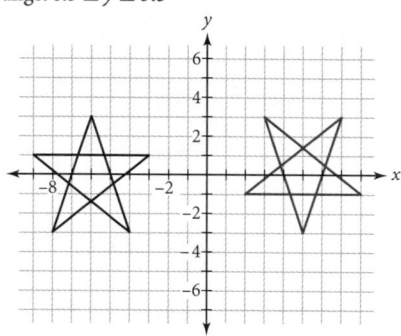

10. For a and b, the graph of a parent function is shown in black. Describe the transformation that creates the red image. Then write a function for the image.

a.

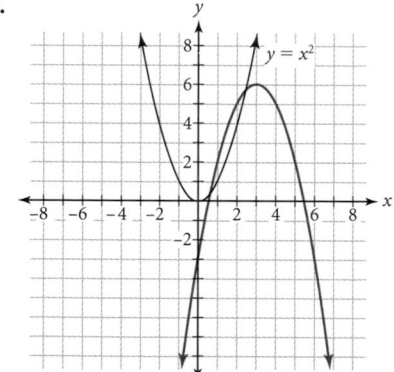

Possible answer: $y = -(x - 3)^2 + 6$

b.

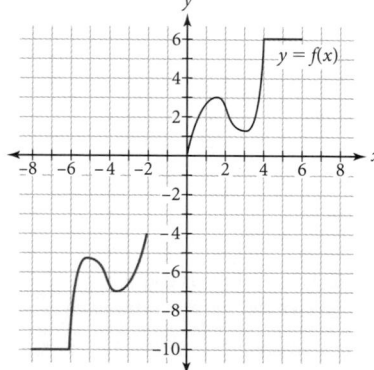

Possible answer: $y = -f(-(x + 2)) - 4$

11. A line of reflection does not have to be the *x*- or *y*-axis. Consider this example in which $y = |x|$ is reflected across the line $x = 4$.

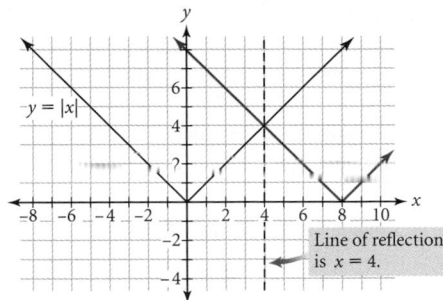

a. Think about each of these transformations as a single reflection. What is the line of reflection?

i.

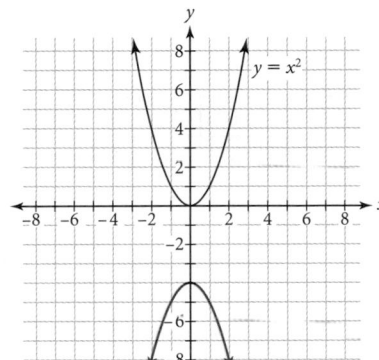

$y = -2$

ii.

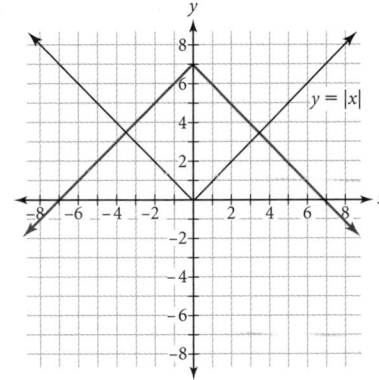

$y = 3.5$

iii.

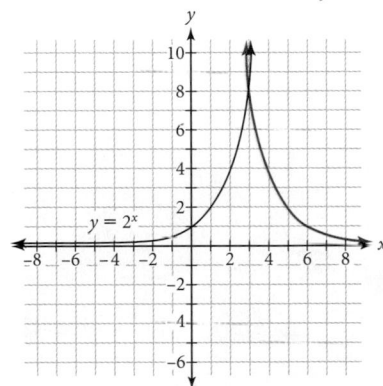

$x = 3$

iv.

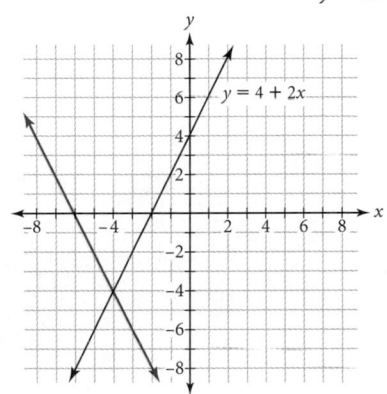

$x = -4$
or $y = -4$

b. Write an equation for the red image in each graph in 11a.

c. The graph of $y = f(x)$ is reflected across the horizontal line $y = b$. What is the equation of the image? $y = -f(x) + 2b$

d. The graph of the function $y = f(x)$ is reflected across the vertical line $x = a$. What is the equation of the image? $y = f(-x + 2a)$

Exercise 11 Graph iv can be thought of in several ways. For 11a, the reflection can be either horizontal or vertical. For 11b graph iv can be thought of in four ways: as a reflection across the *y*-axis followed by a translation of either left 8 units or down 16 units, and as a reflection across the *x*-axis followed by a translation either down 8 units or left 4 units. All are equivalent to $y = -12 - 2x$.

11b. i. $y = -x^2 - 4$

ii. $y = -|x| + 7$

iii. $y = 2^{-(x-6)}$

iv. $y = 2(-(x + 8)) + 4$;
$y = (4 + 2(-x)) - 16$;
$y = -(4 + 2x) - 8$; or
$y = -(4 + 2(x + 4))$

Exercise 12 [ESL] A *reaction* is a chemical change. A *reactant* is the chemical that changes. This exercise reviews Lesson 7.2.

12b. 0 grams; all the reactant will be used.

12d. Both graphs are decreasing exponential graphs. The graph of $y = 500(0.88)^x$ shows a starting amount of 500 grams of reactant; the graph of $y = 500(0.88)^x + 100$ shows a starting amount of 600 grams of reactant. The second graph is a translation of the first graph up 100 units.

Exercise 13 This exercise reviews Lesson 2.3.

13. $47 \text{ T} \cdot \dfrac{1 \text{ cup}}{16 \text{ T}} \cdot \dfrac{1 \text{ quart}}{4 \text{ cups}}$
$= \dfrac{47}{64}$ quarts ≈ 0.734 quart

Exercise 14 This exercise reviews Lesson 5.6.

14a. Possible answers using Q-points: $y = 36 + 3.6(x - 4)$ or $y = 72 + 3.6(x - 14)$

▶ **Review**

12. A chemical reaction consumes 12% of the reactants per minute. A scientist begins with 500 grams of one reactant. So the equation $y = 500(0.88)^x$ gives the amount of reactant remaining, y, after x minutes.

   a. What does the number 0.88 tell you? There is a 12% decrease per minute $(1 - 0.12 = 0.88)$.

   b. What is the long-run value of y? What is the real-world meaning of this value?

   c. What is the long-run value of y for the equation $y = 500(0.88)^x + 100$? What is the real-world meaning of this value? 100 grams; not all the reactant will be used.

   d. Graph $y = 500(0.88)^x$ and $y = 500(0.88)^x + 100$. How are these graphs the same? How are they different?

13. Convert 47 tablespoons to quarts. (16 tablespoons = 1 cup; 1 quart = 4 cups)

14. This table shows the temperature of water in a pan set on a stove.

   a. Find the equation of a line that models this data.

   b. How long will it take for the water to boil (100°C)? approximately 22 minutes

Time (min)	0	2	4	6	8	10	12	14	16	18
Temperature (°C)	22	29	36	44	51	58	65	72	80	87

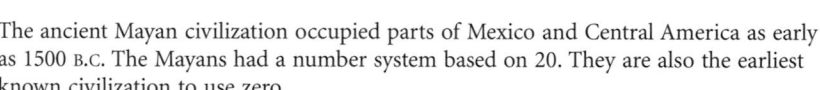

IMPROVING YOUR **REASONING** SKILLS

The ancient Mayan civilization occupied parts of Mexico and Central America as early as 1500 B.C. The Mayans had a number system based on 20. They are also the earliest known civilization to use zero.

Below are the 20 numerals in the Mayan number system. Can you decode the numerals and label them with the numbers 0 to 19? A few are labeled to get you started.

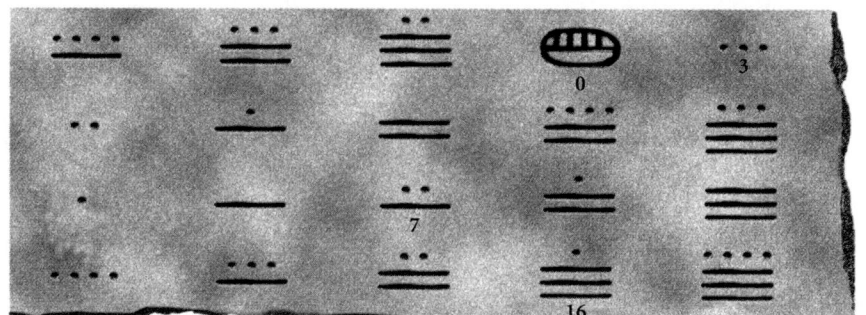

IMPROVING **REASONING** SKILLS

The components of the Mayan number system are often described as *dots* and *rods*. Each rod represents 5 and each dot represents 1. The symbol for zero is often called a *shell*, yet different sources draw it differently.

The base-10 Hindu-Arabic system we use today has 10 digits, 0 through 9. Every other numeral is formed by a placement of these digits. For example, each digit in the numeral 349 represents multiplication by a different power of 10.

In the Mayan number system, the 20 digits, equivalent to our 0 to 19, were used in a top-to-bottom positional number system in which each digit represented a different power of 20. The answers are:

9	13	17	0	3
2	6	10	14	18
1	5	7	11	15
4	8	12	16	19

Stretching and Shrinking Graphs

There is no absolute scale of size in the Universe, for it is boundless towards the great and also boundless towards the small.

OLIVER HEAVISIDE

Imagine what happens to the shape of a picture drawn on a rubber sheet as you **stretch** the sheet vertically.

The width remains the same, but the height changes. You can also **shrink** a picture vertically. This makes the picture appear to have been flattened.

You know how to translate and reflect graphs on the coordinate plane. Now let's see how to change their shape.

PLANNING

LESSON OUTLINE

One day:

25 min	Investigation
5 min	Sharing
10 min	Examples
5 min	Closing
5 min	Exercises

MATERIALS

- graph paper, *optional*
- Calculator Notes 9B, 9C, 9D
- Stretching a Polygon (T or W), *optional*

TEACHING

Our transformation toolbox is expanded to include vertical stretches and shrinks.

NCTM STANDARDS

CONTENT		PROCESS	
	Number		Problem Solving
•	Algebra		Reasoning
•	Geometry	•	Communication
	Measurement	•	Connections
	Data/Probability	•	Representation

LESSON OBJECTIVES

- Discover how to change the shape of polygons and functions
- Explore vertical dilations (stretches and shrinks) of polygons and graphs of equations

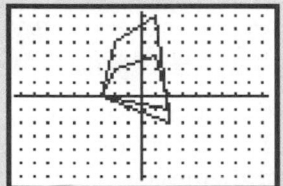

Guiding the Investigation

One step Display the transparency or pass out the Stretching a Polygon worksheet. Ask students how to transform the shape of the original picture to represent the transformed picture. As you watch, suggest as needed that they experiment with specific points.

Step 2 Students may work on graph paper instead of calculators. They should graph both the original quadrilateral and the transformed one.

Step 3 If a group has fewer than four members, the first student to finish with one number should repeat the procedure with an unused number. If there are more than four group members, they can make up additional numbers.

Step 3 For $a = 2$:

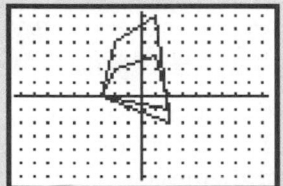

For $a = 3$:

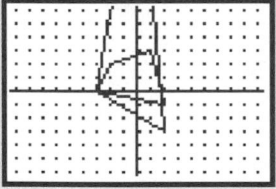

For $a = 0.5$:

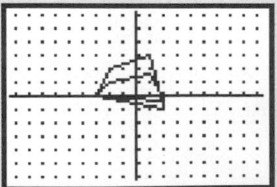

For $a = -2$:

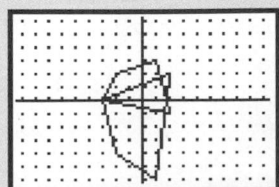

Step 4 Some students may want to use a table to show their results. Sharing can be within groups or with the entire class.

502 CHAPTER 9 Transformations

The German painter Hans Holbein II (1497–1543) used a technique called anamorphosis to hide a stretched skull in his portrait *The Ambassadors* (1533). You can see the skull in the original painting if you look across the page from the lower-left. The painting was originally hung above a doorway so people would notice the skull as they walked through the door. Holbein may have been making a political statement about these two French ambassadors who were members of England's court of King Henry VIII.

Investigation
Changing the Shape of a Graph

In this investigation you will learn how to stretch or shrink a graph vertically.

Step 1 $(1, 3), (2, -1),$ **Step 1** $(-3, 0),$ and $(-2, 2)$

Name the coordinates of the vertices of this quadrilateral.

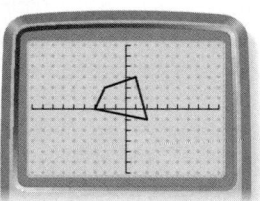

> **Procedure Note**
>
> For this investigation, use a friendly window with a factor of 2.

Step 2 Graph the quadrilateral on your calculator. Use list L1 for the x-coordinates of the vertices and list L2 for the y-coordinates of the vertices.

Step 3 Each member of your group should choose one of these values of a: 2, 3, 0.5, or -2. Use your value of a to define list L3 and list L4 as follows

$$L3 = L1$$
$$L4 = a \cdot L2$$

Graph a second quadrilateral using list L3 for the x-coordinates of the vertices and list L4 for the y-coordinates of the vertices.

Step 4 The y-coordinates of each vertex are multiplied by the factor a; points above, below, and on the x-axis will behave differently as summarized in the table for Step 5.

Step 4 Share your results from Step 3. For each value of a, describe the transformation of the quadrilateral in Step 2. What was the result for each vertex?

Step 5 Organize your results from this first part of the investigation.

Step 5 Possible answer:

	The whole graph	Points above the x-axis	Points below the x-axis	Points on the x-axis
Factor greater than 1	A vertical stretch	Go farther up, away from the x-axis	Go farther down, away from the x-axis	Unchanged
Factor between 0 and 1	A vertical shrink	Go down closer to the x-axis	Go up closer to the x-axis	Unchanged
Factor less than 0	A stretch or a shrink reflected across the x-axis	As above but reflected across the x-axis	As above but reflected across the x-axis	Unchanged

Step 6 Graph should look like that shown on the student page;
$L_1 = \{2, -2, 0, 2\}$,
$L_2 = \{-2, -2, 1, -2\}$.

Step 6 Graph this triangle on your calculator. Use list L_1 for the x-coordinates of the vertices and list L_2 for the y-coordinates of the vertices.

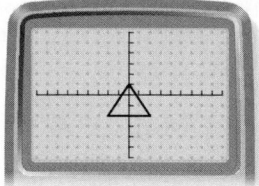

Step 7 Describe how definitions a and b below transform the triangle. Use list L_3 for the x-coordinates of the vertices of the image and list L_4 for the y-coordinates of the vertices of the image. Check your answers by graphing on your calculator.

a. $L_3 = L_1$
   $L_4 = -0.5 \cdot L_2$

b. $L_3 = L_1$
   $L_4 = 2 \cdot L_2 - 2$

Step 8a
$L_3 = L_1$
$L_4 = 3 \cdot L_2$

Step 8b
$L_3 = L_1$
$L_4 = 2 \cdot L_2 + 3$

Step 8 Write definitions for list L_3 and list L_4 in terms of list L_1 and list L_2 to create each image below. Check your definitions by graphing on your calculator.

a.
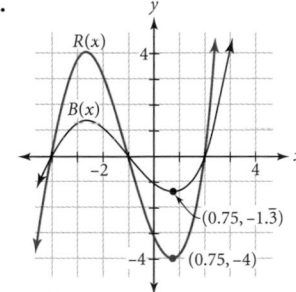

b.
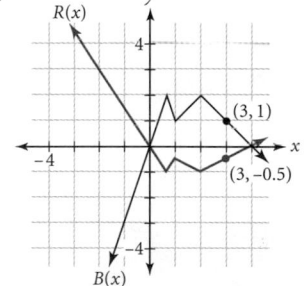

Steps 9 and 10 Graphs depend on equation chosen.

Next, see how you can stretch and shrink the graph of a function.

Step 11 The y-values for Y_2 will be twice the y-values for Y_1. This results in a vertical stretch so that the positive range is two times higher and the negative range is two times lower.

Step 9 Each member of your group should choose an equation from the list below. Enter your equation into Y_1 and graph it on your calculator.

$Y_1(x) = -1 + 0.5x$ ⠀⠀⠀⠀ $Y_1(x) = |x| - 2$
$Y_1(x) = -x^2 + 1$ ⠀⠀⠀⠀ $Y_1(x) = 1.4^x$

Step 10 Enter $Y_2(x) = 2 \cdot Y_1(x)$ and graph it. [▶ 🖵 See **Calculator Note 9B** for specific instructions for your calculator. ◀]

Step 11 Look at a table on your calculator and compare the y-values for Y_1 and Y_2.

Step 12 Repeat Steps 10 and 11, but use these equations for Y_2.

a. $Y_2(x) = 0.5 \cdot Y_1(x)$ ⠀⠀ b. $Y_2(x) = 3 \cdot Y_1(x)$ ⠀⠀ c. $Y_2(x) = -2 \cdot Y_1(x)$

Step 12a The y-values for Y_2 will be one-half the y-values for Y_1; a vertical shrink.

Step 12b The y-values for Y_2 will be three times the y-values for Y_1; a vertical stretch.

Step 12c The values for Y_2 will be two times the y-values for Y_1 and then negated; a vertical stretch and reflection.

Step 13 Write an equation for $R(x)$ in terms of $B(x)$. Then write an equation for $B(x)$ in terms of $R(x)$.

a.

b.

Step 13a $R(x) = 3 \cdot B(x)$;
$B(x) = \frac{1}{3} \cdot R(x)$

Step 13b $R(x) = -\frac{1}{2} \cdot B(x)$;
$B(x) = -2 \cdot R(x)$

Step 7 As written, the equation in part b indicates a stretch followed by a translation. Students entering $L_3 = 2 * (L_2 - 2)$ will see a translation followed by a stretch, with a different result. If any students make this mistake, ask them to present it later so that the class can learn from the idea.

Step 7a a vertical shrink by a factor of 0.5, then a reflection across the x-axis

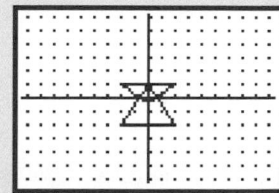

Step 7b a vertical stretch by a factor of 2, then a translation down 2 units

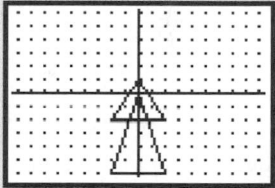

Step 8 [Ask] Each of these images is obtained by stretching and then perhaps translating the figure in Step 6. If students are confused, help them identify the dilation factor. [Ask] "How tall is the original triangle? The transformed triangle is how many times as tall?"

Step 9 If a group has fewer than four members, the first student to finish with one equation should repeat the procedure with an unused equation.

Steps 10 and 12 Calculators vary in syntax. Have Calculator Note 9B handy.

Steps 10 Students may be confused because they're transforming functions that are not parent functions.

SHARING IDEAS

Ask students to report ideas from Steps 5, 8, and 12.

Remind students that, for translations and reflections, the change to a variable in an equation was backward from the change to the corresponding coordinate of a point. [Ask] "Is that the case for a vertical stretch or shrink? How and why?"

As a prelude to Improving Your Reasoning Skills, ask about horizontal stretches and shrinks. [Ask] "How would you change the coefficients of a point? How would you change the equation?"

If you were able to generate some controversy in Lesson 9.2 about whether or not the functions $y = 33(1 + 0.30)^x$ and $y = 92(1 + 0.30)^x$ were in the same family, you might ask the question again. Elicit the idea that they are both dilations of $y = (1 + 0.30)^x$, so they're in the same family.

Watch for students' abilities to name coordinates of a point, graph a polygon, operate on calculator lists, and recognize and describe **translations** and **reflections**.

▶ *EXAMPLE A*

This example describes a vertical shrink in terms of a ratio of distances from the *x*-axis.

MAKING THE CONNECTION

You might mention that stretches and shrinks are called *dilations*. If a figure is dilated by the same factor in two directions, the image is geometrically similar to the original, and the dilation factor is the scale factor of Chapter 3. If you have a projector and a computer with a drawing program, you could illustrate this.

Technology
● ● *CONNECTION* ●

Many computer applications allow you to change the size and shape of clip art. Some applications have commands to change only the horizontal or the vertical scale. If you change only one scale, you distort the picture with a stretch or a shrink. If you change both scales by the same factor, you create a larger or smaller picture that is geometrically similar to the original.

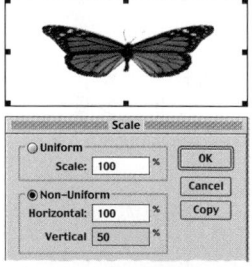

To vertically stretch or shrink a polygon, you multiply the *y*-coordinates of the vertices by a constant factor. To vertically stretch or shrink the graph of a function, you again have to multiply the function by a factor.

EXAMPLE A | Describe how the graph of $y = 0.5|x|$ relates to the graph of $y = |x|$. Then graph both functions.

▶ **Solution** | Tables of values for both functions show that $y = 0.5|x|$ is a vertical shrink. Each *y*-value for $y = 0.5|x|$ is one-half the corresponding *y*-value for $y = |x|$. Multiplying the function by 0.5 has the same effect as multiplying the *y*-coordinate of every point on the graph of $y = |x|$ by 0.5.

x	$y = |x|$	$y = 0.5|x|$
2	2	1
0	0	0
1	1	0.5
5	5	2.5

Graphing the functions together also shows a vertical shrink by a factor of 0.5. Each point on the graph of $y = 0.5|x|$ is one-half the distance from the *x*-axis of the corresponding point on $y = |x|$.

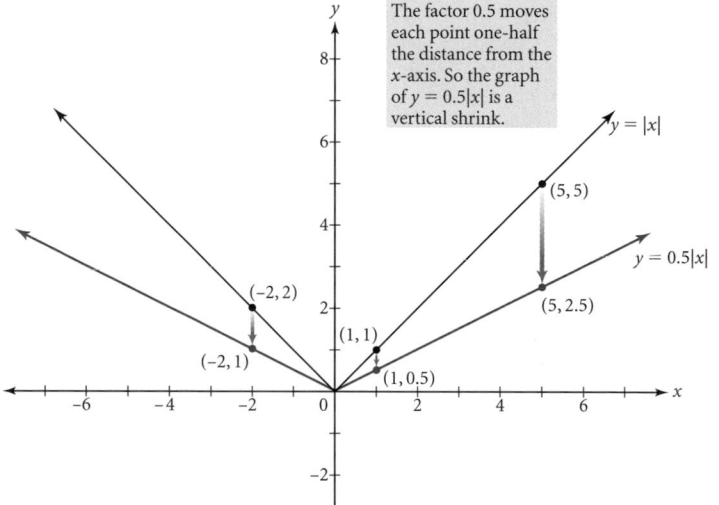

The factor 0.5 moves each point one-half the distance from the *x*-axis. So the graph of $y = 0.5|x|$ is a vertical shrink.

EXAMPLE B

Find an equation for the function shown in this graph.

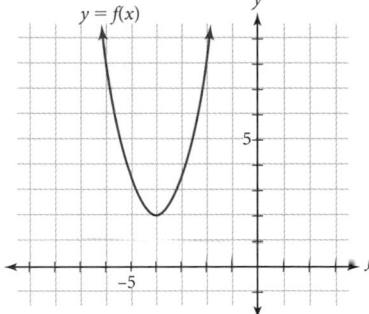

► Solution

The graph is a parabola, so the parent function is $y = x^2$. First determine if a vertical stretch or shrink is necessary. An informal way to do this is to think about corresponding points on the graphs of $y = x^2$ and $y = f(x)$.

The parent function, $y = x^2$
When you move 1 unit left of the vertex, you move 1 unit up to find a point on the graph. When you move 2 units right of the vertex, you move 4 units up to find a point on the graph.

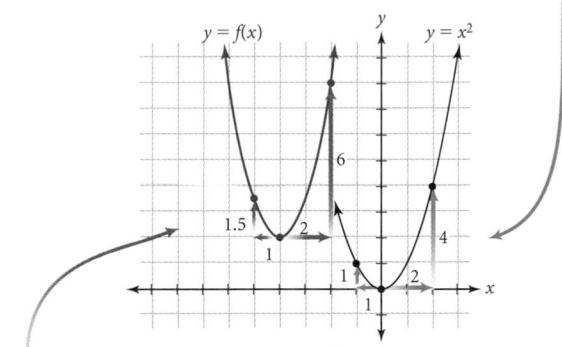

The image function, $y = f(x)$
When you move 1 unit left of the vertex, you move 1.5 units up to find a point on the graph. When you move 2 units right of the vertex, you move 6 units up to find a point on the graph.

For the same x-distances from the vertex on each graph, the corresponding y-distances from the vertex on the image graph $y = f(x)$ are 1.5 times the y-distances on the parent graph $y = x^2$. So the stretch factor is 1.5.

x-distance from vertex	y-distance from vertex of parent function, $y = x^2$	y-distance from vertex of image function, $y = f(x)$	Stretch factor calculation
1	1	1.5	$\frac{1.5}{1} = 1.5$
2	4	6	$\frac{6}{4} = 1.5$

► EXAMPLE B

This example shows how to find the factor by which a graph is stretched vertically. The solution considers the stretch of the graph first, although a translation is most apparent. The reason is twofold. First, like order of operations, it is customary to approach multiplication (stretches, shrinks, and reflections) before addition (translations). Second, doing the stretch factor after translation could erroneously result in
$y = 1.5[(x + 4)^2 + 2]$
$= 1.5(x + 4)^2 + 3$.

One method to check that you have found the correct function is to substitute several points that can be approximated on the graph, such as $(-4, 2)$, $(-2, 8)$, and $(-5, 3.5)$. Another method is to enter the function into the calculator and see whether a table of values corresponds to the graph in the student text.

You can stretch a graph vertically by a factor of *b* in two ways: multiply the *y*-coefficient of each point by *b*, or divide the *y*-variable in the graph's equation by *b*.

The exercises include vertical dilations of polygons and graphs with equations, as well as a few horizontal dilations.

ASSIGNING HOMEWORK

Essential	1–3, 7–10
Performance assessment	6, 9, 13
Portfolio	8
Journal	4, 5, 7, 14
Group	11, 12
Review	15, 16

▶ **Helping with the Exercises**

2a.

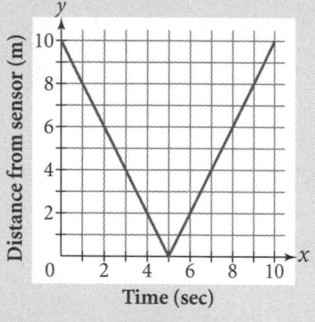

3b. Begin at the motion sensor and walk away at 1.2 m/sec for 5 sec. At 6 m away, turn and walk back at the same speed.

Exercises 4 and 5 The ABS and PARAB calculator programs can give students unlimited practice in writing equations of functions to match given graphs.

$$y = x^2$$ Equation of the parent function.
$$y = 1.5x^2$$ Multiply the parent function, x^2, by a factor of 1.5 for the vertical stretch.

The vertex of the graph of $y = f(x)$ is at $(-4, 2)$. So you must now change the equation to show a translation left 4 units and up 2 units.

$$y = 1.5(x + 4)^2$$ Replace x with $x - (-4)$, or $x + 4$, to translate the graph left 4 units.

$$y - 2 = 1.5(x + 4)^2$$ Replace y with $y - 2$ to translate the graph up 4 units.

$$y = 1.5(x + 4)^2 + 2$$ Solve for y.

The equation for the function is $y = 1.5(x + 4)^2 + 2$.

How can you check that this equation is correct?

Now that you've learned how to translate, reflect, and vertically stretch or shrink a graph, you can transform a function into many forms. This skill gives you a lot of power in mathematics. You can look at a complicated equation and see it as a variation of a simpler function. This skill also allows you to adjust the fit of mathematical models for many situations.

EXERCISES

You will need your calculator for problems **4, 5, 7, 10, 11,** and **12.**

▶ **Practice Your Skills**

1. Ted and Ching-I are using a motion sensor for a "walker" investigation. They find that the graph at right models data for Ted's walk. Write an equation for this graph. $y = |x - 5|$

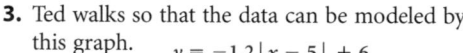

2. Ching-I walks so that her distance from the sensor is always twice Ted's distance from the sensor.
   a. Sketch a graph that models Ching-I's walk.
   b. Write an equation for the graph in 2a. $y = 2|x - 5|$

3. Ted walks so that the data can be modeled by this graph. $y = -1.2|x - 5| + 6$
   a. Write an equation for this graph.
   b. Describe how Ted walked to create this graph.

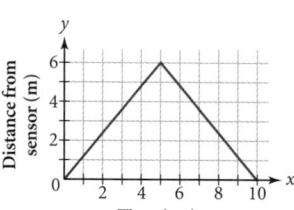

4. Run the ABS program five times. On your own paper, sketch a graph of each randomly generated absolute value function. Find an equation for each graph. [▶ 🖳 See **Calculator Note 9C** to learn how to use the ABS program. ◀] Answers will vary.

5. Run the PARAB program five times. On your own paper, sketch a graph of each randomly generated parabola. Find an equation for each graph. [▶ 🖳 See **Calculator Note 9D** to learn how to use the PARAB program. ◀] Answers will vary.

▶ Reason and Apply

6. This table lists the vertices of a triangle. Name the vertex or vertices that will not be affected by doing a vertical stretch. $(2, 0)$

x	y
2	0
4	2
0	1

7. Graph each function on your calculator. Then describe how each graph relates to the graph of $y = |x|$ or $y = x^2$. Use the words *translation, reflection, vertical stretch,* and *vertical shrink.*
   a. $y = 2x^2$
   b. $y = 0.25|x - 2| + 1$
   c. $y = -(x + 4)^2 - 1$
   d. $y = -2|x - 3| + 4$

8. Each row of the table below describes a single transformation of the parent function $y = |x|$. Copy and complete the table.

Change to the equation $y = |x|$	New equation in $y =$ form	Transformation of the graph of $y = |x|$
Replace x with $x - 3$	$y = |x - 3|$	Translation right 3 units
Replace y with $y + 2$	$y = |x| - 2$	Translation down 2 units
Multiply the right side by -1	$y = -|x|$	Reflection across the y-axis
Replace y with $y - 2$	$y = |x| + 2$	Translation up 2 units
Multiply the right side by $\frac{1}{2}$	$y = \frac{1}{2}|x|$	Vertical shrink by a factor of $\frac{1}{2}$
Replace x with $x + 4$	$y = |x + 4|$	Translation left 4 units
Multiply the right side by 1.5	$y = 1.5|x|$	Vertical stretch by a factor of 1.5
Replace x with $x - 1$	$y = |x - 1|$	Translation right 1 unit
Multiply the right side by 3	$y = 3|x|$	Vertical stretch by a factor of 3

9. Describe the order of transformations of the graph of $y = x^2$ represented by
   a. $y = -(x + 3)^2$
   a reflection across the x-axis and a translation left 3 units
   b. $y = 0.5(x - 2)^2 + 1$
   Possible answer: A vertical shrink by a factor of 0.5; then a translation right 2 units and up 1 unit

10. Draw this **J** on graph paper or on your calculator. Then draw the image defined by each of the definitions in 10a–e. Describe how each image relates to the original figure. (If you use graph paper, give yourself a lot of room or make five individual graphs. If you use a calculator, adjust your friendly window so that you can see both figures at the same time.)

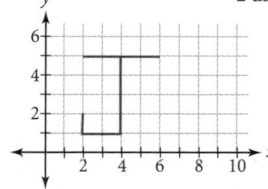

   a. $(x, 3y)$ a vertical stretch by a factor of 3
   b. $(3x, y)$ a horizontal stretch by a factor of 3
   c. $(3x, 3y)$ a horizontal and vertical stretch by a factor of 3
   d. $(0.5x, 0.5y)$ a horizontal and vertical shrink by a factor of 0.5
   e. $(-2x, -2y)$ a horizontal and vertical stretch by a factor of 2; reflected across both the x- and y-axes
   f. Explain why the transformations in 10c, 10d, and 10e are often called "size transformations."
   These transformations proportionately increase or decrease the overall size of the object.

Exercise 10 If students use a calculator, good lists for creating the original figure are $L_1 = \{2, 2, 4, 4, 6, 2\}$ and $L_2 = \{2, 1, 1, 5, 5, 5\}$. Then Lists L_3 and L_4 can be used to define and plot the images.

The horizontal stretch in 10b is the first students have seen. Students will encounter horizontal dilations in Improving Your Reasoning Skills and in the investigation for Lesson 9.7.

In 10c, some students may note that the figures are geometrically similar.

Some students may describe the transformation in 10e as a stretch and a rotation through 180 degrees. Others may try to describe a process of dilating through the origin.

In 10f, a size transformation makes similar figures, whereas a single vertical or horizontal stretch distorts the figure in one direction.

7a. a vertical stretch of $y = x^2$ by a factor of 2

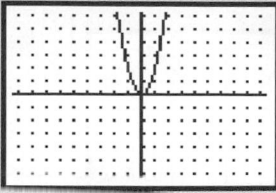

$[-9.4, 9.4, 1, -6.2, 6.2, 1]$

7b. a vertical shrink of $y = |x|$ by a factor of 0.25; then a translation right 2 units and up 1 unit

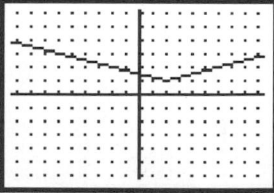

7c. a reflection of $y = x^2$ across the x-axis; then a translation left 4 units and down 1 unit

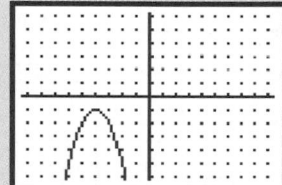

7d. a vertical stretch of $y = |x|$ by a factor of 2 and a reflection across the x-axis; then a translation right 3 units and up 4 units

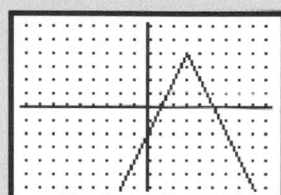

Exercise 8 This is a good exercise to review all of the transformations learned in this chapter.

Exercise 9 Order will not change the answer in 9a. In 9b, the order of the vertical shrink and horizontal translation will not change the answer, but the vertical translation must occur last or the resulting graph will be wrong.

See pages 720–721 for answers to Exercises 10a–10e.

11a. a vertical shrink by a factor of 0.5 and a reflection across the x-axis

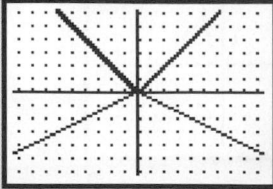

11b. a vertical stretch by a factor of 2 and a translation right 4 units

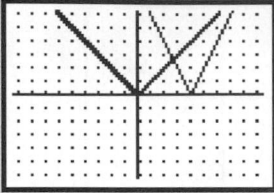

11c. a vertical stretch by a factor of 3 and a reflection across the x-axis; then a translation left 2 units and up 4 units

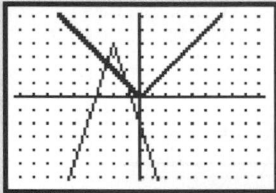

Exercise 12 Here students model a real-world situation with an absolute value function or parabola. Students having difficulties may be advised to identify the vertex first and then estimate the scale factor. By adjusting the scale factor, students can get a better-fitting equation.

In 12a, if students assume the vertex is at January 1, they may get $f(x) = -20|x - 3| + 70$ or $f(x) = -9.3(x\ 5\ 3.3)^2 + 70$.

12a. Possible answer:
$f(x) = -25|x - 3.2| + 80$

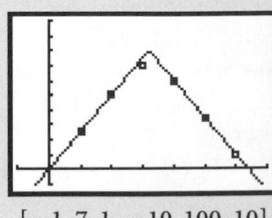

$[-1, 7, 1, -10, 100, 10]$

---

11. Graph $Y_1(x) = abs(x)$ on your calculator. Predict what each graph will look like. Check by comparing the graphs on your calculator. [▶🖥 See **Calculator Note 9B** for specific instructions for your calculator. ◀]

  a. $Y_2(x) = -0.5\,Y_1(x)$
  b. $Y_2(x) = 2\,Y_1(x - 4)$
  c. $Y_2(x) = -3\,Y_1(x + 2) + 4$

12. In Interlochen, Michigan, it begins to snow in early November. The depth of snow increases over the winter. When winter ends, the snow melts and the depth decreases. This table shows data collected in Interlochen.

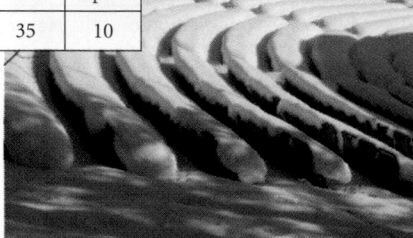

Snow in Interlochen

Date	Nov 1	Dec 1	Jan 1	Feb 1	Mar 1	Apr 1
Depth of snow (cm)	25	50	70	60	35	10

  a. Plot the data. For the dependent variable, let Nov 1 = 1, Dec 1 = 2, and so on. Find a function that models the data.
  b. Use your function to find $f(2.5)$. Explain what this value represents.
  c. Find x if $f(x) = 47$. Explain what this x-value represents.
  d. According to your model, when was the snow the deepest? How deep was it at that time? after 3.2 months (early January); about 80 cm

January snow covers the seats of the outdoor theater at Interlochen Center for the Arts.

13. Deshawn is designing a computer animation program. She has a set of coordinates for the tree shown on the right side. She wants to use 13 frames to move the tree from the right to the left. In each frame, she wants the tree to shrink by 80%. How should she define the coordinates of each image using the coordinates from the previous frame? $(x - 1, 0.8y)$

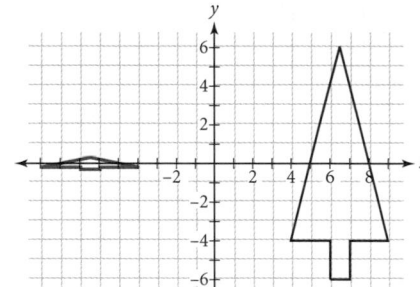

14. Byron says,

If the graph of a function is stretched vertically, but not translated, the factor a is the same as the y-value when x equals 1.

Does Byron's conjecture work for every function in the forms shown below? Tell why or why not.

  a. $y = a \cdot x^2$ Yes; when you substitute 1 for x, you get $y = a \cdot 1^2 = a$.
  b. $y = a \cdot |x|$ Yes; when you substitute 1 for x, you get $y = a \cdot |1| = a$.
  c. $y = a \cdot f(x)$

12b. Using the equation in 12a: $f(2.5) = 62.5$; the depth of the snow after 2.5 months (mid-December) would be about 62.5 cm.

12c. Using the equation in 12a: $x = 1.88$ or $x = 4.52$; the depth of the snow would be 47 cm after 1.88 months (end of November) or after 4.52 months (mid-February).

14c. No; unless $f(1) = 1$, a will not be the same as the y-value. For example, if $f(1) = 3$, then $y = a \cdot f(1) = a \cdot 3$, but $a = \dfrac{y}{3}$.

▶ Review

15. Use the properties of exponents to rewrite each expression without negative exponents.

a. $(2^3)^{-3}$ $\dfrac{1}{2^9}$

b. $(5^2)^5$ 5^{10}

c. $(2^4 \cdot 3^2)^3$ $2^{12} \cdot 3^6$

d. $(3^2 \cdot x^3)^{-4}$ $\dfrac{1}{3^8 \cdot x^{12}}$

Exercise 15 This exercise reviews Lesson 7.6.

16. The equation $y = -52 + 1.6x$ approximates the wind chill temperature in degrees Fahrenheit for a wind speed of 40 miles per hour.

a. Which variable represents the actual temperature? Which variable represents the wind chill temperature? *x* represents actual temperature, and *y* represents wind chill temperature.

b. What *x*-value gives a *y*-value of -15? Explain what your answer means in the context of this problem.
23.125; When the wind chill temperature is $-15°$F with a wind speed of 40 miles per hour, the actual temperature is approximately $23°$F.

Exercise 16 This exercise reviews Lesson 4.6.

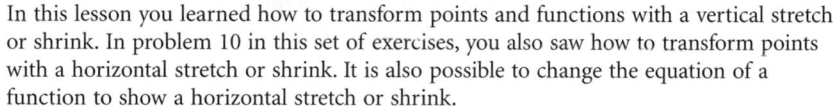

IMPROVING YOUR REASONING SKILLS

In this lesson you learned how to transform points and functions with a vertical stretch or shrink. In problem 10 in this set of exercises, you also saw how to transform points with a horizontal stretch or shrink. It is also possible to change the equation of a function to show a horizontal stretch or shrink.

Consider the graph of $y = x^2$ and its image after a horizontal stretch by a factor of 2. Write an equation for the image.

Describe the image in terms of a vertical stretch or shrink. Write an equation that shows this transformation. Is this equation equivalent to the one that shows a horizontal stretch?

When you vertically stretch or shrink the graph of $y = f(x)$ by a factor of a, you get a graph of $y = a \cdot f(x)$. If you horizontally stretch or shrink the graph of $y = f(x)$ by a factor of b, you will get the graph of what equation?

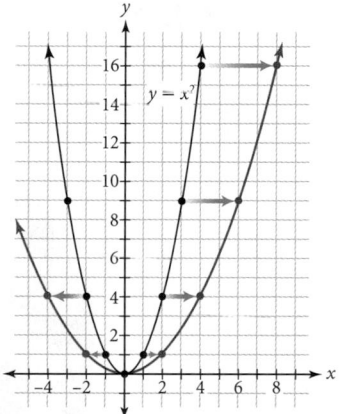

IMPROVING REASONING SKILLS

Possible equations for the horizontal stretch include $y = \left(\dfrac{x}{2}\right)^2$ and $y = (0.5x)^2$. Because these equations are equivalent to $y = 0.25x^2$, the horizontal stretch can also be considered a vertical shrink by a factor of 0.25. Just as a vertical dilation of $f(x)$ by a factor of a is given by $\dfrac{y}{a} = f(x)$, a horizontal stretch or shrink by a factor of b can be represented generically by $y = f\left(\dfrac{x}{b}\right)$. You may want to ask why the variable x is divided by the factor b. [For a point (x, y) on the transformed equation, the point $\left(\dfrac{x}{b}, y\right)$ satisfies the original equation, so the new graph has the equation $y = f\left(\dfrac{x}{b}\right)$.]

PLANNING

LESSON OUTLINE

One day:

30 min   Activity

15 min   Sharing

5 min   Closing

MATERIALS

Experiment 1

- large marbles (one per group)
- poster board (one 8-by-11.5–inch piece for each group. A large sheet of poster board can be cut into quarters.)
- metersticks or yardsticks (one per group)
- tape
- poster board and paper (one each per group)
- paper cups (one per group)
- books (four per group)
- table and chair (one set per group)
- The Rolling Marble (T), *optional*

Experiment 2

- motion sensor
- Calculator Note 4C
- cables for linking calculators

Experiment 3

- stopwatch or watch with second hand

TEACHING

In these activities students see applications of quadratic, absolute value, and linear equations. If you have two days, every group can do each experiment.

Activity Day

Using Transformations to Model Data

In this lesson you'll do experiments to gather data, and then you'll find a function to model to the data. To fit the model, you'll first need to identify a parent function. Then you'll transform the parent function and fit the image function to the data.

There are three experiments to choose from. Your group should choose one experiment. Do the other experiments if time permits.

Activity

Roll, Walk, or Sum

You will need

- a large marble
- tape
- one sheet of paper
- four books
- a sheet of poster board
- a paper cup
- a meterstick or a yardstick
- a table and chair
- a motion sensor
- a stopwatch or a watch with a second hand

Experiment 1: The Rolling Marble

In this experiment you'll write the equation for the path of a falling marble. Then you'll catch the marble at a point you calculate using your equation.

> **Procedure Note**
>
> Use the books and poster board to build a ramp about 30 cm from the edge of the table. Fold the sheet of paper into fan pleats—the smaller the pleats, the better. This paper, when unfolded, will help you locate where the marble hits the floor.

LESSON OBJECTIVES

- Model real-world data with equations that involve vertical transformations of exponential functions
- Model real-world data with equations that describe absolute value or parabolic functions

NCTM STANDARDS

CONTENT		PROCESS	
	Number		Problem Solving
●	Algebra	●	Reasoning
●	Geometry	●	Communication
●	Measurement	●	Connections
●	Data/Probability	●	Representation

Step 1 | Do a trial run. Roll the marble from the top edge of the ramp. Let it roll down the ramp and across the table and drop to the floor. Spot the place where it hits the floor (approximately). Tape the folded paper to the floor in this area.

Step 2 | Now collect data to fix the drop point more precisely. Roll the marble two or three times, and mark the point where it hits the paper. Each roll should be as much like the other rolls as you can make it. So start each marble roll at the same place, and release it the same way each time.

Step 3 Using the sample setup, points are (0, 0), (0, 28), and (14, 0).

Step 4 Using the points in Step 3, an equation is $y = -\frac{1}{7}x^2 + 28$. That is a vertical shrink of the graph of $y = x^2$ by a factor of $\frac{1}{7}$ and a reflection across the x-axis; then a translation up 28 units.

Step 3 | Next, find the coordinates of points for a graph. Let x represent horizontal distance, and let y represent vertical distance. Locate the point on the floor directly below the *edge* of the table. Call this point (0, 0). Measure from (0, 0) up to the point at the edge of the table where the marble rolls off. Name the coordinates of this point. Lastly, measure from (0, 0) to each point where the marble hit the floor. Find the average coordinates for these points on the floor.

Step 4 | As your marble falls, it will follow the path of a parabola. The point where it leaves the table is the vertex of the parabola. Define variables and write an equation that fits your two points.

Next, you'll test your model by using it to calculate a point on the path of the marble. See if you can catch the marble at that point.

Step 5 Students should substitute the seat's height for y in the equation they found in Step 4. The value of x will then be the horizontal distance the cup must be from the table.

Step 5 | Measure the height of the chair seat. Put the chair next to the table and place a small cup on the chair. Use your calculations to adjust the chair so that when you roll the marble, it will land in the cup.

Step 6 | You have only one chance to land in the cup. Release the marble as you did in Step 2. Good luck!

Experiment 2: Walking

In this experiment you'll walk past a motion sensor and model the data you collect.

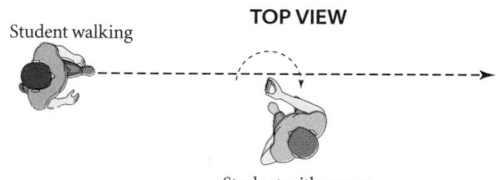

TOP VIEW

Student walking

Student with sensor

> **Procedure Note**
>
> Aim the sensor at the walker. The walker should start about 3 meters away, and then walk quickly toward the sensor, aiming just to the side of it. As the walker passes, the student holding the motion sensor should turn it so that it is always directed at the walker. [▶ 🔲 Use the FREE FORM option of **Calculator Note 4C** to collect your data. ◀]

Step 1 | Walk steadily in the same direction toward the sensor. Pass it and go about 3 meters farther. Record data for the entire walking time.

Step 3 The equation should be in the form $y = a|x - h|$, in which a is the speed of the walkers in m/sec, and h is the starting distance between the two students.

Step 2 | Download the data to each person's calculator. You should expect some erratic data points while the walker is close to the sensor.

Step 3 | Fit the data using function transformations. If the vertex is "missing" from the data, estimate its location.

Another Experiment

An alternative experiment that generates a parabola uses two metersticks or yardsticks and about one meter of string. Holding one end of the string at zero on the meterstick and the other end at various points, students straighten the string to form rectangles. They measure the length and width and calculate the area of each rectangle. After 10 rectangles, they can find an equation that fits the (*length, area*) data they have collected.

Experiment 1

Step 2 The marble will often dent the pleats at the place it hits the floor.

Step 3 The Rolling Marble transparency models the setup with sample coordinates. An open notebook can be used for the ramp instead of poster board.

Step 5 As needed, point out to students that they will get only one chance to try to intercept the marble, so they should be very careful with measurements and calculations.

Step 6 When each group is ready to make its try, you might have the entire class gather around to watch.

Experiment 2

Step 2 Use cables to connect the calculators for distributing the data. If you are using a CBR as your motion sensor, you can use the RANGER program it contains.

Step 3 These data points actually fit a branch of a hyperbola, but an absolute value function (made up of the hyperbola's asymptotes) is a good fit.

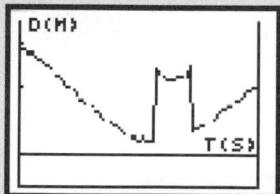

Other walking patterns will produce more clearly visible absolute value graphs. For example, students might walk directly toward the sensor and then turn abruptly and walk directly away.

Experiment 3

Many jobs require testing for speed and accuracy in typing or in numeric keypad entry.

Step 1 This step is for answer checking and for warming up their fingers. If students know the expression $\frac{n(n+1)}{2}$ for the sum of the first n positive integers, they can use it to double-check their answers.

Step 4 Encourage students to use transformations of the line $y = x$ rather than using methods learned in Chapter 5, Fitting a Line to Data.

Step 5 Students may point out that averaging several measurements of the same summation would yield better predictions. It's probably better for groups to spend the time on a different experiment.

SHARING IDEAS

Have a group report results of each experiment, especially if not all groups worked on each one.

Assessing Progress

This is an excellent opportunity to see how well students understand the effect on equations of **translations**, **reflections**, and **dilations**, as well as their abilities to work with the parent **quadratic**, **absolute value**, and **linear** functions.

Closing the Lesson

Knowledge of how transformations affect equations can be very useful in making predictions.

Step 4a Start 1.2 m from the person holding the sensor. Walk at 1.5 m/sec. Pass the sensor at 0.8 sec.

Step 4

Step 4b Start 0.85 m from the person holding the sensor. Walk at 2.1 m/sec. Pass the sensor at about 0.4 sec.

Step 5

Write walking instructions for each of these functions. In your instructions say where to start, how fast to go, and when to pass the sensor.

a. $f(x) = 1.5|x - 1.2|$ **b.** $f(x) = 2.1|x - 0.85|$

If time permits, try following your instructions to see if the graph fits your data.

Experiment 3: Calculating

Who is the fastest calculator operator (CO) in your group? The COs had better warm up their fingers!

Step 1

Step 4 The data is likely to be a direct variation in the form $y = kx$, in which k is the amount of time required to enter one number and the plus sign.

Step 5 Students should substitute 47 for x in their equations from Step 4. The value of y is the time required.

Step 6 If students found a direct variation, then the y-intercept will be 0, which means it takes no time to add no numbers. If students did find a y-intercept, answers will vary.

> **Procedure Note**
>
> You must start with $1 + 2 + 3 \ldots$ each time. It is not fair to use the last result!

Step 1   The CO should carefully calculate sums a–g. Record the answers.

  a. $1 + 2 + 3 + \ldots + 8 + 9 + 10$
  b. $1 + 2 + 3 + \ldots + 13 + 14 + 15$
  c. $1 + 2 + 3 + \ldots + 18 + 19 + 20$
  d. $1 + 2 + 3 + \ldots + 23 + 24 + 25$
  e. $1 + 2 + 3 + \ldots + 28 + 29 + 30$
  f. $1 + 2 + 3 + \ldots + 33 + 34 + 35$
  g. $1 + 2 + 3 + \ldots + 38 + 39 + 40$

Step 2   Next, the CO calculates the first sum, 1 to 10, again *while being timed*. (Record the time only if the CO gets the correct answer. If not, run the trial again.)

Step 3   Repeat Step 2 for sums b–g, that is, 1 to 15, 1 to 20, . . . , 1 to 40. You should have seven data points in the form (*number of numbers added, time*).

Step 4   Find an equation to model the data. Transform it as needed for a better fit.

Step 5   Use your model to predict the time it would take to sum the numbers from 1 to 47. Test your prediction and record the results.

Step 6   What is the y-intercept of your model? Does this value have any real-world meaning? If yes, then what is the meaning? If no, then why not?

Discuss your results with the class. How were the experiments alike? How were they different? How could you recognize the parent function in the data?

Introduction to Rational Functions

In Chapter 3, you learned about inverse variation. The simplest inverse variation equation is $y = \frac{1}{x}$. Look at the graph of this equation.

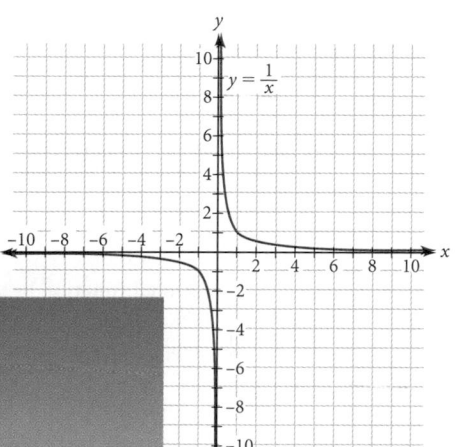

Notice that the graph of $y = \frac{1}{x}$ has two parts. One part is in Quadrant I, and the other is in Quadrant III. In Chapter 3, you wrote inverse variation equations for countable and measurable quantities, such as number of nickels and distance in inches. Since these quantities are always positive, you worked only with the part of the graph in Quadrant I.

Notice that as the x-values get closer and closer to 0, the graph gets closer and closer to the y-axis. As the x-values get farther and farther from 0, the graph gets closer and closer to the x-axis. An **asymptote** is a line that a graph approaches more and more closely. So the graph of $y = \frac{1}{x}$ has two asymptotes: the lines $x = 0$ and $y = 0$. Can you explain why the x- or y-axes are asymptotes for this graph?

Also notice that $y = \frac{1}{x}$ is a function because it passes the vertical line test. You can use the inverse variation function as a parent function to understand many other functions.

Some amusement parks have free-fall rides shaped like a first-quadrant inverse variation graph. This is the Demon Drop at Cedar Point Amusement Park in Ohio.

Investigation
I'm Trying to Be Rational

In the first part of this investigation, you will explore transformations of the parent function $y = \frac{1}{x}$.

> **Procedure Note**
> For this investigation, use a friendly window with a factor of 2.

Step 1 | Graph the parent function $y = \frac{1}{x}$ on your calculator.

Step 2 | Use what you have learned about transformations to predict what the graphs of these functions will look like.

a. $y = \frac{-3}{x}$ **b.** $y = \frac{2}{x} + 3$ **c.** $y = \frac{1}{x - 2}$ **d.** $y = \frac{1}{x + 1} - 2$

NCTM STANDARDS

CONTENT	PROCESS
Number	• Problem Solving
• Algebra	• Reasoning
Geometry	• Communication
Measurement	Connections
Data/Probability	• Representation

LESSON OBJECTIVES

- Investigate basic transformations of the inverse variation function $f(x) = \frac{1}{x}$
- Model real-world data using rational functions

PLANNING

LESSON OUTLINE

One day:

25 min	Investigation
5 min	Sharing
10 min	Example
5 min	Closing
5 min	Exercises

MATERIALS

TEACHING

Graphs of rational functions often have asymptotes or holes. A graph may or may not reach an asymptote. (See page 515.)

 Guiding the Investigation

One step Pose this problem: A bottle contains 1 liter of a 20% salt solution. That is, 0.2 liters of salt is dissolved in the liter of water. How much water should you add to get a 2.5% salt solution? Encourage students to try to understand the situation through a variety of approaches: tables, graphs, equations. They may use as an independent variable the amount of water to be added or the total amount of salt in the solution.

Step 2a Any vertical dilation of an inverse variation can be thought of as a horizontal dilation. That is, a vertical stretch by a factor of a, giving $y = \frac{a}{x}$, is the same as a horizontal stretch by the same factor, with equation $y = \frac{1}{\frac{x}{a}}$.

See page 721 for answers to Steps 1 and 2.

Step 3 [Alert] When entering rational functions into a calculator, a common error is to omit parentheses around the denominator. The function $y = \frac{1}{x+1} - 2$ should be entered as $Y_2 = 1/(x+1) - 2$.

Some calculators (when in continuous mode) will produce a "drag line" in place of the vertical asymptotes. Some calculators will cut off the curve near the vertical asymptote. These are technical shortcomings of the calculator. Suggest that students sketch the actual curve by hand.

Step 4 [Alert] Some students may be confused by the fact that in Lesson 3.4, k was used for the constant of variation. In 4c, a is that constant and k represents the vertical translation. Reiterate that we can use any letters we want, if we say what they represent.

Step 6 to 10 These ideas, not usually topics for first-year algebra, will fascinate some students. The expressions used in a rational expression must both be polynomials, a term introduced in Lesson 10.3.

Step 6 Here the numerator as well as the entire denominator must be enclosed in parentheses. That is, $Y_2 = (x + 3)/((x - 2)(x + 3))$. As an alternative, students might enter $Y_2 = (x + 3)/(x - 2)/(x + 3)$.

Step 8 If students aren't seeing the hole at $x = -3$, be sure they are using the friendly window recommended in the Procedure Note. You might also ask about the domain and range of this function.

Step 9 Again, some calculators may not show the curve going toward infinity near the vertical asymptotes—it may appear to come to a stop. Students may want to sketch a more realistic graph by hand.

See page 721 for answers to Steps 3, 6, 9, and 11.

Step 3   Graph each equation in Step 2 on your calculator along with $y = \frac{1}{x}$. Compare the graph to your prediction in Step 2. How can you tell where asymptotes occur on your calculator screen?

Step 4   Without graphing, describe what the graphs of these functions will look like. Use the words *linear*, *nonlinear*, *increasing*, and *decreasing*. Define the domain and range. Give equations for the asymptotes.

a. $y = \dfrac{5}{x - 4}$

b. $y = \dfrac{-1}{x + 3} - 5$

c. $y = \dfrac{a}{(x - h)} + k$

Step 5 With the exception of $y = -\frac{3}{x}$, these equations are not inverse variations Step 5 because they do not fit the definition of $y = \frac{k}{x}$.

Do you think the equations in Step 2 and Step 4 are inverse variations? Explain why or why not.

Step 7 $y = \dfrac{1}{x - 2}$

Step 8 There is a hole when $x = -3$. Some calculators will show no corresponding y-value when Step 6 $x = -3$. The hole is related to the common factor in the numerator and Step 7 denominator, specifically, $\dfrac{(-3 + 3)}{(-3 + 3)} = \dfrac{0}{0}$. Step 8

The function $y = \frac{5}{x - 4}$ is an example of a **rational function** because it shows a ratio between two expressions, 5 and $x - 4$. Not all rational functions are transformations of $y = \frac{1}{x}$, but you will see similarities in their graphs.

Step 6   Graph this equation on your calculator.

$$y = \frac{(x + 3)}{(x - 2)(x + 3)}$$

Step 7   The graph should look familiar. What graph does it look like?

Step 8   Trace the graph on your calculator. Describe anything unusual that you find. Can you explain your findings?

Step 9   Graph this equation on your calculator.

$$y = \frac{1}{(x + 5)(x - 1)} + 4$$

Step 10 This graph Step 10 has two vertical asymptotes at Step 11 $x = -5$ and $x = 1$; the asymptotes are related to the two factors in the denominator. There is Step 12 also a horizontal asymptote at $y = 4$.

Step 10   Describe the graph. Explain anything unusual that you find.

Step 11   Rational functions often have asymptotes. Some even have holes, that is, x-values for which there are no y-values. Describe how you can tell if a graph has an asymptote or a hole just by looking at the equation and how you can tell where the asymptote or hole will be.

Step 12   Write a rational function that has asymptotes at $x = 2$, $x = -1$, and $y = -3$, and a hole at $x = 8$. Possible answer: $y = \dfrac{(x - 8)}{(x - 8)(x - 2)(x + 1)} - 3$

You can use rational equations to model many real-world applications.

Step 4a A vertical stretch of the graph of $y = \frac{1}{x}$ by a factor of 5 and a translation right 4 units; it will be nonlinear and decreasing. The domain is $x \neq 4$, and the range is $y \neq 0$; there are asymptotes at $x = 4$ and $y = 0$.

Step 4b A reflection of the graph of $y = \frac{1}{x}$ across the x-axis and a translation left 3 units and down 5 units; it will be nonlinear and increasing. The domain is $x \neq -3$, and the range is $y \neq -5$; there are asymptotes at $x = -3$ and $y = -5$.

Step 4c A vertical stretch of the graph of $y = \frac{1}{x}$ by a factor of a (if a is negative, there will be a reflection across the x-axis) and a translation h units horizontally and k units vertically (the signs of h and k will determine whether the translation is right or left, and up or down); it will be nonlinear and decreasing if a is positive, or increasing if a is negative. The domain is $x \neq h$, and the range is $y \neq k$; there are asymptotes at $x = h$ and $y = k$.

EXAMPLE

A salt solution is made from salt and water. A bottle contains 1 liter of a 20% salt solution. This means that the concentration of salt is 20%, or 0.2, of the whole solution.

a. Show what happens to the concentration of salt as you add water to the bottle in half-liter amounts.

b. Find an equation that models the concentration of salt as you add water.

c. How much water should you add to get a 2.5% salt solution?

Mono Lake is a natural saltwater lake located near Lee Vining, California.

▶ **Solution**

a. Use a table to show what happens. The bottle originally contains 20% salt, or 0.2 liter. As you add water, the amount of salt stays the same, but the amount of whole solution increases. Each time you add water, recalculate the concentration of salt by finding the ratio of salt to whole solution.

Added amount of water (L)	0	0.5	1.0	1.5	2.0	2.5	3.0	3.5	4.0	4.5	5.0
Amount of salt (L)	0.2	0.2	0.2	0.2	0.2	0.2	0.2	0.2	0.2	0.2	0.2
Whole solution (L)	1.0	1.5	2.0	2.5	3.0	3.5	4.0	4.5	5.0	5.5	6.0
Concentration of salt	0.2	0.133	0.1	0.08	0.067	0.057	0.05	0.044	0.04	0.036	0.033

b. As the amount of whole solution increases, the concentration of salt decreases, but the *amount* of salt stays the same. This is an inverse variation, and the constant of variation is the amount of salt.

$$\text{concentration} = \frac{salt}{whole\ solution}$$

The equation you need to write should show a relationship between the amount of water you add, x, and the concentration of salt, y. From the table, you can see that the amount of whole solution starts at 1 liter and increases by the amount of water you add. The equation is

Concentration of salt → $y = \dfrac{0.2}{1 + x}$ ← Constant amount of salt

Added amount of water / Starting amount of solution } Whole solution

Ask groups to present their ideas about Steps 3, 4, 5, and 12. As needed, remind students that the product of x and y is constant for an inverse variation.

You may want to have students describe each part of the graph of $y = \frac{1}{x}$ as linear or nonlinear, increasing or decreasing, continuous or discrete. **[Ask]** "Do you think the whole graph is continuous or discrete?" [It is actually called *piecewise continuous* because it has a breaking point between two otherwise continuous parts.] You might point out that the positive quantities in Chapter 3 restricted the domain of $y = \frac{1}{x}$ to $x > 0$, so that all graphs there lay in Quadrant I.

[Ask] "Where have you seen the idea of 'getting closer and closer' before?" [The notion of a mathematical limit arose in Chapter 0 with recursive procedures and strange attractors and in Chapter 2 in which the theoretical probability is the limit of observed probabilities.]

[Ask] "Are there equations whose graphs reach an asymptote?" A graph may or may not reach an asymptote. For example, the graph of $y = \frac{x}{x^2 + 1}$ crosses the x-axis (its horizontal asymptote) at the origin. On the other hand, the graph of the inverse variation $y = \frac{1}{x}$ doesn't reach its horizontal asymptote, the x-axis, because $\frac{1}{x}$ is not 0 for any value of x. Nor does it reach its vertical asymptote, the y-axis, because $\frac{1}{x}$ is undefined when $x = 0$.

▶ **EXAMPLE**

Because many students find ratios a difficult idea, you may want to spend some time on this example, which lets students work further with an inverse variation. You may need to explain that a 20% salt solution (of salt and water) means that the ratio of salt in liters to water also in liters is 20 : 100.

You might want to fill in the table as a class.

[Ask] "What do 0.2 and $1 + x$ represent in this equation? What transformations of the graph of $y = \frac{1}{x}$

result in the graph of $y = \frac{0.2}{1 + x}$?" [A shrink and a shift to the left.]

If you have time, you might suggest that students rework the example with x representing the total amount of water rather than the amount added. In that case they can work with the direct variation $y = \frac{0.2}{x}$ or $xy = 0.2$, and get $x = 8$ when $y = 0.025$. They must then subtract the initial liter of water to find the amount added.

Assessing Progress
Through your observations you can assess students' understanding of the **transformations** studied so far and students' abilities to graph using **friendly windows** and to **trace graphs**.

Closing the Lesson

Any translation of the parent inverse variation $y = \frac{1}{x}$ has **horizontal** and **vertical asymptotes** through the image of the origin under the same transformation. In more general terms, a **rational function** $\frac{f(x)}{g(x)}$ has a horizontal asymptote at $y = L$ if there's a number L to which the values of the function get closer and closer as x gets farther from 0. The graph has a vertical asymptote at $x = L$ if $g(L) = 0$, but $f(L)$ is not 0. If it doesn't have a vertical asymptote at $x = L$ and if $f(L)$ and $g(L)$ are *both* 0, then the graph has a **hole** at a point with x-coordinate L.

BUILDING UNDERSTANDING

Students work with transformations and applications of linear variations and another rational function.

ASSIGNING HOMEWORK

Essential	1–4, 10, 11
Performance assessment	9
Portfolio	8, 12
Journal	11
Group	5–7
Review	13–15

Exercise 1 In 1c, the translation right can be done before the shrink, but the translation down must be done after the shrink. In 1d, the stretch, the reflection, and the translation right can be done in any order, but the translation down must be done after the reflection and stretch.

A graph of the data points and of the equation confirms that this equation is a perfect model.

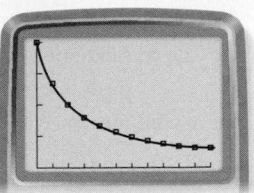

[0, 5.5, 0.5, 0, 0.2, 0.05]

This equation is not an inverse variation because the product of x and y is not constant. It is, however, a transformation of the parent inverse variation function.

c. Use the equation to find the amount of water that you should add. A 2.5% salt solution has a concentration of salt of 0.025.

$$0.025 = \frac{0.2}{1 + x} \qquad \text{Substitute 0.025 for } y.$$

$$0.025 + 0.025x = 0.2 \qquad \text{Multiply both sides by } (1 + x) \text{ and distribute.}$$

$$x = 7 \qquad \text{Solve for } x.$$

You would need to add 7 liters of water to have a 2.5% salt solution.

EXERCISES

You will need your calculator for problems **4, 10, 14,** and **15.**

▶ Practice Your Skills

1. Describe each graph as a transformation of the graph of the parent function $y = |x|$ or $y = x^2$. Then write its equation.

a.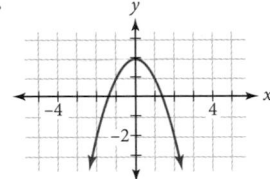

b.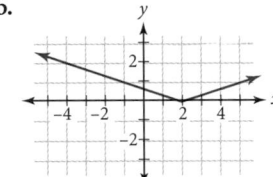

c.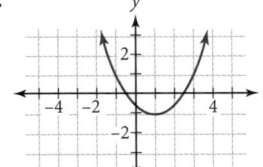

d.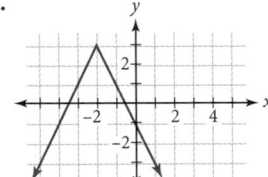

2. Write an equation that generates this table of values. $y = \frac{2}{x}$

x	-4	-3	-2	-1	0	1	2	3	4
y	$-\frac{1}{2}$	$-\frac{2}{3}$	-1	-2	Undefined	2	1	$\frac{2}{3}$	$\frac{1}{2}$

1a. a reflection of the graph of $y = x^2$ across the x-axis; then a translation up 2 units (or translation then reflection); $y = -x^2 + 2$

1b. a vertical shrink of the graph of $y = |x|$ by a factor of $\frac{1}{3}$ and a translation right 2 units; $y = \frac{1}{3}|x - 2|$

1c. Possible answer: a vertical shrink of the graph of $y = x^2$ by a factor of 0.5; then a translation right 1 unit and down 1 unit; $y = 0.5(x - 1)^2 - 1$

1d. Possible answer: a vertical stretch of the graph of $y = |x|$ by a factor of 2 and a reflection across the x-axis; then a translation left 2 units and up 3 units; $y = -2|x + 2| + 3$

3. Write an equation for this graph in the form $y = \frac{a}{x}$. $y = -\frac{5}{x}$

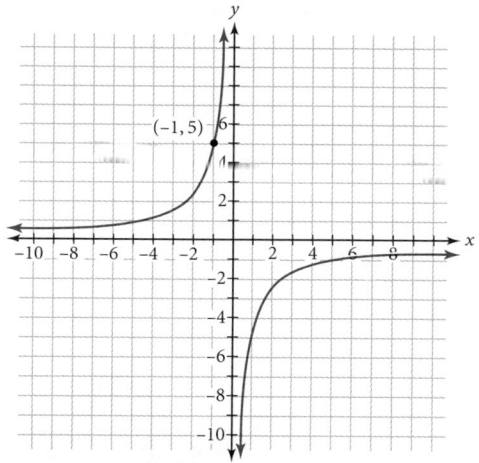

4. Describe each function as a transformation of the graph of the parent function $y = \frac{1}{x}$. Then sketch a graph of each function. Check your answers by graphing on your calculator.

a. $y = \frac{4}{x}$

b. $y = \frac{1}{x-5} - 2$

c. $y = \frac{0.5}{x} + 3$

d. $y = \frac{-3}{x+3}$

5. Describe each of the functions in problem 4 as increasing or decreasing. Give equations for the asymptotes, and give the domain and range.

▶ Reason and Apply

6. Write an equation for each graph. Each calculator screen shows a friendly window with a factor of 1.

a.
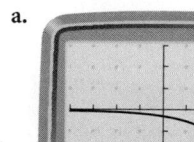
$y = \frac{1}{x-3}$

b.
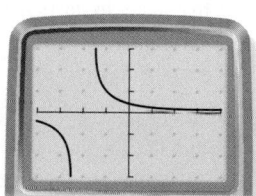
$y = \frac{1}{x+2}$

c.
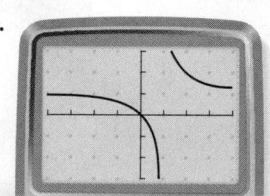
$y = \frac{1}{x-1} + 1$

d.
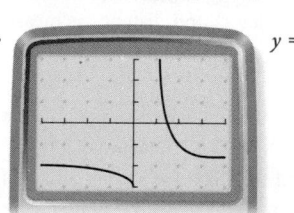
$y = \frac{1}{x-1} - 2$

4a. a vertical stretch by a factor of 4

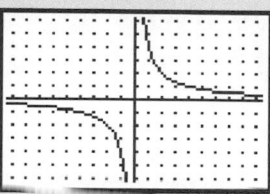

$[-9.4, 9.4, 1, -6.2, 6.2, 1]$

4b. a translation right 5 units and down 2 units

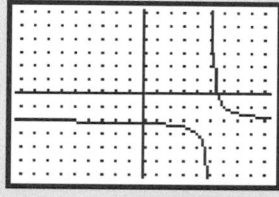

4c. a vertical shrink by a factor of 0.5; then a translation up 3 units

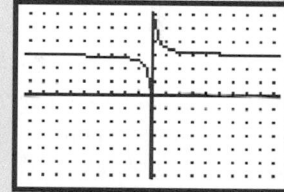

4d. a vertical stretch by a factor of 3, a reflection across the x-axis, and a translation left 3 units

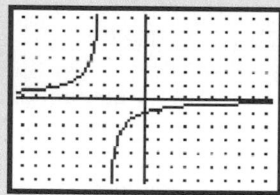

5a. decreasing; asymptotes at $x = 0$ and $y = 0$; domain: $x \neq 0$; range: $y \neq 0$

5b. decreasing; asymptotes at $x = 5$ and $y = -2$; domain: $x \neq 5$; range: $y \neq -2$

5c. decreasing; asymptotes at $x = 0$ and $y = 3$; domain: $x \neq 0$; range: $y \neq 3$

5d. increasing; asymptotes at $x = -3$ and $y = 0$; domain: $x \neq -3$; range: $y \neq 0$

7a. $y = f(-x)$ or $y = \frac{1}{-x}$

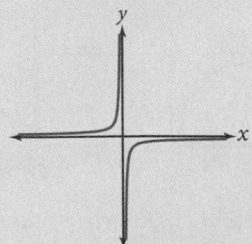

7b. $y = -f(x)$ or $y = -\frac{1}{x}$ (See sketch above.)

7c. Either reflection produces the same graph because $\frac{1}{-x} = -\frac{1}{x}$.

Exercise 8 Many answers are possible to the last three parts. For 8b and 8c, any dilation of the translated graph will have the same asymptotes, as will any reflection across one of those asymptotes. So the number 1 in the numerator may be replaced by some other number. For 8d, multiplying numerator and denominator by $(x + 1)$ will give the desired hole, even if the multiplication is done on the equation for a transformation of $y = \frac{1}{x}$.

8c. Possible answer:
$y = \frac{1}{(x - 2)(x - 3)(x + 4)} + 2$

9. Let x represent the amount of water to add and y represent the concentration of salt. The amount of salt is $0.05(0.5)$.
$y = \frac{0.025}{0.5 + x}$; $0.01 = \frac{0.025}{(0.5 + x)}$;
$x = 2$; 2 liters.

Exercise 11 Some students may use $y = \frac{3500}{x}$. This is okay if they subtract 5 from the value of x that they will get in 11d.

11c. $y = \frac{3500}{5 + x}$, where x is the number of additional businesses that have signed up and y is the cost per business.

Exercise 12 In physics, intensity is the same as brightness. You may want to define this as an "inverse square function."

---

7. Consider the graph of the inverse variation function $f(x) = \frac{1}{x}$. (See page 513.)
   a. Write an equation that reflects the graph across the x-axis. Sketch the image.
   b. Write an equation that reflects the graph across the y-axis. Sketch the image.
   c. Compare your sketches from 7a and 7b. Explain what you find.

8. Write equations for the graphs that meet the descriptions in 8a–d.
   a. A vertical stretch of the graph of $y = \frac{1}{x}$ by a factor of 5. Then a translation right 10 units and down 100 units. $y = \frac{5}{x - 10} - 100$
   b. A transformation of the graph of $y = \frac{1}{x}$ that has asymptotes at $x = -3$ and $y = 0$. The graph shows an increasing function. **Possible answer:** $y = \frac{-1}{x + 3}$
   c. A rational function that has asymptotes at $x = 2$, $x = 3$, $x = -4$, and $y = 2$.
   d. Looks like $y = \frac{1}{x}$ but has a hole where $x = 1$. $y = \frac{(x - 1)}{x(x - 1)}$

9. APPLICATION A nurse needs to treat a patient's eye with a 1% saline solution (salt solution). She finds only a half-liter bottle of 5% saline solution. Write an equation and use it to calculate how much water she should add to create a 1% solution.

10. Solve this equation symbolically.
$$-95 = \frac{5}{x - 10} - 100$$
Check your solution using a calculator graph or table.

11. APPLICATION A business group wants to rent a meeting hall for its job fair during the week of spring break. The rent is $3500, which will be divided among the businesses that agree to participate. So far, only five businesses have signed up.
   a. At this time, what is the cost for each business? $700
   b. Make a table to show what happens to the cost per business as the additional businesses agree to participate.
   c. Write a function for the cost per business related to the number of additional businesses that agree to participate.
   d. How many additional businesses must agree to participate before the cost per business is less than $150? $150 = \frac{3500}{5 + x}$; $x = 18.\overline{3}$; 19 additional businesses.

The saline solution that is used to clean contact lenses is usually a 1% salt solution.

Additional businesses	0	5	10	15	20
Cost per business ($)	700.00	350.00	233.33	175.00	140.00

12. The intensity, I, of a 100-watt light bulb is related to the distance, d, from which it is measured. This rational function shows the relationship when intensity is measured in lux (lumens per square meter) and distance is measured in meters.
$$I = \frac{90}{d^2}$$
   a. Find the intensity of the light 4 meters from the bulb. 5.625 lumens
   b. Find the distance from the bulb if the intensity of the light measures 20 lux. approximately 2.12 meters

10.
$-95 = \frac{5}{x - 10} - 100$	Original equation.
$5 = \frac{5}{x - 10}$	Add 100 to both sides.
$5(x - 10) = 5$	Multiply both sides by $(x - 10)$.
$x - 10 = 1$	Divide both sides by 5.
$x = 11$	Add 10 to both sides.

Solution checking will vary. This graph shows the solution as the intersection of $y = -95$ and $y = \frac{5}{x - 10} - 100$.

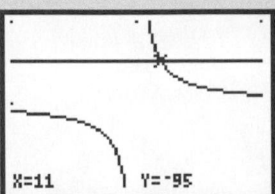
X=11   Y=-95

► **Review**

13. Solve each inequality.
    a. $4 - 2x > 8$ $x < -2$
    b. $-8 + 3(x - 2) \geq -20$ $x \geq -2$
    c. $7 + 2x \leq 3 + 3x$ $x \geq 4$

14. Name the coordinates of the vertex of the graph of $y = 2(x - 3)^2 + 1$. Without graphing, name the points on the parabola whose x-coordinates are 1 unit more or less than that of the vertex. Check your answers by graphing.
    vertex: $(3, 1)$; point to the left: $(2, 3)$; point to the right: $(4, 3)$

15. Name the coordinates of the vertex of the graph of $y = -3|x + 1| + 2$. Without graphing, name the points on the graph whose x-coordinates are 1 unit more or less than that of the vertex. Check your answers by graphing.
    vertex: $(-1, 2)$; point to the left: $(-2, -1)$; point to the right: $(0, -1)$

IMPROVING YOUR **VISUAL THINKING** SKILLS

Describe each striped or plaid fabric pattern as a set of transformations. Which patterns are translations? Which are reflections?

Fabric A

Fabric B

Fabric C

Fabric D
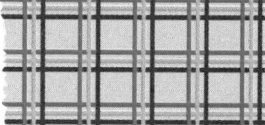

What is the smallest rectangular "unit" that repeats throughout each pattern? Can there be more than one "unit" for a pattern? Suppose a tailor is making a shirt from each fabric pattern. Which shirt should be most expensive? Why?

Exercise 13 This exercise reviews Lesson 6.5.

Exercises 14 and 15 As needed, help students realize that the vertex of $y = (x - h)^2 + k$ or $y = |x - h| + k$ is located at (h, k), an idea informally explored in Lesson 9.2 that will be more formally stated in Chapter 10. As students find the y-coordinates of points whose x-coordinate differs from that of the vertex by 1, they may see that the amount the graph has risen in that unit corresponds to the vertical stretch factor. You might ask why. Encourage students to see that in the original graph the amount of increase over this interval was just 1. These exercises review and extend Lessons 9.2 and 9.4.

IMPROVING **VISUAL THINKING** SKILLS

The stripes in Fabric A exhibit a 123123 . . . pattern. The smallest unit that repeats (through translation) consists of three adjacent stripes.

Fabric B has a 123212321 . . . pattern. You can think of it as generated from reflections, or from the unit of four adjacent stripes repeatedly translated.

Fabric C has the same pattern along its length as Fabric B, as well as a 2121 . . . pattern across its width. You can interpret the pattern as generated by the same transformations as Fabric B or as repeatedly reflecting horizontally.

Fabric D has the same pattern along its length and across its width as Fabric A,

generated by translations. There is no reflection.

It's not possible to cut symmetrical pieces like shirt fronts from a folded piece of Fabric D. Hence there would be more fabric waste as pieces are individually positioned, and more time would be required.

Transformations with Matrices

Say what you know, do what you must, come what may.

SOFIA KOVALEVSKAYA

PLANNING

LESSON OUTLINE

One day:

5 min	Introduction
20 min	Investigation
10 min	Sharing
5 min	Closing
10 min	Exercises

MATERIALS

- graph paper
- Calculator Note 1P

TEACHING

A transformation of points can be represented by putting the coordinates of those points into a matrix and operating on it with another matrix.

One step Show students how the matrix in the introduction represents a quadrilateral. Ask how they might operate on that matrix with another one to represent the quadrilateral after a translation 3 units to the right and 2 units downward. Refer them to Lesson 1.8 as needed. After they've figured out the matrix addition, ask how they might operate on the matrix to achieve reflections and dilations.

INTRODUCTION

In representing the vertices of a polygon, it does not matter which vertex is in the first column or whether vertices are listed clockwise or counterclockwise, as long as the coordinates are listed consecutively.

[ESL] *In consecutive order* means traveling along edges of the polygon listing vertices as you come to them.

You can use a matrix to organize the coordinates of a geometric figure. You can represent this quadrilateral with a 2×4 matrix.

$$\begin{bmatrix} 1 & -2 & -3 & 2 \\ 2 & 1 & -1 & -2 \end{bmatrix}$$

Each column contains the x- and y-coordinates of a vertex. The first row contains all the x-coordinates, and the second row contains all the y-coordinates. All four vertices are in consecutive order in the matrix. When you add or multiply this matrix, the coordinates change. So matrices are useful when you transform coordinates.

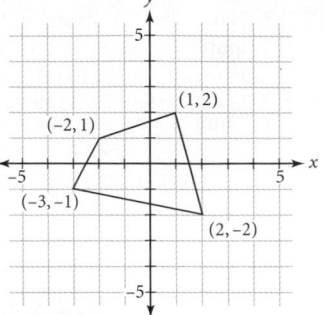

Many textiles, such as this Turkish carpet, use transformations to create interesting patterns.

Investigation
Matrix Transformations

You will need
- graph paper

In this investigation you'll use matrix addition and multiplication to create some familiar transformations.

Step 1 Create a set of coordinate axes on graph paper. Draw the triangle that is represented by this matrix:

$$[A] = \begin{bmatrix} -4 & 3 & 2 \\ -1 & 4 & 0 \end{bmatrix}$$

Step 2 $\begin{bmatrix} 1 & 8 & 7 \\ -1 & 4 & 0 \end{bmatrix}$ **Step 2** Add.

$$[A] + \begin{bmatrix} 5 & 5 & 5 \\ 0 & 0 & 0 \end{bmatrix}$$

LESSON OBJECTIVES

- Review matrix addition and multiplication
- Use matrices to represent transformations of polygons
- Solve real-world problems using matrix models

NCTM STANDARDS

CONTENT		PROCESS	
•	Number	•	Problem Solving
•	Algebra	•	Reasoning
	Geometry		Communication
	Measurement	•	Connections
	Data/Probability	•	Representation

Step 3 | Draw the image represented by your answer in Step 2. Describe the transformation.

Step 4 Possible answer: The transformation adds 5 to each x-coordinate, and the matrix that you added has all 5's in the x-coordinate row.

Step 4 | How is the transformation related to the matrix that you added?

Step 5 | Repeat Steps 1–4, but in Step 2 change what you add to matrix [A] each time. Use a new set of coordinate axes for each transformation.

a. $[A] + \begin{bmatrix} 0 & 0 & 0 \\ -4 & -4 & -4 \end{bmatrix}$

b. $[A] + \begin{bmatrix} 5 & 5 & 5 \\ -4 & -4 & -4 \end{bmatrix}$

c. $[A] + \begin{bmatrix} -6 & -6 & -6 \\ 4 & 4 & 4 \end{bmatrix}$

Next, see if you can work backward.

Step 6 | Write matrix equations to represent these translations.

Step 6a
$\begin{bmatrix} 0 & 0 & 1 & 1 \\ 0 & 1 & 1 & 0 \end{bmatrix}$
$+ \begin{bmatrix} 2 & 2 & 2 & 2 \\ 1.5 & 1.5 & 1.5 & 1.5 \end{bmatrix}$
$= \begin{bmatrix} 2 & 2 & 3 & 3 \\ 1.5 & 2.5 & 2.5 & 1.5 \end{bmatrix}$

a.

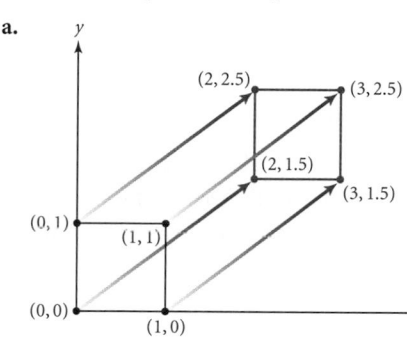

b.

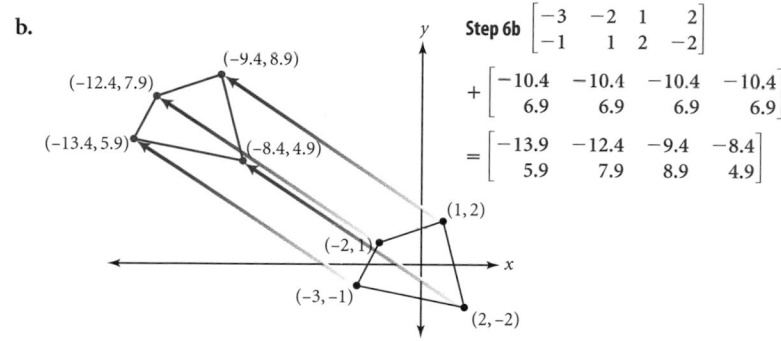

Step 6b
$\begin{bmatrix} -3 & -2 & 1 & 2 \\ -1 & 1 & 2 & -2 \end{bmatrix}$
$+ \begin{bmatrix} -10.4 & -10.4 & -10.4 & -10.4 \\ 6.9 & 6.9 & 6.9 & 6.9 \end{bmatrix}$
$= \begin{bmatrix} -13.9 & -12.4 & -9.4 & -8.4 \\ 5.9 & 7.9 & 8.9 & 4.9 \end{bmatrix}$

Now, see what effect multiplication has.

Step 7 The graph should look like the one shown in the book. Possible answer:
$\begin{bmatrix} 2 & 3 & 6 & 7 & x \\ 2 & 4 & 5 & 1 & y \end{bmatrix}$

Step 7 | Draw this quadrilateral on your own graph paper. Write the coordinates of the vertices in a matrix, [B]. Add a fifth column to your matrix to represent any point of the form (x, y).

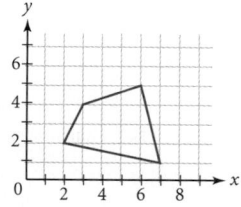

Step 8 | Multiply.

$\begin{bmatrix} 1 & 0 \\ 0 & -1 \end{bmatrix} \cdot [B] \begin{bmatrix} 2 & 3 & 6 & 7 & x \\ -2 & -4 & -5 & -1 & y \end{bmatrix}$

Step 5b $\begin{bmatrix} 1 & 8 & 7 \\ -5 & 0 & -4 \end{bmatrix}$; a translation right 5 units and down 4 units

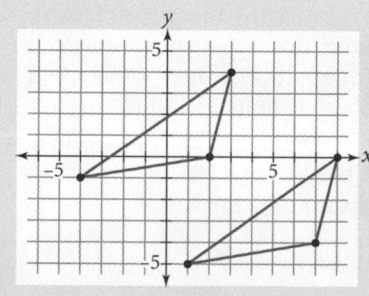

Step 5c $\begin{bmatrix} -10 & -3 & -4 \\ 3 & 8 & 4 \end{bmatrix}$; a translation left 6 units and up 4 units

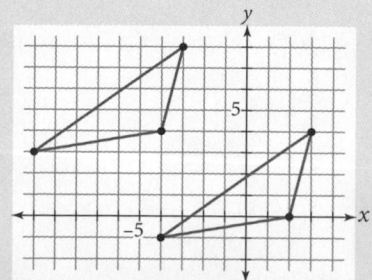

Guiding the Investigation

Steps 1 to 5 Some students might want to see how to do these matrix calculations on a calculator. They could also use the calculators to draw the original and transformed figures. You might challenge some students to use matrices in calculator programs that move figures across the screen, as in the project of Lesson 9.1.

Step 1

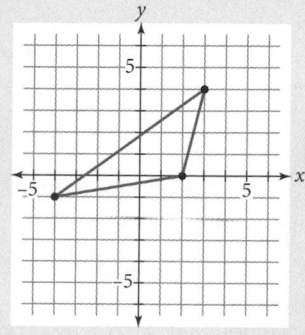

Step 3 a translation right 5 units

Step 5a $\begin{bmatrix} -4 & 3 & 2 \\ -5 & 0 & -4 \end{bmatrix}$; a translation down 4 units

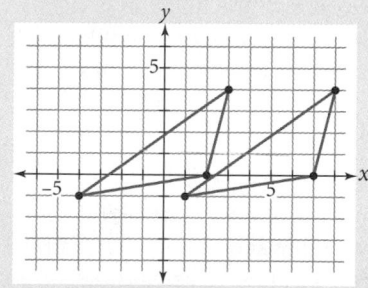

Steps 7 and 8 Some students may not be secure about having letters (x and y) in matrices. Suggest that they just go ahead and see what happens.

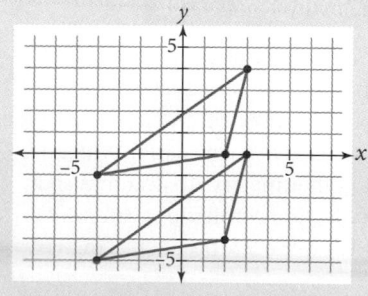

SHARING IDEAS

Have students report their results of Steps 6 and 11.

[Ask] "Does it matter in what order the matrices are multiplied?" [A 2 × 3 matrix representing a triangle must be multiplied by a 2 × 2 matrix on the left.]

"What transformation matrix would produce a reflection over both the *x*- and *y*-axes?" [It would be $\begin{bmatrix} -1 & 0 \\ 0 & -1 \end{bmatrix}$.] This double reflection accomplishes a rotation through 180 degrees.

The opening quotation by Sofia Kovalevskaya (1850–1891), a Russian mathematician, was the motto on her paper "On the Problem of the Rotation of a Solid Body about a Fixed Point." In Exercise 9, students will see rotations through angles other than 180 degrees.

Assessing Progress

You can assess students' abilities at adding and multiplying matrices and at recognizing **translations**, **reflections**, and **dilations** of graphs.

Closing the Lesson

You can represent a polygon by putting the coordinates of its vertices into the columns of a matrix. To achieve a translation, you add another matrix. To accomplish a reflection or dilation, you multiply on the left by a 2 × 2 matrix.

BUILDING UNDERSTANDING

Students work with matrices representing transformations.

See pages 721–722 for answers to Steps 9 and 11a–11c and Exercise 1a.

Step 9 | Draw the image represented by your answer in Step 8. Describe the transformation that happened. **a reflection across the *x*-axis**

Step 10 Possible answer: When $L_4 = -L_2$, all of the *y*-coordinates were negated, creating a reflection across the *x*-axis.

Step 10 | How is the last column of the image matrix related to the transformations you made using lists in this chapter?

Step 11 | Repeat Steps 7–10, but in Step 8 change what you multiply by matrix [*B*] each time. Use a new set of coordinate axes for each transformation.

a. $\begin{bmatrix} -1 & 0 \\ 0 & 1 \end{bmatrix} \cdot [B]$ b. $\begin{bmatrix} 1 & 0 \\ 0 & 0.5 \end{bmatrix} \cdot [B]$ c. $\begin{bmatrix} 0.5 & 0 \\ 0 & 2 \end{bmatrix} \cdot [B]$

EXERCISES

You will need your calculator for problem **5**.

▶ Practice Your Skills

1. The matrix $\begin{bmatrix} -2 & 1 & -2 \\ 2 & 2 & 6 \end{bmatrix}$ represents a triangle.

   a. Name the coordinates and draw the triangle. $(-2, 2), (1, 2), (-2, 6)$

   b. What matrix would you add to translate the triangle down 3 units? $\begin{bmatrix} 0 & 0 & 0 \\ -3 & -3 & -3 \end{bmatrix}$

   c. Calculate the matrix for the image if you translate the triangle down 3 units. $\begin{bmatrix} -2 & 1 & -2 \\ -1 & -1 & 3 \end{bmatrix}$

2. Refer to these triangles.

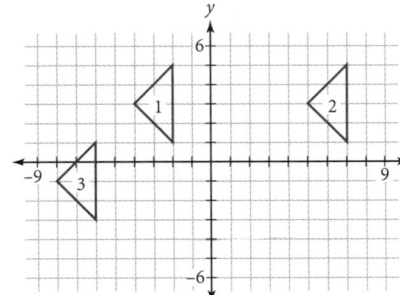

2b. $\begin{bmatrix} -4 & -2 & -2 \\ 3 & 5 & 1 \end{bmatrix} + \begin{bmatrix} 9 & 9 & 9 \\ 0 & 0 & 0 \end{bmatrix} = \begin{bmatrix} 5 & 7 & 7 \\ 3 & 5 & 1 \end{bmatrix}$

2c. $\begin{bmatrix} -4 & -2 & -2 \\ 3 & 5 & 1 \end{bmatrix} + \begin{bmatrix} -4 & -4 & -4 \\ -4 & -4 & -4 \end{bmatrix} = \begin{bmatrix} -8 & -6 & -6 \\ -1 & 1 & -3 \end{bmatrix}$

   a. Write a matrix to represent triangle 1. Possible answer: $\begin{bmatrix} -4 & -2 & -2 \\ 3 & 5 & 1 \end{bmatrix}$

   b. Write the matrix equation to translate from triangle 1 to triangle 2.

   c. Write the matrix equation to translate from triangle 1 to triangle 3.

3. Add or multiply.

   a. $[4 \ \ 7] + [2 \ \ 8]$
   $[6 \ \ 15]$

   b. $\begin{bmatrix} 4 \\ 7 \end{bmatrix} + \begin{bmatrix} 2 \\ 8 \end{bmatrix}$ $\begin{bmatrix} 6 \\ 15 \end{bmatrix}$

   c. $[4 \ \ 7] \cdot \begin{bmatrix} 2 \\ 8 \end{bmatrix}$ [64]

   d. $\begin{bmatrix} 4 \\ 7 \end{bmatrix} \cdot [2 \ \ 8]$ $\begin{bmatrix} 8 & 32 \\ 14 & 56 \end{bmatrix}$

4. In the investigation you saw that matrix multiplication can result in a reflection.

   a. What matrix reflects a figure across the *y*-axis? $\begin{bmatrix} -1 & 0 \\ 0 & 1 \end{bmatrix}$

   b. What matrix reflects a figure across the *x*-axis? $\begin{bmatrix} 1 & 0 \\ 0 & -1 \end{bmatrix}$

ASSIGNING HOMEWORK

Essential	1, 5–7
Performance assessment	4, 5
Portfolio	8
Journal	6, 7
Group	2, 9
Review	3, 10, 11

▶ Helping with the Exercises

Exercise 3 Under the idea that calculators are best used when the focus of an exercise is on solving a problem rather than on learning skills, calculators would not be appropriate here. In fact, you might ask students to write out their calculations in each entry of the solution matrix. In 3d, students may not realize that the product of a 2 × 1 matrix and a 1 × 2 matrix is a 2 × 2 matrix.

▶ Reason and Apply

5. The matrix $\begin{bmatrix} -1 & 2 & 1 & -2 \\ 2 & -1 & -2 & 1 \end{bmatrix}$ represents a quadrilateral.

  a. What kind of quadrilateral is it? a rectangle

  b. Without using your calculator, tell how to find the image of the point $(2, -1)$ when you multiply $\begin{bmatrix} 1 & 0 \\ 2 & 2 \end{bmatrix} \cdot \begin{bmatrix} -1 & 2 & 1 & -2 \\ 2 & -1 & 2 & 1 \end{bmatrix}$. What are the coordinates of this point's image? In what row and column of the image matrix will you find the new x-coordinate? The new y-coordinate?

  c. Multiply $\begin{bmatrix} 1 & 0 \\ 0 & 2 \end{bmatrix} \cdot \begin{bmatrix} -1 & 2 & 1 & -2 \\ 2 & -1 & 2 & 1 \end{bmatrix}$. Check your work with your calculator.

  [▶ 🖵 See **Calculator Note 1P** to review matrix multiplication. ◀] $\begin{bmatrix} -1 & 2 & 1 & -2 \\ 4 & -2 & -4 & 2 \end{bmatrix}$

  d. Draw the image represented by your answer to 5c. What kind of polygon is it? a parallelogram

6. Consider this square.

  a. Write a matrix, $[S]$, to represent it. $[S] = \begin{bmatrix} 0 & 1 & 1 & 0 \\ 0 & 0 & 1 & 1 \end{bmatrix}$

  b. Describe the transformation when you calculate

  i. $\begin{bmatrix} 1 & 0 \\ 0 & 3 \end{bmatrix} \cdot [S]$ **ii.** $\begin{bmatrix} 1 & 0 \\ 0 & -3 \end{bmatrix} \cdot [S]$

  iii. $\begin{bmatrix} 4 & 0 \\ 0 & 2 \end{bmatrix} \cdot [S]$ **iv.** $[S] + \begin{bmatrix} 4 & 4 & 4 & 4 \\ 2 & 2 & 2 & 2 \end{bmatrix}$

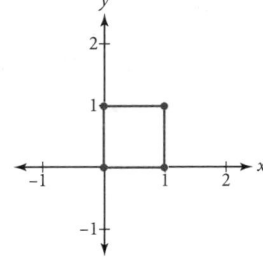

7. Consider this quadrilateral.

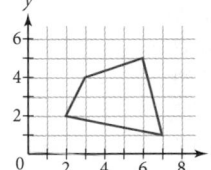

  7b. $\begin{bmatrix} 1 & 0 \\ 0 & 0.5 \end{bmatrix} \cdot [Q] = \begin{bmatrix} 2 & 3 & 6 & 7 \\ 1 & 2 & 2.5 & 0.5 \end{bmatrix}$

  a. Write a matrix, $[Q]$, to represent it. Possible answer: $[Q] = \begin{bmatrix} 2 & 3 & 6 & 7 \\ 2 & 4 & 5 & 1 \end{bmatrix}$

  b. Write a matrix multiplication equation that will vertically shrink the quadrilateral by a factor of 0.5.

  c. Write a matrix multiplication equation that will both vertically and horizontally shrink the quadrilateral by a factor of 0.5. $\begin{bmatrix} 0.5 & 0 \\ 0 & 0.5 \end{bmatrix} \cdot [Q] = \begin{bmatrix} 1 & 1.5 & 3 & 3.5 \\ 1 & 2 & 2.5 & 0.5 \end{bmatrix}$

  d. Multiply $\begin{bmatrix} 1 & 0 \\ 0 & -1 \end{bmatrix} \cdot \begin{bmatrix} -1 & 0 \\ 0 & 1 \end{bmatrix}$. Then multiply the result by matrix $[Q]$. Draw the image of the quadrilateral. Describe the transformation that happened. How is the transformation related to the matrices $\begin{bmatrix} 1 & 0 \\ 0 & -1 \end{bmatrix}$ and $\begin{bmatrix} -1 & 0 \\ 0 & 1 \end{bmatrix}$?

  e. Multiply $\begin{bmatrix} 0 & -1 \\ 1 & 0 \end{bmatrix} \cdot [Q]$. Draw the image. Describe the transformation. $\begin{bmatrix} -2 & -4 & -5 & -1 \\ 2 & 3 & 6 & 7 \end{bmatrix}$; a quarter-turn counterclockwise

7d. $\begin{bmatrix} -1 & 0 \\ 0 & -1 \end{bmatrix}; \begin{bmatrix} -1 & 0 \\ 0 & -1 \end{bmatrix} \cdot [Q]$

$= \begin{bmatrix} -2 & -3 & -6 & -7 \\ -2 & -4 & -5 & -1 \end{bmatrix};$

a reflection across both the x- and y-axes; alone, $\begin{bmatrix} 1 & 0 \\ 0 & -1 \end{bmatrix}$ results in a reflection across the x-axis and

$\begin{bmatrix} -1 & 0 \\ 0 & 1 \end{bmatrix}$ results in a reflection across the y-axis.

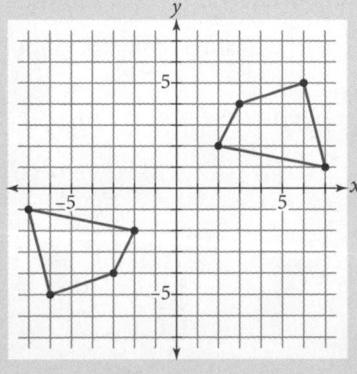

5b. Possible answer: For the x-coordinate, multiply row 1 of the transformation matrix by column 2 of the quadrilateral matrix; $\begin{bmatrix} 1 & 0 \end{bmatrix} \cdot \begin{bmatrix} 2 \\ -1 \end{bmatrix} = 2$; this goes in row 1, column 2, of the image matrix. For the y-coordinate, multiply row 2 of the transformation matrix by column 2 of the quadrilateral matrix; $\begin{bmatrix} 0 & 2 \end{bmatrix} \cdot \begin{bmatrix} 2 \\ -1 \end{bmatrix} = -2$; this goes in row 2, column 2, of the image matrix.

6b. **i.** a vertical stretch by a factor of 3; **ii.** a vertical stretch by a factor of 3 and a reflection across the x-axis; **iii.** a horizontal stretch by a factor of 4 and a vertical stretch by a factor of 2; **iv.** a translation right 4 units and up 2 units

Exercise 7 [Ask] "How is the image in 7c related to the original quadrilateral?" [It will be geometrically similar by a scale factor of 0.5.]

Some students may describe the transformation in 7d as a "half-turn." You might want to take this opportunity to define *rotation*. This is a rotation through 180° about the origin.

You may encourage students to describe the transformation in 7e as a counterclockwise rotation through 90° about the origin. Ask if it, like the 180° rotation, is equivalent to the composite of two reflections. [It can be accomplished by reflecting across the line $y = x$ and then the y-axis.]

7e.

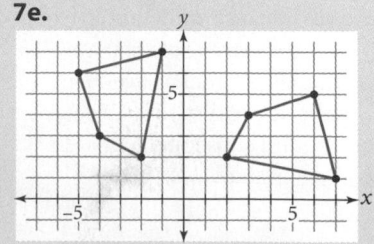

See page 722 for graph to Exercise 5d.

See pages 722–723 for answers to
Exercises 9a–9f, 10a, and 11b.

Exercise 8 **[Ask]** "How are the 3's and 2's in the transformation matrix in 8c related to the transformation and the equation of the image?" [A translation to the right of 3 and a translation up 2.] In the equation, the 3 is subtracted from x before it is squared; the 2 is the constant term.

In part 8d, suggest that students draw the graph of the transformed equation. **[Ask]** "Why does it look like the original graph, even though the image matrix is different from the original matrix?" [The graph of $y = (-x)^2$ is visually identical to that of $y = x^2$.]

Exercise 9 This exercise will challenge many students on different levels. The concept of rotation might be new, and students may have difficulty determining the 60° rotation. Prior experience with angle measurement is useful. It may help some students to ask, "What fraction of a circle was the polygon rotated through?" Students might benefit from using a polygon that has one edge on the x-axis, as in the sample answers, and then determining the rotation of this one edge. **[Alert]** Students may also have trouble accurately calculating with the decimals in this problem. Students' graphing abilities will be tested by having to identify the location of points with decimal coordinates. For example, "How should one approximate $(-0.232, 3.598)$?"

In the transformation matrix, 0.866 is an approximation of $\frac{\sqrt{3}}{2}$, derived from a 30-60-90 triangle with a hypotenuse of 1 unit. Because students have not yet seen square roots that are irrational numbers, have them use the approximation 0.866 and round subsequent answers to 3 decimal places.

---

8. The points $(-2, 4)$, $(-1, 1)$, $(0, 0)$, $(1, 1)$, and $(2, 4)$ are on a parabola.
   a. What is the equation of the parabola that passes through these points? $y = x^2$
   b. Write a matrix, $[P]$, to organize the coordinates of these points. $[P] = \begin{bmatrix} -2 & -1 & 0 & 1 & 2 \\ 4 & 1 & 0 & 1 & 4 \end{bmatrix}$
   c. Add $[P] + \begin{bmatrix} 3 & 3 & 3 & 3 & 3 \\ 2 & 2 & 2 & 2 & 2 \end{bmatrix}$. Write an equation for the parabola that passes through the points organized in the image matrix. $\begin{bmatrix} 1 & 2 & 3 & 4 & 5 \\ 6 & 3 & 2 & 3 & 6 \end{bmatrix}$; $y = (x-3)^2 + 2$
   d. Multiply $\begin{bmatrix} -1 & 0 \\ 0 & 1 \end{bmatrix} \cdot [P]$. Write an equation for the parabola that passes through the points organized in the image matrix. $\begin{bmatrix} 2 & 1 & 0 & -1 & -2 \\ 4 & 1 & 0 & 1 & 4 \end{bmatrix}$; $y = (-x)^2$

9. In problem 7e, you saw a transformation called a **rotation**. A rotation turns a figure about a point called the *center*. The center of a rotation can be inside, outside, or on the figure that is rotated.
   a. Draw a polygon of your own design on graph paper. Represent your polygon with a matrix, $[R]$.
   b. Multiply $\begin{bmatrix} 0.5 & -0.866 \\ 0.866 & 0.5 \end{bmatrix} \cdot [R]$.
      Draw the image of your polygon.
   c. The transformation matrix in 9b rotates the polygon. How many degrees is it rotated? What point is the center of the rotation?
   d. Multiply $\begin{bmatrix} 0.5 & -0.866 \\ 0.866 & 0.5 \end{bmatrix} \cdot \begin{bmatrix} 0.5 & -0.866 \\ 0.866 & 0.5 \end{bmatrix}$ and round the entries in the answer matrix to the nearest thousandth. Then multiply the result by matrix $[R]$. Draw the image of the polygon. Describe the transformation.
   e. Describe how you could rotate your polygon 180° (a half-turn).
   f. Describe how you could rotate your polygon 360° so that the image is the same as the original polygon.

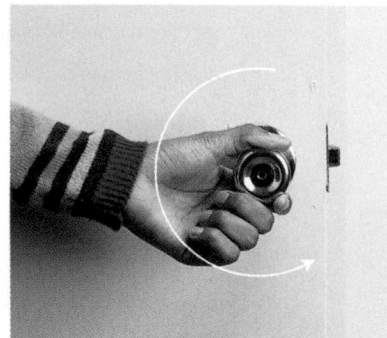

Did you know that you use rotations every day? Opening a door, turning a faucet, tightening a bolt with a wrench—all of these require rotations. The rotational force you use to do these things is called *torque*. Think of other everyday situations that require rotations.

▶ **Review**

10. Tacoma and Jared are doing a "walker" investigation. Tacoma starts 2 m from the motion sensor. He walks away at a rate of 0.5 m/sec for 6 sec. Then he walks back toward the sensor at a rate of 0.5 m/sec for 3 sec.
    a. Sketch a time-distance graph for Tacoma's walk.
    b. Write an equation that fits the graph.
       $y = 5 - 0.5|x - 6|$
11. This table shows the approximate population of the ten most populated countries in 1999.
    a. Give the five-number summary. in millions: 114, 127, 159, 273, 1247
    b. Make a box plot of the data.
    c. Are there any outliers? China and India

Most Populated Countries, 1999

Country	Population (millions)
China	1,247
India	1,001
United States	273
Indonesia	216
Brazil	172
Russia	146
Pakistan	138
Bangladesh	127
Japan	126
Nigeria	114

(2000 World Almanac, pp. 878–879)

For 9e, some students may recall from 7d that $\begin{bmatrix} -1 & 0 \\ 0 & -1 \end{bmatrix} \cdot [R]$ will also accomplish a rotation through 180°.

Exercise 10 This exercise reviews Lesson 9.4.

Exercise 11 This exercise reviews Lesson 1.3.

project

TILES

THE GEOMETER'S SKETCHPAD

The Geometer's Sketchpad was used to create these tiles. Sketchpad has tools to help you create simple polygons and apply transformations. Learn how to use these tools to create your own tiling pattern.

Some floor tiles are simple polygons, like squares. Others have more complex shapes, with curves or unusual angles. But all tiles have one thing in common—they fit together without gaps or overlap.

You can use transformations to create your own tile shape. Start with a polygon that works as a tile. For example, you can start with a rhombus and use transformations to create a complex shape that still works as a tile. In this example, a design drawn on the right side of the rhombus is translated and copied on the left side; a translation is also used for the top and bottom. The result is an interesting shape that still fits together.

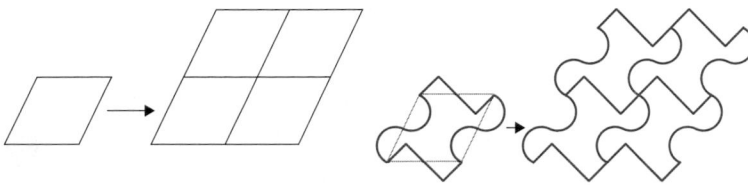

Your project should include

▶ Your tile pattern. Show what a single tile looks like and how several tiles look when they are joined together.

▶ A report of how you created your tile. What polygon did you start with? What transformations did you use?

For an extra challenge, start with a polygon that is not a quadrilateral, like a triangle. Or try other transformations or combinations of transformations. Dot paper, graph paper, a computer drawing program, or The Geometer's Sketchpad software are useful tools for this project.

Learn more about the mathematics of tilings with the Internet links at
www.keymath.com/DA .

Supporting the project

MOTIVATION

Tessellation is the mathematical term for a tiling or a collection of shapes that tile the plane. *Tessellation* comes from *tessera*, the Latin word for tile. Students will have more extensive exposure to creating tessellations with transformations in *Discovering Geometry*.

OUTCOMES

▶ Project includes the required elements and uses correct vocabulary.
• A tile is created from a polygon other than a quadrilateral. (Possibilities: all triangles, hexagons with opposite sides parallel, or a combination of shapes, such as a regular octagon and a square).
• A tile is created from transformations other than translations. (In the example shown in the student text, each side is created by a rotation about the side's midpoint.)
• The student does extensive investigation into which polygons tile the plane and can explain why certain polygons work.
• The student creates a tile that looks like an object, as in the art of M. C. Escher.

PLANNING

LESSON OUTLINE

One day:

5 min   Introduction

15 min   Exercises and helping individuals

15 min   Checking work and helping individuals

15 min   Student self assessment

REVIEWING

Refer students to Exercise 9 of Lesson 9.3, in which Bo is moving a star by translations and reflections. Work through the problem as given. Then add the condition that the star is to dilate vertically by a factor of 0.8 in each frame. Change it to a vertical dilation by a factor of 1.3 in each frame. Then begin with the star centered at the origin and move it vertically the same distance, flipping and dilating horizontally. If you've done Lesson 9.7, review how to accomplish each of these transformations with a matrix. Finally, begin with the graph of $y = |x|$ or $y = \frac{1}{x}$ and write equations for moving it horizontally or vertically 11 units in 11 frames, in each of which it flips and dilates. Depending on the amount of time you have, you or the students might demonstrate these moves through a calculator program.

ASSIGNING HOMEWORK

You might assign evens for homework and have students work alone or in groups on the odds in class. Assign Exercise 10 only if you are covering matrices.

I n this chapter you moved individual points, polygons, and graphs of functions with **transformations.** You learned to **translate, reflect, stretch,** and **shrink** a **parent function** to create a **family of functions** based on it. For example, if you know what the graph of $y = x^2$ looks like, understanding transformations gives you the power to know what the graph of $y = 3(x + 2)^2 - 4$ looks like.

You transformed the graphs of the parent functions $y = |x|$ and $y = x^2$ to create many different absolute value and squaring functions. You can apply the same transformations to the graphs of other parent functions, like $y = x$ or $y = 2^x$, to create many different linear or exponential functions. You can even fit an equation to data by transforming a simple graph into a graph that fits the data better.

You learned that the inverse variation function, $y = \frac{1}{x}$, is one type of **rational function.** The graphs of most rational functions have **asymptotes;** some even have holes. Understanding transformations helps you know where asymptotes and holes will occur.

Finally, you used matrices to organize the coordinates of points and to do transformations. You can use matrices to do translations, reflections, stretches, and shrinks. You can also use them to do more complex transformations, like **rotations.**

EXERCISES

You will need your calculator for problems **4** and **7.**

For problems 1 and 2, consider the black pentagon below as the original figure.

1. The image of the black pentagon after a transformation is shown in red.

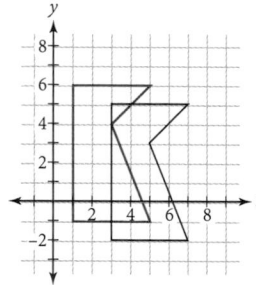

   a. Describe the transformation.   a translation left 2 units and up 1 unit

   b. Define the coordinates of the image using the coordinates of the original figure. $(x - 2, y + 1)$

2. Here are three more transformations of the black pentagon from problem 1.

i.

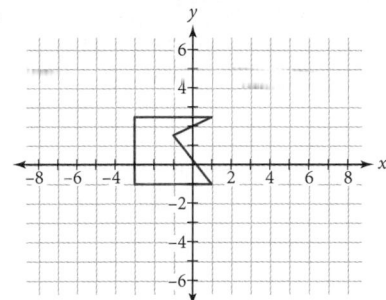

ii.

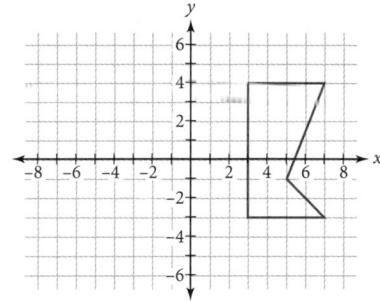

iii.

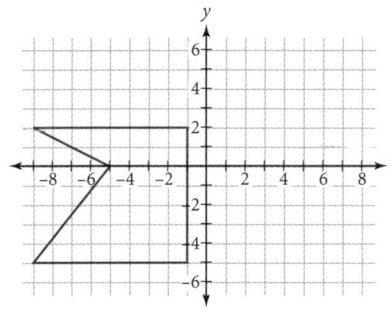

a. Describe the transformations.

b. Patty plots the original pentagon on her calculator. She uses list L_1 for the x-coordinates of the vertices and list L_2 for the y-coordinates. Tell Patty how to define list L_3 and list L_4 for each image shown above.

3. You can create this figure on a calculator by connecting four points. Assume the x-coordinates of each point are entered into list L_1 and the corresponding y-coordinates are entered into list L_2. Explain how to make an image that is

a. A reflection across the x-axis.

b. A reflection across the y-axis.

c. A reflection across the x-axis and a translation right 3 units.

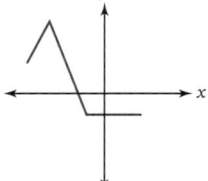

4. Describe each function as a transformation of the graph of the parent function $y = |x|$ or $y = x^2$. Then sketch a graph of each function. Check your answers by graphing on your calculator.

a. $y = 2|x| + 1$ **b.** $y = -|x + 2| + 2$
c. $y = 0.5(-x)^2 - 1$ **d.** $y = -(x - 2)^2 + 1$

5. At right, the graph of $g(x)$ is a transformation of the graph of $f(x)$. Write an equation for $g(x)$ in terms of $f(x)$. $g(x) = f(x - 1) + 2$

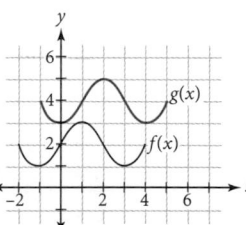

4c. Possible answer: a vertical shrink of the graph of $y = x^2$ by a factor of 0.5; then a reflection across the y-axis; then a translation down 1 unit

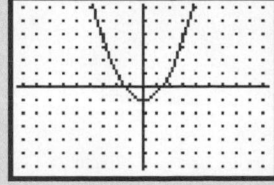

4d. Possible answer: a reflection of the graph of $y = x^2$ across the x-axis; then a translation right 2 units and up 1 unit

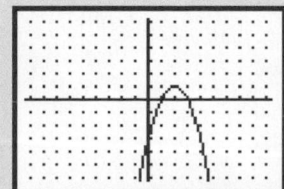

2a. i. a vertical shrink by a factor of 0.5 and a translation left 6 units

2a. ii. Possible answer: a reflection across the x-axis; then a translation up 2 units

2a. iii. Possible answer: a horizontal stretch by a factor of 2 and a reflection across the y-axis; then a translation right 5 units and down 3 units

2b. i. $L_3 = L_1 - 6$; $L_4 = 0.5 \cdot L_2$

2b. ii. Possible answer: $L_3 = L_1$; $L_4 = -L_2 + 2$

2b. iii. Possible answer: $L_3 = -2 \cdot L_1 + 5$; $L_4 = L_2 - 3$

3. Answers will vary. For these possible answers, list L_3 and list L_4 are used for the x- and y-coordinates, respectively, of each image.

3a. $L_3 = L_1, L_4 = -L_2$

3b. $L_3 = -L_1, L_4 = L_2$

3c. $L_3 = L_1 + 3, L_4 = -L_2$

4a. a vertical stretch of the graph of $y = |x|$ by a factor of 2; then a translation up 1 unit

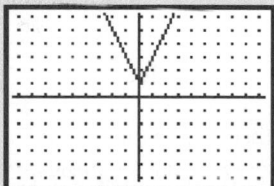

$[-9.4, 9.4, 1, -6.2, 6.2, 1]$

4b. a reflection of the graph of $y = |x|$ across the x-axis; then a translation left 2 units and up 2 units

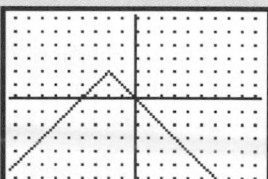

▶ Helping with the Exercises

6a. $y = -|x| + 3$

6b. $y = (x + 4)^2 - 2$

6c. $y = 0.5x^2 - 5$

6d. $y = -2|x - 3| + 1$

7a. The graph should have the same x-intercept as $f(x)$. The y-intercept should be the opposite of that for $f(x)$.

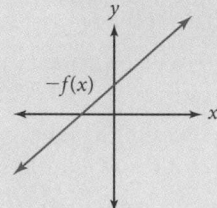

7b. Answers will vary. Possible answer for a friendly window with a factor of 1: If $Y_1 = -x - 2$, $Y_2 = -Y_1$ reflects the graph across the x-axis [because the calculator interprets $-Y_1$ as $-(-x - 2)$, or $x + 2$]; this supports the answer to 7a.

8a. a translation right 3 units; asymptotes: $x = 3, y = 0$

8b. a vertical stretch by a factor of 3 and then a translation left 2 units; asymptotes: $x = -2$, $y = 0$

8c. a translation right 5 units and down 2 units; asymptotes: $x = 5, y = -2$

Exercise 9 Parts b and c may be challenging, since exposure to exponential functions has been limited in this chapter. Help students recognize that transformations can be applied to any function. It may also help show that $y - 2^x$ has an asymptote at $y = 0$ and passes through $(0, 1)$. These facts can then be used to help determine what translations have occurred.

9a. a translation of $y = \frac{1}{x}$ right 3 units and down 2 units; $y = \frac{1}{x - 3} - 2$

9b. a translation of $y = 2^x$ right 4 units and down 2 units; $y = 2^{(x-4)} - 2$

528 CHAPTER 9 Transformations

6. Write the equation for each graph.

a.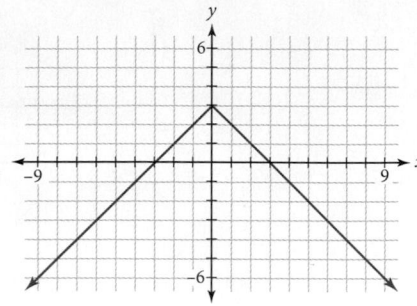

b.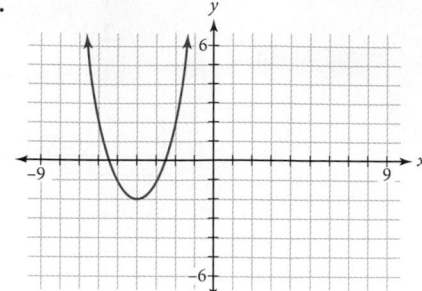

c.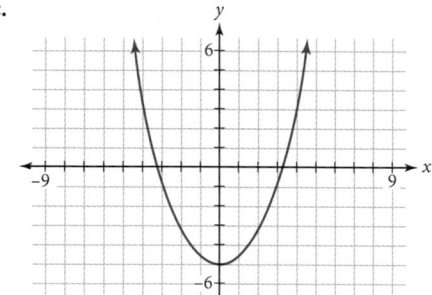

d.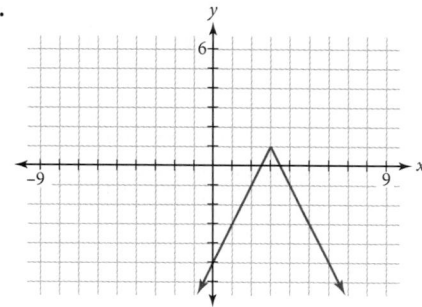

7. Consider the graph of $f(x)$ at right.
 a. Sketch the graph of $-f(x)$.
 b. Enter a linear function into Y_1 on your calculator to create a graph like $f(x)$. Enter $Y_2 = -Y_1$ and graph it too. Describe your results.

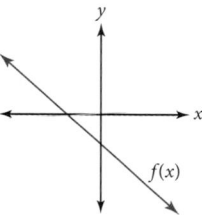

8. Describe each function as a transformation of the graph of the parent function $y = \frac{1}{x}$. Give equations for the asymptotes.

 a. $y = \frac{1}{x - 3}$ b. $y = \frac{3}{x + 2}$ c. $y = \frac{1}{x - 5} - 2$

9. Describe each graph as a transformation of the graph of the parent function $y = 2^x$ or $y = \frac{1}{x}$. Then write an equation for each graph.

a.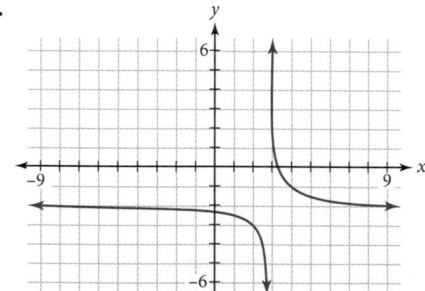

b.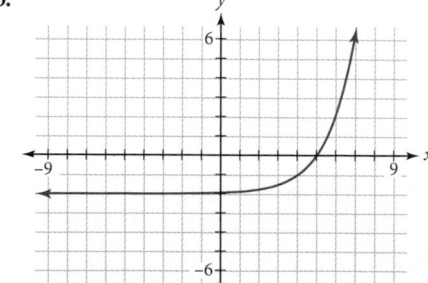

c.

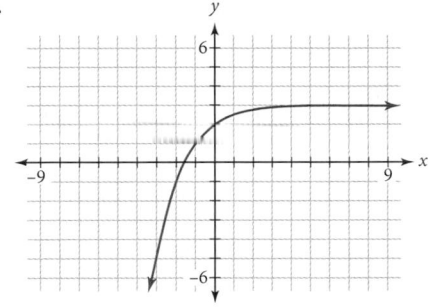

d.

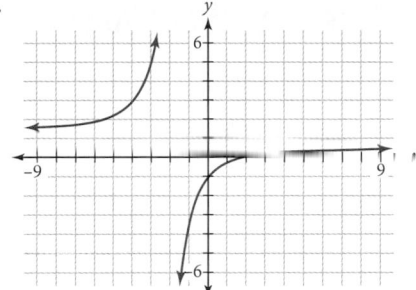

10. Consider this square.

a. Write a matrix, $[A]$, that represents this square.

b. Describe the transformation when you calculate

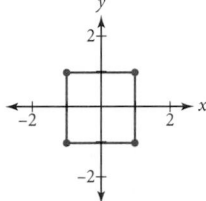

i. $\begin{bmatrix} 1 & 0 \\ 0 & 1 \end{bmatrix} \cdot [A]$ **ii.** $\begin{bmatrix} -1 & 0 \\ 0 & -1 \end{bmatrix} \cdot [A]$

iii. $\begin{bmatrix} 1 & 0 \\ 0 & 3 \end{bmatrix} \cdot [A]$ **iv.** $[A] + \begin{bmatrix} 1 & 1 & 1 & 1 \\ 1 & 1 & 1 & 1 \end{bmatrix}$

TAKE ANOTHER LOOK

In this chapter you saw reflections across the x-axis and across the y-axis. You also saw reflections across other vertical and horizontal lines (see problem 11 in Lesson 9.3). Let's examine another very important line of reflection.

Here is the graph of a function in black. The red image was created by a reflection across the dotted line. What is the equation of the line of reflection?

Identify at least three points on the graph of $y = f(x)$. Then name the image of each point after the reflection. How would you define the coordinates of the image based on the coordinates of the original graph?

The image that results from this type of reflection is called an **inverse**. Is the inverse of a function necessarily a function too? Find an example of a function whose inverse is also a function. Find an example of a function whose inverse is not a function.

Learn more about inverse functions with the links at
www.keymath.com/DA .

Mirrors are used to create reflections. This mirror helps drivers see around a corner on an Italian street.

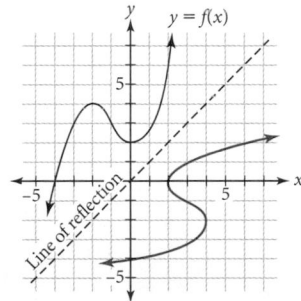

$y = f(x)$

Line of reflection

9c. Possible answer: a reflection of $y = 2^x$ across the x-axis and across the y-axis, followed by a translation up 3 units (or a reflection across the x-axis, followed by a translation up 3 units, followed by a reflection across the y-axis); $y = -2^{(-x)} + 3$

9d. Possible answer: a vertical stretch of $y = \frac{1}{x}$ by a factor of 4 and a reflection across the x-axis, followed by a translation up 1 unit and left 2 units;
$$y = -\frac{4}{x + 2} + 1$$

Exercise 10b If students actually graph the image, they may say that nothing has happened. Because of symmetry, the image will look identical to the original square. However, actually calculating the image matrix will show that the vertices do move, indicating some change.

10a. Possible answer:
$$[A] = \begin{bmatrix} -1 & 1 & 1 & -1 \\ 1 & 1 & -1 & -1 \end{bmatrix}$$

10b. i. Nothing; the image is identical to the original square.

10b. ii. a reflection across the x-axis and across the y-axis, or a rotation of 180°

10b. iii. a vertical stretch by a factor of 3

10b. iv. a translation right 1 unit and up 1 unit

▶ **Take Another Look**

The equation of the line of reflection is $y = x$. Students should find that the coordinates of each point are interchanged after the reflection. For example, the image of $(2, 4)$ is $(4, 2)$. The image of this function is not a function. An example of a function whose inverse is also a function is $y = 2x$. An example of a function whose inverse is not a function is $y = x^2$. The coordinates of the image are defined as (y, x).

[Ask] "Is any function its own inverse?" [Yes, if its graph is symmetric to $y = x$. For example, $y = \frac{1}{x}$ and $y = -x$ are their own inverses, as is $y = x$ itself.]

You may also want to have the students pick an example of a function $f(x)$ whose inverse is also a function. Call the inverse $g(x)$, and ask students to evaluate $f(g(2)), g(f(0))$, and $f(g(x))$.

ASSESSING

For a written test, you can choose the Chapter 9 test from either *Assessment Resources A* or *Assessment Resources B*, use two or three Constructive Assessment items, or select a combination of short answer and deeper questions from the test generator.

FACILITATING SELF ASSESSMENT

To help students complete the portfolio described in Assessing What You've Learned, suggest that they consider for evaluation their work on Lesson 9.1, Exercise 10; Lesson 9.2, Exercises 10 and 11; Lesson 9.3, Exercises 8 and 9; Lesson 9.4, Exercise 8; Lesson 9.6, Exercises 8 and 12; and Lesson 9.7, Exercise 8.

Assessing What You've Learned

ORGANIZE YOUR NOTEBOOK Organize your notes on each type of transformation that you have learned about. Review how each transformation affects individual points and how it changes the equation of a function. Then create a table that summarizes your notes. Use rows for each type of transformation and columns to show effects on points and equations. You can use subrows and subcolumns to further organize the information. For example, you might want to use one row for reflections across the x-axis and another row for reflections across the y-axis. You might want to use one column for changes to $y = f(x)$ and other columns for changes to specific functions like $y = x^2$, $y = |x|$, or $y = \frac{1}{x}$.

UPDATE YOUR PORTFOLIO Choose one piece of work that illustrates each transformation that you have studied in this chapter. Add these to your portfolio. Describe each work in a cover sheet, giving the objective, the result, and what you might have done differently.

PERFORMANCE ASSESSMENT Show a classmate, a family member, or your teacher how you can transform a single parent function into a whole family of functions. Explain how you can write a function for a graph by identifying the transformations. In contrast, show how you can sketch a graph just by looking at the equation.

10

Quadratic Models

Overview

Lesson 10.1 introduces quadratic equations and their graphs in the context of a real-world problem. Students use the equation of a transformed basic parabola to identify the vertex of the parabola. They also solve quadratic equations symbolically and by using graphs and tables. In **Lesson 10.2,** they use a quadratic function to model a real-world problem and learn to find the vertex form from knowing the x-intercepts of a parabola. In **Lesson 10.3,** students learn to change a quadratic equation from vertex form to general form by squaring binomials (using rectangular diagrams). In **Lesson 10.4,** students learn that the roots of a quadratic equation can be quickly found from its factored form, and they explore the relationships among the three forms of a quadratic equation. During Activity Day **Lesson 10.5,** students collect motion data that they then model with quadratic equations. In **Lesson 10.6,** students learn to solve quadratic equations in general form by completing the square. Students study the derivation of the quadratic formula and learn to apply it in **Lesson 10.7. Lesson 10.8** describes cubic functions—their general patterns, characteristics, and graphs.

The Mathematics

A quadratic function is a transformation of $y = x^2$. It has an x^2 term and possibly linear (x) and constant terms.

Three forms

The **vertex form** $y = a(x - h)^2 + k$ appeared in Chapter 9 as the general transformation of the parent function $y = x^2$. This form is most useful for finding the vertex of the parabolic graph. The roots (x-intercepts) of any equation expressed in this form can be found symbolically by undoing or balancing. To find the y-intercept requires some calculation.

The **factored form** $y = a(x - r_1)(x - r_2)$ clearly shows the roots r_1 and r_2. The vertex of the parabola can be found using the average of r_1 and r_2. To find the y-intercept also requires some calculation.

The **general form,** or standard form, is $y = ax^2 + bx + c$. Here the y-intercept is the coefficient c. If the equation is representing the height y of a projectile x seconds after it began moving, the coefficient a is $-\frac{1}{2}$ the gravitational force, and the coefficient b is the initial velocity. Finding the vertex or roots of the general form requires, in effect, changing to one of the other forms.

Changing forms

To change from vertex form to general form requires squaring the binomial $(x - h)$ and combining like terms. Similarly, to get from the factored to the general form, you multiply the binomials $(x - r_1)$ and $(x - r_2)$. Students will understand squaring or multiplying binomials more deeply through diagrams than from memorizing the FOIL (first, outside, inside, last) method.

The primary reason to change quadratic equations from general form is to solve them. In some cases changing to factored form is easy, especially with rectangular diagrams. Then the roots are apparent. In other cases, completing the square (again with a rectangular diagram) can yield vertex form, which can be solved symbolically. The algorithm called the *quadratic formula* is derived this way.

Cubics and general polynomials

The **degree** of a polynomial in variable x is the highest power of x. A polynomial of degree 3 is called a *cubic.*

A polynomial of degree 2 (a quadratic) has three coefficients. If you know three points through which you want a parabola to pass, you can substitute to obtain a system of three equations in three unknowns (the three coefficients) that has a unique

solution, meaning that only one parabola will pass through those three points. If you're given the vertex of the parabola, however, knowing just one other point through which the parabola will pass will determine the parabola, because the figure must also pass through a third point that is a reflection of the second point through the axis of the parabola. Hence the vertex-form equation $y = a(x - h)^2 + k$ can be found from knowing only the vertex (h, k) and one other point, which determines the dilation factor a. Such symmetry is not useful for general polynomials of degree n. Without further conditions, to determine a graph uniquely you need to know $n + 1$ points through which the graph will pass.

Using This Chapter

If you are under time constraints, you might skip Lesson 10.5. If your curriculum doesn't call for factoring, you could omit part of Lesson 10.4. A light treatment of Lessons 10.6 and 10.7 is possible with curricula that don't require completing the square or the quadratic formula. Lesson 10.8 is needed only to be sure students see cube roots and perfect cubes and to revisit the connection between roots and the factored form of an equation.

Resources

Discovering Algebra Resources

Calculator Notes 7B, 10A, 10B

Teaching and Worksheet Masters
Lesson 10.1, 10.2 Coordinate Plane (from Chapter 1)
Lesson 10.5 How High? Sample Data
Lesson 10.5 Rolling Along Sample Data
Lesson 10.6 Fun Training
Lesson 10.7 The Quadratic Formula
Lesson 10.8 Stunt Flying

Assessment Resources A and B
Quiz 1 (Lessons 10.1 and 10.2)
Quiz 2 (Lessons 10.3 to 10.5)
Quiz 3 (Lessons 10.6 to 10.8)
Chapter 10 Test
Chapter 10 Constructive Assessment Options

More Practice Your Skills for Chapter 10

Condensed Lessons for Chapter 10

Other Resources

www.keymath.com/DA

Materials

- graph paper
- rulers or straightedges
- several 24-cm lengths of string
- motion sensors
- empty coffee can
- long table

Pacing Guide

	day 1	day 2	day 3	day 4	day 5	day 6	day 7	day 8	day 9	day 10
standard	10.1	10.2	quiz, 10.3	10.3	10.4	10.5	quiz, 10.6	10.6	10.7	10.8
enriched	10.1	10.2	quiz, 10.3	10.3	10.4	10.5	quiz, project	10.6	10.7	10.8
block	10.1, 10.2	quiz, 10.3	10.4, 10.5	10.6, 10.7	quiz, review	assessment				

	day 11	day 12	day 13	day 14	day 15	day 16	day 17	day 18	day 19	day 20
standard	review	assessment								
enriched	review, TAL	assessment								

Quadratic Models

Buckingham Fountain in Chicago's Grant Park contains 1.5 million gallons of water. When pumped through one of the fountain's 133 jets, the water forms the shape of a parabola as it falls back into the pool. The central spout shoots 135 feet in the air. The relationship between time and the height of free falling objects in the air is described by quadratic equations.

OBJECTIVES

In this chapter you will
- model applications with quadratic functions
- compare features of parabolas to their quadratic equations
- learn strategies for solving quadratic equations
- learn how to combine and factor polynomials
- make connections between some new polynomial functions and their graphs

- Use quadratic functions to model and solve equations based on (*time, height*) relationships for projectiles

- Write quadratic equations that model other real-world data

- Find the *x*-intercepts and vertex of a parabola by graphical and symbolic methods

- Convert quadratic equations among the vertex, factored, and general forms

- Solve quadratic equations using graphs, tables, and symbolic methods

- Learn to multiply binomials and factor trinomials using rectangular diagrams

- Explore cubes, cube roots, and transformations of the parent cubic function

As you and your students look at the many streams of water in the photo of Buckingham Fountain, ask them to describe the parabolas that they see. **[Ask]** "What determines the shape of the parabolas?" [Both the angle and force at which the water is projected.] Compare streams projected straight up to those that leave the fountain at different angles. [If the force behind two fountains is the same, the fountain with the steeper angle will go higher, but the water will not land so

far from the spout as it will for a spout with a smaller angle.] For one of the streams, you know the height. **[Ask]** "What beside the height would you need to know to determine the equation that describes the parabolic path of a drop of water?" [You would also need the point at which the water falls back into the pool. Imagine a coordinate axis with the *y*-axis through the highest point of the stream of water and the *x*-axis through the spout that shoots the water. The equations

could be determined from the three points: the maximum height of the water, the horizontal distance from the spout to the highest point, and the reflection through the *y*-axis of that point.]

The fountain, made of pink Georgia marble, is modeled after the Latona Basin in the gardens of Louis XIV's palace in Versailles, France. Kate Sturges Buckingham gave it to the city of Chicago in 1927.

PLANNING

LESSON OUTLINE

One day:

5 min	Example A
20 min	Investigation
5 min	Sharing
5 min	Example B
5 min	Closing
10 min	Exercises

MATERIALS

• Coordinate Plane (T), one per group, *optional*

• graph paper, *optional*

TEACHING

The height of a projectile is given by a transformation of the squaring function. When the function is written in a form that shows that transformation, an equation involving it can be solved by undoing or balancing.

One step Pose this problem: "A ball is hit straight up. It reaches a height of 68 feet. Four seconds after it is hit, the ball reaches the ground. Sketch a graph of the ball's height in feet in terms of its flight time in seconds." Some students may draw a vertical line, confusing the height function with a path of the ball. Others may conjecture that the graph is that of an absolute value function. **[Ask]** "Does the ball keep the same speed while going up and then suddenly switch to going down at a constant speed?" After students decide the graph should be a parabola, ask when they think it reaches its peak of 68 feet. You may need to ask when they think it will return to the height from which it was hit, giving rise to their making some

LESSON

10.1 Solving Quadratic Equations

We especially need imagination in science; it is not all mathematics, nor all logic, but it is somewhat beauty and poetry.

MARIA MITCHELL

When you throw a ball straight up into the air, its height depends on three major factors—its starting position, the velocity at which it leaves your hand, and the force of **gravity.** The Earth's gravity causes objects to accelerate downward, gathering speed every second. This acceleration due to gravity, called *g*, is 32 ft/sec^2. It means that the object's downward speed increases 32 ft/sec *for each second* in flight. If you plot the height of the ball at each instant of time, the graph of the data is a parabola.

EXAMPLE A

A baseball batter pops a fly ball straight up. The ball reaches a height of 68 feet before falling back down. Roughly 4 seconds after it is hit, the ball bounces off home plate. Sketch a graph that models the ball's height in feet during its flight time in seconds. When is the ball 68 feet high? How many times will it be 20 feet high?

assumptions. As time allows, ask them to write an equation of the graph as a transformation of $y = x^2$. They'll need to find the (negative) dilation factor to make the graph go through (4, 0).

LESSON OBJECTIVES

• Use quadratic functions to model and solve equations based on (*time, height*) relationships for projectiles

• Solve quadratic equations using graphs, tables, and symbolic methods

▶ **Solution**

Height of a Ball

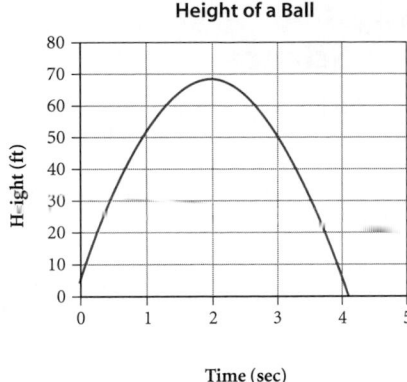

The sketch above pictures the ball's height once it is hit and before it lands on the ground. When the bat hits the ball, it is a few feet above the ground. So the *y*-intercept is just above the origin. The ball's height is 0 when it hits the ground just over 4 seconds later. So the parabola crosses the *x*-axis near the coordinates (4, 0). The ball is at its maximum height of 68 feet after about 2 seconds, or halfway through its flight time. So the vertex of the parabola is near (2, 68).

The ball reaches a height of 20 feet twice—once on its way up and again on its way down. If you sketch $y = 20$ on the same set of axes, you'll see that this line crosses the parabola at two points.

Height of a Ball

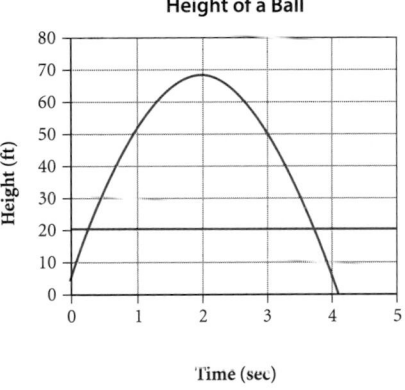

The parabola in Example A is a transformation of the equation $y = x^2$. The function $f(x) = x^2$ and transformations of it are called **quadratic functions,** because the highest power of x is x-squared. The Latin word meaning "to square" is *quadrare*. The function that describes the motion of a ball, and many other projectiles, is a quadratic function. You will learn more about this function in the investigation.

▶ **EXAMPLE A**

This example shows students that the graph of the equation giving the height of a projectile is a parabola and tells them how to interpret such an equation.

Reassure students unfamiliar with baseball terms that the ball is simply rising straight up and then falling back down. The pull of gravity changes the speed of the ball; gravity slows the ball down as it goes up and speeds the ball up as it falls. **[Alert]** Some students may confuse the shape of the graph with the path of the ball. (The path is also parabolic and is discussed in Exercise 7.)

[Language] The word *quadratic* comes from Latin *esquadrare* and means "to make square." In a quadratic equation, the variable is squared; it is raised to the power of 2. The root *quad* means "four" as in *quadrilateral*.

NCTM STANDARDS

CONTENT		PROCESS	
	Number	•	Problem Solving
•	Algebra	•	Reasoning
	Geometry	•	Communication
•	Measurement	•	Connections
	Data/Probability	•	Representation

Step 1 Time is measured from launch. Encourage the use of dimensional analysis: 2.5 is in meters, and 49 is in m/sec. So when the latter is multiplied by t (in seconds), the result is in meters. The gravitational force -9.8 is in m/sec^2; so when it is multiplied by t^2 (in sec^2), the result is again in meters. The $\frac{1}{2}$ has no dimensions.

The equation is not a perfect model for the real world. It is based on the invalid assumptions that there is no air resistance and that the value of g is constant (g decreases with distance from the center of the earth).

Step 3 Some students may think that g should be -32 rather than -9.8. They're thinking of g in the English system, measured in ft/sec^2.

Step 4 Again, watch for confusion between the shape of the graph and the path of the ball.

SHARING IDEAS

Ask students to present their ideas about Steps 3, 6, 7, 8, and 9, perhaps using the Coordinate Plane transparency.

Elicit ideas about what each term in the projectile motion equation represents. In general, the height of a projectile (neglecting air resistance) in meters is given by $h(t) = \frac{1}{2}(-g)t^2 + v_0 t + h_0$, where g is acceleration due to gravity, v_0 is the initial velocity, and h_0 is the initial height. In this case, 2.5 meters is the initial height, and $49t$ is the number of meters gained in t seconds. The initial term, $\frac{1}{2}(-9.8)t^2$, or $-4.9t^2$, represents the downward force of gravity. At first, when t is small, $\frac{1}{2}(-9.8)t^2$ is less than $v_0 t + h_0$, so the object gains in height. When

See page 723 for graphs for Steps 4, 5, 8, and 9.

Investigation
Rocket Science

A model rocket blasts off from a position 2.5 meters above the ground. Its starting velocity is 49 meters per second. Assume that it travels straight up and that the only force acting on it is the downward pull of gravity. In the metric system, the acceleration due to gravity is 9.8 m/sec^2. The quadratic function $h(t) = \frac{1}{2}(-9.8)t^2 + 49t + 2.5$ describes the rocket's **projectile motion.**

Known as the father of rocketry, Robert Hutchings Goddard fired the first successful liquid-fueled rocket in 1926.

Step 1 Let t represent the time in seconds that the rocket is in flight. Let $h(t)$ be the function value that gives the height in meters of the rocket at time t.

Step 2 When $t = 0$ sec, the height of the rocket is 2.5 m.

Step 3 The factor -9.8 is the downward (negative) g.

Step 1 Define the function variables and their units of measurement for this situation.

Step 2 What is the real-world meaning of $h(0) = 2.5$?

Step 3 How is the acceleration due to gravity, or g, represented in the equation? How does the equation show that this force is *downward*?

Next you'll make a graph of the situation.

Step 4 Graph the function $h(t)$. What viewing window shows all the important parts of the parabola?

Step 5 How high does the rocket fly before falling back to Earth? When does it reach this point?

Step 6 How much time passes while the rocket is in flight?

Step 7 Write the equation you must solve to find when $h(t) = 50$.

Step 8 When is the rocket 50 meters above the ground? Use a calculator table to approximate your answers to the nearest tenth of a second.

Step 9 Describe how to answer Step 8 graphically.

Step 5 125 m after 5 sec

Step 6 Just over 10 seconds; the positive x-intercept is at $(10.05, 0)$.

Step 7 $50 = \frac{1}{2}(-9.8)t^2 + 49t + 2.5$; or $50 = -4.9t^2 + 49t + 2.5$

Step 8 at about 1.1 seconds and at about 8.9 seconds

Step 9 Sample answer: Graph $y = h(t)$ and $y = 50$ on the the same set of axes. Find points of intersection.

In the investigation you approximated solutions to a quadratic equation using tables and graphs. Later in this chapter you will learn to solve quadratic equations in the **general form,** $y = ax^2 + bx + c$, using symbolic manipulation. Until then, quadratic equations must be in a certain form for you to solve them symbolically. You will combine the "undo" and "balance" methods on this form in the next example.

t gets large enough, the first term begins to dominate the sum of the other two terms and the object begins falling.

Ask students if they agree with the opening quotation. Mitchell seems to make the assumption that mathematics is calculation and logic, rather than beauty and poetry. Have students seen something other than calculations in this mathematics course? Try to emphasize that mathematics is more a way of thinking than a way of calculating.

To illustrate your point, this might be a good time to have students pose extension problems. Some good ones are "What if the object were dropped from a height instead of beginning on the ground?" "What if it were thrown down from a height?" "Is g the same no matter where you are?" "We know g is, in effect, 0 in space. Does it gradually change from the earth to space, or is there a sudden drop?" The purpose is not to answer these questions but to show that doing mathematics involves posing many problems for future exploration.

EXAMPLE B | Solve $5(x + 2)^2 - 10 = 47$ symbolically. Check your answers with a graph and a table.

▶ **Solution** | Undo each operation as you would when solving a linear equation. To undo the squaring operation, take the square root of both sides. You will get two possible answers.

$$5(x + 2)^2 - 10 = 47$$ The original equation.

$$5(x + 2)^2 - 10 + 10 = 47 + 10$$ Add 10.

$$\frac{5(x + 2)^2}{5} = \frac{57}{5}$$ Divide by 5.

$$(x + 2)^2 = 11.4$$ Reduce.

$$\sqrt{(x + 2)^2} = \sqrt{11.4}$$ Take the square root of both sides.

$$x + 2 = \pm\sqrt{11.4}$$ The $\pm$ symbol shows the two numbers $+\sqrt{11.4}$ and $-\sqrt{11.4}$, whose square is 11.4.

$$x = -2 \pm \sqrt{11.4}$$ Subtract 2 from both sides.

The two solutions are $-2 + \sqrt{11.4}$, or approximately 1.38, and $-2 - \sqrt{11.4}$, or -5.38.

The calculator screens of the graph and the table support each solution.

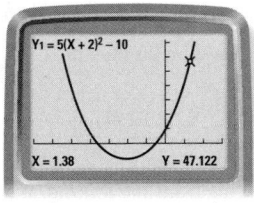 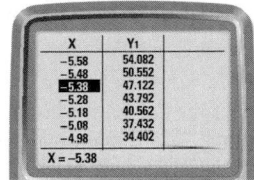

$$[-7, 3, 1, -10, 70, 10]$$

A symbolic approach allows you to find the exact solutions rather than just approximations from a table or a graph. Exact solutions such as $x = -2 \pm \sqrt{11.4}$ are called **radical expressions** because they contain the square root symbol, $\sqrt{}$, and "radical" comes from the Latin word for "root." As you practice solving quadratic equations symbolically, first think about the order of operations. Then concentrate on what each operation does to the equation and how to undo this order. In some situations only one of the solutions you find has a real-world meaning. Always ask yourself whether the answers you find make sense in real-world situations.

EXERCISES

You will need your calculator for problems **1, 2, 5, 6, 8,** and **9.**

▶ **Practice Your Skills**

1. Use a graph to find the number of solutions for each equation. Explain your answer.
   a. $x^2 + 3x - 7 = 11$ 2
   b. $-x^2 + x + 4 = 7$ 0
   c. $x^2 - 6x + 14 = 5$ 1
   d. $-3x^2 - 5x - 2 = -5$ 2

Closing the Lesson

The height of a projectile in meters is given by function $h(t) = -4.9t^2 + v_0 t + h_0$, where t is the number of seconds after launch, v_0 is the initial velocity, and h_0 the initial height. This is a transformation of the squaring function. When the function is written in the familiar form that shows the transformation, an equation involving it can be solved by undoing or balancing. There are often two solutions, and frequently their exact forms are **radical expressions.**

Assessing Progress
Watch for abilities to use **dimensional analysis,** to interpret an expression, to **graph a function,** and to **solve an equation** with graphing and a calculator table.

▶ **EXAMPLE B**
This example shows how to solve a quadratic equation that is written in the form $a(x - h)^2 + k = c$. (In the next lesson students will see that form called the *vertex form.*) They can think either of undoing to get from c back to x (subtract k, divide by a, take square roots, add h) or of balancing, doing the same thing to both sides of the equation. The steps are the same, but the thought process is somewhat different. For solving linear equations, some students might still rely on representing the undoing process by arrow diagrams. Help those students to see that arrow diagrams can be used to represent undoing a quadratic equation of this form, as long as they allow a split to find the two square roots.

Some student may ask, "How big is $\sqrt{11.4}$?" They probably want a decimal name for that number. Even while stressing that $\sqrt{11.4}$ is a perfectly good name—in fact, an accurate name—for the number, point out that 11.4 is between 9 and 16, so its positive square root is between 3 and 4.

Tracing the graph on the window specified in the example gives slightly different values. To obtain the x- and y-values shown, have the calculator evaluate the function from the graph screen for $x = 1.38$.

The number $\sqrt{11.4}$ is sometimes read "radical 11.4," although the word *radical* refers to any root. **[Language]** It comes from the Latin *radix,* for root. You might ask students whether they think a radical, or extreme, opinion always gets to the root, or source, of a problem.

See page 723 for complete answers to Exercises 1a–d.

BUILDING UNDERSTANDING

Students work with graphing and solving quadratic equations, some in real-world contexts.

ASSIGNING HOMEWORK

Essential	1–6, 8
Performance assessment	8
Portfolio	6, 7, 9
Journal	9
Group	5, 7
Review	10, 11

▶ Helping with the Exercises

Exercise 1 Recording sketches of their graphs may help students. They may want to refer to these sketches when drawing the sketches for Exercise 4.

3d. $2(x + 1)^2 = 14$
$(x + 1)^2 = 7$
$x + 1 = \pm\sqrt{7}$
$x = -1 \pm \sqrt{7}$

Exercise 4 As needed, remind students of what the vertex of a parabola is.

Exercise 5 There is no x-term, because the initial velocity is zero. A negative value of t for one time at which the ball has a height of 0 meters is not completely meaningless. If you went back that many seconds before the release and you launched the ball at whatever velocity was needed to reach 147 m at time 0, the situation would be the same.

Exercise 6 As needed, remind students of shifts and vertical dilations.

6d. $h(t) = -4.9(t - 4.7)^2 + 108$

See pages 723–724 for complete answers to Exercises 4a–d and 6d–f.

2. For each equation in problem 1, zoom in on a table to approximate the solutions, if they exist, to the nearest hundredth.
  a. $x = -6, x = 3$ **b.** There are no solutions.   **c.** $x = 3$ **d.** $x \approx -2.14, x \approx 0.47$

3. Use a symbolic method to solve each equation. Show each solution exactly as a radical expression.
  a. $x^2 = 18$ **b.** $x^2 + 3 = 52$ **c.** $(x - 2)^2 = 25$ **d.** $2(x + 1)^2 - 4 = 10$
  $x = \pm\sqrt{18}$ $x^2 = 49, x = \pm 7$ $x - 2 = \pm 5, x = 7$ or $x = -3$

4. Sketch the graph of a quadratic function with
  a. One x-intercept.
  b. Two x-intercepts.
  c. Zero x-intercepts.
  d. The vertex in the first quadrant and two x-intercepts.

▶ Reason and Apply

5. A baseball is dropped from the top of a very tall building. The ball's height, in meters, t seconds after it has been released is $h(t) = -4.9t^2 + 147$.
  a. Find $h(0)$ and give a real-world meaning for this value.
  b. Solve $h(t) = 20$ symbolically and graphically.
  c. Does your answer to 5b mean the ball is 20 meters above the ground twice? Explain your reasoning.
  d. During what interval of time is the ball less than 20 meters high? $t > 5.09$ seconds
  e. When does the ball hit the ground? Justify your answer with a graph. The ball hits the ground when $t \approx 5.48$ seconds because the positive x-intercept is near the point (5.48, 0).

6. **APPLICATION** A small rocket is fired into the air from the ground. It reaches its highest point, 108 meters, at 4.70 seconds. It falls back to the ground at 9.40 seconds.
  a. Name three points that the graph goes through. (0, 0), (4.7, 108), (9.4, 0)
  b. Name a graphing window that lets you see those three points. $[-1, 10, 1, -10, 120, 10]$
  c. What are the coordinates of the vertex of this parabola? (4.7, 108)
  d. Find an equation in the form
    $$y = a(x - h)^2 + k$$
    that fits the three known points. You may need to guess and check to find the value of a.
  e. Find $h(3)$ and give a real-world meaning for this value. 94 meters
  f. Find the t-values for $h(t) = 47$, and describe the real-world meaning for these values.
    $$4.7 \pm \sqrt{\frac{1347.49}{108}}$$

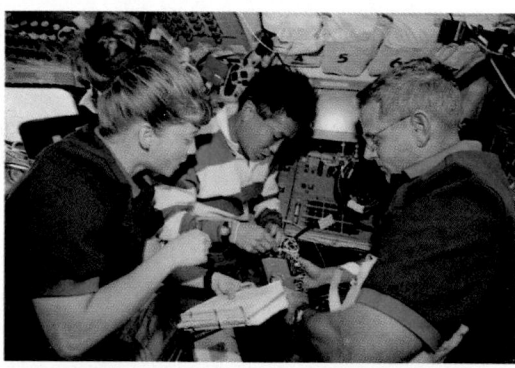

Astronaut trainees Pamela Melroy, Koichi Wakata, and William McArthur are practicing inflight maintenance on a part of a flight deck panel.

5a. $h(0) = -4.9(0)^2 + 147$, or 147. It is the starting height of the ball, when $t = 0$ seconds.

5b. Solve the equation $20 = -4.9t^2 + 147$ symbolically to get $\pm\sqrt{\frac{20 - 147}{-4.9}}$ or approximately ± 5.09. The graph shows two solution points: (5.09, 20) and (−5.09, 20).

5c. No, only the solution (5.09, 20) makes sense because the time t must be positive.

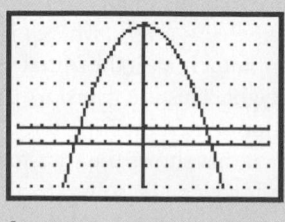

$[-10, 10, 1, -50, 150, 25]$

7. The path of a ball in flight is given by $p(x) = -0.23(x - 3.4)^2 + 4.2$, where x is the horizontal distance in meters and $p(x)$ is the vertical height in meters. Note that in this case the graph is the path of the ball, not the graph of the ball's height over time.

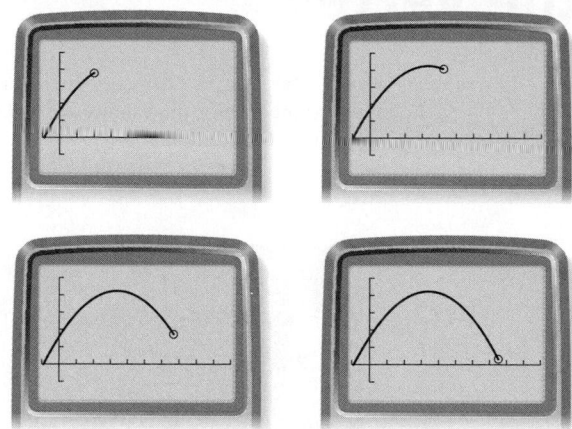

a. Find $p(2)$ and give a real-world meaning for this value.

b. Find the x-values for $p(x) = 2$, and describe their real-world meanings.

c. How high is the ball when it is released? When $x = 0$, the height is 1.5412 m.

d. How far will the ball travel horizontally before it hits the ground?
When $p(x) = 0$, the horizontal distance is $3.4 + \sqrt{\frac{420}{23}}$, or 7.67 m.

8. Solve the equation $4 = -2(x - 3)^2 + 4$ using

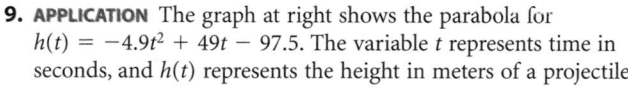

a. A graph.　　　　　b. A table.　　　　　c. Symbolic manipulation.

9. APPLICATION The graph at right shows the parabola for $h(t) = -4.9t^2 + 49t - 97.5$. The variable t represents time in seconds, and $h(t)$ represents the height in meters of a projectile.

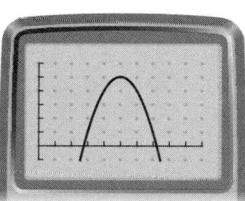

[0, 10, 1, −5, 30, 5]

a. What is a real-world meaning for the x-intercepts in the graph?

b. Find the x-intercepts to the nearest 0.01 second.

c. How can you use 9b to find the vertex of this parabola?

d. What is a real-world meaning for the vertex in the graph?

e. What does $h(3.2)$ tell you?

f. When is the projectile 12.5 m high? Explain how to find these solutions on a graph.

▶ Review

10. Show a step-by-step symbolic solution of the inequality $-3x + 4 > 16$.

11. The solid line in the graph passes through $(0, 6)$ and $(6, 1)$. Write an inequality to describe the shaded region.

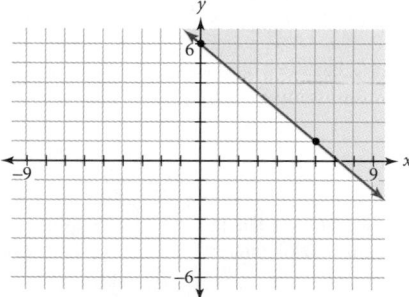

9d. The projectile is 25 meters above the ground, its maximum height, after 5 seconds.

9e. $h(3.2) = 9.124$ meters. This is the height at 3.2 seconds.

Exercises 10 and 11 These exercises review Lessons 6.5 and 6.6.

10. $-3x + 4 > 16$　　The given inequality.
$-3x > 12$　　Subtract 4 from both sides.
$x < -4$　　Divide both sides by −3 and reverse the inequality symbol.

11. The slope is $\frac{1 - 6}{6 - 0} = \frac{-5}{6}$. An intercept form of the inequality is $y \geq 6 - \frac{5}{6}x$.

Exercise 7 [Alert] Be sure students realize that in this case, the graph represents a (*height, distance*) relationship and not (*height, time*). They may wonder why the initial coefficient is not −4.9. The equation for $p(x)$ in this form tells you that the ball reaches its maximum height (the vertex of the parabola) when it is 3.4 m away horizontally from where it started. Substitution allows you to find $p(3.4)$ to be 4.2 m.

7a. After the ball has gone 2 m horizontally, it will be approximately 3.75 m above the ground.

7b. The ball will be 2 m above the ground once it has gone 0.31 m horizontally and again once it has gone 6.49 m horizontally. Exact answers are $3.4 \pm \sqrt{\frac{220}{23}}$.

Exercise 8 There's more than one way to solve this equation using a graph. For example, students might graph the function $Y_1 = -2(x - 3)^2 + 4$ and then trace to find a y-value of 4. Or they might graph Y_1 and then also graph $Y_2 = 4$ and trace to find the point of intersection.

8c. $4 = -2(x - 3)^2 + 4$
$0 = -2(x - 3)^2$
$0 = (x - 3)^2$
$0 = x - 3$
$3 = x$

Exercise 9 This exercise foreshadows work with the general form in Lesson 10.3. If students are stuck on 9c, suggest they use their answer to 9b.

9a. The x-intercepts indicate when the projectile is at ground level.

9b. 2.74 seconds and 7.26 seconds

9c. Find the average of 2.74 and 7.26, which is 5. $h(5) = 25$.

See page 724 for answers to Exercises 8a–b and 9f.

PLANNING

LESSON OUTLINE

One day:

20 min   Investigation

5 min   Sharing

10 min   Examples

5 min   Closing

10 min   Exercises

MATERIALS

- Coordinate Plane (T), one per group, *optional*
- graph paper
- several 24-cm lengths of string

TEACHING

A quadratic equation in vertex form can help optimize real-world situations.

 Guiding the Investigation

The context of this investigation has appeared in Lesson 8.5 and is suggested in the Teacher's Edition as an alternate activity in Lesson 9.5.

One step Pose the problem: "You have 24 meters of fencing material and you want to use all of it to enclose the largest possible rectangular space for your vegetable garden. What dimensions should the rectangle have?" As you circulate among groups, ask them to look for patterns and then to write a quadratic equation fitting the data. Ask how to approximate the *x*-intercepts (the roots of the equation $y = 0$) and wonder aloud how they might find the vertex from knowing the roots.

See page 724 for answers to Steps 1, 2, 3, and 4.

In this lesson you will discover that quadratic functions can model relationships other than projectile motions. You will explore relationships between parabolas and their equations. You will practice writing equations, finding *x*-intercepts, and determining real-world meanings for the *x*-intercepts and the vertex of a parabola.

 ## Investigation
Making the Most of It

You will need

- graph paper

Suppose you have 24 meters of fencing material and you want to use it to enclose a rectangular space for your vegetable garden. Naturally, you want to have the largest area possible for your vegetables. What dimensions should you use for your garden?

Step 1   Find the dimensions of at least eight different rectangular areas, each with a perimeter of 24 meters. You must use all of the fencing material for each garden.

Step 2   Find the area of each garden. Make a table to record the width, length, and area of the possible gardens. It's okay to have widths that are greater than their corresponding lengths.

Width (m)	Length (m)	Area (sq. m)

Step 3   Enter the data for the possible widths into list L1. Enter the area measures into list L2. Which garden width values would give no area? Add these points to your lists.

Step 4   Label a set of axes and plot points in the form (x, y), with *x* representing width in meters and *y* representing area in square meters. Describe as completely as possible what the graph looks like. Does it make sense to connect the points with a smooth curve?

Step 5 The graph reaches its highest point at (6, 36). The 6-by-6 rectangle, a square, has the largest area.

Step 5   Where does your graph reach its highest point? Which rectangular garden has the largest area? What are its dimensions?

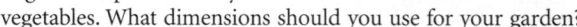

Next you'll write an equation to describe this relationship.

Step 6 The sum of the width and the length is 12. 10 meters; 7.7 meters; $12 - x$ meters

Step 6   Describe a relationship between the values for the garden widths and their corresponding lengths. What is the length of the garden that has a width of 2 meters? A width of 4.3 meters? Write an expression for length in terms of width *x*.

Then ask them to rewrite the equation to show how the parent quadratic function would be transformed to get this one.

Step 1 Students can draw the rectangles if they have trouble visualizing. Use 24-cm lengths of string for kinesthetic learners. A width or length of 0 is not acceptable as a measurement, but they are useful values to list.

Step 2 [Language] *Length* of a rectangle is often defined as its longer dimension. Some students may note that the table begins to repeat, that it will have

symmetry, and they stop because they want to keep the width shorter than the length. If they stop with half the data, their graph won't be a parabola. Say that for this investigation, *width* and *length* refer to the lengths of two edges of the rectangle. As the dimensions change, the width column will sometimes contain the longer measurement. Students who see immediately how to calculate the lengths from the widths may put lengths into list L2 (by formula) and areas into list L3.

Step 7 $y = x(12 - x)$; Step 7 the graph shows that a square has the largest area.

Step 8 x-intercepts at $(0, 0)$ and $(12, 0)$. The rectangle has no area if the width is 0 or 12.

Using your expression for the length from Step 6, write an equation for the area of the garden. Enter this equation into Y₁ and graph it. Does the graph confirm your answer to Step 5?

Step 8 Locate the points where the graph crosses the x-axis. What is the real-world meaning of these points?

Step 9 Do you think the general shape of a garden with a maximum area would change for different perimeters? *Sample answer: A maximum-area garden will always be square.*

In the investigation you found three important points on the graph. The two points on the x-axis are called **x-intercepts.** The x-values of those points are the solutions of the equation $y = f(x)$ when the function value is equal to zero. These solutions give the **roots** to the equation $f(x) = 0$.

In the investigation the roots are the widths that make the garden area equal to zero. The roots help you to find a third important point—the vertex of the parabola.

In Lesson 10.1, you symbolically solved quadratic equations written in the form $y = a(x - h)^2 + k$. In the next example you will learn to approximate roots of the quadratic equation, $0 = ax^2 + bx + c$.

EXAMPLE A

Use a graph and your calculator's table function to approximate the roots of

$$0 = x^2 + 3x - 5$$

▶ Solution

Graph $y = x^2 + 3x - 5$ and find the x-intercepts. On the graph you can see that there are two roots—one appears to be a little less than -4, and the other a little greater than 1.

Search in your calculator's table for the positive x-value that makes the y-value equal to zero. Continue zooming in until you find the positive root, which is about 1.1926. Repeat this process for the negative root, which you'll find to be about -4.1926.

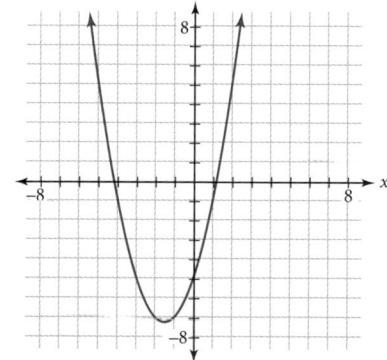

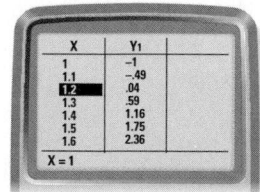
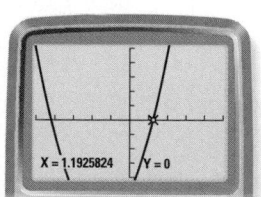

$[-5, 5, 1, -5, 5, 1]$

LESSON OBJECTIVES

- Introduce the general form of a quadratic equation
- Write quadratic equations that model real-world data
- Approximate the x-intercepts and other points on a parabola using graphs, tables, and symbolic methods
- Find the vertex of a parabola from knowing its x-intercepts
- Learn the vertex form of a quadratic equation

NCTM STANDARDS

CONTENT	PROCESS
• Number	• Problem Solving
• Algebra	• Reasoning
• Geometry	Communication
• Measurement	• Connections
Data/Probability	• Representation

Step 4 You might have each group plot its points on a copy of the Coordinate Plane transparency. Then during Sharing you can combine the data points to counter any conjectures that the graph represents an absolute value function. Note any groups that plot (*width, length*) instead of (*width, area*). During Sharing put their plot up with the others and let them critique it with the class.

Step 6 Some students may benefit from solving $2L + 2W = 24$ for L. **[Ask]** "Does the pattern *length* $= 12 -$ *width* make sense?" [Yes, the sum of one length and one width is half the perimeter of 24.]

SHARING IDEAS

If groups plotted their data on transparencies, have them all share those. Then ask for ideas about Steps 7 and 9.

Return to the original question of the investigation. **[Ask]** "What dimensions should you use for your garden?" Probably most will say the dimensions that give maximum area. But others may make suggestions such as the best dimensions are those that allow easier access from the edges or have more length facing the sun. Point out that the utility of a mathematical model depends on assumptions made about the problem.

Real-life problems of finding the largest or smallest of something are called *optimization* problems. Before the advent of graphing calculators, solutions to such problems, especially for functions of degree more than 2, were most easily found through calculus.

Introduce the term *root of an equation* for x-intercepts of its graph. Note that roots in this sense are not necessarily square roots or cube roots.

See page 724 for graph to Step 7.

Observe students' understanding of area and perimeter of a rectangle and of calculator abilities to enter lists, plot points, graph functions, and trace graphs.

▶ EXAMPLE A

This example shows that the roots of an equation $f(x) = 0$ are the x-intercepts of the graph of $f(x)$. Roots of an equation are also called *zeros* of the corresponding function.

The exact roots of this quadratic equation are $\dfrac{-3 \pm \sqrt{29}}{2}$. The calculator may have a feature allowing students to find these x-intercepts (roots) directly from the calculator's graph screen.

▶ EXAMPLE B

This example shows how to find the vertex form of a quadratic by knowing its roots. The procedure requires several steps. Keep reminding students of the goal: to find values for h, k, and a.

[Alert] Some students may not be aware that the number halfway between two numbers is their mean, or average. Averaging the exact roots also results in -1.5.

As needed, point out that $h = -1.5$ implies $x - h = x + 1.5$.

Closing the Lesson

Finding the vertex of the parabolic graph of a quadratic equation can allow solving real-world optimization problems. You can get a quadratic equation into **vertex form** by approximating its **roots,** finding the vertex from the mean of the roots, and adjusting the **dilation factor.**

The line through the vertex that cuts a parabola into two mirror images is called the **line of symmetry.** If you know the roots, you can find the vertex and the line of symmetry.

EXAMPLE B | Find the coordinates (h, k) of the vertex of the parabola $y = x^2 + 3x - 5$. Then write the equation in the form $y = a(x - h)^2 + k$.

▶ **Solution**   This parabola crosses the x-axis twice and has a vertical line of symmetry. The x-coordinate of the vertex lies on the line of symmetry, halfway between the roots. From Example A you know the two roots are approximately 1.1926 and -4.1926. Averaging the two roots gives -1.5. The graph shows that the line of symmetry goes through this x-value.

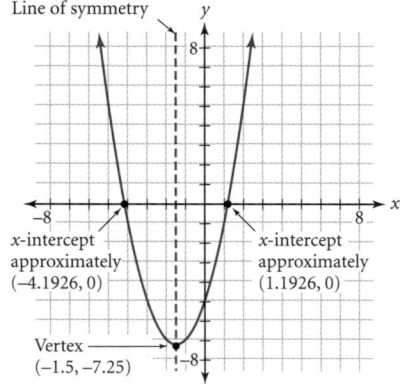

Now use the equation of the parabola, $y = x^2 + 3x - 5$, to find the y-coordinate of the vertex.

$$y = x^2 + 3x - 5 \qquad \text{The equation of the parabola.}$$
$$= (-1.5)^2 + 3(-1.5) - 5 \qquad \text{Substitute } -1.5 \text{ for } x.$$
$$= 2.25 - 4.5 - 5 \qquad \text{Multiply.}$$
$$= -7.25 \qquad \text{Subtract.}$$

So the vertex is $(-1.5, -7.25)$. Sometimes you can find the vertex and see the symmetry in a table of values.

In the table, this point appears to be the lowest point on the parabola.

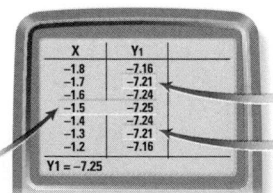

The symmetry of the curve shows up in the repeated y-values on either side of the vertex.

The graph is a transformation of the parent function, $f(x) = x^2$. The vertex (h, k) is $(-1.5, -7.25)$, so there is a translation left 1.5 units and down 7.25 units. Substitute the values h and k into the equation to get $y = (x + 1.5)^2 - 7.25$. Enter the equation into Y2 and graph it.

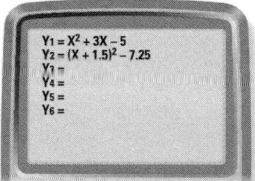 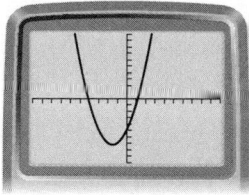 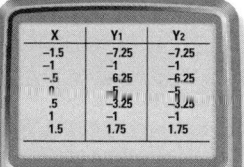

$[-10, 10, 1, -10, 10, 1]$

You can see from the graph and the table that the equations $y = x^2 + 3x - 5$ and $y = (x + 1.5)^2 - 7.25$ are equivalent. So the value of a is 1. The equation $y = 1 \cdot (x + 1.5)^2 - 7.25$ is in the **vertex form,** $y = a(x - h)^2 + k$. It tells you that $(-1.5, -7.25)$ is the vertex.

EXERCISES

You will need your calculator for problems **4, 8,** and **11.**

▶ Practice Your Skills

1. What is the x-coordinate of the vertex of the parabola below? The x-intercepts are at 3 and -2. The window shown is $[-4.7, 4.7, 1, -3.1, 3.1, 1]$.

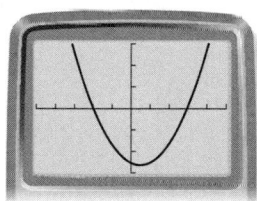

The vertex has an x-value of $\frac{3 + (-2)}{2}$, or 0.5.

2. The equation for the parabola in problem 1 is $y = 0.4x^2 - 0.4x - 2.4$. Explain how to use the x-coordinate you found in problem 1 to find the y-coordinate of the vertex. Substitute 0.5 for x to obtain $0.4(0.5)^2 - 0.4(0.5) - 2.4 = -2.5$, or use a table.

3. Solve $0 = (x + 1.5)^2 - 7.25$ symbolically. Show each step. Compare your solutions with the approximations from Examples A and B.

4. Find the roots of each equation to the nearest thousandth by looking at a graph, zooming in on a table, or both.
   a. $0 = x^2 + 2x - 2$ $x \approx -2.732, 0.732$ b. $0 = -3x^2 - 4x + 3$ $x \approx -1.869, 0.535$

5. Solve each equation symbolically and check your answer.
   a. $(x + 3)^2 = 7$ b. $(x - 2)^2 - 8 = 13$

6. Graph $y = (x + 3)^2$ and $y = 7$. What is the relationship between your solutions to problem 5a and these graphs?

3.
$$0 = (x + 1.5)^2 - 7.25$$
$$7.25 = (x + 1.5)^2$$
$$\pm\sqrt{7.25} = x + 1.5$$
$$-1.5 \pm\sqrt{7.25} = x$$
$$x \approx 1.192582404 \text{ or}$$
$$x \approx -4.192582404$$

BUILDING UNDERSTANDING

The exercises provide practice at approximating roots and at real-world optimization through finding vertices of parabolas.

ASSIGNING HOMEWORK

Essential	1–5, 7, 11
Performance assessment	6, 7, 11
Portfolio	7, 9
Journal	9, 10
Group	8, 9
Review	12

▶ Helping with the Exercises

Exercises 3 and 5 Students can use either undoing or balancing. Accept arrow diagrams.

Exercise 4 Allow students who ask about the root or zero calculator command to use them.

5a. $(x + 3) = \pm\sqrt{7}$
$$x = -3 \pm\sqrt{7}$$
$$x \approx -5.646, -0.354$$

5b. $(x - 2)^2 = 21$
$$x = 2 \pm\sqrt{21}$$
$$x \approx -2.583, 6.583$$

6. Answers will vary. The graph of $y = (x + 3)^2$ intersects the graph of $y = 7$ at $(-5.646, 7)$ and $(-0.354, 7)$.

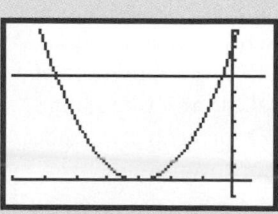

$[-7, 1, 1, -1, 10, 1]$

7a. The ball is on the ground when $h = 0$. This happens when $t = 0$ seconds and $t = 3$ seconds.

7b. The ball is at its highest point when $t = 1.5$ seconds, halfway through its flight.

7c. $h = 36$ when $t = 1.5$, so the ball goes 36 feet high.

Exercise 8 The initial velocity v_0 is 58 ft/sec. The starting position of the ball is 3 feet above the ground.

8a. Between 3.67 seconds when the height is still positive and 3.68 seconds when the height is negative.

8b. Starting the table at 3.67 and setting ΔTbl equal to 0.001 gives an answer of 3.676 seconds.

8c. The maximum height occurs between 1.81 and 1.82 seconds when the height of the ball is at least 55.562 feet.

Exercise 9 Students may be confused by the idea of negative velocity. Velocity is a vector quantity with both magnitude (the speed) and direction. In this case a negative sign indicates that the ball is traveling downward.

9a. Answers will vary. The ball is thrown from an initial height of 5 meters. It reaches a maximum height of about 25 meters in 2 seconds and hits the ground at about 4.3 seconds.

9b. The ball starts at a velocity of 20 meters per second and slows down at a constant rate. At 2 seconds it is not moving. Then it starts falling and is moving down at 22 meters per second when it hits the ground.

9c. As the ball moves up to its maximum height, it is slowing down, moving at 0 meters per second at its peak. It then starts falling and speeding up until it hits the ground.

► **Reason and Apply**

7. The height of a golf ball is given by $h = -16t^2 + 48t$, where t is in seconds and h is in feet.

   a. At what times is the golf ball on the ground?

   b. At what time is the golf ball at its highest point?

   c. How high does the golf ball go?

8. APPLICATION Taylor hits a baseball, and its height in the air at time x is given by the equation $y = -16x^2 + 58x + 3$, where x is in seconds and y is in feet. Use the graph and tables to help you answer these questions.

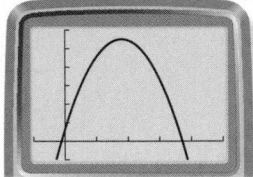 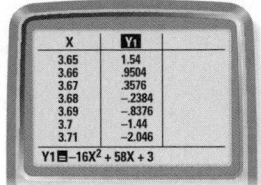 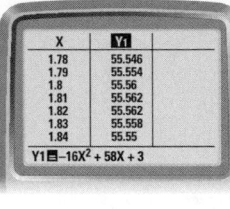

$[-1, 5, 1, -1, 60, 10]$

   a. When does the ball hit the ground?

   b. Use your calculator table to find the answer to three decimal places.

   c. According to the table above, during what time interval is the ball at its highest points? At what time (to the nearest hundredth of a second) is the ball at its highest point, and how high is it?

9. The two graphs at right show aspects of a ball thrown into the air. The first graph shows its height h in meters at any time t in seconds. The second graph shows its velocity v in meters per second at any time t.

   a. What does the first graph tell you about the situation? Use numbers to be as specific as you can.

   b. What does the second graph tell you about the situation? Use numbers to be as specific as you can.

   c. Give a real-world meaning in this context for the negative slope of the lower graph.

   d. What can you say about the ball when the graph of the velocity line intersects the x-axis?

   e. What can you say about the height of the ball when the velocity is 15 meters per second and when it is -15 meters per second?

   f. What are realistic domain and range intervals for the graphs? domain: $0 \le t \le 4.3$; range: $0 \le h(t) \le 26$; $-22 \le v(t) \le 20$

10. Bo and Gale are playing golf. Bo hits his ball, and it is in flight for 3.4 seconds. Gale's ball is in flight for 4.7 seconds.

   a. At what time does each ball reach its highest point?

   b. Can you tell whose ball goes farther or higher? Explain.

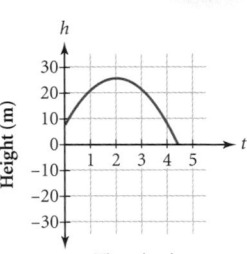

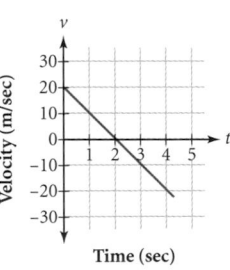

9d. This is when the ball is at its maximum height and not moving. Its velocity is zero.

9e. These are the two times when the ball is about 13 meters high.

Exercise 10 Be sure students understand that they don't have enough information to decide how far or high the ball travels. Gale's ball might have gone higher than Bo's or farther than Bo's.

10a. Bo's ball is at its highest point 0.5(3.4), or 1.7, seconds after it is hit. Gale's is at its highest 0.5(4.7), or 2.35, seconds after it is hit.

10b. You can't tell whose ball goes farther. Even though Gale's is in the air longer, it might just go higher and not as far as Bo's.

11. The table shows the coordinates of a parabola.

   a. On your calculator, plot the points in the table.

   b. Describe the location of the line of symmetry for this graph.

   c. Name the vertex of this graph. (4.5, 19)

   d. Use your knowledge of transformations to write the equation of this parabola in the vertex form, $y = a(x - h)^2 + k$. Check your answer graphically.
   $$y = -3(x - 4.5)^2 + 19$$

x	y
1.5	−8
2.5	7
3.5	16
4.5	19
5.5	16
6.5	7
7.5	−8

▶ **Review**

12. Write equations in the form $y = a + bx$ for each of these graphs. One tick mark represents one unit.

   a.  $y = 2 + \frac{1}{3}x$

   b.  $y = 3.5 - \frac{1}{4}x$

IMPROVING YOUR VISUAL THINKING SKILLS

A parabola is an example of a **conic section.** The Greek geometer Apollonius (255–170 B.C.) defined conic sections by intersecting a double cone with a plane.

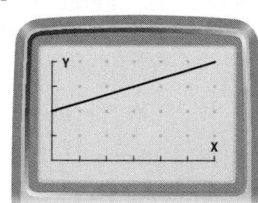

Plane section   Double cone

The plane is a flat surface that extends into infinity. Likewise, both ends of the double cone widen infinitely in opposite directions. To form a parabola, Apollonius sliced the cone with a plane parallel to the cone's edge.

Other examples of conic sections are circles, ellipses, and hyperbolas. How can you intersect a plane with a cone to form these shapes? Are there any other ways that a plane can intersect a double cone?

—Edge

Parabolic section

Circle   Ellipse   Hyperbola

Make a drawing that shows how to form each conic section. Can you form any other shapes?

IMPROVING VISUAL THINKING SKILLS

A plane through the cone's vertex might intersect the cone in a point (if the plane is perpendicular to the axis) or a pair of lines (if the plane contains the axis). These figures are called *degenerate conics.* Planes not through the vertex intersect the cone in the conic sections described. A plane perpendicular to the axis will form a circle, a special case of an ellipse, in which the plane is tilted but not to the extent that it becomes parallel to the edge, when it forms a parabola. Any greater tilt of the plane, including the illustrated hyperbolic section with plane parallel to the cone's axis, will intersect both pieces of the cone and form a hyperbola.

11a.

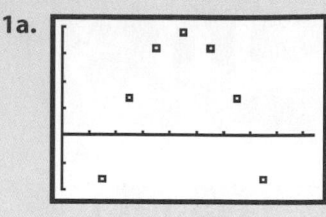

$[0, 9.4, 1, -10, 20, 5]$

11b. It is the vertical line $x = 4.5$.

Exercise 12 This exercise reviews Chapter 4.

Conic Sections

Circular section

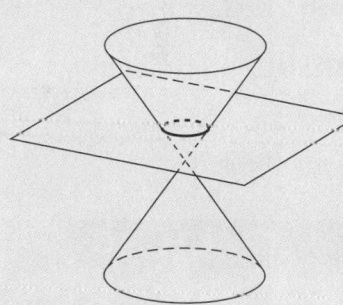

Elliptical section

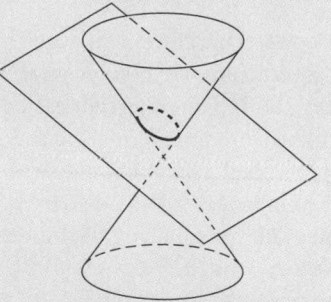

Hyperbolic section

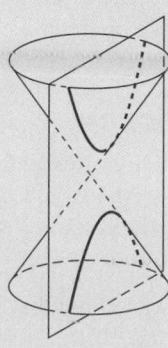

PLANNING

LESSON OUTLINE

First day:

5 min	Introduction
20 min	Example A
25 min	Investigation

Second day:

20 min	Investigation
10 min	Sharing
5 min	Example B
5 min	Closing
10 min	Exercises

MATERIALS

- graph paper
- Calculator Note 7B

TEACHING

To convert a quadratic equation from vertex to general form, you need to square a binomial.

One step Sometimes you want to convert from vertex to general form to find a *y*-intercept easily. Doing so requires squaring the sum, called a *binomial*, into a trinomial. Ask groups to derive methods for squaring the binomials $x + 3$ and $x - 2$. As you observe, challenge students to substitute numbers for *x*, suggest drawing rectangles, and have groups make up their own vertex form to convert to a general form.

INTRODUCTION

Students may confuse letters representing *constants* (a, h, k, b, c) with letters representing *variables* (x, y).

A *term* is something that's added or subtracted. (In contrast, a *factor* is something that's multiplied or divided.) Terms are not necessarily restricted to positive

integer exponents. For example, $4x^{-1}$ might be a term in some expression, but then that expression would not be a polynomial.

[Language] The prefix *poly-* means "many."

▶ EXAMPLE A

This example illustrates the concept of a polynomial by giving examples and non-examples. Note that before counting the number of terms, products are taken and like terms are combined.

Attempt the impossible in order to improve your work.

BETTE DAVIS

From Vertex to General Form

You have learned two forms of a quadratic equation. The vertex form, $y = a(x - h)^2 + k$, gives you information about transformations of the parent function, $y = x^2$. You used the general form, $y = ax^2 + bx + c$, to model many situations of projectile motion. In this lesson you will learn how to convert an equation from the vertex form to the general form.

The general form, $y = ax^2 + bx + c$, is the sum of three terms. A **term** is an algebraic expression that represents only multiplication and division between variables and constants. A sum of terms with positive integer exponents is called a **polynomial.** Variables cannot appear as exponents in a polynomial.

Here are some examples of polynomials.

$$17x \qquad 4.7x^3 + 3x \qquad x^2 + 3x + 7 \qquad 47x^4 - 6x^3 + 0.28x + 7$$

The expression $17x$ has only one term, so it is called a **monomial.** The second expression has two terms and is called a **binomial.** The third expression is a **trinomial** because it has three terms. If there are more than three terms the expression is generally referred to as a polynomial.

EXAMPLE A   Is each algebraic expression a polynomial? If so, how many terms does it have? If not, give a reason why it is not a polynomial.

a. $3x^2 + 4x^{-1} + 7$ **b.** $2^x - 7.5x + 18$

c. $\dfrac{47}{x} + 28$ **d.** $3x + 1 + 2x$

e. $x^2 - x^{10}$ **f.** $-2x^3 \cdot 3x^2$

▶ Solution

Expression	Is it a polynomial?
a. $3x^2 + 4x^{-1} + 7$	No, because the term $4x^{-1}$ has a negative exponent.
b. $2^x - 7.5x + 18$	No, because 2^x has a variable as the exponent.
c. $\dfrac{47}{x} + 28$	No, because the term $\dfrac{47}{x}$ is equivalent to $47x^{-1}$.
d. $3x + 1 + 2x$	Yes, it is a polynomial. It is equivalent to the binomial $1 + 5x$, which has two terms.
e. $x^2 - x^{10}$	Yes. It has two terms and is a binomial.
f. $-2x^3 \cdot 3x^2$	Yes. It involves only multiplication of constants and variables. It is equivalent to the monomial $-6x^5$.

Terms that differ only in their coefficients, such as $3x$ and $2x$, are *like terms*. When you rewrite $3x + 2x$ as $5x$, you are *combining like terms*. In the investigation you will combine like terms when you convert an equation from the vertex form to the general form.

LESSON OBJECTIVES

- Change a quadratic equation from vertex form to polynomial form
- Learn to square a binomial and factor perfect square expressions using rectangular diagrams
- Solve problems using a quadratic equation modeling projectile motion

Investigation
Sneaky Squares

You will need
- graph paper

There are many different, yet equivalent, expressions for a number. For example, 7 is the same as $3 + 4$ and as $10 - 3$. In this investigation you will use these equivalent expressions to model squaring binomials with rectangular diagrams.

Step 1
$(3 + 4)^2 = 9 + 2(12) + 16$
$= 49; 49; 7^2 = (3 + 4)^2$

Step 1 | This diagram shows how to express 7^2 as $(3 + 4)^2$. Find the area of each of the inner rectangles. What is the sum of the rectangular areas? What is the area of the overall square? What conclusions can you make?

Step 2 | For each expression below, draw a diagram on your graph paper like the one in Step 1. Label the area of each rectangle and find the total area of the overall square.

a. $(5 + 3)^2$ 64 **b.** $(4 + 2)^2$ 36 **c.** $(10 + 3)^2$ 169 **d.** $(20 + 5)^2$ 625

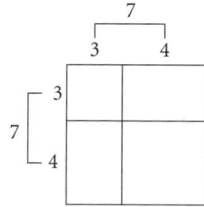

Even though lengths and areas are not negative, you can use the same kind of rectangular diagram to square an expression involving subtraction. You can use different colors, such as red and blue, to distinguish between the negative and the positive numbers. For example, this diagram shows 7^2 as $(10 - 3)^2$.

Step 3 | Draw a rectangular diagram representing each expression. Label each inner rectangle and find the sum.

a. $(5 - 2)^2$ 9 **b.** $(7 - 3)^2$ 16 **c.** $(20 - 2)^2$ 324 **d.** $(50 - 3)^2$ 2209

You can make the same type of rectangular diagram to square an expression involving variables.

Step 4 | Draw a rectangular diagram for each expression. Label each inner rectangle and find the total sum. Combine any like terms you see and express your answer as a trinomial.

a. $(x + 5)^2$ | **b.** $(x - 3)^2$ | **c.** $(x + 11)^2$ | **d.** $(x - 13)^2$
$x^2 + 10x + 25$ | $x^2 - 6x + 9$ | $x^2 + 22x + 121$ | $x^2 - 26x + 169$

NCTM STANDARDS

CONTENT		PROCESS	
	Number		Problem Solving
●	Algebra	●	Reasoning
	Geometry		Communication
	Measurement		Connections
	Data/Probability	●	Representation

Guiding the Investigation

Step 3 [Alert] Some students might need reminding that a positive number multiplied by a negative is negative, and a negative multiplied by a negative is positive. Others might question how an area can be negative. Point out that those rectangles with a −30 inside are representing a negative number, and that the rectangles' actual areas (which, like all areas, are positive) don't matter. Students can check their work by finding the square of each sum. For example, with $(5 - 2)^2$, students probably immediately know $3^2 = 9$, so they can expect the sum of the numbers in their diagram to be 9.

Step 1

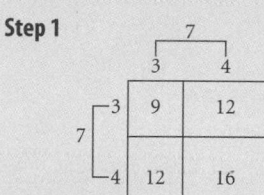

Step 2

Step 3

Step 4

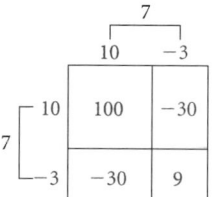

	x	3
x	x^2	$3x$
3	$3x$	9

	x	-5
x	x^2	$-5x$
-5	$-5x$	25

	x	4
x	x^2	$4x$
4	$4x$	16

	x	-6
x	x^2	$-6x$
-6	$-6x$	36

Step 6a $(x + 3)^2 = 49$;
$x + 3 = \pm\sqrt{49}$; $x = -3 \pm 7$;
$x = -10$ or 4

Step 6b $(x - 5)^2 = 81$;
$x - 5 = \pm\sqrt{81}$;
$x = 5 \pm 9$; $x = -4$ or 14

Step 6c $(x + 4)^2 = 121$;
$x + 4 = \pm\sqrt{121}$;
$x = -4 \pm 11$; $x = -15$ or 7

Step 6d $(x - 6)^2 = 64$; $x = 6 \pm 8$;
$x = -2$ or 14

Step 7 Students can make rectangular diagrams to help.

Step 8 You may want to challenge students by asking about a trinomial like $4x^2 + 12x + 9$, which is the square of $2x + 3$.

SHARING IDEAS

Ask for results of Step 6, but, as usual, emphasize the process over the answers.

Have students share their ideas about shortcuts in Step 9. Some may enjoy the game of seeing how fast they can say "The square of a binomial is the square of the first plus twice the product of the first and the second plus the square of the second."

You might point out that the square of a sum is not usually the sum of the squares. **[Ask]** "Under what conditions is it?" [If one term is 0. Here is a simple proof: Assume $(x + y)^2 = x^2 + y^2$ then $x^2 + y^2 = x^2 + 2xy + y^2$, so $2xy = 0$. Either x or y must be zero.

Assessing Progress

Assess general facility with arithmetic and order of operations.

Step 5 You must divide the middle term by 2.

Now use what you have learned to create a rectangular diagram for a trinomial.

Step 5 Make a rectangular diagram for each trinomial. In so doing, what must you do with the middle term? Label each side of the overall square in your diagram, and write the equivalent expression in the form $(x + h)^2$.

a. $x^2 + 6x + 9$ $(x + 3)^2$ **b.** $x^2 - 10x + 25$ $(x - 5)^2$
c. $x^2 + 8x + 16$ $(x + 4)^2$ **d.** $x^2 - 12x + 36$ $(x - 6)^2$

Step 6 Use your results from Step 5 to solve each new equation symbolically. Remember, quadratic equations can have two solutions.

a. $x^2 + 6x + 9 = 49$ **b.** $x^2 - 10x + 25 = 81$
c. $x^2 + 8x + 16 = 121$ **d.** $x^2 - 12x + 36 = 64$

Numbers like 49 are called **perfect squares** because they are the squares of integers, in this case, 7 or -7. The trinomial $x^2 + 6x + 9$ is the square of $x + 3$. So it is also called a perfect square.

Step 7 Which of these trinomials are perfect squares?

a. $x^2 + 14x + 49$ yes; $(x + 7)^2$
b. $x^2 - 18x + 81$ yes; $(x - 9)^2$
c. $x^2 + 20x + 25$ no
d. $x^2 - 12x - 36$ no

Step 8 When the coefficient of x^2 is 1, the last term is the square of half the coefficient of x.

Step 8 Explain how you can recognize a perfect-square trinomial when the coefficient of x^2 is 1. What is the connection between the middle term and the last term?

Step 9 $x^2 + 2hx + h^2$.
Possible answer: double h to get the coefficient of x and square h to get the constant term.

Step 9 Square the expression $(x + h)^2$ by making a rectangular diagram. Then describe a shortcut for this process that makes sense to you.

Knowing how to square a binomial is a useful skill. It allows you to convert equations from vertex form to general form.

EXAMPLE B Rewrite $y = 2(x + 3)^2 - 5$ in the general form, $y = ax^2 + bx + c$.

▶ Solution

			x	3
$y = 2(x + 3)^2 - 5$	The original equation.	x	x^2	$3x$
$y = 2(x^2 + 6x + 9) - 5$	Square the binomial using a rectangular diagram, as shown.			
$y = 2x^2 + 12x + 18 - 5$	Use the distributive property.	3	$3x$	9
$y = 2x^2 + 12x + 13$	Combine like terms.			

You can use a graph or a table on your calculator to verify that the vertex form and the general form of this equation are equivalent. [▶ See **Calculator Note 7B** to review checking different forms of an equation. ◀]

In the example you **expand** $(x + 3)^2$ when you rewrite it as $x^2 + 6x + 9$ in finding the vertex form. The vertex form tells you about translations, reflections, stretches, and shrinks of the graph of the parent function, $y = x^2$. The general form tells you the initial position, velocity, and the force due to gravity in projectile motion applications. Later in the chapter you will learn to convert the general form to the vertex form. Then you'll be able to solve all forms of quadratic equations symbolically.

EXERCISES

You will need your calculator for problems **2**, **4**, and **11**.

▶ Practice Your Skills

1. Is each algebraic expression a polynomial? If so, how many terms does it have? If it is not, give a reason why it is not a polynomial.

a. $x^2 + 3x - 8$ yes; three terms (trinomial)

b. $2x - \dfrac{4}{5}$ yes; two terms (binomial)

c. $5x^{-1} - 2x^2$ no; The first term has a negative exponent.

d. $\dfrac{3}{x^2} - 5x + 2$ no; The first term is equivalent to $3x^{-2}$, which has a negative exponent.

e. $6x$ yes; one term (monomial)

f. $\dfrac{x^2}{3^{-2}} + 5x - 8$ yes; three terms (trinomial)

g. $10x^3 + 5x^2$ yes; two terms (binomial)

h. $3(x - 2)$ Not a polynomial as written but it is equivalent to $3x - 6$, a binomial.

2. Expand each expression. On your calculator, enter the original expression into Y1 and the expanded expression into Y2. With a graph or a table, check that both forms are equivalent.

a. $(x + 5)^2$ $x^2 + 10x + 25$
b. $(x - 7)^2$ $x^2 - 14x + 49$
c. $3(x - 2)^2$ $3(x^2 - 4x + 4)$ or $3x^2 - 12x + 12$

3. Copy each rectangular diagram and fill in the missing values. Then write a squared binomial and an equivalent trinomial that both represent the total area for each diagram.

a.

	?	2
x	x^2	?
?	?	4

b.

	x	?
?	?	$12x$
12	?	144

c.

	x	?
?	x^2	$-7x$
-7	?	?

$(x + 2)^2 = x^2 + 4x + 4$ $(x + 12)^2 = x^2 + 24x + 144$ $(x - 7)^2 = x^2 - 14x + 49$

4. Convert each expression from vertex form to general form. Check your answers by entering the expressions into the Y= screen on your calculator.

a. $(x + 5)^2 + 4$ $x^2 + 10x + 29$
b. $2(x - 7)^2 - 8$ $2x^2 - 28x + 90$
c. $-3(x + 4)^2 + 1$ $-3x^2 - 24x - 47$
d. $0.5(x - 3)^2 - 4.5$ $0.5x^2 - 3x$

▶ Helping with the Exercises

Exercise 1 The expression in 1f is equivalent to the trinomial $9x^2 + 5x - 8$. You may have posted the new words from the lesson introduction. Students can refer to them and give specific names for the polynomials.

3a.

	x	2
x	x^2	$2x$
2	$2x$	4

3b.

	x	12
x	x^2	$12x$
12	$12x$	144

3c.

	x	-7
x	x^2	$-7x$
-7	$-7x$	49

▶ EXAMPLE B

This example shows how to change an equation from vertex form to general form. You could point out that the general form shows the y-intercept easily, although it can be found from the vertex form by substituting 0 for x.

Reinforcing Relationships
A chart can help students summarize and see relationships. Students can make a chart with six columns headed *equation, graph, vertex, line of symmetry, roots,* and *compare with $y = x^2$.* Students then list examples of quadratic equations in both the vertex form and the general form, filling in all six columns for each equation.

Closing the Lesson

To convert a quadratic equation from vertex to general form, you need to square a *binomial.* That can be done with the help of a rectangular diagram, even if the binomial includes subtraction.

BUILDING UNDERSTANDING

Students practice squaring binomials, multiplying one binomial by another, and applying the vertex form of a quadratic equation.

ASSIGNING HOMEWORK

Essential	1–6, 9, 10
Performance assessment	6, 8
Portfolio	11
Journal	7, 13
Group	5, 8, 12
Review	14, 15

Exercise 5 This exercise lays groundwork for the investigation of Lesson 10.4.

5a.

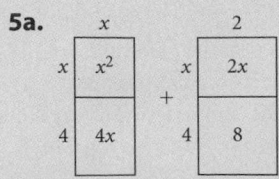

$(x + 2)(x + 4) = x^2 + 6x + 8$

5b.

	x			3
x	x^2		x	$3x$
		$+$		
5	$5x$		5	15

$(x + 3)(x + 5) = x^2 + 8x + 15$

5c.

	x			2
x	x^2		x	$2x$
		$+$		
-5	$-5x$		-5	-10

$(x - 5)(x + 2) = x^2 - 3x - 10$

5d.

	x			0
x	x^2		x	$0x$
		$+$		
-3	$-3x$		-3	0

$(x - 0)(x - 3) = x^2 - 3x$

Exercise 7 Press for an explanation of why it is not correct.

7. No, by squaring the values inside the parentheses, Heather is accounting for only two of the four rectangles in a squaring diagram. She needs to add the two rectangles that sum to the middle term.

Exercise 8 [Alert] Students may be confused by which variable represents speed and which distance in the relationship. We say that an equation "relates y to x." The variable y is the stopping distance and x is the speed. Students can solve the equation in several ways. One method is to approximate the x-coordinate of the rightmost intersection point of $Y_1 = 50$ and $Y_2 = 0.0056x^2 + 0.14x$.

▶ **Reason and Apply**

5. You can use the distributive property to write an equivalent expression for the product of two binomials. For example, you can write $(x + 3)(x + 4)$ as $x(x + 4) + 3(x + 4)$ or $x(x + 3) + 4(x + 3)$.

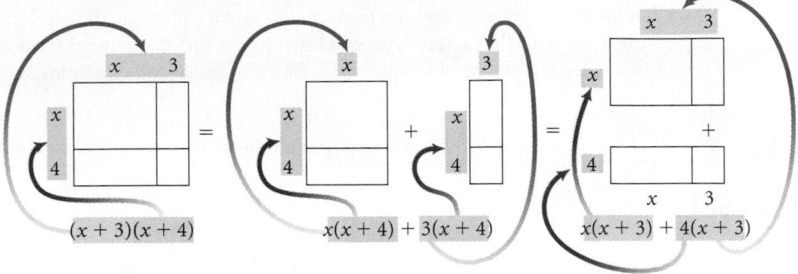

$(x + 3)(x + 4)$　　　$x(x + 4) + 3(x + 4)$　　　$x(x + 3) + 4(x + 3)$

Draw a rectangular diagram to represent each expression. Then write an equation showing the product of the two binomials and the equivalent polynomial in general form.

a. $x(x + 4) + 2(x + 4)$ **b.** $x(x + 5) + 3(x + 5)$

c. $x(x - 5) + 2(x - 5)$ **d.** $x(x - 3) - 0(x - 3)$

6. Consider the graph of the parabola $y = x^2 - 4x + 7$.

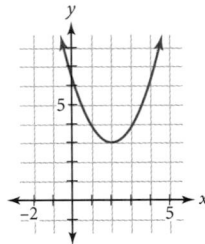

a. What are the coordinates of the vertex? **The coordinates of the vertex are (2, 3).**

b. Write the equation in vertex form. $y = (x - 2)^2 + 3$

c. Check that the equation you wrote in 6b is correct by expanding it to general form. $x^2 - 4x + 4 + 3 = x^2 - 4x + 7$

7. Heather thinks she has found a shortcut to the rectangular diagram method of squaring a binomial. She says that you can just square everything inside the parentheses. That is, $(x + 8)^2$ would be $x^2 + 64$. Is Heather's method correct?

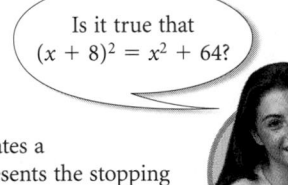

Is it true that $(x + 8)^2 = x^2 + 64$?

8. **APPLICATION** The quadratic equation $y = 0.0056x^2 + 0.14x$ relates a vehicle's stopping distance to its speed. In this equation, y represents the stopping distance in meters and x represents the vehicle's speed in kilometers per hour.

a. Find the stopping distance for a vehicle traveling 100 kph. **70 meters**

b. Write and solve an equation to find the speed of a vehicle that took 50 meters to stop. $50 = 0.0056x^2 + 0.14x$; $x \approx 83$ kph

9. The function $h(t) = -4.9(t - 0.4)^2 + 2.5$ describes the height of a softball thrown by a pitcher.

    a. How high does the ball go? The vertex is at (0.4, 2.5), so the ball reaches a maximum height of 2.5 meters.

    b. What is an equivalent function in general form? $h(t) = -4.9t^2 + 3.92t + 1.716$

    c. At what height did the pitcher release the ball when t was 0 seconds?
    The pitcher released the ball at a height of 1.716 meters.

Lori Harrigan of the USA softball team pitches at the 2000 Olympics in Sydney, Australia.

10. Is the expression on the right equivalent to the expression on the left? If not, correct the right side to make it equivalent.

    a. $(x + 7.5)^2 - 3 \overset{?}{=} x^2 + 15x + 53.25$ yes

    b. $2(x - 4.7)^2 + 2.8 \overset{?}{=} 2x^2 - 9.4x - 41.38$ no; The right side should be $2x^2 - 18.8x + 46.98$.

    c. $-3.5(x + 1.6)^2 - 2.04 \overset{?}{=} -3.5x^2 + 11.2x - 11$ no; The right side should be $-3.5x^2 - 11.2x - 11$.

    d. $-4.9(x - 5.6)^2 + 8.9 \overset{?}{=} -4.9x^2 + 54.88x - 144.764$ yes

11. The Yo-yo Warehouse uses the equation $y = -85x^2 + 552.5x$ to model the relationship between income and price for one of its top-selling yo-yos. In this model, y represents income in dollars and x represents the selling price in dollars of one item.

    a. Graph this relationship on your calculator, and describe a meaningful domain and range for this situation.

    b. Describe a method for finding the vertex of the graph of this relationship. What is the vertex?

    c. What are the real-world meanings of the coordinates of the vertex?

    d. What is the real-world meaning of the two x-intercepts of the graph? The prices produce no income.

    e. Interpret the meaning of this model if $x = 5$.
    If the warehouse charges $5 per item, income will be $637.50.

12. Use a three-by-three rectangular diagram to square each trinomial.

    a. $(x + y + 3)^2$ **b.** $(2x - y + 5)^2$

13. What is the general form of $(x + 4)^2$? Write a paragraph describing several ways to rewrite this expression in general form. $x^2 + 8x + 16$. Answers will vary.

Exercise 10 One way to check the equivalence of two expressions is to enter one as Y₁ and the other as Y₂ and compare their graphs or tables.

Exercise 11 You may need to explain that revenue and income mean the same thing. As the price rises above about $4.25, enough fewer people buy that the revenue goes down. As needed, help students find a good viewing window.

11a. meaningful domain: $0 \leq x \leq 6.5$; meaningful range: $0 \leq y \leq 897.81$

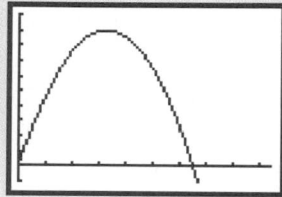

$[0, 9.4, 1, -100, 1000, 100]$

11b. Possible answer: Average the two x-intercepts, then substitute this value into the equation to find the y-coordinate of the vertex. The vertex is (3.25, 897.8125).

11c. The x-coordinate is the price that produces maximum income. The y-coordinate is the maximum income for this relationship.

Exercise 13 Students will probably describe the rectangular diagram method, the use of the distributive property, or the use of the pattern $(a + b)^2 = a^2 + 2ab + b^2$.

12a.

	x	y	3
x	x^2	xy	$3x$
y	xy	y^2	$3y$
3	$3x$	$3y$	9

$x^2 + y^2 + 2xy + 6x + 6y + 9$

12b.

	$2x$	$-y$	5
$2x$	$4x^2$	$-2xy$	$10x$
$-y$	$-2xy$	y^2	$-5y$
5	$10x$	$-5y$	25

$4x^2 + y^2 - 4xy + 20x - 10y + 25$

Exercise 14 This exercise reviews Lesson 8.2.

14. Is the parabola a graph of a function? No, it doesn't pass a vertical line test.

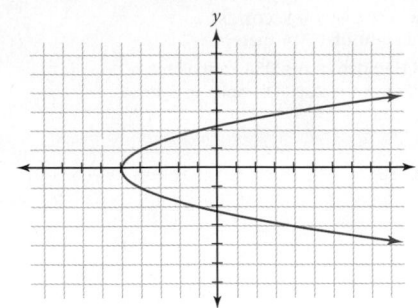

Exercise 15 This exercise reviews Lesson 8.4.

15. Use the graph of $f(x)$ to evaluate each expression. Then think of the numbers 1 through 26 as the letters A through Z to decode a message. The message is POLYNOMIALS.

a. $f(18)$ 16; P

b. $3 \cdot f(3)$ 15; O

c. $f(4^2)$ 12; L

d. $(f(3))^2$ 25; Y

e. $f(17)$ 14; N

f. $f(25)$ 15; O

g. $f(5) + f(15)$ 13; M

h. the greater x-value when $f(x) = 1$ 9; I

i. $f(1) - f(2)$ 1; A

j. $f(4) \cdot f(5)$ 12; L

k. $f(5^2 - 2^2)$ 19; S

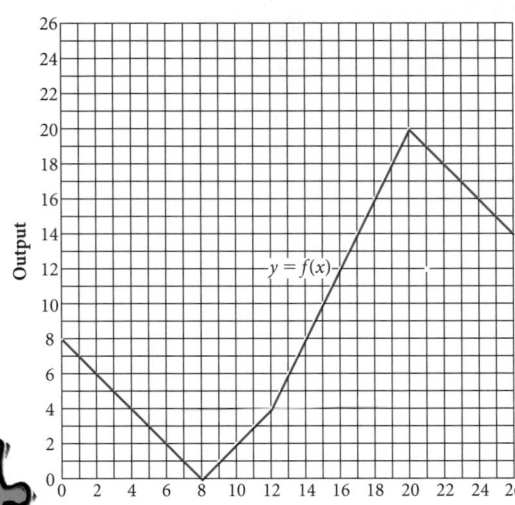

Factored Form

So far you have worked with quadratic equations in vertex form and general form. This lesson will introduce you to another form of quadratic equation, the **factored form:**

$$y = a(x - r_1)(x - r_2)$$

This form helps you identify the roots, r_1 and r_2, of an equation. In the investigation you'll discover connections between the equation in factored form and its graph. You'll also use rectangular diagrams to convert the factored form to the general form and vice versa. Then in the example you'll learn how to use a special property to find the roots of an equation.

Mathematicians assume the right to choose, within the limits of logical contradiction, what path they please in reaching their results.

HENRY ADAMS

Investigation
Getting to the Root of the Matter

You will need

- graph paper

First you'll find the roots of an equation in factored form from its graph.

$$y = a(x - r_1)(x - r_2)$$

Step 1 On your calculator, graph the equations $y = x + 3$ and $y = x - 4$ at the same time.

Step 2 $x = -3$ and $x = 4$

Step 2 What is the x-intercept of each equation you graphed in Step 1?

Step 3 Graph $y = (x + 3)(x - 4)$ on the same set of axes as before. Describe the graph. Where are the x-intercepts of this graph?

Step 4 $y = x^2 - x - 12$; the graph of $y = x^2 - x - 12$ is the same as $y = (x + 3)(x - 4)$ and has identical x-intercepts. So the graph of $y = (x + 3)(x - 4)$ is indeed a parabola.

Step 4 Expand $y = (x + 3)(x - 4)$ to general form. Graph the equation in general form on the same set of axes. What do you notice about this parabola and its x-intercepts? Is the graph of $y = (x + 3)(x - 4)$ a parabola?

NCTM STANDARDS

CONTENT	PROCESS
• Number	Problem Solving
• Algebra	• Reasoning
Geometry	• Communication
Measurement	• Connections
Data/Probability	• Representation

LESSON OBJECTIVES

- Learn that the roots of a quadratic equation can be found quickly from its factored form
- Study factoring of the general form of a quadratic equation
- Explore the relationships among the three forms of a quadratic equation: factored form, vertex form, and general form

LESSON

10.4

PLANNING

LESSON OUTLINE

One day:

20 min	Investigation
5 min	Sharing
5 min	Example
5 min	Closing
15 min	Exercises

MATERIALS

- graph paper

TEACHING

The factored form of a quadratic equation is useful in finding roots (x-intercepts) quickly. It can be converted to and from the other two forms.

Guiding the Investigation

One step **[Ask]** "What are two numbers that have a product of 0?" Entertain all ideas, being sure that products of opposites or reciprocals are rejected. Introduce the term *zero product property*. **[Ask]** "What can you conclude about x if you know that $(x + 3)(x - 4) = 0$?" Point out that if you can get a quadratic equation into this *factored form*, you can find the roots easily. Then challenge the groups to find the factored form of $x^2 + 5x + 6 = 0$. As you circulate, encourage the use of rectangular diagrams and have students check their conjectures with graphs. Some students may also do the factoring by first graphing and finding the x-intercepts.

See pages 724–725 for answers to Steps 1 and 3.

Step 2 Encourage students to determine these numbers without tracing.

Step 3 At this point students have no definitive evidence that the graph of $y = (x + 3)(x - 4)$ is a parabola.

Step 4 This kind of expanding was done in Exercise 5 of Lesson 10.3. Since students know the graph of $y = x^2 - x - 12$ is a parabola and since the equations are equivalent, they are assured that $y = (x + 3)(x - 4)$ does indeed graph as a parabola. They can double-check that $y = x^2 - x - 12$ is equivalent to $y = (x + 3)(x - 4)$ by comparing calculator tables.

Step 6 As needed, remind students of the introduction's definition of "factored form."

Step 6 $(x + 3)(x + 2)$. Enter $x^2 + 5x + 6$ into Y₁ and $(x + 3)(x + 2)$ into Y₂; a graph or table confirms equivalency.

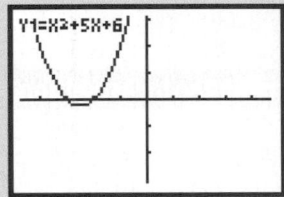

$[-5, 5, 1, -3, 3, 1]$

Step 8 Students can explore the relationships between the factored form of a quadratic equation, its roots, and its graph using the dynamic exploration at www.keymath.com/DA.

Now you'll learn how to find the roots from the general form.

Step 5 Complete the rectangular diagram whose sum is $x^2 + 5x + 6$. A few parts on the diagram have been labeled to get you started.

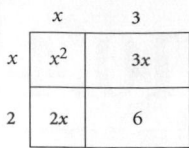

Step 6 Write the multiplication expression of the rectangular diagram in factored form. Use a graph or table to check that this form is equivalent to the original expression.

Step 7 $x = -3$ and $x = -2$

Step 7 Find the roots of the equation $0 = x^2 + 5x + 6$ from its factored form.

Step 8 Rewrite each equation in factored form by completing a rectangular diagram. Then find the roots of each. Check your work by making a graph.

a. $0 = x^2 - 7x + 10 \ (x - 5)(x - 2)$
b. $0 = x^2 + 6x - 16 \ (x - 2)(x + 8)$
c. $0 = x^2 + 2x - 48 \ (x - 6)(x + 8)$
d. $0 = x^2 - 11x + 28 \ (x - 7)(x - 4)$

Now you have learned three forms of a quadratic equation. You can enter each of these forms into your calculator to check that they are equivalent. Here are three equivalent equations that describe the height in meters, y, of an object in motion for x seconds after being thrown upward. Each equation gives different information about the object.

Vertex form	$y = -4.9(x - 1.7)^2 + 15.876$
General form	$y = -4.9x^2 + 16.66x + 1.715$
Factored form	$y = -4.9(x + 0.1)(x - 3.5)$

Which form is best? The answer depends on what you want to know. The vertex form tells you when the maximum height occurs—in this case, 15.876 meters after 1.7 seconds (the vertex). The general form tells you that the object started at a height of 1.715 meters (the y-intercept). The coefficients of x and x^2 give some information about the starting velocity and acceleration. The factored form tells you the times at which the object's height is zero (the roots).

You have already learned how to convert to and from the general form of a quadratic equation. The example will show you how to get the vertex form from the factored form.

EXAMPLE

Write the equation for this parabola in vertex form, general form, and factored form.

▶ Solution

From the graph you can see that the x-intercepts are 3 and -5. So the factored form contains the binomial expressions $(x - 3)$ and $(x + 5)$.

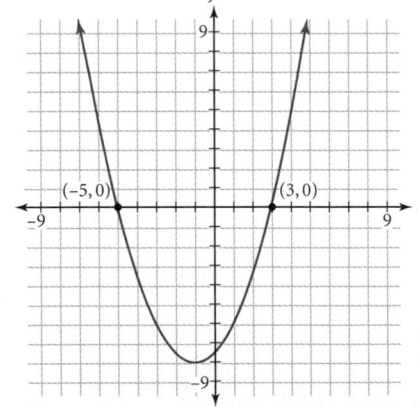

SHARING IDEAS

Have groups present their results from Step 8. Introduce the term *to factor* a trinomial and point out that students should expect to guess and check when factoring. (Although there are algorithmic methods for factoring, students will understand more by checking with rectangular diagrams or distribution.)

If you have time, you might show the two graphs from Step 1 of the investigation and ask whether students could predict where the graph of $y = (x + 3)(x - 4)$ is positive and negative by looking at the two linear graphs.

Three Forms

For the projectile motion illustrated, ask what the roots mean. Only one root, 3.5, is meaningful, because a negative root refers to a time before the object was thrown. Posting examples of these three forms will help many students grasp these concepts.

If you graph $y = (x - 3)(x + 5)$ on your calculator, you'll see it has the same x-intercepts as the graph shown here, but a different vertex. The new vertex is at $(-1, -16)$.

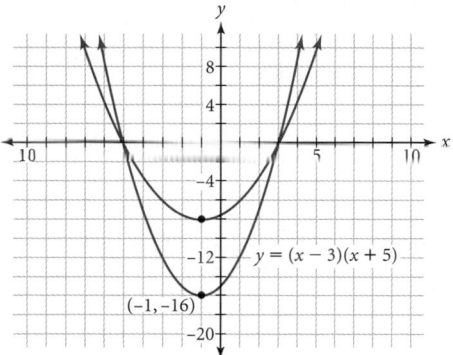

The new vertex needs to be closer to the x-axis, so you need to find the vertical shrink factor a.

The original vertex of the graph shown is $(-1, -8)$. So the graph of the function must have a vertical shrink by a factor of $\frac{-8}{-16}$, or 0.5. The factored form is $y = 0.5(x - 3)(x + 5)$. A calculator graph of this equation looks like the desired parabola.

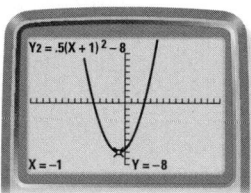

$[-15, 15, 1, -10, 10, 1]$

Now you know that the value of a is 0.5 and that the vertex is at $(-1, -8)$. Substitute this information into the vertex form to get $y = 0.5(x + 1)^2 - 8$.

Expand both forms to find the general form.

$y = 0.5(x - 3)(x + 5)$	The original equations.	$y = 0.5(x + 1)^2 - 8$
$y = 0.5(x^2 - 3x + 5x - 15)$	Expand using rectangular diagrams.	$y = 0.5(x^2 + 1x + 1x + 1) - 8$

	x	-3
x	x^2	$-3x$
5	$5x$	-15

	x	1
x	x^2	$1x$
1	$1x$	1

$y = 0.5(x^2 + 2x - 15)$	Combine like terms.	$y = 0.5(x^2 + 2x + 1) - 8$
$y = 0.5x^2 + x - 7.5$	Distribute and combine.	$y = 0.5x^2 + x - 7.5$

So the three forms of the quadratic equation are

Vertex form	$y = 0.5(x + 1)^2 - 8$
General form	$y = 0.5x^2 + x - 7.5$
Factored form	$y = 0.5(x - 3)(x + 5)$

Assessing Progress

Keep an eye out for abilities to graph equations, solve linear equations, find x-intercepts by tracing, and multiply two binomials.

USING THE QUOTE

The text after the investigation contrasts the usefulness of the three forms of quadratic equations. Draw the class's attention to the quotation opening the lesson and ask whether it applies to the current topic. Point out the three forms of quadratic equations in the student text. Ask about which of these three "paths" they might choose to reach which results.

▶ EXAMPLE

You might ask students what numbers they can multiply together to get 0. Some may suggest opposites until they realize they are thinking of adding. Some may suggest reciprocals, thinking of a product of 1. The goal is to dramatize the *zero product property*—if a product is 0, then at least one of the factors must be 0.

To obtain the vertex coordinates shown on the screen, evaluate the function at $x = -1$ from the graph screen.

Closing the Lesson

The roots r_1 and r_2 of a quadratic equation show up clearly in the factored form $a(x - r_1)(x - r_2) = 0$. The factored form can be converted to the general form by distributing, using a rectangular diagram as needed. The general form can be changed to the factored form by factoring, also using a rectangular diagram.

BUILDING UNDERSTANDING

Students practice solving quadratic equations by factoring, explore how the roots can be combined to get coefficients of the general form, and study the difference of two squares.

ASSIGNING HOMEWORK

Essential	1–7, 11
Performance assessment	6, 9
Portfolio	7, 10
Journal	11f, 12
Group	8, 9, 11
Review	13, 14

▶ Helping with the Exercises

1a. $x + 4 = 0$ or $x + 3.5 = 0$, so $x = -4$ or $x = -3.5$

1b. $x - 2 = 0$ or $x - 6 = 0$, so $x = 2$ or $x = 6$

1c. $x + 3 = 0$ or $x - 7 = 0$ or $x + 8 = 0$, so $x = -3$ or $x = 7$ or $x = -8$

1d. $x = 0$ or $x - 9 = 0$ or $x + 3 = 0$, so $x = 0$ or $x = 9$ or $x = -3$

2a.

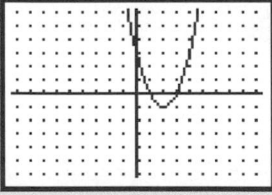

$[-9.4, 9.4, 1, -6.2, 6.2, 1]$

2b.

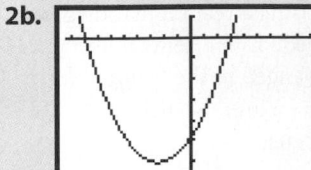

$[-9.4, 9.4, 1, -35, 5, 5]$

2c.

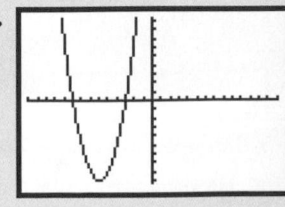

$[-14.1, 14.1, 1, -9.3, 9.3, 1]$

554 CHAPTER 10 Quadratic Models

When finding roots it is helpful to use the factored form. In this example one root is 3 because it makes $(x - 3)$ equal to 0. The other root is -5 because it makes $(x + 5)$ equal to 0. Think of numbers that multiply to zero. If $ab = 0$ or $abc = 0$, the **zero product property** tells you that a, or b, or c must be 0. In an equation like $(x + 3)(x - 5) = 0$, *at least one of the factors must be zero.* The roots of an equation are sometimes called the **zeros** of a function because they make the value of the function equal to zero.

EXERCISES

You will need your calculator for problems **2, 3, 8,** and **11.**

▶ Practice Your Skills

1. Use the zero product property to solve each equation.

   a. $(x + 4)(x + 3.5) = 0$ **b.** $2(x - 2)(x - 6) = 0$

   c. $(x + 3)(x - 7)(x + 8) = 0$ **d.** $x(x - 9)(x + 3) = 0$

2. Graph each equation and then rewrite it in factored form.

   a. $y = x^2 - 4x + 3$ $y = (x - 3)(x - 1)$ **b.** $y = x^2 + 5x - 24$ $y = (x + 8)(x - 3)$

   c. $y = x^2 + 12x + 27$ $y = (x + 3)(x + 9)$ **d.** $y = x^2 - 7x - 30$ $y = (x - 10)(x + 3)$

3. Name the *x*-intercepts for the parabola of each quadratic equation. Then check your answers with a graph.

   a. $y = (x - 7)(x + 2)$ 7 and -2 **b.** $y = 2(x + 1)(x + 8)$ -1 and -8

   c. $y = 3(x - 11)(x + 7)$ 11 and -7 **d.** $y = 0.4(x + 5)(x - 9)$ -5 and 9

4. Write an equation of a quadratic function that corresponds to each pair of *x*-intercepts. Assume there is no vertical stretch or shrink.

   a. 2.5 and -1 $y = (x - 2.5)(x + 1)$ **b.** -4 and -4 $y = (x + 4)(x + 4)$ or $y = (x + 4)^2$

   c. -2 and 2 $y = (x + 2)(x - 2)$ **d.** r_1 and r_2 $y = (x - r_1)(x - r_2)$

5. Consider the equation $y = (x + 1)(x - 3)$.

   a. How many *x*-intercepts does the graph have?   two intercepts; $x = -1$ and $x = 3$

   b. Find the vertex of this parabola. $(1, -4)$

   c. Write the equation in vertex form. Describe the transformations of the parent function, $y = x^2$. $y = (x - 1)^2 - 4$; There is a translation left 1 unit and down 4 units.

▶ Reason and Apply

6. Is the expression on the left equivalent to the expression on the right? If not, change the right side to make it equivalent.

   a. $x^2 + 7x + 12 \overset{?}{=} (x + 3)(x + 4)$ yes

   b. $x^2 - 11x + 30 \overset{?}{=} (x + 6)(x + 5)$ no; change to $(x - 6)(x - 5)$

   c. $2x^2 - 5x - 7 \overset{?}{=} (x - 3.5)(x + 1)$ no; change to $2(x - 3.5)(x + 1)$ or $(2x - 7)(x + 1)$ or $(x - 3.5)(2x + 2)$

   d. $4x^2 + 8x + 4 \overset{?}{=} (x + 1)^2$ no; change to $4(x + 1)^2$

   e. $x^2 - 25 \overset{?}{=} (x + 5)(x - 5)$ yes

   f. $x^2 - 36 \overset{?}{=} (x - 6)^2$ no; change to $(x + 6)(x - 6)$

2d.

$[-20, 20, 2, -50, 10, 2]$

3a.

$[-10, 10, 1, -25, 10, 5]$

3b.

$[-10, 10, 1, -25, 10, 5]$

Exercise 4 Because there's no vertical stretch or shrink, $a = 1$.

7. The sum and product of the roots of a quadratic equation are related to b and c in $y = x^2 + bx + c$. The first row in the table below will help you to recognize this relationship.

   a. Complete the table.

Factored form	Roots	Sum of roots	Product of roots	General form
$y = (x + 3)(x - 4)$	-3 and 4	$-3 + 4 = 1$	$(-3)(4) = -12$	$y = x^2 - 1x - 12$
$y = (x - 5)(x + 2)$	5 and -2	$5 - 2 = 3$	$5(-2) = -10$	$y = x^2 - 3x - 10$
$y = (x + 2)(x + 3)$	-2 and -3	-5	6	$y = x^2 + 5x + 6$
$y = (x - 5)(x + 5)$	5 and -5	0	-25	$y = x^2 - 25$

   b. Use the values of b and c to find the roots of $0 = x^2 + 2x - 8$.
     The sum of the roots needs to be -2, and the product needs to be -8; 2 and -4 satisfy this requirement.

8. In this problem you will discover whether or not knowing the x-intercepts determines a unique quadratic equation. Work through the steps in 8a–e to find an answer. Graph each equation to check your work.

   a. Write an equation for a parabola with x-intercepts at $x = 3$ and $x = 7$. $y = (x - 3)(x - 7)$

   b. Name the vertex of the parabola in 8a. $(5, -4)$

   c. Modify your equation in 8a so that the graph is reflected across the x-axis. Where are the x-intercepts? Where is the vertex? $y = -(x - 3)(x - 7)$; x-intercepts: $x = 3$ and $x = 7$; vertex: $(5, 4)$

   d. Modify your equation in 8a to apply a vertical stretch with a factor of 2. Where are the x-intercepts? Where is the vertex? $y = 2(x - 3)(x - 7)$; x-intercepts: $x = 3$ and $x = 7$; vertex: $(5, -8)$

   e. How many quadratic equations do you think there are with x-intercepts at $x = 3$ and $x = 7$? How are they related to one another?

9. Write a quadratic equation for a parabola with x-intercepts at -3 and 9 and vertex at $(3, -9)$. Express your answer in factored form. $y = 0.25(x + 3)(x - 9)$

10. **APPLICATION** The school ecology club wants to fence in an area along the riverbank to protect some endangered wildflowers that grow there. The club has enough money to buy 200 feet of fencing. It decides to enclose a rectangular space. The fence will form three sides of the rectangle, and the riverbank will form the fourth side.

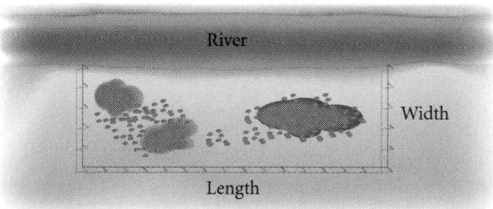

   a. If the width of the enclosure is 30 feet, how much fencing material is available for the length? Sketch this situation. What is the area? length: 140 ft; area: 4200 sq. ft

   b. If the width is w feet, how much fencing material remains for the length, l? $l = 200 - 2w$

   c. Use your answer from 10b to write an equation for the area of the rectangle in factored form. Check your equation with your width and area from 10a.

   d. Which two different widths would give an area equal to 0? $w = 0$ and $w = 100$

   e. Which width will give the maximum area? What is that area? width: 50 ft; maximum area: 5000 sq. ft

Exercise 7 You can extend this exercise by making a game of giving other sums (such as 8) and products (such as 7) and asking students to find the two numbers with that sum and product (1 and 7).

8e. Possible answer: There are infinitely many quadratic equations with roots of 3 and 7. Each is created by substituting a unique value of a into $y = a(x - 3)(x - 7)$. Hence, all are vertical stretches or shrinks and/or reflections across the x-axis. The x-coordinate of each vertex is always 5, but the y-coordinates depend on the value of a.

Exercise 9 Especially if students have worked on Exercise 8, they may wonder whether there's only one quadratic equation satisfying these conditions. Encourage thought about this question. Help them come to realize that any three points lie on only one parabola.

Exercise 10 Students may use guess-and-check to find the maximum, but try to get them all to realize that they can answer the question by finding the vertex of the graph of $A = w(200 - 2w)$.

10c. $A = w \cdot l = w(200 - 2w)$; $30(200 - 2 \cdot 30) = 30(140) = 4200$

3c.

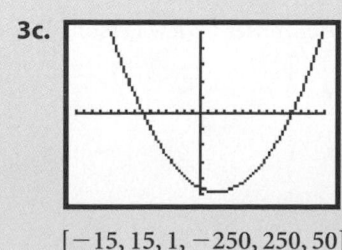

$[-15, 15, 1, -250, 250, 50]$

3d.

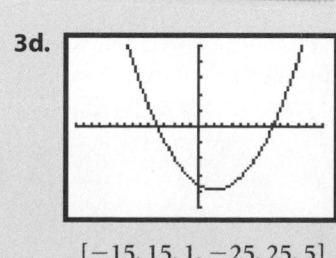

$[-15, 15, 1, -25, 25, 5]$

Exercise 11 This is the only place where the factoring of the differences of squares is introduced, so be sure to assign this problem if it's important to your curriculum.

In 11d, students can think of ii as $y = -1(x^2 - 16)$ and then factor the expression inside the parentheses.

In 11f, the imaginary roots are $2i$ and $-2i$, where $i^2 = -1$. Many students find this idea intriguing, so you may choose to introduce it. (Imaginary numbers are the subject of Take Another Look at the end of the chapter.) Imaginary numbers evolved from solving cubic, not quadratic, equations.

11a. two x-intercepts: $x = 3$ and $x = -3$

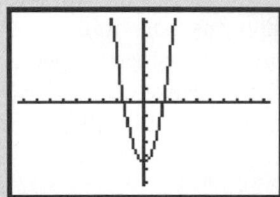

$[-18.8, 18.8, 2, -12.4, 12.4, 2]$

11c. The x-intercepts are the positive and negative square roots of the amount the parabola is translated, or $\pm\sqrt{h}$.

11e. There are no x-intercepts. The graph is above the x-axis.

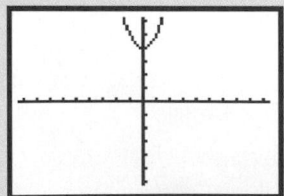

$[-9.4, 9.4, 1, -6.2, 6.2, 1]$

11f. Because there are no x-intercepts, there is no factored form of the equation.

---

11. Consider the equation $y = x^2 - 9$.
   a. Graph the equation. What are the x-intercepts?
   b. Write the factored form of the equation. $y = (x - 3)(x + 3)$
   c. How are the x-intercepts related to the original equation?
   d. Write each equation in factored form. Verify each answer by graphing.
      i. $y = x^2 - 49$ $y = (x + 7)(x - 7)$
      ii. $y = 16 - x^2$ $y = (4 + x)(4 - x)$
      iii. $y = x^2 - 47$ $y = (x + \sqrt{47})(x - \sqrt{47})$
      iv. $y = x^2 - 28$ $y = (x + \sqrt{28})(x - \sqrt{28})$
   e. Graph the equation $y = x^2 + 4$. How many x-intercepts can you see?
   f. Explain the difficulty in trying to write the equation in 11e in factored form.

12. Kayleigh says that the roots of $0 = x^2 + 16$ are 4 and -4 because $(4)^2 = 16$ and $(-4)^2 = 16$. Derek tells Kayleigh that there are no roots for this equation. Who is correct and why?

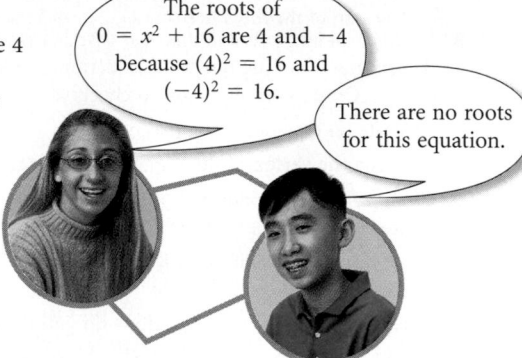

▶ **Review**

13. On graph paper, draw a function that has these properties:
   ▶ Domain of $-4 \leq x \leq 4$
   ▶ $f(-4) = 1$
   ▶ Range of $-3 \leq y \leq 3$
   ▶ $f(3) = 3$

Possible answer:

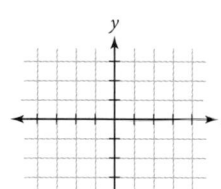

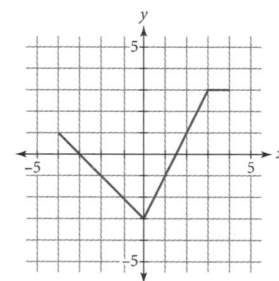

14. On graph paper, draw a graph that is NOT a function with a domain of $-4 \leq x \leq 4$ and a range of $-3 \leq y \leq 3$.

Possible answer:

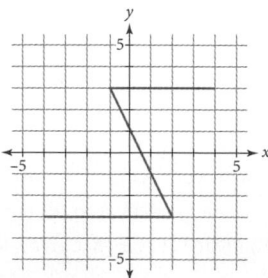

12. Answers will vary. Possible answer: Derek is correct because substituting 4 or -4 for x gives $0 = 16 + 16$, or 32, and not zero; the roots should satisfy the equation.

Exercises 13 and 14 These exercises review Lesson 8.4.

LESSON

10.5

Activity Day

Projectile Motion

You have already learned that quadratic equations model projectile motion. In this lesson you'll do an experiment with projectile motion and find a quadratic function to model the data. If you choose the first experiment, you'll collect data for the *x*-intercepts of a parabola and then find an equation in factored form that matches the graph. If you choose the second experiment, you'll collect parabolic data and then find an equation in vertex form that matches the graph. Read the steps of each experiment and then choose one experiment for your group to do.

Activity
Jump or Roll

You will need
- a motion sensor
- an empty coffee can
- a long table

Each experiment in this activity requires a calculator program. Be sure you have this program in your calculator before you begin. [▶ 🖳 See **Calculator Note 10A** for the required programs. ◀] In the first experiment you will collect data for the zeros of a projectile motion function.

Experiment 1: How High?

The object of this experiment is to find how high you jump.

Step 1 | Set up the program to collect data. Jump straight up without bending your knees in the air. Be sure to land in front of the sensor again. This way, the sensor records the times your feet left the ground and landed.

> **Procedure Note**
>
> Place the motion sensor on the floor. The jumper stands 2 ft or 0.5 m in front of it. There should be a wall or another object about 4 ft or 1 m from the sensor. When the jumper's feet leave the ground, the motion sensor should register a change in distance at a specific instant in time.

PLANNING

LESSON OUTLINE

One day:
45 min   Activities
 5 min   Closing

MATERIALS

- motion sensors
- Calculator Note 10A

Experiment 1
- How High? Sample Data (W), *optional*

Experiment 2
- empty coffee can
- long table
- Rolling Along Sample Data (W), *optional*

TEACHING

Quadratic equations can model motion of various sorts.

🄶 uiding the Activity

If you don't have the resources to collect data for these experiments, provide the students with sample data from the masters How High? and Rolling Along.

If your students are collecting data, try to give each of them a chance both to control the motion (jumping or rolling the can) and to run a motion sensor.

EXPERIMENT 1

The plotted data points will not lie on a parabola, because the motion detector is not following the jump. Instead, the graph will consist of three horizontal line segments from which students determine the root values to substitute, in Step 3, into a quadratic equation for motion.

NCTM STANDARDS

CONTENT	PROCESS
Number	• Problem Solving
• Algebra	• Reasoning
Geometry	• Communication
Measurement	• Connections
Data/Probability	• Representation

LESSON OBJECTIVES

- Use data-collection devices to collect real-world data that can be modeled by quadratic equations
- Write quadratic equations to model real-world data

Distance for the sample data is not height. Step 2 explains what distance is being measured.

The value of $-\frac{1}{2}g$ in Newton's motion equation $h(t) = \frac{1}{2}(-g)t^2 + v_0 t + h_0$ is usually given as -4.9 m/sec^2 or -16 ft/sec^2. For the jumper, however, it's more convenient to work in inches or cm. For inches, this value is $\frac{1}{2} \cdot \frac{-32 \text{ ft}}{\text{sec}^2} \cdot \frac{12 \text{ in.}}{1 \text{ ft}}$, or -192 in./sec^2. For centimeters, it is $\frac{1}{2} \cdot \frac{-9.8 \text{ m}}{\text{sec}^2} \cdot \frac{100 \text{ cm}}{1 \text{ m}}$, or -490 cm/sec^2.

If the motion didn't start when t was 0—as is probably the case for the jumper—then Newton's equation no longer holds. The value of a in the factored form $h = a(t - r_1)(t - r_2)$ equals $-\frac{1}{2}g$, as you can see from expanding $h = a(t - r_1)(t - r_2)$ to get $h = at^2 - a(r_1 + r_2)t + ar_1 r_2$ and comparing to Newton's form. But the other coefficients no longer represent the initial velocity and position. For example, if the roots are $r_1 = 0.09$ and $r_2 = 0.57$, then $-192(t - 0.09)(t - 0.57) = -192t^2 + 126.72t - 9.8496$. You can't conclude that the student jumped from a position 9.8496 inches below the ground with an initial velocity of 126.72 in./sec, because the student jumped from ground level. You can use calculus to find the initial velocity of this sample jump.

EXPERIMENT 2

The motion of the can along the ramp is not projectile motion, so the coefficients will not look familiar. Students can best represent the parabola by an equation in vertex form. By now, some students may have discovered that their calculators will do quadratic regression to find the parabola of best fit. Have them compare the calculator's equation to their own.

Step 2 Sample answer: Feet left the ground at 0.09 seconds and landed at 0.57 seconds.

Step 3 Let t represent the time in seconds and h represent the height in inches (or cm) above the ground. Answer from sample data is below.

Step 2 The data measured by the motion sensor has the form (*time, distance*), where the distance is that between the motion sensor and the nearest object to it. At first this distance is from the sensor to the jumper's feet. Then during the jump the sensor measures the distance to the wall behind the jumper. After the jumper lands the sensor reads the distance to the jumper's feet again. Look at the graph and use the trace feature to determine the instant the feet left the ground. Do this by finding the sharp change in y-values on the graph. Likewise, determine the instant in time when the feet landed back on the ground.

Step 3 If you want to graph the height of your jump over time, what are the variables for the quadratic function in this situation? Substitute the two roots you found in Step 2 for r_1 and r_2 into the equation $h = -192(t - r_1)(t - r_2)$. Use it to calculate the height of your jump in inches. (Or use the equation $h = -490(t - r_1)(t - r_2)$ to find this height in centimeters.) At what time did you reach this height? Explain how you got your answer.

Step 4 Repeat the experiment with each member of your group as a jumper.

Experiment 2: Rolling Along

The object of this experiment is to write a quadratic equation from projectile motion data.

Procedure Note

Prop up one end of the table slightly. Position the motion sensor at the high end of the table and aim it toward the low end.

Step 1 Practice rolling the can up the table directly in front of the sensor. The can should roll up the table, stop about 2 feet from the sensor, and then roll back down. Give the can a short push so that it rolls up the table on its own momentum. Then the force of gravity should cause the can to reverse directions as it rolls back down the slanted table.

Step 2 Set up the program to collect the data. When the sensor begins, gently roll the can up the table. Catch it as it falls off the table.

Step 3 The data collected by the sensor will have the form (*time, distance*). If you do the experiment correctly, the graph should show a parabolic pattern.

Step 4 Find the equation for a parabola that fits your data. Which points did you use to find the equation? In which form is it? Sketch a parabola for this equation onto the graph from Step 3.
Equation for sample data: $y = 0.85(x - 3.2)^2 + 2.74$. Graph is below.

Answers for sample data:
Experiment 1
roots at $t = 0.09$ and $t = 0.57$ sec;
$h = -192(t - 0.09)(t - 0.57)$
(or $h = -490(t - 0.09)(t - 0.57)$).
The vertex is (0.33, 11.06), so the height of the jump is about 11 in. (or 28.1 cm).

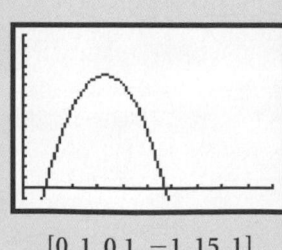

$[0, 1, 0.1, -1, 15, 1]$

Experiment 2

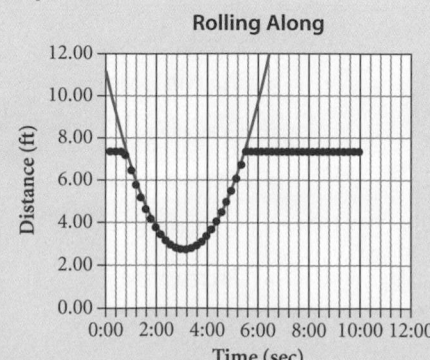

Rolling Along

project

PARABOLA BY DEFINITION

You have learned that the graph of a quadratic equation is a parabola. One definition of a parabola is the set of all points whose distance from a fixed point, the *focus,* is equal to its distance from a fixed line, the *directrix.* (Use the shortest possible distance for the distance between a point and a line.)

You can use The Geometer's Sketchpad or waxed paper to draw a parabola in various ways based on this definition. Start by drawing a line and a point not on the line. Then locate several points equally distant from the focus and the directrix by using the tools in Sketchpad, or by folding waxed paper. On waxed paper fold the focus to lie on the directrix and crease the paper. Repeat to make many creases. The creases will outline a parabola. If you make a similar set of lines in Sketchpad, you can test what happens to the parabola if the focus moves closer to (or farther from) the directrix.

Your project should include

▶ A drawing of the lines with the parabola's focus, directrix, and vertex labeled.

▶ An explanation of how you constructed the lines.

▶ A discussion of how the distance between the focus and the directrix affects the shape of the parabola.

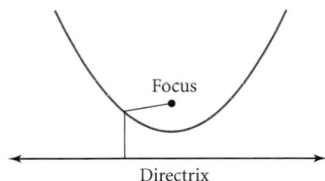

Find more information about parabolas with the helpful Internet links at
www.keymath.com/DA .

THE GEOMETER'S SKETCHPAD

The Geometer's Sketchpad was used to create this parabola. Sketchpad has tools to help you construct points, lines, and the set of points equidistant to both. Learn how to use these tools to create a parabola of your own.

SHARING IDEAS

One group doing each experiment might describe the procedure and another group the results.

Assessing Progress

Look for abilities to find the **vertex form** of a quadratic equation from its **roots** and to adjust an equation to match a graph.

Closing the Lesson

The motion of rising or falling bodies, along ramps as well as in free-fall, can be modeled with quadratic equations.

Using Sketchpad for the Project

A full parabola that will allow you to explore the effect of moving the focus can be drawn in The Geometer's Sketchpad by using this property: When you form an envelope for a parabola by folding waxed paper, the point on the crease that actually lies on the parabola is on the perpendicular to the directrix through the point the focus was placed over to make the crease. After constructing the focus A and directrix, construct a point D on the directrix; construct segment AD and its midpoint E; construct the line (the crease) perpendicular to this segment and through C; construct a perpendicular to the directrix through D; and construct the intersection point F of this line with the crease. Then select F, D, and the directrix and choose Locus from the Construct menu. You get the locus of F as D moves along the directrix—a parabola.

Supporting the project

MOTIVATION

A parabola can be defined as the set of points equidistant from a fixed point (its *focus*) and a fixed line (its *directrix*).

The dynamic sketch at www.keymath.com/DA can help students explore the geometric definition of a parabola.

OUTCOMES

▶ The report includes some representation of the parabola, such as drawings or sheets of folded waxed paper.

▶ There is a clear description of how the representation was created.

▶ The report describes accurately how the distance between the focus and the directrix affects the shape of the parabola.

• The report includes where the vertex is with respect to the focus and directrix.

• The report includes the fact that the distance from the vertex to the focus is $\frac{1}{4a}$, where a is the coefficient of x^2 in the equation of the parabola.

LESSON

10.6

PLANNING

LESSON OUTLINE

One day:

20 min	Investigation
5 min	Sharing
10 min	Examples
5 min	Closing
10 min	Exercises

MATERIALS

- graph paper
- Fun Training (T or W), *optional*

TEACHING

To change a quadratic equation from general to vertex form, or to solve the general form when it's not clear how to factor, you can complete the square.

Guiding the Investigation

One step Show the Fun Training transparency, or distribute it as a handout. Encourage those who approximate the solutions graphically to solve the problem exactly. Some students may want to solve the equation $x(50 - x) = 400$ by saying that $x = 400$ or $50 - x = 400$. Others will struggle to find the vertex form. Suggest that students use rectangular diagrams to help complete the square, getting $(x - 25)^2 = 225$, so that $x = 20$ or $x = 40$. Encourage students to see how the symmetry about $x = 25$ is represented by the expression $25 \pm \sqrt{225}$.

Step 1 If students are having difficulty, ask what number should replace the top question mark, to be multiplied by x to get the result in the top right box.

For every problem there is one solution which is simple, neat, and wrong.

H. L. MENCKEN

Completing the Square

You can always find approximate solutions to quadratic equations by using tables and graphs. If you can convert the equation to the factored form, $y = a(x - r_1)(x - r_2)$, or the vertex form, $y = a(x - h)^2 + k$, then you can use symbolic methods to find exact solutions. In this lesson you'll learn a symbolic method to find exact solutions to equations in the general form, $y = ax^2 + bx + c$.

Recall that rectangular diagrams help you factor some quadratic expressions.

Perfect square trinomial

$x^2 + 6x + 9$

	x	3
x	x^2	$3x$
3	$3x$	9

Factorable trinomial

$x^2 + x - 6$

	x	3
x	x^2	$3x$
-2	$-2x$	-6

In the first diagram the sum of the rectangular areas, $x^2 + 3x + 3x + 9$, is equal to the area of the overall diagram, $(x + 3)^2$. So -3 is the root. In the second diagram the sum is $x^2 + 3x + (-2x) + (-6)$, which equals $(x - 2)(x + 3)$. Both 2 and -3 are the roots.

How do you find the roots of an equation such as $0 = x^2 + x - 1$? It is not a perfect square trinomial, nor is it factorable with integers. For these equations you can use a method called **completing the square.**

Investigation
Searching for Solutions

You will need

- graph paper

To understand how to complete the square with quadratic equations, you'll first work with rectangular diagrams.

Step 1 | Complete each rectangular diagram so that it is a square. How do you know which number to place in the lower right corner?

a.
	x	?
x	x^2	$2x$
?	$2x$	?

$x^2 + 4x + 4 = (x + 2)^2$

b.
	x	?
x	x^2	$-3x$
?	$-3x$	?

$x^2 - 6x + 9 = (x - 3)^2$

c.
	x	?
x	x^2	$-2.5x$
?	$-2.5x$	?

$x^2 - 5x + 6.25 = (x - 2.5)^2$

d.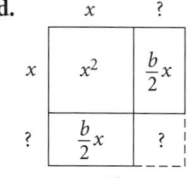
	x	?
x	x^2	$\frac{b}{2}x$
?	$\frac{b}{2}x$	?

$x^2 + bx + \frac{b^2}{4} = \left(x + \frac{b}{2}\right)^2$

Step 1 Sample answer: Take the coefficient of either x-term (half the coefficient of the x-term in expanded form) and square it.

Step 2 | For each diagram in Step 1, write an equation in the form

$x^2 + bx + c = (x + h)^2$

On which side of the equation can you isolate x by undoing the order of operations? the right side, $(x + h)^2$

LESSON OBJECTIVES

- Review how to solve quadratic equations in vertex form
- Solve quadratic equations in polynomial form by completing the square

NCTM STANDARDS

CONTENT		PROCESS	
•	Number		Problem Solving
•	Algebra	•	Reasoning
	Geometry	•	Communication
	Measurement	•	Connections
	Data/Probability	•	Representation

Step 3
a. $(x + 2)^2 = 100$
b. $(x - 3)^2 = 100$
c. $(x - 2.5)^2 = 100$

Step 3 Suppose the area of each diagram in parts a–c is 100 square units. For each square, write an equation that you can solve for x by undoing the order of operations.

Step 4 Solve each equation in Step 3 symbolically. You will get two values for x.

Step 4 a. $x + 2 = \pm 10$,
$x = 8$ or -12
b. $x - 3 = \pm 10$,
$x = 13$ or -7
c. $x - 2.5 = \pm 10$,
$x = 12.5$ or -7.5

All the solutions for x in Step 4 are integers or simple decimals. This means you could have factored the equations with integers. However, the method of completing the square works for other numbers as well. Next you'll consider the solution of an equation that you cannot factor with integers.

Step 5 Consider the equation $x^2 + 6x - 1 = 0$. Describe what's happening in each stage of the solution process.

Stage	Equation	Description
1	$x^2 + 6x - 1 = 0$	The original equation.
2	$x^2 + 6x = 1$	Add 1 to both sides.
3	$x^2 + 6x + 9 = 1 + 9$	Make a perfect-square trinomial by adding 3^2, or 9.

	x	3
x	x^2	$3x$
3	$3x$	9

4	$(x + 3)^2 = 10$	Rewrite the trinomial as a squared binomial.
5	$x + 3 = \pm\sqrt{10}$	Take the square root of both sides.
6	$x = -3 \pm \sqrt{10}$	Subtract 3 from both sides.

Step 6 Use your calculator to find decimal approximations for $-3 + \sqrt{10}$ and $-3 - \sqrt{10}$. Then enter the equation $y = x^2 + 6x - 1$ into Y1. Check your answers with a graph and a table. $x \approx 0.1622776602$ or $x \approx -6.16227766$

Step 7 Repeat the solution stages in Step 5 to find the solutions to $x^2 + 8x - 5 = 0$.
$x = -4 \pm \sqrt{21}$; $x \approx 0.5826$ or $x \approx -8.5826$

The key to solving by completing the square is to express one side of the equation as a perfect-square trinomial. In the investigation the equations are in the form $y = 1x^2 + bx + c$. Note that the coefficient of x^2, called the **leading coefficient**, is 1. However, there are other perfect square trinomials. An example is shown at right.

$4x^2 + 12x + 9$

	$2x$	3
$2x$	$4x^2$	$6x$
3	$6x$	9

$(2x + 3)^2$

In these cases, the leading coefficient is a perfect square number. In Example A you'll learn to complete the square for any quadratic equation in general form.

Step 6

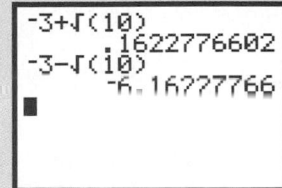

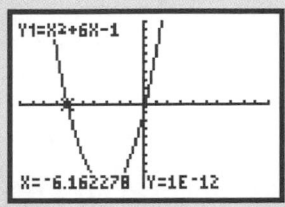

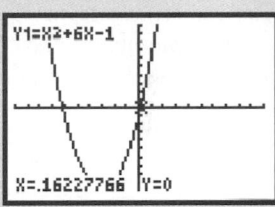

$[-10, 10, 1, -10, 10, 1]$

x-intercepts are approximately 0.16 and -6.16.

X	Y1
.13228	-.1888
.14228	-.1261
.15228	-.0631
.16228	0
.17228	.06335
.18228	.12689
.19228	.19064
X=.1622776601684	

X	Y1
-6.192	.19064
-6.182	.12689
-6.172	.06335
-6.162	1E-12
-6.152	-.0631
-6.142	-.1261
-6.132	-.1888
X=-6.16227766017	

SHARING IDEAS

Have groups share ideas for Steps 4 and 7.

[Ask] "Why is the method called *completing the square*?" Encourage students to draw squares when they can't remember the method.

You might challenge students to solve the equation $x^3 + 12x(x + 4) = 274$ by completing a cube. [A cube can be broken down into a cube with edge length x, a diagonally opposite cube of edge length 4, and three 4-by-x-by-$(x + 4)$ slabs. Completing the cube requires adding the volume of the 4-by-4-by-4 cube to both sides of the equation.]

Assessing Progress

Students will demonstrate their abilities to solve quadratic equations in vertex form symbolically and to graph and trace equations on a calculator.

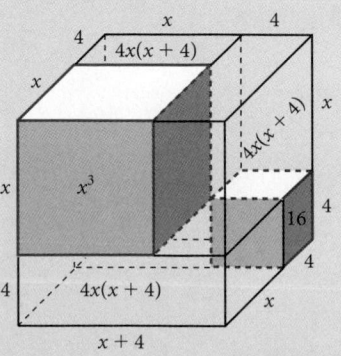

► EXAMPLE A

This example illustrates how to complete the square to solve an equation in which the leading coefficient is not 1.

► EXAMPLE B

In this example, students also complete the square for an equation in which the leading coefficient is not 1. This time they're transforming an equation from general form to vertex form. Because the x-intercepts involve square roots of negative numbers, they are not real numbers, so the parabola doesn't cross the x-axis. Even roots of negative numbers (including square roots, 4th roots, 6th roots, and so on) are examples of imaginary numbers, with no representation on the number line. (Imaginary numbers are the subject of Take Another Look at the end of the chapter.)

EXAMPLE A | Solve the equation $3x^2 + 18x - 8 = 22$ by completing the square.

► Solution | First, transform the equation so that you can write the left side as a perfect-square trinomial in the form $x^2 + 2hx + h^2$.

$3x^2 + 18x - 8 = 22$	The original equation.
$3x^2 + 18x = 30$	Add 8 to both sides of the equation.
$x^2 + 6x = 10$	Divide both sides by 3.

Now you need to decide what number to add to both sides to get a perfect-square trinomial on the left side. Use a rectangular diagram to make a square. When you decide what number to add, you must add it to both sides to balance the equation.

	x	?
x	x^2	$3x$
?	$3x$	?

$x^2 + 6x + 9 = 10 + 9$	Add 9 to both sides to complete the square.
$(x + 3)^2 = 19$	Write the perfect-square trinomial as a squared binomial and combine any like terms.
$x + 3 = \pm\sqrt{19}$	Take the square root of both sides.
$x = -3 \pm\sqrt{19}$	Add -3 to both sides.

The two solutions are $-3 + \sqrt{19}$, or approximately 1.36, and $-3 - \sqrt{19}$, or approximately -7.36.

You can also complete the square to convert the general form of a quadratic equation to the vertex form.

EXAMPLE B | Find the vertex form of the equation $y = 2x^2 + 8x + 11$. Then locate the vertex point and any x-intercepts of the parabola.

► Solution | To convert $y = 2x^2 + 8x + 11$ to the form $y = a(x - h)^2 + k$, complete the square.

$y = 2x^2 + 8x + 11$	The original equation.
$y = 2(x^2 + 4x) + 11$	Factor the 2 from the coefficients.

Now you can complete the square on the expression inside the parentheses. The coefficient of x is 4, so divide that by 2 to get 2. Then add 2^2, or 4, to make a perfect-square trinomial inside the parentheses. You must also subtract 4 inside the parentheses to balance the equation. Note that everything inside the parentheses is multiplied by 2.

$y = 2(x^2 + 4x + 4 - 4) + 11$	Add zero in the form of $4 - 4$.
$y = 2(x^2 + 4x + 4) + 2(-4) + 11$	Rewrite to get a perfect-square trinomial.
$y = 2(x + 2)^2 - 8 + 11$	Express the trinomial as a squared binomial.
$y = 2(x + 2)^2 + 3$	Combine like terms to get the vertex form.

So the vertex is $(-2, 3)$. To find any x-intercept, you can solve the vertex form symbolically.

$$2(x + 2)^2 + 3 = 0 \qquad \text{Substitute 0 for } y \text{ in the original equation.}$$

$$(x + 2)^2 = \frac{-3}{2} \qquad \text{Subtract 3 and then divide both sides by 2.}$$

$$x = -2 \pm \sqrt{\frac{-3}{2}} \qquad \begin{array}{l}\text{Take the square root and then subtract 2}\\ \text{from both sides.}\end{array}$$

If you try to evaluate $-2 \pm \sqrt{\frac{-3}{2}}$, your calculator may give you an error message about a nonreal answer. These roots are not real numbers because the number under the square root sign is negative. **Real numbers** are all numbers except those that involve even roots of negative numbers. Integers, fractions, and any numbers that can be expressed as decimals are real numbers. Every real number is on the x-axis. So this means that $y = 2(x + 2)^2 + 3$ has no x-intercepts. The graph confirms this result.

Note that the vertex is above the x-axis and the parabola opens upward. So the graph does not cross the x-axis.

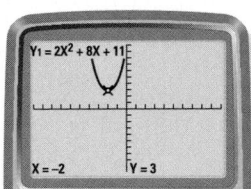

You can now solve any quadratic equation in general form by completing the square. This process leads to a general formula that you will learn in the next lesson.

EXERCISES

You will need your calculator for problems **8**, **11**, and **12**.

▶ **Practice Your Skills**

1. Solve each quadratic equation written in vertex form.
   a. $2(x + 3)^2 - 4 = 0 \quad x = -3 \pm \sqrt{2}$
   b. $-2(x - 5)^2 + 7 = 3 \quad x = 5 \pm \sqrt{2}$
   c. $3(x + 8)^2 - 7 = 0 \quad x = -8 \pm \sqrt{\frac{7}{3}}$
   d. $-5(x + 6)^2 - 3 = -10 \quad x = -6 \pm \sqrt{\frac{7}{5}}$

2. Solve each equation written in factored form.
   a. $(x - 5)(x + 3) = 0 \quad x = 5 \text{ or } x = -3$
   b. $(2x + 6)(x - 7) = 0 \quad x = -3 \text{ or } x = 7$
   c. $(3x + 4)(x + 1) = 0 \quad x = -\frac{4}{3} \text{ or } x = -1$
   d. $x(x + 6)(x + 9) = 0 \quad x = 0 \text{ or } x = -6 \text{ or } x = -9$

3. Decide what number must be added to each expression to make a perfect-square trinomial. Then rewrite the trinomial as a squared binomial.
   a. $x^2 + 18x$
   b. $x^2 - 10x$
   c. $x^2 + 3x$
   d. $x^2 - x$
   e. $x^2 + \frac{2}{3}x$
   f. $x^2 - 1.4x$

4. Solve each quadratic equation by completing the square. Leave your answer in radical form.
   a. $x^2 - 4x - 8 = 0$
   b. $x^2 + 2x - 1 = -5$
   c. $x^2 + 10x - 9 = 0 \quad x = -5 = \pm\sqrt{34}$
   d. $5x^2 + 10x - 7 = 28 \quad x = -1 \pm \sqrt{8}$

4a. $x^2 - 4x - 8 = 0$
$$x^2 - 4x = 8$$
$$x^2 - 4x + 4 = 12$$
$$(x - 2)^2 = 12$$
$$x - 2 = \pm\sqrt{12}$$
$$x = 2 \pm \sqrt{12}$$

4b. $x^2 + 2x - 1 = -5$
$$x^2 + 2x = -4$$
$$x^2 + 2x + 1 = -3$$
$$(x + 1)^2 = -3$$
$$x + 1 = \pm\sqrt{-3}$$
$$x = -1 \pm \sqrt{-3}$$
There are no real roots.

Closing the Lesson

To change a quadratic equation from general to vertex form or to solve the general form when it's not clear how to factor, you can **complete the square.** To do so, you factor, if necessary, to be sure the coefficient of x is 1 and then add the square of half the coefficient of the x term to get a perfect square. Compensate in the rest of the expression or equation for what you have factored and added.

BUILDING UNDERSTANDING

Students practice solving equations by completing the square.

ASSIGNING HOMEWORK

Essential	1, 2, 4, 6, 7
Performance assessment	5, 8, 10
Portfolio	9
Journal	7
Group	3, 9
Review	11, 12

3a. $\left(\frac{18}{2}\right)^2$;
$$x^2 + 18x + 81 = (x + 9)^2$$

3b. $\left(\frac{-10}{2}\right)^2$;
$$x^2 + 10x + 25 = (x - 5)^2$$

3c. $\left(\frac{3}{2}\right)^2; x^2 + 3x + \frac{9}{4} = \left(x + \frac{3}{2}\right)^2$

3d. $\left(\frac{-1}{2}\right)^2; x^2 - x + \frac{1}{4} = \left(x - \frac{1}{2}\right)^2$

3e. $\left(\frac{1}{2} \cdot \frac{2}{3}\right)^2$;
$$x^2 + \frac{2}{3}x + \frac{1}{9} = \left(x + \frac{1}{3}\right)^2$$

3f. $\left(\frac{-1.4}{2}\right)^2$;
$$x^2 - 1.4x + 0.49 = (x - 0.7)^2$$

Exercise 5b Some students may have difficulty understanding that the equation is satisfied by coordinates of any point that lies on the parabola.

5b. Solve the equation $0 = a(5 - 2)^2 - 31.5$; $a = 3.5$.

6a. Let w represent the width in meters. Let l represent the length in meters. Then $l = w + 4$. The area equation is $w(w + 4) = 12$.

6b.
$$w^2 + 4w = 12$$
$$w^2 + 4w + 4 = 12 + 4$$
$$(w + 2)^2 = 16$$
$$w + 2 = \pm 4$$
$$w = -2 \pm 4$$
$$w = -6 \text{ or } 2$$

The width cannot be negative, so it must be 2 meters.

6c. The length is 4 meters more than the width, so the length is 6 meters.

7b. vertex $(-3, 1)$;

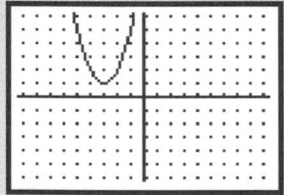

$[-9.4, 9.4, 1, -6.2, 6.2, 1]$

Exercise 7c Students might answer the "why" question in a variety of ways. For example, some may refer to translations (seen from the vertex form of the equation), and others to square roots of negative numbers.

7c.
$$x^2 + 6x + 10 = 0$$
$$x^2 + 6x = -10$$
$$x^2 + 6x + 9 = -10 + 9$$
$$(x + 3)^2 = -1$$
$$x + 3 = \pm\sqrt{-1}$$
$$x = -3 \pm\sqrt{-1}$$

Sample answer: There are no real roots because the graph doesn't cross the x-axis.

► **Reason and Apply**

5. If you know the vertex and one other point on a parabola, you can find its quadratic equation. The vertex (h, k) of this parabola is $(2, -31.5)$, and the other point is $(5, 0)$.

   a. Substitute the values for h and k into the equation $y = a(x - h)^2 + k$. $y = a(x - 2)^2 - 31.5$

   b. To find the value of a, substitute 5 for x and 0 for y. Then solve for a.

   c. Use the a-value you found in 5b to write the equation for the graph in vertex form. $y = 3.5(x - 2)^2 - 31.5$

   d. Use what you learned in 5a–c to write the equation of the graph whose vertex is $(2, 32)$ and that passes through the point $(5, 14)$. $y = -2(x - 2)^2 + 32$

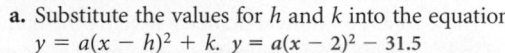

6. The length of a rectangle is 4 meters more than its width. The area is 12 square meters.

   a. Define variables and write an equation for the area of the rectangle in terms of its width.

   b. Solve your equation in 6a by completing the square.

   c. Which solution makes sense for the length of the rectangle?

7. Consider the equation $y = x^2 + 6x + 10$.

   a. Convert this equation to vertex form by completing the square.

   b. Find the vertex. Graph both equations.

   c. Find the roots of the equation $0 = x^2 + 6x + 10$. What happens and why?

7a. $y = x^2 + 6x + \left(\frac{6}{2}\right)^2 - 3^2 + 10$;
$y = x^2 + 6x + 9 - 9 + 10$;
$y = (x + 3)^2 + 1$

8. **APPLICATION** A professional football team uses computers to describe the projectile motion of a football when punted. After compiling data from several games, the computer models the height of an average punt with the equation

$$h(t) = \frac{-16}{3}(t - 2.2)^2 + 26.9$$

where t is the time in seconds and $h(t)$ is the height in yards. The punter's foot makes contact with the ball when $t = 0$.

   a. When does the punt reach its highest point? How high does the football go?

   b. Find the zeros of $h(t) = \frac{-16}{3}(t - 2.2)^2 + 26.9$. Which solution is the hang time—that is, the time it takes until the ball hits the ground? 4.446 seconds

   c. How high is the ball when the punter kicks it?

   d. Graph the equation. What are the real-world meanings of the vertex, the y-intercept, and the x-intercepts?

Brad Maynard punts for the N.Y. Giants.

Exercise 8 The coefficient $-\frac{16}{3}$ results from converting the gravitational constant -16 ft/sec² to yards/sec².

8a. 2.2 seconds; 26.9 yards (80.7 feet)

8b. $t = 2.2 \pm\sqrt{-26.9 \cdot \frac{3}{-16}}$,
$t \approx -0.046$ or $t \approx 4.446$.

8c. The general form is $\frac{-16}{3}t^2 + 23.4\overline{6}t + 1.08\overline{6}$, so the football is about 1 yard high.

8d. The vertex is the maximum height of the ball. The y-intercept is the height of the ball when the punter kicks it. The positive x-intercept is the hang time. The other x-intercept has no real-world meaning.

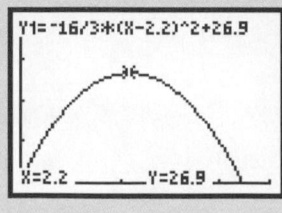

$[0, 5, 1, 0, 40, 10]$

9. **APPLICATION** The Cruisin' Along Company is determining prices for its Caribbean cruise packages. The basic cost is $2500 per person. However, business is slow. To attract corporate clients, the company reduces the cost of each ticket by $5 for each person in the group. The larger the group, the less each person would pay.

   a. Define variables and write an equation for the cost of a single ticket.

   b. Write an equation for the total cost the company charges for a group package. $C = xp = x(2500 - 5x)$

   c. Convert the equation in 9b to vertex form.

   d. What is the total cost of a cruise for a group of 20 people? $C(20) = 20(2500 - 100) = 48,000; \$48,000$

   e. The company accountant reports that the cost of running a cruise is $200,000. Solve the equation
   $$x(2500 - 5x) = 200,000$$
   by completing the square.

   f. What limitations on group size should the cruise company use in order to maximize its profits?

10. **APPLICATION** The rate at which a bear population grows in a park is given by the equation $P(b) = 0.001b(100 - b)$. The function value $P(b)$ represents the rate at which the population is growing in bears per year, and b represents the number of bears.

   a. Find $P(10)$ and provide a real-world meaning for this value.

   b. Solve $P(b) = 0$ and provide real-world meanings for these solutions.

   c. For what size bear population would the population grow fastest?

   d. What is the maximum number of bears the park can support?

   e. What does it mean to say that $P(120) < 0$?

▶ **Review**

11. Find each product. Check your answers by using calculator tables or graphs.

   a. $(x + 1)(2x^2 + 3x + 1)$ $2x^3 + 5x^2 + 4x + 1$ b. $(2x - 5)(3x^2 + 2x - 4)$ $6x^3 - 11x^2 - 18x + 20$

12. Combine like terms in these polynomials. Check your answers by using calculator tables or graphs.

   a. $(x + 1) + (2x^2 + 3x + 1)$ $2x^2 + 4x + 2$ b. $(2x - 5) + (3x^2 + 2x - 4)$ $3x^2 + 4x - 9$
   c. $(x + 1) - (2x^2 + 3x + 1)$ $-2x^2 - 2x$ d. $(2x - 5) - (3x^2 + 2x - 4)$ $-3x^2 - 1$

9e.
$$2500x - 5x^2 = 200,000$$
$$500x - x^2 = 40,000$$
$$x^2 - 500x = -40,000$$
$$x^2 - 500x + 250^2 = 250^2 - 40,000$$
$$(x - 250)^2 = 22,500$$
$$x - 250 = \pm 150$$
$$x = 100 \text{ or } x = 400$$

Exercise 9 Some students might benefit from making a chart with various sizes of groups.

9a. $p = 2500 - 5x$, where p represents the cost in dollars of a single ticket, and x represents the number of tickets sold. Let C represent the total cost of the group package.

9c. $C = -5(x - 250)^2 + 312,500$

9f. Sample answer: The company will lose money if more than 400 people or less than 100 people per group go on the cruise. The vertex is (250, 312500), so the company should limit groups to 250 people to maximize earnings at $312,500.

Exercise 10 Students may miss the fact that $P(b)$ represents a rate of growth rather than the bear population itself, which is given by b. The rate of change is 0 when the population achieves a maximum or minimum. 10d might especially confuse students: They might say, "When there are 101 bears, the growth rate is negative, but that doesn't mean there can't be 101 bears." That argument is valid even though the mathematical model has a maximum at $b = 100$.

10a. $P(10) = 0.9$. This means that when there are 10 bears in the park the population grows at a rate of 0.9 bear per year.

10b. $P(b) = 0$ when $b = 0$ or 100. When there are no bears, the population does not grow. When there are 100 bears, the population does not grow but remains at that level.

10c. The vertex lies halfway between the roots, 0 and 100, so the population is growing fastest when there are 50 bears.

10d. Because the population does not grow when there are 100 bears, this value must be the maximum population.

10e. It means that if 120 bears were brought in, the population would shrink, due to overpopulation.

Exercise 11 Making a 2 × 3 rectangle can help with each of these products. This exercise reviews Lesson 10.4.

Exercise 12 This exercise reviews Lesson 10.3.

PLANNING

LESSON OUTLINE

One day:

25 min	Investigation
5 min	Sharing
5 min	Example
5 min	Closing
10 min	Exercises

MATERIALS

• The Quadratic Formula (T), *optional*

TEACHING

Because completing the square can be labor-intensive, especially when the coefficient of x^2 is not 1, we often use a shortcut called the *quadratic formula*.

Guiding the Investigation

One step Ask students to solve the general-form quadratic equation $ax^2 + bx + c = 0$ by completing the square. For those having difficulties, suggest that they solve a problem with numbers along with the generalization.

Step 1 There is again great potential for students to confuse letters representing constant coefficients (a, b, c) with the letter x representing the independent variable.

Most people are more comfortable with old problems than with new solutions.

ANONYMOUS

The Quadratic Formula

You can solve some quadratic equations symbolically by recognizing their forms:

Vertex form	$-4.9(t - 5)^2 + 75 = 0$
Factored form	$0 = w(200 - 2w)$
Perfect square trinomial	$x^2 + 6x + 9 = 0$

You can also undo the order of operations in other quadratic equations when there is no x-term, as in these:

$$x^2 = 10$$
$$x^2 + 25 = 0$$
$$x^2 - 0.36 = 0$$

If the quadratic expression is in the form $x^2 + bx + c$, you can complete the square by using a rectangular diagram.

	x	$\frac{b}{2}$
x	x^2	$\frac{b}{2}x$
$\frac{b}{2}$	$\frac{b}{2}x$	$\left(\frac{b}{2}\right)^2$

Although it is possible to complete the square for any quadratic equation, it can get messy if your equation is something like $-4.9x^2 + 5x - \frac{16}{3} = 0$.

Let's consider the general case of $ax^2 + bx + c = 0$. The leading coefficient is a. The middle term, bx, is called the **linear term.** The value of c is the **constant term.** Completing the square for the general case gives a formula that solves any quadratic equation. It is called the **quadratic formula.** To use it, all you need to know are the values of a, b, and c.

Investigation
Deriving the Quadratic Formula

You'll solve $2x^2 + 3x - 1 = 0$ and develop the quadratic formula for the general case in the process.

Step 1 $a = 2, b = 3,$
$c = -1$

Identify the values of a, b, and c in the general form, $ax^2 + bx + c = 0$, for the equation $2x^2 + 3x - 1 = 0$.

Step 2 $2x^2 + 3x = 1$

Group all the variable terms on the left side of your equation so that it is in the form

$$ax^2 + bx = -c$$

LESSON OBJECTIVES

• Study the derivation of the quadratic formula
• Learn to use the quadratic formula

NCTM STANDARDS

CONTENT		PROCESS	
	Number		Problem Solving
•	Algebra	•	Reasoning
	Geometry	•	Communication
	Measurement	•	Connections
	Data/Probability	•	Representation

Step 3 $x^2 + \frac{3}{2}x = \frac{1}{2}$ **Step 3**

In order to complete the square, the coefficient of x^2 should be 1. So divide your equation by the value of a. Write it in the form

$$x^2 + \frac{b}{a}x = \frac{-c}{a}$$

Step 4

Use a rectangular diagram to help you complete the square. What number must you add to both sides? Write your new equation in the form

$$x^2 + \frac{b}{a}x + \left(\frac{b}{2a}\right)^2 = \left(\frac{b}{2a}\right)^2 - \frac{c}{a}$$

Step 5 $\left(x + \frac{3}{4}\right)^2 = \frac{9}{16} + \frac{8}{16}$ **Step 5**

Rewrite the trinomial on the left side of your equation as a squared binomial. On the right side, find a common denominator. Write the next stage of your equation in the form

$$\left(x + \frac{b}{2a}\right)^2 = \frac{b^2}{4a^2} - \frac{4ac}{4a^2}$$

Step 6
$x + \frac{3}{4} = \pm\frac{\sqrt{9 + 8}}{\sqrt{16}}$ **Step 6**

Take the square root of both sides of your equation, like this:

$$x + \frac{b}{2a} = \pm\frac{\sqrt{b^2 - 4ac}}{\sqrt{4a^2}}$$

Step 7
$x = -\frac{3}{4} \pm \frac{\sqrt{17}}{4}$ **Step 7**

Get x by itself on the left side, like this:

$$x = -\frac{b}{2a} \pm \frac{\sqrt{b^2 - 4ac}}{2a}$$

Step 8
$x = \frac{-3 + \sqrt{17}}{4}$ or   **Step 8**

$x = \frac{-3 - \sqrt{17}}{4}$

There are two possible solutions given by the equations

$$x = \frac{-b + \sqrt{b^2 - 4ac}}{2a} \quad \text{or} \quad x = \frac{-b - \sqrt{b^2 - 4ac}}{2a}$$

Write your two solutions in radical form.

Step 10 To avoid   **Step 9**
dividing by zero, $a \neq 0$.
To avoid square   **Step 10**
roots of negative
numbers, $b^2 - 4ac \geq 0$.

Write your solutions in decimal form. Check them with a graph and a table.

Consider the expression $\dfrac{-b \pm \sqrt{b^2 - 4ac}}{2a}$. What restrictions should there be so that the solutions exist and are real numbers?

The quadratic formula gives the same solutions that completing the square or factoring does. You don't need to derive the formula each time. All you need to know are the values for a, b, and c. Then you substitute these values into the formula.

> ### Quadratic Formula
>
> If a quadratic equation is written in the general form, $ax^2 + bx + c = 0$,
>
> the roots are given by $x = \dfrac{-b \pm \sqrt{b^2 - 4ac}}{2a}$.

In the next example you'll learn how to use the formula for quadratic equations in general form. You can even use it when the values of a, b, and c are decimals or fractions.

Steps 3 to 5 Some students may want to imitate what they did earlier by factoring out the a, but not dividing through by it, to get $a\left(x^2 + \frac{b}{a}x\right) = -c$ and then $a\left(x^2 + \frac{b}{a}x + \frac{b^2}{4a^2}\right) - \frac{b^2}{4a} = -c$, or $a\left(x + \frac{b}{2a}\right)^2 = -c + \frac{b^2}{4a}$. Continuing with this approach will yield the standard quadratic formula, though with somewhat more difficult computations than those in the student text.

Step 4

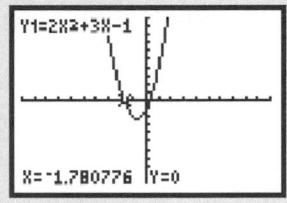

$\frac{9}{16}; x^2 + \frac{3}{2}x + \left(\frac{3}{4}\right)^2 = \left(\frac{3}{4}\right)^2 + \frac{1}{2}$

Step 9 $x \approx -1.7870776406$ and $x \approx 0.2807764064$

Step 10 Besides answering that $b^2 - 4ac$ must not be negative, students may point out that a cannot be 0. Ask when that would be the case, but save the answer for Sharing.

SHARING IDEAS

SHARING IDEAS

Point out that in solving equations with letters instead of numbers, students should develop the habit of being especially careful not to divide by 0. From Step 3 onward in the investigation, a is being divided. Ask if that could lead to difficulties. Indeed, none of these steps hold if $a = 0$, but in that case the equation would be linear, not quadratic.

[Ask] "What would happen if $b^2 - 4ac$ equaled 0." [There would be only one root, a double root. The parabolic graph would have its vertex on the x-axis.]

Ask about the quotation opening the lesson. Are your students more comfortable completing the square than using the quadratic formula?

Assessing Progress

Watch for students' skills at substituting numbers for letters, completing the square, and using calculator graphs and tables.

► **EXAMPLE**

Encourage students to use parentheses as illustrated in the first
step after the formula statement.
Doing so helps them avoid errors
like writing -3^2 for b^2 if $b = -3$.

Closing the Lesson

The **quadratic formula** can be
used instead of completing the
square. It's derived by completing
the square on the general
quadratic equation. It yields two
solutions, $y = \dfrac{-b \pm \sqrt{b^2 - 4ac}}{2a}$,
though they're the same if the
square root has value 0.
[ESL] *Derive* means to obtain a
result from an original true statement by means of step-by-step
reasoning.

BUILDING UNDERSTANDING

Students solve quadratic equations using the quadratic formula.

ASSIGNING HOMEWORK

Essential	1–4, 6, 7
Performance assessment	6, 9
Portfolio	5
Journal	7, 8
Group	1, 5, 8, 10
Review	11, 12

EXAMPLE | Use the quadratic formula to solve $3x^2 + 5x - 7 = 0$.

► **Solution**

The equation is already in general form, so identify the values of a, b, and c.
For this equation, $a = 3$, $b = 5$, and $c = -7$. Here is one way to use the formula:

$$x = \frac{-b \pm \sqrt{b^2 - 4ac}}{2a}$$

The quadratic formula.

$$= \frac{-(\ ) \pm \sqrt{(\ )^2 - 4(\ )(\ )}}{2(\ )}$$

Replace each letter in the formula with
a set of parentheses.

$$= \frac{-(5) \pm \sqrt{(5)^2 - 4(3)(-7)}}{2(3)}$$

Substitute the values of a, b, and c into
the appropriate places.

$$= \frac{-5 \pm \sqrt{25 - (-84)}}{6}$$

Do the operations.

$$= \frac{-5 \pm \sqrt{109}}{6}$$

Subtract.

The two exact roots of the equation are $\dfrac{-5 + \sqrt{109}}{6}$ and $\dfrac{-5 - \sqrt{109}}{6}$.
You can use your calculator to calculate the approximate values, 0.907 and
-2.573, respectively.

To make the formula simpler, think of the expression under the square root sign
as one number. This expression $b^2 - 4ac$ is called the **discriminant.** In the
example the discriminant is 109. So let $d = b^2 - 4ac$. Then the formula becomes

$$x = \frac{-b \pm \sqrt{d}}{2a}$$

If you store these values into your calculator as shown, then you can use the
formula directly on your calculator.

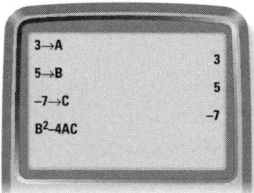
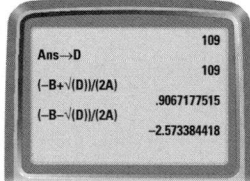

EXERCISES

You will need your calculator for problems **1, 4, 6,** and **12.**

► **Practice Your Skills**

1. Without using a calculator, evaluate each expression in the form $b^2 - 4ac$ for the
   values given. Then check your answers with a calculator.

   a. $a = 3, b = 5, c = 2$ $25 - 24 = 1$

   b. $a = 1, b = -3, c = -3$ $9 - (-12) = 9 + 12 = 21$

   c. $a = -2, b = -6, c = -3$ $36 - 24 = 12$

   d. $a = 9, b = 9, c = 0$ $81 - 0 = 81$

2. Rewrite each quadratic equation in general form if necessary. For each equation, identify the values of a, b, and c.

a. $2x^2 + 3x - 7 = 0$ $a = 2, b = 3, c = -7$

b. $x^2 + 6x = -11$ $x^2 + 6x + 11 = 0; a = 1, b = 6, c = 11$

c. $-3x^2 - 4x + 12 = 0$ $a = -3, b = -4, c = 12$

d. $18 - 4.9x^2 + 47x = 0$

e. $-16x^2 + 28x + 10 = 57$ $-16x^2 + 28x - 47 = 0; a = -16, b = 28, c = -47$

f. $5x^2 - 2x = 7 + 4x$ $5x^2 - 6x - 7 = 0; a = 5, b = -6, c = -7$

3. Solve each quadratic equation. Which equation can you solve readily by completing the square? Which equation has no real solutions?

a. $2x^2 - 3x + 4 = 0$

b. $-2x^2 + 7x = 3$

c. $x^2 - 6x - 8 = 0$

d. $3x^2 + 2x - 1 = 5$

► Reason and Apply

4. Graph the equation $y = x^2 + 3x + 5$. Use the graph and the quadratic formula to answer these questions.

a. How many x-intercepts are on the graph?

b. Use the quadratic formula to try to find the roots of $0 = x^2 + 3x + 5$. What happens when you take the square root?

c. Without looking at a graph, how can you use the quadratic formula to tell if a quadratic equation has any real roots? If the discriminant is negative, there are no real roots. If it is positive or zero, there are real roots.

5. The equation $h = -4.9t^2 + 6.2t + 1.9$ models the height of a soccer ball after Brandi hits it with her head. The t-values represent the time in seconds, and the h-values represent the height in meters. Write an equation that describes each event in 5a–c. Then use the quadratic formula to solve it. Explain the real-world meanings of your solutions.

a. The ball hits the ground.

b. The ball is 3 meters above ground.

c. The ball is 4 meters above ground.

6. Find an equation whose solutions are shown. Evaluate each expression to the nearest 0.1. Make a sketch showing a parabola that has these x-intercepts.

a. $\dfrac{14 \pm \sqrt{(-14)^2 - 4(1)(49)}}{2(1)}$

b. $\dfrac{3 \pm \sqrt{(-3)^2 - 4(2)(2)}}{2(2)}$

c. $\dfrac{3 \pm \sqrt{(-3)^2 - 4(2)(-2)}}{2(2)}$

Brandi Chastain of the USA team heads a soccer ball during the 1999 Women's World Cup at Stanford Stadium.

5a. $-4.9t^2 + 6.2t + 1.9 = 0$; $t \approx -0.255$ sec or $t \approx 1.52$ sec.

5b. $-4.9t^2 + 6.2t + 1.9 = 3$; $t \approx 0.21$ sec or $t \approx 1.05$ sec.

5c. $-4.9t^2 + 6.2t + 1.9 = 4$; $t \approx \dfrac{-6.2 \pm \sqrt{-2.72}}{-9.8}$. This equation has no real solutions, so the ball is never 4 meters high.

6a. Sample answer: $y = x^2 - 14x + 49$. The x-intercept is 7.

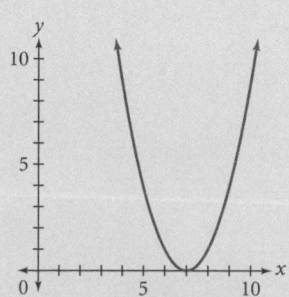

► **Helping with the Exercises**

2d. $-4.9x^2 + 47x + 18 = 0$
$a = -4.9, b = 47, c = 18$

3a. $x = \dfrac{3 \pm \sqrt{-23}}{4}$

There are no real solutions.

3b. Use the quadratic equation or complete the square.
$x = \dfrac{1}{2}$ and $x = 3$

3c. Complete the square.
$$x^2 - 6x - 8 = 0$$
$$x^2 - 6x = 8$$
$$x^2 - 6x + 9 = 17$$
$$(x - 3)^2 = 17$$
$$x - 3 = \pm\sqrt{17}$$
$$x = 3 \pm\sqrt{17}$$

3d. Use the quadratic formula.
$$x = \dfrac{-2 \pm \sqrt{76}}{6} = \dfrac{-1 \pm \sqrt{19}}{3}$$

Exercise 4 Students should conclude that the sign of the value under the square root is the key to this determination. The term *discriminant* to refer to $b^2 - 4ac$ is not mentioned in any exercise. Instead the approach is very informal. This exercise is similar to Exercise 7 in Lesson 10.6, but the solution is different now that students know the quadratic formula.

4a. The graph does not cross the x-axis.

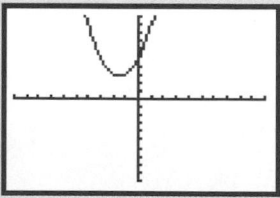

$[-10, 10, 1, -10, 10, 1]$

4b. To use the quadratic formula, $a = 1$, $b = 3$, and $c = 5$.
$$x = \dfrac{-3 \pm \sqrt{-11}}{2}$$

There are no real square roots of negative numbers, so there are no real roots.

See page 725 for answers to Exercises 6b–c.

7. When $b^2 - 4ac$ is negative, there are no real roots. When it's zero, there is one root. When it's positive, there are two roots.

7a. (i) $1^2 - 4(1)(1) = -3$, no x-intercept

7b. (iii) $2^2 - 4(1)(1) = 0$, one x-intercept

7c. (ii) $3^2 - 4(1)(1) = 5$, two x-intercepts

Exercise 8 Many students might find the averaging here to be a challenge.

9. You need to know the t-value when $h = 0$.

Solve $0 = -4.9t^2 + 17t + 2.2$ with $a = -4.9, b = 17, c = 2.2$ to get $t \approx -0.125$ and $t \approx 3.59$. The positive solution of 3.59 seconds makes sense in this situation.

10b. The perimeter does not change.

10c. The area increases and then decreases.

---

7. Match each quadratic equation with its graph. Then explain how to find the number of x-intercepts from the discriminant, $b^2 - 4ac$.

a. $y = x^2 + x + 1$ **b.** $y = x^2 + 2x + 1$ **c.** $y = x^2 + 3x + 1$

i. 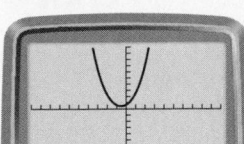   **ii.** 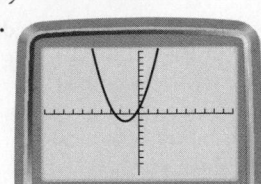   **iii.**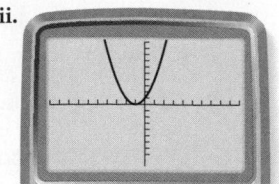

8. The quadratic formula gives two roots for an equation:

$$x = \frac{-b + \sqrt{b^2 - 4ac}}{2a} \quad \text{and} \quad x = \frac{-b - \sqrt{b^2 - 4ac}}{2a}$$

What is the average of these two roots? How does averaging the roots help you find the vertex? *The average of the roots is $\frac{-b}{2a}$. It is the x-coordinate of the vertex of the parabola.*

9. The equation $h = -4.9t^2 + 17t + 2.2$ models the height of a stone thrown into the air, where t is in seconds and h is in meters. Use the quadratic formula to find how long the stone is in the air.

10. APPLICATION A shopkeeper is redesigning the rectangular sign on her store's rooftop. She wants the largest area possible for the sign. When she considers adding an amount to the width, she subtracts that same amount from the length. Her original sign has a width of 4 m and a length of 7 m.

a. Complete the table.

Increase (x) (m)	Width (m)	Length (m)	Area (sq. m)	Perimeter (m)
0	4	7	28	22
0.5	4.5	6.5	29.25	22
1.0	5	6	30	22
1.5	5.5	5.5	30.25	22
2.0	6	5	30	22

b. How do the changes in width and length affect the perimeter?

c. How do the changes in width and length affect the area?

d. Write an equation in factored form for the area A of the rectangle in terms of x, the amount she adds to the width. *$A = (4 + x)(7 - x)$*

e. What are the dimensions of the rectangle with the largest area? *The largest area occurs at the vertex of the parabola, which is at $x = 1.5$. At this point the rectangle has dimensions 5.5 m by 5.5 m, which means the rectangle is a square.*

Review

11. Reduce each rational expression by making a rectangular diagram for the quadratic expression in the numerator. Then cancel one of the binomial factors as shown. For example, to reduce $\frac{x^2 + 3x - 4}{x - 1}$, draw the diagram.

so $\frac{x^2 + 3x - 4}{x - 1} = \frac{(x - 1)(x + 4)}{x - 1} = x + 4$

a. $\dfrac{x^2 - 5x + 6}{x - 3}$ $\quad x - 2$

b. $\dfrac{x^2 + 7x + 6}{x + 1}$ $\quad x + 6$

c. $\dfrac{2x^2 - x - 1}{2x + 1}$ $\quad x - 1$

12. On your graph paper, sketch graphs of these equations. Then use your calculator to check your sketches.

a. $y - 2 = (x - 3)^2$

b. $y - 2 = -2\lvert x - 3 \rvert$

IMPROVING YOUR **REASONING** SKILLS

In Chapter 2, you may have done the project "The Golden Ratio." Now you have the tools to calculate this number. One way to calculate the golden ratio is to add 1 to square it. The symbolic statement of this rule is $x^2 = x + 1$ (or $x = \sqrt{x + 1}$).

You can approximate this value using a recursive routine on your calculator.

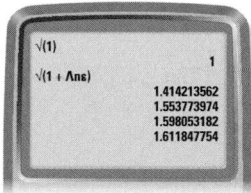

This is the same as calculating $\sqrt{1 + \sqrt{1 + \sqrt{1 + \sqrt{1 + \dots}}}}$

You can also divide both sides of $x^2 = x + 1$ by x to get $x = 1 + \dfrac{1}{x}$.

You can use another recursive routine to approximate x this time.

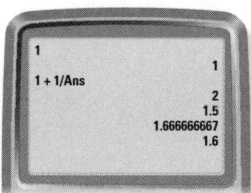

This is the same as calculating $1 + \cfrac{1}{1 + \cfrac{1}{1 + \cfrac{1}{1 + \dots}}}$

Try different starting values for these recursive routines. Do they always result in the same number? Use one of the methods you learned in this chapter to solve $x^2 = x + 1$ symbolically. What are the answers in radical form? Can you write a recursive routine for the negative solution?

IMPROVING **REASONING** SKILLS

Students can begin with any number that avoids dividing by 0 or taking the square root of a negative. For example, for the square root recursion, they might type 17 ENTER. The recursive step is given by $\sqrt{}$ (1 + ANS) ENTER. Repeatedly typing ENTER gives the sequence. For the continued fraction, they could type −3 ENTER,

and then 1 + 1/ANS ENTER. Repeatedly typing ENTER again gives the sequence.

Actually, students had the tools to approximate the value of the Golden Ratio some time ago. Now they can calculate it exactly, using the quadratic formula. The solutions

to the equation $x^2 = x + 1$ are $x = \dfrac{-1 \pm \sqrt{5}}{2}$, with approximations 1.618033989 and −0.6180339887. The negative value can be approximated through a routine whose recursive step is $-1/(1 - \text{ANS})$.

Exercise 11 To help students pay attention to division when there's a possibility of dividing by 0, you might point out that if these expressions were functions, they wouldn't be equal, because their domains are different. This exercise reviews Lesson 9.6.

Exercise 12 This exercise reviews Lessons 8.6 and 8.7.

12a.

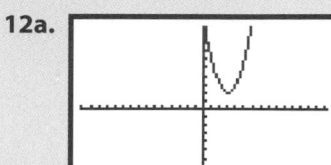

$[-15, 15, 1, -10, 10, 1]$

12b.

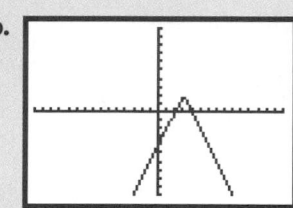

$[-15, 15, 1, -10, 10, 1]$

Cubic Functions

In this lesson you'll learn more about what cubic equations have in common with quadratic equations.

The area of the square at right is 16 square meters (m²), so you can cover the square using 16 smaller squares, each 1 m by 1 m. The sides of a square are equal, so you can write its area formula as *area = side²*. The squaring function, $f(x) = x^2$, models area. Each side length, or input, gives exactly one area, or output.

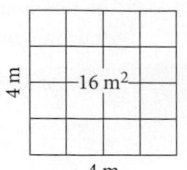

The volume of the cube at right is 64 cubic centimeters (cm³), so you can fill the cube using 64 smaller cubes, each measuring 1 cm by 1 cm by 1 cm. The edges of a cube have equal length, so you can write its volume formula as *volume = (edge length)³*. The cubing function, $f(x) = x^3$, models volume. Each edge length, or input, gives exactly one volume, or output.

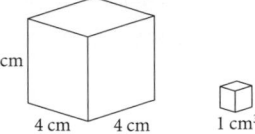

EXAMPLE A

The edge length of a cube with a volume of 64 is 4. So you can write $4^3 = 64$. You call 4 the **cube root** of 64 and the number 64 a **perfect cube** because its cube root is an integer. Then you can express the equation as $4 = \sqrt[3]{64}$. Determine which numbers are perfect cubes.

   a. 59319   **b.** 2197   **c.** 495   **d.** 13824

▶ Solution

Find the cube root of each number: [▶ ▣ See **Calculator Note 10B**. ◀]

   a. $\sqrt[3]{59319} = 39$ **b.** $\sqrt[3]{2197} = 13$

   c. $\sqrt[3]{495} \approx 7.910$ **d.** $\sqrt[3]{13824} = 24$

So 59319, 2197, and 13824 are perfect cubes. The number 495 is not a perfect cube because its cube root is not an integer.

Graphs of cubic functions have different and interesting shapes. In a window $-5 \leq x \leq 5$ and $-4 \leq y \leq 4$, the parent function $y = x^3$ looks like the graph shown.

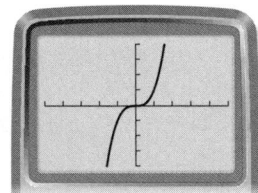

EXAMPLE B

Write an equation for each graph.

a.

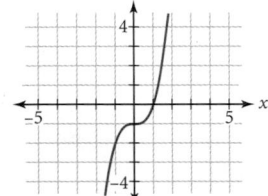

b.

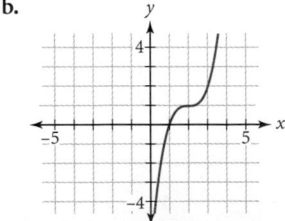

PLANNING

LESSON OUTLINE

One day:

  5 min   Introduction

10 min   Examples A and B

15 min   Investigation

  5 min   Sharing

  5 min   Example C

  5 min   Closing

10 min   Exercises

MATERIALS

- Calculator Note 10B
- Stunt Flying (T or W), *optional*

TEACHING

The relationships between roots, *x*-intercepts, and the factored form of a cubic equation are like those of a quadratic equation.

INTRODUCTION

You might introduce the word *prism*. Just as a square is a special kind of rectangle in which $l = w$, a cube is a special case of a right rectangular prism in which $l = w = h$.

▶ EXAMPLE A

Besides the method described in Calculator Note 10B, students might find cube roots by guess-and-check or by graphing and tracing $y = x^3$.

LESSON OBJECTIVES

- Explore general patterns and characteristics of cubic functions
- Learn formulas that model the areas of squares and the volumes of cubes
- Explore the graphs of power functions and transformations of these graphs

NCTM STANDARDS

CONTENT		PROCESS	
●	Number	●	Problem Solving
●	Algebra	●	Reasoning
●	Geometry	●	Communication
	Measurement		Connections
	Data/Probability	●	Representation

▶ **Solution**

a. Each graph is a transformation of the graph of $y = x^3$. The graph shows a translation of the parent function down 1 unit. So the equation is $y = x^3 - 1$.

b. There is a translation of the graph of $y = x^3$ right 2 units and up 1 unit. So the equation is $y = (x - 2)^3 + 1$.

Investigation
Rooting for Factors

In this investigation you'll discover the connection between the factored form of a cubic equation and its graph.

Step 1 | List the *x*-intercepts for each of these graphs.

Graph A

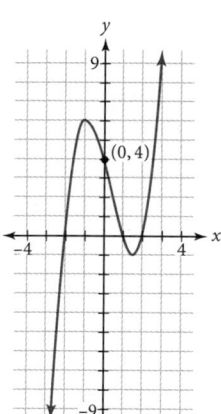

$x = -2, 1,$ and 2

Graph B

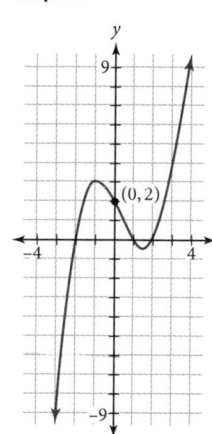

$x = -2, 1,$ and 2

Graph C

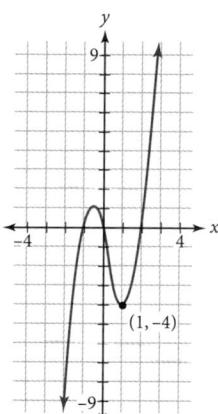

$x = -1, 0,$ and 2

Graph D

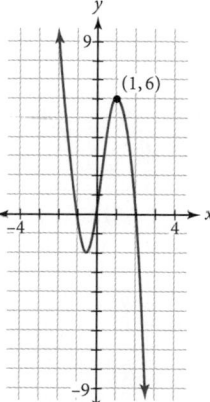

$x = -1, 0,$ and 2

Graph E

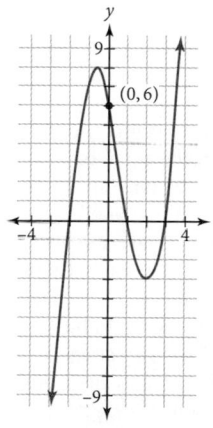

$x = -2, 1,$ and 3

Graph F

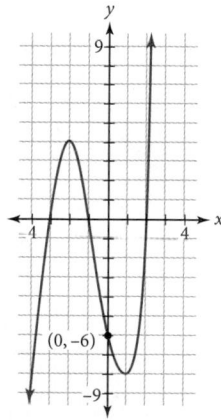

$x = -3, -1,$ and 2

▶ **EXAMPLE B**

This example helps students review transformations. **[Ask]** "What are the similarities between transformations of $y = x^2$ and those of $y = x^3$?" If you discussed the Take Another Look from Chapter 8, you might review inverses by asking whether students think the inverse of $y = x^3$ is a function. They might benefit from solving to get $x = y^{\frac{1}{3}}$ and then seeing its graph.

Guiding the Investigation

One step Pose the problem on the transparency or worksheet Stunt Flying. Students may be baffled at first. Keep assuring them that they have the ability to think through difficult problems. Some will try to graph the path of the stunt plane. Remind them that the question is about the plane's rate equation. (In fact, not enough information is given to determine the flight path.) Focus attention on the main information given: the times at which the plane changed direction—that is, the times at which its velocity was actually 0. Then ask whether a parabola could cross the *x*-axis three times, and encourage students to think of equations with an x^3 term. Finally, if necessary, suggest that they use a factored form. Even after they graph the rate function, they may be confused because it is decreasing at first, when the plane is rising. Keep reminding them that the cubic equation is a rate equation and that when it's positive the height is increasing.

Step 2 If students are expanding the products and then entering them into the Y= for graphing, suggest that they type in the expressions as they are to see what happens.

Step 3 The *x*-intercepts are the roots of the equations. They appear as signed opposites in the factored form. For example, −3 is a root of $y = (x + 3)(x + 2)(x - 2)$.

Step 4 The graph of $y = (x + 3)(x + 2)(x - 2)$ is

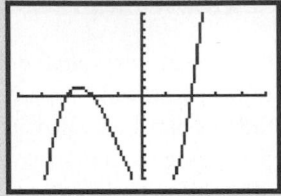

$[-5, 5, 1, -10, 10, 1]$

Reflect across the *x*-axis and apply a vertical shrink with a factor of 0.5 to get $y = -0.5(x + 3)(x + 2)(x - 2)$.

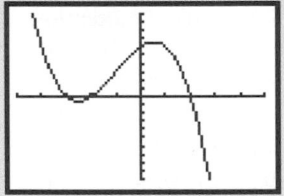

$[-5, 5, 1, -10, 10, 1]$

SHARING IDEAS

For presentations, if possible, choose groups that had different ideas about Step 4.

Ask whether students think there's only one valid equation. If no point on the graph is known, then any vertical dilation factor will give a graph with the same *x*-intercepts, but there should be only one way to adjust the equation to make the graph pass through a given point that's not an *x*-intercept. In fact, there's at most one cubic equation whose graph passes through any given four points.

[Ask] "How are quadratic and cubic equations alike and different?" [The factored forms show the roots for both. They both can

Step 2 a. Graph F
b. Graph C
c. Graph A
d. Graph D
e. Graph B
f. Graph E

Step 2 Each equation below matches exactly one graph in Step 1. Use graphs and tables to find the matches.

a. $y = (x + 3)(x + 1)(x - 2)$ **b.** $y = 2x(x + 1)(x - 2)$
c. $y = (x + 2)(x - 1)(x - 2)$ **d.** $y = -3x(x + 1)(x - 2)$
e. $y = 0.5(x + 2)(x - 1)(x - 2)$ **f.** $y = (x + 2)(x - 1)(x - 3)$

Step 3 Describe how the *x*-intercepts you found in Step 1 relate to the factored forms of the equations in Step 2.

Now you'll write an equation from a graph.

Step 4 Use what you discovered in Steps 1–3 to write an equation with the same *x*-intercepts as the graph shown. Graph your equation; then adjust your equation to match the graph.

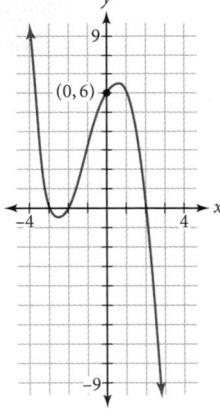

When you can identify the zeros of a function, you can write its equation in factored form. In this lesson you will see cubic equations with integer roots only.

EXAMPLE C   Find an equation for the graph shown.

▶ Solution   There are three *x*-intercepts on the graph. They are at $x = 0, -1$, and 2. Each intercept helps you find a factor in the equation. These factors are $x, x + 1$, and $x - 2$. Graph the equation $y = x(x + 1)(x - 2)$ on your calculator. The shape is correct, but you need to reflect it across the *x*-axis. You also need to vertically stretch the graph. Check the *y*-value of your graph at $x = 1$. The *y*-value is -2. You need it to be 4, so multiply by -2. The correct equation is $y = -2x(x + 1)(x - 2)$. Check this equation by graphing it on your calculator.

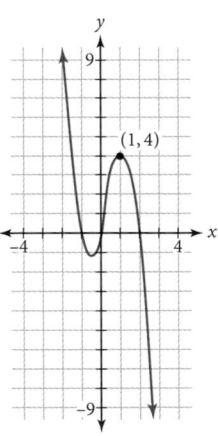

have the same number of roots as the highest power of *x*. Three points determine a quadratic equation and its parabolic graph, whereas four points determine a cubic.]

You might wonder with students about whether or not the highest and lowest values of a cubic occur halfway between roots, as the vertex of a parabola does. If students don't have enough examples to convince them, have them generate more.

▶ EXAMPLE C

This example helps students see how to find the equation of a cubic function given the *x*-intercepts of its graph and a point to determine the stretch factor.

A parabola that touches the *x*-axis at one point also has a double root. A sample equation is $y = (x - 1)^2$.

You can also use what you know about roots to convert cubic equations from general form to factored form. Look at the graph of $y = x^3 - 3x + 2$ at right. It has x-intercepts at $x = -2$ and $x = 1$. However, a cubic equation should have three roots. Where is the third one? Notice that the graph just touches the axis at $x = 1$. It doesn't actually pass through the axis. So the root at $x = 1$ is called a *double root*. The factor $x - 1$ is squared. Graph the factored form $y = (x + 2)(x - 1)^2$. It matches. If it didn't, you would need to look at a specific point to find the scale change required to make it match.

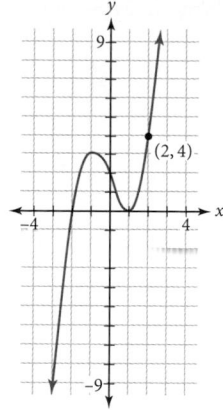

EXERCISES

You will need your calculator for problems **1, 6,** and **7.**

▶ Practice Your Skills

1. Determine whether each number is a perfect square, a perfect cube, or neither.

a. 2,209
perfect square; $47^2 = 2{,}209$

b. 5,832
perfect cube: $18^3 = 5{,}832$

c. 1,224 neither

d. 10,201
perfect square; $101^2 = 10{,}201$

2. Write and solve an equation to find the value of x in each figure.

a.

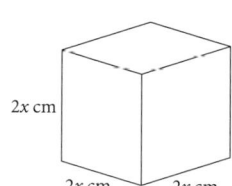

2x cm

2x cm   2x cm

5,832 cm³

b.

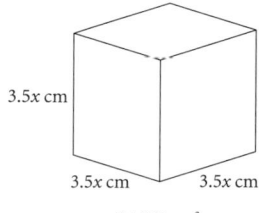

3.5x cm

3.5x cm   3.5x cm

21,952 cm³

c.

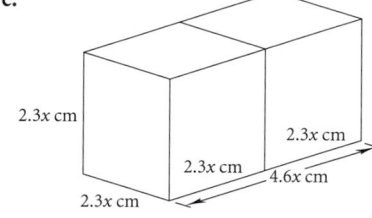

2.3x cm

2.3x cm

2.3x cm   4.6x cm

2.3x cm

3,309 cm³

3. Sometimes you can spot the factors of a polynomial expression without a graph of the equation. The easiest factors to see are those called *common monomial factors*. If you can divide each term by the same expression, then there is a common factor. Factor each expression by removing the largest possible common monomial factor.

a. $4x^2 + 12x$
$4x(x + 3)$

b. $6x^2 - 4x$
$2x(3x - 2)$

c. $14x^4 + 7x^2 - 21x$
$7x(2x^3 + x - 3)$

d. $12x^5 + 6x^3 + 3x^2$
$3x^2(4x^3 + 2x + 1)$

4. Determine whether each table represents a linear function, an exponential function, a cubic function, or a quadratic function.

a.

x	y
2	4
5	25
8	64
11	121
14	196

quadratic

b.

x	y
2	7
5	11
8	15
11	19
14	23

linear

c.

x	y
2	4
5	32
8	256
11	2,048
14	16,384

exponential

d.

x	y
2	8
5	125
8	512
11	1,331
14	2,744

cubic

Assessing Progress
Look for skill at finding x-intercepts of a graph and understanding of factored form and roots.

Closing the Lesson

From the x-intercepts and one more point on the graph of a cubic equation, you can determine the roots of the equation and can write it in factored form. In reverse, if you know the roots or factored form of a cubic equation, you know the x-intercepts of its graph.

BUILDING UNDERSTANDING

Students work with cubes, cube roots, and cubic equations.

ASSIGNING HOMEWORK

Essential	1–5, 8, 9
Performance assessment	4, 5, 8
Portfolio	7, 10
Journal	6
Group	7, 9, 10
Review	11

▶ Helping with the Exercises

Exercise 2c Suggest that students use the formula $V = l \cdot w \cdot h$ or that they imagine the figure as two cubes and write the equation $(2.3x)^3 + (2.3x)^3 = 3309$.

2a. $(2x)^3 = 5{,}832$; $2x = 18$; $x = 9$ cm

2b. $(3.5x)^3 = 21{,}952$; $3.5x = 28$; $x = 8$ cm

2c. $2(2.3x)^3 = 3{,}309$; $2.3x \approx 11.83$; $x \approx 5.14$

Exercise 3 This is the only lesson in the book that deals directly with factoring expressions with common monomial factors. If students are having difficulty understanding what to do, ask, "What factors are common to all of the terms?" A rectangular diagram with just one row might help.

Exercise 6 Any integer n raised to the sixth power will be both a perfect square (of n^3) and a perfect cube (of n^2).

6a. Answers will vary. Three possibilities are $0^2 = 0^3 = 0$, $1^2 = 1^3 = 1$, $8^2 = 4^3 = 64$.

6b. Answers will vary. $(a^3)^2 = (a^2)^3$ for any integer a. For example, $(4^3)^2 = (4^2)^3 = 4096$, which is both a perfect cube and a perfect square. Or enter $Y_1 = x^3$, $Y_2 = x^2$, and $Y_3 = x^6$ into your calculator and look at the table. Y_3 will be Y_1 squared and Y_2 cubed.

7a. If the width is w, the length is $w + 6$ and the height is $w - 2$, so the volume is given by the equation $V = w(w + 6)(w - 2)$.

7b. Three solutions to the equation $47 = w(w + 6)(w - 2)$ are shown on the graph. However, only one solution is a positive value. A table gives the answer: $w \approx 3.4$. Widths greater than about 3.4 cm give volumes greater than 47 cm³.

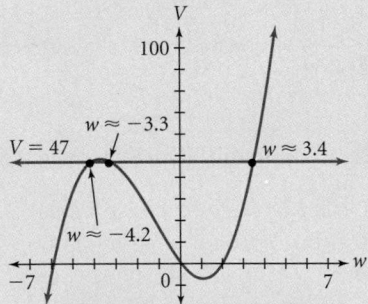

5. Write an equation in factored form for each graph.

a.

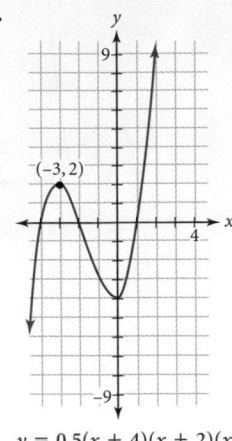

$y = 0.5(x + 4)(x + 2)(x - 1)$

b.

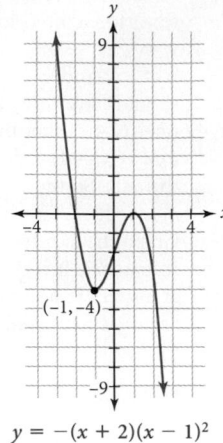

$y = -(x + 2)(x - 1)^2$

Reason and Apply

6. Some numbers are both perfect squares and perfect cubes.

  a. Find at least three numbers that are both a perfect square and a perfect cube.

  b. Define a rule that you can use to find as many numbers as you like that are both perfect squares and perfect cubes.

7. The length of a box is made so that its length is 6 cm more than its width. Its height is 2 cm less than its width.

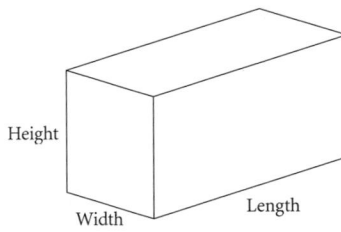

  a. Use the width as the independent variable, and write an equation for the volume of the box.

  b. Suppose you want to ensure that the volume of the box is greater than 47 cm³. Use a graph and a table to describe all possible widths, to the nearest 0.1 cm, of such boxes.

8. Determine whether each statement about the equation $0 = 2x^3 + 4x^2 - 10x$ is true or false.

  a. The equation has three real roots. true

  b. One of the roots is at $x = 2$. false

  c. There is one positive root. true

  d. The graph of $y = 2x^3 + 4x^2 - 10x$ passes through the point $(1, -4)$. true

9. To convert from factored form to general form when there are more than two factors, first multiply any pair of factors. Then multiply the result by the other factor. For example, to rewrite the expression $(x + 1)(x + 3)(x + 4)$ in general form, first multiply the first two factors.

$$(x + 1)(x + 3) = x^2 + 1x + 3x + 3 = x^2 + 4x + 3$$

Then multiply the result by the third factor. You might want to use a rectangular diagram to do this.

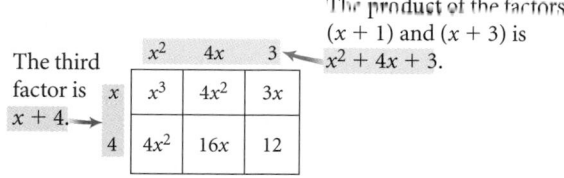

The third factor is $x + 4$.

The product of the factors $(x + 1)$ and $(x + 3)$ is $x^2 + 4x + 3$.

Next, combine like terms to find the sum of the regions.

$$x^3 + 4x^2 + 3x + 4x^2 + 16x + 12 = x^3 + 8x^2 + 19x + 12$$

Convert each expression below from factored form to general form. Use a graph or a table to compare the original factored form to your final general form.

a. $(x + 1)(x + 2)(x + 3)$
$x^3 + 6x^2 + 11x + 6$

b. $(x + 2)(x - 2)(x - 3)$
$x^3 - 3x^2 - 4x + 12$

10. The *girth* of a box is the distance completely around the box in one direction, that is, the length of a string that wraps around the box. Shippers put a maximum limit on the girth of a box rather than trying to limit its length, width, and height. Suppose you must ship a box with a girth of 120 cm in one direction and 160 cm in another direction.

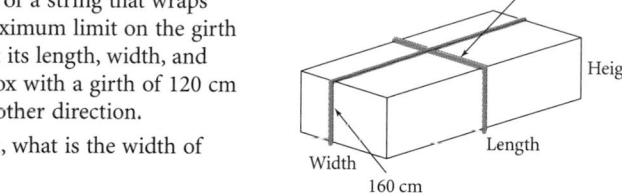

a. If the height of the box is 10 cm, what is the width of the box? 50 cm

b. If the height of the box is 10 cm, what is the length of the box? 70 cm

c. What is the volume of the box described in 10a and b? $10 \cdot 50 \cdot 70 = 35{,}000$ cm³

d. If the height is 15 cm, what are the other two dimensions and what is the volume of the box? The width is 45 cm, and the length is 65 cm. The volume is $15 \cdot 45 \cdot 65 = 43{,}875$ cm³.

e. If the height is x cm, find an expression for the width of the box.

f. If the height is x cm, find an expression for the length of the box.

g. Using your answers to 10e and f, find an equation for the volume of the box. $V = x(60 - x)(80 - x)$

h. What are the roots of the equation you found in 10g, and what do they tell you?

i. Find the dimensions of a box with a volume of 48,488 cm³.

▶ Review

11. Perform the operations, then combine like terms. Check your answers by using tables or graphs.

a. $\left(8x^3 - 5x\right) + \left(3x^3 + 2x^2 + 7x + 12\right)$

b. $\left(8x^3 - 5x\right) - \left(3x^3 + 2x^2 + 7x + 12\right)$

c. $\left(2x^2 - 6x + 11\right) + \left(-8x^2 - 7x + 9\right)$

d. $\left(2x^2 - 6x + 11\right) - \left(-8x^2 - 7x + 9\right)$

e. $\left(2x^2 - 6x + 11\right)\left(-8x^2 - 7x + 9\right)$

Exercise 10 Some students may think that the information given doesn't specify which girth is related to length and which to width. Point out that in this situation, length is assumed to be longer than width.

10e. $w = \dfrac{120 - 2x}{2} = 60 - x$

10f. $l = \dfrac{160 - 2x}{2} = 80 - x$

10h. $x = 0$ cm, $x = 60$ cm, $x = 80$ cm. These x-values make boxes with no volume.

10i. height 22 cm, width 38 cm, and length 58 cm; height 23.265 cm, width 36.735 cm, and length 56.735 cm.

Exercise 11 If students are having difficulty, suggest that they use a 3-by-3 rectangle. This exercise reviews Lesson 10.3.

11a. $11x^3 + 2x^2 + 2x + 12$

11b. $5x^3 - 2x^2 - 12x - 12$

11c. $-6x^2 - 13x + 20$

11d. $10x^2 + x + 2$

11e. $-16x^4 + 34x^3 - 28x^2 - 131x + 99$

PLANNING

LESSON OUTLINE

One day:

5 min	Introduction
30 min	Exercises and helping individuals
15 min	Student self assessment

REVIEWING

Refer students to the equation of the model rocket in the investigation of Lesson 10.1. Ask them to rewrite the equation in vertex form to find the vertex and to find the roots by solving the vertex form equation, by factoring into factored form, and by the quadratic formula. Have them check their factored and vertex forms by expanding back into the general form.

ASSIGNING HOMEWORK

For students who review best in interaction with other students, you might allow students to work in groups to complete parts a and c of problems and require them to do parts b and d on their own.

In this chapter you learned about **quadratic functions.** You learned that they model **projectile motion** and the acceleration due to **gravity.** You discovered important connections between the **roots** and the **x-intercepts** of quadratic equations and graphs. You learned how to use the three different forms of quadratic equations:

General form	$y = ax^2 + bx + c$
Vertex form	$y = a(x - h)^2 + k$
Factored form	$y = a(x - r_1)(x - r_2)$ or $y = ax(x - r_2)$ if $r_1 = 0$

The expression $ax^2 + bx + c$ is a type of **polynomial** because it is the sum of many **terms** or **monomials.** The vertex form gives you information about the **line of symmetry** of the parabola. The factored form shows you the roots of the equation. By the **zero product property,** you know that if the polynomial equals zero, then one of the **binomial** factors, $(x - r_1)$ or $(x - r_2)$, must equal 0. The roots r_1 and r_2 are also called **zeros** of the quadratic function. You learned to expand the vertex and factored forms to the general form by combining like terms.

You first learned to locate solutions to quadratic equations using calculator tables and graphs. You then learned to solve equations symbolically by one of three methods—factor with rectangular diagrams, **complete the square,** or use the **quadratic formula.**

To use the quadratic formula, $x = \frac{-b \pm \sqrt{b^2 - 4ac}}{2a}$, you learned to identify the **leading coefficient,** the **linear term,** and the **constant term** for the **trinomial** $ax^2 + bx + c$. You also learned to calculate the **discriminant,** $b^2 - 4ac$, and saw that it gives information about the number of solutions to the equation.

You saw that solutions to quadratic equations often contain **radical expressions.** You learned that the square root of a negative number does not result in a **real number.** You also learned how to find **cube roots, perfect cubes,** and **perfect squares.** In the last lesson you studied cubic functions.

EXERCISES

You will need your calculator for problems **4** and **9.**

1. Tell whether each statement is true or false. If it is false, change the right side to make it true, but keep it in the same form. That is, if the statement is in factored form, write your corrected version in factored form.

   a. $x^2 + 5x - 24 \stackrel{?}{=} (x + 3)(x - 8)$

   b. $2(x - 1)^2 + 3 \stackrel{?}{=} 2x^2 + x + 1$

   c. $(x + 3)^2 \stackrel{?}{=} x^2 + 9$ false; $x^2 + 6x + 9$

   d. $(x + 2)(2x - 5) \stackrel{?}{=} 2x^2 - x - 10$ true

1a. false; $(x - 3)(x + 8)$

1b. false; $2x^2 - 4x + 5$

2. The equation of the graph at right is

$$y = -2(x + 5)^2 + 4$$

Describe the transformations on the graph of
$y = x^2$ that give this parabola.
Sample response: There is a reflection across the
x-axis $(y = -x^2)$ and a vertical stretch with a factor
of 2 $(y = -2x^2)$. Finally, there is a translation left
5 units and up 4 units $(y = -(x + 5)^2 + 4)$.

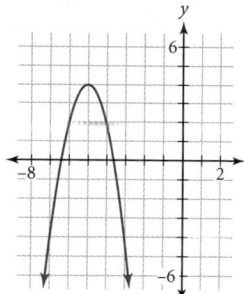

3. Write an equation for each graph below. Choose the form that best fits the
information given.

a.

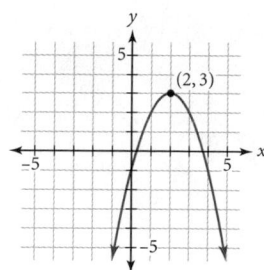

$y = -(x - 2)^2 + 3$; vertex form

b.

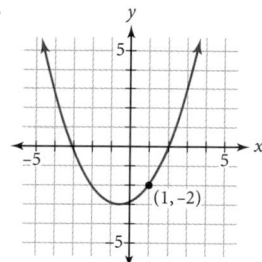

$y = 0.5(x - 2)(x + 3)$; factored form

4. Write an equation in the form $y = a(x - h)^2 + k$ for each graph below. Name an
appropriate viewing window.

a.

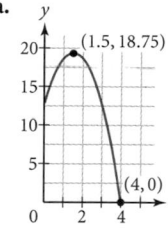

b.

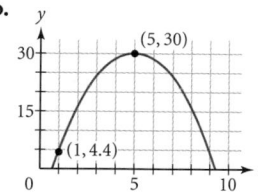

5. Use the zero product property to solve each equation.

a. $(2w + 9)(w - 3) = 0$
$2w + 9 = 0$ or $w - 3 = 0$; $w = -4.5$ or $w = 3$

b. $(2x + 5)(x - 7) = 0$
$2x + 5 = 0$ or $x - 7 = 0$; $x = -2.5$ or $x = 7$

6. Write an equation for a parabola that satisfies the given conditions.

a. The vertex is $(1, -4)$, and one of its x-intercepts is 3. $y = (x - 1)^2 - 4$

b. The x-intercepts are -1.5 and $\frac{1}{3}$.

7. Solve each equation by completing the square. Show each step. Leave your answers
in radical form.

a. $x^2 + 6x - 9 = 13$

b. $3x^2 - 24x + 27 = 0$

▶ **Helping with the Exercises**

4a. $y = -3(x - 1.5)^2 + 18.75$.
Sample window: $[0, 6, 1, 0, 25, 5]$.
4b. $y = -1.6(x - 5)^2 + 30$.
Sample window: $[0, 10, 1, 0, 35, 5]$.

Exercise 6a Students might substitute the coordinates $(3, 0)$ for
x and y in the vertex-form equation $y = a(x - 1)^2 - 4$ and
solve for a. Or they might cite
symmetry to note that the other
x-intercept will be at $(-1, 0)$,
derive the factored-form equation
$y = a(x - 3)(x + 1)$, and solve
for a after substituting the coordinates of the vertex.

6b. Sample answers:
$y = (x + 1.5)\left(x - \dfrac{1}{3}\right)$ and
$y = (2x + 3)(3x - 1)$.

7a. $x^2 + 6x - 9 = 13$
$x^2 + 6x = 22$
$x^2 + 6x + 9 = 22 + 9$
$(x + 3)^2 = 31$
$x + 3 = \pm\sqrt{31}$
$x = -3 \pm\sqrt{31}$

7b. $3x^2 - 24x + 27 = 0$
$3x^2 - 24x = -27$
$x^2 - 8x = -9$
$x^2 - 8x + 16 = -9 + 16$
$(x - 4)^2 = 7$
$x - 4 = \pm\sqrt{7}$
$x = 4 \pm\sqrt{7}$

9a. $f(60) = 8.1$. When there are 60 fish in the tank, the population is growing at a rate of about 8 fish per week.

9b. $f(x) = 0$ for $x = 0$ and $x = 150$. When there are no fish, the population does not grow. When there are 150 fish, the number of fish hatched is equal to the number of fish that die, so the total population does not change.

9c. When there are 75 fish, the population is growing fastest.

9d. The population no longer grows once there are 150 fish, so this is the maximum number of fish the tank has to support.

9e.

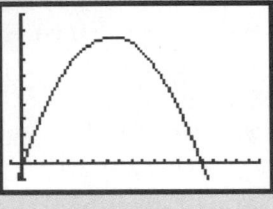

$[-10, 200, 10, -1, 10, 1]$

10. The roots are at 0 and 1.6 seconds, so start with the equation $y = x(x - 1.6)$. Then reflect the graph across the x-axis. When $x = 0.5$, $y = 0.55$, and we need the value of y to be 8.8. So apply a vertical stretch with a factor of $\frac{8.8}{.55}$, or 16. The final equation is $y = -16x(x - 1.6)$.

11a. No x-intercepts means taking the square root of a negative number. So $(-6)^2 - 4(1)(c) < 0$; $-4c < -36$; $c > 9$. Or translate the graph of $y = x^2 - 6x$ vertically to see that for values of $c > 9$, the parabola does not cross the x-axis.

11b. One x-intercept implies a double root, so $x^2 - 6x + c$ must be a perfect-square trinomial. Make a rectangular diagram to find $\left(\frac{-6}{2}\right)^2 = 9$, so $x^2 - 6x + 9$ is a perfect-square trinomial and $c = 9$. The graph touches the x-axis once. You can solve $b^2 - 4ac = 36 - 4c = 0$ to get $c = 9$.

11c. For $c < 9$, $b^2 - 4ac > 0$, so the discriminant gives two real roots. The parabola $y = x^2 - 6x + c$ crosses the x-axis twice for values of c that are less than 9.

---

8. Solve each equation by using the quadratic formula. Determine whether there are real number solutions. Leave your answer in radical form.

   a. $5x^2 - 13x + 18 = 0$ **b.** $-3x^2 + 7x + 9 = 0$

8a. $x = \dfrac{13 \pm \sqrt{-191}}{10}$, no real-number solutions.

8b. $x = \dfrac{-7 \pm \sqrt{157}}{-6}$

9. The function $f(x) = 0.0015x(150 - x)$ models the rate at which the population of fish grows in a large aquarium. The x-value is the number of fish, and the $f(x)$-value is the rate of increase in the number of fish per week.

   a. Find $f(60)$, and give a real-world meaning for this value.

   b. For what values of x does $f(x) = 0$? What do these values represent?

   c. How many fish are there when the population is growing fastest?

   d. What is the maximum number of fish the aquarium has to support?

   e. Graph this function.

10. A toy rocket blasts off from ground level. After 0.5 second it is 8.8 feet high. It hits the ground after 1.6 seconds. Write an equation in factored form to model the height of the rocket as a function of time.

11. Name values of c so that $y = x^2 - 6x + c$ satisfies each condition below. Use the discriminant, $b^2 - 4ac$, or translate the graph of $y = x^2 - 6x$ to help you.

   a. The graph of the equation has no x-intercepts.

   b. The graph of the equation has exactly one x-intercept.

   c. The graph of the equation has two x-intercepts.

12. Use the quadratic formula to find the roots of each equation.

   a. $x^2 + 10x - 6 = 0$ **b.** $3x^2 - 8x + 5 = 0$ $x = 1$ and $x = \frac{5}{3}$

   $x = -5 + \sqrt{31}$ and $x = -5 - \sqrt{31}$

13. For each graph, identify the x-intercepts and write an equation in factored form.

   a.

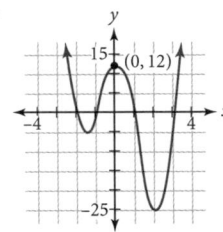

$x = -2, -1, 1,$ and 3;
$y = 2(x - 3)(x + 2)(x + 1)(x - 1)$

   b.

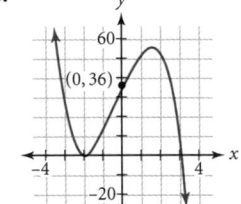

$x = -2$ (double root) and 3;
$y = -3(x + 2)^2(x - 3)$

14. Make a rectangular diagram to factor each expression.

   a. $x^2 + 7x + 12$ $(x + 3)(x + 4)$

   b. $x^2 - 14x + 49$ $(x - 7)^2$

   c. $x^2 + 3x - 28$ $(x + 7)(x - 4)$

   d. $x^2 - 81$ $(x - 9)(x + 9)$

14a.

	x	3
x	x^2	$3x$
4	$4x$	12

14b.

	x	-7
x	x^2	$-7x$
-7	$-7x$	49

14c.

	x	-4
x	x^2	$-4x$
7	$7x$	-28

14d.

	x	-9
x	x^2	$-9x$
9	$9x$	-81

TAKE ANOTHER LOOK

In this chapter you have encountered many equations, such as $x^2 = -4$, that have no real solutions. The solutions to these equations exist in another set of numbers called **imaginary numbers.** To find the solution to $x^2 = -4$, mathematicians write $x = 2i$ or $x = -2i$. The symbol i represents the imaginary unit.

Express i as a square root of a negative number. (Hint: $2i = \sqrt{-4}$, so factor the 4 out of the radical to see what i is.) What happens if you multiply i by itself to find i^2? Use this result to find i^3 and i^4. What happens if you keep multiplying by i?

Use the pattern you discovered to calculate i^{10}, i^{25}, and i^{100}.

Learn more about imaginary numbers with the links at www.keymath.com/DA .

Assessing What You've Learned

ORGANIZE YOUR NOTEBOOK Choose your best graph of a parabola from this chapter. Label the vertex, roots, line of symmetry, and y-intercept of the graph. Show the quadratic equation for the graph in each quadratic form—general, vertex, and factored.

WRITE IN YOUR JOURNAL Add to your journal by answering one of these prompts.
▶ There are many ways to solve quadratic equations—calculator tables and graphs, factoring, completing the square, and the quadratic formula. Which method do you like best? Do you always use the same method, or does the method depend on the problem?
▶ Compare each form of a quadratic equation—general, vertex, and factored. What information does each form tell you? How can you convert an equation from one form to another?

PERFORMANCE ASSESSMENT Show a classmate, a family member, or your teacher that you can solve any quadratic equation. Demonstrate how to find solutions with a calculator (graph or table) and by hand (factoring, completing the square, or using the quadratic formula).

GIVE A PRESENTATION Work with a partner or in a group to create your own problem about projectile motion. It can be about the height of a ball, the path of a rocket, or some other object. If possible, conduct an experiment to collect data. Decide which information will be given and which form of quadratic equation to use. Make up a question about your problem. Solve the problem using one of the methods you learned in this chapter. Present the problem and its solution to the class on a poster.

▶ **Take Another Look**

Many graphing calculators can operate on imaginary numbers, at least enough to generate the patterns here. $i^2 = -1$; $i^3 = (-1)i = -i$; $i^4 = 1$; $i^5 = i$. This pattern repeats every four powers so that $i^{10} = i^8 \cdot i^2$ (or -1); $i^{25} = i$; and $i^{100} = 1$.

ASSESSING

Use the test generator or Assessment Resources A or B to create a written test that is appropriate for your class. Combine the test results with the informal assessments you have made of students' group process skills and the knowledge and skills they have demonstrated in presentations, journals, and class participation.

FACILITATING SELF ASSESSMENT

To help students complete the portfolio described in Assessing What You've Learned, suggest that they consider for evaluation their work on Lesson 10.1, Exercises 6, 7, 9; Lesson 10.2, Exercises 7, 9; Lesson 10.3, Exercise 11; Lesson 10.4, Exercises 7, 10; Lesson 10.6, Exercise 9; Lesson 10.7, Exercise 5; and Lesson 10.8, Exercises 7, 10.

11

Introduction to Geometry

Overview

In this chapter algebra is applied to geometric concepts as students are introduced to analytic geometry and trigonometric ratios. Through a study of different polygons, **Lesson 11.1** reviews the slopes of parallel lines and shows how the slopes of perpendicular lines are related. In **Lesson 11.2,** students see how to find the coordinates of a segment's midpoint. In **Lesson 11.3,** students prepare for deriving the Pythagorean theorem by learning how to draw line segments whose lengths represent radical quantities. The Pythagorean theorem and its applications are the focus of **Lesson 11.4. Lesson 11.5** has students rewrite radical expressions in different forms, using rules for operating on radicals. In **Lesson 11.6,** students derive and learn to apply the distance formula. Trigonometry is the focus of the last two lessons. In **Lesson 11.7,** students study similar triangles and learn the basic trigonometric ratios. Applications of inverse trigonometric ratios are featured in **Lesson 11.8.**

The Mathematics

Approaches to geometry are usually classified as analytic (coordinate) and synthetic. Most geometry associated with algebra is analytic, as is the case in the student text.

Synthetic geometry

In about 300 B.C., the Greek mathematician Euclid wrote 13 books called the *Elements,* the culmination of all mathematics known in the Western world at that time. Because of paradoxes, Euclid was wary of measurement, so Euclid worked only with synthetic geometry. As a whole, the *Elements* encompasses number theory, ratios, and geometry. The climax of the first book, which lays the groundwork for them all and focuses on triangles, is the Pythagorean theorem.

Euclid states this theorem in terms of square figures: The square on the hypotenuse of a right triangle equals the sum of the squares on the other two sides. We, more comfortable with associating numbers with shapes, would insert the word *area:* the area of the square on the hypotenuse equals the sum of the areas of the squares on the other two sides. Even more commonly, in this algebraic age, we tend to abandon the consideration of areas of squares altogether and work with the lengths of the triangle's sides. We usually write $c^2 = a^2 + b^2$, although there's no reason we can't use other letters for the three lengths. Indeed, we often need to do so in applying the theorem.

Analytic geometry

In this text students have already worked with analytic geometry. They have plotted points and graphed equations. In this chapter they consider polygons, which are not graphs of equations $y = f(x)$.

They already know that two lines that have the same slope are parallel. Here they learn that two lines whose slopes are opposite reciprocals $\left(\text{such as } \frac{3}{4} \text{ and } -\frac{4}{3}\right)$ are perpendicular.

The idea of midpoint of a line segment is also familiar. What's new in this chapter is that the coordinates of the midpoint can be found by averaging (finding the mean of) the coordinates of the segment's endpoints.

The reason the Pythagorean theorem, which is a part of synthetic geometry, is introduced is to derive the formula for the distance between two points whose coordinates are known.

Trigonometry

Trigonometry, which comes from the Greek meaning "triangle measure," relates angles of triangles to the lengths of their sides. The student uses ratios of sides of right triangles to introduce trigonometry

for acute angles only. Trigonometry can be extended to obtuse angles, angles with a measure of more than 180 degrees, and even negative angles, but the right triangle context must be stretched significantly to do so. The extensions tend to be more effective if approached through periodic (circular) functions, in which angles are measured by radians, the number or radii they sweep out on a circle. Periodic functions are useful in modeling various kinds of waves as well as phenomena such as mean daily temperatures.

Using This Chapter

Check your district's standards to see how much analytic geometry and trigonometry must be included in the course. This chapter is good for courses that have some time for exploration at the end. For an abbreviated approach, consider only the highlights of Lessons 11.1, 11.3, 11.4, and 11.6. If students are familiar with right triangles, you might do only the trigonometry in Lessons 11.7 and 11.8.

Resources

Discovering Algebra Resources

Calculator Notes 7B, 11A, 11B

Teaching and Worksheet Masters
  Lessons 11.1, 11.2, 11.3, 11.6, 11.8 Surveying
  Lesson 11.1 Quadrilaterals
  Lesson 11.3 What's My Area?
  Lesson 11.5 Pyramid Net
  Lesson 11.6 Amusement Park Map
  Lesson 11.7 Earth and Moon
  Lesson 11.8 Contour Map

Assessment Resources A and B
  Quiz 1 (Lessons 11.1 and 11.2)
  Quiz 2 (Lessons 11.3 to 11.5)
  Quiz 3 (Lessons 11.6 to 11.8)
  Chapter 11 Test
  Chapter 11 Constructive Assessment Options
  Chapters 8 to 11 Exam
  Final Exam

More Practice Your Skills for Chapter 11

Condensed Lessons for Chapter 11

Other Resources

Dan Bennett. *Pythagoras Plugged In.* Emeryville, CA: Key Curriculum Press, 1995.

Lawrence W. Swienciki. *The Ambitious Horse.* Emeryville, CA: Key Curriculum Press, 2001.

Diane Venters, Elaine Krajenke Ellison. *Mathematical Quilts.* Emeryville, CA: Key Curriculum Press, 1999.

www.keymath.com/DA

Materials

- graph paper
- straightedges
- centimeter rulers
- protractors

Pacing Guide

	day 1	day 2	day 3	day 4	day 5	day 6	day 7	day 8	day 9	day 10
standard	11.1	11.2	quiz, 11.3	11.4	11.5	11.5	quiz, 11.6	11.6	11.7	11.7
enriched	11.1	11.2	quiz, 11.3	11.4, project	11.5	11.5, project	quiz, 11.6	11.6	11.7	11.7
block	11.1, 11.2	quiz, 11.3, 11.4	11.5	quiz, 11.6	11.7	11.8	quiz, review	assessment		

	day 11	day 12	day 13	day 14	day 15	day 16	day 17	day 18	day 19	day 20
standard	11.8	review	assessment							
enriched	11.8	review, TAL	assessment							

Introduction to Geometry

- Extend knowledge about slopes of parallel lines to perpendicular lines and extend knowledge of midpoints to coordinates

- Learn the definitions of various quadrilaterals defined by their parallel and perpendicular sides

- Learn to find the areas of polygons by decomposing the figures into triangles and rectangles and by removing triangles from larger rectangles

- See the Pythagorean theorem and its applications both geometrically and algebraically

- Work with radical expressions: squaring, distributing, simplifying, and applying to quadratic equations and parabolas

- See how the Pythagorean theorem leads to the distance formula and its applications in coordinate geometry

- Use similar triangles to express the definitions of the sine, cosine, and tangent as ratios

- Learn to find an acute angle in a real-world situation when one of its trigonometric ratios is known

These brightly colored wall paintings are a traditional art form of South Africa's Ndebele tribe. Ndebele women paint murals like these to celebrate special occasions such as weddings and harvests. Learning and continuing the ancient art form is an important part of training for young girls. The painters' use of universally recognized geometric shapes help these murals transcend time and cultural boundaries.

OBJECTIVES

In this chapter you will
- learn concepts, definitions, and symbols important in geometry
- use algebra to describe geometric relationships
- discover some properties of parallel and perpendicular lines
- find the coordinates of a line segment's midpoint
- calculate the distance between two points
- learn more about square roots
- explore important relationships between the sides of a right triangle

As you look at these designs with students, you can talk about patterns and transformations. **[Ask]** "What polygons do you see?" [triangles, trapezoids, rectangles, hexagons, dodecagons, . . .] "How are the shapes of the painting related to the shape of the fence?" [Both contain parts that are symmetric, lines are parallel, angles are repeated, designs are aligned, . . .] "Where do you see transformations?" [translations: along the top borders, shapes under these borders repeat; reflection: any of the elements that have bilateral symmetry; 2-fold rotation]

The Ndebele art style has come into international recognition through the work of the South African artist Esther Mahlangu. In 1991, she was the first woman to contribute to BMW's Art Car Collection—a series of automobiles that have been painted by notable artists (including Alexander Calder and Andy Warhol). In 1994, she was commissioned by the National Museum of Women in the Arts (Washington, D.C.) to paint a mural for an exhibition in her honor.

Parallel and Perpendicular

When you draw geometric figures on scaled coordinate axes, you are doing **analytic geometry.** You describe relationships and properties of the figures using the axes to identify points and write the equations of lines. In this lesson you will discover some interesting connections between algebra and geometry.

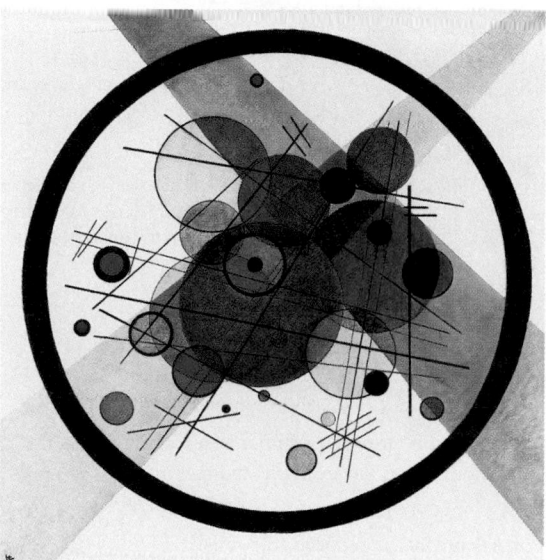

The Russian artist Wassily Kandinsky (1866–1944) used parallel and perpendicular line segments in his 1923 work titled *Circles in a Circle.*

Parallel lines are lines in the same plane that never intersect. They are always the same distance apart. You draw arrowheads on each line to show that they are parallel.

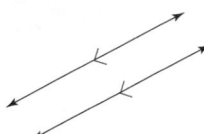

Perpendicular lines are lines that meet at a right angle, that is, at an angle that measures 90°. In fact, four right angles are formed where perpendicular lines intersect. You draw a small box in one of the angles to show that the lines are perpendicular.

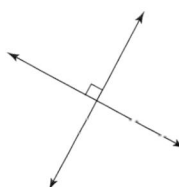

Investigation
Slopes

You will need
- graph paper
- a straightedge

A rectangle has two pairs of parallel line segments and four right angles. When you draw a rectangle on the coordinate plane and notice the slopes of its sides, you will discover how the slopes of parallel and perpendicular lines are related.

Step 1 Draw coordinate axes centered on graph paper. Each member of your group should choose one of the following sets of points. Plot the points and connect them, in order, to form a closed polygon. You should have a rectangle.

a. $A(6, 20)$, $B(13, 11)$, $C(-5, -3)$, $D(-12, 6)$
b. $A(3, -1)$, $B(-3, 7)$, $C(9, 16)$, $D(15, 8)$
c. $A(-11, 21)$, $B(17, 11)$, $C(12, -3)$, $D(-16, 7)$
d. $A(3, -10)$, $B(-5, 22)$, $C(7, 25)$, $D(15, -7)$

Step 2
slope of $\overline{AD}$ and $\overline{BC}$:
$\frac{7}{9}$
$\frac{3}{4}$
$\frac{14}{5}$
$\frac{1}{4}$

Step 3
slope of $\overline{AB}$ and $\overline{DC}$:
$-\frac{9}{7}$
$-\frac{4}{3}$
$-\frac{5}{14}$
-4

NCTM STANDARDS

CONTENT	PROCESS
Number	Problem Solving
• Algebra	• Reasoning
• Geometry	Communication
Measurement	Connections
Data/Probability	• Representation

LESSON OBJECTIVES

- Discover how slopes of parallel and perpendicular lines are related
- Learn the definitions of various quadrilaterals defined by their parallel and perpendicular sides

PLANNING

LESSON OUTLINE

One day:
5 min	Introduction
20 min	Investigation
5 min	Sharing
10 min	Examples
5 min	Closing
5 min	Exercises

MATERIALS

- graph paper
- straightedges
- Surveying (T or W), *optional*
- Quadrilaterals (T)

TEACHING

Further study of analytic (coordinate) geometry begins by extending students' knowledge of slopes of parallel lines to geometric figures and learning about slopes of perpendicular lines in those figures.

INTRODUCTION

Remind students what a *plane* is. Work so far with a coordinate plane has included plotting individual points or graphing lines or other functions. Now we turn to other geometric figures composed of line segments.

Any pair of parallel lines will lie in a single plane, but in space three lines each parallel to the other two might not be in the same plane. **[Ask]** "Must a pair of lines either meet or be parallel?" [No, there are lines in space that never meet and are not parallel; they are called *skew.*]

See page 725 for answers to Steps 1a–d.

For a tool to draw line segments, the name *straightedge* (rather than *ruler*) is used to point out that any measurement markings on the tool are irrelevant.

One step Display the Surveying transparency and ask, "What can you say about the shape of this plot of land? Are any of its edges parallel or perpendicular?" As you circulate, keep asking students how they can know for certain whether two lines are parallel or perpendicular. As needed, suggest that they check the slopes of known cases. If they check only the case of horizontal and vertical lines, suggest that they try a rotation of the lines about the origin. In Sharing, discuss the assumptions needed to have two of the edges parallel. (Is the measurement really 637.5 feet?) Students might solve the problem with any of the four vertices at the origin and may begin a discussion of whether or not a plane is a good representation of this part of the earth.

Step 1 Each student in the group can choose a different set of points. As needed, point out that the notation $A(6, 20)$ means that the point named A has coordinates $(6, 20)$. As needed, help students see that the rectangle is slanted. Students can add "perpendicular boxes."

[Language] A line over the top of other symbols, as used to designate a line segment, is called a *vinculum*. Technically, part of the standard square root symbol is a vinculum. Vincula used to be employed in many places where we now use parentheses.

Step 4 The idea that slopes of parallel lines are equal will be

See page 725 for answers to Steps 5, 6, and 8.

Steps 2 and 3 See answers on page 583.

Step 4 Parallel lines have the same slopes.

The slope of a line segment is the same as the slope of the line containing the segment. You can write the segment between A and D as $\overline{AD}$.

Step 2 Find the slopes of $\overline{AD}$ and $\overline{BC}$.

Step 3 Find the slopes of $\overline{AB}$ and $\overline{DC}$.

Step 4 What conclusion can you make about the slopes of parallel lines based on your answers to Steps 2 and 3?

The ties underneath these railroad tracks in British Columbia are a real-world example of parallel segments.

The tracks are also parallel as long as they don't curve; they only seem to converge as they recede from view

To find the **reciprocal** of a number, you write the number as a fraction and then invert the numerator and the denominator. For example, the reciprocal of $\frac{2}{3}$ is $\frac{3}{2}$. The product of reciprocals is 1.

Step 5 Express the slope values of $\overline{AB}$ and $\overline{BC}$ as reduced fractions.

Step 6 Express the slope values of $\overline{AD}$ and $\overline{DC}$ as reduced fractions.

Step 7 Perpendicular lines have opposite reciprocal slopes; their product is −1. Any pair of perpendicular lines should reinforce this conclusion.

These street intersections in New York City are a real-world example of perpendicular lines.

Step 7 What conclusion can you make about the slopes of perpendicular lines? What is their product? Check your conclusion by finding the slopes of any other pair of perpendicular sides in your rectangle.

Step 8 On the coordinate plane, draw two new pairs of parallel lines that have the slope relationship you discovered in Step 7. What figure is formed where the two pairs of lines intersect?

remembered by many students from the several times it appeared in Chapter 5.

Step 7 [Alert] Students might need help in seeing the relationship here, because both "taking a reciprocal" and "finding the opposite" are needed. If students are having trouble, have them graph another set of lines with slope -2 and $+\frac{1}{2}$. Or ask them to draw perpendicular lines and find the slope.

Step 8 Try to help students realize that the relationship works both ways. **[Alert]** For students having difficulties understanding the slope relationships, you might suggest that they draw slope triangles on the rectangles' sides.

You can draw any polygon on a graph and assign coordinate pairs to its vertices. Then you can use these points to calculate slopes, lengths of sides, perimeters, areas, and even the sizes of angles. You can use this information to draw conclusions about the polygon.

A **right triangle** has one right angle. The sides that form the right angle are called **legs,** and the side opposite the right angle is called the **hypotenuse.**

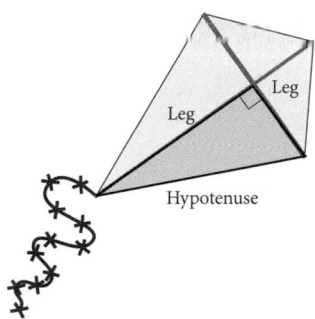

Leg
Leg
Hypotenuse

EXAMPLE A

Triangle *ABC* (written as △*ABC*) is formed by connecting the points (1, 3), (9, 5), and (10, 1). Is it a right triangle?

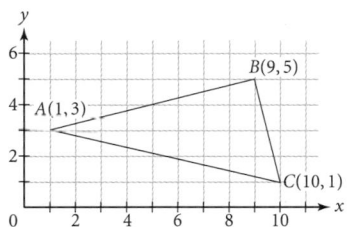

► **Solution**

The slope of $\overline{AB}$ is $\frac{1}{4}$, the slope of $\overline{AC}$ is $\frac{-2}{9}$, and the slope of $\overline{BC}$ is -4. The slopes $\frac{1}{4}$ and -4 are negative reciprocals of each other, so the sides with these slopes are perpendicular. That means angle *B* is a right angle. So these three points define a right triangle. Did you notice that the product of the two slopes, $\frac{1}{4}$ and -4, is -1?

A **trapezoid** is a quadrilateral with one pair of opposite sides that are parallel and one pair of opposite sides that are not parallel. A trapezoid with one of the nonparallel sides perpendicular to both parallel sides is a **right trapezoid.**

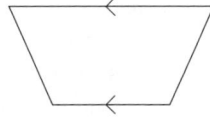
Trapezoid

Right trapezoid

► **EXAMPLE A**

This example shows how the relationship between slopes of perpendicular lines can be applied to find out if a triangle has a right angle. As needed, remind students that -4 can be written as $-\frac{4}{1}$.

The △ symbol is customarily and frequently used in geometry to mean *triangle.* Students should not confuse this with delta, introduced in Lesson 5.1 to help represent a line's slope $\left(\frac{\Delta y}{\Delta x}\right)$. Quadrilaterals and other polygons do not generally have symbols, although some textbooks will use symbols for frequently occurring quadrilaterals such as the square or parallelogram.

You may want to introduce the symbol ∠ as "angle." That is ∠B is read as "angle B." "Angle B" means the angle ∠ABC (or ∠CBA).

Have students present ideas about Step 4 only if there was confusion as they worked. You may want to introduce the symbol | | to mean "is parallel to." For example, writing $\overline{AD}\,||\,\overline{BC}$ means segment *AD* is parallel to segment *BC*.

Focus class discussion on Steps 7 and 8. Although the term *negative reciprocals* is often used to describe the relationship between, say, $\frac{2}{3}$ and $-\frac{3}{2}$, the term *opposite reciprocals* might be preferable because it better captures the fact that each is related to the other in the same way. (That is $\frac{2}{3}$ is the opposite reciprocal of $-\frac{3}{2}$. But only $-\frac{3}{2}$ is a negative number.) Students encountered opposites in the introduction of absolute values in Lesson 8.6.

Ask whether the slopes of a horizontal and a vertical line are opposite reciprocals of each other, or, equivalently, if their product is -1. Because the slope of a vertical line is undefined, these kinds of lines, though perpendicular, contradict the claim "if two lines are perpendicular, then their slopes are opposite reciprocals." The converse of that claim is true, however: If their slopes are opposite reciprocals, then two lines are perpendicular.

You may want to introduce the symbol ⊥ to mean "is perpendicular to." For example, $\overline{AD} \perp \overline{BC}$ means segment *AD* is perpendicular to segment *BC*.

Assessing Progress
Through your observations of group work and presentations, you can assess students' abilities to **plot points,** find **slopes** of line segments, and **reduce fractions.**

► **EXAMPLE B**

This example illustrates how the relationships between both parallel and perpendicular lines can be applied to learn something about geometric figures.

Be sure to sketch pictures when introducing the term *trapezoid*. With this definition of trapezoid, no parallelogram is a trapezoid. Therefore, to verify that the given figure is a trapezoid, the student must verify that the slope of $\overline{CD}$ is indeed different from that of $\overline{AB}$. (The term *trapezoid* is often defined without the condition that one pair of edges is nonparallel. With that definition, every parallelogram is a trapezoid, and this additional step wouldn't be needed.)

Closing the Lesson

Slopes of parallel lines are equal. Slopes of **perpendicular** lines are **opposite reciprocals.** These two facts can be applied to learn something about geometric figures whose vertices are given by coordinates. For example, a triangle can be checked to determine if it's a **right triangle,** and a quadrilateral can be checked to determine whether it's a **trapezoid,** parallelogram, or rectangle.

BUILDING UNDERSTANDING

The exercises give practice in working with slopes of parallel and perpendicular lines.

ASSIGNING HOMEWORK

EXAMPLE B | Classify as specifically as possible the polygon formed by the points $A(-4, 1)$, $B(-2, 4)$, $C(4, 0)$, and $D(-1, -1)$.

► **Solution** | Plot the vertices on a coordinate plane and connect them to form a quadrilateral.

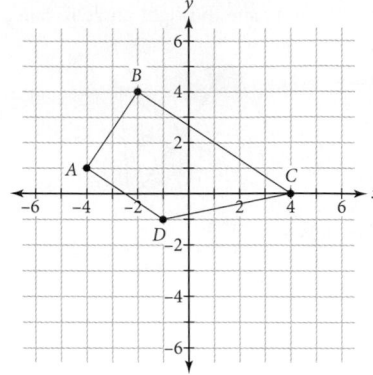

Calculate the slopes of the sides. Notice equal slopes (parallel sides) and negative reciprocal slopes (perpendicular sides).

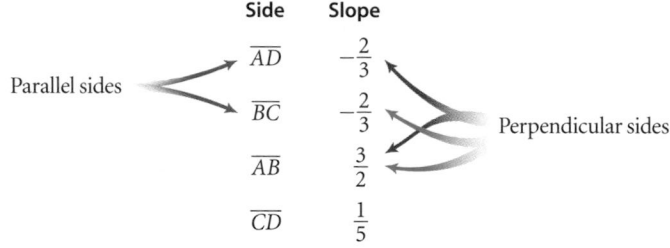

Side	Slope
$\overline{AD}$	$-\frac{2}{3}$
$\overline{BC}$	$-\frac{2}{3}$
$\overline{AB}$	$\frac{3}{2}$
$\overline{CD}$	$\frac{1}{5}$

Quadrilateral $ABCD$ has one set of parallel sides and one side perpendicular to that pair. So $ABCD$ is a right trapezoid.

EXERCISES

▶ Practice Your Skills

1. Find the slope of each line.

   a. $y = 0.8(x - 4) + 7$ 0.8   **b.** $y = 5 - 2x$ -2

   c. $y = -1.25(x - 3) + 1$ -1.25 **d.** $y = -4 + 2x$ 2

   e. $6x - 4y = 11$ $\frac{3}{2}$ **f.** $3x + 2y = 12$ $-\frac{3}{2}$

   g. $-9x + 6y = -4$ $\frac{3}{2}$ **h.** $10x - 15y = 7$ $\frac{2}{3}$

2. Write the equation of a line with a slope of $\frac{3}{4}$ that passes through the point $(-2, 5)$. $y = 5 + \frac{3}{4}(x + 2)$

▶ Helping with the Exercises

Exercise 1 [Ask] "Is there a quick way to find the slope from the three coefficients of $Ax + By = C$?" Help students solve the general equation once to see that the slope will be $-\frac{A}{B}$. Then, for example, they can see quickly that the slope in 1e is $-\frac{6}{-4} = \frac{3}{2}$.

3. Determine whether each pair of lines is parallel, perpendicular, or neither.

 a. $y = 0.8(x - 4) + 7$
  $y = -1.25(x - 3) + 1$ **perpendicular**

 b. $y = 5 - 2x$
  $y = -4 + 2x$ **neither**

 c. $6x - 4y = 11$
  $-9x + 6y = -4$ **parallel**

 d. $3x + 2y = 12$
  $10x - 15y = 7$ **perpendicular**

Exercise 3 This exercise uses the same equations as Exercise 1, so students can use their answers to complete this problem.

▶ Reason and Apply

For problems 4–11, plot each set of points on graph paper and connect them in order to form a polygon. Classify each polygon using the most specific term that describes it. Justify your answers by finding the slopes of the sides of the polygons.

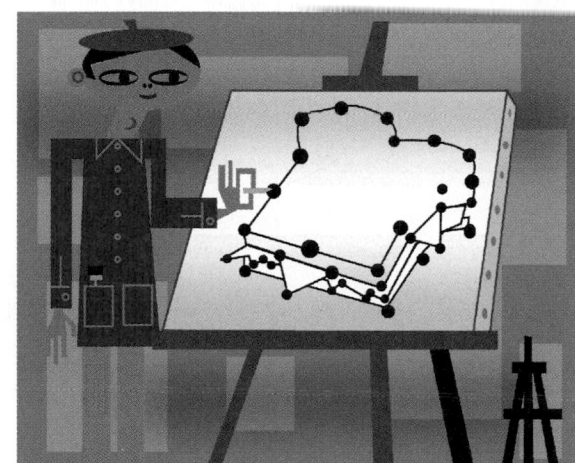

 4. $(-5, 0), (1, 4), (6, 3), (-3, -3)$

 5. $(-3, -2), (3, 1), (5, -3), (-1, -6)$

 6. $(-3, 4), (0, 4), (3, 0), (3, -4)$

 7. $(-1, 4), (2, 7), (5, -2), (2, -5)$

 8. $(-4, -1), (-2, 7), (2, 6), (3, 3)$

 9. $(0, 4), (2, 8), (6, -2), (2, -1)$

 10. $(-8, -2), (-4, 4), (5, -2), (1, -8)$

 11. $(-2, 2), (1, 5), (4, 2), (1, -3)$

Exercises 4 to 11 If students need a review of terms, you might go over the Quadrilaterals transparency before they begin work on the assignment. If students are working individually, you might assign only 4 through 8. Or you might assign two of problems 4 through 11 to each member of a group of four.

 12. Name coordinates for the vertices of a quadrilateral that has two right angles and no parallel sides.

Exercise 12 Encourage students to plot their quadrilaterals on graph paper.

▶ Review

 13. Find the solution for each system, if there is one.

 a. $\begin{cases} y = 0 \\ y = 2 + 3x \end{cases}$ $\left(-\frac{2}{3}, 0\right)$

 b. $\begin{cases} y = 0.25x - 0.25 \\ y = 0.75 + x \end{cases}$ $(-1.\overline{3}, -0.58\overline{3})$

 c. $\begin{cases} 2y = x - 2 \\ 3y = x - 3 \end{cases}$ $(0, -1)$

Exercise 13 This exercise reviews Lesson 6.3.

 14. Multiply and combine like terms.

 a. $x(x + 2)(2x - 1)$ $2x^3 + 3x^2 - 2x$

 b. $(0.1x - 2.1)(0.1x + 2.1)$ $0.01x^2 - 4.41$

Exercise 14 Allow students to use rectangular diagrams. This exercise reviews Lesson 10.3.

 15. At right, what is the ratio of the total area of shaded triangles to the area of the largest triangle? $\frac{6}{16}$ or $\frac{3}{8}$

Exercise 15 This exercise reviews Lesson 0.1 but could be done even if Chapter 0 was omitted.

 16. Find the value halfway between

 a. 3 and 11 7

 b. −4 and 7 1.5

 c. −12 and −1 −6.5

 d. 2 and 47 24.5

Exercise 16 This exercise reviews Lesson 1.2, because students find the mean of the two values. It will help prepare students for calculating midpoints in Lesson 11.2.

4. right trapezoid; slopes: $\frac{2}{3}, -\frac{1}{5}, \frac{2}{3}, -\frac{3}{2}$

5. rectangle; slopes: $\frac{1}{2}, -2, \frac{1}{2}, -2$

6. trapezoid; slopes: 0, $-\frac{4}{3}$, undefined, $-\frac{4}{3}$

7. parallelogram; slopes: 1, −3, 1, −3

8. quadrilateral; slopes: 4, $-\frac{1}{4}, -3, \frac{4}{7}$

9. trapezoid; slopes: 2, $-\frac{5}{2}, -\frac{1}{4}, -\frac{5}{2}$

10. rectangle; slopes: $-\frac{2}{3}, \frac{3}{2}, -\frac{2}{3}, \frac{3}{2}$

11. quadrilateral; slopes: 1, −1, $\frac{5}{3}, -\frac{5}{3}$

12. Possible answer:

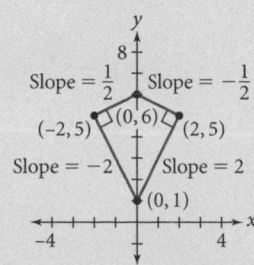

LESSON OUTLINE

One day:

5 min	Introduction
20 min	Investigation
5 min	Sharing
10 min	Example
5 min	Closing
5 min	Exercises

MATERIALS

- graph paper
- straightedges
- Surveying (T), *optional*

TEACHING

The midpoint of a line segment was reviewed in Chapter 0. Using analytic geometry, its coordinates can be found from those of the segment's endpoints.

INTRODUCTION

You may want to ask students why they think the same term, *median,* is used for both data analysis and geometry. The median of a data set divides the set into two halves with the same number of data points, and the median of a triangle divides it into two halves with equal areas.

Guiding the Investigation

One step Show the Surveying transparency used in Lesson 11.1 and pose this problem: "A fellow surveyor has suggested that you can divide the plot of land in half by connecting the midpoints of two of the edges. Find the midpoints of the four edges. What do you think of the suggestion?" As students work, suggest as needed that they try a

variety of approaches to finding the midpoints. Some students may use graph paper and find areas by counting squares.

Step 1

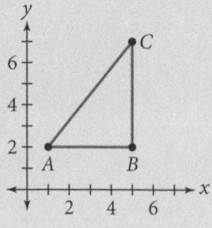

LESSON
11.2

Finding the Midpoint

I n analytic geometry you can use the algebraic concept of slope to identify parallel and perpendicular lines. That helps you recognize and draw geometric figures like rectangles and right triangles. Another useful skill is to find a **midpoint,** the middle point, of a line segment. Midpoints are used, for example, to draw these two geometric figures.

Balance is beautiful.
MIYOKO OHNO

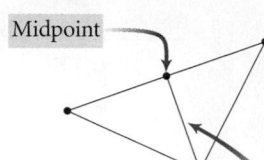

Midpoint

A **median** of a triangle is a segment that connects a vertex to the midpoint of the opposite side.

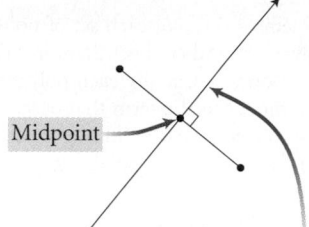

Midpoint

A **perpendicular bisector** is a line that divides a segment in half and that passes through the segment at a right angle.

Investigation
In the Middle

In this investigation you will discover a method for finding the coordinates of the midpoint of a segment. As you work through the steps, think about which algebra concepts help you find a midpoint.

You will need
- graph paper
- a straightedge

Step 2 $(3, 2)$; some students will find this by visual inspection; others may average the x-coordinates of the endpoints.

Step 3 $(5, 4.5)$; some students will find this by inspection; others may average the y-coordinates of the endpoints.

Step 4 $(3, 4.5)$; the x-coordinate is halfway between the x-coordinates of A and C. The y-coordinate is halfway between the y-coordinates of A and C.

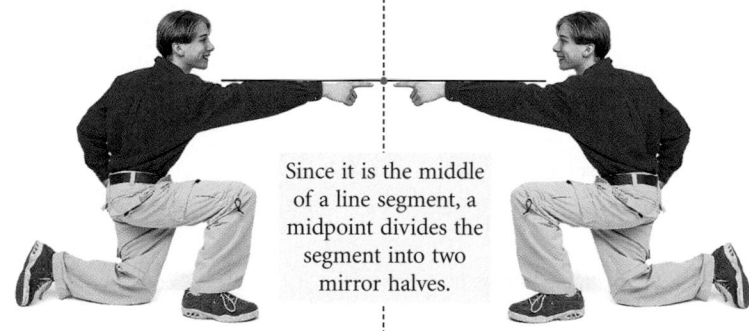

Since it is the middle of a line segment, a midpoint divides the segment into two mirror halves.

Step 1 Plot the points $A(1, 2)$, $B(5, 2)$, and $C(5, 7)$ and connect them.

Step 2 Find the midpoint of $\overline{AB}$. How did you find this point?

Step 3 Find the midpoint of $\overline{BC}$. How did you find this point?

Step 4 Find the midpoint of $\overline{AC}$. How does the midpoint's x-coordinate compare to the x-coordinates of A and C? How does its y-coordinate compare to the y-coordinates of A and C?

Step 2 You may want to define the points at the ends of a segment as *endpoints*. The term is used in the student text without definition. A definition does appear in the glossary.

Step 4 If students have been answering the questions by measuring, estimating, or counting grid lines, they may find this task difficult. Encourage them to draw horizontal and vertical lines through the midpoint of $\overline{AC}$.

Step 6 One answer: Find the distance between the *x*-coordinates and add half of this distance to the leftmost *x*-coordinate; then find the distance between the *y*-coordinates and add half of this distance to the lowest *y*-coordinate. Another answer: Average the *x*-coordinates and average the *y*-coordinates.

Step 7 If an answer in Step 6 involved counting grid squares or looking at the graph, the students should state a generalized answer similar to the possible answers given for Step 6.

Step 5

Step 6

Step 7

Step 8

Consider the points $D(2, 5)$ and $E(7, 11)$. Find the midpoint of $\overline{DE}$. $(4.5, 8)$

Explain how to find the coordinates of the midpoint of a line segment between any two points.

Does your method in Step 6 work even if you don't graph the points? If not, change your method so that it will work.

Find the midpoint of the segment between each pair of points.

a. $F(-7, 42)$ and $G(2, 14)$ $(-2.5, 28)$
b. $H(2.4, -1.8)$ and $J(-4.4, -2.2)$ $(-1, -2)$

There are several ways to find the midpoint of a segment. However, the midpoint is always halfway between the two endpoints, so its *x*-coordinate will be the average of the *x*-coordinates of the endpoints. Likewise, its *y*-coordinate will be the average of the *y*-coordinates of the endpoints.

In the next example you'll combine your knowledge of midpoints and slopes.

EXAMPLE

A triangle has vertices $A(-4, 3)$, $B(5, 9)$, and $C(0, -3)$.

a. Write the equation of the median from vertex B.
b. Write the equation of the perpendicular bisector of $\overline{AB}$.

▶ **Solution**

First, plot $\triangle ABC$.

a. The median from vertex B connects to the midpoint of $\overline{AC}$. Find the midpoint of $\overline{AC}$, then sketch the median.

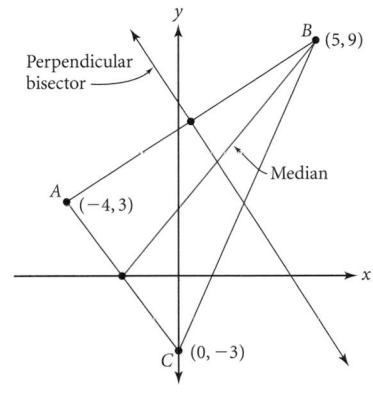

Find the average of the *x*-coordinates of A and C.

Find the average of the *y*-coordinates of A and C.

The midpoint of $\overline{AC}$.

$$\left(\frac{-4+0}{2}, \frac{3+-3}{2}\right) \text{ or } (-2, 0)$$

Step 6 As needed, suggest that students think about how finding the midpoint of a line segment relates to previous concepts studied in this book. Specifically, a midpoint can be found by calculating the mean of the *x*- and *y*-coordinates of the endpoints of the segment.

SHARING IDEAS

Have students share their methods from Step 7. Then ask them to state the rule symbolically, aiming for the midpoint formula. Encourage all students to assist with this derivation, which many might find difficult to articulate. Keep a balance between supporting students and challenging them to think.

Point out the quotation opening the lesson. *Design News* called Miyoko Ohno "the first ranking bridge designer in Japan." Ask students what *balance* means. They may discuss the physical, the visual, the emotional, and other realms. **[Ask]** "Is physical balance always achieved at the midpoint?" [Thinking of a beam whose mass is uniformly distributed, they may say yes. Or they may think of the beam balance of Chapter 3, with weights placed in various locations, and say no.]

Assessing Progress

Watch for students' ability to plot points and find the **mean** of two numbers. Also check for their understanding of **midpoint.**

▶ **EXAMPLE**

In this example students see an application of the midpoint formula to medians of triangles and perpendicular bisectors of line segments. Have students draw the triangle before going through the solution. As necessary, refer students to the pictures of a median and a perpendicular bisector on page 588. You may want to post a copy of the midpoint formula.

LESSON OBJECTIVES

- Discover the coordinates of the midpoint of a segment in terms of those of its endpoints
- Use coordinates of the midpoint of a segment to write equations of lines in polygons

NCTM STANDARDS

CONTENT		PROCESS	
	Number	●	Problem Solving
●	Algebra	●	Reasoning
●	Geometry	●	Communication
●	Measurement		Connections
	Data/Probability	●	Representation

Closing the Lesson

The coordinates of the **midpoint** of a line segment are the means (averages) of the respective coordinates of the segment's endpoints. The midpoint can be used to find **medians** of triangles and **perpendicular bisectors** of line segments.

BUILDING UNDERSTANDING

Students practice finding coordinates of midpoints.

ASSIGNING HOMEWORK

Essential	1–3, 6, 8
Performance assessment	6, 8, 9
Portfolio	8, 10
Journal	4, 6, 7, 9
Group	4, 5, 10, 11
Review	1, 2, 4, 12, 13

Helping with the Exercises

Exercises 1, 2, and 4 This is the first use of single letters as names for lines. If students are confused, they may write out equations for the lines.

Now use the coordinates of vertex B and the midpoint of $\overline{AC}$ to find the slope of the median.

Find the slope between vertex B and the midpoint.

$$\text{Slope} = \frac{9 - 0}{5 - (-2)} = \frac{9}{7}$$

Use the coordinates of the midpoint and the slope to write the equation of the median in point-slope form.

$$y = 0 + \frac{9}{7}(x - (-2)) \quad \text{or} \quad y = \frac{9}{7}(x + 2)$$

b. Look back at the sketch on page 589. The perpendicular bisector goes through the midpoint of $\overline{AB}$ and is perpendicular to $\overline{AB}$. First find the midpoint of $\overline{AB}$.

$$\left(\frac{-4 + 5}{2}, \frac{3 + 9}{2}\right) \text{ or } \left(\frac{1}{2}, 6\right)$$

The slope of $\overline{AB}$ is $\frac{9 - 3}{5 - (-4)}$ which equals $\frac{6}{9}$ or $\frac{2}{3}$. The slope of the perpendicular bisector is the negative reciprocal, or $\frac{-3}{2}$.

The equation of the perpendicular bisector of $\overline{AB}$ in point-slope form is

$$y = 6 + \frac{-3}{2}\left(x - \frac{1}{2}\right)$$

What you have learned about finding the midpoint of a segment is summarized by this formula.

Midpoint Formula

If the endpoints of a segment have the coordinates (x_1, y_1) and (x_2, y_2), the midpoint of the segment has the coordinates

$$\left(\frac{x_1 + x_2}{2}, \frac{y_1 + y_2}{2}\right)$$

EXERCISES

Practice Your Skills

1. Line ℓ has a slope of 1.2. What is the slope of line p that is parallel to line ℓ? 1.2

2. Line ℓ has a slope of 1.2. Line m is perpendicular to line ℓ.
   a. What is the slope of line m? $\frac{-1}{1.2} = -\frac{5}{6} = -0.8\overline{3}$
   b. What is the product of the slopes of line ℓ and line m? -1

3. Find the midpoint of the segment between each pair of points.
   a. $(4, 5)$ and $(-3, -2)$ $(0.5, 1.5)$
   b. $(7, -1)$ and $(5, -8)$ $(6, -4.5)$

4. The equation of line ℓ has the form $Ax + By = C$. What is the slope of a line

a. Perpendicular to line ℓ? $\frac{B}{A}$ **b.** Parallel to line ℓ? $\frac{-A}{B}$

5. Find the midpoint of a segment with the endpoints (a, b) and (c, d). $\left(\frac{a+c}{2}, \frac{b+d}{2}\right)$

▶ Reason and Apply

6. The vertices of $\triangle ABC$ are $A(0, 0)$, $B(1, 5)$, and $C(0, 4)$. Is it a right triangle? Explain how you know.

7. There is a situation in which two lines are perpendicular but the product of their slopes is not -1. Describe this situation.

8. The points $A(2, 1)$ and $B(4, 6)$ are the endpoints of a segment.

a. Find the midpoint of $\overline{AB}$. $(3, 3.5)$

b. Write the equation of the perpendicular bisector of $\overline{AB}$. $y = 3.5 + \frac{-2}{5}(x - 3)$

9. Sketch this quadrilateral on your own paper.

a. Find the midpoint of each side.

b. Connect the midpoints in order. What polygon is formed? How do you know?

c. Draw the diagonals of the polygon formed in 9b. Are the diagonals perpendicular? Explain how you know.

10. On graph paper, draw a triangle with vertices $A(11, 6)$, $B(4, -8)$, and $C(-6, 6)$.

a. Find the midpoint of each side. Label the midpoint of $\overline{AB}$ point D, the midpoint of $\overline{BC}$ point E, and the midpoint of $\overline{CA}$ point F.

b. Find the slopes of $\overline{AB}$ and $\overline{FE}$. What is special about these two line segments?

c. Do your findings in 10b hold true between $\overline{BC}$ and $\overline{DF}$ and between $\overline{CA}$ and $\overline{ED}$? Explain why or why not. Yes; the slopes of $\overline{BC}$ and $\overline{DF}$ are both $\frac{-7}{5}$, and the slopes of $\overline{CA}$ and $\overline{ED}$ are both 0.

d. Compare the length of $\overline{ED}$ to the length of $\overline{CA}$. The length of $\overline{ED}$ is $\frac{1}{2}$ the length of $\overline{CA}$.

e. Without calculating, compare the area of $\triangle ABC$ to the area of $\triangle DEF$.

f. If you connect the midpoints of the sides of $\triangle DEF$, how does the area of the new triangle compare to the area of $\triangle ABC$? The area of the new triangle would be $\frac{1}{16}$ the area of $\triangle ABC$.

11. In 11a–c, you are given the midpoint of a segment and one endpoint. Find the other endpoint.

a. midpoint: $(7, 4)$ endpoint: $(2, 4)$ $(12, 4)$

b. midpoint: $(9, 7)$ endpoint: $(15, 9)$ $(3, 5)$

c. midpoint: $(-1, -2)$ endpoint: $(3, -7.5)$ $(-5, 3.5)$

The starting position of the game Cat's Cradle shows triangles, parallel lines, and midpoints. What other geometric shapes do you see?

[Graph showing quadrilateral with vertices $C(13, 12)$, $D(5, 8)$, $A(3, 2)$, $B(17, 4)$]

10a. $D(7.5, -1)$, $E(-1, -1)$, $F(2.5, 6)$

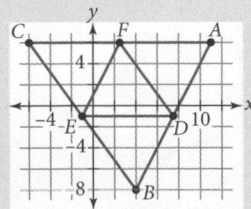

10b. The slopes of $\overline{AB}$ and $\overline{FE}$ are both 2; they are parallel.

10e. The area of $\triangle DEF$ is $\frac{1}{4}$ the area of $\triangle ABC$.

Exercise 4 This exercise relies on the insight obtained from Exercise 1e–1h in Lesson 11.1. Students may need to re-derive that slope from the standard form $Ax + By = C$.

Exercise 6 Even if students can answer the question by calculation, encourage them to sketch the triangle.

6. Yes. Possible answer: The slope of $\overline{AB}$ is 5 and the slope of $\overline{BC}$ is $\frac{-1}{5}$, so angle B is a right angle.

Exercise 7 If students aren't thinking of horizontal and vertical lines, you might suggest that they try the standard problem-solving strategy of considering extreme cases. Because the slope of a vertical line is undefined, the product of its slope with that of a horizontal line (0) is also undefined.

7. One line is horizontal and the other is vertical. A horizontal line has a slope of 0 and a vertical line has an undefined slope, so the product is also undefined.

9a. midpoint of $\overline{AB}$: $(10, 3)$
midpoint of $\overline{BC}$: $(15, 8)$
midpoint of $\overline{CD}$: $(9, 10)$
midpoint of $\overline{DA}$: $(4, 5)$

9b. parallelogram: The opposite sides are parallel because the slopes are 1, $\frac{-1}{3}$, 1, and $\frac{-1}{3}$.

9c. No; the slopes of the diagonals are $\frac{3}{11}$ and -7.

Exercise 10 You may want to introduce the notation ED to mean "the length of $\overline{ED}$." Hence, 10d could be reworded as "Compare ED to CA." **[Alert]** 10e might be a challenge, especially for students who are trying to use formulas to find areas. Encourage them to think about which of the smaller triangles have the same shape and size. (All four are congruent, so they all have the same area.) You might actually use the word *congruent*, which was introduced in Chapters 0 and 3. The idea that congruent triangles have the same area is used again in Exercise 13.

Exercise 12 [Ask] "Is the student text correct in saying that these are equations of lines?" Students may not have realized that equations of lines can be written like this, especially with x isolated. This exercise reviews Chapter 6 and Lesson 8.7.

12b. Possible answer: $x + y = 2$ and $y = 2x + 5$ or any line of the form $y = 3 + m(x + 1)$, where m is any number, or $x = -1$

12c. Possible answer: $y = (x + 1)^2 + 3$; any parabola of the form $y = a(x + 1)^2 + 3$ will have its vertex at this point.

Exercise 13 This exercise reviews Lesson 11.1.

13a. Possible answers:

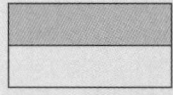 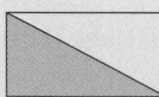

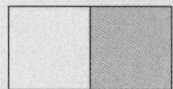 

13b. Answers will vary. Any method that uses only one line segment will form congruent polygons. Some other methods may also produce congruent polygons.

13c. Yes. If you imagine a vertical segment through the upper vertex of the triangle, you can see that the triangles on either side of the vertical line are congruent.

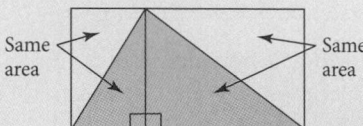
Same area — Same area

▶ **Review**

12. Two intersecting lines have the equations $2x - 3y + 12 = 1$ and $x = 2y - 7$.
    a. Find the coordinates of the point of intersection. $(-1, 3)$
    b. Write the equations of two different lines that intersect at this same point.
    c. Write the equation of a parabola that passes through this same point.

13. Draw four congruent rectangles—that is, all the same size and shape.
    a. Shade half the area in each rectangle. Use a different way of dividing the rectangle each time.
    b. Which of your methods in 13a divide the rectangle into congruent polygons?
    c. Ripley divided one of her rectangles like this. Is the area divided in half? Explain.

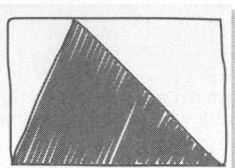

IMPROVING YOUR GEOMETRY SKILLS

This puzzle was created by the English mathematician Charles Dodgson (1832–1898). You may know him better as Lewis Carroll, the author of *Alice's Adventures in Wonderland*.

Cut an 8-by-8 square into pieces like this:

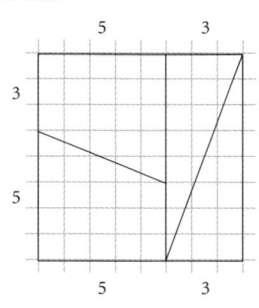

Reassemble them like this.

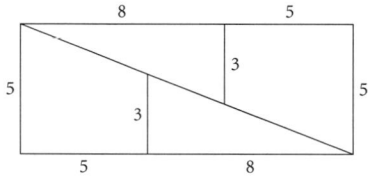

What is the area of the square? What is the area of the rectangle? Why aren't they equal?

IMPROVING GEOMETRY SKILLS

The area of the square is 64 square units, and the area of the rectangle is 65 square units. The slope of each triangle's hypotenuse is $\frac{-2}{5}$ or -0.4. The slope of each right trapezoid's nonperpendicular side is $\frac{-3}{8}$ or -0.375. The slopes of these segments are close enough to fool the eye, but the segments do not lie on the same line. Instead, they create a parallelogram whose area is 1 square unit. This and 59 other puzzles and paradoxes are collected in *One Equals Zero and Other Mathematical Surprises* by Nitsa Movshovitz-Hadar and John Webb (Emeryville, CA: Key Curriculum Press, 1998).

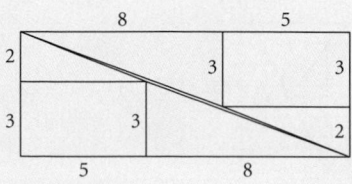

Squares, Right Triangles, and Areas

Triangles, squares, rectangles, and other polygons have been important throughout the history of design and construction. Finding the areas of farms, lots, floors, and walls is important for city planners, architects, building contractors, interior designers, and people in building trades and other occupations. An architect designs space for the people who will use a building. A contractor must be able to determine an approximate price per square foot to bid a job.

In this lesson you will practice finding the area of squares and the lengths of their sides on graph paper. You'll use this relationship to find the lengths of the sides of a right triangle.

Framing a house requires many parallels, perpendiculars, and area calculations.

EXAMPLE A | Find the area of each shape on the grid at right.

▶ **Solution** | The rectangle has an area of 3 square units. The area of the triangle is half the area of the rectangle, so it has an area of 1.5 square units.

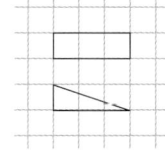

You can often draw or visualize a rectangle or square related to an area to help you find the area.

EXAMPLE B | Find the area of square *ABCD*.

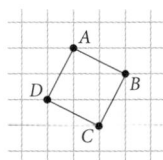

▶ **Solution** | Using the grid lines, draw a square around square *ABCD*. The outer square, *MNOP*, has an area of 9 square units. Each triangle has an area equal to half of 2 square units, or 1 square unit.

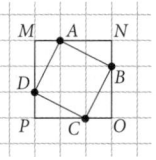

$$Area\ of\ square\ ABCD = Area\ of\ square\ MNOP - 4(1)$$

So the area of square *ABCD* is $9 - 4$, or 5 square units.

NCTM STANDARDS

CONTENT	PROCESS
• Number	Problem Solving
• Algebra	• Reasoning
• Geometry	Communication
Measurement	Connections
Data/Probability	• Representation

LESSON OBJECTIVES

• Learn how areas of figures can be found by decomposing the figures into triangles and rectangles or by removing triangles from larger rectangles

• Lay the groundwork for understanding the Pythagorean theorem by learning that if the area of a square is s, then its side length is $\sqrt{s}$

PLANNING

LESSON OUTLINE

One day:

10 min	Examples A, B
15 min	Investigation
5 min	Sharing
5 min	Example C
5 min	Closing
10 min	Exercises

MATERIALS

• graph paper
• straightedges
• Surveying (T), *optional*
• What's My Area? (W), *optional*

TEACHING

Finding areas of figures with vertices on a grid can lead to drawing segments with length the square root of an integer.

One step Display the Surveying transparency used first in Lesson 11.1. Ask what the area of the plot is. Encourage students to make a scale drawing on graph paper and to find the area both by adding up areas of smaller pieces and by subtracting off triangular areas from a larger rectangle. As needed, suggest that students consider right triangles as halves of rectangles. Ask that differing answers be presented during Sharing so that students can critique the various approaches.

▶ **EXAMPLE A**

This example reviews the meaning of area while reminding students that right triangles can be seen as half of rectangles. Students may have seen this idea in Exercise 13 of Lesson 11.2.

► **EXAMPLE B**

Here students see how to find the area of a figure by subtracting off the areas of pieces around the figure. The area can be found in other ways as well. Some students may be able to "count" the area of ABCD by piecing together whole grid squares.

Guiding the Investigation

This investigation is written for students working in pairs. Copying the figures onto their own graph paper, or perhaps even a geoboard, provides excellent visualization practice for students. If you are short on time, however, you can give each pair a copy of the worksheet What's My Area?

Step 1 Encourage students to use both addition and subtraction in finding these areas.

Step 2 Lengths are always positive, so we do not use $\pm\sqrt{}$ when expressing them.

Step 3 [Ask] "Is the second figure really a square? How can you tell?" [Justifications should mention that the slopes of adjacent edges are opposite reciprocals.] Two ways students can find the area are by adding areas of an interior square and four triangles or by subtracting areas of four triangles from the area of a larger square.

SHARING IDEAS

Students might report results of Step 1 (especially if there is disagreement) and show their ideas from Step 4.

[Ask] "How many different squares can be drawn with vertices at grid points on a 5-by-5 grid?" If "different" is interpreted as "different in size," then help the students find a systematic method to show that there are 11 (6 tilted). If "different" includes the placement, then again try to derive a systematic method to count all 105 (50 tilted).

Investigation
What's My Area?

You will need
• graph paper
• a straightedge

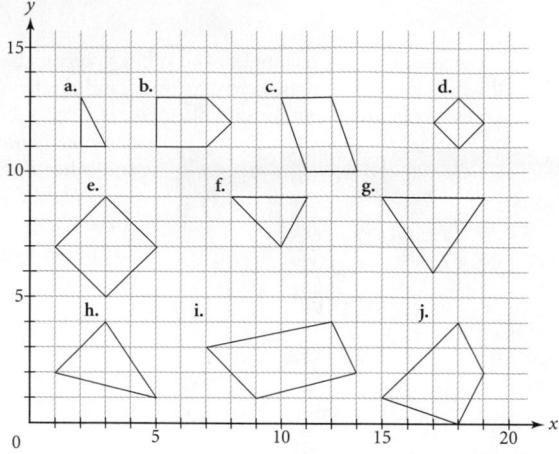

Step 1a 1 square unit
Step 1b 5 square units
Step 1c 6 square units
Step 1d 2 square units
Step 1e 8 square units
Step 1f 3 square units
Step 1g 6 square units
Step 1h 5 square units
Step 1i 10.5 square units
Step 1j 8 square units Step 1

Copy these shapes onto your graph paper. Work with a partner to find the area of each figure.

If you know the side length, s, of a square, then the area of the square is s^2. Likewise, if you know that the area of a square is s^2, then the side length is $\sqrt{s^2}$, or s. So the square labeled d in Step 1, which has an area of 2, has a side length of $\sqrt{2}$ units.

Step 2 area: 8 square Step 2
units; side length:
$\sqrt{8}$ units   Step 3

Step 3a area: 9 square units; side length: 3 units

Step 3b area: 10 square units; side length: $\sqrt{10}$ units

What are the area and side length of the square labeled e in Step 1?

What are the area and side length of each of these squares?

a.
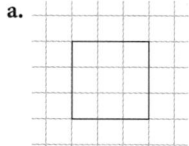

b.
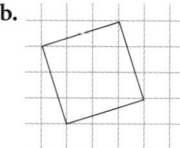

Step 4 Squares with Step 4
area 2, 5, 8, 10, 13, and 17 can be constructed by tilting the square relative to the grid; squares with area 1, 4, 9, 16, and 25 are possible by following grid lines. Side lengths will be the square root of the area.

Shown below are the smallest and largest squares with grid points for vertices that can be drawn on a 5-by-5 grid. Draw at least five other different-size squares on a 5-by-5 grid. They may be tilted, but they must be square, and their vertices must be on the grid. Find the area and side length of each square.

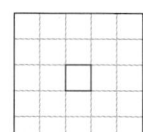

EXAMPLE C | Draw a line segment that is exactly $\sqrt{10}$ units long.

► **Solution** A square with an area of 10 square units has a side length of $\sqrt{10}$ units. Ten is not a perfect square, so you will have to draw this square tilted. Start with the next largest perfect square, that is 16 square units (4-by-4), and subtract the areas of the four triangles to get 10. Here are two ways to draw a square tilted in a 4-by-4 square. Only the square on the left has an area of 10 square units.

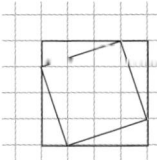 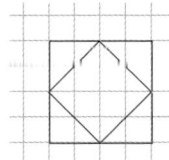

So a line segment with a length of $\sqrt{10}$ units looks like this:

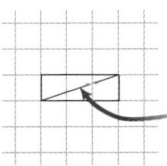

This red segment has a length of $\sqrt{10}$ units.

If a 4-by-4 square had not worked, you could have tried a larger square.

You can draw segments on graph paper with lengths equal to many square root values, but you may have to guess and check!

This fabric quilt, *Spiraling Pythagorean Triples,* shows several tilted squares. It was made by Diana Venters, a mathematician who uses mathematical themes in her quilts. You will learn about the Pythagorean theorem in Lesson 11.4.

See more mathematical quilts with the links at **www.keymath.com/DA** .

EXERCISES

You will need your calculator for problems **2** and **10**.

► **Practice Your Skills**

1. Find an exact solution for each equation. Leave your answers in radical form.
   a. $x^2 = 47$ $\pm\sqrt{47}$
   b. $(x - 4)^2 = 28$ $4\pm\sqrt{28}$
   c. $(x + 2)^2 - 3 = 11$ $-2\pm\sqrt{14}$
   d. $2(x - 1)^2 + 4 = 18$ $1\pm\sqrt{7}$

2. Calculate decimal approximations for your solutions to problem 1. Round your answers to the nearest thousandth. Check each answer by substituting it into the original equation.

► **Helping with the Exercises**

Exercise 2 When checking with decimal approximations, students will not get the exact value on the right. For instance, for 2a, $(6.856)^2 = 47.004736$, not 47. You may need to remind students of the difference between exact values and decimal approximations.

2a. ±6.856

2b. $-1.292, 9.292$

2c. $-5.742, 1.742$

2d. $-1.646, 3.646$

► **EXAMPLE C**

This example shows how to construct a segment whose length is the (positive) square root of an integer. Students may recognize this particular square from Step 3 of the investigation, but being able to think though the process in reverse will be very important for the next lesson.

If you have not done so in earlier chapters, you may want to show the shorthand for "square units" as "units²." Specifically, you can show ft², in², m², and so on.

Assessing Progress
Watch for abilities to see shapes as what's left over when something else is removed as well as a composite of smaller shapes. You can also assess understanding of **squares** of numbers and their **square roots.**

Closing the Lesson

Areas of many figures can be found by considering the figures as composites of smaller rectangles and triangles or as the remainder when triangles are removed from a larger rectangle. For many integers, a segment whose length is the square root of that integer can be constructed by making a square whose area is that integer.

BUILDING UNDERSTANDING

Students work with finding areas and constructing square roots.

ASSIGNING HOMEWORK

Essential	1, 3, 4, 6–9
Performance assessment	7, 8
Portfolio	5, 9
Group	2, 9
Review	10–11

Exercise 3 Some students will find it difficult to find areas by subtracting from the area of a circumscribed rectangle, the preferred method for 3e and perhaps 3d. Encourage them by saying that once they can find area by this method, they will find it easier than trying to piece together parts of little squares.

3a. 4 square units

3b. 12 square units

3c. 2 square units

3d. 6 square units

3e. 20 square units

3f. 18 square units

Exercises 4, 5a, and 5b In these exercises you have a chance to preview reducing radicals. You might encourage students to view the side lengths as being made up of short segments, each $\sqrt{2}$ units in length. For example, the square in Exercise 4 has 3 short $\sqrt{2}$ segments on each side. From that students can see that $\sqrt{18} = 3\sqrt{2}$.

Exercise 4 Students are not expected to reduce radical form at this point. For instance, they may leave the answer as $\sqrt{18}$ rather than rewriting it as $3\sqrt{2}$. Reducing radicals will be addressed in Lesson 11.5.

5. polygon 3a: $\sqrt{8}$ units and $\sqrt{2}$ units

polygon 3b: $\sqrt{8}$ units and $\sqrt{18}$ units

polygon 3e: $\sqrt{50}$ units, $\sqrt{50}$ units, and $\sqrt{40}$ units

Exercise 6 Encourage students to generalize that triangles with equal bases and equal heights have the same area.

Exercise 7 Again, encourage generalization to a pattern.

Exercise 8 Encourage creative thinking. **[Alert]** Suggest to any students having difficulty to use their knowledge of slope.

8. (6, 11) and (12, 7) or (−2, −1) and (4, −5)

3. Find the area of each figure at right.

4. Find the side length of the square in 3f.
$\sqrt{18}$ units

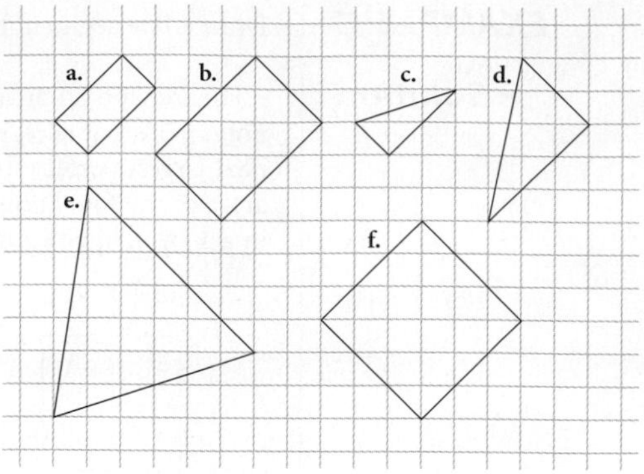

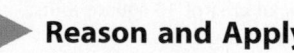

Reason and Apply

5. Find the side lengths of the polygons in 3a, 3b, and 3e.

6. Find the area of each triangle below. You may want to draw each triangle separately on graph paper. Each triangle has an area of 18 square units.

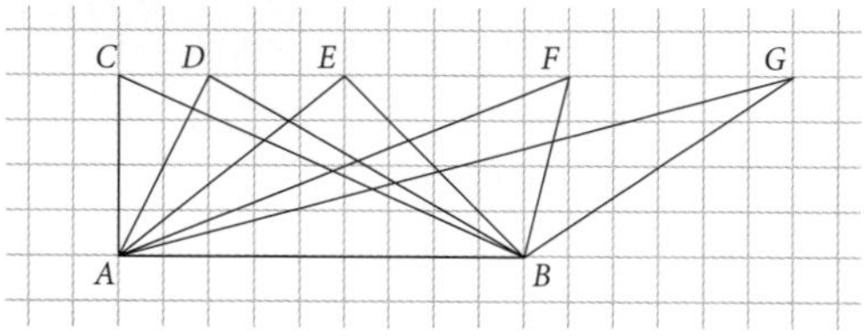

a. $\triangle ABC$ **b.** $\triangle ABD$ **c.** $\triangle ABE$ **d.** $\triangle ABF$ **e.** $\triangle ABG$

7. In the figure at right $\triangle ABC$ is a right triangle.

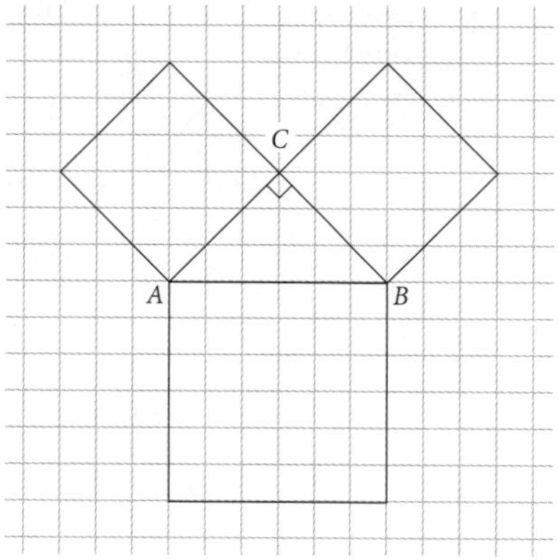

a. Find the area of each of the squares built on the sides of this triangle. 36 square units, 18 square units, 18 square units

b. Find the lengths of $\overline{AB}$, $\overline{BC}$, and $\overline{AC}$.
length of $\overline{AB}$: 6 units; length of $\overline{BC}$: $\sqrt{18}$ units; length of $\overline{AC}$: $\sqrt{18}$ units

8. A square is drawn on graph paper. One side of the square is the segment with endpoints (2, 5) and (8, 1). Find the other two vertices of the square. There are two possible solutions. Can you find both?

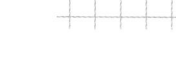

9. Below is a right triangle.

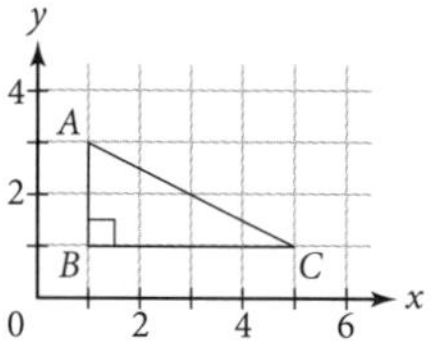

a. Which side is the hypotenuse of $\triangle ABC$? Which sides are the legs? $\overline{AC}$ is the hypotenuse; $\overline{AB}$ and $\overline{BC}$ are the legs.

b. Draw this triangle on graph paper and draw a square on each side, as in problem 7.

c. Find the area of each square you drew in 9b. 20 square units, 16 square units, 4 square units

d. Find the lengths of $\overline{AB}$, $\overline{BC}$, and $\overline{AC}$. length of $\overline{AB}$: 2 units; length of $\overline{BC}$: 4 units; length of $\overline{AC}$: $\sqrt{20}$ units

e. What is the relationship between the areas of the three squares?
The areas of the two smaller squares add up to the area of the larger square.

Exercise 9 Legs and hypotenuse were defined in Lesson 11.1. If you did not do so in Lesson 11.2, you might want to define AB as the length of $\overline{AB}$ and restate this pattern in terms of lengths: $(AB)^2 + (CB)^2 = (AC)^2$.

9b.

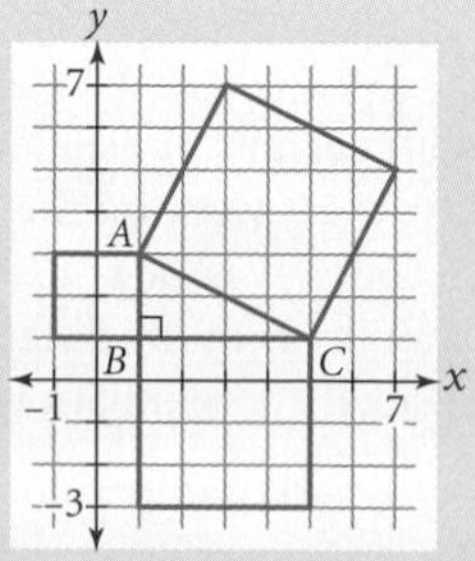

▶ Review

10. The population of City A is currently 47,000 and is growing at a rate of 4.5% per year. The population of City B is currently 56,000 and is decreasing at a rate of 1.2% per year.

   a. What will the population of the two cities be in 5 years? City A: 58,571; City B: 52,720

   b. When will the population of City A first exceed 150,000? in 27 years

   c. If the population decrease in City B began 10 years ago, how large was the population before the decline started? 63,106

11. Use all these clues to find the equation of the one function that they describe.

   ▶ The graph of the equation is a parabola that crosses the *x*-axis twice.

   ▶ If you write the equation in factored form, one of the factors is $x + 7$.

   ▶ The graph of the equation has *y*-intercept 14.

   ▶ The axis of symmetry of the graph passes through the point $(-4, -2)$.
   $y = 2(x + 7)(x + 1)$

Exercise 10 This exercise reviews Lesson 2.4.

Exercise 11 This exercise reviews Lesson 10.3.

IMPROVING YOUR **VISUAL THINKING** SKILLS

The Chokwe people of northeastern Angola, Africa, are respected for their mat weaving designs. They weave horizontal white strands with vertical brown strands. In the design below, the first brown strand passes over one white strand and then under four white strands; the next brown strand to the right repeats the weaving pattern, but the design is translated down two units.

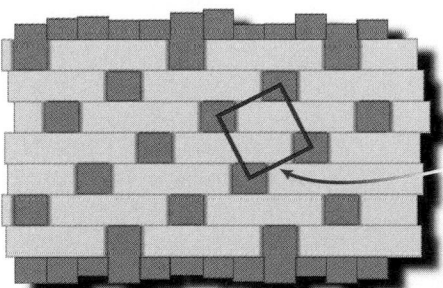

> The exposed brown strands could be connected to create tilted squares throughout the design.

Notice that the Chokwe design creates tilted squares similar to those you saw in this lesson. These tilted squares are repeated throughout the design.

Paulus Gerdes, a Mozambican mathematician, calls this design a "(1, −2)-solution" for finding a pattern of tilted squares. (*Geometry from Africa*, 1999, p. 75) That means if you move 1 unit right and 2 units down from any brown square, you hit another brown square.

Is this the only design that could be called a (1, −2)-solution? For example, can passing each brown strand over one white strand and then under three white strands create a pattern of tilted squares? How about over one, under two? How about over one, under five? Describe your results.

What under-over design creates a (1, −3)-solution?

IMPROVING **VISUAL THINKING** SKILLS

Weaving under four, over one is the only way to get a (1, −2)-solution. (It might be called a (1, 3) solution as well.) Under three, over one either creates diagonal stripes, or it creates rhombuses leaving alternating white strands without any brown strands over them, so the mat would fall apart. Under two, over one creates diagonal stripes, not squares. Under five, over one creates stripes or rectangles that fall apart. Under nine, over one is the (1, −3)-solution. Students may conjecture that the (1, *k*)-solution must go under k^2, over one. This is indeed true. The (1, −4)-solution goes under sixteen, over one. The (1, −1)-solution goes under one, over one. In general, imagine one square *A* in which a brown strand appears. In the next column to the right, the brown strand to which the first one is connected in a tilted square is *k* white strands lower. So the appearance of the brown strand below *A* must be connected to a point *k* units to the right and one unit up. This will be a brown strand only if each brown strand passes under k^2 white strands.

LESSON
11.4

PLANNING

LESSON OUTLINE

One day:

5 min	Introduction
20 min	Investigation
10 min	Sharing
5 min	Example
5 min	Closing
5 min	Exercises

MATERIALS

- graph paper
- straightedges

TEACHING

The Pythagorean theorem is a statement about areas of geometric figures as well as an algebraic claim about lengths of line segments.

One step Pose the problem of the example, insisting on an exact answer. As needed, encourage students to experiment with measurements of smaller right triangles and look for patterns.

INTRODUCTION

In earlier classes, students may have derived the formula for area of a triangle by dividing a rectangle in half along the diagonal. In this book, they have seen this as the basis for area of a triangle in Example A in Lesson 11.3, and they may have chosen this as a way to divide a rectangle in Exercise 13a in Lesson 11.2. The fact that some triangles do make up half a rectangle and others do not could be illustrated with Exercise 6 in Lesson 11.3. Obtuse triangles do not form half-rectangles. They (and other triangles) do, however, form half-parallelograms, which themselves can be modified into rectangles.

There can be no mystery in a result you have discovered for yourself.

W. W. SAWYER

The Pythagorean Theorem

You've seen that the area of a right triangle is half the area of the rectangle drawn around it. This is true for many other triangles too, but not all triangles.

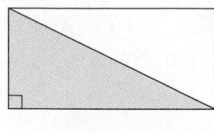

The area of this right triangle is half the area of the rectangle.

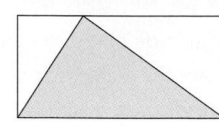

The area of this right triangle is also half the area of the rectangle.

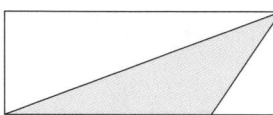

The area of this triangle is actually less than half the area of the rectangle around it.

You might have figured out that triangles with the same base and the same height have the same area. This is true whether or not the triangles all fit inside the same rectangle. The area formula for a triangle is

$$Area = \frac{base \cdot height}{2} \quad \text{or} \quad A = \frac{1}{2}bh$$

You can use the formula to find the area of a triangle without adding grid squares or subtracting areas.

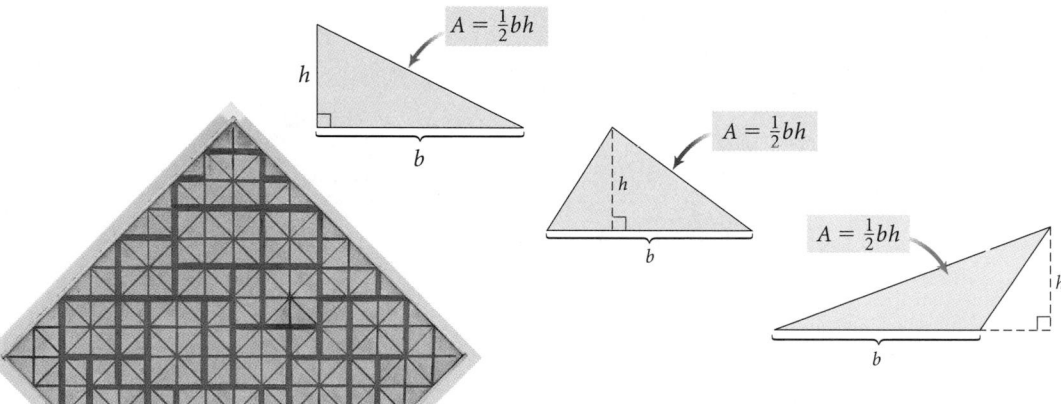

You'll remember a formula more easily if you discover it yourself. In this lesson you will discover a right triangle formula that planners and builders have used for thousands of years.

The Dutch artist Piet Mondrian (1872–1944) is famous for his use of straight lines, right angles, and geometric shapes. *Composition in Black and Gray* (1919) shows many right triangles and quadrilaterals.

Exercise 6 in Lesson 11.3 illustrated that triangles with the same base and height have equal areas. You may want to revisit this idea.

LESSON OBJECTIVES

- Discover the Pythagorean theorem by exploring right triangles and the squares built on each side
- Apply the Pythagorean theorem to real-world problems

Investigation
The Sides of a Right Triangle

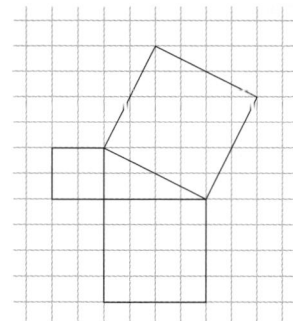
Guiding the Investigation

You will need
- graph paper
- a straightedge

This investigation will help you discover a very useful formula that relates the lengths of the sides of a right triangle.

This is only a sample. Your right triangle should be larger or smaller.

Step 1 Draw a right triangle on graph paper with its legs on the grid lines and its vertices at grid intersections.

Step 2 Draw a square on each side of your triangle.

Step 3 Find the area of each square and record it.

Step 4 As a group or as a class, combine your results in a large table like this one. Look for a relationship between the numbers in each row of the table.

Step 3 Using the sample triangle in the book:

	Area of square on leg 1	Area of square on leg 2	Area of square on hypotenuse
Trisha's triangle	4	16	20
Joe's triangle			

Step 5 Calculate the lengths of the legs and the hypotenuse for each triangle based on the areas you calculated in Step 3.

Step 5 Using the sample triangle in the book:

	Length of leg 1	Length of leg 2	Length of hypotenuse
Trisha's triangle	2	4	$\sqrt{20}$
Joe's triangle			

Step 6 Use what you discovered about the areas of the squares to write a rule relating the lengths of the legs to the length of the hypotenuse.
$$(\text{length of leg 1})^2 + (\text{length of leg 2})^2 = (\text{length of hypotenuse})^2$$

Step 1 Be sure each member of a group draws a different right triangle.

Step 2 The full extent of each leg or hypotenuse should be used as an edge of a square.

Step 3 If time permits, have each student repeat Steps 1, 2, and 3 for a second, different right triangle.

Step 4 If students find that some data points don't fit their conjecture, have them go back and check for mistakes in their drawings or calculations.

Step 6 You may want to encourage students to repeat the investigation with their own triangles.

SHARING

Have students share whatever variety of conjectures they derived. Some may refer to areas and others to lengths or squares of numbers.

Point out the student text's statement of the Pythagorean theorem. Ask how students' own conjectures are the same or different. Following history, the investigation led to a statement about areas: "The square *on* the hypotenuse equals (in area) the sum of the squares *on* the other two sides." The statement in Step 6 and the box is more numerical and algebraic: "The square *of* the (length of the) hypotenuse equals the sum of the squares *of* the (lengths of) the other two sides."

Although the theorem is often stated using the letters *a, b,* and *c,* there's no reason the edges of the triangle should have those names. In fact, students will be working with right triangles whose edges have other names.

NCTM STANDARDS

CONTENT		PROCESS	
	Number		Problem Solving
•	Algebra	•	Reasoning
•	Geometry		Communication
•	Measurement	•	Connections
	Data/Probability	•	Representation

Assessing Progress
During the investigation and presentations, make note of students' abilities to calculate areas of tilted squares and to record data systematically.

The history connection diagram indicates an algebraic proof starting with $(a - b)^2 + 4ab = c^2$. Area proofs like the one in this diagram appear in an ancient Chinese text called (with various spellings and translations) the *Chou Pe*, probably written by different authors over several centuries. The original diagram showed only a 3-4-5 right triangle embedded in a 7-by-7 square and indicated a geometric rather than algebraic proof. Other civilizations, including the Babylonians and Egyptians, had even earlier knowledge of the Pythagorean relationship, though perhaps without proofs. Centuries later, the Hindu mathematician Bhaskara (b A.D. 1114) used the Chinese diagram as the basis of his proof of the Pythagorean theorem.

PYTHAGOREAN THEOREM

[Ask] "Is the hypotenuse always the longest side of a right triangle?" Students may think about this geometrically or algebraically. [The square root of the sum of two squares of numbers must be bigger than either of the numbers.] **[Ask]** "Does the Pythagorean theorem hold even if the lengths of sides are fractions?"

Using the Quote

Warwick Sawyer is an English mathematician who developed a method by which engineering apprentices learned mathematics by handling physical objects.

The many proofs of the Pythagorean theorem give different insights into the related ideas. The process of deriving theorems requires creative, intuitive, even messy conjecturing. The role of proof is not so much to create truth but to test conjectures and to gain insight into underlying concepts. Making students aware of this process may help those who prefer creativity and messy thinking to be more interested in mathematics. Ask students if they feel they have discovered the

What you discovered in the investigation is the famous **Pythagorean theorem.** A theorem is a mathematical formula or statement that has been proven to be true. This theorem is named after Pythagoras, a Greek mathematician who lived around 500 B.C. This relationship was discovered and used by people in cultures before Pythagoras, but the theorem is usually given his name.

The Pythagorean Theorem

The sum of the squares of the lengths of the legs a and b of a right triangle equals the square of the length of the hypotenuse c.

$$a^2 + b^2 = c^2$$

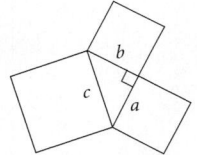

The next example shows how you can use what you learned in the investigation to find the missing length of a side of a right triangle.

EXAMPLE

A baseball diamond is a square with 90 ft between first and second base. What is the distance from home plate to second base?

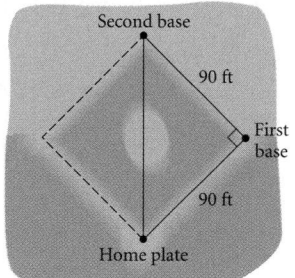

▶ **Solution**

The distance between home plate and second base is the hypotenuse of a right triangle. Call it c. This means that the area of a square on c equals the sum of the areas of the squares on each leg. The legs a and b are equal in this case.

History
• CONNECTION •

The earliest known proof of the "Pythagorean" theorem came from China over 2500 years ago. Can you use this version of the Chinese proof to show $a^2 + b^2 = c^2$?

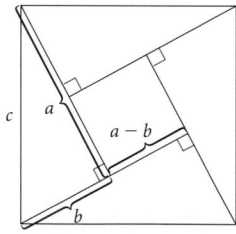

$$
\begin{aligned}
c^2 &= a^2 + b^2 & \text{The Pythagorean theorem.} \\
&= 90^2 + 90^2 & \text{Each leg is 90 ft.} \\
&= 8{,}100 + 8{,}100 & \text{Square each leg length.} \\
c^2 &= 16{,}200 & \text{Add.} \\
c &= \sqrt{16{,}200} \approx 127.3 & \text{Find the square root.}
\end{aligned}
$$

So the distance from home plate to second base is approximately 127.3 ft.

Pythagorean theorem themselves and if they feel, as the quote implies, there's no mystery in it.

You might also ask students if they think that non-right triangles obey the Pythagorean theorem. If you don't answer the question, some students may pursue research on their own. In Take Another Look at the end of this chapter, students are asked to describe these patterns using inequalities.
[Link] The more general relationship for all triangles is called the Law of Cosines and is usually taught in trigonometry courses.

▶ **EXAMPLE**

This example gives a real-world application of the Pythagorean theorem.

Symbolically, $|c| = \sqrt{16{,}200}$ or $c \approx \pm 127.3$. Praise students who recall that there are two square roots.
[Ask] "Why does the solution omit one of them?" [When the negative square root is not a meaningful solution, we often skip the step of considering both roots.]

EXERCISES

*You will need your calculator for problems **2, 5, 7, 9**, and **11**.*

▶ Practice Your Skills

In problems 1–4, a and b are the legs of a right triangle and c is the hypotenuse.

1. Suppose the square on side c has an area of 2601 cm² and the square on side b has an area of 2025 cm². What is the area of the square on side a? 576 cm²

2. Using the areas from problem 1, find each side length, a, b, and c. $a = 24$ cm, $b = 45$ cm, $c = 51$ cm

3. Suppose $a = 10$ cm and $c = 20$ cm. Find the exact length of side b in radical form. $c = \sqrt{300}$ cm

4. Suppose the right triangle is isosceles (two equal sides).

   a. Which two sides are the same length: the two legs or a leg and the hypotenuse? the two legs

   b. If the two equal sides are each 8 cm in length, what is the exact length of the third side in radical form? $c = \sqrt{128}$ cm

▶ Reason and Apply

5. APPLICATION Triangles that are similar to a right triangle with sides 3, 4, and 5 are often used in construction. The roof shown here is 36 ft wide. The two halves of the roof are congruent. Each half is a right triangle with sides proportional to 3, 4, and 5. (The shorter leg is the vertical leg.)

   a. How high above the attic floor should the roof peak be? 13.5 ft

   b. How far is the roof peak from the roof edge? 22.5 ft

   c. What is the shingled area of the roof if the building is 48 feet long? 2160 ft²

6. Cal and Al are trying to solve the problem $\sqrt{x + 4} = 5$. Cal says that $\sqrt{x + 4} = \sqrt{x} + \sqrt{4}$. Al disagrees but can't explain why. Who is right? Explain your reasoning.

$\sqrt{x + 4} = \sqrt{x} + \sqrt{4}$

I disagree.

7. You will need a centimeter ruler for this problem.

   a. Measure the length and width of your textbook cover in centimeters. approximately 21.6 cm-by-27.6 cm

   b. Use the Pythagorean theorem to calculate the diagonal length using the length and width you measured in 7a. approximately 35.0 cm

   c. Measure one of the diagonals of the cover.

   d. How close are the values you found in 7b and c? Should they be approximately the same? Answers will vary. The two results should be approximately the same.

8. Miya was trying to solve the problem $x^2 + 4^2 = 5^2$. She took the square root of both sides and got $x + 4 = 5$, which means x equals 1. Explain why her answer is wrong, and show how to find the correct answer. Answers will vary. If $x = 1$, then Miya is claiming that $1^2 + 4^2 = 5^2$, but $17 \neq 25$. You have to isolate x before you take the square root: $x^2 + 16 = 25$, $x^2 = 9$, $x = \pm 3$.

---

6. Al is right. Answers will vary. As an example, $\sqrt{9 + 4} \neq \sqrt{9} + \sqrt{4}$.

Exercise 7 Students can also use notebook paper or perhaps their desktops.

7c. Answers will vary but should be close to 35 cm.

Closing the Lesson

The **Pythagorean theorem** claims that the square drawn on the hypotenuse of a right triangle is equal in area to the sum of the squares drawn on the legs. In algebraic terms, $a^2 + b^2 = c^2$, where a and b are the lengths of the legs and c is the length of the hypotenuse.

BUILDING UNDERSTANDING

Students work with the Pythagorean theorem and its converse.

ASSIGNING HOMEWORK

Essential	1–5, 6 or 8, 7 or 9, 10
Performance assessment	7, 9
Portfolio	10, 13
Journal	6, 8, 12
Group	10–13
Review	14

▶ Helping with the Exercises

Exercise 3 At this point, students may leave their answers unsimplified, because reducing radicals won't be introduced until Lesson 11.5.

Exercise 4 As needed, point out that the student text says that an isosceles triangle has two sides with the same length. A decimal name for the solution doesn't give an "exact length" in this case.

Exercise 5 [Alert] Be ready to help students having difficulty with the proportions involved.

Exercise 6 Ask for good justifications. Students may give examples in which the equality fails or may show that squaring both sides of the equation leads to $4\sqrt{x} = 0$, implying that $x = 0$.

Exercise 9 As needed, remind students that 1 mile = 5280 feet.

As an extension, ask students how much a mile of road pavement will buckle if it expands 1 foot, all pushed to the same place. The hypotenuse is barely longer than one leg, yet the other leg is almost 103 feet.

Exercise 10 The reverse of a theorem is called its *converse*. Evidence by measurement shows that the theorem's converse *seems* to be true, but it does not prove it. **[Ask]** "Does the truth of a theorem imply the truth of its converse?" Have students examine statements like "If it's a dog, then it has a tail," whose converse is "If it has a tail, then it's a dog." Although converses differ in meaning and often in truth, the converse of the Pythagorean theorem can be proved to be true.

10a. right triangles: i, ii, iv; not a right triangle: iii

Exercise 11 This exercise is long and challenging. You might want to use it as a group problem or for extra credit. TVs are not actually manufactured to maximize area as indicated in 11g. Instead, there are two standard aspect ratios—4-by-3 for a standard TV and 16-by-9 for a widescreen TV. A standard TV is therefore based on a 3-4-5 triangle, so the actual dimensions of a 27-inch TV would be 21.6 by 16.2 inches. Theatrical movies generally have one of three aspect ratios: 1.33 (4-by-3), 1.85, or 2.35. Students may be familiar with these ratios from DVDs that maintain the widescreen aspect ratios.

9. The launching pad for a hot air balloon is 1.2 miles away from where you're standing. If the balloon rises vertically 3000 feet into the air, how far (in feet) will it be from you?
approximately 7010 ft

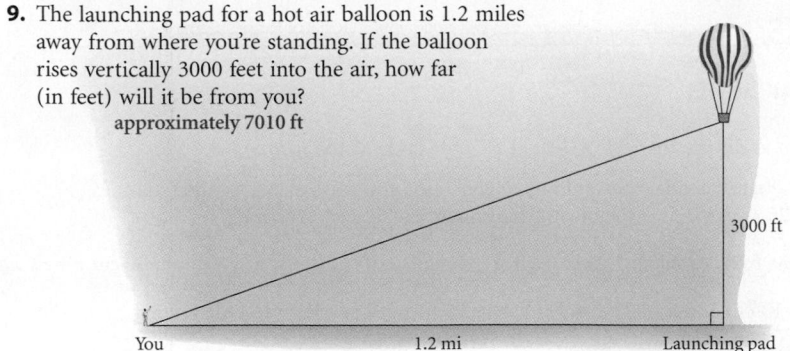

You　　　　　1.2 mi　　　　Launching pad

3000 ft

10. Strips of graph paper may help in 10a.

　a. Draw or make triangles with the side lengths that are given in i–iv. Then use a protractor or the corner of a sheet of paper to find whether or not each triangle is a right triangle.

　　i. 5, 12, 13　　　　　**ii.** 7, 24, 25
　　iii. 8, 10, 12　　　　**iv.** 6, 8, 10

　b. Based on your results from 10a does the Pythagorean theorem work in reverse? In other words, if the relationship $a^2 + b^2 = c^2$ is true, is the triangle necessarily a right triangle?
　　　　Yes, the theorem works in reverse.

11. A 27-inch TV has a screen that measures 27 inches on its diagonal. Complete the following steps and use the Pythagorean theorem to find the **dimensions** of the screen with maximum area for a 27-inch TV.

　a. Enter the positive integers from 1 to 26 into list L₁ on your calculator to represent the possible screen widths. $L_1 = \{1, 2, 3, \ldots, 26\}$

　b. Imagine a screen 1 inch wide. Calculate the length of a 27-inch TV screen with a width of 1 inch, and enter your answer into the first row of list L₂. approximately 26.98 in.

　c. Define list L₂ to calculate all the possible screen lengths. $L_2 = \sqrt{(27^2 - L_1^2)}$

　d. What is the area of a 27-inch screen with a width of 2 inches? approximately 53.85 in.²

　e. Define list L₃ to calculate all possible screen areas. $L_3 = L_1 \cdot L_2$, or $L_3 = L_1 \cdot \sqrt{(27^2 - L_1^2)}$

　f. Plot points in the form (*width, area*) and find an equation that fits these points.

　g. What screen dimensions give the largest area for a 27-inch TV?

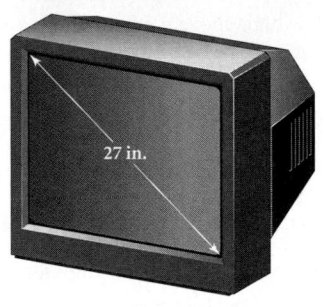
27 in.

The size of a television is measured on its diagonal.

27-inch TV

Width (L₁)	Length (L₂)	Area (L₃)
1		
2		
3		
⋮		
26		

11f. A model that works is $y = x\sqrt{27^2 - x^2}$, where x is the width and y is the area.

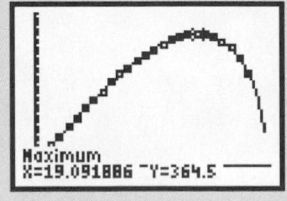

Maximum
X=19.091886　Y=364.5

$[0, 27, 5, -50, 450, 50]$

11g. Trace the graph or use a table to find that a 19-by-19-in. square gives maximum area.

12a. Possible answer: The ratios $\frac{10}{5}$, $\frac{24}{12}$, and $\frac{26}{13}$ are all equal to 2, so the sides are proportional and the triangles are similar.

12b. Yes. Possible answer: Similar figures have congruent angles. Another possible answer: The numbers satisfy the Pythagorean theorem.

12. In problem 10, you showed that a triangle with side lengths of 5, 12, and 13 units is a right triangle. See page 602.

   a. Explain why a triangle with side lengths of 10, 24, and 26 units is similar to this triangle.

   b. Is the triangle in 12a a right triangle? Explain.

13. APPLICATION Nadia Ferrell wants to build an awning over her porch. She wants the slope of the awning to be $\frac{5}{12}$. The porch is 8 feet deep, and the roof line is 14 feet above the porch. She draws this sketch to help her plan.

   a. How long will the awning be from the roof line to the porch support posts? Show your work.

   b. How tall will the posts be that hold up the front of the awning? Show your work.

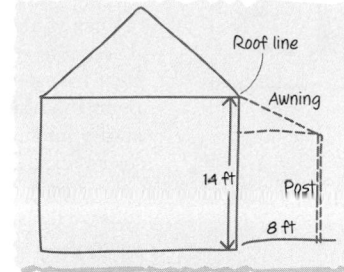

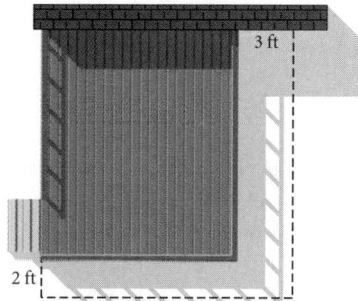

Review

14. Ibrahim Patterson is planning to expand his square deck. He will add 3 feet to the width and 2 feet to the length to get a total area of 210 square feet. Find the dimensions of his original deck. Show your work.
$(x + 3)(x + 2) = 210, x^2 + 5x + 6 = 210, x^2 + 5x - 204 = 0,$
$x = -17$ or $x = 12$; the original deck is 12 feet on a side.

project

PYTHAGORAS REVISITED

You know that the Pythagorean theorem says that the sum of the areas of the squares on the two legs of a right triangle is equal to the area of the square on the hypotenuse.

But if the shapes aren't squares, but are similar, does the sum of the areas on the legs still equal the area on the hypotenuse? Does a Pythagorean-like relationship still hold for the shape in these drawings?

Design your own similar shapes on the sides of a right triangle. Carefully measure or calculate their areas. Then examine and report on your results. Your project should include

▶ Your right triangle drawing with similar shapes on each side.

▶ Your measurements and calculations.

▶ A written explanation of how you drew the similar shapes, how you calculated the areas, and a conclusion whether or not a Pythagorean-like relationship holds for any shape.

Dot paper, graph paper, a computer drawing program, or The Geometer's Sketchpad software are useful tools for this project.

THE GEOMETER'S SKETCHPAD

In The Geometer's Sketchpad, you can drag parts of these figures and the triangles will remain right triangles and the three shapes will remain similar. Sketchpad also measures areas. Learn how to use its tools to create your own Pythagorean drawing!

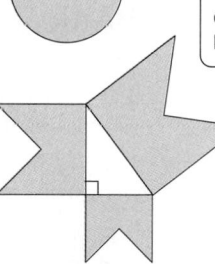

Exercise 13 In 13a, you may need to point out that the angle at the peak of the roof is a right angle. Students may also need to be told that a slope of $\frac{5}{12}$ creates a right triangle with legs 5 and 12. Therefore, the hypotenuse is 13. The awning triangle will be similar to this 5-12-13 triangle.

[Alert] If students are having difficulties with 13b, help them see the length of the posts when projected against the house and ask how to find that length by subtracting.

13a. $\frac{13}{12} = \frac{a}{8}, a = 8.\overline{6}$;

8 ft 8 in. long

13b. $\frac{5}{12} = \frac{14 - b}{8}, b = 10\frac{2}{3}$;

10 ft 8 in. tall

Exercise 14 Students might use a variety of approaches, including graphical, to solve the equation $x^2 + 5x - 204 = 0$. This exercise reviews Lesson 10.4.

Proof for the Project
Students might give a proof following this outline: If you drop a perpendicular from the right angle to the hypotenuse, you divide the right triangle into two smaller right triangles. All three of these triangles are similar, and their hypotenuses are the edges of the original triangle, so their scale factors are in the ratio of $a : b : c$. Hence, any similar figures with these scale factors will have area ratios $a^2 : b^2 : c^2$. Because the two smaller triangles partition the original triangle, we can show that $a^2 + b^2 = c^2$.

Supporting the **project**

MOTIVATION

The Pythagorean theorem is actually a special case of a more general theorem: If similar figures are built on the legs and hypotenuse, then the sum of the areas of the figures on the legs equals the area of the figure on the hypotenuse.

OUTCOMES

▶ An explanation of how the similar shapes on three edges of a right triangle were drawn and how their areas were calculated.

▶ A supported conclusion about a Pythagorean-like relationship among the shapes. (The conclusion may or may not be true, but it must be supported by the calculations.)

• One or more drawings with accurately drawn similar shapes.

• Varied methods of calculating area as a self-check.

• A thorough conclusion including reasons that a Pythagorean-like relationship does indeed hold for any shape.

LESSON

11.5

Operations with Roots

When you use the Pythagorean theorem to find the length of a right triangle's side, the result is often a radical expression. You can always find an approximate value for a radical expression with a calculator, but sometimes it's better to leave the answer as an exact value, or in radical form. But there is more than one way to write an exact value or a radical expression. In this lesson you'll discover ways to rewrite radical expressions so that you can recognize solutions in a variety of forms.

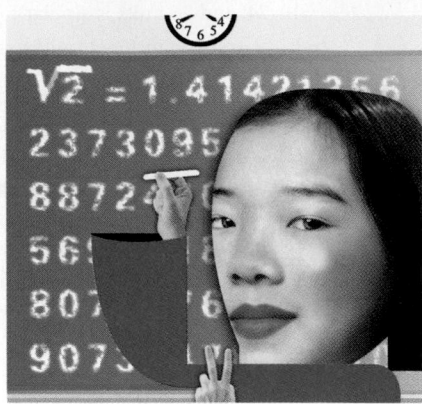

EXAMPLE A | Draw a segment that is $\sqrt{13} + \sqrt{13}$ units long.

▶ **Solution**

First think of two perfect squares whose sum is 13, such as $4 + 9 = 13$. Then draw a right triangle on graph paper using the square roots of your numbers, 2 and 3, for the leg lengths. By the Pythagorean theorem, the hypotenuse of your triangle is $\sqrt{13}$.

Now draw a second congruent triangle so that the hypotenuses form a single segment. This pair of hypotenuses is $\sqrt{13} + \sqrt{13}$, or $2\sqrt{13}$, units long.

The combined length is $\sqrt{13} + \sqrt{13}$ or $2\sqrt{13}$.

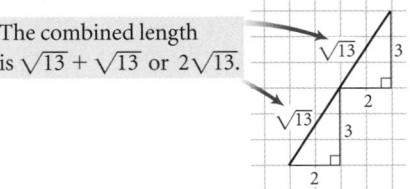

Investigation
Radical Expressions

How can you tell if two different radical expressions are equivalent? Is it possible to add, subtract, multiply, or divide radical expressions? You'll answer these questions as you work through this investigation.

You will need

- graph paper

Step 1 | On graph paper, draw line segments for each length given below. You may need more than one triangle to create some of the lengths.

Step 1g same as f
Step 1h same as e
Step 1i same as d

a. $\sqrt{18}$ b. $\sqrt{40}$ c. $\sqrt{20}$
d. $2\sqrt{5}$ e. $3\sqrt{2}$ f. $2\sqrt{10}$
g. $\sqrt{10} + \sqrt{10}$ h. $\sqrt{2} + \sqrt{2} + \sqrt{2}$ i. $\sqrt{5} + \sqrt{5}$

Guiding the Investigation

Because radical expressions are often irrational numbers (whose decimal names are infinite and nonrepeating), students might think it is impossible to add, subtract, multiply, or divide with them.

Step 1 If students are having difficulties, suggest that they look back at the triangles drawn in the previous investigation.

Step 1a
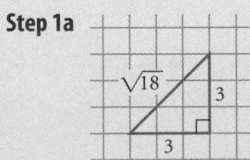

Step 1c
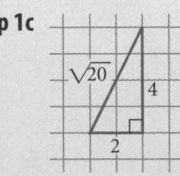

Step 1b
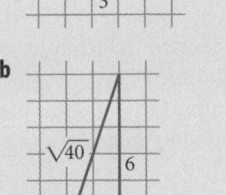

Step 1d
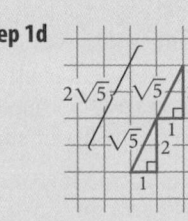

PLANNING

LESSON OUTLINE

First day:

10 min   Example A
40 min   Investigation

Second day:

15 min   Investigation
10 min   Sharing
10 min   Examples B, C
5 min   Closing
10 min   Exercises

MATERIALS

- graph paper
- Calculator Note 7B
- Pyramid Net (W), *optional*

TEACHING

In some cases radical expressions can be rewritten with fewer occurrences of the square root sign.

▶ EXAMPLE A

This example introduces the idea of different radical names for the same number. Before students consider the solution, ask if they can think of another way to write $\sqrt{13} + \sqrt{13}$. Some students might claim that the sum equals $\sqrt{26}$. Have them test that conjecture on their calculators. It may come as a slight surprise to some students that the sum equals $2\sqrt{13}$. They may believe easily enough that "1 apple plus 1 apple equals 2 apples," but they might understandably see the multiplication by 2 in $2\sqrt{13}$ as different from the use of 2 as an adjective in "2 apples."

See page 726 for answers to Steps 1e and f.

Steps 2–3 a, e, and h are equivalent; b, f, and g are equivalent; c, d, and i are equivalent.

Step 2 Do any of the segments seem to be the same length?

Step 3 Use your calculator to find a decimal approximation to the nearest ten thousandth for each expression in Step 1. Which expressions are equivalent?

Step 4a $4\sqrt{x}$

Step 4b $\sqrt{xy}$

Step 4c $x\sqrt{y}$ or $\sqrt{n^2 y}$

Step 4d x or $\sqrt{x^2}$

Step 4e $\sqrt{x}$

Step 4 Find another way to write each expression below. Choose positive values for the variables, and use decimal approximations to check that your expression is equivalent to the original expression.

    a. $\sqrt{x} + \sqrt{x} + \sqrt{x} + \sqrt{x}$ **b.** $\sqrt{x} \cdot \sqrt{y}$ **c.** $\sqrt{x \cdot x \cdot y}$

    d. $\left(\sqrt{x}\right)^2$ **e.** $\dfrac{\sqrt{xy}}{\sqrt{y}}$

Step 5 Possible answers: Radical expressions with the same numbers under the square root symbol may be added by adding the coefficients; radical expressions are multiplied by multiplying the numbers under the square root symbols; you can divide one radical expression by another by dividing the numbers under the square root symbols and then putting the answer under a single square root symbol.

Step 5 Summarize what you discovered about adding, multiplying, and dividing radical expressions.

Step 6 Use what you've learned to find the area of each quadrilateral below. Give each answer in radical form as well as a decimal approximation to the nearest hundredth.

a.

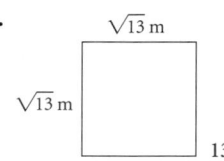

$\sqrt{13}$ m, $\sqrt{13}$ m, 13 m²

b.

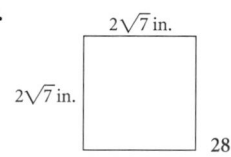

$2\sqrt{7}$ in., $2\sqrt{7}$ in., 28 in.²

c.

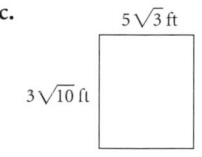

$5\sqrt{3}$ ft, $3\sqrt{10}$ ft, $15\sqrt{30}$ ft²; approximately 82.16 ft²

d.

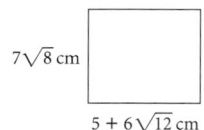

$7\sqrt{8}$ cm, $5 + 6\sqrt{12}$ cm

$\left(35\sqrt{8} + 42\sqrt{96}\right)$ cm², or $\left(70\sqrt{2} + 168\sqrt{6}\right)$ cm²; approximately 510.51 cm²

EXAMPLE B

Rewrite each expression with as few square root symbols as possible and no parentheses.

    a. $2\sqrt{3} + 5\sqrt{3}$ **b.** $3\sqrt{5} \cdot 2\sqrt{7}$

    c. $\dfrac{\sqrt{15}}{\sqrt{3}}$ **d.** $\sqrt{3}\left(5\sqrt{2} + 3\sqrt{3}\right)$

▶ Solution

a. Add the terms using the distributive property.

$$2\sqrt{3} + 5\sqrt{3} = (2 + 5)\sqrt{3} = 7\sqrt{3}$$

The distributive property allows you to factor out $\sqrt{3}$.

LESSON OBJECTIVES

- Recognize equivalent radical expressions
- Square binomials with radical terms
- Use the distributive property with radicals
- Simplify expressions under a radical by factoring out a perfect square
- Revisit quadratic equations and parabolas

NCTM STANDARDS

CONTENT	PROCESS
Number	Problem Solving
• Algebra	• Reasoning
• Geometry	Communication
Measurement	Connections
Data/Probability	• Representation

Step 3 Emphasize that although the decimal approximations are helpful for determining equivalency, they are not exact answers. Hence, 4.1234567891 and 4.1234567892 are both 4.1235 when rounded to the ten-thousandths place, yet the original numbers are *not* equal.

Step 4 Students may want to refer to Calculator Note 7B. As needed, remind students of the meaning of square roots.

Step 6 Remind students that rectangles have four right angles (as originally stated in the investigation in Lesson 11.1). Hence, all of these figures are rectangles, perhaps squares.

SHARING IDEAS

Ask students to present, in writing, their rules from Step 5. After they've presented a variety of rules, you might say that the square root symbol is a *radical* and the number or expression under a radical is the *radicand*. Then help them rephrase their rules using these terms. (Symbols for cube roots and other roots are also radicals.)

Point out that the numbers in Step 6 are irrational: They can't be written as ratios of integers. Ask whether these irrational numbers are "infinite" or "go on forever." Some students will say so, thinking of the decimal names rather than the numbers themselves. They all have very simple names that involve radicals.

Assessing Progress

Students can demonstrate their abilities to draw line segments of **radical lengths** using appropriate **right triangles,** to approximate values of **square roots,** and to find areas of squares and other rectangles.

► **EXAMPLE B**

This example illustrates simplifying radical names for numbers.
In part d, some students may
still need to see or write the
intermediary step that
$\sqrt{2}\sqrt{3} = \sqrt{2 \cdot 3} = \sqrt{6}$.

[Ask] "Can you simplify $5\sqrt{6} + 9$
any further?" These unlike terms
can't be combined. You might
draw an analogy to $5x + 9$.

The decimal approximations that your calculator gives support the idea that
$2\sqrt{3} + 5\sqrt{3}$ is equivalent to $7\sqrt{3}$.

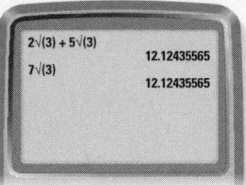

b. All of the numbers are multiplied. So use the commutative property of
multiplication to group coefficients together and radical expressions together.

First the commutative property
allows you to swap $\sqrt{5}$ and 2.

$$3\sqrt{5} \cdot 2\sqrt{7} = 3 \cdot 2 \cdot \sqrt{5} \cdot \sqrt{7}$$

Then to multiply two radical expressions, you multiply
the numbers under the square root symbols.

$$3 \cdot 2 \cdot \sqrt{5} \cdot \sqrt{7} = 6\sqrt{5 \cdot 7} = 6\sqrt{35}$$

Use your calculator. Do the decimal approximations support the idea that
$3\sqrt{5} \cdot 2\sqrt{7}$ is equivalent to $6\sqrt{35}$?

c. You can combine the numbers under one square root symbol and divide.

$$\frac{\sqrt{15}}{\sqrt{3}} = \sqrt{\frac{15}{3}} = \sqrt{5}$$

To divide radical expressions,
you can rewrite the numbers
under one square root symbol.

d. $\sqrt{3}\left(5\sqrt{2} + 3\sqrt{3}\right) = 5\sqrt{2} \cdot \sqrt{3} + 3\sqrt{3} \cdot \sqrt{3}$ Distribute $\sqrt{3}$.

$\qquad\qquad\qquad = 5\sqrt{6} + 3\sqrt{9}$ Multiply the radical expressions.

$\qquad\qquad\qquad = 5\sqrt{6} + 3 \cdot 3$ $\sqrt{9}$ is equal to 3.

$\qquad\qquad\qquad = 5\sqrt{6} + 9$ Multiply.

The investigation and Example B have illustrated several rules for rewriting radical expressions.

Rules for Rewriting Radical Expressions

For $x \geq 0$ and $y \geq 0$, and any values of a or b, these rules are true:

Addition of Radical Expressions

$$a\sqrt{x} + b\sqrt{x} = (a + b)\sqrt{x}$$

Multiplication of Radical Expressions

$$a\sqrt{x} \cdot b\sqrt{y} = a \cdot b\sqrt{x \cdot y}$$

For $x \geq 0$ and $y > 0$, this rule is true:

Division of Radical Expressions

$$\frac{\sqrt{x}}{\sqrt{y}} = \sqrt{\frac{x}{y}}$$

EXAMPLE C | Show that $6 + \sqrt{20}$ is a solution to the equation $0 = 0.5x^2 - 6x + 8$.

▶ **Solution** | If $6 + \sqrt{20}$ is a solution, then you will get a true statement when you substitute it into the equation and evaluate both sides.

$0 = 0.5x^2 - 6x + 8$ — The original equation.

$0 \overset{?}{=} 0.5\left(6 + \sqrt{20}\right)^2 - 6\left(6 + \sqrt{20}\right) + 8$ — Substitute $-6 + \sqrt{20}$ for x.

$0 \overset{?}{=} 0.5\left(6 + \sqrt{20}\right)^2 - 36 - 6\sqrt{20} + 8$ — Distribute the 6 over $6 + \sqrt{20}$.

$0 \overset{?}{=} 0.5\left(36 + 6\sqrt{20} + 6\sqrt{20} + 20\right) - 36 - 6\sqrt{20} + 8$ — Use a rectangular diagram to square the expression $6 + \sqrt{20}$.

	6	$\sqrt{20}$
6	36	$6\sqrt{20}$
$\sqrt{20}$	$6\sqrt{20}$	20

$0 \overset{?}{=} 18 + 3\sqrt{20} + 3\sqrt{20} + 10 - 36 - 6\sqrt{20} + 8$ — Distribute the 0.5 over the expression in parentheses.

$0 \overset{?}{=} 18 + 10 - 36 + 8$ — Combine the radical expressions.

$0 = 0$ — Add and subtract.

The right side of the equation does evalute to 0, so $6 + \sqrt{20}$ is a solution to the equation $0 = 0.5x^2 - 6x + 8$.

▶ **EXAMPLE C**

In this example, students see how to check that a number given by a radical name is the solution to a quadratic equation. Point out that the proposed solution may have come from applying the quadratic formula to the equation.

Note that in squaring $6 + \sqrt{20}$, the middle terms are written $6\sqrt{20}$ rather than $\sqrt{20}\,6$ to make it clear that the 6 is not under the radical.

Some students may be confused that in the step called "Combine the radical expressions" there are no radical expressions. You may need to highlight that $3\sqrt{20} + 3\sqrt{20} - 6\sqrt{20} = 0\sqrt{20} = 0$, so no square root symbols remain at this step.

Closing the Lesson

Three important rules allow rewriting of **radical expressions.** For non-negative (or positive for divisors) values of x and y:

$$a\sqrt{x} + b\sqrt{x} = (a + b)\sqrt{x}$$

$$\sqrt{x} \cdot \sqrt{y} = \sqrt{xy}$$

$$\frac{\sqrt{x}}{\sqrt{y}} = \sqrt{\frac{x}{y}}$$

You can use these rules from left to right to reduce the number of radicals. You can also use them from right to left as well. For example, $\sqrt{20} = \sqrt{4} \cdot \sqrt{5} = 2\sqrt{5}$ (This idea appears in Exercise 10.)

BUILDING UNDERSTANDING

Many of the exercises review the Pythagorean theorem or quadratic equations while providing practice in operating on radicals.

ASSIGNING HOMEWORK

Essential	1–6, 9–11
Performance assessment	6, 11
Portfolio	12–14
Journal	12
Group	10, 12–14
Review	7, 8, 15, 16

▶ Helping with the Exercises

Exercise 3 [Alert] Students may need to be advised that they are being asked to change quadratic equations from factored to general form. If they still need to, allow them to use rectangular diagrams for the distributivity. As in Example C, they should write $x\sqrt{3}$ rather than $\sqrt{3}x$ and $x\sqrt{5}$ instead of $\sqrt{5}x$ to make it easier to see that the x is not under the radical.

3a. $y = x^2 - 3$

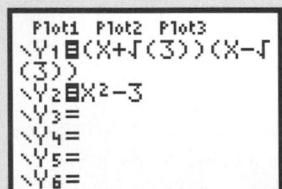

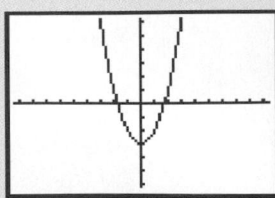

$[-9.4, 9.4, 1, -6.2, 6.2, 1]$

See page 726 for answers to Exercises 3b, 6a, 6b, and 6c.

608 CHAPTER 11 Introduction to Geometry

EXERCISES

You will need your calculator for problems **1, 3, 4, 5, 6, 7, 8,** and **12.**

▶ Practice Your Skills

1. Rewrite each expression with as few square root symbols as possible and no parentheses. Use your calculator to support your answers with decimal approximations.

   a. $2\sqrt{3} + \sqrt{3}$ $3\sqrt{3}$
   b. $\sqrt{5} \cdot \sqrt{2} \cdot \sqrt{5}$ $5\sqrt{2}$
   c. $\sqrt{2}\left(\sqrt{2} + \sqrt{3}\right)$ $2 + \sqrt{6}$
   d. $\sqrt{5} - \sqrt{2} + 3\sqrt{5} + 6\sqrt{2}$ $4\sqrt{5} + 5\sqrt{2}$
   e. $\sqrt{3}\left(\sqrt{2}\right) + 5\sqrt{6}$ $6\sqrt{6}$
   f. $\sqrt{2}\left(\sqrt{21}\right) + \sqrt{3}\left(\sqrt{14}\right)$ $2\sqrt{42}$
   g. $\dfrac{\sqrt{35}}{\sqrt{7}}$ $\sqrt{5}$
   h. $\sqrt{5}\left(4\sqrt{5}\right)$ 20

2. Find the exact length of the missing side for each right triangle.

   a.
   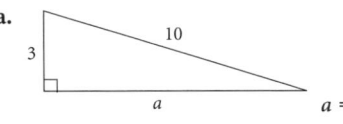
   $a = \sqrt{91}$

   b.
   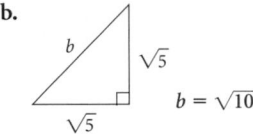
   $b = \sqrt{10}$

   c.
   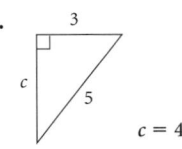
   $c = 4$

   d.
   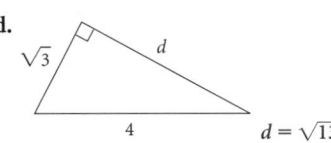
   $d = \sqrt{13}$

3. Write the equation for each parabola in general form. Use your calculator to check that both forms give the same graph or table.

   a. $y = \left(x + \sqrt{3}\right)\left(x - \sqrt{3}\right)$
   b. $y = \left(x + \sqrt{5}\right)\left(x + \sqrt{5}\right)$

4. Name the x-intercepts for each parabola in problem 3. Give both the exact value and a decimal approximation to the nearest thousandth for each x-value.
   a. $x = \pm\sqrt{3} \approx \pm1.732$ **b.** $x = -\sqrt{5} \approx -2.236$

5. Name the vertex for each parabola in problem 3. Give both exact values and decimal approximations to the nearest thousandth for the coordinates of each vertex.
   a. $(0, -3)$ **b.** $\left(-\sqrt{5}, 0\right) \approx (-2.236, 0)$

▶ Reason and Apply

6. Write the equation for each parabola in general form. Use your calculator to check that both forms have the same graph or table.

   a. $y = \left(x + 4\sqrt{7}\right)\left(x - 4\sqrt{7}\right)$ Vertex: $(0, 112)$
   b. $y = 2\left(x - 2\sqrt{6}\right)\left(x + 3\sqrt{6}\right)$ $\left(\dfrac{-\sqrt{6}}{2}, -75\right) \approx (-1.225, -75)$
   c. $y = \left(x + 3 + \sqrt{2}\right)\left(x + 3 - \sqrt{2}\right)$ $(-3, -2)$

7. Name the x-intercepts for each parabola in problem 6. Give both the exact values and a decimal approximation to the nearest thousandth for each x-value.
   a. $\pm4\sqrt{7} \approx \pm10.583$ **b.** $2\sqrt{6} \approx 4.899$ and $-3\sqrt{6} \approx -7.348$

8. Name the vertex for each parabola in problem 6. Give both exact values and decimal approximations to the nearest thousandth for the coordinates of each vertex.

X	Y1	Y2
-3	6	6
-2	1	1
-1	-2	-2
0	-3	-3
1	-2	-2
2	1	1
3	6	6

X = -3

Exercises 3b and 4b There is no positive root because each factor is $(x + \sqrt{5})$.

Exercises 7 and 8 These exercises use radical expressions and review Chapter 10.

7c. $-3 - \sqrt{2} \approx -4.414$ and $-3 + \sqrt{2} \approx -1.586$

9. A radical expression with a coefficient can be rewritten without a coefficient. Here's an example:

$2\sqrt{5}$ The original expression.

$\sqrt{4} \cdot \sqrt{5}$ $\sqrt{4}$ is equivalent to 2.

$\sqrt{20}$ Multiply.

Use this method to rewrite each radical expression.

a. $4\sqrt{7}$ $\sqrt{112}$ **b.** $5\sqrt{22}$ $\sqrt{550}$ **c.** $18\sqrt{3}$ $\sqrt{972}$ **d.** $30\sqrt{5}$ $\sqrt{4500}$

10. You can rewrite some radical expressions using the fact that they contain perfect square factors. Here's an example:

$\sqrt{125}$ The original expression.

$\sqrt{25 \cdot 5}$ 25 is a perfect-square factor of 125.

$\sqrt{25} \cdot \sqrt{5}$ Rewrite the expression as two radical expressions.

$5\sqrt{5}$ Find the square root of 25.

Use this method to rewrite each radical expression.

a. $\sqrt{72}$ $6\sqrt{2}$ **b.** $\sqrt{27}$ $3\sqrt{3}$ **c.** $\sqrt{1800}$ $30\sqrt{2}$ **d.** $\sqrt{147}$ $7\sqrt{3}$

11. Show that $6 - \sqrt{20}$ is a solution to the equation $0 = 0.5x^2 - 6x + 8$.

12. APPLICATION The Great Pyramid of Cheops in Egypt has a square base with a side length of 800 feet. Its triangular faces are almost equilateral. These diagrams show an unfolded and a folded scale model of the pyramid.

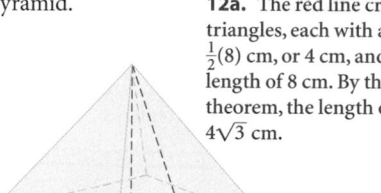

12a. The red line creates two right triangles, each with a leg length of $\frac{1}{2}(8)$ cm, or 4 cm, and a hypotenuse length of 8 cm. By the Pythagorean theorem, the length of the other leg is $4\sqrt{3}$ cm.

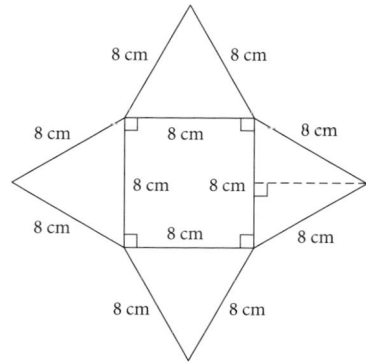

a. Note the model's measurements. Explain how to find the distance from the centers of the sides of the base to the tip of each triangle. (That is, find the length of the segment shown in red.)

b. Find the height of the model. (That is, find the length of the segment shown in blue.) $4\sqrt{2}$ cm

c. Use your result from 12b to find the approximate height of the Great Pyramid.
approximately 566 ft

The Great Pyramid of Cheops was built using over 2,300,000 blocks of stone weighing 2.5 tons each.

11. $0 = 0.5x^2 - 6x + 8$ The original equation.

$0 \overset{?}{=} 0.5\left(6 - \sqrt{20}\right)^2 - 6\left(6 - \sqrt{20}\right) + 8$ Substitute $6 - \sqrt{20}$ for x.

$0 \overset{?}{=} 0.5\left(6 - \sqrt{20}\right)^2 - 36 + 6\sqrt{20} + 8$ Distribute the -6 over $6 - \sqrt{20}$.

$0 \overset{?}{=} 0.5(36 - 6\sqrt{20} - 6\sqrt{20} + 20) - 36 + 6\sqrt{20} + 8$ Square the expression.

$0 \overset{?}{=} 18 - 3\sqrt{20} - 3\sqrt{20} + 10 - 36 + 6\sqrt{20} + 8$ Distribute the 0.5 over the expression in parentheses.

$0 \overset{?}{=} 18 + 10 - 36 + 8$ Combine the radical expressions.

$0 = 0$ Add and subtract.

Exercise 9 Some students may still need to see or write the intermediary step
$$\left(\sqrt{4} \cdot \sqrt{5} = \sqrt{4 \cdot 5} = \sqrt{20}\right).$$
Students can use decimal approximations to check their answers roughly.

Exercise 10 Ideally, students will become comfortable moving between the two types of radical expressions in Exercises 9 and 10.

Students need to identify the largest perfect square. Some students can do this more easily in several steps. For example, in 10c,
$$\begin{aligned} \sqrt{1800} &= \sqrt{18 \cdot 100} \\ &= \sqrt{9 \cdot 2 \cdot 100} \\ &= \sqrt{9} \cdot \sqrt{100} \cdot \sqrt{2} \\ &= 3 \cdot 10 \cdot \sqrt{2} \\ &= 30\sqrt{2} \end{aligned}$$
Because simplifying expressions under a radical by factoring out a perfect square is a new skill for students, you may want them to work through this problem in groups.

Exercise 11 This exercise is slightly different from Example C. Insist on valid logic in checking.

Exercises 12–14 These challenging applications of the Pythagorean theorem may be assigned for extra credit or review or as group work.

Exercise 12 You might want to ask students to create a net for a scale model of the Great Pyramid of Cheops, or they could cut out the model provided on the worksheet Pyramid Net and make a pyramid. The length being asked for, an altitude of a triangular face, is called the *slant height* or *lateral height* of the pyramid.

To give students practice in simplifying radical expressions, the text approximates the faces with equilateral triangles. The actual sides have lengths 800 ft, 743 ft, and 743 ft, making the height 481 ft instead of 510 ft.

Exercise 13 This exercise assumes that the pole will not bend. As needed, encourage students to draw an auxiliary line.

Solutions should involve two triangles, as in Exercise 12. You might want to have a box available as well as a dowel or straw that students can place inside the box to help visualize the situation.

Exercise 14 The answer for side c contains a square root symbol within a square root symbol, which students may not have seen before.

Exercise 15 This exercise reviews Lesson 9.4.

Exercise 16 This exercise reviews Lesson 10.1.

13. The longest pole that fits in a rectangular box goes from one corner to the corner farthest from it. Find the longest pole that fits a 30-cm-by-50-cm-by-20-cm box. Show all your work. Give the answer as an exact value. $10\sqrt{38}$ **cm**

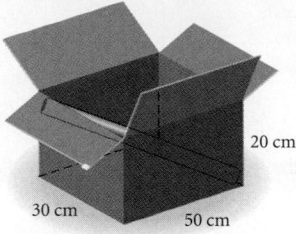

14. Find the exact lengths of sides a, b, and c in the figure at right.
$a = 2\sqrt{2}$ cm, $b = 2\sqrt{3}$ cm, $c = \sqrt{8\sqrt{3} + 12}$ cm

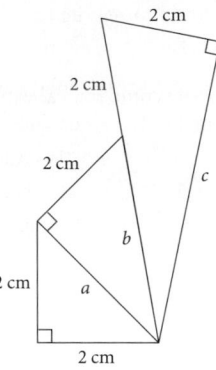

▶ **Review**

15. Write an equation for each transformation of the graph of $y = x^2$.
   a. a translation up 3 units and right 2 units $y = (x - 2)^2 + 3$
   b. a reflection across the x-axis and then a translation up 4 units $y = -x^2 + 4$
   c. a vertical stretch by a factor of 3 and then a translation right 1 unit $y = 3(x - 1)^2$

16. How many x-intercepts does the graph of each equation in 15a–c have?
   a. no x-intercepts   **b.** two x-intercepts   **c.** one x-intercept

project

SHOW ME PROOF

Throughout history, many different civilizations have used the right triangle relationship $a^2 + b^2 = c^2$. The people of Babylonia, Egypt, China, Greece, and India all found this relationship useful and fascinating.

Along the way, there also have been many different proofs of this theorem. Drawing squares on each side of a right triangle is only one of them. Research the history of the Pythagorean theorem and locate a proof you find interesting. Prepare a paper or a presentation of the proof.

Your project should include

▶ A clear and accurate presentation of the proof. Include diagrams and mathematical equations when appropriate.

▶ A written or verbal explanation of why the proof works. (You may need to do some research to fully understand what a proof is.)

▶ The history associated with the proof.

▶ A list of the resources you used.

Supporting the project

MOTIVATION

Students learn something about the structure of a proof and that a theorem can have more than one proof.

OUTCOMES

▶ Diagrams and mathematics support and explain the proof.
▶ The history of the proof is discussed.
▶ Resources are cited accurately.
• More than one proof is presented and proofs are compared and contrasted.

• The presentation shows deep understanding of what constitutes a proof, as opposed to isolated examples.
• Ancient peoples (such as the Babylonians) are mentioned; they had knowledge of the Pythagorean relation but, as far as we know, did not have a formal proof.

A Distance Formula

If you hike 2 kilometers east and 1 kilometer north from your campsite, do you know how far you are from camp? In this lesson you will use coordinate geometry and the Pythagorean theorem to find the distance between any two points.

The shortest distance between two points is under construction.

NOELIE ALTITO

EXAMPLE A

If you start at your campsite at the point $(0, 0)$ and walk 2 km east and 1 km north to the point $(2, 1)$, how far are you from your campsite?

▶ **Solution**

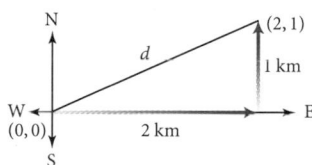

The east-west leg of the right triangle is $2 - 0$ or 2 km in length, and the north-south leg of the triangle is $1 - 0$ or 1 km in length. Let d be your distance from camp, the distance between points $(0, 0)$ and $(2, 1)$. By the Pythagorean theorem, $d^2 = 2^2 + 1^2$, so $d = \sqrt{2^2 + 1^2}$. So your distance from camp is $\sqrt{5}$ km, or approximately 2.24 km.

The four thunderbirds in the center of this Eastern Sioux buckskin pouch (ca. 1820) stand for the four cardinal directions— north, east, south, and west.

NCTM STANDARDS

CONTENT	PROCESS
Number	• Problem Solving
• Algebra	• Reasoning
• Geometry	Communication
• Measurement	• Connections
• Data/Probability	• Representation

LESSON OBJECTIVES

- Learn how to calculate the distance between two points on a grid using right triangles and thus to generate the distance formula
- Learn to apply the symbolic form of the distance formula

PLANNING

LESSON OUTLINE

First day:

10 min   Example A

40 min   Investigation

Second day:

15 min   Investigation

10 min   Sharing

10 min   Examples B, C

5 min   Closing

10 min   Exercises

MATERIALS

- Amusement Park (W), *optional*
- Surveying (W), *optional*

TEACHING

The Pythagorean theorem with analytic geometry gives a formula for calculating the distance between two points.

One step Display the Surveying master from Lesson 11.1. Say that to write up a description of the property, you need to find the length of each edge of the plot. Encourage students to draw right triangles and, as they finish, to derive a general formula. As they find the distance formula, ask them to determine the lengths of the plot's diagonals and to check all results by measuring and adjusting for the scale of the drawing.

▶ **EXAMPLE A**

This example shows how to use the Pythagorean theorem to calculate a distance. Students should notice the subtraction of 0 in calculating horizontal and vertical distances. It is shown to help students generalize to cases in which neither point is the origin. Because d is a distance, only the positive square root is considered.

Step 2 You might pass out copies of the Amusement Park worksheet so that students aren't tempted to write in their books.

Some students may have difficulties deciding what right triangle to draw. You might suggest that they think of slope triangles. And when the triangle is drawn, they may not see how to find the lengths of the legs. Suggest that they drop perpendiculars to the axes.

Students also may find it challenging to work with general units rather than feet or meters. As needed, encourage them to think about appropriate length in feet or meters for each unit (in Step 3 they're asked to assume that each unit is 0.1 mi), but try to get results in units first.

Steps 5–9 Some students may have difficulties working in the abstract. You might ask them to write out the steps for finding one specific distance in a column and then write the general approach in an adjacent column, using variables.

SHARING IDEAS

Have students present any variety of ideas they came up with in Step 8. Work with the class to derive the distance formula. In the process, ask if the order of subtraction of the x- and y-coordinates matters. Elicit the idea that the squaring operation will yield the same positive radicand either way.

Ask students to calculate distances, assuming that the grid lines are roads along which they must travel. Ask what the distance formula would be in that case. The use of absolute values makes the distance simple to express: $|x_1 - x_2| + |y_1 - y_2|$. Ask if this is the same as the standard distance formula. Some students may think so, believing that the square root of a sum is the sum of

the square roots. If you can challenge them to determine why the data conflict with their belief, they may correct their misconception in a deeper way than if you just tell them they're mistaken.

Investigation
Amusement Park

You will need
- graph paper

In this investigation you will discover a general formula for the distance between two points.

Step 1 Copy the map of amusement park attractions and assign coordinates to each attraction on the map.

Step 2 Find the distance between each pair of attractions in a–e. When appropriate, draw a right triangle. Use what you know about right triangles and the Pythagorean theorem to find the exact distance between each pair of attractions.

Step 2a 6 units
Step 2b $\sqrt{10}$ units
Step 2c 2 units
Step 2d 5 units
Step 2e $\sqrt{85}$ units

Step 3 Roller Coaster and Sledge Hammer; 10 units or 1 mile

Step 4 $11\sqrt{5}$ units or approximately 2.5 miles

    a. Bumper Cars and Sledge Hammer
    b. Ferris Wheel and Hall of Mirrors
    c. Mime Tent and Hall of Mirrors
    d. Refreshment Stand and Ball Toss
    e. Bumper Cars and Mime Tent

Step 3 Which pairs of attractions are farthest apart? If each grid unit represents 0.1 mile, how far apart are they?

Step 4 Chris parked his car at the coordinates $(17, -9)$. If each grid unit represents 0.1 mile, how far is it from the Refreshment Stand to his car? (Try to do this without plotting the location of his car.)

Two new attractions are being considered. The first attraction, designated P_1, will be located at the coordinates (x_1, y_1) as shown at right, and the second building, designated P_2, will be located at the coordinates (x_2, y_2).

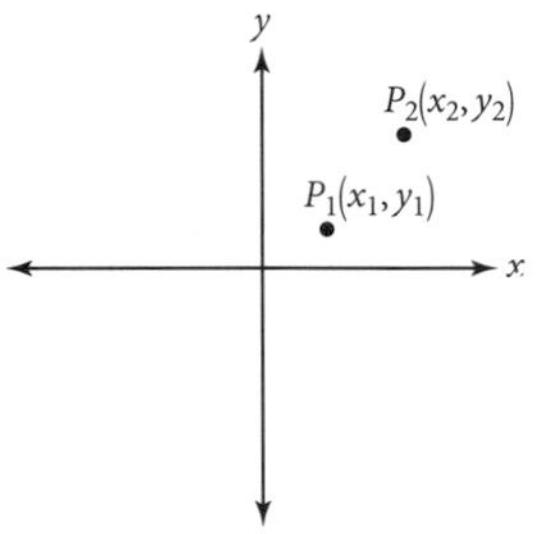

Step 6 $y_2 - y_1$

Step 7 $x_2 - x_1$

Step 5 Sketch a right triangle with horizontal and vertical legs and hypotenuse $\overline{P_1P_2}$.

Step 6 Write an expression for the vertical distance between P_1 and P_2.

Step 7 Write an expression for the horizontal distance between P_1 and P_2.

Step 8 Write an expression for the distance between these two points. (This formula should work for any two points.)

Step 9 Verify that your formula works by using it to find the distance between the Bumper Cars and the Mime Tent.

Assessing Progress

Look for abilities to assign coordinates to points, draw right triangles, and use notation to write a general expression representing a sequence of steps.

Step 8 *distance between P_1 and P_2* $= \sqrt{(x_2 - x_1)^2 + (y_2 - y_1)^2}$

Step 9 *distance* $= \sqrt{(-4 - 3)^2 + (-3 - 3)^2} = \sqrt{(-7)^2 + (-6)^2} = \sqrt{49 + 36} = \sqrt{85}$ units

The Pythagorean theorem is an efficient way to solve many problems involving distance.

EXAMPLE B

If you walk 6 kilometers northeast from (2, 1), at a compass reading of 45°, what is your new location?

▶ **Solution**

First you find the horizontal and vertical change from your starting position. If you walk northeast, you walk just as far to the east as you walk to the north. Your path creates an isosceles right triangle. Use the Pythagorean theorem to find a in the sketch below.

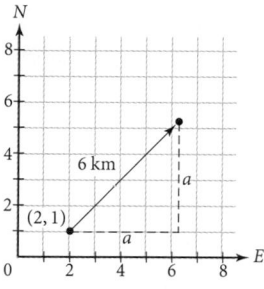

$a^2 + b^2 = c^2$ Pythagorean theorem.

$a^2 + a^2 = 6^2$ Substitute 6 for c, and substitute a for b because both legs are the same length.

$2a^2 = 36$ Add $a^2 + a^2$.

$a = \sqrt{18}$ Divide by 2 and take the square root of both sides.

To get the coordinates of your new location, add $\sqrt{18}$ to each coordinate of your starting location, (2 , 1). Your new location is exactly $\left(2 + \sqrt{18}, 1 + \sqrt{18}\right)$, or about 6.2 kilometers east and 5.2 kilometers north of the campsite.

You can also use similar trianges to find a. The triangle created by your path is similar to the triangle created when one grid square is cut in half. The smaller triangle has legs of 1 unit each and a hypotenuse of $\sqrt{2}$ units. Set up a proportion and solve for a.

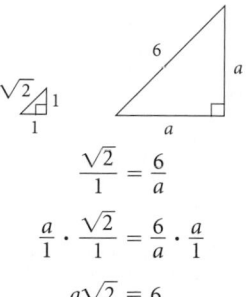

$\dfrac{\sqrt{2}}{1} = \dfrac{6}{a}$ Set up a proportion that compares the hypotenuses to the legs.

$\dfrac{a}{1} \cdot \dfrac{\sqrt{2}}{1} = \dfrac{6}{a} \cdot \dfrac{a}{1}$ Multiply both sides by a.

$a\sqrt{2} = 6$ Reduce.

▶ **EXAMPLE B**

Whereas Example A shows how to find a distance given some points, this example shows how to find a point given a distance.

For this triangle you might want to introduce the terminology of a 1-1-$\sqrt{2}$ triangle or a $45°$-$45°$-$90°$ triangle.

This example shows how solving an equation in radicals may produce a number that does not check as a solution. Such a root is called an *extraneous root*. Here, -3.4 is an extraneous root.

Closing the Lesson

The **distance formula** for calculating the distance between points (x_1, y_1) and (x_2, y_2) is derived by applying the Pythagorean theorem to a triangle with a right angle at a point in line horizontally with one of the given points and vertically with the other. The result is that the distance is $\sqrt{(x_2 - x_1)^2 + (y_2 - y_1)^2}$.

$$a = \frac{6}{\sqrt{2}}$$ Divide both sides by $\sqrt{2}$.

$$a = \frac{\sqrt{36}}{\sqrt{2}}$$ Rewrite 6 as $\sqrt{36}$.

$$a = \sqrt{\frac{36}{2}}$$ Rewrite $\frac{\sqrt{36}}{\sqrt{2}}$ as $\sqrt{\frac{36}{2}}$.

$$a = \sqrt{18}$$ Rewrite $\sqrt{\frac{36}{2}}$ as $\sqrt{18}$.

Your result is the same as above.

Sometimes equations from distance problems have variables within a square root. Example C shows how to work with these situations.

EXAMPLE C | Solve the equation $\sqrt{15 + x} = x$.

► **Solution**

$$\sqrt{15 + x} = x$$ The original equation.

$$\left(\sqrt{15 + x}\right)^2 = x^2$$ Square both sides to undo the square root.

$$15 + x = x^2$$ The result of squaring.

$$0 = x^2 - x - 15$$ Subtract 15 and x from both sides to get a trinomial set equal to 0.

$$x \approx -3.4 \text{ and } x \approx 4.4$$ Use the quadratic formula, a graph, or a table to find two possible solutions.

Check:

$$\sqrt{15 + (-3.4)} \neq -3.4$$ The square root of a number can't be negative. So 3.4 is not a solution.

$$\sqrt{15 + 4.4} = \sqrt{19.4} \approx 4.4$$ This solution checks.

Whenever you solve a square root equation, be careful to check whether or not each solution satisfies the original equation.

In the second part of the investigation you used the Pythagorean theorem to derive the **distance formula.** When you know the coordinates of two points, this formula allows you to find the distance between the points even without plotting them.

Distance Formula

The distance d between $P_1(x_1, y_1)$ and $P_2(x_2, y_2)$ is given by the formula

$$d = \sqrt{(x_2 - x_1)^2 + (y_2 - y_1)^2}$$

Look back at Example A. Can you show how to use the distance formula to solve this problem without graphing?

The length of a segment is the same as the distance between the endpoints of the segment. So, the distance formula has many applications in analytic geometry.

EXERCISES

You will need your calculator for problems **8** and **10**.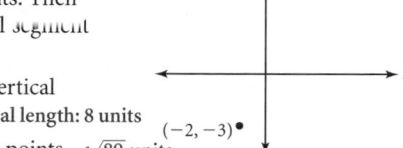

Practice Your Skills

1. Is a triangle with side lengths of 9 cm, 16 cm, and 25 cm a right triangle? Explain. **No, because $9^2 + 16^2 \neq 25^2$.**

2. Plot these points on graph paper.
   a. Draw the segment between the two points. Then draw a horizontal segment and a vertical segment to create a right triangle.
   b. Find the lengths of the horizontal and vertical segments. **horizontal length: 5 units; vertical length: 8 units**
   c. Find the exact distance between the two points. **$\sqrt{89}$ units**

   (3, 5)

   (−2, −3)

3. On his homework, Matt wrote that the distance between two points was

   $$\sqrt{(6-1)^2 + (3-7)^2}$$

   What two points was Matt working with? **Possible answer: (6, 3) and (1, 7)**

Reason and Apply

4. Quadrilateral *ABCD* is pictured at right.
   a. What is the slope of each side?
   b. What type of quadrilateral is it? **parallelogram**
   c. Find the length of each side.

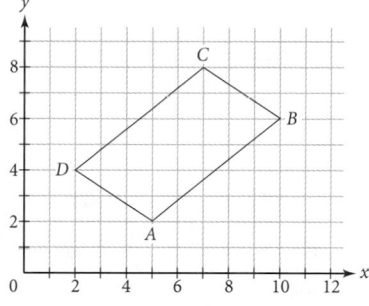

For problems 5 and 6, refer to the Amusement Park investigation on page 612.

5. Use the distance formula to find the distance between each pair of attractions.
   a. Refreshment Stand and Bumper Cars **$\sqrt{26}$ units or approximately 0.5 mile**
   b. Acrobats and Hall of Mirrors **5 units or 0.5 mile**

6. Jake's sawdust spreader breaks down halfway between the Roller Coaster and the Sledge Hammer. At what point on the map on page 612 does the breakdown occur? **(−1, 1)**

For problems 5 and 6, refer to the Amusement Park investigation on page 612.

Students work with the distance formula.

ASSIGNING HOMEWORK

Essential	1, 2, 4–7
Performance assessment	6
Portfolio	7
Journal	3, 8
Group	8
Review	9, 10

▶ Helping with the Exercises

Exercise 2b If students are having difficulty finding the lengths of these segments, suggest that they drop perpendiculars from the endpoints to the axes.

2a. Possible answer:

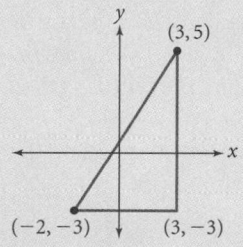

(3, 5)

(−2, −3)   (3, −3)

Exercise 3 Be open to two possible solutions here, because the *x*- and *y*-coordinates can be subtracted in either order.

Exercise 4 You may want to ask students to make a conjecture about the lengths of opposite sides of any parallelogram and to test it on other parallelograms.

4a. slope of $\overline{AB}$ and $\overline{DC}$: $\frac{4}{5}$; slope of $\overline{AD}$ and $\overline{BC}$: $-\frac{2}{3}$

4c. length of $\overline{AB}$ and $\overline{DC}$: $\sqrt{41}$ units; length of $\overline{AD}$ and $\overline{BC}$: $\sqrt{13}$ units

Exercise 6 If students solve this problem using the midpoint formula of Lesson 11.2, have them check with the distance formula.

Exercise 8c As students compare the solutions they get using these two methods, they see again that some apparent solutions to a square root equation may be extraneous. If students are able to explain that the extraneous solution results when the original equation is squared, they are more likely to remember how important it is to verify solutions. If the symbolic solution is difficult for too many students for you to help individually, you might rely on groups.

Exercise 9 This exercise reviews skills learned in Lesson 11.5.

Exercise 10 This is technically a two-step exercise. Considering a right triangle, students will first have to find the length of the other leg based on the slope. Students should ultimately use the Pythagorean theorem to find the length of the ramp itself, or the hypotenuse of the right triangle. If students give the answer 60 ft (not 60 ft 1 in.), they may have found only the length of the other leg. (The slope of 0.05 was determined based on the information at the web site http://hometime.com/projects/howto/access/pc2aces2.htm.)

[Ask] "Do you think Martin will be able to build the ramp as a straight ramp?" [Probably not, because it would need to reach 60 ft horizontally from the building. Students may have seen wheelchair ramps that wind around a building or switch back.]

As an extension, turn the problem around. **[Ask]** "How much will a horizontal 60-foot walkway that is fixed at one end rise at the other end if one inch were added to its length?" [The answer of 3 feet might be surprising.] If you leave the problem open, students may want to check and recheck the arithmetic. Be sensitive to how, for many of them, changing their intuition is threatening because they see it as changing who they are.

7. A rectangular box has the dimensions shown in the diagram.
   a. What is the length of the diagonal $\overline{BD}$? 5 cm
   b. What is the length of the diagonal $\overline{BH}$? 13 cm

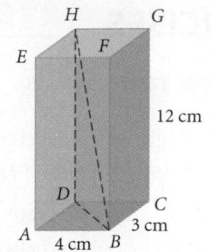

8. Consider the equation
   $$\sqrt{20 - x} = x$$
   a. Solve the equation symbolically.
   b. Solve the equation using a graph or a table.
   c. Explain why you get two possible solutions when you solve the equation symbolically and only one solution when you look at a graph or table. Substitute both possible solutions into the original equation, and describe what happens.

   $x = 4$ is a solution; $x = -5$ is not a realistic solution because a square root cannot be negative.

▶ **Review**

9. Solve each equation.
   a. $\dfrac{3}{5} = \dfrac{a}{105}$ $a = 63$ b. $\dfrac{1}{\sqrt{2}} = \dfrac{b}{7\sqrt{2}}$ $b = 7$ c. $\dfrac{\sqrt{3}}{2} = \dfrac{c}{\sqrt{12}}$ $c = 3$

10. **APPLICATION** Martin Weber is building a wheelchair ramp at the Town Hall. The ramp will meet a door that is 3 feet off the ground. Building codes in his area require an exterior ramp to have a slope of 0.05. How long will the ramp need to be? Give your answer in exact form and as an approximation to the nearest inch.
    $3\sqrt{401}$ ft or approximately 60 ft 1 in.

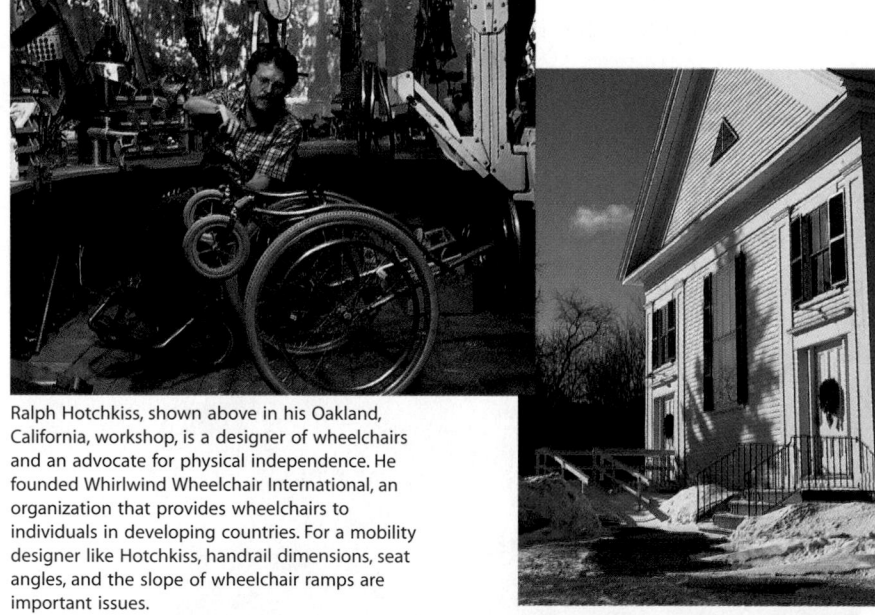

Ralph Hotchkiss, shown above in his Oakland, California, workshop, is a designer of wheelchairs and an advocate for physical independence. He founded Whirlwind Wheelchair International, an organization that provides wheelchairs to individuals in developing countries. For a mobility designer like Hotchkiss, handrail dimensions, seat angles, and the slope of wheelchair ramps are important issues.

8a. $\sqrt{20 - x} = x$
$$20 - x = x^2$$
$$0 = x^2 + x - 20$$
$$0 = (x + 5)(x - 4)$$
$$x = -5, \text{ or } x = 4$$

8b. Possible answer: The intersection of the graphs of $Y_1 = \sqrt{(20 - x)}$ and $Y_2 = x$ occurs once at $x = 4$.

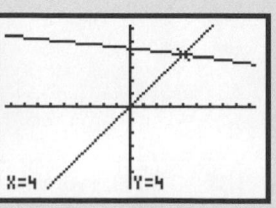

$[-9.4, 9.4, 1, -6.2, 6.2, 1]$

LESSON 11.7

Similar Triangles and Trigonometric Functions

Similar figures have corresponding angles that are equal and corresponding side lengths that are proportional. So the figures have the same shape, but one is an enlargement of the other. You can use ratios and proportions to compare and calculate length measurements for similar figures.

These Japanese cat figurines, called Maneki Neko, are near examples of three-dimensional similar figures. In what ways are they not mathematically similar?

EXAMPLE A

Elene is walking to school. From where she stands, she can see the 5-meter flagpole on top of her school. She holds her centimeter ruler approximately 50 centimeters from her eye. Against the ruler, the flagpole looks 25 centimeters tall. How far is she from school?

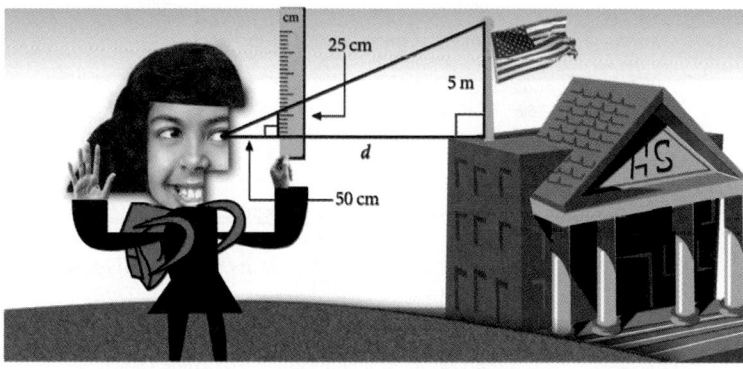

▶ **Solution**

This solution creates two similar triangles. One triangle is formed by Elene's eye and the 0-cm and 25-cm marks on her ruler. The other triangle is formed by Elene's eye and the ends of the flagpole. Elene's eye is a common vertex.

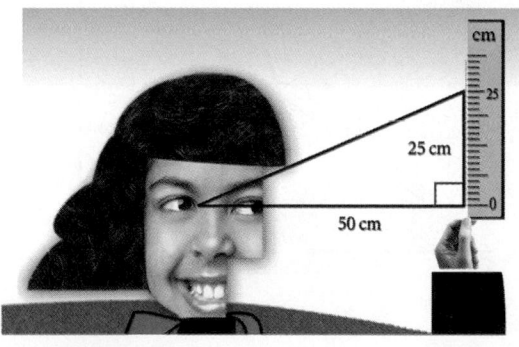
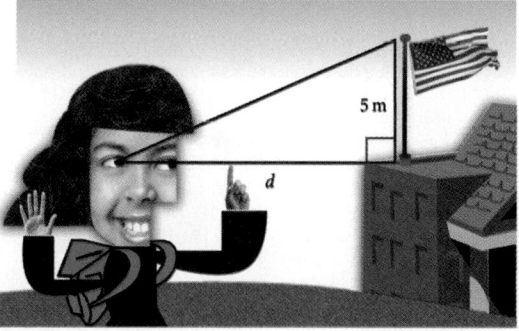

NCTM STANDARDS

CONTENT	PROCESS
Number	• Problem Solving
• Algebra	• Reasoning
• Geometry	• Communication
• Measurement	Connections
Data/Probability	• Representation

LESSON OBJECTIVES

- Review the basic properties of similar triangles
- Review (or learn) how to measure acute angles using a protractor
- Discover the definitions of the sine, cosine, and tangent ratios

PLANNING

LESSON OUTLINE

First day:
10 min   Example A
40 min   Investigation

Second day:
15 min   Investigation
10 min   Sharing
10 min   Example B
5 min   Closing
10 min   Exercises

MATERIALS

- graph paper
- straightedges
- protractors
- Calculator Note 11A
- Earth and Moon (T), *optional*

TEACHING

Trigonometry can replace similar triangles in finding distances. The concepts of ratio, proportion, and similarity were first introduced in Chapters 2 and 3.

One step Display the Earth and Moon transparency. Challenge students to calculate the radius of the moon (*RM*) given that the distance to the moon (*EM*) is 240,000 miles, that the angle at *R* is a right angle, and that the angle at *E* is 0.26 degrees. As you circulate, encourage them to draw scale models and look for ratios. During Sharing, lead students to the idea that the ratios of corresponding side lengths of all similar right triangles (in which corresponding angles have the same measure) are the same.

This example shows how to determine a length using similar triangles and points out that corresponding angles of similar triangles have the same measure. Students may need reminding that similar figures are dilations of each other. Saying that "similar" means "same shape" leads some students to think that, for example, all rectangles are similar to each other but not to any triangles.

Use the ratios of adjacent sides to write the proportion

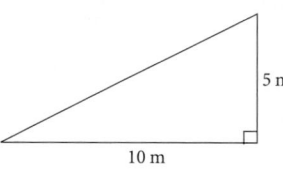

Here, $d = 10$, so Elene is 10 meters from school.

Look back at the similar right triangles in Example A.

Notice the ratios compare the vertical legs to the horizontal legs. Both ratios, $\frac{25 \text{ cm}}{50 \text{ cm}}$ and $\frac{5 \text{ m}}{10 \text{ m}}$, equal 0.5.

If another triangle similar to those has a horizontal leg of 7 meters, then the vertical leg would be 3.5 meters in length. The angles in all of these triangles are about 26.6°, 63.4°, and 90°. Any other right triangle with a value of 0.5 for the ratio of its vertical leg to its horizontal leg also has angles with these measures.

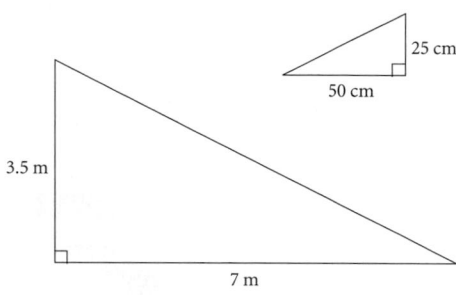

In each of these similar triangles, the ratio of the vertical leg to the horizontal leg is 0.5. All three have angles measuring approximately 26.6°, 63.4°, and 90°.

Likewise, if a right triangle has these angle measures, the ratio of its vertical leg to its horizontal leg is 0.5. There is a connection between the angle measures of a triangle and the ratios of its sides. You'll explore this connection in the investigation.

The angle below measures 36°. You already know that an angle that measures 90° is called a right angle. An angle that measures less than 90° is called an **acute angle.** An angle that measures more than 90° but less than 180° is called an **obtuse angle.**

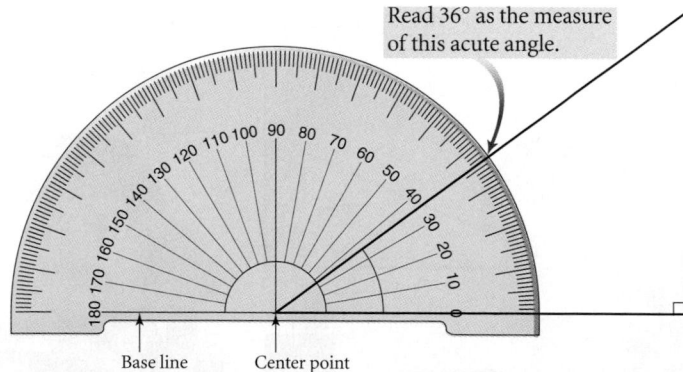

Read 36° as the measure of this acute angle.

Base line   Center point

Investigation
Ratio, Ratio, Ratio

You will need

- graph paper
- a straightedge
- a protractor

In this investigation you'll learn about some very important ratios in right triangles.

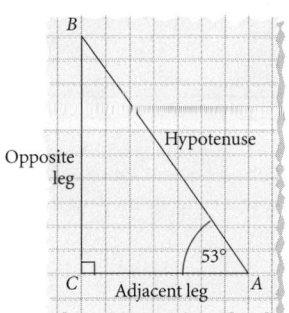

This is only a sample. Your triangle should look different.

Procedure Note

Measuring Angles
Place the center point of your protractor on the vertex of the angle. Line up the base line with one side of the angle. Notice the mark that the other side of the angle passes through. See the example at the bottom of page 610.

Step 1 On graph paper, use a straightedge to draw a right triangle. Use the grid lines of the graph paper to make sure the legs are perpendicular. Make the triangle large enough for you to measure its angles accurately.

Step 2 Label one of the acute angles *A*. Measure it.

If you look at one acute angle in a right triangle, the **adjacent leg** is the leg of the triangle that is part of the measured angle. The **opposite leg** is the leg of the triangle that is not part of the angle you are looking at.

Step 3 Make a table like this one and record the information for each triangle drawn by a member of your group.

	Tony's triangle	Alice's triangle
Measure of angle A		
Length of adjacent leg (a)		
Length of opposite leg (o)		
Length of hypotenuse (h)		

Step 4 Make a new table like this one and calculate the ratios for each triangle drawn by a member of your group.

	Tony's triangle	Alice's triangle
Measure of angle A		
$\frac{o}{h}$		
$\frac{a}{h}$		
$\frac{o}{a}$		

Guiding the Investigation

Step 1 Each member of the group should draw a different right triangle.

Step 3 As needed, remind students that a *leg* of a right triangle is a side that isn't the hypotenuse. Whereas in stating the Pythagorean theorem, the leg called *a* was opposite to the angle called *A*, in trigonometry the leg called *a* is adjacent to angle *A*. The letter *a* stands for *adjacent*, just as *o* stands for *opposite* and *h* for *hypotenuse*.

Students should use radicals to express lengths exactly.

Have at least one group share its results from Step 6, and summarize the results by pointing out the box on page 620 of the student text. You might draw a large triangle and then a bunch of line segments within it parallel to one of its legs to point out the equality of the ratios corresponding to the angle the triangles formed share.

Students might be helped by the mnemonic SOHCAHTOA (pronounced "soh-cah-toh-ah") standing for "sine: opposite over hypotenuse, cosine: adjacent over hypotenuse, tangent: opposite over adjacent."

[Ask] "Are there any other ratios of the three sides of a right triangle?" [Sine, cosine, and tangent are only three of six trigonometric functions. The other three are *secant* $\frac{h}{a}$, *cosecant* $\frac{h}{o}$ and *cotangent* $\frac{a}{o}$].

[Ask] "Do you think that trigonometric ratios apply to non-right triangles?" Challenge students to find two non-right triangles that have an angle in common but different ratios of, say, opposite to adjacent. [Like the Pythagorean theorem, trigonometric ratios are valid only for right triangles.]

Some students might reasonably ask why we need trigonometry. Similar triangles seem to suffice, as in Elene's finding the distance to the flagpole. Often it's impossible to draw similar triangles, as when you're trying to measure distance in space (as in the one-step problem) or surveying. Point out the History Connection. Pitiscus was born in Silesia, now part of Poland. Trigonometric ratios (and tables for many angles) were developed 1700 years before Pitiscus for the purpose of measuring such distances.

Step 5 With your calculator in degree mode, find the value of the **sine,** the **cosine,** and the **tangent** of angle A. [▶ ☐ See **Calculator Note 11A** to learn about these functions on your calculator. ◀] Record these values to the nearest 0.01 in a table like this one.

	Tony's triangle	Alice's triangle
Measure of angle A		
sine (A)		
cosine (A)		
tangent (A)		

Step 6 The results for Steps 4 and 5 should be about the same.

$$sine = \frac{opposite\ leg}{hypotenuse},$$
$$cosine = \frac{adjacent\ leg}{hypotenuse},$$
$$tangent = \frac{opposite\ leg}{adjacent\ leg}$$

Step 6 Compare your results for Steps 4 and 5. Define each function—sine, cosine, and tangent—in terms of a ratio of the lengths of the adjacent leg, the opposite leg, and the hypotenuse.

Step 8 Students should find that the ratios are the same for any pair of similar right triangles.

Step 7 Draw a larger right triangle with an acute angle D equal to your original angle A.

Step 8 Measure the side lengths and calculate the sine, the cosine, and the tangent of angle D. What do you find?

In the investigation you discovered that some ratios of the sides of a right triangle have special names: sine, cosine, and tangent. Sine, cosine, and tangent are all called **trigonometric functions.** They are fundamental to the branch of mathematics called **trigonometry.** Learning to identify the parts of a right triangle and to evaluate these functions for particular angle measures is an important problem-solving tool.

History
• CONNECTION •

The word *trigonometry* comes from the Greek words for triangle and measurement. Its first use in English was in a 1614 translation of *Trigonometry: Doctrine of Triangles* by the Silesian mathematician Bartholmeo Pitiscus (1561–1613).

Trigonometric Functions

For acute angle A in a right angle, the trigonometric functions are

$$sine\ of\ angle\ A = \frac{length\ of\ opposite\ leg}{length\ of\ hypotenuse} \quad or \quad \sin A = \frac{o}{h}$$

$$cosine\ of\ angle\ A = \frac{length\ of\ adjacent\ leg}{length\ of\ hypotenuse} \quad or \quad \cos A = \frac{a}{h}$$

$$tangent\ of\ angle\ A = \frac{length\ of\ opposite\ leg}{length\ of\ adjacent\ leg} \quad or \quad \tan A = \frac{o}{a}$$

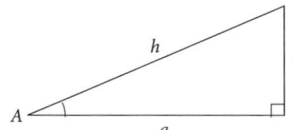

In this lesson you will get practice writing ratios associated with the trigonometric functions. You will also get more practice with ratio, proportion, and similarity. In the next lesson you will apply the ratios to solve problems.

EXAMPLE B

Find these ratios for this triangle.

a. sin A

b. cos A

c. tan B

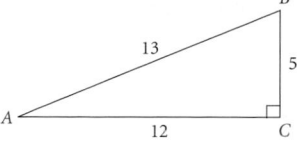

▶ **Solution**

For angle A, the side length values are $a = 12$, $o = 5$, and $h = 13$.

a. $\sin A = \dfrac{o}{h} = \dfrac{5}{13}$

b. $\cos A = \dfrac{a}{h} = \dfrac{12}{5}$

Using angle B changes which leg is opposite and which is adjacent. For angle B, the side length values are $a = 5$, $o = 12$, and $h = 13$.

c. $\tan B = \dfrac{o}{a} = \dfrac{12}{5}$

Note that identifying the opposite and adjacent legs depends on which angle you are using. Be careful to identify the correct sides and angles when using trigonometric functions.

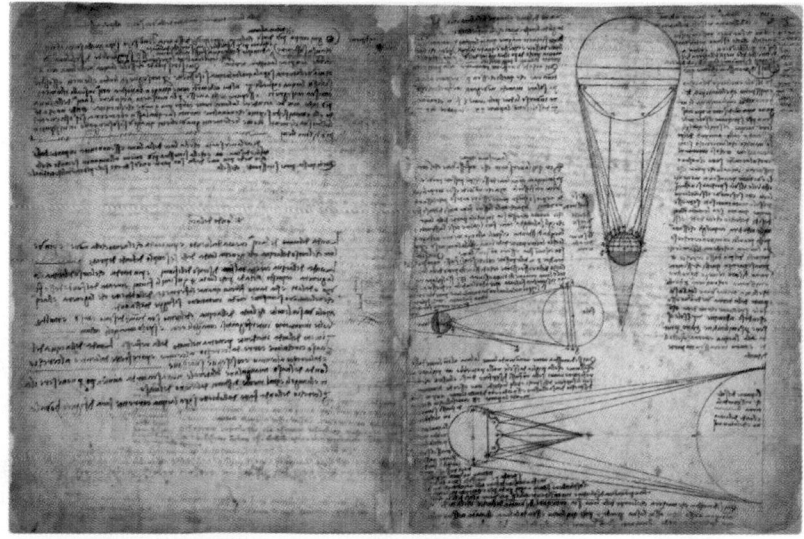

This sketch by Leonardo da Vinci (Italian, 1452–1519) explains why moonlight is less bright than sunlight. A diverse genius, Leonardo was a painter, draftsman, sculptor, architect, and engineer. The triangles used in this sketch show that he was also knowledgeable about geometry and trigonometry.

Assessing Progress

Check for students' understanding of right triangles and abilities to collect data systematically.

▶ **EXAMPLE B**

You might have students find all three ratios for both angles A and B and look for relationships between the ratios. The angles A and B are complements because together they make up a right angle. The word *cosine* stands for *complement's sine*, so cos A = sin B and vice versa.

[Ask] "Do the ratios hold for the right angle?" [Although the sine, as opposite over hypotenuse, will be 1, it's not clear which leg is adjacent. Trigonometric ratios are extended for right angles by other means.]

Closing the Lesson

Trigonometry is useful in finding distances that are lengths of sides of right triangles, especially when you know an angle and another length. The major trigonometry ratios are **sine**, **cosine**, and **tangent**.

Students work with similar triangles and trigonometric ratios.

ASSIGNING HOMEWORK

Essential	1, 3, 5–7
Performance assessment	6
Portfolio	7
Journal	4
Group	8–10
Review	2, 11, 12

▶ **Helping with the Exercises**

Exercise 1 If students are having difficulties with 1a or 1c, remind them to invert both sides of the equation and then solve for x (rather than encouraging cross-multiplying).

Exercise 2 This exercise reviews Chapter 3.

Exercise 3 You can use comparison of 3a and 3b to discuss trigonometric ratios for complementary angles.

Exercise 4 Students may not be able to answer the question in 4a if they don't realize that the acute angles at C and E are equal in measure. They may be convinced if you point out that the sum of angles of all triangles is the same, or if you talk about stretching. In Chapter 3, the ratio of corresponding side lengths (referred to in 4b) was called the *scale factor*. **[Alert]** If students are having trouble with this exercise, encourage them to draw two separate triangles and label the side lengths of each.

4a. Yes. Possible answer: Angle A is common to both triangles and angles B and D are both right angles, so angles C and D must also be congruent. Because all three angles are congruent, the triangles are similar.

EXERCISES

You will need your calculator for problems **7, 8, 9, 10,** and **12.**

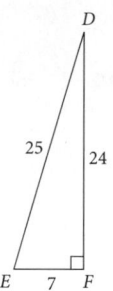

▶ **Practice Your Skills**

1. Solve each equation for x.

  a. $\frac{2}{3} = \frac{18}{x}$ $x = 27$ **b.** $\frac{7}{8} = \frac{x}{40}$ $x = 35$ **c.** $\frac{1}{4} = \frac{\sqrt{10}}{\sqrt{x}}$ $x = 160$ **d.** $\frac{2}{x} = \frac{x}{8}$ $x = \pm 4$

2. APPLICATION One inch on a road map represents 50 miles on the ground. Two cities are 3.6 inches apart on a map. What is the actual distance between the cities? 180 miles

3. Find these ratios for the triangle at right.

  a. $\sin D$ $\sin D = \frac{7}{25}$
  b. $\cos E$ $\cos E = \frac{7}{25}$
  c. $\tan D$ $\tan D = \frac{7}{24}$

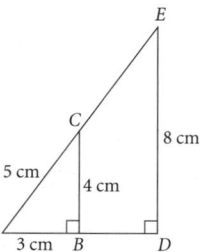

4. The diagram at right shows $\triangle ABC$ and $\triangle ADE$.

  a. Are the triangles similar? Why or why not? (Hint: The angles in any triangle sum to 180°.)
  b. Find the ratio of corresponding side lengths of $\triangle ADE$ to $\triangle ABC$. $\frac{8}{4} = 2$
  c. Find the lengths of $\overline{AD}$ and $\overline{AE}$. 6 cm and 10 cm
  d. Find the areas of $\triangle ADE$ and $\triangle ABC$. 24 cm² and 6 cm²
  e. What is the ratio of the area of $\triangle ADE$ to the area of $\triangle ABC$? $\frac{24}{6} = 4 = (2)^2$

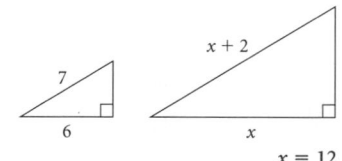

▶ **Reason and Apply**

5. Find x in each pair of similar triangles.

  a.

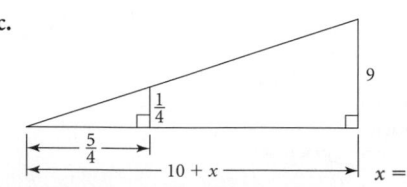

$x = 12$

  b.

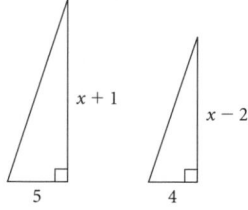

$x = 14$

  c.

$x = 35$

Exercise 5 The proportions in 5a and 5b are easily solved when x is in the numerator of each side.

For 5a use the proportion
$$\frac{x}{6} = \frac{x + 2}{7}$$

For 5b use the proportion
$$\frac{x + 1}{5} = \frac{x - 2}{4}$$

6. An 8-meter ladder is leaning against a building. The bottom of the ladder is 2 meters from the building.

a. How high on the building does the ladder reach?

b. A windowsill is 7 meters high on the building. How far from the building should the bottom of the ladder be placed to meet the windowsill?

6a. $\sqrt{60}$ m or approximately 7.75 m

6b. $\sqrt{15}$ m or approximately 3.87 m

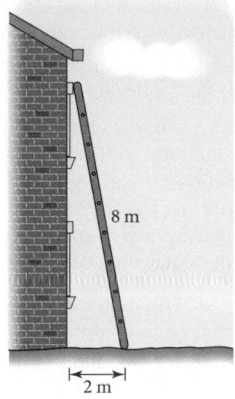

7. The wire attached to the top of a telephone pole makes a 65° angle with the level ground. The distance from the base of the pole to where the wire is attached to the ground is d. The height of the pole is h. The length of the wire is w.

a. What trigonometric function of 65° is the same as $\frac{d}{w}$? cosine

b. What trigonometric function of 65° is the same as $\frac{h}{w}$? sine

c. What trigonometric function of 65° is the same as $\frac{h}{d}$? tangent

d. Use your calculator to approximate the values in 7a–c to the nearest ten thousandth.

e. If the wire is attached to the ground 2.6 meters from the pole, how high is the pole? approximately 5.6 m

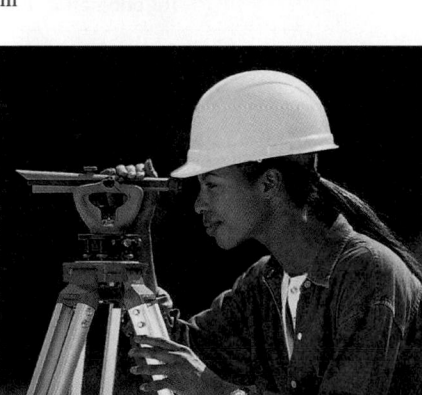

8. Consider this right triangle with a 28° angle.

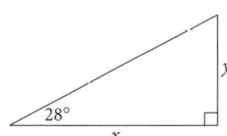

a. Write an equation that relates x and y, the legs of the right triangle. $\tan 28° = \frac{y}{x}$ or $y = x \cdot \tan 28°$

b. Graph your equation on your calculator. Make a sketch of the graph on your paper. Describe the graph.

c. Find y if $x = 100$. $y \approx 53.17$

d. If $y = 80$, find x. $x \approx 151$

Surveyors use a tool called a theodolite, or transit, to measure angles. The angles are sometimes used with trigonometry to measure distances.

Exercise 7 Here's an application of trigonometry. You may point out how the angle and one leg of a right triangle are being used in 7e to find a length that's not otherwise easily available. If students ask, however, be ready to acknowledge that indeed the height could be found by using similar triangles, such as those cast by shadows.

7d. $\cos 65° \approx 0.4226$; $\sin 65° \approx 0.9063$; $\tan 65° \approx 2.1445$

Exercise 8 This exercise could be saved for a later review. Students might graph either $y = x \tan 28$ or $y = \frac{x}{\tan 28}$. **[Alert]** If they are having difficulties answering the questions, remind them of the trace function on their calculators.

8b. The graph is a direct variation.

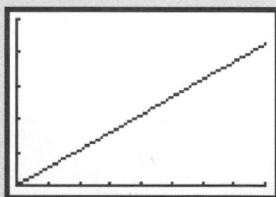

[0, 200, 25, 0, 125, 25]

Exercises 9 and 10 Students are not asked to rationalize denominators, but this might be a good time to approach the subject if you want your students to learn this skill. These are good group problems and are also suitable for review.

9a–b.

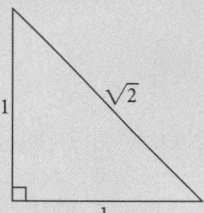

10a–b.

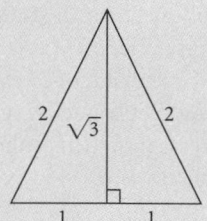

9. You need a protractor and a straightedge for this problem.
   a. Draw an isosceles right triangle. Each acute angle should be 45°.
   b. Label one of the legs of your isosceles right triangle "1 unit." Find the exact lengths of the other two sides.
   c. Make a table like this one on your own paper. First write each ratio using the lengths you found in 9b. Then use your calculator to find a decimal approximation for each exact value to the nearest ten thousandth. Finally, find each ratio using the trigonometric function keys on your calculator. Check that your decimal approximations and the values using the trigonometric function keys are the same.

Trigonometric Functions for a 45° Angle

	Sine	Cosine	Tangent
Exact value of ratio	$\frac{1}{\sqrt{2}}$	$\frac{1}{\sqrt{2}}$	$\frac{1}{1}$
Decimal approximation of exact value	0.7071	0.7071	1.0000
Value by trigonometric function keys	0.7071	0.7071	1.0000

10. You need a protractor and a straightedge for this problem.
    a. Draw an equilateral triangle. Each angle should be 60°. Draw a segment from one vertex to the midpoint of the opposite side. You should have two triangles with angles measuring 30°, 60°, and 90°.
    b. Label one side of your equilateral triangle "2 units." Find the exact lengths of the other two sides of your 30°-60°-90° triangles.
    c. Make tables like these on your own paper. First write each ratio using the lengths you found in 10b. Then use your calculator to find a decimal approximation for each exact value to the nearest ten thousandth. Finally, find each ratio using the trigonometric function keys on your calculator. Check that your decimal approximations and the values using the trigonometric function keys are the same.

Trigonometric Functions for a 30° Angle

	Sine	Cosine	Tangent
Exact value of ratio	$\frac{1}{2}$	$\frac{\sqrt{3}}{2}$	$\frac{1}{\sqrt{3}}$
Decimal approximation of exact value	0.5000	0.8660	0.5774
Value by trigonometric function keys	0.5000	0.8660	0.5774

Trigonometric Functions for a 60° Angle

	Sine	Cosine	Tangent
Exact value of ratio	$\frac{\sqrt{3}}{2}$	$\frac{1}{2}$	$\frac{\sqrt{3}}{1}$
Decimal approximation of exact value	0.8660	0.5000	1.7321
Value by trigonometric function keys	0.8660	0.5000	1.7321

▶ Review

11. Here are four linear equations.

$$y = 2x - 1 \qquad x + 2y = 4$$
$$y = 3 + 2x \qquad x = -2y + 10$$

a. Graph the four lines. What polygon is formed?

b. Find the coordinates of the vertices of the polygon. $(1.2, 1.4), (2.4, 3.8), (0.8, 4.6), (-0.4, 2.2)$

c. Find the linear equations for the diagonals of the polygon. $y = 11 \quad 0.15 \, y \quad \frac{17}{7} \, | \, \frac{4}{7} \, u$

d. Find the coordinates of the point where the diagonals intersect. $(1, 3)$

12. Find the missing side lengths in this figure.

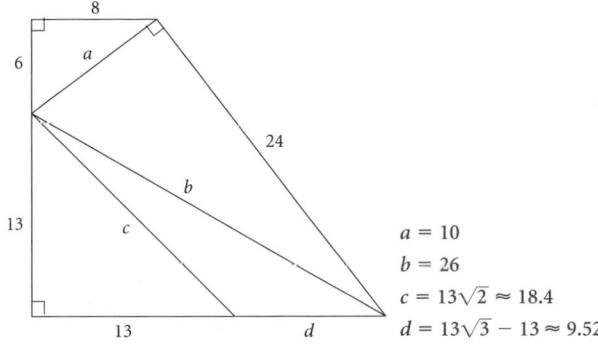

$a = 10$
$b = 26$
$c = 13\sqrt{2} \approx 18.4$
$d = 13\sqrt{3} - 13 \approx 9.52$

Two Tibetan Buddhist monks create a mandala from colored sand. The delicate geometric design will take days to create and will then be dismantled in a special ceremony.

Learn more about the mathematics and meaning of mandalas with the Internet links at **www.keymath.com/DA**.

Exercise 11 This exercise reviews Lesson 6.3.

11a. rectangle

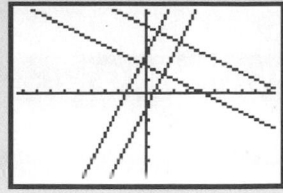

$[-9.4, 9.4, 1, 6.2, 6.2, 1]$

Exercise 12 This exercise reviews Lesson 11.4.

PLANNING

LESSON OUTLINE

One day:

10 min	Examples
20 min	Investigation
5 min	Sharing
5 min	Closing
10 min	Exercises

MATERIALS

- centimeter rulers
- protractors
- Calculator Note 11B
- Surveying (T), *optional*
- Contour Map (W), *optional*

TEACHING

To find angles when given the side lengths of a right triangle, you can work backward from finding trigonometric ratios.

One step Display the Surveying transparency. Remind students that to describe the plot of land, you need to determine the angle at each corner. Ask students to figure out ways of doing that. As you circulate, remind them as needed that if they know an angle and one edge length of a right triangle, they can find the other edge lengths from trigonometric ratios. They may find the angles using guess-and-check. Some may realize that their calculators have ways of going backward. Have students present a variety of approaches during Sharing.

▶ **EXAMPLE A**

This example reviews the use of trigonometric ratios for finding lengths given the measure of an angle.

LESSON
11.8

Trigonometry

In Lesson 11.7, you learned about trigonometric ratios in right triangles. The trigonometric functions allow you to find the ratios of side lengths when you know the measure of an acute angle. So if you know the length of one side and the measure of one acute angle, you can solve for the lengths of the other sides.

EXAMPLE A   Consider this triangle.

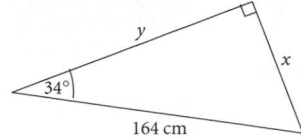

a. Find the length of the side labeled x.
b. Find the length of the side labeled y.

▶ **Solution**

a. The variable x represents the length of the side opposite the 34° angle. The length of the hypotenuse is 164 cm.

$\sin A = \dfrac{o}{h}$	Definition of sine.
$\sin 34° = \dfrac{x}{164}$	Substitute 34° for the measure of the angle and 164 for the hypotenuse.
$164 \sin 34° = x$	Multiply both sides by 164.
$91.7 \approx x$	Multiply and round to the nearest tenth.

The side labeled x is approximately 91.7 cm.

b. The variable y represents the length of the side adjacent to the 34° angle. The length of the hypotenuse remains 164 cm.

$\cos A = \dfrac{a}{h}$	Definition of cosine.
$\cos 34° = \dfrac{y}{164}$	Substitute the measure of the angle and hypotenuse.
$164 \cos 34° = y$	Multiply both sides by 164.
$136.0 \approx y$	Multiply and round to the nearest tenth.

The side labeled y is approximately 136.0 cm.

What if you know the lengths of the sides but want to know the measure of an acute angle? You can use the inverses of the trigonometric functions to find the angle measure when you know the ratio. The inverses of the trigonometric functions are inverse sine, inverse cosine, and inverse tangent. They are written $\sin^{-1}$, $\cos^{-1}$, and $\tan^{-1}$.

LESSON OBJECTIVES

- Become familiar with the concept of inverse trigonometric functions
- Learn to interpret topographic maps by using inverse tangent ratios
- Solve real-world problems using trigonometric ratios

NCTM STANDARDS

CONTENT		PROCESS	
	Number	•	Problem Solving
•	Algebra	•	Reasoning
•	Geometry	•	Communication
•	Measurement	•	Connections
	Data/Probability	•	Representation

EXAMPLE B

Find the measure of angle A.

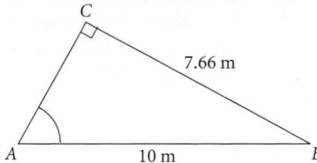

Solution

Because you know the length of the side opposite angle A and the length of the hypotenuse, you can find the sine ratio.

$$\sin A = \frac{7.66}{10} = 0.766$$

You find the measure of the angle with the inverse sine of 0.766.

[▶ 💻 See **Calculator Note 11B** to learn about the inverses of the trigonometric functions. ◀]

$$A = \sin^{-1}(0.766) \approx 50°$$

So the measure of angle A is approximately 50°. Check your answer using the sine function.

$$\sin 50° \approx 0.766$$

You use an inverse to undo a function. You may have noticed in Example B that sin and $\sin^{-1}$ undo each other the same way that squaring and finding the square root undo each other.

Investigation
Reading Topographic Maps

You will need
- a centimeter ruler
- a protractor

A **topographic map,** or contour map, reveals the shape of the land surface by showing different levels of elevation. The simple map below shows the elevation of a hill. You can use the map to determine the size of this hill and how steep it is. In this investigation you will take an imaginary hike over the summit and calculate the steepness of the hill at different points along the way.

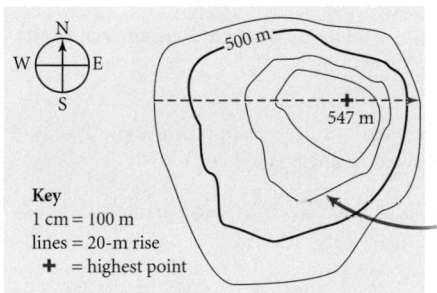

Key
1 cm = 100 m
lines = 20-m rise
✛ = highest point

Each of these rings is called a **contour line,** or isometric line. There is a 20-m rise between each contour line.

Step 1 The climb would be easier from the west. From the east the same elevation changes take place in a shorter horizontal distance so the climb is steeper. Some students may prefer a steep climb to a steep descent and choose to   **Step 1** start at the east.

If you want to hike over the summit, will it be easier to hike up from the west or the east? How can you tell?

EXAMPLE B

This example introduces the notion of the inverse sine. Point out that the −1 is not an exponent but part of the symbol for the inverse trig ratio.

You might ask whether this approach will always work. Elicit the idea that $\sin^{-1} x$ is not defined for values of x more than 1 or less than −1. Indeed, it's not a function, because many angles have the same sine. For this reason, mathematicians have defined $\text{Sin}^{-1} x$, $\text{Cos}^{-1} x$, and $\text{Tan}^{-1} x$, which have restricted domains and are always functions.

 Guiding the Investigation

Step 1 Students who are not familiar with topographic maps (often called *topo sheets*) may find this very difficult. Stress that if you walk around the hill on a contour line, your elevation doesn't change.

You may want to distribute the optional worksheet Contour Map.

Step 3

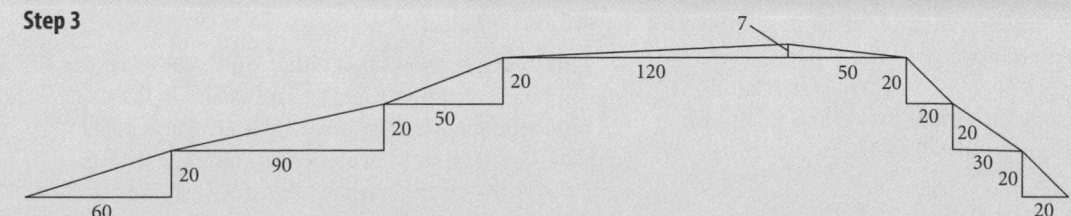

Step 2 Again, students without experience may find it difficult to determine the elevations involved. Remind them that there's a 20 m rise between contour lines. It is a positive 20 m between contour lines in sections 1 though 3 and negative 20 m in sections 6 though 8. You might need to point out that the topo sheet is a view from above. The distances are horizontal, not along the surface. Help students use the key to determine horizontal distances.

Step 5 The "angle of the climb" is the acute angle between the horizontal leg (the base) and the hypotenuse. Encourage students to discuss which inverse function they should use. It doesn't matter. Students should not forget to use parentheses around the fractions on their calculators. The answer chart, using a negative rise for downward slopes, results in a negative ratio and a negative angle.

Step 6 Help students visualize the negative angles as measured down from the horizontal.

SHARING IDEAS

If student answers or approaches vary, have them share results of Steps 5 and 6. Some students may be confused, even after a presentation by either you or other students. People's filters may keep them from seeing things that are clear to others.

You might say that in the days before calculators, people would use a table to find the angle that had a given trigonometric ratio.

[Ask] How does the overall angle of climb upward relate to the angles over the sections in which you're climbing upward. Let students discuss whether or not it's the average.

Using the Quote
Direct students' attention to the quotation that opens this last lesson of the book. Marie Curie

See page 726 for answers to Steps 2–5.

Step 2 Suppose you want to go from the west side of the hill, over the summit, and down the east side along the dotted-line trail shown on the map. You begin your hike at the edge of the hill, which has an elevation of 480 m above sea level, and travel east. By the contour lines and the peak on the map, your hike will be divided into 8 sections. Find the horizontal and vertical distance traveled for each section of your hike.

Step 3 See bottom of page 627.

Step 3 Draw a slope triangle representing each section of the hike. On graph paper, draw a right triangle with a base representing the horizontal distance and a leg representing the vertical distance. Find the slope of each hypotenuse.

Step 4 Use the Pythagorean theorem to calculate the actual distance you hiked in each section.

Step 5 Find the angle of the climb for each section of the hike. Use an inverse trigonometric function.

Step 6 Answers should be close to the answers for Step 5.

Step 6 Use a protractor to measure the angle of the climb in each of your slope triangles in Step 3. How do these answers compare to your answers in Step 5?

Step 7 If you have a calculator, trigonometry might be more convenient. If you only have a protractor, measuring the angle would be best.

Step 7 You have used two methods for finding the angle of the climb:

1. Drawing triangles and using a protractor to find angle measures.
2. Using trigonometry to calculate angle measures.

Are there times when one method of finding angle measures is more convenient than the other? Explain your thinking.

EXAMPLE C Consider this triangle.

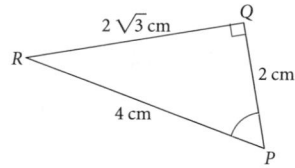

a. Name the lengths of the sides opposite angle P and adjacent to angle P.

b. Use the side lengths to find exact ratios for $\sin P$, $\cos P$, and $\tan P$.

c. Find the measure of angle P using each of the inverse trigonometric functions.

▶ **Solution**

a. The length of the side opposite angle P is $2\sqrt{3}$ cm, and the length of the side adjacent to angle P is 2 cm.

b. $\sin P = \dfrac{2\sqrt{3}}{4} = \dfrac{\sqrt{3}}{2}$ $\cos P = \dfrac{2}{4} = \dfrac{1}{2}$ $\tan P = \dfrac{2\sqrt{3}}{2} = \sqrt{3}$

c. $\sin^{-1}\left(\dfrac{\sqrt{3}}{2}\right) = 60°$ $\cos^{-1}\left(\dfrac{1}{2}\right) = 60°$ $\tan^{-1}\left(\sqrt{3}\right) = 60°$

Each of the inverse trigonometric functions gives the measure of angle P as 60°.

(1867–1934) was a French physicist best known for her work with radioactivity. Ask what students see that remains to be done. In particular, what don't they understand as well as they'd like to, and what other questions do they have?

Assessing Progress
Through your observations of group work and presentations, you can assess students' abilities to draw **slope triangles,** calculate slopes, apply the **Pythagorean theorem,** and use **trigonometric ratios.**

▶ **EXAMPLE C**

This example shows that either $\sin^{-1}$, $\cos^{-1}$, or $\tan^{-1}$ can be used to find the measure of an angle of a right triangle when all three sides are known.
[Link] Students will become very familiar with the 30°-60°-90° triangle when they study geometry.

EXERCISES

You will need your calculator for problems **2, 3, 4, 5, 6, 7, 8, 9, 10,** and **11.**

▶ Practice Your Skills

1. Use the triangle at right as a guide. For 1a–f, fill in the correct angle, side, or ratio.

   a. $c^2 - a^2 = \boxed{d}^2$

   b. $\tan \boxed{A} = \dfrac{a}{d}$

   c. $\cos \boxed{A} = \dfrac{d}{c}$

   d. $\sin^{-1}\boxed{} = D \quad \dfrac{d}{c}$

   e. $\sin D = \cos \boxed{A}$

   f. $\sin \boxed{A} = \dfrac{a}{c}$

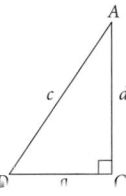

2. You will need a straightedge and a protractor for this problem.

   a. On graph paper, draw a right triangle with legs exactly 3 and 4 units long.

   b. Write trigonometric ratios and use inverse functions to find the angle measures.

   c. Measure the angles to check your answers to 2b. The angles should measure approximately 37° and 53°.

3. Use a trigonometric ratio to find the length of side x in the triangle at right.
   $x \approx 44.6$ m

4. Sketch a right triangle to illustrate the ratio

   $$\tan 25° = \frac{6.8}{b}$$

   Then find the length of side b.

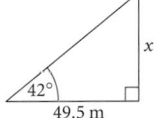

▶ Reason and Apply

5. Find the measure of each labeled angle or side length to the nearest tenth of a degree or centimeter.

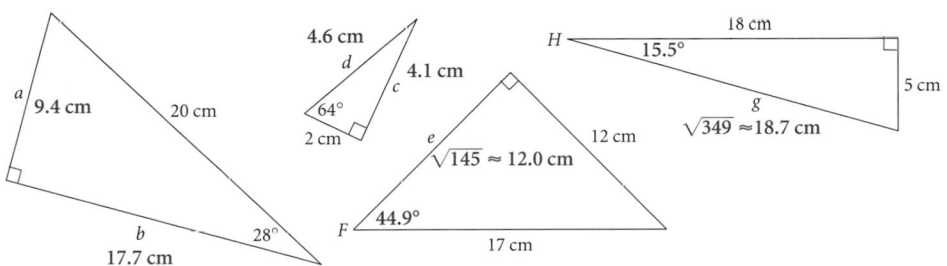

6. The legs of $\triangle PQR$ measure 8 cm and 15 cm.

   a. Find the length of the hypotenuse. 17 cm

   b. Find the area of the triangle. 60 cm²

   c. Find the measure of angle P. approximately 28°

   d. Find the measure of angle Q. approximately 62°

   e. What is the sum of the measures of angles P, Q, and R? 180°

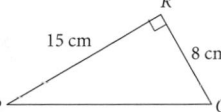

Closing the Lesson

You can work backward from finding trigonometric ratios of angles by using the $\sin^{-1}$, $\cos^{-1}$, and $\tan^{-1}$ buttons on a calculator to find angles corresponding to ratios of a right triangle's side lengths.

BUILDING UNDERSTANDING

These exercises involve trigonometric ratios and their inverses.

ASSIGNING HOMEWORK

Essential	**1–6**
Performance assessment	**6**
Portfolio	**8, 10**
Journal	**9**
Group	**7–11**
Review	**12, 13**

▶ Helping with the Exercises

Exercise 1 You may want to mention that the side marked with a lowercase letter is usually opposite the angle marked by the corresponding uppercase letter.

2a.

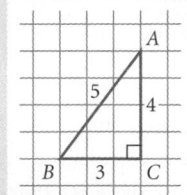

2b. Possible answer:

$$A = \sin^{-1}\left(\frac{3}{5}\right) = 37°,$$

$$B = \sin^{-1}\left(\frac{4}{5}\right) \approx 53°$$

4. $b \approx 14.6$

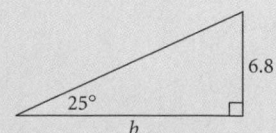

Exercise 7 Students may already know that the sum of the angles in any triangle is 180°. This fact will arise in a geometry course.

Exercise 8 If your school has stairways, your students might enjoy getting out of the classroom. Assign groups to different stairways around the school. Some groups may also want to investigate a stairway at home or in a public area such as a shopping center. You might provide metersticks to help with the estimation in 8a. To answer 8e, students must find $\tan\left(\tan^{-1}\left(\frac{7}{12}\right)\right)$. Because tan and $\tan^{-1}$ are inverse functions, the result is simply $\frac{7}{12}$.

8d. $\tan^{-1}\left(\frac{7}{12}\right) \approx 30.26°$

Exercise 9 The road grade is the slope, expressed as a percent. Students might well be surprised that roads that feel steep while driving or walking may have grades of no more than 10° or 12°.

Some students may give the answer of 150 to 9b, taking 15% of 1000. This approach ignores the fact that the 1000 feet of driving surface is along the slant, the hypotenuse of a triangle, whereas the 15% grade refers to the ratio of the legs. Two steps are needed. First find the angle using $\tan^{-1}$ and then find the vertical leg using sine. Some students might find it interesting that the two steps can be combined with a single calculation:

$x = 1000 \sin\left(\tan^{-1}\left(\frac{15}{100}\right)\right)$

≈ 148.34

---

7. APPLICATION The **angle of elevation** is the angle between the horizontal and the line of sight. The angle of elevation of the roof of this building is 31°. Use a trigonometric ratio to find the height of the building. **approximately 81.1 m**

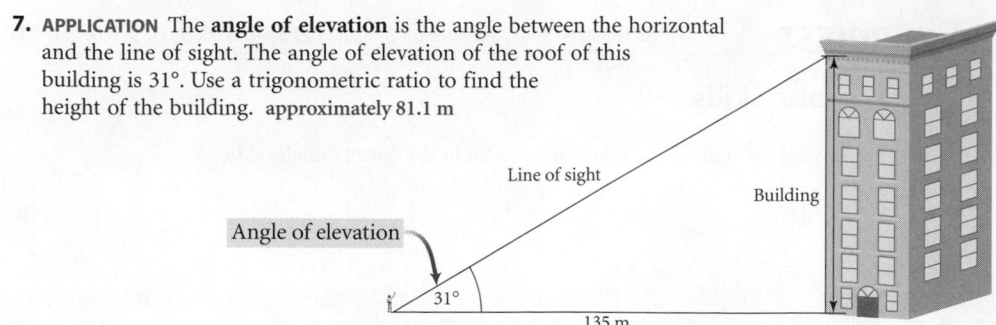

Line of sight

Building

Angle of elevation

31°

135 m

8. You will need to find an actual stairway to do this problem. Use the diagram below as a guide.

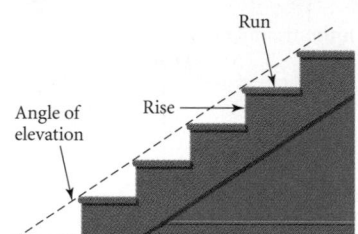

Run

Angle of elevation

Rise

Answers will vary. An average rise-to-run ratio for stairs is $\frac{7}{12}$. The answers provided are based on this ratio.

a. Estimate the angle of elevation of the stairs. **approximately 30°**

b. Measure the rise and run of several steps. **rise: 7; run: 12**

c. Find the slope of the line going up the stairs. $\frac{7}{12}$

d. Calculate the angle of elevation of the stairs.

e. Find the tangent of the angle of elevation. $0.58\overline{3}$ or $\frac{7}{12}$

9. APPLICATION The grade of a road is a percent calculated from the ratio

$$\frac{vertical\ distance\ traveled}{horizontal\ distance\ traveled}$$

The road in the sketch below has a 5% grade.

A sextant is a tool used to measure the angle of elevation of the sun or a star, thereby allowing one to determine latitude on the earth's surface. Here, Richard Byrd (1888–1957) checks his sextant before a historical 1926 flight over the North Pole.

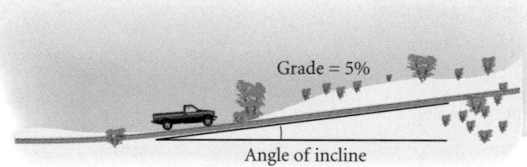

Grade = 5%

Angle of incline

a. Find the angle of incline of the road. **approximately 2.86°**

b. A very steep street has a grade of 15%. If you drive 1000 feet on this street, how much has your elevation changed? **about 148 ft**

10. Find the area of each figure to the nearest 0.1 cm².

a.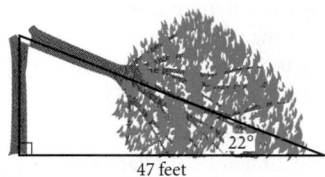

5 cm

20°

68.7 cm²

b.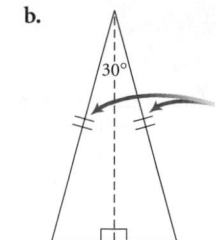

30°

These marks mean these sides are congruent, or have equal lengths.

8 cm

59.7 cm²

11. A tree is struck by lightning and breaks as shown. The tip of the tree touches the ground 47 feet from the stump and makes a 22° angle with the ground.

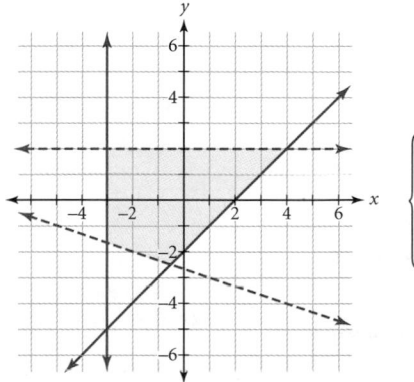

22°

47 feet

a. How high is the part of the trunk that is still standing? **approximately 19 ft**

b. How long is the portion of the tree that is bent over? **approximately 51 ft**

c. How tall was the tree originally? **approximately 70 feet**

Review

12. Write the system of inequalities whose solution is shown here.

$$\begin{cases} y < 2 \\ x \geq -3 \\ y > -\frac{8}{3} - \frac{1}{3}x \\ y \geq -2 + x \end{cases}$$

13. Give the equation for a rational function with asymptotes at $y = -3$ and $x = 2$ and a hole where $x = -7$.

Possible answer: $y = -3 + \dfrac{x + 7}{(x - 2)(x + 7)}$

Exercise 10 There are several ways to find the areas. Encourage students to share their methods. For 10b, students might find the altitude of the triangle by drawing a median from the 30° angle and then determining its length.

Exercise 12 This exercise reviews Lesson 6.7. If you did not cover systems of inequalities in Chapter 6, do not assign this exercise.

Exercise 13 This exercise reviews the rational functions of Lesson 9.6, an optional lesson.

CHAPTER 11 REVIEW

LESSON OUTLINE

One day:

5 min	Introduction
15 min	Exercises and helping individuals
15 min	Checking work and helping individuals
15 min	Student self assessment

REVIEWING

Direct students' attention to Exercise 4 in Lesson 11.1. Ask which pairs of sides, if any, are parallel or perpendicular. Have students find the lengths of all four sides and the diagonals. (Because two edges are perpendicular, you can review the Pythagorean theorem to find the lengths of the diagonals.) Ask if the diagonals cross at their midpoints. You can also ask students to find the angle measures, which can be found using inverse trigonometric ratios.

ASSIGNING HOMEWORK

If students can complete Exercises 6–9 on their own, they will have a good understanding of the chapter.

▶ Helping with the Exercises

Exercise 1 Decimal approximations may be used to check the exact answers in radical form.

2. The area is 5 square units. Here are two possible strategies:

i. Draw a square around the tilted square using the grid lines. Subtract the outer triangles from the area of the larger square: $9 - 4(1) = 5$.

You began this chapter by exploring relationships between algebra and geometry. You used **analytic geometry** to discover properties related to the slopes of **parallel** and **perpendicular lines.** You also used analytic geometry to find the **midpoint** of a segment.

Then you explored area and side relationships for squares drawn on graph paper. You learned how to draw segments whose lengths are equal to many different square roots. You also found ways to rewrite radical expressions.

You discovered the **Pythagorean theorem,** an important relationship between the lengths of the **legs** and **hypotenuse** of a **right triangle.** This relationship is useful in many professions and has been valuable to many civilizations for thousands of years. You used analytic geometry and the Pythagorean theorem to find a formula for the distance between any two points.

Finally, you reviewed ratio, proportion, and similarity. Similar right triangles introduced you to **trigonometric functions**—sine, cosine, and tangent. For an **acute angle** in a right triangle, these functions are defined by ratios between the **opposite leg, adjacent leg,** and hypotenuse.

EXERCISES

You will need your calculator for problems **6, 7, 8,** and **18.**

1. Rewrite each expression with as few square root symbols as possible.

a. $4\sqrt{5} + 4\sqrt{5}$ $8\sqrt{5}$ b. $10\sqrt{17} - 6\sqrt{17}$ $4\sqrt{17}$ c. $138\sqrt{3} + 21\sqrt{3} - 36\sqrt{3}$ $123\sqrt{3}$

d. $\sqrt{5} \cdot \sqrt{3}$ $\sqrt{15}$ e. $4\sqrt{5} \cdot 4\sqrt{5}$ 80 f. $(10\sqrt{17})^2$ 1700

g. $\sqrt{6} \cdot \sqrt{15}$ $3\sqrt{10}$ h. $4\sqrt{25} \cdot 4\sqrt{5}$ $80\sqrt{5}$ i. $\sqrt{2} + \sqrt{3}$ $\sqrt{2} + \sqrt{3}$

j. $\sqrt{2} + \sqrt{8}$ $3\sqrt{2}$ k. $\dfrac{\sqrt{18}}{\sqrt{3}}$ $\sqrt{6}$ l. $\sqrt{3} + \sqrt{27}$ $4\sqrt{3}$

2. Find the area of the tilted square. Use two different strategies to check your answers.

3. Use analytic geometry to show that the sides of the square in problem 2 are perpendicular. The slopes of the sides are $\frac{1}{2}$, -2, $\frac{1}{2}$, and -2. The product of the slopes of each pair of adjacent sides is -1, so the sides are perpendicular.

4. Explain how to draw a square with a side length of $\sqrt{29}$ units.

5. APPLICATION Is a triangle with side lengths 5 feet, 12 feet, and 13 feet a right triangle? Explain how you know. Then explain how a construction worker could use a 60-foot piece of rope to make sure that the corners of a building foundation are right angles.

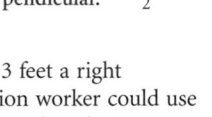

ii. Find the length of the side between $(1, 0)$ and $(3, 1)$: $\sqrt{(3-1)^2 + (1-0)^2} = \sqrt{5}$. Square the side length to find the area: $\left(\sqrt{5}\right)^2 = 5$.

4. Possible answer: Draw a 7-by-7 square on graph paper and remove triangles with areas of 5 square units (legs 2 units and 5 units) from each corner. The area of the remaining square is $49 - 4 \cdot 5 = 29$ square units.

Exercise 5 This procedure in construction is sometimes called "squaring up."

5. Possible answer: Sides of 5 ft, 12 ft, and 13 ft satisfy the Pythagorean theorem and form a right triangle. Side lengths of 10 ft, 24 ft, and 26 ft, which sum to 60 ft, also form a right triangle. Stretch 10 ft of the rope along one wall and 24 ft of the rope along the adjacent wall; the remaining 26 ft of rope should exactly fit along the hypotenuse if the foundation corners are right angles.

6. Draw this quadrilateral on graph paper.

   a. Name the coordinates of the vertices of this quadrilateral.

   b. Find the slope of each side.

   c. What kind of quadrilateral is this? Explain how you know.

   d. Find the coordinates of the midpoint of each side. Mark the midpoints on your drawing. Connect the midpoints in order.

   e. Use the distance formula to find the length of each side of the figure formed by connecting the midpoints in 6d.

   f. Find the slope of each side of the figure formed in 6d. The slopes are $-0.5, -5.5, -0.5,$ and $-5.5.$

   g. What kind of figure is formed in 6d? Explain how you know. It is a rhombus. The sides are all the same length and opposite sides have equal slope, so they are parallel.

7. Find the approximate lengths of the legs of this right triangle.

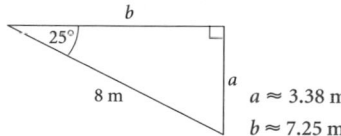

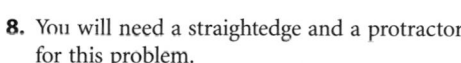

$a \approx 3.38$ m

$b \approx 7.25$ m

8. You will need a straightedge and a protractor for this problem.

   a. Carefully draw a right triangle with sides 5, 12, and 13 units on graph paper.

   b. Measure the angle opposite the 5-unit side. approximately 67°

   c. Find the measure of the angle opposite the 5-unit side using $\sin^{-1}, \cos^{-1},$ and $\tan^{-1}.$ $\sin^{-1}\left(\frac{12}{13}\right) \approx 67°, \cos^{-1}\left(\frac{5}{13}\right) \approx 67°,$ and $\tan^{-1}\left(\frac{12}{5}\right) \approx 67°$

   d. Explain how you can find the measure of the angle opposite the 12-unit side. What is the measure of this angle? Possible answer: Subtract 67° from 90°; approximately 23°.

9. A rectangular box has the dimensions shown in the diagram at right.

   a. What is the length of the diagonal $\overline{AC}$?

   b. What is the length of the diagonal $\overline{AG}$?

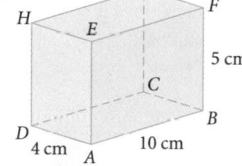

10. If the sides of a triangle are enlarged by a factor of $k,$ then its area is enlarged by a factor of $k^2.$ Check that this is true using an example. Then explain why it will be true for any triangle. Possible answer: If a triangle has base 8 cm and height 4 cm, its area is 16 cm². If the triangle is enlarged by a factor of 3, its base will be 24 cm, its height will be 12 cm, and its area will be 144 cm², which equals $3^2 \times 16$ cm³. Another possible answer: For any triangle, the area is $A = \frac{1}{2}bh.$ If the sides are enlarged by a factor of $k,$ the area is enlarged by k^2: $A = \frac{1}{2}(kb)(kh)$ or $A = \frac{1}{2}bh \cdot k^2.$

6a. $A(-4, 2), B(0, 5), C(6, -3), D(2, -6)$

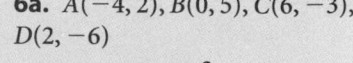

6b. slope of $\overline{AB}$: $\frac{3}{4}$;

slope of $\overline{BC}$: $\frac{-4}{3}$;

slope of $\overline{CD}$: $\frac{3}{4}$;

slope of $\overline{AD}$: $\frac{-4}{3}$

6c. It is a rectangle. The product of the slopes of adjacent sides is $-1,$ so each pair is perpendicular.

6d.

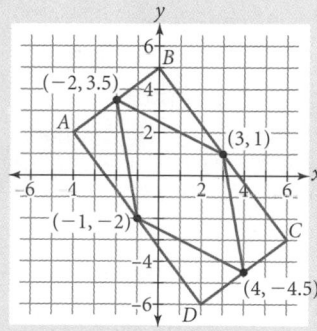

6e. Each side length is $\sqrt{31.25}$ units or approximately 5.59 units.

8a. Sample answer:

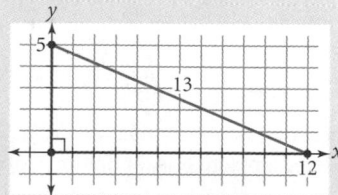

Exercise 9 This is the third diagonal exercise of this sort in the student text. You may encourage students to extend the Pythagorean theorem to three dimensions: $a^2 + b^2 + c^2 = d^2$ or $d = \sqrt{a^2 + b^2 + c^2}.$

9a. $\sqrt{116}$ cm or approximately 10.77 cm

9b. $\sqrt{141}$ cm or approximately 11.87 cm

Mixed Review

This mixed review serves as a review of the whole book. There are two exercises each from the most recent chapters (Chapters 7–10), and one exercise each from the other chapters except Chapter 0.

Exercise 11 This exercise reviews Chapter 5. These "normal" temperatures are based on records over the 30-year period 1961–1990. The *2001 World Almanac* does not specify whether normal means the mean, median, or mode. Your class might discuss which measure of center would be best.

11a. $y = 60 + 1.08(x - 40)$ or $y = 34 + 1.08(x - 16)$

Exercise 12 This exercise reviews Chapter 6.

12a. $\begin{cases} 3a + 1.5p = 13.74 \\ 2a + 3p = 16.32 \end{cases}$,

where a is the price per pound for dried apricots and p is the price per pound for dried papaya.

MIXED REVIEW

▶ **11.** This table gives the normal minimum and maximum January temperatures for 18 U.S. cities.

January Temperatures Across the United States

City	Minimum temp. (°F)	Maximum temp. (°F)	City	Minimum temp. (°F)	Maximum temp. (°F)
Mobile, AL	40	60	Helena, MT	10	30
Little Rock, AK	29	49	Atlantic City, NJ	21	40
Denver, CO	16	43	New York, NY	26	37
Jacksonville, FL	41	64	Cleveland, OH	18	32
Honolulu, HI	66	80	Pittsburgh, PA	19	34
Indianapolis, IN	17	34	Rapid City, SD	11	34
New Orleans, LA	42	61	Houston, TX	40	61
Boston, MA	22	36	Richmond, VA	26	46
Minneapolis, MN	3	21	Lander, WY	8	31

(*2001 World Almanac*, p. 243)

a. Let x represent the normal minimum temperature, and let y represent the normal maximum temperature. Use the Q-point method to find an equation for a line of fit for the data. (Round the slope to the nearest hundredth.)

b. The normal minimum January temperature for Memphis, Tennessee, is 31°F. Use your equation to predict the normal maximum January temperature. approximately 50°

c. The normal maximum January temperature for Charleston, South Carolina, is 58°F. Use your equation to predict the normal minimum January temperature. approximately 38°

12. The HealthyFood Market sells dried fruit by the pound. Jan bought 3 pounds of dried apricots and 1.5 pounds of dried papaya for $13.74. Yoshi bought 2 pounds of dried apricots and 3 pounds of dried papaya for $16.32.

a. Write a system of equations to represent this situation.

b. How much does a pound of dried apricots cost? How much does a pound of dried papaya cost?
apricots: $2.79; papaya: $3.58

13. Tell whether the relationship between x and y is a direct variation, an inverse variation, or neither, and explain how you know. If the relationship is a direct or inverse variation, write its equation.

a.

x	y
0.2	10
0.8	2.5
1	2
4	0.5
5	0.4

b.

x	y
0.3	6
0	3
1	13
3	33
10.0	103

c.

x	y
0.8	0.2
1	0.25
3	0.75
12	3
28.0	7

d.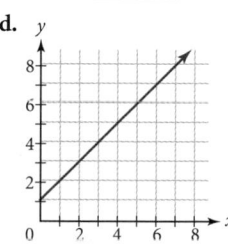

e.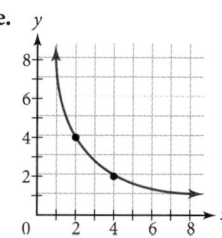

f.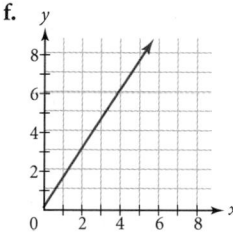

14. Here is a graph of a function f.

   a. Use words such as *linear, nonlinear, increasing,* and *decreasing* to describe the behavior of the function.

   b. What is the range of this function? $0 \le y \le 5$

   c. What is $f(3)$? 3

   d. For what x-values does $f(x) = 2$? 1, 5, 12

   e. For what x-values does $f(x) = 5$? $7 \le x \le 9$

15. Use the properties of exponents to rewrite each expression. Your answers should have only positive exponents.

   a. $(3x^2y)^3$ $27x^6y^3$ **b.** $\dfrac{5^2p^7q^3}{5p^3q}$ $5p^4q^2$ **c.** $x^{-4}y^{-2}x^5$ $\dfrac{x}{y^2}$ **d.** $m^2(n^{-4} + m^{-6})$ $\dfrac{m^2}{n^4} + \dfrac{1}{m^4}$

16. Here are the running times in minutes of 22 movies in the new-release section of a video store.

120	116	93	108	134	90	112	99	93	104	110
105	97	115	100	82	102	105	104	105	112	179

   a. Find the mean, median, and mode of the data. mean: 108.4; median: 105; mode: 105

   b. Find the five-number summary of the data and create a box plot.

   c. Create a histogram of the data. Use an appropriate bin width.

   d. Make at least three observations about the data based on your results from 16a–c.

16b. five-number summary: 82, 99, 105, 112, 179

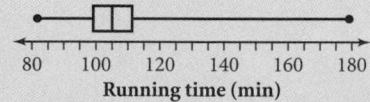

16c. Bin widths may vary.

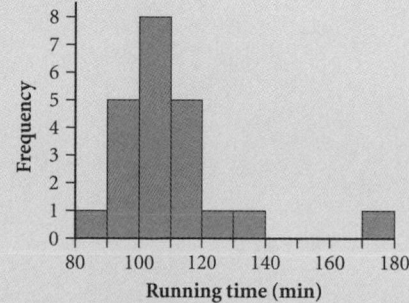

13a. Inverse variation. Possible explanation: The product of x and y is constant; $xy = 2$ or $y = \frac{2}{x}$.

13b. Neither. Possible explanation: The product is not constant, so it is not an inverse variation. The y-value for $x = 0$ is not 0, so it is not a direct variation.

13c. Direct variation. Possible explanation: The ratio of y to x is constant; $y = 0.25x$.

13d. Neither. Possible explanation: The graph is not a curve, so the relationship is not an inverse variation. The line does not pass through the origin, so it is not a direct variation.

13e. Inverse variation. Possible explanation: The product of the x- and y-coordinates for any point on the curve is 8; $xy = 8$ or $y = \frac{8}{x}$.

13f. Direct variation. Possible explanation: The graph is a straight line through the origin; $y = 1.5x$.

14a. Possible answer: For $0 < x < 3$, f is nonlinear and increasing at a slower and slower rate. For $3 < x < 5$, f is linear and decreasing. For $5 < x < 7$, f is linear and increasing. For $7 < x < 9$, f is linear and constant (neither increasing nor decreasing). For $9 < x < 12$, f is nonlinear and decreasing at a slower and slower rate.

16d. Sample answers: (1) About 75% of the new releases have running times of 112 minutes or less. (2) None of the new releases have running times between 140 and 169 minutes. (3) Most (18) of the running times are between 90 and 119 minutes.

Exercise 17 This exercise reviews Chapter 10.

Exercise 18 This exercise reviews Chapter 7.

Exercise 19 This exercise reviews Chapter 8.

Exercise 20 This exercise reviews Chapter 2.

Exercise 21 This exercise reviews Chapter 9. Students might see this as a translation right and a vertical shrink, described by $(x + 8, 0.5y)$.

17. Write the equation for this parabola in
  a. Factored form. $y = (x + 3)(x - 1)$
  b. Vertex form. $y = (x + 1)^2 - 4$
  c. General form. $y = x^2 + 2x - 3$

18. Six years ago, Maya's grandfather gave her his baseball card collection. Since then, the value of the collection has increased by 8% each year. The collection is now worth $1900.

  a. How much was the collection worth when Maya first received it? approximately $1197

  b. If the value of the collection continues to grow at the same rate, how much will it be worth 10 years from now? approximately $4102

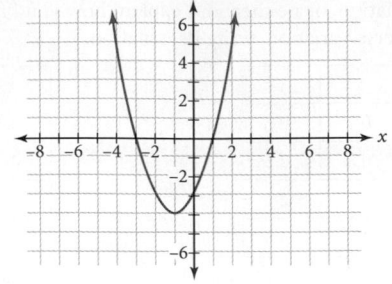

19. If $f(x) = x^2 + |x| - 4$, find
  a. $f(-5)$ 26
  b. $f(2)$ 2
  c. $f(-7) - f(4)$ 36
  d. $f(-7 - 4)$ 128
  e. $-3 \cdot f(3)$ -24

20. **APPLICATION** The Galaxy of Shoes store is having a 22nd anniversary sale. Everything in the store is reduced by 22%.

  a. Anita buys a pair of steel-toed boots originally priced at $79.99. What is the discounted price of the boots? $62.39

  b. The sales tax on Anita's boots is 5%. What total price will Anita pay for the boots? $65.51

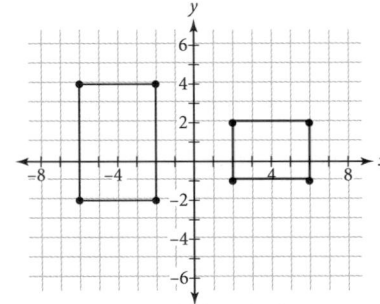

21. The image of the black rectangle after a transformation is shown in red.

  a. Describe the transformation. Possible answer: a reflection across the y-axis and a vertical shrink by a factor of 0.5

  b. Define the coordinates of any point in the image using (x, y) as the coordinates of any point in the original figure. $(-x, 0.5y)$

22. Solve for *x*.

    a. $0 = (x + 5)(x - 2)$ $x = -5$ or $x = 2$

    b. $0 = x^2 + 8x + 16$ $x = 4$

    c. $x^2 - 5x = 2x + 30$ $x = -3$ or $x = 10$

    d. $x^2 = 5x$ $\pm\sqrt{5}$

23. Give the equation for each graph.

    a.

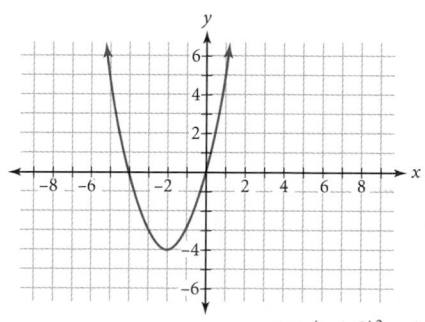

$y = (x + 2)^2 - 4$

    b.

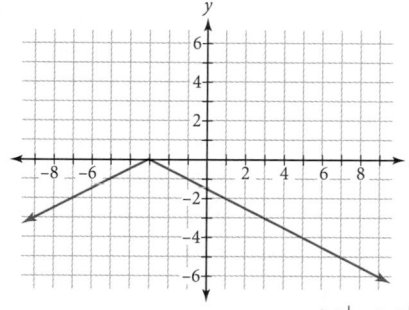

$y = -0.5|x + 3|$

24. Zoe Kovalesky visits companies and teaches the employees how to use their computers and software. She charges a fixed fee to visit a company, plus an amount for each employee in the training class. This table shows the number of employees trained and the total bill for the last five companies she visited.

Computer Training

Employees trained	Total bill ($)
5	400
11	610
17	820
3	330
25	1100

    a. How much does Zoe charge for each employee in a training class? $35

    b. What fixed fee does Zoe charge to visit a company? $225

    c. Write a recursive routine to find the total bill for any number of employees.

    d. Write an equation to calculate the total bill, *y*, for any number of employees, *x*. $y = 225 + 35x$

    e. ACME, Inc., has hired Zoe to train 12 employees. How much will the total bill be? $645

    f. Last week, Zoe taught a training class at the Widget Company. The total bill was $505. How many employees were in the class?   8

25. The vertices of $\triangle ABC$ are $A(-6, 1)$, $B(2, 7)$, and $C(10, 1)$. Do 25a–d before you graph the triangle.

    a. Find the length and slope of each side.

    b. What kind of triangle is $\triangle ABC$? Explain how you know.

    c. Find the midpoint of $\overline{AC}$ and call it *D*. $D(2, 1)$

    d. If points *B* and *D* are connected, they form $\overline{BD}$. That creates two triangles. What kind of triangles are $\triangle ABD$ and $\triangle BCD$? Explain how you know.

    e. Draw $\triangle ABC$ on graph paper and draw $\overline{BD}$. Does your drawing support your results from 25a–d?

Exercise 22 This exercise reviews Chapter 10.

Exercise 23 This exercise reviews Chapter 9.

Exercise 24 This exercise reviews Chapter 4.

24c. $\{0, 225\}$ (ENTER); $\{\text{Ans}(1) + 1, \text{Ans}(2) + 35\}$ (ENTER), (ENTER), ...

Exercise 25 This exercise reviews Chapter 11. Some students may conjecture that the triangle has a right angle, but though slopes of $\overline{AB}$ and $\overline{BC}$ are opposite, they are not reciprocals.

25a.

segment	length	slope
$\overline{AB}$	10	$\frac{3}{4}$
$\overline{BC}$	10	$-\frac{3}{4}$
$\overline{AC}$	16	0

25b. isosceles triangle; two sides have equal length

25d. Right triangles. Possible explanation: $\overline{BD}$ has an undefined slope, so it is vertical; $\overline{AC}$ has a slope of 0, so it is horizontal.

25e. A drawing should confirm 25a–d.

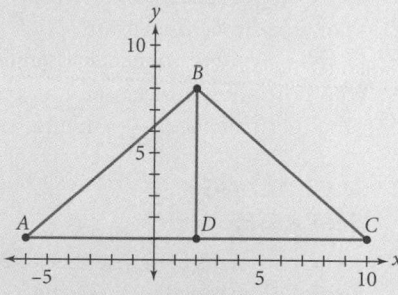

▶ Take Another Look

The Pythagorean relationship doesn't hold for sides opposite and adjacent to non-right angles in triangles, but students can find inequalities by measurement. If c is opposite an angle that's less than 90 degrees (an acute angle), then c^2 is less than $a^2 + b^2$. And if c is opposite an angle that's more than 90 degrees (an obtuse angle), then c^2 is more than $a^2 + b^2$.

You may want to mention that the difference between $a^2 + b^2$ and c^2 is actually $2ab \cos C$. This relationship is called the *Law of Cosines*. When angle C is acute, this difference is positive, so c^2 is less than $a^2 + b^2$. When C is obtuse, you're subtracting a negative number from $a^2 + b^2$ to get a larger c^2. **[Link]** Students will see why the Law of Cosines holds when they take advanced algebra.

ASSESSING

Use the form of written assessment you have found best for your class. For your end-of-year written assessment, you can use the unit exam or final exam from Assessment Resources A or B. You might choose to use some of the Constructive Assessment items for a take-home final exam. Combine your written assessment with portfolios, projects, and other performance assessments.

FACILITATING SELF ASSESSMENT

To help students complete the portfolio described in Assessing What You've Learned, suggest that they consider for evaluation their work on Lesson 11.1, Exercise 12; Lesson 11.2, Exercises 8, 10; Lesson 11.3, Exercises 5, 9; Lesson 11.4, Exercises 10, 13; Lesson 11.5, Exercises 12–14; Lesson 11.6, Exercise 7; Lesson 11.7, Exercise 7; and Lesson 11.8, Exercises 8, 10.

TAKE ANOTHER LOOK

▶ You have seen that the Pythagorean theorem, $a^2 + b^2 = c^2$, holds true for right triangles. What about triangles that don't have a right angle? Is there a relationship between the side lengths of any triangle?

If a triangle is not a right triangle, you can classify it as one of two other types of triangles, based on its angles. An **acute triangle** has three angles that are all acute. An **obtuse triangle** has one angle that is obtuse.

Use your straightedge and protractor to draw an acute triangle. Label the longest side c, and label the shorter sides a and b. Find the square of the length of each side and compare them. Does the relationship $a^2 + b^2 = c^2$ still hold true? If not, state an equation or inequality that does hold true.

Make a conjecture about the squares of the side lengths for an obtuse triangle. Then draw an obtuse triangle and measure the sides. What relationship do you find this time?

Summarize your results.

You can learn how trigonometry is used in acute and obtuse triangles with the Internet links at **www.keymath.com/DA** .

Assessing What You've Learned

 ORGANIZE YOUR NOTEBOOK Make sure your notebook contains notes and examples of analytic geometry. Your notes should give you quick reference to important concepts like the slope of parallel and perpendicular lines, the Pythagorean theorem, finding a midpoint of a segment, and finding the distance between two points. In your math class next year you may be studying more advanced concepts of geometry, so your notes can help you in the future too.

 UPDATE YOUR PORTFOLIO Geometry is the study of points, lines, angles, and shapes, so you have drawn and graphed a lot of figures for this chapter. Choose several pieces of work that illustrate what you have learned, and show how you can use algebra and geometry together. For each piece of work, make a cover sheet that gives the objective, the result, and what you might have done differently.

 PERFORMANCE ASSESSMENT Show a classmate, a family member, or your teacher that you understand how analytic geometry combines algebra and geometry. Show both the geometric and algebraic method of finding the midpoint of a segment, or the distance between two points. You may also want to show a geometric proof of the Pythagorean theorem and the algebraic formula that results. Be sure to compare and contrast the geometric method and the algebraic method. Explain which method you prefer and why.

Selected Answers

This section contains answers for the odd-numbered problems in each set of Exercises. When a problem has many possible answers, you are given only one sample solution or a hint on how to begin.

LESSON 0.1

1a. $\frac{1}{8}$; $\frac{1}{16} + \frac{1}{16}$ or $2 \times \frac{1}{16}$

1b. $\frac{3}{64}$; $\frac{1}{64} + \frac{1}{64} + \frac{1}{64}$ or $3 \times \frac{1}{64}$

1c. $\frac{3}{5}$; $\frac{1}{25} + \frac{1}{25} + \frac{1}{25} + \frac{1}{25} + \frac{1}{25} + \frac{1}{25} + \frac{1}{25} + \frac{1}{25} + \frac{1}{25} + \frac{1}{25} + \frac{1}{25} + \frac{1}{25} + \frac{1}{25} + \frac{1}{25} + \frac{1}{25}$ or $15 \times \frac{1}{25}$ or $1 - \frac{10}{25} = \frac{3}{5}$

1d. $\frac{7}{625}$; $\frac{1}{625} + \frac{1}{625} + \frac{1}{625} + \frac{1}{625} + \frac{1}{625} + \frac{1}{625} + \frac{1}{625}$ or $7 \times \frac{1}{625}$

3a.

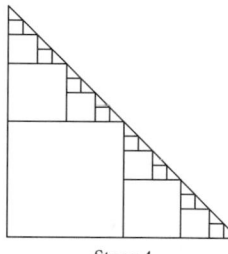

Stage 4

3b. 32   **3c.** 48   **3d.** 56

5. Sample answers:

5a.    **5b.**

5c.    **5d.**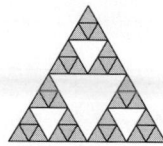

7a. Sample answer: Divide each side of the square into thirds and connect those points with lines parallel to the sides. A square is formed in the middle. Erase everything except the center square. To get the next stage, do the same thing in all eight squares formed around the middle square.

7b.

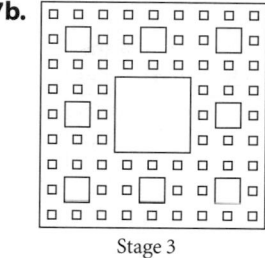

Stage 3

7c. $\frac{1}{9}$; $\frac{17}{81}$; $\frac{217}{729}$ **7d.** $\frac{8}{9}$; $\frac{64}{81}$; $\frac{512}{729}$

9a. $\frac{9}{16}$

9b. Sample answers: $12 \times \left(\frac{9}{16}\right) = 6\frac{3}{4}$; $(12 \div 16) \times 9 = 6.75$

11a. $\frac{1}{4} \times \frac{1}{4} \times 32 = \frac{32}{16} = 2$

Sample answer:

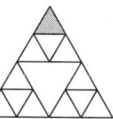

11b. $\frac{3}{4} \times \frac{1}{4} \times \frac{1}{4} \times 32 = \frac{96}{64} = \frac{3}{2} = 1\frac{1}{2}$

Sample answer:

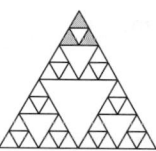

11c. $\frac{1}{2} \times \frac{1}{2} \times \frac{1}{4} \times 32 = \frac{32}{16} = 2$

Sample answer:

13. $1 - \frac{11}{32} = \frac{21}{32}$

Selected Answers

1a. 5^4 **1b.** 7^5 **1c.** 3^7 **1d.** 2^3

3a. 3^3 **3b.** 2^5 **3c.** 5^4 or 25^2 **3d.** 7^3

5a. 4 or 2^2; 8 or 2^3 **5b.** 2^5

7a. 64 **7b.** 512 **7c.** 8^2; 8^3 **7d.** 262,144 or 8^6

7e. The exponent is always one less than the stage number.

7f. Yes, because $8^0 = 1$.

9b.

Stage number	Area of one shaded triangle	Total area of the shaded triangles
0	1	1
1	$\frac{1}{4}$	$\frac{3}{4}$
2	$\frac{1}{16}$	$9 \cdot \frac{1}{16}$ or $\frac{3}{4} \cdot \frac{3}{4} = \frac{9}{16}$
3	$\frac{1}{64}$	$\frac{3}{4} \cdot \frac{9}{16} = \frac{27}{64}$

9c. The area of one shaded triangle is $\frac{1}{4}$ the area of one of the previous shaded triangles. The total area of the shaded triangles in each figure is $\frac{3}{4}$ the shaded area in the previous figure.

11. Hint: Think of a situation in which $\frac{3}{4}$ of something is divided into 5 pieces.

1a. $\frac{125}{8}$; 15.63 **1b.** $\frac{25}{9}$; 2.78

1c. $\frac{2401}{81}$; 29.64 **1d.** $\frac{729}{64}$; 11.39

3. $\left(\frac{5}{3}\right)^4 \approx 7.72$; $\left(\frac{5}{3}\right)^5 \approx 12.86$; Stage 5

5a. See below.

5b. Stage 5: $\left(\frac{5}{4}\right)^5 \approx 3.05$; Stage 11: $\left(\frac{5}{4}\right)^{11} \approx 11.64$

7a. See below.

7b. Stage 5

7c. No. Stage 6 has a length of slightly more than 161, and Stage 7 has a length of over 376.

5a. (*Lesson 0.3*)

Stage number	Total length		
	Multiplication form	Exponent form	Decimal form
2	$5 \cdot 5 \cdot \frac{1}{4} \cdot \frac{1}{4} = \frac{25}{16}$	$5^2 \cdot \left(\frac{1}{4}\right)^2 = \left(\frac{5}{4}\right)^2$	1.56
3	$5 \cdot 5 \cdot 5 \cdot \frac{1}{4} \cdot \frac{1}{4} \cdot \frac{1}{4} = \frac{125}{64}$	$5^3 \cdot \left(\frac{1}{4}\right)^3 = \left(\frac{5}{4}\right)^3$	1.95

7a. (*Lesson 0.3*)

Stage number	Total length		
	Expanded form	Exponent form	Decimal form
2	$7 \cdot 7 \cdot \frac{1}{3} \cdot \frac{1}{3} = \frac{49}{9}$	$7^2 \cdot \left(\frac{1}{3}\right)^2 = \left(\frac{7}{3}\right)^2$	5.44
3	$7 \cdot 7 \cdot 7 \cdot \frac{1}{3} \cdot \frac{1}{3} \cdot \frac{1}{3} = \frac{343}{27}$	$7^3 \cdot \left(\frac{1}{3}\right)^3 = \left(\frac{7}{3}\right)^3$	12.70
4	$7 \cdot 7 \cdot 7 \cdot 7 \cdot \frac{1}{3} \cdot \frac{1}{3} \cdot \frac{1}{3} \cdot \frac{1}{3} = \frac{2401}{81}$	$7^4 \cdot \left(\frac{1}{3}\right)^4 = \left(\frac{7}{3}\right)^4$	29.64

Selected Answers

9. 2.8

11a. 4   **11b.** 16   **11c.** 64   **11d.** $4^1, 4^2, 4^3$

11e. The exponent is one less than the stage number. For Stage 1, $4^0 = 1$

LESSON 0.4

1a. 3   **1b.** -3 **1c.** -7 **1d.** -3

3a. -8 **3b.** -25 **3c.** -17 **3d.** -11

5a. In the first recursion, he should get $-0.2 \cdot 2 = -0.4$, not $+0.4$. His arithmetic when evaluating $0.4 - 4$ was correct. In the second recursion, he used the wrong value (-3.6 instead of -4.4) because of his previous error. His arithmetic was also incorrect because $-0.2 \cdot -3.6 = +0.72$, not -0.72. His arithmetic when evaluating $-0.72 - 4$ was correct.

5b. $-4.4, -3.12, -3.376, -3.3248$

5c. $-3.8, -3.24, -3.352$

5d. Yes. The calculations seem to be approaching a value close to -3.3.

7a.

Starting value	2	-1	10
First recursion	-3	3	-19
Second recursion	7	-5	39
Third recursion	-13	11	-77

7b. No, the values get farther and farther apart.

9a. i. 7.5 or $\frac{15}{2}$

9a. ii. -10

9a. iii. 6.25

9b. When the coefficient is 0.2, the attractor value is 1.25 times the constant. In general the attractor value is $\dfrac{\text{constant term}}{1 - \text{coefficient of the box}}$

9c. Sample answer: $0.2 \cdot \square + 1.8$.

11. -22.5

LESSON 0.5

1a. 8.0 cm   **1b.** 4.3 cm   **1c.** 7.2 cm

3a–c.

3d. points D and B; no

5d. The resulting figure should resemble a right-angle Sierpiński triangle.

7a. This game fills the entire square.

7b. This game creates a small Sierpiński triangle at each corner of the triangle.

7c. This game creates four small Sierpiński carpets, one at each corner of the square.

7d. This game creates an overlapping pattern like the Sierpiński triangle.

9. Point should divide segment into an 8-cm and a 4-cm segment.

11. See below.

11. (*Lesson 0.5*)

Stage number	Total length		
	Multiplication form	**Exponent form**	**Decimal form**
1	$6 \cdot \dfrac{1}{4}$	$6^1 \cdot \left(\dfrac{1}{4}\right)^1$	1.5
2	$6 \cdot 6 \dfrac{1}{4} \cdot \dfrac{1}{4}$	$6^2 \cdot \left(\dfrac{1}{4}\right)^2 = \left(\dfrac{6}{4}\right)^2 = \dfrac{9}{4}$	2.25
3	$6 \cdot 6 \cdot 6 \cdot \left(\dfrac{1}{4}\right) \cdot \left(\dfrac{1}{4}\right) \cdot \left(\dfrac{1}{4}\right)$	$6^3 \cdot \left(\dfrac{1}{4}\right)^3 = \left(\dfrac{6}{4}\right)^3 = \dfrac{27}{8}$	3.38
4	$6 \cdot 6 \cdot 6 \cdot 6 \cdot \left(\dfrac{1}{4}\right) \cdot \left(\dfrac{1}{4}\right) \cdot \left(\dfrac{1}{4}\right) \cdot \left(\dfrac{1}{4}\right)$	$6^4 \cdot \left(\dfrac{1}{4}\right)^4 = \left(\dfrac{6}{4}\right)^4 = \dfrac{81}{16}$	5.06

1a. 1 **1b.** 3 **1c.** 9 **1d.** 27

1e. 81 **1f.** 6561 **1g.** 531,441 **1h.** 1

1i. $\frac{1}{3}$ **1j.** $\frac{1}{9}$ **1k.** $\frac{1}{27}$

1l. $\frac{1}{81}$ **1m.** $\frac{1}{6561}$ **1n.** $\frac{1}{531,441}$

3a. 72 **3b.** 290 **3c.** -10

3d. 312 **3e.** $2.1\bar{6}$ **3f.** -34

5a. $\frac{1}{16} + \frac{1}{16} + \frac{1}{16} = \frac{3}{16}$ **5b.** $\frac{1}{16} + \frac{1}{16} + \frac{1}{4} = \frac{3}{8}$

5c. $\frac{1}{9} + \frac{1}{9} + \frac{1}{81} + \frac{1}{81} = \frac{20}{81}$

5d. $\frac{1}{4} + \frac{1}{16} + \frac{1}{64} + \frac{1}{64} = \frac{11}{32}$

7a. See below.

7b. $\left(\frac{7}{5}\right)^{20} \approx 836.68$

CHAPTER 1 · CHAPTER **1** CHAPTER 1 · CHAPTER

LESSON 1.1

1. maximum: 93 bpm; minimum: 64 bpm; range: 93 bpm $-$ 64 bpm $=$ 29 bpm

3a. Uranus

3b. Mercury and Venus have no satellites.

3c. 9 **3d.** nine times

5a. 80 bpm **5b.** 29 bpm

5c. She counted her pulse rates for one full minute.

5d. Any whole number could occur, not just multiples of four.

5e. A full minute, sometimes longer, to ensure accuracy.

7. 3 in Classical

9a. 10^4 **9b.** $2^3 \cdot 5^6$ **9c.** $\frac{3^6}{8^3}$

11a. 18;

Doubles of 225	**450**	900	1800	**3600**	7200
Doubles of 1	**2**	4	8	**16**	32

11b. $9\frac{1}{2}$;

Doubles of 6	6	12	24	**48**	96
Doubles of 1	1	2	4	**8**	16

Half of 6	3
Half of 1	$\frac{1}{2}$

LESSON 1.2

1a. mean and median are 6; mode 5

1b. mean 5.1; median 5; modes 3, 8

1c. mean 10.25; median 9; no mode

1d. mean 17.5; median 20; mode 20

3.

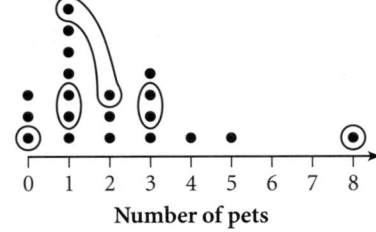

Number of pets

3a. 20 **3b.** 8 **3c.** 1

5a. $18.24. If you multiply the mean by the number of items, you get the sum of the items.

7. 45.5 seconds

9a. Multiply the mean by 10; together they weigh approximately 15,274 pounds.

9b. Five of the fish caught weigh 1449 lbs or less, and five weigh 1449 lbs or more.

9c. 2664 pounds.

7a. (*Chapter 0 Review*)

Stage number	Total length		
	Multiplication form	Exponent form	Decimal form
0	1	1	1
1	$7\left(\frac{1}{5}\right)$	$7^1\left(\frac{1}{5}\right)^1 = \left(\frac{7}{5}\right)^1 = \frac{7}{5}$	1.4
2	$7 \cdot 7 \cdot \left(\frac{1}{5}\right) \cdot \left(\frac{1}{5}\right)$	$7^2\left(\frac{1}{5}\right)^2 = \left(\frac{7}{5}\right)^2 = \frac{49}{25}$	1.96

Selected Answers

11a. See below.

11b. mean: 33.95; median: 30; mode: 29

11c. Either the mode or the median is probably best. The mean is distorted by one extremely high value.

13a. 12 cm   **13b.** 8 cm   **13c.** 2.4 cm

LESSON 1.2

1a. 5, 10, 23, 37, 50   **1b.** 10, 22, 31.5, 37, 50

1c. 14, 22.5, 26, 41, 47   **1d.** 5, 10, 19, 34.5, 47

3. 1, 4, 6, 7, 9

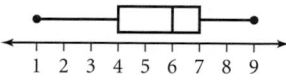

5a. Quartiles are the boundaries dividing a data set into four groups, or quarters, with approximately the same number of values.

5b. The range.

5c. The interquartile range, or IQR.

5d. Outliers are at or near the minimun and maximum values, which are the endpoints of the whiskers.

7a. 151, 426, 644, 1020, 1305

7b. The mean for the Bulls is about 675 points, and the mean for the Raptors is about 715 points; the medians are 416 points and 644 points, respectively. The Chicago mean is much higher than its median because of Michael Jordan's total points. On the average, individual Toronto players scored more points than individual Chicago players.

7c. The median probably best represents the total-points-scored data for the Bulls. Students can justify choosing either the mean or the median for the Raptors. As a team owner, you might think the mean better reflects your team's talents.

7d. The lengths of the boxes are about the same, and the medians seem to divide the boxes into regions that look about the same length for the two teams.

7e. See the graph below. Without Jordan, the range of the data is much smaller, and the length of the box is a little shorter. If Jordan's points scored are eliminated, the Raptors have the higher-scoring players. There is more variation in the number of points scored by individual Raptors than for individual Bulls.

9a. 35 feet

9b. More information is needed. The length is between 11 and 17.5 feet.

9c. 65 mi/hr

9d. The ten longest snakes vary in length from about 8 ft to 35 ft. About half of the snakes range in length from about 11 ft to 25 ft. Running speeds of the ten fastest mammals range from 42 mi/hr to 65 mi/hr. About half of the speeds are between 43 mi/hr and 50 mi/hr. The cheetah appears to run much faster than most other mammals.

9e. No, the units of these data sets are different.

9f. about 47 mi/hr

11. Hint: For the median age to be 14, the middle age must be 14. For the mean age to be 22, the sum of the ages must be 5 · 22, or 110.

LESSON 1.4

1a. matinee: 29; evening: 30

1b. matinee   **1c.** None

1d. The number is less than or equal to 4.

3a. 51

3b. Approximately $\frac{1}{4}$ of the countries had a female life expectancy between approximately 81 and 83 years.

3c. 2

3d. There are no bins to the right of 85 in the histogram. Also, the maximum point in the box plot is located at approximately 83 years.

5a. minimum: 6.0 cm; maximum: 8.5 cm; range: 2.5 cm

11a. (*Lesson 1.2*)   **Highest-Paid Athletes**

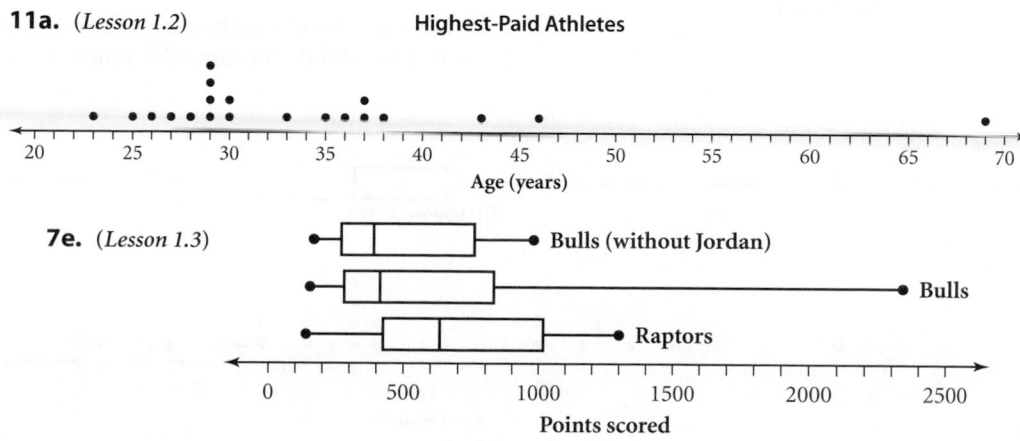

7e. (*Lesson 1.3*)

5b. Ring Finger Length

```
6 │ 0  5  5
7 │ 0  0  0  5
8 │ 5
```

Key
```
6 │ 0  means 6.0 cm
```

7a. Hint: When you roll a die, there are 6 possible outcomes, and each outcome has an equal chance of being rolled.

7b. Hint: Estimates are likely to be clumped around the actual value.

7c. Hint: What is the range of ages of people in your school? Approximately how many students are in each grade?

7d. Fairly flat bins on the left getting taller as you go to the right, with the last two bins (8–9 and 9–10) both being 25 units tall.

9a. Hint: There is a definite clustering near the center. Assign a grade for this "average" performance and then work your way out.

9b. The outlier should get an A.

9c. Hint: Try to describe the thinking process you used while answering 9a.

11. See below.

LESSON 1.6

1.

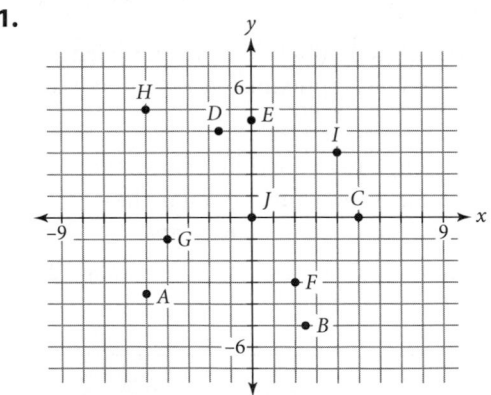

3. Your calculator will tell you the correct answers.

5a. $A(-7, -4)$, $B(0, -2)$, $C(3, 6)$, $D(0, 3)$, $E(4, -4)$, $F(-2, -6)$, $G(4, 0)$, $H(7, 1)$, $I(-5, 2)$, and $J(-5, 4)$

5b.

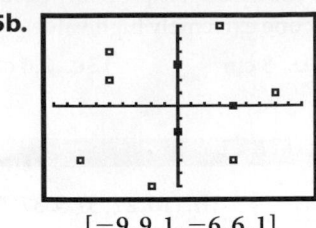

$[-9, 9, 1, -6, 6, 1]$

5c. Points B, D, and G

5d. I: C and H II: I and J III: A and F IV: E

7a. Approximate answers (the second coordinates are in millions): $(1984, 280)$, $(1985, 320)$, $(1986, 340)$, $(1987, 415)$, $(1988, 450)$, $(1989, 445)$, $(1990, 440)$, $(1991, 360)$, $(1992, 370)$, $(1993, 340)$, $(1994, 345)$, $(1995, 273)$, $(1996, 225)$, $(1997, 173)$, $(1998, 159)$, $(1999, 142)$.

7b.

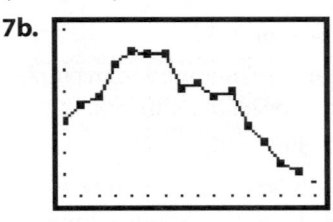

$[1984, 1999, 1, 100, 500, 50]$

7c. Hint: During what years did shipments increase? During what years did shipments decrease?

9a. $(-1.5, 2.6)$, $(-3, 0)$, $(-1.5, -2.6)$, $(1.5, -2.6)$

11a. Hint: With an odd number of values, the middle value must be the median, 15. The minimum and maximum must be 5 and 47, respectively. Now find two values between the minimum and median that give a first quartile of 12; and find two values between the median and maximum that give a third quartile of 30.

11b. Hint: With 10 data values, the median will divide the data set into two groups of 5 data values. The middle value of the lower half must be the first quartile, 12; and the middle value of the upper half must be the third quartile, 30.

11c. Hint: Try adding two values to your data set from 11b, but maintain the same five-number summary.

---

11. (*Lesson 1.4*)

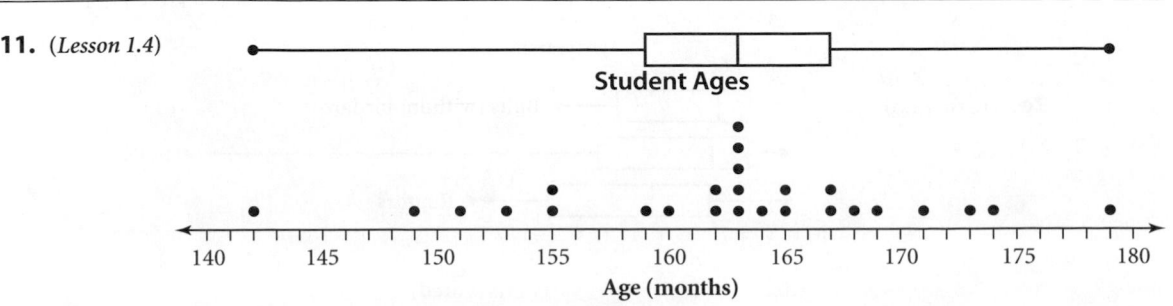

1a. Actual number of dinosaurs; estimated number of dinosaurs.

1b.
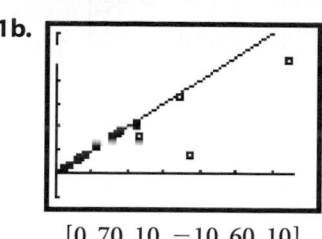
$$[0, 70, 10, -10, 60, 10]$$

1c. $y = x$

1d. No. In no case is the estimated number of dinosaurs more than the actual number.

1e. Yes. The estimated count of five different species was less than the actual count.

3. Overestimates are *B, C, D, E,* and *G.* Underestimates are *A, F, H,* and *I.*

5a. The more the rubber band is stretched, the farther it flies.

5b. Hint: Find 15 on the *x*-axis and move up into the graph. What *y*-coordinate would give you a point "in line" with the other points?

5c. Hint: Find 400 on the *y*-axis and move right into the graph. What *y*-coordinate would give you a point "in line" with the other points?

7a. It provides the differences between the estimated number of each species and the actual count. This helps identify over- and underestimates.

7b.
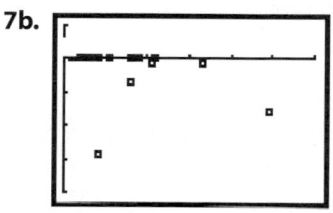
$$[0, 60, 10, -40, 10, 10]$$

7c. 5; underestimates

7d. $(8, -29)$; the 8 is the estimated number of velociraptors; the number of velociraptors was underestimated by 29.

9a. More states had students with higher verbal scores than mathematics scores.

9b. Those states in which students scored higher in mathematics had larger student populations taking the test. The national average is an average of all students, not an average of all state averages.

11a. Hint: If the median is not in the box, it must be located on one of the ends of the box.

11b. Hint: The difference between Q1 and Q3 will be zero.

11c. Hint: The minimum needs to be far removed from the other data values.

11d. Hint: If there is no right whisker, the maximum will be on the right end of the box.

1. Randall Cunningham threw 19 touchdown passes in 1992.

3. 3×4

5. $\begin{bmatrix} 788 & 489 & 35 & 19 \\ 809 & 492 & 53 & 21 \\ 919 & 590 & 61 & 19 \end{bmatrix}$ This matrix gives the totals from the two years.

7. $\begin{bmatrix} -158 & -115 & -11 & -9 \\ 41 & 26 & 15 & -1 \\ 115 & 54 & 11 & 5 \end{bmatrix}$

This matrix gives the difference between the 1998 totals and the 1992 totals.

9a. $\begin{bmatrix} 8 & -5 & 4.5 \\ -6 & 9.5 & 5 \end{bmatrix}$ **9b.** $\begin{bmatrix} -3 & 4 & -2.5 \\ 2 & -6 & -4 \end{bmatrix}$

9c. $\begin{bmatrix} 15 & -3 & 6 \\ -12 & 10.5 & 3 \end{bmatrix}$ **9d.** $\begin{bmatrix} 4 & -2.5 & 2.25 \\ -3 & 4.75 & 2.5 \end{bmatrix}$

11. Hint: Think about the first matrix representing cost per unit of two items, and the second matrix as units of each item.

13a. Hint: The minimum in the data set must be 0, and the maximum must be 7. Also, the data value 2 must occur more frequently than any other.

13b. Hint: The minimum must be 22.2, and the maximum must be 30.4. No values in the list should be repeated.

1a. Mean: 41.5; divide the sum of the numbers by 14. Median: 40; list the numbers in ascending order and find the mean of the two middle numbers. Mode: 36; find the most frequently occurring number.

1b. 27, 36, 40, 46, 58

Battery Life
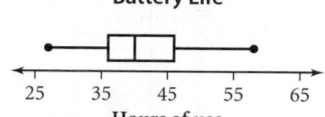
Hours of use

3a.

Mean Annual Wages, 1998

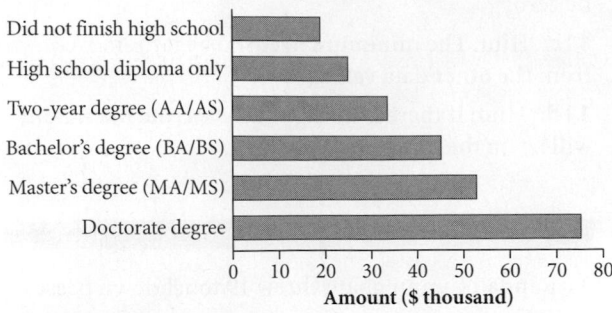

3b. Greatest jump: from a master's degree to a doctorate; smallest difference: from not finishing high school to a high school diploma.

5a. Mean: approximately 154; median: 121; there is no mode.

5b. Bin widths may vary.

Pages Read in Current Book

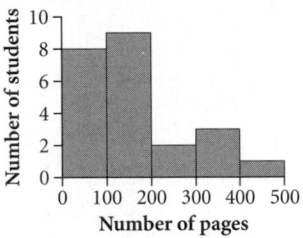

Pages Read in Current Book

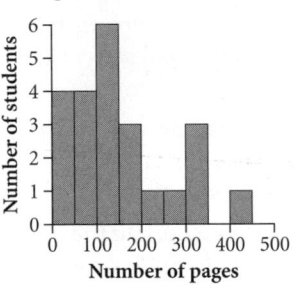

5c. **Pages Read in Current Book**

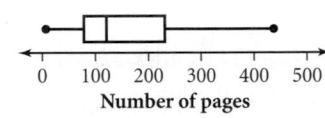

5d. Sample answer: Most of the students questioned had read fewer than 200 pages, with a fairly even distribution between 0 and 200.

7a. $[A] = \begin{bmatrix} 5.00 & 8.00 \\ 3.50 & 4.75 \\ 3.50 & 4.00 \end{bmatrix}$, $[B] = \begin{bmatrix} 0.50 & 0.75 \\ 0.50 & 0.25 \\ 0.50 & 0.25 \end{bmatrix}$

$[C] = [43 \ 81 \ 37]$

7b. $[A] + [B] = \begin{bmatrix} 5.50 & 8.75 \\ 4.00 & 5.00 \\ 4.00 & 4.25 \end{bmatrix}$

7c. $[C] \cdot ([A] + [B]) = [708.5 \ \ 938.5]$
matinee: \$708.50
evening: \$938.50

9a. 2,820,000

9b. See below.

9c. **The Ten Most Populated U.S. Cities, 1998**

0	97
1	08 11 20 22 44 79
2	82
3	60
4	
5	
6	
7	42

Key	
2	82 means 2.82 million

9d. **The Ten Most Populated U.S. Cities, 1998**

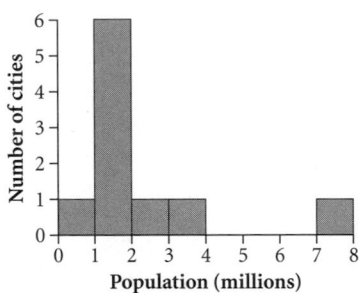

9b. (*Chapter 1 Review*) **The Ten Most Populated U.S. Cities, 1998**

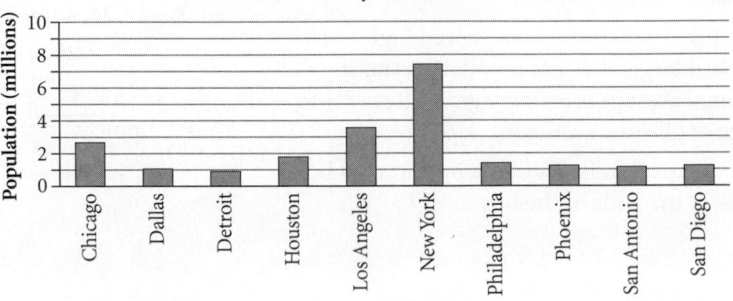

9e. The bar graph helps show how each city compares with the others, since they remain identified by name. The stem plot shows distribution but also shows actual values. The histogram shows distribution and a definite clustering between 1 and 2 million, but it does not show individual cities or populations.

LESSON 2.1

1a. 0.875 **1b.** 0.65 **1c.** 2.6 **1d.** 2.08

3a. $\dfrac{240 \text{ miles}}{1 \text{ hour}}$

3b. $\dfrac{10 \text{ parts capsaicin}}{1,000,000 \text{ parts water}}$ or $\dfrac{1 \text{ part capsaicin}}{100,000 \text{ parts water}}$

3c. $\dfrac{350 \text{ women-owned firms}}{1000 \text{ firms}}$ or

$\dfrac{7 \text{ women-owned firms}}{20 \text{ firms}}$

3d. $\dfrac{35,500 \text{ dollars}}{1 \text{ person}}$

5a. $T = 18$ **5b.** $R = 28$ **5c.** $S = 73.5$
5d. $x = 2.1$ **5e.** $M = 6$ **5f.** $n = 21$
5g. $c = 31.2$ **5h.** $W = 9$

7a. 85%

7b. $\dfrac{t}{7.38} = \dfrac{85}{100}, t = \$6.27$

9. $2\frac{2}{3}$ cups of water and $\frac{2}{3}$ cup of oatmeal; $6\frac{2}{3}$ cups water and $1\frac{2}{3}$ cups oatmeal

11.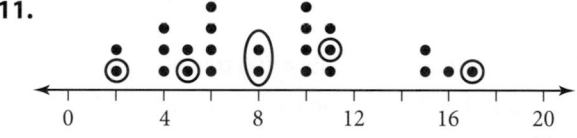

13a. -12 **13b.** -4 **13c.** -8 **13d.** $-\frac{2}{3}$

LESSON 2.2

1a. 32% of what number is 24?

1b. 48% of 450 is what number?

1c. What percent of 117 is 98? or 98 is what percent of 117?

3. 269

5a. Marie should win over half the games.

5b. $\dfrac{28 \text{ games won by Marie}}{28 + 19 \text{ total games}} = \dfrac{M}{12}; M = 7.15$ or 7 games

5c. $\dfrac{19}{47} = \dfrac{30}{G}; G \approx 74$ games

7. approximately 76 errors

9a. 1 sulfur atom, 2 hydrogen atoms, and 4 oxygen atoms

9b. 200 hydrogen atoms would combine with 100 sulfur atoms and 400 oxygen atoms.

9c. Use all 400 atoms of oxygen, 200 atoms of hydrogen, and 100 atoms of sulfur to make 100 molecules of sulfuric acid.

11. Matt; by the order of operations, you multiply before you subtract.

LESSON 2.3

1a. $x = 49.4$ **1b.** $n = 10$
1c. $x \approx 216$ **1d.** $x = 583.\overline{3}$

3a. $\dfrac{50 \text{ m}}{1 \text{ sec}} \cdot \dfrac{1 \text{ km}}{1000 \text{ m}} \cdot \dfrac{60 \text{ sec}}{1 \text{ min}} \cdot \dfrac{60 \text{ min}}{1 \text{ hr}} = 180 \text{ km/hr}$

3b. $0.025 \text{ day} \cdot \dfrac{24 \text{ hr}}{1 \text{ day}} \cdot \dfrac{60 \text{ min}}{1 \text{ hr}} \cdot \dfrac{60 \text{ sec}}{1 \text{ min}} = 2160 \text{ sec}$

3c. $1200 \text{ oz} \cdot \dfrac{1 \text{ lb}}{16 \text{ oz}} \cdot \dfrac{1 \text{ ton}}{2000 \text{ lb}} = 0.0375 \text{ ton}$

5a. 158.8 cm **5b.** 244 cm
5c. 4.72 in. **5d.** 1.28 in.

7a. Measurement in Yards and Feet

Yards	1	2	3	4	5
Feet	3	6	9	12	15

7b. For each additional yard there are 3 more feet.

7c. $\dfrac{f}{y} = \dfrac{3}{1}$

7d. i. 450 feet **7d. ii.** 128 yards

9a. Fifteen 12-oz cans to make 960 oz.

9b. $\dfrac{12}{64}$ or 0.1875 oz

9c. $\dfrac{\text{number of ounces of concentrate}}{\text{number of ounces of lemonade}} = \dfrac{12}{64}$

9d. $\dfrac{16}{L} = \dfrac{12}{64}; L \approx 85$ oz

11. If the profits are divided in proportion to the number of students in the clubs, the Math Club would get $288, leaving $192 for the Chess Club.

13. laughing kookaburra: 46 cm, green kingfisher: 22 cm, belted kingfisher: 33 cm, pygmy kingfisher: 10 cm, ringed kingfisher: 41 cm

LESSON 2.4

1a. F **1b.** T **1c.** F **1d.** T

3. 2001: 277,722,000; 2002: 280,777,000
2003: 283,865,000

5a. approximately 49% **5b.** approximately 19.2%

5c. approximately 1,398,110,000

7a. $7.76 **7b.** $7.49

7c. Her wage has dropped by $0.01 per hour because the increase was calculated as 3.5% of $7.50, but the decrease was based on $7.76.

9. Hint: Each bin contains data values in a range. For example, the first bin contains two heights between 120 and 129, so you can choose any two heights in this range for your data set.

11a.

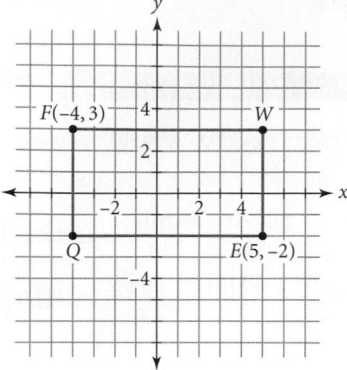

11b. Hint: W and Q do not need to be diagonally opposite to make a rectangle.

LESSON 2.5

1. Type AB = 3750; Type B = 9000; Type A = 30,000; Type O = 32,250

3. No, the total height of all the bars must be 100%.

5a.

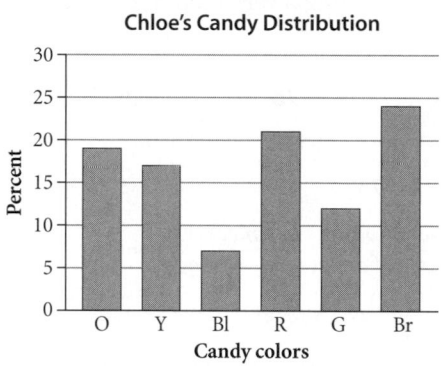

5b.

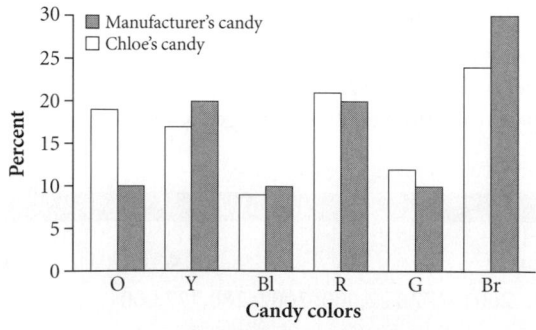

Chloe's bag of candy had the same dominant color as the graph from the manufacturer, and her least color was one of the least manufactured. But the distributions are not very close.

7a. 9th: 27%; 10th: 26%; 11th: 25%; 12th: 22%

7b. 9th: 189; 10th: 172; 11th: 170; 12th: 147; 0.3%; 2 students

7c.

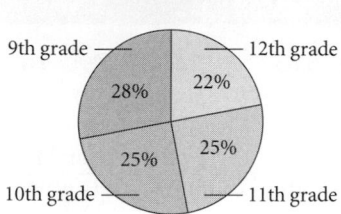

Proportionally, the ninth grade increased 1% and the tenth grade decreased 1%.

9a. 180 pulses per second

9b. $18.\overline{6}$ meters per second

LESSON 2.6

1a. heads or tails

1b. 1, 2, 3, 4, 5, or 6

1c. 2, 3, 4, 5, 6, 7, 8, 9, 10, 11, 12

1d. A, B, C, D, or E

3a. $\frac{27}{100} = 0.27$

3b. P(landing in shaded area) = 0.25 because the shaded area is $\frac{1}{4}$ of the circle.

5f. P(every teacher will give 100 free points) = 0

5g. P(earth will rotate on its axis in 24 hours) = 1

7a. Finding and counting a litter is a trial; an outcome may be having one cub (or two or three or four).

7b. No. If outcomes were equally likely, then the number of litters of each size would have been almost the same, with about nine litters of each size.

7c. $\frac{22}{35} \approx 0.63$

9a. $\frac{1}{6} \approx 17\%$

9b. $\frac{90 + 165}{360} = \frac{255}{360} \approx 71\%$

9c. $\frac{45 + 60}{360} = \frac{105}{360} \approx 29\%$

11. Sample answer: $(-4, 1), (-1, 3), (4, 3), (1, 1)$.

13. 3

1. Probability: $\frac{74}{180} \approx 0.411$; observed probability: $\frac{15}{50} = 0.30$. Sample answer: Perhaps your method of selecting students was not random. For example, your results could be biased because you talked only to students who were participating in after-school activities or only to students in a particular class.

3a. $\frac{15}{250} = 0.06$

3b. $\frac{235}{250} = 0.94$

3c. 0.06

5a. H, H, T, H, H, T

5b. Find the cumulative sum of list L1.

```
2randInt(0,1,100
)-1→L1
{1 -1 1 1 -1 -1…
cumSum(L1)
{1 0 1 2 1 0 -1…
```

5c. After many steps, you may be close to 0.

5d. With many trials you might be closer to 0.

7. Hint: See **Calculator Note 2A** to review random integers.

9. 229

11. Hint: Look back at problem 3 in Lesson 2.6 (page 125).

1a. $n = 8.75$

1b. $w = 84.6$

1c. $k = 5\frac{1}{6}$ or $5.1\overline{6}$

3. Hint: One sample answer is $\frac{5 \text{ hours}}{7 \text{ birdhouses}} = \frac{x \text{ hours}}{30 \text{ birdhouses}}$. Find two more ways to write this proportion.

5a. Hint: Pick any x-coordinate and divide it in half to get the y-coordinate.

5b. All points appear to lie on a line.

7a. 12.5 cm²; $\frac{12.5}{40} = 0.3125$

7b. 32.5 cm²; $\frac{32.5}{45} = 0.7\overline{2}$

9. Sample answer: If the person is talking about the entire state, this cannot occur. All scores cannot be greater than the middle score. The probability is 0.

11. 1365 shih rice; 169 shih millet

1a. $\frac{3 \text{ pounds}}{30 \text{ days}} = 0.1$ pound per day

1b. $\frac{5 \text{ pounds}}{45 \text{ days}} = 0.\overline{1}$ pound per day

1c. Crystal's cat

3a. 0.1 gallon per mile

3b. 22 gallons

3c. 150 miles

5a. $\frac{24901.55 \text{ miles}}{(2 \cdot 365 + 2 \cdot 30.4 + 2) \text{ days}} \approx 31.4$ miles per day

5b. $\frac{31.4 \text{ miles}}{1 \text{ day}} \cdot \frac{(1.5 \cdot 365) \text{ days}}{1} \approx 17191.5$ miles

5c. $\frac{31.4 \text{ miles}}{1 \text{ day}} = \frac{60,000 \text{ miles}}{t}$; $t \approx 1911$ days, or more than 5 years

7a. $2.49 per box, 42¢ per bar, $2.99 per box, 25¢ per ounce

7b. yes

7c. $\frac{1.495 \text{ ounces}}{1 \text{ bar}}$

7d. approximately 25¢

7e. Hint: Compare the price per box, price per bar, price per ounce, and ounces per bar for both brands.

9a. Hint: One sample answer is $\frac{20 \text{ pounds of food}}{1 \text{ week}}$. Find another ratio stated in the problem.

9b. $936

9c. $\frac{20}{85} = \frac{f}{60}$; 14 pounds of food per week

9d. $280.80

11a. $x = \frac{21}{5}$ or 4.2

11b. $x = \frac{22}{9}$ or $2.\overline{4}$

11c. $x = \frac{cd}{e}$

13a.

Jelly Beans in Small Bag

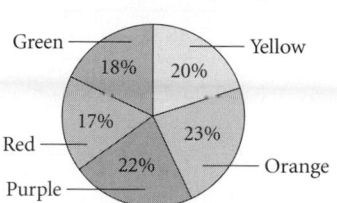

Selected Answers

13b.

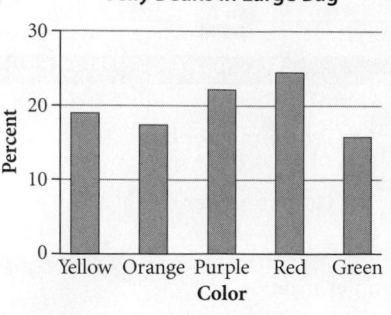

Jelly Beans in Large Bag

[Bar graph showing Percent (0–30) on vertical axis and Color on horizontal axis, with bars for Yellow, Orange, Purple, Red, Green.]

13c. In the small bag, orange occurs most frequently and red least frequently. In the large bag, red occurs most often and green least often.

13d. 85 pieces. No, because each color does not occur equally often.

LESSON 3.2

1a. 40 **1b.** 75

3.

Distance (miles)	Distance (kilometers)
2.8	4.5
7.8	12.5
650.0	1040.0
937.5	1500.0

5a. $3.13

5b. 4 yards

5c. $1.25

7a. Bernard Lavery's Vegetables

Vegetable	Weight (kilograms)	Weight (pounds)
Cabbage	56	123
Summer squash	49	108
Zucchini	29	64
Kohlrabi	28	62
Celery	21	46
Radish	13	28
Cucumber	9	20
Brussels sprout	8	18
Carrot	5	11

(*The Top 10 of Everything 1998*, p. 98)

7b. $y = 2.2x$

7c. 2.95 kilograms

7d. 7920 pounds

7e. 100 lb = 45.4$\overline{5}$ kg; 100 kg = 220 lb

9a. Sample answer: {100, 50, 20, 10, 5, 1, 0.50, 0.25, 0.10, 0.05, 0.01}

9b. Sample answer: To convert to Japanese yen, multiply the list by 108.770. {10877, 5438.5, 2175.4, 1087.7, 543.85, 108.77, 54.385, 27.19, 10.88, 5.44, 1.09}

9c. Divide list L2 by the exchange rate to obtain the original values.

9d. Use dimensional analysis.

$$\frac{2119.150 \text{ liras}}{1 \text{ dollar}} \cdot \frac{1 \text{ dollar}}{2.140 \text{ marks}} = 990.257 \text{ liras per mark.}$$

11a. 81.25 mph

11b. 40 kilometers per hour

11c. 65 mph is 104 kilometers per hour. A speed limit sign might post 100 kilometers per hour.

13a. $\frac{2}{10} = \frac{1}{5}$

13b. No. All the tables are equally likely to have their table number spun; 9 or 10 being spun affects all tables equally.

LESSON 3.3

1a. $y = \frac{3}{4} = 0.75$

1b. $y = 2$

1c. $x = 100$

1d. $x = 108$

3a. Sample answer: $l = \frac{8}{6}s$, where l is a length on the larger polygon an s is a length on the smaller polygon.

3b. $w = 9.\overline{3}$ cm; $x = 2.25$ cm; $y = 6$ cm; $z = 4$ cm

5a. 42 miles

5b. approximately 1.5 inches

5c. a little over 3 inches

5d. 52.5 miles

7. Rhombuses i and iii are similar, and rhombuses iv and v are similar. In each similar pair, corresponding sides are proportional and the angle measures are the same.

9a.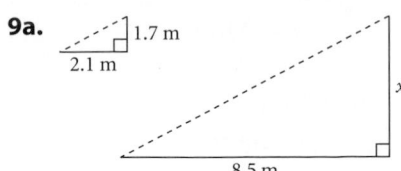

9b. $\frac{1.7}{2.1} = \frac{x}{8.5}$; $x \approx 6.88$ or 6.9 meters high

9c. You could measure the length of the tree's shadow and write a proportion using a person's height and the length of his or her shadow.

11a. Sample answer: 1 cm represents 250 km

11b. Sample answer: Using the scale of 1 cm = 250 km, an equation would be $y = \frac{1}{250}x$, where x is the actual river length in kilometers and y is the scale drawing length in centimeters. Approximate lengths are around 26.7 cm, 25.8 cm, 25.2 cm, 23.9 cm, and 22.2 cm, respectively.

13a. $\frac{30}{100} = \frac{x}{24}$; $x \approx 7$

13b. $\frac{30}{100} = \frac{p}{s}$

13c. $p = \frac{30}{100}s$ or $p = 0.3s$

13d. part = percent · total

15a. $0.31 per day

15b. $60\frac{2}{3}$ pounds of food per cat per year

15c. It costs about $113, or 8.7 fourteen-pound bags. The owner will have to buy 9 bags and spend $116.82.

17. P(mango flavored) = $\frac{4}{48}$ or $\frac{1}{12}$ or 0.083

LESSON 3.4

1a. $y = \frac{15}{x}$

1b. $y = \frac{35}{x}$

1c. $y = \frac{3}{x}$

3. Hint: Pick any value for x, put it into the equation, and evaluate to find y. You could do all 5 points at once using list calculations on your calculator.

5. This is not an inverse variation. The product of the quantities

(*time spent watching TV, time spent doing homework*)

is not a constant. Instead, the sum is a constant. This is a relationship of the form $x + y = k$ or $y = k - x$, not an inverse variation $xy = k$ or $y = \frac{k}{x}$.

7a. $62.\overline{3}$ N, 93.5 N, and 187 N.

7b. As you move closer to the hinge, it takes more force to open the door. Moving from 15 cm to 10 cm needs an increase of about 31.2 N. Moving 5 cm closer requires an increase of 93.5 N. As you move closer, the force needed increases more rapidly. When you get very close to the hinge, the force needed becomes extremely large.

7c. On the graph, the curve goes up very steeply near the y-axis.

9a. If the balance point is at the center, then the weight of an unknown object will be exactly the same as the weight that balances it on the other side. If the balance point is off-center, you must know the two distances and do some calculation.

9b. $15 \cdot M = 20 \cdot 7$; $M \approx 9.3$ kg

11a. On this graph, x represents frequency and y represents tube length.

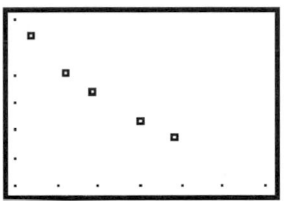

[400, 1000, 100, 30, 90, 10]

11b. Sample answer: $y = \frac{37,227.1}{x}$ where 37,227.1 is the mean of the products of the frequencies and tube lengths.

11c. $y = \frac{37,227.1}{x}$; $y \approx 42.3$ cm

13. $s = 0.85p$. The sale price is $11.86.

CHAPTER 3 REVIEW

1a. approximately 13 miles per gallon or 0.077 gallon per mile

1b. 65 miles   **1c.** 7.7 gallons

3a. If x represents the weight in kilograms and y represents weight in pounds, one equation is $y = 2.2x$ where 2.2 is the data set's mean ratio of pounds to kilograms.

3b. approximately 13.6 kilograms

3c. 55 pounds

5a. approximately 2.2 inches

5b. 15.75 miles

7a. Because the product of the x- and y-values is approximately constant, it is an inverse relationship.

7b. Sample answer: $y = \dfrac{45.5}{x}$, where 45.5 is the mean of the products.

7c. $y = \dfrac{45.5}{32}, y \approx 1.4$

9a. 4 feet

9b. Yes, Robbie can balance by sitting 2.9 feet from the center.

11. No, they won't fit. 210 centimeters is 6.89 feet.

13. **Algebra Grades**

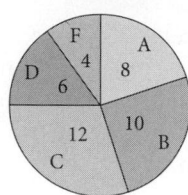

The size and the angle of each piece of the circle graph vary directly with the percent of students who earned each grade.

15a. approximately 1100 thousand visitors

15b. 354, 465, 740, 1272, 3494

15c.

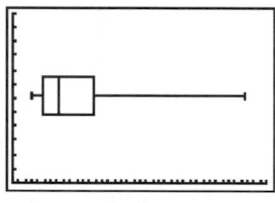

[0, 4000, 100, 1, 12, 1]

15d. Yosemite; the number of visitors exceeds 1272 by more than 1.5 (1272 − 465).

17a. 9 amperes

17b. 6 ohms

CHAPTER 4 · CHAPTER **4** CHAPTER 4 · CHAPTER

LESSON 4.1

1a. 3

1b. −10

1c. $\dfrac{1}{4}$

1d. −6

3a. $2L + 2W$

3b. $2(15 + 4) = 38; 2 \cdot 15 + 2 \cdot 4 = 38$

5a. $84

5b. You can multiply 7 by each day's hours and then add those products to find the week's total. Or you can find the total hours for the week first and multiply this result by $7 \cdot 4 + 7 \cdot 3 + 7 \cdot 5 = 7(4 + 3 + 5)$.

7a. $5.40   **7b.** $3.15

7c. $7.28

9a. First multiply 16 by 4.5, then add 9.

9b. First divide 18 by 3, then add 15.

9c. First square 6, then add −5, then multiply by 4 and subtract 124 from 3.

11a. Sample answer: $(3 + 2)(5) − 7 = 18$

11b. Sample answer: $8 − 5(6 − 7) = 13$

13. $\dfrac{(6 + 3) \cdot 4^2}{8 + 2} − 9 = 5.4$

15a. **Distance Traveled**

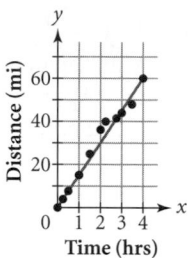

15b. 15 mph

15c. $y = 15x$

15d. Downhill during the intervals of 1 to 2.25 hours away from home and 3.5 to 4 hours (the bike's velocity is above average) and uphill during the interval of 2.25 to 3.5 hours away from home (the bike's speed is slower than average).

17. The 20% discount is $2.59. The 8% tax on 10.36 is $0.83. The tip amount on $12.95 is $1.94, so the total is $13.13.

LESSON 4.2

1a. −12   **1b.** 32

1c. −24   **1d.** 35

1e. 13   **1f.** 3

1g. −19   **1h.** −6

1i. −4

3a. 15   **3b.** 18

3c. 17   **3d.** 5

5a. Juwan forgot the parentheses.

5b. $(2 + 3)4 − 5$ or $4(2 + 3) − 5$.

7. See below.

9a.

Description	Jack's sequence	Nina's sequence
Pick the starting number.	5	3
Multiply by 2.	10	6
Multiply by 3.	30	10
Add 6.	36	24
Divide by 3.	12	8
Subtract your original number.	7	5
Subtract your original number again.	2	2

9b.

Description	Jack's sequence	Nina's sequence
Pick the starting number.	−10	10
Add 2.	−8	12
Multiply by 3.	−24	36
Add 9.	−15	45
Subtract 15.	−30	30
Multiply by 2.	−60	60
Divide by 6 (you should have your original number).	−10	10

11a–c. Hint: Look back at Example A (page 191). The number trick in that example always results in −18.

13. Hint: Look back at Example B (page 192). The number trick in that example always results in the starting number.

15a. value: 25; order of operations: add 7, multiply by 5, divide by 3.

15b.

Equation: $\dfrac{5(x+7)}{3} = -18$ Work backward

Operation on x	Undo operation	$x = -17.8$
+ (7)	− (7)	−10.8
· (5)	÷ (5)	−54
÷ (3)	· (3)	−18

17a. 202.3 ft/minute   **17b.** 102.8 cm/sec

19a. $1\dfrac{11}{12}$ cups   **19b.** $13.05

LESSON 4.3

1a. 15   **1b.** −16   **1c.** −5

3a. 5 (ENTER) Ans + 3 (ENTER) , (ENTER) , . . .

3b.

Figure #	Perimeter
1	5
2	8
3	11
4	14
5	17

3c. 32   **3d.** Figure 15

5a. Start with 3, then apply the rule Ans + 6; 10th term = 57.

5b. Start with 1.7, then apply the rule Ans − 0.5; 10th term = −2.8.

5c. Start with −3, then apply the rule Ans · −2; 10th term = 1536.

5d. Start with 384, then apply the rule Ans/2 or Ans · 0.5; 10th term = 0.75.

---

7. (*Lesson 4.2*)

Description	Daxun's sequence	Lacy's sequence	Claudia's sequence	Al's sequence
Pick the starting number.	14	−5	−8.6	x
Add 5.	19	0	−3.6	$x + 5$
Multiply by 4.	76	0	−14.4	$4(x + 5)$
Subtract 12.	64	−12	−26.4	$4(x + 5) - 12$
Divide by 4.	16	−3	−6.6	$\dfrac{4(x + 5) - 12}{4}$
Subtract the original number.	2	2	2	$\dfrac{4(x + 5) - 12}{4} - x$

7a. Sample answer: The smallest square has an area of 1. The next larger white square has an area of 4, which is 3 more than the smallest square. The next larger gray square has an area of 9, which is 5 more than the 4-unit white square.

7b. 1 (ENTER) Ans + 2 (ENTER) , (ENTER) , . . .

7c. 17, the value of the ninth term in the sequence

7d. 39

7e. The 48th term is 95. Students might press (ENTER) 48 times or compute 2(48) − 1.

9a. 4 meters

9b. Press 101 and then Ans − 4. The 19th term represents the height of the 7th floor. The height is 29 meters.

9c. 26 terms

9d. 19 meters

11a. $17 \cdot 7 = 119$ **11b.** 14

11c. Hint: Be sure to try other intervals of 100, such as 300 to 400 and 400 to 500.

11d. Hint: Try to describe a method you can calculate with pencil and paper, and another method using a recursive routine on your calculator.

13a. 1 (ENTER) , Ans · 3 (ENTER) , (ENTER) . . .
The 9th term is 6561.

13b. 5 (ENTER) , Ans · (−1) (ENTER) , (ENTER) . . .
The 123rd term is 5.

13c. −16.2 (ENTER) , Ans + 1.4 (ENTER) , (ENTER) . . .
The 13th term is the first positive term.

13d. −1 (ENTER) , Ans · (−2) (ENTER) , (ENTER) . . .
The 8th term is the first to be greater than 100.

15a. $297.25

15b. $7.25 (4 \cdot 8 + 6 \cdot 1.5) = 7.25(41) = 297.25$

LESSON 4.4

1a. negative; −1517   **1b.** positive; 472

1c. positive; $12.\overline{3}$ **1d.** positive; 326

1e. negative; $-3.\overline{3}$ **1f.** negative; −1464

3.

x	y
0	4
1	3
2	2
3	1
4	0

x	y
0	−1.5
1	0
2	1.5
3	3

5a. $-9.\overline{3}$

5b. Equation:
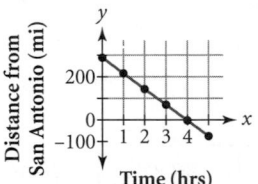

$$\frac{4 - 5(x + 3)}{6} = 12$$

Work backward $x = -16.6$

Operation on x	Undo operation	
+ (3)	− (3)	← −13.6
· (−5)	÷ (−5)	← 68
+ (4)	− (4)	← 72
÷ (6)	· (6)	← 12

7a. {0, 272} (ENTER) {Ans(1) + 1, Ans(2) − 68}
(ENTER) , (ENTER) , (ENTER) , (ENTER) , (ENTER)

7b.
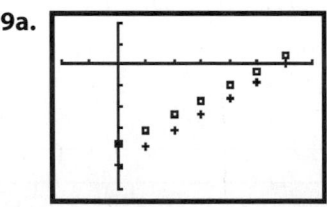

7c. The starting value is the point (0, 272) on the graph.

7d. On the graph, you move right 1 unit and down 68 units to get from one point to the next. In the recursive routine, you add 1 to the first number and subtract 68 from the second number.

7e. This is a linear graph relating a distance to any time between 0 and 5 hours. The line represents the distances at all possible times; points only represent distances at certain times.

7f. The car is within 100 miles of San Antonio after 2.53 hours have elapsed. Explanations will vary. Graphically, it is the time after which the line crosses the horizontal line $y = 100$.

7g. The car takes 4 hours to reach San Antonio. Answers will vary. The answer is the fourth entry in the table. Graphically, it is where the line crosses the x-axis.

9a.

[−10, 35, 5, −60, 20, 10]

9b. The points for each submarine appear to lie on a line. The USS *Dallas* surfaces at a faster rate.

9c. Yes, each line means that any time in this range corresponds to depth below the surface.

9d. The submarine's nose rises slightly above the water when surfacing.

11a. Hint: Since the bicyclist pedals at two different rates, the graph will be made of two distinct sections. How will these sections compare? Where will one section end and the other start?

11b.

Bicyclist

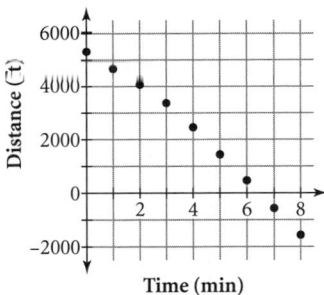

11c. Sample answer: Where on the graph does the bicyclist pass you? The answer is on the x-axis between 6 and 7 minutes.

13a. Subtract 32 from the temperature in Fahrenheit, multiply the difference by 5, and then divide by 9.

13b. $F = \dfrac{9C}{5} + 32$; $C = \dfrac{5(F - 32)}{9}$

15a.

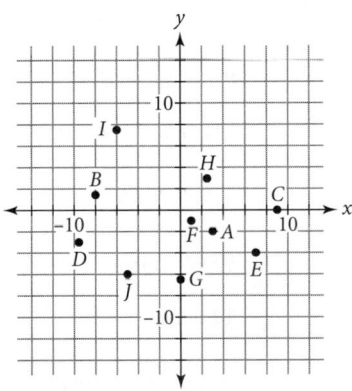

15b. Quad I: H; Quad II: B, I; Quad III: D, J; Quad IV: A, E, F; x-axis: C; y-axis: G.

15c. Hint: One sample answer is, "If both the x-coordinate and the y-coordinate are positive, the point will be in Quadrant I." Modify this statement to describe criteria for the other quadrants and the axes.

LESSON 4.6

1a. ii   **1b.** iv   **1c.** iii   **1d.** i

3a. $d \approx 38.3$ ft   **3b.** $d \approx 25.42$ ft

3c. The walker started 4.7 feet away from the motion sensor.

3d. The walker was walking at a rate of 2.8 feet per second.

5a. $-2\,\text{L}_1 + 10$ **5b.** $4\text{L}_1 - 4\text{L}_2$

7a. Sample answer: Jo has an initial start-up cost of $300 for equipment and expenses. She makes $15 for every lawn she mows, N.

7b. Sample answers: How many lawns will Jo have to mow to break even? [Solve $-300 + 15N = 0$. Jo must mow 20 lawns.] How much profit will Jo earn if she mows 40 lawns? [Substitute 40 for N. $300.]

7c. Subtracting 300 from $15N$ is the same as adding $15N$ to -300.

7d. The input variable is N for number of lawns, and the output variable is P for profit.

9a. $y = 45 + 0.12x$, where x represents dollar amounts customers spend and y represents Manny's daily income in dollars.

9b.

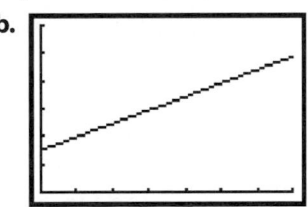

[0, 840, 120, 0, 180, 30]

9c. $45 + 0.12 \cdot 312 = \$82.44$

9d. between $500 and $625

11a. $\dfrac{8}{n} = \dfrac{15}{100}$, $n \approx 53.3$ **11b.** $\dfrac{15}{100} = \dfrac{n}{18.95}$, $n = 2.8$

11c. $\dfrac{p}{100} = \dfrac{326}{64}$, $p \approx 509.4$ **11d.** $\dfrac{10}{100} = \dfrac{40}{n}$, $n = 400$

13a. ii   **13b.** iv   **13c.** iii   **13d.** i

15a. The expression equals -4.

Ans $-$ 8	-3
Ans $\cdot$ 4	-12
Ans/3	-4

15b. $y = 14$

LESSON 4.7

1a.

Input	Output
x	y
20	100
-30	-25
16	90
15	87.5
-12.5	18.75

1b.

L₁	L₂
x	y
0	-5.2
-8	74.8
24	-245.2
-35	344.8
-5.2	46.8

3a. The rate is negative, so the line goes from the upper left to the lower right.

3b. The rate is not negative or positive. It is zero. The line is a horizontal line.

3c. The rate is positive, so the line goes from the lower left to the upper right.

3d. The rate for the speedier walker will be greater than the rate for the person walking more slowly, so the graph for the speedier walker will be steeper than the graph for the slower walker.

5a. i. 3.5; ii. 8; iii. -1.4

5b. i. -6; ii. 1; iii. 23; the y-intercept

5c. i. $y = -6 + 3.5x$; ii. $y = 1 + 8x$;
iii. $y = 23 - 1.4x$

7a. $35 + 0.8(25) = 55$ miles

7b. 50 min

9a. 990 square units

9b. Sample answer: $33x = 990$; $x = \dfrac{990}{33}$

9c. 30 units

11a.

-15	-15
Ans + 52	37
Ans/1.6	23.125
$52 + 1.6(23.125) = -15$	Check

11b.

52	52
Ans $- 7$	45
Ans/-3	-15
$7 - 3(-15) = 52$	Check

13a. 70.4 lengths   **13b.** 2.2 feet per second

13c. 43.7 lengths

13d. 40 lengths for a kilometer, 64 lengths for a mile

LESSON 4.8

1a. $2x = 6$ **1b.** $x + 2 = 5$

1c. $2x - 1 = 3$ **1d.** $2 = 2x - 3$

3a.

$0.1x + 12 = 2.2$	Original equation.
$0.1x + 12 - 12 = 2.2 - 12$	Subtract 12 from both sides.
$0.1x = -9.8$	Remove the 0 and subtract.
$x = -98$	Divide both sides by 0.1.

3b.

$\dfrac{12 + 3.12x}{3} = -100$	Original equation.
$12 + 3.12x = -300$	Multiply both sides by 3.
$-12 + 12 + 3.12x = -12 - 300$	Subtract 12 from both sides.
$3.12x = -312$	Remove the 0.
$x = -100$	Divide both sides by 3.12.

5a. $x = \dfrac{1}{12}$ **5b.** $x = 36$

7a.

$4 - 1.2x = 12.4$	Original equation.
$4 - 4 - 1.2x = 12.4 - 4$	Subtract 4 from both sides.
$-1.2x = 8.4$	Subtract.
$\dfrac{-1.2x}{-1.2} = \dfrac{8.4}{-1.2}$	Divide both sides by -1.2.
$x = -7$	Reduce.

7b.

Start with 12.4.	12.4
Ans $- 4$	8.4
Ans/-1.2	-7

7c.

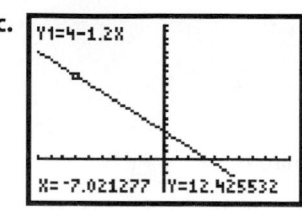

$[-10, 10, 1, -5, 20, 1]$

7d.

X	Y1	
-3	7.6	
-4	8.8	
-5	10	
-6	11.2	
-7	12.4	
-8	13.6	
-9	14.8	
X=-7		

9a. $r = \dfrac{C}{2\pi}$ **9b.** $h = \dfrac{2A}{b}$

9c. $l = \dfrac{P}{2} - w$ **9d.** $s = \dfrac{P}{4}$

9e. $t = \dfrac{d}{r}$ **9f.** $h = \dfrac{2A}{a + b}$

11a. $-\dfrac{1}{5}$ **11b.** -17

11c. 2.3   **11d.** x

13. $120

15. Your equation is correct when the graph of your line exactly matches the program's line.

CHAPTER 4 REVIEW

1a. $x = -7$ **1b.** $x = -23.4$

3a. iii   **3b.** i

3c. ii

5a. $y = x$ **5b.** $y = -3 + x$

5c. $y = -4.3 + 2.3x$ **5d.** $y = 1$

7a. 0 represents no bookcases sold; -850 represents fixed overhead, such as startup costs; Ans(1) represents the previously calculated number of bookcases sold; Ans(1) + 1 represents the current number of bookcases sold, one more than the previous; Ans(2) represents the profit for the previous number of bookcases; Ans(2) + 70 represents the profit for the current number of bookcases—the company makes $70 more profit for each additional bookcase.

7b.

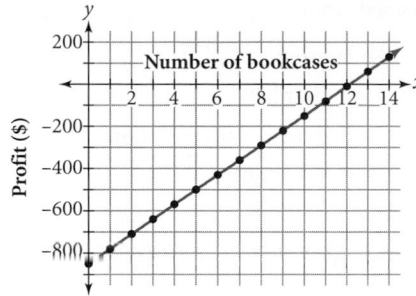

7c. Sample answer: The graph crosses the *x*-axis at approximately 12.1 and is positive after that. It shows you need to make at least 13 bookcases to make a profit.

7d. −850, the profit if the company makes 0 bookcases, is the *y*-intercept; 70, the amount of additional profit for each additional bookcase, is the rate of change.

7e. No, partial bookcases cannot be sold.

9a. 4 seconds

9b. Away; the distance is increasing.

9c. approximately 0.5 meters

9d. $\dfrac{2.9 - 0.5}{4} = 0.6$ meter per second

9e. $\dfrac{5.5 \text{ meters}}{0.6 \text{ meters/second}} = 9.1\overline{6}$ seconds. Approximately 9 seconds.

9f. The graph is a straight line.

11a. $L_2 = -5.7 + 2.3 \cdot L_1$

11b. $L_2 = -5 - 8 \cdot L_1$

11c. $L_2 = 12 + 0.5 \cdot L_1$

13a. Hint: You could use lists, tables, graphs, guessing and checking, working backward, or the balancing method.

13b. $2(3.5 - 6) = 2(-2.5) = -5$

15a. $t \approx -1.2°$ **15b.** $t \approx 38.1°$

15c. $t \approx 38.1°$ **15d.** $t \approx 27.4°$

CHAPTER 5 · CHAPTER **5** CHAPTER 5 · CHAPTER

LESSON 5.1

1a. 2   **1b.** $\dfrac{2}{3}$ **1c.** $-\dfrac{4}{3}$

3. Sample answers:

3a. $(1, 7), (-1, 1)$ **3b.** $(3, 3), (1, 13)$

3c. $(12, 3), (4, 9)$ **3d.** $(6, 7.2), (4, 6.8)$

5a. i. The *x*-values don't change, so the slope is undefined.

5a. ii. The *y*-values decrease as the *x*-values increase, so the slope is negative.

5a. iii. The *y*-values don't change, so the slope is 0.

5a. iv. The *y*-values increase as the *x*-values increase, so the slope is positive.

5b. i. Using the points $(4, 0)$ and $(4, 3)$, the slope is $\dfrac{3 - 0}{4 - 4} = \dfrac{3}{0}$. Since you can't divide by 0, the slope is undefined.

5b. ii. Using the points $(1, 3)$ and $(4, -3)$, the slope is $\dfrac{-3 - 3}{4 - 1} = \dfrac{-6}{3} = -2$.

5b. iii. Using the points $(-4, -5)$ and $(-3, -5)$, the slope is $\dfrac{-5 - (-5)}{-3 - (-4)} = \dfrac{-5 + 5}{-3 + 4} = \dfrac{0}{1} = 0$.

5b. iv. Using the points $(0, -2)$ and $(4, 1)$, the slope is $\dfrac{1 - (-2)}{4 - 0} = \dfrac{3}{4}$.

5c. i. $x = 4$; **ii.** $y = 5 - 2x$; **iii.** $y = -5$; **iv.** $y = -2 + \dfrac{3}{4}x$

7a. Use the slope to move backward from $(4, 16.75)$; $(4 - 1, 16.75 - 2.95) = (3, 13.80)$, or \$13.80 for 3 hours; $(3 - 1, 13.80 - 2.95) = (2, 10.85)$, or \$10.85 for 2 hours.

7b. Continuing the process in 7a leads to $(0, 4.95)$, or \$4.95 for 0 hours. This is the flat monthly rate for Hector's Internet service.

7c. $y = 4.95 + 2.95x$, where *x* is time in hours and *y* is total fee in dollars.

7d. Substitute 28 for *x* and solve for *y*: $y = 4.95 + 2.95(28) = 87.55.$ \$87.55 for 28 hours.

9. $y = e - \dfrac{a}{c}x$

11a. i. Line 2 is a better choice.

11a. ii. Either line 3 or line 4 is a reasonable fit.

11b. Hint: Remember that not all data is linear.

13. Hint: One sample answer is $\{3, 3, 6, 16, 22\}$. Try to find another.

15a. 85%   **15b.** 150%   **15c.** 6.5%   **15d.** 107%

LESSON 5.2

1a. No. Although this line goes through four points, too many points are below the line.

1b. No. Although the slope of the line shows the general direction of the data, too many points are below the line.

1c. Yes. About the same number of points are above the line as below the line, and the slope of the line shows the general direction of the data.

1d. No. Although the same number of points are above the line as below the line, the slope of the line doesn't show the direction of the data.

3a. $y = -2 + \dfrac{2}{3}x$

3b. $y = 2 - \frac{2}{3}x$

3c. $y = -2 - \frac{2}{5}x$

3d. $y = 3$

5a. The number of representatives depends on the population.

5b. Let x represent population in millions, and let y represent the number of representatives.

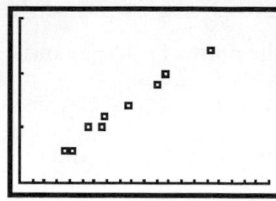

$[0, 10, 0.5, 0, 15, 5]$

5c. Sample answer: $y = 1.6x$. The slope represents the number of representatives per 1 million people. The y-intercept means that a state with no population would have no representatives.

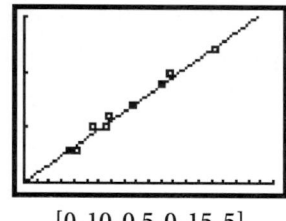

$[0, 10, 0.5, 0, 15, 5]$

5d. The equation $y = 1.6x$ gives $y = 1.6(33) = 52.8$ or 53 representatives. On July 19, 2000, California had 52 representatives.

5e. The equation $y = 1.6x$ gives $8 = 1.6x$;
$x = \frac{8}{1.6} = 5$; 5 million.
The estimated population of Minnesota in July 1999 was 4.8 million.

5f. The relationship should be a direct variation because it should go through the point $(0, 0)$. A state with no population would have no representatives.

7a. $\frac{y_2 - y_1}{x_2 - x_1} = \frac{4.4 - 3.4}{4.5 - 2}$; the slope is
0.4 meter per second.

7b. The y-intercept is 2.6 meters; students can find this by working backward with the slope or by estimating from a graph.

7c. $y = 2.6 + 0.4x$

9a. Sample answer: $y = -8 + 4x$

9b. Sample answer: $y = -2x$

9c. $y = 6 + x$ **9d.** $y = 10$

11a. neither

11b. inverse variation; $y = \frac{100}{x}$

11c. direct variation; $y = -2.5x$

11d. direct variation; $y = \frac{1}{13}x$

13a. mean: $24.8\overline{6}$; median: 21

13b. mean: 44.5; median: 40

13c. mean: approximately 140.1; median: 145

13d. mean: 85.75; median: 86.5

LESSON 5.3

1a. $4; (5, 3)$ **1b.** $2; (-3.1, 1.9)$

1c. $-3.47; (7, -2)$ **1d.** $-1.38; (2.5, 5)$

3a. 2

3b. $y = -1 + 2(x + 2)$

3c. $y = 13 + 2(x - 5)$

3d. The graphs coincide and the tables are identical.

5. There are no "correct" answers for this game.

7a. AD: $y = 2 + 0.2(x + 1)$ or $y = 3 + 0.2(x - 4)$
BC: $y = -2 + 0.2(x + 3)$ or $y = -1 + 0.2(x - 2)$
AB: $y = 2 + 2(x + 1)$ or $y = -2 + 2(x + 3)$
DC: $y = 3 + 2(x - 4)$ or $y = -1 + 2(x - 2)$

7b. The slopes are the same; the coordinates of the points are different.

7c. $ABCD$ appears to be a parallelogram because each pair of opposite sides is parallel. The equal slopes in 7b mean that the opposite sides are parallel.

9a. The data appears linear.

$[10, 50, 5, 250, 800, 50]$

9b. Sample answer: Using the points $(21, 360)$ and $(43, 620)$, an equation is $y = 620 + 11.82(x - 43)$.

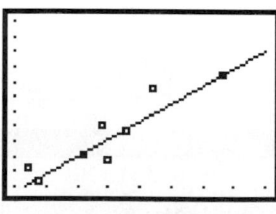

$[10, 50, 5, 250, 800, 50]$

9c. In the graph of $y = 620 + 11.82(x - 43)$ approximately 490 calories.

9d. Compared to the graph of $y = 620 + 11.82(x - 43)$, the point lies above the line. A point above the line means the sandwich has more calories than the model predicts.

9e. Using $y = 620 + 11.82(x - 43)$, four points are above the line, two are on the line, and two are below the line

9f. Hint: Remember that a good model usually shows the general direction of the data and has about the same number of points above the line as below.

9g. Using $y = 620 + 11.82(x - 43)$, approximately 112 calories; this makes sense because not all calories in food come from fat.

11a. 4.125 liters　　　　**11b.** 180 K

13a. $-1; (5, -1)$

13b. undefined; $(2, 3)$

13c. $-\frac{5}{2}; (0, -3)$

LESSON 5.4

1a. not equivalent; $-3x - 9$

1b. equivalent　　　　**1c.** equivalent

1d. not equivalent; $-2(x + 4)$ or $2(-x - 4)$

3a. $x = 4$; division property

3b. $-x = 92$; addition property
$x = -92$; multiplication property

3c. $x = -7$; subtraction property

3d. $x = 112$; multiplication property

5a. $(-5, 25)$　　　　**5b.** $x = 0$

7a. $y = 5(2 + x)$　　　　**7b.** $y = 5(x + 2)$

7c. The y_1 value is missing, which means it is zero; $y = 0 + 5(x + 2)$.

7d. $(-2, 0)$; this is the x-intercept.

9a. $x = 2$; the point $(2, 0)$ is the x-intercept.

9b. $y = 3$; the point $(0, 3)$ is the y-intercept.

9c.

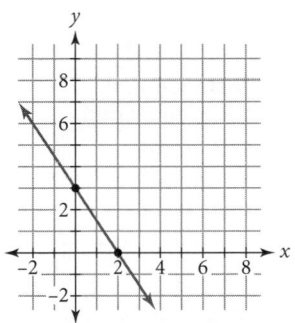

9d. The slope is $-\frac{3}{2}; y = 3 - \frac{3}{2}x$.

9e.

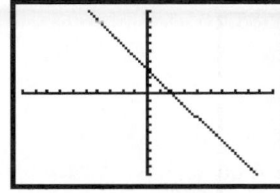

$$[-10, 10, 1, -10, 10, 1]$$

The two lines are the same; hence the equations are equivalent.

9f. $3x + 2y = 6$　　Original equation.
$2y = 6 - 3x$　　Subtraction property with $3x$.
$y = 3 - \frac{3}{2}x$　　Division property with 2.

11a. $y = 15.20 + 0.85(x - 20)$

11b. $19.45

11c. The equation is used to model the bill only when Dorine is logged on for more than 15 hours. Substituting 15 for x gives the flat rate of $10.95 for all amounts of time less than 15.

11d. 30 hours

13a. Hint: It seems logical that compensation should increase over time because of inflation. Are there any countries for which compensation behaved contrary to this assumption?

13b. Germany is the country with the greatest increase ($25.91), and Mexico is the country with the least increase ($0.04).

13c. Top box is 1975, middle 1985, lowest 1995.

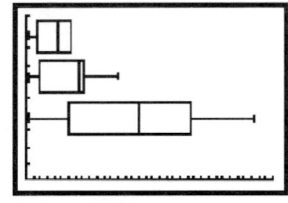

$$[0, 35, 5, 0, 10, 1]$$

Notice that there is a much larger range in data for 1995 than in 1975. The lowest compensation has not changed much, whereas the top end has moved considerably. The median has moved upward as well.

15. $z = \frac{3.8 + 5.4}{0.2} - 6.2; z = 39.8$

LESSON 5.5

1a. $y = 1 + 2(x - 1)$ or $y = 5 + 2(x - 3)$

1b. $y = 3 + \frac{2}{3}(x - 1)$ or $y = 5 + \frac{2}{3}(x - 4)$

1c. $y = 6 - \frac{4}{3}(x - 1)$ or $y = 2 - \frac{4}{3}(x - 4)$

3. The x-intercept of $y = b(x - x_1)$ is at $x = x_1$.

3a. 3　　　　**3b.** -4　　　　**3c.** 6

5a.

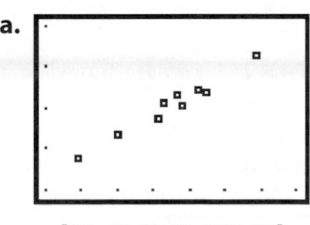

$$[10, 45, 5, 40, 120, 10]$$

5b. Sample answer: Using the points $(20, 67)$ and $(31.2, 88.6)$, an equation is $y = 67 + 1.9(x - 20)$.

5c. Using $y = 67 + 1.9(x - 20)$:

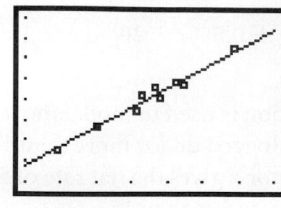

$[10, 45, 5, 40, 120, 10]$

5d. $y = 32 + 1.8(x - 0)$ or $y = 212 + 1.8(x - 100)$

5e. The sample equation in 5b gives $y = 29 + 1.9x$; the equations in 5d both give $y = 32 + 1.8x$.

5f. The difference could be a result of measurement error or faulty procedures.

7a. $y = -11 + 2x$

7b. $y = 17 + 41x$

7c. $y = 59 - 6x$

7d. $y = 9 - 4x$

9a. -3

9b. $y = 106 - 3(x - 10)$

9c. After 45 full days, there will be only one biscuit left, so it will be empty in the middle of the 46th day.

9d. When the box was new, before Anchor had any biscuits, there were 136 biscuits.

LESSON 5.6

1a. 166, 405, 623, 1052, 1483

1b. 204, 514, 756, 1194, 1991

1c.

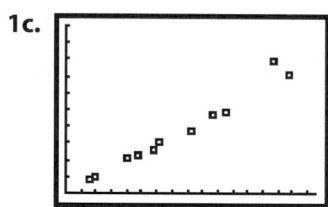

$[0, 1650, 100, 0, 2500, 250]$

1d. The slope will be positive because as the flying distance increases so does the driving distance.

1e. $(405, 514), (1052, 1194)$

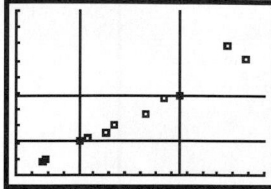

$[0, 1650, 100, 0, 2500, 250]$

1f.

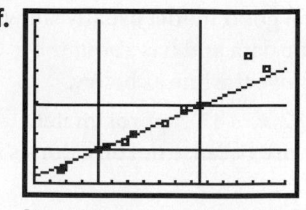

$[0, 1650, 100, 0, 2500, 250]$

The slope is approximately 1.05;
$y = 1194 + 1.05(x - 1052)$ or $y = 514 + 1.05(x - 405)$.

1g. approximately 1054 miles

1h. approximately 535 or 536 miles

3a. Sample answer: The slope will be positive and you will choose the lower-left and upper-right corners of the rectangle for the Q-points.

3b. Sample answer: The slope will be negative and you will choose the upper-left and lower-right corners of the rectangle.

5. Hint: Look back at the graph on page 291. Based on the scattering of data points, why was the Q-point $(6, 15)$ not one of the data points?

7a. $y = 1.3 + 0.625(x - 4)$ or $y = 6.3 + 0.625(x - 12)$

7b. The elevator is rising at a rate of 0.625 second per floor.

7c. 36.3 seconds after 2:00, or approximately 2:00:36

7d. almost at the 74th floor

9a. Hint: Since the elevators are traveling in opposite directions at equal rates, a good method of estimation is to find the average of the elevators' starting positions.

9b. At 28.8 sec, or at about 2:00:29, the elevators will pass at the 48th floor.

11a. Start with 370, then use the rule Ans $- 54$.

Time (hr)	Distance from Mt. Rushmore (mi)
0	370
1	316
2	262
3	208
4	154
5	100
6	46

11b.

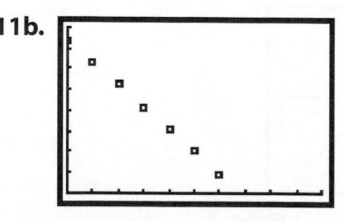

$[0, 10, 1, 0, 400, 50]$

11c.

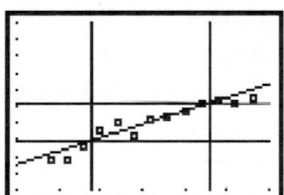

[0, 10, 1, 0, 400, 50]

The line represents the distance remaining at any time during the trip. With the line, you can see how far you are at any time, instead of just at the top of each hour.

11d. −54; the real-world meaning of the slope is that your distance from Mt. Rushmore decreases by 54 miles each hour.

11e. The car will reach the Wall Drug Store in the first half hour of the 5th hour of the trip. You can see this on the graph if you look at the line where it has a y-value of about 80.

11f. The car will reach Mt. Rushmore after almost 7 hours of travel. You can see this on the graph or in the table since after 7 hours, the car would be 8 miles too far if it kept going.

LESSON 5.7

1a. $(6, 6)$

1b. $(5, 9)$

1c. $y = 9 - 3(x - 5)$

1d. $y = 24 - 3x$

1e. $(8, 0)$

3a. $y = \dfrac{18 - 2x}{5}$ or $y = 3.6 - 0.4x$

3b. $y = \dfrac{-12 - 5x}{-2}$ or $y = 6 + 2.5x$

5. Let x represent distance from Los Angeles in miles, and let y represent elapsed time from Seattle in minutes.

5a. $y = 1385.5 - 1.49(x - 411.5)$ or $y = 240 - 1.49(x - 1181.5)$. The slope means the distance from Los Angeles decreases by 1 mile each 1.49 minutes.

5b. approximately 1701 min, or 28 hrs 21 min, by the first equation or approximately 1702 min, or 28 hrs 22 min, by the second equation

5c. approximately 939 miles by the first equation or 940 miles by the second equation

7. 15.4321 grains per gram.

1.
$$-3 = \frac{4 - 10}{x_2 - 2}$$
$$-3(x_2 - 2) = -6$$
$$x_2 - 2 = 2$$
$$x_2 = 4$$

3. Line a has slope -1, y-intercept 1, and equation $y - 1 = x$. Line b has slope 2, y-intercept -2, and equation $y = -2 + 2x$.

5a. $(-4.5, -3.5)$

5b. $y = 2x + 5.5$

5c. $y = 2(x + 2.75)$; the x-intercept is -2.75.

5d. The x-coordinate is 5.5; $y = 16.5 + 2(x - 5.5)$.

5e. Hint: You can try graphing, using a calculator table, or putting all equations in intercept form.

7a. $y = 12600 - 1350x$

7b. -1350; the car's value decreases by $1,350 each year.

7c. 12,600; Karl paid $12,600 for the car.

7d. $9\frac{1}{3}$; in $9\frac{1}{3}$ years the car will have no monetary value.

9a. 1952, 1956, 1976, 1990, 2000; 1.67, 1.835, 1.93, 2.02, 2.05

9b. The Q-points are (1962, 1.835) and (1990, 2.020).

9c. $y = 1.835 + 0.007(x - 1962)$ or $y = 2.020 + 0.007(x - 1990)$

9d. Hint: Remember that a good model usually shows the general direction of the data and has about the same number of points above the line as below.

[1940, 2000, 10, 1.5, 2.5, 0.1]

9e. Using $y = 1.835 + 0.007(x - 1962)$, the prediction is 2.19 m.

11a. Written as $y = a + bx$, b is the slope and a is the y-intercept.

11b. If the points are (x_1, y_1) and (x_2, y_2), then the slope of the line is $\dfrac{y_2 - y_1}{x_2 - x_1} = b$. The equation is $y = y_1 + b(x - x_1)$.

Selected Answers

LESSON 6.1

1a. Yes, because $47 + 3(-15.6) = 0.2$ and $8 + 0.5(-15.6) = 0.2$.

1b. No, because $23 \neq 12 + (-4)$. The point satisfies only one of the equations.

1c. No, $12.3 \neq 4.5 + 5(2)$. Furthermore, the lines are parallel, so there is no solution to the system.

3. In both cases, the calculator gives exact solutions that satisfy each system.

3a. $(8, 7)$

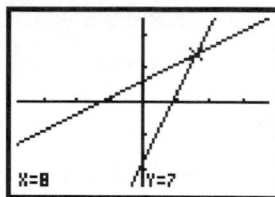

3b. $(1.5, 0.5)$

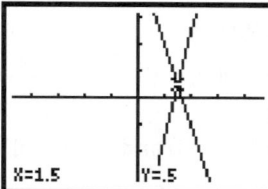

5. The results of substituting the x- and y-values into the original equation tell you that each coordinate satisfies both the standard form and the intercept form of the equation.

5a. $y = 3 - 2x$; $(1, 1)$: $4(1) + 2(1) = 6$

5b. $y = -4 + 0.4x$; $(1, -3.6)$: $2(1) - 5(-3.6) = 20$
A coordinate satisfies both forms of a linear equation.

7a. $P = -5000 + 2.5N$

7b.

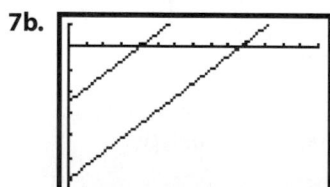

$[0, 7000, 500, -13000, 20000, 2000]$

7c. Sally will always profit more than Gizmo.kom for the same number of web site hits. Because their lines never intersect, there is no solution to the system of equations, and their profits will never be equal.

7d. Sally pays less to start Gadget.kom, but she profits at the same rate as Gizmo.kom. Her profit will always be $7,000 more for the same N-values.

9a. $d = 9 - t$ where d is drill team member's distance from end zone; $d = 3 + 0.5t$ where d is tuba player's distance from end zone

9b. $(4, 5)$. After 4 seconds, the tuba player bumps into the drill team member at the 5-yard mark.

11a. Because lines with different slopes always intersect, the y-intercept a can equal any number and b can be any number except -5.

11b. $a \neq 2$ and $b = -5$, same slope, different y-intercept, lines do not intersect.

11c. $a = 2$ and $b = -5$, same slope and y-intercept, lines overlap.

13a. 85

13b. -8.2

13c. 3

13d. 3.5

13e. 1.5

15a. $\begin{bmatrix} 1 & -11 \\ -6 & 8 \end{bmatrix}$

15b. $\begin{bmatrix} 13 & -1 \\ 7 & 8 \end{bmatrix}$

LESSON 6.2

1. 3. Add $2.5t$ to both sides.
  5. Divide both sides by 4.

3a. $x = -2$

3b. $y = 10$

3c. $d = 3$

3d. $t = 4$

5a. $-x + 8$

5b. $13x - 8$

7a. $N = 7,777\frac{7}{9}, P = 7,444\frac{4}{9}$

7b. The approximate solution, $N \approx 7778$ and $P \approx 7444$, is more meaningful because there cannot be a fractional number of web site hits.

9a. $A + C = 200$

9b. $8A + 4C = 1304$

9c. $A = 126$ and $C = 74$, so the theater sold 126 adult tickets and 74 child tickets.

11a. $\begin{cases} d = 35 + 0.8t \\ d = 1.1t \end{cases}, \left(116\frac{2}{3}, 128\frac{1}{3}\right)$.

The pickup passes the sports car roughly 128 miles from Flint after approximately 117 minutes.

11b. $\begin{cases} d = 220 - 1.2\,t \\ d = 1.1t \end{cases}$,

$\left(\dfrac{2200}{23}, \dfrac{2420}{23}\right) \approx (95.7, 105.2)$. The minivan meets the pickup truck about 105 miles from Flint after approximately 96 minutes.

11c. $\begin{cases} d = 220 - 1.2t \\ d = 35 + 0.8t \end{cases}$,

$(92.5, 109)$. The minivan meets the sports car 109 miles from Flint after 92.5 minutes.

11d. $220 - 1.2t = 2(35 + 0.8t);\ t \approx 53.6$ min, minivan is about 156 mi, sports car is about 78 mi.

11e. The solutions found using substitution are exact (if not rounded off). Recursive routines sometimes give approximate answers because of their discreteness.

13a. 12.1 ft/sec

13b. 50 seconds

13c. $y = 100 + 12.1x$, where x represents the time in seconds and y represents her height above ground level. To find out how long her ride to the observation deck is, solve the equation $520 = 100 + 12.1x$.

15a. i

15b. iii

15c. ii

LESSON 6.3

1a. $y = \dfrac{10 - 5x}{2}$ or $y = \dfrac{10}{2} - \dfrac{5x}{2}$

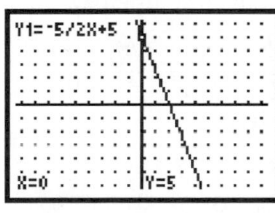

$[-9.4, 9.4, 1, -6.2, 6.2, 1]$

1b. $y = \dfrac{30 - 15x}{6}$ or $y = \dfrac{30}{6} - \dfrac{15x}{6}$

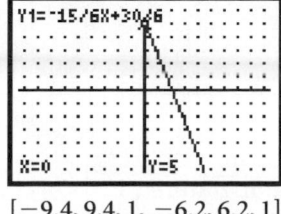

$[-9.4, 9.4, 1, -6.2, 6.2, 1]$

The graph is the same as the graph for 1a. Both equations are equivalent to $y = 5 - \dfrac{5}{2}x$.

3a. $(-2.5, -1)$

3b. $(3, -2)$

5a. Multiply the first equation by -5 and the second equation by 3, or multiply the first equation by 5 and the second equation by -3.

5b. Multiply the first equation by -8 and the second equation by 7, or multiply the first equation by 8 and the second equation by -7.

7a. $(4, 2)$

7b. $(3, -1)$

7c. $(-3, -1)$

9a. $y = 163 - x$ and $y = -33 + x$

9b. $y = 65$

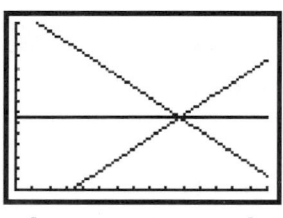

$[0, 150, 10, 0, 140, 10]$

9c. $x = 98$

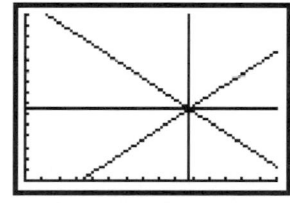

$[0, 150, 10, 0, 140, 10]$

9d. The four lines intersect at the same point, $(98, 65)$. The solution to the system must satisfy all the equations—the original equations in the system and any new equations created by combining pairs of equations.

11a. substitution

11b. Her solution is half right. Anisha didn't find the value $y = -1$ when $x = 4$.

13a. Let c represent gallons burned in the city and h represent gallons burned on the highway.

$\begin{cases} c + h = 11 \\ 17c + 25h = 220 \end{cases}$

13b. $(6.875, 4.125)$; 6.875 gallons in the city, 4.125 gallons on the highway

13c. $\dfrac{17 \text{ mi}}{\text{gal}} \cdot 6.875 \text{ gal} \approx 117$ city mi,

$\dfrac{25 \text{ mi}}{\text{gal}} \cdot 4.125 \text{ gal} \approx 103$ hwy mi

13d. Check: $\begin{cases} 6.875 + 4.125 = 11 \\ 17(6.875) + 25(4.125) = 220 \end{cases}$

and $117 + 103 = 220$.

15a. See below.

15b. $T = 95.2 - 0.004E$. The slope is the rate of change in temperature with elevation. The y-intercept (in this case, T-intercept) is the temperature that day at sea level in the same area.

15c. At the summit the temperature was 13.9 degrees Fahrenheit.

17a. P (top) $= \dfrac{400}{769}$ or 0.52 roughly.

P (bottom) $= \dfrac{369}{769}$ or 0.48 roughly.

17b. A trial is assigning a locker. An outcome is whether a locker is in the top row or the bottom row.

LESSON 6.4

1a. $\begin{cases} 2x + 1.5y = 12.75 \\ -3x + 4y = 9 \end{cases}$

1b. $\begin{cases} \dfrac{1}{2}x = \dfrac{1}{2} \\ -x + 2y = 0 \end{cases}$

1c. $\begin{cases} 2x + 3y = 1 \\ 2y = 0 \end{cases}$

3a. $(8.5, 2.8)$

3b. $\left(\dfrac{1}{2}, \dfrac{13}{16} \right)$

3c. $(0, 0)$

5a. $\begin{cases} 3x + y = 7 \\ 2x + y = 21 \end{cases}$

5b. $\begin{bmatrix} 3 & 1 & 7 \\ 2 & 1 & 21 \end{bmatrix}$

7a.

	Adults	Children	Total (kg)
Monday	40	15	10.80
Tuesday	35	22	12.29

7b. Let x represent the average weight of chips an adult eats and y represent the average weight of chips a child eats. The system is $\begin{cases} 40x + 15y = 10.8 \\ 35x + 22y = 12.29 \end{cases}$

7c. $\begin{bmatrix} 40 & 15 & 10.8 \\ 35 & 22 & 12.29 \end{bmatrix}$

7d. $\begin{bmatrix} 1 & 0 & 0.15 \\ 0 & 1 & 0.32 \end{bmatrix}$

7e. Each adult ate an average of about 0.15 kg (150 g) of chips, and each child ate an average of 0.32 kg (320 g) of chips.

9a. Let x represent the number of small trucks and y represent the number of large trucks. The system is $\begin{cases} 5x + 12y = 532 \\ 7x + 4y = 284 \end{cases}$

9b. $\begin{bmatrix} 5 & 12 & 532 \\ 7 & 4 & 284 \end{bmatrix}$

9c. $\begin{bmatrix} 1 & 0 & 20 \\ 0 & 1 & 36 \end{bmatrix}$

9d. Zoe should order 20 small trucks and 36 large trucks.

11a. $\begin{cases} m + t + w = 286 \\ m - t = 7 \\ t - w = 24 \end{cases}$

11b. $\begin{bmatrix} 1 & 1 & 1 & 286 \\ 1 & -1 & 0 & 7 \\ 0 & 1 & -1 & 24 \end{bmatrix}$

The rows represent each equation. The columns represent the coefficients of each variable and the constants.

11c. $\begin{bmatrix} 1 & 0 & 0 & 108 \\ 0 & 1 & 0 & 101 \\ 0 & 0 & 1 & 77 \end{bmatrix}$

11d. They cycled 108 km on Monday, 101 km on Tuesday, and 77 km on Wednesday.

13a. $4, \text{Ans} - 0.5$

13b. $-3, \text{Ans} + 2$

13c. $1/2, \text{Ans} - 1$

13d. $0, \text{Ans} + 1$

15. $\begin{bmatrix} 1 & 3 \\ -2 & 1 \\ 3 & 23 \end{bmatrix} \rightarrow \begin{matrix} 3 & -3 & 0 \\ -6 & -1 & -7 \\ 9 & -23 & -14 \end{matrix}$

$-7y = -14, y = 2$ and $x = 7$

15a. *(Lesson 6.3)*

Marsha's Climb

	Elevation (ft)	Temperature (°F)
Start	4,300	78
Rest station	7,800	64
Highest point	11,900	47.6

1a. Multiply by 4; $12 < 28$

1b. Multiply by -3; $-15 \geq -36$

1c. Add -10; $-14 \geq x - 10$

1d. Subtract 8; $b - 5 > 7$

1e. Divide by 3; $8d < 10\frac{2}{3}$

1f. Divide by -3; $-8x \geq -10\frac{2}{3}$

3a. $x \leq -1$

3b. $x > 0$

3c. $x \geq -2$

3d. $-2 < x < 1$

3e. $0 < x \leq 2$

5a. $y = \dfrac{5.2 - 3x}{4} = 1.3 - 0.75x$

5b. $y = \dfrac{2x}{3} + 5$

7a. $x \leq 3$

7b. $x < -2$

7c. $x \geq -3$

7d. $0 \geq x$ or $x \leq 0$

9a. Add 3 to both sides; $4 < 5$.

9b. Divide both sides by 2 (or multiply by 0.5); $3 > 1$.

9c. Multiply both sides by -3; $3 > -3$.

9d. Multiply both sides by 2; $0 < 6$.

11a. The variable x drops out of the inequality, leaving $-3 > 3$, which is never true. So the original inequality is not true for any number x. You can't draw a graph to represent this situation on a number line.

11b. The variable x drops out of the inequality, leaving $-6.6 \geq -15$, which is always true. So the original inequality is true for any number x. The graph would be a line with arrows on both ends.

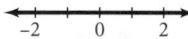

13a. Multiply 12 by 3.2 to get 38.4. Subtract 38.4 from 72 to get 33.6.

13b. Square 5 to get 25. Subtract 25 from 3 to get -22. Multiply -22 by 1.5 to get -33. Add -33 to 2 to get -31.

13c. Divide 21 by 7 to get 3 and divide 6 by 2 to get 3. Subtract 3 from 3 to get 0.

15a. $-2x - 16$

15b. $3 - 4y$

15c. $-z + 5$

1a. iii   **1b.** ii   **1c.** i   **1d.** iv

3a. 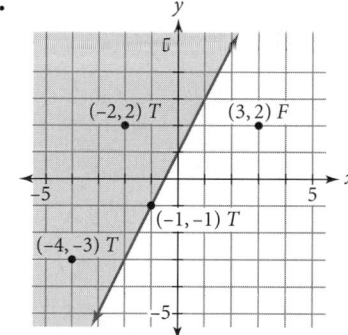 **3b.**

3c. **3d.**

5a–c.

7a. $y \leq 1 - 2x$ **7b.** $y < -2 + \frac{2}{3}x$

7c. $y > 1 - 0.5x$ **7d.** $y \geq -2 + \frac{1}{3}x$

9a.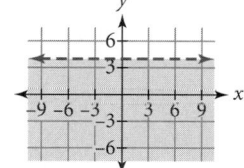

9b.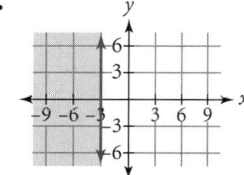

9c.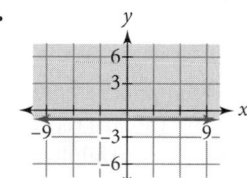

9d.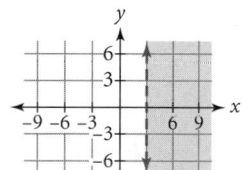

11.

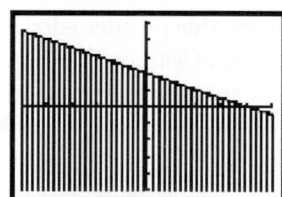

$[-5, 5, 1, -5, 5, 1]$

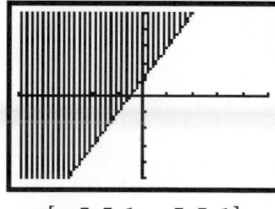

$[-5, 5, 1, -5, 5, 1]$

13a. about 27 mph

13b. Since $d = r \cdot t$, and the distance was the same for both Ellie and her grandmother, you can set these products equal to each other. If you let $r =$ Ellie's grandmother's speed, then $2.5(65) = 6r$.

Selected Answers

1a. iii   **1b.** i   **1c.** ii

3a. $y \geq -x + 2; y \geq x - 2$

3b.

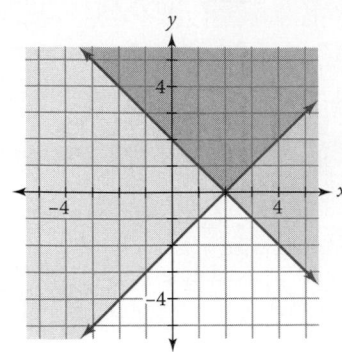

5. $\begin{cases} y > 2 - x \\ y < 2 \\ x < 3 \end{cases}$

7a. $\begin{cases} A \leq C \\ A + C \leq 75 \\ A \geq 0 \\ C \geq 0 \end{cases}$

7b.

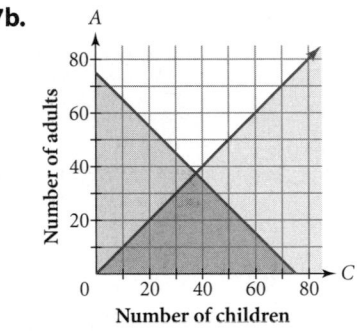

All the points in the dark-shaded triangular region satisfy the two inequalities. The point (50, 10) represents the situation in which 50 children escort 10 adults.

7c. Sample answer: It is possible to have all children and no adults at the restaurant. One possible additional constraint is that there must be at least one adult per five children, or $A \geq \frac{1}{5}C$. The solution for this set of constraints is the triangular region bounded by $A \leq C$, $A + C \leq 75$, and $A \geq 0.2C$.

9. $x \geq 3$ and $y \geq -2 + \frac{1}{2}x$

11. The region is a pentagon.

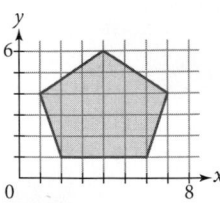

13a. $713.15   **13b.** $957.80

15a. 4   **15b.** 10

15c. $\dfrac{10\left(\dfrac{3x+12}{5} - 1.4\right) - 10}{6}$, which simplifies to x

1. line a: $y = 1 - x$; line b: $y = 3 + \frac{5}{2}x$; intersection: $\left(-\frac{4}{7}, \frac{11}{7}\right)$

3. $(3.75, 4.625)$

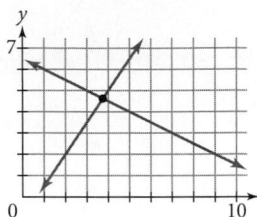

5a. ... the slopes are the same, but the intercepts are different (the lines are parallel).

5b. ... the slopes are the same and the intercepts are the same (the lines overlap).

5c. ... the slopes are different (the lines intersect in a single point).

7. $x \leq -1$

9a. 10 sq. m/min

9b. 7 sq. m/min

9c. No. He will cut 156 sq. m, and the lawn measures 396 sq. m.

9d. $10h + 7l = 396$

9e. $\frac{1}{30}$ liters per min

9f. $\frac{3}{200}$ liters per min

9g. Yes. He will use $\frac{34}{75}$ liter, and the tank holds 1.2 liters.

9h. $\frac{h}{30} + \frac{3l}{200} = 1.2$

9i. $l = 14.4$ min, $h = 29.52$ min

9j. If Harold cuts for 29.52 minutes at the higher speed and 14.4 minutes at the lower speed, he will finish Mr. Fleming's lawn and use one full tank of gas.

CHAPTER 7 · CHAPTER **7** CHAPTER 7 · CHAPTER

1a. starting value: 16; multiplier: 1.25; 7th term: 61.035

1b. starting value: 27; multiplier: $\frac{2}{3}$ or $0.\overline{6}$; 7th term: $2.\overline{370}$ or $\frac{64}{27}$

3. Start with 72, then apply the rule Ans · $(1 + 0.40)$; first five terms are 72, 100.8, 141.12, 197.568, and 276.5952.

5. Start with 32, then apply the rule Ans · 0.75; Stage 2 has a shaded area of 18 square units.

7a. Start with 115, then apply the rule Ans · $(1 - 0.03)$.

Selected Answers

7b. 12 minutes

9a. Start with 6.9, then apply the rule
Ans · (1 + 0.142).

9b. See below.

9c.

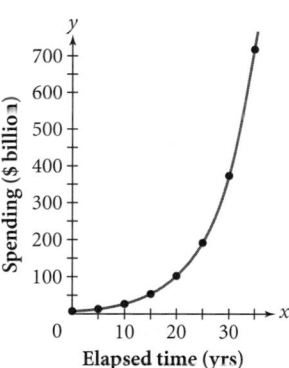

9d. Sample answer: The graph implies a smooth, ever-increasing amount of Medicare spending, which is probably not realistic.

11a. See below.

11b. See below.

11c. The graph of the first plan is linear. The graph of the second is not; its slope increases between consecutive points.

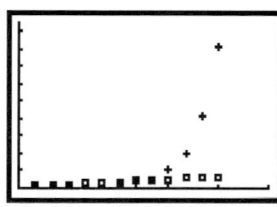

[0, 15, 3, 0, 4750, 500]

11d. Possible answer: Grace's charity will receive more total money with option 2. It will also give the charity more income later in the year when the budget may be tighter.

13a. i   **13b.** iii   **13c.** ii   **13d.** iv

LESSON 7.2

1a. 7^8 **1b.** $3^4 \cdot 5^5$ **1c.** $(1 + 0.12)^4$

3a. ii   **3b.** iii   **3c.** iv   **3d.** i

5a. $y = 1.2 \cdot 2^x$ **5b.** $y = 500 \cdot 0.2^x$

5c. $y = 125 \cdot 0.4^x$

7. Your calculator will tell you when you have the correct answer.

9a. $9,200

9b. Start with 11,500, then apply the rule
Ans · (1 − 0.2).

9c.

Time elapsed (yrs)	0	1	2	3	4
Value ($)	11,500	9,200	7,360	5,888	4,710.40

9d. $y = 11,500(1 − 0.2)^x$

9e.

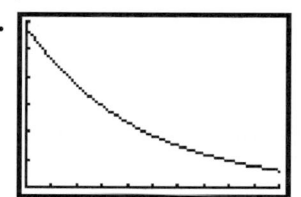

[0, 10, 1, 0, 12000, 2000]

9b. (*Lesson 7.1*)

Medicare Spending

Year	1970	1975	1980	1985	1990	1995	2000	2005
Elapsed time (yrs) x	0	5	10	15	20	25	30	35
Spending ($ billion) y	6.9	13.4	26.0	50.6	98.2	190.8	370.5	719.7

11a. (*Lesson 7.1*)

	Jan	Feb	Mar	Apr	May	June	July	Aug	Sep	Oct	Nov	Dec
Option 1	$50	$25	$25	$25	$25	$25	$25	$25	$25	$25	$25	$25
Option 2	$1	$2	$4	$8	$16	$32	$64	$128	$256	$512	$1,024	$2,048

11b. (*Lesson 7.1*)

	Jan	Feb	Mar	Apr	May	June	July	Aug	Sep	Oct	Nov	Dec
Option 1	$50	$75	$100	$125	$150	$175	$200	$225	$250	$275	$300	$325
Option 2	$1	$3	$7	$15	$31	$63	$127	$255	$511	$1,023	$2,047	$4,095

11. Your calculator will tell you when you have the correct answer.

13a. $y = 5000(1 + 0.05)^x$

13b.

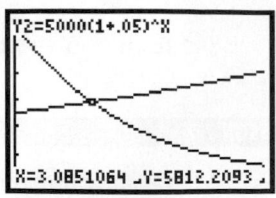

[0, 10, 1, 0, 12000, 2000]

The intersection point represents the time and the value of both cars when their value will be the same. By tracing the graph shown, students should see that both cars will be worth approximately $5,806 after a little less than 3 years 1 month.

15. Hint: Since you multiply by $(1 - 0.035)$, your situation needs to describe something that decreases over time. The equation $y = 100(1 - 0.035)^x$ will help you solve the problem you create in 15b.

LESSON 7.3

1a. $5x^4$ **1b.** $15x^{10}$ **1c.** $8x^{10}$ **1d.** $-2x^4 - 2x^6$

3a. 3^{40} **3b.** 7^{12} **3c.** x^{12} **3d.** y^{40}

5. Student 2 was correct. $2x^2 = 2 \cdot x \cdot x$; substitute $x = 3$: $2 \cdot 3 \cdot 3 = 18$

7. a, d, and g: 27,521.40084; b and f: 11,711.2344; c and h: 1060.32; e: 129,350.5839

9. Enclose the -5 in parentheses.

11. Sample answers:

11a. $x^3 \cdot x^5 = x^8$ **11b.** $(x^3)^5 = x^{15}$

11c. $(3x)^5 = 3^5 x^5 = 243 x^5$

11d. Exponents are added when you multiply two exponential expressions with the same base. Exponents are multiplied when an exponential expression is raised to a power. An exponent is distributed when a product is raised to a power.

13a. $x^4 + x^5$ **13b.** $-2x^4 - 2x^6$

13c. $17x^{7.5} + 8.5x^{8.4}$

15a. $4.5x - 47$

15b. $4.5x - 47 > 0$; $x > 10.\overline{4}$; he must shovel 11 sidewalks to pay for his equipment.

15c. $4.5x - 47 > 100$; $x > 32.\overline{6}$; he must shovel 33 sidewalks to pay for his expenses and buy a lawn mower.

17a. $(5.3625, 0.70625)$

17b. approximately $(3.095, 0.762)$

LESSON 7.4

1a. 3.4×10^{10} **1b.** -2.1×10^6 **1c.** 1.006×10^4

3a. $12x^6$ **3b.** $7y^{16}$ **3c.** $2b^6 + b^5$ **3d.** $10x^4 - 6x^2$

5. $3.5 \times 10^7 = 3.5 \cdot 10 \cdot 10 \cdot 10 \cdot 10 \cdot 10 \cdot 10 \cdot 10$
$= 35,000,000$;

$3.5^7 = 3.5 \cdot 3.5 \cdot 3.5 \cdot 3.5 \cdot 3.5 \cdot 3.5 \cdot 3.5$
$= 6433.9296875$

7a. 3.3411×10^{25} **7b.** approximately 3.6×10^{47}

9a. Yes, because they are both equal to 51,800,000,000.

9b. Al's answer   **9c.** Possible answer: 518×10^8

9d. Rewrite the digits before the 10 in scientific notation, then use the multiplication property of exponents to add the exponents on the 10's. In this case, $4.325 \times 10^2 \times 10^3 = 4.325 \times 10^5$.

11a. 2,000,000,000; 2×10^9

11b. 7.3×10^{11} calls per year

13a. 9.46×10^{12} km; 1.0×10^5 light-years

13b. 9.46×10^{17} km

13c. $\dfrac{(9.46 \times 10^{17})}{(1.27 \times 10^4)} = 7.45 \times 10^{13}$

15.

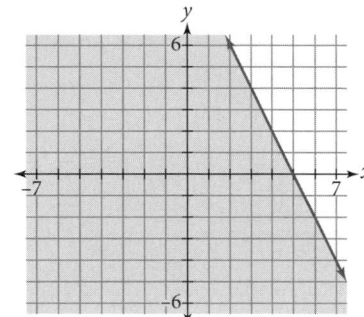

LESSON 7.5

1. $x^3 y$

3. Sample answer: $\dfrac{3^6}{3^2}$ means there are 6 factors of 3 in the numerator and 2 factors of 3 in the denominator. So there are 2 factors of 1, or $\frac{3}{3}$, in the entire expression, leaving 4 factors of 3 in the numerator, or 3^4.

5a. $648x^{11}$ **5b.** $25x^2$ **5c.** $-4x^2$ **5d.** $-108x^5 y^{14}$

7a. about 132 people per square mile

7b. about 867 people per square mile

7c. The population of Japan was about 6.6 times denser than that of Mexico.

9. four days earlier

Method 1: Use a recursive routine

864 (ENTER) , Ans/3 (ENTER) , (ENTER) , (ENTER) , (ENTER)

Method 2: Use an equation: $y = 864\left(\frac{1}{3}\right)^x$;
look at the table to find x when y is less than 20.

11a. approximately 61 years

11b. approximately 1.3 years

11c. approximately 25.4 years

13. approximately 2.272×10^5 flowers

15a. (Answers recorded to tenths.) Mercury: 4.1 cm;
Venus: 10 cm; Earth: 10.5 cm; Mars: 5.6 cm;
Jupiter: 117.3 cm; Saturn: 94.7 cm; Uranus: 69.3 cm;
Neptune: 41.3 cm; Pluto: 2 cm; Sun: 1152 cm

15b. Hint: Convert the largest model diameter to
meters. Will Halley be able to make this model?
Consider a different size for Pluto that gives a
reasonable diameter for the largest model.

LESSON 7.6

1a. $\dfrac{1}{2^3}$ **1b.** $\dfrac{1}{5^2}$ **1c.** $\dfrac{1.35}{10^4}$

3a. -5 **3b.** -4 **3c.** -4

5. Sample answer: Negative exponents mean to use a
reciprocal with the exponent positive.
$6^{-3} = \dfrac{1}{6^3} = \dfrac{1}{216}, -6^3 = -216$

7a. $3500(1 + 0.04)^{-4}$; approximately $2,992

7b. $250(1 + 0.04)^{-3}$; approximately $222

7c. $25(1 + 0.04)^{-5}$; approximately $21

7d. $126,800(1 + 0.04)^{-30}$; approximately $39,095

9a. true; $(2^3)^2 = 2^3 \cdot 2^3 = 2 \cdot 2 \cdot 2 \cdot 2 \cdot 2 \cdot 2 = 2^6$

9b. false; $(3^3)^4 = 3^3 \cdot 3^3 \cdot 3^3 \cdot 3^3$
$= 3 \cdot 3 \cdot 3 \cdot 3 \cdot 3 \cdot 3 \cdot 3 \cdot 3 \cdot 3 \cdot 3 \cdot 3 \cdot 3 = 3^{12}$

9c. false; $(10^{-2})^4 = \left(\dfrac{1}{10^2}\right)^4$
$= \left(\dfrac{1}{10 \cdot 10}\right)\left(\dfrac{1}{10 \cdot 10}\right)\left(\dfrac{1}{10 \cdot 10}\right)\left(\dfrac{1}{10 \cdot 10}\right) = \dfrac{1}{10^8}$
$= 10^{-8}$

9d. true; $(5^{-3})^{-4} = \left(\dfrac{1}{5^3}\right)^{-4} = \dfrac{1}{\left(\dfrac{1}{5^3}\right)^4}$
$= \dfrac{1}{\left(\dfrac{1}{5 \cdot 5 \cdot 5}\right)\left(\dfrac{1}{5 \cdot 5 \cdot 5}\right)\left(\dfrac{1}{5 \cdot 5 \cdot 5}\right)\left(\dfrac{1}{5 \cdot 5 \cdot 5}\right)} = \dfrac{1}{\dfrac{1}{5^{12}}}$
$= \dfrac{1}{5^{-12}} = 5^{12}$

11. $36(1 + 0.5)^{(4-6)} = 36(1 + 0.5)^{-2}$

13. 1×10^{-1} or $\dfrac{1}{10}$ of a chocolate bar per person

15a. i. 4×10^5 **15a.** ii. 3.1×10^{10}

15a. iii. 1.21×10^5 **15a.** iv. 2×10^{-4}

15b. Sample answer: Divide the coefficients of the
powers of 10, and then divide the powers of 10
(subtract the exponents).

15c. 0.6×10^5 or 6×10^4 in scientific notation

LESSON 7.7

1a. $1 + 0.15$; rate of increase: 15%

1b. $1 + 0.08$; rate of increase: 8%

1c. $1 - 0.24$; rate of decrease: 24%

1d. $1 - 0.002$; rate of decrease: 0.2%

1e. $1 + 1.5$; rate of increase: 150%

3. $B = 250(1 + 0.0425)^t$

5a. The ratios are 0.957, 0.956, 0.965, 0.964, 0.963,
0.961, 0.959, 0.958, 0.971, and 0.955.

5b. $0.9609 \approx 0.96$

5c. $1 - 0.04$ **5d.** $y = 47(1 - 0.04)^x$

5e.

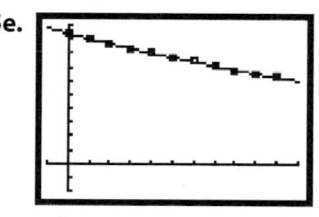

$[-1, 11, 1, -10, 50, 5]$

Adjustments to A or r are not necessary—the fit
is good as is.

5f. 55 minutes

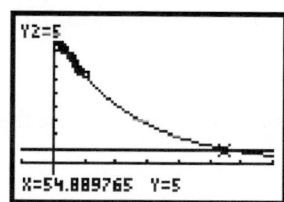

$[-10, 70, 10, -10, 50, 5]$

7a. Sample answer: Let x represent years since 1990,
and let y represent median price in dollars. An
equation is $y = 85,000(1 + 0.06)^x$, where 0.06 is
derived from the mean ratio of about 1.06.

7b. Hint: Add 5 to the current year and then
subtract 1990 to find the number of years since 1990.
Substitute this value for the appropriate variable
in your equation.

7c. Hint: Multiply your answer from 7b by 0.10.

7d. Hint: 5 years is the same as 60 months. So divide
your answer from 7c by 60.

9. 1000 nanoseconds per microsecond

11a. $y = 2(1 + 0.5)^x$

11b. $y = 2(1 + 0.5)^{0.5} \approx 2.45$; approximately 2.45 liters

11c. approximately 115 liters

11d. After about 30.4, or 30 minutes 24 seconds

13. $x = 3$ cm; $y = 7.2$ cm; $z = 9$ cm

CHAPTER 7 REVIEW

1a. 3^4 **1b.** 3^3 **1c.** 3^2

1d. 3^{-1} **1e.** 3^{-2} **1f.** 3^0

3. Hint: Since you multiply by $(1 - 0.15)$, your situation needs to describe something that decreases over time.

5a. $y = 200(1 + 0.4)^x$ **5b.** $y = 850(1 - 0.15)^x$

x	y
0	200
1	280
2	392
3	548.8
4	768.32
5	1075.648
6	1505.9072

x	y
−2	1176.4706
−1	1000.0000
0	850
1	722.5
2	614.125
3	522.00625
4	443.7053

7. approximately 1.17×10^0 years

9a. False; 3 to the power of 3 is not 9; $27x^6$

9b. False; you can't use the multiplication property of exponents if the bases are different; $3^2 \cdot 2^3$ or 72

9c. False; the exponent −2 applies only to the x; $\dfrac{2}{x^2}$

9d. False; the power property of exponents says to multiply exponents; $\dfrac{x^6}{y^9}$

11a. Let t be the number of T-shirts, and let s be the number of sweatshirts.
$t + s = 12$
$6t + 10s = 88$

11b. 8 T-shirts and 4 sweatshirts

13a. Start with 21, then apply the rule Ans − 4; 10th term = −15.

13b. Start with −5, then apply the rule Ans · (−3); 10th term = 98,415.

13c. Start with 2, then apply the rule Ans + 7; 10th term = 65.

15a. $y = 1.6x$, where x is a measurement in miles and y is a measurement in kilometers

15b. 400 km   **15c.** 3.2 km

15d. approximately 168 m

17a. −4   **17b.** 13   **17c.** −12

19. The following sample answers use the Q-point method and a decimal approximation of the slope.

19a. $y = 96 - 5.7(x - 5)$

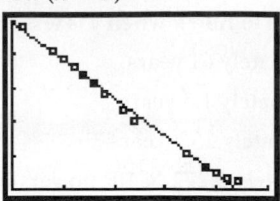

$[0, 25, 5, 0, 125, 25]$

19b. $y = 124.5 - 5.7x$ **19c.** approximately 16 days

19d. The y-intercept would become 200;
$y = 200 - 5.7x$.

19e. 86 grams

CHAPTER 8 · CHAPTER **8** CHAPTER 8 · CHAPTER

LESSON 8.1

1a. SBOHF   **1b.** EPNBJO

1c. UBCMF   **1d.** HSBQI

3a. SECRET CODES

3b. The coding scheme is a letter-shift of +1.

5a. {1:00, 2:00, 3:00, 4:00, 5:00, 6:00, 7:00, 8:00, 9:00, 10:00, 11:00, 12:00}

5b. {0100, 0200, 0300, 0400, 0500, 0600, 0700, 0800, 0900, 1000, 1100, 1200, 1300, 1400, 1500, 1600, 1700, 1800, 1900, 2000, 2100, 2200, 2300, 2400}

5c. It is not a function because each standard time designation has two military time designations. If students distinguish A.M. from P.M. times, then it is a function.

7a. L_1 = {6, 21, 14, 3, 20, 9, 15, 14, 19}
L_2 = {15, 30, 23, 12, 29, 18, 24, 23, 28}

7b. You have to subtract 26 from the numbers greater than 26. The new list L_2 is {15, 4, 23, 12, 3, 18, 24, 23, 2}.

7c. ODWLCRXWB

7d.

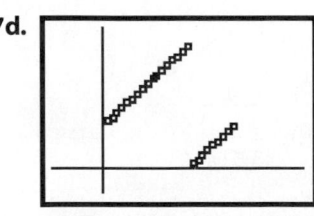

$[-10, 37, 0, -5, 31, 0]$

7e. Avoid any multiple of 26, such as 0, ±26, ±52, and so on.

9. ALGEBRA

Selected Answers

11a. Input: $\{0, 1, -1, 2, -2\}$; output: $\{0, 1, 2\}$; the relationship is a function.

11b. Input: $\{1, 4, 9\}$; output: $\{1, -1, 2, -2, 3, -3\}$; the relationship is not a function.

13a. Subtract the input letter's position from 27 to get the output letter's position.

13b. Yes, each input matches no more than one output.

13c. Sample answer: LISA codes into QRHZ.

15a. $28b^6$	**15b.** $11a^3$	**15c.** 1.75
15d. $1372d^3$	**15e.** Not possible	**15f.** $4x^5$

LESSON 8.2

1a.

Input x	Output y
-4	1
-1	3.4
1.5	5.4
6.4	9.32
9	11.4

1b.

Domain x	Range y
-4	4.4
-1	2
2.4	-0.72
11	-7.6
14	-10

3.

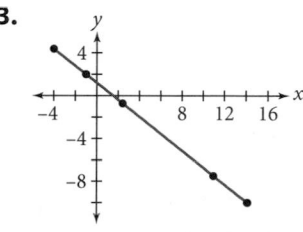

5. Start at the 2-m mark and stand still for 2 sec; walk toward the 4-m mark at 2 m/sec for 1 sec. Stand still for another second; walk toward the 8-m mark at 4 m/sec for 1 sec; then stand still for 3 sec. Yes, the graph represents a function.

7a. No; Los Angeles, for example, has more than one area code $(213, 310, \ldots)$.

7b. Yes; each person has only one birth date.

7c. No; the same last name will correspond to many different first names.

7d. Yes; each state has only one capital.

9a. Not a function; the input 3 has two different output values, 10 and 8.

9b. A function; each x-value corresponds to only one y-value.

9c. A function; each x-value corresponds to only one y-value.

11. Hint: Your graph will not pass the vertical line test.

13a. domain: $\{-5, -4, -3, -2, -1, 0, 1, 2, 3, 4, 5\}$; range: $\{0, 1, 2, 3, 4, 5\}$

13b. domain: $0 \leq x \leq 360$; range: $-1 \leq y \leq 1$

13c. domain: all numbers x; range: all numbers y

15a. When $x = 23.125, y = -15$. Answers will vary. Zoom in on the table by changing the start values and the table increments (ΔTbl).

15b. The lines intersect at the solution point.

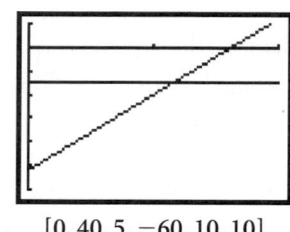

$[0, 40, 5, -60, 10, 10]$

LESSON 8.3

1a. The level of medication drops quickly at first and then decreases more slowly over time. After 10 hours, about 30 milligrams are still in the bloodstream.

1b. The independent variable is time. The dependent variable is the amount of medication in milligrams in the bloodstream.

1c. The domain is the time in hours from 0 through 10. The range is the amount of medication in milligrams from about 30 through 500.

1d. Sample answer: Assuming $(1, 400)$ is a point on the curve, the equation is $y = 500(1 - 0.20)^x$.

3a. The reading on the scale depends on the weight of the dog, so the dog's weight is the independent variable and the reading on the scale is the dependent variable.

3b. The amount of time you spend in the plane depends on the distance you fly, so the distance between the cities is the independent variable and the amount of time in the plane is the dependent variable.

3c. The wax sticks to the candle wick each time you dip it, so the number of dips is the independent variable and the diameter of the candle is the dependent variable.

5. $x \to y$; independent $\to$ dependent; input $\to$ output; time $\to$ distance (or distance $\to$ time)

7. Student graphs should consist of three line segments, each less steep than the one before it.

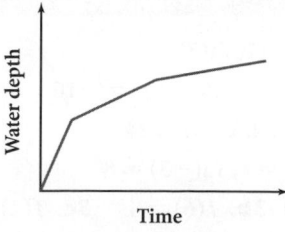

9a. If the *x*-axis is labeled in days, a reasonable domain is from 0 to 120, with 0 representing May 1. If the *x*-axis is labeled in months or another unit of time, the domain will differ accordingly. The range should go from 0 to about 5 inches, or 13 centimeters, depending on how high grass grows before it's cut.

9b. The domain should go from 0 to a large number such as 500 or more, depending on the number of students in the school. The range is a set of integer values from 1 to a maximum that depends on the number of students in the school.

9c. The domain should go from 0 to about 5 seconds. The range should go from 0 to about 300 feet, or 100 meters.

9d. The number of dice is the domain; it is any non-negative integer. The probability is the range, and it is any number between 0 and 1 for observed outcomes. The theoretical probability for a large number of dice is close to 1.

9e. The angle measure is the domain, and it goes from 0° to 180°. The area is the range, and it could be as small as 0 and as large as about 60 sq. in. or 400 sq. cm, depending on the dimensions of the coat hanger.

11a. Hint: The graph should be made up of at least three horizontal segments.

11b. Hint: How will your speed change when the school bus makes a stop? How will your speed change when the school bus starts driving again?

11c. Hint: Your graph might look like an upside-down U.

13a. decreases. **13b.** increases.

13c. $h = \dfrac{1000}{(\pi r^2)}$

13d.

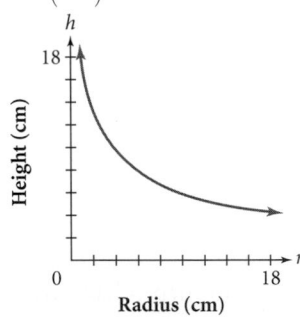

15a. -2.5 **15b.** -4 **15c.** $\dfrac{5}{3}$

LESSON 8.4

1a. $3(3) + 2 = 11, Y_2(3) = 11$

1b. $3(-4) + 2 = -10, Y_1(3) = -10$

1c. $(5)^2 - 1 = 24, Y_2(5) = 24$

1d. $(-3)^2 - 1 = 8, Y_2(-3) = 8$

3a. $f(4) = 0$ **3b.** $f(6) = 4$ **3c.** $f(2) = 2, f(5) = 2$

3d. $f(0.5) = 1, f(3) = 1, f(4.5) = 1$

3e. Three **3f.** $x > 5$

3g. domain: $-1 \le x \le 7$; range: $0 \le y \le 6$

5a. $f(x) = 7x + 5$ **5b.** $f(x) = 5 + 7(x - 1)$

7a. amount of medication in milligrams

7b. time in hours **7c.** $0 \le x \le 10$; all real numbers x

7d. $53 < y \le 500, y > 0$

7e. 500 **7f.** about 4 hours

9a. Hint: Be sure to try positive and negative numbers, fractions, and zero.

9b. Hint: Be sure to try positive and negative numbers, fractions, and zero.

9c. Hint: You will need to enter numbers into a few lists to use this command.

9d. Hint: There are two different possible answers depending on how you use the random command.

11a. $f(x)$: independent variable x is time in seconds; dependent variable y is height in meters.
$g(x)$: independent variable x is time in seconds; dependent variable y is velocity in meters per second.

11b. $f(x)$: domain $0 \le x < 3.2$; range $0 \le y \le 50$.
$g(x)$: domain $0 \le x < 3.2$; range $-31 \le y \le 0$.

11c. Sample answer: For the graph of $f(x)$, the ball is dropped from an initial height of 50 meters. It hits the ground after about 3.2 seconds. At the moment the ball is dropped, its velocity is 0 m/sec. For the graph of $g(x)$, the velocity starts at 0 m/sec and changes at a constant rate, becoming more and more negative.

11d. In the 1st second, the ball falls about 5 m, from 50 m at $x = 0$ to about 45 m at $x = 1$.

11e. In the 2nd second, the ball falls about 15 m, from about 45 m at $x = 1$ to about 30 m at $x = 2$.

11f. For the graph of $f(x)$, the ball hits the ground after about 3.2 seconds. For the graph of $g(x)$, at $x \approx 3.2$ the velocity is about -31 m/sec, or $g(x) \approx -31$.

13. No, the input -2 has two different output values, 3 and 0, and the input 3 has two different outputs, 1 and -2.

15a. $\dfrac{1}{2^9}$ **15b.** 5^{10}

15c. $2^{12}3^6$ **15d.** $\dfrac{1}{3^8 5^{12}}$

17a. $x = -4$ **17b.** $x = -0.75$ **17c.** $x = 4$

LESSON 8.5

1. Sample answer:

1a.

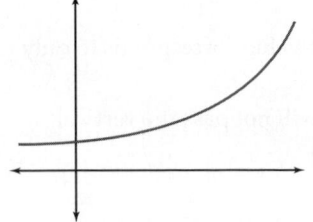

1b.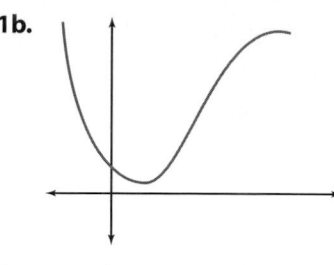

1c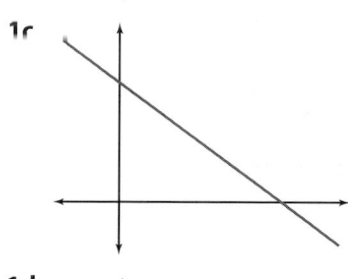

1d.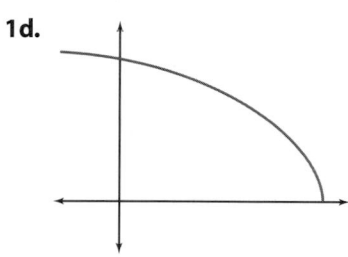

3a. Linear, increasing, constant rate of change

3b. Nonlinear, increasing, fast rate of change, then slowing down

3c. Nonlinear, decreasing, fast rate of change, then slowing down

3d. Linear, neither increasing nor decreasing, zero rate of change

5. Graph B

7. Hint: Think about some actual data values. For example, if it takes 1 volunteer 10 minutes to clean up 1 square yard of campus, how long should it take 2 volunteers? How about 3 volunteers or 4 volunteers? What kind of relationship do you see in the data?

9a. about 11:00 A.M.

9b. between 10:10 and 10:40 A.M.

9c. between 9:00 and 9:45 A.M. and then again after 11:00 A.M.

11a. licks, independent; mass, dependent

11b. scoops, independent; cost, dependent

11c. amount of stretch, independent; flight distance, dependent

11d. number of coins, independent; number of heads, dependent

13a. $y = -25.9 - 2.5x$

13b. The slope is -2.5 and the y-intercept is -25.9.

15. The five-number summary is 293, 307, 343.5, 431, 601.

1a. 7 **1b.** 0.5

1c. 5 **1d.** 9

1e. -5 **1f.** -5

1g. 12 **1h.** 3

3a. $12 = 12$ **3b.** $40 = 40$

3c. $15 > 9$ **3d.** $9 < 13$

3e. $4 = 4$ **3f.** $16 > \dfrac{1}{16}$

5.

x	-4	-1.5	0	1.2	3	4.75
y	4	1.5	0	1.2	3	4.75

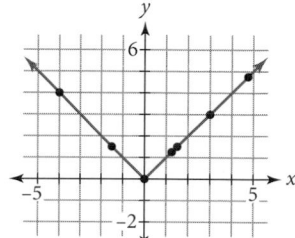

7. The solutions are $(2.85, 2.85)$ and $(-2.85, 2.85)$.

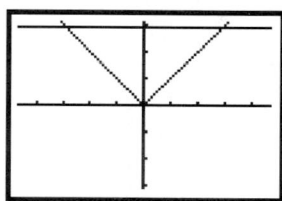

$[-4.7, 4.7, 1, -3.2, 3.2, 1]$

9. The walker starts 5 meters away from the motion sensor and walks toward the motion sensor at a rate of 1 meter per second for 4 seconds and then walks away from the motion sensor at the same rate.

11a. $g(5) = |5| + 6 = 11$

11b. $h(1) = 18(1.5) = 27$

11c. $g(-2) = |-2| + 6 = 8; h(-2) = 18(1.5)^{-2} = 8$

11d. $f(3) = 7 + 4 \cdot 3 = 19$

13. Find the mean of the absolute values of the deviations for each lake. Spider Lake is 0.675 lbs, and Doll Lake is 0.375 lbs. The fish weights varied more in Spider Lake.

15a. $-1\dfrac{2}{3} < x$ or $x > -1\dfrac{2}{3}$

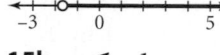

15b. $x \leq -1$

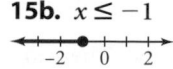

15c. $15 \geq x$ or $x \leq 15$

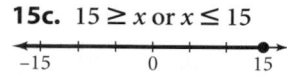

Selected Answers

1.

Side (cm)	Perimeter (cm)	Area (sq cm)
1	4	1
2	8	4
3	12	9
4	16	16
14	56	196
15	60	225
21	84	441
25.2	100.8	635.04
47	188	2209

3a. $x = \pm 6$

3b. $x = \pm 6$

3c. $x = \pm 3.8$

3d. $x = \pm 3.8$

5. $-1 \leq x \leq 1$. Good graphing window answers will vary. One possibility is a TI-83 friendly window like $[-2.35, 2.35, 1, 0, 3.1, 1]$.

7.

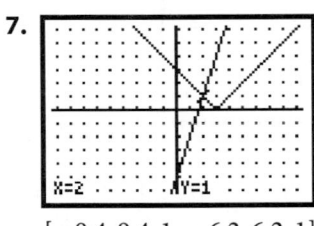

$[-9.4, 9.4, 1, -6.2, 6.2, 1]$

Sample answer: They tell us that there is exactly one solution and that the x-value that makes the equation true is $x = 2$.

9a. $1 + 3 + 5 + 7 + 9 = 25$, or 5^2.

9b. $1 + 3 + 5 + 7 + 9 + \ldots + 29 = 225$, or 15^2.

9c. The sum of the first n positive odd integers is n^2.

9d. Each larger square is created by adding a border of small squares on two sides. The number of little squares added on each time is the next odd number. The resulting figure is the next square number.

11a. 16: 1-by-1, 9: 2-by-2, 4: 3-by-3, and 1: 4-by-4

11b. A 3-by-3 grid has 14 squares: 9: 1-by-1, 4: 2-by-2, and 1: 3-by-3. A 2-by-2 grid has 5 squares: 4: 1-by-1 and 1: 2-by-2. A 1-by-1 grid has 1 square.

11c. Sample answer: One possible response: There are n^2 1-by-1 squares, $(n - 1)^2$ 2-by-2 squares, $(n - 2)^2$ 3-by-3 squares, and so on. There would be 55 squares in a 5-by-5 grid.

13a. $y = 400(0.75)^x$

13b.

x	y
0	400
4	126.5625
3	168.75
1	300
≈ -3.19	1000

1a. $-2 \leq x \leq 4$

1b. $1 \leq f(x) \leq 3$

1c. 1

1d. at $x = -1$ or 3

3. The graph is a horizontal line segment at 0.5 meters per second.

5a. Hint: Keep in mind that the graph shows time and velocity. It does not show relative position.

5b. No. Because Abby starts out moving faster than both Bea and Caitlin, even when she slows down to their speed, she stays ahead.

7a. Hint: Your graph will pass the vertical line test.

7b. Hint: Your graph will not pass the vertical line test.

9a. $f(-3) = |-3| = 3$

9b. $f(2) = |2| = 2$

9c. $x = 10$ or -10

1a. $(-2, 3), (4, 1), (2, -5)$

1b. $(-5, 1), (3, 4), (6, -3), (-3, -5)$

3a. a translation up 4 units

3b. The x-coordinates are unchanged.

3c. The y-coordinates are increased by 4.

5a. a translation right 10 units and down 8 units

5b. $L_3 = L_1 + 10, L_4 = L_2 - 8$

5c. The signs would change: $L_3 = L_1 - 10, L_4 = L_2 + 8$.

Selected Answers

7a.

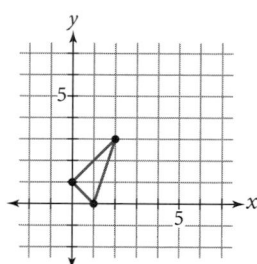

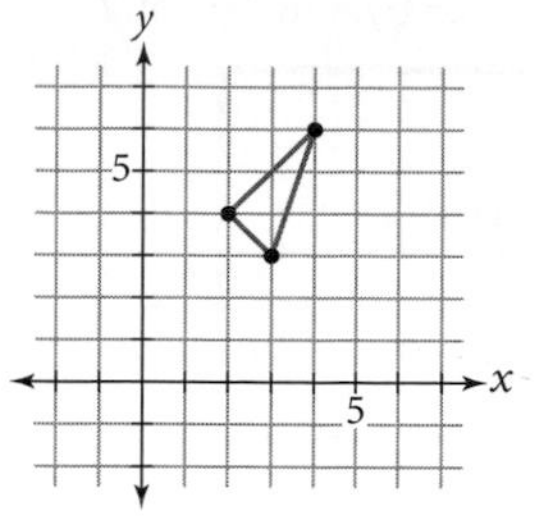

7b.

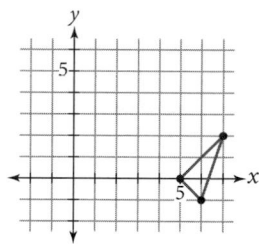

7c.

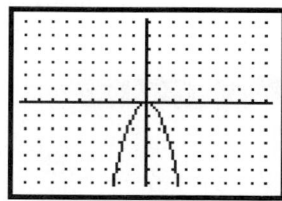

9a. a translation right 12 units and up 7 units

9b. $(x + 12, y + 7)$

9c. $\left(x + \dfrac{12}{20}, y + \dfrac{7}{20}\right) = (x + 0.6, y + 0.35)$

11a. 17   **11b.** -10

11c. $8 + 3x$ **11d.** $-1 + 6x$

13a. $y = -2 + x$ **13b.** $y = 1 - 0.5(x - 1)$

13c. $y = |x|$ **13d.** $y = 1.5^x$

LESSON 9.2

1a. 19   **1b.** 5

1c. 8   **1d.** $2|x + 6| + 1$

3a. $(1, -3)$ **3b.** $(-5, -3)$

3c. $(6, 4)$ **3d.** $(-1, 0)$

5a. $y = x^2 - 2$ **5b.** $y = 4^{(x-5)}$

5c. $y = |x + 4| + 1$

7a. The input variable is t, time, the output variable is d, distance.

7b. Time is in seconds and distance is in meters.

7c. domain: $0 \le t \le 5$; range: $1 \le d \le 4$

7d. Sample answer: Beth starts 3 m from her teacher and walks toward the teacher at 1 m/sec for 2 sec. When she turns in the test, she is 1 m from the teacher. Beth then turns and walks away from the teacher at 1 m/sec for 3 sec.

7e. $d = |t - 2| + 1$

9a. a translation right 3 units

9b. a translation left 2 units

9c. a translation down 2 units

9d. a translation up 3 units

11a. 20°C   **11b.** 221°C; 188°C

11c. approximately 15% per hour

11d. Let x represent time in hours, and let y represent temperature in degrees Celsius, $y = 221(1 - 0.15)^x$.

11e. $y = 221(1 - 0.15)^x + 20$; the vertical translation is needed to make the long-run value 20°C.

11f. $y = 221(1 - 0.15)^{(x-5)} + 20$; the horizontal translation is needed because the first temperature was measured 5 hours after the bowl was removed from the kiln.

11g. The temperature of the bowl immediately after it was removed from the kiln was approximately 518°C.

11h. To make sure the temperature is 25°C or less, round up to 29 hours.

13a. $(4, 8)$ **13b.** $y = b(x - 4) + 8$

13c. (H, V) **13d.** $y = b(x - H) + V$

15a. $(0.75, 6.5)$ **15b.** $(-3.5, -24.5)$

15c. $(3, 1)$

LESSON 9.3

1a. -1 **1b.** 37.5   **1c.** -10

1d. $0.5(-x - 3)^2 - 3$ **1e.** $-0.5(x - 3)^2 + 3$

3a. a reflection across the x-axis

3b. a translation right 6 units or a reflection across the y-axis

3c. a translation left 2 units

3d. a translation left 2 units and a reflection across the x-axis

5a. a reflection across the x-axis

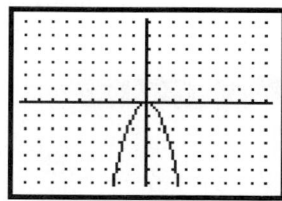

$[-9.4, 9.4, 1, -6.2, 6.2, 1]$

5b. a translation left 3 units and a reflection across the x-axis

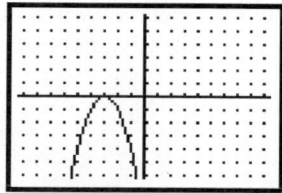

$[-9.4, 9.4, 1, -6.2, 6.2, 1]$

5c. a reflection across the *x*-axis followed by a translation up 3 units

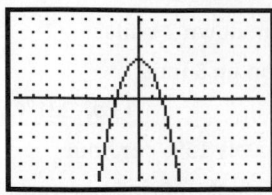

$[-9.4, 9.4, 1, -6.2, 6.2, 1]$

5d. a reflection across the *y*-axis and a translation up 3 units

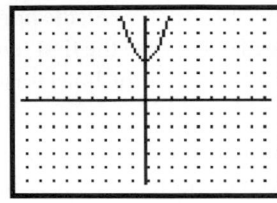

$[-9.4, 9.4, 1, -6.2, 6.2, 1]$

7a. a reflection across the *y*-axis

7b. a translation left 8 units and a reflection across the *x*-axis

7c. a translation right 2 units and down 4 units

7d. a reflection across the line $y = x$

9. $(x + 1, -y)$

11a. i. $y = -2$ **11a.** ii. $y = 3.5$

11a. iii. $x = 3$ **11a.** iv. $x = -4$ or $y = -4$

11b. i. $y = -x^2 - 4$ **11b.** ii. $y = -|x| + 7$

11b. iii. $y = 2^{-(x-6)}$

11b. iv. $y = 2(-(x + 8)) + 4$;
    $y = (4 + 2(-x)) - 16$;
    $y = -(4 + 2x) - 8$; or
    $y = -(4 + 2(x + 4))$

11c. $y = -f(x) + 2b$

11d. $y = f(-x + 2a)$

13. $47\,\text{T} \cdot \dfrac{1\,\text{cup}}{16\,\text{T}} \cdot \dfrac{1\,\text{quart}}{4\,\text{cups}} = \dfrac{47}{64}\,\text{quart} \approx 0.734\,\text{quart}$

LESSON 9.4

1. $y = |x - 5|$

3a. $y = -1.2|x - 5| + 6$

3b. Begin at the motion sensor and walk away at 1.2 m/sec for 5 sec. At 6 m away, turn and walk back at the same speed.

5. Your equation will be correct when the graphs coincide and when a table of values for Y_1 and Y_0 are identical.

7a. a vertical stretch of $y = x^2$ by a factor of 2

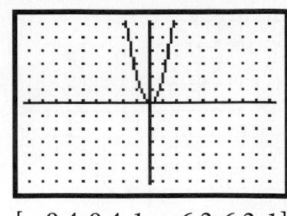

$[-9.4, 9.4, 1, -6.2, 6.2, 1]$

7b. a vertical shrink of $y = |x|$ by a factor of 0.25; then a translation right 2 units and up 1 unit

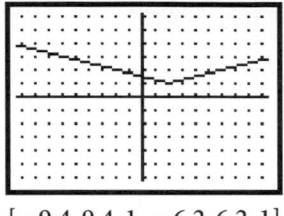

$[-9.4, 9.4, 1, -6.2, 6.2, 1]$

7c. a reflection of $y = x^2$ across the *x*-axis; then a translation left 4 units and down 1 unit

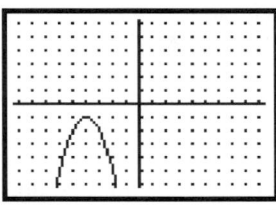

$[-9.4, 9.4, 1, -6.2, 6.2, 1]$

7d. a vertical stretch of $y = |x|$ by a factor of 2 and a reflection across the *x*-axis; then a translation right 3 units and up 4 units

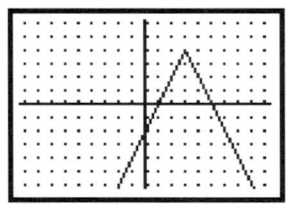

$[-9.4, 9.4, 1, -6.2, 6.2, 1]$

9a. a reflection across the *x*-axis and a translation left 3 units

9b. a vertical shrink by a factor of 0.5; then a translation right 2 units and up 1 unit

11a. a vertical shrink by a factor of 0.5 and a reflection across the *x*-axis

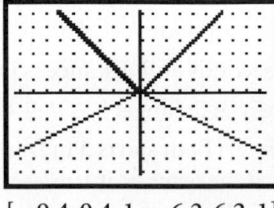

$[-9.4, 9.4, 1, -6.2, 6.2, 1]$

11b. a vertical stretch by a factor of 2 and a translation right 4 units

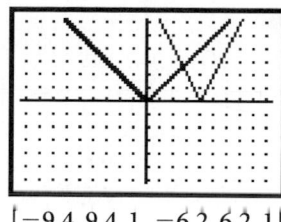

$[-9.4, 9.4, 1, -6.2, 6.2, 1]$

11c. a vertical stretch by a factor of 3 and a reflection across the x-axis; then a translation left 2 units and up 4 units

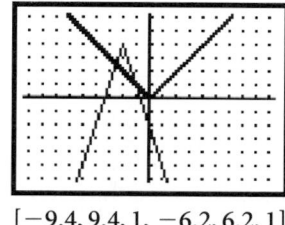

$[-9.4, 9.4, 1, -6.2, 6.2, 1]$

13. $(x - 1, 0.8y)$

15a. $\dfrac{1}{2^9}$　　　　**15b.** 5^{10}

15c. $2^{12} \cdot 3^6$　　　　**15d.** $\dfrac{1}{3^8 \cdot x^{12}}$

LESSON 9.6

1a. a reflection of the graph of $y = x^2$ across the x-axis; then a translation up 2 units (or translation then reflection); $y = -x^2 + 2$

1b. a vertical shrink of the graph of $y = |x|$ by a factor of $\frac{1}{3}$ and a translation right 2 units; $y = \frac{1}{3}|x - 2|$

1c. a vertical shrink of the graph of $y = x^2$ by a factor of 0.5; then a translation right 1 unit and down 1 unit; $y = 0.5(x - 1)^2 - 1$

1d. a vertical stretch of the graph of $y = |x|$ by a factor of 2 and a reflection across the x-axis; then a translation left 2 units and up 3 units; $y = -2|x + 2| + 3$

3. $y = -\dfrac{5}{x}$

5a. decreasing; asymptotes at $x = 0$ and $y = 0$; domain: $x \neq 0$; range: $y \neq 0$

5b. decreasing; asymptotes at $x = 5$ and $y = -2$; domain: $x \neq 5$; range: $y \neq -2$

5c. decreasing; asymptotes at $x = 0$ and $y = 3$; domain: $x \neq 0$; range: $y \neq 3$

5d. increasing; asymptotes at $x = -3$ and $y = 0$; domain: $x \neq -3$; range: $y \neq 0$

7a. $y = f(-x)$ or $y = \dfrac{1}{-x}$

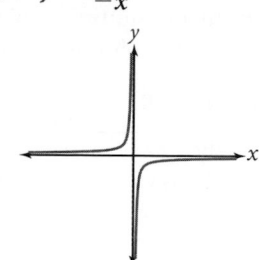

7b. $y = -f(x)$ or $y = -\dfrac{1}{x}$ (See sketch above.)

7c. Either reflection produces the same graph because $\dfrac{1}{-x} = -\dfrac{1}{x}$.

9. Let x represent the amount of water to add and y represent the concentration of salt. The amount of salt is $0.05(0.5)$.

$y = \dfrac{0.025}{0.5 + x}$; $0.01 = \dfrac{0.025}{(0.5 + x)}$; $x = 2$; 2 liters.

11a. \$700

11b.

Additional businesses	0	5	10	15	20
Cost per business (\$)	700.00	350.00	233.33	175.00	140.00

11c. $y = \dfrac{3500}{5 + x}$, where x is the number of additional businesses that have signed up and y is the cost per business.

11d. $150 = \dfrac{3500}{5 + x}$; $x = 18.\overline{3}$; 19 additional businesses.

13a. $x < -2$

13b. $x \geq -2$

13c. $x \geq 4$

15. vertex: $(-1, 2)$; point to the left: $(-2, -1)$; point to the right: $(0, -1)$

LESSON 9.7

1a. $(-2, 2), (1, 2), (-2, 6)$

1b. $\begin{bmatrix} 0 & 0 & 0 \\ -3 & -3 & -3 \end{bmatrix}$　　**1c.** $\begin{bmatrix} -2 & 1 & -2 \\ -1 & -1 & 3 \end{bmatrix}$

3a. $\begin{bmatrix} 6 & 15 \end{bmatrix}$　　**3b.** $\begin{bmatrix} 6 \\ 15 \end{bmatrix}$

3c. $\begin{bmatrix} 64 \end{bmatrix}$　　**3d.** $\begin{bmatrix} 8 & 32 \\ 14 & 56 \end{bmatrix}$

5a. a rectangle

5b. For the *x*-coordinate, multiply row 1 of the transformation matrix by column 2 of the quadrilateral matrix; $[1 \quad 0] \cdot \begin{bmatrix} 2 \\ -1 \end{bmatrix} = 2$; this goes in row 1, column 2, of the image matrix. For the *y*-coordinate, multiply row 2 of the transformation matrix by column 2 of the quadrilateral matrix;

$[0 \quad 2] \cdot \begin{bmatrix} 2 \\ -1 \end{bmatrix} = -2$; this goes in row 2, column 2, of the image matrix.

5c. $\begin{bmatrix} -1 & 2 & 1 & -2 \\ 4 & -2 & -4 & 2 \end{bmatrix}$

5d. a parallelogram

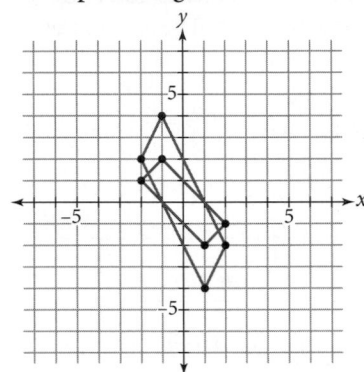

7a. $[Q] = \begin{bmatrix} 2 & 3 & 6 & 7 \\ 2 & 4 & 5 & 1 \end{bmatrix}$

7b. $\begin{bmatrix} 1 & 0 \\ 0 & 0.5 \end{bmatrix} \cdot [Q] = \begin{bmatrix} 2 & 3 & 6 & 7 \\ 1 & 2 & 2.5 & 0.5 \end{bmatrix}$

7c. $\begin{bmatrix} 0.5 & 0 \\ 0 & 0.5 \end{bmatrix} \cdot [Q] = \begin{bmatrix} 1 & 1.5 & 3 & 3.5 \\ 1 & 2 & 2.5 & 0.5 \end{bmatrix}$

7d. $\begin{bmatrix} -1 & 0 \\ 0 & -1 \end{bmatrix}; \begin{bmatrix} -1 & 0 \\ 0 & -1 \end{bmatrix} \cdot [Q]$

$= \begin{bmatrix} -2 & -3 & -6 & -7 \\ -2 & -4 & -5 & -1 \end{bmatrix}$; a reflection across both the

x- and *y*-axes; alone, $\begin{bmatrix} 1 & 0 \\ 0 & -1 \end{bmatrix}$ results in a reflection

across the *x*-axis and $\begin{bmatrix} -1 & 0 \\ 0 & 1 \end{bmatrix}$ results in a reflection

across the *y*-axis.

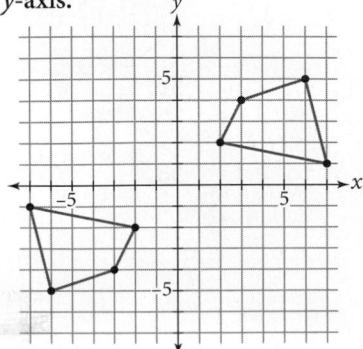

7e. $\begin{bmatrix} -2 & -4 & -5 & -1 \\ 2 & 3 & 6 & 7 \end{bmatrix}$; a quarter-turn counterclockwise

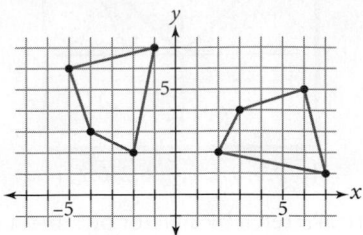

9a. Sample answer: $\begin{bmatrix} 0 & 3 & 4 \\ 0 & 2 & 0 \end{bmatrix}$

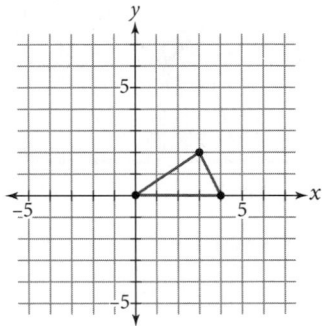

9b.

$\begin{bmatrix} 0.5 & -0.866 \\ 0.866 & -0.5 \end{bmatrix} \cdot \begin{bmatrix} 0 & 3 & 4 \\ 0 & 2 & 0 \end{bmatrix} = \begin{bmatrix} 0 & -0.232 & 2 \\ 0 & 3.598 & 3.464 \end{bmatrix}$

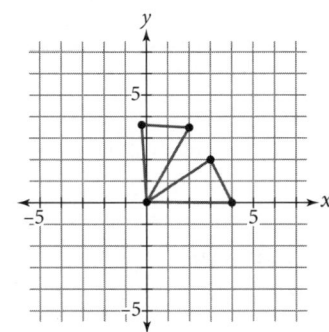

9c. a rotation 60° about the origin, or the point $(0, 0)$

9d. $\begin{bmatrix} -0.500 & -0.866 \\ 0.866 & -0.5 \end{bmatrix}$;

$\begin{bmatrix} -0.500 & -0.866 \\ 0.866 & -0.5 \end{bmatrix} \cdot \begin{bmatrix} 0 & 3 & 4 \\ 0 & 2 & 0 \end{bmatrix} = \begin{bmatrix} 0 & -3.232 & -2 \\ 0 & 1.598 & 3.464 \end{bmatrix}$;

a rotation 120° about the origin

(See the graph at the top of page 679.)

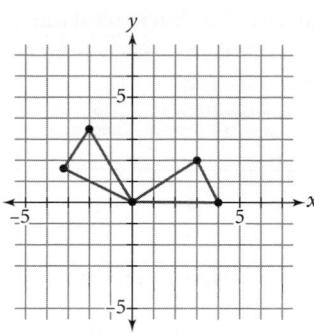

9e. Multiply $\begin{bmatrix} 0.5 & -0.866 \\ 0.866 & -0.5 \end{bmatrix}$ by itself three times

$(3 \cdot 60° = 180°)$. Then multiply the result by matrix $[R]$.

9f. Multiply $\begin{bmatrix} 0.5 & -0.866 \\ 0.866 & -0.5 \end{bmatrix}$ by itself six times

$(6 \cdot 60° = 360°)$. Then multiply the result by matrix

$[R]$ or just multiply $[R]$ by $\begin{bmatrix} 1 & 0 \\ 0 & 1 \end{bmatrix}$.

11a. in millions: 114, 127, 159, 273, 1247

11b. **Most Populated Countries, 1999**

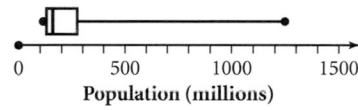

Population (millions)

11c. China and India

CHAPTER 9 REVIEW

1a. a translation left 2 units and up 1 unit

1b. $(x - 2, y + 1)$

3. For these sample answers, list L₃ and list L₄ are used for the x- and y-coordinates, respectively, of each image.

3a. $L_3 = L_1, L_4 = -L_2$

3b. $L_3 = -L_1, L_4 = L_2$

3c. $L_3 = L_1 + 3, L_4 = -L_2$

5. $g(x) = f(x - 1) + 2$

7a. The graph should have the same x-intercept as $f(x)$. The y-intercept should be the opposite of that for $f(x)$.

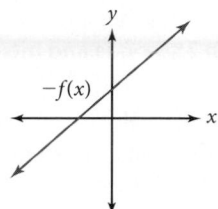

7b. Hint: Y₂ = −Y₁ is the same as $y = -f(x)$

9a. a translation of $y = \dfrac{1}{x}$ right 3 units and down 2 units; $y = \dfrac{1}{x - 3} - 2$

9b. a translation of $y = 2^x$ right 4 units and down 2 units; $y = 2^{(x-4)} - 2$

9c. a reflection of $y = 2^x$ across the x-axis and across the y-axis, followed by a translation up 3 units (or a reflection across the x-axis, followed by a translation up 3 units, followed by a reflection across the y-axis); $y = -2^{(-x)} + 3$

9d. a vertical stretch of $y = \dfrac{1}{x}$ by a factor of 4 and a reflection across the x-axis, followed by a translation up 1 unit and left 2 units; $y = -\dfrac{4}{x + 2} + 1$

LESSON 10.1

1. Enter one side of the equation into Y₁ and the other into Y₂ on the calculator. Then find the number of intersection points.

1a. two solutions

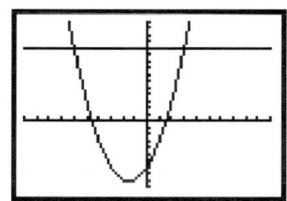

$[-10, 10, 1, -10, 15, 1]$

1b. no solutions

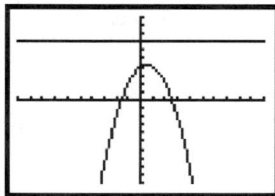

$[-10, 10, 1, -10, 10, 1]$

1c. one solution

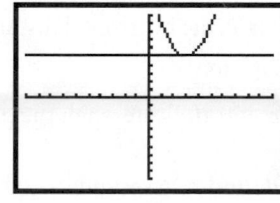

$[-10, 10, 1, -10, 10, 1]$

Selected Answers

1d. two solutions

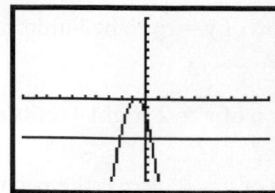

$$[-10, 10, 1, -10, 10, 1]$$

3a. $x = \pm\sqrt{18}$

3b. $x^2 = 49, x = \pm 7$

3c. $x - 2 = \pm 5, x = 7$ or $x = -3$

3d. $2(x + 1)^2 = 14, (x + 1)^2 = 7, x + 1 = \pm\sqrt{7},$
$x = -1 \pm\sqrt{7}$

5a. $h(0) = -4.9(0)^2 + 147$, or 147. It is the starting height of the ball, when $t = 0$ seconds.

5b. Solve the equation $20 = -4.9t^2 + 147$ symbolically to get $\pm\sqrt{\dfrac{20 - 147}{-4.9}}$ or approximately ± 5.09. The graph shows two solution points: $(5.09, 20)$ and $(-5.09, 20)$.

5c. No, only the solution $(5.09, 20)$ makes sense because the time t must be positive.

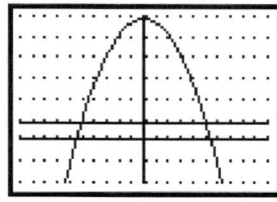

$$[-10, 10, 1, -50, 150, 25]$$

5d. $t > 5.09$ seconds

5e. The ball hits the ground when $t \approx 5.48$ seconds because the positive x-intercept is near the point $(5.48, 0)$.

7a. After the ball has gone 2 m horizontally, it will be approximately 3.75 m above the ground.

7b. The ball will be 2 m above the ground once it has gone 0.31 m horizontally and again once it has gone 6.49 m horizontally. Exact answers are $3.4 \pm\sqrt{\dfrac{220}{23}}$.

7c. When $x = 0$, the height is 1.5412 m.

7d. When $p(x) = 0$, the horizontal distance is $3.4 + \sqrt{\dfrac{420}{23}}$, or 7.67 m.

9a. The x-intercepts indicate when the projectile is at ground level.

9b. 2.74 seconds and 7.26 seconds

9c. Find the average of 2.74 and 7.26, which is 5. $h(5) = 25$.

9d. The projectile is 25 meters above the ground, its maximum height, after 5 seconds.

9e. $h(3.2) = 9.124$ meters. This is the height at 3.2 seconds.

9f. Sample answer: The horizontal line $y = 12.5$ intersects the parabola twice—when $x \approx 3.4$ seconds and when $x \approx 6.6$ seconds.

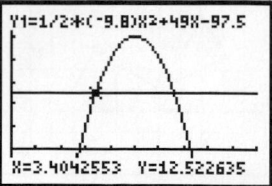

$$[0, 10, 1, -5, 30, 5]$$

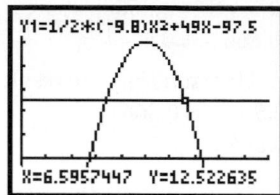

$$[0, 10, 1, -5, 30, 5]$$

LESSON 10.2

1. The vertex has an x-value of $\dfrac{3 + (-2)}{2}$, or 0.5.

3.
$$0 = (x + 1.5)^2 - 7.25$$
$$7.25 = (x + 1.5)^2$$
$$\pm\sqrt{7.25} = x + 1.5$$
$$-1.5 \pm\sqrt{7.25} = x$$
$$x \approx 1.92582404 \text{ or }$$
$$x \approx -4.192582404$$

5a.
$$(x + 3) = \pm\sqrt{7}$$
$$x = -3 \pm\sqrt{7}$$
$$x \approx -5.646, -0.354$$

5b.
$$(x - 2)^2 = 21$$
$$x = 2 \pm\sqrt{21}$$
$$x \approx -2.583, 6.583$$

7a. The ball is on the ground when $h = 0$. This happens when $t = 0$ seconds and $t = 3$ seconds.

7b. The ball is at its highest point when $t = 1.5$ seconds, halfway through its flight.

7c. $h = 36$ when $t = 1.5$, so the ball goes 36 feet high.

9a. Sample answer: The ball is thrown from an initial height of 5 meters. It reaches a maximum height of about 25 meters in 2 seconds and hits the ground at about 4.3 seconds.

9b. The ball starts at a velocity of 20 meters per second and slows down at a constant rate. At 2 seconds it is not moving. Then it starts falling and is moving down at 22 meters per second when it hits the ground.

9c. As the ball moves up to its maximum height, it is slowing down, moving at 0 meters per second at its peak. It then starts falling and speeding up until it hits the ground.

9d. This is when the ball is at its maximum height and not moving. Its velocity is zero.

9e. These are the two times when the ball is about 13 meters high.

9f. domain: $0 \le t \le 4.3$; range: $0 \le h(t) \le 26$; $-22 < v(t) \le 20$

11a.

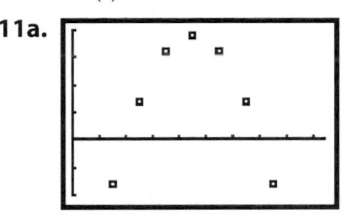

$[0, 9.4, 1, -10, 20, 5]$

11b. It is the vertical line $x = 4.5$.

11c. $(4.5, 19)$

11d. $y = -3(x - 4.5)^2 + 19$

LESSON 10.3

1a. yes; three terms (trinomial)

1b. yes; two terms (binomial)

1c. no; The first term has a negative exponent.

1d. no; The first term is equivalent to $3x^{-2}$, which has a negative exponent.

1e. yes; one term (monomial)

1f. yes; three terms (trinomial)

1g. yes; two terms (binomial)

1h. Not a polynomial as written but it is equivalent to $3x - 6$, a binomial.

3a.

	x	2
x	x^2	$2x$
2	$2x$	4

$(x + 2)^2 = x^2 + 4x + 4$

3b.

	x	12
x	x^2	$12x$
12	$12x$	144

$(x + 12)^2 = x^2 + 24x + 144$

3c.

	x	-7
x	x^2	$-7x$
-7	$-7x$	49

$(x - 7)^2 = x^2 - 14x + 49$

5a.

	x	
x	x^2
4	$4x$

$+$

	2
x	$2x$
4	8

$(x + 2)(x + 4) = x^2 + 6x + 8$

5b.

	x
x	x^2
5	$5x$

$+$

	3
x	$3x$
5	15

$(x + 3)(x + 5) = x^2 + 8x + 15$

5c.

	x
x	x^2
-5	$-5x$

$+$

	2
x	$2x$
-5	-10

$(x - 5)(x + 2) = x^2 - 3x - 10$

5d.

	x
x	x^2
-3	$-3x$

$+$

	0
x	$0x$
-3	0

$(x - 0)(x - 3) = x^2 - 3x$

7. No, by squaring the values inside the parentheses, Heather is accounting for only two of the four rectangles in a squaring diagram. She needs to add the two rectangles that sum to the middle term.

9a. The vertex is at $(0.4, 2.5)$, so the ball reaches a maximum height of 2.5 meters.

9b. $h(t) = -4.9t^2 + 3.92t + 1.716$

9c. The pitcher released the ball at a height of 1.716 meters.

11a. meaningful domain: $0 \le x \le 6.5$; meaningful range: $0 \le y \le 897.81$

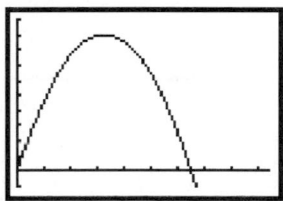

$[0, 9.4, 1, -100, 1000, 100]$

11b. Sample answer: Average the two x-intercepts, then substitute this value into the equation to find the y-coordinate of the vertex. The vertex is $(3.25, 897.8125)$.

11c. The x-coordinate is the price that produces maximum income. The y-coordinate is the maximum income for this relationship.

11d. The prices produce no income.

11e. If the warehouse charges $5 per item, income will be $637.50.

13. $x^2 + 8x + 16$

15. The message is POLYNOMIALS.

15a. 16; P **15b.** 15; O **15c.** 12; L

15d. 25; Y **15e.** 14; N **15f.** 15; O

15g. 13; M **15h.** 9; I **15i.** 1; A

15j. 12; L **15k.** 19; S

LESSON 10.4

1a. $x + 4 = 0$ or $x + 3.5 = 0$, so $x = -4$ or $x = -3.5$

1b. $x - 2 = 0$ or $x - 6 = 0$, so $x = 2$ or $x = 6$

1c. $x + 3 = 0$ or $x - 7 = 0$ or $x + 8 = 0$, so $x = -3$ or $x = 7$ or $x = -8$

1d. $x = 0$ or $x - 9 = 0$ or $x + 3 = 0$, so $x = 0$ or $x = 9$ or $x = -3$

3a. 7 and -2

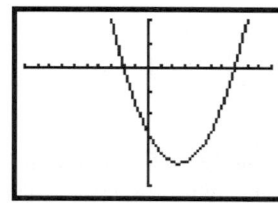

$[-10, 10, 1, -25, 10, 5]$

3b. -1 and -8

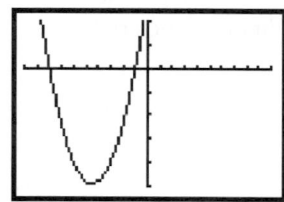

$[-10, 10, 1, -25, 10, 5]$

3c. 11 and -7

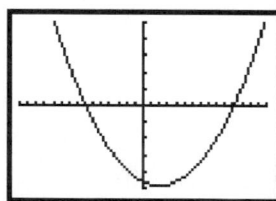

$[-15, 15, 1, -250, 250, 50]$

3d. -5 and 9

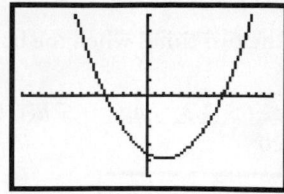

$[-15, 15, 1, -25, 25, 5]$

5a. two intercepts; $x = -1$ and $x = 3$

5b. $(1, -4)$

5c. $y = (x - 1)^2 - 4$; There is a translation left 1 unit and down 4 units.

7a. See below.

7b. The sum of the roots needs to be -2, and the product needs to be -8; 2 and -4 satisfy this requirement.

9. $y = 0.25(x + 3)(x - 9)$

11a. two x-intercepts: $x = 3$ and $x = -3$

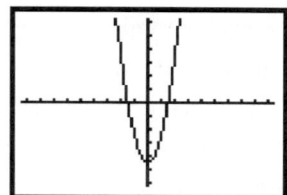

$[-18.8, 18.8, 2, -12.4, 12.4, 2]$

11b. $y = (x - 3)(x + 3)$

11c. The x-intercepts are the positive and negative square roots of the amount the parabola is translated, or $\pm\sqrt{h}$.

11d. i. $y = (x + 7)(x - 7)$

11d. ii. $y = (4 + x)(4 - x)$

11d. iii. $y = (x + \sqrt{47})(x - \sqrt{47})$

11d. iv. $y = (x + \sqrt{28})(x - \sqrt{28})$

7a. *(Lesson 10.4)*

Factored form	Roots	Sum of roots	Product of roots	General form
$y = (x + 3)(x - 4)$	-3 and 4	$-3 + 4 = 1$	$(-3)(4) = -12$	$y = x^2 - 1x - 12$
$y = (x - 5)(x + 2)$	5 and -2	$5 - 2 = 3$	$5(-2) = -10$	$y = x^2 - 3x - 10$
$y = (x + 2)(x + 3)$	-2 and -3	-5	6	$y = x^2 + 5x + 6$
$y = (x - 5)(x + 5)$	5 and -5	0	-25	$y = x^2 - 25$

11e. There are no x-intercepts. The graph is above the x-axis.

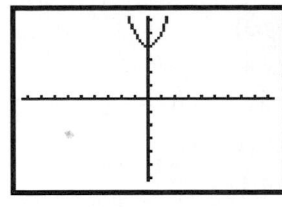

$[-9.4, 9.4, 1, -6.2, 6.2, 1]$

11f. Because there are no x-intercepts, there is no factored form of the equation.

13. Sample answer:

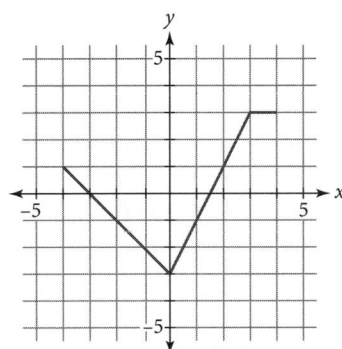

LESSON 10.6

1a. $x = -3 \pm \sqrt{2}$ **1b.** $x = 5 \pm \sqrt{2}$

1c. $x = -8 \pm \sqrt{\frac{7}{3}}$ **1d.** $x = -6 \pm \sqrt{\frac{7}{5}}$

3a. $\left(\frac{18}{2}\right)^2; x^2 + 18x + 81 = (x + 9)^2$

3b. $\left(\frac{-10}{2}\right)^2; x^2 + 10x + 25 = (x - 5)^2$

3c. $\left(\frac{3}{2}\right)^2; x^2 + 3x + \frac{9}{4} = \left(x + \frac{3}{2}\right)^2$

3d. $\left(\frac{-1}{2}\right)^2; x^2 - x + \frac{1}{4} = \left(x - \frac{1}{2}\right)^2$

3e. $\left(\frac{1}{2} \cdot \frac{2}{3}\right)^2; x^2 + \frac{2}{3}x + \frac{1}{9} = \left(x + \frac{1}{3}\right)^2$

3f. $\left(\frac{-1.4}{2}\right)^2; x^2 - 1.4x + 0.49 = (x - 0.7)^2$

5a. $y = a(x - 2)^2 - 31.5$

5b. Solve the equation $0 = a(5 - 2)^2 - 31.5; a = 3.5$.

5c. $y = 3.5(x - 2)^2 - 31.5$

5d. $y = -2(x - 2)^2 + 32$

7a. $y = x^2 + 6x + \left(\frac{6}{2}\right)^2 - 3^2 + 10;$
$y = x^2 + 6x + 9 - 9 + 10; y = (x + 3)^2 + 1$

7b. vertex $(-3, 1)$;

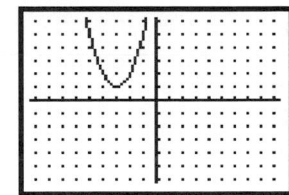

$[-9.4, 9.4, 1, -6.2, 6.2, 1]$

7c. $x^2 + 6x + 10 = 0$

$$x^2 + 6x = -10$$
$$x^2 + 6x + 9 = -10 + 9$$
$$(x + 3)^2 = -1$$
$$x + 3 = \pm\sqrt{-1}$$
$$x = -3 \pm\sqrt{-1}$$

Sample answer: There are no real roots because the graph doesn't cross the x-axis.

9a. $p = 2500 - 5x$, where p represents the cost in dollars of a single ticket, and x represents the number of tickets sold. Let C represent the total cost of the group package.

9b. $C = xp = x(2500 - 5x)$

9c. $C = -5(x - 250)^2 + 312,500$

9d. $C(20) = 20(2500 - 100) = 48,000; \$48,000$

9e.
$$2500x - 5x^2 = 200,000$$
$$500x - x = 40,000$$
$$x^2 - 500x = -40,000$$
$$x^2 - 500x + 250^2 = 250^2 - 40,000$$
$$(x - 250)^2 = 22,500$$
$$x - 250 = \pm 150$$
$$x = 100 \text{ or } x = 400$$

9f. Sample answer: The company will lose money if more than 400 people or less than 100 people per group go on the cruise. The vertex is $(250, 312500)$, so the company should limit groups to 250 people to maximize earnings at $312,500.

11a. $2x^3 + 5x^2 + 4x + 1$

11b. $6x^3 - 11x^2 - 18x + 20$

LESSON 10.7

1a. $25 - 24 = 1$

1b. $9 - (-12) = 9 + 12 = 21$

1c. $36 - 24 = 12$

1d. $81 - 0 = 81$

3a. $x = \dfrac{3 \pm \sqrt{-23}}{4}$. There are no real solutions.

3b. Use the quadratic equation or complete the square.
$x = \dfrac{1}{2}$ and $x = 3$

3c. Complete the square.
$$x^2 - 6x - 8 = 0$$
$$x^2 - 6x = 8$$
$$x^2 - 6x + 9 = 17$$
$$(x - 3)^2 = 17$$
$$x - 3 = \pm\sqrt{17}$$
$$x = 3 \pm \sqrt{17}$$

3d. Use the quadratic formula.
$x = \dfrac{-2 \pm \sqrt{76}}{6} = \dfrac{-1 \pm \sqrt{19}}{3}$

5a. $-4.9t^2 + 6.2t + 1.9 = 0$; $t \approx -0.255$ sec or $t \approx 1.52$ sec. The ball hits the ground after 1.52 seconds.

5b. $-4.9t^2 + 6.2t + 1.9 = 3$; $t \approx 0.21$ sec or $t \approx 1.05$ sec. The ball is 3 meters high when $t \approx 0.21$ seconds and when $t \approx 1.05$ seconds.

5c. $-4.9t^2 + 6.2t + 1.9 = 4$;

$t \approx \dfrac{-6.2 \pm \sqrt{-2.72}}{-9.8}$. This equation has no real

solutions, so the ball is never 4 meters high.

7. When $b^2 - 4ac$ is negative, there are no real roots. When it's zero, there is one real root. When it's positive, there are two roots.

7a. i; $1^2 - 4(1)(1) = -3$, no x-intercept

7b. iii; $2^2 - 4(1)(1) = 0$, one x-intercept

7c. ii; $3^2 - 4(1)(1) = 5$, two x-intercepts

9. You need to know the t-value when $h = 0$.

Solve $0 = -4.9t^2 + 17t + 2.2$ with $a = -4.9$, $b = 17$, $c = 2.2$ to get $t \approx -0.125$ and $t \approx 3.59$. The positive solution of 3.59 seconds makes sense in this situation.

11a. $x - 2$

11b. $x + 6$

11c. $x - 1$

LESSON 10.8

1a. perfect square; $47^2 = 2{,}209$

1b. perfect cube: $18^3 = 5{,}832$

1c. neither

1d. perfect square; $101^2 = 10{,}201$

3a. $4x(x + 3)$

3b. $2x(3x - 2)$

3c. $7x(2x^3 + x - 3)$

3d. $3x^2(4x^3 + 2x + 1)$

5a. $y = 0.5(x + 4)(x + 2)(x - 1)$

5b. $y = -(x + 2)(x - 1)^2$

7a. $V = w(w + 6)(w - 2)$

7b. Three solutions to the equation $47 = w(w + 6)(w - 2)$ are shown on the graph. However, only one solution is a positive value. A table gives the answer $w \approx 3.4$. Widths greater than about 3.4 cm give volumes greater than 47 cm³.

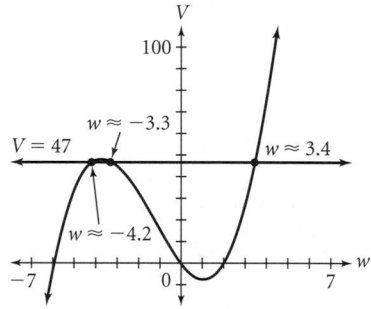

9a. $x^3 + 6x^2 + 11x + 6$

9b. $x^3 - 3x^2 - 4x + 12$

11a. $11x^3 + 2x^2 + 2x + 12$

11b. $5x^3 - 2x^2 - 12x - 12$

11c. $-6x^2 - 13x + 20$

11d. $10x^2 + x + 2$

11e. $-16x^4 + 34x^3 - 28x^2 - 131x + 99$

CHAPTER 10 REVIEW

1a. false; $(x - 3)(x + 8)$

1b. false; $2x^2 - 4x + 5$

1c. false; $x^2 + 6x + 9$

1d. true

3a. $y = -(x - 2)^2 + 3$; vertex form

3b. $y = 0.5(x - 2)(x + 3)$; factored form

5a. $2w + 9 = 0$ or $w - 3 = 0$; $w = -4.5$ or $w = 3$

5b. $2x + 5 = 0$ or $x - 7 = 0$; $x = -2.5$ or $x = 7$

7a. $x^2 + 6x - 9 = 13$

$$x^2 + 6x = 22$$
$$x^2 + 6x + 9 = 22 + 9$$
$$(x + 3)^2 = 31$$
$$x + 3 = \pm\sqrt{31}$$
$$x = -3 \pm\sqrt{31}$$

7b. $3x^2 - 24x + 27 = 0$

$$3x^2 - 24x = -27$$
$$x^2 - 8x = -9$$
$$x^2 - 8x + 16 = -9 + 16$$
$$(x - 4)^2 = 7$$
$$x - 4 = \pm\sqrt{7}$$
$$x = 4 \pm\sqrt{7}$$

9a. $f(60) = 8.1$. When there are 60 fish in the tank, the population is growing at a rate of about 8 fish per week.

9b. $f(x) = 0$ for $x = 0$ and $x = 150$. When there are no fish, the population does not grow. When there are 150 fish, the number of fish hatched is equal to the number of fish that die so the total population does not change.

9c. When there are 75 fish, the population is growing fastest.

9d. The population no longer grows once there are 150 fish, so this is the maximum number of fish the tank has to support.

9e.

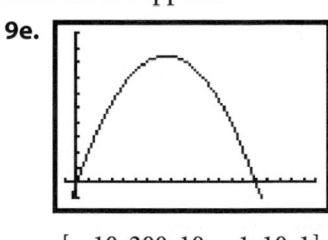

$[-10, 200, 10, -1, 10, 1]$

11a. No x-intercepts means taking the square root of a negative number. So $(-6)^2 - 4(1)(c) < 0$; $-4c -36$; $c > 9$. Or translate the graph of $y = x^2 - 6x$ vertically to see that for values of $c > 9$, the parabola does not cross the x-axis.

11b. One x-intercept implies a double root, so $x^2 - 6x + c$ must be a perfect-square trinomial. Make a rectangular diagram to find $\left(\frac{-6}{2}\right)^2 = 9$, so $x^2 - 6x + 9$ is a perfect-square trinomial and $c = 9$. The graph touches the x-axis once. You can solve $b^2 - 4ac = 36 - 4c = 0$ to get $c = 9$.

11c. For $c < 9$, $b^2 - 4ac > 0$, so the discriminant gives two real roots. Also, the parabola $y = x^2 - 6x + c$ crosses the x-axis twice for values of c that are less than 9.

13a. $x = -2, -1, 1$, and 3; $y = 2(x - 3)(x + 2)(x + 1)(x - 1)$

13b. $x = -2$ (double root) and 3; $y = -3(x + 2)^2(x - 3)$

LESSON 11.1

1a. 0.8   **1b.** -2 **1c.** -1.25 **1d.** 2

1e. $\frac{3}{2}$ **1f.** $-\frac{3}{2}$ **1g.** $\frac{3}{2}$ **1h.** $\frac{2}{3}$

3a. perpendicular   **3b.** neither

3c. parallel   **3d.** perpendicular

5. rectangle; slopes: $\frac{1}{2}, -2, \frac{1}{2}, -2$

7. parallelogram; slopes: $1, -3, 1, -3$

9. trapezoid; slopes: $2, -\frac{5}{2}, -\frac{1}{4}, -\frac{5}{2}$

11. quadrilateral; slopes: $1, -1, \frac{5}{3}, -\frac{5}{3}$

13a. $\left(-\frac{2}{3}, 0\right)$ **13b.** $(-1.\overline{3}, -0.58\overline{3})$

13c. $(0, -1)$ **15.** $\frac{6}{16}$ or $\frac{3}{8}$

LESSON 11.2

1. 1.2

3a. $(0.5, 1.5)$

3b. $(6, -4.5)$

5. $\left(\dfrac{a + c}{2}, \dfrac{b + d}{2}\right)$

7. One line is horizontal and the other is vertical. A horizontal line has a slope of 0 and a vertical line has an undefined slope, so the product is also undefined.

9a. midpoint of $\overline{AB}$: $(10, 3)$
midpoint of $\overline{BC}$: $(15, 8)$
midpoint of $\overline{CD}$: $(9, 10)$
midpoint of $\overline{DA}$: $(4, 5)$

9b. parallelogram: The opposite sides are parallel because the slopes are $1, \frac{-1}{3}, 1$, and $\frac{-1}{3}$.

9c. No; the slopes of the diagonals are $\frac{3}{11}$ and -7.

11a. $(12, 4)$ **11b.** $(3, 5)$ **11c.** $(-5, 3.5)$

13a. Sample answers:

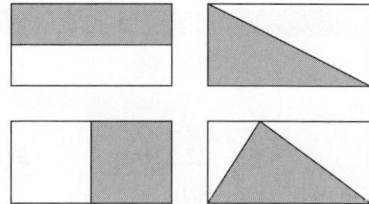

Selected Answers

13b. Any method that uses only one line segment will form congruent polygons. Some other methods may also produce congruent polygons.

13c. Yes. If you imagine a vertical segment through the upper vertex of the triangle, you can see that the triangles on either side of the vertical line are congruent.

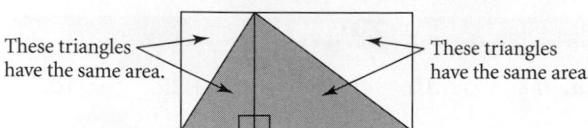

These triangles have the same area.

These triangles have the same area.

So the area of the rectangle is divided in half.

LESSON 11.3

1a. $\pm\sqrt{47}$ **1b.** $4 \pm \sqrt{28}$

1c. $-2 \pm \sqrt{14}$ **1d.** $1 \pm \sqrt{7}$

3a. 4 square units **3b.** 12 square units

3c. 2 square units **3d.** 6 square units

3e. 20 square units **3f.** 18 square units

5. polygon 3a: $\sqrt{8}$ units and $\sqrt{2}$ units

polygon 3b: $\sqrt{8}$ units and $\sqrt{18}$ units

polygon 3e: $\sqrt{50}$ units, $\sqrt{50}$ units, and $\sqrt{40}$ units

7a. 36 square units, 18 square units, 18 square units

7b. Length of $\overline{AB}$: 6 units; length of $\overline{BC}$: $\sqrt{18}$ units; length of $\overline{AC}$: $\sqrt{18}$ units.

9a. $\overline{AC}$ is the hypotenuse; $\overline{AB}$ and BC are the legs.

9b.

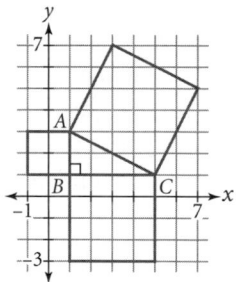

9c. 20 square units, 16 square units, 4 square units

9d. Length of $\overline{AB}$: 2 units; length of $\overline{BC}$: 4 units; length of $\overline{AC}$: $\sqrt{20}$ units

9e. The areas of the two smaller squares add up to the area of the larger square.

11. $y = 2(x + 7)(x + 1)$

LESSON 11.4

1. 576 cm²

3. $c = \sqrt{300}$ cm

5a. 13.5 ft **5b.** 22.5 ft **5c.** 2160 ft²

7a. approximately 21.6 cm-by-27.6 cm

7b. approximately 35.0 cm

7c. Answers will vary but should be close to 35 cm.

7d. The two results should be approximately the same.

9. approximately 7010 ft

11a. $L_1 = \{1, 2, 3, \ldots, 26\}$

11b. approximately 26.98 in.

11c. $L_2 = \sqrt{(27^2 - L_1{}^2)}$

11d. approximately 53.85 in²

11e. $L_3 = L_1 \cdot L_2$, or $L_3 = L_1 \cdot \sqrt{(27^2 - L_1{}^2)}$

11f. A model that works is $y = x\sqrt{27^2 - x^2}$, where x is the width and y is the area.

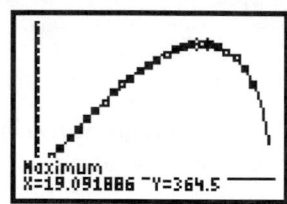

$[0, 27, 5, -50, 450, 50]$

11g. Trace the graph or use a table to find that a 19-by-19-in. square gives maximum area.

13a. $\frac{13}{12} = \frac{a}{8}$, $a = 8.\overline{6}$; 8 ft 8 in. long

13b. $\frac{5}{12} = \frac{14 - b}{8}$, $b = 10\frac{2}{3}$; 10 ft 8 in. tall.

LESSON 11.5

1a. $3\sqrt{3}$ **1b.** $5\sqrt{2}$

1c. $2 + \sqrt{6}$ **1d.** $4\sqrt{5} + 5\sqrt{2}$

1e. $6\sqrt{6}$ **1f.** $2\sqrt{42}$

1g. $\sqrt{5}$ **1h.** 20

3a. $y = x^2 - 3$

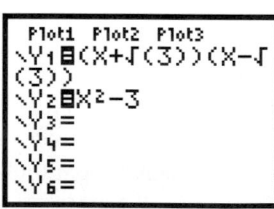

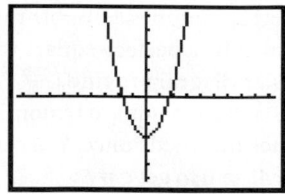

$[-9.4, 9.4, 1, -6.2, 6.2, 1]$

3b. $y = x^2 + 2x\sqrt{5} + 5$

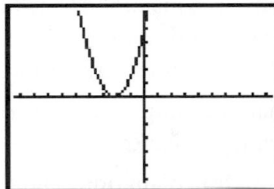

$[-9.4, 9.4, 1, -6.2, 6.2, 1]$

5a. $(0, -3)$

5b. $\left(-\sqrt{5}, 0\right) \approx (-2.236, 0)$

7a. $\pm 4\sqrt{7} \approx 10.583$

7b. $2\sqrt{6} \approx 4.899$ and $-3\sqrt{6} \approx -7.348$

7c. $-3 - \sqrt{2} \approx -4.414$ and $-3 + \sqrt{2} \approx -1.586$

9a. $\sqrt{112}$ **9b.** $\sqrt{550}$ **9c.** $\sqrt{972}$ **9d.** $\sqrt{4500}$

11. See below.

13. $10\sqrt{38}$ cm

15a. $y = (x - 2)^2 + 3$

15b. $y = -x^2 + 4$

15c. $y = 3(x - 1)^2$

LESSON 11.6

1. No, because $9^2 + 16^2 \neq 25^2$.

3. Sample answer: $(6, 3)$ and $(1, 7)$.

5a. $\sqrt{26}$ units or approximately 0.5 mile

5b. 5 units or 0.5 mile

7a. 5 cm

7b. 13 cm

9a. $a = 63$

9b. $b = 7$

9c. $c = 3$

LESSON 11.7

1a. $x = 27$

1b. $x = 35$

1c. $x = 160$

1d. $x = \pm 4$

3a. $\sin D = \dfrac{7}{25}$

3b. $\cos E = \dfrac{7}{25}$

3c. $\tan D = \dfrac{7}{24}$

5a. $x = 12$

5b. $x = 14$

5c. $x = 35$

7a. cosine

7b. sine

7c. tangent

7d. $\cos 65° \approx 0.4226$; $\sin 65° \approx 0.9063$; $\tan 65° \approx 2.1445$

7e. approximately 5.6 m

---

11. (*Lesson 11.5*)

$0 = 0.5x^2 - 6x + 8$ The original equation.

$0 \overset{?}{=} 0.5\left(6 - \sqrt{20}\right)^2 - 6\left(6 - \sqrt{20}\right) + 8$ Substitute $6 - \sqrt{20}$ for x.

$0 \overset{?}{=} 0.5\left(6 - \sqrt{20}\right)^2 - 36 + 6\sqrt{20} + 8$ Distribute the -6 over $6 - \sqrt{20}$.

$0 \overset{?}{=} 0.5(36 - 6\sqrt{20} - 6\sqrt{20} + 20) - 36 + 6\sqrt{20} + 8$ Square the expression.

$0 \overset{?}{=} 18 - 3\sqrt{20} - 3\sqrt{20} + 10 - 36 + 6\sqrt{20} + 8$ Distribute the 0.5 over the expression in parentheses.

$0 \overset{?}{=} 18 + 10 - 36 + 8$ Combine the radical expressions.

$0 = 0$ Add and subtract.

11a. rectangle

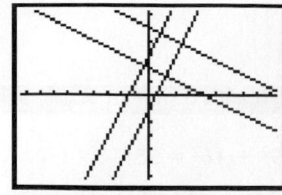

$[-9.4, 9.4, 1, -6.2, 6.2, 1]$

11b. $(1.2, 1.4), (2.4, 3.8), (0.8, 4.6), (-0.4, 2.2)$

11c. $y = 11 - 8x, y = \dfrac{17}{7} + \dfrac{4}{7}x$

11d. $(1, 3)$

LESSON 11.8

1a. d **1b.** A **1c.** A

1d. $\dfrac{d}{c}$ **1e.** A **1f.** A

3. $x \approx 44.6$ m

5. $a \approx 9.4$ cm $b \approx 17.7$ cm
$c \approx 4.1$ cm $d \approx 4.6$ cm
$e = \sqrt{145} \approx 12.0$ cm $F \approx 44.9°$
$g = \sqrt{349} \approx 18.7$ cm $H \approx 15.5°$

7. approximately 81.1 m

9a. approximately 2.86°

9b. about 148 ft

11a. approximately 19 ft

11b. approximately 51 ft

11c. approximately 70 feet

13. Sample answer: $y = -3 + \dfrac{x + 7}{(x - 2)(x + 7)}$

CHAPTER 11 REVIEW

1a. $8\sqrt{5}$ **1b.** $4\sqrt{17}$ **1c.** $123\sqrt{3}$

1d. $\sqrt{15}$ **1e.** 80 **1f.** 1700

1g. $3\sqrt{10}$ **1h.** $80\sqrt{5}$ **1i.** $\sqrt{2} + \sqrt{3}$

1j. $3\sqrt{2}$ **1k.** $\sqrt{6}$ **1l.** $4\sqrt{3}$

3. The slopes of the sides are $\dfrac{1}{2}, -2, \dfrac{1}{2}$ and -2.
The product of the slopes of each pair of adjacent sides is -1, so the sides are perpendicular.

5. Hint: Find a similar right triangle that has a perimeter of 60 feet.

7. $a \approx 3.38$ m
$b \approx 7.75$ m

9a. $\sqrt{116}$ cm or approximately 10.77 cm

9b. $\sqrt{141}$ cm or approximately 11.87 cm

11a. $y = 60 + 1.08(x - 40)$ or $y = 34 + 1.08(x - 16)$

11b. approximately 50°

11c. approximately 38°

13a. Inverse variation. Sample explanation: The product of x and y is constant; $xy = 2$ or $y = \dfrac{2}{x}$.

13b. Neither. Sample explanation: The product is not constant, so it is not an inverse variation. The y-value for $x = 0$ is not 0, so it is not a direct variation.

13c. Direct variation. Sample explanation: The ratio of y to x is constant; $y = 0.25x$.

13d. Neither. Sample explanation: The graph is not a curve, so the relationship is not an inverse variation. The line does not pass through the origin, so it is not a direct variation.

13e. Inverse variation. Sample explanation: The product of the x- and y-coordinates for any point on the curve is 8; $xy = 8$ or $y = \dfrac{8}{x}$.

13f. Direct variation. Sample explanation: The graph is a straight line through the origin; $y = 1.5x$.

15a. $27x6y^3$ **15b.** $5p^4q^2$

15c. $\dfrac{x}{y^2}$ **15d.** $\dfrac{m^2}{n^4} + \dfrac{1}{m^4}$

17a. $y = (x + 3)(x - 1)$

17b. $y = (x + 1)^2 - 4$

17c. $y = x^2 + 2x - 3$

19a. 26 **19b.** 2 **19c.** 36

19d. 128 **19e.** -24

21. Hint: There is more than one way to describe this transformation. Your answer to 21b will rely on your answer to 21a.

23a. $y = (x + 2)^2 - 4$

23b. $y = -0.5|x + 3|$

9c. (*Lesson 11.7*)

Trigonometric Functions for a 45° Angle

	Sine	Cosine	Tangent
Exact value of ratio	$\dfrac{1}{\sqrt{2}}$	$\dfrac{1}{\sqrt{2}}$	$\dfrac{1}{1}$
Decimal approximation of exact value	0.7071	0.7071	1.0000
Value by trigonometric function keys	0.7071	0.7071	1.0000

Selected Answers

segment	length	slope
$\overline{AB}$	10	$\frac{3}{4}$
$\overline{BC}$	10	$-\frac{3}{4}$
$\overline{AC}$	16	0

25b. isosceles triangle; two sides have equal length

25c. $D(2, 1)$

25d. Right triangles. Sample explanation: $\overline{BD}$ has an undefined slope, so it is vertical; $\overline{AC}$ has a slope of 0, so it is horizontal.

25e. A drawing does confirm 25a–d.

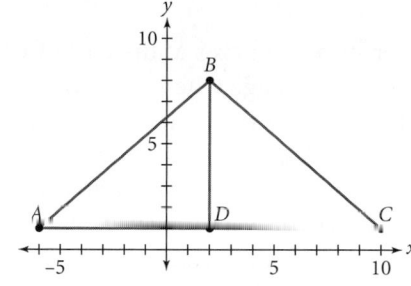

Glossary

The number in parentheses at the end of each definition gives the page where each word or phrase is first used in the text. Some words and phrases are introduced more than once, either because they have different application in different chapters or because they first appeared within features such as Project or Take Another Look; in these cases, there may be multiple page numbers listed.

A

absolute value A number's distance from 0 on the number line. The absolute value of a number gives its size, or magnitude, whether the number is positive or negative. The absolute value of a number x is shown as $|x|$. For example, $|-9| = 9$ and $|4| = 4$. (460)

absolute value function The function $f(x) = |x|$, which gives the absolute value of a number. The absolute value function is defined by two rules: If $x \geq 0$, then $f(x) = x$. If $x < 0$, then $f(x) = -x$. (462)

acute angle An angle that measures less than 90°. (618)

acute triangle A triangle with three acute angles. (638)

addition expression An expression whose only operation is addition. There are also subtraction expressions, multiplication expressions, and division expressions. See **algebraic expression.** (3)

addition property of equality If $a = b$, then $a + c = b + c$ for any number c. (278)

adjacent leg If you consider an acute angle of a right triangle, the adjacent leg is the leg that is part of the angle. (619)

algebraic expression A symbolic representation of mathematical operations that can involve both numbers and variables. (185)

analytic geometry The study of geometry using coordinate axes and algebra. (583)

angle of elevation The angle between a horizontal line and the line of sight. (630)

appreciation An increase in monetary value over time. (380)

associative property of addition For any values of a, b, and c, $a + (b + c) = (a + b) + c$. (278)

associative property of multiplication For any values of a, b, and c, $a(bc) = (ab)c$. (278)

asymptote A line that a graph approaches more and more closely, but never actually reaches. (513)

attractor A number that the results get closer and closer to when an expression is evaluated recursively. (24)

average The number obtained by dividing the sum of the values in a data set by the number of values. Formally called the mean. (44)

axis One of two perpendicular number lines used to locate points in the coordinate plane. The horizontal axis is often called the x-axis, and the vertical axis is often called the y-axis. The plural of axis is axes. (67)

B

balancing method A method of solving an equation that involves performing the same operation on both sides until the variable is isolated on one side. (233)

bar graph A data display in which bars are used to show measures or counts for various categories. (38)

base A number or an expression that is raised to a power. For example, $x + 2$ is the base in the expression $(x + 2)^3$, and 5 is the base in the expression 5^y. (376)

bimodal Used to describe a data set that has two modes. (44)

binary number A number written in base 2. Binary numbers consist only of the digits 0 and 1. Computers store information in binary form. (431)

binomial A polynomial with exactly two terms. Examples of binomials include $-3x + x^4$, $x - 12$, and $x^3 - x^{12}$. (544)

bins Intervals on the horizontal axis of a histogram that data values are grouped into. Boundary values fall into the bin to the right. (57)

box plot A one-variable data display that shows the five-number summary of a data set. A box plot is drawn over a horizontal number line. The ends of the box indicate the first and third quartiles. A vertical segment inside the box indicates the median. Horizontal segments, called whiskers, extend from the left end of the box to the minimum value and from the right end of the box to the maximum value. (51)

box-and-whisker plot See **box plot.**

category A group of data with the same attribute. For example, data about people's eye color could be grouped into three categories: blue, brown, and green. (39)

center (of rotation) The point that a figure turns about during a rotation. (524)

chaotic Systematic and nonrandom, yet producing results that look random. Small changes to the input value of a chaotic process can result in large changes to the output value. (29)

coefficient A number that is multiplied by a variable. For example, in a linear equation in intercept form $y = a + bx$, b is the coefficient of x. (217)

combining like terms Adding terms that have exactly the same variables raised to the same exponents. For example, in the expression $4x + 2x^2 - x + 5 + 7x^2$ you can combine the like terms $4x$ and $-x$ and the like terms $2x^2$ and $7x^2$ to get $3x + 9x^2 + 5$. (544)

common denominator A common multiple of the denominators of two or more fractions. For example, 30 is a common denominator of $\frac{7}{10}$ and $\frac{4}{15}$. (6)

common monomial factor A monomial that is a factor of every term in an expression. For example, $3x$ is a common monomial factor of $12x^3 - 6x^2 + 9x$. (575)

commutative property of addition For any values of a and b, $a + b = b + a$. (278)

commutative property of multiplication For any values of a and b, $ab = ba$. (278)

completing the square Adding a constant term to an expression of the form $x^2 + bx$ to form a perfect-square trinomial. For example, to complete the square in the expression $x^2 + 12x$, add 36. This gives $x^2 + 12x + 36$, which is equivalent to $(x + 6)^2$. To solve a quadratic equation by completing the square, write it in the form $x^2 + bx = c$, complete the square on the left side (adding the same number to the right side), rewrite the left side as a binomial squared, and then take the square root of both sides. (560)

compound inequality A combination of two inequalities. For example, $-5 < x \leq 1$ is a compound inequality that combines the inequalities $x > -5$ and $x \leq 1$. (342)

congruent Having the same shape and size. Two angles are congruent if they have the same measure. Two segments are congruent if they have the same length. Two figures are congruent if you can move one to fit exactly on top of the other. (3, 156)

conic section Any curve that can be formed by the intersection of a plane and a double cone. Parabolas, circles, ellipses, and hyperbolas are all examples of conic sections. (543)

conjecture A statement that might be true but that has not been proven. Conjectures are usually based on data patterns or on experience. (65)

constant A value that does not change. (147)

constant multiplier In a sequence that grows or decreases exponentially, the number each term is multiplied by to get the next term. The value of $1 + r$ in the exponential equation $y = A(1 + r)^x$. (368)

constant of variation The constant ratio in a direct variation or the constant product in an inverse variation. The value of k in the direct variation equation $y = kx$ or in the inverse variation equation $y = \frac{k}{x}$. (148, 166)

constant term A term that includes no variable. In the expression $ax^2 + bx + c$, the constant term is c. (566)

constraint A limitation on the values of the variables in a situation. A system of inequalities can model the constraints in many real-world situations. (356)

continuous function A function that has no breaks in the domain or range. The graph of a continuous function is a line or curve with no holes or gaps. (455)

contour line See **isometric line.**

contour map A map that uses isometric lines to show elevations above sea level, revealing the character of the terrain. Also called a topographic map. (248, 627)

conversion factor A ratio used to convert measurements from one unit to another. (106)

coordinate plane A plane with a pair of scaled, perpendicular axes allowing you to locate points with ordered pairs and to represent lines and curves by equations. (67)

coordinates An ordered pair of numbers in the form (x, y) that describes the location of a point on the coordinate plane. The x-coordinate describes the point's horizontal distance and direction from the origin, and the y-coordinate describes its vertical distance and direction from the origin. (67)

cosine If A is an acute angle in a right triangle, $cosine\ of\ angle\ A = \frac{length\ of\ adjacent\ leg}{length\ of\ hypotenuse}$, or $\cos A = \frac{a}{h}$. (620)

cryptography The study of coding and decoding messages. (424)

cube (of a number) A number raised to the third power. The cube of a number x is "x cubed" and is written x^3. For example, the cube of 4 is 4^3, which is equal to 64. (572)

cube root The cube root of a number a is the number b such that $a = b^3$. The cube root of a is denoted $\sqrt[3]{a}$. For example, $\sqrt[3]{64} = 4$ and $\sqrt[3]{-125} = -5$. (572)

cubing function The function $f(x) = x^3$, which gives the cube of a number. (572)

data A collection of information, numbers, or pairs of numbers, usually measurements for a real-world situation. (38)

data analysis The process of calculating statistics and making graphs to summarize a data set. (65)

decreasing A term used to describe the behavior of a function. A function is decreasing on an interval of its domain if the y-values decrease as the x-values increase. Visually, the graph of the function goes down as you read from left to right for that part of the domain. (455)

decreasing function A function that is always decreasing. (453)

dependent variable A variable whose values depend on the values of another variable (called the independent variable). In a graph of the relationship between two variables, the values on the vertical axis usually represent values of the dependent variable. (441)

depreciation A decrease in monetary value over time. (379)

deviation from the mean A data value minus the mean of its data set. The deviations of the data values from the mean give an idea of the spread of the data values. (460)

dimensional analysis A strategy for converting measurements from one unit to another by multiplying by a string of conversion factors. (106)

dimensions (of a matrix) The number of rows and columns in a matrix. If a matrix has 2 rows and 4 columns, its dimensions are 2 × 4. (80)

dimensions (of a rectangle) The width and length of a rectangle. If a rectangle is 2 units wide and 4 units long, its dimensions are 2-by-4. (602)

direct variation A relationship in which the ratio of two variables is constant. That is, a relationship in which two variables are directly proportional. A direct variation has an equation in the form $y = kx$, where x and y are the variables and k is a number called the constant of variation. (148)

directly proportional Used to describe two variables whose values have a constant ratio. (148)

directrix See **parabola.**

discrete function A function whose domain and range are made up of distinct values rather than intervals of real numbers. The graph of a discrete function is made up of distinct points. (455)

discriminant The expression under the square root symbol in the quadratic formula. If a quadratic equation is written in the form $ax^2 + bx + c = 0$, then the discriminant is $b^2 - 4ac$. If the discriminant is greater than 0, the quadratic equation has two solutions. If the discriminant equals 0, the equation has one real solution. If the discriminant is less than 0, the equation has no real solutions. (568)

distance formula The distance, d, between points (x_1, y_1) and (x_2, y_2) is given by the formula $d = \sqrt{(x_2 - x_1)^2 + (y_2 - y_1)^2}$. (614)

distributive property For any values of a, b, and c, $a(b + c) = a(b) + a(c)$. (184, 278)

division property of equality If $a = b$, then $\frac{a}{c} = \frac{b}{c}$ for any nonzero number c. (278)

division property of exponents For any nonzero value of b and any values of m and n, $\frac{b^m}{b^n} = b^{m-n}$. (395)

domain The set of input values for a function. (426)

dot plot A one-variable data display in which each data value is represented by a dot above that value on a horizontal number line. (39)

double root A value r is a double root of an equation $f(x) = 0$ if $(x - r)^2$ is a factor of $f(x)$. The graph of $y = f(x)$ will touch, but not cross, the x-axis at $x = r$. For example, 3 is a double root of the equation $0 = (x - 3)^2$. The graph of $y = (x - 3)^2$ touches the x-axis at $x = 3$. (575)

E

elimination method A method for solving a system of equations that involves adding or subtracting the equations to eliminate a variable. In some cases, both sides of one or both equations must be multiplied by a number before the equations are added or subtracted. For example, to solve $\begin{cases} 3x - 2y = 5 \\ -6x + y = 11 \end{cases}$, you could multiply the first equation by 2 and then add the equations to eliminate x. (324)

engineering notation A notation in which a number is written as a number greater than or equal to 1 but less than 1000, multiplied by 10 to a power that is a multiple of 3. For example, in engineering notation, the number 10,800,000 is written 10.8×10^6. (405)

equally likely Used to describe outcomes that have the same probability of occurring. For example, when you toss a coin, heads and tails are equally likely. (124)

equation A statement that says the value of one number or algebraic expression is equal to the value of another number or algebraic expression. (193)

equilateral triangle A triangle with three sides of the same length. (29, 283)

evaluate (an expression) To find the value of an expression. If an expression contains variables, values must be substituted for the variables before the expression can be evaluated. For example, if $3x^2 - 4$ is evaluated for $x = 2$, the result is $3(2)^2 - 4$, or 8. (22)

even temperament A method of tuning an instrument based on an equal tuning ratio between adjacent notes (that is, an exponential equation). (409)

expand (an algebraic expression) To rewrite an expression by multiplying factors and combining like terms. For example, to expand $(x + 8)(x - 2)$, rewrite it as $x^2 + 6x - 16$. (547)

expanded form (of a repeated multiplication expression) The form of a repeated multiplication expression in which every occurrence of each factor is shown. For example, the expanded form of the expression $3^2 \cdot 5^4$ is $3 \cdot 3 \cdot 5 \cdot 5 \cdot 5 \cdot 5$. (376)

experimental frequency The number of times a particular outcome occurred during the trials of an experiment. (123)

exponent A number or variable written as a small superscript of a number or variable, called the base, that indicates how many times the base is being used as a factor. For example, in the expression y^4, the exponent 4 means four factors of y, so $y^4 = y \cdot y \cdot y \cdot y$. (10)

exponential equation An equation in which a variable appears in the exponent. (376)

exponential form The form of an expression in which repeated multiplication is written using exponents. For example, the exponential form of $3 \cdot 3 \cdot 5 \cdot 5 \cdot 5 \cdot 5$ is $3^2 \cdot 5^4$. (376)

exponential growth A growth pattern in which amounts increase by a constant percent. Exponential growth can be modeled by the equation $y = A(1 + r)^x$, where A is the starting value, r is the rate of growth written as a decimal or fraction, x is the number of time periods elapsed, and y is the final value. (377)

 F

factor One of the numbers, variables, or expressions multiplied to obtain a product. (10)

factored form An expression written as a product of expressions, rather than as a sum or difference. For example, $3(x + 2)$ and $y(4 - w)$ are in factored form. See **factoring**. (369)

factored form (of a quadratic equation) The form $y = a(x - r_1)(x - r_2)$, where $a \neq 0$. The values r_1 and r_2 are the zeros of the quadratic function. (551, 578)

factoring The process of rewriting an expression as a product of factors. For example, to factor $7x - 28$, rewrite it as $7(x - 4)$. To factor $x^2 + x - 2$, rewrite it as $(x - 1)(x + 2)$. (280)

family of functions A group of functions with the same parent function. For example, $y = |x - 5|$ and $y = -2|x| + 3$ are both members of the family of functions with parent function $y = |x|$. (485)

fault A break in a rock formation caused by the movement of the earth's crust, in which the rocks on opposite sides of the break move in different directions. (485)

feasible region In a linear programming problem, the set of points that satisfy all the constraints. If the constraints are given as a system of inequalities, the feasible region is the solution to the system. (364)

first quartile (Q1) The median of the values below the median of a data set. (50)

first-quadrant graph A coordinate graph in which all the points are in the first quadrant. (69)

five-number summary The minimum, first quartile, median, third quartile, and maximum of a data set. The five-number summary helps show how the data values are spread. (50)

focus See **parabola**.

fractal The result of infinitely many applications of a recursive procedure to a geometric figure. The resulting figure has self-similarity. From the Latin word *fractus,* meaning broken or irregular. (2, 4, 15)

frequency The number of times a value appears in a data set. (57)

function A rule or relationship in which there is exactly one output value for each input value. (426)

function notation A notation in which a function is named with a letter and the input is shown in parentheses after the function name. For example, $f(x) = x^2 + 1$ represents the function $y = x^2 + 1$. The letter f is the name of the function, and $f(x)$ (read "f of x") stands for the output for the input x. The output of this function for $x = 2$ is written $f(2)$, so $f(2) = 5$. (446)

 G

general equation An equation that represents a whole family of equations. For example, the general equation $y = kx$ represents the family of equations that includes $y = 4x$ and $y = -3.4x$. (179)

general form (of a quadratic equation) The form $y = ax^2 + bx + c$, in which $a \neq 0$. (534, 578)

girth The distance around an object in one direction. The girth of a box is the length of string that wraps around the box. (577)

glyph A symbol that presents information nonverbally. (73)

golden ratio The ratio $\frac{1 + \sqrt{5}}{2}$, often considered an aesthetically "ideal" ratio. Examples of the golden ratio can be found in the environment, in art, and in architecture. (99)

gradient The inclination of a roadway. Also called the grade of the road. (232, 630)

graph sketch A rough graph in which the axes are labeled with variable names but not with specific scale values. (213)

gravity The force of attraction between two objects. Gravity causes objects to accelerate toward Earth at a rate of 32 ft/sec^2 or 9.8 m/sec^2. (532)

Glossary

half-life The time needed for an amount of a substance to decrease by one-half. (414)

half-plane The points on a plane that fall on one side of a boundary line. The solution of a linear inequality in two variables is a half-plane. (348)

hexagon A polygon with exactly six sides. (72)

histogram A one-variable data display that uses bins to show the distribution of values in a data set. Each bin corresponds to an interval of data values; the height of a bin indicates the number, or frequency, of values in that interval. (57)

horizontal axis The horizontal number line on a coordinate graph or data display. Also called the x-axis. (39, 67)

hypotenuse The side of a right triangle opposite the right angle. (585)

image The figure or graph of a function that is the result of a transformation of an original figure or graph of a function. (478)

imaginary number A number in the set of numbers that includes square roots of negative numbers. In the set of imaginary numbers, $\sqrt{-1}$ is represented by the letter i. For example, the solution to $x^2 = -4$ is $\sqrt{-4}$ or $2i$. (581)

increasing Used to describe the behavior of a function. A function is increasing on an interval of its domain if the y-values increase as the x-values increase. Visually, the graph of the function goes up as you read from left to right for that part of the domain. (455)

increasing function A function that is always increasing. (453)

independent variable A variable whose values affect the values of another variable (called the dependent variable). In a graph of the relationship between two variables, values on the horizontal axis usually represent values of the independent variable. (441)

inequality A statement that one quantity is less than or greater than another. For example, $x + 7 \geq -3$ and $6 + 2 < 11$ are inequalities. (339)

intercept form The form $y = a + bx$ of a linear equation. The value of a is the y-intercept, and the value of b, the coefficient of x, is the slope of the line. (217)

interest A percent of the balance added to an account at regular time intervals. (369)

interquartile range (IQR) The difference between the third quartile and the first quartile of a data set. (52)

interval The set of numbers between two given numbers, or the distance between two numbers on a number line or axis. (39)

inverse Reversed in order or effect. In an inverse mathematical relationship, as one quantity increases, the other decreases. (163)

inverse (of a function) The relationship that reverses the inputs and outputs of a function. For example, the inverse of the function $y = x + 2$ is $y = x - 2$. (473, 529)

inverse variation A relationship in which the product of two variables is constant. That is, a relationship in which two variables are inversely proportional. An inverse variation has an equation in the form $xy = k$, or $y = \frac{k}{x}$, in which x and y are the variables and k is a number called the constant of variation. (166)

inversely proportional Used to describe two variables whose values have a constant product. (166)

invert To switch the positions of two objects. For example, to invert the fraction $\frac{3}{4}$, switch the numerator and the denominator to get $\frac{4}{3}$. When you invert a fraction, the result is the reciprocal of the fraction. (94)

irrational number A number that cannot be expressed as the ratio of two integers. In decimal form, an irrational number has an infinite number of digits and doesn't show a repeating pattern. Examples of irrational numbers include π and $\sqrt{2}$. (96)

isometric line A line on a contour map that shows elevation above sea level. All the points on an isometric line have the same elevation. Also called a contour line. (248, 627)

isosceles triangle A triangle with two sides of the same length. (283)

 K

key A guide for interpreting the values in a data display. For example, a stem plot has a key that shows how to read the stem and leaf values. (59)

Koch curve A fractal generated recursively by beginning with a line segment and, at each stage, constructing an equilateral triangle on the middle third of each line segment and removing the edge of that triangle on the line segment. (14)

 L

leading coefficient In a polynomial, the coefficient of the term with the highest power of the variable. For example, in the polynomial $3x^2 - 7x + 4$, the leading coefficient is 3. (561)

leg One of the perpendicular sides of a right triangle. (585)

light-year The distance light travels in one year. About 9460 billion kilometers. (392)

line of fit A line used to model a set of data. A line of fit shows the general direction of the data and has about the same number of data points above and below it. (261)

line of symmetry A line that divides a figure into mirror-image halves. In a parabola that opens up or down, the line of symmetry is the vertical line through the vertex. (540)

linear In the shape of a line or represented by a line. In mathematics, a linear equation or expression has variables raised only to the power of 1. For example, $y = 3x + 1$ is a linear equation. (207)

linear function A function characterized by a constant rate of change—that is, as the x-values change by a constant amount, the y-values also change by a constant amount. The graph of a linear function is a straight line. (452)

linear programming A process that applies the concepts of constraints, points of intersection, and algebraic expressions to solve application problems. (364)

linear relationship A relationship that you can represent with a straight-line graph. A linear relationship is characterized by a constant rate of change—that is, as the value of one variable changes by a constant amount, the value of the other variable also changes by a constant amount. (207)

linear term A term in which a constant is multiplied by a variable to the first power. In the expression $ax^2 + bx + c$, the linear term is bx. (566)

long-run value The value that the y-values approach as the x-values increase. (489)

lowest terms The form of a fraction in which the numerator and denominator have no common factors except 1. (6)

 M

matrix A rectangular array of numbers or expressions, enclosed in brackets. (80)

maximum The greatest value in a data set or the greatest value of a function. (39, 454)

mean The number obtained by dividing the sum of the values in a data set by the number of values. Often called the average. (44)

measure of center A single number used to summarize a one-variable data set. The mean, median, and mode are measures of center. (44)

median (of a data set) If a data set contains an odd number of values, the median is the middle value when the values are listed in order. If a data set contains an even number of values, the median is the mean of the two middle values when the values are listed in order. (44)

median (of a triangle) A segment from the vertex of a triangle to the midpoint of the opposite side. (588)

midpoint The point on a line segment halfway between the endpoints. If a segment is drawn on a coordinate grid, you can use the midpoint formula to find the coordinates of its midpoint. (2, 588)

midpoint formula If the endpoints of a segment are (x_1, y_1) and (x_2, y_2), then the midpoint of the segment is $\left(\frac{x_1 + x_2}{2}, \frac{y_1 + y_2}{2}\right)$. (590)

Glossary

minimum The least value in a data set or the least value of a function. (39, 454)

mode The value or values that occur most often in a data set. A data set may have more than one mode or no mode. (44)

monomial A polynomial with only one term. Examples of monomials include $-3x$, r^4, and $7x^2$. (544)

multiplication property of equality If $a = b$, then $ac = bc$ for any number c. (278)

multiplication property of exponents For any values of b, m, and n, $b^m \cdot b^n = b^{m+n}$. (384)

nonlinear Not in the shape of a line or not able to be represented by a line. In mathematics, a nonlinear equation or expression has variables raised to powers other than 1. For example, $x^2 + 5x$ is a nonlinear expression. (452)

nonlinear function A function characterized by a nonconstant rate of change—that is, as the x-values change by a constant amount, the y-values change by varying amounts. (452)

observed probability A probability based on experience or collected data. Also called relative frequency. (123)

obtuse angle An angle that measures more than 90°. (618)

obtuse triangle A triangle with an obtuse angle. (283, 638)

one-variable data Data that measures only one trait or quantity. A one-variable data set consists of single values, not pairs of data values. (67)

opposite leg If you consider an acute angle of a right triangle, the opposite leg is the leg that is *not* part of the angle. (619)

order of magnitude A way of expressing the size of an extremely large or extremely small number by giving the power of 10 associated with the number. For example, the number 6.01×10^{26} is on the order of 10^{26} and the number 2.43×10^{-11} is on the order of 10^{-11}. (421)

order of operations The agreed-upon order in which operations are carried out when evaluating an expression: (1) evaluate all expressions within parentheses or other grouping symbols, (2) evaluate all powers, (3) multiply and divide from left to right, and (4) add and subtract from left to right. (5, 182)

ordered pair A pair of numbers named in an order that matters. For example, $(3, 5)$ is different from $(5, 3)$. The coordinates of a point are given as an ordered pair in which the first number is the x-coordinate and the second number is the y-coordinate. (67)

origin The point on the coordinate plane where the x- and y-axes intersect. The origin has coordinates $(0, 0)$. (67)

outcome A possible result of one trial of an experiment. (123)

outlier A value that is far outside the range of most of the other values in a data set. As a general rule, a data value is considered an outlier if the distance from the value to the first quartile or third quartile (whichever is nearest) is more than 1.5 times the interquartile range. (46)

parabola The graph of a function in the family of functions with parent function $y = x^2$. The set of all points whose distance from a fixed point, the focus, is equal to the distance from a fixed line, the directrix. (467, 559)

parallel lines Lines in the same plane that never intersect. They are always the same distance apart. (583)

parallelogram A quadrilateral with opposite sides that are parallel. (159)

parent function The most basic form of a function. A parent function can be transformed to create a family of functions. For example, $y = x^2$ is a parent function that can be transformed to create a family of functions that includes $y = x^2 + 2$ and $y = 3(x - 4)^2$. (485)

pentagon A polygon with exactly five sides. (32)

perfect cube A number that is equal to the cube of an integer. For example, -125 is a perfect cube because $-125 = (-5)^3$. (572)

perfect square A number that is equal to the square of an integer, or a polynomial that is equal to the square of another polynomial. For example, 64 is a perfect square because it is equal to 8^2, and $x^2 - 10x + 25$ is a perfect-square trinomial because it is equal to $(x - 5)^2$. (546)

perpendicular bisector A line that passes through the midpoint of a segment and is perpendicular to the segment. (588)

perpendicular lines Lines that meet at a right angle. (583)

pictograph A data display with symbols showing the number of data items in each category. Each symbol in a pictograph stands for a specific number of data items. (38)

point-slope form The form $y = y_1 + b(x - x_1)$ of a linear equation, in which (x_1, y_1) is a point on the line and b is the slope. (271)

polygon A closed figure made up of segments that do not cross each other. (19)

polynomial A sum of terms that have positive integer exponents. For example, $-4x^2 + x$ and $x^3 - 6x^2 + 9$ are polynomials. (544)

population density The number of people per square mile. (397)

power properties of exponents For any values a, b, m, and n, $(b^m)^n = b^{mn}$ and $(ab)^n = a^n b^n$. (385)

predict To make an educated guess, usually based on a pattern. (9)

probability A number between 0 and 1 that gives the chance that an outcome will happen. An outcome with a probability of 0 is impossible. An outcome with a probability of 1 is certain to happen. (122)

product The result of multiplication. (83)

projectile motion The motion of a thrown, kicked, fired, or launched object—such as a ball—that has no means of propelling itself. (534)

proportion An equation stating that two ratios are equal. For example, $\frac{34}{72} = \frac{x}{18}$ is a proportion. (94)

Pythagorean theorem The sum of the squares of the lengths of the legs a and b of a right triangle equals the square of the length of the hypotenuse c—that is, $a^2 + b^2 = c^2$. (600)

Q

Q-points On a scatter plot, the vertices of the rectangle formed by drawing vertical lines through the first and third quartiles of the x-values and horizontal lines through the first and third quartiles of the y-values. If the points show an increasing linear trend, then the line through the lower-left and upper-right Q-points is a line of fit. If the points show a decreasing linear trend, then the line through the upper-left and lower-right Q-points is a line of fit. (289)

quadrant One of the four regions that a coordinate plane is divided into by the two axes. The quadrants are numbered I, II, III, and IV, starting in the upper right and moving counterclockwise. (67)

quadratic formula If a quadratic equation is written in the form $ax^2 + bx + c = 0$, then the solutions of the equation are given by $x = \frac{-b \pm \sqrt{b^2 - 4ac}}{2a}$. (566, 567)

quadratic function Any function in the family with parent function $f(x) = x^2$. Examples of quadratic functions are $f(x) = 1.5x^2 + 2$, $f(x) = (x - 4)^2$, and $f(x) = 5x^2 - 3x + 12$. (533)

quadrilateral A polygon with exactly four sides. (273)

R

radical expression An expression containing a square root symbol, $\sqrt{}$. Examples of radical expressions are $\sqrt{x + 4}$ and $3 \pm \sqrt{19}$. (535)

radioactive decay The process by which an unstable chemical element loses mass or energy, transforming it into a different element or isotope. (406)

raised to the power A term used to connect the base and the exponent in an exponential expression. For example, in the expression 7^4, the base 7 is raised to the power of 4. (385)

random Not ordered, unpredictable. (28, 129)

range (of a data set) The difference between the maximum and minimum values in a data set. (40)

Glossary

range (of a function) The set of output values for a function. (426)

rate A ratio, often with 1 in the denominator. (140)

rate of change The difference between two output values divided by the difference between the corresponding input values. For a linear relationship, the rate of change is constant. (226)

ratio A comparison of two quantities, often written in fraction form. (93)

rational function A function, such as $f(x) = \frac{3}{x + 2}$ or $f(x) = \frac{x - 1}{(x + 3)(x - 1)}$, that is expressed as the ratio of two polynomial expressions. (514)

rational number A number that can be written as a ratio of two integers. (96)

real number Any number that can be represented on the number line. The real numbers include integers, rational numbers, and irrational numbers. The real numbers do *not* include imaginary numbers. (563)

reciprocal The multiplicative inverse. The reciprocal of a given number is the number you multiply it by to get 1. To find the reciprocal of a number, you can write the number as a fraction and then invert the fraction. For example, the reciprocal of $\frac{3}{4}$ is $\frac{4}{3}$. (584)

rectangle A quadrilateral with four right angles. In a rectangle, opposite sides are parallel. (31)

recursive Describes a procedure that is applied over and over again, starting with a number or geometric figure, to produce a sequence of numbers or figures. Each stage of a recursive procedure builds on the previous stage. The resulting sequence is said to be generated recursively, and the procedure is called recursion. (2)

recursive routine A starting value and a recursive rule for generating a recursive sequence. (199)

recursive rule The instructions for producing each stage of a recursive sequence from the previous stage. (3)

recursive sequence An ordered list of numbers defined by a starting value and a recursive rule. You generate a recursive sequence by applying the rule to the starting value, then applying the rule to the resulting value, and so on. (199)

reflection A transformation that flips a figure or graph over a line, creating a mirror image. (493)

reflection across the x-axis A transformation that flips a figure or graph across the x-axis. Reflecting a point across the x-axis changes the sign of its y-coordinate. (493)

reflection across the y-axis A transformation that flips a figure or graph across the y-axis. Reflecting a point across the y-axis changes the sign of its x-coordinate. (493)

relative frequency The ratio of the number of times a particular outcome occurred to the total number of trials. Also called observed probability. (123)

relative frequency graph A data display (usually a bar graph or a circle graph) that compares the number in each category to the total for all the categories. Relative frequency graphs show fractions or percents, rather than actual values. (116)

repeating decimal A decimal number with a digit or group of digits after the decimal point that repeats infinitely. (93)

rhombus A quadrilateral with opposite sides parallel and all sides the same length. (159)

right angle An angle that measures 90°. (31, 583)

right trapezoid A trapezoid with two right angles. In a right trapezoid, one of the nonparallel sides is perpendicular to both parallel sides. (585)

right triangle A triangle with a right angle. (31, 252, 283, 585)

roots The solutions of an equation. When the equation is written in the form $f(x) = 0$, the roots are the x-intercepts of the graph of $y = f(x)$. For example, the roots of $(x - 2)(x + 1) = 0$ are 2 and -1. These roots are the x-intercepts of the graph of $y = (x - 2)(x + 1)$. (539)

rotation A transformation that turns a figure about a point called the center of rotation. (524)

row operations Operations performed on the rows of a matrix in order to transform it into a matrix with a diagonal of 1's with 0's above and below, creating a solution matrix. These are allowable row operations: multiply (or divide) all the numbers in a row by a nonzero number, add (or subtract) all the numbers in a row to (or from)

corresponding numbers in another row, add (or subtract) a multiple of the numbers in one row to (or from) the corresponding numbers in another row. (332)

sample A part of a population selected to represent the entire population. Sampling is the process of selecting and studying a sample from a population in order to make conjectures about the whole population. (100)

scale (of an axis or a number line) The values that correspond to the intervals of a coordinate axis or number line. (39)

scale factor A rate that relates the measurements in a scaled figure to the measurements in the original figure. (154)

scatter plot A two-variable data display in which values on a horizontal axis represent values of one variable and values on a vertical axis represent values of the other variable. The coordinates of each point represent a pair of data values. (67)

scientific notation A notation in which a number is written as a number greater than or equal to 1 but less than 10, multiplied by an integer power of 10. For example, in scientific notation, the number 32,000 is written 3.2×10^4. (388)

segment Two points on a line (endpoints) and all the points between them on the line. Also called a line segment. (3)

self-similar Describes a figure in which part of the figure is similar to—that is, has the same shape as—the whole figure. (6)

shrink A transformation that decreases the height or width of a figure. A vertical shrink decreases the height but leaves the width unchanged. A horizontal shrink decreases the width but leaves the height unchanged. A vertical shrink by a factor of a multiplies the y-coordinate of each point on a figure or graph by a. A horizontal shrink by a factor of b multiplies the x-coordinate of each point on a figure by b. (501)

Sierpiński triangle A fractal created by Waclaw Sierpiński by starting with a filled-in equilateral triangle and recursively removing every triangle

whose vertices are midpoints of triangles remaining from the previous stage. You can create a Sierpiński-like fractal design by starting with an equilateral triangle and recursively connecting the midpoints of the sides of each upward-pointing triangle. (3)

similar figures Figures that have the same shape. Similar polygons have proportional sides and congruent angles. (156)

simulate To model an experiment with another experiment, called a *simulation,* so that the outcomes of the simulation have the same probabilities as the corresponding outcomes of the original experiment. For example, you can simulate tossing a coin by randomly generating a string of 0's and 1's on your calculator. (100)

sine If A is an acute angle in a right triangle, $sine\ of\ angle\ A = \frac{length\ of\ opposite\ leg}{length\ of\ hypotenuse}$, or $\sin A = \frac{o}{h}$. (620)

slope The steepness of a line or the rate of change of a linear relationship. If (x_1, y_1) and (x_2, y_2) are two points on a line, then the slope of the line is $\frac{y_2 - y_1}{x_2 - x_1}$. The slope is the value of b when the equation of the line is written in intercept form $y = a + bx$, and it is the value of m when the equation of the line is written in slope-intercept form $y = mx + b$. (251, 254, 265)

slope triangle A right triangle formed by drawing arrows to show the vertical and horizontal change from one point to another point on a line. (252)

slope-intercept form The form $y = mx + b$ of a linear equation. The value of m is the slope and the value of b is the y-intercept. (265)

solution The value(s) of the variable(s) that make an equation or inequality true. (193)

spread A property of one-variable data that indicates how the data values are distributed from least to greatest, and where gaps or clusters occur. Statistics such as the range, the interquartile range, and the five-number summary can help describe the spread of data. (39)

square A quadrilateral in which all four angles are right angles and all four sides have the same length. (7)

square (of a number) The product of a number and itself. The square of a number x is "x squared" and is written x^2. For example, the square of 6 is 6^2, which is equal to 36. (466)

square root The square root of a number a is a number b so that $a = b^2$. Every positive number has two square roots. For example, the square roots of 36 are -6 and 6 because $6^2 = 36$ and $(-6)^2 = 36$. The square root symbol, $\sqrt{\ }$, means the positive square root of a number. So, $\sqrt{36} = 6$. (468)

square root function The function that undoes squaring, giving only the positive square root (that is, the positive number that, when multiplied by itself, gives the input). The square root function is written $f(x) = \sqrt{x}$. For example, $\sqrt{144} = 12$. (468)

squaring The process of multiplying a number by itself. See **square** of a number. (466)

squaring function The function $f(x) = x^2$, which gives the square of a number. (471)

standard form The form $ax + by = c$ of a linear equation, in which a and b are not both 0. (277)

statistics Numbers, such as the mean, median, and range, used to summarize or represent a data set. Statistics also refers to the science of collecting, organizing, and interpreting information. (40)

stem plot A one-variable data display used to show the distribution of a fairly small set of data values. Generally, the left digit(s) of the data values, called the stems, are listed in a column on the left side of the plot. The remaining digits, called the leaves, are listed in order to the right of the corresponding stem. A key is usually included. (59)

stem-and-leaf plot See **stem plot.**

strange attractor A figure that the stages generated by a random recursive procedure get closer and closer to. (29)

stretch A transformation that increases the height or width of a figure. A vertical stretch increases the height but leaves the width unchanged. A horizontal stretch increases the width but leaves the height unchanged. A vertical stretch by a factor of a multiplies the y-coordinate of each point on a figure or graph by a. A horizontal stretch by a factor of b multiplies the x-coordinate of each point on a figure by b. (501)

substitution method A method for solving a system of equations that involves solving one of the equations for one variable and substituting the resulting expression into the other equation. For example, to find the solution of $\begin{cases} y + 2 = 3x \\ y - 1 = x + 3 \end{cases}$, you can solve the first equation for y to get $y = 3x - 2$ and then substitute $3x - 2$ for y in the second equation. (316)

subtraction property of equality If $a = b$, then $a - c = b - c$ for any number c. (278)

symbolic manipulation Applying mathematical properties to rewrite an equation or expression in equivalent form. (320)

symmetric Having a sense of balance, or symmetry. Symmetric is most often used to describe figures with mirror symmetry, or line symmetry—that is, figures that you can fold in half so that one half matches exactly with the other half. (52)

system of equations A set of two or more equations with the same variables. (308)

system of inequalities A set of two or more inequalities with the same variables. (354)

T

tangent If A is an acute angle in a right triangle, $tangent\ of\ angle\ A = \dfrac{length\ of\ opposite\ leg}{length\ of\ adjacent\ leg}$, or $\tan A = \frac{o}{h}$. (620)

term An algebraic expression that represents only multiplication and division between variables and constants. For example, in the polynomial $x^3 - 6x^2 + 9$, the terms are x^3, $-6x^2$, and 9. (544)

terminating decimal A decimal number with only a finite number of nonzero digits after the decimal point. (93)

theoretical probability A probability calculated by analyzing a situation, rather than by performing an experiment. If the outcomes are equally likely, then the theoretical probability of a particular group of outcomes is the ratio of the number of outcomes in that group to the total number of possible outcomes. For example, when you roll a die, one of the six possible outcomes is a 2, so the theoretical probability of rolling a 2 is $\frac{1}{6}$. (123)

third quartile (Q3) The median of the values above the median of a data set. (50)

topographic map See **contour map.**

transformation A change in the size or position of a figure or graph. Translations, reflections, stretches, shrinks, and rotations are types of transformations. (476)

translation A transformation that slides a figure or graph to a new position. (478)

trapezoid A quadrilateral with one pair of opposite sides that are parallel and one pair of opposite sides that are not parallel. (585)

trial One round of an experiment. (123)

trigonometric functions The sine, cosine, and tangent functions, which express relationships among the measures of the acute angles in a right triangle and the ratios of the side lengths. (620)

trigonometry The study of the relationships among sides and angles of right triangles. (620)

trinomial A polynomial with exactly three terms. Examples of trinomials include $x + 2x^3 + 4$, $x^2 - 6x + 9$, and $3x^3 + 2x^2 + x$. (544)

two-variable data set A collection of data that measures two traits or quantities. A two-variable data set consists of pairs of values. (67)

undoing method A method of solving an equation that involves working backward to reverse each operation until the variable is isolated on one side of the equation. (194)

value of an expression The numerical result of evaluating an expression. (22)

variable A trait or quantity whose value can change, or vary. In algebra, letters often represent variables. (67, 94)

vertex (of an absolute value graph) The point where the graph changes direction from increasing to decreasing or from decreasing to increasing. (483)

vertex (of a parabola) The point where the graph changes direction from increasing to decreasing or

from decreasing to increasing. (486)

vertex (of a polygon) A "corner" of a polygon. An endpoint of one of the polygon's sides. The plural of vertex is vertices. (28)

vertex form (of a quadratic equation) The form $y = a(x - h)^2 + k$, where $a \neq 0$. The point (h, k) is the vertex of the parabola. (541, 578)

vertical axis The vertical number line on a coordinate graph or data display. Also called the y-axis. (39)

vertical line test A method for determining whether a graph on the xy-coordinate plane represents a function. If all possible vertical lines cross the graph only once or not at all, the graph represents a function. If even one vertical line crosses the graph in more than one point, the graph does not represent a function. (433)

x-intercept The x-coordinate of a point where a graph meets the x-axis. For example, the graph of $y = (x + 2)(x - 4)$ has two x-intercepts, -2 and 4. (539)

y-intercept The y-coordinate of the point where a graph crosses the y-axis. The value of y when x is 0. The y-intercept of a line is the value of a when the equation for the line is written in intercept form $y = a + bx$, and it is the value of b when the equation for the line is written in slope-intercept form $y = mx + b$. (217)

zero product property If the product of two or more factors equals zero, then at least one of the factors equals zero. For example, if $x(x + 2)(x - 3) = 0$, then $x = 0$ or $x + 2 = 0$ or $x - 3 = 0$. (554)

zeros (of a function) The values of the independent variable (the x-values) that make the corresponding values of the function (the $f(x)$-values) equal to zero. For example, the zeros of the function $f(x) = (x - 1)(x + 7)$ are 1 and -7 because $f(1) = 0$ and $f(-7) = 0$. See **roots.** (554)

Glossary

Index

A

abacus, 269
Abbey, Edward, 382
absolute value, 460
    and deviation from the mean, 460–461
absolute-value equations, solving, 462–463
absolute-value function, 462
    family of, 485
    transformation of, 483–484
acceleration, 532
Activities. *See* Investigations
acute angle, 618
acute triangle, 238, 638
Adams, Henry, 551
addition
    associative property of, 278
    commutative property of, 278
    of matrices, 81–82
    order of operations for, 5, 182
    property of equality, 278
    of radical expressions, 607
adjacent leg, 619, 621
Africa, 125, 582, 597
Aikman, Troy, 85
Albers, Anni, 485
algebraic expressions, 185
Alther, Lisa, 324
Altito, Noelie, 611
analytic geometry, 583
    distance formula and, 611–614
    midpoint and, 588–590
    slope and, 583–586
anemometer, 121
Angelou, Maya, 476
angle(s), 618
angle of elevation, 630
Angola, 597
animation, 476, 481, 482, 498, 508
Apollonius, 543
application
    animation, 476, 481, 482, 498, 508
    appreciation and depreciation, 304, 379, 380, 381, 395–396, 401–402, 636
    archaeology, 407
    architecture, 154–156, 257, 260, 419, 475, 593, 616
    art, 92, 111–112, 157, 181, 211, 222, 250, 260, 366, 475, 502, 531, 582, 597, 611
    astronomy, 41, 389–390, 392, 398, 421

automobiles, 62, 72, 141, 143, 152, 175, 207–209, 223, 300, 304, 329, 371, 379, 380, 381, 395–396, 401–402
biology, wildlife, 49, 56, 109, 398
biology, human, 11, 70, 96, 114, 119, 128, 134, 171, 204, 296–297, 388, 392, 417, 458
business, 71, 86, 89, 114, 153, 221, 240, 245, 246, 313, 321, 329, 336, 355–356, 359, 363, 364, 371, 387, 518, 549, 565, 637
census data, 64, 382
chemistry, 40, 98, 104, 106, 286–287, 290–291, 318–319, 500, 515
computers, 155, 246, 413, 427, 431, 494, 504
consumer awareness, 47, 63, 76, 82–83, 110–111, 113, 135, 141, 143, 144, 148–149, 150, 161, 171, 187, 189, 198, 210, 231, 232, 240, 241, 251–252, 257, 281, 294, 337, 338, 373, 418, 472, 634, 636
cooking, 98, 108, 115, 212, 500
design, 155, 158, 189, 357, 570, 593
education, 113, 131, 135, 313, 419
electronics, 167, 178
energy, 72, 87, 141, 142, 143, 175, 223, 329
engineering, 199, 203, 204, 206, 232, 262–263, 294, 372, 405, 630
entertainment, 45–46, 47, 60, 186
entomology, 97, 129, 383, 398, 412, 418
environment, 90, 286
film making, 114, 459
fitness, 216–219, 221, 222, 224, 232, 282, 337, 359
fundraising, 109, 136, 169, 205, 372
garbage, 275
geography, 116, 160
government, 104, 267
health, 263–265, 274, 293, 336
horticulture, 151, 176
income, 55, 88, 97, 99, 113, 139–141, 178, 187, 205, 222, 282, 305, 353, 360, 421
inflation, 403, 417
interest, 368–369, 377, 379, 386–387, 397, 404
librarianship, 103, 117–118
life expectancy, 61, 284–285

manufacturing, 134, 166, 282, 364, 372, 398, 404, 490
maps, 159, 160, 162, 248, 612, 627–628
medicine, 39, 41, 371, 397, 442, 518
meteorology, 73, 122, 153, 205, 225–229, 230, 247, 273, 330, 346, 439, 508, 509, 634
monetary systems, 152
music, 38, 42, 138, 167, 170, 299, 408–409, 411, 423
oceanography, 76
painting, 175
pets, 47, 56, 144, 161, 287
physics, 165–166, 169, 170, 177, 222, 275, 389, 391, 392, 402, 411, 413, 414–415, 418, 518, 524, 534, 536
population, bacterial, 378, 421, 449, 487, 580
population, human, 64, 90, 98, 112, 120, 121, 144, 267, 270–271, 295, 382, 392, 397, 403, 404, 490, 524, 597
population, wildlife, 100–102, 103, 122–123, 126, 131, 133, 464, 565
projectiles, 534, 536, 536–537, 542, 557–558, 564
publishing, 103
savings, 178, 240, 266, 344, 411
scale models, 158, 198, 315
seismology, 48, 485
shipping, 146–147, 274, 336, 346, 357, 577
sports, 48, 50–51, 54–55, 56, 70, 79, 85, 89, 128, 141–143, 174, 247, 293, 298, 300, 304, 322, 353, 444, 532, 542, 549, 564, 569
surveying, 160, 361, 623, 630
technology, 89, 121, 413, 602, 623, 630
testing/assessments, 63, 73, 78, 136
travel, 144, 152, 168, 189, 207–209, 210, 258, 292, 294, 299, 314–315, 322, 548, 602
work, 55, 80, 88, 132, 282, 458
appreciation and depreciation, 304, 379, 380, 381, 395–396, 401–402, 636
Arab mathematics, 198
archaeology, 407
architecture, 154–156, 257, 260, 419, 475, 593, 616

Index

area
    fractals and, 3–6
    of polygons, 593–594
    squaring function as modeling, 572
    of triangles, 593, 598–600
Armstrong, Lance, 174
art, 92, 111–112, 157, 181, 211, 222, 250, 260, 366, 475, 502, 531, 582, 597, 611
Assessing What You've Learned
    Giving a Presentation, 249, 306, 365, 422, 474, 581
    Keeping a Portfolio, 36, 91, 137, 180, 206, 249, 306, 365, 474, 530, 638
    Organize Your Notebook, 6, 137, 180, 249, 306, 365, 474, 530, 581, 638
    Performance Assessment, 180, 306, 422, 530, 581, 638
    Write in Your Journal, 91, 137, 180, 249, 306, 365, 422, 474, 581
associative property, 278
astronomy, 41, 389–390, 392, 398, 421
asymptote, 513
attractors, 22–25
    See also strange attractors
automobiles, 62, 72, 141, 143, 152, 175, 207–209, 223, 300, 304, 329, 371, 379, 380, 381, 395–396, 401–402
average. *See* mean
Avogadro, Amadeo, 391
Avogadro's number, 391
axis. *See* horizontal axis; vertical axis

B

Babylonia, 610
baker's dozen, 150
balancing method, 233–237, 278, 535
Balet Folklórico, 186
bar graphs, 38–39, 116–118
base, 376
baseball statistics, 79
bicycles, 161, 172–174, 189, 224
bimodal data, 44–45
binary numbers, 431
binomials, 544, 548
bins, 57–60
biology, human, 41, 79, 96, 114, 119, 128, 134, 171, 204, 296–297, 388, 392, 417, 458
biology, wildlife 49, 56, 109, 398
Bohr, Niels, 129
Bouguer, Pierre, 339
boundary values, 57

box-and-whisker plots. *See* box plots
box plots, 51–52, 57
Braille, Louis, 206
Brooks, Romaine, 250
Bullock, James, 93
business, 71, 86, 89, 114, 153, 221, 240, 245, 246, 313, 321, 329, 336, 355–356, 359, 363, 364, 371, 387, 518, 549, 565, 637
Byrd, Richard, 630

C

capture-recapture method, 100–103
carbon dating, 407
Carroll, Lewis, 592
cars. *See* automobiles
Cassat, Mary, 104
categories, 39
Cat's Cradle, 591
cause and effect, 442
Celsius, conversion of, 153, 448, 473–474
census data, 64, 382
center, measures of, 44–46
    See also mean; median; mode
center of rotation, 524
centimeters, conversion of, 105–106
Central America, 500
chaotic processes, 29
Chastain, Brandi, 569
Chatelet, Emilie du, 452
chemistry, 40, 98, 104, 106, 286–287, 290–291, 318–319, 500, 515
China, 136, 145, 161, 215, 338, 600, 610
Christo, artist, 150
circle graphs, 116–118
Clay, Henry, 74
coefficients, 217
    leading, 561, 566
    like terms and, 544
    nonzero, 225
    radical expressions and, 609
    zero, elimination method and, 326
combining like terms, 544
common denominator, 6
common monomial factors, 575
commutative property, 278
completing the square, 560–563, 566
compound inequalities, 342
computers, 155, 246, 413, 427, 431, 494, 504
congruent, 3, 156
conic sections, 543
conjectures, 65–66
Connections
    careers, 39, 155, 157
    consumer, 141
    health, 264

    history, 40, 96, 147, 254, 333, 339, 390, 427, 600, 620
    music, 409
    nature, 6, 15
    science, 29, 106, 166, 479, 485
    social science, 382
    technology, 84, 494, 504
    See also applications; cultural connections
constant, 147
    effect on graphs, 179
    in recursive routines, 206
constant multipliers
    expanded form of expressions with, 376
    modeling data and, 408
    in recursive routines, 367–370, 374, 376, 408
    See also exponents
constant of variation, 166
constant term, 566
constraints, 356–357
construction, 175, 538–539, 555, 570, 593, 601, 603, 616, 632
consumer awareness, 47, 63, 76, 82–83, 110–111, 113, 135, 141, 143, 144, 148–149, 150, 161, 171, 187, 189, 198, 210, 231, 232, 240, 241, 251–252, 257, 281, 294, 337, 338, 373, 418, 472, 634, 636
continental drift, 479
continuous functions, 455
contour lines. *See* isometric lines
contour maps, 248, 627–628
conversion factors, 106, 148
conversion of units, 105–106
    conversion factors, 106, 148
    cups/liters, 212
    cups/quarts, 500
    dimensional analysis, 106
    feet/ells, 135
    feet/rods, 108
    inches/centimeters, 105–106
    miles/kilometers, 146–147
    moles/milliliters, 106
    moles/molecules, 391
    monetary, 152
    pounds/grams, 212
    pounds/kilograms, 151
    scale drawings and, 154–157
    temperature, 153, 448
cooking, 98, 108, 115, 212, 500
Coolidge, Susan, 367
coordinate plane, 67
    half-plane, 348
    horizontal axis, 67
    ordered pairs, 67, 68
    origin, 67
    quadrants, 67
    transformation and, 476–477
    vertical axis, 67

coordinates, 67
cosine, 620–621, 624
Crichton, Michael, 74
cryptography, 424–427, 431
cube(s), 572
cube root, 572
cubic equations, 572–575
cubic functions, 572
cubic numbers, 470
cultural connections
    Africa, 125, 582, 597
    Babylonians, 610
    China, 136, 145, 161, 215, 338,
      600, 610
    Egypt, 43, 424, 609, 610
    England, 592, 630
    France, 366
    Germany, 260
    Greece, 409, 543, 610
    Hebrew, 431
    India, 162, 610
    Mayans, 252, 500
    Mexico, 252, 500
    Myanmar, 181
    Native Americans, 181, 611
    Silesia, 620
    Spain, 475
    Thailand, 311
    Tibet, 625
    See also Connections
Cunningham, Randall, 85
Curie, Marie, 626
cylinders, 445

data, 38
    bimodal, 44–45
    collection of, 67–68, 69, 301–302
    estimations of, 74–75
    first-quartile, 50
    five-number summary of, 50–52
    frequency of, 57, 58
    interquartile range (IQR), 52
    maximum values in, 39
    measures of center of. See mean;
      median; mode
    minimum values in, 39
    modeling of. See modeling data
    one-variable, 67, 87
    outliers, 46, 56
    range of, 40, 46
    spreads of, 39, 40, 51, 52
    third-quartile, 50
    two-variable, 67–69
    See also graphs; matrices
data analysis, 65
Davis, Bette, 544
decimals
    conversion of fractions into, 93
    probability, 122

    repeating, 93
    terminating, 93
decreasing functions, 455
dependent variable, 441, 442
depreciation. See appreciation and
    depreciation
design, 155, 158, 189, 357, 570, 593
deviation from the mean, 460–461
Dickinson, Emily, 406
dimension, matrix, 80–84
dimension, rectangle, 595
dimensional analysis, 106
directly proportional, 148
directrix, 559
direct variation, 146–149
    equation for, 148
    and intercept form, 218
    percentages as, 161
    scale drawings and, 154
disabilities, people with, 142, 174,
    353, 616
discrete functions, 455
discriminant, 568
distance formula, 611–614
distance/velocity formula, 176
distributive property, 183–186, 278
    and product of binomials, 548
    reversal of (factoring), 280
division
    Egyptian doubling method for, 43
    order of operations for, 5, 182
    property of equality, 278
    property of exponents, 393
    rewriting radical expressions, 607
    symbols for, 93
division property of exponents, 395
Dodgson, Charles, 592
domain, 426–427, 441, 453, 455
dot plots, 39–40, 44–46, 57
double root, 575

education, 113, 131, 135, 313, 419
Egypt, 43, 424, 609, 610
Einstein, Albert, 446
Einstein's problem, 188
electronics, 167, 178
elimination method, 324–327
ell, 135
energy, 72, 87, 141, 142, 143, 175,
    223, 329
engineering, 199, 203, 204, 206, 232,
    262–263, 294, 372, 405, 630
engineering notation, 405
England, 592, 630
enlarging and reducing, 111–112
    See also scale drawings
entertainment, 45–46, 47, 60, 186
entomology, 97, 129, 383, 398, 412, 418
environment, 90, 286

equality, properties of, 278
equally likely outcomes, 124
equations, 193
    coefficients. See coefficients
    cubic, 572–575
    direct variation, 148
    equivalent, 276–279
    exponential. See exponential
      equations
    functions expressed as, 446, 448
    general, 179
    inverse variation, 166
    linear. See linear equations
    and number tricks, 193
    quadratic. See quadratic
      equations
    solutions to, defined, 193
    See also functions
equations, solving
    absolute value, 462–463
    balancing method, 233–237, 278,
      535
    calculator methods, 233, 237,
      308–311, 317, 462, 534
    completing the square, 560–563,
      566
    linear equations. See equation
      systems, solving; linear
      equations, solving
    parabola, 467–468
    properties used in, 278
    quadratic equations. See
      quadratic equations, solving
    square root, 614
    squares, 467–468
    symbolic methods of, 233, 237,
      320, 462–463, 468, 534–535,
      560–563, 566
    by undoing operations, 193–194,
      233, 237, 467, 535, 566
equilateral triangle, 29, 283
estimating, 74–75
evaluation of expressions, 22,
    190–194
even temperament, 409
expanded form of repeated
    multiplication, 376
expanding an expression, 547
exponential equations, 374–377
    for decreasing patterns, 401–402,
      489–490
    expanded form of, 376, 394
    exponential form of, 376
    for growth, 377
    long-run value of y in, 489
    modeling data with, 406–409,
      414–415
exponential functions, 485, 486–487
exponential growth, 377
exponents, 10
    base of, 376

Index

J

K

L

M

money systems, 152
monomials, 544
Moore, Gordon, 413
Moore's Law, 413
multiplication
    Arab and Persian method, 198
    associative property of, 278
    of binomials, 548
    commutative property of, 278
    formula of ancient India, 162
    of matrices, 82–84
    order of operations for, 5
    property of equality, 278
    property of exponents, 384
    of radical expressions, 607
    of reciprocals, 584
    as recursive routine, 367–370
    symbols for, 10
multiplication property of
    exponents, 384
music, 38, 42, 138, 167, 170, 299,
    408–409, 411, 423
Myanmar, 181

scientific notation for. See
    scientific notation
    See also negative numbers
number tricks, 190–193

O

observed probability, 123, 129–131
obtuse angle, 618
obtuse triangle, 283, 638
oceanography, 76
Ohno, Miyoko, 588
one-variable data, 67, 87
operations. See addition, division,
    multiplication, subtraction
    See also order of operations
opposite leg, 619, 621
ordered pairs, 67, 68
order of magnitude, 421
order of operations, 5, 182
    and distributive property,
        182–186
    and number tricks, 190–193
    and squaring, 466
    undoing, 193–194, 233, 237, 467,
        535, 566
origin, 67
orthodontic treatment, 397
outcomes, 123, 124
outliers, 46, 56

P

painting, 175
Panama Canal, 147
Pangaea, 479
parabolas, 467, 532
    definition of, 559
    graphing, 466–467
    line of symmetry of, 540
    translation of, 486
    vertex, 486, 539, 540
    See also quadratic equations
parallel lines, 262, 583
    slope of, 583–586
    symbol for, 583
parallelogram, 159
parent function, 485, 513
Pascal's triangle, 215
patterns, of Sierpiński triangle, 9
pentominoes, 109
percentages
    as direct variation, 161
    finding an unknown part, 102
    finding an unknown percent, 101
    finding an unknown total,
        101–102
    increasing and decreasing,
        110–112
    probability, 122

perfect cubes, 572
perfect squares, 546, 560
    completing the square and,
        560–562
    radical expressions and, 609
perpendicular bisector, 588
perpendicular lines, 583
    slope of, 583–586
    symbol for, 583
Persia, 198
pets, 47, 56, 144, 161, 287
Philolaus, 388
physics, 165–166, 169, 170, 177, 222,
    275, 389, 391, 392, 402, 411, 413,
    414–415, 418, 518, 524, 534, 536
pi, 96
Picasso, Pablo, 432
pictographs, 38
Pitiscus, Bartholmeo, 620
plate tectonics, 479
points, transformations of, 478–479,
    507
point-slope form, 270–272
    equation for, 271
    equivalent equations and,
        276–279
    modeling data using, 272,
        284–285, 296–297
polygon(s)
    area of, 593–594
    families of, 171
    graphing of, 585–586
    matrix transformation of,
        520–522, 524
    reflection of, 492
    stretching and shrinking,
        502–503, 504
    translation of, 476–479
polynomials, 544
    common monomial factors of,
        575
    multiplication of, 548
population, bacterial, 378, 421, 449,
    487, 580
population density, 397
population, human, 64, 90, 98, 112,
    120, 121, 144, 267, 270–271, 295,
    382, 392, 397, 403, 404, 490, 524,
    597
population, wildlife, 100–102, 103,
    122–123, 126, 131, 133, 464, 565
pounds, conversion of, 151
power properties of exponents, 385
power, raising to a, 385
Presley, Elvis, 461
probability, 122
    equally likely outcomes, 124
    observed, 123, 129–131
    outcomes, 123, 124
    random outcomes, 129–131
    as ratio, 122

Index

median of, 588, 589–590
obtuse, 283, 638
right. *See* right triangle(s)
slope and, 283
vertex of, 28
See also Pythagorean theorem
trigonometric functions, 619–621, 624
    inverse, 626–628
trigonometry, 620
trinomials, 544, 560
tuning forks, 170
tuning instruments, 408–409
Turing, Alan, 427
two-variable data, 67–69

undoing method, 190–194
unit conversion. *See* conversion of units

value, absolute. *See* absolute value
value of the expression, 22
variables, 67
    cause and effect and, 442
    dependent, 441, 442
    elimination of, 324–327
    independent, 441, 442, 446
    input, 220
    one-variable data, 67, 87
    output, 220
    in proportions, 94–96
    two-variable data, 67–69

variation
    constant of, 166
    direct. *See* direct variation
    inverse. *See* inverse variation
Venters, Diana, 595
vertex (of parabola), 486, 539, 540
vertex (of polygon), 28
vertex form of quadratic equations, 541
    conversion of factored form to, 552–553
    conversion of general form to, 562
    conversion to general form, 544–547
    information given by, 547, 552
    transformations and, 541, 544, 547
vertical axis (*y*-axis), 39, 67
    dependent variable shown on, 441
    reflection across, 493
vertical change over horizontal change. *See* slope
vertical lines
    as failing vertical line test, 435
    slope of, as undefined, 255
vertical line test, 433–435
Volterra, Vito, 446
volume, 572

Wakata, Koichi, 536
Washington Monument, 419

weed fractals, 12
Wegener, Alfred, 479
Weigand, Charmion Von, 466
Whitehead, Alfred North, 50
wildlife. *See* population, wildlife
Willis, Suzanne, 389
Woods, Tiger, 99
work, 55, 80, 88, 132, 282, 458

x-axis. *See* horizontal axis
x-intercepts
    of cubic equations, 573–574
    of linear equations, 243
    of quadratic equations, 539

y-axis. *See* vertical axis
Yelesina, Yelena, 304
y-intercept, 217
Young, Steve, 85

zero
    coefficients as, 326
    as exponent, 399–401
    Mayans and, 500
    product property of, 554
zeros. *See* roots
zero product property, 554
Zhu Shijie, 215

Photo Credits

Abbreviations: top (*t*), middle (*m*), bottom (*b*), left (*l*), right (*r*)

Cover
Background image: Pat O'Hara/DRK Photo; boat image: Marc Epstein/DRK Photo; all other images: Ken Karp Photography

Front Matter
v (*t*): Ken Karp Photography; **v** (*b*): George Holton/ Photo Researchers; **xix** (*t*): Ken Karp Photography; **xix** (*b*): Ken Karp Photography; **xx**: Ken Karp Photography; **xxi**: Ken Karp Photography; **xxiii**: Ken Karp Photography; **xxiv**: Ken Karp Photography; **xxv**: Ken Karp Photography; **xxvi** (*l*): Ken Karp Photography; **xxvi** (*r*): Ken Karp Photography

Chapter 0
1: Copyright 2000 Lifesmith Classic Fractals, Palmdale, CA, USA. All rights reserved. http://www.lifesmith.com; **4** (*r*): Ken Karp Photography; **5** (*l*): Roger Ressmeyer/Corbis; **5** (*r*): Ken Karp Photography; **6**: Cheryl Fenton; **12**: Copyright 2000 Lifesmith Classic Fractals, Palmdale, CA, USA. All rights reserved. http://www.lifesmith.com; **15** (*l*): Yann-Arthus-Bertrand/Corbis; **15** (*r*): Ken Karp Photography; **22**: Ken Karp Photography; **29** (*l*): AFP/Corbis; **29** (*r*): Gary Braasch/Corbis; **33** (*t*): Copyright 2000 Lifesmith Classic Fractals, Palmdale, CA, USA. All rights reserved. http://www.lifesmith.com; **33** (*m*): Ken Karp Photography

Chapter 1
37: The Stock Market; **38**: Ken Karp Photography; **39** (*t*): Corbis; **39** (*b*): Ken Karp Photography; **42**: Ted Horowitz/The Stock Market; **44**: Cheryl Fenton; **45**: FPG; **47**: Cedar Point photo by Dan Feight; **48** (*l*): *The Hollow of the Deep-Sea Wave Off Kanagawa* by Katsushika Hokusai/Minneapolis Institute of Art Acc. No. 74.1.230; **48** (*r*): Ken Karp Photography; **49**: Catherine Noren/Stock Boston; **50**: Jonathan Daniel/Allsport; **51** (*b*): Jim Amos/Photo Researchers; **55**: Roy Pinney/FPG; **56**: Ken Karp Photography; **58**: Ken Karp Photography; **61** (*b*): Alison Wright/Corbis; **61** (*mr*): Patricio Robles Gil/Bruce Coleman, Inc.; **61** (*mbl*): Betty Press/Woodfin Camp & Associates; **61** (*ml*): Sharon Smith/Bruce Coleman, Inc.; **61** (*mbr*): Catherine Karnow/Woodfin Camp & Associates; **65**: Ken Karp Photography; **66**: Ken Karp Photography; **69**: Joseph Sohm ChromoSohm, Inc./Corbis; **70**: Gregg Mancuso/Stock Boston; **72**: John Collier/FPG; **74**: Michael Yamashita/Woodfin Camp & Associates; **80**: Library of Congress; **81**: Ken Karp Photography; **84**: Christian Michaels/FPG; **85**: Joseph Sohm/ChromoSohm, Inc./Corbis; **86**: Roger Ball/The Stock Market; **87** (*t*): Ken Karp Photography; **88**: Doug Pensinger/Allsport

Chapter 2
92: Morton Beebe/Corbis; **97**: Ken Karp Photography; **99**: Corbis; **100**: Michael Heron/Woodfin Camp & Associates; **101** (*l*): Steve & Dave Maslowski/Photo Researchers, Inc.; **101** (*m*): Mark Stouffer/ Animals Animals; **101** (*r*): Gary Meszaros/Photo Researchers, Inc.; **103** (*t*): Ken Karp Photography; **103** (*b*): John Henley/The Stock Market; **104**: Library of Congress; **105**: Robert Fried/Stock Boston; **106** (*r*): Cheryl Fenton; **106** (*l*): Robert Holmes/Corbis; **107**: Corbis; **108**: Cheryl Fenton; **109** (*l*): M. Harvery/DRK Photo; **109** (*r*): Peter & Beverly Pickford/DRK Photo; **109** (*m*): Maslowski/Photo Researchers; **109** (*ml*): Anthony Mercieca/Photo Researchers, Inc.; **109** (*mt*): Kevin Schafer/Corbis; **110**: Ken Karp Photography; **112**: U.S. Postal Service; **113**: Ken Karp Photography; **114**: Ken Karp Photography; **115**: Cheryl Fenton; **117**: Kelly-Mooney Photography/Corbis; **119**: Larry Mulvehill/Photo Researchers; **121** (*t*): David L. Brown/Tom Stack & Associates; **121** (*b*): Cary Wolinsky/Stock Boston; **123**: Cheryl Fenton; **125**: Cheryl Fenton; **126**: Kennan Ward/The Stock Market; **129**: S. Dalton/Animals Animals; **133**: Tom Lazar/Animals Animals; **134** (*t*): Steve & Dave Maslowski/Photo Researchers, Inc.; **135**: Jim Harrison/Stock Boston

Chapter 3
138 (*l*): Rob Hann/Retna; **139** (*r*): R. W. Jones/Corbis; **139** (*l*): Steve Chenn/Corbis; **138** (*br*): Steve Double/Retna; **138** (*tr*): Rob Hann/ Retna; **141**: Cheryl Fenton; **142**: Rick Hansen Institute; **144**: Ken Karp Photography; **145**: Burstein Collection/Corbis; **147**: Will and Deni McIntyre/Photo Researchers, Inc.; **150** (*t*): Manfred Vollmer/ Corbis; **150** (*b*): Archivo Iconografico, SA/Corbis; **152**: John Carter/ Photo Researchers, Inc.; **153**: Ken Karp Photography; **155** (*l*): Stephen Simpson/FPG; **155** (*r*): Telegraph Colour Library/ FPG; **157** (*l*): Bettmann/Corbis; **157** (*r*): James Blank/Bruce Coleman, Inc.; **157** (*m*): Library of Congress; **158**: Cheryl Fenton; **160**: Robert Caputo/Stock Boston; **161**: Andrew Holbrooke/ The Stock Market; **162**: Ken Karp Photography; **167**: Ted Horowitz/ The Stock Market; **170**: Richard Megna/Fundamental Photographs; **172**: Ken Karp Photography; **174**: AFP/Corbis; **175**: Ken Karp Photography; **176**: Judith Canty/Stock Boston; **178** (*r*): Tom Bean/ DRK Photo; **178** (*l*): Marc Muench/David Muench Photography

Chapter 4
181 (*t*): Alison Wright/Corbis; **181** (*m*): Christie's Images; **185**: Cheryl Fenton; **186**: Danny Lehman/Corbis; **189**: Ken Karp Photography; **198**: Cheryl Fenton; **199**: Rafael Macia/Photo Researchers, Inc.; **200**: Cheryl Fenton; **200** (*m*): Cheryl Fenton; **203**: Photofest; **204**: Jeffry Myers/Stock Boston; **205**: David Falconer/Bruce Coleman, Inc.; **211**: D. Burnett/Woodfin Camp & Associates; **213**: Ken Karp Photography; **217** (*t*): Duomo/Corbis; **217** (*b*): Ann McCarthy/The Stock Market; **218** (*b*): Anderson/The Image Works; **220**: Hertz Rent-A-Car; **223**: Claude Charlier/The Stock Market; **224**: George D. Lepp/Photo Researchers, Inc.; **225**: Bettmann/Corbis; **227**: John Eastcott/YVA Momatiuk/Woodfin Camp & Associates; **228**: Corbis; **232**: Douglas Peebles/Corbis; **233**: James P. Blair/Corbis; **242**: Ken Karp Photography; **243**: Cheryl Fenton; **247**: Ken Karp Photography; **248**: Bernard Soutrit/Woodfin Camp & Associates

Chapter 5
250: Smithsonian American Art Museum, Washington, D.C./Art Resource, NY; **251**: Collection of Gretchen and John Berggruen, San Francisco; **252**: Robert Frerck/Woodfin Camp & Associates; **255** (*t*): Ken Karp Photography; **255** (*m*): Ken Karp Photography; **255** (*mr*): Ken Karp Photography; **257** (*t*): Ken Karp Photography; **257** (*b*): Philip Gould/Corbis; **258**: James Marshall/The Stock Market; **260**: The Museum of Modern Art, New York. gift of Philip Johnson. Photograph © 2000 The Museum of Modern Art, New York; **261**: S. Turner/Animals, Animals; **263** (*t*): Peter Menzel/Stock Boston; **263** (*b*): *Hamburger* (1983) by David Gilhooly, Collection of Harry W. and Mary Margaret Anderson, Photo by M. Lee Fatheree; **267**: AFP/Corbis; **268**: Ken Karp Photography; **274**: Ken Karp Photography; **275**: John DeWaele/Stock Boston; **276**: Bettmann/ Corbis © 2001 Andy Warhol Foundation for the Visual Arts/ARS, New York; **279**: Interlochen Center for the Arts; **282** (*t*): David Weintraub/Stock Boson; **282** (*b*): Roger Ball/The Stock Market; **284**: Steven Rubin/The Image Works; **285**: Morton Beebe, S. F./ Corbis; **286**: Michael Sedam/Corbis; **288**: Ken Karp Photography; **290**: Bob Daemmrich/Stock Boston; **295**: Tom Bean/DRK Photo; **299**: Amtrak; **301**: Ken Karp Photography; **303** (*b*): Robert Frerck/ Woodfin Camp & Associates; **303** (*b*): Bob Daemmrich/Stock Boston; **304**: Mike Powell/Allsport

Chapter 6
307: George Holton/Photo Researchers, Inc.; **308** (*l*): Tom Bean/ Corbis; **308** (*r*): Tom Bean/Corbis; **320** (*l*): Tom Bean/Corbis; **320** (*r*): Tom Bean/Corbis; **310** (*l*): Ken Karp Photography; **310** (*r*): Ken Karp Photography; **311**: George Chan/Photo Researchers, Inc.; **314**: Gerard Smith/Photo Researchers, Inc.; **315** (*t*): Corbis; **315** (*b*): Ken Karp Photography; **316**: Philip James;

Photo Credits

317: Cheryl Fenton; 321 (*t*): Photofest; 321 (*l*): Photofest; 321 (*r*): Photofest; 322: Mike Powell/Allsport; 323: Douglas Mesney/ The Stock Market; 329 (*l*): A. Ramey/Stock Boston; 329 (*r*): Ronnie Kaufman/The Stock Market; 330: Don Mason/The Stock Market; 331: Dan McCoy/Rainbow; 334: Milton Rand/Tom Stack & Associates; 341: George Olson/Woodfin Camp & Associates; 344: Ken Karp Photography; 345: Ken Karp Photography; 346 (*t*): U. S. Postal Service; 346 (*b*): NASA; 347: Charles Mann/ The Stock Market; 353: Ezra Shaw/Allsport; 354: Cheryl Fenton; 355: Ken Karp Photography; 355: Cheryl Fenton; 356: Cheryl Fenton; 357: Bob Daemmrich/Stock Boston; 361: Tom Walker/Stock Boston; 362: Ken Karp Photography; 364: Tony Hawk/Allsport

Chapter 7

366: Francis G. Mayer/Corbis; 367: Ken Karp Photography; 367 (*b*): Cheryl Fenton; 371: Jose Pelaez/The Stock Market; 372: Cheryl Fenton; 374: Courtesy of José Tence Ruiz; 378: Geoff Tompkinson/Photo Researchers; 379: Timothy Eagan/Woodfin Camp & Associates; 380: Cheryl Fenton; 382: Tom & DeeAnn McCarthy/The Stock Market; 383: Runk-Schoenberger/Grant Heilman Photography; 388: Paul Thiessen/Tom Stack & Associates; 389 (*t*): Mark Godfrey/The Image Works; 389 (*b*): Bettmann/ Corbis; 390: Cheryl Fenton; 391: Bettmann/Corbis; 392 (*t*): Cheryl Fenton; 397: Lee White/Corbis; 398: Stephen Dalton/Photo Researchers, Inc.; 402: Peter Arnold, Inc.; 404 (*t*): Ken Karp Photography; 404 (*b*): Diane Enkells/Stock Boston; 405: Dick Luria/ Photo Researchers; 406 (*l*): Chuck Savage/The Stock Market; 406 (*m*): Cydney Conger/Corbis; 406 (*r*): Hulton-Deutsch Collection/Corbis; 407: AP/Wide World; 408: Tom & Therisa Stack/ Tom Stack & Associates; 411 (*t*): Ken Karp Photography; 411 (*b*): Alexander Tsiaras/Photo Researchers, Inc.; 413 (*t*): Hulton-Deutsch Collection/Corbis; 413 (*b*): Manfred Cage/Peter Arnold, Inc.; 414: Ken Karp Photography; 417: Image by Man Ray © Man Ray Trust-ADAGP/ARS, 2001; 418: Michael Lustbader/Photo Researchers, Inc.; 419: T. J. Florian/Rainbow; 420: Cheryl Fenton

Chapter 8

423: Scala/Art Resource, NY; 424 (*l*): Archivo Ifonografico, SA/ Corbis; 424 (*r*): Stuart Craig/Bruce Coleman, Inc.; 424 (*b*): Ken Karp Photography; 427: Archive Photos; 432: *Guitare Et Journal* by Pablo Picasso/Christie's Images; 435: J. Pickerell/The Image Works; 439: Tom Stewart/The Stock Market; 440 (*l*): Schalkwijk/Art Resource, NY; 440 (*r*): Schalkwijk/Art Resource, NY; 441: Corbis; 446: Bettmann/Corbis; 452: Ed Young/Photo Researchers, Inc.; 459: J. Barry O'Rourke/The Stock Market; 461: A. Ramey/Woodfin Camp & Associates; 464: Bob & Clara Calhoun/Bruce Coleman, Inc.; 466 (*t*): National Museum of Women in Art/Gift of Wallace and Wilhelmina Holladay; 466 (*b*): Frank Siteman/Stock Boston; 471 (*t*): Archive Photos; 471 (*m*): J. Pickerell/The Image Works; 471 (*b*): Bettmann/Corbis; 472: Richard Hamilton Smith/Corbis

Chapter 9

475: J. C. Carton/Bruce Coleman, Inc.; 476: © 2001 Alias/Wavefront, a division of Silicon Graphics Limited; 483: Ken Karp Photography; 485 (*t*): National Museum of Women in the Arts/Gift of Wallace and Wilhelmina Holladay; 485 (*b*): Corbis; 490 (*t*): Kevin R. Morris/ Corbis; 490 (*b*): Ken Karp Photography; 501 (*l*): Ken Karp Photography; 501 (*r*): Ken Karp Photography; 502: National Gallery, London/Art Resource; 508: Interlochen Center for the Arts; 510: Ken Karp Photography; 513: Cedar Point photo by Dan Feicht; 515: David D. Keaton/The Stock Market; 518 (*t*): Ken Karp Photography; 518 (*b*): Norman Tomalin/Bruce Coleman, Inc.; 519: Cheryl Fenton; 520: Phyllis Picardi/Stock Boston; 524: Ken Karp Photography; 526: Phyllis Picardi/Stock Boston; 529: Galen Rowell/Corbis

Chapter 10

531: T. J. Florian/Rainbow; 532: Ken Karp Photography; 534: Bettmann/Corbis; 536: NASA; 538: Lee Foster/Bruce Coleman, Inc.; 542: Ken Karp Photography; 549 (*t*): Doug Pensinger/Allsport; 549 (*b*): Rick Gayle/The Stock Market; 551: Ken Karp Photography; 557: Ken Karp Photography; 560: Ken Karp Photography; 564: Ezra Shaw/Allsport; 565: Stuart Westmorland/Photo Researchers, Inc.; 569: Tom Hauck/Allsport

Chapter 11

582: Lindsay Hebberd/Corbis; 583: ©Philadelphia Museum of Art/ Corbis; 584 (*t*): Jeff Greenberg/Rainbow; 584 (*b*): Richard Berenholtz/ The Stock Market; 588: Ken Karp Photography; 591: Ken Karp Photography; 593: Coco McCoy/Rainbow; 595: Quilt by Diana Venters/*Mathematical Quilts* by Diana Venters and Elaine Krajenke Ellison; 598: Philadelphia Museum of Art/Corbis; 602: Ken Karp Photography; 609: Will & Deni McIntyre/Photo Researchers, Inc.; 611: Founders Society Purchase with funds from Flint Ink Corporation/Photograph © 1991 The Detroit Institute of Art, Accession Number 81.488; 613: Cheryl Fenton; 616 (*r*): William Johnson/Stock Boston; 616 (*l*): Peter Menzel/Stock Boston; 617: Robert Holmes/Corbis; 621: Corbis; 623: Michael Keller/ The Stock Market; 625: Daniel E. Wray/The Image Works; 630: Bettmann/Corbis; 632: Michael Keller/The Stock Market; 634: Cheryl Fenton

Additional Answers

Step 5 Sample data results

Fifty-eight white beans were removed and replaced with red beans.

Number of tagged fish	Total number of fish	Fraction of tagged fish
3	60	$\frac{3}{60}$ or $\frac{1}{20}$
6	67	$\frac{6}{67}$
3	73	$\frac{3}{73}$
4	66	$\frac{4}{66}$ or $\frac{2}{33}$
4	88	$\frac{4}{88}$ or $\frac{1}{22}$
2	72	$\frac{2}{72}$ or $\frac{1}{36}$

The mean of the ratios of tagged to total fish in a sample is 0.052 and the median is 0.048. Because 0.05 is the average of the mean and the median, you might use the fraction $\frac{3}{60}$. Solve the proportion: $\frac{58}{x} = \frac{3}{60}$; $x = 1160$.

Step 2

Continental Land Areas (millions of km²)

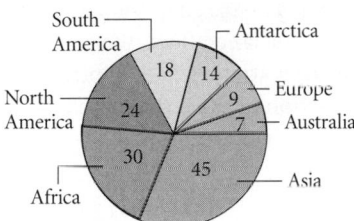

Step 4

Continental Land Areas

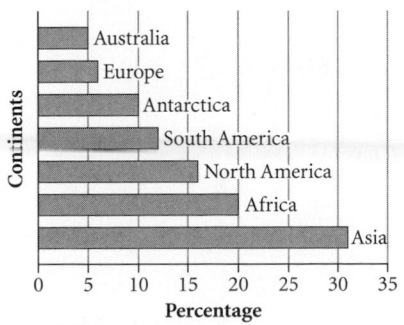

Step 1

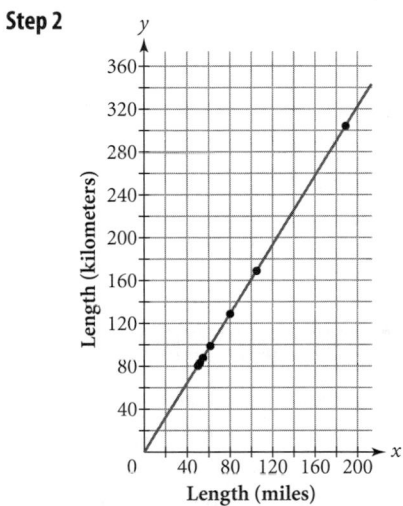

Step 2

Step 3

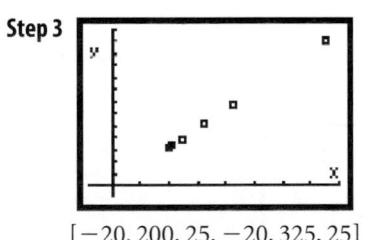

$[-20, 200, 25, -20, 325, 25]$

Additional Answers

Additional Answers

Step 3

Time (min)	Minivan (mi)	Sports car (mi)	Pickup (mi)
100	100	115	110
120	076	131	132
180	004	179	198
190	−8	187	209
200	−20	195	220
230	−56	219	253
240	−68	227	264

4a.

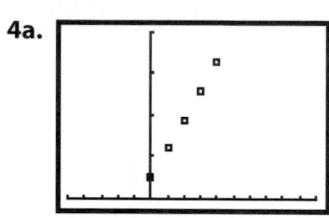

$[-5, 10, 1, 0, 40, 10]$

4b.

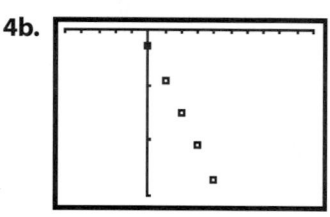

$[-5, 10, 1, 0, -30, 10]$

Step 1c Start at the 3-m mark and walk toward the 4-m mark at $\frac{1}{4}$ m per sec for 4 sec. Then walk toward the 0-mark at 1 m per sec.

Step 1d Start at the 0.5-m mark and walk toward the 4-m mark at 1 m per sec for 2 sec. Then walk toward the 0-m mark at $\frac{3}{4}$ m per sec for 2 sec. Then walk toward the 4-m mark at 1 m per sec for 2 sec.

Step 3a

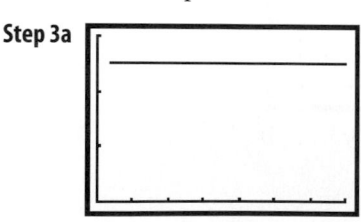

$[0, 7, 1, 0, 3, 1]$

Step 3b

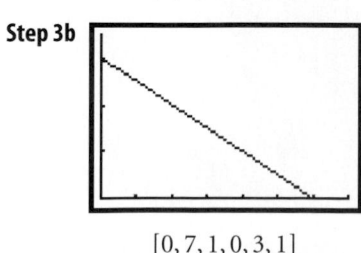

$[0, 7, 1, 0, 3, 1]$

Step 3c

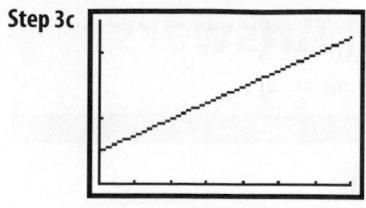

$[0, 7, 1, 0, 3, 1]$

Step 4 Table a: Start at the 0.8-m mark and walk toward the 4-m mark at 0.2 m per sec.

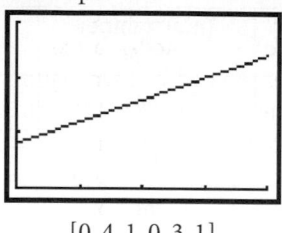

$[0, 4, 1, 0, 3, 1]$

Table b: Start at the 4-m mark and walk toward the 0-m mark at 0.4 m per sec.

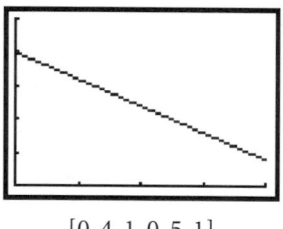

$[0, 4, 1, 0, 5, 1]$

Step 5 Table a: Starting value is 0.8, rule is add 0.2.

Table b: Starting value is 4, rule is subtract 0.4.

The starting position in the walking instructions matches the routine's starting value and corresponds to where the graph crosses the y-axis. The speed of the walker matches the change value of the rule and affects the steepness of the line graphed. The operation (add or subtract) of the rule relates to the direction of the walker and affects whether the line goes up or down.

Pascal's Triangle Project

The next 4 rows are:

1, 6, 15, 15, 6, 1
1, 7, 21, 35, 35, 21, 7, 1
1, 8, 28, 56, 70, 56, 28, 8, 1
1, 9, 36, 84, 126, 126, 84, 36, 9, 1

Fibonacci numbers

```
       1  3  3  1        21
    1  4  6  4  1
  1  5 10 10  5  1
1  6 15 20 15  6  1
1  7 21 35 35 21  7  1
```

Sierpiński's triangle

▸ If the odd numbers in Pascal's triangle are colored in, the even numbers are left uncolored, and the triangle is extended infinitely, then it becomes a Sierpiński triangle.

- This implies that the number of odd numbers in the (infinite) triangle is relatively so small that the probability of choosing one at random is close to 0.

LESSON 5.5, PAGE 285

Step 7 It is reasonable. Although the combined data is not a simple average of male and female, the line should still lie between the two.

Step 8 Answers will vary. The intercept-form method first gives a parallel line that has to be raised or lowered based on estimation (weakness); however, that method involves adjusting the line to a fit (strength). This point-slope method immediately gives you a line (strength); but the line must go through points and could possibly benefit from adjusting (weakness). The point-slope method also increases the chance that different people will get the same equation of best fit.

LESSON 5.7, PAGE 296

Step 1

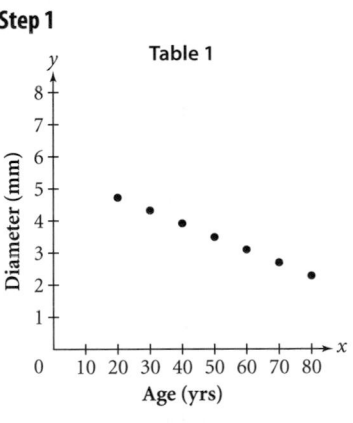

Table 1

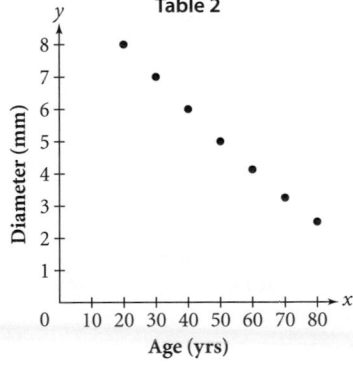

Table 2

LESSON 6.1, PAGE 310

Step 1

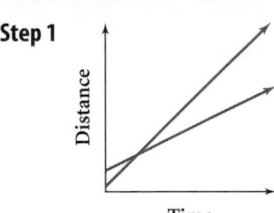

The steeper line represents the faster walker.

Step 4

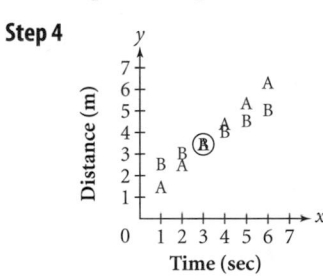

Step 6 Where Will They Meet?

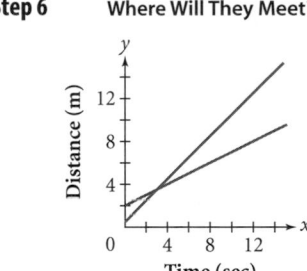

LESSON 6.1, PAGE 313

6d. Graphing windows will vary. The ones shown are $[0, 12000, 1000, -15000, 15000, 6000]$.

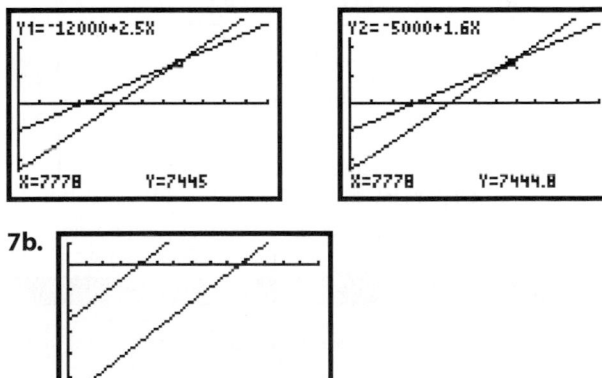

7b.

$[0, 7000, 500, -13000, 20000, 2000]$

14b.

Postage Costs

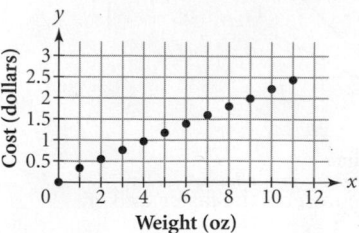

Step 2 See table. The rate of change does not indicate a linear pattern because it is not constant.

Step 3

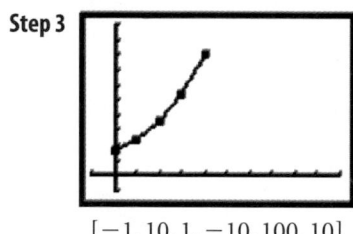

$[-1, 10, 1, -10, 100, 10]$

The slope of the line segments from point to point increases.

Steps 4 and 5

	Standard notation	Scientific notation
	5,000	5×10^3
a.	250	2.5×10^2
b.	$-5,530$	-5.53×10^3
c.	14,000	1.4×10^4
d.	7,000,000	7×10^6
e.	18	1.8×10^1
f.	$-470,000$	-4.7×10^5

Step 1 a. $\dfrac{5^9}{5^6} = \dfrac{5 \cdot 5 \cdot 5 \cdot \cancel{5} \cdot \cancel{5} \cdot \cancel{5} \cdot \cancel{5} \cdot \cancel{5} \cdot \cancel{5}}{\cancel{5} \cdot \cancel{5} \cdot \cancel{5} \cdot \cancel{5} \cdot \cancel{5} \cdot \cancel{5}} = 5^3$

b. $\dfrac{3^3 \cdot 5^3}{3 \cdot 5^2} = \dfrac{3 \cdot 3 \cdot \cancel{3} \cdot 5 \cdot \cancel{5} \cdot \cancel{5}}{\cancel{3} \cdot \cancel{5} \cdot \cancel{5}} = 3^2 \cdot 5^1$

c. $\dfrac{4^4 x^6}{4^2 x^3} = \dfrac{4 \cdot 4 \cdot \cancel{4} \cdot \cancel{4} \cdot x \cdot x \cdot x \cdot \cancel{x} \cdot \cancel{x} \cdot \cancel{x}}{\cancel{4} \cdot \cancel{4} \cdot \cancel{x} \cdot \cancel{x} \cdot \cancel{x}} = 4^2 x^3$

Steps 1–3 Data will vary. The following is a sample set of data recorded using 201 counters and a 68° angle on the plate.

"Years" elapsed	"Atoms" remaining	Successive ratios
0	201	
1	147	0.7313
2	120	0.8163
3	94	0.7833
4	71	0.7553
5	52	0.7324
6	42	0.8077
7	32	0.7619
8	28	0.8750
9	22	0.7857
10	18	0.8182
11	15	0.8333
11	12	0.8000
13	10	0.8333
14	9	0.9000

Step 4 Students should notice an exponential pattern. Here is a graph of the sample data (W):

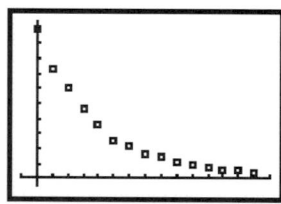

$[0, 15, 1, 0, 200, 20]$

Step 9 The equation from Step 8 does not fit very well.

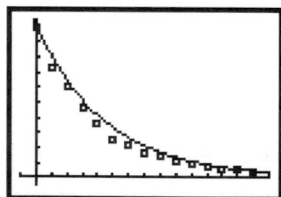

Step 10 Students may find that A seems to adjust the curve's position vertically and that r seems to change the steepness of the curve. The equation $y = 201(1 - 0.22)^x$ fits the sample data better.

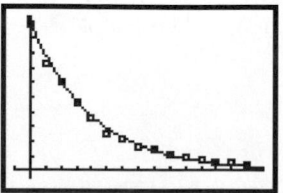

Step 12 The calculator table probably will vary from the collected data; the value at year 0 may be the only equivalent value. Explanations should include contrasting actual data to theoretical models.

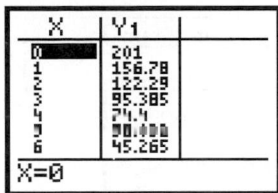

Step 1

Graph A

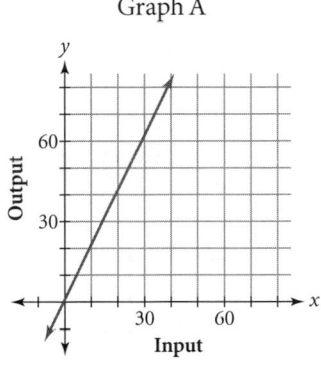

Step 6

Graph B

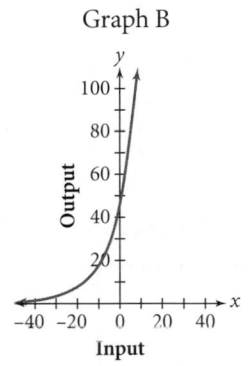

Input: all real numbers
Output: all positive numbers

8c. The independent variable is time. The dependent variable is the height of the ball because the height of the ball depends on the amount of time that has passed since it was hit. The graph could resemble a curve that climbs and then falls.

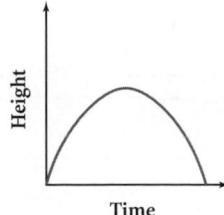

8d. The independent variable is the number of dice thrown, and the dependent variable is the probability. The graph is points starting at $(0, 0)$ and getting close to 1 as the number of dice becomes more than 20.

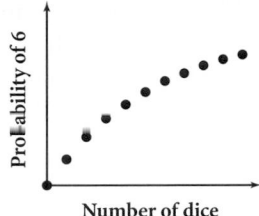

8e. The independent variable is the angle, and the area depends on the measure of this angle. The graph representing the area will increase and then fall to 0.

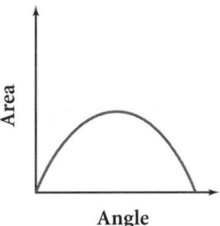

9b. The domain should go from 0 to a large number such as 500 or more, depending on the number of students in the school. The range is a set of integer values from 1 to a maximum that depends on the number of students in the school.

9c. The domain should go from 0 to about 5 seconds. The range should go from 0 to about 300 feet, or 100 meters.

9d. The number of dice is the domain; it is any non-negative integer. The probability is the range, and it is any number between 0 and 1 for observed outcomes. The theoretical probability for a large number of dice is close to 1.

9e. The angle measure is the domain, and it goes from 0° to 180°. The area is the range, and it could be as small as 0 and as large as about 60 sq. in. or 400 sq. cm, depending on the dimensions of the coat hanger.

Step 4 This graph shows the relationship between any number and its square.

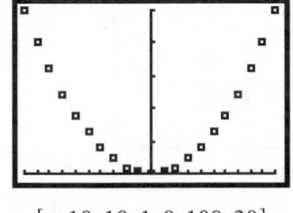

$[-10, 10, 1, 0, 100, 20]$

Additional Answers

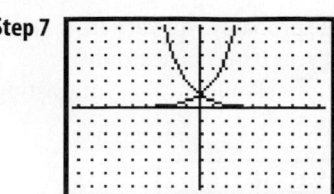

Step 7

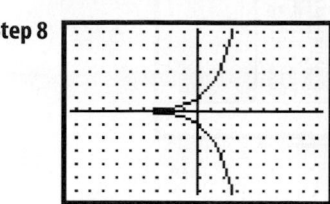

Step 8

Step 8 This line is in the middle of the parabola and is on top of the *y*-axis. The left and right sides of the parabola are reflection images of each other over this "mirror" line, also called the "line of symmetry."

Step 9 They are alike in that they are both continuous and symmetric. The absolute value graph looks like a V and the parabola is smoothly curved, almost like a U.

Step 10 Only points in the first quadrant could represent squares because both side lengths and areas are positive numbers.

Step 10 Changing the sign of the *x*-coordinates of points, or negating the variable *x* in a function, produces a flip across the *y*-axis; likewise, changing the sign of the *y*-coordinates of points, or negating the variable *y* in a function, produces a flip across the *x*-axis.

12. Answers will vary. It's impossible for x^2 to be negative because if *x* is a negative number, x^2 is a negative times a negative, which is positive. If *x* is a positive number, x^2 is a positive times a positive, which is also positive. So x^2 must be positive, or at least equal to 0, no matter what *x* is.

CHAPTER 8 REVIEW, PAGE 472

6b.

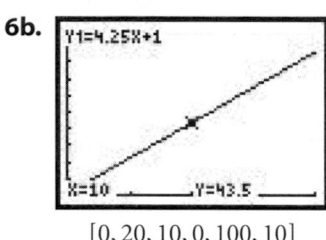

Y1=4.25X+1
X=10   Y=43.5

$[0, 20, 10, 0, 100, 10]$

7a. Answers will vary. The graph will pass the vertical line test.

7b. Answers will vary. The graph will fail the vertical line test.

LESSON 9.2, PAGE 483

Steps 1 and 2

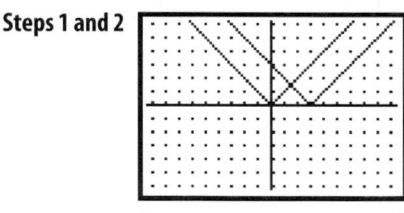

$[-9.4, 9.4, 1, -6.2, 6.2, 1]$

LESSON 9.3, PAGES 492–493

Step 6

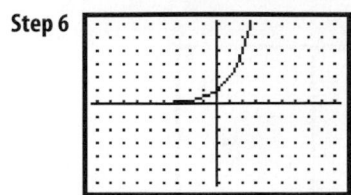

LESSON 9.4, PAGE 507

10a.

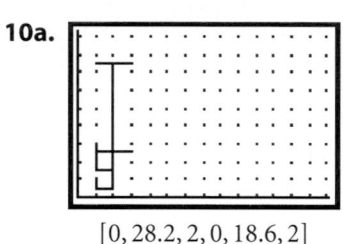

$[0, 28.2, 2, 0, 18.6, 2]$

10b.

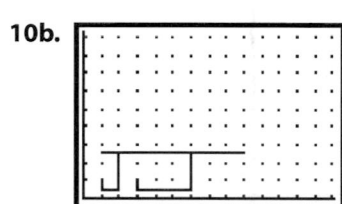

10c.

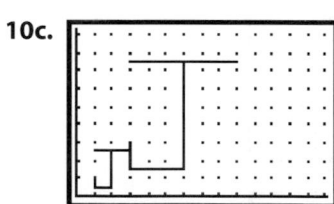

10d.

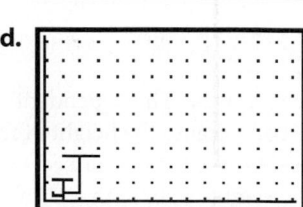

10e.

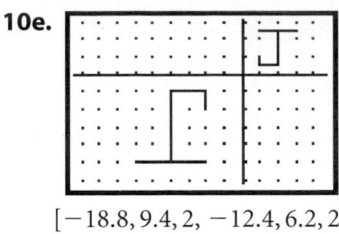

$[-18.8, 9.4, 2, -12.4, 6.2, 2]$

Step 1

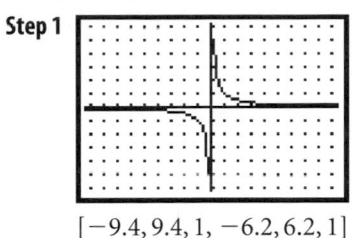

$[-9.4, 9.4, 1, -6.2, 6.2, 1]$

Step 2 Predictions will vary.

a. a vertical stretch of the graph of $y = \frac{1}{x}$ by a factor of 3 and a reflection across the x-axis

b. a vertical stretch of the graph of $y = \frac{1}{x}$ by a factor of 2; then a translation up 3 units

c. a translation of the graph of $y = \frac{1}{x}$ right 2 units

d. a translation of the graph of $y = \frac{1}{x}$ left 1 unit and down 2 units

Step 3a

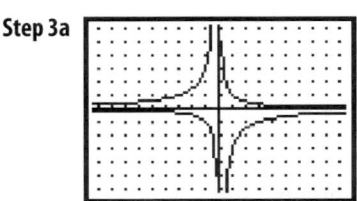

Step 3b

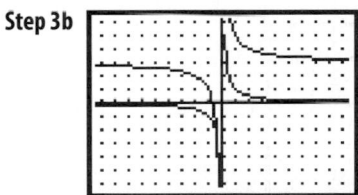

Step 3c

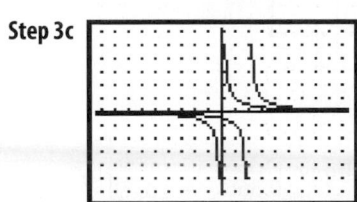

Step 3d

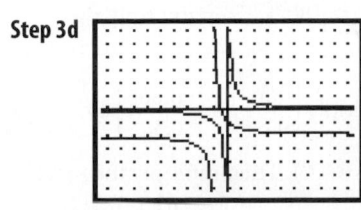

Different calculators will show the asymptotes in different ways.

Step 6

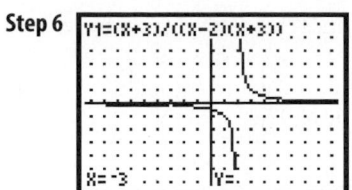

The hole at $x = -3$ is barely visible, but students can see that there is no y-value when $x = -3$.

Step 9

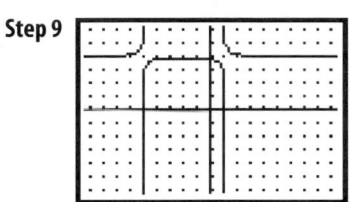

Step 11 A hole occurs at $x = h$ when the same factor, $(x - h)$, occurs in the numerator and denominator. A vertical asymptote occurs at $x = h$ for each factor $(x - h)$ in the denominator that does not appear in the numerator. A horizontal asymptote occurs at $y = k$ when k is the constant of the equation.

Step 9

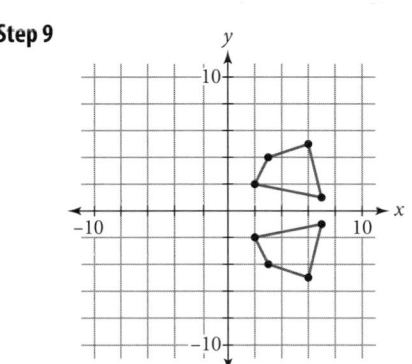

Step 11a

$$\begin{bmatrix} 2 & -3 & -6 & -7 & -x \\ 2 & 4 & 5 & 1 & y \end{bmatrix};$$

a reflection across the y-axis

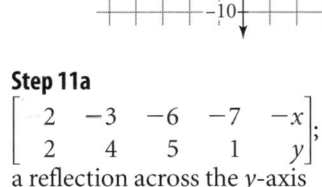

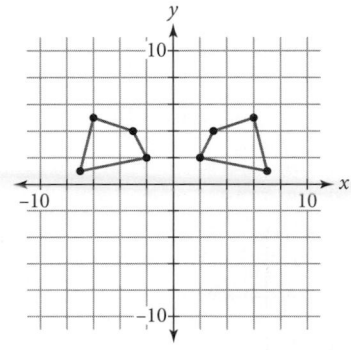

Additional Answers

Step 11b

$$\begin{bmatrix} 2 & 3 & 6 & 7 & x \\ 1 & 2 & 2.5 & 0.5 & 0.5y \end{bmatrix};$$

a vertical shrink by a factor of 0.5

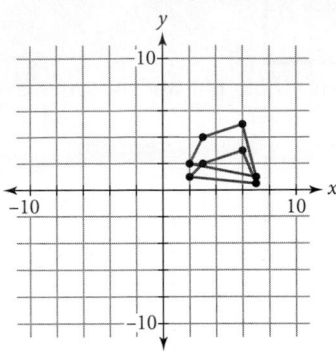

Step 11c

$$\begin{bmatrix} 1 & 1.5 & 3 & 3.5 & 0.5x \\ 4 & 8 & 10 & 2 & 2y \end{bmatrix};$$

a horizontal shrink by a factor of 0.5 and a vertical stretch by a factor of 2

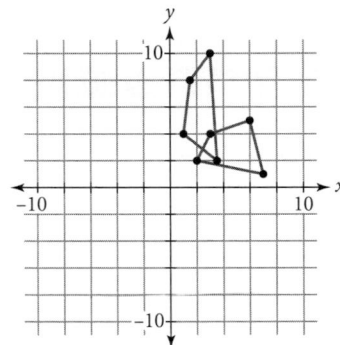

1a.

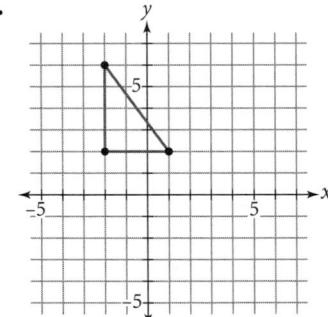

5d.

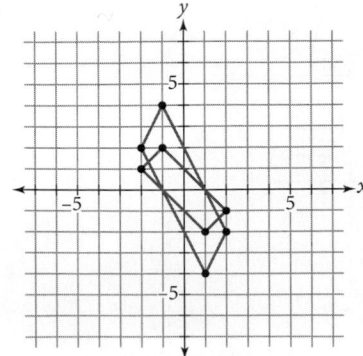

9a. Answers will vary. Sample answer: $\begin{bmatrix} 0 & 3 & 4 \\ 0 & 2 & 0 \end{bmatrix}$

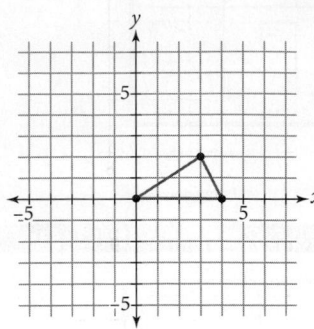

9b. Answers will vary. Sample answer:

$$\begin{bmatrix} 0.5 & -0.866 \\ 0.866 & -0.5 \end{bmatrix} \cdot \begin{bmatrix} 0 & 3 & 4 \\ 0 & 2 & 0 \end{bmatrix} = \begin{bmatrix} 0 & -0.232 & 2 \\ 0 & 3.598 & 3.464 \end{bmatrix}$$

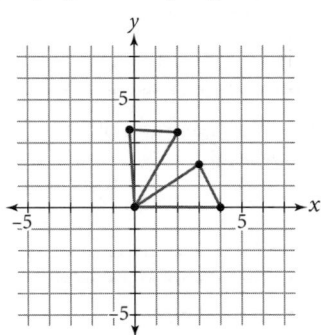

9c. a rotation 60° about the origin, or the point (0, 0)

9d. $\begin{bmatrix} -0.500 & -0.866 \\ 0.866 & -0.5 \end{bmatrix}$; sample answer:

$$\begin{bmatrix} -0.500 & -0.866 \\ 0.866 & -0.5 \end{bmatrix} \cdot \begin{bmatrix} 0 & 3 & 4 \\ 0 & 2 & 0 \end{bmatrix} = \begin{bmatrix} 0 & -3.232 & -2 \\ 0 & 1.598 & 3.464 \end{bmatrix};$$

a rotation 120° about the origin

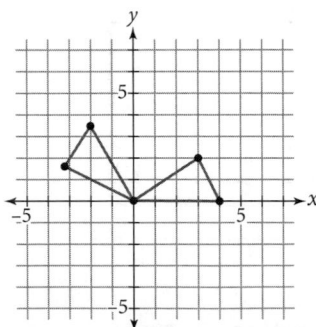

9e. Possible answer: Multiply $\begin{bmatrix} 0.5 & -0.866 \\ 0.866 & -0.5 \end{bmatrix}$ by

itself three times (3 · 60° = 180°). Then multiply the result by matrix [R].

9f. Possible answer: Multiply $\begin{bmatrix} 0.5 & -0.866 \\ 0.866 & -0.5 \end{bmatrix}$ by

itself six times (6 · 60° = 360°). Then multiply the result

by matrix [R] or just multiply [R] by $\begin{bmatrix} 1 & 0 \\ 0 & 1 \end{bmatrix}$.

10a.

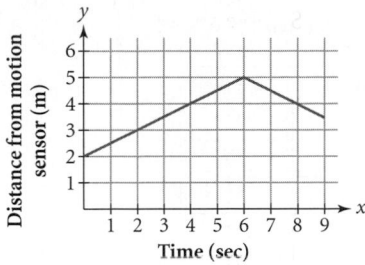

11b. Most Populated Countries, 1999

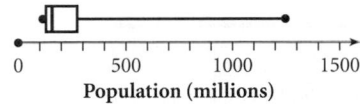

LESSON 10.1, PAGES 534–537

Step 4 Sample using window $[-1, 12, 1, -10, 150, 10]$

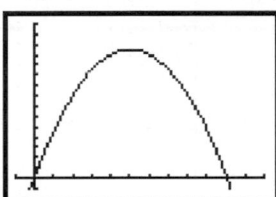

Step 5

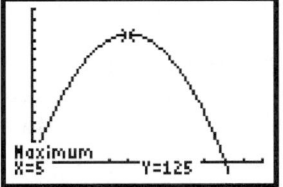

Step 8

X	Y₁	
1.04	48.16	
1.05	48.548	
1.06	48.934	
1.07	49.32	
1.08	49.705	
1.09	50.088	
1.1	50.471	

X=1.09

X	Y₁	Y₂
8.87	51.613	50
8.88	51.233	50
8.89	50.853	50
8.9	50.471	50
8.91	50.088	50
8.92	49.705	50
8.93	49.32	50

X=8.91

Step 9

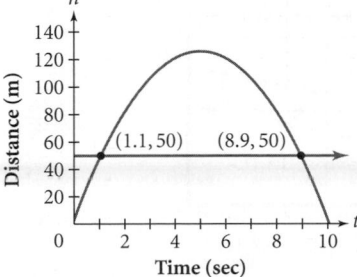

1. Enter one side of the equation into Y₁ and the other into Y₂ on the calculator. Then find the number of intersection points.

1a. two solutions

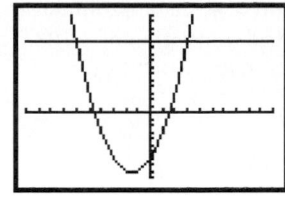

$[-10, 10, 1, -10, 15, 1]$

1b. no solutions

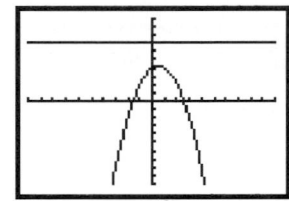

$[-10, 10, 1, -10, 10, 1]$

1c. one solution

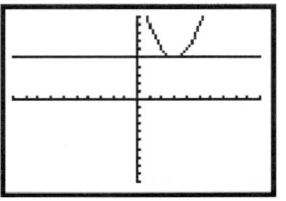

$[-10, 10, 1, -10, 10, 1]$

1d. two solutions

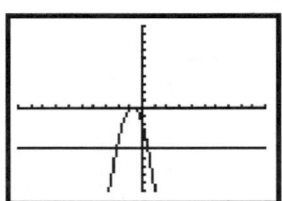

$[-10, 10, 1, -10, 10, 1]$

4. Sample answers:

4a.

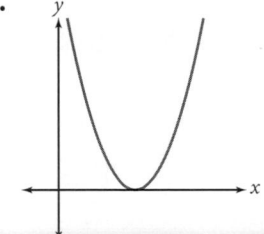

4b.

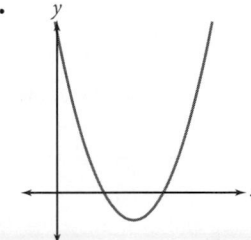

4c.

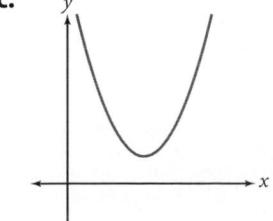

4d.

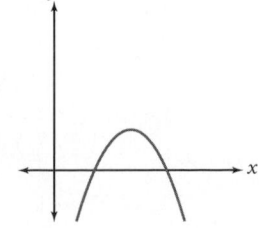

Additional Answers

6d. Start with $h(t) = -t^2$ and a vertical dilation of 108, a horizontal dilation of 4.7, a horizontal translation of $+4.7$, and a vertical translation of $+108$. The equation is $h(t) = -108\left(\frac{1}{4.7}(t - 4.7)\right)^2 + 108$ $= -4.9(t - 4.7)^2 + 108$.

6f. Answers are $4.7 \pm \sqrt{\dfrac{1347.49}{108}}$.

At about 1.17 sec and 8.23 sec, the rocket will be 47 m above the ground.

8a. The graph shows that only one solution exists.

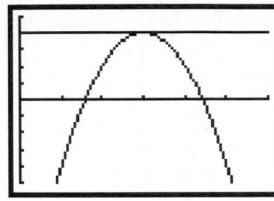

$[0, 6, 1, -5, 5, 1]$

8b. The table shows the solution to be at $x = 3$.

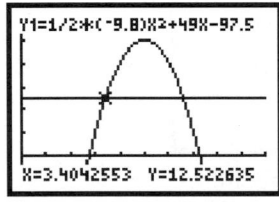

9f. Answers will vary. The horizontal line $y = 12.5$ intersects the parabola twice—when $x \approx 3.4$ seconds and when $x \approx 6.6$ seconds.

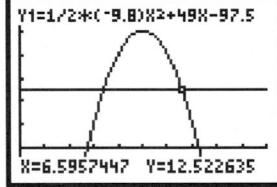

$[0, 10, 1, -5, 30, 5]$

Step 2

Width	Length	Area
1	11	11
2	10	20
3	9	27
4	8	32
5	7	35
6	6	36

Step 3 Sample answers:

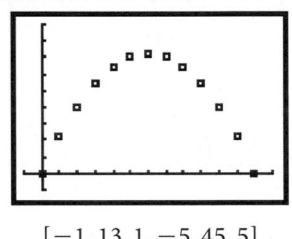

Step 4 Sample graph:

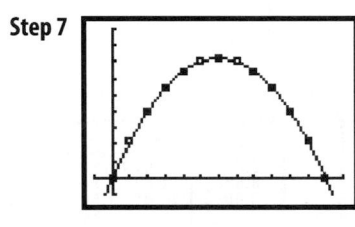

$[-1, 13, 1, -5, 45, 5]$

The area increases rapidly at first until the width becomes 6, then it decreases. Yes, it makes sense to draw a curve through points to represent decimal widths and areas.

Step 7

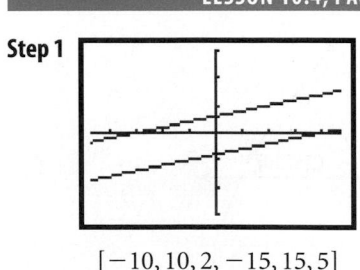

$[-1, 13, 1, -1, 9, 1]$

LESSON 10.2, PAGES 538–539

Step 1 Six sample answers:

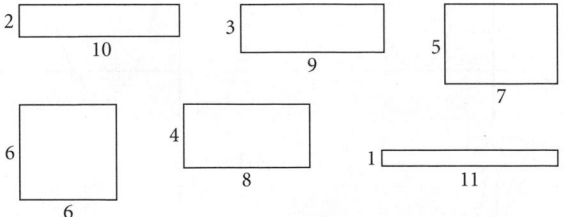

LESSON 10.4, PAGE 551

Step 1

$[-10, 10, 2, -15, 15, 5]$

Step 3

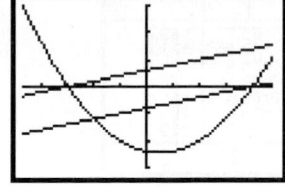

$[-10, 10, 2, -15, 15, 5]$

The graph looks like a parabola. The x-intercepts are where $y = x + 3$ and $y = x - 4$ cross the x-axis.

LESSON 10.7, PAGE 569

6. Sample answers:

6b. $y = 2x^2 - 3x + 2$. There are no x-intercepts.

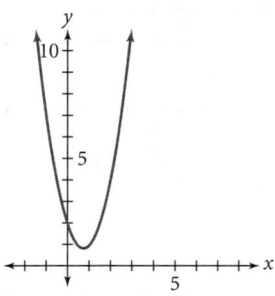

6c. $y = 2x^2 - 3x - 2$. The x-intercepts are -0.5 and 2.

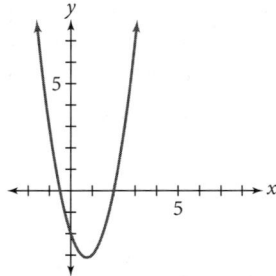

LESSON 11.1, PAGE 583

Step 1a

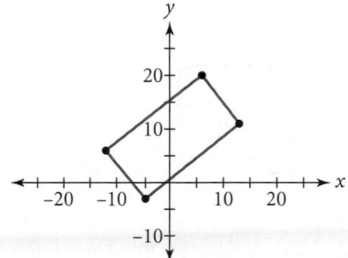

Step 1b

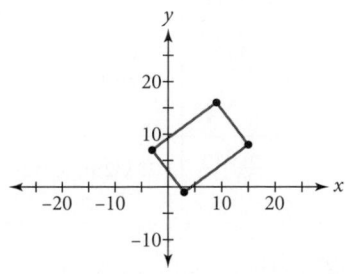

Step 1c

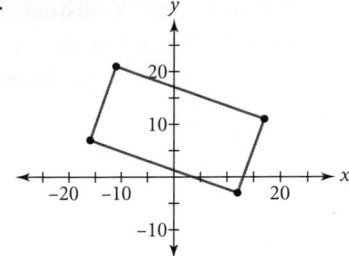

Step 1d

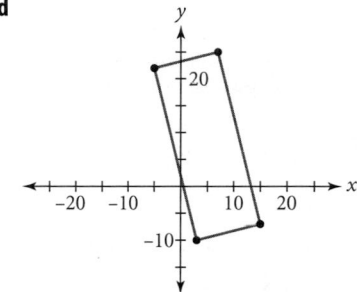

LESSON 11.1, PAGE 584

Step 5a slope of $\overline{AB}$: $-\frac{9}{7}$; slope of $\overline{BC}$: $\frac{7}{9}$

Step 5b slope of $\overline{AB}$: $-\frac{4}{3}$; slope of $\overline{BC}$: $\frac{3}{4}$

Step 5c slope of $\overline{AB}$: $-\frac{5}{14}$; slope of $\overline{BC}$: $\frac{14}{5}$

Step 5d slope of $\overline{AB}$: $-\frac{4}{1}$; slope of $\overline{BC}$: $\frac{1}{4}$

Step 6a slope of $\overline{AD}$: $\frac{7}{9}$; slope of $\overline{DC}$: $-\frac{9}{7}$

Step 6b slope of $\overline{AD}$: $\frac{3}{4}$; slope of $\overline{DC}$: $-\frac{4}{3}$

Step 6c slope of $\overline{AD}$: $\frac{14}{5}$; slope of $\overline{DC}$: $-\frac{5}{14}$

Step 6d slope of $\overline{AD}$: $\frac{1}{4}$; slope of $\overline{DC}$: $-\frac{4}{1}$

Step 8 Sample answer:

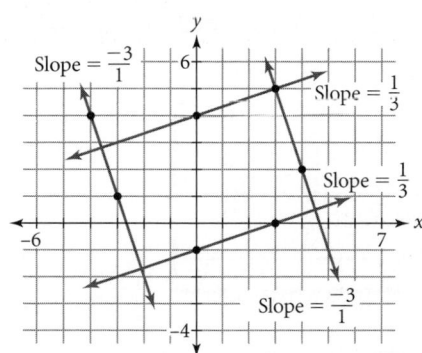

A rectangle is formed.

Step 1f

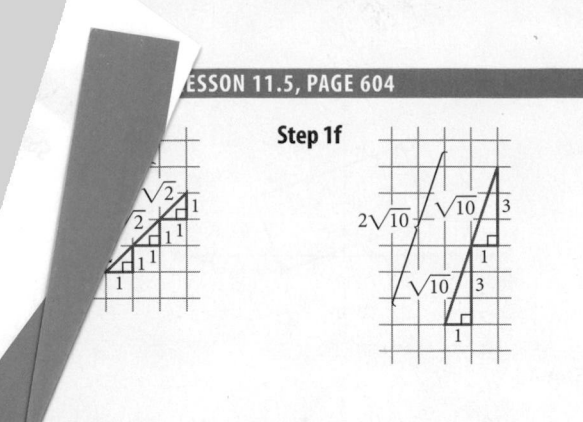

5. $y = x^2 + 2x\sqrt{5} + 5$

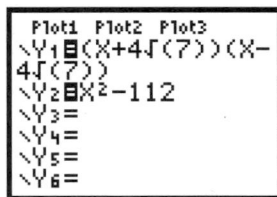

Plot1 Plot2 Plot3
\Y1⊟(X+√(5))(X+√(3))
\Y2⊟X²+2X√(5)+5
\Y3=
\Y4=
\Y5=
\Y6=

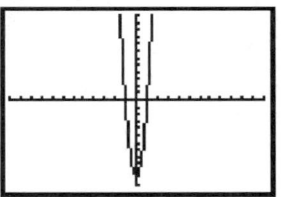

$[-9.4, 9.4, 1, -6.2, 6.2, 1]$

X	Y₁	Y₂
-3	.50359	.50359
-2	.05573	.05573
-1	1.5279	1.5279
0	5	5
1	10.472	10.472
2	17.944	17.944
3	27.416	27.416
X=-3		

6a. $y = x^2 - 112$

Plot1 Plot2 Plot3
\Y1⊟(X+4√(7))(X-4√(7))
\Y2⊟X²-112
\Y3=
\Y4=
\Y5=
\Y6=

$[-120, 120, 10, -120, 120, 10]$

X	Y₁	Y₂
-3	-103	-103
-2	-108	-108
-1	-111	-111
0	-112	-112
1	-111	-111
2	-108	-108
3	-103	-103
X=-3		

6b. $y = 2x^2 + 2x\sqrt{6} - 72$

Plot1 Plot2 Plot3
\Y1⊟2(X-2√(6))(X+3√(6))
\Y2⊟2X²+2X√(6)-72
\Y3=
\Y4=
\Y5=

$[-100, 100, 10, -100, 100, 10]$

X	Y₁	Y₂
-3	-68.7	-68.7
-2	-73.8	-73.8
-1	-74.9	-74.9
0	-72	-72
1	-65.1	-65.1
2	-54.2	-54.2
3	-39.3	-39.3
X=-3		

6c. $y = x^2 + 6x + 7$

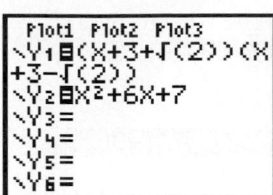

Plot1 Plot2 Plot3
\Y1⊟(X+3+√(2))(X+3-√(2))
\Y2⊟X²+6X+7
\Y3=
\Y4=
\Y5=
\Y6=

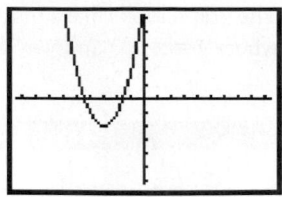

$[-9.4, 9.4, 1, -6.2, 6.2, 1]$

X	Y₁	Y₂
-3	-2	-2
-2	-1	-1
-1	2	2
0	7	7
1	14	14
2	23	23
3	34	34
X=-3		

Steps 2–5

Section	Horizontal distance traveled (Step 2)	Slope (Step 3)	Diagonal distance (Step 4)	Angle of climb (Step 5)
1	60	$\frac{20\text{ m}}{60\text{ m}} = \frac{1}{3}$	$20\sqrt{10}$ m ≈ 63.2 m	$\tan^{-1}\left(\frac{1}{3}\right)$ $\approx 18.4°$
2	90	$\frac{20\text{ m}}{90\text{ m}} = \frac{2}{9}$	$10\sqrt{85}$ m ≈ 92.2 m	$\tan^{-1}\left(\frac{2}{9}\right)$ $\approx 12.5°$
3	50	$\frac{20\text{ m}}{50\text{ m}} = \frac{2}{5}$	$10\sqrt{29}$ m ≈ 53.9 m	$\tan^{-1}\left(\frac{2}{5}\right)$ $\approx 21.8°$
4	120	$\frac{7\text{ m}}{120\text{ m}}$ $= \frac{7}{120}$	$\sqrt{14,449}$ m ≈ 120.2 m	$\tan^{-1}\left(\frac{7}{120}\right)$ $\approx 3.3°$
5	50	$\frac{-7\text{ m}}{50\text{ m}}$ $= -\frac{7}{50}$	$\sqrt{2,549}$ m ≈ 50.5 m	$\tan^{-1}\left(-\frac{7}{50}\right)$ $\approx -8.0°$
6	20	$\frac{-20\text{ m}}{20\text{ m}}$ $= -1$	$20\sqrt{2}$ m ≈ 28.3 m	$\tan^{-1}(-1)$ $= -45°$
7	30	$\frac{-20\text{ m}}{30\text{ m}}$ $= -\frac{2}{3}$	$10\sqrt{13}$ m ≈ 36.1 m	$\tan^{-1}\left(-\frac{2}{3}\right)$ $\approx -33.7°$
8	20	$\frac{-20\text{ m}}{20\text{ m}}$ $= -1$	$20\sqrt{2}$ m ≈ 28.3 m	$\tan^{-1}(-1)$ $\approx -45°$

Additional Answers